Progress in
Fourier Transform
Spectroscopy

Proceedings of the
10th International Conference,
August 27 – September 1, 1995,
Budapest, Hungary

Edited by
J. Mink, G. Keresztury,
R. Kellner

Mikrochimica Acta
Supplement 14

SpringerWienNewYork

Prof. Dr. János Mink
Department of Analytical Chemistry, University of Veszprém, Veszprém, Hungary
and Institute of Isotopes, Hungarian Academy of Sciences, Budapest, Hungary

Dr. Gábor Keresztury
Central Research Institute of Chemistry, Hungarian Academy of Sciences, Budapest, Hungary

Prof. Dr. Robert Kellner
Institut für Analytische Chemie, Technische Universität Wien, Wien, Austria

© 1997 Springer-Verlag Wien
Printed in Austria

Product Liability: The publisher can give no guarantee for information about drug dosage and
application thereof contained in this book. In every individual case the respective user must check its
accuracy by consulting other pharmaceutical literature. The use of registered names, trademarks, etc.
in this publication does not imply, even in the absence of a specific statement, that such names are
exempt from the relevant protective laws and regulations and therefore free for general use.

Typesetting: Thomson Press, New Delhi, India
Printing: MANZ, A-1050 Wien
Graphic design: Ecke Bonk
Printed on acid-free and chlorine free bleached paper

With 611 Figures

Die Deutsche Bibliothek – CIP-Einheitsaufnahme

Progress in Fourier transform spectroscopy : proceedings
of the 10th international conference, August 27 – September 1,
1995, Budapest, Hungary / ed. by J. Mink ... – Wien ; New
York : Springer, 1997
   (Mikrochimica acta : Supplementum ; 14)
   ISBN 3-211-82931-8
NE: Mink, Janos [Hrsg.]; Mikrochimica acta / Supplementum

ISSN 0026-3672
ISBN 3-211-82931-8 Springer-Verlag Wien New York

Preface

The 10th International Conference on Fourier Transform Spectroscopy (ICOFTS) was held in the Budapest Convention Centre, Budapest, Hungary, between August 27 and September 1, 1995.

To highlight the 10th event of the highly successful ICOFTS-series with a special opening session, the conference started with a lecture by Nobel Laureate Richard R. Ernst on "The Complementarity of Frequency, Time and Space Domains in Spectroscopy" which provided a glimpse into our prominent "neighbour": FT-NMR.

Also during the opening session, the Fritz-Pregl Medal was awarded to Professor Peter R. Griffiths, while the Friedrich-Emich Medal was given to Professor János Mink by Prof. M. Grasserbauer, president of the Austrian Society for Analytical Chemistry.

The regular sessions were opened by Jeanette G. Grasselli-Brown with a short overview of the history of ICOFTS from its start in 1970. The 19 invited plenary lectures given in the order of their appearance in this volume have provided a framework to the conference that covered advances in the most important areas of Fourier-transform spectroscopy. More than 350 further communications have presented the results of recent research done in 30 countries from five continents, in the poster sessions. In addition, round table discussions have been organized in three parallel sessions around prominent topics of current interest, namely on Catalysis, Environmental and Life Sciences, and Quantitative Accuracy of FTIR Spectrometry.

There were more than 400 registered participants at the meeting, e.g. from Germany (75), Hungary (50), USA (46), Austria (28), UK (26), the Netherlands (22), Russia (16), France (15), Japan (13), Switzerland (11), Italy (10), etc.

The scientific content of the Conference clearly demonstrated that Fourier-transform spectroscopy is already a well established method used in a great variety of scientific fields. Progress has been reported in new and quickly developing fields like FT-Raman, 2D-FTS, step-scan, photoacoustic, time-resolved and emission spectroscopy, FT-IR and FT-Raman microscopy, vibrational circular dichroism and a broad variety of applications in analytical chemistry, biology, catalysis, semiconductors, polymers, coupled techniques, chemometrics, high-pressure, low temperature, matrix-isolation, high-resolution, environmental and atmospheric studies, etc. The very broad and practically unlimited field of applications gathered scientists from very different areas, indicating the high degree of interdisciplinarity of Fourier transform spectroscopy.

Despite the many challenges that Budapest offers to visitors, most participants preferred the spacious, air-conditioned quarters of the Convention Centre using the opportunity for lively discussions. The attendance was very high at all scientific sessions from the Monday morning opening till the end of the last lecture on Friday afternoon.

The conference was sponsored and supported by the following organizations:

– Federation of European Chemical Societies (under Event no. 203)
– Hungarian Acadamy of Sciences (HAS)
– Working Committee on Laser Physics and Spectroscopy of the HAS
– Ministry of Culture and Education of Hungary
– Austrian Society for Analytical Chemistry
– International Union of Pure and Applied Chemistry
– Hungarian National Committee for Technological Development
– Institute of Isotopes of the HAS
– University of Veszprém
– Central Research Institute for Chemistry of the HAS.

The organizers thank Bio-Rad, Bruker and Perkin-Elmer who generously sponsored the social events.

The conference was highlighted by technical exhibitions from instrument manufacturers, software companies and science publishers:

AABSPEC International Ltd. ABL&E – JASCO Hungary Ltd., Academic Press Ltd., ATI Unicam Analytical Technology, Inc., Bio-Rad Laboratories Ltd., Sadtler Division, Bomem/Hartman & Brown, Bühler AG/ANSATEC, CIC Photonics, Inc., Galactic Industries Corporation, Harrick Scientific Corporation, High Pressure Diamond Optics Inc., Inovex GmbH/MIDAC Corporation, Jasco Europe srl., Marcel Dekker, Inc., Marco Polo Bt., MTEC Photoacoustics, Inc., M. Theiss, Nicolet Instrument Corporation, Renishaw plc Transducer Systems Division, Sentech Gesellschaft für Sensortechnik GmbH, Shimadzu Europa GmbH, Soft Science, Spectra-Tech Ltd and Springer-Verlag Wien.

The editors of this volume would like to thank Silvia Schilgerius, Elisabeth Hunger and the technical staff of Springer-Verlag Wien, for their cooperation and technical assistance in the preparation of this volume. Well deserved thanks are due to the participants of the conference and all authors of these articles for patiently waiting for the appearance of the Proceedings.

In accordance with our tradition to maintain a high scientific standard of the Proceedings, all papers appearing in this volume have been refereed. The editors are indebted to the experts who served on the referees' panel during and after the conference for their extremely valuable help.

The 11th International Conference on Fourier Transform Spectroscopy is being organized by James A. de Haseth and Richard A. Dluhy, and will be held in Athens, Georgia, USA, August 10–15, 1997.

J. Mink, G. Keresztury, R. Kellner

Editorial note:
All papers in this volume have undergone the refereeing procedure required by *Mikrochimica Acta*. The accepted papers – corresponding to the 19 plenary lectures and 203 of the presented 350 posters – are arranged in the following way:

1. Plenary Lectures, L1 to L19.
2. Poster papers grouped in 17 topical sections.

Contents

Analytical Applications, Problem Solving – I

Analytical Applications, Problem Solving – II, General Areas

Chemometrics, Data Treatment, 2D-FTS

Microscopy, High Pressure FTS

Low Temperature FTS, Matrix Isolation

Polymers

Biological Samples

Theoretical Studies – II

High Resolution and Atmospheric Studies

Step-Scan, 2D-FTS, PAS, Time Resolved FTS

Surfaces, Films, ATR, IRRAS, DRIFTS

Instrumentation, Emission

Double Modulation, VCD

IR-Sensors and Fibers

Mikrochim. Acta [Suppl.] 14, 1–7 (1997)

Raman Spectroscopy of Polymeric Fibers

Bruce Chase

Corporate Center for Analytical Sciences, Dupont Experimental Station, Wilmington, DE 19880-0328, U.S.A.

Abstract. Raman spectroscopy has many advantages for the study of polymeric fibers. The scattering geometry is uniquely suited to the sampling problem, and the ability to obtain both second and fourth order orientation functions from polarized data makes the Raman measurement an ideal candidate for fiber characterization. Raman measurements using visible excitation on stationary fibers are found to have lower fluorescence interference than that found for bulk polymer. However, the low background levels are associated with a rapid photo-bleaching process which is ineffective for moving fibers. FT-Raman measurements alleviate this problem, but a fiber optic probe suitable for measurements in the near infrared is needed. Preliminary results, obtained by using such a probe are demonstrated.

Key words: Raman spectroscopy, polymeric fibers, fiber optic probe, photo-bleaching.

The major driving force in the development of FT-Raman was the possibility of significant fluorescence suppression, which has now been well demonstrated [1]. This aspect of the technique is particularly important in the study of polymers, where conventional Raman spectra are often severely limited by fluorescence background. This signal can arise from impurities which are either intentionally part of the composition, such as anti-oxidants or UV absorbers, or included as random contamination. For many polymers it is now quite feasible to obtain good Raman data by using near infrared excitation. However, the source and behavior of the fluorescence under visible excitation is still not well understood. On occasion, this background can be bleached out by prolonged visible excitation and quite good spectra can be obtained without resorting to near infrared excitation. Additionally, the processing which the polymer has undergone can also make a difference in the level of background fluorescence. Since Raman measurements have many advantages for the study of fibers, including the capability for determining orientation as well as composition, it is fundamentally important to understand how the background changes as a function of sample preparation and form.

Experimental

All FT-Raman spectra were obtained on a Nicolet 910 equipped with a germanium detector. A Quantronix 416 Nd/YAG laser was used for excitation with power levels ranging from 100 to 500 mw. Visible Raman spectra were taken with a Kaiser Instruments Holospec® f/1.8 monochromator and a Photometrics CH210 CCD camera equipped with an EEV 1511 chip operated at $-45\,°C$. A Kaiser Holoprobe® fiber optic probe head was used along with a Coherent DPSS532 diode laser operating at 532 nm with a sample power of 45 mw.

Results

Many polymers which from their electronic absorption spectrum would not be expected to fluoresce, show significant background signals when excited in the visible. Raman spectra obtained at 532 nm for bulk samples of 6,6 nylon, a polyurethane and polyethylene terephthalate are shown in Figs. 1–3. In all cases large background signals were observed and there was no indication of significant photo-bleaching. Irradiation times of one hour resulted in at most a 50% reduction in background. As shown in Fig. 3, even a well studied sample such as polyethylene terephthalate can have a fluorescence background if a commercial grade material is examined. However, all of these samples show extremely good spectra when examined at 1064 nm with FT-Raman as shown in Figs. 4–6. Even a material such as polyphenylene terephthalamide, which is high-

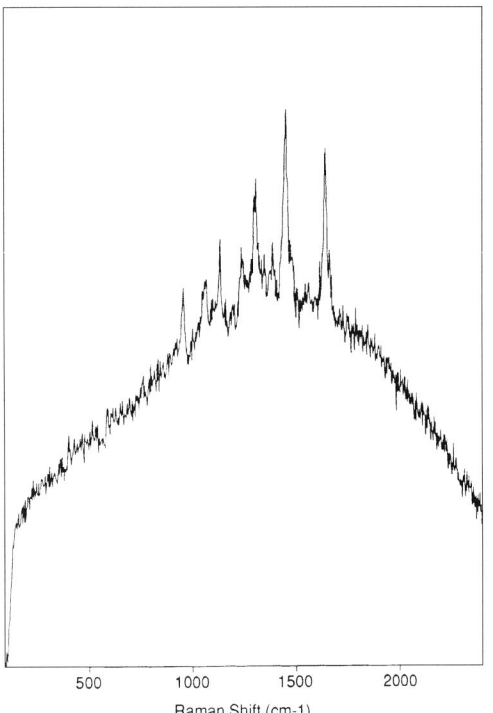

Fig. 1. Raman spectrum of bulk 6,6 nylon, at 532 nm, 45 mw

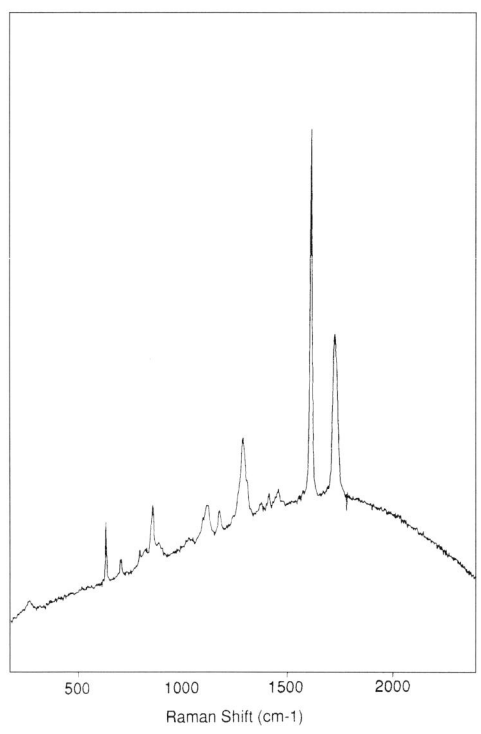

Fig. 3. Raman spectrum of bulk polyethylene terephthalate, at 532 nm, 45 mw

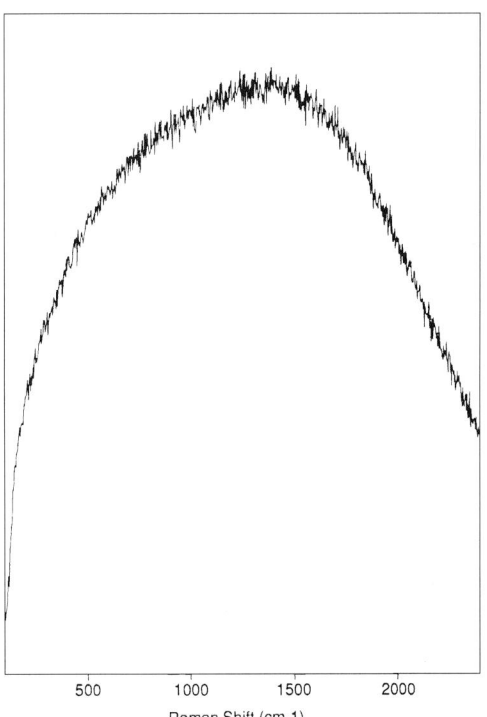

Fig. 2. Raman spectrum of bulk polyurethane, at 532 nm, 45 mw

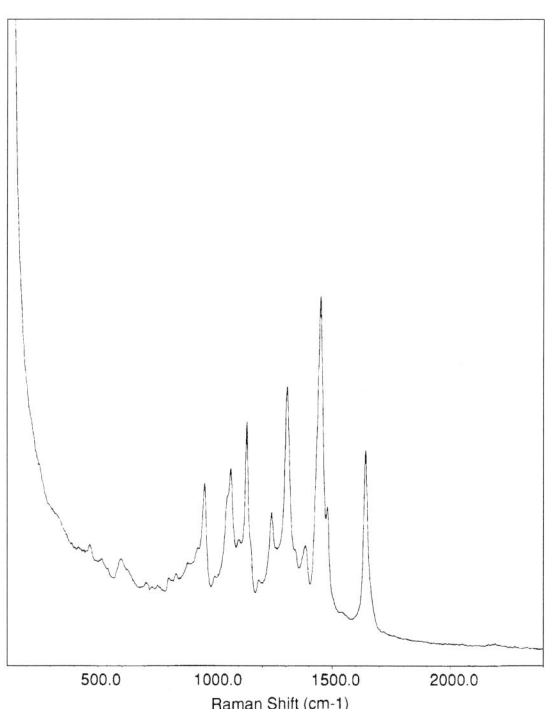

Fig. 4. FT-Raman spectrum of bulk 6,6 nylon, at 1064 nm, 200 mw

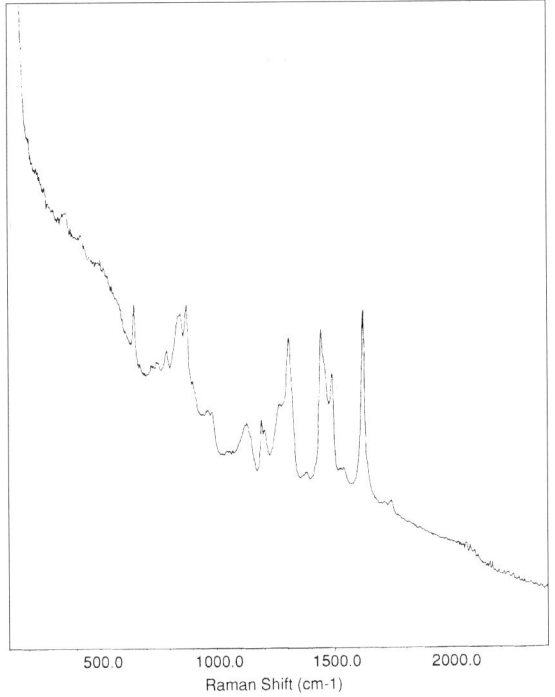

Fig. 5. FT-Raman spectrum of bulk polyurethane, at 1064 nm, 250 mw

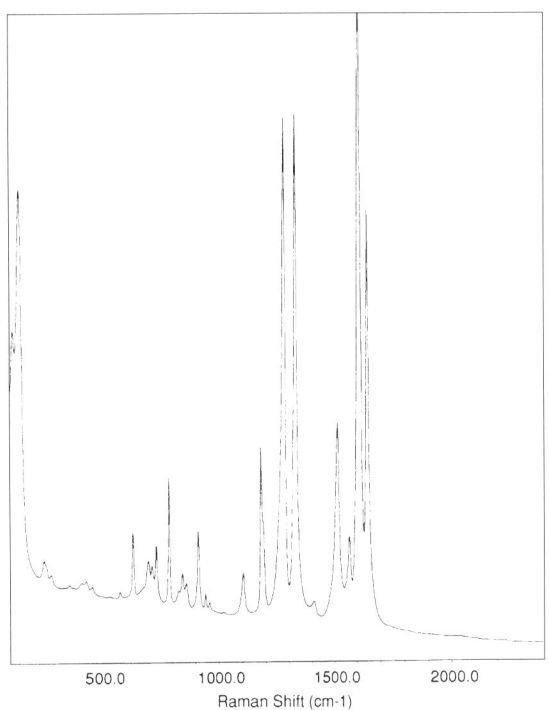

Fig. 7. FT-Raman spectrum of bulk polyphenylene terephthalamide, at 1064 nm, 100 mw

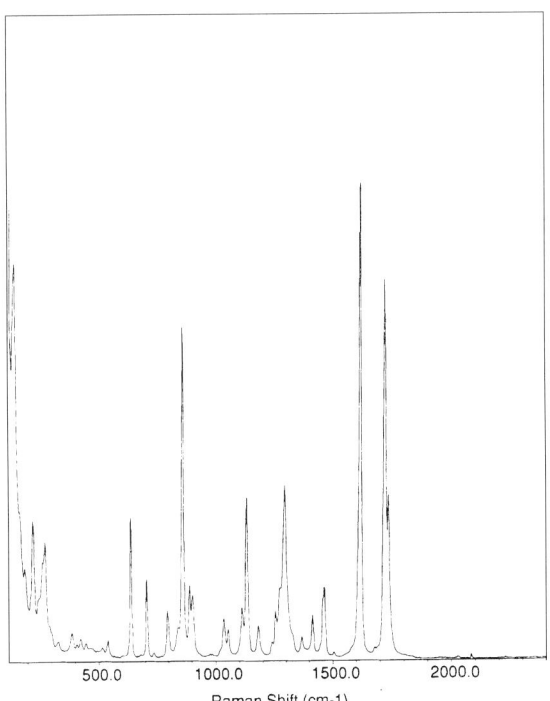

Fig. 6. FT-Raman spectrum of bulk polyethylene terephthalate, at 1064 nm, 150 mw

ly fluorescent in the visible and subject to thermal degradation when excited in the visible, yields excellent data in the near infrared, as shown in Fig. 7. This ability to obtain Raman data by using near infrared excitation with a wide range of polymers has been documented by many workers in the last five years [2–4], and has allowed significant structural studies to be accomplished. However, in many cases, the real interest is in fibers made from these bulk polymers. Both structural and segmental orientation information can be obtained from the Raman spectra [5], which would enhance our capability for structure–property correlations. Initially, one might expect the fibers made from these bulk polymers to show similar fluorescence backgrounds when excited in the visible. Figures 8–10 show visible Raman spectra of fibers made from 6,6 nylon, polyurethane, and polyethylene terephthalate respectively. In all cases, the fluorescence background is much less severe than that found for the bulk polymer. It is possible that the fiber processing removed fluorescent impurities from the bulk polymer by a type of zone refining process occurring during extrusion, or that the fluorescence is much more efficiently bleached in the fiber geometry. The latter explanation is supported by the observation that multiple fiber bundles yield Raman spectra with higher background signals. These results indicate that on-line Raman spectra of fibers may well be obtainable by use of visible excitation. The

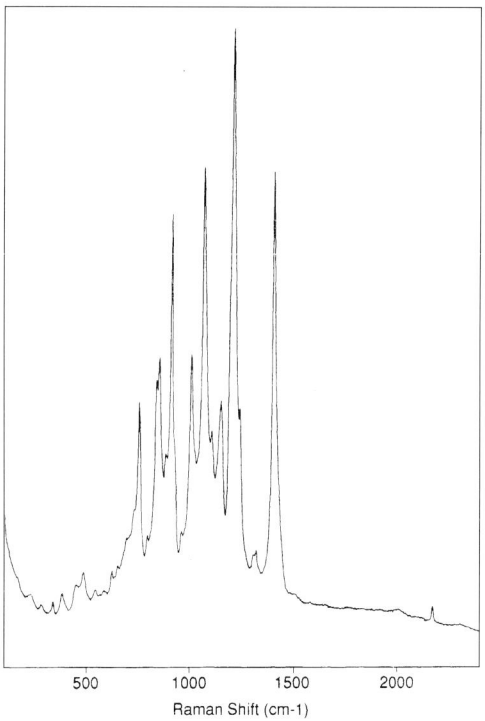

Raman Shift (cm-1)

Fig. 8. Raman spectrum of 6,6 nylon fibers, at 532 nm, 45 mw

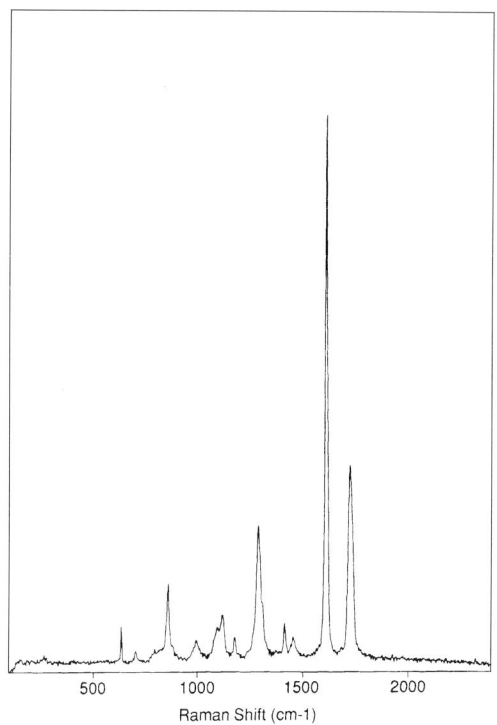

Raman Shift (cm-1)

Fig. 10. Raman spectrum of polyethylene terephthalate fibers, at 532 nm, 45 mw

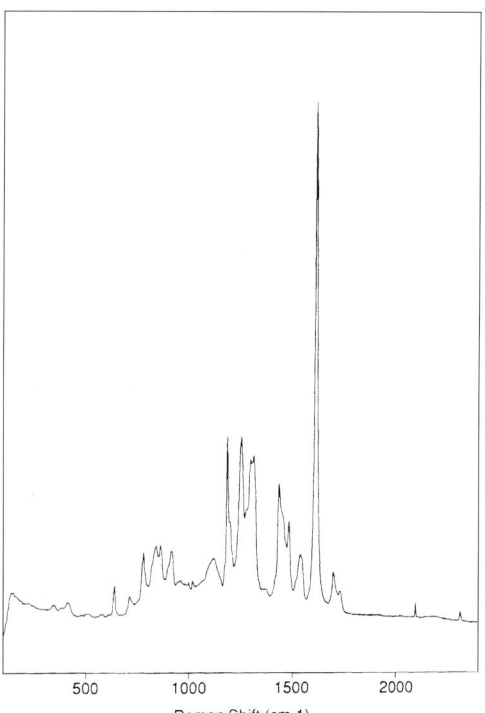

Raman Shift (cm-1)

Fig. 9. Raman spectrum of polyurethane fibers, at 532 nm, 45 mw

development of the Holoprobe® fiber optic probe, by Kaiser Optical Systems, will allow monitoring of the fibers at various points in the spinning process, with a monochromator located at a significant distance from the process, since only the fiber probe head needs to be physically located near the process. As a first approximation to the spinning process, this technique was used to examine PET fibers being drawn on a fiber draw machine. The fiber bundle is passed through two roll bars operating at different angular velocities, imparting a draw to the fibers. If desired, the fibers can be heated during the draw process. Unfortunately, it was found the Raman spectra of moving PET fibers exhibited extremely high fluorescence backgrounds, as shown in Fig. 11. This supports the hypothesis that photo-bleaching in stationary fibers is the mechanism by which the fluorescence background is reduced. For the moving fibers, new material is continually brought into the beam, and the photo-bleaching process never reaches completion. For the bulk polymer there is so much material present in or near the scattering volume that the fluorescence is never entirely photo-bleached out.

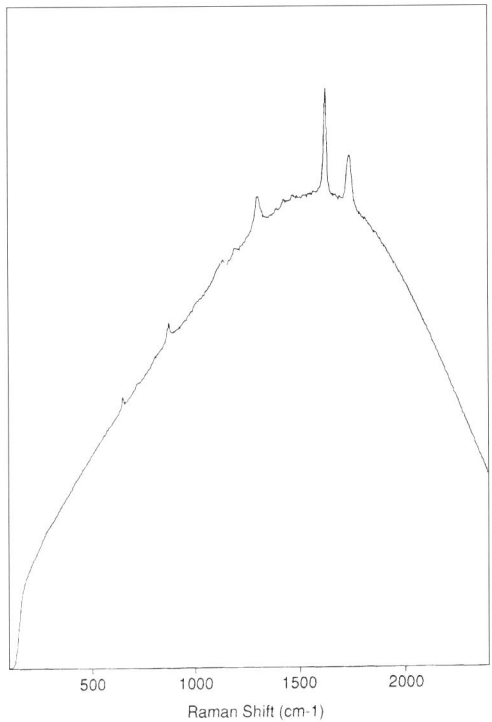

Fig. 11. Raman spectrum of polyethylene terephthalate fibers moving at 5 cm/sec past the focal point, 532 nm, 45 mw

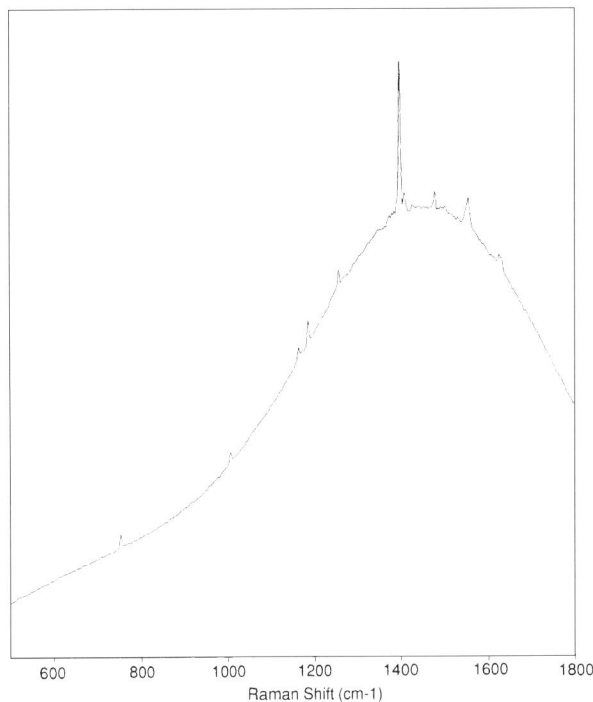

Fig. 12. Raman spectrum of anthracene, at 532 nm, 45 mw

The use of CCD detection, coupled with visible excitation, allows the acquisition of extremely low noise spectra even in the presence of a background. Under these circumstances, it should be possible to remove the background by curve fitting or some similar procedure to produce usable spectra. However, it was found that increasing the measurement time beyond a certain point (100 seconds by summing 100 block averages of 1-second integrations) did not improve the S/N ratio. For such a high background, the pixel to pixel (or binned pixel block to binned pixel block) variation in quantum efficiency limits the observed S/N. It should be possible to remove this effect by use of flat fielding, or dividing the observed Raman spectrum by a spectrum of a continuum source in the same frequency range. This was attempted by using a tungsten source to illuminate the sample, but the pixel to pixel variation differed between the Raman spectrum and the white light spectrum. The most likely cause for this difference was slightly different illumination of the sample by the laser and the white light, resulting in different illumination and optical filling of the fiber optic. In order for flat fielding to operate correctly, the continuum source must illuminate the sample in an identical fashion to the laser illumination,

giving identical illumination distribution of all the pixels which are being binned on the CCD. This was accomplished by using a highly fluorescent material illuminated by the focused 1064 nm laser source in place of the fiber sample. A portion of a yellow cardboard sheet which showed no signs of photo-bleaching (a Kodak lens tissue package) was found to meet these requirements. Figure 12 shows the spectrum of a photodegraded anthracene sample which has a high fluorescence background. Figure 13 shows the same spectrum which has been baseline corrected (a) or flat fielded (b). There is a clear reduction in the high-frequency noise which was due to pixel to pixel variation in quantum efficiency.

While this approach can be used to obtain reasonable Raman data, it would be more advantageous to be able to do remote FT-Raman spectroscopy by using a fiber optic probe similar in design to the Holoprobe®. Figure 14 shows a schematic of such a device. The incident laser radiation is made collinear with the optical axis defined by the sample and the collection fiber optic, by reflecting it off a holographic notch filter. With this arrangement, laser intensities of several hundred milliwatts at the sample were achieved. A critical component of the design of such a system is the provi-

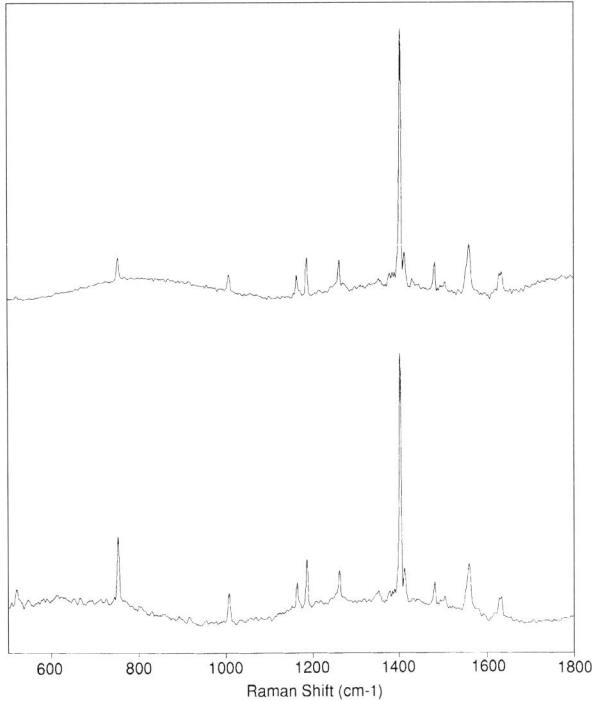

Fig. 13. Spectrum from Fig. 12 upper: baseline corrected, lower flat fielded

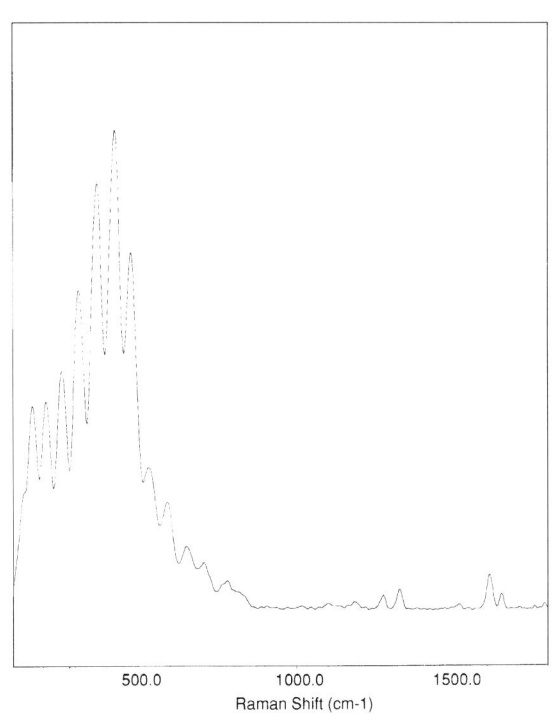

Fig. 15. FT-Raman spectrum when input fiber optic is unfiltered, showing features associated with silica scattering and interferences

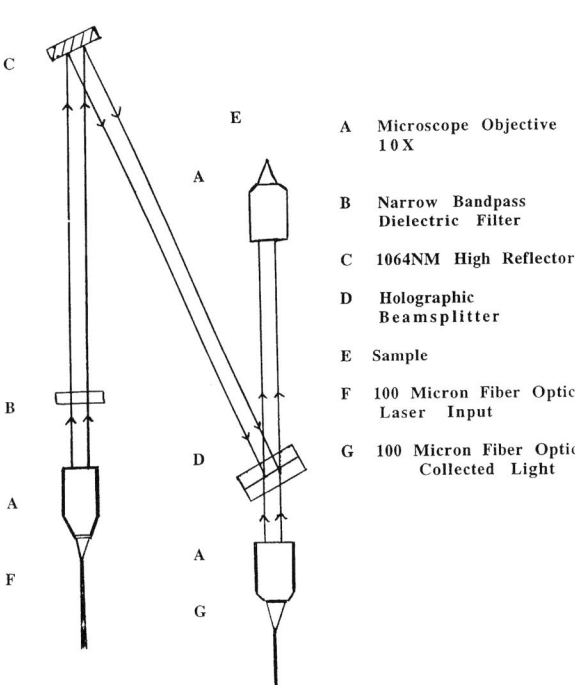

A   Microscope Objective
    10 X

B   Narrow Bandpass
    Dielectric Filter

C   1064NM High Reflector

D   Holographic
    Beamsplitter

E   Sample

F   100 Micron Fiber Optic
    Laser Input

G   100 Micron Fiber Optic
    Collected Light

Fig. 14. Optical layout for 1064 fiber optic collection system

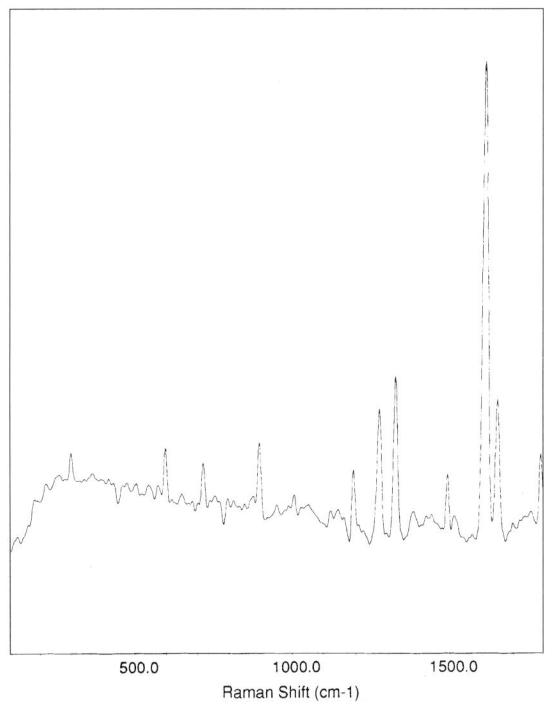

Fig. 16. FT-Raman spectrum obtained through fiber optic sampling of polyphenylene terephthalamide fibers, at 1064 nm, 200 mw

sion for filtering the Raman scattering from silica, which can occur in both the excitation and collection fibers. If either fiber is unfiltered the Raman spectrum is dominated by silica features as shown in Fig. 15. Preliminary results on fibers are demonstrated in Fig. 16 which shows the FT-Raman spectrum of a poly-phenylene terephthalamide fiber, obtained with this unit. The scattered intensity is quite low, and the overlap between illuminated area and collection volume needs to be further optimized. However, it should now be quite feasible to obtain FT-Raman spectra remotely over relatively long runs of fiber optic, without interference from scattering within the fiber optic cable.

References

[1] D. B. Chase, J. F. Rabolt (eds.), *Fourier Transform Raman Spectroscopy: From Concept to Experiment*, Academic Press, New York, 1994.
[2] J. R. Durig (ed.), *Proceedings of the XII International Conference on Raman Spectroscopy*, Wiley, New York, 1990.
[3] W. Kiefer (ed.), *Proceedings of the XIII International Conference on Raman Spectroscopy*, Wiley, New York, 1992.
[4] N. T. Yu, (ed.), *Proceedings of the XIV International Conference on Raman Spectroscopy*, Wiley, New York, 1994.
[5] M. J. Citra, D. B. Chase, R. M. Ikeda, K. H. Gardener, *Macromolecules* **1995**, *28*, 4007.

Mikrochim. Acta [Suppl.] 14, 9–13 (1997)

High-Resolution Fourier-Transform Spectroscopy in Solid State Physics

Marina N. Popova

Institute of Spectroscopy, Russian Academy of Sciences, 142092 Troitsk, Moscow Region, Russia

Abstract. Some problems in solid state physics are outlined where the use of high-resolution Fourier-transform spectroscopy offers new and unique possibilities. The work of the author's group is summarized, mainly. Three specific problems are discussed in more detail, namely, hyperfine interactions, isotope effects, and magnetic properties of insulators and semiconductors.

Key words: hyperfine and isotope structure, magnetic properties.

The advantages of the Fourier-transform (FT) spectrometers over the grating ones are well known. These are the advantages of Jacquinot (large throughput), Fellgett (better signal-to-noise ratio due to the simultaneous registration of all the spectral elements), Connes (precise wavenumber scale), and Michelson (high resolution and broad spectral range). One of the main advantages, the Fellgett or multiplex advantage, is proportional to $\sqrt{M}$, where M is the number of spectral elements. It has the greatest value in the case of intensity-independent noise of infrared detectors, but still exists in the case of photon noise in the visible, provided line-emmission spectra are registered.

The spectra of many atoms and molecules consist of narrow lines spread over a broad spectral range (M is large). The use of high-resolution Fourier-transform spectroscopy (FTS) enabled collection of vast amounts of information, inaccessible for conventional spectroscopy, on vibrational–rotational spectra of different molecules, determination of molecular constants with high precision, making atlases of atomic lines, and a new higher level of research in atomic and, especially, molecular physics. High-resolution laboratory FT spectrometers were constructed and used in the main spectroscopic centres, starting from the pioneer works of Pierre and Janine Connes and their collaborators at the Laboratoire Amé Cotton in the sixties and early seventies. Commercial instruments are now available.

In 1983, the large laboratory high-resolution (0.005 cm^{-1}) Fourier-transform spectrometer UFS-02 was installed in the Institute of Spectroscopy. This instrument [1], meant for both gas phase and solid state research, was designed and constructed at the Central Construction Bureau of Laboratory Scientific Instruments of the USSR Academy of Sciences, in close collaboration with G. N. Zhizhin's department of solid state spectroscopy of the Institute of Spectroscopy. Not long before, G. N. Zhizhin had invited me to review those problems of solid state physics where high spectral resolution is essential, and I wrote the paper "Solid state physics and high resolution spectroscopy" (unpublished). Among the problems where not only high resolution but also the wide spectral band offered by FTS is important I considered the following: (a) Rydberg series of shallow donors and acceptors in semiconductors; (b) FIR transitions between sublevels of the impurity ground level; (c) spectroscopy of laser crystals; (d) optical analogue of the Mössbauer effect; (e) excitonic series in semiconductors

From this list, only the first two topics could yield, at that time, examples of research done with the help of FTS (see, e.g. [2–6]). The levels of shallow donors and acceptors form hydrogen-like sequences, and transitions between them lie in the region 80–500 μm for germanium, 20–50 μm for silicon. In ultrapure semiconducting materials the spectral lines of residual shallow donors and acceptors are narrow (less than 0.03 cm^{-1} in some cases [2]). Very high sensitivity in detecting residual impurities in ultrapure materials has been achieved by combining the photoconductivity method, first proposed by T. M. Lifshits, and high-resolution FTS [2, 3].

The ground level of the transition metal impurity in a crystal is split into several sublevels by spin–orbit interaction, and that of the rare-earth (RE) impurity is split by a crystal field. FIR transitions between these sublevels have been registered and stuided by FIR FT spectrometers with 1.0–0.1 cm^{-1} spectral resolution [7–9].

In this presentation, I shall give a brief summary of high-resolution FTS solid state research performed by my group. Most of the problems considered could not be solved without high-resolution FTS. The instruments we used are briefly described in the experimental section. The following sections deal with our high-precision measurements of laser crystals, the study of hyperfine structure in the optical spectra of Ho^{3+} in LiYF$_{4,5}$ and molecular clusters of Ho^{6}Li$_x$^{7}Li$_{8-x}$F$_8$ in LiYF$_4$:Ho^{3+}. The final section deals with the magnetic properties of insulators and semiconductors, studied by the rare-earth spectral probe method.

Experimental

Two Fourier-transform spectrometers were used in our studies, namely, the high-resolution (0.005 cm^{-1}) infrared (0.8–100 μm) rapid scanning laboratory FT spectrometer UFS-02 [1] already mentioned in the introduction, and the Bomem DA3.002 high resolution (0.0026 cm^{-1}) FT spectrometer. The samples were in an optical cryostat, either in liquid helium or in helium vapour, at a temperature of 1.5–300 K. A given temperature could be controlled with a precision of ±0.02 K. We worked out a special technique for taking the spectra of polycrystalline opaque samples [10, 11]. The powder specimens were carefully mixed with ethanol and put on the sapphire platelet directly before the window of a detector. The whole assembly was installed in a cryostat.

High-Precision Spectroscopy of Laser Crystals

On our new high-resolution instrument UFS-02, we started with the work on laser crystals activated with RE ions. Precise measurments of spectral line positions, widths, shapes, and of oscillator strengths have been made for the important YAG–Er laser system Y$_{3-x}$Er$_x$Al$_5$O$_{12}$; the mechanisms of spectral line broadening were investigated [12–14]. A number of new laser materials were studied [15, 16].

Interesting results were obtained in the course of high-precision spectral measurements of the efficient multifrequency laser crystal LiYF$_4$:Ho^{3+}. They are briefly described in the next section.

Hyperfine Interactions

Nuclear hyperfine structure (HFS) due to the interaction between the electronic core and the nucleus of an atom has been intensively studied in atomic spectroscopy. There are few examples of a resolved HFS in crystals (see, e.g. [17]). Recently, the use of selective laser excitation and other high-resolution techniques enabled observation of HFS that is otherwise masked by inhomogeneous broadening. Only a few separate lines appropriate for these techniques have been studied up till now [17, 18].

We reported the first observation of hyperfine structure in the optical spectra of LiYF$_4$:Ho^{3+} [19] and studied this HFS in detail [19–22]. It was the first observation of the HFS in the infrared spectral region, the first broad band observation of HFS for a number of multiplets (the $^5I_8 \rightarrow$ $^5I_7$, $^5I_6$, $^5I_5$, $^5I_4$ optical transitions in the Ho^{3+} ion were studied). Examples of the spectra are presented in Fig. 1.

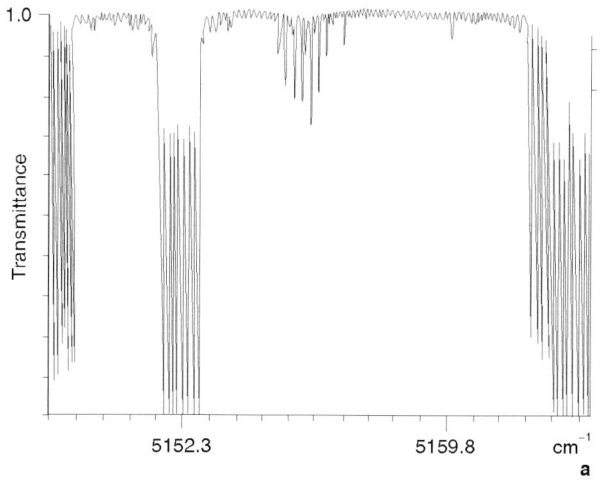

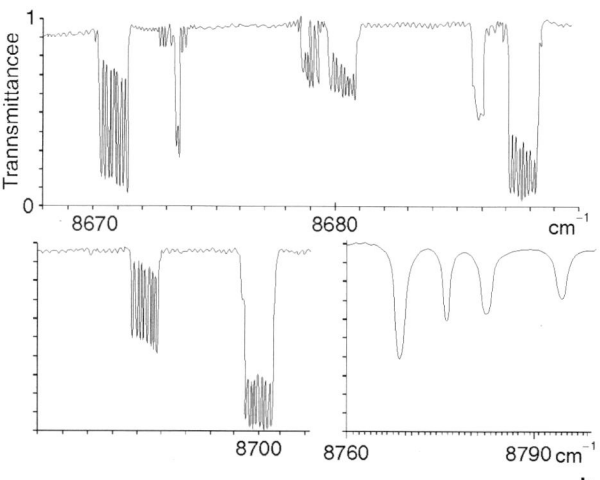

Fig. 1. Absorption of LiYF$_4$:Ho^{3+} crystal in the regions of the transitions **a** $^5I_8 \rightarrow ^5I_7$ and **b** $^5I_8 \rightarrow ^5I_6$

The diagonal part of the magneto-dipole hyperfine (HF) interaction results in eight equidistant hyperfine components in the spectrum, while the non-diagonal part and the quadrupole HF interaction contribute to a non-equidistance. From the intervals, we found magnetic g-factors for the excited states, without magnetic measurements, determined the strength of a magnetic field created at the Ho^{3+} ion's nucleus by electrons in a given state, and estimated the electric field gradient at the nucleus. From the broadening of high-frequency lines (which is clearly seen in Fig. 1b), we estimated the intramultiplet relaxation rates due to non-radiative phonon transitions.

We observed for the first time new peculiarities of the HFS in optical spectra, namely, irregular HF intervals, complex intensity distribution, forbidden HF lines. Calculations of the HFS, in the framework of the crystal field theory, have been performed for the first time for several multiplets (originated from the $^5I_{8,7,6,5,4}$ free ion levels). Good agreement between the calculated and measured HFS made it possible to understand all the peculiarities in the observed HFS and also revealed the good quality of the calculated wave functions. These wave functions can be used to determine various parameters of the Ho^{3+} ion in the important laser material $LiYF_4$.

Isotope Effects

Isotope effects provide a good test of physical theories. In activated crystals, the isotope spectral effects were attributed to the interaction of optical electrons with locally perturbed lattice vibrations that depend on the mass of impurity ion, while local static changes of the lattice structure near the different isotopes were not taken into account [23–25]. We have shown experimentally that the change of static crystal field due to isotope substitution contributes essentially to the observed isotope shifts [26].

Recently, we have found and studied the isotope structure in the optical spectra of $LiYF_4$:Ho^{3+} (0.1 at.%) associated with the lithium isotopes [26–29]. Figure 2 presents an example of such a structure. The singlet–doublet transition with hyperfine structure is shown. As can be seen, eight hyperfine components in the "pure" form are present only in the $^7LiYF_4$ crystal with homogeneous isotopic composition, and in the samples of mixed composition, containing isotopes ^7Li and ^6Li, the hyperfine components are split into individual narrow lines (with half-width down to 0.007

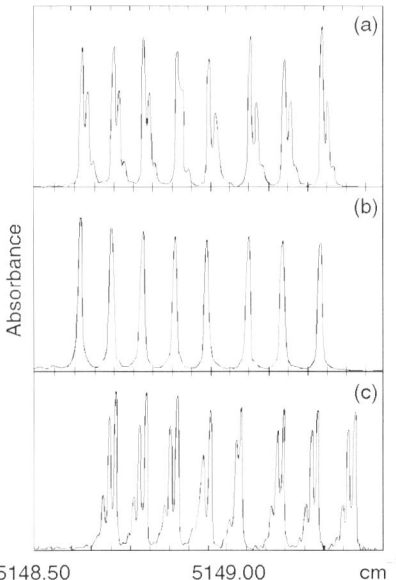

Fig. 2. Hyperfine components of the transition $\Gamma_2^{(1)}(^5I_8) \to \Gamma_{34}^{(1)}(^5I_7)$ in the crystals $^7Li_{1-x}$ ^6Li_x YF_4:Ho (0.1 at.%). **a** $x = 0.0742$ (natural abundance of lithium isotopes), **b** $x = 0.005$, **c** $x = 0.9$

cm^{-1}). This isotopic structure of the hyperfine components is equidistant in most cases (with period of 0.01–0.03 cm^{-1}).

It follows from the relative intensities of isotope components that the strongest line comes from the Ho^{3+} impurity ions with homogeneous isotopic composition in the closest surroundings, which includes eight lithium sites ($Ho^7Li_8F_8$ or $Ho^6Li_8F_8$ clusters), and the next strongest line represents the centres with one foreign isotope ($Ho^7Li_7{}^6LiF_8$ or $Ho^6Li_7{}^7LiF_8$ clusters). It is clearly seen from Fig. 2 that the middle interval in the HFS is broader for these last centres in comparison with that for the first ones. The explanation is that the isotope substitution lowers the symmetry of the local static crystal field and lifts the degeneracy of the Γ_{34} level.

The reason for the dependence of the static crystal field upon the isotope composition lies in the anharmonicity of zero-field vibrations. Because of the difference in masses of the lithium isotopes the amplitudes of their zero-field vibrations differ, and this leads, through the anharmonisity, to different equilibrium positions of the nearest fluorine ions. Such a qualitative picture [26] has been later confirmed by calculations, and the quantitative microscopic theory of isotope spectral effects has been developed [29, 30].

The isotope structure in the spectra can be used for precise determination of the unknown relative abundance of isotopes in a crystal [28, 31].

Megnetic Phenomena

The spectral probe method is widely used for studying magnetic phase transitions. The internal magnetic field that appears from magnetic ordering in a system splits and shifts the energy levels of a probe ion. The detection of appropriate spectral line splittings or shifts delivers information on a magnetic state. Typically, the spectral resolution of grating spectrometers used for such studies is 0.5–1 cm^{-1} in the visible. As we have shown recently [29], the inhomogeneous width of some rare-earth spectral lines in a crystal may be as small as 0.007 cm^{-1}. The most intense lines allowed in free rare-earth ion optical transitions lie in the infrared and occupy broad spectral regions. It is advantageous to register such spectra with a Fourier-transform spectrometer rather than a classical one.

We have demonstrated by the example of the rare-earth/transition-metal magnetic compounds related to high-T_c superconductors, that high-resolution FT spectroscopy considerably broadens the scope of the spectral probe method and offers new possibilities in studying magnetics. The high-resolution spectrum of a rare-earth ion in a magnetic crystal is extremely sensitive to the changes of a local magnetic field. The splittings, linewidths and line-shapes of spectral lines provide information on the magnetic phase transitions and their nature, magnetic structure in an ordered state, short range order.

As an example, the spectra at different temperatures of the Er^{3+} probe in the so-called "green phase" of Yb–Ba–Cu–O superconducting ceramics are shown in Fig. 3a [32]. The spectral lines at 6532 cm^{-1} and 6540 cm^{-1} correspond to two non-equivalent structural positions. The splitting and the narrowing of the line at 6532 cm^{-1} indicates a magnetic ordering at $T_c = 16.5 \pm 0.5$ K. The temperature of magnetic ordering follows from the point of inflection in the splitting *vs.* temperature curve [33] – see Fig. 3b.

The splitting does not vanish at $T = T_c$; a "tail" is observed due to short-range order at $T > T_c$. Such a "tail" is the longer the lower the dimensionality of a magnetic system; low-dimensional magnetism can be studied in such a way [33–35].

In the temperature interval 5.5–3.5 K the spectrum of Er^{3+} in Yb_2BaCuO_5 changes and can be represen-

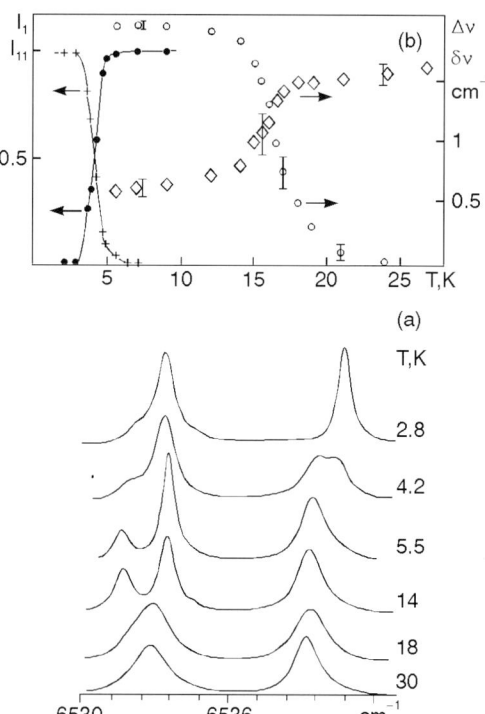

Fig. 3. Temperature dependences of **a** the Er^{3+} probe spectrum in Yb_2BaCuO_5, **b** the splitting Δv of the line at 6532 cm^{-1} (circles) and the half-width of its high-frequency component (diamonds) in the high-temperature magnetic phase; the relative integral intensities I_I and I_{II} of two subspectra in the line at 6540 cm^{-1} (crosses and filled circles)

ted as a superposition of two different spectra with temperature-dependent relative intensities. These spectral changes are more pronounced for the line at 6540 cm^{-1} (see Fig. 3a). Such a behaviour points to a first order spin-reorientation transition (at $T_R = 4.0$ K) when two magnetic phases coexist.

While this phase transition in Yb_2BaCuO_5 follows unambiguously from the spectra, other methods failed to detect it. I would like to mention here also our work on Dy_2BaCuO_5 [36] where high-resolution FT spectra enabled us to detect a very delicate splitting of the low-temperature magnetic transition. It was found that an ordering of the dysprosium subsystem at 11.26 K is followed by spin reorientation at 11.08 K.

More comprehensive surveys of our high-resolution FTS studies of magnetics can be found in [37, 38].

Conclusions

In conclusion, our work has revealed some more problems to add to the list given in the introduction, where the advantages of high resolution FTS are applicable to solid state research: (a) hyperfine interactions; (b) isotope effects; (c) magnetic phase transitions; (d) low-dimensional magnetism.

It should be emphasized that either the wide spectral band offered by FTS, or high sensitivity, or both are essential for these studies. If we compare other high-resolution spectroscopic techniques, we find that tunable lasers usually scan over about one wave number in one mode, and in few selected spectral regions. The Fabry-Perot interferometer, if working with the resolution appropriate for the study of hyperfine and isotope structure of Ho^{3+}, would not cover even one spectral line with hyperfine structure, because of its narrow free spectral range.

Acknowledgements. I am grateful to G. N. Zhizhin, who initiated and encouraged high-resolution FTS studies in the Institute of Spectroscopy. I would like to thank my collaborators and students N. I. Agladze, I. V. Paukov, Yu. A. Hadjiiskii, G. G. Chepurko who contributed to this work. This work was made possible in part by Grant No. 95-02-03796 from the Russian Fund for Fundamental Research and by Grant No. JEJ100 from the International Science Foundation and Russian Government.

References

[1] N. I. Agladze, A. A. Balashov, V. S. Bukreev, N. G. Kultepin, M. N. Popova, V. A. Vagin, E. A. Vinogradov, G. N. Zhizhin, *Proc. SPIE* **1985**, *553*, 452.

[2] S. D. Seccombe, D. M. Korn, *Solid State Comm.* **1972**, *11*, 1539.

[3] E. E. Haller, W. L. Hansen, *Solid State Comm.* **1974**, *15*, 687.

[4] E. E. Haller, *Phys. Rev. Lett.* **1978**, *40*, 584.

[5] B. Pajot, J. Kauppinen, R. Anttila, *Solid State Comm.* **1979**, *31*, 759.

[6] C. Jagannath, Z. W. Grabowski, A. K. Ramdas, *Phys. Rev. B* **1981**, *23*, 2082.

[7] A. Hadni, *L'infrarouge Lointain*, Presses Universitaires de France, Paris, 1969, p. 219.

[8] R. R. Joyce, P. L. Richards, *Phys. Rev.* **1969**, *179*, 375.

[9] R. G. Howell, D. J. Newman, *Phys. Stat. Solidi* **1968**, *29*, 697.

[10] N. I. Agladze, G. G. Chepurko, E. P. Khlybov, M. N. Popova, *Proc. SPIE* **1989**, *1145*, 321.

[11] G. G. Chepurko, I. V. Paukov, M. N. Popova *Proc. SPIE* **1992**, *1575*, 564.

[12] N. I. Agladze, A. A. Balashov, G. N. Zhizhin, M. N. Popova *Opt. Spektrosk.* **1984**, *57*, 379.

[13] N. I. Agladze, M. N. Popova, E. A. Vinogradov, T. M. Murina, V. I. Zhekov, *Optics Commun.* **1988**, *65*, 351.

[14] N. I. Agladze, E. A. Vinogradov, Kh. S. Bagdasarov, V. I. Zhekov, T. M. Murina, M. N. Popova, E. A. Fedorov, *Kristallografiya* **1988**, *33*, 912.

[15] N. I. Agladze, V. A. Antonov, P. A. Arseñev, Kh. S. Bagdasarov, V. F. Zolin, V. M. Markushev, I. T. Makhmudov, M. N. Popova, *Zh. Prikla. Spektrosk.* **1985**, *43*, 798.

[16] N. I. Agladze, V. A. Antonov, P. A. Arseñev, Ch. M. Briskina, V. F. Zolin, V. M. Markushev, M. N. Popova, D. S. Kholodnyi, *Zh. Prikla. Spektrosk.* **1988**, *48*, 613.

[17] R. M. Macfarlane, R. M. Shelby, in *Spectroscopy of Solids Containing Rare Earth Ions*, (A. A. Kaplyanskii, R. M. Macfarlane eds. North-Holland, Amsterdam, 1987, Chap. 3.

[18] T. Boonyarith, J. P. D. Martin, N. B. Manson, *Phys. Rev. B* **1993**, *47*, 14696.

[19] N. I. Agladze, M. N. Popova *Solid State Commun.* **1985**, *55*, 1097.

[20] N. I. Agladze, E. A. Vinogradov, M. N. Popova, *Zh. Eksp. Teor. Fiz.* **1986**, *91*, 1210 [*JETP* **1986**, *64*, 716].

[21] N. I. Agladze, E. A. Vinogradov, M. N. Popova, *Opt. Spektrosk.* **1986**, *61*, 3.

[22] N. I. Agladze, *Ph.D. Thesis*, Troitsk, 1991.

[23] G. F. Imbusch, W. M. Yen, A. L. Schawlow, G. E. Devlin, J. P. Remeika, *Phys. Rev.* **1964**, *136*, A481.

[24] N. Pelletier-Allard, R. Pelletier. *J. Phys. C* **1984**, *17*, 2129.

[25] C.-Y. Huang, *Phys. Rev.* **1968**, *168*, 334.

[26] N. I. Agladze, M. N. Popova, G. N. Zhizhin, V. J. Egorov, M. A. Petrova, *Phys. Rev. Lett.* **1991**, *66*, 477.

[27] N. I. Agladze, M. N. Popova, *II International Symposium on Rare Earths Spectroscopy*, 9–14 Sept. 1989, Changchung, China, *Abstracts*, p. 1.

[28] N. I. Agladze, M. N. Popova, G. N. Zhizhin, M. Becucci, S. Califano, M. Inguscio, F. S. Pavone, *Zh. Eksp. Teor. Fiz.* **1993**, *103*, 2215 [*JETP* **1993**, *76*, 1110].

[29] N. I. Agladze, M. N. Popova, M. A. Koreiba, B. Z. Malkin, V. R. Pekurovskii, *Zh. Eksp. Teor. Fiz.* **1993** *104*, 4171 [*JETP* **1993**, *77*, 1021].

[30] B. Z. Malkin, S. K. Saikin, *Proc. SPIE* **1996**, *2706*, 193.

[31] N. I. Agladze, M. N. Popova, in *Proceedings of the 12th Int. Conf. on Defects in Insulating Materials* (Germany 1992), (O. Kanert, J.-M. Spaeth, eds.) World Scientific, Singapore, 1993, Vol 1, p. 583.

[32] I. V. Paukov, M. N. Popova, B. V. Mill, *Phys. Lett. A* **1992**, *169*, 301.

[33] M. N. Popova, I. V. Paukov, *Opt. Spektrosk.* **1994**, *76*, 285.

[34] I. V. Paukov, M. N. Popova, B. V. Mill, *Phys. Lett. A* **1991**, *157*, 306 (and Correction p. 442).

[35] I. V. Paukov, M. N. Popova, J. Klamut, *Phys. Lett. A* **1994**, *189*, 103.

[36] M. N. Popova, G. G. Chepurko, *Pis'ma Zh. Eksp. Teor. Fiz.* **1990**, *52*, 1157 [*JETP Lett* **1990**, *52*, 562].

[37] M. N. Popova, *J. Appl. Phys.* **1994**, *76*, 7105.

[38] M. N. Popova, *Proc. SPIE* **1996**, *2706*, 182.

Mikrochim. Acta [Suppl.] 14, 15–22 (1997)
© Springer-Verlag 1997

Quantitative Infrared Intensities of Neat Liquids: Their Measurement and Use

John E. Bertie

Department of Chemistry, University of Alberta, Edmonton T6G 2G2, Canada

Abstract. This paper presents a summary of some of the work done in this laboratory over the past 6 years. It has become clear that well-aligned modern Fourier transform spectrometers are sufficiently photometrically accurate and sufficiently free from major differences between users that absolute infrared absorption intensities can be determined and transferred between well-aligned instruments with an error not exceeding about 2%. It is essential, however, to correct the baselines of transmission spectra, on the basis of measurements in cells with very long path-lengths. Such developments have led to the establishment of intensity standards for infrared spectroscopy of liquids. These have been accepted by the International Union of Pure and Applied Chemistry, and published. Such developments have also led to the reliable determination of relatively small intensity differences, such as those caused by isotopic substitution or by change in the intermolecular environment in binary liquid mixtures. Knowledge of correct intensities has allowed the development of an approximate procedure that greatly simplifies the computations. On the theoretical side, the availability of accuracy to within 2% has created the need to develop methods for obtaining molecular information from absorption intensities which introduce errors no larger than $\sim 0.1\%$, so that the experimental accuracy is not degraded by avoidable theoretical approximations. A major issue is the determination of the integrated intensities of overlapping bands and the assignment of the intensity under the baseline. The most appropriate intensity quantity to use is the imaginary molar polarizability, $\alpha_m''(\tilde{\nu})$. The spectrum of this quantity has been little used, but can be obtained easily with modern computerized spectrometers from transmission spectra, attenuated total reflection spectra, or specular reflection spectra. Most bands in the α_m'' spectrum of non-hydrogen-bonded liquids have essentially the Classical Damped Harmonic oscillator shape, and carefully fitting such bands to the spectrum is currently the most objective way to determine the integrated intensities. All the intensity above zero ordinate is usually explained by such fits if the α_m'' spectra were calculated from baseline-corrected experimental absorbance spectra. These points are illustrated in this paper by spectra taken from recent work in this laboratory.

Key words: infrared intensities, optical constants, methanol, chlorobenzene, acetonitrile.

The development and improvement of Fourier transform infrared spectrometers over the past 25 years has allowed the measurement of absolute infrared absorption intensities of neat liquids with a simplicity, wavenumber range, precision, and accuracy that could only be dreamed of by those who developed this field through work on dispersive instruments. This fact, together with the major advances in theoretical computations in the same period, have made the comparison of the best results with the results of theoretical calculation a rewarding exercise rather than the frustrating exercise of earlier days. This paper will illustrate these statements with a summary of recent work in this laboratory.

It is desirable to discuss explicitly what is meant by the word "quantitative" in the title of this paper. It is quantitative spectroscopy when an analyst takes great care to keep all experimental factors constant during the measurement of a set of standard samples to create an analytical calibration set, and during the measurement of the set of unknown samples which are then quantitatively analysed by comparing the measurements against the calibration set. The important point in such an experiment is that the different measurements must be extremely reproducible, i.e. precise, in order that the comparison is accurate. If the comparison is accurate, the analysis will be accurate, unless other factors intervene. But this is not the meaning of "quantitative" intended in this paper. There is no requirement that the individual measurements in the experiment above should be accurate, i.e. should give the correct answer, as long as they give the same answer each time they are made. That is, the measurements must be precise but need not be accurate.

In this paper, the term "quantitative" indicates that the correct value of the intensity is sought. Accurate intensities are sought, intensities that are the same no matter where they are measured. We cannot know when the correct intensity has been obtained, so the best that can be done is to provide a defensible estimate of how close to the correct intensity the measured intensities lie.

The procedures for achieving this are exactly the opposite of the procedures for the analytical experiment above. The highest precision possible is needed, but it cannot be obtained by keeping experimental factors constant which simply ensures that systematic errors are constant and do not influence the precision. The constant systematic errors are likely to mean that the measured intensities are wrong but that the error or inaccuracy is not revealed. To estimate the accuracy of the measured intensities, it is necessary to make measurements under different conditions, varying as many factors as possible in order to reveal and estimate the systematic experimental errors. The precision of the average values which result from such a procedure is worse than the precision from the analytical experiments discussed above, but a defensible estimate of the accuracy of the measurements is possible. Note, however, that it is always possible that an important systematic error has been overlooked, so the search for accuracy has no end.

Absolute intensities of liquids can be calculated from transmission, reflection, or attenuated total reflection measurements. This paper describes transmission measurements almost exclusively. The paper contains a brief discussion of the problems involved in the measurement of intensities of liquids by transmission, and how they have been overcome sufficiently to allow the acceptance of a published set of secondary intensity standards that are believed accurate to about 2%. The computations involved yield spectra of the real and imaginary refractive indices. The computations are sufficiently complex in detail to discourage their use, although an established computer program is available, so the paper continues with a report of a much simpler calculation that, although approximate, is more accurate then current measurements for all but the most intense bands.

The paper then turns to the interpretation of the accurate intensities in terms of physical properties of the liquids, and recommends the use of the imaginary

molar polarizability spectrum, which is readily calculated from the refractive index spectra.

In most cases, the major limitation to the accuracy with which the intensity of a particular transition, or group of transitions which contribute to a single band, can be determined is the separation of the area under the spectrum into the contributions from the different bands. The cautious use of curve fitting for this purpose is illustrated. The paper concludes with a brief account of the analysis of the measured intensities of liquid methanol-d_0, -d_1, -d_3 and -d_4, which illustrates the reliability of the molecular parameters we can currently obtain from accurate intensity measurements.

A summary [1] and a discussion [2] of the intensity quantities used in this paper are given elsewhere.

Absorption Intensities of Liquids by Transmission Measurements

The "absorbance" spectrum obtained from the transmission of infrared radiation through a cell full of liquid is shown in Fig. 1. The windows were potassium bromide and the sample liquid was benzene. Superimposed on this spectrum is a near-sine wave, which is the "absorbance" spectrum of the same cell empty. The reflection of radiation at the inner surfaces of the windows changes markedly when the cell is emptied, because the refractive index of the windows is very similar to that of the sample but is very different from that of air. Consequently, interference fringes are seen in the spectrum of the empty cell, which clearly cannot be used to correct the spectrum of the cell full of liquid for the intensity lost by reflection from the windows.

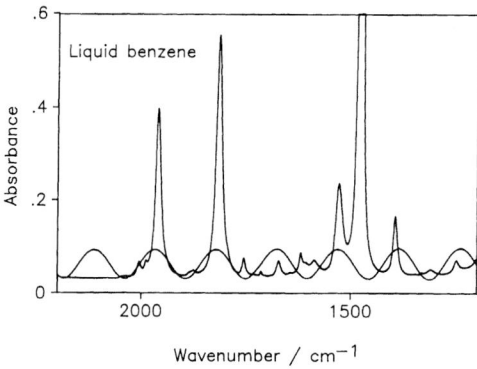

Fig. 1. The "absorbance" spectrum of a KBr-windowed cell full of liquid benzene-h_6, superimposed on that of the empty cell. The path-length is $33 \, \mu m$. The spectra shown are termed *experimental absorbance spectra* because they are influenced by factors other than absorption

The spectrum of the empty cell is the necessary source of the cell path-length, d, which is calculated from the spacing of the fringes, $\Delta\tilde{\nu}$, by $d = 1/(2\Delta\tilde{\nu})$. The path-length obtained in this way is usually accurate to about 1%.

In order to obtain the absorption intensity of a liquid sample from the "absorbance" spectrum of a cell full of liquid, it must be recognized that what the instrument calls the absorbance spectrum differs from what IUPAC calls an absorbance spectrum and what all scientists call absorbance when they use the Beer–Lambert law. IUPAC defines [3] absorbance as due solely to loss of energy by absorption. In comparison, the instrument's "absorbance" spectrum of a cell full of liquid is influenced by three factors [4]: the absorption of light by the liquid, the loss of light due to reflection, and the loss of light due to variable processes that, empirically, cause unpredictable fluctuations in the level of the baseline. The last two factors affect the instrument's "absorbance" spectrum but they are not due to absorption so cannot contribute to the true absorbance. Thus, we call the instrument's "absorbance" spectrum the *experimental absorbance* spectrum and say that the last two factors cause *apparent absorbance* [4]. Thus, the three factors which contribute to the experimental absorbance, EA, spectrum of the cell full of liquid are the *absorbance*, A, the apparent absorbance due to reflection, AA_R, and the apparent absorbance due to baseline errors, AA_B. Thus,

$$EA = A + AA_R + AA_B. \tag{1}$$

In order to determine the absorption by the liquid, AA_R and AA_B must be removed.

The baseline errors, AA_B, are removed by calculating where the baseline should be at *anchor points* in the baseline to either side of an absorption peak, and shifting the EA spectrum to put it there. To calculate where the baseline should be, the reflection losses must be calculated and added to the absorbance of the sample at the anchor points. The reflection losses are calculated as described in [4] and [5], from the known real refractive index spectrum of the windows and an approximate, constant refractive index of the sample; the result is not sensitive to the latter. The absorbance by the sample at the anchor points is calculated as the path-length times the linear absorption coefficient, K, (the absorbance per unit path-length [3]), the latter having been determined to about 1% accuracy from the spectra of the liquid in cells with extremely long paths. In practice the computations are done by the FORTRAN programs BASELINE and ANCHORPT.

Once the baseline has been corrected, the EA spectrum is converted into the real and imaginary refractive index spectra of the sample liquid. The procedure was developed and used in the 1970s by Norman Jones and his colleagues at the National Research Council of Canada [5, 6], and was modified by C. Dale Keefe in this laboratory [4, 7] into the FORTRAN program RNJ46A. An overview of this method has been given elsewhere [8].

In this laboratory it was found that variation of many instrumental parameters had little effect on the results obtained. Thus Si-on-CaF_2 and Ge-on-KBr beam-splitters were used, the interferometer modulator was realigned once, the detector was realigned several times, the optical path through the cell was realigned and the cell holder was repositioned, the cell was taken out of the instrument and dried between each sample, many different cells with different lengths and window materials were used, different apodization functions were used, different optical retardation velocities were used, *etc.* In spite of all of these variations, remarkable consistency in the imaginary refractive index spectra was obtained. The top box of Fig. 2 shows 15 imaginary refractive index, k, spectra com-

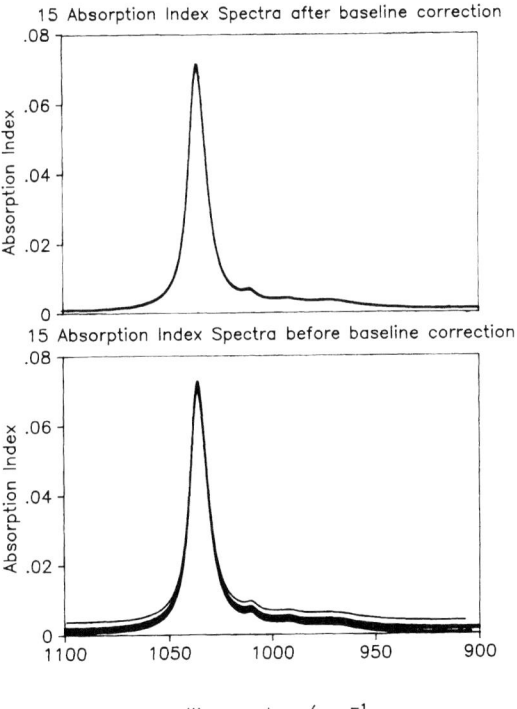

Fig. 2. Fifteen imaginary refractive index spectra of benzene calculated from 15 independent experimental absorbance spectra. Top: with baseline correction. Bottom: without baseline correction

puted from 15 independent experimental absorbance spectra. The lower box shows the k spectra calculated without baseline correction. Comparison of the two boxes clearly shows the importance of baseline correction. The agreement when baseline correction is used is impressive, and suggests that the k spectra have very good accuracy, to the extent that the accuracy can be determined from measurements on just one instrument.

It should be noted that the molar absorption coefficient spectrum is more familiar to chemists, and is calculated from the k spectrum by

$$E_m = \frac{4\pi \tilde{v} k}{2.303\, C} \qquad (2)$$

where C is the molar concentration.

To explore the accuracy further, scientists in other laboratories were asked to run spectra of empty cells and cells full of standard liquids with the use of good analytical technique and cells of good analytical quality. Spectra were obtained from about 6 different laboratories. One or two showed clear signs of instrumental misalignment and were not used, but the remainder were converted into real and imaginary refractive index spectra in this laboratory. The agreement between the intensities measured in different laboratories for benzene [7], toluene [9], chlorobenzene [10], and dichloromethane [11] was of the order of 2%. These results indicate the intrinsic intensity accuracy of well-aligned Fourier transform spectrometers, mainly with room-temperature detectors, though one laboratory used an HgCdTe detector at 77 K.

The IUPAC Intensity Standards

This work has led to the acceptance by IUPAC of secondary intensity standards for the calibration of infrared spectra of liquids. The standards are based on 43 bands of benzene, toluene, chlorobenzene, and dichloromethane, and have been published by Blackwell [12]. They include areas under the molar absorption coefficient spectrum between specified wavenumbers, both areas above zero ordinate and areas above the linear baseline drawn through the integration limits. They also include values of E_m, k, and the real refractive index, n, tabulated at $0.5\,\mathrm{cm}^{-1}$ intervals. The areas are believed accurate to 2% or better. The E_m and k values are believed accurate to 3% or better near absorption bands and 10% or better in the baseline.

The IUPAC standards include a diskette which contains a program IRYTRUE which provides a re-

liable procedure for calibrating the effective pathlength of transmission cells by use of the standards. The program's associated files contain the standard spectra in readable digital form in ASCII format.

The standards have been used in this laboratory [13] to calibrate transmission cells and to calibrate the CIRCLE, cylindrical multiple ATR, cell for the measurement of absolute intensities [1, 14–16]. Use of several different bands of the four liquids gives transmission cell path-lengths that agree with those measured from interference fringes to about 3% for 11 μm cells and 1% or better for path lengths of 100 μm or greater.

There is no doubt that many uses will develop for these intensity standards as their availability becomes more widely known. There is also no doubt that additional studies will lead to the improvement of the standards. These first infrared intensity standards should find many uses and simulate much new work.

Simpler Computations

The computations required to obtain real, n, and imaginary, k, refractive index spectra from experimental absorbance spectra are exact and are easily done with computer program RNJ46A. However, they are sufficiently complex that people are discouraged from doing them. To this end a much simpler, approximate, calculation has been explored and found to be sufficiently accurate for most work [17].

The exact method calculates the refractive index spectra by iteration [4–7]. Very approximate n and k spectra are computed and then used in the Fresnel equations of physical optics to compute the transmission and reflection at each interface in the cell. In this way the experimental absorbance spectrum is calculated. The calculated EA spectrum is then compared with that observed, the k spectrum is adjusted to improve the fit, the n spectrum is recalculated by Kramers–Kronig transform of the new k spectrum, and these new spectra are used in Fresnel's equations in another cycle of refinement. When the refinement has converged, the E_m spectrum is calculated from the k spectrum through Eq. (2).

The key feature of the exact method for the present discussion is that AA_R, the apparent absorbance due to reflection losses, is calculated by applying Fresnel's equations to each interface, as part of an iterative procedure. In the simpler approximate method AA_R is calculated by treating the cell full of liquid as though it were a single window. Thus $AA_R = -\log_{10} T$, where $T = 2n_w/(n_w^2 + 1)$ and n_w is the refractive index of the windows. The latter changes slightly across the infrared, so AA_R changes with wavenumber.

This simplification eliminates the need for an iterative calculation. To determine the linear absorption coefficient, K, at an anchor point, AA_R calculated in this way is simply subtracted from the experimental absorbance, EA, at the anchor point of the sample in a very thick cell, and the K value is obtained by dividing the result by the cell path-length. The assumption is made that baseline errors can be neglected provided the EA exceeds 0.3, and is justified by the agreement between values of K obtained from cells of different path-lengths. To calculate the imaginary refractive index, k, spectrum from the EA spectrum of the sample in a cell of normal length, AA_R and the baseline correction AA_B are subtracted from EA at each wavenumber to give the absorbance, A, directly, and E_m is calculated as A/Cd, where C and d are the molar concentration and path-length: k is related to E_m through Eq. (2) and is also directly calculated from the absorbance.

This method is clearly much simpler but is also approximate. The error introduced by the approximation is expected to increase from zero as the difference between the real refractive index of the liquid and windows increases from zero. This implies that absorption bands might be distorted by the approximate method, because of the anomalous dispersion that accompanies absorption, and that the distortion should increase with the strength of the band.

The effectiveness of the method is shown for chlorobenzene between KBr windows in Fig. 3. The figure shows two spectra calculated from the same EA spectrum with baseline correction. The two spectra, one calculated by the exact method and the second by the approximate method, are indistinguishable on the scale of the graph. In fact, the approximate spectrum lies within 0.4% of the exact one except that the difference increases to 1% near 1480 cm^{-1}, on the high-wavenumber side of the strongest band before the shoulder at 1500 cm^{-1}.

The strongest band in Fig. 3 has 2/3 the intensity of the strongest band of chlorobenzene, which is at 740 cm^{-1} and is very strong. For the 740 cm^{-1} band, the difference between the approximate and exact methods is up to 6%, with the maximum deviation again occurring on the high-wavenumber side of the band. For bands weaker than those shown, the agreement of the approximate and exact methods is even

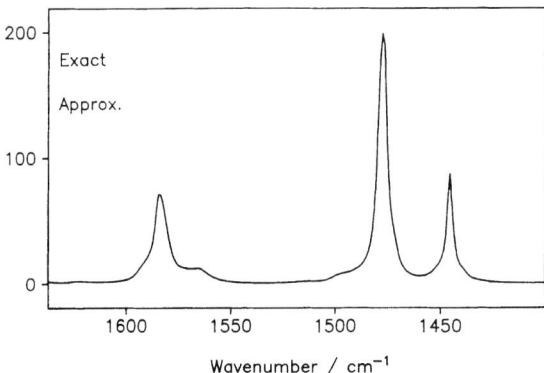

Fig. 3. Molar absorption coefficient spectra of chlorobenzene calculated from experimental absorbance spectra between KBr windows by the exact and approximate methods. The two spectra in the figure are indistinguishable on the scale shown. The units of E_m are L mole^{-1} cm^{-1}

better than that shown in Fig. 3, being about 0.1% for bands with maximum E_m values near 5 L mole^{-1} cm^{-1}.

Clearly, the approximate method is accurate to better than the accuracy of the experimental data for all but the strongest band of chlorobenzene. It should be used with caution until further study of the effect of the refractive index of the liquid has been made, but it seems conservative to suggest that it can be used with good accuracy for bands of organic liquids between alkali metal halide windows that have peak values of E_m less than about 100.

Intensity Quantities for Physico-chemical Analysis

To analyse absorption intensities in terms of physico-chemical properties it is necessary to transform the real and imaginary refractive index spectra into imaginary molar polarizability spectra under the Lorentz local field. Use of the Lorentz local field corrects the spectra for predictable macroscopic dielectric effects, which can be misleading as noted below. It accounts correctly only for isotropic effects, so does not disguise the chemically interesting spectral changes due to short-range anisotropic interactions. The equations (1) and details (2) have been given elsewhere. Briefly, complex dielectric constant spectra are first calculated as the square of the complex refractive index spectra, ($\varepsilon' = n^2 - k^2, \varepsilon'' = 2nk$) and are then used in the Lorentz–Lorenz relation

$$\frac{\hat{\varepsilon} - 1}{\hat{\varepsilon} + 2} = \frac{4\pi}{3 V_m} \hat{\alpha}_m \qquad (3)$$

where V_m is the molar volume, to yield the complex molar polarizability spectra, $\hat{\alpha}_m(\tilde{v}) = \alpha'_m(\tilde{v}) + i\,\alpha''_m(\tilde{v})$.

It is appropriate to summarize here the main reasons for using the imaginary molar polarizability spectrum, $\alpha''_m(\tilde{v})$, instead of the spectrum of a more familiar quantity, in order to obtain information about the physico-chemical properties of the liquids.

(1) According to theory, the α''_m spectrum is the sum of individual bands. According to the most commonly used theory, these bands have the classical damped harmonic oscillator (CDHO) shape. In contrast, the E_m, k, or ε'' spectrum is not just the sum of individual bands, because the bands interact if they are intense.

(2) The peak wavenumber of a band in the $\tilde{v}\alpha''_m$ spectrum is the wavenumber of the mechanical oscillator that causes the band. In contrast, the peak wavenumber of a band in the E_m, k, or ε'' spectrum depends on the intensity of the band as well as the mechanical oscillator wavenumber.

(3) The shape of a band in the α''_m spectrum is the shape of the oscillator function. Under the CDHO theory the α''_m band is symmetric. In contrast, the shape of a band in the E_m, k, or ε'' spectrum is skewed by dielectric effects. For weak bands this skewing is small but for intense bands it can be serious and misleading [2].

Points (2) and (3) are illustrated in Fig. 4, which shows parts of the E_m, ε'', α''_m and k spectra of methanol, scaled for convenient presentation. The top box shows the broad, intense OH stretching band and the lower box shows the sharp and intense CO stretching band.

(4) The integrated intensity C_j of band j is the area under the band in the $\tilde{v}\alpha''_m$ spectrum. Under the double harmonic approximation C_j is directly proportional to the square of the dipole moment derivative with respect to the normal coordinate [1, 2]; i.e. $C_j \propto |\delta\mu/\delta Q|^2$. In contrast, the calculation of $|\delta\mu/\delta Q|^2$ from the E_m, ε'' or k spectrum requires a value, $\underline{n}$ or $\underline{\varepsilon}$, of the real refractive index or dielectric constant that is the hypothetical value that would exist at the wavenumber of the band if the band were absent. It is not usually possible to determine this accurately [2].

(5) The integrated intensity C_j and the integrated intensity $A_j = \int E_m\,d\tilde{v}$, that is frequently used for gases, provide a convenient relation between the intrinsic intensities of corresponding transitions in the liquid

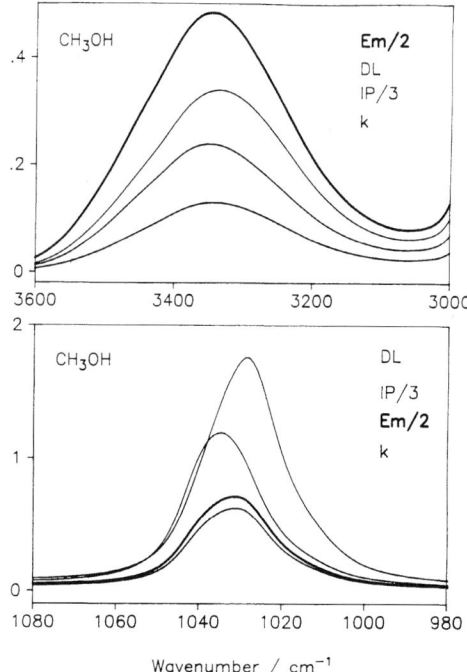

Fig. 4. The OH stretching (top box) and CO stretching (bottom box) bands in the E_m, ε'', α''_m and k spectra of methanol. The symbols in the top right of each box shown the scaling factor that was used for convenience and show the order of the peak maxima from top to bottom. The units of E_m are 10^5 cm^2 mol^{-1} and of α''_m (indicated by IP for imaginary polarizability) cm^3 mol^{-1}. ε'' (indicated by DL for dielectric loss) and k are dimensionless

and gas phases. The equation

$$8\pi^2 \frac{C_{\text{liq}}}{A_{\text{gas}}} = \frac{|\delta\mu/\delta Q|^2_{\text{liq}}}{|\delta\mu/\delta Q|^2_{\text{gas}}}$$

applies to all bands of all compounds. Thus, if the dipole moment derivative is the same in the liquid and gas phases, the ratio $8\pi^2 \dfrac{C_{\text{liq}}}{A_{\text{gas}}}$ equals 1 for all bands of all compounds. This contrasts with the widely used Polo–Wilson equation

$$\frac{A_{\text{liq}}}{A_{\text{gas}}} = \frac{|\delta\mu/\delta Q|^2_{\text{liq}}}{|\delta\mu/\delta Q|^2_{\text{gas}}} \left\{ \frac{n^2+2}{3} \right\}^2 \frac{1}{n}$$

which not only requires the value of $\underline{n}$ that is difficult to determine accurately but also gives a ratio that is different for each band of one compound, as well as different for different compounds, because the value of n changes across the spectrum.

(6) The transition probability of a band is proportional to the area under the α''_m spectrum.

Analysis of CD₃OD Spectrum by Curve Fitting

A very important advantage of the α_m'' spectrum is realized when curve fitting is used to separate the contributions from the different bands to the area under the spectrum. The statistical errors of the fit are far smaller when the α_m'' spectrum is fitted than when the k or E_m spectrum is fitted. This is, of course, of vital importance when choosing between different fits.

In the case of strongly overlapping bands, and even in the case of mildly overlapping bands, we know of no method that is more objective than *cautious* curve fitting of the α_m'' spectrum. We have found that most bands can be fitted very well by CDHO bands, and that bands that involve hydrogen-bonded hydrogen atoms require Gaussian bands, usually more than one, to fit them satisfactorily. One must, of course, be very cautious to avoid the use of more bands than are absolutely necessary. We have found that only occasionally is it necessary to use more bands than are visible in the experimental α_m'' spectrum.

Figure 5 illustrates a difficult example of such curve fitting. It shows the 1200–400 cm⁻¹ region of the α_m'' spectrum of CD₃OD. The region contains strongly overlapping CD₃ rocking and deformation bands near 1080 cm⁻¹ strongly coupled to the intense CO stretching band at 990 cm⁻¹. Below 990 cm⁻¹ are a weak band due to a CD₃ rock, the in-plane C–O–D bend and, near 500 cm⁻¹, the intense out-of-plane motion of the deuterium-bonded deuterium atom. The wavenumber of this latter band does not change with the isotopic composition of the methyl group, which shows that it is an out-of-plane O–D–O motion and is not the torsion about the C–O bond that occurs in the gas phase.

The top curve in Fig. 5 is actually two curves, the experimental spectrum and the sum of the fitted bands. They are difficult to distinguish, which shows that an excellent fit was obtained. The bands used to obtain the fit are shown beneath the top curve. All are CDHO bands except the four broad bands near 800 and 500 cm⁻¹ which fitted the bands due to deuterium atom motion. Of particular note is that very few bands were used that are not visible as peaks, shoulders or asymmetry in the spectrum. It is clear in some cases, that the band does not have exactly the shape of the band used to fit it, so that more than one band must be used to obtain the fit, but it is equally clear that more than one transition may contribute to the band. It is also fairly clear, although with significant uncertainty,

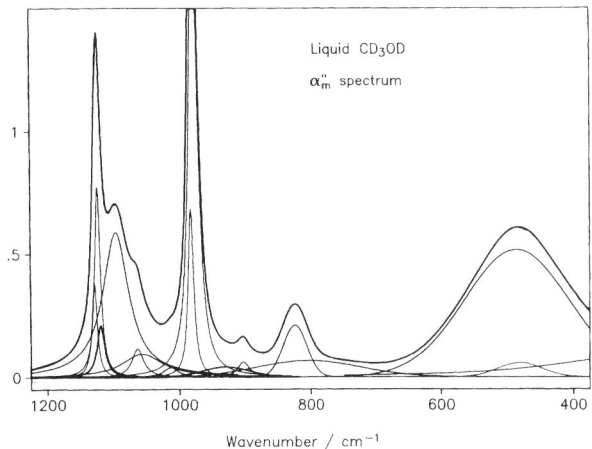

Fig. 5. The top curve is the experimental α_m'' spectrum of CD₃OD, essentially coincident with the sum of the bands fitted to it. The lower curves show the bands required for the fit

which of the fitted bands contribute to the intensity of each observed band.

The integrated intensities were determined [18] for essentially all of the fundamentals of CH₃OH [19], CH₃OD [19], CD₃OH [20] and CD₃OD [20] by such methods, and were analysed [18] by the eigenvectors of a normal coordinate calculation with the aid of the directions of the dipole moment derivative vectors calculated by Torii and Tasumi by *ab initio* methods [21]. The results indicate the current level of our ability to obtain physico-chemical information from absolute intensity spectra, and can be summarized as follows.

The α_m'' spectra are believed accurate to about 3% [19, 20]. The integrated intensities were obtained by curve fitting and also by the more chemical method of comparison of the spectra of different isotopomers [19, 20]. Comparison of the results of the two methods allowed us to estimate the accuracy of the integrated intensities as 2% for the OH and OD stretches, 5% for the CO stretches, 3% for the sum of the three CH₃ and CD₃ stretches, but only about 20% for the intensities of the individual CH and CD stretching vibrations, between 1% and 50% for the C–O–H and C–O–D in-plane deformations, 6% for the out-of-plane O–H–O vibration, 5% for the sum of the CH₃ deformations and 30% for the sum of the CD₃ deformations, and 40% for the weak CH₃ and CD₃ rocking bands. The experimental integrated intensities were analysed under the double harmonic theory. This involved fitting the intensities to $\delta\mu/\delta R$ parameters, the molecular dipole moment derivatives with respect to internal coordinates. These parameters were held invariant for the four isotopomers. The experimental intensities were fitted by these parameters to approximately their estimated accuracy noted above.

Acknowledgement. The author thanks the Natural Sciences and Engineering Research Council of Canada for financial support of this work, and thanks the recent members of his group, S. L. Zhang, C. D. Keefe, Z. Lan, Y. Apelblat and R. Norman Jones, without whom this lecture and paper would not have been possible.

References

[1] J. E. Bertie, S. L. Zhang, H. H. Eysel, S. Baluja, M. K. Ahmed, *Appl. Spectrosc.* **1993**, *47*, 1100.

[2] J. E. Bertie, S. L. Zhang, C. D. Keefe, *J. Mol. Struct.* **1994**, *324*, 157.

[3] I. Mills, T. Cvitaš, K. Homann, N. Kallay, K. Kuchitsu, *Quantities, Units and Symbols in Physical Chemistry, 2nd Ed.*, International Union of Pure and Applied Chemistry, Blackwell, Oxford, 1988, p. 32.

[4] J. E. Bertie, C. D. Keefe, R. N. Jones, *Can. J. Chem.* **1991**, *69*, 1609.

[5] J. P. Hawranek, P. Neelakantan, R. P. Young, R. N. Jones, *Spectrochim. Acta* **1976**, *32A*, 75.

[6] T. G. Goplen, D. G. Cameron, R. N. Jones, *Appl. Spectrosc.* **1980**, *34*, 657.

[7] J. E. Bertie, R. N. Jones, C. D. Keefe, *Appl. Spectrosc.* **1993**, *47*, 891.

[8] J. E. Bertie, S. L. Zhang, C. D. Keefe, *Vib. Spectrosc.* **1995**, *8*, 215.

[9] J. E. Bertie, R. N. Jones, Y. Apelblat, C. D. Keefe, *Appl. Spectrosc.* **1994**, *48*, 127.

[10] J. E. Bertie, R. N. Jones, Y. Apelblat, *Appl. Spectrosc.* **1994**, *48*, 144.

[11] J. E. Bertie, Z. Lan, R. N. Jones, Y. Apelblat, *Appl. Spectrosc.* **1995**, *49*, 840.

[12] J. E. Bertie, C. D. Keefe, R. N. Jones, *Tables of Intensities for the Calibration of Infrared Spectroscopic Measurements in the Liquid Phase*, International Union of Pure and Applied Chemistry, Blackwell, Oxford, 1995.

[13] J. E. Bertie, Y. Apelblat, R. N. Jones, C. D. Keefe, S. L. Zhang, *Appl. Spectrosc.* **1995**, *49*, 1821.

[14] J. E. Bertie, H. H. Eysel, *Appl. Spectrosc.* **1985**, *39*, 392.

[15] J. E. Bertie, H. Harke, M. K. Ahmed, H. H. Eysel, *Croat. Chem. Acta* **1988**, *61*, 391.

[16] J. E. Bertie, S. L. Zhang, R. Manji, *Appl. Spectrosc.* **1992**, *46*, 1660.

[17] S. E. Bertie, Y. Apelblat, *Appl. Spectrosc.* **1996**, *50*, 1039.

[18] S. L. Zhang, *J. Chem. Phys.* submitted.

[19] J. E. Bertie, S. L. Zhang, *Appl. Spectrosc.* **1994**, *48*, 176.

[20] J. E. Bertie, S. L. Zhang, *J. Chem. Phys.* **1994**, *101*, 8364.

[21] H. Torii, M. Tasumi, *J. Chem. Phys.* **1993**, *99*, 8459.

Mikrochim. Acta [Suppl.] 14, 23–31 (1997)

Digging for Light in Semiconductor Mines – Infrared Views of Luminescent Porous Silicon

Wolfgang Theiß

I. Physikalisches Institut, Aachen University of Technology (RWTH), D-52056 Aachen, Federal Republic of Germany

Abstract. Infrared reflection spectroscopy is presented as a tool for the investigation of porous silicon. The simulation approach used for spectrum interpretation is discussed. The porous silicon aging behaviour, a relation between silicon–hydrogen vibrational modes and luminescence as well as the structuring of porous silicon required for devices are discussed.

Key words: infrared spectroscopy, silicon, porous silicon, optics.

Porous silicon, which is very easily produced by an electrochemical etching process, has turned out to be a much more efficient visible light emitter than bulk silicon [1]. Very much effort has been put into research on this new material since an extension of silicon chip technology to silicon optoelectronics has been a dream for a long time.

Infrared spectroscopy is a widely used tool for the investigation of semiconductors. In this paper its application to obtain useful results in porous silicon science is demonstrated. The focus is on two aspects: spectrum analysis by "simulation spectroscopy", which has been put forward significantly in recent years, and a summary of the infrared properties of porous silicon that were obtained by using this method.

After a short general introduction to porous silicon (preparation, microstructure, luminescence properties) experimental considerations concerning infrared spectroscopy are given that lead to the choice of external reflection as the most suitable technique. The applied data analysis by a simulation approach is discussed in some detail and finally results are given covering the aging behaviour, a correlation of infrared and luminescence properties as well as an IR characterization of structured porous silicon layer systems. The use of porous silicon superlattices as easily designable infrared filters is proposed in the final section.

Preparation, Microstructure and Luminescence Properties of Porous Silicon

This section summarizes some general properties of porous silicon which are good to know before doing optical spectroscopy. The focus is on the remarkably simple preparation and the extraordinary microstructure of the material obtained, which is responsible for the enhanced light emission.

Porous silicon layers are prepared by electrochemical etching of silicon wafers in hydrofluoric acid [2]. A typical set-up is shown in Fig. 1. Part of the wafer is covered by the acid, which provides the fluoride ions required for the dissolution of silicon atoms as SiF_6^{2-}-complexes. The wafer must have an electrode on its other side since positive charges (holes) are also necessary for the process and are obtained from this electrode. Of course, in the case of n-type material holes are not available and must be created by visible or UV illumination (usually from the back of the wafer) during the etching process. The HF solution in most cases consists of HF, water and a considerable fraction of ethanol to ensure easy penetration of the fluid into the already etched pores.

Under certain conditions (which mainly depend on the HF concentration, the doping level of the substrate and the applied current density) the contrary of electropolishing occurs: on the "hills" of surface inhomogeneities a slightly smaller hole concentration exists as compared to the "valleys" and hence the silicon dissolution proceeds faster in the valleys. This mechanism

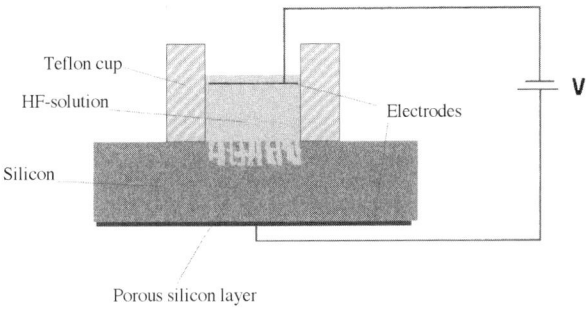

Fig. 1. Typical set-up for the preparation of porous silicon layers on silicon wafers

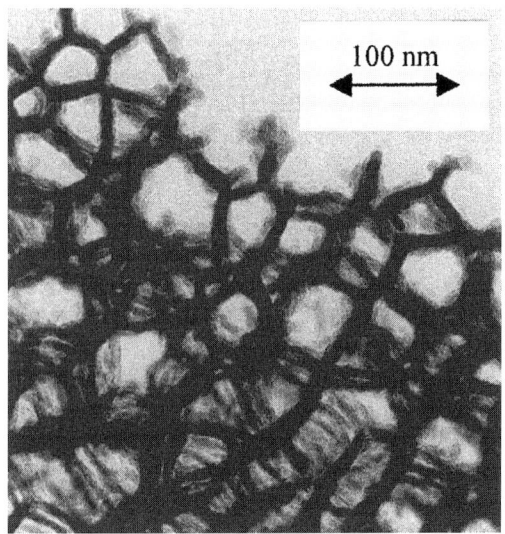

Fig. 2. TEM picture of a mesoporous sample (taken from [3])

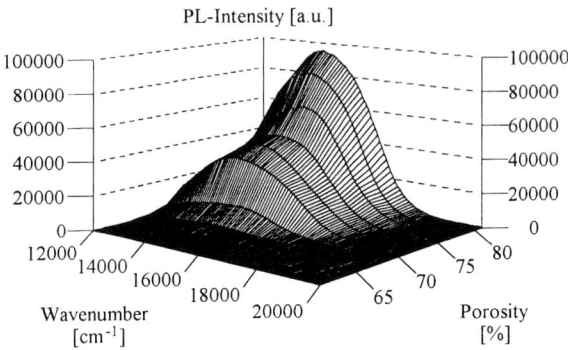

Fig. 3. Photoluminescence spectra (in the visible spectral range) of porous silicon prepared on $0.2\,\Omega\,cm$ p-doped substrates. As can be seen, the intensity strongly depends on porosity

leads to pore growth with hole depletion in the remaining silicon structures.

The average size of the pores (and the remaining silicon pore walls) can vary in a wide range depending on the preparation conditions mentioned above – a classification into macroporous (pores in the micrometre range), mesoporous (pores in the 10–100 nm range) and micro- or nanoporous silicon (feature sizes below 10 nm) is widely used. Figure 2 shows a mesoporous sample (solid phase dark), which is a network of connected silicon "wires" with diameters in the 10 nm range. Nanoporous silicon can be considered as a loosely connected system of microcrystals with diameters less than 5 nm. This type of porous silicon is characterized by a huge specific internal surface in the range 200–1000 m^2/g.

Significant light emission occurs only in nanoporous silicon. The broad luminescence spectra usually peak in the red visible spectral range (see Fig. 3), and time-resolved measurements indicate a very slow decay in the 1–100 µs range. Although there is no general agreement on the mechanism of the light emission the following rough explanation is satisfying in many respects. The lack of light emission in bulk silicon is due to the indirect band gap in silicon. For radiative recombination of electrons in the conduction band minimum with holes in the valence band maximum, additional particles (usually phonons) are needed to provide the required momentum for the transition. In consequence this process is very unlikely to occur and radiative recombination rates are very small. Electron-hole pairs waiting for radiative recombination move long distances and scan the crystal for non-radiative recombination centres such as lattice or surface defects or doping atoms. These are usually reached before the light emission process, which is thus efficiently quenched.

In small crystallites the band structure of silicon changes and – as is the case for many other materials – the band gap widens with decreasing size. Nanoporous silicon can be considered as a collection of crystallites with band gaps in the 1.5–2 eV range. The very narrow interconnections have even larger gaps. This leads to an efficient localization of electron–hole pairs in the largest crystallites (with the smallest band gap) which cannot escape by crossing the large band gap interconnections. If these largest crystallites are small enough to be free of any volume and surface defects, non-radiative transitions cannot occur any more and light emission in the visible spectral range is favoured. This model of light emission in porous silicon will be supported by infrared results below.

Doing Infrared Spectroscopy on Porous Silicon

Now I turn to the application of IR spectroscopy to porous silicon investigation. After choice of an appropriate experimental technique, the data analysis by a simulation approach is discussed in some detail since it is not a widespread method but is well suited for many spectroscopic problems.

Experimental Techniques

The experimental spectroscopic techniques to be used for the inspection of typical porous silicon layers must be chosen with respect to the special design of such samples.

Porous silicon is often produced on standard silicon wafers with only one side polished (the side on which the porous silicon is etched). Because the silicon wafer serves as an electrode in the etching process a contact on the back of the wafer must be provided, which is often realized by depositing a metal layer. The porous layer has a thickness of typically a few microns and has interfaces which are sharp enough to lead to pronounced interference structures in infrared spectra. At higher frequencies in the visible, interferences can still be observed, with amplitudes decaying with increasing frequency since porous silicon becomes opaque as the frequencies approach the electronic interband transitions in the ultraviolet. Owing to the very small penetration depth of electromagnetic radiation in the UV, only reflection techniques can be used in this spectral range.

Also, transmission measurements cannot be performed directly in the infrared region without removing the metal layer from the back of the wafer. In addition, a quantitative analysis of the transmission spectra is not straightforward, because of uncontrolled scattering of radiation at the rough back surface. The direct conversion of a transmission spectrum into absorbance, which certainly is the greatest advantage of transmission spectroscopy, is not possible anyway in the case of porous silicon, because the conditions necessary for this operation are not fulfilled: the reflectivity of the sample and reference (which is the bare silicon wafer in most cases) is not the same and the interference structures in the spectra indicate that radiation contributing to the transmission is passing through the thin porous layer more than just once.

Hence, external reflection should be the method of choice for routine spectroscopy on porous silicon layers. Of course, a suitable data analysis must be provided which takes into account the intermixture of interference patterns and absorption bands. This can be achieved – as will be shown below – by performing realistic spectrum simulations and adjusting model parameters to fit the measured data.

It should be mentioned that ATR (attenuated total reflection) spectroscopy with variable angle of incidence has been used successfully as a depth profiling technique for porous silicon layers [4]. Due to the low thermal conductivity of highly porous materials photoacoustic spectroscopy has also been used as a characterizing method – in combination with quantitative analysis (once more based on spectrum simulation, see [5]) the thermal conductivity of porous silicon can be obtained by optical means [6]. Beyond routine spectroscopy in situ studies where the porous layer was etched directly into a silicon ATR prism (multiple reflections) have been reported [7].

Data Evaluation by Simulation

As mentioned above, the analysis of the optical spectra is done in this work by a comparison of simulated and measured data. The model calculations are based on three steps. First the dielectric function of the solid phase of the porous layer must be defined (which may exhibit differences from that of bulk silicon). Secondly, the macroscopic response of the porous system to electric fields (as applied by the incident infrared light beam) has to be determined, which must be done by using appropriate effective medium theories. Finally, wave propagation through the sample must be followed and all partial waves contributing to the reflected or transmitted radiation must be summed.

All calculations have been done by using the following dielectric function model (given in wavenumbers $\tilde{v}$) for the solid component of porous silicon:

$$\varepsilon(\tilde{v}) = \varepsilon_\infty - \frac{\tilde{v}_P^2}{\tilde{v}^2 + i\tilde{v}\tilde{v}_\tau}$$

$$+ \sum_i \frac{1}{\sqrt{2\pi}\sigma_i} \int_{-\infty}^{\infty} \exp\left(-\frac{(x - \Omega_{0,i})^2}{2\sigma_i^2}\right) \frac{\Omega_{P,i}^2}{x^2 - \tilde{v}^2 - i\tilde{v}\Omega_{\tau,i}} dx$$

$$(1)$$

The first term ε_∞ is just a real and constant contribution representing the dielectric function in the high frequency limit where all other contributions vanish. The second is the Drude model for free carriers. The third term in Eq. (1) is a sum of extended harmonic oscillator contributions representing vibrational modes in the infrared, but is also used for electronic transitions in the visible and UV. Brendel and Bormann suggested [8] replacing the classical harmonic oscillator model (characterized by the resonance frequency Ω_0, the oscillator strength Ω_P^2 and the damping constant Ω_τ) by an integration of harmonic oscillator terms with Gauss-distributed resonance frequencies. This accounts for small local variations leading to slightly varying resonance frequencies and has been developed for disordered systems such as amorphous substances. The fourth parameter σ (the width of the Gauss distribution) adds to this oscillator model the freedom to switch from a Lorentzian to a Gaussian line shape continuously.

After the solid phase dielectric function ε has been set up, the averaging to the macroscopic observable effective dielectric function ε_{eff} has to be done. It should be mentioned at this stage that the assignment of one dielectric function to the solid phase is a crude approximation, since the band structure varies locally, as pointed out above. On the other hand, satisfying effective medium concepts (as given below) are available at present only for two-phase composites, which prevents a more detailed description of the microstructure.

A general expression for ε_{eff} is known as the Bergman representation [9, 10]:

$$\varepsilon_{\text{eff}} = \varepsilon_M \left(1 - (1-p) \int_0^1 \frac{g(n,p)}{\dfrac{\varepsilon_M}{\varepsilon_M - \varepsilon} - n} \, dn \right). \quad (2)$$

Here ε_M is the dielectric function of the host material (the empty pores, i.e. $\varepsilon_M = 1$ for porous silicon) p is the porosity and the function $g(n,p)$ – the so-called spectral density – takes into account all topological details. Different topologies should be described by different spectral densities, but unfortunately not very much is known about the relation between $g(n,p)$ and its corresponding topology. Some experience in the choice of spectral densities, obtained by studying the optical properties of many inhomogeneous systems, has been summarized in [10] and [11]. For the case of the silicon network that makes up porous silicon the following parameterization for $g(n,p)$

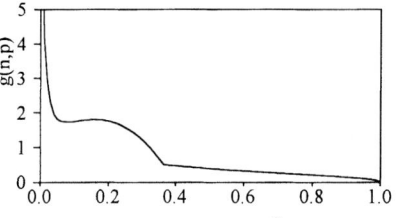

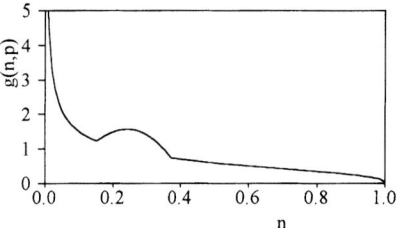

Fig. 4. Examples of spectral densities generated by parameterization given in Eq. (3). The top graph corresponds to $p = 0.61$, $g_0 = 0.12$ and $d = 0.16$, the bottom one was obtained by setting $p = 0.79$, $g_0 = 0.04$ and $d = 0.11$

has been used:

$$g(n,p) = g_0 \delta(n) + \frac{3\sqrt{3g_0}}{2\pi n(1-p)^2}$$
$$\cdot \left[p^2 \left| \frac{1-n}{n} \right|^{1/3} + p(1-p) \left| \frac{1-n}{n} \right|^{2/3} \right]$$
$$\cdot + \left(1 - \frac{g_0}{(1-p)^2} \right) \left[-\frac{3}{4d^3} \left(n - \frac{p}{3} \right)^2 + \frac{3}{4d} \right]$$
$$\cdot \theta \left(\frac{p}{3} - d + n \right) \theta \left(\frac{p}{3} + d - n \right). \quad (3)$$

The three parameters in Eq. (3) used to describe the microtopology of porous silicon are the porosity p, the so-called percolation strength g_0 and the "broadening parameter" d. Spectral densities generated according to Eq. (3) (examples are given in Fig. 4) are much more realistic than oversimplified effective medium treatments such as those due to Maxwell Garnett or Bruggeman, as discussed in [10].

Once the decision on the effective medium treatment has been made and ε_{eff} is calculated, all quantities regarding wave propagation in a given arrangement such as reflection and transmission coefficients or absorption losses, can be calculated. In this work the algorithm described in [12] and some refinements given in [13] are used.

Finally, to adjust the simulated spectra to measured ones a least-squares fit is employed, leading to optimized values of the dielectric function parameters and layer thicknesses. It should be noted that from the dielectric functions obtained other quantities such as complex refractive indices or absorption coefficients can be calculated easily.

Discussion of a Typical Reflectance Spectrum

In this section, typical features of external reflectance spectra are discussed in order to demonstrate the use of the simulation approach. The top graph of Fig. 5 shows the reflectivity of a single porous layer of 70% porosity on a silicon wafer (resistivity $0.2\,\Omega\,cm$) and the result of the first try to simulate the spectrum by using just a dielectric constant for the "silicon" that remains after the etching. The main interference patterns are described well in the model which gives a layer thickness of $4.7\,\mu m$. Nevertheless, the sharp "dips" at around 630, 910 and $2100\,cm^{-1}$ that are not present in the simulation demand a modification of the model in terms of the addition of several vibrational modes. An improved version of the parameter fit shown in the bottom part of Fig. 5 now includes most details. The

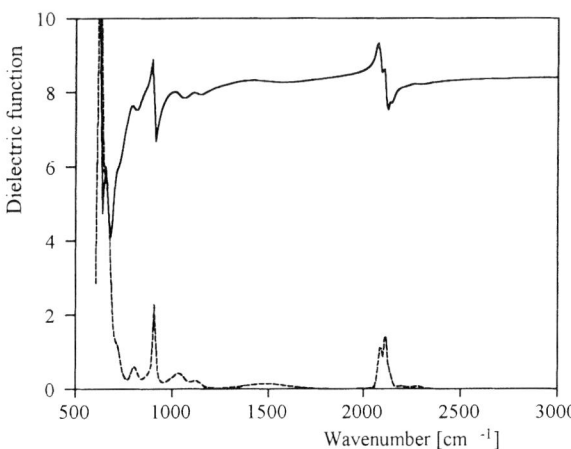

Fig. 6. Dielectric function of the pore wall material as obtained from the "fit" shown in Fig. 5 (bottom curve). The real part is displayed as solid line, the imaginary part is shown dashed

dielectric function that has been used in Fig. 6. The strongest modes are Si_3–Si–H and Si_2–Si–H_2 wagging modes (625 and $660\,cm^{-1}$), the Si_2–Si–H_2 scissor mode ($905\,cm^{-1}$) and various –Si–H_x stretching modes (2080, 2115 and $2140\,cm^{-1}$). Band assignments can be found in [14 and 15]. After HF etching the huge internal surface of porous silicon is covered by hydrogen, which saturates the dangling silicon bonds. Also some Si–O–Si vibrational modes are usually needed to describe porous silicon. The oxygen uptake depends on the age and the treatment of the sample. Aging under ambient conditions will be discussed below.

Results

Results are given in the following for porous silicon layers prepared on boron-doped substrates of various dopant concentrations, thicknesses and porosities. All have been obtained from reflectance spectra and parameter fits as discussed above.

Aging of Porous Silicon

To learn what happens to porous silicon stored under ambient conditions the time-dependence of the optical properties of some samples has been investigated [13–16]. In time intervals ranging from a few minutes (shortly after exposure to ambient air) to many weeks (after more than a year) the reflectivity has been re-

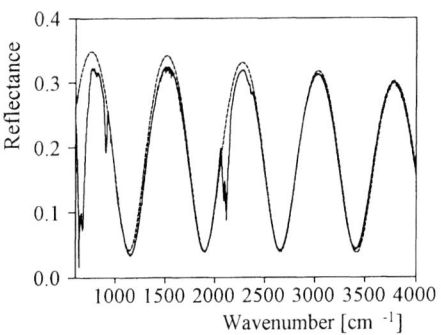

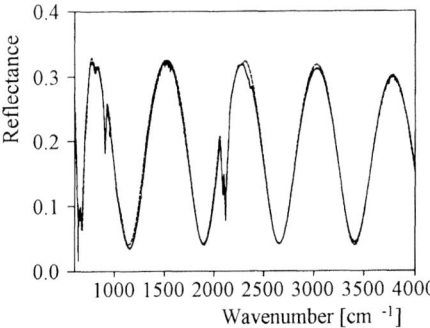

Fig. 5. Comparison of measured (solid lines) and simulated (dashed) reflectance spectra of a typical porous silicon layer (70% porosity) on a p-doped substrate. Top: constant and real dielectric functions used for the pore walls. Bottom: improved fit by adding vibrational modes

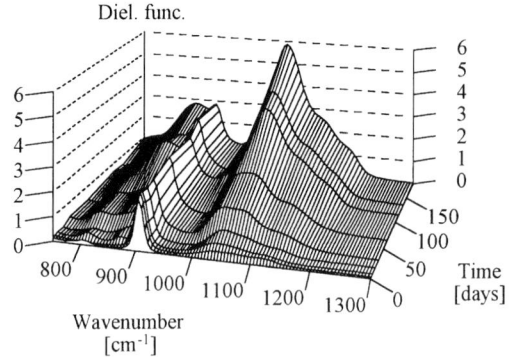

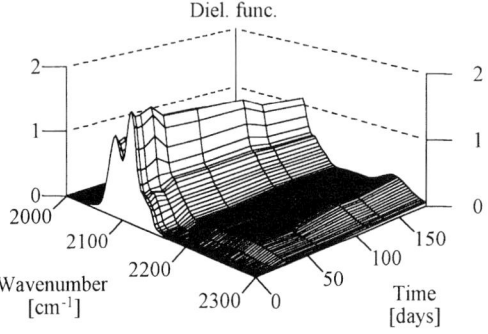

Fig. 7. The dielectric function of the pore wall material *vs.* time for a 75% porosity layer on a *p*-doped substrate. The top graph shows, besides the quite stable Si–H scissor mode, growing silicon–oxygen modes, whereas the bottom graph reflects the changes of the Si–H stretching modes due to the attack on the silicon back-bonds by oxygen atoms

corded, always for exactly the same sample position. The time-dependent dielectric functions obtained this way are summarized in Fig. 7 for a 75% porosity sample. Increasing Si–O–Si modes (1000–1200 cm^{-1}) can be used to monitor the oxygen concentration, which after 6 months reaches about 2.8×10^{22} cm^{-3} which is already more than half the value of SiO$_2$. This oxidation, of course, must be taken into account in device considerations. Figure 7 (bottom) also shows the Si–H stretching region where a decrease of the strong Si–H bands at 2080, 2115 and 2140 cm^{-1} and an increase of the modes at around 2200 and 2250 cm^{-1} is observed. The latter are up-shifted versions of the former, with the frequency shift caused by the presence of oxygen in the silicon back-bonds (i.e. O–Si–H configurations). The ratio of the strengths of these two types of stretching modes can be used to monitor the presence of oxygen in the back-bond configuration. More details on aging effects can be found in [13].

Correlation of Infrared Properties to Photoluminescence

As can be seen from Fig. 7, freshly prepared samples do show only Si–H vibrations, which is consistent with the common assumption that after HF treatment the silicon surface is completely covered by hydrogen. This ensures a good passivation of silicon bonds, which is necessary for light emission. Unsaturated dangling silicon bonds are known as efficient non-radiative recombination centres. One dangling bond on a nanometer-sized crystallite can quench the luminescence completely.

As mentioned above, luminescent porous silicon can be considered as a collection of more or less separated particles with dimensions in the nanometer range. With the (very) crude approximation that there are isolated spheres the strength of the Si–H stretching modes (being proportional to the concentration of hydrogen in the solid phase) can be used to measure the size of the average sphere since every hydrogen atom corresponds to one silicon surface atom. Hence the Si–H mode strength is proportional to the ratio of the crystallite surface area to its volume which is inversely proportional to the radius r of the sphere.

On the other hand, for the photoluminescence the intensity relation $\sim W_R/(W_R + W_{NR})$ holds, where W_R is the radiative and W_{NR} the non-radiative recombination rate, respectively. If $W_R \ll W_{NR}$ (which holds for porous silicon) this simplifies to intensity $\sim W_R/W_{NR}$. If we now assume that non-radiative recombinations

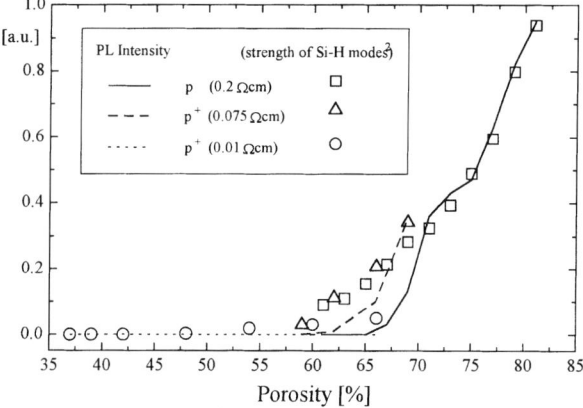

Fig. 8. Photoluminescence intensity *vs.* the square of the hydrogen concentration for samples of various porosities and substrate doping levels (See text for discussion)

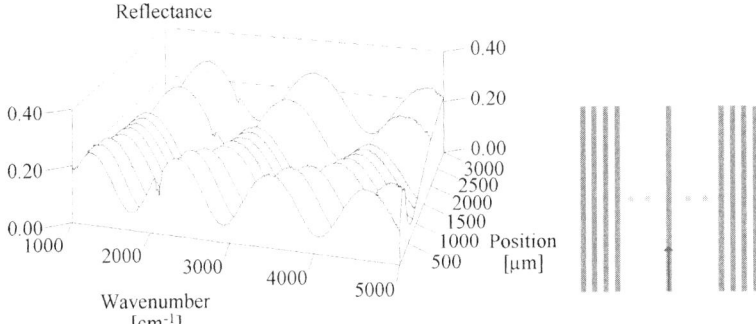

Fig. 9. IR microscopy scan along the central strip as shown in the right-hand sketch. The shifting interference patterns indicate an inhomogeneous thickness of the porous layer

are mainly due to surface defects (e.g. dangling silicon bonds) $W_{NR} \sim r^2$ should be a good approximation, reflecting that the chance of having a defect-free silicon crystallite increases with decreasing size. With W_R constant (i.e. independent of the average sphere radius) we get luminescence intensity $\sim 1/r^2 \sim$ (Si–H mode strength)^2 which establishes a correlation of photoluminescence intensity and IR spectra.

The photoluminescence peak intensity and the quantity $\sum \Omega_P^2$ (where the summation is over all Si–H stretching modes), as obtained for many samples of different doping level and porosities, are plotted in Fig. 8. The expected relation is indeed found, although there are some discrepancies for medium porosities (60–70%). In this range neglecting the interconnections between the crystallites is not a good approach and the photoluminescence is probably smaller than expected, since surface defects on neighbouring crystallites can be reached for non-radiative recombination.

Towards Devices: Structured Porous Silicon

One advantage of porous silicon is that it is still silicon: no material is better known and many techniques developed for chip technology can be applied to porous silicon, too. Photolithography methods, for example, have been used to prepare laterally structured porous silicon layers such as small pixels or strips which would be needed for "luminescent silicon displays" or optical waveguides on silicon chips. Photoresist on the wafer prevents silicon dissolution, which only takes place where the photoresist has been removed.

IR microspectroscopy was used to investigate the properties of small porous silicon structures prepared by standard photolithography techniques. Various lithography masks have been applied – here results are given for 40 strips of 10 mm length and 50 µm width.

The spacing between the strips was also 50 µm, making a total width of 4 mm. Figure 9 shows reflectivity measurements along the central strip (as indicated in the figure) which show (by the shifting interference patterns) layer thickness inhomogeneities typical for structured porous silicon.

Analysing the microscope spectra in detail we obtain thickness profiles as shown in Fig. 10. Apparently, at the edges of the structured area larger thicknesses occur (which are accompanied by slightly higher porosities). This can be explained as due to enhanced etching currents at the boundaries, where the holes needed for silicon dissolution are collected from

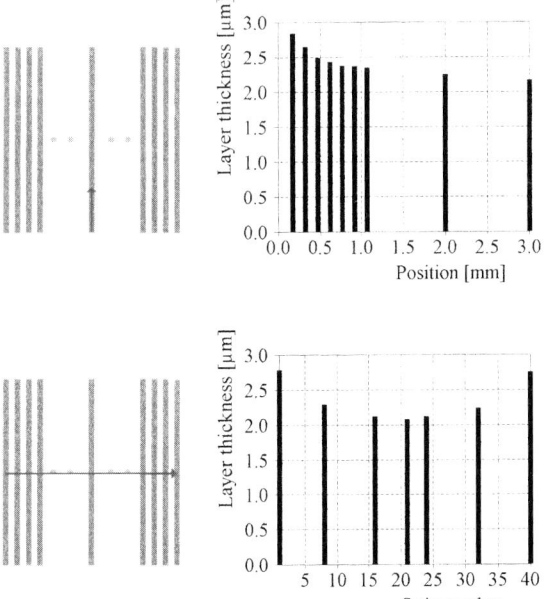

Fig. 10. Thickness profiles along the indicated directions, as obtained from "simulation analysis" of IR microscope reflectivity spectra like the ones shown in Fig. 9

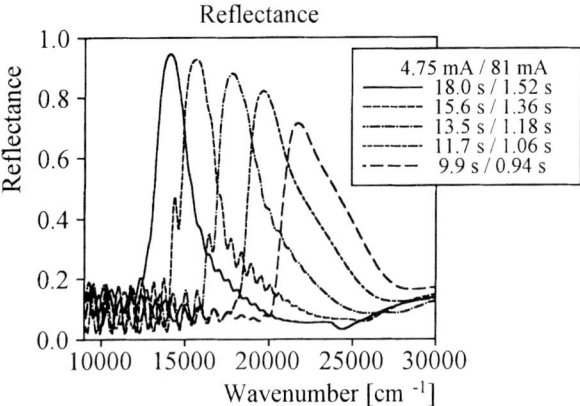

Fig. 11. Reflectivity of various superlattices in the visible spectral range. After the determination of the porosity-dependence of the refractive index of porous layers on *p*-doped material (0.2 Ω cm) it was possible to predict the reflectivity of a system of 20 double layers of high and low porosity before production. Here a set of reflectance filters covering the visible spectral range has been etched by switching the anodization current between 4.75 and 81 mA for the indicated time intervals

during etching. Since the porosity depends almost linearly on the current density and the etch front moves very homogeneously "into the substrate" an arbitrary (effective) refractive index profile can be produced by using a programmable current source [17]. Examples of "porosity superlattices" realized in this way are shown in Figs. 11 and 12. The latter demonstrates the possible use of such multilayer systems as filters in the infrared spectral region. Until now no mechanically stable transmission filters have been produced, although porous silicon layer systems can be removed quite easily from the substrate by a "current shock" (switching to high current densities leads to electro-polishing instead of pore growth and lifts off the porous layers).

A combination of horizontal (photolithography) and vertical structuring to obtain small porosity super-lattice structures is currently under investigation. This would be needed for display applications or silicon colour-sensitive detector arrays.

a much larger area than in the centre. From the thickness profiles a typical scale of 1 mm for such edge effects can be deduced.

Structuring perpendicular to the wafer surface (vertical structuring) can be achieved in a fascinatingly simple way, namely by varying the current density

Acknowledgements. I would like to thank my co-workers M. Arntzen, S. Hilbrich and M. Wernke for the investigation of a huge amount of samples, that were prepared by R. Arens-Fischer, M. G. Berger, M. Krüger and M. Thönissen at the Institut für Schicht- und Ionentechnik (Forschungszentrum Jülich GmbH). The latter also provided photoluminescence spectra. Part of this work was financed by the Bundesministerium für Forschung und Technologie (BMFT) under contract No. 01BM401/3.

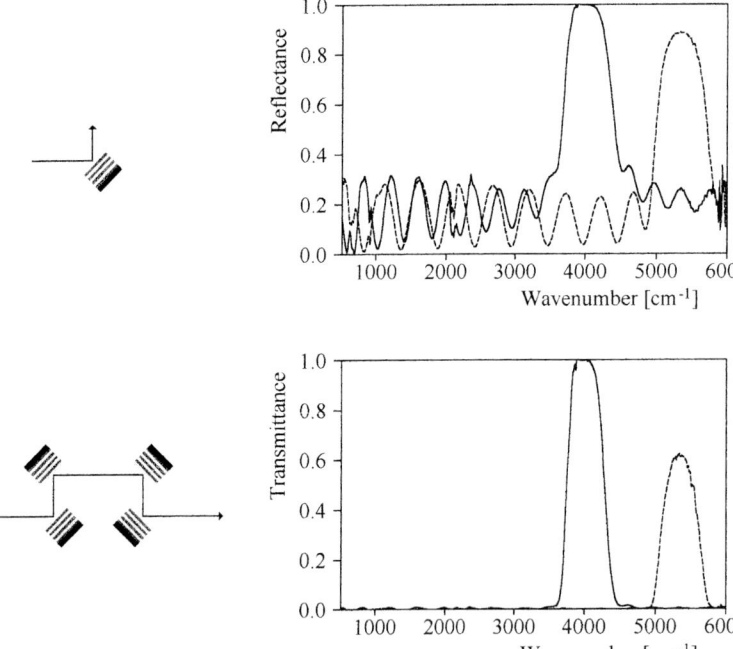

Fig. 12. The top curve shows the IR reflectivity of 10 porous double layers (solid line: 0.4 μm 65% porosity, 0.4 μm 75% porosity; dashed line: 0.4 μm 65% porosity, 0.2 μm 75% porosity) on a *p*-doped 0.2 Ω cm wafer. A series of 4 reflections leads to a transmission band-pass behaviour as shown in the bottom spectra. Note the large reflectivity at 4000 cm^{-1}, even larger than that of the gold reference mirror

References

[1] L. T. Canham, *Appl. Phys. Lett.* **1990**, 57, 1046.

[2] J.-N. Chazalviel, in: *Porous Silicon Science and Technology* (J. C. Vial, J. Derrien, eds.), Springer, Berlin Heidelberg New York Tokyo, Les Editions de Physique, 1995, p. 17.

[3] A. G. Cullis, *Unpublished*, similar pictures can be found in A. G. Cullis, *Studies of Porous Silicon by Electron Microscopy* in: *Optical Properties of Low Dimensional Silicon Structures* (D. C. Bensahel, L. T. Canham, S. Ossicini, eds.), Kluwer, Dordrecht, 1993, p. 147.

[4] W. Theiß, M. Wernke, V. Offermann, *Thin Solid Films* **1995**, 255, 181.

[5] P. Grosse, R. Wynands, *Appl. Phys. B* **1989**, 48, 59.

[6] W. Theiß, in: *Porous Silicon Science and Technology* (J. C. Vial, J. Derrien, eds.), Springer, Berlin Heidelberg New York Tokyo, Les Editions de Physique, 1995, p. 189.

[7] V. M. Dubin, F. Ozanam, J.-N. Chazalviel, *Vib. Spectrosc.* **1995**, 8, 159.

[8] R. Brendel, D. Bormann, *J. Appl. Phys.* **1992**, 71, 1.

[9] D. J. Bergman, *Phys. Rep. Phys. Lett. Sect. C* **1978**, 43, 377.

[10] W. Theiß, in: *Festkörperprobleme/Advances in Solid State Physics, Vol. 33* (R. Helbig, ed.), Vieweg, Braunschweig, 1994, p. 149.

[11] W. Theiß, S. Henkel, M. Arntzen, *Thin Solid Films* **1995**, 255, 177.

[12] B. Harbecke, *Appl. Phys. B* **1986**, 39, 165.

[13] W. Theiß, *Surf. Sci. Rep.* in press.

[14] K. H. Beckmann, *Surf. Sci.* **1965**, 3, 314.

[15] P. Gupta, V. L. Colvin, S. M. George, *Phys. Rev. B* **1988**, 37, 8234.

[16] W. Theiß, M. Arntzen, S. Hilbrich, M. Wernke, R. Arens-Fischer, M. G. Berger, *Phys. States Solidi B* **1995**, 190, 15.

[17] M. G. Berger, C. Dieker, M. Thönissen, L. Vescan, H. Lüth, H. Münder, W. Theiß, M. Wernke, P. Grosse., *J. Phys. D: Appl. Phys.* **1994**, 27, 1333.

Mikrochim. Acta [Suppl.] 14, 33–42 (1997)

Infrared Reflection Absorption Spectroscopic Study on Photochemical Processes in Langmuir–Blodgett Films

Koichi Itoh*, **Masato Yamamoto**, **Naoki Furuyama**, and **Atushi Saito**

Department of Chemistry, School of Science and Engineering, Waseda University, Shinjuku-ku, Tokyo 169, Japan

Abstract. Infrared reflection absorption (IRA) spectroscopy was applied to study photodimerization processes induced by UV-light irradiation of Langmuir–Blodgett (LB) films of a stilbazole derivative ($C_{18}S$) and a diaryl-1,3-butadiene ($C_{18}B$) embedded in fully deuterated arachidic acid (DA) on a silver substrate. Structures of dimerization products were proposed by analysing the IRA spectra. The dimerization processes of both $C_{18}S$ and $C_{18}B$ were found to consist of two steps of second order reactions, the first step proceeding much faster than the second. These kinetic features correspond to the time-course of the orientation change of the alkyl side chains of $C_{18}S$ and $C_{18}B$. The result indicates that a relaxation process of the repulsive interaction between the side chains and the matrix molecules, which is accumulated in the first step, proceeds in the second step. The IRA spectroscopy was also applied to study structures and photochemical processes induced by UV-irradiation of the LB films of a cinnamic acid derivative ($C_{22}CA$) on a silver substrate. The LB films give three sets of $\nu_{C=O}$ and $\nu_{C=C}$ bands at 1730 and 1641, 1678 and 1624, and 1714 and 1637 cm^{-1}. The first set of bands was assigned to the $C_{22}CA$ molecules in a non-hydrogen-bonded state, and the second and third sets were assigned to the molecules forming hydrogen-bonded dimers with *trans* (i.e. the C=O and C=C groups takes on a *trans* conformation around the C–C bond connecting the two groups) and *cis* conformations, respectively. The non-hydrogen-bonded state exists only in the first monolayer, interacting directly with the substrate surface. The *trans* hydrogen-bonded dimer is formed within each LB monolayers, while the *cis* dimer is formed between neighboring LB monolayers. Upon UV-irradiation, the $C_{22}CA$ molecules in the non-hydrogen-bonded state do not show any photochemical process, while the molecules forming the *trans* hydrogen-bonded dimer undergo photodimerization, and the molecules forming the *cis* dimer are converted into a non-hydrogen-bonded state.

Key words: photochemical processes, infrared reflection absorption, Langmuir–Blogett films.

The Langmuir–Blodgett (LB) technique is an efficient method for fabrication of monomolecular layers with purposely designed molecular arrangements, and this has been used for modelling various phenomena in physics, chemistry and biology which depend on precise structures in molecular levels. Fourier-transform infrared reflection absorption (IRA) spectroscopy is a method to obtain grazing incident angle p-polarized reflection spectra of molecular monolayers and multilayers adsorbed on metal surfaces. The high sensitivity of this technique allows us to observe high-quality vibrational spectra of even submonolayer quantities of adsorbates.

In the present paper, we have applied IRA spectroscopy to investigate photochemical processes induced by UV-light irradiation of the LB films containing the N-(1-octadecyl)-4-stilbazolium cation ($C_{18}S$) [1,2], 1-phenyl-4-[4-(N-octadecylpyridiniumyl]-1,3-butadiene ($C_{18}B$) [3], and *trans-p*-docosyloxycinnamic acid ($C_{22}CA$) (see Fig. 1). Photochemical processes in well-organized molecular systems such as crystalline states were first studied by Schmidt [4] who found that, upon UV irradiation, cinnamic acid and its derivatives in the crystalline state undergo photodimerization, as illustrated in Fig. 1C, and established a relationship between the configurations of the dimerization products and the molecular arrangements of the starting materials in the crystal. Quina and Whitten [5] studied a similar photodimerization process of $C_{18}S$ incorporated into the LB films of arachidic acid by using absorption and emission spectroscopy; they indicated that the process leads to the formation of structures such as those depicted in Fig. 1A. Quina and Whitten [5] also gave a brief discussion of the photodimerization process of $C_{18}B$ and proposed possible dimerization products, as shown in Fig. 1B. In the present work we prepared the LB films of fully deuterated arachidic acid, $CD_3(CD_2)_{18}COOH$, which is abbreviated to DA, containing $C_{18}S$ and $C_{18}B$ with incorporation ratios ($C_{18}S$ or $C_{18}B$:DA) of (1:4) and (1:9)

---

* To whom correspondence should be addressed

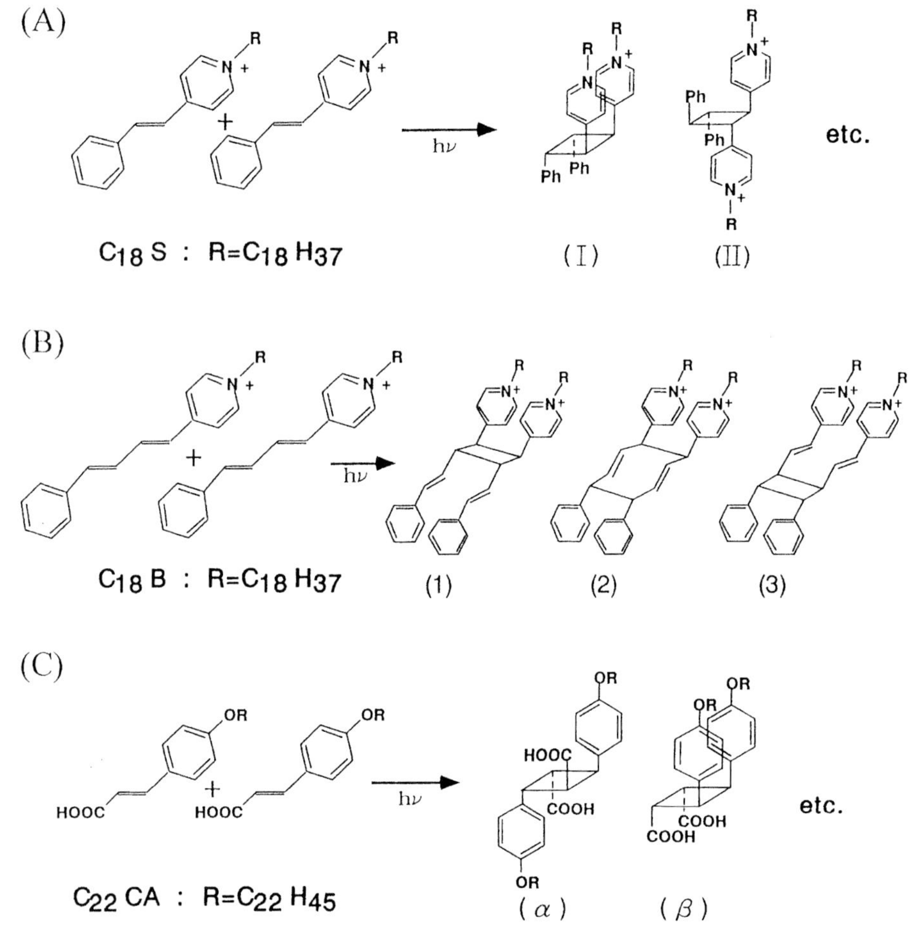

(A)

C$_{18}$S : R=C$_{18}$H$_{37}$ (I)   (II)   etc.

(B)

C$_{18}$B : R=C$_{18}$H$_{37}$ (1)   (2)   (3)

(C)

C$_{22}$CA : R=C$_{22}$H$_{45}$ (α)   (β)   etc.

Fig. 1. A–C Structures of C$_{18}$S, C$_{18}$B and C$_{22}$CA and their possible photodimerization products

and applied IRA spectroscopy to investigate how C$_{18}$S and C$_{18}$B change their orientation and structures during photodimerization in the matrix molecules (DA). A series of LB films of C$_{22}$CA on a silver substrate was also prepared by changing both the number of mono-layers (1–9) and the surface pressure (5–30 mN/m). The structures of the LB films were studied by means of IRA spectroscopy, and the correlation between the structures and photochemical processes induced by UV-irradiation was investigated.

Experimental

Materials

C$_{18}$S, C$_{18}$B and C$_{22}$CA were prepared by the procedures already reported [5, 6]. Fully deuterated arachidic acid (DA) was purchased from Cambridge Isotope Laboratories and used without further purification.

Fabrication of LB Films

LB film fabrication was performed by a usual method, already reported [1–3]. The pH of the trough water was kept at 4.5 for the

preparation of LB films of the matrix molecule (DA) and C$_{22}$CA in the acid form. CdCl$_2$ was added to the trough water (pH 7.0) to the level of 2×10^{-4} M for preparation of LB films of DA in the salt form. The temperature of the trough water was kept at 20 °C for the fabrication of the mixed LB films and at 13 °C for the fabrication of the LB films of C$_{22}$CA. A silver film with a thickness of about 100 nm, which was deposited on a slide glass by an evaporation method, was used as a substrate.

Photochemical Reactions

Photodimerization of the LB film samples containing C$_{18}$S was induced by irradiation at 340 nm and that of C$_{18}$B by irradiation at 370 nm. The 340- and 370 nm excitation light was obtained from a 150 W xenon lamp by using Hoya U340 and U370 filters, as appropriate. The photochemical process of the LB films of C$_{22}$CA was induced by irradiation with a 100 W high-pressure Hg lamp (Oreal, Model 66033).

IRA Measurements

IRA spectra were obtained by using JEOR JIR-5500 and Bio-Rad FTS-45A Fourier-transform infrared spectrometers equipped with a liquid-nitrogen cooled MCT detector with resolution of 4 cm^{-1}.

Results and Discussion

Photodimerization of $C_{18}S$ and $C_{18}B$ Embedded in the LB Films of Fully Deuterated Arachidic Acid

IRA Spectral Changes during Photodimerization of $C_{18}S$ and $C_{18}B$. Figure 2 summarizes the IRA spectral changes observed for $C_{18}S$ and $C_{18}B$, respectively, embedded in the LB films of fully deuterated arachidic acid, [DA,($C_{18}S$:DA = 1:4 and $C_{18}B$:DA = 1:4)] in the acid form. Figure 2A indicates that the 1626-cm^{-1} band due to an olefinic C=C stretching vibration (v(C=C)) and the 969-cm^{-1} band associated with a CH out-of-plane bending vibration show a precipitous intensity decrease with the onset of irradiation. In addi-

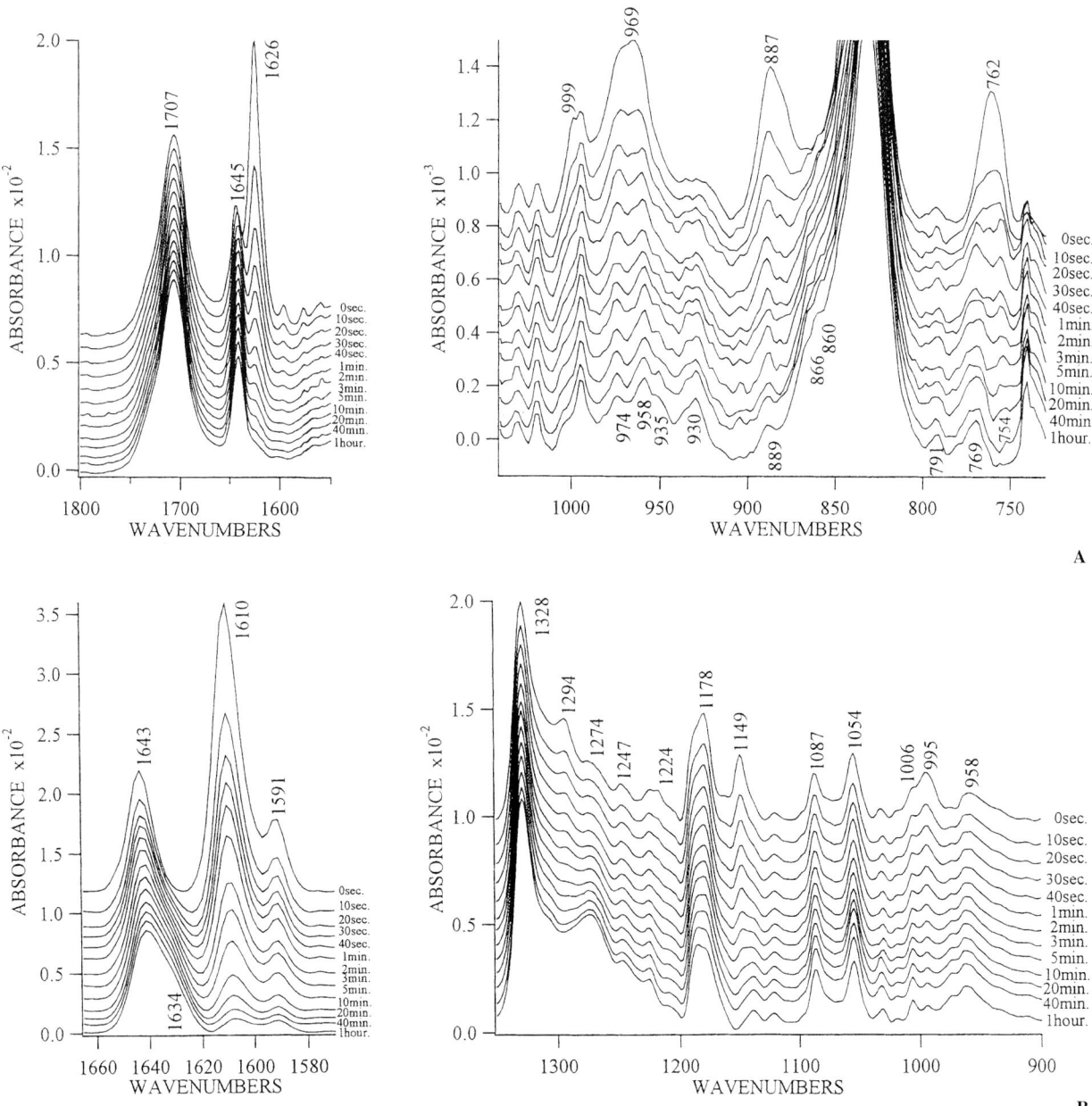

Fig. 2. IRA spectral changes induced by UV-irradiation observed for $C_{18}S$ **A** and $C_{18}B$ **B** embedded in the LB films of DA ($C_{18}A$ or $C_{18}B$:DA = 1:4) in the acid form. [The numbers of the mixed monolayers of the samples are 15 **A** and 31 **B**, and the irradiation wavelengths are 340 nm **A** 370 nm **B**]. Each number at the right-hand side of each spectrum indicates the irradiation time

tion, there appear IR bands at 974, 930, 866, 860 and 754 cm^{-1} at the beginning of irradiation (0–ca. 30 s), and after about 30 s of irradiation another set of IR bands appears at 958 and 769 cm^{-1}. IR bands corresponding to these bands are also observed for the photodimerization products of the *N*-methylstyl-bazolium cation, (not shown in this paper), indicating that photodimerization takes place in the LB film. Based on the correlation between the configurations of the photodimerization products of *N*-methylstyl-bazolium cation and their IR spectra, we concluded that the first set of bands can be ascribed to the configuration similar to that of type I in Fig. 1A and the second set to the configuration similar to that of type II. Thus, the IRA spectra proved that formation of a cyclobutane ring with type I configuration takes place at the beginning of the photodimerization and that formation of the ring with type II starts after a certain period of irradiation (about 30 s).

The 1610-cm^{-1} band in Fig. 2B is due to an out-of-phase mode of the coupled stretching modes of the neighbouring C=C groups of $C_{18}B$ [7]. On irradiation at 370 nm the intensity of the 1610-cm^{-1} band decreases precipitously. In addition, the 995-cm^{-1} band due to a CH out-of-plane bending vibration [8] also decreases in intensity. These results corroborate that a photochemical reaction takes place, resulting in the disappearance of the butadiene group. The 1643-cm^{-1} band in Fig. 2B is due to a ring stretching vibration of the pyridiniumyl group of $C_{18}B$. As the photochemical reaction proceeds, there appear new bands at 1634 cm^{-1} (a shoulder of the 1643-cm^{-1} band) and at 1274 cm^{-1}; the former band is ascribed to a v(C=C) band and the latter to one of the characteristic vibrations of a cyclobutane ring [9]. These results suggests that the photochemical reaction product takes one of the proposed structures 1 and 3 of Fig. 1B.

Kinetic features of the photodimerization of $C_{18}S$ and $C_{18}B$. In addition to the LB films discussed above, the IRA study was extended to the LB films with the matrix molecule (DA) in the salt form and containing $C_{18}S$ and $C_{18}B$ with an incorporation ratio of 1:9 in the matrix molecule in both the acid and salt forms. The rate and extent of photodimerization were monitored by means of the v(C=C) band intensity at 1626 cm^{-1} for $C_{18}S$ and at 1610 cm^{-1} for $C_{18}B$. Figure 3 plots the reciprocal of the relative intensity of the v(C=C) band against irradiation time for the photodimerization of

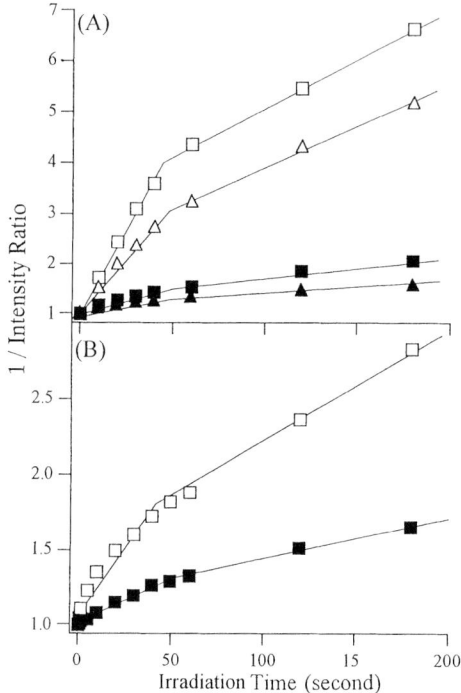

Fig. 3. The plots of the reciprocal of the relative intensity of the v(C=C) band [1626 cm^{-1} **A** and 1610 cm^{-1} **B**] against irradiation time for the LB films containing $C_{18}S$ (**A**) and C_{18} (**B**). **A** (□) The 15-monolayer LB film of the 1:4 mixture in the matrix with the acid form. (△) The 15-monolayer LB film of the 1:9 mixture in the matrix with the acid form. (■) The 15-monolayer LB film of the 1:4 mixture in the matrix with the salt form. (▲) The 15-monolayer LB film of the 1:9 mixture in the matrix with the salt form. **B** (□) The 31-monolayer LB film of the 1:4 mixture in the matrix with the acid form. (■) The 31-monolayer LB film of the 1:4 mixture in the matrix with the salt form

$C_{18}S$ and of $C_{18}B$. The plots for each sample conform to two sets of straight lines, proving that the photodimerization in all the LB films follows two steps of second-order reactions, the initial step (from 0 to ca. 40–70 s) proceeding faster than the second (from ca. 40–70 s to 200 s).

Figure 4A summarizes the relative intensity change of the CH$_3$ asymmetric stretching [v_{as}(CH$_3$), 2956 cm^{-1}], CH$_3$ symmetric stretching [v_s(CH$_3$), 2873 cm^{-1}] and CH$_2$ symmetric stretching [v_s(CH$_2$), 2850 cm^{-1}] vibrations [10] of the alkyl group of $C_{18}S$ during the photodimerization in the 15-monolayer LB film of the (1:4) mixture with the matrix in the salt form: Fig. 4B shows the corresponding intensity changes observed for $C_{18}B$ during the photodimerization in the 31-monolayer LB film of the (1:4) mixture with the matrix in the salt form. The relative intensity of each band is normalized to the intensity measured at $t = 0$ s.

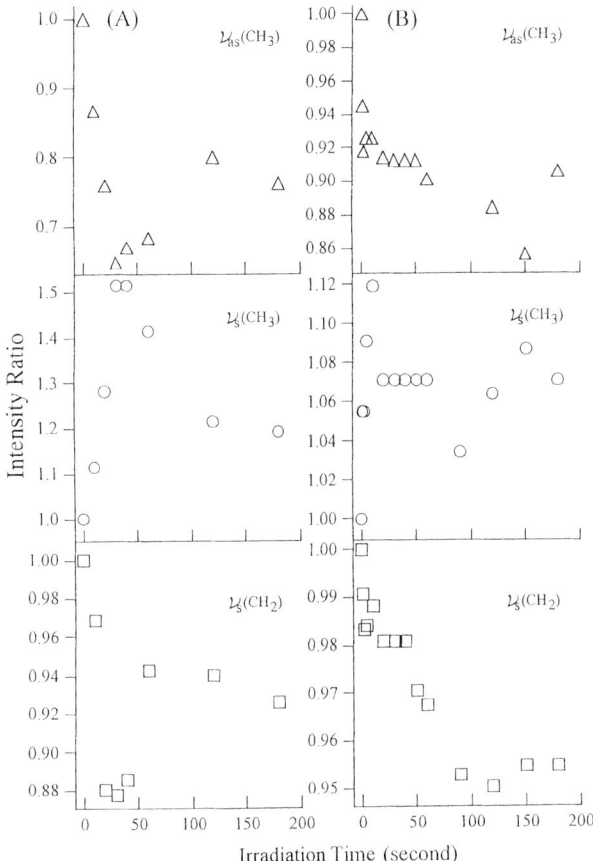

Fig. 4. Relative intensity changes in the $v_{as}(CH_3)$ (2956 cm^{-1}), $v_s(CH_3)$ (2873 cm^{-1}) and $v_s(CH_2)$ (2850 cm^{-1}) bands of **A** the 15-monolayer LB film of the $C_{18}S$ and DA mixture ($C_{18}S:DA = 1:4$) in the salt form of the matrix and **B** the 31-monolayer LB film of the $C_{18}B$ and DA mixture ($C_{18}B:DA = 1:4$) in the salt form of the matrix

The intensity changes of the CH_3 and CH_2 vibrations observed for the photodimerization of $C_{18}S$ and $C_{18}B$ in other LB film samples show trends similar to those of Fig. 4.

From Fig. 4A it is clear that the relative intensities of all the CH_3 and CH_2 stretching vibrations of $C_{18}S$ change in two steps; e.g., the intensity of $v_{as}(CH_3)$ shows a rapid decrease in an initial irradiation period (0–ca. 40 s) and then a slight increase in the following period (40–180 s); in contrast, the $v_s(CH_3)$ band exhibits an opposite trend of the intensity change. The direction of the transition moment of the $v_{as}(CH_3)$ vibration is perpendicular to the symmetry axis of the methyl group and that of the $v_s(CH_3)$ vibration is parallel to the axis. According to the surface selection rule of IRA spectroscopy [11], the above-mentioned intensity changes indicate that, in the initial step of photodimer-

ization, the symmetry axis of the methyl group changes its orientation from a tilted state to a less tilted one, and in the second step, the methyl group changes its orientation in the direction of the initial orientation. The intensity change of the $v_s(CH_2)$ vibration, which has its transmission moment perpendicular to the axis of the alkyl chain, shows a trend similar to that of the $v_{as}(CH_3)$ vibration, indicating that the alkyl group also orients its axis from a tilted state to a less tilted one during the first step and, then back towards the direction of the original orientation.

As Fig. 4B shows, the relative intensity change of the $v_s(CH_2)$ vibration of $C_{18}B$ is similar to that observed for the corresponding band of $C_{18}S$, i.e. in the first step (0–ca. 100 s) it shows a rapid decrease, indicating a rapid average orientation change of the alkyl groups from a tilted state to a less titled one; in the second step (100–300 s) the intensity shows a slight intensity increase, suggesting that the average orientation of the alkyl group changes towards the original direction. The intensity changes of the $v_{as}(CH_3)$ and $v_s(CH_3)$ bands in Fig. 5B give scattered features, indicating that, in contrast to the case of $C_{18}S$, the axis of the methyl group of $C_{18}B$ changes its orientation in a random fashion during the photodimerization process.

The two-step orientation changes of the alkyl group observed, especially for $C_{18}S$, correspond to the kinetic features observed for the photodimerization processes in Fig. 3. Most of the $C_{18}S$ molecules assume a stacking state, probably in a *syn* head-to-head arrangement, which is favourable for dimerization in the LB film matrices prior to irradiation. With the onset of irradiation, these molecules readily undergo dimerization to form the type I configuration, resulting in both the fast step of the reaction and the rapid orientation changes of the alkyl groups. These processes should accumulate repulsive interaction between the alkyl chains and the matrix molecules. The second slower orientation change of the alkyl groups can be interpreted as a relaxation process of the repulsive interaction. During the relaxation, some of $C_{18}S$ molecules, which are in an unfavourable stacking state for dimerization, are rearranged to dimerize, resulting in the formation of the type II configuration.

Structures and Photochemical Processes of the LB Films of $C_{22}CA$

Photodimerization of $C_{22}CA$ in the crystalline state. Upon UV irradiation by a 100 W high-pressure

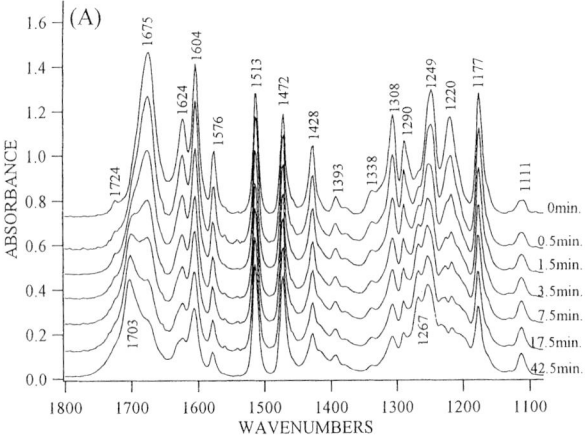

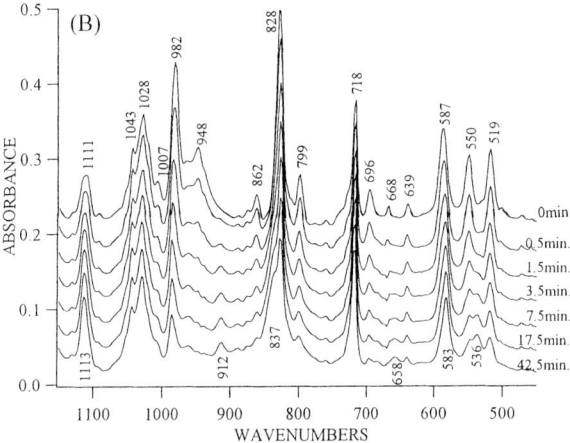

Fig. 5. Infrared spectral changes in the 1800–1100-cm^{-1} **A** and 1100–500-cm^{-1} **B** regions during photodimerization of C_{22}CA in the crystalline state. The number at the right-hand end of each spectrum indicates the irradiation time

Hg lamp, C_{22}CA in the KBr disk shows infrared (IR) spectral changes, as summarized in Fig. 5. The 1675-cm^{-1} band due to a C=O stretching vibration $[v(C=O)]$ of the carboxyl group shifts to 1703 cm^{-1}, and the bands at 1624 and 982 cm^{-1}, which are assigned to $v(C=C)$ and CH out-of-plane bending vibrations of the olefinic group, respectively, reduce their intensity, and there appear new bands at 1267, 912 and 536 cm^{-1}. According to Schmidt [4] cinnamic acid in the α crystalline state is converted into α-truxillic acid upon UV-irradiation. IR spectral changes associated with the conversion (not shown in this paper) are similar to those in Fig. 5. From these results we can conclude that, upon UV-irradiation, C_{22}CA in the crystalline state undergoes dimerization, resulting in the formation of a cyclobutane ring with a configuration similar to that of (α) in Fig. 1C. The similarity

between the frequencies of the IR bands associated with $v(C=O)$ and $v(C=C)$ observed for the α-cinnamic acid, which forms a hydrogen-bonded dimer [4, 13], and the corresponding frequencies observed for C_{22}CA in the crystalline state, indicates that the 1675- and 1624- cm^{-1} bands observed before irradiation (Fig. 5) can be assigned to the $v(C=O)$ and $v(C=C)$ vibrations of the carboxyl groups forming a hydrogen-bonded dimer. A resonance interaction between the C=O and C=C groups of C_{22}CA is removed by photo-dimerization; this is one of the reasons for the higher frequency shift of the 1675-cm^{-1} band to 1703 cm^{-1}.

Structure and photochemical process of one-monolayer LB films of C_{22}CA. Figure 6A is the surface pressure *vs.* area plot observed for C_{22}CA at 20 °C and pH 4.5: Fig. 6B shows the IRA spectra of one-monolayer LB films of C_{22}CA on a silver substrate prepared at surface pressures in the 10–30 mN/m region. The LB film prepared at 10 mN/m gives rise to bands at 1730 and 1641 cm^{-1}, which are ascribable to $v(C=O)$ and $v(C=C)$, respectively. Upon increasing the surface pressure there appear bands at 1684 and 1624 cm^{-1} in addition to the former set of bands; the intensities of the 1684 and 1624-cm^{-1} bands increase with the surface pressure. The 1684- and 1624-cm^{-1} bands correspond to the 1678- and 1624-cm^{-1} bands observed for C_{22}CA in the crystalline state (Fig. 5), which are assigned to the $v(C=O)$ and $v(C=C)$ vibrations, respectively, of the carboxyl group forming the hydrogen-bonded dimer. Presumably, the 1730- and 1641-cm^{-1} bands are due to a carboxyl group in a free state or non-hydrogen-bonded state. When the one-monolayer LB films are prepared at the lower surface pressure, the C_{22}CA molecules are loosely packed with each other, resulting in the formation of the non-hydrogen-bonded state; on the other hand, use of a higher surface pressure causes a densely packed state of the C_{22}CA molecules, which leads to the formation of the hydrogen-bonded state similar to that in the crystalline state.

Figure 7 illustrates the spectral changes induced by UV-irradiation (with a 100 W high-pressure Hg lamp) of the one-monolayer LB films of C_{22}CA. The irradiation does not cause appreciable spectral change of the LB film prepared at 10 mN/m (Fig. 7A) except for slight frequency lowering of the $v(C=O)$ band from 1730 to 1728 cm^{-1}. On the other hand, the LB films prepared at 30 and 20 mN/m show appreciable change, i.e. with the onset of irradiation the 1684- and 1624-

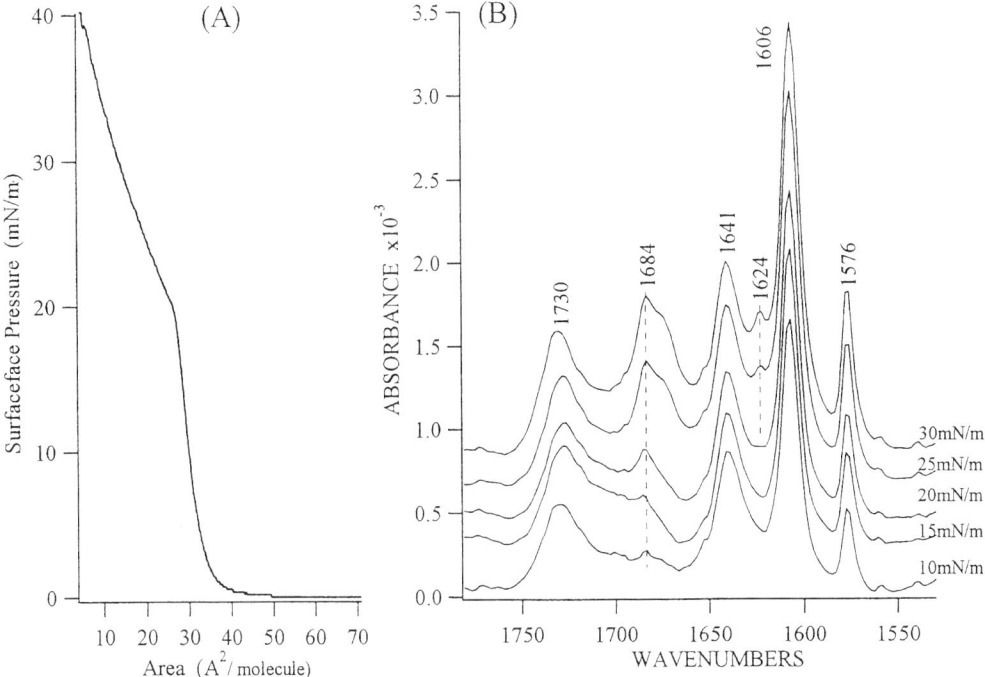

Fig. 6. A Surface pressure–area curve of $C_{22}CA$ at 20°C and pH 4.5 (see text). **B** IRA spectra of one-monolayer LB films of $C_{22}CA$ prepared at various surface pressures

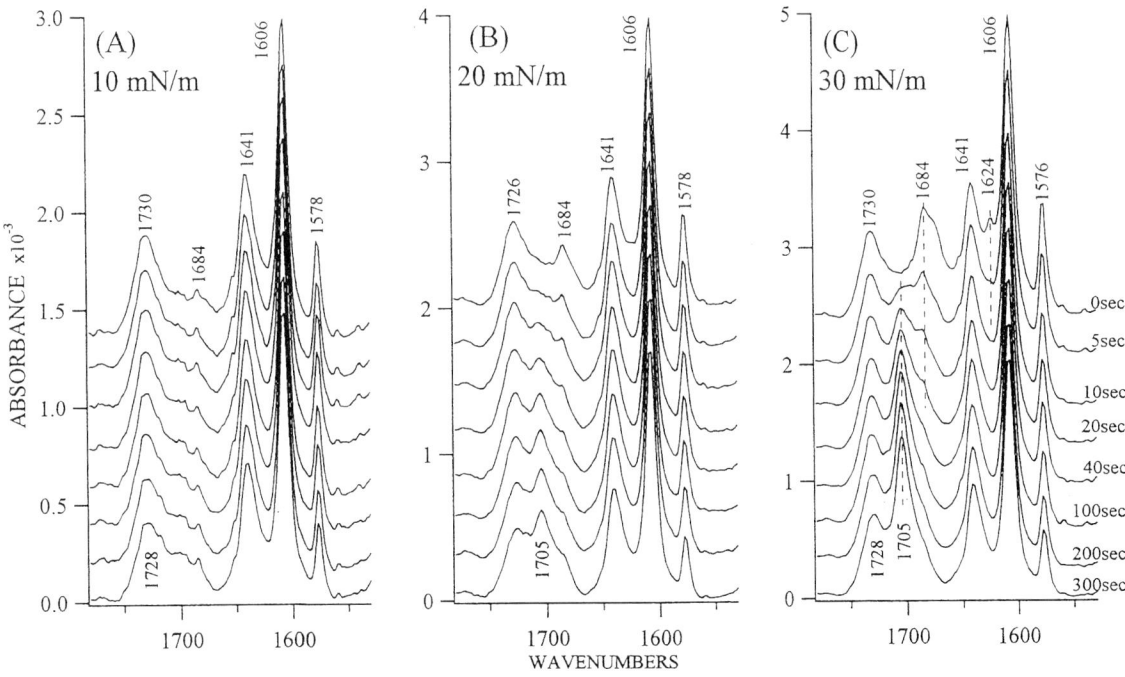

Fig. 7. IRA spectral changes during UV-irradiation of one-monolayer LB films of $C_{22}CA$ prepared at surface pressures of 10 mN/m **A**, 20 mN/m **B** and 30 mN/m **C**. The number at the right-hand end of each spectrum indicates the irradiation time

cm^{-1} bands disappear and there appears a new band at 1705 cm^{-1} (Fig. 7, B and C). These spectral changes are similar to those observed for the photodimerization process of $C_{22}CA$ in the crystalline state (Fig. 5).

The 1730- and 1641-cm^{-1} bands in Fig. 7, B and C, do not show appreciable change upon irradiation, as in the case of the one-monolayer LB film prepared at 10 mN/m. From these results we can conclude that, upon irradiation, the $C_{22}CA$ molecules forming the hydrogen-bonded dimer in the one-monolayer LB films undergo a photodimerization, while those in the non-hydrogen-bonded (or free) state do not show any photochemical process.

Structures and photochemical processes of multilayer LB films of $C_{22}CA$. Figure 8 exhibits the IRA spectra of the LB films of $C_{22}CA$ with the number of monolayers (L), 1, 3, 5, 7 and 9, which were prepared at 25 mN/m. The films with $L \geqslant 3$ give $\nu(C=O)$ and $\nu(C=C)$ bands at 1714 and 1637 cm^{-1}, respectively, in addition to the 1732-, 1678- and 1641-cm^{-1} bands observed for the monolayer LB films. The intensity of the 1714-, 1678- and 1637-cm^{-1} bands increase with L, while that of the 1732-cm^{-1} band remains virtually constant. The 1732- and 1641-cm^{-1} bands are assigned to a non-hydrogen-bonded state; the existence of these bands only in the one-monolayer film suggests that an interaction of the $C_{22}CA$ molecules with the substrate cause the formation of the non-hydrogen-bonded state. The 1678-cm^{-1} band corresponds to the 1684-cm^{-1} band shown in Fig. 6B and assigned to the $C_{22}CA$ molecules forming the hydrogen-bonded dimer within each monolayer in the LB films. [The reason for the discrepancy of the $\nu(C=O)$ frequencies (1678 and 1684 cm^{-1}) observed for the one-monolayer LB film in Figs. 6B and 8 has not been clarified yet; the spectral feature in the $\nu(C=O)$ region depends critically on preparation conditions such as temperature, pH, surface pressure and dipping speed etc.) The $\nu(C=C)$ band at 1624 cm^{-1}, which is associated with the 1678- (or 1684-) cm^{-1} band, is buried between the 1641 and 1606 cm^{-1} bands in the spectrum of the one-monolayer LB film in Fig. 8.) The 1714- and 1637-cm^{-1} bands in Fig. 8 indicate that the LB films with $L \geqslant 3$ contain a structure in addition to those found in the monolayer LB films. The frequency of the $\nu(C=O)$ band at 1714 cm^{-1}, which is appreciably lower than that of the non-hydrogen-bonded state (1732 cm^{-1}), suggests that the third structure in the LB films with $L \geqslant 3$ also contains a hydrogen-bonded state between the carboxyl groups. There are at least two possible structures for hydrogen-bonded dimers, i.e. a dimer with the $C=O$ and $C=C$ bonds taking a *cis* conformation (*cis* dimer) and that with the $C=O$ and $C=C$ bonds taking a *trans* conformation (*trans* dimer): Fig. 9 shows these two structures for the case of acrylic acid. The X-ray analysis could not determine whether cinnamic acid in the α-crystalline state assumes the *trans* or the *cis* conformation, since the carboxyl moiety is disordered by interchange of carbonyl and hydroxy groups [13]. According to Umemura and Hayashi [14], acrylic acid forming the *cis* dimer (Fig. 9A) gives $\nu(C=O)$ and $\nu(C=C)$ infrared bands at 1725 and 1636 cm^{-1}, respectively, and acrylic acid forming the *trans* dimer gives the corresponding bands at 1705 and 1617 cm^{-1}. Based on these results, we tentatively conclude that the IRA bands at 1678 and 1624 cm^{-1} observed for the LB films are ascribable to the $C_{22}CA$ molecules forming the *trans* dimer and the IRA bands at 1714 and 1637 cm^{-1} to the $C_{22}CA$ molecules forming the *cis* dimer.

Figure 10A shows the IRA spectral changes of the LB film (L = 3) of $C_{22}CA$ induced by irradiation with a 100 W high-pressure Hg lamp. Upon irradiation the

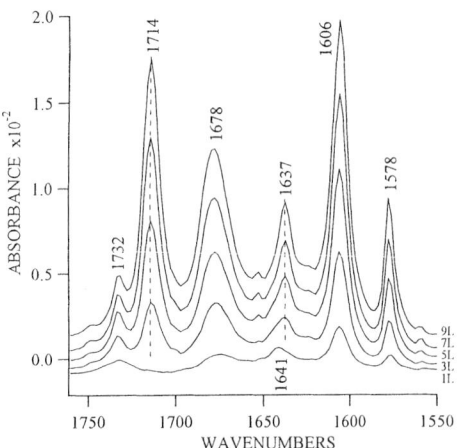

Fig. 8. IRA spectra of the LB films of $C_{22}CA$ with 1, 3, 5, 7 and 9 monolayers (prepared at 25 mN/m)

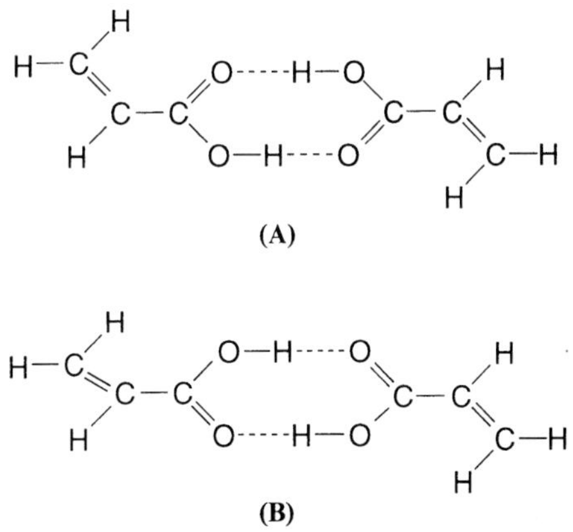

Fig. 9. Structures of hydrogen-bonded dimers of acrylic acid with the *cis* **A** and *trans* **B** conformations

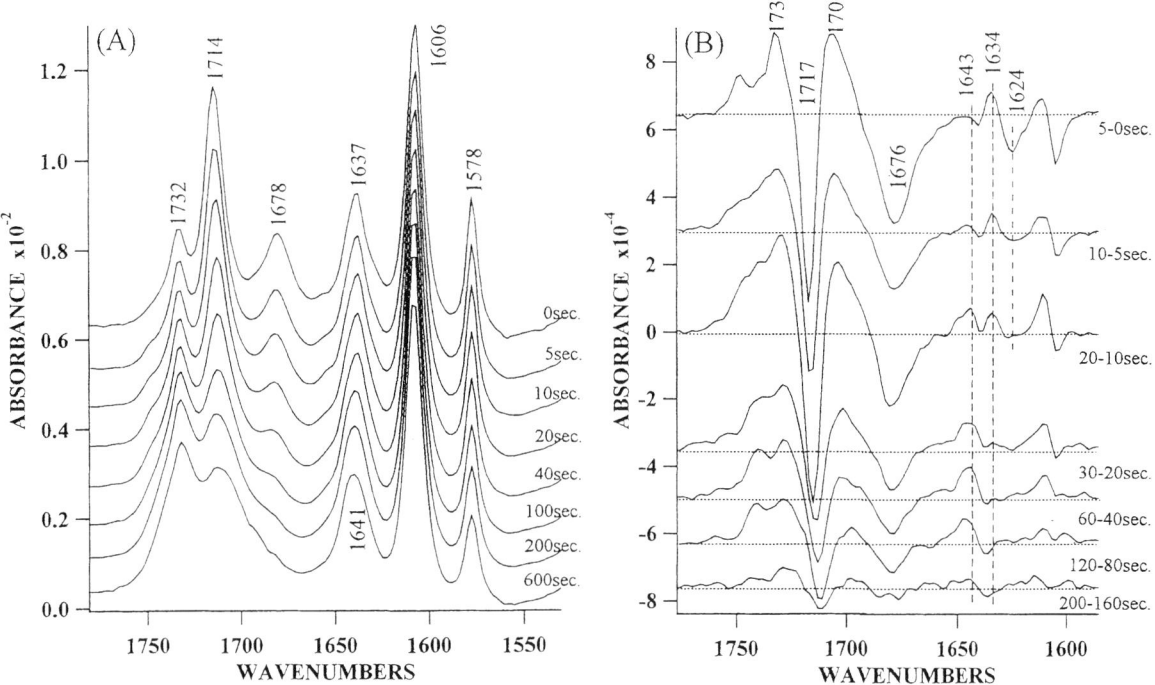

Fig. 10. A IRA spectral changes during UV-irradiation of a three-monolayer LB film of $C_{22}CA$ (prepared at 25 mN/m). The number at the right-hand end of each spectrum indicates the irradiation time. **B** Difference spectra calculated from the IRA spectra in **A** (see text)

1678-cm^{-1} band reduces its intensity, and there appears a shoulder near 1705 cm^{-1}. These spectral changes are similar to those observed in Fig. 7; thus, as in the case of the one-monolayer film, the $C_{22}CA$ molecules forming the *trans* dimer undergo photodimerization. The 1732-cm^{-1} band does not show any change, indicating that the non-hydrogen-bonded state remains unchanged. From Fig. 10A it is also clear that the irradiation causes intensity reduction of the 1714-cm^{-1} band. This result cannot be interpreted as due to photodimerization, because the $\nu(C=C)$ band at 1637 cm^{-1} does not show intensity decrease comparable to that of the 1714-cm^{-1} band. More detailed insight into the spectral changes can be obtained from the series of difference spectra in Fig. 10B, which are calculated by subtracting a spectrum measured at an irradiation time, t_1, from that measured at t_2 ($t_1 < t_2$). Positive and negative peaks in the figure indicate intensity increase and decrease, respectively, caused by the irradiation. The negative peaks at 1676 and 1624 cm^{-1} and the positive one at 1706 cm^{-1} observed in a initial step of irradiation (0–120 s) correspond to the photodimerization of the $C_{22}CA$ molecules forming the *trans* dimer. A negative peak at 1717 cm^{-1} seems to correspond to a positive peak in the 1730–1740-cm^{-1} region; this result suggests that the $C_{22}CA$ molecules

forming the *cis* dimer are converted into a non-hydrogen-bonded state. A positive peak at around 1643 cm^{-1}, which is ascribable to the non-hydrogen-bonded state, is observed for $t_2 \geqslant 20$ s, corroborating the conversion.

Thus, the LB films of $C_{22}CA$ with L $\geqslant 3$ consists of three structures, which show different photochemical behavior. That is, the non-hydrogen-bonded state, which exists mainly in the first monolayer on the silver substrate, does not show any photochemical reaction; the hydrogen-bonded dimer with the *trans* conformation, which is formed within each monolayer, undergoes photodimerization; the hydrogen-bonded dimer with the *cis* conformation, which is formed between neighbouring monolayers, is converted into a non-hydrogen-bonded state. We are now investigating the effect of the wavelength of the irradiation light on the photochemical processes, in order to get more detailed insight into these processes.

References

[1] M. Yamamoto, T. Wajima, A. Kameyama, K. Itoh, *J. Phys. Chem.* **1992**, *96*, 10365.
[2] M. Yamamoto, K. Itoh, A. Nishigaki, S. Ohshima, *J. Phys. Chem.* **1995**, *99*, 3655.

[3] A. Saito, T. Wajima, M. Yamamoto, K. Itoh, *Langmuir* **1995**, *11*, 1277.

[4] G. M. J. Schmidt, *J. Chem. Soc.* **1964**, 2014.

[5] F. H. Quina, D. G. Whitten, *J. Am. Chem. Soc.* **1977**, *99*, 877.

[6] G. W. Gray, B. Jones, *J. Chem. Soc.* **1954**, 1467.

[7] Y. Furukawa, H. Takeuchi, I. Harada, M. Tasumi, *Bull. Chem. Soc. Jpn.* **1983**, *56*, 392.

[8] R. J. Hemley, B. R. Brooks, M. Karplus, *J. Chem. Phys.* **1986**, *85*, 6550.

[9] A. Annamalai, T. A. Keiderling, *J. Mol. Spectrosc.* **1985**, *109*, 46.

[10] R. G. Snyder, *J. Chem. Phys.* **1967**, *47*, 1316.

[11] R. G. Greenler, *J. Chem. Phys.* **1966**, *44*, 310.

[12] R. R. Chance, A. Prock, R. Silbey, in: *Advances in Chemical Physics* (I. Prigogine, S. A. Rice, eds.), Wiley, New York, 1978, Vol. 37, p.1.

[13] R. F. Bryan, D. P. Freyberg, *J. Chem. Soc. Perkin II* **1975**, 1835.

[14] J. Umemura, S. Hayashi, *Bull. Inst. Chem. Res. Kyoto Univ.* **1974**, *52*, 585.

Mikrochim. Acta [Suppl.] 14, 43–50 (1997)

Monitoring the Mechanism of Biological Reactions by Infrared Spectroscopy

Friedrich Siebert

Institut für Biophysik und Strahlenbiologie, Albert-Ludwigs-Universität, Albertstr. 23, D-79104 Freiburg, Federal Republic of Germany

Abstract. Reasons are presented for regarding infrared spectroscopy as an extremely useful tool for studying the molecular mechanism of biological reactions. Different methods are described as well as their applications to different biological systems. Time-resolved infrared techniques are emphasized. It is shown that the method has become especially powerful if combined with the modern techniques of molecular biology.

Key words: infrared difference spectroscopy, time-resolved, isotope-labelling, site-directed mutagenesis, caged compounds, stopped-flow.

The elucidation of the molecular mechanism of biological reactions is one of the main goals in the field of biochemistry. Several questions are addressed, such as what are the molecular interactions of the enzyme with the substrate that are eventually responsible for the proper function. Enzymes can be controlled by cofactors, and knowledge of the respective interactions is of importance; in many cases, protein–protein and/or protein–peptide (hormone) interactions play a major role. A special class of enzymes, the so-called photobiological systems, contains as active centres chromophores embedded within the protein. The absorption of light initiates the molecular changes in the chromophore and hence subsequently in the protein that are required for function of the enzyme. In order to understand the molecular mechanisms, the unveiling of these processes is of great importance, as well as the determination of the chromophore–protein interaction. In order to perform their functions, enzymes must adopt a well-defined structure, which is finally determined by the amino-acid sequence of the peptide chain. Therefore, a basic problem of biophysical chemistry is the monitoring of the folding pathway, i.e. the reaction of the peptide chain from the non-ordered coil to three-dimensional structure of the protein.

All biochemical reactions are processes in time. Therefore, the usual methods of structural biology, i.e. X-ray crystallography, electron microscopy and NMR spectroscopy, which represent essentially static methods, are difficult to apply. (However, it should be mentioned that the first method has recently also been extended to slow time-resolved studies [1]). On the contrary, optical methods are well suited for time-resolved studies. Among the methods of optical spectroscopy, vibrational spectroscopy appears especially promising for the investigation of biochemical reactions, since vibrational spectra are sensitive to the structure of molecules and their interactions. Resonance Raman spectroscopy has been widely employed for the study of chromophores and their light-induced reactions in photobiological systems [2–4]. Infrared spectroscopy, the complementary method of vibrational spectroscopy, is not selective for particular components of an enzyme: all the constituents contribute to the absorption spectrum of a protein. Therefore, infrared spectroscopy could, in principle, provide the information required for the study of biological reactions. However, the non-selectivity prohibits the direct application: the thousands of atoms in a protein cause a correspondingly large number of normal modes and infrared absorption bands. Their overlap renders it difficult to interpret such an absorption spectrum in molecular terms. In addition, biological reactions in most cases require the presence of water, which is a very strong infrared absorber and therefore complicates infrared measurements. Thus, one could think that infrared spectroscopy is not a useful tool for the study of biological processes. However, it will be shown in this article that, on the contrary, infrared spectroscopy represents a very elegant method, and that deep insights into biological reactions can be achieved which would be difficult to obtain with other methods. In the recent past, several articles related to this topic have appeared [5–9].

Description of the Various Methods and Their Applications

The problem of non-selectivity has been overcome by the development of the method of infrared difference spectroscopy. The idea is simple: it appears reasonable that only a small part of the complex enzyme actually participates in the reaction. If it is possible to stabilize the enzyme in well defined states of the reaction pathway for a time long enough for the spectra to be recorded with sufficient accuracy, then it is possible to form the difference spectrum between two states of interest. This difference spectrum contains information about only those molecular groups that have changed in the transition from the one state to the other. Thus, selectivity is introduced with respect to those parts of the enzyme which are of special interest. The possibility of stabilization of reaction intermediates even allows the application of static methods. If, however, such stabilization is difficult to realize, time-resolved techniques have to be employed. The drastic reduction of absorption bands will facilitate the interpretation of the spectra. From the goals mentioned it is clear that the vibrations of a single amino-acid and their changes should be reflected in the spectra. A rough estimate (and the actual results) will demonstrate that the spectra should be recorded with sufficient accuracy to resolve absorbance changes of the order of 10^{-5} in an average absorbance of 0.7. The high sensitivity of modern FT-IR instruments has greatly contributed to achievement of this high amplitude resolution. In addition, it has also helped to overcome the strong absorbance of water, if highly concentrated samples (20–100 μg of protein per μl of water) and thin cuvettes (4–6 μm) are used.

Under these sample conditions, starting the reaction without interfering with the transmission properties imposes several problems. Photobiological systems the reactions of which can be triggered by light, are most convenient to investigate by infrared spectroscopy. The trigger is very specific and does not alter the physical appearance of the sample. In addition, intermediates of the photoreaction can often be stabilized by lowering the temperature. Therefore, FT-IR difference spectroscopy was first applied to such systems. The photoreactions of the light-driven proton-pump bacteriorhodopsin [10–12] and of the visual pigment rhodopsin [12, 13] have been studied and the results have confirmed the high accuracy of the difference spectra.

From the results it became evident that the difference spectra contain a large number of well resolved bands, that await further interpretation, about which three basic questions can be asked. What are the molecular groups causing the bands? Which molecular changes are reflected in the spectra? Are these groups important for the enzyme function, or do they merely represent a type of reporter groups undergoing molecular changes in response to functionally important alterations of other groups? It also became clear that for the interpretation of the spectra additional tools are required.

Owing to the non-selectivity of infrared spectroscopy, for photobiological systems it is important to discriminate between the bands caused by the chromophore and those caused by the protein. Many systems allow the removal of the native chromophore and its replacement by an artificial one. In this way, isotopically labelled chromophores can be introduced and the corresponding shifts of bands in the difference spectra will then allow the assignment of these bands to chromophore vibrations. In addition, the observed shifts are prerequisites for normal mode description of the bands necessary for a molecular interpretation of the spectral changes. This method of isotope-labelling has been successfully applied to the retinal chromophores in rhodopsin (11-*cis*-retinal) [14–16] and bacteriorhodopsin (all-*trans*-retinal) [17, 18]. The C=N stretching vibrations of the protonated retinal Schiff base, by which the chromophore is bound to the protein via a lysine group, could be identified in the initial states and the photoproducts, and from the spectral changes and isotope shifts important conclusions could be drawn about the hydrogen-bonding interaction of the Schiff base with its environment. By assigning C–C stretching modes and hydrogen out-of-plane bending modes, information on retinal isomerization and on twists of the polyene chain could be obtained. For some bacterial systems labelled amino-acids can be introduced into the protein, and in this way protein bands can also be identified [19–22]. An example of this will be presented later when another method of infrared spectroscopy is discussed.

Modern molecular biology has also contributed to the interpretation of the difference spectra. If recombinant enzymes, i.e. enzymes synthesized in an appropriate expression system can be prepared, one or several amino-acids of the sequence can be exchanged. The effects of the mutations can be followed both functionally as well as spectroscopically and great progress has been made in the elucidation of the pro-ton-pumping mechanism of bacteriorhodopsin by such investigations [23–25]. An application for the visual system rhodopsin is shown in Fig. 1. Here, the counter-ion of the protonated retinylidene Schiff base, glutamic acid 113, has been replaced by an aspartic acid (E113D) or by an alanine (E113A) group. Depicted are the difference spectra for the metarhodopsin II state, the state responsible for triggering the enzymatic cascade in the visual process [26]. It is known that in this state the Schiff base becomes deprotonated. The three spectra are very much alike, indicating that no larger structural alterations are caused by the muta-tions. However, a closer look reveals that the positive band of the photoproduct metarhodopsin II at around $1710\,\text{cm}^{-1}$ is missing in the spectrum of the E113A mutant. This shows that the positive band is caused by

glutamic acid 113, and that this group becomes proto-nated with the deprotonation of the Schiff base. Thus, the counter-ion Glu 113 functions as a proton acceptor for Schiff base deprotonation. This step (neutralization of the charge at position 113) has been shown to be crucial for the visual process. The alterations in the fingerprint region (C–C stretching region between 1300 and $1100\,\text{cm}^{-1}$) that are evident in the E113A spectrum indicate that the counter-ion exerts a special interaction not only with the Schiff base but also with the C-12–C-14 region of the retinal.

The replacement of the native chromophore with an artificial one enables the incorporation of chemically modified retinals and the consequences of this replace-ment can be studied functionally as well as spectro-scopically, similarly to the case of amino-acid replace-ment. This has successfully been employed for retinal proteins. In the case of rhodopsin, it has been shown that the methyl group of the chromophore at position 9 of the polyene chain is of utmost importance for the visual excitation process: removal of this group greatly diminishes the capability of the illuminated pigment to trigger the enzymatic cascade [27]: the molecular events during the photoreaction are drastically altered. Also, an intact cyclohexene ring of the retinal is re-quired for a correct coupling of the chromophore isomerization to the protein conformational changes: replacement of the ring by two ethyl groups abolishes the visual excitation and strongly alters the molecular events of the photoreaction [28]. Also, in bacterior-hodopsin, the interaction of the 9-methyl group of the retinal with the protein is important for an efficient pump. Removal of this group has a strong influence both on the early as well as on the later part of the photoreaction. In particular, the photocycle is slowed down by a factor of 250, rendering the pump much less effective. In combination with infrared experiments on mutants, evidence has been provided that the 9-methyl group interacts with tryptophan 182 [29].

So far, static methods could be applied, since inter-mediates of the photoreaction could be stabilized by lowering the temperature or by adjusting the pH of the solvent. In many cases, however, intermediates cannot be stabilized, e.g. the late photoproducts called "N" and "O" of the photocycle of bacteriorhodopsin. Therefore, time-resolved infrared spectroscopic tech-niques are required. A very efficient extension of static FT-IR difference spectroscopy is the rapid-scan tech-nique. Here, advantage is taken of the property of modern FT-IR instruments that the time needed to

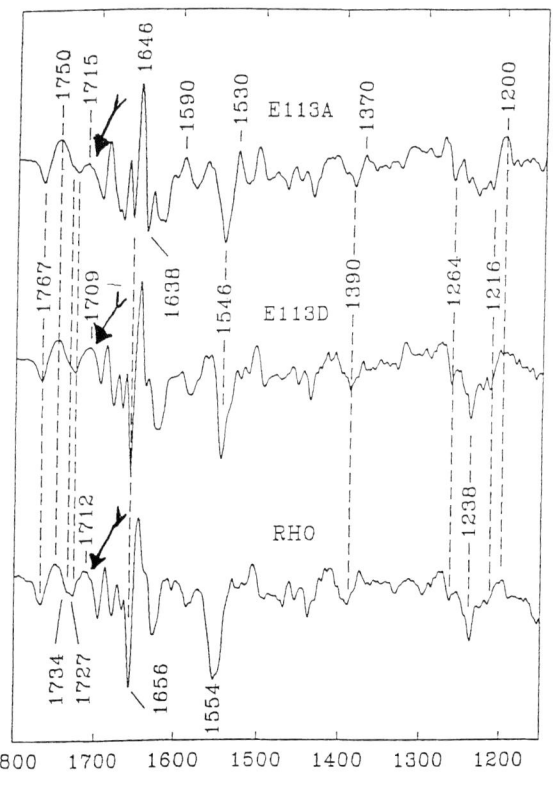

Wavenumber (cm⁻¹)

Fig. 1. FT-IR difference spectra of the metarhodopsin II photoreac-tion of the mutants Glu-113->Ala (E113A), Glu-113->Asp (E113D), and wild type rhodopsin (RHO) expressed in COS cells. Experiments were done at 0 °C with detergent-solubilized pigments. The arrows indicate the band due to protonated glutamic acid 113 and aspartic acid 113 and the lack of this band. *A* alanine, *E* glutamic acid, *D* aspartic acid (Adapted from [26])

record a single interferogram at a resolution of $4\,cm^{-1}$ can be as short as 4 ms. Thus, by synchronizing the trigger process with the movement of the interferometer mirror, difference spectra can be formed between states at various times after the trigger and the state before the trigger. By introducing appropriate delays, a wide time-scale can be covered. To achieve a sufficient signal/noise ratio, several measurements have to be averaged. For reactions that can be triggered many times (e.g. those of bacteriorhodopsin), all the measurements can be performed with the same sample. For other systems, triggerable only once, for each measurement a new sample has to be used. In instruments capable of recording a two-sided interferogram for the forward and backward movement of the mirror, it is possible to split the interferograms into four, thereby increasing the time resolution. The rapid-scan technique has very successfully been applied to the study of the late intermediates of bacteriorhodopsin [24, 30–32]. The time-resolution of this method is primarily limited by the movement of the mirror. In order to overcome at least partially this restriction, the stroboscopic technique has been developed. Here, only a certain part of the interferogram is recorded with one scan. By adjustment of an appropriate delay between the trigger and the movement of the mirror, in the subsequent measurements the remainder of the interferogram is recovered. The time-resolution is increased according to the fraction of the interferogram recorded with a single scan. Of course, the number of reactions to be triggered is increased accordingly. For this method it is necessary that the reaction can be triggered many thousand times with the same sample. The time-resolution could be improved to approx. $20\,\mu s$, and therefore, the earlier intermediates of the photoreaction of bacteriorhodopsin became accessible [33–35]. The technique is called stroboscopic time-resolved FT-IR spectroscopy. The ultimate (theoretical) time-resolution is obtained with this technique if only a single interferogram point is measured with a single scan. Then, the conversion time of the AD converter, which has to have an accuracy of at least 16 bit, imposes a new limitation.

In order to overcome this new limitation, the broadband FT-IR technique was first given up and monochromatic techniques were introduced. Since the amplitude of the spectral changes is very small, it is sufficient to record them with a resolution of 8 bit. Much faster AD converters are available and the time-resolution can be extended into the 2 ns region. In order to achieve this time-resolution, a correspondingly fast detector has to be employed. Photovoltaic MCT detectors offer such possibilities. The complete time-resolved difference spectrum is obtained by tuning the measuring beam to the respective wavenumber and repeating the measurement. As a tunable monochromatic beam either a globar with a monochromator [19, 36] or tunable lasers (CO lasers [37, 38] or lead salt diodes [39, 40]) can be employed. The high intensity of the laser beams considerably reduces the number of signals to be averaged, although, especially for the diode lasers, the tuning process may sometimes be tedious. The single-wavelength method appears especially suitable if only a small spectral range needs to be covered. A typical example is shown in Fig. 2. Here, the spectral changes of bacteriorhodopsin between 1800 and $1700\,cm^{-1}$ are shown, together with representative traces of the time-resolved changes [19]. Two spectra are shown, one of native bacteriorhodopsin, the other of bacteriorhodopsin in which all the aspartic acid groups have been replaced by ones ^{13}C-labelled at position 4 (i.e. the side-chain carboxyl group). A spectral shift of the band at $1760\,cm^{-1}$ by $40\,cm^{-1}$ is evident, indicating that the corresponding band is caused by an aspartic acid group. This shows that an aspartic acid group becomes protonated in this time-range, and it can be shown that this proton transfer represents an essential step in the proton-pumping mechanism. The slower rise time of the trace at $1755\,cm^{-1}$ indicates that the same aspartic acid group later undergoes an environmental change resulting in increasing hydrogen bonding.

The success of the single wavelength technique prompted us to ask whether it is possible to combine this data-acquisition method with the FT-IR method. The result was the development of time-resolved step-scan FT-IR spectroscopy. This technique became available when manufacturers provided interferometers in which the movable mirror is capable of a precise stepwise movement. The idea of step-scan FT-IR spectroscopy is simple: at each sampling point of the interferometer the mirror is held fixed and the reaction is triggered. In this way, at each sampling point the temporal change of the interferogram is recorded. After rearranging of the data in the computer and the FT operation, the spectral changes can be obtained at any time covered by the temporal trace. The method can only be applied to systems which can be triggered very many times. For a spectral resolution of $4\,cm^{-1}$ and a free spectral range of $2000\,cm^{-1}$, some 560 sampling

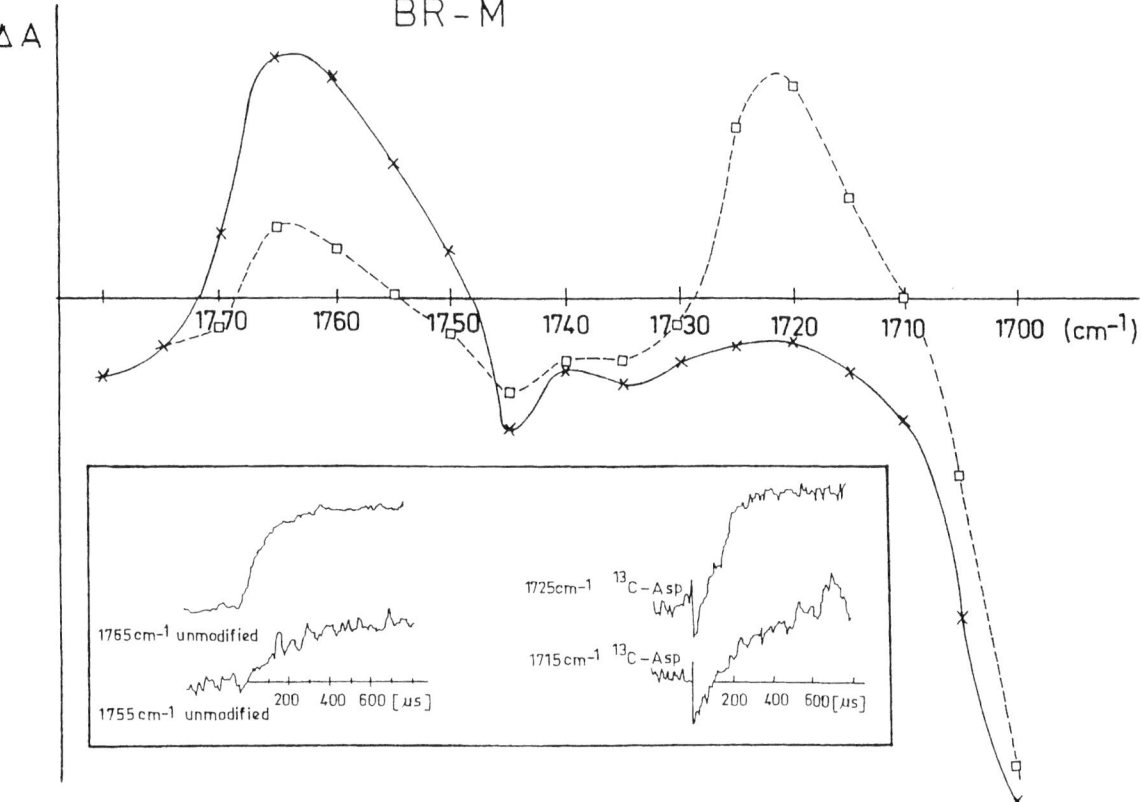

Fig. 2. Time-resolved difference spectrum of bacteriorhodopsin obtained by the single-wavelength method, demonstrating protonation of an aspartic acid. Crosses correspond to the unmodified sample, open squares to a sample labelled with (4-^{13}C)Asp. Lower part shows corresponding time courses of the two samples at the specified wavenumbers. (Adapted from [19])

points are needed. This number represents the minimum number of reactions to be triggered. In most cases, depending on the time resolution, signals have to be averaged, thereby increasing the number of trigger processes accordingly. As in the case of the single-wavelength method, the time resolution can be extended into the 2 ns range. The advantages of this method were first demonstrated by applying it to the investigation of the photoreaction of bacteriorhodopsin [41]. Later, by increasing the time resolution into the 100 ns range, information on the dynamics of the chromophore isomerization could be obtained [42]. Figure 3 demonstrates how the complete dynamics of the photoreaction of bacteriorhodopsin can be covered by a single measurement, involving about 60000 excitations.

Hitherto it has only been described how reactions of photobiological samples can be investigated. The advantage of these samples is that their reactions can be triggered by light. By far the largest number of enzymes, however, comprise non-photobiological systems. Therefore methods have to be developed that allow the reactions of these systems to be started in a well-defined way. It was advantageous that a similar problem arose in the field of physiology. For kinetic studies it was important to deliver special compounds to a tissue at precisely defined times, overcoming the problem of long diffusion. Here, the so-called caged compounds have been developed. These are substrates or cofactors of enzymes, which are made inactive by a dye molecule covalently linked to the site of interaction with the enzyme. The dye absorbs in the UV, and by a corresponding flash it can be cleaved off, thereby enabling the interaction. Caged ATP, GTP, cGMP, Ca^{2+}, H^+, sugars, and so on, are now commercially available, and kits are available that allow the synthesis of special compounds according to the needs of the experiment. Some reviews have recently appeared [43–45]. It is obvious that this method of starting a reaction can also be applied for infrared studies. Figure 4 demonstrates the application of caged Ca^{2+} for the study of the binding of Ca^{2+} to the Ca-ATPase of sarcoplas-

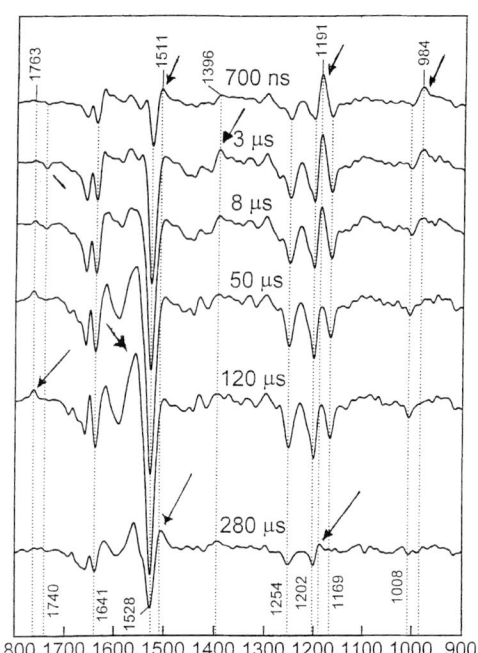

Fig. 3. Step-scan experiments on the photoreaction of bacteriorhodopsin. In order to increase the yield of the O-intermediate, the experiment was done at 40 °C. 128 signals were averaged at each sampling point of the interferogram. Spectral resolution is 4 cm⁻¹. The arrows indicate the important features of the KL (700 ns), L (3 μs), M (120 μs) O (280 μs) intermediates. (O. Weidlich and F. Siebert, unpublished)

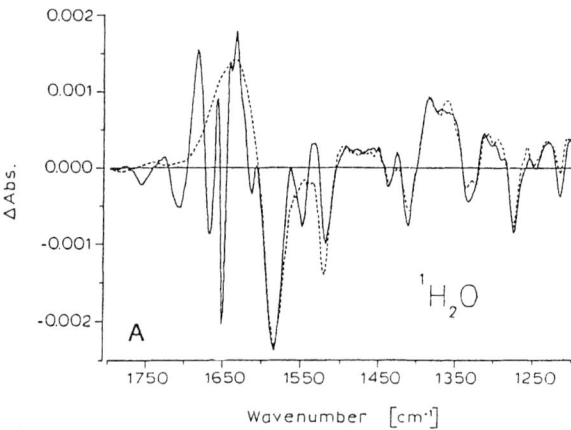

Fig. 4. The effect of Ca^{2+}-binding to the Ca-ATPase of sarcoplasmatic reticulum, using caged Ca^{2+} DM-Nitrophen. Dashed line: difference spectrum of the caged compound; solid line: difference spectrum of ATPase plus caged compound. (Adapted from [46])

matic reticulum [46]. To cage the Ca^{2+} the photolysable chelating agent DM-nitrophen has been used. For a detailed evaluation, several corrections have to be made: the photolysis of the caged compound produces

infrared difference bands that have to be taken into account. The interaction of the caged compound with the enzyme, both in the "caged" and in the photolysed state, may cause band shifts relative to the compound in enzyme-free solution. In addition, the difference bands of the compound without Ca^{2+} have to be considered. These corrections can be made but are not shown in the figure. From a direct inspection two conclusions can already be made: the two negative bands at above 1700 cm⁻¹ indicate that the binding involves two carboxyl groups which become deprotonated; the bands between 1680 and 1620 cm⁻¹ (amide I), corrected for the bands of the cage, indicate that structural changes of the protein occur upon Ca^{2+} binding. The reactions of this system with its natural substrates ATP and ADP have also been studied with the corresponding caged compounds [47, 48].

The classical methods of triggering biochemical reactions are the stopped-flow and rapid-mixing techniques. Until recently, it has been thought that they would be difficult to implement for infrared sudies, since the very thin sample cuvettes and the rather viscous protein solutions used (high protein concentration) impede the necessary rapid flow and mixing process. However, for solutions in D_2O the feasibility of a stopped-flow apparatus for infrared studies has recently been demonstrated [49]. The mixing chamber is directly mounted on the CaF_2-window of the cuvette, allowing a dead time of only about 5 ms. Thus, it is now relatively easy to apply substrates to the enzyme, vary the pH, and start the folding of proteins by dilution of the denaturant (e.g. urea or guanidinium chloride). It can be foreseen that this method will find many applications in biochemical research.

The last method to be described can be especially applied to enzymes located in biological membranes. It uses an ATR cell as sample compartment. It can be shown that, by drying the membranes on the surface of the ATR crystal, the membranes become immobilized even if afterwards an aqueous solution is flowed across the surface. Thus, substrates and cofactors can be added without causing major disturbances of the sample. In comparison to the use of caged compounds or the stopped-flow method, the mixing time is much longer (several seconds or even minutes). The advantage of this method is that normal aqueous solutions can be applied. It has been used to study the influence of pH on the proton-pump bacteriorhodopsin [50, 51] and to monitor conformational changes of the acetylcholine receptor upon interaction with agonists [52].

Since the membranes are usually oriented on the crystal surface, infrared dichroic studies can also conveniently be made.

Conclusions

The ever increasing number of publications using infrared spectroscopy for monitoring biological reactions demonstrates that the various methods described in this article have found many applications. It can be said that with this general technique molecular information has become available which would have been difficult, if not impossible, to obtain with other techniques. In particular, the combination with the technique of site-directed mutagenesis has dramatically increased the information deduced from the infrared difference spectra. In several cases, this spectroscopic method could be used to describe the effects of mutation at a molecular level. It is not possible here to review all the applications. The large field of photosynthesis research has been omitted, e.g. [53, 54], as well as the class of redox-induced reactions, using an electrochemical cell, e.g. [55]. Also, the class of time-resolved experiments using pico- and femtosecond lasers (e.g. see references in [56, 57]) have not been covered, mainly because they do not, in most cases, apply FT-IR techniques. Infrared spectroscopy, as applied to biochemical research, has changed from an exotic technique used by few experts, to a technique which can be applied almost as a routine. One further development appears crucial: in order to deduce even more molecular information from the difference spectra, more theoretical work is necessary to describe the bands in the difference spectra in terms of precise molecular structure and interaction.

Acknowledgement. I would like to thank all my co-workers in the Institut für Biophysik und Strahlenbiologie for their excellent collaboration. Part of the work mentioned in this article is based on their efforts. The work has been made possible by several grants from the Deutsche Forschungsgemeinschaft.

References

[1] I. Schlichting, S. C. Almo, G. Rapp, K. Wilson, K. Petratos, A. Lentfer, A. Wittinghofer, W. Kabsch, E. F. Pai, G. A. Petsko, R. S. Goody, *Nature* **1990**, *345*, 309.

[2] R. A. Mathies, *Methods in Enzymology, Vol. 246: Biochemical Spectroscopy* (K. Sauer ed.), Academic Press, San Diego, 1995, pp. 377–389.

[3] T. G. Spiro, R. S. Czernuszewski, in: *Methods in Enzymology, Vol. 246, Biochemical Spectroscopy* (K. Sauer ed.), Academic Press, San Diego, 1995, pp. 416–460.

[4] J. R. Kincaid, in: *Methods in Enzymology, Vol. 246, Biochemical Spectroscopy* (K. Sauer, ed.), Academic Press, San Diego, 1995, pp. 460–501.

[5] F. Siebert, in: *Methods in Enzymology, Vol. 246, Biochemical Spectroscopy* (K. Sauer, ed.), Academic Press, Orlando, 1995, pp. 501–526.

[6] F. Siebert, in: *Infrared Spectroscopy of Biomolecules* (H. H. Mantsch, D. Chapman, eds.), Wiley-Liss, 1996, pp. 83–106.

[7] W. Mäntele, *Trends Biochem. Sciences* **1993**, *18*, 197.

[8] K. Gerwert, *Curr. Opin. Struct. Biol.* **1993**, *3*, 769.

[9] M. S. Braiman, K. J. Rothschild, *Ann. Rev. Biophys. Biophys. Chem.* **1988**, *17*, 541.

[10] K. J. Rothschild, H. Marrero, *Proc. Natl. Acad. Sci. USA* **1982**, *79*, 4045.

[11] K. Bagley, G. Dollinger, L. Eisenstein, A. K. Singh, L. Zimányi, *Proc. Natl. Acad. Sci. USA* **1982**, *79*, 4972.

[12] F. Siebert, W. Mäntele, *Eur. J. Biochem.* **1983**, *130*, 565.

[13] K. J. Rothschild, W. A. Cantore, H. Marrero, *Science* **1983**, *219*, 1333.

[14] K. A. Bagley, V. Balogh-Nair, A. A. Croteau, G. Dollinger, T. G. Ebrey, L. Eisenstein, M. K. Hong, K. Nakanishi, J. Vittitow, *Biochemistry* **1985**, *24*, 6055.

[15] U. M. Ganter, W. Gärtner, F. Siebert, *Biochemistry* **1988**, *27*, 7480.

[16] Y. J. Ohkita, J. Sasaki, A. Maeda, T. Yoshizawa, M. Groesbeek, P. Verdegem, J. Lugtenburg, *Biophys. Chem.* **1995**, *56*, 71.

[17] K. Gerwert, F. Siebert, *EMBO Ji.* **1986**, *5*, 805.

[18] A. Maeda, J. Sasaki, J. M. Pfefferlé, Y. Shichida, T. Yoshizawa, *Photochem. Photobiol.* **1991**, *54*, 911.

[19] M. Engelhard, K. Gerwert, B. Hess, W. Kreutz, F. Siebert, *Biochemistry* **1985**, *24*, 400.

[20] S.-L. Lin, P. Ormos, L. Eisenstein, R. Govindjee, K. Konno, K. Nakanishi, *Biochemistry* **1987**, *26*, 8327.

[21] P. Roepe, D. Gray, J. Lugtenburg, E. M. M. van den Berg, J. Herzfeld, K. J. Rothschild, *J. Am. Chem. Soc.* **1988**, *110*, 7223.

[22] K. J. Rothschild, P. Roepe, P. L. Ahl, T. N. Earnest, R. A. Bogomolni, S. K. Das Gupta, C. M. Mulliken, J. Herzfeld, *Proc. Natl. Acad. Sci. USA* **1986**, *83*, 347.

[23] K. J. Rothschild, *J. Bioenerg. Biomembr.* **1992**, *24*, 147.

[24] K. Fahmy, O. Weidlich, M. Engelhard, J. Tittor, D. Oesterhelt, F. Siebert, *Photochem. Photobiol.* **1992**, *56*, 1073.

[25] K. Gerwert, B. Hess, J. Soppa, D. Oesterhelt, *Proc. Natl. Acad. Sci. USA* **1989**, *86*, 4943.

[26] F. Jäger, K. Fahmy, T. P. Sakmar, F. Siebert, *Biochemistry* **1994**, *33*, 10878

[27] U. M. Ganter, E. D. Schmid, D. Perez-Sala, R. R. Rando, F. Siebert, *Biochemistry* **1989**, *28*, 5954.

[28] F. Jäger, S. Jäger, O. Kräutle, N. Friedman, M. Sheves, K. P. Hofmann, F. Siebert, *Biochemistry* **1994**, *33*, 7389.

[29] O. Weidlich, N. Friedman, M. Sheves, F. Siebert, *Biochemistry*, in press.

[30] M. S. Braiman, P. L. Ahl, K. J. Rothschild, *Proc. Natl. Acad. Sci. USA* **1987**, *84*, 5221.

[31] K. Gerwert, G. Souvignier, B. Hess, *Proc. Natl. Acad. Sci. USA* **1990**, *87*, 9774.

[32] R. Kräutle, W. Gärtner, U. M. Ganter, C. Longstaff, R. R. Rando, F. Siebert, *Biochemistry* **1990**, *29*, 3915.

[33] M. S. Braiman, O. Bousché, K. J. Rothschild, *Proc. Natl. Acad. Sci. USA* **1991**, *88*, 2388.

[34] O. Bousché, M. S. Braiman, Y-W. He, T. Marti, H. G. Khorana, K. J. Rothschild, *J. Biol. Chem.* **1991**, *266*, 11063.

[35] G. Souvignier, K. Gerwert, *Biophys. J.* **1992**, *63*, 1393.

[36] T. Yuzawa, C. Kato, M. W. George, H. Hamaguchi, *Appl. Spectrosc.* **1994**, *48*, 684.

[37] A. J. Dixon, P. Glyn, M. A. Healy, P. M. Hodges, T. Jenkins, M. Poliakoff, J. J. Turner, *Spectrochim. Acta* **1988**, *44A*, 1309.

[38] P. Glyn, M. W. George, P. M. Hodges, J. J. Turner, *J. Chem. Soc. Chem. Commun.* **1989**, 1655.

[39] R. Hienerwadel, D. Thibodeau, F. Lenz, E. Nabedryk, J. Breton, W. Kreutz, W. Mäntele, *Biochemistry* **1992**, *31*, 5799.

[40] R. Hienerwadel, S. Grzybek, C. Fogel, W. Kreutz, M. Y. Okamura, M. L. Paddock, J. Breton, E. Nabedryk, W. Mäntele, *Biochemistry* **1995**, *34*, 2832.

[41] W. Uhmann, A. Becker, C. Taran, F. Siebert, *Appl. Spectrosc.* **1991**, *45*, 390.

[42] O. Weidlich, F. Siebert, *Appl. Spectrosc.* **1993**, *47*, 1394.

[43] J. H. Kaplan, A. P. Somlyo, *Trends Neurosci.* **1989**, *12*, 54.

[44] J. H. Kaplan, G. C. R. Ellis-Davies, *Proc. Natl. Acad. Sci. USA* **1988**, *85*, 6571.

[45] J. A. McCray, D. R. Trentham, *Annu. Rev. Biophys. Biophys. Chem.* **1989**, *18*, 239.

[46] H. Georg, A. Barth, W. Kreutz, F. Siebert, W. Mäntele, *Biochim. Biophys. Acta* **1994**, *1188*, 139.

[47] A. Barth, W. Mäntele, W. Kreutz, *Biochim. Biophys. Acta* **1991**, *1057*, 115.

[48] A. Barth, W. Kreutz, W. Mäntele, *Biochim. Biophys. Acta* **1994**, *1194*, 75.

[49] A. J. White, K. Drabble, C. W. Wharton, *Biochem. J.* **1995**, *306*, 843.

[50] H. Marrero, K. J. Rothschild, *Biophys. J.* **1987**, *52*, 629.

[51] S. Száraz, D. Oesterhelt, P. Ormos, *Biophys. J.* **1994**, *67*, 1706.

[52] J. E. Baenziger, K. W. Miller, K. J. Rothschild, *Biophys. J.* **1992**, *61*, 983.

[53] R. Hienerwadel, D. Thibodeau, F. Lenz, J. Breton, E. Nabedryk, W. Kreutz, W. Mäntele, *Time-Resolved Vibrational Spectroscopy V.* (H. Fakahashi, ed.), Berlin, Heidelberg New York Tokyo, 1992.

[54] J. Breton, J.-R. Burie, C. Boullais, G. Berger, E. Nabedryk, *Biochemistry* **1994**, *33*, 12405.

[55] C. Berthomieu, A. Boussac, W. Mäntele, J. Breton, E. Nabedryk, *Biochemistry* **1992**, *31*, 11460.

[56] *Time-resolved Vibrational Spectroscopy VI* Springer, Berlin, Heidelberg New York Tokyo, 1994.

[57] *Time-resolved Vibrational Spectroscopy V*, Springer, Berlin, Heidelberg New York Tokyo, 1992.

Mikrochim. Acta [Suppl.] 14, 51–56 (1997)

In situ Fourier Transform Infrared Studies of Active Sites and Reaction Mechanisms in Heterogeneous Catalysis: Hydrocarbon Conversion on H-Zeolites

Jean-Claude Lavalley*, **Sophie Jolly-Feaugas**, **André Janin**, and **Jacques Saussey**

Laboratoire Catalyse et Spectrochimie, URA CNRS 414, ISMRA, Université de Caen, 6, Bd du Maréchal Juin, 14050 Caen Cedex, France

Abstract. Adsorption of CO at $-180\,°C$ and pyridine at $150\,°C$ shows that a dealuminated HY zeolite prepared by hydrothermal treatment at $550\,°C$ followed by mild acid leaching presents Lewis acid sites, due to extra-framework aluminium debris, and very strong Brønsted acid sites ($H_0 \sim -12$), characterized by an IR band at 3600 cm^{-1}. Use of an IR cell as a flow reactor shows that these hydroxyl groups are active sites for n-hexane cracking at $400\,°C$, whereas an HY zeolite de-aluminated by isomorphous substitution is almost inactive in the same conditions. GC analysis of the products formed indicates that cracking occurs through a protolytic mechanism. All the results obtained show that the IR cell designed for flow experiments can be used as a conventional reactor.

Key words: hydrocarbon cracking, HY-zeolites, flow-cell reactor, dealumination.

Heterogeneous catalysis is crucial to chemical technology. In particular, the ability of catalysts to crack long-chain hydrocarbons to form smaller chain hydrocarbons was critical for the emerging automobile industry. Cracking catalysts are solid acids. The first solid acid catalysts commercially used were amorphous silica – alumina; however, the most important advance in this technology in the last three decades has been the development of zeolite catalysts.

Zeolites are three-dimensional crystalline aluminosilicates with uniform molecular-scale pores. HY zeolites are generally used in cracking and related reactions since their framework structure (pore opening $7.4\,Å$ connecting supercages with a diameter of $12\,Å$) is large enough to admit large molecules.

The formula for the unit cell of the dehydrated HY zeolite is $H_{56}(AlO_2)_{56}(SiO)_{136}$ giving rise to an Si/Al ratio of 2.4. However this solid is not thermally stable. Dealumination increases its stability. Amongst the techniques used to increase the Si/Al ratio, steaming followed by mild acid leaching seems the most convenient [1]. However, it leads to a complex solid presenting not only Brønsted acid sites, but also Lewis acid ones [2]. Moreover, Brønsted acid sites have been found to be heterogeneous [3], so a main question arises: which are the acid sites that are catalytically active for a given reaction?

The aim of the present study is to characterize, by infrared spectroscopy, using adsorption of different probe molecules, the acidity of two dealuminated HY zeolites with a similar Si/Al_{total} ratio. One denoted by HY_{SA}, has been prepared by conventional treatment (steaming followed by mild acid leaching); the other, named HY_i, has been obtained by treatment with ammonium hexafluosilicate, by isomorphous substitution, leading therefore to a more homogeneous material [4]. The activity of both samples for n-hexane cracking, a demanding reaction [5], has been compared by using an infrared cell as a reactor. The amount of the different products formed has been carefully analysed in order to elucidate the reaction mechanism.

Experimental

The two dealuminated samples were prepared from a low-sodium (280 ppm) form of an NH_4Y zeolite (Zeocat E 2268, Si/Al = 2.72). It was modified either by treatment with $(NH_4)_2SiF_6$ (HY_i:$Si/Al_{total} = 5.53$; $Si/Al^{IV} = 5.4$; $a_0 = 24.47\,Å$) or hydrothermal treatment at $550\,°C$ followed by acid leaching with $1\,N$ HCl (HY_{SA}:$Si/Al_{total} = 6.41$; $Si/Al^{IV} = 11.2$; $a_0 = 24.38\,Å$; Al^{IV} represents Al in AlO_4 tetrahedra).

Catalytic measurements and pyridine adsorption were performed in a special IR cell used as a reactor. Its design has already been reported [6]. The cell is installed in a continuous-flow system directly built on a Nicolet 5 SX FT-IR spectrometer, which allowed the dead volume to be minimized. The pressure can be increased up to 20 bars and the temperature up to $400\,°C$. This system will be referred as the in situ dynamic one. Each zeolite sample (self-supported wafer, 16 mm diameter, 15 mg weight and 0.2 mm thick) was ramped in the reactor cell over a 3-hour period to $400\,°C$ under a nitrogen stream (15 ml/min) and held at this temperature for 1 hour. Spectra could be recorded every 4 s by averaging 5 interferograms. The wavenumbers of the different bands particularly those of the hydroxyl groups, were

---

* To whom correspondence should be addressed

lower at higher temperature than those generally reported for room temperature, the differences reaching ca. 7 cm^{-1} at 400 °C. The GC collection was performed every 30 s and products were analysed with a Delsi Nermag chromatograph equipped with an FID detector.

CO adsorption was studied by using a conventional cell allowing recording spectra at −180 °C.

Results

In the HY$_i$ sample, the Si/Al$_{total}$ and Si/Al^{IV} ratios are similar, indicating the lack of framework debris. Accordingly, its IR spectrum (Fig. 1a) mainly shows two strong bands near 3622 and 3547 cm^{-1} characteristic of HF OH (high frequency, OH located in the supercages) and LF OH (low frequency, OH located in sodalite cages or hexagonal prisms), respectively.

In contrast, the steaming followed by mild acid leaching gives rise to extra-framework materials, explaining the strong intensity of the silanol band, near 3732 cm^{-1}, and the occurrence of additional hydroxyls characterized by a band at ca. 3593 cm^{-1} (Fig. 1b). The latter was assigned to framework OH groups perturbed by extra-framework aluminium species [7], and is situated at 3600 cm^{-1} when the spectra are recorded at room temperature.

Acidity: Probe Molecule Adsorption

Acidity is defined relative to a base, used as the probe molecule, in acid–base interaction. The Brønsted acidity of a solid is defined as its ability to donate, or at least partially transfer, a proton. Its Lewis acidity is defined as its ability to accept an electron pair through the formation of a coordinate bond. The probes used in the present study are carbon monoxide (adsorbed at −180 °C) and pyridine.

Carbon monoxide. Figure 2 shows the results relative to CO adsorption at −180 °C. Subtracted spectra show that CO specifically interacts with the HF OH groups of HY$_i$, through a hydrogen bond, explaining the appearance of a broad band at 3333 cm^{-1}. The bond strength can be estimated by the frequency difference:

$$\Delta v OH = v - v_p$$

where $v = vOH$ frequency before CO adsorption, $v_p = vOH$ frequency for the perturbed OH groups.

On HY$_{SA}$ (Fig. 2), CO interacts with HF OH groups but also with those characterized by the 3600 cm^{-1} band. Comparison of the results shows that the acidity of the 3600 cm^{-1} hydroxyl groups ($\Delta vOH = 390$ cm^{-1}) is higher than that of the HF OH of HY$_{SA}$ ($\Delta vOH = 330$ cm^{-1}), itself higher than that of the HF OH of HY$_i$ ($\Delta vOH = 300$ cm^{-1}). These results are in nice agreement with those reported earlier, for CO [8] or ethylene adsorption [9].

Pyridine. Pyridine has been chosen since it well characterizes Brønsted acid sites through the formation of pyridinium species [bands at 1631, 1618 and 1542 cm^{-1}] and Lewis acid sites, the coordinated species giving rise to bands near 1450 and 1600–1620 cm^{-1} [10].

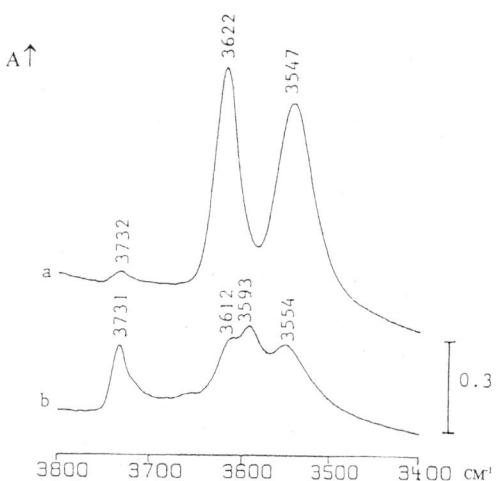

Fig. 1. IR spectra of HY$_i$ (*a*) and HY$_{SA}$ (*b*) activated under flow of nitrogen at 400 °C (spectra recorded at 400 °C)

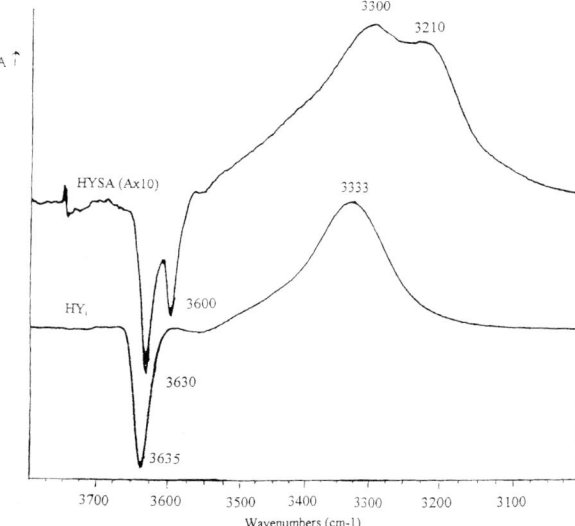

Fig. 2. CO adsorption on HY$_i$ and HY$_{SA}$ (the spectrum of the activated sample before CO adsosrption has been subtracted, spectra recorded at −180 °C)

On the zeolites activated at 400 °C under a dry nitrogen flow, 2 μl of pyridine were introduced at 150 °C, then the temperature of the cell was raised to 400 °C under nitrogen in order to study the stability of the species formed and in particular to eliminate physisorbed species.

The spectra in the 1650–1400 cm^{-1} frequency range are reported in Fig. 3. It clearly appears that both samples present Brønsted acid sites (1540–1537 cm^{-1} band). On the other hand, only HY$_{SA}$ presents Lewis acid sites (band at 1452 cm^{-1}, persisting under nitrogen flow at 400 °C).

Reactivity: n-Hexane Cracking

The FT-IR results obtained in working conditions have already been published [11, 12]. Let us recall that, in the conditions used, the HY$_i$ sample was almost inactive. On the other hand, the initial conversion of the HY$_{SA}$ sample was about 25% at 400 °C after 1 min of time on stream. The products formed mainly arose from cracking. Deactivation occurred rapidly due to coke formation, characterized by bands at 1586 and 1348 cm^{-1}. Subtracted spectra showed that the 3600 cm^{-1} band (3593 cm^{-1} in working conditions) due to highly acidic hydroxyls, was the most affected during the reaction. Specific poisoning of these hydroxyl groups by 2,6-lutidine (2,6-dimethylpyridine) clearly demonstrated that the activity in n-hexane cracking of HY$_{SA}$ was due to the strong Brønsted acid sites characterized by the ν(OH) band at ca. 3600 cm^{-1}. On the other hand, Lewis acidity without these strong Brønsted acid sites present did not generate any activity.

In order to determine the reaction mechanism, the activity and selectivity of both zeolites were studied, with the IR cell as reactor. Over the activated sample, N$_2$ + hexane (partial pressure of n-hexane 0.06 bar)

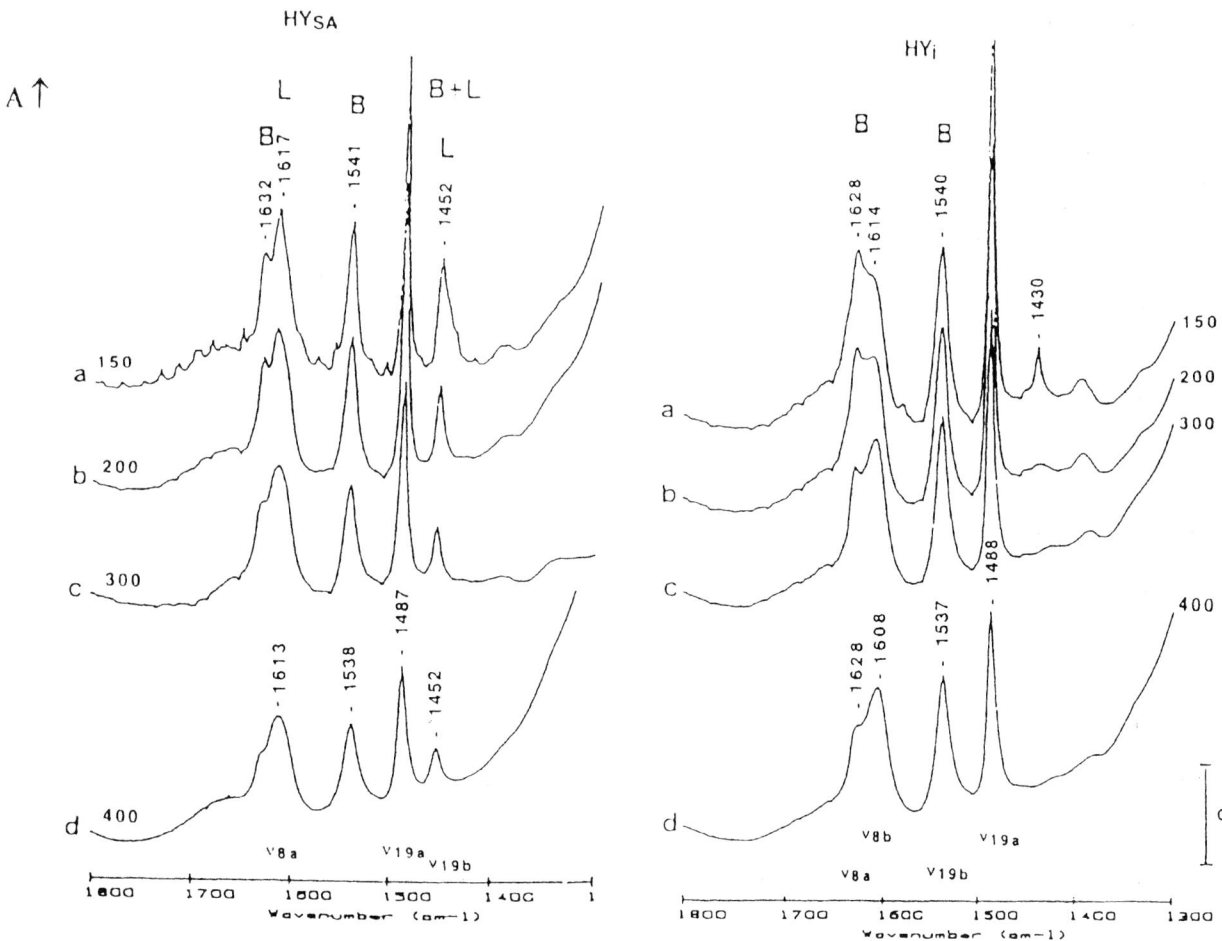

Fig. 3. IR spectra of HY$_i$ and HY$_{SA}$ activated in dynamic conditions at 400 °C, then saturated by pyridine at 150 °C, followed by N$_2$ flow at 150, 200, 300 and 400 °C (*a, b, c, d*, respectively)

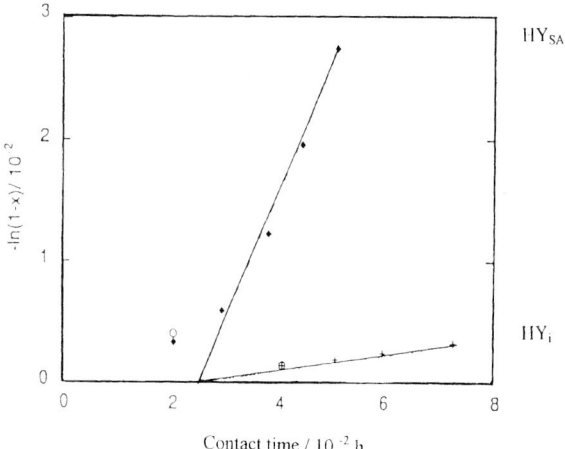

Fig. 4. Rate of n-hexane cracking over HY_i and HY_{SA} at 400 °C

be formed according to the literature results obtained for different zeolites [13, 14]. On the other hand, in the conditions used, no deactivation occurred (lack of any IR band at 1586 cm^{-1}).

Discussion

As expected, the HY_i sample does not present any Lewis acidity, indicating the lack of extra-framework debris. Its Brønsted acidity, as shown by CO adsorption results, is mainly due to HF OH groups, located in the supercages. By contrast, the steaming followed by mild acid leaching gives rise to extra-framework materials, explaining the strong intensity of the silanol band, near 3732 cm^{-1}, the occurrence of additional hydroxyls characterized by a band at ca. 3600 cm^{-1} (3593 cm^{-1} at 400 °C) and the presence of Lewis acid sites as shown by pyridine adsorption. The 3600 cm^{-1} hydroxyl groups are strongly acidic as shown by CO adsorption, confirming previous results obtained either by pyridine desorption [2] or ethylene adsorption [9] followed by IR spectroscopy. Several hypotheses have been proposed to explain the high acidity of the 3600 cm^{-1} OH groups. It has been suggested that the framework Brønsted acid strength was increased by Lewis acid sites [14, 15]. Alternatively, Garralon et al. [2] proposed that it was due to the formation of amorphous silica–alumina inside the pores. Carvajal et al. [7] explained it as involving aluminium cations located in the sodalite cages. There would withdraw electrons from the framework hydroxyl groups, thus making the proton more acidic. It is difficult to distinguish between these hypotheses, although the observations [3] of an extra band at 3525 cm^{-1}, also corresponding to strong acidic sites and which could result from the shift of the LF OH groups, supports the third one. Moreover X-ray diffraction analysis confirmed that steaming generated extra-framework Al species, some of them being indeed located in the centre of the sodalite cages [16].

The acidity of the HF groups of HY_{SA} is higher than that of those in HY_i. This confirms previous results [17] showing that the strength of framework hydroxyls increases with the framework dealumination level.

In order to determine an acidity scale of Brønsted acid sites, we have studied adsorption of CO (at -180 °C) and C_2H_4 (at -60 °C) on different materials such as silica, alumina, phosphated alumina, silica-alumina, and different zeolites, including modified HY and β-zeolites [18]. Figure 5 shows that the v(OH) frequency shift due to CO adsorption is linearly related to that obtained using C_2H_4 as probe. This indicates that the two molecules used probe the same sites, without effects such as steric hindrance.

was intermittently flowed for 3 min, in order to reach the stationary state, at different flow-rates. Then the products formed were analysed by gas chromatography while the catalyst was maintained under pure N_2 flow at 400 °C.

In Fig. 4, we report the variation of $-\ln(1-x)$ (where x = degree of conversion) with contact time for each catalyst studied. Linear variations are observed, showing that the cracking rate is first order in n-hexane. As already reported [12], the activity of HY_i is very low in the conditions used, being lower than that of HY_{SA} by a factor of 10–15.

Extrapolation to very low contact times, of the amount of each product formed allows us to reach the selectivity at zero conversion. The results are reported in Table 1, in which $C_n^=$, C_n and iC_n signify alkenes, alkanes and iso-alkanes, respectively. Note that the device used did not allow us to detect H_2, expected to

Table 1. Selectivity of HY_i and HY_{SA} sample at initial conversion

Selectivity	HY_i	HY_{SA}
Methane C_1	4.1	3.1
Pentenes $C_5^=$	1.5	0.9
Pentanes C_5	0	0.6
iC_5	0.7	1.5
Ethane C_2	5.1	8
Butenes $C_4^=$	6.7	4.1
Ethylene $C_2^=$	17	5.6
Butanes C_4	2.3	3.7
iC_4	0.7	2.1
Propane C_3	5.7	20.5
Propene $C_3^=$	30	34
Hexenes $C_6^=$	22	6.3
2-methylpentane	2.4	5.2
3-methylpentane	1.8	4.5

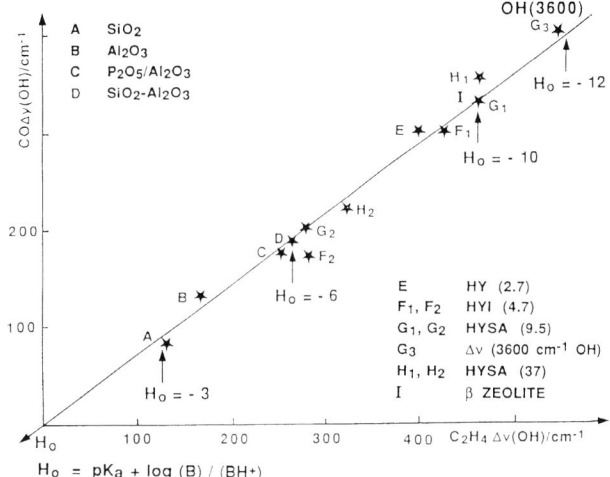

Fig. 5. Scale for the determination of Brønsted acidity of different materials: $\nu(OH)$ shift observed from CO adsorption ($-180\,^\circ C$) vs. that obtained from C_2H_4 adsorption ($-60\,^\circ C$). H_0 values are those reported in [19]

Recently Umansky et al. [19], using adsorbed H_0 indicators, calibrated the acidities of different solid acids by means of those of sulphuric acid solutions having the same H_0 values. They found $H_0 = -3, -6$ and -10 for silica gel, silica–alumina and a β-zeolite, respectively.

Taking into account these results and those reported in Fig. 5, the H_0 value of the $3600\ cm^{-1}$ hydroxyl groups can be estimated as -12, at the limit of superacidity, whereas the H_0 value for HF OH groups is close to -10.

All the IR results obtained by using the IR cell as reactor showed that the $3600\ cm^{-1}$ hydroxyl groups are mainly active for n-hexane cracking at $400\,^\circ C$. In particular, specific poisoning by 2,6-lutidine showed that Lewis acid sites alone are inactive toward n-hexane cracking [11, 12]. A study of 3-methylpentane and 2,3-dimethylbutane on HY, amorphous silica-alumina and H-ZSM5 has also shown that Lewis acid sites do not appear to participate in cracking [20].

Two mechanisms have been proposed in the literature for acid-catalysed paraffin cracking [21]:
(i) mechanism A, involving as the essential feature a bimolecular hydride transfer in the rate-determining step:

$$R-\underset{H}{\underset{|}{C}}-R + R_1-\overset{+}{\underset{|}{C}}-R_1 \xrightarrow{\text{H-transfer}} R-\overset{+}{\underset{|}{C}}-R + R_1-\underset{H}{\underset{|}{C}}-R_1$$

$$R-\overset{+}{\underset{|}{C}}-R \xrightarrow{\beta\text{-scission}} \text{olefin} + R_2-\overset{+}{\underset{|}{C}}-R_2$$

(ii) mechanism B, involving a monomolecular reaction which proceeds via a penta-coordinated carbonium ion intermediate:

$$H^+ + R-\underset{H}{\underset{|}{C}}-R \longrightarrow R-\overset{+}{C}-R$$

$$RH + R-\overset{+}{C}-H \quad\quad H_2 + R-\overset{+}{C}-R$$

The two mechanisms can be distinguished according to the nature of the major cracking products formed. Mechanism B predominates at high temperature, low hydrocarbon partial pressure and low conversion [13, 21, 22].

In the case of n-hexane as the alkane, the $C_6H_{15}^+$ carbonium can lead to $H_2 + C_6H_{13}^+$, $CH_4 + C_5H_{11}^+$, $C_2H_6 + C_4H_9^+$, $C_3H_8 + C_3H_7^+$ or/and $C_4H_{10} + C_2H_5^+$. The carbenium formed may (i) lose a proton and form the corresponding alkene (hexene, pentene, butene, propene and/or ethene) or (ii) be involved in secondary reactions (isomerization, alkylation or H-transfer reactions) [13].

In order to indicate the relative contributions of the two different cracking mechanisms, Wielers et al. [13] used a "cracking mechanism ratio" (CMR), which is defined by the ratio (in molar selectivities):

$$\frac{CH_4 + C_2H_6 + C_2H_4}{iC_4}$$

CH_4, C_2H_6 and C_2H_4 are typical products of the protolytic cracking route [mechanism B], whereas isobutane is typical of the classical cracking route [mechanism B]. A higher value for the CMR (>1) reflects a significant contribution of the protolytic cracking route, while a low value ($0 < CMR < 1$) is indicative of classical carbenium ion chemistry. From data reported in Table 1, the CMR value is 70 and 5.6 for HY_i and HY_{SA} respectively, clearly indicating that, for both materials, the protolytic mechanism is mainly operative. This is in agreement with literature results [13, 23].

Conclusion

This study demonstrates the application of infrared spectrocopy in catalysis studies. Adsorption of probe molecules such as carbon monoxide leads to the determination, on the H_0 scale, of the strength of the Brønsted acid sites of dealuminated HY zeolites, whereas that of pyridine shows the presence or absence of Lewis acid sites. Use of an IR cell as a reactor permits characterization of active sites for n-hexane cracking, while an on-line gas chromatograph allows determination of the amount of the different products formed, in order to establish the reaction mechanism. The results show that the cell used indeed acts as a conventional reactor.

Acknowledgements. The authors are grateful to Institut Français du Pétrole (IFP) for financial support and Drs. E. Benazzi and N. Zanier for helpful discussions.

References

[1] P. K. Maher, F. D. Hunter, J. Scherzer, in: *Adv. Chem. Ser.* (*Molecular Sieve Zeolites*–I) **1971**, *101*, 266.

[2] G. Garralon, A. Corma, V. Fornes, *Zeolites* **1989**, *9*, 84.

[3] S. Khabtou, T. Chevreau, J. C. Lavalley, *Microporous Mater.* **1994**, *3*, 133.

[4] G. W. Skeels, D. W. Breck, *Proc. 6th Int. Zeolite Conf. 1983*, **1984**, 87.

[5] J. N. Miale, N. Y. Chen, N. Y. Weisz, *J. Catal.* **1966**, *6*, 278.

[6] J. F. Joly, N. Zanier-Szydlowski, S. Colin, F. Raatz, J. Saussey, J. C. Lavalley, *Catal. Today* **1991**, *9*, 31.

[7] R. Carvajal, P. J. Chu, J. H. Lunsford, *J. Catal.* **1990**, *125*, 123.

[8] M. A. Makarova, K. M. Al-Ghefaili, J. Dwyer, *J. Chem. Soc. Farad. Trans.* **1994**, *90*, 383.

[9] A. Chambellan, T. Chevreau, S. Khabtou, M. Marzin, J. C. Lavalley, *Zeolites* **1992**, *12*, 306.

[10] E. P. Parry, *J. Catal.* **1963**, *2*, 371.

[11] S. Jolly, J. Saussey, J. C. Lavalley, N. Zanier, E. Benazzi, J. F. Joly, *Ber. Bunsenges Phys. Chem.* **1993**, *97*, 313.

[12] S. Jolly, J. Saussey, J. C. Lavalley, *J. Mol. Catal..* **1994**, *86* 401.

[13] A. F. H. Wielers, M. Vaarkamp, M. F. M. Post, *J. Catal.* **1991**, *127*, 51.

[14] R. A. Beyerlein, G. B. Mcvicker, L. N. Yacullo, J. J. Ziemiak, *J. Phys. Chem.* **1988**, *92*, 1967.

[15] C. Mirodatos, D. Barthomeuf, *J. Chem. Soc., Chem. Commun.* **1981**, 39.

[16] E. Merlen, *Ph. D. Thesis*, University of Paris VI, 1989.

[17] A. Janin, J. C. Lavalley, A. Macedo, F. Raatz in: *Perspective in Molecular Sieve Science ACS Symp. Ser.* **1988**, *368*, 117.

[18] M. Maache, *Ph. D. Thesis*, Caen, 1992.

[19] B. Umansky, J. Engelhardt, W. K. Hall, *J. Catal.* **1991**, *127*, 128.

[20] J. Abbot, B. W. Wojciechowski, *J. Catal.* **1989**, 115, 1.

[21] W. O. Haag, R. M. Dessau, *Proc. 8th Int. Congr. Catal. 1984* **1985**, II-305.

[22] R. Shigeishi, A. Garforth, I. Harris, I. Dwyer, *J. Catal.* **1991**, *130*, 423.

[23] J. Dwyer, J. Dewing, K. Karim, S. Holmes, A. F. Ojo, A. A. Garforth, D. Rawlence, in: *Zeolite Chemistry and Catalysis.* (P. A. Jacobs et al., eds.) Elsevier, Amsterdam, 1991, p. 1.

Mikrochim. Acta [Suppl.] 14, 57–66 (1997)

FT-IR Spectra of Nanoparticles: Surface and Adsorbate Modes

J. P. Devlin[1,*] and **V. Buch**[2]

[1] Department of Chemistry, Oklahoma State University, Stillwater, OK, 74078, U.S.A.
[2] Department of Physical Chemistry, The Hebrew University, Jerusalem, Israel

Abstract. Nanocrystals and amorphous nanoparticles (~ 20 nm diameter) have been prepared in the vapor phase by the rapid expansion of a dilute mixture of volatile substances (H_2O, CO_2, mesitylene, etc.) in N_2/He into a precooled (30–130 K) static infrared cell. The nanoparticles can be studied spectroscopically while suspended in the inert carrier gas, or as deposits on an infrared window, formed by use of multiple load–pump cycles. Since the particles are of a size for which $\sim 10\%$ of the molecules are at the surface, it is possible to obtain high quality spectra of the surface-localized vibrational modes and/or of monolayer quantities of adsorbate molecules, the latter quite generally, whereas direct observation of the surface-localized modes has been limited to H-bonded samples for which the frequencies differ significantly from interior mode frequencies. For example, from the difference between spectra for "large" *vs* "small" nanocrystals, the absorbance from bulk ice modes can be largely eliminated so that it is possible to observe several O–H(D) stretch and bend modes of surface molecules. With guidance from simulated spectra of an ice cluster and a slab of cubic ice, it has been possible to assign bands to vibrations of three classes of surface water molecules. The simulations show that a relaxed crystalline ice surface has three types of surface water molecules: 3-coordinated with a dangling H(D), 3-coordinated with a dangling O, and surface 4-coordinated with a distorted tetrahedral structure. With mode assignments for the surface water molecules, it becomes possible to interpret the interactions between small adsorbate molecules and the ice surface through the magnitude of the shifts and intensity variations of specific modes. Of course, interpretations based on these effects can be complemented by simultaneous observations on the modes of the adsorbate molecules. In the case of *monolayer* CF_4 on surfaces of a variety of nanoparticles, the antisymmetric stretch mode appears as an intense TO-LO doublet with a large (~ 80 cm^{-1}) splitting that is a sensitive probe of the surface structure and tecture.

Key words: nanoparticles, infrared spectra, surface-localized, vibrational modes, ice and ice adsorbates, surface structure and texture.

This paper considers the FT-IR spectra of networks of self-supporting amorphous or crystalline molecular nanoparticles suspended on an infrared-transparent plate. Following their formation in the gas phase by use of a bulb (or static) process first described by Ewing and Sheng [1], a network of particles is formed as $\sim 5\%$ of each batch becomes attached to the end plates of the infrared cluster cell. Typically the nanoparticles have $\sim 10\%$ of their molecules at the surface, so, combined with the absence of limitations on either the optical thickness or lifetime of such samples, infrared observations of particle surface-localized vibrations and of modes of sub-monolayer quantities of adsorbate molecules are greatly facilitated [2].

Once a network of nanoparticles has been formed, by repetitive load–pump cycles, one can usually determine pressure and temperature conditions under which small molecules will adsorb in monolayer amounts, so the potential for study of adsorbates on solid molecular surfaces at cryogenic temperatures by this approach can be viewed as a general one. Further, it is often a simple matter to choose the phase of the nanoparticles by selection of the formation conditions or an appropriate annealing temperature.

The study of the surface-localized modes of nanoparticles appears to have a less general potential. The spectrum of a network of nanoparticles, like the spectrum of the corresponding gas phase nanoparticles, typically varies substantially from that of the corresponding thin film, particularly *for vibrational modes with large dipole oscillator strengths*. However, these differences are related to dielectric constant and particle size/shape effects rather than surface-localized particle vibrations [3–5]. To date, surface-localized modes of nanoparticles have been observed only for hydrogen-bonded systems for which variations of modal frequencies with extent of H-bonding are notoriously large. In fact, a preponderance of the data have been obtained for cubic ice [2], so this material is, necessarily, a primary subject of this paper.

Water ice is a particularly important material in science and nature, so an in-depth study of its surface is

---

* To whom correspondence should be addressed

warranted. The interaction of molecules with the surface of crystalline and amorphous ice has emerged as a prime concern of diverse areas of scientific study. For example, ice is generally recognized as a key player in the chemistry of the stratosphere, with icy particle surfaces, sities of concentration and activation of otherwise stable molecules, of particular interest [6–8]. Similarly, on the available evidence, one can legitimately refer to the surface of amorphous ice as a prime chemical factory of interstellar space, particularly in dense interstellar clouds, where molecule formation and reaction is thought to often depend on the concentration of atoms and molecules on/within "dirty" amorphous ice that has slowly accreted on refractory dust particles [9, 10]. Supporting evidence is accumulating, such as the recognition of the three-micron ice band [11] and the vibrational fundamental of H_2 adsorbed on amorphous ice [12], both observed in absorption towards stellar objects. Closely related is the chemistry of comets, which are thought to be aggregates of relatively unprocessed ice-covered dust particles of the pre-solar cloud from which our solar system was formed. The surface of ice also clearly has a role in the icing of materials on earth, and its interplay with crystalline point defects may be involved in the generation and interparticle transfer of charge that ultimately leads to electrical displays in storms [13].

Because of the great sensitivity of the H_2O stretching modes to H-bond strengths, vibrational spectra are a versatile probe of the molecular environment at ice surfaces. A full appreciation of the corresponding spectroscopic data has emerged in recent years because of advances in the computational simulation of structure [14] and spectra [15–17] of complex extended H-bonded systems. The particular surface-localized mode identified as the stretch of unbonded O–H (or O–D) groups produces a band that is well separated from the bulk-ice absorption bands and which has, therefore, been the object of the most frequent and intensive study. As a result, there is little uncertainty in the observation or assignment of the dangling-H(D) mode band of 2-coordinate/3-coordinate surface water molecules at 3720/3696 (2748/2728) cm^{-1} for amorphous ice and 3692 (2725) cm^{-1} for 3-coordinate cubic-ice surface molecules [18]. Further, the response of this mode to the presence of small molecules at ice surfaces has led to its use as a sensitive probe of the ice–adsorbate interaction [19–22].

Infrared bands of other surface-localized stretching modes occur at somewhat lower frequencies [23, 24],

with more overlap by bands of the bulk-ice modes, so their observation is dependent on indirect methods, such as infrared difference spectroscopy. Experimental difference spectra, (particularly comparing spectra of small and large nanocrystals so as to mute the interior spectrum of ice), combined with predictions of band positions and intensities from simulated spectra of reconstructed ice surfaces [2], will be used in this paper to identify six surface-localized modes that arise from the three distinct classes of surface water molecules. These categories of water molecules within the reconstructed (111) surface bilayer of cubic ice are identified, from the simulated structures, as 3-coordinated molecules with dangling-H(D) or d-H(D), 3-coordinated molecules with a dangling-O or d-O, and distorted 4-coordinated or s-4 surface molecules. Similar spectroscopic and simulation results will also be used to distinguish, somewhat less clearly, between surface and subsurface modes of the ice nanocrystals.

The second major goal of this paper is to emphasize the importance of the surface texture for heterogeneous processes, such as the ionization of acids on ice particles, and to show that a great deal can be learned about surface texture through the behaviour of the spectrum of surface-adsorbed CF_4. CF_4 has been shown to be a special molecule, uniquely useful as a powerful probe of surface structures, primarily because the antisymmetric stretch mode, being triply degenerate, projects an exceptionally strong dipole oscillation independent of direction. Consequently, strong dipole coupling within a monolayer of CF_4 gives rise to an absorption band complex with well-defined intense features corresponding to transverse (~ 1240 cm^{-1}; T-mode; dipole parallel to the surface) and longitudinal (~ 1320 cm^{-1}; L-mode; dipole perpendicular to the surface) mode frequencies [25]. Both modes are infrared active because of the "spherical" shape of the nanoparticle substrates under study.

The precise from of this band complex varies sensitively with the degree of smoothness and lateral regularity of the CF_4 monolayer in a manner that has been thoroughly simulated [26]. The predictions of the CF_4 spectral response to smoothness and/or lateral regularity have been semiquantitatively verified by observations of CF_4 monolayers on surfaces varying from exceedingly rough microporous amorphous ice to smooth and laterally regular nanocrystals of CO_2 and mesitylene (symmetrical trimethyl benzene). The conclusions from these comparisons of theory and data will be used to characterize the surface of cubic ice nanocrystals.

Experimental

FT-IR spectra were measured for nanoparticles formed in the rapid expansion of 0.2–1.0% mixtures of a vapor in N_2 or He medium into a precooled cluster cell of high heat capacity at temperatures ranging from 30 to 130 K. The cylindrical cluster cell is mounted within a vacuum chamber on the cold end of an APD closed-cycle cooler which provides a minimum sampling temperature of ~ 25 K (Fig. 1). Rather than the nanoparticles being scanned in the gas phase, they are allowed to diffuse to the end windows of the cluster cell, where approximately 5% of the "new" clusters remain attached after each load-and-evacuation cycle. The "deposition" process selects small clusters which, during repetitive cycles, can accumulate on the end windows to effective thicknesses of a few microns. This produces a stable sample of nanoparticles, for which, in the case of ice nanocrystals, the bands of the surface-isolated modes are as prominent as for the corresponding gas-phase samples.

If held at appropriately low temperatures, these nanoparticles can be used over indefinite periods of time and numerous adsorption/desorption cycles for observation of adsorbate effects. From the stability of their spectra, it is clear that the individual nanoparticles retain a high degree of integrity during adsorption/desorption cycles with various small molecules. Since the lifetime of the nanoparticles is not a limitation, extended FT-IR scan times permit measurements of surface-localized modes/adsorbate bands with exceptional signal-to-noise levels. Further, if desired, the average size of the nanoparticles can be adjusted by annealing in an appropriate temperature range (e.g., 120–150 K for ice nanocrystals). Molecules of the smaller particles, which have a higher vapor pressure, are transferred to the more stable larger particles of the network of suspended nanoparticles.

Particularly useful data are obtained from the difference of spectra of the same sample taken before and after an adjustment of average particle size. Such comparisons have been made either between two spectra without adsorbate or between two spectra of ice saturated with the same adsorbate [2]. Such spectra reflect the conversion of surface/subsurface molecules into interior ice molecules, with the surface molecules complexed with adsorbate molecules in the latter

case. Difference spectra for nanocrystals, bare and with surface adsorbates under equilibrium gas pressures, were also determined: since adsorbate-induced shifts are limited primarily to the top surface molecules of the nanoparticles, these spectra allow differentiation between surface and subsurface ice-molecule modes.

Spectra have been measured at nominally 2 or $4 \, cm^{-1}$ resolution, by using typically 200–400 scans with a Bio-Rad FTS-20 or FTS-40 spectrometer. Spectra have been obtained for nanoparticles of H_2O (D_2O), CO_2, and symmetrical trimethylated benzene (mesitylene), both pure and covered with small molecule adsorbates. With typical surface-localized band peak absorbances of less than 0.01, a large fraction of the data for ice nanoparticles has been obtained for D_2O samples so as to minimize atmospheric interferences.

Results and Discussion

This section divides into two main parts. First we will consider the spectroscopy of the ice surface (and subsurface), making assignments of individual absorption features to specific modes of particular classes of surface molecules. Using this assignment, we will look briefly at the magnitudes and significance of the response of these modes to monolayer uptake of several small molecule adsorbates. In the second part, the basis of the usefulness of the spectrum of CF_4 as a probe of the ice surface, which lies in a comparison between the predicted spectrum for model "films" and the observed patterns for CF_4 adsorbed on a variety of nanoparticles, will be examined. The CF_4 adsorbate spectrum will be revealed as a sensitive probe of both the smoothness and regularity (crystallinity) of the surfaces of molecular solids in general [26].

Assignment of the Surface-Localized Modes of Cubic Ice

It is difficult to envision an approach to the spectroscopy of the ice surface without an image of that surface. Such an image has been developed by starting with the perfect surface of ice cleaved perpendicularly to the c axis (Fig. 2a). The unrelaxed exposed surface consists of a bilayer, the top layer of which contains 3-coordinated molecules with an even distribution between the d-H(D) and d-O classes. The bottom layer contains only fully coordinated s-4 molecules. A top view of the hexagonally regular bilayer, with proton disorder characteristic of cubic and hexagonal ice, is given in Fig. 2b.

The cleaved surface is expected to relax (or reconstruct) to some degree with time. The nature of that reconstruction is of considerable current interest and is an important consideration in analysis of the surface and adsorbate spectra. We have simulated the recon-

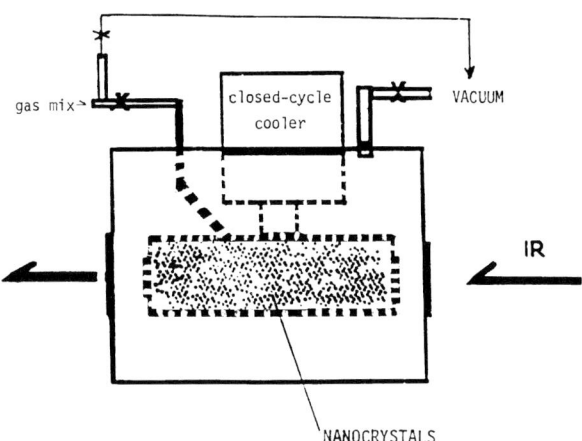

gas mix

closed-cycle cooler

VACUUM

IR

NANOCRYSTALS

Fig. 1. A schematic of the infrared static cold-cluster cell

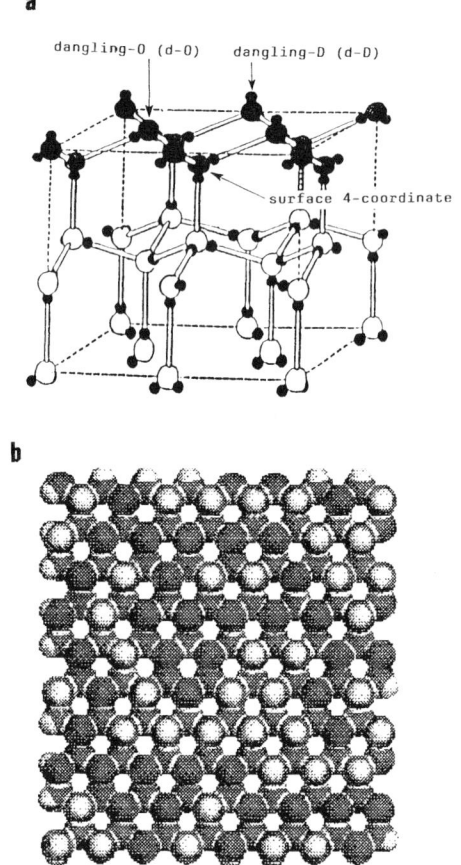

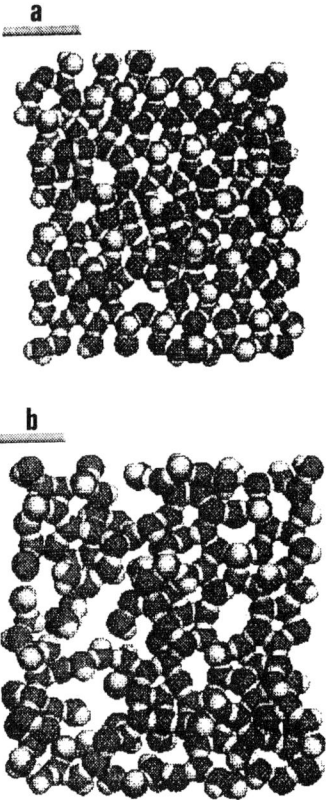

Fig. 3. Reconstructed topmost bilayer of cubic ice slab annealed at **a** 200 K, **b** 250 K

Fig. 2. **a** Crystalline ice. Surface bilayer molecular groups for freshly cleaved (hexagonal) ice, **b** topmost bilayer of a model crystalline (cubic) ice surface (111)

struction, particularly to represent the surfaces of annealed ice nanocrystals, using an ice slab of six bilayers, with a TIPS2 intermolecular potential [14] and cyclic boundary conditions. Simulated warming of this slab was found to result in surface relaxation that maximizes the H-bonding of the surface water molecules. This occurs with introduction of considerable lateral irregularity and even vertical surface roughness at the higher temperatures (> 200 K) of reconstruction [26]. This progression can be viewed in Fig. 3 which compares the top bilayers for slabs annealed at 200 and 250 K.

The ultimate in surface roughness for ice is attained in the vapor deposition of microporous amorphous ice. This structure has similarly been simulated in earlier studies, by using a 450-molecule water cluster formed under energy conditions representative of 10 K [19]. However, the vertical as well as much of the lateral irregularity of such a cluster is lost during thorough

annealing/recooling above 250 K (Fig. 4), producing a nearly spherical cluster $(D_2O)_{450}$ with a distribution of surface molecules over the d-D, d-O and s-4 classes similar to that of the relaxed slabs shown in Fig. 3. Using a hybridized intra/intermolecular potential represented by Morse oscillators. Buch has calculated the infrared spectrum of such a cluster [27]. Bands that can reasonably be attributed to modes "localized" on particular types of surface molecules have been identified and are displayed in Fig. 5.

As noted earlier, with the exception of the dangling-H(D) mode of 3-coordinated surface molecules, difference spectra are used to enhance the experimental view of the several surface-molecule bands of ice nanocrystals. Though the band intensities of the other surface-localized modes, which involve H-bonded O–H(D) groups, are greater, the lower frequencies and greater breadth of the bands requires an approach that discriminates against the intense bands of the interior modes of the nanocrystals. We will examine two approaches to the experimental spectra, based on difference spectroscopy, one of which tends to reveal the

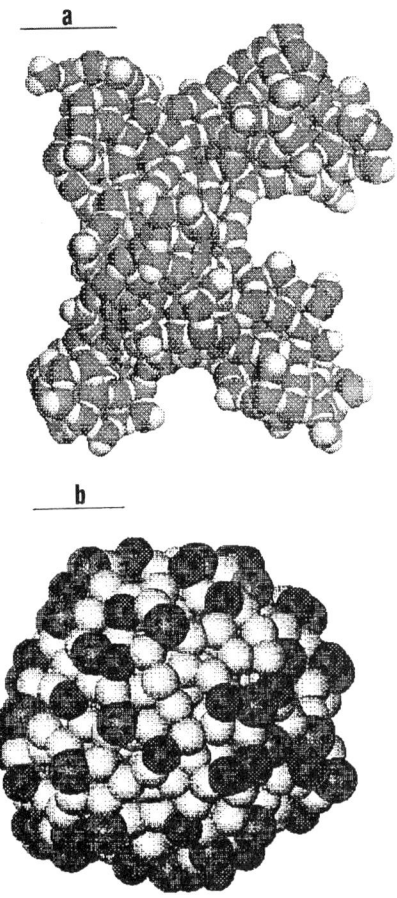

Fig. 4. Model amorphous ice clusters (H$_2$O) 250: **a** prepared by molecular condensation at 10 K, **b** after annealing at >250 K

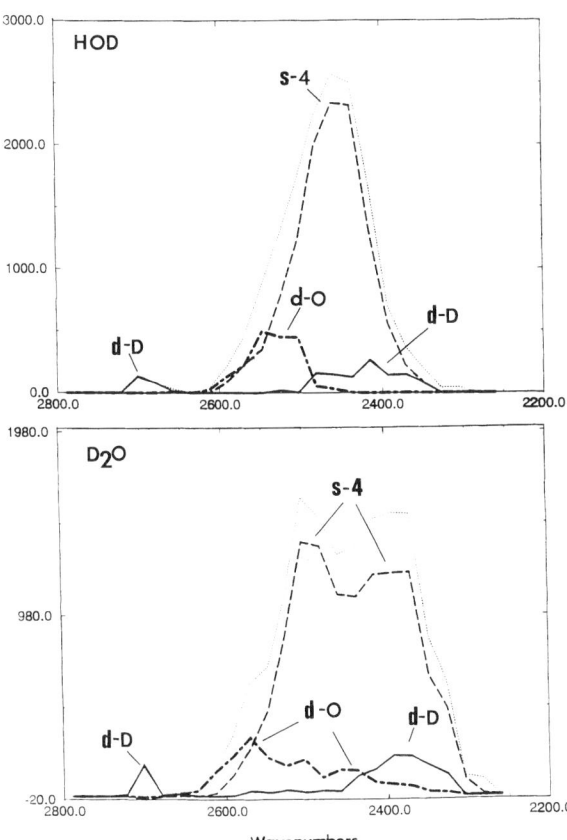

Fig. 5. Simulated O–D stretch spectra for surface molecules of cluster of Fig. 4b: dotted, all external molecules; dashed, s-4 molecules; dot–dashed, d-O molecules; solid line, d-D molecules. Bottom figure is for pure D$_2$O cluster; the top figure is for isolated HOD modeled with no coupling between bonds

combined surface- and subsurface-localized mode spectra and one that places the focus more directly on the molecular modes of the first surface bilayer.

Difference spectra between small and large nanocrystals. The result of subtracting the spectrum of larger nanocrystals of cubic ice from smaller nanocrystals is presented in Fig. 6. The positive bands above 2500 cm^{-1}, reflecting lost intensity from annealing, must originate from O–D modes localized near the surface, while the three dominant negative bands below 2500 cm^{-1} are readily recognized as produced by the increased amount of interior ice molecules (see Fig. 8e for spectrum of the interior ice modes). The simulated spectrum in Fig. 5 suggests assignment of the three surface-isolated mode bands to out-of-phase vibrations of surface D$_2$O molecules, as follows: d-D at 2725 cm^{-1}; d-O at 2640 cm^{-1}; s-4 at 2580 cm^{-1}.

That these bands are each produced by surface-isolated modes of cubic ice is made clear by their

response to the presence of a monolayer of adsorbed small molecules. Many examples have been investigated [2], of which Fig. 7 shows the results for adsorbed CO and ethylene molecules at 83 K. These difference spectra, comparing adsorbate-covered nanocrystals with larger crystals, also covered with the same adsorbate, show that major shifts to 20–60 cm^{-1} are induced in the d-D and d-O modes by these particular adsorbates. However, observation of the shift of the s-4 mode is hindered by the onset of the negative interior-mode bands near 2500 cm^{-1}.

The simulated spectra (Fig. 5) also include features for the three in-phase D$_2$O surface-molecule modes but each lies in the region of the intense interior-mode bands below 2500 cm^{-1}. One approach to the observation of these bands might be to "add" out the negative intensity of the interior-mode bands of Fig. 6a, using the spectrum of the annealed nanocrystals for which the surface-isolated mode bands are relatively insignificant. However, annealing converts subsurface as well

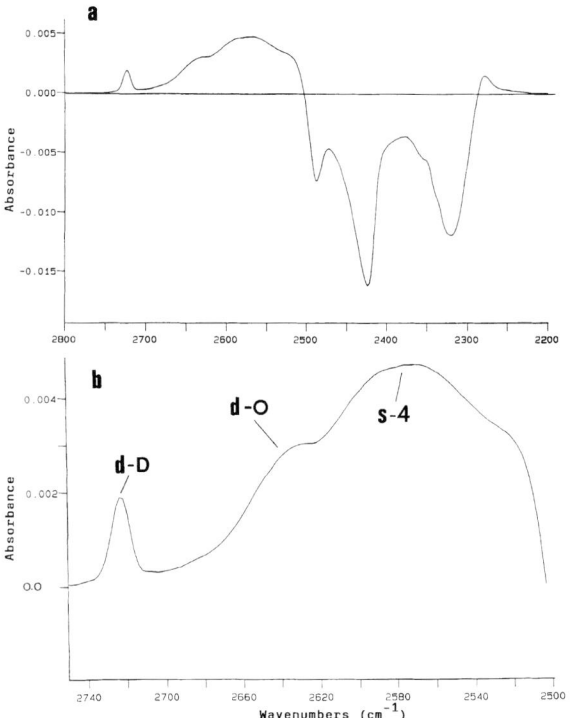

Fig. 6. Infrared difference spectrum (83 K) from spectrum of ice nanocrystals annealed at 120 K, minus the spectrum for the same ice sample annealed at 140 K. The bottom curve is an enlargement of part of the difference spectrum in (**a**).

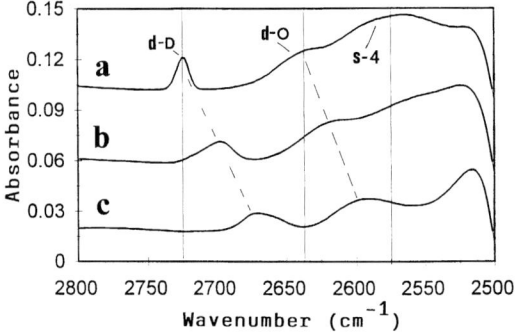

Fig. 7. Comparison of difference spectra (83 K), obtained by subtracting spectrum of larger from that of smaller nanocrystals, with the nanocrystals *a* bare, *b* covered with CO and *c* covered with ethylene.

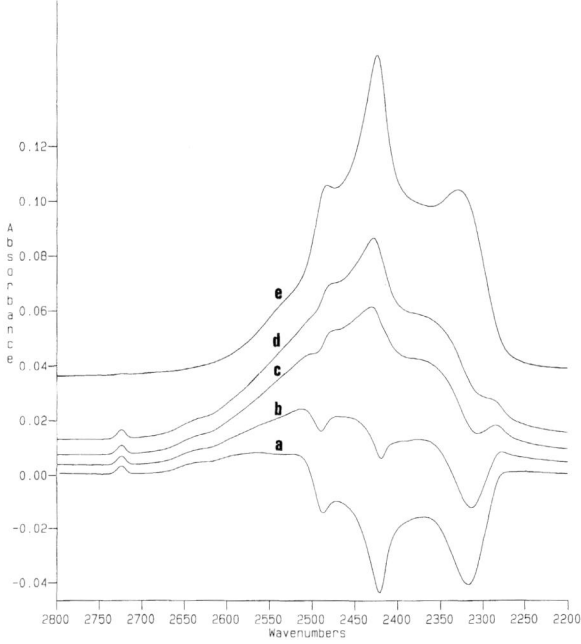

Fig. 8. Infrared difference spectrum (*a*) as in Fig. 6 but with annealing temperatures of 83 and 141 K. In curves *b*, *c* and *d* a percentage of the interior mode spectrum *e* has been added to the difference spectrum *a* to (partially) account for the surface/sub-surface molecules converted into interior molecules during annealing: *a* 10%; *b* 20%; *c* 25%

Fig. 8 corresponds to subtracting 100% of the spectrum of ice annealed at 141 K from that of fresh ice nanocrystals, with both spectra obtained at 83 K. Curves (b), (c) and (d) correspond to the addition of 10, 20 and 25% of the band intensity of the annealed ice (Fig. 8e) to the difference spectrum (a). It is clear that more than 20% of the interior band intensity must be added to eliminate the negative interior ice bands from curve (a). The inference is that $\sim 25\%$ of the annealed ice interior-mode intensity (Fig. 8e) develops during conversion of small into larger nanocrystals. That is, molecules with spectrum (d) are converted into molecules with spectrum (e) by the annealing at 141 K.

Since, for the fresh 20 nm nanocrystals, only $\sim 10\%$ of the molecules are in the surface bilayer, the implication is that $\sim 60\%$ of the increase in interior-mode intensity results from conversion of *subsurface* molecules into interior molecules. This result is consistent with the view that ~ 3 bilayers (i.e., six layers) of water molecules make up the combined surface and subsurface region [2].

An unfortunate consequence of the combined surface and subsurface contributions to the spectrum of Fig. 8d is that not all of the surface-isolated modes are

as surface molecules into interior molecules, and the simulations of the ice slabs with reconstructed surfaces indicate that disorder is significant for one or more sublayers of the surface bilayer.

For these reasons, when the interior ice-mode spectrum is added to a spectrum such as Fig. 6a, a complex result is obtained. Like Fig. 6a, the bottom curve of

apparent. Though the d-D, d-O and, to a lesser degree, the s-4 bands identified in Fig. 6 remain visible, the most prominent features of Fig. 8d, near 2480, 2425 and 2360 cm^{-1}, are attributed to subsurface-localized modes since they are not shifted by the presence of adsorbates, and are close in position, if not form, to the interior mode bands of Fig. 8e. Consequently, the combined surface and subsurface spectrum of Fig. 8d is best analyzed through comparison to simulated spectra for the first three surface layers of the relaxed slab of cubic ice [2].

Difference spectra between bare and coated nanocrystals. The identification of the remaining ice surface-isolated modes is nevertheless possible with reasonable confidence. One approach is to adsorb HCl at temperatures in the 40 K range. The acid slowly attacks the surface, converting much of the surface bilayer into broad-banded hydronium ions complexed with H$_2$O and Cl$^-$, but with relatively less influence on the ice subsurface layers. The difference spectrum, between the original and the surface after exposure to HCl, reveals six fairly distinct bands, three of which are the bands of the surface-isolated modes assigned in Figs. 5 and 6 [28]. One might conjecture that the other three bands, at much lower frequencies, are produced by the in-phase surface-isolated modes.

A second approach, the one demonstrated here, is to observe the difference spectrum between the ice surface when bare and when covered with a layer of molecular

adsorbate [2]. The adsorbate presumably shifts the surface-isolated mode bands while having little influence on the subsurface modes. Thus, in Fig. 9 we see that such difference spectra, for H$_2$, N$_2$, CO and C$_2$H$_4$ adsorbates, each have six maxima with similar frequencies. From these maxima and the simulation results (Fig. 5), the remaining three surface-isolated in-phase O–D stretch modes are assigned the approximate frequencies (cm^{-1}): d-D 2300; D–O 2480; s-4 2430. It can be noted that the adsorbate-induced shifts, as seen in the negative bands of Fig. 9, increase through the series from H$_2$ to ethylene, interfering progressively more seriously with the unshifted (positive) bands. The new assignments are consistent with the results for ice nanocrystals based on the effects of acid exposure.

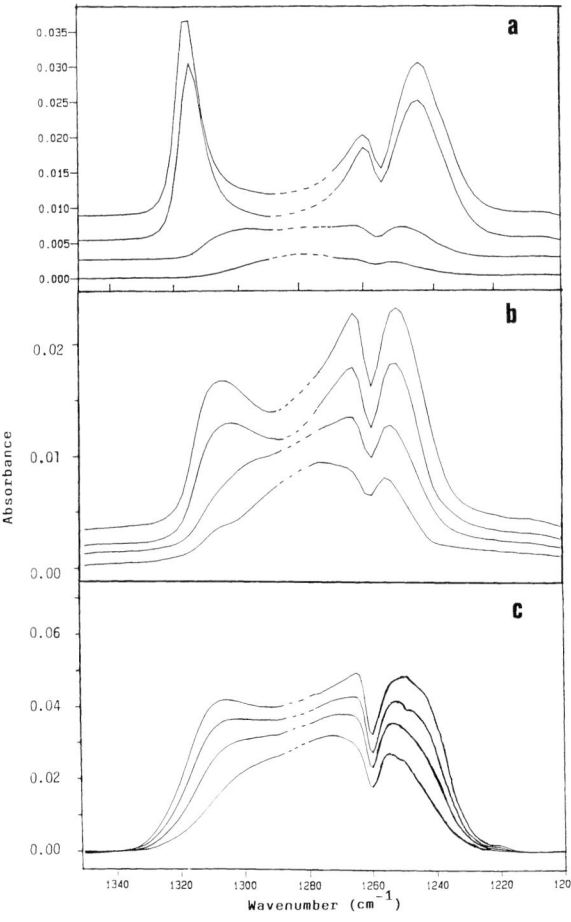

Fig. 10. The infrared band complex of the antisymmetric stretch mode of CF$_4$ adsorbed on ice surfaces at 83 K: from bottom to top of each series the CF$_4$ (g) pressures were 0.01, 0.02, 0.04 and 0.06 Torr. Samples were **a** 140 K annealed nanocrystals, **b** partially amorphous nanoparticles, **c** amorphous film

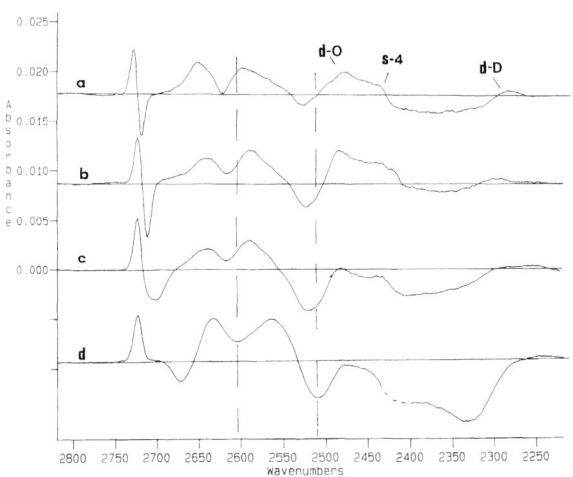

Fig. 9. Infrared difference spectra from spectrum of bare ice nanocrystals minus spectra of nanocrystals covered with adsorbate: adsorbate and scanning temperatures were *a* H$_2$, 30 K; *b* N$_2$, 83 K; *c* CO, 83 K; *d* ethylene, 90 K

Recognizing Surface Texture From the $CF_4\ v_3$ Band

The nature of the surface sites and the texture of a surface may represent the most, important factors in a heterogeneous process. For example, by entering molecule-sized crevices and holes of the ice surface, an HCl molecule may encounter the "wet" environment required for ionization. We have examined the identification of the surface sites and seen how occupation of those sites can be monitored from the shifting of the bands of the associated surface-localized vibrational modes. The texture of a surface is largely a separate question, requiring a different approach. Here we examine that question through the unique and sensitive response of the T-L band complex of adsorbed CF_4 to both the lateral regularity (crystallinity) and the vertical roughness of the CF_4 film and the supporting surface [25, 26].

The antisymmetric-stretch band of adsorbed CF_4, with increasing coverage, is shown for three different types of ice surfaces in Fig. 10. The curves of Fig. 10a reveal the emergence of dominant T (1243 cm^{-1}) and L (1320 cm^{-1}) bands as the CF_4 coverage of *annealed* ice nanocrystals approaches monolayer amounts at 83 K and 0.06 Torr. Broader and less well-defined T and L features are produced by CF_4 adsorbed on ice nanoparticles known to contain an amorphous fraction (based on the O–D stretch mode of isolated HOD) (Fig. 10b). When adsorbed on the micropore walls of an amorphous ice film (Fig. 10c) the CF_4 band continues to span the range from T to L, but the band structure is largely lost.

These series of spectra hint at the sensitivity of the band complex to the nature of the underlying surface and might suggest that the annealed nanocrystalline ice of Fig. 10a has a hexagonally regular surface. However, simulated spectra of adsorbed CF_4 show that in fact the spectrum of Fig. 10a is indicative of a vertically smooth but laterally irregular surface. Presented in Fig. 11 are the expected series for CF_4 on (a) the somewhat regular ice surface shown in Fig. 3a and (b) the strongly reconstructed surface in Fig. 3b. The latter series matches well the results of Fig. 10a, whereas the series of Fig. 11a differs qualitatively from the observed series. The conclusion is that the ice nanocrystals annealed at 140 K have a laterally irregular surface. Identifying factors are the considerable breadth of the T-mode band, the rapid reduction in the T-L splitting that accompanies reduced coverage and the broad centrally positioned band for CF_4 adsorbed at 0.02 Torr.

By contrast, spectra with greater intensity between the T- and L-mode peak positions, such as in Fig. 10 (band c) cannot be simulated without assumption of significant vertical surface irregularity. For example,

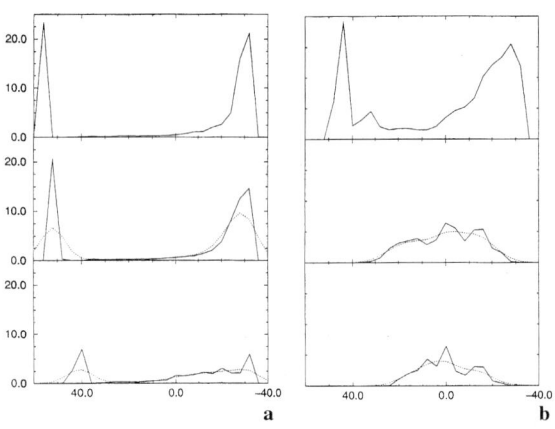

Fig. 11. Simulated spectra (83 K) of the antisymmetric stretch of CF_4 adsorbed on the model surfaces (**a**) of Fig. 3a and (**b**) of Fig. 3b with a gas overpressure of 0.01, 0.02 and 0.06 Torr, from bottom to top

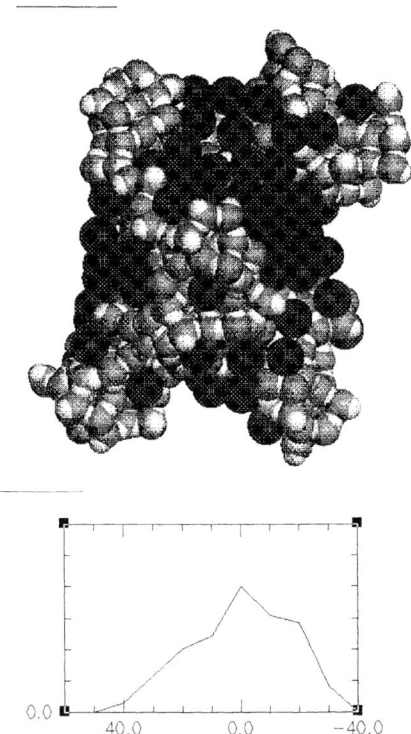

Fig. 12. Simulated image and spectrum of CF_4 adsorbed on a model amorphous ice cluster. Molecular graphics program: MOIL-VIEW as cited in [29]

the breadth and shape of the band complex, for CF_4 on microporous amorphous ice (Fig. 10c) is reasonably matched by using the cluster model of Fig. 12. The pools of adsorbed CF_4, within nm-sized hemipores, result in the broad simulated band at the bottom of the figure, with the band intensity centered near the decoupled frequency ($\sim 1275\,cm^{-1}$). At the other extreme, spectra with intensity only at the T and L positions, with the T-mode peak greater than that of the L (and with sharp T and L features, even at very low coverage), characterize a regular (crystalline) CF_4 layer (Fig. 13a) [26].

The experimental spectra for CF_4 on crystalline mesitylene and crystalline CO_2 nanoparticles, as given in Fig. 13 (b and c), show the T-L band characteristics expected from simulations for a smooth *regular* surface (Fig. 11a). The T and L features remain dominant and nearly unshifted for surface coverage as low as 40%, the T and L peaks are both sharp, and, for near

monolayer coverage, very little band intensity occurs at intermediate frequencies. Also, the peak intensity of the T-mode band is greater than for the L mode, as expected for a crystalline monolayer film (Fig. 13a). These are not the characteristics of the series of the Fig. 10a for CF_4 adsorbed on well annealed crystalline ice, confirming the view that the equilibrium ice surface is reasonably smooth but with significant lateral disorder (much as in Fig. 3b). A comparison of the spectra in Figs. 10–13 also indicates that the T-L band complex of the antisymmetric stretching mode of adsorbed CF_4 is a powerful indicator of surface texture that has general applicability.

Acknowledgements. JPD and VB acknowledge the support of NSF grant CHE-931 9176, PRF grant #29560-AC6, and BSF grant CHE-9022055.

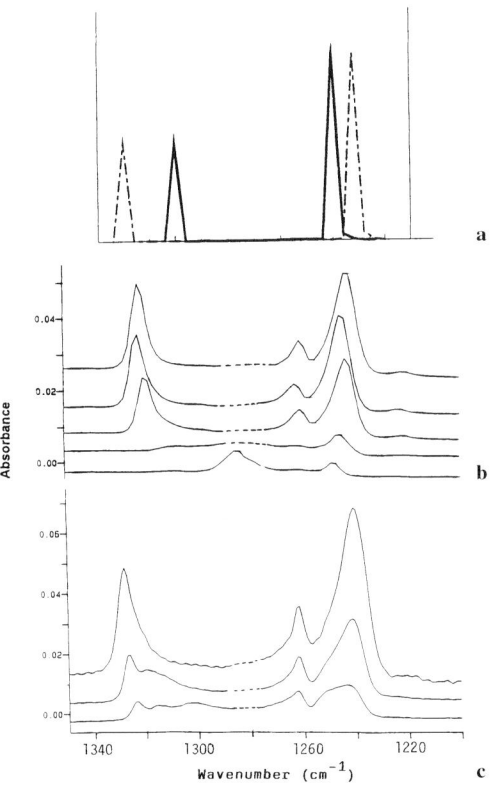

Fig. 13. Simulated spectra of CF_4 in a monolayer film (**a**) compared with observed spectra of CF_4 adsorbed at monolayer and sub-monolayer amounts on **b** mesitylene nanocrystals, **c** CO_2 nanocrystals. Simulations were for cubic (dot–dashed) and hexagonal (solid line) crystalline monolayers.

References

[1] G. E. Ewing, D. T. Sheng, *J. Phys. Chem.* **1988**, *92*, 4063.

[2] B. Rowland, N. S. Kadagathur, J. P. Devlin, V. Buch, T. Feldman, M. J. Wojcik, *J. Chem. Phys.* **1995**, *102*, 8328.

[3] R. Disselkamp, G. E. Ewing, *J. Chem. Soc. Faraday Trans.* **1990**, *86*, 2369.

[4] M. A. Ovchinnikov, C. A. Wight, *J. Chem. Phys.* **1994**, *100*, 972.

[5] J. A. Barnes, T. E. Gough, M. Stoer, *J. Chem. Phys.* **1991**, *95*, 4840.

[6] B. J. Koehler, L. S. McNeil, A. M. Middlebrook, M. A. Tolbert, *J. Geophys. Res.* **1993**, *98*, 10563.

[7] L. Delzeit, B. Rowland, J. P. Devlin, *J. Phys. Chem.* **1993**, *97*, 10312.

[8] A. B. Horn, T. Koch, M. A. Chesters, M. R. S. McCoustra, J. R. Sodeau, *J. Phys. Chem.* **1994**, *98*, 946.

[9] A. G. G. M. Tielens, L. J. Allamandola, in: *Physical Processes in Interstellar Clouds* (G. E. Morfill, M. Scholer, eds.), Reidel, Dordrecht, 1987.

[10] V. Buch, in: *Molecular Astrophysics* (T. W. Hartquist, ed.), Cambridge University Press, Cambridge, 1990.

[11] W. Hagens, A. G. G. M. Tielens, J. M. Greenberg, *Astron. Astrophys.* **1983**, *117*, 132.

[12] S. A. Sandford, L. J. Allamandola, T. R. Geballe, *Science* **1993**, *262*, 400.

[13] G. W. Gross, *J. Geophys. Res.* **1982**, *87*, 7170.

[14] W. L. Jorgensen, *J. Chem. Phys.* **1982**, *77*, 4156.

[15] S. A. Rice, M. S. Bergren, A. C. Belch, G. J. Nielson, *J. Phys. Chem.* **1983**, *87*, 4295.

[16] J. R. Reimers, R. O. Watts, *Chem. Phys. Lett.* **1983**, *94*, 222.

[17] M. J. Wojcik, V. Buch, J. P. Devlin, *J. Chem. Phys.* **1993**, *99*, 2332.

[18] V. Buch, J. P. Devlin, *J. Chem. Phys.* **1991**, *94*, 4091.

[19] H. G. Hixson, M. J. Wojcik, M. S. Devlin, J. P. Devlin, V. Buch, *J. Chem. Phys.* **1992**, *97*, 753.

[20] V. Buch, J. P. Devlin, *J. Chem. Phys.* **1993**, *98*, 4195.

[21] J. P. Devlin, *J. Phys. Chem.* **1992**, *96*, 6185.

[22] A. B. Horn, M. A. Chesters, M. R. S. McCoustra, J. R. Sodeau, *J. Chem. Soc. Faraday Trans.* **1992**, *88*, 1077.

[23] B. Rowland, M. Fisher, J. P. Devlin, *J. Phys. Chem.* **1993**, *97*, 2485.

[24] E. Honegger, S. Leutwyler, *J. Chem. Phys.* **1988**, *88*, 2582.

[25] B. Rowland, N. S. Kadagathur, J. P. Devlin, *J. Chem. Phys.* **1995**, *102*, 13.

[26] V. Buch, L. Delzeit, C. Blackledge, J. P. Devlin, *J. Phys. Chem.* **1996**, *100*, 3732.

[27] V. Buch, *Unpublished work.*

[28] L. Delzeit, J. P. Devlin, *J. Phys. Chem.* submitted.

[29] Drawings in this figure and in Figs. 2, 3 and 4 used the molecular graphics program: MOIL-VIEW as described in C. Simmerling, R. Elber and J. Zhang in *Modeling of Biomolecular Structures and Mechanisms*, A. Pullman (ed.), Kluwer, Netherlands, 1995.

Mikrochim. Acta [Suppl.] 14, 67–77 (1997)
© Springer-Verlag 1997

Medical Applications of Infrared Spectroscopy

H. M. Heise

Institut für Spektrochemie und Angewandte Spektroskopie, Bunsen-Kirchhoff-Str. 11, D-44139 Dortmund, Federal Republic of Germany

Abstract. Interest in the infrared spectral range for medical applications increases rapidly due to improvements in instrumentation and data processing. The use of lasers is fairly common for different tissue treatments, e.g. surgery. Optical imaging of tissue over greater distances, based on the relative transparency of biological tissue to short-wave near infrared radiation, is still in its infancy. Near infrared spectrometry is being investigated as a non-invasive clinical tool for improved understanding of in vivo processes. Some of the hottest research is carried out to obtain therapy-relevant biological parameters in human tumours. Of great interest is cerebral haemodynamics, as well as the intracellular reduction–oxidation (redox) level of cytochrome aa$_3$, e.g in the brain of human neonates. Further studies cover the transcutaneous determination of important metabolites such as blood glucose. Mid and near infrared spectroscopy can also be exploited in combination with discrete sampling. Samples can be from the great variety of biofluids, such as whole blood or derived fluids, or for example, urinary calculi or tissue samples even at cellular level. IR spectroscopy can deliver reagent-free multicomponent assays. In particular, the quantitative analysis of blood substrates and its evaluation for application in the clinical laboratory have been undertaken by us. Extra-corporal sensors can be based on such technology. The prognosis for future development is that IR spectroscopy will play an important role in medical diagnostic instrumentation close to the patient.

Key words: infrared spectroscopy, in vivo monitoring, clinical chemistry, metabolites, oxygen metabolism, histopathology.

The potential of infrared (IR) spectroscopy for clinical chemistry and biomedical applications has been increased by improvements in instrumentation and sample processing. There are two main fields, in vitro and in vivo studies, the latter of which can be carried out invasively, non-invasively or ex vivo, as in extra-corporal measurements. Recently a review on this broad subject of medical applications has been published [1].

IR spectroscopy plays an important role in the elucidation of structures and identification of organic and inorganic compounds and is one of the most important physical methods in the chemical laboratory, because of the substantial information content of mid and near infrared spectra. However, in the past, applications for medical investigations have been rather scarce. One field of early application was the analysis of urinary calculi by the KB pellet technique [2]. This is an example of the qualitative and quantitative analysis of principal or mediumly concentrated compounds. More refined data evaluation reveals this technique as a promising tool for routine quantification in the clinical laboratory [3]. An obstacle in the analysis of physiological samples was the high water content of these specimens, so that new emerging techniques such as attenuated total reflectance (ATR) or exploitation of the near infrared (NIR) were necessary to tackle the aqueous bioenvironment.

The principle of IR spectrometry for in vivo studies was mainly developed by using in vitro samples. It seems reasonable that such an approach needs further refinements for attempting quantitative in vivo measurements, as the boundary conditions due to physiological variations are more difficult to be observed than for in vitro studies. In 1977 Jöbsis published an important paper on the use of NIR spectroscopy for measurements of intact tissue of centimetre thickness [4]. Since then an upsurge has been noticed in biochemical and medical applications.

In order to monitor oxygen metabolism in biological systems scientists extended their observation to the near infrared, observing changes in absorbance between 760 and 950 nm, where haemoglobin shows significant absorption in both its reduced and oxygenated states. A similar component is myoglobin in muscle tissue. Another chromophore of interest is cytochrome aa$_3$ in the cell mitochondrial membrane, the terminal compound in the respiratory chain, and study of its redox state has been based on its spectral absorption features at around 840 nm. For these species, quantitative clinical monitoring studies have been successful. Cerebral monitoring, e.g. for neonates, has been achieved by monitoring a number of parameters, such as blood flow, oxygen delivery and blood volume and the response to changing the arterial CO_2 tension. For a recent review on the in vivo NIR applications see [5]; in another paper the various pitfalls and problems associated with the NIR measurement of mitochondrial cytochrome oxidase are discussed [6].

Other medical parameters, such as metabolites, are of great interest. Non-invasive glucose measurement is important for critically ill patients. Another enormous demand is for a sensor for self-monitoring of diabetic patients, particularly those who are insulin-dependent and undergoing a so-called intensified insulin therapy. Such a measurement is a challenge, as there is no "reporter" molecule available correlated to the blood glucose concentration and carrying a chromophore for the favoured short-wave near infrared (due to the silicon technology at the spectroscopists' disposal), similar to the compounds mentioned above. A logical decision was to exploit the conventional long-wave near infrared, where glucose absorptivities with low combination and overtone bands could provide a basis for direct spectrometry. Owing to the rather low concentrations, competition with other compounds is evident, so that sophisticated calibration techniques are compulsory if a reliable non-invasive glucose assay is to be achieved. My intention is to report on the status of mid and near IR spectroscopy in the medical field, and one part will cover our applications of different measurement techniques for in vitro and in vivo spectroscopy.

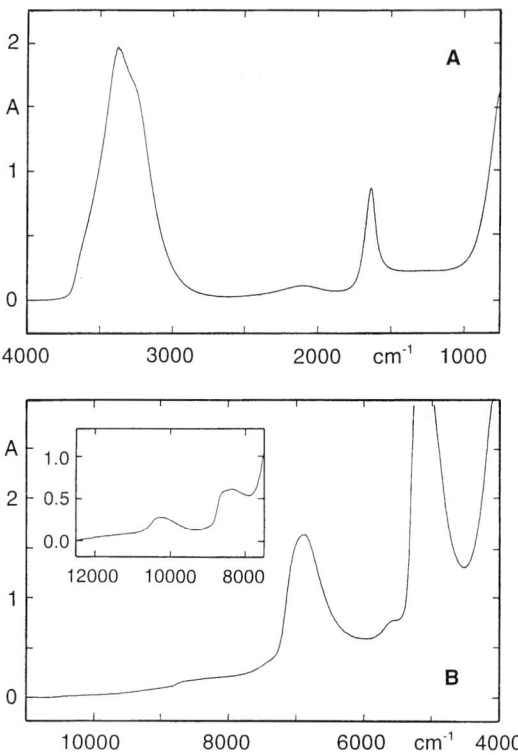

Fig. 1. Absorbance IR-spectra of water. **A** ATR-spectrum measured by using a micro-Circle cell (temperature 37 °C). **B** Spectra measured by transmittance, temperature 25 °C, cell path-length 1 mm and for short-wave NIR 10 mm (see inset)

In vitro Studies

Infrared spectroscopy is a widely applied technique for obtaining qualitative and quantitative results for a great variety of samples. Advances in direct spectrometry due to improvements in measuring equipment and refinements in data evaluation by chemometrics provide a significant impact on analytical chemistry in the biomedical field. In general, the mid infrared spectral range contains more selective information than does the near infrared, where mainly broad overtone and combination bands of CH–, OH– and NH– vibrations are seen. Structural changes can still be easily distinguished, as has been shown by using near IR photoacoustic spectra of homo-polypeptides [7]. Since the intensity of fundamental bands in the mid infrared is high, short optical path-lengths of less than 50 µm are required, in particular for physiological samples of high water content. ATR is an appropriate technique and typically uses optical materials compatible with aqueous biofluids. Optical path-lengths in the near infrared are in the range of millimetres, or even centimetres if the short-wave near infrared is considered for spectroscopic analysis (see Fig. 1).

Studies of dissected tissues have been presented, helping the pathologist in identifying tissue status. In the past, such studies had been hampered because of the strong absorption of tissue water and difficulties with sample preparation. Thin slices obtained with a microtome can now be investigated [8, 9]. Microspectroscopic studies have also been undertaken to probe the sub-cellular chemical composition of atherosclerotic arterial walls [10]. Another possibility for preparation is provided by cell smears, as performed with exfoliated human cervical cells [11]. The non-invasive characterization of intact microbial cells was recently described by Naumann et al. [12].

Mid IR ATR spectroscopy has been frequently applied to study of the secondary structure of polypeptides, proteins and membrane proteins in biological systems. Conformational changes, e.g. in bladder mucosa, are displayed in the spectra [13]. The same applies to other disease-induced tissue changes, which are preceded by alterations at the molecular level. An informative review on the IR spectroscopic tools in medical diagnosis and pathology was given by Mantsch and Jackson [14]. Diseases such as multiple sclerosis, Alzheimer's disease and cancer have attracted spectroscopists and physicians aiming at rapid diagnostic tests; for further references see [14].

Another measurement technique is photoacoustics, which offers the possibility for non-destructive depth profiling and the IR spectroscopic characterization of tissue biopsy samples such as brain tissue or calcified tissues like bone and teeth [15]. Additionally, physiological fluids have also been studied, in particular for their glucose concentrations [16, 17].

Recently, near infrared studies were reported on the identification of air-dried breast carcinomatous tissue by reflectance spectroscopy [18]. Four wavelength intervals were found to differentiate between normal and carcinomatous tissues. With the use of principal component and discriminant analysis, near infrared spectra of pap smears ($5000–4000\,cm^{-1}$) could be classified for screening purposes [19]. A different multivariate approach for the analysis of near IR spectra obtained by a fibre-optic probe has been used for arterial analysis [20].

A wide variety of fluids can be measured, for example blood, interstitial fluid, aqueous humour, lymph, urine, sweat, saliva, tears, cerebrospinal fluid. Often the identification of extravasal fluids is necessary, or their quantification for testing clinical significance. Dry films from synovial fluid have been investigated for CO_2 clathrates as a marker for arthritis [21]. Another fluid studied by the same technique was saliva as reported by Ahmed and Mantsch [22]. We prefer, instead, the ATR technique in combination with scaled water absorbance subtraction. In Fig. 2 such a difference spectrum is shown. Saliva contains many glycoproteins rich in sialine acid. Saliva has received much attention in monitoring blood glucose, because it is easily accessible (the latest unsuccessful example is given in [23]). As demonstrated by many studies, there is no correlation between the glucose concentrations in blood and that found in other accessible fluids.

The horizontal ATR accessory is very convenient for the study of fluids and tissues, as it also allows in vivo measurements (other techniques for this field, that are possible with transmittance or diffuse reflectance measurements, are discussed in the following section). On the basis of the ATR technique, Kaiser had proposed a non-invasive blood glucose assay using mid IR spectrometry [24]. In a recent paper by Kajiwara et al., results were reported from transcutaneous measurements of oral mucosa [25]. We also have studied this tissue extensively, but with negative results, owing to the layered skin structure [26]; see also Fig. 2 for spectra of dry inner and outer lip tissue. One must be aware of the minute radiation penetration depth inherent to the ATR technique. This is quite evident from Fig. 3, where the spectra of dry and wet inner lip tissue, and saliva, of the same person are shown, with compensation for water absorbance. Figure 4 shows the penetration depth of the electrical field at which its amplitude is exponentially damped to $1/e$ of the incident field strength, dependent on the reflection angle for the absorbing medium of water at $1000\,cm^{-1}$ (the ATR crystal material was ZnSe); also given is the reflectance for s- and p-polarized radiation. If we consider the depth *vs.* radiation power as being proportional to the square of the field strength, the values must be halved! This gives an idea of the layer thickness and, eventually, the mass of substance needed for spectrum recording.

Since our first papers on the ATR spectroscopy study of whole blood and blood plasma for quantitative analysis of several substrates such as glucose, total protein, total cholesterol, triglycerides, urea and uric acid [27, 28], significant improvements were made, so

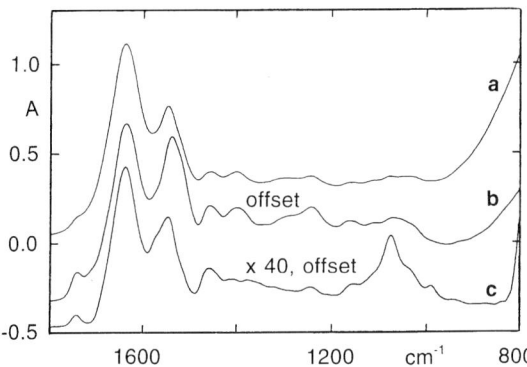

Fig. 2. ATR-spectra of different biological samples measured by a horizontal ATR accessory (ZnSe-crystal for 45° reflection). *a* Dry oral mucous tissue of the inner lip. *b* Skin from outer lip. *c* Saliva components after scaled water absorbance subtraction

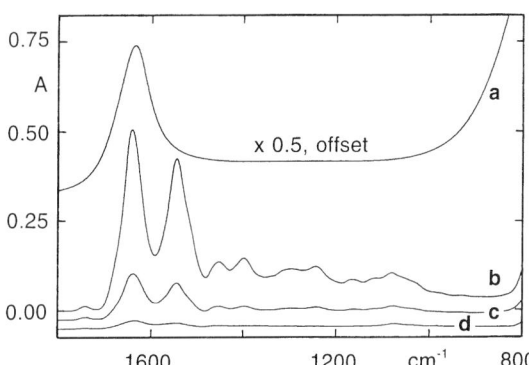

Fig. 3. ATR-experiment demonstrating limited optical penetration depth. *a* Water absorbance with horizontal ATR accessory. *b* Inner lip tissue after scaled water absorbance subtration. *c* Inner lip with a film of saliva after scaled water absorbance subtraction. *d* Saliva components after water absorbance compensation

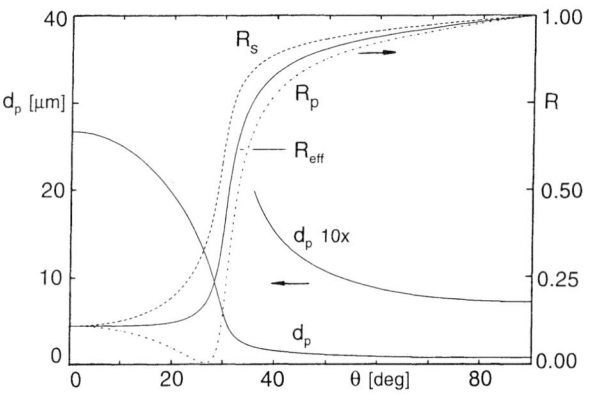

Fig. 4. Penetration depth d_p of the electrical radiation field and Fresnel-reflectances R_s and R_p for the interface ZnSe and water at $1000\,cm^{-1}$ plotted against the angle of reflection ($n_{ZnSe} = 2.41$ and $n_{H2O} = 1.212 - i \times 0.0552$); effective reflectance for unpolarized radiation is $R_{eff} = 0.5\,(R_s + R_p)$

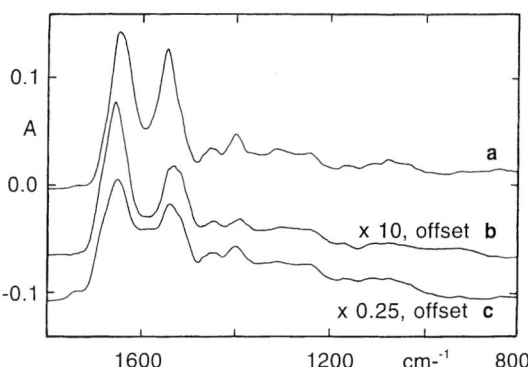

Fig. 5. Measurement techniques for blood substrates in blood plasma. *a* ATR-spectrum of a blood plasma sample, using a micro-Circle cell. *b* Protein residue on ATR-crystal after plasma measurement and water rinsing. *c* DR-spectrum of dry blood plasma film

that clinically acceptable performance can now be reached [29, 30]; see also Table 1. For routine analysis, it is necessary to time the measurements and to clean the flow-through cell (micro-Circle cell) with detergents after recording plasma samples, owing to the protein adsorption kinetics experienced with such biotic fluids. In Fig. 5 the problem is clearly demonstrated for an experiment where the plasma resided for two minutes in the cell and was subsequently purged with water only. The adsorbed protein shows much smaller degree of glycosilation. There are other limitations as shown in [31]. An extensive investigation on spectral interferences in use of blood serum for glucose assay has been presented by Bhandare et al. [32].

Investigations were made by the author, of different measurement techniques which could favourably be used in a reagent-free clinical multicomponent analyser. The same plasma sample was transferred onto a diffusely reflecting substrate as a film and dried at room temperature for a few minutes. The diffuse reflectance (DR) spectra were recorded by using a device developed by Korte and Otto [33], which discriminates for Fresnel-reflectance. Spectra taken after different time intervals showed that crystallization and other effects have to be considered in the calibration stage (see Fig. 6). The time-dependent changes are noticeably smaller with serum samples.

For continuous monitoring of blood glucose, however, another technique, the coupling of microdialysis with an ATR flow-through cell, provides a sensing device that is an alternative to enzymatic amperometric biosensors. The dialysis membrane filters out com-

Table 1. Mean square prediction errors of various partial least-squares calibration models for different blood substrates based on plasma spectra from 125 patients (optimum results for each spectral range in italics; see also text)

Model		Total protein [g/l]	Total cholestrol [mg/dl]	Triglycerides [mg/dl]	Glucose [mg/dl]
SW-NIR[a]	H₂O comp. absorbance	1.16	*14.9*	23.5	
SW-NIR[a]	-log (single beam)	*1.08*	15.4	*23.4*	47.3
NIR[b]	H₂O comp. absorbance	0.98	8.3	13.7	18.0
NIR[b]	-log (single beam)	1.07	*7.7*	*12.1*	*16.2*
MIR[c]	H₂O comp. absorbance	1.22	*8.1*	10.3	10.4
MIR[c]	-log (single beam)	*0.90*	8.2	10.3	*9.8*

[a] Spectral range for all substrates 11015–7621 cm⁻¹, sample path-length 10 mm.
[a] Spectral range for protein 6001–5508 cm⁻¹, for cholesterol and triglycerides 6001–5508 and 4520–4212 cm⁻¹, for glucose 6788–5461 and 4736–4212 cm⁻¹, sample path-length 1 mm.
[c] Spectral range for protein 1700–1351 cm⁻¹, for cholesterol 3001–2800, 1800–1701 and 1500–1099 cm⁻¹, for triglycerides 1800–1701 cm⁻¹ and 1500–1099 cm⁻¹, for glucose 1200–950 cm⁻¹, measured by micro-Circle cell.

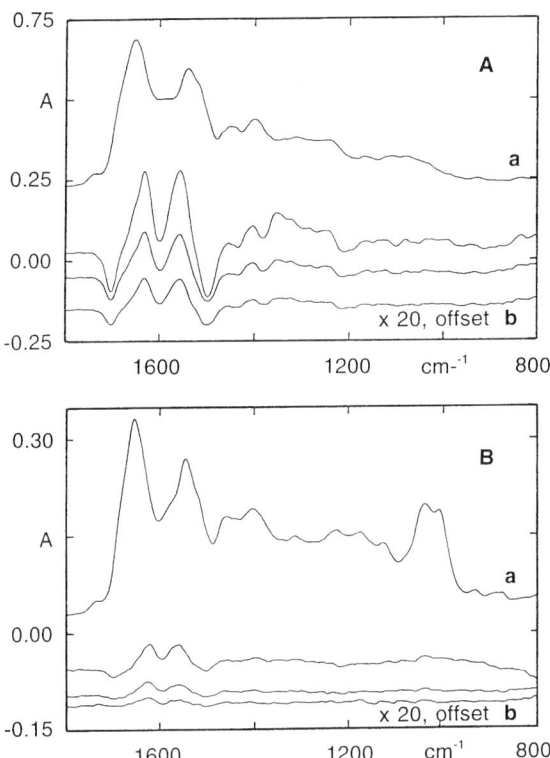

Fig. 6. DR-spectra of dry films from blood plasma (**A**) and serum (**B**). *a* Spectrum after 5 min of fluid deposit. *b* Difference spectra for additional time lapse of 3, 7 and 60 min, respectively (from bottom to top)

pounds of high relative molar masses, at the cost of analyte dilution. In Fig. 7, a comparison of the spectra from the original blood plasma and the microdialysate (perfusion flow-rate 3 μl/min), recorded by using the micro-Circle cell, is shown. The aqueous glucose absorption bands at above 1000 cm^{-1} are evident. Quantitative results are reported in a separate paper of this issue [34].

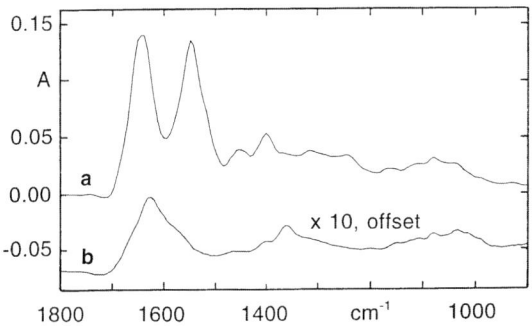

Fig. 7. Measurement techniques for glucose in blood plasma. *a* ATR-spectrum of blood plasma. *b* ATR-spectrum of microdialysate from the same plasma obtained at perfusion flow-rate of 3 μl/min, using a micro-Circle cell

So far, we have mainly discussed mid infrared measurements for substrate analysis. The near infrared has repeatedly been chosen for in vitro measurements, because of the ease of sample handling. Transmittance cells with quartz or glass windows having optical path-lengths in the millimetre range can be used. Another reason for choosing this range is that similar in vivo measurements rely on a sufficient penetration depth into tissue, which is achieved by using near infrared radiation. An idea of the different spectral features to be expected in multivariate analysis is provided by Fig. 8. Different strategies for the data evaluation have been developed, in particular for glucose measurements, see e.g. [35]. With our approach, multivariate spectral analysis, based for example on partial least-squares, allows us to monitor low concentrations of physiological components against an extremely large and varying background. One prerequisite is certainly the selection of appropriate wavelength intervals to obtain optimum prediction models. We have investigated the information content of different spectral regions, in particular the mid, near and short-wave near infrared, and the calibration performance based on such data. These studies were also of interest for potential in vivo applications. In Table 1 a comparison of prediction results is given for total protein, total cholesterol, triglycerides and glucose based on the same population of plasma samples from a hospital population [26, 29–31, 36, 37]. Average concentrations found were as follows: total protein 70.5 g/l, total

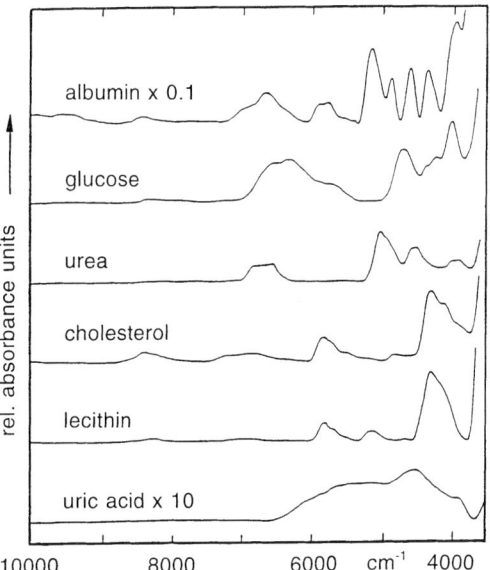

Fig. 8. Blood substrate spectra in the near-infrared. For proteins, albumin was chosen as a model compound and lecithin for triglycerides; the absorbances were scaled to the mean substrate concentrations in blood plasma

cholesterol 219 mg/dl, triglycerides 163.1 mg/dl and glucose 207.5 mg/dl. For the highly concentrated protein, the calibration performance is not very much affected by picking different spectral ranges. This is not the case for total cholesterol and triglycerides, where the short-wave NIR provides prediction results worse by a factor of two than those obtained with the mid IR and long-wave NIR. Completely different results are achieved for glucose, where we find best selectivity with mid infrared data. The NIR calibrations are certainly affected by the variances arising from the aqueous matrix. Water absorption bands here are known to be very sensitive to changes in temperature, electrolyte and protein content. Nevertheless, in vivo measurements do not seem to be totally out of reach.

Other quantitative measurements, in particular for whole blood, are also reported. The measurement of haemoglobin in unlysed blood could be realized with single-term second-derivative ratio of log $1/T$ data at 1740 and 1346 nm [38]. An important parameter, especially for blood used in transfusion during surgery, is blood durability. By measuring the blood samples in their original plastic bags, it was shown that blood changes during storage could be followed by NIR spectroscopy [39].

Quantitative measurements of discrete samples, e.g. of body fluids such as blood and derived fluids such as plasma and serum can be made by using multivariate calibration techniques. IR spectroscopy can also be a powerful adjunct to histopathology. IR spectroscopy shows great promise as a tool in medical diagnosis and pathology, in particular because of its non-destructive analysis capabilities. More applications such as, for instance, the characterization of inclusions bonded to tissues in implants, and their transformation into material with properties close to biological tissues [40], will be seen in the future.

In vivo Measurements

In vivo studies require measurement techniques which do not interfere with the integrity of the living object. Spectroscopy in the visible and infrared range has been used for some time to quantify endogenous chromophores in tissues, such as haemoglobin, myoglobin and cytochromes, and to differentiate their chemical states that are important for oxygen metabolism. The basis for this are changes in the spectral absorption coefficients, which can be related, for example, to blood oxygenation and tissue metabolites. Diagnostic techniques are also desired for discriminating between healthy and malignant tissues, e.g. those originating from cancer.

Non-invasive optical studies compete with other measurement techniques such as X-ray or positron emission tomography. Although these optical methods are non-invasive, cell damage on the molecular level might occur when high energy radiation is involved. Sonography or magnetic resonance spectroscopy are other techniques for non-invasive measurement; the latter, in particular, is of great importance in the medical field, as it renders good possibilities for patient imaging, as well as for the in vivo analysis of physiological processes. Frequencies up to 400 MHz are applied with magnetic field intensities up to 4 T. The instrumental apparatus and the computer capacity needed, however, are quite enormous (for a review on the current status of in vivo MR spectroscopy see [41]).

Optical tomography identifying hidden objects non-invasively also belongs to the category of in vivo measurement. Optical spectroscopy has shown a promising potential (see for example [42]). It is in particular discussed as an alternative to X-ray mammography for studying the female breast for cancer detection. In the literature, extensive time-resolved studies have been published, e.g. [43]. Ultimately, it may be possible to localize tissue abnormalities by using near infrared spectroscopy analogously to computed tomographic X-ray imaging; for an overview on the basics of optical tomography see Wilson et al. [44].

The increasing use of infrared radiation in diagnostic and therapeutic medicine has created a need to understand radiation transport in tissue, especially since fluence–depth profiles and diffuse transmittance and reflectance are important for the quantitative analysis of diagnostic measurements aimed at drug and metabolite concentrations or the degree of blood oxygenation. This is the broadening field of diagnostic spectroscopy, whereas photochemical interaction is the goal in photodynamic therapy, e.g. of cancer, where the local effect is certainly dependent on the intensity of visible radiation, which is absorbed by specific molecules, so-called photosensitizers, added to the tissue. Thermal interaction is the primary effect needed in laser surgery, where absorption of the energy in the form of infrared radiation by the main tissue constituent, i.e. water, occurs, so that tissue coagulation and ablation is the result. Photophysical processes, such as thermal and photochemical interaction which need to be considered in medical laser applications have been reviewed by Boulnois [45].

Knowledge of the spatial distribution of photon-visit probability is not only important for imaging, but also for quantitative diagnostics, since it defines the effective tissue volume which is probed by in vivo spectroscopy. For such a purpose, modelling of the radiation transport in tissue is required. Additionally, internal fluence rates can be derived, necessary for medical treatment, where the radiation dose is important, e.g. for photochemical and thermal reactions. Another question concerns the average depth from which the information is derived. An example with simulation and measurement of signals in laser flowmetry on human skin can be provided [46]. Estimation of the wavelength-dependent optical path-length through tissue is required in order to quantify absolute analyte concentration values [47]. The macroscopic tissue volumes for medical measurements by diffuse transmittance or reflectance are in the order of some mm^3 up to cm^3, when radiation from around 600 to 1300 nm, the so-called therapeutic window, and longer wavelengths are used. Here we find rather low absorption by water, which clearly dominates the in vivo spectra in the long-wave NIR; absorption at shorter wavelength is due to tissue chromophores such as haemoglobin, melanin and others [45].

For biological tissues we find anisotropic scattering, which is highly forward-directed due to discontinuities in the refractive index at the cellular level [48]. This is the reason why simple theories such as that derived by Kubelka and Munk [49], which are very popular for the mid infrared, are not valid, so that more sophisticated modelling is required (see, e.g. [50]). For that purpose reliable optical constants of body tissues are also necessary.

The radiation can be delivered by optical fibres or reflection optics and measurements may be made by utilizing the diffusely transmitted or reflected intensity. The spectra do not depend only on the optical parameters of the tissue studied, but also depend on the collection geometry used for collecting the photons interacting with the in vivo tissue. To explain our spectra and for a better understanding of the radiation transport, e.g. by means of maximum penetration depths and integral photon path-lengths, we started extensive investigations using Monte Carlo simulations of the "photon random walk" through tissue [26, 51, 52]. The optical parameters needed are shown in Fig. 9A, which presents the schematics of the Monte Carlo simulation. An optimized custom-made accessory based on mirrors (see Fig. 9B) for measuring diffuse reflected radiation, and a fibre optic probe, were used for in vivo skin measurements. The achievable spectral performance (noise level and practicability aspects under the constraint of the acceptable measurement time of one minute) obtained by using these accessories and the results from Monte Carlo simulations are compared and presented in another paper in this issue [52]. In Fig. 10 the DR spectra of two different skin tissues, viz, oral mucosa and thumb skin, measured by means of these accessories are displayed.

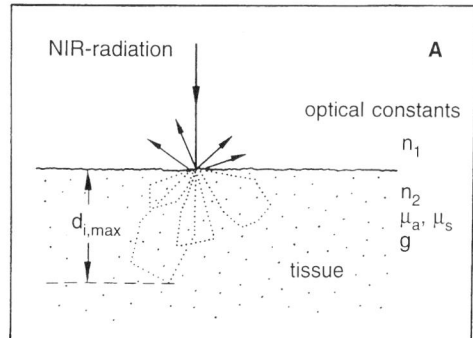

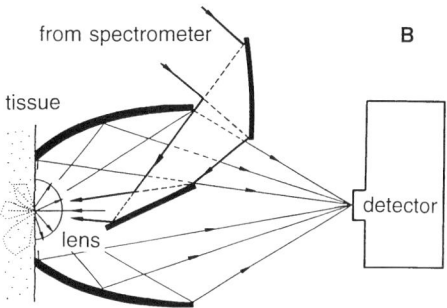

Fig. 9. Schematics of the skin tissue DR-experiment. **A** The following optical constants are necessary to model radiation transport: refractive index n_1 for the material in contact with skin, n_2 refractive index, absorption and scattering coefficients μ_a and μ_s, and anisotropy parameter for the scattering phase function g for skin tissue. **B** Optimized DR-accessory based on mirrors

As a different solid angle is seen by each of the devices, a different average penetration depth for IR radiation is achieved. The numerical aperture of the fibres allows only for an inner cone with half opening angle of $9°$, whereas the reflection optics accumulates radiation from the hemisphere above the sample, except for the illumination cone with half opening angle of $35°$. The difference in penetration depth is clearly seen in the doublet band below $6000 \, cm^{-1}$, which indicates that subcutaneous fatty tissue is also probed by the optic fibre (see also [51]). Additionally, the water band intensities differ significantly, which can be correlated to the mean integral photon path-length. An even greater gap is observed between the spectra of thumb tissue, for which a top layer of highly scattering stratum corneum exists. The spectra measured by the DR accessory are much affected by this scattering, whereas the spectrum recorded by the fibre optic probe is similar to that of the mucous tissue except for keratin spectral features at around $5600 \, cm^{-1}$.

We are mainly interested in the metabolic status of the body, i.e. studying the concentration of meta-

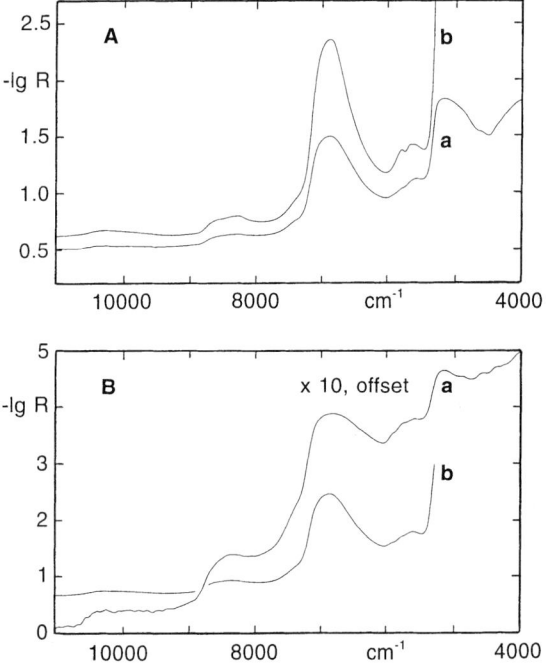

Fig. 10. DR-spectra of mucous tissue of the inner lip (**A**) and thumb skin (**B**) recorded by using different accessories: reflection optics (*a*), and fibre optic probe (*b*)

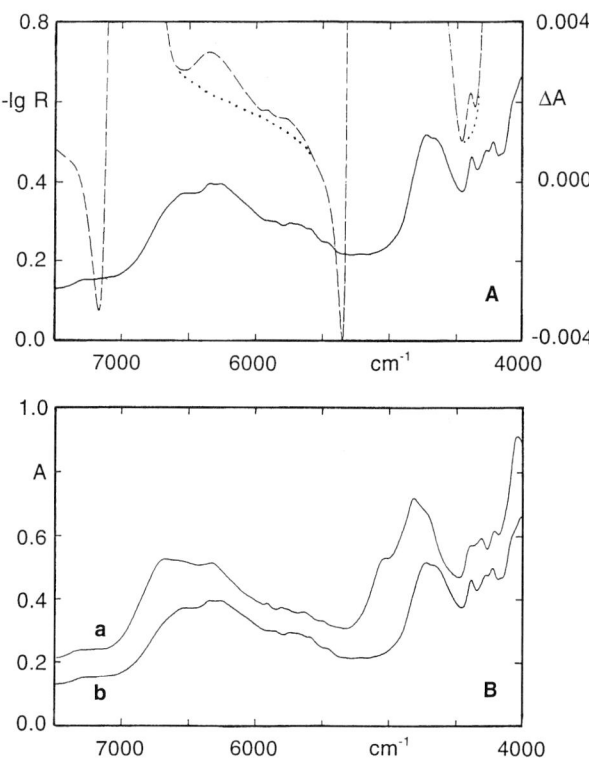

Fig. 11. Near-IR spectra of glucose. **A** Absorbance difference spectrum of aqueous glucose (c = 800 mg/dl, path-length 1.5 mm; water partly compensated, dashed curve) and DR-spectrum of anhydrous crystalline glucose (solid curve). **B** Comparison of DR-spectra of glucosemonohydrate (*a*) and crystalline anhydrous glucose (*b*)

bolites, in particular glucose in blood, by spectroscopy of skin tissue, which contains a high density of blood vessels. Referring to the results of our in vitro studies (see above), we were led to the use of diffuse reflectance spectrometry at around 6400 cm^{-1}. We conducted many experiments with a single diabetic person, in addition to a population of over 130 different patients. The results from our non-invasive glucose assay are well described in previous publications (see [26, 53–55]. The mean square prediction errors from multivariate partial least-squares calibration were at best around 45 mg/dl with single person calibration. This performance is so far not acceptable for monitoring in the normal and hypoglycemic concentration range (about 120 mg/dl and below). Further references on non-invasive glucose assays reported in the literature can be found in [55].

The boundary conditions for glucose monitoring can be estimated from Fig. 11A, where a difference spectrum of aqueous glucose with water absorption partly compensated is shown. For clarity the DR spectrum of anhydrous crystalline glucose is also shown. This demonstrates clearly the magnitude and availability of sugar absorptivities for a non-invasive glucose assay. To quantify glucose at a concentration of 10 mg/dl with an optical path-length around 1 mm the spectral noise level should be preferably an absorbance

of below 10^{-5}, which can at the moment only be achieved by using reflection optics, as long as we have to rely on conventional FT-NIR spectrometers. It is quite interesting to see which glucose bands are influenced, when a spectrum of glucose monohydrate is compared with that of the anhydrous form (Fig. 11B).

We have also studied the variances involved in a one-person experiment with repositioning of the DR equipment. In Fig. 12 the mean lip spectrum and its standard deviation from eight measurements with low and constant blood glucose concentration are shown. The spectral variation experienced is best seen by plotting the difference spectra against the mean. A factor analysis provides us with the principal components, against which other spectra can be orthogonalized [56]. Some difference spectra are presented in Fig. 12C and 12D, the residues of which clearly make evident how important the modelling of physiological variations is, if classical calibration strategy is to be used for a glucose assay.

A few more advanced applications of in vivo NIR spectroscopy will be reported to conclude this review. There is a qualitative interpretation of skin DR spectra

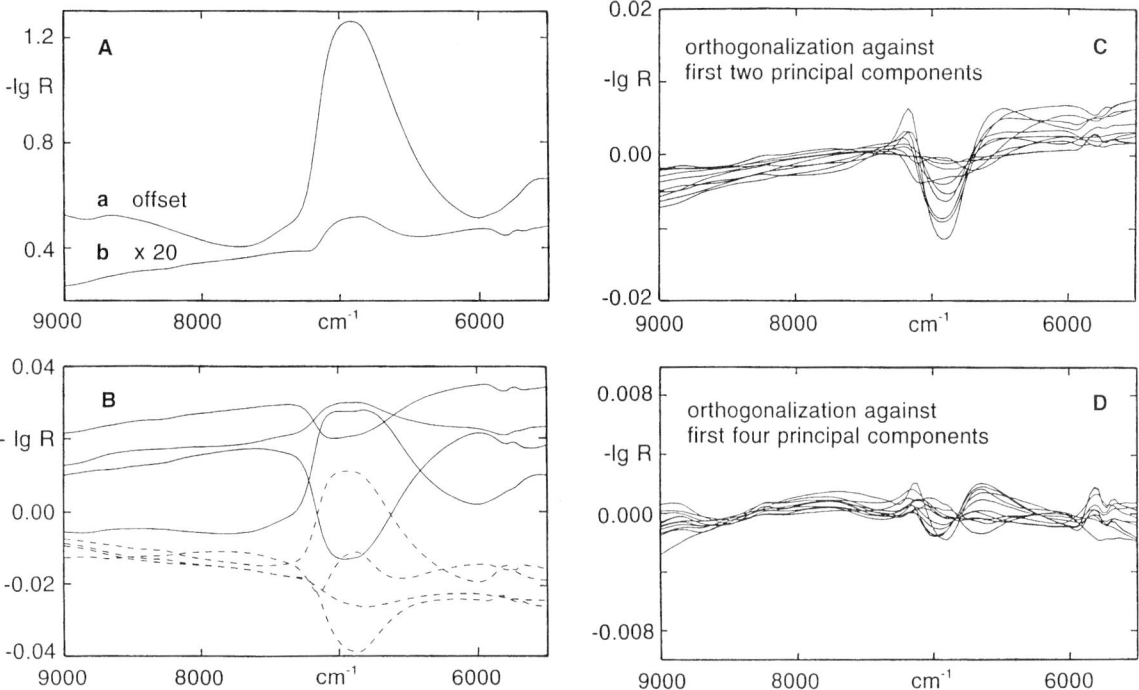

Fig. 12. Variations in the logarithmic single-beam DR-spectra of the inner lip of one test person. **A** Mean of eight spectra (*a*) and their standard deviation (*b*). **B** Difference spectra versus mean spectrum. **C, D** Residuum spectra for a different data set after preprocessing with factor spectra obtained from the data set shown in **A** and **B**

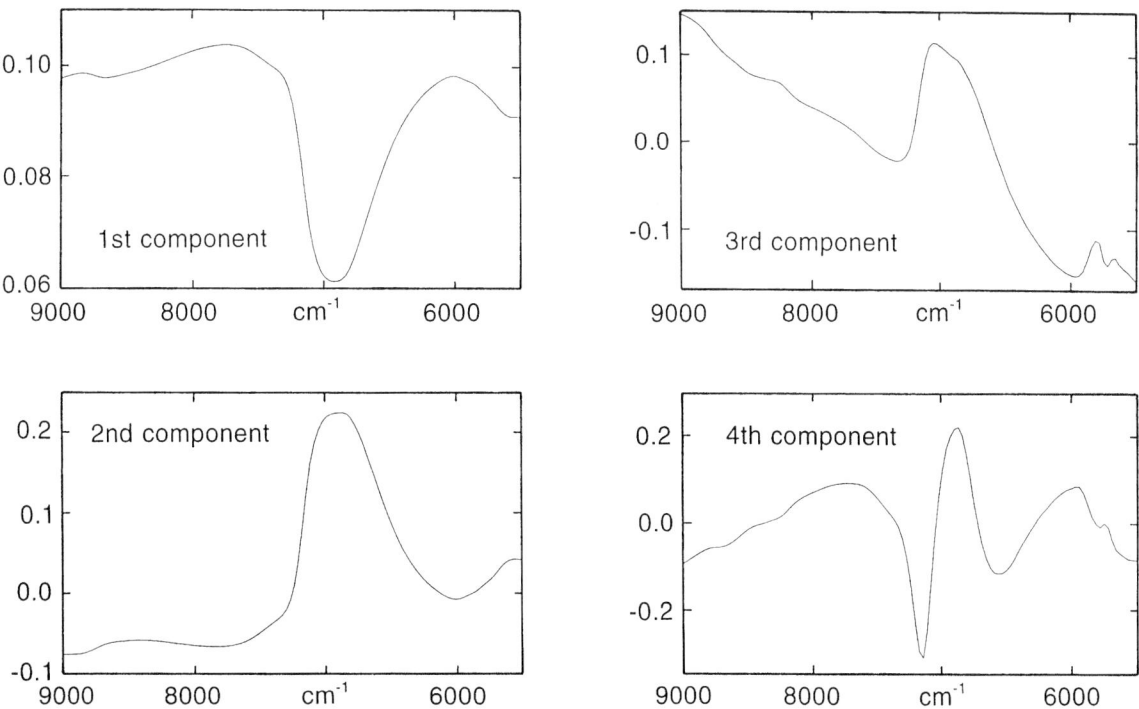

Fig. 13. First factor spectra obtained from a principal component analysis of background lip spectra (see also Fig. 12)

to determine moisturization [57]. Some other near IR studies, on aging and stroke causing brain-tissue damage in gerbils, were more than skin-deep. Non-invasively recorded DR spectra between 1.1 and 2.5 µm were evaluated to reveal age effects and the amount of brain-tissue oedema, from the changes in protein and lipid composition. The results could be used for therapies to reduce ischaemic and post-ischaemic damage [58].

There is increasing interest in NIR spectroscopy to monitor signals related to neural activity [59]. Other measurements were made to monitor changes in cerebral haemodynamics during inspiration and expiration. Pulsed laser diodes giving four wavelengths between 780 and 900 nm were used as the NIR radiation source [60]. Further investigations studied the effect of different pharmaceutical agents, for instance indomethacin or aminophylline, on the cerebral oxygen metabolism in premature infants, see for example [61]. With adaptation of this technique the intravasal oxygen content in tumours of nude mice has been measured [62]. A recent study on Hb, HbO_2 and $Cytaa_3$ by a total least-squares technique, shows clearly the influence of the number of wavelengths used, as well as the wavelength subset considered. Optimization of the linear equation system by minimizing its condition number leads to much improved prediction performance [63].

Conclusions

One trend is certainly towards decentralization in the hospital, which means that the analytical instruments move closer to the patient where the answer is needed, thus avoiding central laboratories. This move is parallel to the development of non-invasive measurement techniques, which are, by implication, non-destructive, so monitoring of physiological variations becomes possible.

Due to the complexity of the biological systems and the analytes to be studied, the amount of data from spectroscopic techniques, i.e. multivariate spectra of high dimensionality, will increase tremendously, partly because faster measurements are possible, so that sophisticated data reduction becomes vital. This is the domain of chemometrics, which will play a more important role, in particular as the power of computer facilities is steadily growing.

For a better understanding of in vivo biological systems, including their physiological variances, correct spectral modelling is absolutely essential. Best measurements of high reproducibility with consideration of tissue variability must be taken into account to reach the accuracy necessary for practical application for the patient at home or in the hospital, and for safe

operation providing warnings and trends. NIR spectroscopy can yield clinically relevant information, in particular, on the energy metabolism in tissues of patients with arteriovascular disease, metabolic disorders, tumours and acute or chronic hypoxia.

Acknowledgements. Financial support by the Ministerium für Wissenschaft und Forschung des Landes Nordrhein-Westfalen, the Bundesminister für Bildung und Forschung, the Deutsche Forschungsgemeinschaft and Boehringer Mannheim GmbH (Mannheim, Germany) is gratefully acknowledged. The author also thanks Profs. Th. Koschinsky, F.A. Gries and H. Reinauer from the Diabetes-Forschungsinstitut (Düsseldorf, Germany) for support and discussions. I would also like to thank Dr.-Ing. R. Marbach, Mr. A. Bittner and Mrs. M. Hillig for their invaluable assistance.

References

[1] H. M. Heise, *Laboratoriumsmedizin* **1991**, *15*, 470.
[2] A. Hesse, G. Schrumpf, I. Schilling, *Z. Urol. Nephrol.* **1974**, *67*, 367.
[3] M. Volmer, A. Bolck, B. G. Wolthers, A. J. de Ruiter, D. A. Doornbos, W. van der Slik, *Clin. Chem.* **1993**, *39*, 948.
[4] F. F. Jöbsis, *Science* **1977**, *198*, 1264.
[5] J. S. Wyatt, D. T. Delpy, in: *Imaging Techniques of the CNS of the Neonates* Heidelberg New York Tokyo (J. Haddad, D. Christmann, J. Messer, eds.), Springer, Berlin Heidelberg, New York Tokyo, 1991.
[6] C. E. Cooper, S. J. Matcher, J. S. Wyatt, M. Cope, G. C. Brown, E. M. Nemoto, D. T. Delpy, *Biochem. Soc. Trans.* **1994**, *22*, 974.
[7] J. Wang, M. K. Ahmed, M. G. Sowa, H. H. Mantsch, *Proc. SPIE* **1994**, *2089*, 492.
[8] T. J. O'Leary, W. F. Engler, K. M. Ventre, *Appl. Spectrosc.* **1989** *43*, 1095
[9] P. T. T. Wong, B. Rigas, *Appl. Spectrosc.* **1990**, *44*, 1715
[10] D. R. Kodali, D. M. Small, J. Powell, K. Krishnan, *Appl. Spectrosc.* **1991** *45*, 1310.
[11] P. T. T. Wong, R. K. Wong, T. A. Caputo, T. A. Godwin, B. Rigas, *Proc. Natl. Acad. Sci. USA* **1991**, *88* 10988.
[12] D. Naumann, S. Keller, D. Helm, Ch. Schultz, B. Schrader, *J. Mol. Struct.* **1995**, *347* 399.
[13] S.-Y. Lin, R.-C. Liang, C.-H. Yang, H.-S. Hsu, A. T.-L. Lin, *Anal. Meth. Instrum.* **1993**, *1*, 191.
[14] H. H. Mantsch, M. Jackson, *J. Mol. Struct.* **1995**, *347*, 187.
[15] M. G. Sowa, H. H. Mantsch, *J. Mol. Struct.* **1993**, *300*, 239.
[16] G. B. Christison, H. A. MacKenzie, *Med. Biol. Eng. Comput.* **1993**, *31*, 284.
[17] H. A. Mackenzie, G. B. Christison, P. Hodgson, D. Blanc, *Sens. Actuators* **1993**, *B 11*, 213.
[18] J. Wallon, S. H. Yan, J. Tong, M. Meurens, J. Haot, *Appl. Spectrosc.* **1994**, *48*, 190.
[19] Z. Ge, C. W. Brown, H. J. Kisner, *Appl. Spectrosc.* **1995**, *49*, 432.
[20] L. A. Cassis, R. A. Lodder, *Anal. Chem.* **1993**, *65*, 1247.
[21] H. H. Eysel, M. Jackson, H. H. Mantsch, G. T. D. Thomson, *Appl. Spectrosc.* **1993**, *47*, 1519.
[22] M. K. Ahmed, H. H. Mantsch, *Proc. SPIE.* **1994**, *2089*, 520.
[23] G. G. Guilbault, G. Palleschi, G. Lubrano, *Biosens. Bioelectron.* **1995**, *10*, 379.
[24] N. Kaiser, *IEEE Trans. Biomed. Eng.* **1979**, *26*, 597.
[25] K. Kajiwara, T. Uemura, H. Kishikawa, K. Nishida, Y. Hashiguchi, M. Uehara, M. Sakakida, K. Ichinose, M. Shichiri, *Med. Biol. Eng. Comput.* **1993**, *31*, S17.

[26] R. Marbach, *Meßverfahren zur IR-spektrometrischen Blut-glucosebestimmung, Fortschr. Ber. VDI Ser. 8, Vol. 346,* VDI Düsseldorf 1993.

[27] H. M. Heise, R. Marbach, G. Janatsch, J. D. Kruse-Jarres, *Anal. Chem.* **1989**, *61*, 2009.

[28] G. Janatsch, J. D. Kruse-Jarres, R. Marbach, H. M. Heise, *Anal. Chem.* **1989**, *61*, 2016.

[29] H. M. Heise, R. Marbach, Th. Koschinsky, F. A. Gries, *Appl. Spectrosc.* **1994**, *48*, 85.

[30] H. M. Heise, A. Bittner, *J. Mol. Struct.* **1995**, *348*, 127.

[31] H. M. Heise, A. Bittner, *J. Mol. Struct.* **1995**, *348*, 21.

[32] P. Bhandare, Y. Mendelson, E. Stohr, R. A. Peura, *Vib. Spectrosc.* **1994**, *6*, 363.

[33] E. H. Korte A. Otto, *Appl. Spectrosc.* **1988**, *42*, 38.

[34] A. Bittner H. M. Heise Th. Koschinsky, F. A. Gries *Mikrochim. Acta [Suppl].* **1997**, *14*, 827.

[35] K. H. Hazen, M. A. Arnold, G. W. Small, *Appl. Spectrosc.* **1994**, *48*, 477.

[36] A. Bittner, R. Marbach H. M. Heise, *J. Mol. Struct.* **1995**, *349*, 341.

[37] A. Bittner, Ph. D. Thesis, University of Essen, in preparation.

[38] J. T. Kuenstner, K. H. Norris, W. F. McCarthy, *Appl. Spectrosc.* **1994**, *48*, 484.

[39] J. L. Gonczy, L. L. Gyarmati, in: *Making Light Work: Advances in Near Infrared Spectroscopy* (I. Murray, I. A. Cowe, eds.), VCH, Weinheim 1992.

[40] M. Blazewicz, C. Paluszkiewicz, *Proc. SPIE.* **1992**, *1575*, 424.

[41] M. Hajek, Quat. Magn. Reson. *Biol. Med.* **1995**, *2*, 165.

[42] G. Müller, (ed.), *Medical Optical Tomography: Functional Imaging and Monitoring, Vol. 11*, Society of Photo-Optical Instrumentation Engineers, Bellingham, Washington, 1993.

[43] G. Mitic, J. Kölzer, J. Otto, E. Plies, G. Sölkner, W. Zinth, *Appl. Optics* **1994**, *33*, 6699.

[44] B. C. Wilson, E. M. Sevick, M. S. Patterson, B. Chance, *Proc. IEEE* **1992**, *80*, 918.

[45] J. -L. Boulnois, *Lasers Med. Science* **1986**, *1*, 47.

[46] M. H. Koelink, F. F. M. de Mul, J. Greve, R. Graaf, A. C. M. Dassel, J. A. Aarnoudse, *Proc. SPIE* **1991**, *1431*, 63.

[47] S. J. Matcher, M. Cope, D. T. Delpy, *Phys. Med. Biol.* **1994**, *39*, 177.

[48] B. C. Wilson, S. L. Jacques, *IEEE J. Quant. Electron.* **1990**, *26*, 2186.

[49] P. Kubelka, F. Munk, *Z. Techn. Phys.* **1931**, *12*, 593.

[50] M. S. Patterson, B. C. Wilson, D. R. Wyman, *Lasers Med. Science* **1991**, *6*, 155.

[51] R. Marbach, H. M. Heise, *Appl. Optics* **1995**, *34*, 610.

[52] A. Bittner, S. Thomaßen, H. M. Heise, *Mikrochim. Acta [Suppl.]* **1997**, *14*, 429.

[53] R. Marbach, Th. Koschinsky, F. A. Gries, H. M. Heise, *Appl. Spectrosc.* **1993**, *47*, 875.

[54] H. M. Heise, R. Marbach, *Proc. SPIE* **1994**, *2089*, 114.

[55] H. M. Heise, R. Marbach, Th. Koschinsky, F. A. Gries, *Artif. Organs* **1994**, *18*, 439.

[56] H. M. Heise, A. Bittner, in preparation.

[57] K. A. Martin, in: *Making Light Work: Advances in Near Infrared Spectroscopy* (I. Murray, I. A. Cowe, eds.), VCH, Weinheim, 1992.

[58] J. M. Carney, W. Landrum, L. Mayes, Y. Zou, R. A. Lodder, *Anal. Chem.* **1993**, *65*, 1305.

[59] A. Villringer, J. Planck, C. Hock, L. Schleinkofer, U. Dirnagl, *Neurosci. Lett.* **1993**, *154*, 101.

[60] C. E. Elwell, H. Owen-Reece, M. Cope, A. D. Edwards, J. S. Wyatt, E. O. R. Reynolds, D. T. Delpy, *Adv. Exp. Med. Biol.* **1994**, *345*, 619.

[61] H.-U. Bucher, M. Wolf, M. Keel, G. von Siebenthal, G. Duc, *Eur. J. Pediatr.* **1994**, *153*, 123.

[62] F. Steinberg, M. Leßmann, C. Streffer, in: *Oxygen Transport to Tissue XV* (P. Vaupel, R. Zander, D. F. Bruley, eds.), Plenum, New York, 1994.

[63] S. van Huffel, P. Casaer, P. van Mele, G. Willems, *Proc. SPIE* **1995**, *2389*, 743.

Mikrochim. Acta [Suppl.] 14, 79–88 (1997)
© Springer-Verlag 1997

The Colorful World of Quasilinearity as Revealed by High Resolution Molecular Spectroscopy

Sieghard Albert, Manfred Winnewisser*, and **Brenda P. Winnewisser**

Physikalisch-Chemisches Institut der Justus-Liebig-Universität, Heinrich-Buff-Ring 58, D-35392 Gießen, Federal Republic of Germany

Abstract. The analysis of the mid infrared spectrum (1200–4000 cm^{-1}) of fulminic acid yielded the identification of more than 110 new vibrational levels. Each assigned vibrational level shows interactions enhanced by the quasilinearity of the molecule HCNO. Most of the resonances can be classified into a network of resonance systems. This network can be built up with the help of the basic resonance systems 00002/00010, 00004/00100 and 00015/01000 that include Coriolis-type and Fermi-type resonances. In this paper the 00200/00104/00007/[00033] and 00114/00210/00202/01003/[00017] resonance systems are discussed in detail.

Key words: quasilinear molecule, high resolution spectroscopy, resonance.

Fulminic acid, HCNO, has been the subject of many experimental and theoretical investigations because of its quasilinear HCN-bending vibration (v_5) which is located at 224 cm^{-1}. This value is in sharp contrast to the v_2 bending fundamental of HCN at 711 cm^{-1}. The other vibrational modes of HCNO are the CH-stretching (v_1), CN-stretching (v_2), NO-stretching (v_3) and CNO-bending (v_4) modes.

The spectrum of HCNO has been measured from the microwave [1, 2] to the ultraviolet spectral regions [3], including studies in the millimetre wave [4–6], and infrared regions [7–11]. In spite of this long research history it is still not possible to reproduce well the experimental data by *ab initio* calculations, as can be seen from Table 1. The development of the semirigid-bender model [12, 13] for the HCN-bending motion made a great contribution towards understanding the rovibrational dynamics of HCNO. As a result of the anharmonic potential for v_5, two basic resonances were identified during the analysis of the HCNO rovibra-

tional spectrum in the far-infrared region [14]: a Coriolis-type interaction between the states 00002 and 00010, and an anharmonic or Fermi-type interaction between the states 00004 and 00100, as can be seen in Fig. 1 where the vibrational term values up to 1250 cm^{-1} are shown.

The analysis of the rovibrational spectrum of HCNO in the mid infrared region up to 4000 cm^{-1} allowed the identification of more than 110 new vibrational levels. Each level shows interactions of local or global character [15]. Most of the resonances caused by these interactions can be classified into a grand network of resonance systems as is illustrated in Fig. 2.

Table 1. The five normal modes of HCNO, calculated *ab initio*, in comparison to the experimental values

Year	v_1	v_2	v_3	v_4	v_5	Method/Basis-set
1980[a]	3715	2435	982	750	516	SCF/4–31G
1982[b]	3709	2427	991	550	549	SCF/4–31G
1982[b]	3710	2530	1333	550	549	MP3/6–31G**
1989[c]	3537	2282	1332	582	496	MP2/6–31G**
1993[d]	3361	2194	1302	461	498	MP2/TZ2P/DZP
1993[d]	3240	2181	1287	538	178	Variational method
1994[e]	3316	2160	1244	527	437	CCSD(T)/TZ2P
1994[e]	3280	2285	1310	549	230	S-VWN/TZ2P
1994[e]	3290	2190	1230	538	237	B-LYP/TZ2Pf
	3336	2196	1254	537	224	experiment

[a] A. Komornicki, J.D. Goddard, H.F. Schäffer, *J. Am. Chem. Soc.* **1980**, *102*, 1763; [b] L. Farnell, R.H. Nobes, L. Radom, *J. Mol. Spectrosc.* **1980**, *93*, 271; [c] J.H. Teles, G. Maier, B.A. Hess, Jr., L.J. Schaad, M. Winnewisser, B.P. Winnewisser, *Chem. Ber.* **1989**, *122*, 753; [d] N. Pinnavaia, M.J. Bramley, M.-D. Su, W.H. Green, N.C. Handy, *Mol. Phys.* **1993**, *78*, 319; [e] N.C. Handy, Ch.W. Murray, R.D. Amos, *Phil. Mag.* **1994**, *69B*, 755.

---

* To whom correspondence should be addressed

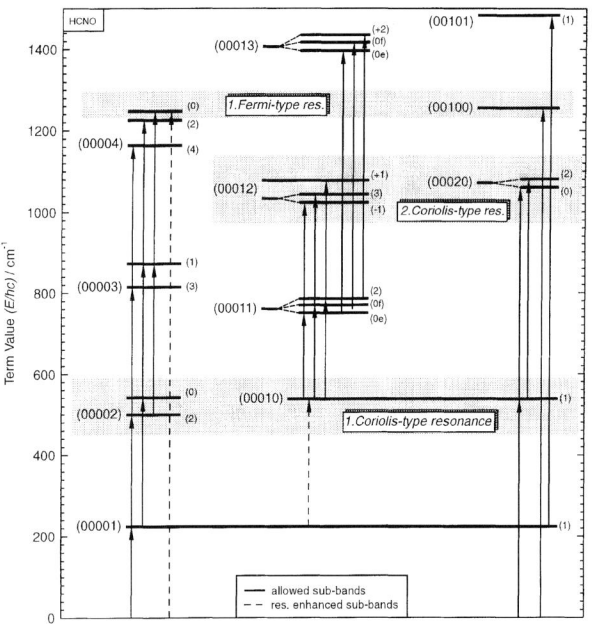

Fig. 1. Vibrational term value diagram of HCNO up to 1250 cm^{-1}; the different resonance systems are shaded

ed term values, intensity information, and data from various regions of the infrared spectrum to identify and classify the levels and resonances.

In the following discussion the vibrational states of the five normal modes are characterized by the quantum numbers v_1, v_2, v_3, v_4, and v_5. The vibrational levels within the excited bending states v_4 and v_5 are labeled by the vibrational angular momentum quantum numbers l_4 and l_5, where $k = l_4 + l_5$. The full notation is thus $v_1 v_2 v_3 (v_4^{l_4} v_5^{l_5})^k$ together with the symmetry labels e and f. Where there are two or more vibrational levels with the same k within one vibrational state, they are distinguished by the characterization $(+)$ and $(-)$, giving the sign of $k = l_4 + l_5$ resulting when $l_4 > 0$. A resonance system consisting of two or more states is indicated by slashes as in 00002/00010. Vibrational states which are written in brackets have not yet been observed. Their vibrational term values are estimated. The J values within a vibrational level are referred to as the rovibrational levels.

The various resonance systems will be discussed in separate publications. In this paper we present as examples the 00200/00104/00007/[00033] and the 00114/00210/00202/01003/[00017] resonance systems, which are discussed in detail. Their analysis illustrates the extensive use of graphical manipulations of reduc-

Spectrum and Assignment

The assignment of transitions belonging to a sub-band consisting of P- and R-branches was carried out efficiently with an interactive Loomis–Wood assignment program [16]. The correct J assignment was checked by combination differences. The subsequent initial data reduction was achieved with a program [17]

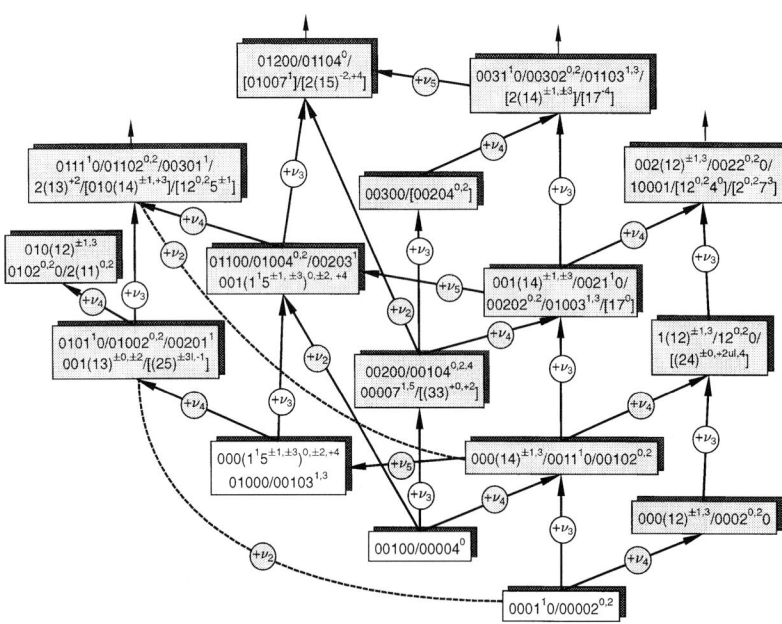

Fig. 2. The complete network of resonance systems of HCNO for excitation up to 5000 cm^{-1}

which calculates band centres and coefficients in a power series (ps) expansion of the rovibrational term value in $J(J + 1)$

$$G(J) = G_c + B_{ps}J(J+1) - D_{ps}[J(J+1)]^2 + H_{ps}[J(J+1)]^3$$
$$+ L_{ps}[J(J+1)]^4 \quad (1)$$

for the upper and the lower levels. The band centre $\tilde{v}_c$ is defined as $\tilde{v}_c = G'_c - G''_c$. Table 2 lists the power series coefficients of the assigned sub-bands of HCNO which are used in the analysis of the resonance systems described here. These power series coefficients are only physically significant if no interactions such as l-type or accidental resonances are present. The accidental interactions are described as Fermi-type resonances if they are J-independent and as Coriolis-type resonances if they are J-dependent. Furthermore $\Delta l = 0$ in a Fermi-type resonance pair and $\Delta l \neq 0$ in a Coriolis-type resonance pair in a linear molecule.

At the moment it is impossible to describe the complex interaction phenomena of the assigned rovibrational levels of HCNO with a comprehensive Hamiltonian, as described in [18] for a truly linear molecule. Therefore, in order to make vibrational assignments and identify resonance partners, types of resonances and cross-over points, reduced term values defined according to three models T_{RED1}, T_{RED2}, and T_{RED3} have been used to obtain an initial analysis of the resonance systems. These models are defined as:

model 1 $\quad T_{RED1} = T_{meas} - [B^{gs}J(J+1) - D^{gs}[J(J+1)]^2]$

model 2 $\quad T^{(1,2)}_{RED2} = T_{meas} - \left[\dfrac{B^{(1)} + B^{(2)}}{2} J(J+1) \right.$

$$\left. - \dfrac{D^{(1)} + D^{(2)}}{2}[J(J+1)]^2 \right]$$

model 3 $\quad T_{RED3} = T_{meas} - [G_c + B_{J\leqslant 20}J(J+1) - D_{J\leqslant 20}[J(J+1)]^2]$
$$(2)$$

T_{meas} represents the experimentally determined term value, B^{gs} and D^{gs} are the power series coefficients of the ground state and B, D are the power series coefficients of the rovibrational levels under consideration. $T^{(1,2)}_{RED2}$ is specific for the vibrational levels 1 and 2. The subscript $J \leqslant 20$ means that the power series coefficients are determined only with regard to the first twenty J values, so that T_{RED3} highlights J-dependent deviations from these values at higher J.

The 00200/00104/00007/[00033] Resonance System of HCNO

Addition of a quantum of v_3 to the Fermi-type basic resonance pair $00100/00004^0$ in Fig. 1 leads to the 00200/00104/00007/[00033] resonance system. The anharmonicity of the v_5 bending mode brings the 00007 and 00033 vibrational states into resonance. This system can be divided into two resonance groups. Group 1 consists of the interacting rovibrational levels 00200, $00104^{0,2e,f}$, $00007^{1,3,e,f}$ and the estimated 00033 rovibrational levels. Group 2 contains the 00104^4 and 00007^5 rovibrational levels. The observed interactions in resonance group 2 are not strong, however, because of the great difference in energy between the vibrational levels involved. The assigned rovibrational levels are identified by the $00200 \leftarrow 00000$, $00007^{1,3,5} \leftarrow 00005^{1,3,5}$, $00104^{0,2} \leftarrow 00002^{0,2}$ and $00104^4 \leftarrow 00004^4$ transitions which are shown in the term value diagram in Fig. 3. As can be seen from this figure, the combined information from various regions of the HCNO spectrum was necessary in order to identify the vibrational levels and interactions in this resonance system.

An overview of the complex interactions in the 00200/00104/00007/[00033] resonance system is shown in terms of the reduced term values of the group 1 rovibrational levels in Fig. 4 (upper left). As the expanded reduced term values of the $00007^{1e,f}$ and the $00104^{2e,f}$ rovibrational levels (upper right) illustrate, there is a crossing of the e rovibrational levels at $J \approx 16/17 [J(J+1) \approx 272/306]$ and of the f rovibrational levels at $J \approx 28/29 [J(J+1) \approx 812/870]$. The crossing points are clearly confirmed by observation of the intensities of the $00104^{2e,f} \leftarrow 00002^{2e,f}$ sub-band (Fig. 5). This interaction is a Coriolis-type resonance ($\Delta l = 1$) and therefore J-dependent. In a first approximation, the J-dependence can be expressed as $C\sqrt{J(J+1)}$, where C is the Coriolis-coupling constant. In consideration of this approximation and the estimated crossing points mentioned above the Coriolis-coupling constant can be determined to be $C \approx 8 \times 10^{-3} \, \text{cm}^{-1}$.

As the reduced term values T_{RED1} (Fig. 4, upper right) and the intensities of the $00104^{2e,f} \leftarrow 00002^{2e,f}$ sub-band (Fig. 5) show, the crossing point of the f rovibrational levels is at a lower J value than that of the e rovibrational levels which is in agreement with the fact that $\Delta B^e_{ps} = 1.3 \times 10^{-3} \, \text{cm}^{-1} < \Delta B^f_{ps} = 4.3 \times 10^{-3} \, \text{cm}^{-1}$. Because it is a Coriolis-type reson-

Table 2. Power series constants of sub–bands in the spectrum of HCNO analysed in the present work

Sub–band	$\tilde{\nu}_c$	B''_{ps} / B'_{ps}	D''_{ps} / D'_{ps} $\times 10^{-6}$	H''_{ps} / H'_{ps} $\times 10^{-12}$	L''_{ps} / L'_{ps} $\times 10^{-15}$	σ $\times 10^{-4}$	#
$07^{1e} \leftarrow 05^{1e}$	846.06123(25)	0.3843520(54)	0.1112(70)	–	–		
		0.3848300(64)	−0.0356(17)	−154(16)	–	3.684	34
$07^{1f} \leftarrow 05^{1f}$	846.0571(11)	0.386795(35)	0.128(25)	–	–		
		0.387936(35)	0.112(25)	–	–	34.068	46
$07^3 \leftarrow 05^3$	858.155678(81)	0.385616(26)	0.1770(30)	–	–		
		0.3863712(25)	0.2147(28)	–	–	1.869	32
$07^5 \leftarrow 05^5$	878.805084(59)	0.3856358(12)	0.15700(70)	–	–		
		0.3863366(12)	0.17155(67)	–	–	1.654	53
$00104^0 \leftarrow 02^0$	1964.043509(38)	0.3841213(10)	0.13575(94)	1.58(26)	–		
		0.3832909(11)	−0.0321(12)	−26.75(59)	3.65(12)	1.174	84
$00104^0 \leftarrow 00$	2505.80763(16)	0.3825626(47)	0.1407(79)	–	–		
		0.3832907(53)	−0.0148(98)	–	–	2.132	20
$00104^{2e} \leftarrow 02^{2e}$	1985.40896(21)	0.3842254(48)	0.1666(51)	–	–		
		0.3835329(68)	0.119(32)	−416(70)	539(51)	4.334	48
$00104^{2f} \leftarrow 02^{2f}$	1985.40827(13)	0.3842259(48)	0.144(15)	–	–		
		0.3835808(68)	0.910(53)	3763(140)	–	1.913	24
$00104^4 \leftarrow 04^4$	1266.05350(11)	0.3852957(26)	0.1576(23)	–	–		
		0.3837104(24)	0.1589(20)	–	–	2.039	36
$001(14)^{-1e} \leftarrow (12)^{-1e}$	1984.0541(89)	0.384683(20)	0.215(19)	–	–		
		0.383707(18)	0.234(16)	–	–	9.896	20
$001(14)^{-1e} \leftarrow 03^{1e}$	2144.7299(10)	0.383739(17)	0.124(14)	–	–		
		0.383781(18)	0.427(33)	140(27)	–	9.342	23
$001(14)^{-1f} \leftarrow (12)^{-1f}$	1984.0446(48)	0.384831(14)	0.232(25)	–	–		
		0.384159(13)	0.266(20)	–	–	5.600	21
$001(14)^{-1f} \leftarrow 03^{1f}$	2144.7314(20)	0.3855675(53)	0.1486(38)	–	–		
		0.3841841(56)	0.2871(64)	34.9(31)	–	4.394	31
$001(14)^{+1e} \leftarrow 03^{1e}$	2171.5067(10)	0.3837461(27)	0.1319(21)	–	–		
		0.3826491(27)	−0.0851(40)	−47.1(28)	–	2.249	36
$001(14)^{+1f} \leftarrow 03^{1f}$	2171.5060(11)	0.38554997(99)	0.13717(59)	–	–		
		0.3840032(13)	−0.0157(33)	−26.2(40)	4.67(85)	0.917	30
$001(14)^{+3} \leftarrow (12)^3$	1985.7340(18)	0.3845594(27)	0.1269(28)	–	–		
		0.3839629(32)	0.1445(70)	−22.7(59)	–	2.246	37
$001(14)^{-3} \leftarrow 03^3$	2143.39734(13)	0.3848950(27)	0.1562(16)	–	–		
		0.3835463(29)	0.2423(24)	24.90(61)	–	3.522	56
$001(14)^{-3} \leftarrow 02^2$	2458.38119(14)	0.3842358(52)	0.1806(88)	–	–		
		0.3835480(54)	0.2451(85)	–	–	2.231	21
$00200 \leftarrow 00$	2498.28729(44)	0.3825649(60)	0.1418(12)	–	–		
		0.3793010(66)	0.2053(34)	16.5(14)	−1.93(17)	18.380	89
$00202^0 \leftarrow 02^0$	2507.242311(56)	0.3841223(12)	0.13486(72)	–	–		
		0.3808311(15)	0.05364(24)	−40.0(22)	0.3236(73)	1.730	64
$00202^{2e} \leftarrow 02^{2e}$	2515.156094(81)	0.3842404(20)	0.1796(17)	3.28(41)	–		
		0.3808824(22)	0.1466(20)	−11.05(74)	1.86(10)	2.363	79
$00202^{2f} \leftarrow 02^{2f}$	2515.156193(99)	0.3842141(25)	0.1371(22)	−1.17(52)	–		
		0.3808540(26)	0.0992(25)	−15.67(96)	2.01(15)	3.01	79
$0021^{1e}0 \leftarrow 1^{1e}0$	2500.476914(93)	0.3826598(13)	0.13493(36)	–	–		
		0.3794190(14)	0.1945(10)	14.39(59)	−1.27(11)	2.919	80
$0021^{1f}0 \leftarrow 1^{1f}0$	2500.47608(45)	0.3834427(97)	0.1444(26)	–	–		
		0.380252(10)	0.2746(72)	55.1(41)	−8.17(81)	16.759	69
$01003^{1e} \leftarrow 03^{1e}$	2162.45179(13)	0.3837432(26)	0.1298(11)	–	–		
		0.3823175(26)	0.2012(12)	6.07(25)	–	4.183	60
$01003^{1f} \leftarrow 03^{1f}$	2162.45378(51)	0.385558(11)	0.1390(27)	–	–		
		0.383177(11)	0.0644(68)	−35.2(41)	5.64(78)	19.210	69
$01003^3 \leftarrow 03^3$	2156.54696(11)	0.38489355(97)	0.15587(36)	–	–		
		0.3831700(11)	0.1254(13)	6.66(92)	−1.26(21)	1.780	66

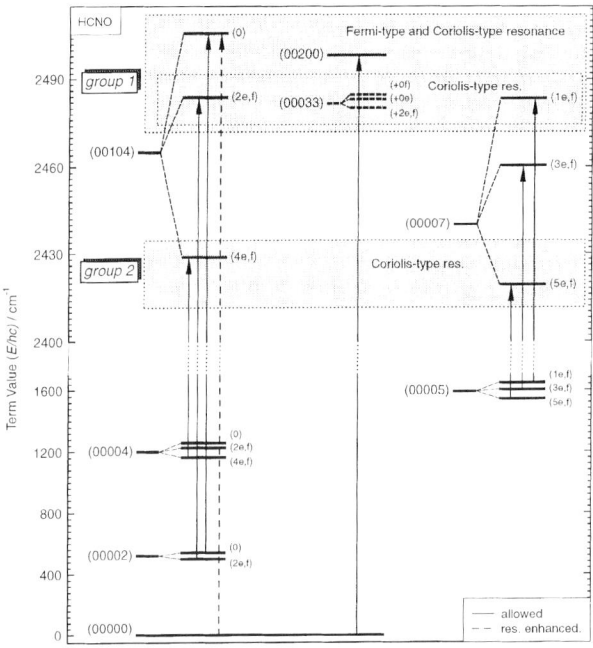

Fig. 3. Vibrational term value diagram of the 00200/00104/00007/[00033] resonance system of HCNO and the assigned transitions; the different resonance groups are shaded

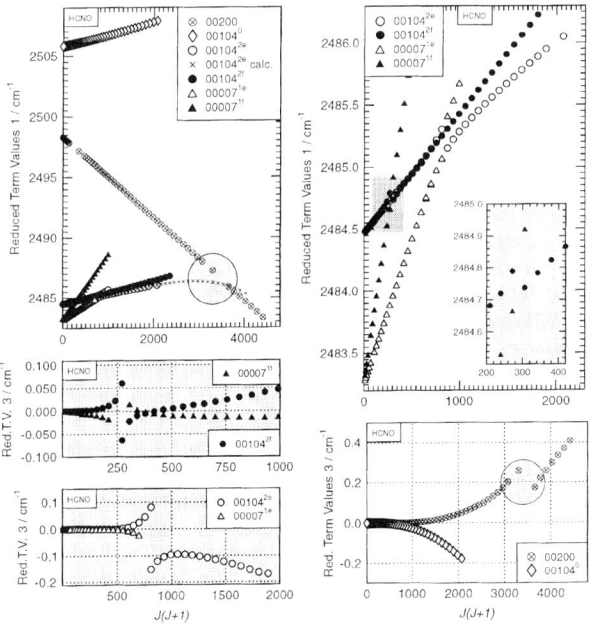

Fig. 4. Upper left: reduced term values of the $00104^{0,2e,f}$, $00007^{1e,f}$ and 00200 rovibrational levels of HCNO calculated by model 1. Upper right: enlargement of the crossing region of the $00104^{2e,f}$ and $00007^{1e,f}$ rovibrational levels (model 1). Middle and lower left: reduced term values of the $00104^{2e,f}$ and $00007^{1e,f}$ rovibrational levels (model 3). Lower right: reduced term values of the 00200 and 00104^0 rovibrational levels (model 3)

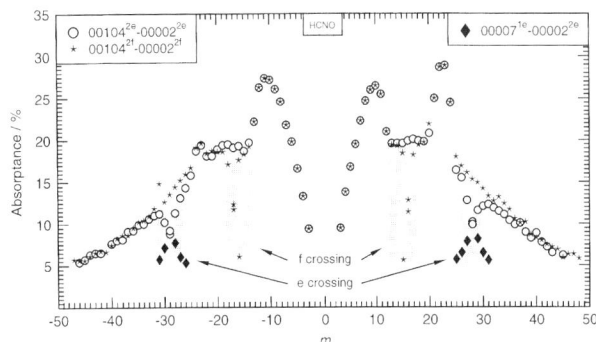

Fig. 5. Intensities of the $00104^{2e,f} \leftarrow 00002^{2e,f}$ sub-band; the crossing regions are shaded

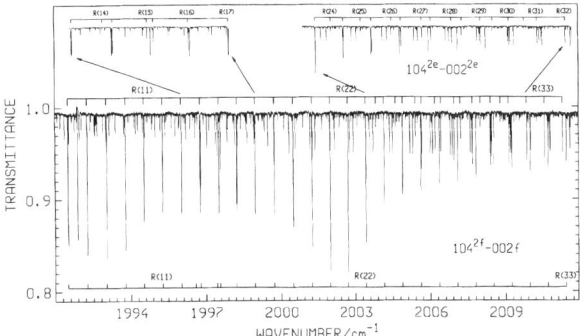

Fig. 6. Part of the R-branches of the $00104^{2e,f} \leftarrow 00002^{2e,f}$ sub-band of HCNO; the crossing regions are expanded

ance, the interaction of the $00007^{1e}/00104^{2e}$ rovibrational levels is stronger at the crossing point than that of the $00007^{1f}/00104^{2f}$ levels. For that reason a few resonance-enhanced transitions of the $00007^{1e} \leftarrow 00002^{2e}$ sub-band could be assigned in the crossing region, as is shown by the intensities in Fig. 5 and part of the R-branches in Fig. 6. The different strength of the e and f crossings is very well shown by using model 3 of the reduced term values (Fig. 4, middle and lower left). The $00007^{1e,f}/00104^{2e,f}$ resonance pair is an excellent example of a transition from a local perturbation (f-rovibrational levels) to a stronger and thus more global interaction (e-rovibrational levels).

In addition, the 00007^{1e} rovibrational levels should cross the 00200 rovibrational levels at $J = 51/52[J(J + 1) = 2652/2756]$ as predictions show. Indeed, the 00200 rovibrational levels are very weakly perturbed at these J values. In comparison to the above-mentioned Coriolis-type resonance, the Coriolis-type interaction between the 00007^{1e} and 00200 rovibrational levels is much weaker. An explanation for this may be the larger change in two quantum numbers ($\Delta v_3 = 2$ and $\Delta v_5 = 5$).

The strongest interaction in the 00200/00104/00007/[00033] resonance system is the Fermi-type interaction between the 00104^0 and 00200 rovibrational levels, as the reduced term values in Fig. 4 (lower right) illustrate. As a result of the interaction a resonance-enhanced sub-band, $00104^0 \leftarrow 00000$, could be assigned. This Fermi-type resonance pair can be deduced from the basic resonance pair 00004^0/00100 (Fig. 1) by addition of one further quantum of v_3. In contrast to the crossing observed for the basic resonance 00004^0/00100, there is no crossing of the 00200 and 00104^0 rovibrational levels. As a consequence of the differences in B_{ps} values, however, there is a crossing of the 00200 and 00104^{2e} rovibrational levels at $J \approx 62/65[J(J+1) \approx 3906/4290]$. This crossing of the 00200 rovibrational levels is shown by use of the reduced term values T_{RED3} in the grey circle of Fig. 4 (lower right). A prediction of the 00104^{2e} rovibrational levels up to $J = 65[J(J+1) = 4290]$ based on the experimentally determined power series constants B_{ps} and D_{ps} illustrates clearly the crossing of the involved rovibrational levels in terms of the reduced term values T_{RED1} (Fig. 4, upper left). In addition, the 00104^{2e} rovibrational levels show weak perturbations at $J = 42/43$ $[J(J+1) = 1806/1892]$ which we attribute to the $0003^{3,1}3^1$ rovibrational levels, which can only be roughly predicted with the current data.

The resonance pairs of the resonance system discussed, together with the term value differences for each pair, are listed in Table 3. The indication in the table of a local resonance implies in each case an avoided crossing. Because of the relatively large vibrational term value separations, the interactions between the rovibrational levels $00200/00007^{1e}$, $00104^0/00007^{1e}$ and $00104^{2e,f}/00007^{3e,f}$ can be neglected in the initial analysis. As opposed to these resonance pairs, the observed interaction in the $00104^4/00007^5$ resonance pair increases with increasing J value, because the separation in energy decreases and the off-diagonal-element (Coriolis-type) increases with increasing J value. At $J > 60[J(J+1) > 3600]$ the 00104^4 and 00007^5 rovibrational levels should cross. As mentioned above, it is at the moment impossible to describe the discussed interactions by a Hamiltonian suitable for a quasilinear molecule.

The 00114/00210/00202/01003/[00017] Resonance System of HCNO

Addition of a further quantum of v_4 to the 00200/00104/00007/[00033] resonance system leads to the 00114/00210/00202/01003/[00017] system. This

Table 3. The resonance pairs of the 00200/00104/00007/[00033] resonance system.

Pair	Partner 1/Partner 2	Δl	Type		ΔG_c
Coriolis-type and Fermi-type resonances					
1.	00104^0/00200	0	Ft[a]	gl[b]	7.49
2.	00104^0/00007^{1e}	1	Ct	lo	22.52
3.	00200/00104^{2e}	2	Ct	lo	13.83
4.	00200/00007^{1e}	1	Ct	lo	15.03
5.	$00104^{2e,f}$/$00007^{1e,f}$	1	Ct	lo	1.20
6.	$00104^{2e,f}$/$00007^{3e,f}$	1	Ct	-	24.13
7.	$00104^{4e,f}$/$00007^{5e,f}$	1	Ct	lo	9.51
8.	$00104^{2e,f}$/$000(3^33^{-1})^{+2e,f}$	0	Ft	lc	2[c]
l-type resonances					
9.	00104^0/00104^{2e}	2	l	gl	21.32
10.	$00104^{2e,f}$/$00104^{4e,f}$	2	l	gl	56.58
11.	$00007^{1e,f}$/$00007^{3e,f}$	2	l	gl	22.95
12.	$00007^{3e,f}$/$00007^{5e,f}$	2	l	gl	45.59

[a] Ft: Fermi-type resonance; Ct: Coriolis-type resonance; l: l-type resonance. [b] gl: global perturbation; lo: local perturbation; [c] estimated value.

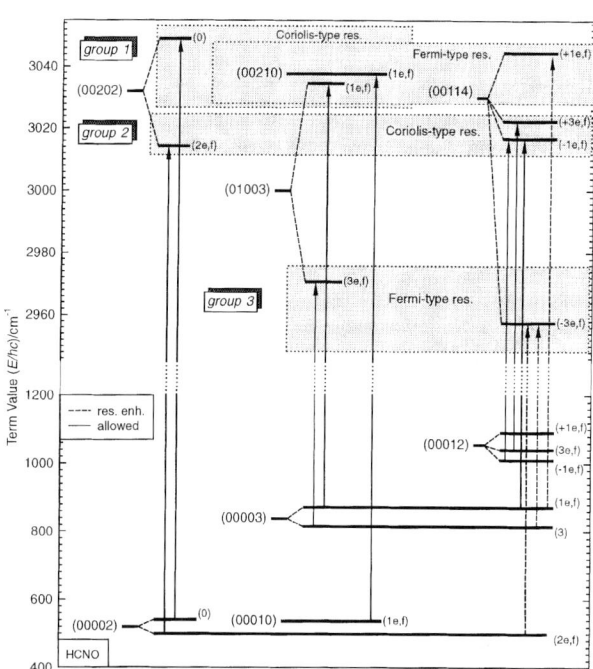

Fig. 7. Vibrational term value diagram of the 00114/00210/00202/01003 resonance system of HCNO and the assigned transitions; the different resonance groups are shaded

resonance system is most conveniently divided into three resonance groups, as shown in Fig. 7. Group 1 consists of the $001(1^14^0)^{+1e,f}$, 00202°, $0021^{1e,f}0$ and $01003^{1e,f}$ rovibrational levels. Group 2 contains the $001(1^14^{\pm2})^{+3,-1e,f}$ and $00202^{2e,f}$ rovibrational levels and estimated $000(1^17^1)^{0,+2}$ vibrational levels. Finally, group 3 consists of the $001(1^14^{-4})^{-3}$ and 01003^3 rovibrational levels and the estimated $000(1^17^{-5})^{-4}$ rovibrational levels. Figure 7 shows the vibrational term value diagram including the assigned transitions; the 00017 vibrational levels are omitted here for simplicity. The $0021^{1e,f}0$, $00202^{0,2e,f}$ and $01003^{1e,f}$ rovibrational levels were assigned, using the $\Delta l = 0$ transitions $0021^{1e,f}0 \leftarrow 0001^{1e,f}0$, $00202^{0,2e,f} \leftarrow 00002^{0,2e,f}$ and $01003^{1e,f} \leftarrow 00003^{1e,f}$. The $001(1^14^{\pm2})^{-1,+3,e,f}$ rovibrational levels were characterized by the $001(1^14^{\pm2})^{-1,+3,e,f} \leftarrow 000(1^12^2)^{-1,3e,f}$ transitions whereas the $001(1^14^0)^{+1e,f}$ rovibrational levels could be located through the $001(1^14^0)^{+1e,f} \leftarrow 00003^{1e,f}$ resonance-enhanced sub-band.

Resonance group 1 consists of a Fermi triplet $01003^{1e,f}/0021^{1e,f}0/001(1^14^0)^{+1e,f}$ and the Coriolis-type resonance $001(1^14^0)^{+1e}/00202^0$. Figure 8 (upper left) provides an overview of all involved resonance partners of group 1 in terms of their reduced term values following model 1. As shown in this figure (upper left), there is a crossing of the $01003^{1e,f}$ and $0021^{1e,f}0$ rovibrational levels at $J = 31/34[J(J+1) = 992/1190]$. The crossing point is shown in more detail in Fig. 8 (upper right). This illustrates the weak local interaction due to the large change in four quantum numbers.

As mentioned above, the resonance-enhanced $001(1^14^0)^{+1e,f} \leftarrow 00003^{1e,f}$ sub-band could be observed. Its intensity is a result of the Fermi-type interaction between the $001(1^14^0)^{+1e,f}$ and $01003^{1e,f}$ rovibrational levels. Another possible Fermi-type partner $001(1^14^{-2})^{-1e,f}$ can be neglected because of the larger vibrational energy separation ($\Delta G_c = 17.72\,\mathrm{cm}^{-1}$) compared to the vibrational energy separation ($\Delta G_c = 9.06\,\mathrm{cm}^{-1}$) of $001(1^14^0)^{+1e,f}$ and $01003^{1e,f}$. As the reduced term values in Fig. 8 (upper left) illustrate, there is no crossing of the $001(1^14^0)^{+1e,f}$ and $01003^{1e,f}$ rovibrational levels because the lower lying $01003^{1e,f}$ rovibrational levels have the smaller rotational constant B_{ps}.

The rovibrational levels of the third pair in the Fermi triad, $0021^{1e,f}0/001(1^14^0)^{+1e,f}$, do not cross either, as the reduced term values in Fig. 8 (upper left) illustrate. The fact that they interact can be deduced from the observed interaction pair $0011^{1e,f}0/000(1^14^0)^{+1e,f}$ by addition of one quantum of v_3 [15]. The reduced term values of the $0021^{1e,f}0$ and $01003^{1e,f}$ rovibrational levels, calculated according to model 3, provide a comparison (Fig. 8, lower left and right) between the local Fermi-type perturbation due to the $0021^{1e,f}0$ and $01003^{1e,f}$ interaction and the global Fermi-type interaction caused by the $0021^{1e,f}0/001(1^14^0)^{+1e,f}$ and $01003^{1e,f}0/001(1^14^0)^{+1e,f}$ Fermi-type pairs of levels as the deviations in Fig. 8 (middle and lower left and right) illustrate.

The last important interaction in group 1 is caused by the Coriolis-type resonance involving the rovibrational levels $00202^0/001(1^14^0)^{+1e}$. As the reduced term values (model 1) in Fig. 8 (upper left) show, there should be a crossing of the involved rovibrational levels at $J > 50[J(J+1) > 2550]$. The perturbation can clearly be demonstrated by the reduced term values calculated by model 3. In Fig. 8 (middle right) the positive and negative deviations of the 00202^0 and $001(1^14^0)^{+1e}$ rovibrational levels are shown. The negative deviations of the $001(1^14^0)^{+1f}$ rovibrational levels (Fig. 8, middle left) are also caused by the above mentioned Fermi-type resonance pairs.

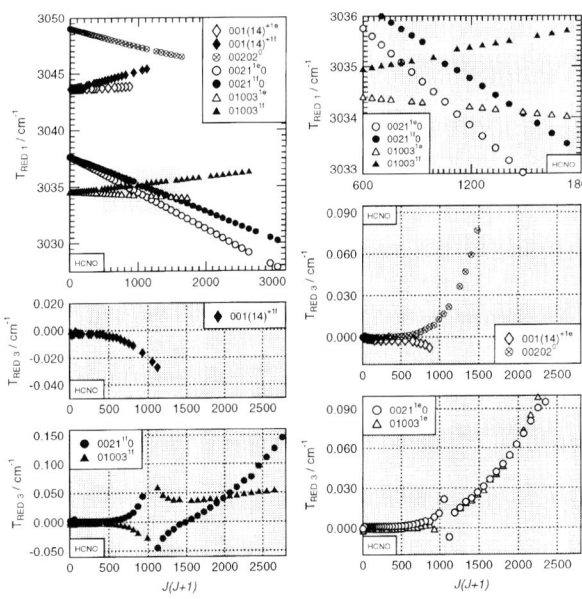

Fig. 8. Upper left: reduced term values of the $001(1^14^0)^{+1e,f}$, $01003^{1e,f}$, 00202^0 and $0021^{1e,f}$ rovibrational levels of HCNO calculated by model 1. Upper right: enlargement of the crossing region of the $01003^{1e,f}$ and $0021^{1e,f}0$ rovibrational levels (model 1). Middle and lower left: reduced term values of the $001(1^14^0)^{+1f}$, 0021^{1f} and 01003^{1f} rovibrational levels (model 3). Middle and lower right: reduced term values of the $001(1^14^0)^{+1e}$, 00202^0, $0021^{1e}0$ and 01003^{1e} rovibrational levels (model 3)

The $001(1^14^{-2})^{-1e,f}/00202^{2e,f}$ Coriolis-type resonance is the most important interaction in resonance group 2. Because the lower lying $00202^{2e,f}$ rovibrational levels have the smaller rotational constant B_{ps}, there is no crossing of the involved rovibrational levels. In the $000(1^14^{-2})^{-1e,f}/00102^{2e,f}$ pair [15], from which the $001(1^14^{-2})^{-1e,f}/00202^{2e,f}$ resonance pair can be inferred by addition of one further quantum of ν_3, a crossing is observed. The observable interaction due to the Coriolis-type resonance in the resonance pair

$001(1^14^2)^{3e,f}/00202^{2e,f}$ is much smaller because of the greater vibrational energy separation ($\Delta G_c = 8.28$ cm^{-1}). In addition, the $001(1^14^{-2})^{-1e,f}$ rovibrational levels show local perturbations at $J = 16/18$ [$J(J+1) = 272/342$] (e-levels) and at $J = 19/21$ [$J(J+1) = 380/462$] (f-levels). According to term value estimation and with regard to the extrapolation from the $00200/00104/00007$ resonance system, the interaction should be caused by the $000(1^17^{-1})^{0e,f}$ rovibrational levels. A comparison of the local pertur-

Table 4. The resonance pairs of the 00114/00210/00202/01003/[00017] resonance system

Pair	Partner 1/Partner 2	Δl	Type		ΔG_c	IME^a
Coriolis-type and Fermi-type resonances of the resonance groups 1 and 2						
1.	$00202^0/001(1^14^0)^{+1e}$	1	Ct^b	gl^c	5.39	C_A
2.	$00202^0/0021^1e0$	1	Ct	gl	11.35	C_B
3.	$00202^0/01003^{1e}$	1	Ct	gl	14.45	C_C
4.	$001(1^14^0)^{+1e,f}/0021^{1e,f}0$	0	Ft	gl	5.96	F_A
5.	$0021^{1e,f}0/01003^{1e,f}$	0	Ft	lo	3.10	F_B
6.	$001(1^14^0)^{+1e,f}/01003^{1e,f}$	0	Ft	gl	9.06	F_C
7.	$001(1^14^{-2})^{-1e,f}/00202^{2e,f}$	1	Ct	gl	2.63	C_D
8.	$001(1^14^2)^{+3e,f}/00202^{2e,f}$	1	Ct	gl	8.28	C_E
9.	$000(1^17^1)^{+2e,f}/001(1^14^{-2})^{-1e,f}$	1	Ct	lo	4^d	C_F
10.	$0001(1^17^{-1})^{0e,f}/001(1^14^{-2})^{-1e,f}$	1	Ct	lo	-2^d	C_G
11.	$001(1^17^1)^{+2e,f}/001(1^14^2)^{3e,f}$	1	Ct	lo	-2^d	C_H
12.	$00202^0/001(1^14^{-2})^{-1e}$	1	Ct	-	32.17	C_I
13.	$0021^{1e,f}0/001(1^14^{-2})^{-1e,f}$	0	Ft	-	20.82	F_D
14.	$0021^{1e,f}0/00202^{2e,f}$	1	Ct	-	23.45	C_J
15.	$01003^{1e,f}/001(1^14^{-2})^{-1e,f}$	0	Ft	gl	17.72	F_E
16.	$01003^{1e,f}/00202^{2e,f}$	0	Ct	gl	20.35	C_K
17.	$001(1^14^0)^{+1e,f}/00202^{2e,f}$	1	Ct	-	29.41	C_L
l-type resonances						
18.	$001(1^14^0)^{+1e,f}/001(1^14^{-2})^{-1e,f}$	2	*l*	gl	26.78	$U_{+1,-1}$
19.	$001(1^14^0)^{+1e,f}/001(1^14^2)^{+3e,f}$	2	*l*	gl	21.13	$U_{+1,+3}$
20.	$001(1^14^2)^{+3e,f}/001(1^14^{-2})^{-1e,f}$	2	*l*	gl	5.65	$U_{-1,+3}$
21.	$001(1^14^{-2})^{-1}/001(1^14^{-4})^{-3}$	2	*l*	gl	59.40	$U_{-1,-3}$
22.	$00202^0/00202^{2e}$	2	*l*	gl	34.80	$U_{0,2}$
23.	$01003^{1e,f}/01003^{3e,f}$	2	*l*	gl	63.97	$U_{1,3}$
Fermi-type and Coriolis-type resonances of the resonance group 3						
24.	$01003^3/001(1^14^{-4})^{-3}$	0	Ft	gl	13.15	F_F
25.	$01003^3/000(1^17^{-5})^{-4}$	1	Ct	lo	22^d	C_M
26.	$001(1^14^{-4})^{-3}/000(1^17^{-5})^{-4}$	1	Ct	lo	9^d	C_N
Fermi-type and Coriolis-type resonances of the resonance groups 2 and 3						
27.	$001(1^14^2)^{+3}/01003^3$	0	Ft	-	51.90	F_G
28.	$00202^2/01003^3$	1	Ct	-	43.62	C_O
29.	$00202^2/001(1^14^{-4})^{-3}$	1	Ct	-	56.77	C_P

^{a} Interaction matrix element.

^{b} Ct: Coriolis-type resonance, Ft: Fermi-type resonance, l: l-type resonance.

^{c} gl: global, lo: local.

^{d} estimated value.

bation of $001(1^14^{-2})^{-1e,f}$ and $00104^{2e,f}$ illustrates that the interaction of the latter level is stronger. This observation agrees with the observation, made in other cases, that excitation by an additional quantum of v_4 decreases the interaction matrix element [15].

In resonance group 3 (Fig. 7), the two sub-bands $001(1^14^{-4})^{-3} \leftarrow 00003^3$ and $01003^3 \leftarrow 00003^3$ could be assigned. Their observation can only be explained if a Fermi-type resonance between the $001(1^14^{-4})^{-3}$ and 01003^3 rovibrational levels is assumed. The identification of the $001(1^14^{-4})^{-3}$ rovibrational levels was confirmed by extrapolating the term value and the rotational constant. It is surprising that both sub-bands have similar intensities, considering the vibrational term value separation of $\Delta G_c = 13.15\,\mathrm{cm}^{-1}$. A comparison with the other Fermi-type resonance enhanced sub-band $001(1^14^0)^{+1e,f} \leftarrow 00003^{1e,f}$, $01003^{1e,f} \leftarrow 00003^{1e,f}$ and its allowed partners shows that in this case, the intensity of the allowed sub-band is much stronger than that of the resonance-enhanced sub-band. In addition, the $001(1^14^{-4})^{-3}$ rovibrational levels are strongly perturbed at $J > 33[J(J+1) > 1122]$, probably by the $000(1^17^{-5})^{-4}$ rovibrational levels. This identification is based on term value extrapolation.

All resonance pairs of the $00114/00210/00202/01003/[00017]$ resonance system are listed in Table 4. The complete local network of the system, which may be considered as a mapping of the interaction matrix linking the relevant levels, is presented in Fig. 9.

Discussion

The resonance systems discussed above are part of the extensive network of resonance systems shown in Fig. 2. Some of the perturbations in the $00114/00210/00202/01003/[00017]$ resonance system, in particular, due to the dark states 00017, could be identified on the basis of extrapolation from the $00200/00104/00007/[00033]$ resonance system. In addition to the basic resonances $\Delta v_3 = 1$, $\Delta v_4 = -4$ and $\Delta v_4 = 1$, $\Delta v_5 = 2$, another important resonance emerged from the network:

$$0011v_5^{|l|}/0100(v_5-1)^{|l|}$$

$$\text{with} \quad v_5 = 3,4,5 \text{ and } |l| = 0,1,2,3 \tag{3}$$

This is the set of interactions with $\Delta v_2 = 1$, $\Delta v_3 = -1$, $\Delta v_4 = -1$, $\Delta v_5 = -1$ which first appears in the $00015/01000/00103$ resonance system which will be discussed in detail in another publication. It is characterized most simply by the quartic force constant k_{2345}.

The $001(1^14^0)^{+1e,f}/01003^{1e,f}$ and $001(1^14^{-4})^{-3}/01003^3$ resonances, two of the Fermi-type pairs discussed above, belong to this group. Figure 10 illus-

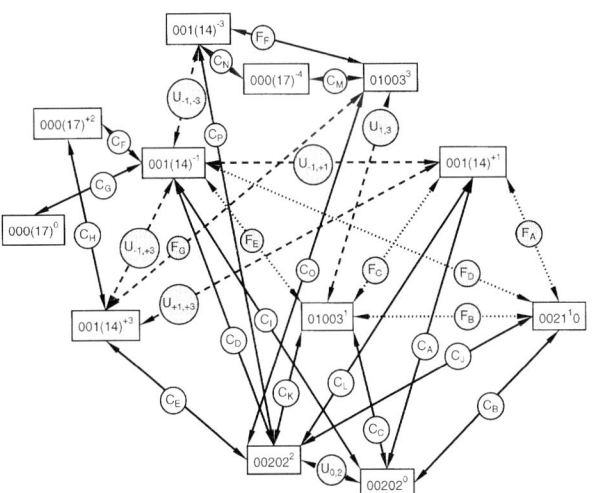

Fig. 9. The local network of the specific interactions in the $000114/00210/00202/01003/[00017]$ resonance system of HCNO; the interactions are identified as in Table 4

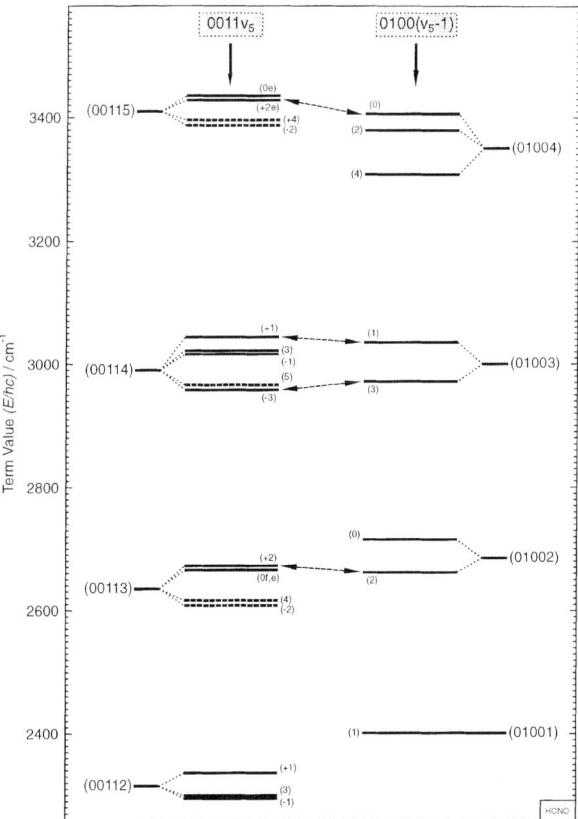

Fig. 10. Vibrational term value diagram of the Fermi-type resonance pairs $0011v_5/0100(v_5-1)$ of HCNO; the interactions are indicated by arrows

trates the evolution of the interacting set $\Delta v_2 = 1$, $\Delta v_3 = -1$, $\Delta v_4 = -1$, $\Delta v_5 = -1$ in terms of a term value diagram. Due to the strong anharmonicity of the v_5 mode the term value difference between the different l values of the two polyads $00v_3[1v_5]^{|l|}$ and $01(v_3 - 1)[0(v_5 - 1)]^{|l|}$ shifts significantly with increasing v_5 value, so the interactions can be detected. This is indicated in Fig. 10 by arrows. The effect of these Fermi-type interactions is observed in the spectrum primarily through the presence of resonance-enhanced sub-bands.

Acknowledgements. We thank Dr. Stefan Klee, Georg Mellau and Klaus Lattner for their help in recording the spectra, and Dr. Rolf Wilmes for the preparation of the precursor for HCNO. The laboratory and computational work was supported in part by the Deutsche Forschungsgemeinschaft, Fonds der chemischen Industrie and a Max-Planck Research Award.

References

[1] M. Winnewisser, H. K. Bodenseh, *Z. Naturforsch.* **1967**, *22a*, 1724.

[2] H. K. Bodenseh, M. Winnewisser, *Z. Naturforsch.* **1969**, *24a*, 1966

[3] W. D. Sheasley, C. W. Mathews, *J. Mol. Spectrosc.* **1972**, *43*, 467.

[4] M. Winnewisser, B. P. Winnewisser, *Z. Naturforsch.* **1971**, *26a*, 128.

[5] M. Winnewisser, B. P. Winnewisser, *J. Mol. Spectrosc.* **1972**, *41*, 143.

[6] B. P. Winnewisser, M. Winnewisser, C. W. Mathews, K.M.T. Yamada, *J. Mol. Spectrosc.* **1987**, *126*, 460.

[7] B. P. Winnewisser, *J. Mol. Spectrosc.* **1971**, *40*, 164.

[8] B. P. Winnewisser, M. Winnewisser, F. Winther, *J. Mol. Spectrosc.* **1974**, *51*, 65.

[9] F. Winther, *J. Mol. Spectrosc.* **1976**, *62*, 232.

[10] E. L. Ferretti, K. Narahari Rao, *J. Mol. Spectrosc.* **1975**, *56*, 494.

[11] B. P. Winnewisser, P. Jensen, *J. Mol. Spectrosc.* **1983**, *101*, 408.

[12] P. R. Bunker, B. M. Landsberg, B. P. Winnewisser, *J. Mol. Spectrosc.* **1979**, *74*, 9.

[13] P. Jensen, *J. Mol. Spectrosc.* **1983**, *101*, 422.

[14] K. Yamada, B. P. Winnewisser, M. Winnewisser, *J. Mol. Spectrosc.* **1975**, *56*, 449.

[15] S. Albert, *Ph.D. Thesis*, Justus-Liebig-Universtät, Gießen, 1995.

[16] B. P. Winnewisser, J. Reinstädtler, K. M. T. Yamada, J. Behrend, *J. Mol. Spectrosc.* **1989**, *136*, 12.

[17] F. Stroh, *Ph.D. Thesis*, Justus-Liebig-Universität, Gießen, 1991.

[18] K. M. T. Yamada, F. W. Birss, M. R. Aliev, *J. Mol. Spectrosc.* **1985**, *112*, 347.

Mikrochim. Acta [Suppl.] 14, 89–101 (1997)
© Springer-Verlag 1997

Stratospheric FTS

Crofton B. Farmer

Jet Propulsion Laboratory, California Institute of Technology, Pasadena, CA 91109, U.S.A.

Abstract. Fourier-transform spectroscopy has played a prominent role over the past quarter century in furthering our understanding of the properties of the upper atmosphere and, in particular, in elucidating the processes that control the chemical and dynamical stability of the stratosphere. The first detections of HCl and HF in the early 1970s led to the discovery of the discrimination between anthropogenic and natural emissions of halogens which the changing ratio of these two trace compounds provided. More recently, the comparison, against model predictions of refined and revised measurements of the altitude-dependent budgets of the source and reservoir components of the chlorine, fluorine and nitrogen families of trace molecular species, has formed the basis of well-constrained tests against which the effectiveness of international protocols to control the release of industrial chemicals can be assessed. The contribution of the FTS methods to these and other global atmospheric environmental problems has been invaluable. Experimentalists have responded to the challenge of adapting Fourier-transform techniques to the often difficult environments of spacecraft, balloon, aircraft or remote ground-based sites in order to exploit the advantages afforded by making observations from these platforms. These developments and the current status of the results obtained from them are described.

Key words: atmospheric environment, atmospheric contaminants, ozone loss, halogen compounds, nitrogen oxides, FT spectroscopy.

The last time I had the pleasure of addressing this body was in Vienna, eight years ago. On that occasion my task was relatively easy; I spoke about FTS from space, the ATMOS instrument having had its first flight abroad the Space shuttle and we had an exciting array of new and unexpected results to present. It is rather more difficult this time, however, for the reason that in the decade since the first flight of ATMOS the level of research activity on the Earth's atmosphere by use of Fourier-transform instruments has increased dramatically, reflecting public and political awareness of a problem and a need and, of course, the particular value of the advantages of FTS applied to these problems. So my difficulty has been how to extract from all of the results of this busy period a representative sample. I have chosen what I see as the salient points of what we have learned about atmospheric properties and processes that can be directly attributed to the application of FTS techniques to the global environmental problems of the upper atmosphere. In doing this I will inevitably offend some researchers by leaving them out and I can only hope they will forgive me.

Background

Fourier transform spectroscopy has played a prominent role in atmospheric research since about 1970 – a 25 year period which has been motivated by two basic topics of interest: (*i*) investigation of the minor component and trace composition of the atmosphere, in order to understand the interaction of radiation in the context of its effects on the physical state and dynamics of the atmosphere – a central feature of climate studies, and (*ii*) investigation of the effects of man's activities in altering the compositional structure of the atmosphere and of the effects of these changes on the stability of the ozone layer in particular – a study of processes.

Table 1 summarizes the events of the past quarter century. The initial period, which might be called the period of *realization*, was characterized by a paucity of information about the stratosphere – even to the point that the natural production and loss mechanisms that governed the unperturbed ozone layer were not understood – and by the progressive realization that the effects of reactive forms of nitrogen, hydrogen and chlorine introduced into the atmosphere by fleets of high altitude aircraft, solid-fuel rocket motors, fertilizers and finally, industrial halocarbons, would increase the rate of destruction of ozone to a dangerous level. This period saw the development of several aircraft and balloon-borne infrared FT instruments, the

Table 1. FTS in the upper atmosphere. Historical perspective

1970–1975 Realization

Impact of:	High alt. aircraft	NO_x, HO_x, aerosol
	Solid fuel rocket motors	Cl_x
	Fertilizers	NO_x
	Industrial halocarbons	Cl_x

Early measurements; NO and HCl; Concorde, U-2, balloon flights

---

1976–1985 Exploration

Development of new balloon and aircraft instruments
Balloon intercomparison campaigns
First spaces flight of high resolution FTS
Discovery of Antarctic ozone loss
Simultaneity; detection of new trace constituents

1986–1995 Processes

Measurement of NO_x and Cl_x budgets
Antarctic and Arctic expeditions
NDSC established
Pinatubo eruption
Ubiquitous role of heterogeneous chemistry

---

first detections of two important trace species in this context, NO and HCl, and, interestingly from a historical viewpoint, a consensus among chemical kineticists that heterogeneous catalytic reactions on the surfaces of stratospheric aerosol particles would not be significant in comparison with the destruction of ozone through gas phase reactions. The decade that followed was essentially, a period of *exploration*, characterized by simultaneity of measurements of trace species (a feature of FT spectroscopy that renders it specially valuable to upper atmosphere studies) and by a number of new detections. This productive period also saw the scientifically interesting but practically frustrating and expensive balloon intercomparison campaigns and, at its culmination, a coincidence of two significant events that were to influence the direction of upper atmospheric research for the decade to follow. These were the first spaceborne flight of a high resolution atmospheric FT spectrometer and the publication at the same time (May 1985) of the seminal paper describing the discovery of the Antarctic "ozone hole". During the decade up to the present time we have seen the clarification of the atmospheric budgets of the N, H, and halogen families of trace species, the success of a number of ground-based and aircraft expeditions to study in detail the mechanisms of polar and mid-latitude ozone loss, and the establishment of the Network

for the Detection of Stratospheric Change (NDSC). In all of these activities, Fourier-transform spectroscopy has played a vital part. The period, characterized by the elucidation of *processes*, also saw the enactment of international protocols for the regulation of industrial chemicals; this collaboration between science, industry and policy makers can undoubtedly be viewed as a milestone in the evolution of international environmental cooperation.

In this overview I shall concentrate mainly on the results that have established our present understanding of upper atmospheric processes. But first I would like to put this into context by summarizing the state of knowledge of the composition of the upper atmosphere prior to the intense effort to unravel the puzzle of the Antarctic ozone loss, and then briefly review the different methods which have been used in applying FTS techniques to these problems.

Figure 1 shows the range of temperature and pressure for the region of the atmosphere of interest here. The important reactions occur from the upper troposphere through the stratosphere to the lower meso-

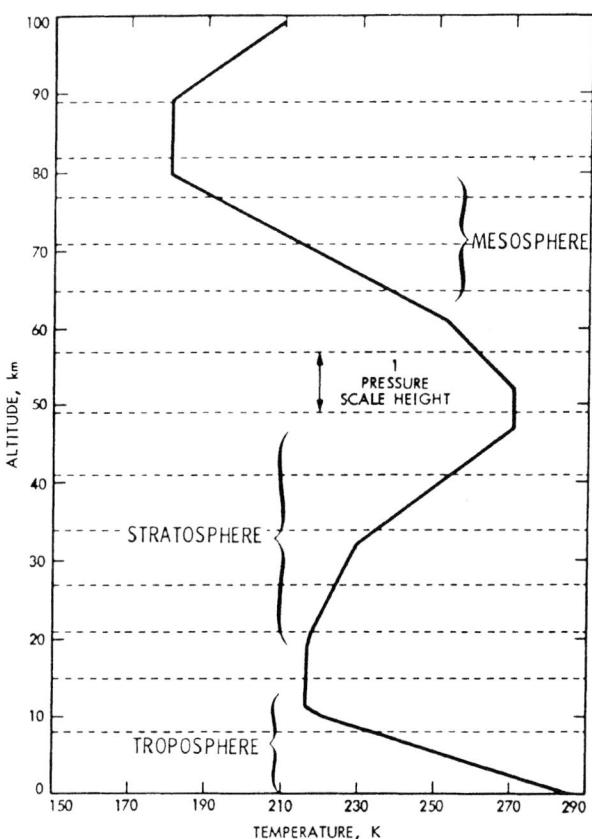

Fig. 1. Atmospheric temperature as a function of altitude. The region of interest here is from the surface to the lower mesosphere

sphere. The *scale height* is the factor which defines the exponential rate of decrease – in this case of pressure: the pressure falls by $1/e$ in about 8 km, so we're dealing with a pressure or density range of some 10 scale heights, or about 5 orders of magnitude. The degree to which the scale heights of non-uniformly mixed gases are different from the atmospheric density or pressure scale, provides a measure of their residence time in that region of the atmosphere. A property of the pressure scale height is that the integrated column density above a given pressure level is equal to that in a uniform layer having a thickness of one scale height, at the base pressure of the column. Thus we can easily see that the volume of the atmosphere is equal to the scale height times the surface area of the Earth – which turns out to have a value of about $4 \times 10^9 \, km^3$. This is an easy number to remember, since it is about the same as the global population at the end of the 1970s. So we each are custodians now of somewhat less than a 1 km cube of atmosphere; this fact becomes useful when we consider how easily we are able to affect fundamental properties of the atmosphere.

When Chapman first proposed (in the 1930s) the mechanism to explain the dynamic equilibrium between production and loss of ozone in the stratosphere, he was unable to account fully for the observed loss rate. It was not until much later that additional natural loss processes were identified; in fact, the first measurement of NO in the stratosphere in 1971 was behind the earliest concerns that were voiced regarding the stability of the ozone layer. The most insidious of these "potential threats", however, was revealed in 1974 when Rowland and Molina made headlines with their prediction [1] concerning chlorine, which they felt was accumulating at an alarming rate in the lower atmosphere. They pointed out that the tropospheric loading of chlorine was mainly in the form of stable species which would not be removed by the normal processes, such as rain-out; they would therefore diffuse slowly into the upper atmosphere where photodissociation would release reactive atomic chlorine. These species were the chlorofluorocarbons, of course, and the alarm generated sufficient concern for one of the sources of CFCs – aerosol spray cans – to be outlawed almost immediately. A concerted efforts was made to search in the stratosphere for the expected chemical sink for chlorine – HCl – and it was finally detected (in 1975) by our group at JPL, using an FTS with a resolution of $0.1 \, cm^{-1}$ [2]. The instrument, mounted in a U-2 aircraft flying at 19 km altitude, measured

the atmospheric absorption of sunlight at low solar elevations.

Figure 2 shows schematically the relationships between the natural and anthropogenic sources of the reactive species involved in the catalytic cycles that destroy ozone. These are all gas phase reactions. Of particular interest to the present topic are the three branches – interfering reactions – that remove chlorine and nitrogen from the catalytic cycles. Note that the cycles require sunlight in order to generate atomic oxygen. On average, the ClO radical cycles through the $O_3 + ClO$ reaction some 10^4 to 10^5 times before it meets an NO_2 molecule to form chlorine nitrate, a "reservoir" gas that temporarily takes it out of the loop. With this efficiency of recycling, coupled with ozone having a relative concentration of parts per million, it is not surprising that chlorine need only be present in the upper atmosphere in amounts measured in parts per billion for it to have a significant effect on the ozone concentration. The formation of chlorine nitrate removes both chlorine and nitrogen from the action. A key to the development of a major perturbation to stratospheric ozone lies in the balance between available chlorine and nitrogen; for example, if the chlorine is in excess, removal of nitrogen allows the destruction of ozone by chlorine to proceed unchecked.

At the time of the beginning of intensified theoretical and experimental studies of the upper atmosphere in the mid-1970s, many of these gases in the atmosphere had not even been detected, and the distributions of many others were unknown. Since the reactions repre-

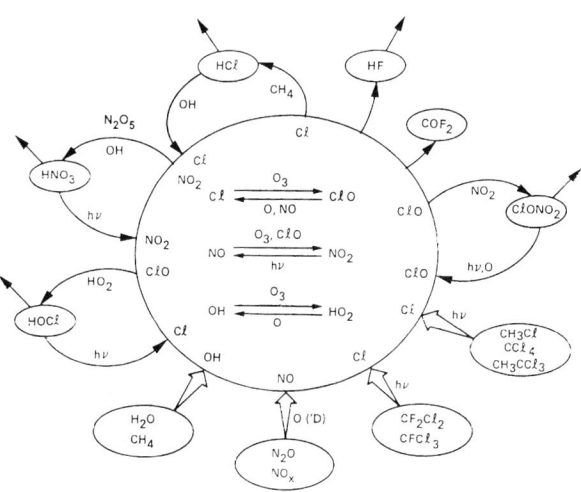

Fig. 2. Schematic representation of the sources and sinks of the reactive species which are involved in catalytic cycles with O_3

sented in this diagram are occurring simultaneously, but at rates that depend on the local temperature, pressure and solar insolation, the ability to measure their distributions *simultaneously* is essential in any attempt to test the validity of photochemical models. This is where the multiplex and speed advantages of Fourier-transform spectroscopy have been particularly beneficial.

Observations have been made in both the absorption mode, using the sun as the radiation source (this includes reflection from the moon), and in emission. They have been made from ground-based sites, from a variety of high altitude aircraft, from stratospheric balloons and from satellites in Earth orbit. Each of the observation techniques has its own advantages and disadvantages; as the results show, researchers have been able to successfully minimize the disadvantages in order to exploit a necessary feature of their particular chosen method. Without attempting a detailed comparison between the absorption and emission techniques, it will suffice in the present context to say that, whereas solar absorption spectroscopy has the advantage of high signal levels – a great help when measurements have to be made rapidly – they can only be made during daytime, they occasionally run into difficulties with diurnally varying species which have short photochemical lifetimes, and limb occultation observations made from orbit are restrictive in their access to full global and seasonal coverage. Against this, limb emission measurements made from aircraft, balloon and space platforms enjoy the advantage of full temporal and spatial coverage, but are subject to some loss of accuracy resulting from the need for good knowledge of the thermal profile along the line of sight in order to extract line opacities.

Observations from Space

The ATMOS (a convenient acronym for Atmospheric Trace Molecule Spectroscopy) instrument was developed at JPL for the purpose of recording the atmospheric absorption spectrum by observing the sun during sunrise and sunset as seen from the orbiting Space Shuttle [3]. The tangent point of the sun–spacecraft line of sight moves vertically through the atmosphere typically at about 2 km/s, so that spectra must be recorded rapidly under these conditions if good vertical resolution is to be achieved. ATMOS is capable of sampling one million points every two seconds. At the heart of this superb instrument, which

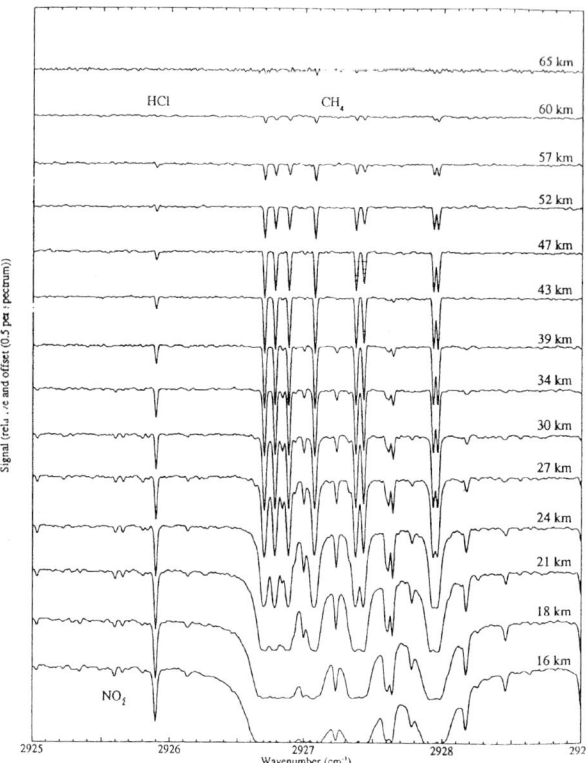

Fig. 3. A 4 cm^{-1} segment of a sequence of spectra by ATMOS during a sunset occultation from spacelab-3. The tangent altitudes corresponding to each successive zpd (zero path difference) are shown on the right

was built by Honeywell, is a pair of retroreflectors moving in contra-motion. The spectral range of ATMOS is from 600 to 4000 cm^{-1}; the optical configuration is double-passed, and the unapodized resolution is 0.01 cm^{-1}.

Figure 3 illustrates a small portion of a sequence of sunset spectra, showing the growth of the HCl R1 line, together with a group of CH$_4$ lines, from the top of the stratosphere down to the tropopause. Several new detections, including those of N$_2$O$_5$, ClNO$_3$ (sometimes written as ClONO$_2$) and HNO$_4$, were made from these initial ATMOS observations; Figure 4 summarizes the results of the retrievals of the vertical distributions of minor and trace molecular constituents made from this flight. Notice the very large dynamic range of the measurements – volume mixing ratios spanning almost 7 orders of magnitude and covering an altitude range that reaches into the lower thermosphere in some cases.

It is convenient to comment here on two aspects of what was learned from these results. The first concerns the halogen budgets, shown in Fig. 5. Here we see

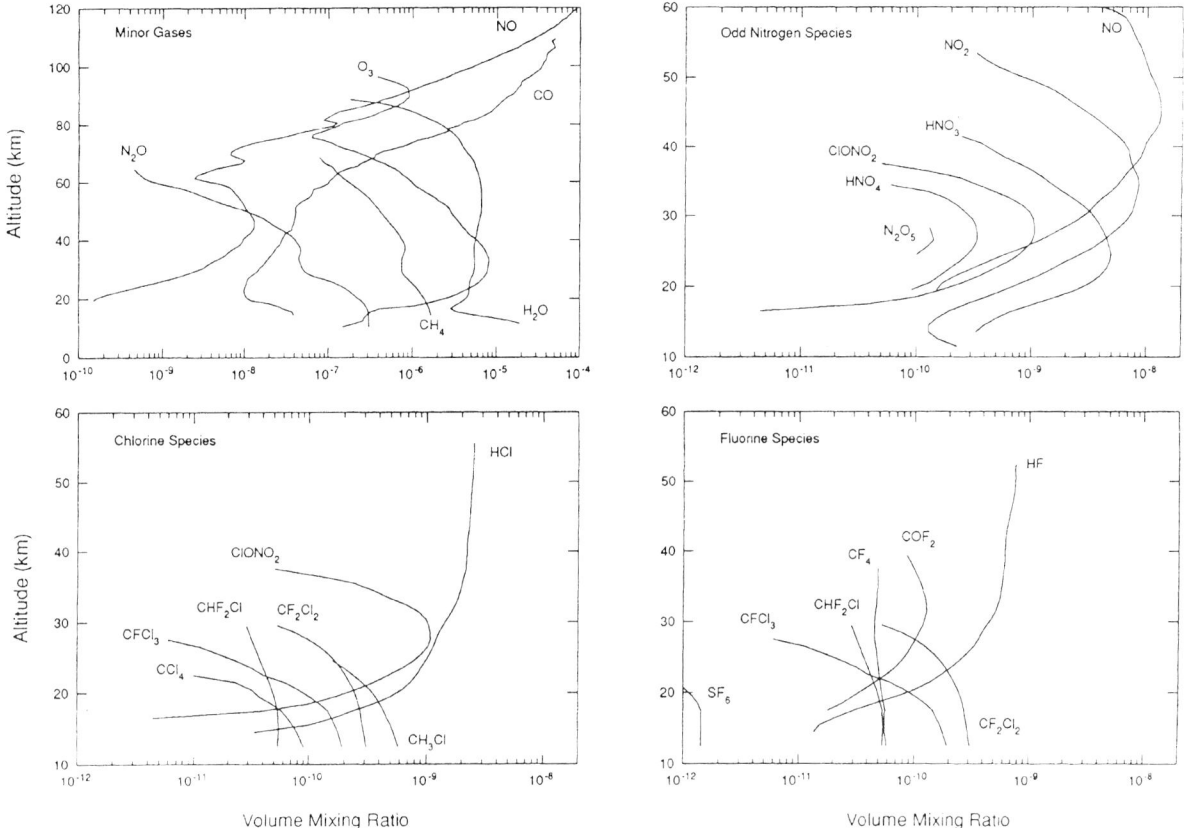

Fig. 4. Volume mixing-ratio profiles for 30°N latitude retrieved from ATMOS

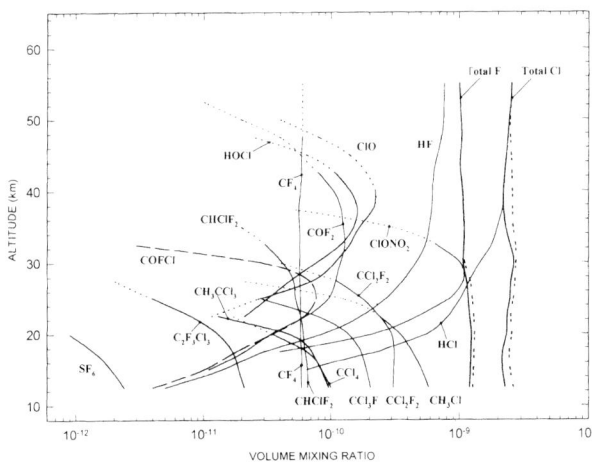

Fig. 5. The altitude distribution of chlorine-containing species; from Zander et al. [4]. (Reprinted by permission of the copyright holders. Copyright 1992, Kluwer Academic Press, Dordrecht.)

clearly the transition from the source gases at low altitudes, through the temporary reservoirs, to the final sinks, HCl and HF. The sum of chlorine or fluorine atoms remains almost constant throughout the entire altitude range, suggesting that there are no missing species of significance [4]. The slight decrease with altitude of the sum of total Cl and F reflects the finite mixing time (about 5 years) required for gases injected into the atmosphere at the surface to reach the top of the stratosphere. From this result alone there could be no doubt that chlorine in the upper atmosphere was there as a result of man's actions. This figure also serves to illustrate the fact that in order to monitor the chlorine loading of the atmosphere it is only necessary to measure HCl at the top of the stratosphere – a relatively simple and inexpensive monitoring task. Even the total column abundances of HCl and HF, which can be determined from the ground, are adequate indicators of the total atmospheric halogen burden.

The second important result concerns the implications of the measurements of the distributions of nitrogen oxides, in particular N_2O_5, and nitric acid. N_2O_5 is a difficult gas to measure because of the line density of its spectrum, coupled with interference from CH_4, O_3 HNO_3 and N_2O bands; its presence had been conjec-

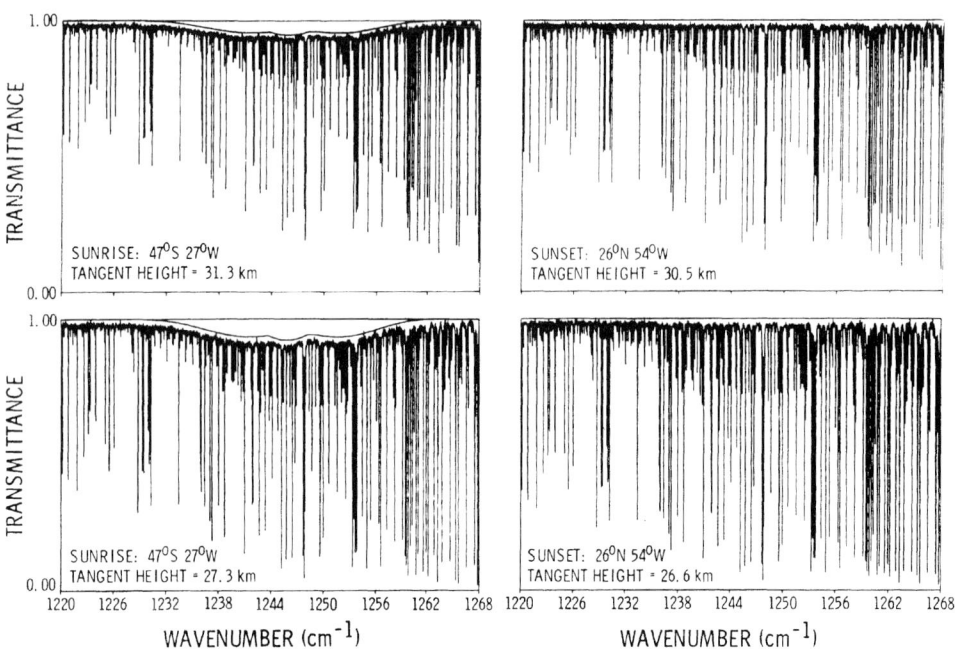

Fig. 6. Comparison between sunrise and sunset spectra in the region of the v_{12} N_2O_5 band (from [5])

tured for some time and attempts to detect it by differencing radiometric emission measurements were inconclusive. The opportunity to compare high resolution sunrise and sunset spectra and show the difference in the continuum which matched the unresolved shape of the band (N_2O_5 is formed at night and is therefore expected in the sunrise spectrum only) allowed an unambiguous identification to be made [5]. Figure 6 illustrates the case of the v_{12} band at $1250\,cm^{-1}$.

A number of attempts were made to model the sets of simultaneous vertical profiles obtained from the Spacelab-3 and succeeding ATLAS flights of the instrument. It was found that the distributions of NO, NO_2 and N_2O_5 were not compatible with the observed HNO_3 when only gas phase chemistry was considered. Using the measured background aerosol distribution, however, to compute the added effect of the hydrolysis of N_2O_5 on particle surfaces ($N_2O_5 + H_2O \rightarrow 2HNO_3$), gave good agreement between the expected and observed distributions [6]. This can be seen best in the ratios of NO_2/HNO_3 and N_2O_5/HNO_3, which are illustrated in Fig. 7a and b. Thus the heterogeneous production of nitric acid on aerosol particles at midlatitudes provided another potential mechanism for denitrification of the stratosphere, and this result took on added significance with the appearance of the large

and persistent sulphuric acid cloud from the eruption of Mount Pinatubo in 1991 (see later).

Before leaving the topic of FTS measurements from space I would just comment on the status of future plans and expectations. To date, ATMOS is the only high resolution FT spectrometer to have made observations of the Earth atmosphere from space. A number of new instruments are in the wings, in preparation for flights scheduled to occur in the coming decade. These include MIPAS/ENVISAT (1998) and TES/EOS (2002), which are emission instruments, and ILAS/ADEOS (1996) and DOPI (1996), which operate in the absorption mode. The last flight of ATMOS was on the ATLAS-3 mission, in November 1994. As things stand at present the future of ATMOS is uncertain; there are no firm plans to refly the instrument in its present form and I fear that the continuity of measurements, especially those of the halogen budget, is in jeopardy. We are all subject to the vicissitudes of the other kind of budget and of political fickleness.

Polar Campaigns

The unequivocal presence of $ClONO_2$ in the stratosphere revealed by the ATMOS spectra obtained in May of 1985 underscored the coincidence of the appearance of the paper announcing the discovery of the

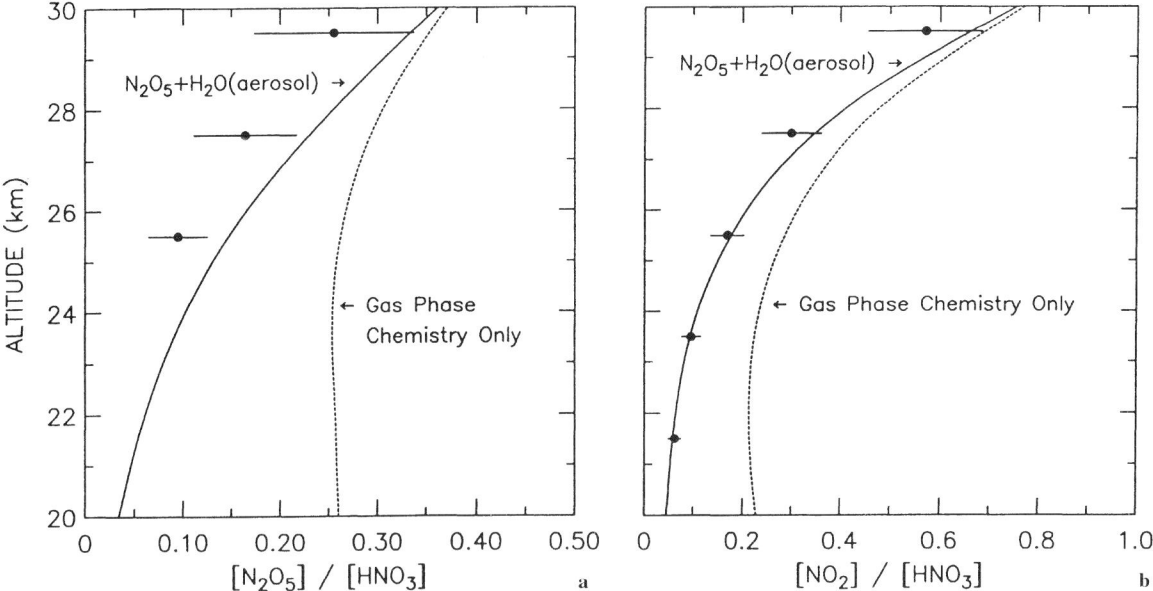

Fig. 7. The altitude variation of the ratio N_2O_5/HNO_3 (**a**) and NO_2 (**b**) as derived from ATMOS results, compared with model distributions with (solid) and without (dotted) the inclusion of the hydrolysis of N_2O_5 on particle surfaces. From [6]

Antarctic ozone "hole" by Farman et al. [7] in Nature that same month. While theoreticians put forward competing theories to explain this strange seasonal loss of ozone above Antarctica, a first expedition to the ice was hastily organized for the following (1986) austral spring. One of the four instruments taken to McMurdo for this exercise was an FT spectrometer, the JPL Mark 4 interferometer. This instrument was originally designed for balloon flights, but it was used for the ground-based measurements in Antarctica as well as for the aircraft flights into the Antarctic and Arctic vortices and of course the balloon flights that occurred in the intervening periods.

The spectra were obtained from a position near the edge of the Antarctic vortex so that, as the vortex moved, the instruments were able to sample the air successively inside and outside the region of rapidly decreasing column ozone [8, 9]. Measurements of the column densities of the tropospheric source gases N_2O and CH_4, which act as tracers of atmospheric motions, showed that air within the vortex was *descending*, rather than rising, a result which discounted the suggestion that a reversal of the normal circulation could be bringing ozone-poor air from lower latitudes into the vortex. Similarly, the levels of NO and NO_2 were seen to be much lower inside the vortex than outside, a result which dispelled the notion that the ozone loss was the result of increased levels of nitrogen oxides in

the middle and upper stratosphere caused by a change in solar activity. And finally, the partitioning of chlorine between HCl and $ClNO_3$ was reversed, and total chlorine amounts were low, during the period of erosion of the ozone column at the beginning of spring, followed by a return to the normal partitioning and abundances as the ozone column recovered in November. These results, taken together with UV and microwave measurements of ClO_2 and ClO respectively, pointed conclusively to perturbed chlorine chemistry as the cause of the annual springtime ozone reduction. However, the details of the mechanism which could cause such rapid and efficient ozone destruction remained a puzzle for some time and were the subject of a series of well organized and well instrumented aircraft flights during the succeeding years.

These flights exploited a combination of in situ instruments, mounted in NASA ER-2 aircraft (the research version of the reconnaisance U-2 of the cold-war era), and remote sensing instruments (which included two solar absorption FT spectrometers) carried by the NASA DC-8. The meteorologists were strongly influential in these compaigns and they taught the spectroscopists about conservation of volume mixing ratio, how air parcels mix along isentropic surfaces, and how to reduce the confusion that otherwise results from the effects of atmospheric dynamics on the altitude distribution of trace gases, by using the local number density of the long-lived tracers (such as N_2O, CH_4 and HF) to normalize the local and column values. This alternative way of looking at the compositional data was invaluable in that it provided a surrogate for the exposure of an air parcel to solar UV radiation. The polar aircraft flights which have been made over the past 8 years, together with aircraft and balloon flights at mid-latitudes, have advanced tremendously our understanding of upper atmospheric processes. In the space available here it is not possible to do more than show some examples of the impressive quality of the spectra that are currently being obtained and of the progress that has been made in their analysis.

To successfully record high resolution spectra from a vibrating aircraft or from a gondola swinging beneath a stratospheric balloon, careful attention must be paid to several engineering aspects of the instrument design, in particular, the control of the position of the moving reflector. The scheme that we adopted some years ago for the JPL Mark 4 balloon FT spectrometer is to use a long plastic nut machined to provide a near interference fit to the threaded bar that provides the drive for the moving carriage. The nut contacts about 100 of the grooves of the bar, integrating out any random errors of the thread and providing complete freedom from back-lash – an essential requirement for an unstable platform under gravity. This instrument is also double-passed and has a maximum unapodized resolution of $0.003\,cm^{-1}$. The system has performed excellently in both the aircraft and balloon environments, as well as in the difficult thermal and wind conditions of the Antarctic ice shelf. Examples from spectra obtained during these flights, showing also the quality of the synthetic spectral fits and the resultant residuals, are illustrated in Fig 8a and b. The aircraft instruments provided column measurements above the flight altitude for a large number of source and reservoir species as functions of latitude, as the flight trajectories penetrated the "wall" of the vortex to its interior. In particular, these included the key species O_3, NO, NO_2, HNO_3, $ClNO_3$, HCl and HF. Figure 9 shows these distributions for the case of the first Antarctic aircraft campaign [10], and illustrates dramatically the denitrification and removal of inorganic chlorine reservoirs that has occurred within the "chemical processor", as it has been called. The most striking demonstration of the direct action of chlorine on ozone, however, came from the in situ fluorescence measurements from the ER-2 Fig. 10 as it flew through the wall of the vortex [11].

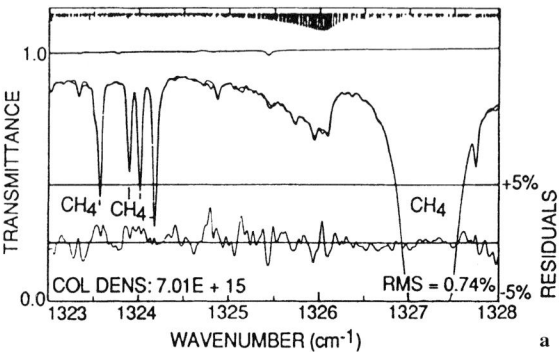

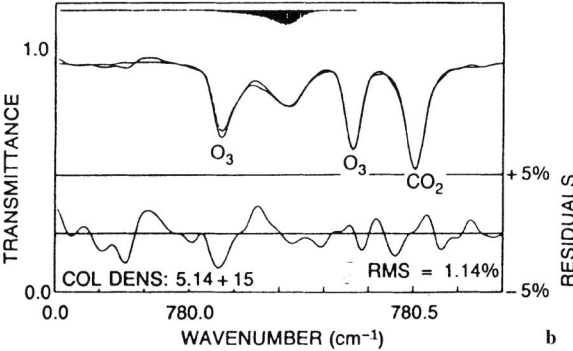

Fig. 8. Examples of spectra obtained from Antarctic aircraft flights in Sept. 1987. The lower trace in each case is the difference between the calculated and observed spectra; the upper trace shows the solar spectrum and the tick marks represent the positions and strengths of lines used in the simulations. Upper panel HNO_3, solar zenith angle 85.3°; lower panel $ClNO_3$, solar zenith angle 89.1°. From [10]. (Reprinted by permission of the copyright holders. Copyright 1989 American Geophysical Union)

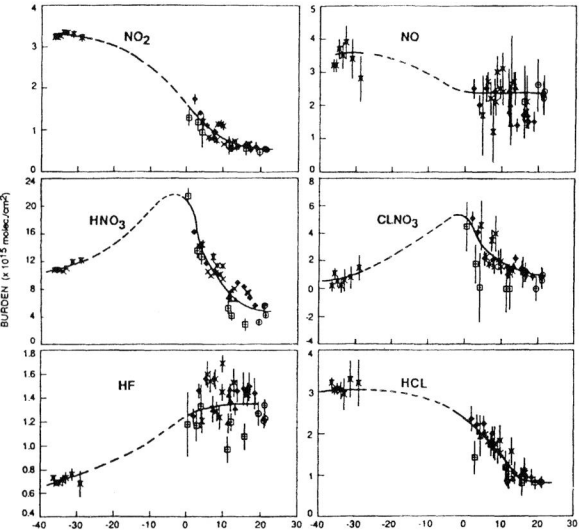

Fig. 9. Latitude dependences of NO, NO_2, HNO_3, $ClNO_3$, HCl and HF referred to the edge of the polar vortex. From Toon et al. [10]. (Reprinter: permission as for Fig. 8)

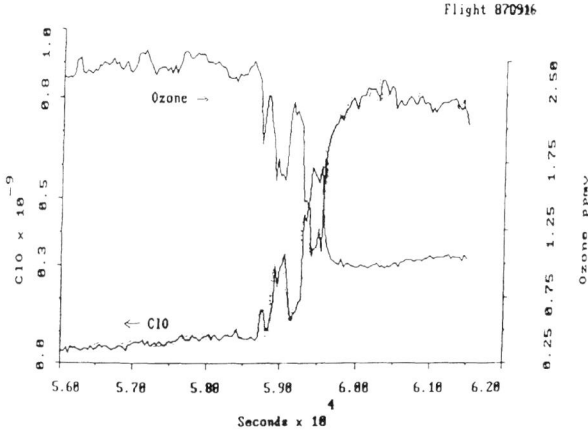

Fig. 10. Anti-correlation between ClO and O_3 from in situ, measurements made from the EF-2 aircraft as it penetrated the wall of the Antarctic vortex. Adapted from [11]

A qualitative explanation of the chemical and dynamical processes that cause the polar ozone loss suggested by these and, of course, the many other in situ measurements made during these intense exercises, as well as by supporting laboratory measurements, is as follows. Starting with the fall season, the polar ozone distribution has its characteristic circumpolar maximum. During winter (the polar night) the stratosphere cools to space; because the S. Pole is isolated, the circulation – the polar vortex – is persistent, the air inside having a slow sinking motion. (The N. Pole, by contrast, is not isolated and the northern polar vortex is fragmented.) Polar stratospheric clouds (PSCs) form at very low temperatures – they are ubiquitous in the south and sporadic in the north. The more common of these aerosols are nitric acid trihydrate, which forms at about 195 K; PSCs of this type ("type 1") consists of many small particles. If temperatures fall to 190 K, larger, pure ice, particles ("type 2") form. These particles grow and precipitate out. Heterogeneous reactions begin; HCl diffuses into the ice; $ClNO_3$ reacts on the surface with HCl or H_2O; HNO_3 sticks on the particle surfaces; Cl_2 is ejected. When the sun rises at the beginning of spring, Cl_2 immediately dissociates, Cl forms ClO and Cl_2O_2, NO_x is locked up as HNO_3 on the particles, ClO and Cl_2O_2 (and BrO) destroy ozone. This highly perturbed chemistry is confined to the "processor". As ozone is reduced the stratosphere absorbs less sunlight and cools even further, causing more dehydration. The heterogeneous reactions are rapid enough to remove as much as half

of the ozone column in the contained region within a period of three weeks.

The Role of Balloon-Borne Measurements

Observations from balloons have played an important part in upper atmosphere research. In the early days, balloons provided access to altitudes above those which could be reached with aircraft and from which limb geometries with tangent heights up to 40 km were attainable. Today, observations from balloons continue to play a valuable role in stratospheric research, particularly in validating in situ and spacecraft remote-sensing observations at intermediate spatial scales, and in testing specific aspects of the chemistry and dynamics of the atmosphere, by using chosen combinations of instruments, relatively quickly and economically. In this context I have tried to select results that illustrate the remarkable precision that is currently achieved with FTS balloon observations as well as their having made a clear contribution to progress in stratospheric research.

Because of the importance of HCl as a direct indicator of global chlorine loading it is vital that confidence in the precision and accuracy of its measurements should be high. How well do ground-based column measurements and local mixing ratio measurements derived from spectral fitting algorithms applied to an occultation sequence of spectra, compare? Figure 11 shows an example of results from balloon flight in 1993 in which column measurements from high-sun spectra recorded during the ascent of the balloon are

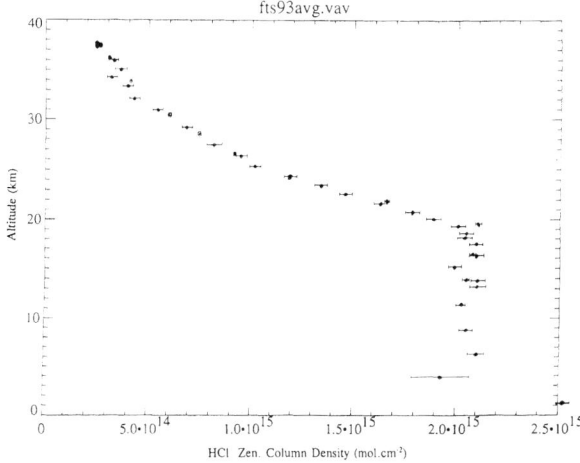

Fig. 11. Zenith column densities of HCl from balloon ascent and sunset occultation spectra. Courtesy G. C. Toon

compared with column values above successive tangent altitudes retrieved from the spectra that were recorded at the balloon float altitude. The two profiles of zenith column *vs* base altitude are indistinguishable to within relative standard deviations of about 5%, except at the surface where the ascent results show the added burden of the tropospheric component of HCl, which is confined to a layer of about 2 or 3 km thickness.

There are many examples of the high quality of not only the balloon spectra themselves, but of careful laboratory work to improve molecular spectral parameters, together with further development of radiative transfer and spectral fitting algorithms that are essential to the analysis of the observational data. In this context must be cited the work of the Denver University group [12], whose long history of successful balloon flights with high resolution FT spectrometers has made a considerable impact on the spectroscopy of the upper atmosphere.

The groups at SAO, IROE and LaRC [13, 14] have used the far infrared region between 50 and 160 cm^{-1} to exploit the strong rotational transitions of several important stratospheric species that are not accessible at near and mid infrared frequencies. These include members of the HO$_x$ family (HO$_2$, H$_2$O$_2$ and OH) as well as HOCl and HBr. An example of a day–night pair of spectra [15] in the region of the 118 cm^{-1} transition of OH is shown in Fig. 12.

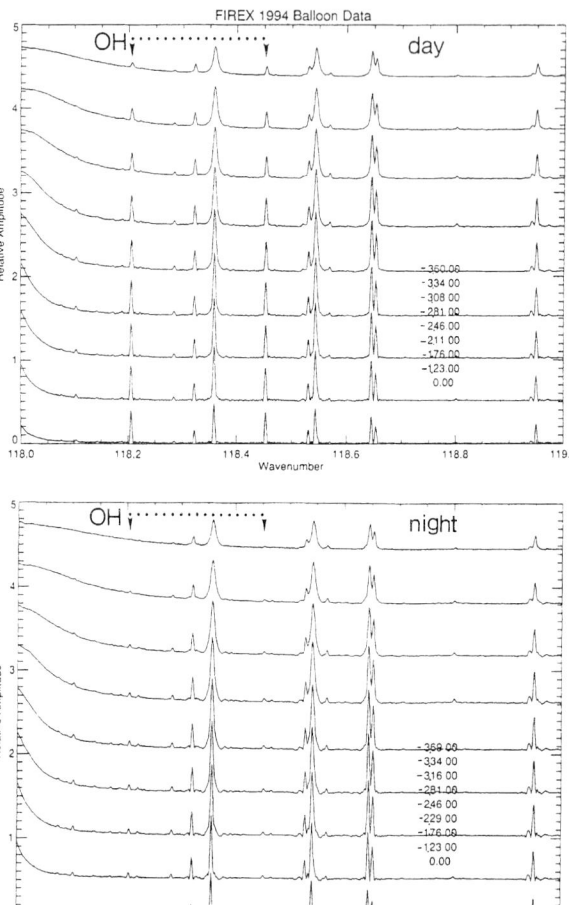

Fig. 12. Far IR emission spectrum of the atmosphere in the region of the OH rotational transition at 118.2 cm^{-1}. Courtesy B. Carli, M. Carlotti, B. M. Dinelli, F. Mencaraglia, J. H. Park

The Network for the Detection of Stratospheric Change

At about the time of the start of the polar aircraft flights, NASA established the Network for the Detection of Stratospheric Change (NDSC), now an international collaboration with stations in the U.S. Europe, the Arctic, New Zealand, Hawaii, and elsewhere. The sites are equipped with microwave, LIDAR, and FTIR instruments to monitor the important source, radical and reservoir molecular species that are diagnostic of changes in the state of the stratosphere. This network was set up as a result of the early realization that the chlorine and NO$_x$ budgets at least, and any episodes involving these principal chemical families, can in fact be monitored from the surface of the earth, given appropriate instrumentation. An example is appropriate here of the consistently high quality work that has been carried out at the ISSJ in the Bernese Alps, one of the combination of observatories that make up the European Alpine station of the NDSC. Among the

many scientific contributions based on the archive of beautiful spectra recorded at the Jungfraujoch is the record of HCl and HF columns. This record, which was started in the mid-1970s [16], is shown in Fig. 13. Despite the large seasonal and secular variability in the column amounts, the long-term trend in the abundances of both gases, as well as the rapid change in the HF/HCl ratio (another clear diagnostic of their anthropogenic origin) are clearly seen. Following are two further examples from the many important results of recent ground-based studies.

It was mentioned earlier that the measured mid-latitude distributions of NO$_x$ species could only be explained by the inclusion of the heterogeneous reaction of N$_2$O$_5$ with water vapor on stratospheric aerosol particles. The eruption of Pinatubo in June 1991 provided a striking demonstration of this. Figure 14

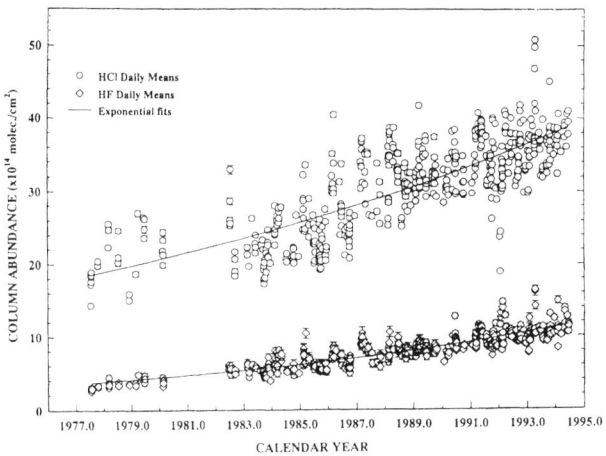

Fig. 13. Record of total atmospheric column measurements of HCl and HF from the International Scientific Station of the Jungfraujoch; courtesy of R. Zander

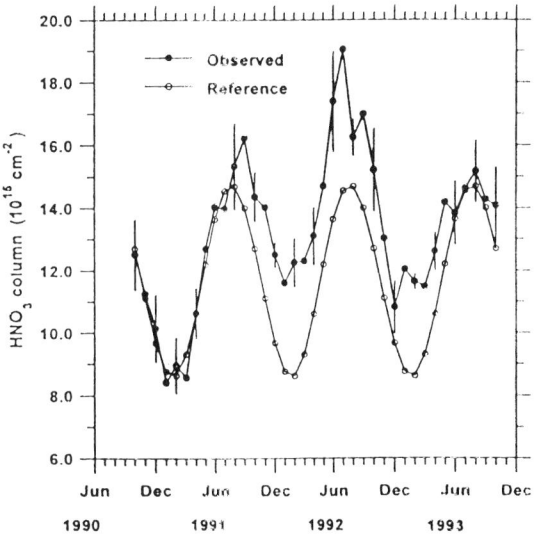

Fig. 15. Monthly averaged HNO_3 column amounts observed at Lauder. The reference values established from the data points prior to the Mt. Pinatubo eruption are shown as open circles. From Koike et al. [17]. (Reprinted by permission of the copyright holders. Copyright 1994, American Geophysical Union)

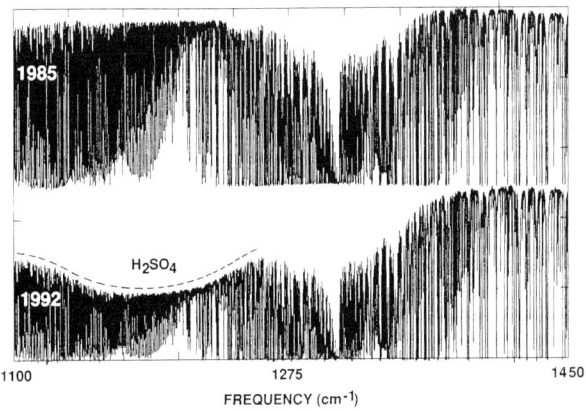

Fig. 14. ATMOS spectra for 24 km altitude before and after the Mt. Pinatubo eruption of June 1991, showing the strong sulphuric acid aerosol absorption

dust cloud, or an asteroid impact, for example) may not have been so far-fetched after all. In any event, this piece of work seems to me to capture the essence of what the NDSC is all about.

The second example of high quality ground-based FT spectroscopy that I want to cite is taken from the work of Notholt and his colleagues, who have used the moon (to be strict, I should say sunlight reflected from the moon) as the radiation source for observations (made with a Bruker 120M FT spectrometer) from the NDSC site at Spitzbergen (at 79 °N) during the Arctic winter. For these measurements Notholt et al. [18] averaged typically 15 spectra recorded over a period of 3 hours to achieve the required signal/noise ratio at 0.02 cm^{-1} resolution. Figure 16a illustrates the results for $ClONO_2$, with a synthetic fit giving residuals of less than 1%; Fig. 16b shows a Moon transmission spectrum for the region of the HCl R1 line. Using this technique Notholt et al. were able to show clearly the passage over their site of air-masses cold enough for PSCs to have formed and which had been strongly depleted in HCl and $ClONO_2$ – air–masses, in other words, which had been primed very early in the winter for catalytic ozone destruction as soon as they emerged into sunlight. These results achieved the difficult demonstration of the winter preconditioning of the northern polar containment vessel, and have contributed to

shows the H_2SO_4 aerosol band seen in spectra over the 24–28 km altitude range from the May 1992 ATMOS flight on the ATLAS-1 mission. Long-term monitoring of stratospheric gases from the NDSC site at Lauder, New Zealand had established the normal seasonal variation of HNO_3 above the site. After the Pinatubo aerosol cloud had spread to the latitude of Lauder, observers there [17] were able to detect easily a decrease in stratospheric NO_2 and a concomitant increase in HNO_3 (Fig. 15). So the scenario in which unchecked addition of chlorine to the atmosphere could cause a runaway destruction of ozone in the event of extensive removal of NO_x resulting from a widespread increase in aerosol loading (a nuclear

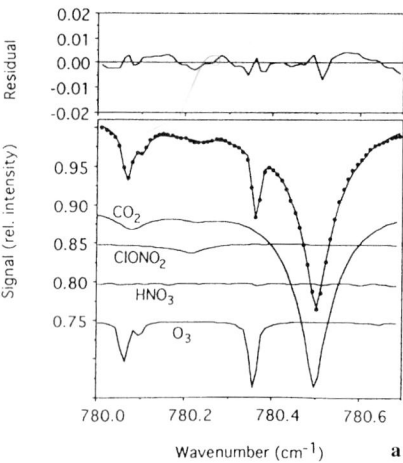

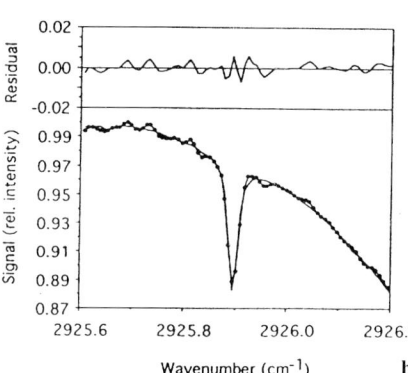

Fig. 16. Measured and computed spectra of **a** $ClNO_3$ and **b** HCl. The measured spectra were obtained by Notholt et al. [18] at Spitzbergen (79 deg. N) in December 1992 using moonlight as the radiation source. (Measured spectra Reprinted by permission of the copyright holders. Copyright 1995, American Geophysical Union)

the general belief that the same processes that cause the Antarctic stratospheric ozone loss also operate in the Arctic, with the major difference that the fragmentary nature of the vortex, coupled with the much less frequent occurrence of temperatures low enough for PSC formation, results in far less severe episodes of seasonal ozone loss in the north.

The Effect of Regulation of Industrial Halocarbons

After 20 years of measurements of stratospheric HCl, and some 6 years since the first restrictions on the release of CFCs into the troposphere went into effect, it is interesting to compare a current assessment of the projected Cl and F loading of the atmosphere with the HCl and HF mixing ratios at the top of the stratosphere. This is shown in Fig. 17. The result seems surprisingly accurate. The solid lines show the actual and projected loading derived from industry figures, and represent the mixing ratios for the atmosphere at the surface; the measured values refer to the uniformly mixed region at the top of the stratosphere, the delay between them reflecting the approximately 5-year time for vertical mixing. If the predictions are correct, we will have to wait until some 4 or 5 years from now before we can expect to see a reversal of the growth of the stratospheric halogen sinks.

Conclusions

In concluding, FTS has clearly played a significant role in environmental studies in the upper atmosphere over the past quarter of a century, more so perhaps than any other technique. It must be remembered, however, that the successes are the result of the combined efforts of laboratory spectroscopists, chemical kineticists, modelers and instrument engineerings, as well as those of us who enjoy looking at and analyzing spectra.

Fig. 17. Past and projected atmospheric chlorine and fluorine loading (taken from industry estimates). The HCl and HF values are from ATMOS measurements at the top of the stratosphere (see text)

References

[1] F. S. Rowland, M. J. Molina, *Rev. Geophys. Space Phys.* **1975**, *13*, 1.

[2] C. B. Farmer, O. F. Raper, R. H. Norton, *Geophys. Res. Lett.* **1976**, *3*, 13.

[3] C. B. Farmer, *Mikrochim. Acta* **1987**, *III*, 189.

[4] R. Zander, M. R. Gunson, C. B. Farmer, C. P. Rinsland, F. W. Irion, E. Mahieu, *J. Atmos. Chem.* **1992**, *15*, 171.

[5] G. C. Toon, C. B. Farmer, R. H. Norton, *Nature* **1986**, *319*, 570.

[6] M. B. McElroy, R. J. Salawitch, K. Minschwaner, *Planet. Space Sci.* **1992**, *40*, 373.

[7] J. C. Farman, B. G. Gardiner, J. D. Shanklin, *Nature* **1985**, *315*, 207.

[8] C. B. Farmer, G. C. Toon, P. W. Schaper, J.-F. Blavier, L. L. Lowes, *Nature* **1987**, *329*, 126.

[9] G. C. Toon, C. B. Farmer, P. W. Schaper, J.-F. Blavier, L. L. Lowes, *J. Geophys. Res.* **1989**, *94* (*11*), 613.

[10] G. C. Toon, C. B. Farmer, L. L. Lowes, P. W. Schaper, J.-F. Blavier, R. H. Norton, *J. Geophys. Res.* **1989**, *94* (*16*), 571.

[11] J. G. Anderson, W. H. Brune, M. H. Proffitt, *J. Geophys. Res.* **1989**, *94*, 11465.

[12] A. Goldman, F. J. Murcray, R. D. Blatherwick, J. J. Kosters, F. H. Murcray, D. G. Murcray, *J. Geophys. Res.* **1989**, *94*, 14945.

[13] B. Carli, J. H. Park, *J. Geophys. Res.* **1988**, *93*, 3851.

[14] D. G. Johnson, K. W. Jucks, W. A. Traub, K. V. Chance, *J. Geophys. Res.* **1995**, *100*, 3091.

[15] B. Carli, M. Carlotti, B. M. Dinelli, F. Mencaraglia, J. H. Park, *J. Geophys. Res.* **1989**, *94*, 11049.

[16] R. Zander, *Private Communication.*

[17] M. Koike, N. B. Jones, W. A. Matthews, P. V. Johnston, R. L. McKenzie, D. Kinnison, J. Rodriguez, *Geophys. Res. Lett.* **1994**, *21*, 597.

[18] J. Notholt, P. Von der Gathen, S. Poil, *J. Geophys. Res.* **1995**, *100* (*11*), 269.

Mikrochim. Acta [Suppl.] 14, 103–108 (1997)

From Vibrational Intensity Spectroscopy to Non-linear Optical Properties of Organic Molecules in Electronics and Photonics

Mirella Del Zoppo*, **Chiara Castiglioni**, **Paola Zuliani**, and **Giuseppe Zerbi**

Dipartimento di Chimica Industriale e Ingegneria Chimica, Politecnico di Milano, P. L. Da Vinci, 32, I-20133 Milano, Italy

Abstract. A new use of vibrational intensity spectroscopy is proposed for the characterization of non-linear optical responses of polyconjugated organic molecules. The experimental and theoretical results obtained for several molecules are reported and discussed. These data prove that it is possible to use the vibrational approach as a new method for measuring molecular hyperpolarizabilities, which in some cases can be competitive with the more traditional ones. A theoretical justification of the method is also presented.

Key words: polyconjugated organic molecules, hyperpolarizability, non-linear optical properties, vibrational spectroscopy.

In the recent past non-linear optical processes have been increasingly exploited in a variety of optoelectronic and photonic applications. The basic phenomenon of an optically induced intensity-dependent change of the refractive index is of fundamental importance in all-optical switching and computing. Optical switching allows much faster data-processing than electronic switching does, and parallel processing also becomes feasible. Of course the efficiency of these non-linear optical processes relies on the material employed. Most of the materials currently used in the fabrication of photonic devices are inorganic crystals (e.g. potassium dideuterium phosphate KDP, lithium niobate $LiNbO_3$, barium titanate $BaTiO_3$). Although the technology for these materials is highly developed and their non-linear optical susceptibilities are large enough for most current photonic applications, they also have some drawbacks. A particularly severe limitation is the relatively slow optical switching time characteristic of photorefractive ferroelectric inorganic crystals. For this reason new non-linear optical materials are needed to extend the range of photonic applications. Organic materials and inorganic semiconductors are the main candidates as new non-linear optical media. Organic materials [1–5] are of major interest because of their relatively low cost, ease of fabrication and integration into devices, tailorability (which allows fine tuning of the chemical structure and properties for a given non-linear optical process), high laser-damage thresholds, low dielectric constants, fast non-linear optical response times, and off-resonance non-linear optical susceptibilities comparable to or exceeding those of inorganic crystals. These features justify the enormous interest which has recently grown in the study and characterization of these materials. Several techniques, both experimental and theoretical, have been employed in efforts to understand the mechanisms which lead to an improved performance of the material.

With this work we wish to point out how vibrational spectroscopy can help in the non-linear optical characterization of the molecules under study. More precisely, our final goal is to present a new and simple method of measuring the molecular non-linearities, based on use of the vibrational spectra. In contrast to the traditional experimental techniques [second and third harmonic generation (SHG and THG), electric field induced second harmonic generation (EFISH), etc.] which require the use of expensive laser equipment with a carefully designed set-up, our method of measurement can be routinely applied in any spectroscopy laboratory.

The method relies on a completely new use of vibrational intensity spectroscopy. Frequency spectroscopy has always been preferred to intensity spectroscopy.

---

* To whom correspondence should be addressed

Frequency seems a much more easily accessible datum, but from experience gained over the years, it appears that the intensity datum is much richer in information, especially when the electronic nature and properties of the molecule are considered.

Infrared intensities have already been successfully used to obtain quantitative estimates of equilibrium fixed atomic charges q_{α}° and charge fluxes (or charge redistribution induced by the vibration) $\partial q_{\alpha}^{\circ}/\partial Q_k$ [6, 7]. Within the Equilibrium Charges and Charge Fluxes model (ECCF) [8, 9] infrared intensities can be expressed as functions of (q_{α}°, $\partial q_{\alpha}^{\circ}/\partial R_t$ and $\partial R_t/\partial Q_k$).

Since the organic molecules to be used in non-linear optical applications are low band-gap conjugated systems, here we will consider only the intensity behaviour of this class of compounds. The prototype polyconjugated molecule is *trans*-polyacetylene. Analysis of its relative intensities suggests that charge fluxes along the chain and hence the charge mobility are very large [10]. Moreover the extent of charge mobility seems to be connected with the degree of conjugation along the chain. Indeed, charge fluxes are smaller in the shortest polyacetylene oligomers.

Also, the spectrum of doped polyacetylene [11] can be interpreted on the basis of extremely large charge fluxes along the backbone chain. The enhancement of conductivity obtained by doping is related to the occurrence of highly mobile charge defects. The enormous intrinsic activity of the doping-induced infrared modes is essentially due to the occurrence of huge charge fluxes along the chain.

Raman intensities are even more important in the study of polyconjugated molecules. For this class of molecules, Raman spectra, both in frequency and intensity, can be nicely rationalized in terms of a two-state model [12] in which only one relevant excited state is taken into account. Under this hypothesis the Raman scattering cross-section of ith normal mode is proportional to $|R(-\omega_1, \omega_2, \omega_i)|^2$, given by

$$|R(-\omega_1, \omega_2, \omega_3)| = \frac{2[E_{eg}^2 + \hbar^2\omega_1\omega_2]}{[E_{eg}^2 - (\hbar\omega_1)^2][E_{eg}^2 - (\hbar\omega_2)^2]}$$
$$\times [\langle g'|\mathbf{e}_1\cdot\boldsymbol{\mu}|e'\rangle\langle e'|(\partial H/\partial Q_i)|e'\rangle\langle e'|\mathbf{e}_2\cdot\boldsymbol{\mu}|g'\rangle] \quad (1)$$

where ω_1 is the frequency (in s^{-1}) of the exciting laser, $\omega_2 = \omega_1 - \omega_i$ is the Raman scattered frequency, ω_i being the vibrational frequency of the ith normal mode, μ is the dipole moment operator, $\mathbf{e}_1$ and $\mathbf{e}_2$ are the polarization vectors of the incident and scattered light respectively, g' and e' are the eigenfunctions of the

purely electronic ground and excited states and E_{eg} is the energy difference between the excited and electronic states.

Equation (1) clearly indicates that the Raman intensities are functions of the electronic excited states. It is particularly instructive to consider in greater detail the second factor, which contains what we can call the electron–phonon coupling term. According to the Hellman–Feynmann theorem it is possible to write [13]:

$$\langle e|(\partial H/\partial Q_k)^{\circ}|e\rangle = (\partial E^e/\partial Q_k)^{\circ} \cong \omega_k(\Delta Q_k)^{eg} \quad (2)$$

Where E_e is the energy of the electronic excited state and $(\Delta Q_k)^{eg}$ is the displacement of the equilibrium position of the normal coordinate in going from the ground to the excited state. This equation relates the energy of the excited state (electron) with a particular totally symmetric normal mode (phonon) for which the equilibrium geometry in the ground and excited states is different.

It follows that both infrared and Raman intensities are directly related to the molecular electronic structure; this observation is the key for establishing a relationship between nuclear motions (vibrations) and electronic properties. This is not yet enough. A more fundamental property of polyconjugated systems is that electron–phonon coupling is large and this allows accurate estimation of the non-linear optical response on the basis of the vibrational spectra. Electron–phonon coupling means that as a consequence of the low band-gap, the electronic and vibrational energies are not so far apart, thus making it possible to develop some kind of coupling between the two. As a consequence, whenever the electronic cloud is excited, the nuclei are also affected, and vice versa particular motions of the nuclei can induce some charge rearrangement. This kind of coupling is most effective only along certain particular directions in the vibrational space, which are determined by the electronic nature (i.e. charge distribution and mobility) of the excited states [14]. It is only for vibrations along this direction that the duality between nuclear motions and electronic properties can be found. Recalling Eq. (2), the direction we look for is given by $(\Delta Q_k)^{eg}$. In the case of polyene systems it is intuitive that this direction is along the molecular backbone chain where the π electrons can "freely" move.

Results and Discussion

We can now discuss in greater detail the vibrational method we have proposed for the determination of molecular polarizabilities [15]. When a molecule is exposed to the action of a strong static field the electronic charge distribution is suddenly distorted. As a consequence of the new equilibrium distribution, infinitesimal nuclear shifts are induced. The starting point in our model is to find an analytic expression which describes the effect of this nuclear rearrangement on molecular polarizabilities. The final equations expressing this kind of contribution turn out to be:

$$\alpha_{nm}^r = \frac{1}{4\pi^2 c^2} \sum_k \left(\frac{1}{\nu_k^2}\right) \left[\left(\frac{\partial\mu_n}{\partial Q_k}\right)\left(\frac{\partial\mu_m}{\partial Q_k}\right)\right] \quad (3)$$

$$\beta_{nmp}^r = \frac{1}{4\pi^2 c^2} \sum_k \left(\frac{1}{\nu_k^2}\right) \left[\left(\frac{\partial\mu_n}{\partial Q_k}\right)\left(\frac{\partial\alpha_{mp}}{\partial Q_k}\right)\right.$$
$$\left. + \left(\frac{\partial\mu_m}{\partial Q_k}\right)\left(\frac{\partial\alpha_{np}}{\partial Q_k}\right) + \left(\frac{\partial\mu_p}{\partial Q_k}\right)\left(\frac{\partial\alpha_{nm}}{\partial Q_k}\right)\right] \quad (4)$$

$$\gamma_{nmps}^r = \frac{1}{4\pi^2 c^2} \sum_k \left(\frac{1}{\nu_k^2}\right) \left[\left(\frac{\partial\mu_n}{\partial Q_k}\right)\left(\frac{\partial\beta_{mps}}{\partial Q_k}\right)\right.$$
$$+ \left(\frac{\partial\mu_m}{\partial Q_k}\right)\left(\frac{\partial\beta_{nps}}{\partial Q_k}\right) + \left(\frac{\partial\mu_p}{\partial Q_k}\right)\left(\frac{\partial\beta_{nms}}{\partial Q_k}\right)$$
$$+ \left(\frac{\partial\mu_s}{\partial Q_k}\right)\left(\frac{\partial\beta_{nmp}}{\partial Q_k}\right) + \left(\frac{\partial\alpha_{nm}}{\partial Q_k}\right)\left(\frac{\partial\alpha_{ps}}{\partial Q_k}\right)$$
$$\left. + \left(\frac{\partial\alpha_{np}}{\partial Q_k}\right)\left(\frac{\partial\alpha_{ms}}{\partial Q_k}\right) + \left(\frac{\partial\alpha_{ns}}{\partial Q_k}\right)\left(\frac{\partial\alpha_{mp}}{\partial Q_k}\right)\right] \quad (5)$$

The quantities $\partial\mu_n/\partial Q_k$, $\partial\alpha_{nm}/\partial Q_k$, $\partial\beta_{nms}/\partial Q_k$, where n, m, s indicate the Cartesian components, can be obtained from infrared intensities, Raman and hyper Raman cross-sections, respectively; ν_k is the vibrational frequency of the kth normal mode Q_k. Eqs. (3)–(5) are derived under the hypothesis of both mechanical and electrical harmonicity.

The reliability of the method has been tested on a large variety of molecules [16–19] for which independent data were available, both for second and third order non-linearities. The problem could be analysed both theoretically and experimentally. We pursued both ways and the methodology adopted was on the one hand to calculate the electronic molecular properties together with the electronic hyperpolarizabilities (β^e and γ^e) with quantum mechanical *ab initio* calculations and on the other to calculate the infrared and Raman spectra to be used together with Eqs. (4) and (5) to evaluate the vibrational hyperpolarizabilities (β^r

and γ^r)*. The next step was the comparison of the values obtained for the molecular electronic hyperpolarizabilities and their vibrational counterparts. The final result was surprising: for all the systems considered, calculations indicated that $\beta^e \cong \beta^r$ and $\gamma^e \cong \gamma^r$. An analogous approach can be taken experimentally. In this case the comparison was made between our measured vibrational contributions (infrared and Raman absolute intensities) and other experimental determinations obtained independently in other laboratories by the use of traditional techniques (such as EFISH and THG). Once again the general result was: $\beta^e \cong \beta^r$ and $\gamma^e \cong \gamma^r$. These results are summarized in Table 1 where β values for several molecules are reported. The molecules were chosen as prototypes representative of the different classes of molecules generally used as non-linear optical chromophores for second-order polarizabilities. Data for γ are shown in Fig. 1 for polyene systems and in Fig. 2 for oligo alkyl thiophenes.

A more accurate analysis of the vibrational spectra can offer some new insight into the physics of these compounds. From Eqs. (4) and (5) it appears that the largest contributions to vibrational hyperpolarizabilities come from those normal modes which have the largest intensities. It must be noted that in the case of β a simultaneous strong infrared and Raman activity is required. The analysis of the vibrational spectra show that the normal modes which mostly contribute are skeletal modes associated with CC stretchings. This result is not surprising in view of the discussion reported in the introduction. Only the normal modes oscillating in the direction of the preferential electron–phonon coupling can have an extremely large Raman intensity. In a similar way the most intense infrared modes are those for which charge fluxes along the chain are the largest. Indeed, calculations of these flux contributions to infrared intensities for these normal modes show that these contributions are one order of magnitude larger than the charge contributions.

Another interesting effect that can be analysed is the influence of the surrounding medium on the non-linear optical behaviour. The non-linear optical properties of

---

* It must be noted that in Eq. (5) hyper Raman cross-sections also appear. Actually, for the molecule used in this work to evaluate γ^r, the Raman intensities are so large that the contributions coming from the last four terms can certainly be neglected without introducing any appreciable error. This simplification is no longer valid for other classes of molecules where the infrared/hyper Raman contribution becomes competitive or even dominant.

Table 1. Comparison between β and β^r values for some selected organic molecules obtained both theoretically (with ab-initio 3–21 G basis set) and experimentally

Molecule	β_r^μ (3–21 G)	β^μ (3–21 G)	β_r^μ (exp)	β^μ (exp)
I	10.67	9.55	12.6	10 (a)
II	28.24	30.96	–	24 (a)
III	8.27	11.08	–	–
IV	–	–	50 (b)	46 (c)
V n = 1	11.6	10.2	10.2	3.8 (d)
V n = 3	34.7	32.0	34.6	39.8 (d)
V n = 4	–	–	50.9	42.1 (d)

β values are in units of 10^{-30} esu.

(a) from EFISH experiments [20].
(b) β_{yyy}.
(c) from HLS experiments [21].
(d) from EFISH experiments [22].

some molecules can be fine tuned by interactions with the environment. Particularly amenable for this kind of study is the use of various solvents of increasing polarity [28].

Some molecules which proved to be very sensitive to this kind of effect are push–pull polyenes. These are polyene systems with terminal electron donor and acceptor groups:

$$D\text{———}A_n$$

Such molecules possess a low-energy charge-transfer state and exhibit large second and third order non-linearities. The rationale for understanding this effect is the fact that the solvent acts as a stabilizer of the charge-transfer form. The increase in charge-transfer character in the molecular ground state alters the chain structure, making the carbon backbone less alternated. It is the fine balance between solvent polarity on the one hand and molecular structural parameters (i.e. donor and acceptor strength and chain length) on the other that determines the extent of modulation of the structural parameter (bond length alternation). The induced modulation of the chain structure has a great influence on the non-linear behaviour (both electronic and vibrational) since, as we have already seen, it is along the molecular axis that π electrons form a de-

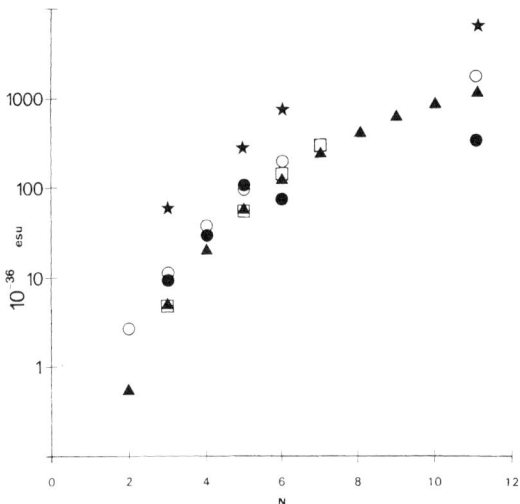

Fig. 1. Comparison among γ values obtained from different methods for polyene systems of increasing chain length: ($\bigcirc$) γ^r from calculated (*ab initio* 6–31 G) Raman intensities [17]. ($\blacktriangle$) γ from *ab initio* calculations (6–31 G) [23]. ($\bullet$) γ^r from experimental Raman cross sections [18]. ($\bigstar$) γ from THG measurements [24]. ($\square$) γ^r from experimental Raman cross sections of polyenovanillines [19]

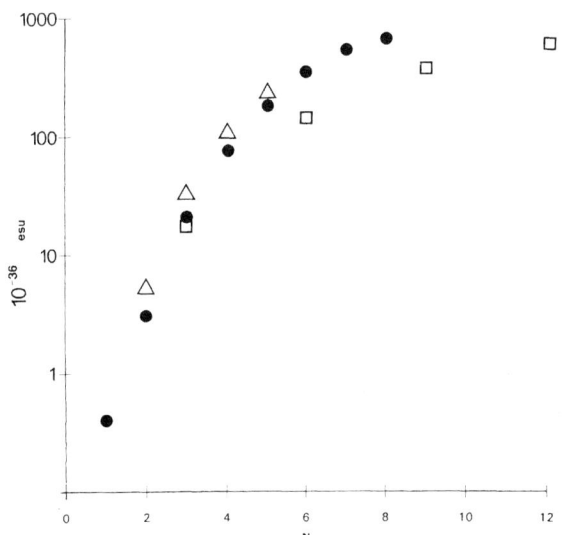

Fig. 2. Comparison between γ^r values obtained from experimental ($\square$) [25] and computed ($\triangle$) [26] Raman cross sections and γ values ($\bullet$) calculated with SOS method [27] for oligothiophenes with increasing chain length

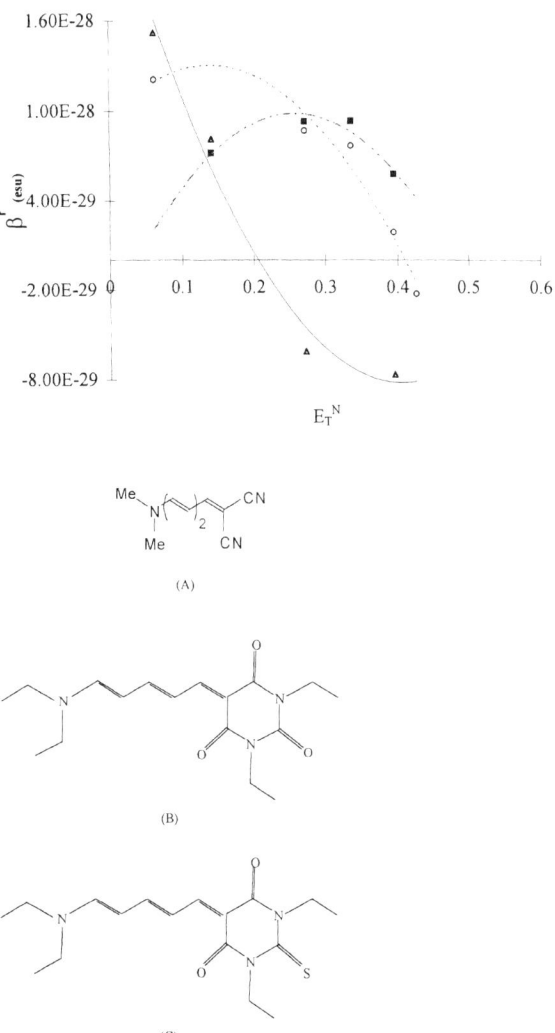

Fig. 3. Experimental β^r values for molecules A ($\blacksquare$), B ($\bigcirc$) and C ($\triangle$) in various solvents (respectively: C_6H_6, CCl_4, $CHCl_3$, CH_2Cl_2, CH_3CN, CH_3NO_2 in order of increasing polarity). Lines are only meant as a guide to the eye

localized and easily polarizable network, thus allowing coupling with the nuclei. In Fig. 3 we show the solvent-dependent behaviour of β^r for three different push–pull polyenes dissolved in various solvents, of increasing polarity. Depending on the kind of unperturbed chain structure (when in an apolar solvent), different kinds of modulation are obtained. Particularly interesting is the sign inversion of β. From theoretical considerations it is known that in correspondence to a vanishing value of β the chain structure must be equalized or "metallic".

In this case also, all our experimental findings are in extremely good agreement with results obtained by means of other experimental techniques [29, 30] and also with theoretical calculations [31].

Conclusions

In this paper we have presented only a few selected examples which support our belief that the coincidence between electronic and vibrational hyperpolarizabilities cannot be fortuitous. The physical reasons which justify our belief have also first been qualitatively discussed: we have referred to the peculiar delocalized nature of the electronic cloud characteristic of polyconjugated systems.

However, to be more rigorous in our conclusions, we have recently shown [32] with a theoretical model that β^r of push–pull polyenes can be expressed in terms of exactly the same electronic quantities which appear in the analytic expression of β^e. This result has been obtained with a suitable parametrization of the molecular wavefunction by introducing an explicit dependence from the vibrational coordinate.

We are thus able to provide, in addition to the large amount of experimental evidence, a theoretical justification for the vibrational approach. The essence of the method relies on the possibility of inducing the same kind of molecular polarization both by direct electronic excitation and by a particular nuclear oscillation.

The experimental results obtained and the theoretical justification presented allow proposal of the vibrational method as an alternative method of measurement of the optical non-linearities of organic polyconjugated materials.

References

[1] D. S. Chemla, J. Zyss (eds.), *Nonlinear Optical Properties of Organic Molecules and Crystals*, Vols. 1 and 2, Academic Press, New York, 1987.

[2] D. J. Williams, *Angew. Chem. Int. Ed. Engl.* **1984**, *23*, 690.

[3] P. N. Prasad, D. J. Williams, *Introduction to Nonlinear Optical Effects in Molecules and Polymers*, Wiley, New York, 1991.

[4] S. R. Marder, J. E. Sohn, G. D. Stucky (eds.), *Materials for Nonlinear Optics*, American Chemical Society, Washington, D. C., 1991.

[5] R. A. Hann, D. Bloor (eds.), *Organic Materials for Nonlinear Optics*, Royal Society of Chemistry, London, 1989.

[6] C. Castiglioni, M. Gussoni, G. Zerbi, *J. Mol. Struct.* **1989**, *198*, 475.

[7] M. Gussoni, C. Castiglioni, M. N. Ramos, M. Rui, G. Zerbi, *J. Mol. Struct.* **1990**, *224*, 445.

[8] J. C. Decius, *J. Mol. Spectrosc.* **1975**, *57*, 348.

[9] A. J. van Straten, W. M. A. Smit, *J. Mol. Spectrosc.* **1976**, *62*, 297.

[10] M. Gussoni, C. Castiglioni, M. Miragoli, G. Lugli, G. Zerbi, *Spectrochim. Acta* **1985**, *41A*, 371.

[11] G. Zerbi, M. Gussoni, C. Castiglioni, *Springer Series in Solid State Science 63*, 1985, 156.

[12] C. Castiglioni, M. Del Zoppo, G. Zerbi, *J. Raman Spectrosc.* **1993**, *24*, 485.

[13] W. L. Peticolas, C. Blazej, *Chem. Phys. Lett.* **1979**, *63*, 604.

[14] M. Gussoni, C. Castiglioni, G. Zerbi, in: *Advances in Material Science Spectroscopy* (R. H. Clark, R. E. Hester, eds.), Wiley, Chichester, 1991, Chap. 5.

[15] C. Castiglioni, M. Gussoni, M. Del Zoppo, G. Zerbi, *Solid State Commun.* **1992**, *82*, 13.

[16] C. Castiglioni, M. Del Zoppo, P. Zuliani, G. Zerbi, *Synth. Met.* **1995**, *74*, 171.

[17] M. Del Zoppo, C. Castiglioni, G. Zerbi, M. Rui, M. Gussoni, *Synth. Met.* **1992**, *51*, 135.

[18] M. Del Zoppo, C. Castiglioni, M. Veronelli, G. Zerbi, *Synth. Met.* **1993**, *57*, 3919.

[19] P. Zuliani, M. Del Zoppo, C. Castiglioni, G. Zerbi, C. Andraud, T. Brotin, A. Collet, *J. Phys. Chem.* **1995**, *99*, 16242.

[20] L. T. Cheng, W. Tam, S. H. Marder, A. E. Stiegman, G. Rikken, C. W. Spangler, *J. Phys. Chem.* **1991**, *95*, 10643.

[21] J. Zyss, T. Chau Van, C. Dhenaut, I. Ledoux, *Chem. Phys.* **1993**, *177*, 281.

[22] C. Andraud, A. Collet, *Private communication.*

[23] G. J. B. Hurst, M. Dupuis, E. Clementi, *J. Chem. Phys.* **1988**, *89*, 385.

[24] J. P. Hermann, J. Ducuing, *J. Appl. Phys.* **1974**, *45*, 5100.

[25] M. Veronelli, *Thesis*, Politecnico di Milano, 1994.

[26] V. Hernandez, G. Zerbi, to be published.

[27] D. Beljonne, Z. Shuai, J. L. Brédas, *J. Chem. Phys.* **1993**, *98*, 8819.

[28] P. Zuliani, M. Del Zoppo, C. Castiglioni, G. Zerbi, S. R. Marder, J. W. Perry, *J. Chem. Phys.* **1995**, *103*, 9935.

[29] S. R. Marder, D. N. Beratan, L.-T. Cheng, *Science* **1991**, *252*, 103.

[30] S. R. Marder, J. W. Perry, G. Bourhill, C. B. Gorman, B. G. Tiemann, K. Mansour, *Science* **1993**, *261*, 186.

[31] F. Meyers, S. R. Marder, B. M. Pierce, J. L. Brédas, *Chem. Phys. Lett.* **1994**, *228*, 171.

[32] C. Castiglioni, M. Del Zoppo, G. Zerbi, *Phys. Rev. B* **1996**, *53*, 13319.

Mikrochim. Acta [Suppl.] 14, 109–119 (1997)

Determination of Conformation of Biopolymers by Particle Beam/Infrared Spectrometry

James A. de Haseth* and **Vincent E. Turula**

Department of Chemistry, University of Georigia, Athens, Georgia 30602-2556, U.S.A.

Abstract. The "Particle Beam" is an aerosol apparatus developed as an HPLC high-presure liquid (chromatography) interface for both mass spectrometry and infrared spectrometry. The effluent can either be the output of a liquid chromatographic column, or a constant concentration solution supplied by a simple liquid chromatography pump. The operating features of the particle beam LC/FT-IR apparatus make it highly suitable for use in investigations of dynamic biopolymer solution structure. The more volatile solvent (mobile phase) is rapidly evaporated from the analyte molecules, and removed to vacuum (*desolvation*). The desolvation process is endothermic and occurs in a few milliseconds, so the low-energy conformers in solution are trapped as the aerosol particles cool and eventually freeze. Internal rotations, restricted by desolvation, are not likely to occur. The desolvated protein molecules are deposited on an infrared-transparent substrate and analyzed by off-line FT-IR microscopy. Therefore, this rapid sampling of a protein in solution produces a deposit of dried particles that can be spectroscopically analyzed, yielding a solid-state spectrum highly representative of the protein's structure in solution. Preliminary experiments have shown that the structural integrity and biological activity of globular proteins were maintained during use of the particle beam. The results presented here illustrate applications of the particle beam in obtaining secondary structure information from dynamic protein structure experiments. The particle beam interface was used to collect IR spectra of several enzymes separated by reversed-phase HPLC. These spectra illustrate the effects of the chromatographic conditions (e.g. mobile phase composition, elution procedure, and stationary phase) upon the secondary structure, and provided insight into conformational effects on the retention behavior of proteins.

Key words: FT-IR spectrometry, proteins, conformation.

One area in which Fourier-transform infrared (FT-IR) spectrometry has found application is kinetic studies. With the revival of step-scanning interferometry has come a resurgence in time-resolved spectrometry (TRS). There are two main domains for the application of TRS. In one, the duration of the experiment is much shorter than the interferometer scan-time; but only if the experiment is reproducible and can be repeated continuously and rapidly with data collection synchronized with the experiment time scale can all the information be collected. In the other, the interferometer scan-time is insignificant in comparison with the experiment duration, so a single run consisting of many separate scan sets provides all the information. Either of these two conditions can be met in the study of many systems, but a very large number of experiments do not meet these requirements.

Time-resolved spectrometry can also be accomplished if transient species are trapped in such a fashion that their structure is maintained. If the structure can be maintained, the spectra of the transient species can be examined at leisure. The trade-off is that the experiment may have to be run more than once so that different species can be trapped. This is condition far different from the one for step-scan time-resolved spectrometry; the experiment must be reproduced, but it does not have to be repeated continuously and rapidly. Such is the case with protein folding and denaturation studies; proteins can be triggered to unfold or fold, and can be trapped in an intermediate state.

An approach has been undertaken in this laboratory that is aimed at the development of a technique with particle beam/FT-IR spectrometry that will allow for estimation of protein solution structure without masking water absorptions being present in the spectrum. As the particle beam can prepare samples in such a state rapidly, it can be used not only for the IR characterization of proteins separated by HPLC, but also to study in vitro folding mechanisms for globular proteins. This manuscript presents preliminary studies to verify that protein folding mechanisms can be examined with this technique.

---

* To whom correspondence should be addressed

Proteins are biopolymers that are composed of amino-acids joined together by amide linkages known as peptide bonds [1]. There are four levels of protein structure, defined as primary, secondary, tertiary, and quaternary [2]. The primary structure is the sequence of amino-acid residues that constitute the protein. The *conformation* of a biopolymer is its detailed three-dimensional structure, and is determined by interactions both within the molecule and with the aqueous solvent. Protein three-dimensional structure and function have been related, in part, to the physical and chemical properties of the amino-acids, the amino-acid sequence, chain length, and in some regard the presence of stereoisomers [3]. *Secondary structure* encompasses the three-dimensional spatial arrangement of the peptide backbone without regard to the conformation of the side-chains. The geometry of the amide planes as well as the secondary structure order is a result of hydrogen-bonding. These structures include α-helices, β-sheets, β-turns, and random coils. It is in the area of secondary structure determination that vibrational spectrometry has been applied.

Infrared spectrometry has been used extensively to examine protein structure in solution. The position and intensity of infrared (IR) amide absorption bands are sensitive to globular protein secondary structure content [4]. Information about the protein secondary structure is contained in the amide I, II, and III regions. These bands arise from delocalized vibrations of the peptide linkages. *N*-Methylacetamide (CH$_3$CONH-CH$_3$) is a molecule that closely resembles the common repeat unit of the polypeptide chain, and has therefore been used to describe the amide vibrations [4]. The amide I, II, and III vibrations are seen at 1650, 1560, and 1300 cm^{-1}, respectively. Because the majority of energy distribution of the amide I band is associated with the C=O stretch, it is recognized as the most informative for structure correlation. The amide bands are sensitive to hydrogen bonding, as in the vast hydrogen-bonded network of biological macromolecules, modes become coupled. The spectra of biopolymers are complex and the conformationally sensitive amide absorption bands, which are intrinsically overlapped, must be separated from each other. For this, the spectra must undergo mathematical manipulations. These include *deconvolution* to resolve underlying component bands, *differentiation* to determine band centers, and *curve-fitting* to obtain band areas that correspond to the percentage contributions to the structure, of the species giving rise to the bands.

The use of IR spectroscopy for the analysis of proteins and polypeptides dates back to the 1950 s [5]. The desire to obtain protein spectra in an environment close to the native aqueous state led workers to employ thin transmission cells to minimize the effects of the solvent, as the strong absorption bands of water resulted in the swamping of large regions of the spectrum, particularly in the amide I region. To circumvent this situation, deuterium oxide was used as a medium for IR protein structure investigations [6]. Transmission experiments with dispersive spectrophotometers and D$_2$O were difficult, owing to the inability to remove contaminant HOD, and had questionable background compensation as it was difficult to match cell pathlengths. Byler and Susi found good agreement for the secondary structure estimates of 17 proteins between their D$_2$O FT-IR spectra and corresponding X-ray crystallographic determinations [7]. Jakobsen and co-workers have used attenuated total reflectance (ATR) FT-IR spectrometry to study globular protein structural changes induced by non-aqueous solvents [8–10]. The proteins were adsorbed from aqueous solution onto a germanium ATR element. The resultant spectral changes led to conformational assignments. The condensed phase spectra of proteins obtained in these analyses yield intrinsically broad band contours. In both the transmission and ATR techniques, spectral subtraction of the water absorption bands was required to reveal the amide absorptions.

The particle beam system is an aerosol apparatus originally developed as an HPLC interface for mass spectrometry [11], and later infrared spectrometry [12]. The apparatus eliminates the more volatile solvent (mobile phase) by evaporation, leaving the analyte molecules. In normal operation the protein solution is pumped through a small orifice and exits as a liquid jet. This column of liquid is then nebulized with a stream of helium. The resultant protein/water droplets are directed into a cylindrical glass chamber where the water molecules are stripped by evaporation and removed (*desolvation*). After the desolvation process, the protein molecules are dried to an extent that can be controlled by certain parameters. The protein molecules are subsequently impacted against an infrared-transparent substrate, which is later removed and analyzed by off-line FT-IR microscopy. The desolvation process is endothermic and occurs in a few milliseconds, so the low-energy conformers in solution are trapped as the aerosol particles cool and eventually freeze; internal rotations, restricted by desolvation, are not likely to occur. Therefore, this rapid sampling of a protein in solution produces a dried particle beam deposit which can be spectroscopically analyzed and thus yield a solid-state spectrum highly representative of the protein's structure in solution.

Preliminary studies were aimed at determining the ability of the particle beam to prepare protein films without causing structural degradation. The three major processes in particle beam operation were isolated, so that if any structural degradation occurred, the process responsible for it could be identified. The three processes in particle beam operation are: production of a liquid jet in monodisperse aerosol generation, desolvation, and deposition. The protein chosen for this work was β-lactoglobulin (βLG, MW approximately 35000) a dimeric milk protein that has a high β-sheet content with some α-helices. Comparisons of reference IR data for native and denatured βLG have led to its conformational elucidation. The amide I and II IR bands of βLG solutions are similar to the same bands in the βLG solid-film spectra [13]. Susi and co-workers examined native βLG in various pH-denatured states in D_2O solutions and found that a mixture of β-sheets and α-helix, with some disordered segments, comprised the native structure [14, 15]. The conclusion of each of our experiments was based on comparisons of βLG IR spectra with reference data generated in this laboratory by parallel sampling techniques (i.e. ATR CIRcle® cell, and thin solid-films). As structural changes can be discerned by means of frequency shifts and intensity changes in the IR amide I, II, and III bands, spectral comparisons were made with the original band contours. Deconvolution and curve-fitting were used only as a supplement. Band assignments were made first by the use of literature data as a benchmark, then more exact assignments were made by following band contour changes that were induced by the exposure of βLG to structure-altering solvents (e.g. methanol or ethylene glycol).

Another study undertaken in our laboratory was the separation of various protein and polypeptide mixtures by HPLC. These complex biological mixtures can possess a range of molecular weights and diverse amino-acid compositions, yet can be separated effectively and rapidly by reversed-phase HPLC (RPC) [16–18]. The retention of proteins in RPC is complex, as the interaction between the stationary phase and the protein in governed by the protein amino-acid constitution, distribution of hydrophobic residues within the protein, and the 3-D structure or conformation of the biomolecules. Highly acidic mobile phase conditions and non-viscous organic modifiers in the mobile phase are essential for effective separation [16]. These solvents induce alteration to the tertiary and quaternary levels of structure, and the organic modifier, depending upon its nature, permeates into the compact protein. This causes unfolding, which exposes hydrophobic chains previously buried in the core of the compact protein. A mobile phase with a substantial amount of organic modifier stabilizes the unfolded protein, with its hydrophobic regions in direct contact with the solvent. Optimal interaction occurs when the protein is denatured, because there is then maximum surface contact between the stationary phase ligands and a majority of the peptide residues [16]. The conformational state of the protein, therefore, has a direct bearing on its retention behavior [17, 18].

Separation efficiency and sample resolution, which can be manipulated by gradient parameters, reflect how the separation affects conformational order. In some instances the overall conversion time from native to unfolded forms is comparable to the time of separation. The elution of either form may be obtained by use of a steep gradient such that the separation time is comparable to the time needel to attain a particular form [19–21]. In addition to the secondary equilibria, conformational interconversion, protein–protein aggregation, metastable adsorption, and multimonomer dissociation degradation may also occur in RPC. In these situations it becomes difficult to differentiate between sample compounds and artifacts [22].

Native protein structure can be preserved in RPC, but, the retention process is responsible for production of secondary conformational forms which give either asymmetric or multiple peaks, which depend upon the chromatographic conditions. Most of the literature that concerns conformational effects in RPC is directed to the dynamic state of protein tertiary and quaternary structure, because these structures drastically influence separations. In separate RPC studies of pure papain [23], and ribonuclease A [24] this effect was illustrated.

Hindered by steric restriction, compact proteins with ordered structure do not diffuse into the chromatographic stationary phase as well as they do when unfolded. During RPC, proteins unravel as conditions facilitate disorganization of the tertiary and quaternary levels of structure. The contribution to retention from these levels is then a minor factor. It is not certain, however, how pH, solvents, pressure, and elution conditions affect secondary structure. Accepted retention models do not consider secondary structure. In this study the state of protein secondary structure following RPC separation is determined. Presented here are several experiments that address the question of how the secondary structural integrity of globular proteins is affected by analytical RPC.

Experimental

Particle Beam

The particle beam operation is described elsewhere [12], but a brief description follows. A 10 µL injection loop was used to introduce the protein solution into the flowing stream, in these preliminary studies. The liquid was pumped at 0.3 mL/min into the apparatus to produce a Rayleigh jet which was dispersed by helium at approximately 20 psig pressure (Fig. 1). The jet was impacted onto either potassium chloride, zinc selenide, or calcium fluoride windows. The deposit diameter varied and was dependent upon the horizontal position of the substrate collector. Particle beam deposit spectra were acquired with a Spectra-Tech IR Plan™ microscope (Spectra-Tech, Shelton, CT) enclosed in a laboratory-designed Plexiglass™ purge compartment, and interfaced to the Perkin–Elimer 1725X FT-IR spectrometer. The microscope was equipped with a mercury cadmium telluride (MCT) detector, and 1.5 mm upper and 1.0 mm lower circular apertures were used, giving a 100 µm diameter field of view. Spectral manipulations such as digital subtraction, deconvolution and curve-fitting were calculated with GRAMS™ software (Galactic Industries, Salem, NH).

Spectra were recorded with use of a variety of sampling methods. A sandwich cell constructed with polished calcium fluoride windows (25 mm dia. × 4 mm, International Crystal Inc., Garfield, N. J., cat #0002D-181) with a 6 µm thick Teflon™ spacer (Harrick Scientific, Ossining, N. Y., cat#PSS-M25) was used to record transmission spectra of solutions. Reference spectra of solid films were prepared by the deposition of protein solution onto a CaF_2 window and evaporation to dryness in an ambient temperature vacuum desiccator over dryrite. All transmission spectra were collected with a Perkin–Elmer 1725X Fourier transform infrared (FT-IR) spectrometer (DTGS detector) with 1000–2000 single-beam scans co-added, with a resolution of 4 cm^{-1}. Attenuated total reflectance measurements were made with a sealed ATR micro CIRcle® cell (200 µL volume) and a zinc selenide (element (spectra-Tech, Shelton, CT). The cell was

mounted in the sample compartment of a Bio-Red FTS-60 FT-IR spectrometer or in a Bio-Rad FTS-60A/896 FT-IR spectrometer (Bio-Red Laboratories, Digilab Division, Cambridge, MA) such that the buffer or protein solution could be loaded externally, while the purge integrity of the instrument was maintained. A Lure-Lok® syringe with a capacity of 3 mL was used to load the circle cell. Spectra were collected at regular time intervals so that protein adsorption onto the element could be followed. An MCT detector was used for the circle cell measurements, with 2000 scans at 4 cm^{-1} resolution, with boxcar truncation. The ZnSe element was cleaned by a soap solution scrub, sonication, and finally solvent rinses. The empty cell was scanned prior to analysis so that any residual protein was detected and recorded.

Chemicals and Reagents

β-Lactoglobulin and lysozyme were purchased from the Sigma Chemical Company, (St. Louis, MO), and used without further purification. Protein solutions were prepared in an aqueous solution of 18 MΩ demineralized water, 0.15 M sodium chloride, and a phosphate buffer that consisted of 5.75 mM potassium dihydrogen phosphate and 31.4 mM anhydrous disodium hydrogen phosphate. Although it was difficult to substract the phosphate buffer bands from thin film spectra completely, and hence the structurally important amide III band was occasionally masked, the presence of the buffer eliminated the potential for structural changes induced by ambient pH fluctuations, and enabled close simulation of physiological pH. The concentration of the protein solutions used was varied, and depended upon the sensitivity of the experiment: for ATR 10 mg/mL; for solution transmission 100 mg/mL; for particle beam < 10 mg/mL.

HPLC grade acetonitrile was obtained from J. T. Baker, Phillipsburg, NJ and 2-propanol from Fisher Chemical, Fairlawn, NJ. Water was demineralized to 18 MΩ resistivity with a Barnstead NANO ultrapure water system. The HPLC solvents were degassed continuously with helium in their reservoirs. Two buffers were employed to control sample pH: phosphate buffer, which depending on the buffer strength desired, contained appropriate amounts of potassium diphosphate monobasic, and anhydrous sodium phosphate dibasic supplied by J. T. Baker, and a TRIS buffer consisting of TRIZMA® Hydrochloride, (tris[hydroxymethyl]aminomethane hydrochloride) and TRIZMA® base (tris[hydroxymethyl] aminomethane) (Sigma Chemical Company). Trifluoroacetic acid was supplied by Aldrich Chemical Co. The four globular proteins (ribonuclease A (RNase A), α-chymotrypsin, lysozyme, trypsin) analyzed in this study were of the highest available purity, from Sigma Chemical, and were used without further purification. All reagents for the stop-rate activity assay of ribonuclease A were purchased from Sigma Chemical, and included type VI ribonucleic acid from Torula yeast, and sodium acetate trihydrate.

Chromatography

Reversed-phase HPLC separations were performed on a Hewlett–Packard 1090L binary gradient HPLC system (San Jose, CA) with a 6-port Rheodyne injection valve (Cotati, CA) and 10 µm frit-filter (Upchurch Scientific, Oak Harbor, WA). Separations were performed with n-alkyl silica support colums: narrow-bore (250 × 2.1 mm) Macrosphere 300 Å n-butyl 7 µm (Alltech Associates, Deerfield, IL) and analytical (250 × 4.6 mm ID) 300 Å n-butyl, and n-octadecyl 5 µm (Vydac, Hesperia, CA). Two organic modifiers were used: pure acetonitrile, and a 1:1 2-propanol/acetonitrile mixture. Detection was done with a Kratos Spectroflow 783G variable wavelength detector

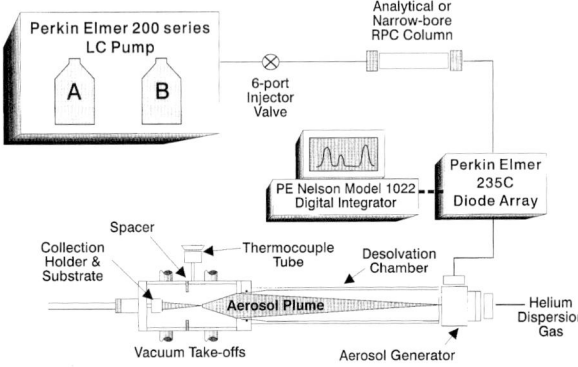

Fig. 1. Particle beam HPLC instrumental set-up

set for 10 mV output at 0.25 absorbance. Signal output was captured by a PE Nelson Model 1022 digital integrator (Perkin Elmer Norwalk, CT) with a single channel and scanned at 600 points per minute for these experiments. Post-run processing was performed with both PE Nelson integrator software and Grams 386™ software (Galactic Industries Corp, Salem, NH). An array basic program prepared by the authors was used to convert the data-point abscissa into minutes. Depending on the experiment, several sizes of injection loop were used (20, 10, and 5 µl), but for most of the separations and particle beam collections the sample concentrations were 5–10 mg/mL, used with a 10 µL loop, which gave 50–100 µg on-column injection. In determinations of the capacity factor (k') a solution of uracil (Aldrich), monitored at 214 nm, was injected to mark the unretained peak time t_0.

Results and Discussion

Particle Beam Considerations

For operation of the particle beam apparatus, the aerosol formed must be close to monodisperse; that is, it must be an aerosol with a narrow distribution of particle sizes. This ensures identical desolvation for each particle. The particle beam apparatus uses an inert polyimide-coated fused-silica tube with an inner diameter of 25 µm to produce the aerosol. As globular proteins with molecular weights of tens of thousands have average radii of gyration that may, depending upon the state of folding, approach the inner diameter of the fused-silica tube, it was considered important to determine whether shear degradation occurred, and whether the structure was affected [25]. To detect any shear degradation, the jet formation process was isolated from the particle beam apparatus. Instead of passing βLG through the entire particle beam apparatus it was passed through only the 25 µm fused-silica tube, and collected for subsequent analysis. A previous timing experiment with a UV-Visible detector and an organic dye (Erythorisin B, f.w. 879.9) determined the optimal collection time after injection. The absorption pattern of the dye indicated a laminar flow profile at this flow-rate. βLG did not appear to move through the fused silica-tube in a compact slug as the dye did. It is speculated that this was caused by some protein adhering to the stainless steel tubing. The collected βLG solution was analyzed by both solid film transmission and ATR circle cell FT-IR spectrometry. Both solid and liquid state experimental spectra were nearly identical to their corresponding reference spectra.

The glass desolvation chamber of the particle beam apparatus is wrapped with heating tape so that the chamber is maintained at ambient temperature during aerosol evaporation. As evaporation is an endothermic process, the desolvation chamber cools during solvent evaporation; if the chamber is warmed to 25–30°C, condensation of the solvent vapor is prevented. To what degree, if any, over-heating affects protein structure, was investigated by operation of the particle beam at a drastically elevated temperature. A test was made with the desolvation temperature raised and maintained at 160°C, at which a βLG deposit onto a ZnSe window was made. The resulting spectra were identical to the other particle beam spectra. Based on velocity measurements made with identical operating parameters, the time a protein molecule is resident in the 30 cm long desolvation chamber is approximately 1–5 milliseconds. Since no spectral change was observed in the βLG deposit collected through the desolvation chamber at 160°C, it was concluded that excessive interface heat did not transfer to the aerosol, nor did it alter the structure of the travesing protein.

During the solvent elimination process in the particle beam apparatus, aqueous protein aerosol particle cool as the ambient water evaporates. Towards the later stages of the desolvation process, the temperature of the particles drops further and causes the small amount of residual water, trapped with the protein, to freeze, so that a low-temperature protein/water "comet" or particle is produced. The low temperature beam of particles then impacts on an IR-transparent substrate. The substrate is later removed from the interface and placed in an FT-IR microscope for spectral measurement. No deleterious effects on structure are believed to occur as a result of temperature reduction, but the massive protein particle moves rapidly and strikes the subsrate with considerable kinetic energy or momentum. If denaturation were to occur it is very probable that it would be a result of this high-energy impact. Two spectral comparisons, again of one liquid and one solid state, were made to determine whether impact with the substrate causes protein conformation changes. Immediately after a buffered βLG solution was run through the particle beam apparatus and deposited onto a CaF_2 window, it was redissolved with 10 µL of buffer/saline solution directly on the window. An identical CaF_2 window was placed above this βLG solution with a 6 µm thick teflon spacer to make the cell. Spectra of the redissolved deposit (original, deconvoluted, 2nd derivative) are shown in Fig. 2, B, D, F. As isolation from solvent vapor is based on momentum separation, throughput is high for massive macromolecules [26]. However, it was difficult to determine this solution concentration, as the transport

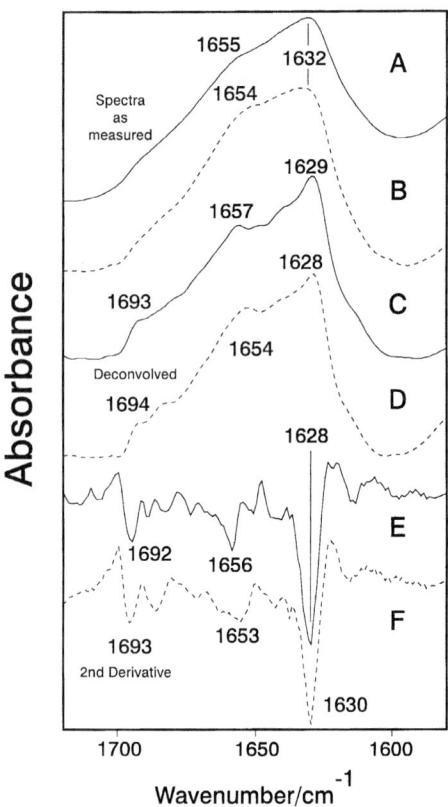

Fig. 2. Amide I region of a reference spectrum of βLG (solid line): *A* spectrum as measured, *C* deconvolved, *E* second derivative, compared to a particle beam deposit redissolved; (dashed line): *B* spectrum as measured, *D* deconvolved, and *F* second derivative

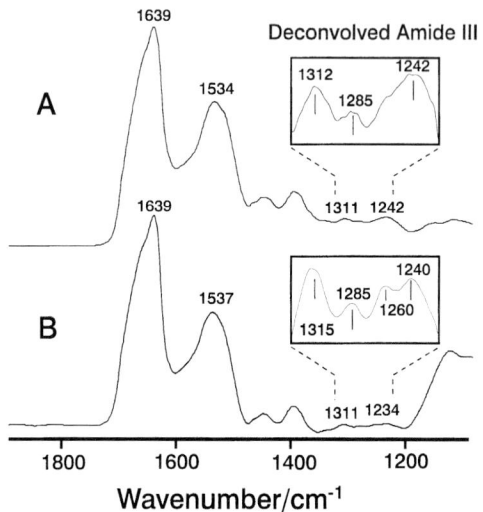

Fig. 3. *A* βLG particle beam deposit, and *B* cast film of βLG; the amide III region has been deconvolved and expanded

efficiency of the particle beam varies slightly with aerosol generator alignment. The reference solution, 100 mg/mL (Fig. 2, A, C, E), was originally believed to be more concentrated than the deposition solution. The original contours of the amide I region, as well as the deconvoluted and 2nd derivative spectra are identical for these βLG solution transmission spectra (Fig. 2). Both specta exhibit amide I maxima at 1632 cm^{-1}, corresponding directly to literature values of βLG in H$_2$O [7, 15]. The absorbance measured for the particle beam rinsed deposit was greater than that of the reference solution (the absorbance scale is not graduated in Fig. 2, but the intensity for the reference deposit was 65% of that for the rinsed deposit). This indicated that the transport efficiency (quantity injected/quantity recovered) was somewhat greater than 20%. Slight difference in the intensity of the component bands illustrate the difficulty inherent in quantitative measurements with the use of this sampling technique.

In the solid state, similar results for βLG are observed (Fig. 3A and B) as the amide I band maximum is

at around 1639 cm^{-1} for both the vacuum-evaporated reference and particle beam deposits. Deconvolution and curve-fitting revealed a slight difference between the two sets of spectra. The amide I absorption maximum at 1639 cm^{-1} was definitively assigned to the β-sheet. Beta structure is known to be exhibited as several component bands in IR spectra [2]. An amide band at 1639 cm^{-1} has been previously assigned to a disordered conformation in solution spectra of alkaline denatured βLG (in 2-propanol-d$_1$, pD = 12) [27]. Protein spectra acquired in deuterium oxide environments, however, cannot be related directly to the corresponding H$_2$O spectra, and often wavenumber shifts of several bands are observed [28, 29]. No concomitant amide II or III disordered band intensity increases were observed in the raw data; the amide III disordered 1261 cm^{-1} band was decreased in intensity in the particle beam spectrum. Therefore, structural disordering as a result of impact was ruled out.

From the excellent comparison (both of wavenumber and relative intensity) between the contours of particle beam deposit, evaporated thin film and solution spectra, it is clear that the secondary structure does not change upon high-energy impact.

Much information about protein structure can be obtained from the spectra of proteins that have been exposed to structure-altering non-aqueous solvents; these were introduced into the particle beam by means of a gradient elution HPLC program. The resulting spectral changes were followed by IR spectrometry, and provided reference for band assignments. Although non-aqueous solvent systems often irreversibly perturb protein structural integrity, the worth of this experiment is demonstrated and the IR band assignment within the complex protein spectrum is shown to be definitive. Timasheff et al. measured the effect of the intensity change of amide I IR bands as the structure transition from predominant β-sheet to α-helix structure occurred with a series of D_2O/CH_3OD mixtures [15]. A trend of increased amide I α-helix intensity was observed with increased CH_3OD concentration. In similar fashion, an experiment was performed in this study to determine whether the particle beam could be operated in this mode, and successfully desolvate proteins in changing solvent environments. Solvents that cause known structural changes were used, e.g. methanol for the formation of α-helices, ethylene glycol for the separation of β-sheets [30]. This experiment was performed by a continuous particle beam acquisition of 5 injections of 10 mg/mL βLG solutions each in a different solvent environment, starting with 100% water, and changing to 100% methanol in four equal steps. In this manner, βLG was exposed to greater concentrations of methanol with each successive injection, which caused more β-sheet to α-helix conversion. A large 32 mm diameter CaF_2 window was used as the collection substrate and mounted onto the holder so that if the shaft was rotated by approximately 70° after each deposit, a clean portion of the window was moved into the beam path. Although irregular, a single collection track of 5 deposits was easily made with this configuration. A separate experiment to observe βLG amide III bands was made with only three different solvent concentrations, and without the phosphate buffer as it masks the amide III region. The original contours of the particle beam spectra show the intensity changes that occur in the amide I and II bands with increasing methanol concentration. Representative spectra are illustrated in Fig. 4. Again the most noticeable change occurs with the amide I bands. Clearly, the intensity of the α-helix component (at approximately 1660 cm^{-1}) increases, with a simultaneous decrease in the intensity of the β-sheet component (at approximately 1635 cm^{-1}). Band positions were confirmed by deconvolution. Also in Fig. 4 are the deconvoluted and curve-fit contours of these spectra, which illustrate, to a greater extent than the original spectra, the trends with increasing methanol concentration. Evaporated films of the methanol/water mixtures that correspond to those used in the particle beam acquisition were

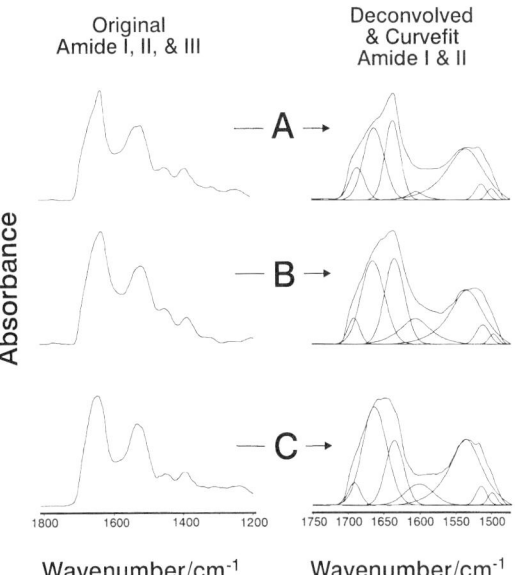

Fig. 4. β-Lactoglobulin beam deposits from different mobile phases: *A* 100% water, *B* 50% water/50% methanol, *C* 25% water/75% methanol

measured and no significant spectral intensity changes were observed. Apparently, as the methanol evaporated in the vacuum desiccator, the protein crystallized on the window in a conformation very close to its native conformation. With the particle beam deposits, the dynamic conformation was maintained upon impact.

Reversed-Phase HPLC/Particle Beam

The particle beam interface can operate marginally at flow-rates optimal for analytical scale separations (0.5–1.5 mL/min), so methods were developed at flow-rates permissible for operation of the particle beam. Operational limitations are intrinsic to solvent elimination devices. The particle beam operates effectively in the range 0.15–0.6 mL/min. To achieve complete desolvation, the appropriate fused-silica capillary ID needs to be used at different flow-rates: the 25 μm bore spans the largest mobile phase flow-rate range, 0.3–0.6 mL/min; 20 μm bore 0.2–0.3 mL/min; 16 μm bore, 0.15–0.2 mL/min [31].

A mixture of three enzymes (ribonuclease A – M.W. 13700, an α/β protein; lysozyme – M.W. 14000, predominantly α-helix; α-chymotrypsin, – M.W. 21600, high β-sheet + low α-helix) [32], all of modest size and representative of the two major types of secondary structure, were chosen for the initial separation. Gradient elution chromatographic methods were developed

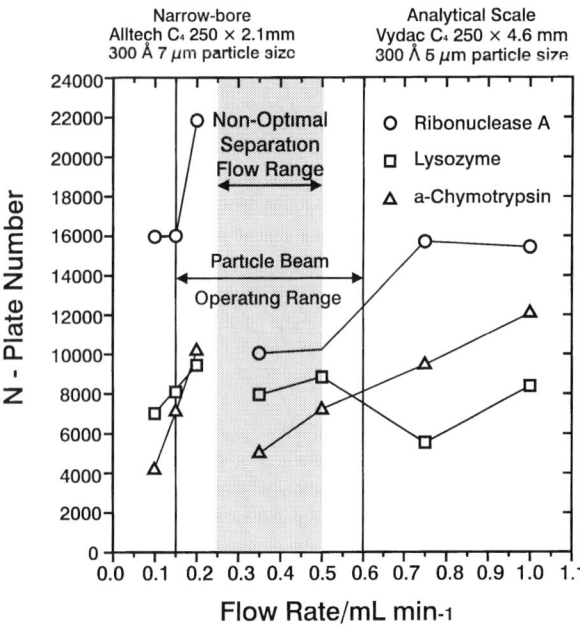

Fig. 5. Column efficiency for the separation of ribonuclease A, lysozyme, and α-chymotrypsin; the operational range of the particle beam apparatus is shown

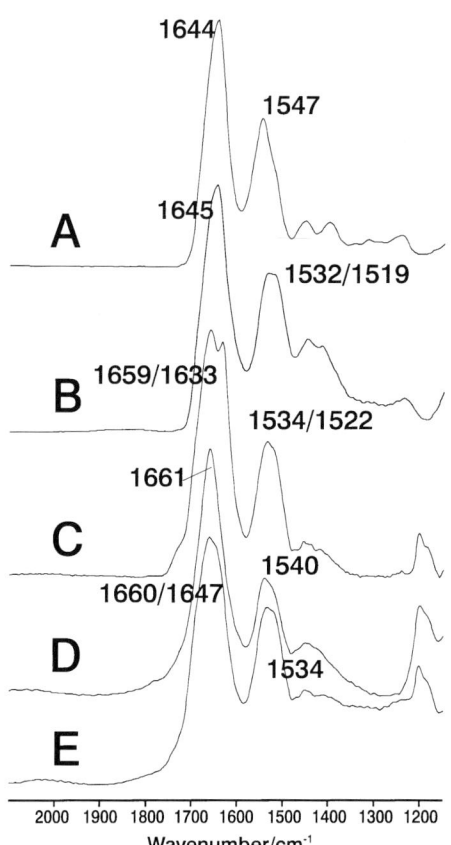

Fig. 6. Spectra of ribonuclease A: *A* reference spectrum of 5% Rnase A (w/v) in H_2O *B* reference particle beam deposit from pure H_2O (no RPC column), *C* particle beam deposit from RPC C_{18} gradient elution, *D* particle beam deposit from RPC C_4 gradient elution, and, *E* deposit from D redissolved and measured as a cast film on CaF_2

at flow-rates permissible for operation of the particle beam without compromise of chromatographic resolution. The gradient slope was optimized and baseline resolution achieved at these lower flow-rates. Separation efficiency (*N*) for the n-butyl analytical column suffered with decreasing flow for the three enzymes, but resolution of adjacent peaks was still good (see Fig. 5).

Infrared spectra obtained with the particle beam illustrate the alteration to protein secondary structure in RPC. Original contour IR spectra RNase A are shown in Fig. 6. The reference spectra (A and B) are directly comparable to those of an external reference (D_2O solution: 21% α-Helix, 50% β-sheet [7]; the particle beam gives the following estimates: 30% α-helix, 51% β-sheet). Amide I is conformationally convolved, but the peak maximum is indicative of the presence of both α-helix and β-sheet and can be deconvolved further to accentuate this. Amide II positions are slightly different between the reference and particle beam spectra, with a 10–15 cm^{-1} shift, because of the difference in the hydrogen bonding between solid and liquid phases. In the n-octadecyl gradient elution spectrum (C) some alteration of both α-helix and β-sheet to disorder occurs. This manifests itself by way of a peak maximum shift relative to that of the reference spectra and indicative of a random coil structure (the 1659 cm^{-1}

center frequency is not disordered *per se*, but shows how large an amount has changed). The peak at 1633 cm^{-1} shows that significant β-sheet content remains after elution and suggests that on the n-octadecyl column more α-helix is converted into a random coil; given the geometrical similarity between the α-helix and the disordered state this seems highly plausible. Unfortunately, the highly conformationally sensitive amide III was too weak to provide any further insight into the conformation. In contrast to the n-octadecyl elution of RNase A, the n-butyl showed greater loss of β-sheet (spectrum D). After the IR spectrum had been acquired, the deposit was removed from the CaF_2 substrate, dissolved in pure water, dessolvated, and rescanned to give spectrum E, which shows that a large portion of the native secondary structure had been reorganized. Similar results occurred with gradient elution when a 1:1 2-propanol/acetonitrile modifier was used. Apparently, the protein with disulfide cysteine linkages intact was not reversibly denatured.

Full mid-IR spectra of α-chymotrypsin are illustrated in Fig. 7. The liquid phase spectrum B was obtained by the ATR technique and shows much structural detail, including the amide III; the amide III pattern observed is consistent with the high β-sheet content. Previous IR studies produced estimations of secondary structure consistent with these results. (D_2O solution: 12% α-helix, 51% β-sheet [7]; particle beam H_2O: 23% α-helix, 63% β-sheet [33]). The resultant IR spectra of α-chymotrypsin show RPC trends identical to those observed for RNase A. The n-octadecyl spectrum (C) shows randomization of the α-helix as well. Here, however, the split peak maximum favors the β-sheet region (1636 cm^{-1}) as expected. This, com-

pared to the n-butyl spectrum (D), shows a vast reduction in ordered structure. Recasting from reconstituted RPC n-butyl deposits (with both modifiers, spectra E–G) showed again a conversion back to a structure very close to that of the native, with an intensity indicative of the high β-sheet content. Lastly, it was clear that less β-sheet structure traversed the n-butyl columns intact than did the n-octadecyl. α-Chymotrypsin is folded into two anti-parallel β-barrels (cylindrical arrangement of six β-strands each) [32]. The rigid n-butyl ligand network forces disorganization of these surface β-sheet frames during adsorption.

Because the elution process employed in RPC governs the rate at which a protein passes through an RPC column, it was considered important to investigate how isocratic elution affects secondary structure, compared to gradient elution. In gradient elution the protein velocity is dynamic, changing with the mobile phase, whereas in isocratic elution it is constant. In isocratic elution the mobile phase strength determines the rate at which a protein moves; it is the equilibrium or partition between mobile/stationary phases that changes. The capacity factor (k') is the retention parameter that provides insight regarding rate. Small changes in the percentage organic modifier (acetonitrile, acetonitrile/2-propanol) have drastic effects upon k' [20, 34]. At very low values of k', protein bands move through the column at an appreciable rate and there is little stationary phase interaction.

Lysozyme was also included in these gradient studies. The spectra were consistent in being highly symmetric, with the amide I peak maximum centered at 1660 cm^{-1}. The solution spectrum of lysozome prepared in the mobile phase (44% acetonitrile:56% water) was acquired (Fig. 8, spectrum A). A modest shoulder at 1622 cm^{-1} indicated that this solvent system might have altered the hydrogen bonding or hydrophobic environment such that some β-sheets formed in a protein that has few (and little proclivity to do so). Spectrum B shows the particle beam collection of the 44% acetonitrile elution. The definiteness of the 1635 cm^{-1} shoulder clearly indicates the presence of β-sheet. Although the bands at 1622 and 1635 cm^{-1} differ by 13 cm^{-1}, they are representative of β-structure in the liquid and solid states respectively. When the particle beam deposit was dissolved in pure water, desolvated, and scanned it produced a spectrum similar to that of the native lysozyme. The 1450 cm^{-1} band is not for a protein but is a surface artifact – it does not change in intensity from B to C. An evaporated film

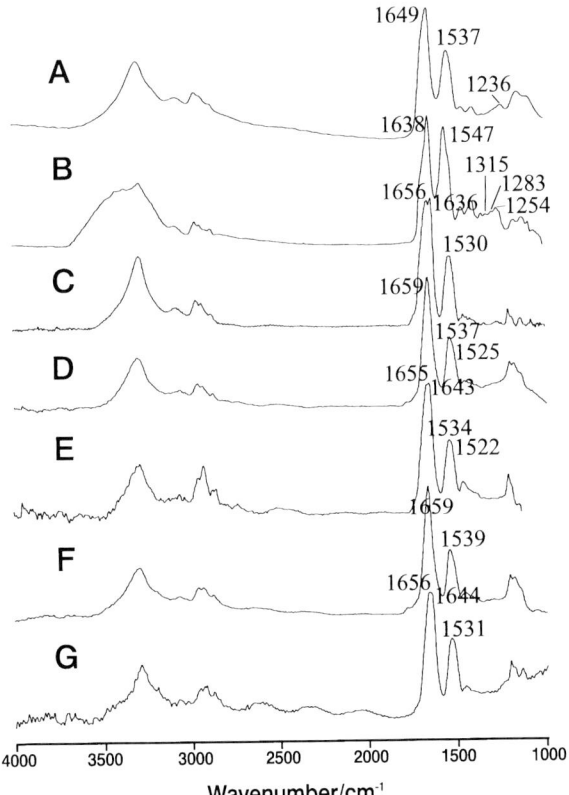

Fig. 7. α-Chymotrypsin spectra: *A* reference particle beam deposit from pure H_2O (no column), *B* reference ATR solution spectrum, 2% (w/v) in H_2O *C* particle beam deposit from RPC C_{18} acetonitrile modifier gradient elution, *D* particle beam deposit from RPC C_4 acetonitrile modifier gradient elution, *E* deposit from D redissolved and measured as a cast film on CaF_2 *F* particle beam deposit from RPC C_4 1:1 2-propanol/acetonitrile modifier gradient elution, and, *G* deposit from F redissolved and measured as a cast film on CaF_2

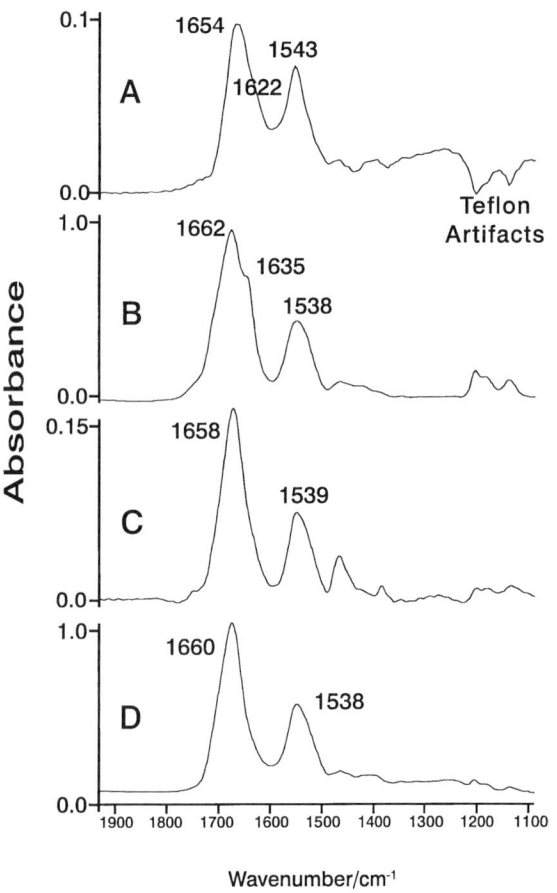

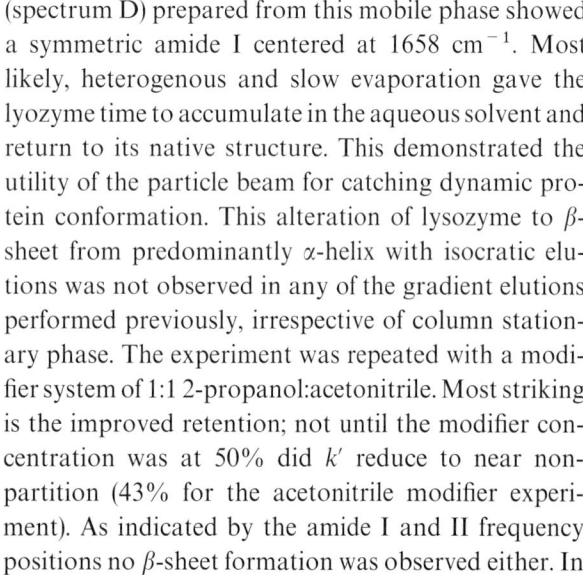

Fig. 8. Lysozyme spectra: *A* reference ATR solution spectrum, 2% (w/v) in 44% acetonitrile/56% H$_2$O/TFA, *B* particle beam deposit from 44% acetonitrile/56% H$_2$O RPC C$_4$ isocratic elution, *C* deposit from B redissolved and measured as a cast film on CaF$_2$, and *D* cast film from 44% acetonitrile/56% H$_2$O

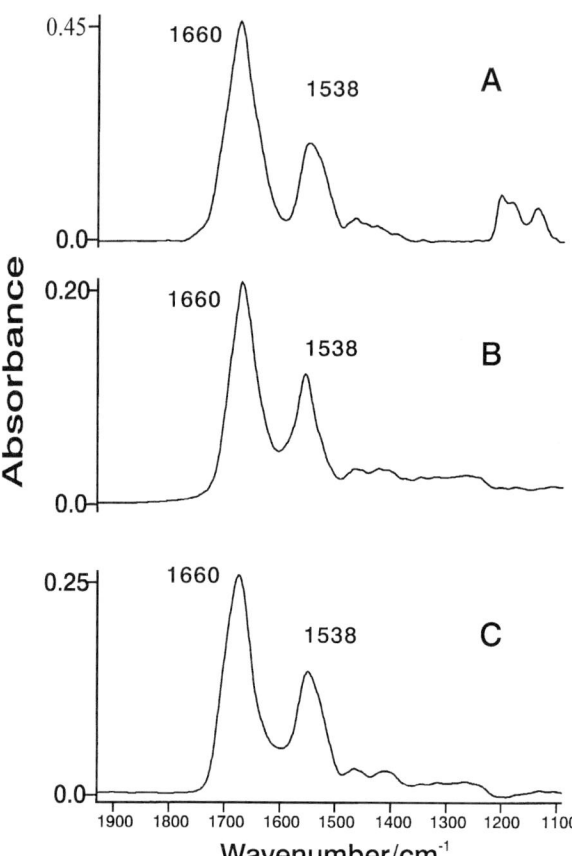

Fig. 9. Lysozyme spectra: *A* particle beam deposit from 45% 1:1 2-propanol/acetonitrile/55% H$_2$O, *B* reference spectrum 5% (w/v) in H$_2$O, *C* reference particle beam deposit from pure H$_2$O (no column)

(spectrum D) prepared from this mobile phase showed a symmetric amide I centered at 1658 cm^{-1}. Most likely, heterogenous and slow evaporation gave the lyozyme time to accumulate in the aqueous solvent and return to its native structure. This demonstrated the utility of the particle beam for catching dynamic protein conformation. This alteration of lysozyme to β-sheet from predominantly α-helix with isocratic elutions was not observed in any of the gradient elutions performed previously, irrespective of column stationary phase. The experiment was repeated with a modifier system of 1:1 2-propanol:acetonitrile. Most striking is the improved retention; not until the modifier concentration was at 50% did *k'* reduce to near non-partition (43% for the acetonitrile modifier experiment). As indicated by the amide I and II frequency positions no β-sheet formation was observed either. In

Fig. 9 the IR spectra of a particle beam collection at 45% modifier (spectrum A) are nearly identical to the reference spectrum of lysozyme in pure water solution (B) and a particle beam collection from pure water (C). Each spectrum has amide I and II bands in the α-helix positions. For these studies it was concluded that secondary structure alteration can occur with strong mobile phases in isocratic elution, and 2-propanol can preserve 3-D structure better than acetonitrile does. This may not be the case for proteins with high β-sheet content, as alcohols are known to stablize α-helices [30, 35, 36]. It was deduced that surface-degradation was not a factor in these situations, and it is speculated that a combination of organic modifier, pressure, and velocity are responsible for the changes. These changes would most likely be different for different proteins and sets of RPC conditions.

Conclusions

The ability to desolvate proteins from complex solutions with the particle beam system has been demonstrated. These studies have shown that the protein structural integrity is maintained. The processes of liquid jet formation and desolvation and deposition cause no degradation of any protein studied. Furthermore, through biological activity assays it was found that the tertiary and quaternary structural integrity of enzymes is largely maintained. It is presently under investigation whether the given conformational make-up of proteins determines their stability in passage through the particle beam apparatus. Both βLG and lysozyme particle beam spectra produced secondary structure estimates similar to their solution estimates. It has also been shown that the particle beam apparatus can prepare proteins for IR analysis for very dilute solutions (nanogram quantities injected). Particle beam spectra were obtained at concentrations well below those required by any ancillary liquid sampling technique used in this study. It was demonstrated that particle beam spectra can be used with HPLC programs to introduce structure-altering substances which enable assignment of the numerous protein IR amide bands. The differences in the original particle beam band contours of the βLG spectra caused by methanol were discernable without deconvolution. Lastly, it was also demonstrated that the particle beam could be used to detect elution conformation of globular proteins in reversed-phase HPLC experiments. Results from this work clearly showed that a high degree of ordered secondary structure was randomized upon elution. For several enzymes examined, however, the native structure reorganized to a large degree after the particle beam deposits were redissolved. With utilization of the rapid desolvation feature of the particle beam, future work will involve the trapping of folding/refolding intermediates for spectroscopic analysis.

References

[1] R. E. Dickerson, I. Geis, *The Structure and Action of Proteins*, Benjamin, Menlo Park, CA, 1969, pp. 1–43.

[2] K. U. Linderstrøm-Lang, in: *Lane Medical Lectures, No. VI*, Stanford University Press, Stanford, CA, 1952, p. 58.

[3] C. Ghélis, J. Yon *Protein Folding*, Academic Press, New York, 1982, pp. 37–120.

[4] H. Susi, in: *Structure and Stability of Biological Molecules* (S. N. Timasheff, G. D. Fasman, eds.), M. Dekker, New York, 1969, pp. 575–633.

[5] G. B. B. M. Sutherland, *Adv. Protein Chem.* **1952**, 7, 291.

[6] P. Doty, A. Wada, J. T. Yang, E. R. Blout, *J. Polymer Sci.* **1957**, 23, 851.

[7] D. M. Byler, H. Susi, *Biopolymers* **1986**, 25, 469.

[8] R. J. Jakobsen, F. M. Wasacz, J. W. Brasch, K. B. Smith, *Biopolymers* **1986**, 25, 639.

[9] F. M. Wasacz, J. M. Olinger, R. J. Jakobsen, *Biochemistry* **1987**, 26, 1464.

[10] R. J. Jakobsen, F. M. Wasacz, in: *Proteins at Interfaces: Physiochemical and Biochemical Studies* (J. L. Brash, T. A. Horbett, eds.), American Chemical Society, Washington, D. C., 1987, p. 339.

[11] R. C. Willoughby, R. F. Browner, *Anal. Chem.* **1984**, 56, 2626.

[12] J. A. de Haseth, R. M. Robertson, *Microchem. J.* **1989**, 40, 77.

[13] S. N. Timasheff, H. Susi, *J. Biol. Chem.* **1966**, 241, 249.

[14] H. Susi, S. N. Timasheff, L. Stevens, J. Biol. Chem. **1967**, 242, 5460.

[15] S. N. Timasheff, H. Susi, L. Stevens, *J. Biol. Chem.* **1967**, 242, 5467.

[16] F. E. Regnier, *Science* **1987**, 238, 319.

[17] X. Geng, F. E. Regnier, *J. Chromatogr.* **1984**, 296, 15.

[18] S. Lin, B. L. Karger, *J. Chromatogr.* **1990**, 499, 89.

[19] M. T. W. Hearn, A. N. Hodder, M.-I. Aguilar, *J. Chromatogr.* **1985**, 327, 47.

[20] L. R. Synder, K. M. Gooding, F. E. Regneir (eds.), in: *HPLC of Biological Macromolecules: Methods and Applications, Vol. 51*, Dekker, New York, 1990, pp. 231–257.

[21] S. Wicar, M. G. Mulkerrin, G. Bathory, L. H. Khundkar, B. L. Karger, *Anal. Chem.* **1994**, 66, 3908.

[22] A. Sadana, *Bioseparations* **1992**, 3, 145.

[23] S. A. Cohen, K. Benedek, S. Dong, Y. Tapuhi, B. L. Karger, *Anal. Chem.* **1984**, 56, 217.

[24] S. A. Cohen. K. Benedek, Y. Tapuhi, J. C. Ford, B. L. Karger, *Anal. Biochem.* **1985**, 144, 275.

[25] V. E. Turula, J. A. de Haseth, *Appl. Spectrosc.* **1994**, 48, 1255.

[26] V. E. Turula, J. A. de Haseth, R. F. Browner, *19th Annual Meeting of the Federation of Analytical Chemistry and Spectroscopy Societies (FACSS)*, Philadephia, Pennsylvania, September 1992.

[27] J. M. Purcell, H. Susi, *I. Biochem. Biophys. Methods* **1984**, 9, 193.

[28] P. W. Holloway, *Proc. SPIE* **1989**, 1145, 266.

[29] P. W. Holloway, H. H. Mantsch, *Biochemistry* **1989**, 28, 931.

[30] J. S. Singer, in: *Advances in Protein Chemistry* (C. B. Anfinsen, M. L. Anson, K. Bailey, J. T. Edsall, eds.), Academic Press, New York, 1962, p. 1.

[31] V. E. Turula, J. A. de Haseth, *Anal. Chem.* **1996**, 68, 629.

[32] C. Branden, J. Tooze, *Introduction to Protein Structure*, Garland, New York, 1991.

[33] J. Koyama, J. Nomura, Y. Shiojima, Y. Ohtsu, I. Horii, *J. Chromatogr.* **1992**, 625, 217.

[34] J. W. Nelson, N. R. Kallenbach, *Biochemistry* **1989**, 28, 5256.

[35] A. L. Fink, B. Painter, *Biochemistry* **1987**, 26, 1665.

[36] V. E. Turula, J. A. de Haseth, *46th Pittsburgh Conference and Exposition on Analytical Chemistry*, New Orleans, Louisiana, March 6–10, 1995.

Mikrochim. Acta [Suppl.] 14, 121–124 (1997)

New Developments in FT Spectrometer Design

James W. Brault

University of Colorado, C.I.R.E.S; NOAA R/E/AL5, 325 Broadway, Boulder, CO 80303, U.S.A.

Abstract. Laser fringes have long been used to establish the x-axis in interferometric spectrometry, but solutions for the intensity axis have been far less satisfactory. Now we are seeing the rapid commercial development of low-cost, medium speed sigma–delta analog-to-digital converters, developed for stereo audio applications. A single chip provides two channels of 20-bit precision at 50 kHz, a significant improvement over many current systems of much greater cost and complexity. However, while the laser works in the *spatial* domain, this converter operates strictly in the *time* domain; it cannot be triggered. We have developed a bridge between these two domains: the adaptive digital filter (ADF), which not only permits us to use this converter to obtain measurements at arbitrary times, but as a bonus shows us how to move much of the complexity of an interferometric control and data acquisition system from hardware to software. For example, flexible subdivision (to increase the free spectral range) is easily obtained via a simple and efficient algorithm, free of laser ghosts. Compensation for drive velocity variability is also possible, requiring only a modest amount of computer memory.

Key words: FT Spectrometer design, fringe subdivision, digital filter, velocity compensation.

In principle, Fourier-transform spectrometry is appealingly simple: we divide a light beam into two parts, pass the two beams through paths which differ in length by x, recombine them and direct the resulting beam to a detector, where $I(x_i)$ is measured at a number of equally-spaced values of x_i. This function, the *interferogram*, is known to be the Fourier transform of the desired spectrum, which is then derived in a matter of seconds on modern computers by the calculation of a discrete fourier transform (DFT). (There are, of course, some subtle considerations involved in this computational process, but they are now reasonably well understood and will be ignored here.)

The task of an instrument designer preparing to produce an FT Spectrometer is thus to create an appropriate set of stationary and moving optics (collimators, beam-splitters, moving reflector optics, recombiners, etc.) and then drive the moving optics in such a way that we can (1) recognize the positions of

equal intervals x_i (2) accurately measure $I(x_i)$ at those positions.

In general, neither of these is a trivial task for precision Fourier-transform spectrometry. Measurements in the visible and UV require that x_i be accurate to 1 nm or better, and broad-band absorption measurements in the near IR can reach 10^{12} detected photons per measurements, giving an inherent S/N of 10^6 in a single data point – a formidable job for an analog-to-digital (A/D) converter.

These were some of our thoughts in early 1993 when we were given the opportunity of building an FT Spectrometer for long-path atmospheric work, subject to some rather severe constraints on budget, manpower (~ 1 man-year) and time (9 months).

Review of Existing Techniques

The detection of equal intervals is easy in the IR, where the fringe crossings of a He–Ne laser provide a free spectral range of $\sim 8000 \, \mathrm{cm}^{-1}$ (or any integral division thereof), but any work further toward the visible would require fringe *sub*division, which is generally rather expensive, troublesome, and subject to systematic errors which generate ghosts. Some of the systems now in use are as follows:

1. Use both positive and negative-going fringe crossings. This is certainly the simplest solution, but it is inflexible, raises the free spectral range (FSR) to only $16000 \, \mathrm{cm}^{-1}$, and produces "laser ghosts" whenever the spacing between the two pairs of crossings is not identical, due to discriminator drift.

2. Use the times of the last two fringe crossings to *extrapolate* to the time at which the next fringe subdivision is needed. This can be used to give any desired subdivision, but its weakness lies in the extrapolation; any problems, especially periodic, in the drive are amplified, again leading to ghosts or excess noise.

3. Using a phase-locked loop, lock an oscillator to some high multiple of the basic fringe rate, and derive the subdivision from this oscillator. This sounds very different from system 2, but is in fact simply one of the ways of implementing the same extrapolation, and is subject to the same problems.

4. If polarizing optics are used in the laser path, it is possible to derive two separate laser fringe signals 90° out of phase. By use of sophisticated circuitry, these signals may be analysed to derive a much finer grid, but any systematic error in the derived phase position is thus periodic, and hence laser ghosts are difficult to eliminate.

5. If a Zeeman-split two frequency laser is used, position information is available in the phase of the Zeeman frequency signal, typically in the low MHz ranges, so that extrapolation is effectively eliminated. Such systems are technically complex and expensive and are still subject to laser ghosts when the two frequencies inevitably become slightly mixed.

The meaurement of intensities shows somewhat less variation, with almost all systems employing a sample-and-hold amplifier followed by a triggered A/D converter. Some approaches in current use are the following: (1) A straight 15–16 bit converter with S/H. (2) As for 1, but preceded by a switchable gain amplifier, with 1–4 preset gain changes of 2,4 or 8 across the interferogram. (3) As for 2, but with 8 grains in powers of 2, automatically selected for every sample.

The 16-bit converter is useful for weak sources, narrow band absorption spectra or emission spectra, all of which have lower photon rates, but fails for high photon rates where the photon noise falls below the digitization limit (or, more likely, below the noise of the anti-aliasing filter used before the converter). The gain changes in system 2 are adequate for most problems, but share with system 3 the problem that the gains, offset and phase characteristics (time delays) of all the amplifiers must be perfectly matched at the switching points or a form of intermodulation distortion will result.

An Alternative Approach

Although solutions to these two major measurement tasks do exist, they are in general complex, expensive and somewhat temperamental, and a much simpler technique had to be found if we were to produce an instrument in time. The first tantalizing glimpse of a possible alternative came with the appearance of an 18-bit A/D chip designed for the consumer stereo audio market, soon upgraded to 20 bits (Crystal CS-5390); 20 bits; 2 channels at 30–50 kHz; noise < 1 lsb;

superb linearity; no sample-and-hold or anti-aliasing filters needed; negligible cost due to mass production. But (of course) there was a catch: this converter was designed to run continuously, locked to a clock ($f_{samp} = F_{clock}/1024$) – it cannot be triggered. Indeed, many of its best features are only *possible* because it deals with a continuous stream of information.

With this converter and a simple interval timer gated by the fringe crossings, however, we are getting very close to a useful solution. It is clear that we can easily get the times $t(x_n)$, where the x_n are the fringe crossings, and the intensities $I(t_j)$, where the t_j are the sample times; all that is missing is some way to use this information to derive $I[x_i(t)]$, where the x_i are in general some subdivision of the fringes. The first step is to invoke the sampling theorem: if (1) the fringe time $t(x)$ is a band-limited function of x, and (2) the band limit is $< 1/2$ the laser fringe rate, then knowing the times of fringe crossings is enough to recover $t(x)$ for any x.

It is likely that most systems will easily meet this criterion, since velocity variations are usually confined to the low audio range, which includes the power line frequency and its first few harmonics as well as typical mechanical resonance frequencies. As an example, let us assume that sampling takes place at 50 kHz with reduction by a factor of 4, so that the final data output rate is 12.5 kHz (note that this adds 1 bit to the A/D dynamic range). For the sampling interval, assume a free spectral range of 33000 cm^{-1} (about 300 nm, near to atmospheric cut-off); this will require one of the slowest speeds and hence lowest fringe frequencies. The laser itself will then have a fringe frequency of 3 kHz, so the band limit must be less than 1500 Hz – quite comfortable.

The simplest way to recover $t(x)$ for any given x is to find the polynomial that passes through an even number of fringe times on either side of the desired x_i, and evaluate this polynomial at x_i to find the desired $t(x_i)$; generally a cubic will be found sufficient. As shown in Fig. 1, fringe subdivision is thus reduced to a trival arithmetic problem.

It only remains to find a way to derive a sample at a given time from a series of samples obtained at known fixed times. The solution to this part of the problem turns out to be equally simple: it is a minor variant of the common digital filter now found in almost every FT Spectrometer. Let us just for a moment go back to the continuous functions which underlie the digital filters that we usually apply to sampled data. If the functions are properly band-limited, then the digital samples contain everything we need to know to reconstruct the associated continuous function; we simply convolute the digital data with the appropriate

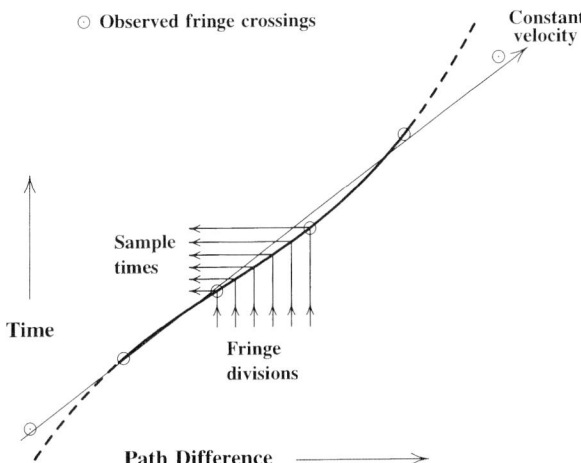

Fig. 1. Using a cubic to interpolate subdivisions between two fringe crossings

continuous sinc function to find its value at any desired sample position.

In practice, if we wish to be able to interpolate a maximum of m possible values between existing samples, using a digital filter of n taps, we simply create a long digital filter with mn taps. We then parse these taps into m separate filters, with each filter obtained by using only each mth point starting at point $1, 2, 3 \cdots, m$, as in Fig. 2.

Thus, within the granularity determined only by the number of kernels, we can obtain a sample at the desired time by a simple and computationally efficient process. We have chosen $m = 1024$, which is also the

ratio of the master clock frequency to the raw sample frequency, and measure all times in clock counts ("ticks", 1 tick = 20 nanosec). Once we have determined the time t_N of the Nth desired data point, then $j = t_N / 1024$ (integer arithmetic) gives the index of the next lower raw sample, and the remainder $k = t_N - 1024j$ gives the index of the desired kernel in a properly parsed array. The vector product of this kernel and the appropriate portion of the raw sample array then yields the desired data point. Because it can be used in this way for interpolation, as well as other more sophisticated operations, we have used the name *Adaptive Digital Filter* for this generalized approach to digital filtering.

Again, if we use n taps and take $m = 1024$, the memory requirement for the kernel array is only $n \times 4 \times 1024 = 4096 \times n = \sim 200$ kilobytes for $n = 50$ taps. On the better current (mid-1995) PC's, a multiply, replace – add process requires only $\sim 0.2\,\mu sec$, so very long kernels may be used even at 10 kHz output rates, without the need for any dedicated digital signal processing equipment.

Hardware Implementation

The simplicity of the resulting hardware can be appreciated by reference to Fig. 3, which shows a diagram of the system as it has been implemented in the NOAA FTSpectrometer. Outside the PC, the hardware is minimal: to handle the interferogram signal and position information we need only the detectors and their usual associated preamplifiers (including any necessary gain adjustments required to deal with sources of widely differing intensity); the laser detectors and their preamplifiers; discriminators to convert the laser sine waves into square waves; and the 20 bit converter chip itself. Communcation between these external compo-

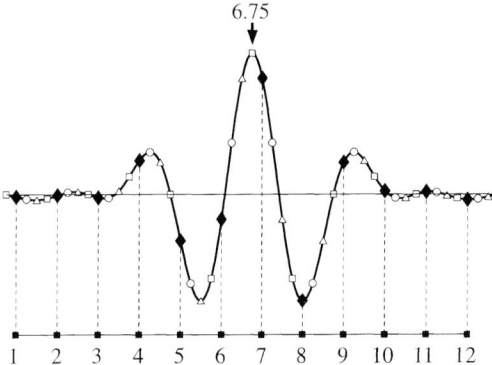

Fig. 2. A crude example of interpolating kernels, showing for clarity a 12 tap filter with only 4 subdivisions per sample (actual numbers are more like 50 taps and 1000 subdivisions). In this example, the taps represented by filled diamonds are used to weight 12 samples, shown schematically below; the sum of the products gives the interpolated value at 6.75 on this scale. Moving the curve one subdivision to the left (circles) would yield a point at 6.50, and a two subdivision shift (triangles) would give 6.25

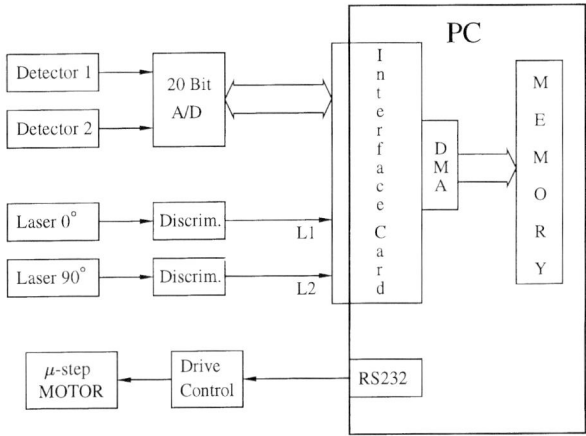

Fig. 3. The system diagram of the NOAA FT spectrometer

nents and the computer is completely handled by 7 optical fibers, providing interference-free signals. Positioning of the moving reflectors is handled open-loop by a commercial micro-stepping motor and indexer which accepts commands from an RS232 link. Inside the PC there is a single custom interface card which: (a) supplies the 50 MHz clock, (b) converts serial data from the A/D to parallel and sends it to the PC's DMA handler, (c) contains an interval timer chip for measuring fringe times, using the 50 MHz clock (also sent to DMA), and (d) contains an up–down counter used with the L1/L2 signals to keep position information. That is all there is – the rest of the functions are in software.

This approach thus simultaneously offers reduced complexity and maintenance, higher precision, and lower cost than conventional designs. Some of the obvious advantages are:

- no gain changes are necessary during a scan, so no amplifier matching is required; yet a dynamic range near 21 bits is obtained;
- no anti-aliasing analog filters are required, so system noise is reduced;
- no sample-and-hold needed;
- no complicated servo control needed; considerable velocity variation permitted;
- subdivision always obtained by interpolation, not extrapolation;
- no special digital processors needed;
- DC term included in measurement: allows monitoring of modulation efficiency; source ratioing (I/I_0) with a single detector;
- $t(x)$ provides a useful drive diagnostic;
- significant design changes possible through software, not hardware.

To illustrate the last point, note that linear combinations of digital filters may be used with a single detector for special problems, or the output of two different detectors can be processed with individual filters and the results summed, thus mixing the signals without mixing their noise components. It must be emphasized that the generation of these filters is a very simple computational process requiring only a few seconds of time at set-up.

A final example of powerful design changes easily carried out is the full compensation of drive velocity variation (to be implemented soon on the NOAA FT Spectrometer). In this application, we recognize that if there are significant velocity variations and if any components in the signal path show a frequency-dependent response, then the interferogram will be contaminated by the velocity variations, through a combination of a change in the amplitude of the fringes and a shift in their phase, even if sampling is rigidly tied to fringe crossings. *Any* system is subject to these problems to the extent that drive velocity variations occur. The only system inherently free of this problem would be one which either had perfect velocity control or infinite bandwidth for all signal paths (and this includes the digital filter!). In fact, the digital filter itself is a major source of frequency-dependent response, though being non-causal, the dependence appears only in the amplitude.

Fortunately, proper application of the ADF permits us to greatly reduce the effects of velocity variation in real time, even if the variation is appreciable. A complete discussion of this procedure is beyond the scope of this paper; we will only note here that the procedure depends on the creation of a series of, say, 10–20 filter arrays, each appropriate to a slightly different velocity. Since the fringe times give us the instantaneous velocity directly, we need only refer to the last fringe time to decide which array to enter – again, a computationally efficient process. This next level of compensation thus requires no new hardware, and needs only of the order of 2 megabytes of the computer's memory for enough arrays to reduce the velocity effects by more than an order of magnitude.

Mikrochim. Acta [Suppl.] 14, 125–131 (1997)

The Application of FT-IR Spectroelectrochemistry to the Study of Conductive Polymers

Helmut Neugebauer*,** and Zhao Ping

Institute of Physical Chemistry, University of Vienna, Währinger Strasse 42, A-1090 Wien, Austria

Abstract. The application of in situ Fourier-transform infrared spectroscopy to the investigation of the properties of conductive polymers and to the study of their electrochemical reaction processes is described. Two different experimental methods, external reflection absorption spectroscopy and internal reflection spectroscopy, are discussed. Results of the study of electronic effects, absorption band intensity enhancement, structural changes of the polymer chain, and changes in the content of counter-ions in the polymer matrix are presented.

Key words: spectroelectrochemistry, conductive polymers, conducting polymers, in situ spectroscopy, FT-IR spectroscopy.

The interest in conductive polymers started at the end of the 1970s, when an increase in the conductivity of polyacetylene exposed to oxidizing agents was observed. In the meantime, a great number of substances belonging to this novel class of materials have been found. Because of their extraordinary properties and their promising application potentialities, interest in conductive polymers is still very high. In particular their electrochemical properties are the subject of intense scientific work (see e.g. [1]). Besides the classical electrochemical investigation techniques (e.g. current/voltage curves), in situ techniques are increasingly applied during electrochemical reactions. Amongst these techniques are quartz microbalance investigations [2], EPR spectroscopy [3], and spectroscopic techniques, especially infrared spectroscopy [4–6], which is a powerful method for the determination not only of structural properties of the polymer chain, but also of electronic properties, which are related to the conductivity of the material. Owing to its sensitivity and short measurement time, Fourier-transform infrared (FT-IR) spectroscopy is commonly applied.

In the present paper, the principles and application possibilities of in situ FT-IR techniques for the study of the properties and reaction mechanisms of conductive polymers during electrochemical transitions, are given.

Experimental

The main problem for the in situ IR investigation of electrode surfaces during electrochemical reactions is the strong absorption of the IR light in the electrolyte solution. In order to minimize this absorption, the path-length of the IR radiation in the electrolyte has to be kept very short (a few μm). Two experimental set-ups can be applied: external reflection absorption spectroscopy and internal reflection spectroscopy. Surveys of the methods are given in [7–10].

External Reflection Absorption Spectroscopy

Figure 1 shows the principal set-up for external reflection absorption spectroscopy. A platinum disc working electrode, covered with the polymer, is located at a distance of a few μm from an IR-transparent

* Present address: Institute of Physical Chemistry, University of Linz, Altenberger strasse 69, A-4040 Linz, Austria
** To whom correspondence should be addressed

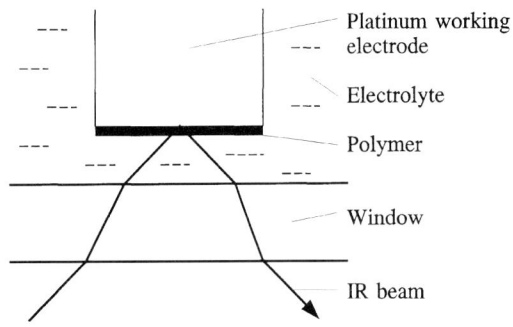

Fig. 1. Set-up for external reflection absorption spectroscopy

window. A counter-electrode and a reference electrode (not shown in Fig. 1) are mounted besides the working electrode. The IR beam passes through the window, the thin electroyte layer and the polymer. After reflection at the platinum surface the radiation reaches the detector. Transmission-like spectra can be obtained, if disturbing effects (reflection at the window, reflection at the polymer surface) are avoided or corrected.

A major problem is posed by changes in the composition of the electrolyte solution in the thin electrolyte layer, due to the current of the electrochemical reaction. The influence of this effect on the IR spectra hinders the determination of the amount of counter-ions in the polymer matrix, which is an important parameter for the investigation of the polymer reactions. A description of the specific problems concerning in situ external reflection absorption spectroscopy of conducting polymers is given in [11].

Internal Reflection Spectroscopy

By use of internal reflection spectroscopy, the electrochemical difficulties of thin electrolyte layer systems can be avoided. Figure 2 shows the set-up for this method.

An IR-transparent reflection element with high refractive index and sufficient electrical conductivity acts as the working electrode. The polymer covers the surface of the reflection element. The IR beam is totally reflected at that electrode surface. Owing to the interactions of the evanescent wave with absorbing substances near the surface, IR spectra of the polymer can be obtained without spectral disturbances due to absorption in the electrolyte solution at greater distances from the surface [12]. The main problem of this method lies in the stability of the reflection element in the electrolyte solution. The most common material used in internal reflection spectroelectrochemistry, germanium, cannot be used in aqueous electrolytes at anodic potentials. Recently, a new reflection element system especially suited for the study of the electrochemical reactions of conductive polymers in aqueous electrolytes, was developed [13]. The reflection element consists of an inert material, e.g. zinc selenide, covered with a thin evaporated metal grid on the surface to act as an electrode.

Measurement Procedure

Infrared spectra are recorded consecutively during slow cyclovoltammetric experiments. In cyclovoltammetry, oxidation and/or reduction reaction processes can be detected by the appearance of current peaks in the current/voltage diagram. An IR spectrum recorded just before the reaction investigated is chosen as the reference spectrum. The IR spectra measured during the reaction are related to the reference spectrum. In this way, difference spectra are obtained, which show the spectral changes during the reaction. Upward pointing bands belong to substances being produced, downward pointing bands to substances that are being removed.

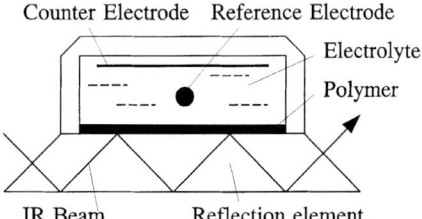

Fig. 2. Set-up for internal reflection spectroscopy

Polymer Film Preparation

The polymer films were prepared electrochemically on the electrode surfaces in electrolytes containing the respective monomer. Details of the polymerization procedure are given in [9, 10, 13–18].

Results and Discussion

The properties of a conductive polymer change dramatically when it is coverted from the insulating form into the conducting form. A number of remarkable effects accompany the property changes. The most interesting, which can be studied by using IR spectroelectrochemistry, are electronic effects, absorption band intensity enhancement, structural changes of the polymer chain, and changes in the content of counterions in the polymer matrix.

Electronic Effects

The electronic effects can be described with band structure models as shown in Fig. 3. In this figure, the electronic structure of a polymer with a non-degenerate ground state is shown (for a detailed description of the electronic properties of conductive polymers with non-degenerate ground state and with degenerate ground state, especially polyacetylene, see [19, 20]).

In the non-conducting neutral form, conductive polymers are semiconductors. During conversion into the conducting form, new electronic states appear in the band gap, related to the formation of charge carriers (polarons and bipolarons in the case of polymers with a non-degenerate ground state). Together with the new electronic states, new optical transitions occur. In the case of polarons, three new transitions can be found: valence band → lower polaron state, valence band → upper polaron state, and lower polaron state → upper polaron state. Bipolarons, where the lower state in the band gap is unoccupied (in the case of positively charged bipolarons), show two new transi-

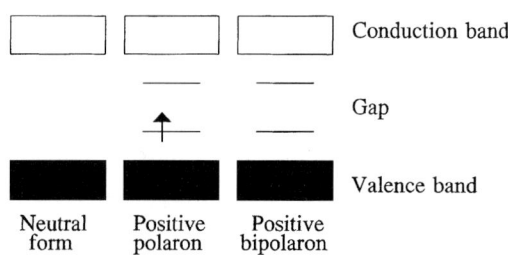

Fig. 3. Band structure model for conductive polymers with non-degenerate ground state

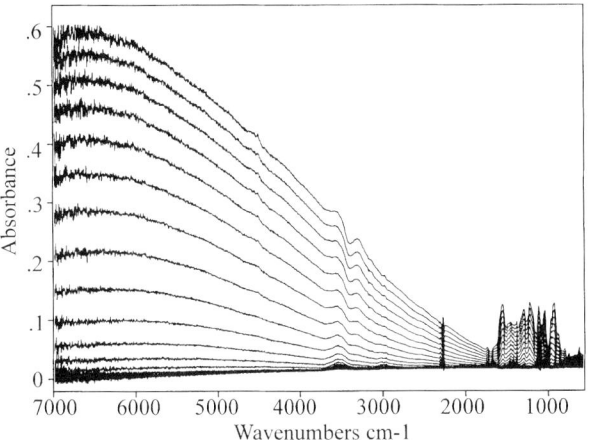

Fig. 4. Difference spectra during the oxidation reaction of polypyrrole; Reference spectrum – neutral form

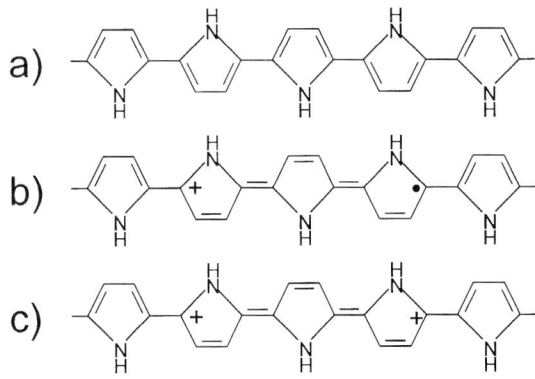

Fig. 5. Structures in polypyrrole: **a** neutral form, **b** polaron, **c** bipolaron

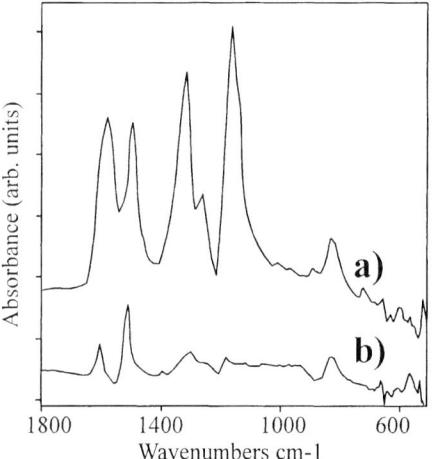

Fig. 6. Transmission spectra of polyaniline: *a* conducting middle oxidized salt form, *b* non-conducting middle oxidized base form

tions: valence band → lower bipolaron state, and valence band → upper bipolaron state. In both cases, the electronic transition with the lowest energy (valence band → lower state) is visible in IR spectroscopy. In the mid IR region (up to 7000 cm^{-1}), the transition occurs in the high energy region of the spectra. Figure 4 shows difference spectra obtained during the oxidation of polypyrrole in $CH_3CN/LiClO_4$ electrolyte, by using external reflection absorption spectroscopy.

Besides changes in the vibrational part of the spectrum (below 1600 cm^{-1}), the spectra are dominated by a strong increase in absorption in the high energy range. The absorption is related to the increasing conductivity of polypyrole and can be attributed to the electronic transition into the lower (polaron or bipolaron) state as mentioned above. The peak maximum of this absorption is seen in optical spectroscopy at 1eV (ca. 8000 cm^{-1}) [20]. A broad and intense electronic absorption in the energy range above ca. 2000 cm^{-1} is found for most conductive polymers in the conducting form, although the interpretation depends on the nature of the substance. For example, in polyaniline the absorption is attributed to the occurrence of a "polaron lattice" [21].

Absorption Band Intensity Enhancement

The charge carriers in conducting polymers with a non-degenerate ground state are polarons and bipolarons. A schematic representation of these quasiparticles in polypyrrole is given in Fig. 5.

Polarons are singly charged and have one unpaired electron, whereas bipolarons have two charges and no unpaired electrons. The charges on the polymer chain lead to a dramatic increase in the changes of the dipole moment of specific polymer vibrations, mainly vibrations which are related to the conjugated system [22, 23]. The increasing changes in dipole moment lead to an increasing interaction with the electromagnetic radiation and therefore to a strong increase in absorption band intensity in the IR spectra. Again, the increase in absorption intensity during the conversion from the non-conducting into the conducting form is a common feature of most conductive polymers. A scheme similar to that shown in Fig. 5 for polypyrrole can easily be devised for other polymers, e.g. polyaniline [24]. In Fig. 6, the absorption spectrum of polyaniline in a non-conducting state (Fig. 6b), obtained ex situ by transmission spectroscopy, is compared with a spectrum of the polymer in the conducting form (Fig. 6a). The strong increase of the band intensities is clearly visible.

The strong absorption increase has the consequence that in IR difference spectroscopy during the conversion of a polymer from the non-conducting into the conducting form, the negative bands of the vanishing non-conducting form are hardly visible. Difference spectra of those processes are dominated by the absorption spectrum of the conducting form.

Structural Changes of the Polymer Chain

As can be seen from Fig. 5, aromatic rings in the benzenoid form are converted into quinoid rings during the formation of polarons and bipolarons in conductive polymers with aromatic ring systems. Since the infrared absorption bands of quinoid and benzenoid rings are different, the conversion can be seen by IR spectroscopy. Figure 7 shows IR difference spectra during the oxidation of polyaniline in an aqueous acidic electrolyte, obtained by internal reflection absorption spectroscopy using a ZnSe reflection element with an evaporated metal grid as the working electrode system [13]. In Fig. 7, the polymer is converted from the non-conducting reduced form into the conducting middle oxidized form (for details of the electrochemistry of conducting polymers see e.g. [25]).

The negative bands at 1502 and 1605 cm^{-1}, labelled B in Fig. 7, are related to the decreasing amount of

benzenoid rings, whereas the positive bands at 1475 and 1562 cm^{-1} (Q) indicate the formation of rings with more quinoid structure (in examination of the spectral effects in detail, the aromatic ring system of the conducting middle oxidized form of polyaniline is described more precisely as a partly semi-quinoid structure [14]. In accordance with the strong absorption intensities of bands of the conducting form of polyaniline, the intensities of the positive bands of the partly quinoid system generated are greater than those of the negative bands for the vanishing benzenoid ring system (see previous paragraph). Details of the vibrational spectra and the structure of polyaniline and related compounds are given in [26]. A theoretical study of the vibrational spectroscopy of the non-conducting reduced form of polyaniline can be found in [27].

Changes in the Content of Counter-Ions in the Polymer Matrix

The conversion of conductive polymers from the non-conducting into the conducting form is accompanied by oxidation or reduction of the polymer matrix. In order to balance the electrical charges, counter-ions have to diffuse into the polymer (or ions with the same charge have to leave the polymer). The determination of the ion content, preferentially performed in situ, is therefore an important requirement for the characterization of the substance and for the formulation of the reaction mechanisms. Polyaniline in particular shows a very complex behaviour, depending not only on the oxidation state of the polymer (redox transitions), but also on the pH value of the surrounding electrolyte solution (salt–base transitions) [24, 28, 29]. The conversion into the conducting form can be performed either electrochemically by changing the potential of the electrode, or by decreasing the pH value, when the polymer is in its middle oxidized form. A great number of publications describing the various insulator-to-metal transitions have appeared in the literature (for IR spectroscopic characterization see [14–17]).

The in situ IR determination of the ion content of polyaniline in aqueous electrolytes has several difficulties: (1) the most commonly used anions either have no IR absorption (e.g. Cl$^-$) or have bands which overlap with polymer bands (e.g. ClO$_4^-$, BF$_4^-$, PF$_6^-$); (2) in external reflection absorption spectroscopy, the determination of the ion content of the polymer is hindered by the spectral influence of the thin electrolyte layer (see experimental part); (3) in internal reflection spec-

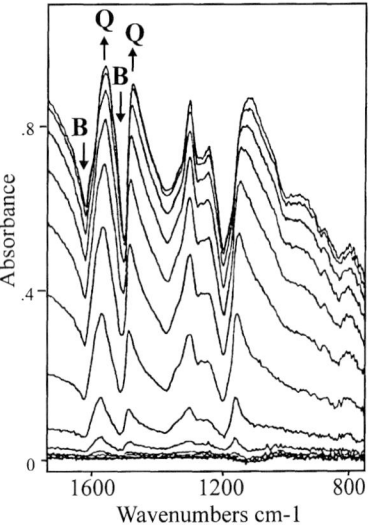

Fig. 7. Difference spectra during the oxidation reaction of polyaniline in acidic electrolyte; reference spectrum – non-conducting reduced form

troscopy, the reflection element must be stable in the acidic electrolyte. To avoid these problems, a special ex situ method, using the same spectroelectrochemical cell as used in external reflection absorption spectroscopy, was developed: the ammonia flush conversion technique (AFC) [16]. As an example, Fig. 8 shows the AFC determination of the ClO_4^- content of the middle oxidized conducting form of polyaniline.

A polyaniline-coated platinum electrode is equilibrated in an external reflection absorption spectroelectrochemical cell in the appropriate electrolyte. The conducting state of the polymer is established electrochemically. After washing and drying, an ex situ IR spectrum of the polymer is recorded [spectrum (a) in Fig. 8]. The cell is then flushed with gaseous NH_3, converting the salt form of polyaniline into the base form with the formation of NH_4ClO_4 [spectrum (b) in Fig. 8]. The cell is washed with H_2O to remove NH_4ClO_4 and dried again. The resulting spectrum (c) in Fig. 8 is the spectrum of the base form of polyaniline. A spectral subtraction (b)−(c) yields spectrum (d), which is the calculated spectrum of NH_4ClO_4. From the band intensity at around 1100 cm^{-1} the amount of ClO_4^-, which was originally in the polymer, can be determined. The results of the AFC method in acidic electrolytes at different electrode potentials are shown in Fig. 9.

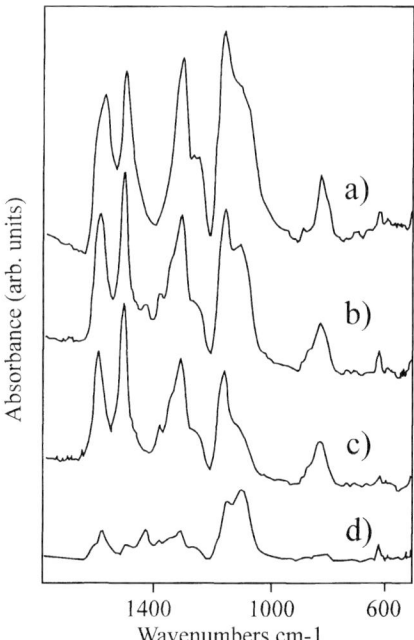

Fig. 8. Determination of the ClO_4^- content in polyaniline by the AFC method: *a* conducting middle oxidized form, *b* after treatment with NH_3 gas, *c* after washing with H_2O, *d* spectral subtraction *b* − *c*

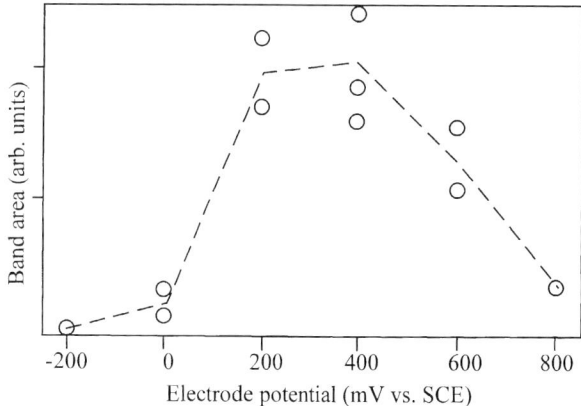

Fig. 9. Amount of ClO_4^- in polyaniline at various electrode potentials, acidic electrolyte, AFC method

The maximum amount of ClO_4^- anions is found in the middle oxidized conducting form of polyaniline at around +400 mV *vs.* a saturated calomel electrode (SCE). Taking into account the published ion content (one anion per two nitrogen atoms) of this form of polyaniline [24], the amount of anions in the other oxidation states can be calculated.

Since changes in the ion content due to the washing procedures which are necessary for the ex situ experiments cannot be excluded, in situ methods are favourable for the determination of the amount of counterions in the polymer. Recently, two developments have improved the possibility of applying IR in situ techniques: first, the development of metal grid electrodes by evaporation on the surface of inert reflection elements allows the application of internal reflection spectroelectrochemistry [13], secondly, ReO_4^- has been found to be a well suited counter-ion for IR purposes, since it has an absorption band which does not interfere with polyaniline bands [18, 30]. Figure 10 shows difference spectra obtained during the first oxidation process of polyaniline in $HReO_4$ solution at pH 1, obtained by internal reflection spectroscopy using a ZnSe reflection element with a platinum–gold metal grid on the surface.

Since the ReO_4^- band at around 900 cm^{-1} is well separated from the polymer bands, and no interferences due to a thin electrolyte layer occur, the intensity of the band is a direct measure of the changes in the ion content of the polymer during the oxidation process. By use of a flow cell, not only the redox transitions of polyaniline, but also transitions due to different pH values of the electrolyte (salt–base transitions) can be

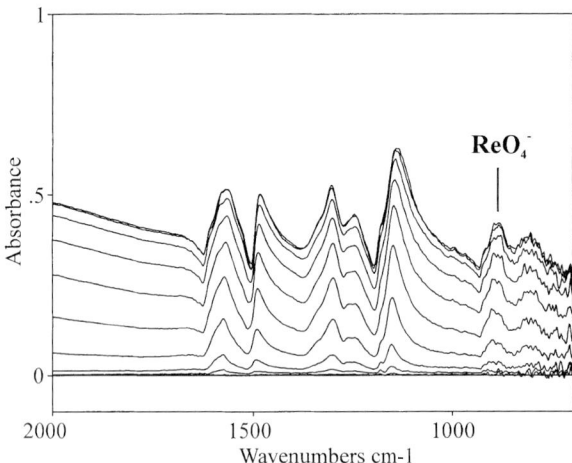

Fig. 10. Difference spectra during the first oxidation reaction of polyaniline: reference spectrum–reduced form; $HReO_4$ electrolyte, pH 1, internal reflection spectroscopy

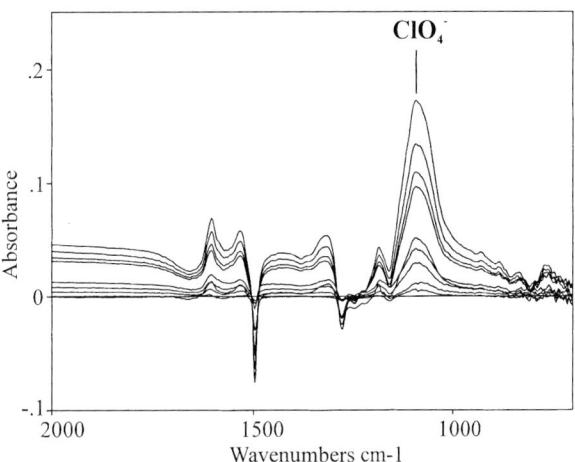

Fig. 11. Difference spectra during the base→salt transition of the reduced form of polyaniline: reference spectrum at pH 8; $HClO_4/NaClO_4$ electrolyte; uppermost curve at pH 1; internal reflection spectroscopy

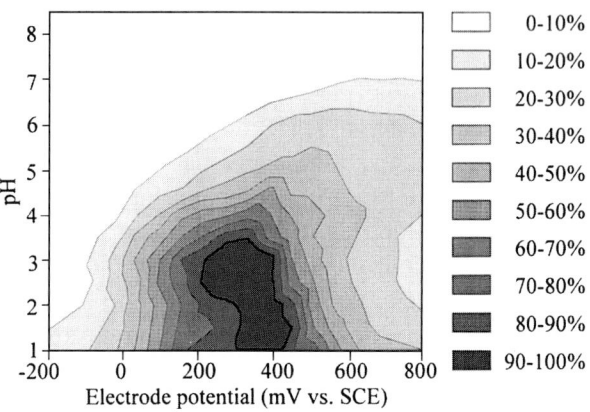

Fig. 12. Anion content in polyaniline *vs.* electrode potential *vs.* pH; 100% = maximum amount of anions in the conducting middle oxidized form (one anion per two nitrogen atoms)

studied. In Fig. 11, the spectral changes during the base→salt conversion of polyaniline at −200 mV *vs.* SCE are displayed.

In Fig. 11, the polymer remains in the reduced, non-conducting state, and the pH is changed from 8, where the reference spectrum is taken, to 1. Since the changes of the polymer absorptions during the base→salt transition of the reduced form are much smaller than the transitions where the conducting middle oxidized form is involved, the determination of the anion content also can be obtained in this case with ClO_4^- anions (in the middle oxidized form, the absorp-

tion band of ClO_4^- interferes strongly with a polymer band). The intensity of the ClO_4^- band at around 1100 cm^{-1} is a direct measure of the ion content, which increases during the transition of the base form of reduced polyaniline into the salt form (maximum change at around pH 2.5).

From a number of experiments during electrochemical redox reactions at different pH values and during salt–base transitions at different electrode potentials, a three-dimensional diagram describing the anion content of polyaniline in various states can be derived. The diagram is shown in Fig. 12.

Similarly to the results obtained by using the AFC method, the amount of anions in different oxidation states and at various pH values can be calculated, taking into account the known amount of anions in the middle oxidized salt form of polyaniline. From the data, detailed information on the contribution of the anions to the different reaction processes of polyaniline can be obtained.

Conclusion

FT-IR spectroelectrochemistry is a valuable tool for studying the properties and reactions of conductive polymers. With two experimental set-ups, external reflection absorption spectroscopy and internal reflection spectroscopy, information on electronic effects as well as on the structural properties of these highly interesting substances can be obtained.

Acknowledgements. The work was partly supported by the " Fonds zur Förderung der wissenschaftlichen Forschung" of Austria, Project 7771. The authors thank N.S. Sariciftci, A. Moser, J. Theiner and A. Neckel for their support, cooperation and valuable discussions.

References

[1] G. P. Evans, in: *Advances in Electrochemical Science and Engineering, Vol. 1* (H. Gerischer, C. W. Tobias, eds.), VCH, Weinheim, 1990, p.1.

[2] D. Orata, D. A. Buttry, *J. Am. Chem. Soc.* **1987**, *109*, 3574.

[3] E. M. Geniès, M. Lapkowski, *J. Electroanal. Chem.* **1987**, *236*, 199.

[4] H. Neugebauer, G. Nauer, A. Neckel, G. Tourillon, F. Garnier, P. Lang, *J. Phys. Chem.* **1984**, *88*, 652.

[5] H. Neugebauer, A. Neckel, N. Brinda-Konopik, in: *Electronic Properties of Polymers and Related Compounds* (H. Kuzmany, M. Mehring, S. Roth, eds.), Springer, Berlin Heidelberg New York Tokyo, 1985, p.227.

[6] P. A. Christensen, A. Hamnett, D. C. Read, *Electrochim. Acta* **1994**, *39*, 187.

[7] R. Kellner, H. Neugebauer, G. Nauer, A. Neckel, *Proc. SPIE* **1985**, *553*, 12.

[8] A. Neckel, *Mikrochim. Acta* **1987**, *III*, 263.

[9] H. Neugebauer, N. S. Sariciftci, in: *Lower-Dimensional Systems and Molecular Electronics* (R. M. Metzger, P. Day, G. C. Papavassiliou, eds.), Plenum, New York, 1991, p. 401.

[10] H. Neugebauer, *Macromol. Symp.* **1995**, *94*, 61.

[11] H. Neugebauer, N. S. Sariciftci, H. Kuzmany, A. Neckel, in: *Electronic Properties of Conjugated Polymers III* (H. Kuzmany, M. Mehring, S. Roth, eds.), Springer, Berlin Heidelberg New York Tokyo, 1989, p.137

[12] N. J. Harrick, *Internal Reflection Spectroscopy*, Interscience, New York, 1967.

[13] A. Moser, H. Neugebauer, K. Maurer, J. Theiner, A. Neckel, in: *Electronic Properties of Polymers* (H. Kuzmany, M. Mehring,

S. Roth, eds.), Springer, Berlin Heidelberg New York Tokyo, 1992, p.276.

[14] N. S. Sariciftci, H. Kuzmany, H. Neugebauer, A. Neckel, *J. Chem. Phys.* **1990**, *92*, 4530.

[15] H. Kuzmany, N. S. Sariciftci, N. Neugebauer, A. Neckel, *Phys. Rev. Lett.* **1988**, *60*, 212.

[16] H. Neugebauer, A. Neckel, N. S. Sariciftci, H. Kuzmany, *Synth. Met.* **1989**, *29*, E185.

[17] N. S. Sariciftci, M. Bartonek, H. Kuzmany, H. Neugebauer, A. Neckel, *Synth. Met.* **1989**, *29*, E193.

[18] Z. Ping, H. Neugebauer, A. Neckel, *Electrochim. Acta,* **1996**, *41*, 767.

[19] W. P. Su, in: *Handbook of Conducting Polymers* (T. A. Skotheim, ed.), Dekker, New York, 1986, p. 757.

[20] J.-L. Brédas, in: *Handbook of Conducting Polymers* (T. A. Skotheim, ed.), Dekker New York, 1986, p. 859.

[21] S. Stafström, J. L. Brédas, A. J. Epstein, H. S. Woo, D. B. Tanner, W. S. Huang, A. G. MacDiarmid, *Phys. Rev. Lett.* **1987**, *59*, 1464.

[22] G. Zerbi, M. Gussoni, C. Castiglioni, in: *Conjugated Polymers* (J. L. Brédas, R. Silbey, eds.), Kluwer, Dordrecht, 1991, 435.

[23] M. Gussoni, C. Castiglioni, G. Zerbi, in: *Spectroscopy of Advanced Materials* (R. J. H. Clark, R. E. Hester, eds.), Wiley, Chichester. **1991**, p. 251.

[24] A. G. MacDiarmid, A. J. Epstein, *Faraday Discuss. Chem. Soc.* **1989**, *88*, 317.

[25] J. Heinze, *Synth. Met.* **1991**, *41–43*, 2805.

[26] I. Harada, Y. Furukawa, F. Ueda, *Synth. Met.* **1989**, *29*, E303.

[27] R. Kostić, D. Raković, I. E. Davidova, L. A. Gribov, *Phys. Rev. B* **1992**, *45*, 728.

[28] A. G. MacDiarmid, J. C. Chiang, A. F. Richter, A. J. Epstein, *Synth. Met.* **1987**, *18*, 285.

[29] A. J. Epstein, J. M. Ginder, F. Zuo, R. W. Bigelow, H.-S. Woo, D. B. Tanner, A. F. Richter, W.-S. Huang, A. G. MacDiarmid, *Synth. Met.* **1987**, *18*, 303.

[30] Z. Ping, H. Neugebauer, A. Neckel, *Synth. Met.* **1995**, *69*, 161.

Mikrochim. Acta [Suppl.] 14, 133–141 (1997)

Astronomical Fourier-Transform Spectroscopy of the 1990s

J. P. Maillard

Institut d'Astrophysique de Paris, CNRS, 98 bis Blvd Arago, 75014 Paris, France

Abstract. The number of FT spectrometers mounted behind optical telescopes has decreased with the development of modern cold-grating infrared spectrometers, the instruments of choice for the study of faint astronomical objects. However, Fourier-transform spectroscopy remains unrivalled for high resolution work in the infrared. The instrument in operation behind the Canada–France–Hawaii telescope is presented. The sources commonly under study, strong infrared sources, are those objects of the solar system with an atmosphere (planets and satellites) and various types of stars. Of particular interest are the late-type stars surrounded by cold circumstellar envelopes or very young objects embedded in their parent molecular clouds. The latter sources are used to probe the interior of giant molecular complexes. The astronomical sources offer conditions difficult to obtain in the laboratory. The most significant results obtained with an FT spectrometer in the last few years include the detection of the molecular ion H_3^+ on Jupiter and the radical C_5 in the carbon-rich envelope of the evolved star IRC + 10216. Results will be also mentioned from the millimetric range, a new window of exploration from the ground, where FT spectrometers are starting to be used, to benefit from their broad spectral coverage. New applications have also been tested. For example, with the development of 2-D infrared arrays, spectro-imaging becomes possible and an imaging FT spectrometer is in operation behind the Canada–France–Hawaii telescope. In another field, an FT spectrometer is used as seismometer to detect the acoustic oscillation spectrum of giant planets. The next challenge for Fourier-transform spectroscopy is the era of new technology telescopes. The possibilities of an FT spectrometer behind an 8-m class telescope will serve as a conclusion.

Key words: infrared, planetary atmospheres, molecular clouds, seismology.

Astronomical Fourier-transform spectroscopy in the infrared (FT-IR) is considered to have really started with the first recording of high-resolution spectra of Venus and Mars, in the near infrared, at the Observatoire de Haute-Provence in 1965 [1]. These spectra of the two planets covered the range 4000–9000 cm^{-1}, unexplored up to that time, and had a resolution of 0.08 cm^{-1}, roughly three times better than the resolution which had been reached previously in this spectral range, and then only for the Sun. They were published in 1969 in an Atlas des Spectres dans le Proche Infrarouge de Vènus, Mars, Jupiter et Saturne [2]. One year later, the most prominent results obtained at Laboratoire Aimè Cotton, where the method had been developed, were presented at the Aspen International Conference on Fourier Spectroscopy, 1970 [3], the first meeting devoted exclusively to this new spectroscopic tool. In the following decade, many astronomical FT spectro-meters were built and mounted behind ground-based telescopes, flown on balloons, and embarked on airplanes and spacecraft. In a review of 1978 [4], at least twenty groups in the world, were engaged in a program of astronomical FT-IR. In the following years, many results reviewed in 1984 by Ridgway and Brault [5] were accumulated for a variety of sources. For the main objects of the solar system alone, excluding the Sun, it can be considered that most of the information retrieved from high-resolution spectroscopic data (abundances, chemical composition, isotopic ratios) has been obtained from FT spectra, as shown by a review in 1987 [6].

What is the current situation of FT-IR for astronomical applications? The FT spectrometer called IRIS (InfraRed Interferometric Spectrometer), on board Voyager 2 launched in 1977, sent to Earth its last data, spectra of the surface of Neptune and Triton at the time of the encounter, in 1989. The instruments at Kitt Peak Observatory [7, 8] and on the Kuiper Airborne Observatory [9] have been decommissioned. The last ground-based infrared instrument to be put into operation was the FT spectrometer on the Canada–France–Hawaii telescope in 1982 [10].

Does this mean that astronomical FT-IR has become completely obsolete? After the enthusiasm of the first era, the complexity of the technique has limited the number of instruments maintained in operation. A further reason why FT spectrometers did not become a common instrument in observatories is that high-resolution spectroscopy in astronomy is a field in which only a small number of people are interested. CCD cameras and low-resolution spectrographs are simpler and give access to the majority of intrinsically faint celestial objects. They are preferred for basic astronomical instrumentation. However, FT-IR has remained so far without serious competitors in high-resolution infrared spectroscopy. The CFH-FT spectrometer part of the telescope instrumentation, remains active. Some of the most significant results obtained with it are illustrated in the present review, with a few results from the Kitt Peak instruments.

Brief Description of the CFH-FT Spectrometer

Quoted by Beer [11] as "state of the art in astronomical FTSs", the CFH-FT spectrometer is briefly described below. The 3.6-m telescope to which it is attached opened in 1980. It is located on the top of the Mauna Kea volcano in Hawaii, at an altitude of 4200 m, the highest astronomical site. This exceptional place now attracts the new generation of 8- and 10-m telescopes. The design of the instrument derives directly from the previous generations of FT spectrometers developed in France [12, 13]. Several improvements and modifications were introduced because an instrument specialized for astronomy differs in many aspects from a laboratory instrument. Compared with the instruments of previous generations, it had to be able to work not only on bright sources like Venus and few red giant stars, but on much fainter sources and also at longer wave-lengths, in the thermal infrared, where almost no high-resolution spectroscopic work had been done so far. To realize these goals and take advantage of the exceptional infrared qualities of the site, the main choice made was to build an instrument for the infrared Cassegrain focus, instead of the Coudé focus. Other instruments of this type had been restricted to operation in a coudé room because of the stable and protected environment it offers. An infrared focus presents a much lower emissivity and a gain in luminosity by 50%, because there are no mirrors between the telescope and the instrument. However, using a Cassegrain focus meant a need to compensate for the effects of variable flexures and changing temperature in order to keep the interferometric alignments correctly

adjusted all the time. Provided these difficulties are overcome, a gain in signal-to-noise ratio by a factor of ~ 2.5 at beyond 2.5 µm, where the thermal background radiation from the sky, the telescope and the instrument becomes the major source of noise, is obtained, compared to a similar instrument at the Coudé focus [14]. Owing to this potential gain and the properties of InSb detectors, the spectral range covered was extended from 1 to 5.4 µm, giving access to six atmospheric windows. Selectable, cooled, narrow-band filters are chosen to have a band-pass in this range, well-adjusted to the regions of interest for the sources to be observed during a run, in a constant effort to optimize the detectivity. A choice of circular apertures makes it possible to work either on stellar or on extended objects. Projected on the sky, they correspond respectively to 2.5, 5, 8 and 12 arcsec. The smallest aperture can be used on stellar sources, without loss of light, because of the image quality at the CFH telescope (image spread $\leqslant 0.6$ arcsec FWHM), yielding an efficient limitation of the thermal background. The broad spectral range is covered with two interchangeable beam-splitters, permanently mounted, coated for each half of the domain respectively.

The optical design uses cat's-eye retroreflectors in the interferometer arms to collect all the energy on two outputs, instead of flat mirrors, often accepted in laboratory instruments. This design has the advantage also of offering two inputs, which allows cancellation of the strong thermal background radiation beyond 2.5 µm (which can be several hundred times the flux level of the sources to be observed) by matching on the sky the two entrance apertures. A switching of the telescope corrects for the residual imbalance. A 60-cm maximum path-difference has been chosen (0.01 cm^{-1} FWHM), suitable for most of the astrophysical conditions, provided that in many cases the limitation does not come from the instrument but from the brightness of the source. The optical layout of the instrument is given in Fig. 1.

The recording method is based on the step-by-step technique with an internal modulation, consisting in a modulation of the path-difference, introduced in the first generation of instruments [15], which has proved to be an efficient method for cancelling atmospheric scintillation noise. The various commands, including selection of the parameters, exploration of the zero path-difference and data acquisition, are computer-controlled. A real-time processing is incorporated into the software, and full data processing is available on-

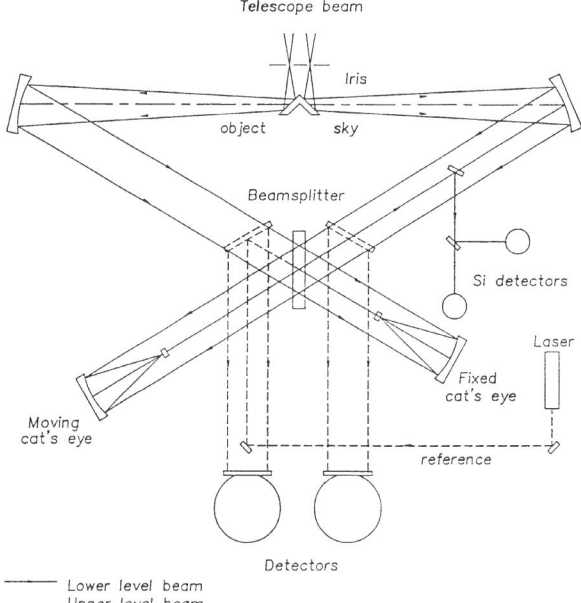

Fig. 1. Optical layout of the CFH-FTS. The instrument is a dual-input dual-output system. The instrument is mounted at the infrared Cassegrain focus ($f/35$). The two entrance apertures are located in the focal plane of the telescope, 52 arcsec apart on the sky. In presence of strong thermal background emission, the source is alternately switched from one entrance to the other every 5–10 min. A stabilized monomode He–Ne laser controls the stepping, the positioning, and the square-wave modulation of the path difference. A dewar with an InSb photodiode (0.5×0.5) is mounted on each output beam. In imaging mode an interface forms the two images of the entrance focal plane on an HgCdTe array with a plate scale of 0.33 arcsec/pixel

line at the summit, which gives the possibility of a first look at the data during recording and immediately after an integration, an important aspect for making the best use of the precious observing time. More details of the instrument can be found in [10] and [14].

The instrument was put into full service in 1984, when the infrared focus became operational, and has since been maintained and upgraded by the Telescope Corporation. The FT spectrometer is available for the three communities which share the telescope time, with scientific programs from planets to active nuclei of galaxies. A selection of recent results illustrates some current topics in astrophysics, and shows the permanent advantages of FT-IR in this field.

Selected Astrophysical Results

A Workshop held at Brussels in 1994, devoted to Laboratory and Astronomical High Resolution Spectra gave a good opportunity to review many

astronomical results obtained by FTS, in particular in solar and stellar spectroscopy [16–18] to produce atlases. The Sun is obviously a source for high-resolution spectroscopy. For example, spectacular solar spectra covering the range 4800–600 cm^{-1} have been obtained with an FT spectrometer on board the Space Shuttle at a spectral resolution of 0.013 cm^{-1} [16]. In the present review, results for three types of less classical sources are illustrated, dealing with the chemical composition of planetary atmospheres, which have been traditionally an active domain for FT-IR, circumstellar envelopes of evolved stars and the environment of young stellar objects. The latter represent a new class of sources, detected by infrared surveys because they are totally obscured in the core of molecular clouds. In each field, particular results placed in a larger context are evoked, but no attempt is made to review all the results obtained in these various fields.

Planetary Atmospheres

At the workshop mentioned above, a review devoted to planetary atmospheres was presented [19]. We limit this account to some original results on Venus, Mars and Jupiter. Though Venus and Mars, the brightest sources in the near infrared, were the first two astronomical sources to be observed at high resolution by an FT spectrometer [1], with for Venus the resolution later pushed up to 500,000 [13], the extension of the spectral range to longer wavelengths and the increase in sensitivity, has brought several new results for these telluric planets. The understanding of their evolution compared with that of the Earth, stimulated by new space missions (Phobos, Galileo, Magellan), has renewed the interest in a detailed study of their atmosphere. Deuterated molecules have been detected on Mars and Venus [20, 21] and new molecular compounds were revealed by the observation of the dark side of Venus [22]. The complex chemistry of giant-planet atmospheres remains an object of interest, leading to the first spectroscopic detection of the molecular ion H_3^+ [23] in the polar zones of Jupiter.

Abundance of deuterium in the solar system. Deuterium is an element of cosmological interest, which is destroyed in the stellar photospheres. Hence, it can be detected only in media which have remained cold enough. A trace of the deuterium which was present in the solar nebula at the formation of the solar system can be found in the planetary atmospheres. The comparison of the solar D/H ratio, 4.6 Gyr old, with its value in different locations of the Galaxy informs us about the evolution of the deuterium abundance, so we can retrieve its primordial abundance. On the other hand, variations of the D/H ratio between the different objects of the Solar System is indicative of the particular evolution of each body. The ratio was totally unknown for Mars and controversial for Venus, being based on an uncertain determination from an experiment on board the Pioneer Venus probe. The measurement on Mars was secured from the observation with the CFH-FT spectrometer of 22 lines belonging to the v_1 band of HDO at 3.67 μm, in a spectrum at a resolution of 90,000, taken with a 5 arcsec aperture on the planet [20]. The dryness of the site, (0.3 mm of precipitable water) was another important factor in separating the lines from the absorptions of the deuterated water present in the Earth's atmosphere and to remeasure the water vapour abundance on Mars at the same time, important for determining correctly the D/H ratio. A mean ratio of 6 times the telluric value was deduced, which implies an enrichment of deuterium with respect to hydrogen by a gravitational selection, due to a rapid escape of hydrogen from Mars in the planet's early history. This deduction is consistent with a much wetter and warmer atmosphere than today's suggested by geological features on the surface.

On Venus, another method was devised by observing its dark side. Infrared imagery at 1.7 and 2.2 μm of the night-side had shown hot spots, interpreted as holes in the cloud cover through which emission from warm lower layers escapes. The spectroscopy of these regions made it possible to probe much deeper into the Venusian atmosphere, considerably increasing the column density of minor constituents observed. A strong deuterium enrichment by a factor of 100 compared with the telluric value was measured from many absorptions of HDO and H_2O between 4120 and 4275 cm^{-1}, belonging respectively to the v_1 and v_3 bands for H_2O and the $v_1 + v_2$ and $3v_3$ bands for HDO [21]. Removal of an original large water reservoir through dissociation and hydrogen escape, which was faster than on Mars, appears a plausible explanation of the data. This type of observation was very productive, making possible, beside the deuterium measurement, the determination, previously inaccessible from the Earth, of the water vapour distribution with altitude. The sulphur compound OCS, important for the chemistry of the atmosphere of Venus, was detected for the first time [22], and new insights into the oxygen air-

glow excitation, by detecting localized emission of the forbidden band $^1\Delta$ at 1.27 μm [24], was obtained. The most recent laboratory data on faint combination bands of CO_2 were required to interpret the spectra correctly [25].

Jovian aurorae, the H_3^+ ion. Strong H_2 emissions concentrated in an oval near the poles of Jupiter were detected first by the UV spectrometer on board Voyager 1 [26], and were interpreted as a signature of an auroral emission excited by high-energy particles trapped by the Jovian magnetic field, comparable with the phenomenon observed on Earth. To better understand the phenomenon, an observation of the quadrupole lines of H_2 in the 2.2 μm region, which should also be excited in the polar regions, was undertaken with the CFH-FT spectrometer. The 1–0 $S_1(1)$ line in emission, and two other quadrupole lines, were in fact detected [23], with an aperture size of 5 arcsec on the South pole of the planet, at a resolution of 20,000, a compromise to obtain a high enough signal-to-noise ratio in the limited amount of time available (10 min per spectrum, owing to the fast rotation of Jupiter). Because the spectra were recorded over the whole K band, the basic advantage of an FT spectrometer, almost 30 totally unexpected emission lines appeared in the spectra. After several months of theoretical efforts to improve ab initio calculations, they were identified as the $2v_2$ band of the molecular ion H_3^+, which was not yet known in the laboratory, making this detection the first spectroscopic observation of this ion in space [23]. Further laboratory work was undertaken on this band [27], a good illustration of the importance of a fruitful collaboration between the observers and the specialists in molecular spectroscopy.

H_3^+ is the result of the reaction of molecular hydrogen with H_2^+. This ion on Jupiter is produced by the precipitation of the electrons at the poles. With the overtone clearly identified, the detection of the fundamental band v_2 at 4 μm was expected to be easy. The observation provided a pure emission spectrum of this band of H_3^+ (Fig. 2), with 42 lines. Most of the spectrum falls on the top of the P-branch of the v_3 band of CH_4, which is totally saturated on Jupiter, and hence, completely dark, except for the higher J lines. A rotational temperature was derived for the southern and northern zones, of 1000 ± 40 K and 835 ± 50 K respectively [28], consistent with excitation taking place in the range 1–100 nbar pressure in the stratosphere. A thermal population of the band was assumed. To check this

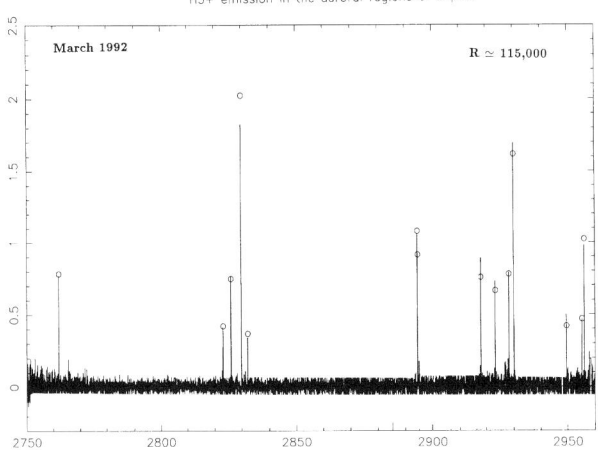

H3+ emission in the auroral regions of Jupiter

March 1992   R $\simeq$ 115,000

Fig. 2. Spectrum of H_3^+ recorded on the southern hemisphere of Jupiter at a resolution of 115000 in order to fully resolve the line profiles. To display the full spectral range the linewidth (FWHM) of $0.037 \, cm^{-1}$ is not appreciable. A fit of the lineshape on the strongest lines gave a kinetic temperature of 1150 K. The open circles represents the intensities of the lines from the best fit obtained for a rotational temperature of 1250 K

assumption an attempt was made to fully resolve the line profiles, which was successful. Again, only the CFH-FT spectrometer was able to provide the 10^5 resolution at $2900 \, cm^{-1}$ over the several hundreds of cm^{-1} needed to resolve the lines [29], allowing direct derivation of the kinetic temperature, which fell in agreement with the rotational temperature. The shape of the H_3^+ spectrum (Fig. 2) also suggested trying a direct imaging of the emission zones by isolating several strong emission lines with a narrow band filter at 3.4 μm, using an infrared camera, an instrument which has become currently available in observatories. The changing aspect of the emission related to the rotation of the planet could be studied. By the association of spectroscopy and imaging, a full appraisal of the Jovian aurorae is emerging [30].

The auroral phenomenon has many implications which test our understanding of the Jovian atmosphere. All the related effects depend on the structure of the magnetic field, the energy of the surrounding particles, the chain of reactions resulting from the precipitation of particles and the stratospheric circulation. H_3^+ appears a crucial probe to study all these parameters.

Circumstellar Shells

The dusty envelopes around late-type stars, resulting from mass loss in cool giants, are abundant sources of

molecules. A good review on the spectroscopy of molecules in this environment was given in the workshop quoted above [18]. The carbon star IRC + 10216 is favourite object for both microwave and infrared spectroscopy. It also represents a good illustration of the fruitful interplay between astronomical and laboratory work. When the high-resolution 13 μm spectra of IRC + 10216 were first recorded with the FT spectrometer on the 4-m Mayall telescope, no laboratory spectra of SiS were available. This detection prompted the production of the vibration–rotation emission spectrum of SiS obtained by reaction of SiS_2 and Si at 1000 °C [31], and its recording with the MacMath FT spectrometer. This also happened with the 4.6 μm absorption spectrum of C_3 [32] which was identified in advance of the laboratory work. The C_3 lines, remarkably strong in the IRC + 10126 spectrum, suggested search for the next member of the carbon chain series, C_5 [33]. All these spectra required very high resolution, only feasible with an FT spectrometer in this spectral range. The C_5 lines are weak (maximum central depth of 3.7%) and very narrow ($< 0.010 \, cm^{-1}$) because they form at the cold edge (40 K) of the circumstellar envelope. However, the astronomical measurements were in excellent agreement with the subsequent laboratory measurements.

These detections of new molecules in the infrared, combined with the detections in the millimetric range, help to enrich chemical models of circumstellar envelopes to make the link with the composition of the interstellar medium, and in particular of molecular clouds, where new stars form.

Young Stellar Objects

The process of star formation is an important topic in modern astrophysics. Detailed studies can be conducted in our own galaxy. Two types of site are distinguished, the Giant Molecular Clouds (the archetype is Orion) and the Dark Cloud Complexes (e.g. Taurus-Auriga). The latter forms exclusively stars with a mass comparable with that of the Sun. High-mass stars ($\geqslant 5 M \odot$) are grouped in clusters of a few units in the core of giant clouds. They are estimated to be of spectral type O or B, with a total luminosity of 10^4–10^5 times the solar luminosity. They are the brightest infrared sources, deeply embedded in molecular clouds, and therefore only accessible in the thermal infrared. From this point of view, they present a special interest for probing the gas and dust content on their line of sight and for studying the interaction of massive stars with their parent cloud in their early phase of evolution. By use of high-resolution absorption spectroscopy with the CFH-FT spectrometer, new results have been obtained on the environment of ten high-mass young stellar objects [34] during several compaigns of observation, associated when possible with sub-millimetric CO mapping [35]. The main part of the vibration–rotation band ($\Delta v = 1$) of ^{12}CO and ^{13}CO (2080–2180 cm^{-1}) was covered with a limit of resolution of 0.05 cm^{-1}, giving the possibility of resolving the line profiles of more than 50 lines simultaneously. With all these data, two regions can be discriminated on the line of sight: the extended molecular cloud, and the stellar neighbourhood, which itself divides into two components, a quiescent high-density shell and fast outflowing gas. Density, temperature and velocity are determined for each region. The results obtained for the molecular clouds are consistent with other determinations, based essentially on millimetric observations. On the contrary, the results on the close environment of the massive sources (radius between 100 and 1000 AU) are completely new, because accessible only by observation in the infrared. A warm quiescent gas, probably heated by gas–grain collisions, is detected for all the sources, with temperatures ranging from 120 to 1000 K, while the average temperature of the clouds is only 45 K. The gas of the molecular cloud blown away in the vicinity of the central star is piled up in a shell around the young object.

The velocity structure of the outflows detected in absorption, reachable by high resolution, implies that the absorbing gas is not the same as that seen in CO millimetre emission lines, which has a bipolar structure centred on the infrared source, with extended lobes. On the contrary, the infrared outflows are close to the source. The relationship between the two classes of flow is a new question which has been opened by high-resolution spectroscopy.

Starting with CO, which is the most common molecule, these sources can be used to search by the same method for other molecules such as NH_3, OCS, HCN, C_2H_2, and the H_3^+ ion. As for the circumstellar envelopes, a complete picture of the chemical composition in the core of the molecular clouds remains to be built by absorption spectroscopy, which is complementary to the millimetric emission observations. A new field has opened in which FT-IR can play an essential role.

New Modes of Operation

Besides the observational activity evoked above, new modes of operation have emerged. With the discovery of the submillimetric and millimetric windows, FT spectrometers for this range have started to be considered. In the infrared, the multiplex properties of these instruments, combined with the frequency stability of the reference laser prompted attempts at detection of pressure oscillations of objects such as planets and stars. With the advent of infrared array detectors another mode exploits the so-called throughput advantage to combine spectroscopy and imagery.

Extension to the Submillimetric Range

In the laboratory, FT spectrometers exist that cover the range from the UV to the far infrared. In astronomy, the domain has been mainly the infrared from 0.9 to 13 μm for the ground-based instruments. Beyond this, a large spectral region is totally blocked and can only be studied from space. However, in the driest sites, new windows at 350, 450, 850 μm and 1.2 and 3 mm have become exploitable. Telescopes devoted to this intermediate range between the infrared and radiowaves have been built in the last decade, on Mauna Kea, in Chile, and in the south of Spain. For the instrumentation of these telescopes the properties of FTS have been rediscovered. The standard instrumentation consist of heterodyne receivers. Ultra-high resolution is possible (10^6) but on very small spectral frequency ranges adapted to study narrow emission lines of well-known frequency one at a time. A survey of a large domain requires another type of instrument. The one most used is an FT spectrometer. The resolution is necessarily much lower but full coverage of the window becomes possible. Already, two new FT spectrometers have been recently installed, one at the Caltech Submillimeter Observatory (CSO) [36] with a limit of resolution of 0.0136 cm^{-1} and the other on the James Clerk Maxwell Telescope (JCMT) with a resolution of 0.0125 cm^{-1} [37].

Seismology of Giant Planets and Solar-type Stars

Thirty years ago, a new branch of astrophysics opened with the detection of the 5-minute oscillations of the Sun. Later, these were correctly interpreted as acoustic oscillations driven by convection in the solar atmosphere. The detailed study of these oscillations, which have been resolved into millions of modes, is a unique tool to determine the internal structure of the Sun. Obviously, it would be of considerable interest to be able to obtain the same kind of information on stars. The giant planets, made essentially of hydrogen and helium, are supposed also to generate pressure oscillations. The dramatic difference of energy between the Sun and the brightest stars or Jupiter has made this objective out of reach. The oscillations correspond to vertical motions of the surface which translate into a Doppler effect on the photosphere lines for a star, and on the absorption lines in solar light diffused by the planetary atmosphere. This effect is very small (20 cm/s per mode for the Sun). To have some chance of positively detecting the pressure oscillations in objects other than the Sun, many modes have to be added; in other words, only a low frequency resolution can be expected. To improve the sensitivity the Doppler effect on a maximum of lines has to be summed. An FT spectrometer, with its multiplex properties, can provide an instrumental solution. By accurate stabilization of path-difference on the slope of fringe feature created by all the stellar lines present in the spectral range admitted in the instrument, a Doppler signal can be generated as a function of time. Its FT contains the oscillation spectrum of the object. Such an experiment was tested on Jupiter in January 1990 with the CFH-FT spectrometer at the time of opposition, and repeated one year later to confirm the results [38]. The R-branch of the $3v_3$ band of CH_4 at 9100 cm^{-1}, which is the only accessible unsaturated band of methane in Jupiter's spectrum, was used. A signal of Jovian origin was positively detected over 4 nights of observation. Modes of l-degree 1 and 2 have been identified, the size of the aperture on the planetary disk (12 arcsec) filtering the higher degrees, which simplifies a spectrum that is already very complex. An analysis of this spectrum has shown a characteristic frequency of 136 μHz, a signature of the discontinuity between a solid nucleus and the envelope of hydrogen and helium, which forms the major part of the planet. Through a model, this value can be related to the depth of the envelope, its density, and the mixing ratio of hydrogen and helium. Current models are unable to explain correctly the observed value, calling for a revision of the internal structure of the planet.

Encouraged by this positive result, the method has been tried also on Saturn, which is 25 times fainter and with the rings hiding a large portion of the planetary surface, making the detection more delicate. There are also projects to test solar-type stars for which the amplitude of oscillations to detect is predictable. It can be noted that the CFH-FT spectrometer was directly usable for this application because of its very accurate positioning servo-system required by the step-by-step data acquisition mode. The interest in this method may justify the building of specialized interferometers, for the new field of asteroseismology, portable to other telescopes to make a network, in order to record without interruption, the Doppler signal on a source over several days.

Spectro-imaging in the 1–2.5 μm Range

The Michelson interferometer and the Fabry–Perot etalon are well-known for their throughput advantage. With the advent of modern array detectors, an image of the input aperture (the focal plane of the telescope) can be made behind the FT spectrometer, on a two-dimensional array, replacing the pair of single detectors normally used. Again, benefiting from the step-by-step mode, a frame is recorded at each step, this being an image of the entrance field at a given path-difference. All the frames form a cube of data, with x and y the coordinates of each point of the field and z the path-difference. The intensities of all the points with the same x and y coordinates in the cube form the interferogram of this source on the sky, which can be transformed to create its spectrum. By this method, the spectra of all the points are recorded simultaneously. This mounting offers an efficient manner of working in the infrared spectro-imaging of extended objects. Many astronomical sources are seen extended at the focus of a telescope – planets (Venus, Mars, Jupiter, Saturn), galactic planetary nebulae, and galaxies – or are made of several close sources, such as star-forming regions, the Galactic Centre and globular clusters. A camera, an HgCdTe array (256 × 256 pixels) – a low noise device that is sensitive in the range 1–2.5 μm – has been coupled to the CFH-FT spectrometer, offering a circular field of 20 arcsec [39]. Among the programs already conducted have been study of the dark side of Venus in the oxygen emission band (see above) and the study of several typical planetary nebulae (PN) [40]. Studies of classical PN (NGC 7027, GL 2688, BD + 30 3639 and CRL 618) were made through a K' filter (2–2.3 μm), covering a region which contains several 1–0, 2–1 H_2 lines, the Brγ line (2.16 μm), HeI and HeII lines. With these lines and the continuum observed in a single scan, the imaging of the various components – the molecular envelope (H_2 lines), the ionized region (Brγ, HeI and HeII), the dust (continuum) – was possible. Through a narrow band filter centred on the strongest H_2 line (2.12 μm) kinematics studies of the molecular envelope were possible at a resolution of 30 km/s. This technique is very promising and fully exploits all the properties of an FT spectrometer.

Conclusion: New Challenges

The few results reported here, and the new applications under development, have shown that an FT spectrometer behind a large telescope, in a good infrared astronomical site can produce unique results in many area of modern astrophysics, provided it is optimized for faint sources and continuously upgraded with new detectors and new computing facilities. However, a new generation of telescopes will bring a revolution in the observational capabilities before the turn of the century. With use of active technologies to control the shape of the primary mirror, telescopes of 8- and 10-m class are already under construction. The Very Large Telescope (VLT) project started by the European Southern Observatory consists of four telescopes of 8-m class in Chile. The Gemini project with the U.S., U.K. and Canada as the principal members of the consortium is based on one 8-m telescope on Mauna Kea and one in Chile. The Subaru telescope is the Japanese 8-m project, also in Hawaii. With increased telescope diameters, a grating spectrometer becomes more and more difficult to adapt. It requires a 0.2 arcsec slit for reaching a resolution of 10^5, which means a dramatic loss of light on a stellar image. Image slicers can solve the problem in the visible, but not in the thermal infrared. New techniques called adaptive optics, which compensate in real-time the effect of turbulence on the shape of the image, are under study, which should be able to approach diffraction-limited images in the infrared. However, to work properly, the corrective system which is absolutely required to reach a high spectral resolution needs a close reference star (a few arcseconds apart) of sufficient brightness. It also contributes noticeably to the emissivity of the focus, which does not favour work in the thermal infrared. At any rate, resolutions higher than 10^5 become impracticable because of the size of the grating to be mounted and of the pieces needing to be cooled to liquid-nitrogen temperature. As a result, no high-resolution infrared echelle spectrograph is currently under construction

for the new generation telescopes. With an FT spectrometers, moderately increasing the diameter of the beams solves the problem of the optical matching, and resolutions beyond the 10^5 limit are easily feasible. With the grating spectrograph, high resolution must be limited to stellar objects. Only the FT spectrometer can combine very high resolution and an angular diameter of several arcsec, which is of interest for many extended objects, from planets to galactic nuclei. Finally, the sensitivity of the FT spectrometer, which is under question, can be improved by going to a multi-channel and cryogenic instrument [41]. This perspective represents a new exciting challenge for astronomical Fourier-transform spectroscopy. That can also be the complete end, since as stated in the introduction, simpler instruments are preferred for the basic instrumentation, especially in a time of tight budgets.

References

[1] J. Connes, P. Connes, *J. Opt. Soc. Am.* **1966**, *56*, 896.
[2] J. Connes, P. Connes, J. P. Maillard, *Atlas des Spectres dans le Proche Infrarouge de Venus, Mars, Jupiter et Saturne*, Editions du Centre National de la Recherche Scientifique, 1969.
[3] G. A. Vanasse, A. T. Stair, D. J. Baker (eds.), *Aspen International Conference on Fourier Spectroscopy, 1970*, AFCRL 71-0019, Special Reports No. 114, 1971.
[4] J. P. Maillard, in: *High Resolution Spectrometry* (M. Hack, ed.), 4th *Trieste International Colloquium in Astrophysics*, 1978, p. 108.
[5] S. T. Ridgway, J. W. Brault, *Ann. Rev. Astron. Astrophys.* **1984** *22*, 291.
[6] C. de Bergh, *Mikrochim. Acta* **1987**, *III*, 141.
[7] J. W. Brault, *Ossni. Mem. Astrofis. Arcetri* **1979**, *106*, 33.
[8] D. N. B. Hall, S. T. Ridgway, E. A. Bell, J. M. Yarborough, *Proc. SPIE*, **1979**, *172*, 121.
[9] D. S. Davis, H. P. Larson, M. Williams, G. Michel, P. Connes, *Appl. Opt.* **1980**, *19*, 4138.
[10] J. P. Maillard, G. Michel, in: *Instrumentation for Astronomy with Large Optical Telescopes* (C. M. Humphries, ed.), IAU Colloquium No. 67, Reidel, Dordrecht, 1982, p. 213.
[11] R. Beer, *Remote Sensing by Fourier Transform Spectroscopy*, Wiley–Interscience, New York, 1991, p. 104.
[12] G. Guelachvili, J. P. Maillard, in: *Aspen International Conference on Spectroscopy, 1970* (G. A. Vanasse, A. T. Stair, D. J. Baker, eds.), AFCRL 71–0019, Special Reports No. 114, 1971, p. 151.
[13] P. Connes, G. Michel, *Appl. Opt.* **1975**, *14*, 2067.
[14] J. P. Maillard, in: *The Impact of Very High S/N Spectroscopy on Stellar Physics* (G. Cayrel de Strobel, M. Spite, eds.), IAU Symposium No. 132, Kluwer, Dordrecht, 1988, p. 71.
[15] J. Connes, P. Connes, J. P. Maillard, *J. Physique* **1967**, *28* Colloque C2, 136.
[16] M. Geller, in: *Laboratory and Astronomical High Resolution Spectra* (A. J. Sauval, R. Blomme, N. Grevesse, eds.), ASP Series, Vol. 81, 1995, p. 88.
[17] R. L. Kurucz, in: *Laboratory and Astronomical High Resolution Spectra* (A. J. Sauval, R. Blomme, N. Grevesse, eds.), ASP Series Vol. 81, 1995, p. 17.

[18] K. H. Hinkle, in: *Laboratory and Astronomical High Resolution Spectra* (A. J. Sauval, R. Blomme, N. Grevesse, eds.), ASP Series Vol. 81, 1995, p. 482.

[19] J. P. Maillard, in: *Laboratory and Astronomical High Resolution Spectra* (A. J. Sauval, R. Blomme, N. Grevesse, eds.), ASP Series Vol. 81, 1995, p. 343.

[20] T. Owen, J. P. Maillard, C. de Bergh, B. L. Lutz, *Science* **1988**, *240*, 1767.

[21] C. de Bergh, B. Bézard, T. Owen, D. Crisp, J. P. Maillard, B. L. Lutz, *Science* **1991**, *251*, 547.

[22] B. Bézard, C. de Bergh, D. Crisp, J. P. Maillard, *Nature* **1990** *345*, 508.

[23] P. Drossart, J. P. Maillard, J. Caldwell, S. J. Kim, J. K. G. Watson, W. A. Majewski, J. Tennyson, S. Miller, S. K. Atreya, J. T. Clarke, J. H. White, Jr., R. Wagener, *Nature* **1989**, *340*, 539.

[24] J. P. Maillard, B. Bézard, L. Domisse, D. Crisp, D. Simons, *Bull. Am. Astron. Soc.* **1993**, *25*, 1095.

[25] J. B. Pollack, J. B. Dalton, D. Grinspoon, R. B. Wattson, R. Freedman, D. Crisp, B. Bézard, C. de Bergh, L. P. Giver, Q. Ma, R. Tipping, *Icarus* **1993**, *103*, 1.

[26] A. L. Broadfoot, M. J. S. Belton, P. Z. Takacs, B. R. Sandel, D. E. Shemansky, J. B. Holberg, J. M. Ajello, S. K. Atreya, T. M. Donahue, H. W. Moos, J. L. Bertaux, J. E. Blamont, D. F. Strobel, J. C. McConnell, A. Dalgarno, R. Goody, M. B. McElroy, *Science* **1979**, *204*, 979.

[27] W. A. Majewski, P. A. Feldman, J. K. G. Watson, S. Miller, J. Tennyson, *Ap. J.* **1989**, *347*, L51.

[28] J. P. Maillard, P. Drossart, J. K. G. Watson, S. J. Kim, J. Caldwell, *Ap. J.* **1990**, *363*, L37.

[29] P. Drossart, J. P. Maillard, J. Caldwell, J. Rosenqvist, *Ap. J.* **1993**, *402*, L25.

[30] S. J. Kim, P. Drossart, J. Caldwell, J. P. Maillard, T. Herbst, M. Shure, *Nature* **1991**, *353*, 536.

[31] C. I. Frum, R. Engleman, Jr., P. F. Bernath, *J. Chem. Phys.* **1990**, *93*, 5457.

[32] K. W. Hinkle, J. J. Keady, P. F. Bernath, *Science* **1988**, *241*, 1319.

[33] P. F. Bernath, K. H. Hinkle, J. J. Keady, *Science* **1989**, *244*, 562.

[34] J. P. Maillard, G. F. Mitchell, in: *Astrochemistry of Cosmic Phenomena* (P. Singh, ed.), IAU Symposium No. 150, Kluwer, Dordrecht, 1992, p. 259.

[35] G. F. Mitchell, S. W. Lee, J. P. Maillard, H. Matthews, T. I. Hasegawa, A. I. Harris, *Ap. J.* **1995**, *438*, 794.

[36] E. W. Weisstein, E. Serabyn, *Icarus* **1994**, *109*, 367.

[37] D. A. Naylor, T. A. Clark, G. R. Davis, W. D. Duncan, G. J. Tompkins, *MNRAS* **1993**, *260*, 875.

[38] B. Mosser, D. Mekarnia, J. P. Maillard, J. Gay, D. Gautier, P. Delache, *Astron. Astrophys.* **1993**, *267*, 604.

[39] J. P. Maillard, in: *Tridimensional Optical Spectroscopic Methods in Astrophysics* (G. Comte, M. Marcelin, eds.), IAU Colloquium No. 149, ASP Conf. Series Vol. 71, 1995, p. 316.

[40] P. Cox, J. P. Maillard, P. J. Huggins, T. Forveille, D. A. Simons, F. Rigaut, R. Bachiller, S. Guilloteau, A. Omont, *Astron. Astrophys.* 1996, accepted.

[41] J. P. Maillard, in: *High Resolution Spectroscopy with the VLT* (M. H. Ulrich, ed.), ESO Conference No. 40, 1992, p. 239.

Mikrochim. Acta [Suppl.] 14, 143–147 (1997)
© Springer-Verlag 1997

Automatic Interpretation of Infrared Spectra

Christian Affolter[1,*], Knut Baumann[1], Jean-Thomas Clerc[1], Hans Schriber[2], and Ernö Pretsch[2]

[1] Pharmaceutical Institute, University of Bern, Baltzerstr. 5, CH-3000 Bern, Switzerland
[2] Department of Organic Chemistry, Swiss Federal Institute of Technology (ETH), Universitätstr. 16, CH-8092 Zürich, Switzerland

Abstract. Normal frequency ranges tabulated for large fragments of organic molecules have been tested against IR spectra collected in a spectroscopic database. The results show that the correlation tables cannot be reliably used for automatic spectra interpretation. On the other hand, vibrational spectra of an organic compound can be predicted on the basis of a spectroscopic database by making a structure similarity search.

Key words: spectra prediction, database, computer-aided interpretation.

There is an ever increasing need for automatically processing spectral data for the elucidation or confirmation of chemical structures. If the unknown compound is recorded in a spectroscopic database, a simple library search might be sufficient for its identification, but other cases have to be solved by spectra interpretation. One approach is to generate, with the help of a program, all possible isomers consistent with the molecular formula [1, 2]. Those structures that are compatible with all experimental spectra are then potential solutions to the problem. To reduce the huge search space, substructures and other constraints can be applied in the structure generation process, using both substructures required to be present ("goodlist") and known to be absent ("badlist") [3, 4]. The power of these two kinds of information in reducing the search space depends on the size of the fragments, the number of their free valences and the kind of atoms and bonds included. Goodlist entries are especially efficient in reducing the search space if they consist of large fragments that contain only a small number of free valences and atoms of low occurrence in the molecular formula. The opposite holds for badlist items, in that small

fragments with many free valences and atoms of high occurrence are most useful (see Fig. 1). Since IR spectroscopy has a special power for recognizing larger substructures with few free valences, the chances of identifying relevant fragments could be expected to be better for goodlists than for badlists.

It is tempting to use spectra interpretation programs to automatically generate entries for goodlists and/or badlists. Tabulated IR frequency ranges are available for a series of functional groups [5–7] and are widely employed as interpretation aids. Computer programs attempting to automatically interpret spectra on the

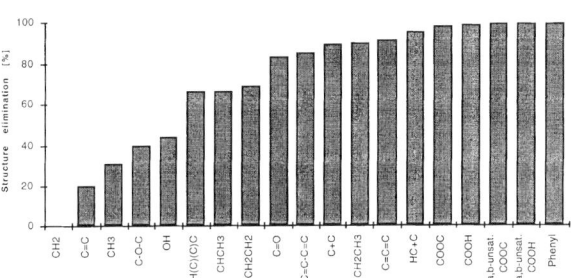

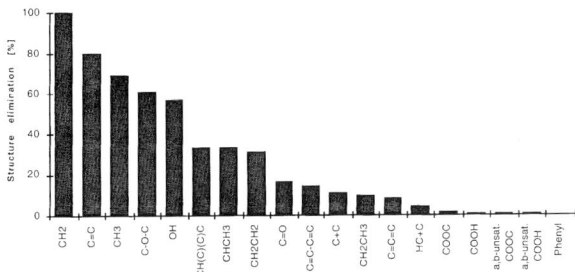

Fig. 1. Power of various goodlist (top) and badlist (bottom) fragments in reducing the number of generated constitutions of molecular formula $C_8H_{10}O_2$ (total number, 607,376 [16]). Note that the goodlist and badlist are complementary

---

* To whom correspondence should be addressed

basis of such ranges have already been developed [8]. However, the practical applicability of such expert systems, which try to mimic the thinking of a chemist, has remained very limited. For an automatic interpretation, all possible effects of surrounding groups should be considered in order not to lose valid structures. This would entail very broad and therefore non-selective frequency ranges.

Recently, a collection of normal vibrations of a series of large, mostly univalent fragments has become available [9] on the basis of which the prediction of IR spectra of whole molecules by combining the vibration frequencies of their substructure units is proposed. Therefore, rather large terminal groups of up to 14 atoms were selected for which it could be assumed that their IR frequencies are but slightly affected by the rest of the molecule and hence may be considered as generally valid. One could expect that such groups are ideally suited as goodlist entries. Since for each functional group, typically 5–10 frequency ranges are tabulated, correlations based on them can be assumed to be much more selective than those using just one or two frequency intervals. One purpose of this work was to test the applicability of these correlation tables for an automatic compilation of goodlists and/or badlists as input for structure generators.

A different strategy is applied when the structure generation process has ended. The complete structures are used to predict the spectra, which are then compared with the experimental ones and ranked according to their similarity. Though quantum mechanical methods can predict vibrational spectra, the actual computational demands for reliable estimates are still prohibitively large in realistic cases [10, 11]. In the second part of the paper, a simple method for predicting IR spectra is proposed. It is based on a previously developed structure descriptor [12, 13] further optimized for this application. The fundamental idea is to encode a chemical structure as a vector, search for structurally similar entries in a reference library and use their spectra as a basis for the estimation. Spectrum prediction thus becomes a kind of library search.

Results and Discussion

Applicability of Tabulated Frequency Ranges

In order to investigate the applicability of tabulated normal vibration frequency ranges [9] to a great number of compounds, 62 terminal fragments with 3–8

atoms attached to a defined node were selected and tested against the corresponding database entries. Altogether 20, 257 structures containing one of the selected fragments were studied. The SpecInfo Database, Version 2.1 [14] consisting of 20,075 IR spectra of 18,356 compounds served as reference. The automatic peak-picking algorithm of this database was used together with substructure searches applying standard SQL commands of the underlying database program Sybase [15].

The 62 selected fragments are characterized by 5–12 frequency ranges. In 51% of all cases, at least one of the predicted ranges proved to be empty. Since for 23 of the fragments fewer than 20 reference structures were available, the study was continued with the remaining 39. For 26 of these, over 50% lacked the predicted band in one or more regions (cf. Table 1). Usually, these bands were found missing quite evenly from all the regions involved (cf. Fig. 2). In only one case (SO_2), all 126 entries satisfied the tabulated requirements. As an illustration, the spectra of a selection of aldehydes are shown in Fig. 3 together with the predicted ranges.

Table 1. Substructure fragments, number of occurrences in the database and of cases lacking at least one IR band in the expected range of normal vibrations

Fragment	Occurrence	Rule violated	Percentage
$BR-CH_2-(CO)$	26	12	46.3
$Cl-CH_2-(\langle sp^3 \rangle)$	178	132	74.2
$Cl-CH_2-(C=C)$	20	19	95.0
$Cl-CH_2-(CO)$	79	41	51.9
$H-\overset{\overset{O}{\|\|}}{C}-(\langle\cdot\rangle)$	153	21	13.7
$H-\overset{\overset{O}{\|\|}}{C}-(O)$	39	27	69.2
$HO-CH_2-$	620	193	31.1
$H_2C=C\langle^{(\langle sp^3 \rangle)}_{(\langle sp^3 \rangle)}$	70	60	85.7
$H_2C=CH-(\langle sp^3 \rangle)$	455	382	84.0
$H_2C=CH-(O\ or\ S)$	67	41	61.2
$H_2N-(CO)$	190	53	27.9
$H_2N-(CS)$	32	10	31.3
$H_2N-(\langle\cdot\rangle)$	1291	563	43.6
H_2N-CH_2-	95	73	78.6
$H_2N-\overset{\overset{O}{\|\|}}{C}-(\langle\cdot\rangle)$	56	43	76.8

(*continued*)

Table 1. (*continued*)

Fragment	Occurrence	Rule violated	Percentage
$H_3C-(CH_2)$	3271	1211	37.0
$H_3C-((CH_2CH_2)$	1421	1108	78.0
$H_3C-(CH=)$	2466	974	39.5
H_3C—	1857	1147	61.8
$H_3C-(CO-\langle sp^3\rangle)$	635	442	69.6
$H_3C-CO-\langle sp\rangle$ or $\langle sp^2\rangle)$	232	174	75.0
$H_3C-(CON\langle sp^3\rangle)$	210	203	96.7
$H_3C-(CO——)$	115	100	87.0
H_3C (COO)	270	207	76.7
$H_3C-(NH)$	138	90	65.2
$H_3C-(NHCO)$	108	51	47.2
$H_3C-(O-\langle sp^3\rangle)$	414	252	60.9
$H_3C-(O-\langle sp\rangle$ or $\langle sp^2\rangle)$	1430	886	62.0
$H_3C-(O——)$	996	701	70.4
$H_3C-(OCO)$	1131	471	41.6
$H_3C-(S)$	123	73	59.3
$H_3C-(SO_2)$	65	32	49.2
$H_3C-\overset{O}{\overset{\|}{C}}-NH-$	184	161	87.5
H_3C-NH-	66	61	92.4
$H_3C-NH-(CO)$	108	67	62.0
$H_3C-NH-\overset{O}{\overset{\|}{C}}-$	108	68	63.0
$H_3C-O-CH_2-$	131	116	88.5
O_2N-	1131	166	14.7
$-SO_2-$	126	0	0.0

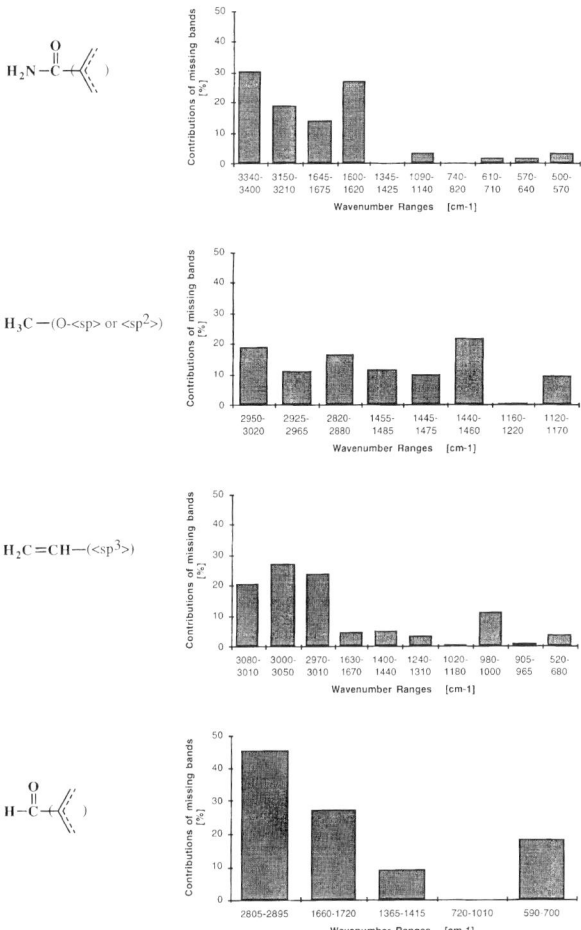

Fig. 2. Percentage of compounds that do not show IR bands in the ranges predicted for four selected fragments

There could be several reasons for this surprising result. (1) The signal is present in the spectrum but lies outside the predicted wavenumber range. This was the most common case ($>94\%$) except for OH- and NH-containing compounds with broad absorption bands. (2) The signal overlaps with a broad neighbouring band. Generally, this occurred in around 4% of the cases investigated, but for compounds with an OH or NH group, band overlaps were responsible for 30% of the failures. (3) The signal is missing owing to errors in the database. Fewer than 2% of the failures were due to this effect. The percentages given have been estimated on the basis of manually investigating some 200 cases.

It has to be noted that bands outside the predicted regions cause a fragment to be mistaken as a badlist item, thereby preventing the structure of the unknown to be assembled. This error would not have such fatal consequences when a goodlist rule is applied, because no statement is then made about the corresponding fragment. On the other hand, for the automatic generation of goodlist entries from IR spectra, it is important to know how many components exhibit the required bands in all regions predicted for a given group, without it being present in their structure. This question, however, has not been investigated here because IR spectroscopy is not suitable on its own for reliably predicting the presence of substructures.

Spectra Prediction

In order to predict IR spectra, chemical structures are represented as vectors of length n. These vectors are

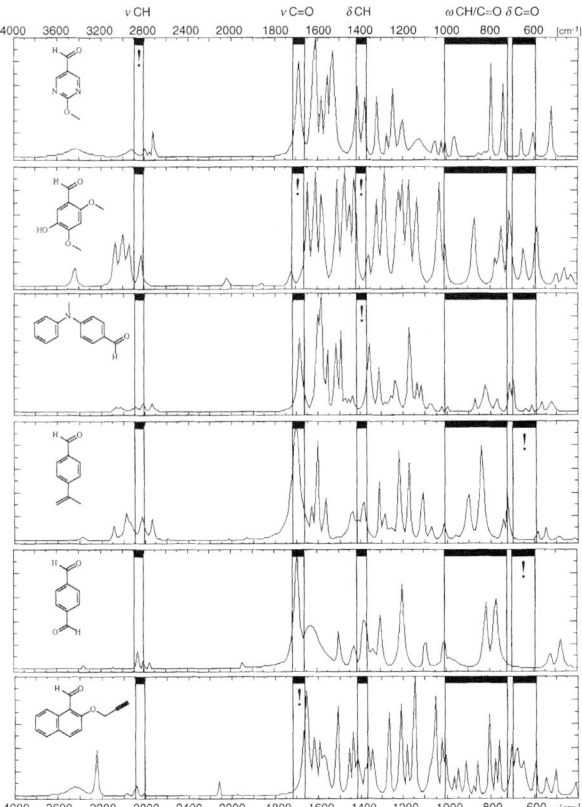

Fig. 3. Predicted frequency ranges (broad bars) in the IR spectra of six compounds with an aldehyde group. Exclamation marks indicate lack of a predicted band

based on fragments, topology and geometry and span an *n*-dimensional space, the so-called structure space. All structures contained in a spectroscopic reference databse are mapped into this space (cf. Fig. 4, left). Likewise, the spectra encoded as *m* data points are mapped into an *m*-dimensional spectra space (cf. Fig. 4, right). To predict a spectrum, the structure in question is projected into the structure space and its nearest neighbours are selected. Then, the spectra of these compounds are identified in the spectra space and used as a basis for predicting the spectrum of interest. Thus, spectrum prediction is basically performed by interpolating between selected spectra, taking into account the distances between the corresponding structures in the structure space. A series of examples for predicted and observed spectra, together with the respective distances (D) in the structure space, are shown in Fig. 5. In the present preliminary version, the structure descriptor is based on topology only, and instead of interpolation, the spectrum of the nearest neighbour is shown without any processing.

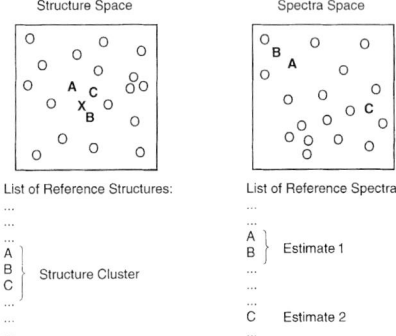

Fig. 4. Schematic representation of the structure (left) and spectra (right) spaces. In the former, the database entries A, B and C are the closest ones to the target structure X, whereas in the spectra space they form two clusters, each of them corresponding to an estimate of the spectrum of X

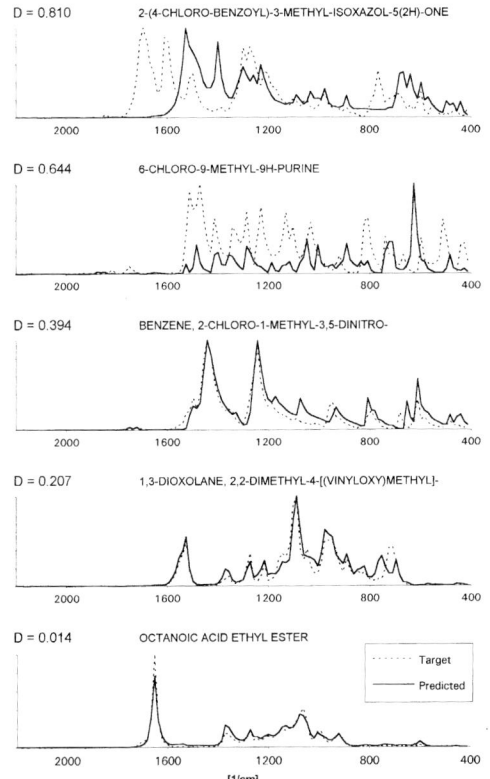

Fig. 5. Comparison of predicted and observed spectra for different distances D between target and reference structures in the structure space. Unreliable predictions (extrapolations) are signalled by large values of D

The main advantage of the approach described is its ability to provide information about the reliability of the estimate. Cases of extrapolation are immediately recognized by an excessive distance in the structure space between the target and its nearest neighbour (cf. Fig. 5, top). Neighbours in the structure space may map into distinct groups in the spectra space (cf. Fig. 4), each group generating another solution. If the structure neighbours map into widely scattered points of the spectra space, the target is not adequately represented and no reliable prediction is possible. Thus, the system is aware of its limitations and no meaningless results are produced.

Various improvements of the current implementation are envisaged. In particular, the structure representation will be further developed by including three-dimensional geometric descriptors and taking into account the peculiarities of IR spectroscopy. There is also room for improving the interpolation algorithms.

Conclusion

It has been shown that for the automatic interpretation of IR spectra, it is rather risky to use tabulated substructure–subspectra correlations, even if they are based on a series of normal frequencies of large terminal fragments. The applicability of spectra predictions based on a structure similarity search has been demonstrated, in a preliminary implementation.

Acknowledgements. This work was partly supported by the Swiss National Science Foundation. The authors thank Dr. D. Wegmann for careful reading of the manuscript.

References

[1] N. A. B. Gray, *Computer-Assisted Structure Elucidation*, Wiley, New York, 1986.

[2] B. D. Christie, M. E. Munk, *J. Chem. Inf. Comput. Sci.* **1988**, *28*, 87.

[3] C. A. Shelley, M. E. Munk, *Anal. Chim. Acta* **1981**, *133*, 507.

[4] S. Sasaki, I. Fujiwara, H. Abe, T. Yamasaki, *Anal. Chim. Acta* **1980**, *122*, 87.

[5] N. B. Colthup, L. H. Daly, S. E. Wiberley, *Introduction to Infrared and Raman Spectroscopy*, 3rd Ed., Academic Press, Boston, 1990.

[6] D. Lin-Vien, N. B. Colthup, W. G. Fateley, J. G. Grasselli, *The Handbook of Infrared and Raman Characteristic Frequencies of Organic Molecules*, Academic Press, Boston, 1991.

[7] A. Gloor, M. Cadisch, R. Bürgin Schaller, M. Farkas, T. Kocsis, J. T. Clerc, E. Pretsch, R. Aeschimann, M. Badertscher, T. Brodmeier, A. Fürst, H.-J. Hediger, M. Junghans, M. Kubinyi, M. E. Munk, H. Schriber, D. Wegmann, *SpecTool: A Hypermedia Book for Structure Elucidation of Organic Compounds with Spectroscopic Methods*, Chemical Concepts, D-69442 Weinheim, 1994.

[8] H. B. Woodruff, G. M. Smith, *Anal. Chem.* **1980**, *52*, 2321.

[9] N. P. G. Roeges, *A. Guide to the Complete Interpretation of Infrared Spectra of Organic Structures*, Wiley, Chichester, 1994.

[10] P. Pulay, G. Fogarasi, X. Zhou, P. W. Taylor, *Vib. Spectrosc.* **1990**, *1*, 159.

[11] R. Herges, U. Weigel, *Anal. Chim. Acta*, submitted.

[12] J. T. Clerc, A. L. Terkovics, *Anal. Chim. Acta* **1990**, *235*, 93.

[13] C. Affolter, J. T. Clerc, *Chemom. Intell. Lab. Syst.* **1993**, *21*, 151.

[14] SpecInfo, Chemical Concepts GmbH, P.O. Box 100202, D-69442 Weinheim.

[15] Sybase, Sybase Inc., 6475 Christie Ave., Emeryville, CA 94608, USA.

[16] A. Kerber, R. Laue, D. Moser, *Anal. Chim. Acta* **1990**, *235*, 221.

Mikrochim. Acta [Suppl.] 14, 149–156 (1997)

Time-Resolved FT Emission Spectroscopy and its Application in Molecular Spectroscopy and Kinetics

Hai-Lung Dai

Department of Chemistry, University of Pennsylvania, Philadelphia, PA 19104–6323, U.S.A.

Abstract. With a 10-cm interferometer running in the step-scan mode, the following characteristics can be achieved for a time- and frequency-resolved emission spectrum: 25 ns and 0.25 cm^{-1} in the visible/UV and 500 ns (detector response-time limited) and 0.1 cm^{-1} in the IR. The time resolution is achieved by arresting the movable mirror and recording the time-dependent interferometric signal with a transient digitizer, followed by the same procedure at each point along the interferometer axis. The signal at a selected time slice will then be assembled for Fourier transform into a spectrum. The ability to record an emission spectrum with time resolution is useful in many studies in molecular spectroscopy and kinetics. In our laboratory, the following advances have recently been made: (1) The FT emission technique in the visible/UV can be used as an effective means for dispersing fluorescence. This technique is particularly advantageous in detecting weak fluorescence from small molecules and transient species. (2) Time-resolved dispersed fluorescence spectra following the initial preparation of a single excited rovibronic level allow straightforward deduction of state-to-state energy transfer rates and mechanism. (3) The emission in the IR from highly vibrationally excited molecules can be used to deduce the energy content of the emitting molecules. Following the preparation of molecules excited with a specific high vibrational energy by a laser pulse, the energy content of the excited molecules during a collisional deactivation process can be monitored by using time-resolved FT-IR emission spectroscopy.

Key words: time-resolved, IR emission, dispersed fluorescence, energy transfer.

Temporally and spectrally resolved fluorescence spectra from samples excited by a pulsed light source can generate much valuable information such as excited-state lifetimes, energy transfer and reaction rates and mechanisms, and ground-state molecular rotational and vibrational constants that allow deduction of molecular structures and structure-related chemical properties. Acquisition of such information, particularly for transient species such as radicals, reactive intermediates, molecular ions, weakly bound molecular complexes, and highly excited molecules, constitutes the bulk of present day studies in molecular spectroscopy and kinetics.

Many approaches, using various kinds of dispersion instruments or multiple resonance laser spectroscopic techniques, have been developed for recording time- and frequency-resolved emission spectra. An efficient and straightforward technique based on Fourier-transform spectrometers has been proven useful for such purposes [1–22]. This time-resolved Fourier-transform emission spectroscopy (TR-FTES) technique is easily accessible for emission ranging from the infrared to the visible and ultraviolet. As a time-resolved spectroscopic technique, TR-FTES has many attributes: a large portion of the emission spectrum can be recorded in one experiment, high spectral resolution is obtainable in all spectral regions, it is a zero-background technique, the complete time behavior of the system can be measured in one experiment, accurate absolute frequencies are measured without the need for calibration, it has a throughput advantage compared to grating spectrometers with similar optics, and a multiplex advantage when the signal-to-noise ratio of the system is limited by the noise of the detector.

There are two different experimental approaches to performing time-resolved Fourier-transform spectroscopy. In the "record-on-the-flight" approach the movable mirror of the interferometer is scanned at a constant rate. The production of the transient event, and the subsequent data collection, are synchronized to the mirror position by monitoring with a photodiode the interference pattern of an He–Ne reference laser that passes along the axis of the interferometer. As the mirror moves, the photodiode signal varies cosinusoidally, and when it is zero – the "zero-crossing" of the He–Ne laser – a trigger pulse is electronically generated to trigger the excitation system and the data collection. When the interferometer mirror can be made to move slowly relative to the transient process being investigated, then the transient process can be initiated and digitized at each He–Ne zero-crossing, and a complete interferogram can be obtained in a single mirror sweep. However, if the time-scale of the transient process is comparable to, or longer than, the time taken for the interferometer mirror to move between He–Ne zero-crossings, then a more sophisticated interleaved sampling technique must be used. A detailed discussion of this procedure can be found in [10, 12, 13].

The second approach to time-resolved Fourier transform spectroscopy, which we adopt, is to use a step-scan interferometer [4, 5, 14, 15, 18]. In this "arrest-and-record" scheme the interferometer mirror is moved to a set position – referenced by the zero-crossings of an He–Ne laser – and kept stationary while the transient event is initiated and data are collected and averaged. The mirror is then moved to the next position and the process is repeated. In this way a complete interferogram, with averaging, is obtained in one mirror sweep. This technique is less prone to artifacts, as any variation in the signal over the course of the scan will only affect the signal-to-noise ratio of the resulting spectrum, and will not introduce spurious or shifted peaks. The step-scan approach also allows more flexibility in the time-scale of the experiment; specifically, because the interferometer mirror can be held indefinitely at an He–Ne zero-crossing there is no upper limit on the time-scale of the transient process that can be investigated. The lower limit for the time-scale is determined by the detector, the digitizing electronics of the spectrometer, and the pulsed system used to initiate the transient event.

The operation of our step-scan FT spectrometer is described in the next section. The subsequent sections contain examples of its application in spectroscopic and kinetic studies of a transient radical species and highly vibrationally excited molecules.

Step-scan, Time-resolved Fourier-transform Emission Spectroscopy

The apparatus and the operating procedure of the TR-FTES technique have been described elsewhere [18]. The operating principles and necessary details are laid out here. The Fourier-transform spectrometer

used in our experiments was a Bruker IFS-88 equipped with step-scan and time-resolved capabilities. The interferometer can be run in both step-scan operation, for time-resolved measurements, and continuous scan operation for steady-state measurements. A diagram of the experimental apparatus is shown in Fig. 1.

Light from the gas cell is collected with a 2″ diameter, 4″ focal length lens (L1) and then focused into the IFS-88 spectrometer with a 2″ diameter, 8″ focal length lens (L2), to match the f/4 focussing characteristics of the spectrometer. Before passing through the interferometer, the collected light is focused onto an aperture (A) in the spectrometer to reduce the divergence of the light beam through the interferometer optics (such divergence ultimately limits the maximum obtainable resolution).

The spectrometer contains a classical Michelson interferometer, which can be switched under software control from the conventional rapid scan mode into the step-scan mode. The interferometer consists of a quartz beam-splitter (BS), a fixed mirror (FM), and a movable mirror (MM) mounted on the slide of a rectangular air-bearing with approximately 4 cm distance of travel (the largest achievable optical path difference is twice this). The movable mirror is driven by a linear motor attached to the air-bearing slide. The movement of the movable mirror is tracked by

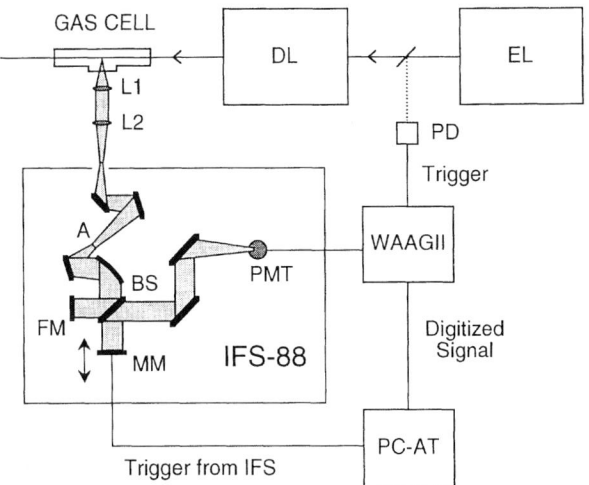

Fig. 1. Schematic of the experimental apparatus: DL and EL are the dye and the excimer lasers, respectively; L1 and L2 are lenses; A is an adjustable aperture; BS is a beam splitter; FM and MM are the fixed and the moving mirror of the interferometer, respectively; PMT is a photomultiplier tube; PD is a fast photodiode; WAAGII is the transient digitizer; PC-AT is an AT compatible personal computer. (Reproduced from [18] by permission of the copyright holders. Copyright 1992, American Institute of Physics.)

monitoring the interference signals from an He–Ne laser that is passed down the axis of the interferometer.

In step-scan mode the movable mirror can be positioned at optical path differences of multiples of one-half the He–Ne reference laser wavelength ($0.316\,\mu m$) by locating the zero-crossings in the laser interference pattern: this is achieved by microprocessor control of the linear motor that drives the movable mirror of the interferometer. The positioning accuracy of this interferometer system has been measured as ± 1.1 nm. The time taken for the mirror to stabilize at a zero-crossing – the "settling time" – is approximately 25 msec; however, the actual time between steps also depends on the distance travelled, so is slightly greater than the settling time.

The emission from the molecular system, triggered by a laser pulse, is passed through the spectrometer and focused onto a photomultiplier (for visible or UV light), an MCT detector (for IR light), or a silicon photodiode (for visible and near IR light). The transient signal from the detector is then treated by a transient digitizer. The operation of the IFS-88 spectrometer and the data collection are controlled by a Dell 325D, AT-compatible, personal computer (PC-AT) equipped with a Bruker acquisition processor board and OPUS™ software. Data acquired with the transient digitizer are subsequently rearranged and Fourier-transformed by the acquisition processor and displayed.

During a TR-FTES experiment, when the interferometer mirror is stepped to a new position (a zero-crossing of the He–Ne reference laser) the control electronics of the IFS-88 spectrometer check the mirror position. Once the mirror has settled to the He–Ne zero-crossing a trigger is sent to the PC-AT to start data acquisition. The digitized signal from the transient digitizer is then collected for a given number of laser shots (for signal averaging) while the mirror position is held constant, and is stored on disk.

After averaging of the laser shots, an $n \times m$ array is obtained in which the columns correspond to the times $t_1, t_2, \ldots t_m$, and the rows to the mirror positions $x_1, x_2, \ldots x_n$ – see Fig. 2. For example, the ith row of the array – $I(x_i, t_1), I(x_i, t_2), \ldots I(x_i, t_m)$ – contains the transient signal at times $t_1, t_2, \ldots t_m$, for mirror position x_i. Fourier-transforming the kth column of the array, $I(x_1, t_k), \ldots I(x_n, t_k)$, gives the emission spectrum at time $t_k, I(\nu_1, t_k), I(\nu_2, t_k), \ldots I(\nu_p, t_k)$. The spacing of the frequency elements, $\nu_1, \nu_2, \ldots \nu_p$, is determined by the length of the interferogram and the amount of zero-

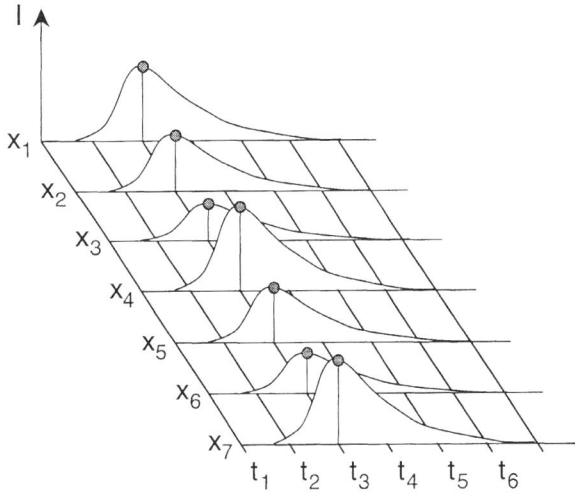

Fig. 2. Representation of the data array collected during a time-resolved Fourier transform spectroscopy experiment. At each mirror position, x_i, the transient signal is digitized at times $t_1, t_2, t_3, \ldots t_m$. The signal can be averaged over many events. After the scan is completed, the signal as a function of mirror position ($x_1, x_2, x_3, \ldots x_n$) at each time slice, represented here by the filled circles at t_3, is Fourier-transformed to give the spectrum at that time. (Reproduced from [18] by permission of the copyright holders. Copyright 1992, American Institute of Physics.)

filling used in the Fourier-transformation; the number of frequency elements is determined by the required spectral range.

Fourier-Transform Dispersed Fluorescence Spectroscopy of Singlet Methylene

It was widely recognized in the last decade that by examining the downward transitions from an electronically excited state, highly excited vibrational levels of the electronic ground state can be studied. These downward transitions can be detected by dispersing the spontaneous emission from the electronically excited state. However, conventional detection of dispersed fluorescence by using grating spectrometers is hard pressed to provide the kind of resolution and sensitivity that is desirable.

Grating spectrometers allow only photons within the selected bandwidth to be detected. On the other hand, by use of a Fourier-transform spectrometer to resolve the spectrum, all the photons are detected in the interferogram. This "multiplex" advantage has been demonstrated in IR emission studies where the detector noise is dominated by the background. For emission studies in the UV or visible, when a photomultiplier tube (PMT) is used, photon shot-to-shot noise, PMT dark current, and laser fluctuations all contribute to the signal noise. There is no multiplex advantage to using a Fourier-transform spectrometer for the shot-to-shot noise and laser intensity fluctuations, but a multiplex advantage for the PMT dark-current noise still exists. This is significant for low intensity fluorescence detection. In addition, the throughput advantage of FT spectrometers over grating spectrometers means that dispersed fluorescence can be detected with much higher sensitivity.

The first demonstration of these advantages was reported in a study of the high vibrational levels of CH_2 in the lowest singlet $\tilde{a}^1A_1$ state [23]. Spectra obtained showed that several previously unknown vibrational levels in the 5000–7500 cm^{-1} region of $\tilde{a}^1A_1$ CH_2, including the $v_2 = 5$ level, can be detected with excellent signal/noise ratios for a sample with < 10 mTorr CH_2 partial pressure, with only 10^5 laser pulses.

In the experiment, $\tilde{a}^1A_1$ CH_2 was produced by pulsed photolysis of ketene at 308 nm. The photolysis pulse was supplied by half the output of a Lambda Physik 210i excimer laser (360 mJ pulse energy, 30 Hz repetition rate). A single rotational transition in the CH_2 $\tilde{b} \leftarrow \tilde{a}$ 2_0^{16}, 2_0^{18}, or 2_0^{19} bands was excited by a Lambda Physik 3002 dye laser (0.3 cm^{-1} bandwidth, 30 mJ pulse energy at 500 nm, and 20 ns pulse duration). The dye laser was pumped by the other half of the excimer pulse and optically delayed by ca. 20 ns to allow for partial thermalization of the rotationally and vibrationally hot $\tilde{a}^1A_1$ CH_2 produced from ketene photolysis. Ketene was flowed continuously through the gas cell to avoid build-up of reaction products during an experiment, and the total pressure was normally maintained at approximately 0.8–0.9 Torr.

FTES spectra obtained for the CH_2 $\tilde{b} \rightarrow \tilde{a}$ band are shown in Fig. 3. In Fig. 3a emission from the CH_2 $\tilde{b}^1B_1(0, 18^0, 0)0_{00}$ level was recorded with a 3949 cm^{-1} bandwidth and 2 cm^{-1} resolution. This level was excited through the $\tilde{b} \leftarrow \tilde{a} 2_0^{18}$ $^PP_{1,0}(1)$ transition. Emission to the (0, 4, 0), (1, 2, 0), (2, 0, 0), (0, 5, 0) and (1, 3, 0) vibrational levels of the CH_2 $\tilde{a}^1A_1$ state can be seen and assigned in the spectra. Each of these bands consists of a single rotational transition: $0_{00} \rightarrow 1_{10}$. In Fig. 3b an FTES spectrum obtained after excitation of the CH_2 $\tilde{b}^1B_1 (0, 19^1, 0)$ 4_{13} level is shown. This spectrum was recorded with 1 cm^{-1} resolution and a bandwidth of 1974 cm^{-1}. Emission to the $\tilde{a}^1A_1$ (0, 5, 0) and (1, 3, 0) vibrational levels can be seen and the rotational levels in the $\tilde{a}^1A_1$ state are labelled in the figure. Similar spectra were obtained for the $\tilde{b}^1B_1$ (0, 19^1, 0)

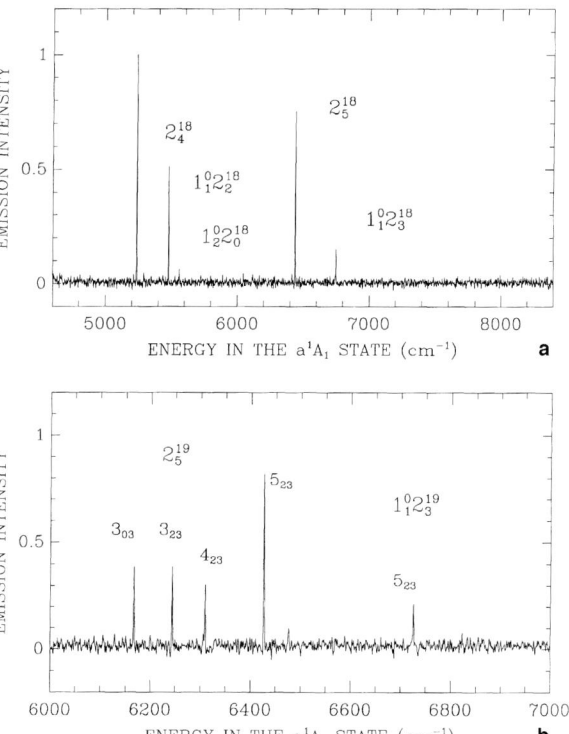

Fig. 3. a Fourier-transform dispersed fluorescence spectra recorded after excitation of the $CH_2 \tilde{b}^1B_1 (0. 18^0, 0) J_{K_aK_c} = 0_{00}$ level. Five $\tilde{b} \rightarrow \tilde{a}$ vibrational bands can be seen and are labelled in the spectra. Each band consists of a single rotational transition $0_{00} \rightarrow 1_{10}$. **b** Fourier-transform dispersed fluorescence spectra recorded after excitation of the CH_2 $\tilde{b}^1B_1(0, 18^0, 0)J_{K_aK_c} = 4_{13}$ level. Two $\tilde{b} \rightarrow \tilde{a}$ bands can be seen, 2_5^{19} and $1_1^02_3^{19}$ The rotational levels in the $\tilde{a}^1A_1$ state are labelled in the figure. (Reproduced from [23] by permission of the copyright holders. Copyright 1993, American Institute of Physics.)

$1_{11}, 2_{12}, 3_{13}$ and 5_{15} levels. The $\tilde{a}^1A_1$ (0, 4, 0) and (1, 2, 0) levels have also been observed in emission from the $\tilde{b}^1B_1$ (0, 16^0, 0) and (0, 17^1, 0) states. With the present laser repetition rate, these spectra with wavenumbers of a few thousands and ~ 1 cm^{-1} resolution are obtained in only *a couple of hours* of laser operation.

This first Fourier-transform detection of dispersed fluorescence spectra in a pulsed laser-induced fluorescence experiment has shown much better resolution and sensitivity than can be obtained with grating spectrometers. Dispersed fluorescence spectra over broad wavelength ranges can be efficiently obtained with modest resolution and high sensitivity, and with even better than 10^{-7} sec time-resolution. The newly acquired FTES spectra have allowed us to detect many highly excited vibrational levels in the lowest singlet $\tilde{a}$ state as well as in the excited singlet $\tilde{b}$ state [23, 24].

The information allowed us to understand the origin of many previously unassignable (due to large ΔK selection rules) $\tilde{a} \leftrightarrow \tilde{b}$ transitions [25], to deduce the barrier to linearity in the $\tilde{a}$ state and the singlet–singlet splitting, and to observe the Renner–Teller effect on the high vibrational levels [26].

State-to-state Energy Transfer and Reaction Rate of Excited Rovibronic Levels of Singlet Methylene

Despite intensive research on the spectroscopy and dynamics of highly vibrationally excited molecules in recent years, very little information is available about the state-to-state energy transfer kinetics of high vibrational levels. Using the TR-FTES technique, we have determined the cross-sections and propensity rules for state-to-state rotational energy transfer in the fifteenth bending overtone of the $\tilde{b}^1 B_1$ state of CH_2, induced by collisions with ketene [19].

In its $\tilde{b}^1 B_1$ state CH_2 is a quasilinear molecule and the rotational and bending degrees of freedom are strongly coupled. Thus, the $\tilde{b}^1 B_1$ $(0, 16^0, 0)$ level of CH_2 provides an excellent example of a non-rigid and highly vibrationally excited molecule. In fact, no rotationally state-resolved kinetic measurements have been performed on this molecule to date, for the $\tilde{a}^1 A_1$ or $\tilde{b}^1 B_1$ states. The existence of numerous perturbations in the $CH_2 \tilde{b}^1 B_1$ state (singlet–triplet, Renner–Teller, and Fermi resonance interactions have all been observed) makes the question of how energy relaxation occurs for individual $\tilde{b}^1 B_1$ levels an interesting one.

The reaction of CH_2 is also a very interesting subject. CH_2 is extremely reactive in its $\tilde{a}^1 A_1$ state. The products and kinetics of reaction have been studied for a wide variety of partners [27–29]. In the $\tilde{a}^1 A_1$ state CH_2 has a pair of non-bonding electrons. In the excited $\tilde{b}^1 B_1$ state, however, the two electrons have paired spins but occupy different molecular orbitals. Thus, the reactivity of the $\tilde{b}^1 B_1$ state, in contrast to that of the $\tilde{a}^1 A_1$ state, is an interesting issue.

In the experiments, pulsed photolysis of ketene at 308 nm was used to produce $\tilde{a}^1 A_1$ CH_2, which was subsequently excited to a single rotational level of the $\tilde{b}^1 B_1$ $(0, 16^0, 0)$ state by a pulsed dye laser. The dispersed fluorescence from that state was then recorded as a function of time. The fluorescence spectrum shows peaks due to the initially excited rotational level and "daughter" peaks – which grow in with time – from nearby rotational levels. These levels are populated as a result of rotational energy changing collisions with

ketene. Measurement of the fluorescence lifetime of the initially excited level as a function of ketene pressure yields the total cross-section for collisional removal of $\tilde{b}^1 B_1$ CH_2, and analysis of the daughter peak intensity gives the state-to-state RET (rotational energy transfer) cross-sections. Subtraction of the RET contribution from the total collisional cross-section yields the cross-section for the removal of $\tilde{b}^1 B_1$ CH_2 by reactive collisions with ketene.

An example of the time-resolved spectra obtained is shown in Fig. 4. In this experiment the $\tilde{b}^1 B_1 (0, 16^0, 0) 0_{00}$ level was initially excited; transitions can be clearly seen from the 0_{00} level, and from the 2_{02} level which is populated by RET from the 0_{00} level. In Fig. 5, the relative intensities of the $\tilde{b} \to \tilde{a} 2_1^{16}$ $0_{00} \to 1_{10}$ and $2_{02} \to 3_{12}$ transitions are plotted against time. Note that after a short instrument response rise time (ca. 70 ns) the $\tilde{b} \to \tilde{a} 2_1^{16} 0_{00} \to 1_{10}$ transition simply decays, indicating the decrease of the 0_{00} population with time. In contrast, the $2_{02} \to 3_{12}$ "daughter" transition first grows in and then decays. This behaviour

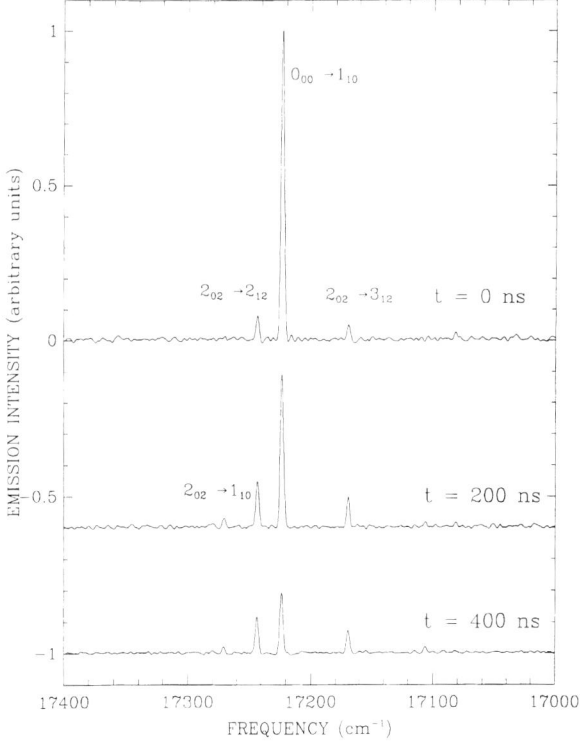

Fig. 4. TR-FTES spectra of the $CH_2 \tilde{b} \to \tilde{a} 2_1^{16}$ band at $t = 0, 200,$ and 400 ns after excitation of the $J_{K_a K_c} = 0_{00}$ level. The 2_{02} level is populated by RET from the 0_{00} level. The ketene pressure was 200 m Torr. (Reproduced from [19] by permission of the copyright holders, Copyright 1993, American Institute of Physics)

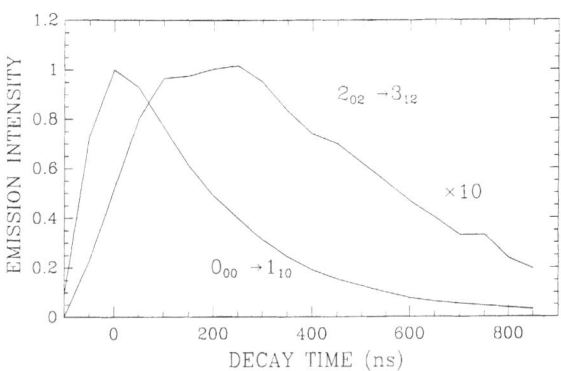

Fig. 5. Intensity of the $\tilde{b} \rightarrow \tilde{a} 2_1^{16}\, 0_{00} \rightarrow 1_{10}$ and $2_{02} \rightarrow 3_{12}$ transitions as a function of time (ns) after excitation of the $CH_2\,\tilde{b}B_1\,(0,\,16^0,\,0)0_{00}$ level at 200 mTorr of ketene. The intensities have been normalized to the maximum intensity of the $0_{00} \rightarrow 1_{10}$ transition, and the $2_{02} \rightarrow 3_{12}$ transition has been multiplied by 10 for display purposes. (Reproduced from [19] by permission of the copyright holders, Copyright 1993, American Institute of Physics)

confirms that the 2_{02} level is populated by an energy transfer process.

The propensity rules observed for RET show that the highly vibrationally excited $\tilde{b}^1B_1\,(0,\,16^0,\,0)$ CH_2 interacts with ketene through a quadrupole–dipole interaction. Interaction through a quadrupole moment implies that there is no permanent dipole moment for $\tilde{b}^1B_1\,CH_2$ undergoing large amplitude bending motion above the barrier to linearity. RET from this non-rigid species is determined by the time-averaged behavior of the vibrational motion. Both RET and reaction occur rapidly for $\tilde{b}^1B_1\,CH_2$ in collisions with ketene. The cross-sections for RET range from approximately 1 to 5 times the hard-sphere gas kinetic cross-section, which was calculated to be approximately $10\,\text{Å}^2$, and those for reactive scattering range from 2 to 6 times the hard-sphere rate. The reactive cross-sections, σ_j^{rxn}, of $\tilde{b}^1B_1$ $(0,\,16^0,\,0)$ CH_2 show that CH_2 is more reactive in its $\tilde{b}^1B_1$ state than in its $\tilde{a}^1A_1$ state. Furthermore, the σ_j^{rxn} values of the rotational levels studied appear to increase linearly with CH_2 rotational energy.

Collisional Deactivation of Highly Vibrationally Excited Molecules

For decades, collisional energy transfer between molecules has been extensively studied to understand the role of energy in chemical processes. Techniques for studying collisional deactivation of highly excited polyatomic molecules can be grouped into two types: experiments that monitor the initially excited mol-

ecules or ones that probe the species that receive energy during the collision. Experimental techniques used to directly monitor the excited molecules are transient optical pumping [30–32] and IR emission [20, 21, 33, 34].

The IR emission technique relies on the principle that IR emission spectra from a single rovibrational level reflect the vibrational wavefunction in terms of the IR-active normal modes. When such a population distribution of molecules is prepared, time-resolved IR intensity measurements at a single vibrational transition yield the average energy of that vibrational mode during collisional deactivation [33, 34]. By resolving the IR emission band-shape, or by simultaneous monitoring of several vibrational transitions, information about the width of the population distribution can be obtained. By using TR-FTES the complete time-evolution of all IR-active vibrational modes in the system can be monitored at high spectral resolution in a single experiment. In particular, we have shown that TR-FTES spectra from an ensemble of highly vibrationally excited molecules can give the average energy and width of the evolving population distribution [20, 21].

TR-FTES studies of collisional deactivation of vibrationally excited NO_2 by a variety of gases are illustrated here as an example. The vibrationally excited NO_2 is prepared by pulsed laser excitation of the $\tilde{A}/\tilde{B} \leftarrow \tilde{X}$ transition at 475 nm followed by fast internal conversion. At $t = 0$, the excited NO_2 molecules have an energy of $21050\,\text{cm}^{-1}$ set by the photon energy. TR-FTIRES spectra recorded with a band-width of $15\,798\,\text{cm}^{-1}$ and a resolution of $16\,\text{cm}^{-1}$ are shown in Fig. 6. The spectra are arranged from top-to-bottom in order of increasing time from $0\,\mu s$ (defined by the arrival of the laser pulse). Spectra were collected every $0.5\,\mu s$, but only those at $2\,\mu s$ intervals are shown in Fig. 6. The NO_2 pressure was 190 mTorr and the spectra shown are the results of averaging three scans, with 30 laser shots collected per mirror position for each scan. The data presented in Fig. 6 took ca. 2 hours of laser operation time to collect.

As collisional deactivation progresses, two sharp peaks centered at $2600\,\text{cm}^{-1}$ and $1450\,\text{cm}^{-1}$ appear, which are the NO_2 $\nu_1 + \nu_3$ and ν_3 transitions, respectively. Both these peaks are considerably broader than the room temperature absorption bands of NO_2 and are significantly red-shifted from their fundamental positions. This can clearly be seen in Fig. 7, where spectra of the ν_3 band recorded at $1\,\text{cm}^{-1}$ resolution

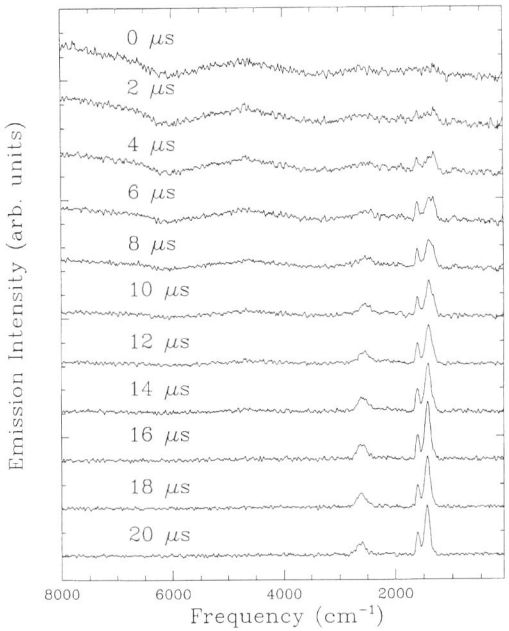

Fig. 6. NO_2 TR-FTIRES spectra recorded after pulsed laser excitation at 475 nm. The time after the laser pulse is labelled in the figure. The NO_2 pressure was 190 mTorr. (Reproduced from [20] by permission of the copyright holders. Copyright 1994, American Institute of Physics.)

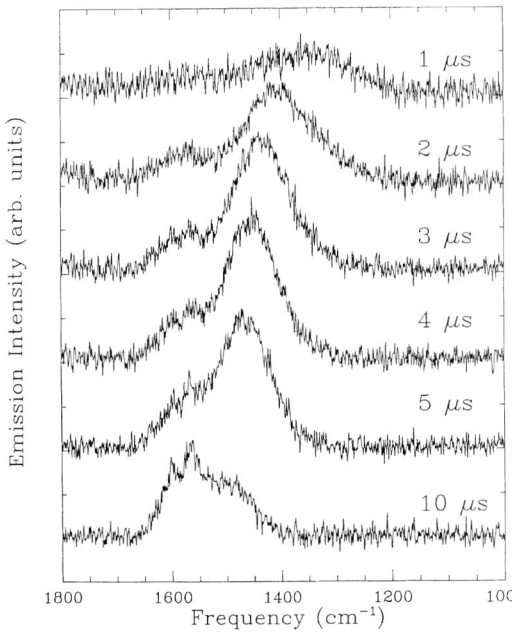

Fig. 7. TR-FTIRES spectra of the $NO_2 \nu_3$ transition recorded with 1 cm^{-1} resolution and at 1.70 Torr pressure. (Reproduced from [20] by permission of the copyright holders. Copyright 1994, American Institute of Physics.)

are presented. Early in the process ν_3 band is shifted by almost 300 cm^{-1} from the fundamental transition (1617 cm^{-1}). This indicates that emission originates from highly anharmonic, excited vibrational levels of the $\tilde{X}^2 A_1$ state. As time progresses the ν_3 and $\nu_1 + \nu_3$ frequencies are blue-shifted because the vibrationally hot NO_2 molecules are collisionally deactivated to lower-energy, less anharmonic vibrational levels.

Quantitative information about collisional deactivation can be extracted from the TR-FTIRES spectra by modelling the ν_3 and $\nu_1 + \nu_3$ bands: the average energy, $\langle E \rangle$, and width, σ^2, of the population distribution of excited NO_2 molecules undergoing collisional deactivation, and the average energy lost per collision, $\langle \Delta E \rangle$, for the vibrationally hot NO_2 molecules. The energy distribution of vibrationally excited NO_2 during collisional deactivation, extracted from the emission spectra, shows that the energy loss per collision with ambient NO_2 increases dramatically from < 50 cm^{-1} below 13000 cm^{-1} energy to 1300 cm^{-1} at 20000 cm^{-1} energy.

Results for collisional deactivation of highly vibrationally excited NO_2 and CS_2 by a variety of buffer gases, examined by TR-FTES, show that there is a dramatic increase in the average energy removed per collision for NO_2 excited above ~ 10000 cm^{-1} and for CS_2 above ~ 26000 cm^{-1}. These energies correspond to the origins of the lowest excited $\tilde{A}^2 B_2 / \tilde{B}^2 B_1$ states of NO_2 and the lowest excited $R^3 A_2$ state of CS_2. Mixing between these excited electronic states with the ground electronic state enhances collisional relaxation by allowing the electronic transition dipole to contribute to collisional energy transfer [35].

TR-FTES can also be used to monitor the excited population of the bath molecules in collision with the highly excited molecules prepared by laser excitation. In a study of highly excited NO_2 by cold N_2O or CO_2, we have observed IR emission from multiply-excited bath molecules. The study found single collisions that transfer more than 3500 cm^{-1} energy in vibration-to-vibration energy transfer from NO_2 to N_2O or CO_2 [21]. This observation provides evidence supporting the concept of supercollisions.

Acknowledgements. This work is supported by the Basic Energy Sciences of the US Department of Energy through Grant No. DE-FG02-86ER 134584 and a US Department of Energy University Instrumentation Grant. The author is grateful to collaborators in his laboratory who did all the measurements and analysis: Prof. Gregory Hartland, Dr. Wei Xie, and Dr. Dong Qin.

References

[1] J. J. Sloan, E. J. Kruus, in: *Time-Resolved Spectroscopy* (R. J. H. Clark, R. E. Hester, eds.), Wiley, New York, 1989, pp. 219–253.

[2] P. M. Aker, J. J. Sloan, *J. Chem. Phys.* **1986**, *85*, 1412.

[3] E. J. Krus, B. I. Niefer, J. J. Sloan, *J. Chem. Phys.* **1988**, *88*, 985.

[4] R. E. Murphy, F. H. Cook, H. Sakai, *J. Opt. Soc. Am.* **1975**, *65*, 600.

[5] H. Sakai, R. E. Murphy, *Appl. Opt.* **1978**, *17*, 1342.

[6] G. E. Caledonia, B. D. Green, R. E. Murphy, *J. Chem. Phys.* **1979**, *71*, 4369.

[7] B. D. Green, G. E. Caledonia, R. E. Murphy, F. X. Robert, *J. Chem. Phys.* **1982**, *76*, 2441.

[8] A. W. Mantz, *Appl. Spectrosc.* **1976**, *30*, 459.

[9] A. W. Mantz, *Appl. Opt.* **1978**, *17*, 1347.

[10] A. A. Garrison, R. A. Crocombe, G. Mamantov, J. A. de Haseth, *Appl. Spectrosc.* **1980**, *34*, 399.

[11] B. D. Moore, M. Poliakoff, M. B. Simpson, J. J. Turner, *J. Phys. Chem.* **1985**, *89*, 850.

[12] T. R. Fletcher, S. R. Leone, *J. Chem. Phys.* **1988**, *88*, 4720.

[13] E. L. Woodbridge, M. N. R. Ashfold, S. R. Leone, *J. Chem. Phys.* **1991**, *94*, 4195.

[14] G. Hancock, D. E. Heard, *Chem. Phys. Lett.* **1989**, *158*, 167.

[15] W. Uhmann, A. Becker, C. Taran, F. Siebert, *Appl. Spectrosc.* **1991**, *45*, 390.

[16] R. A. Palmer, C. J. Manning, J. A. Rzepiela, J. M. Widder, J. L. Chao, *Appl. Spectrosc.* **1989**, *49*, 193.

[17] J. A. Dodd, S. J. Lipson, W. A. M. Blumberg, *J. Chem. Phys.* **1991**, *95*, 5752.

[18] G. V. Hartland, W. Xie, H.-L. Dai, A. Simon, M. J. Anderson, *Rev. Sci. Instrum.* **1992**, *63*, 3261.

[19] G. V. Hartland, D. Qin, H.-L. Dai, *J. Chem. Phys.* **1993**, *98*, 6906.

[20] G. V. Hartland, D. Qin, H.-L. Dai, *J. Chem. Phys.* **1994**, *100*, 7832.

[21] G. V. Hartland, D. Qin, H.-L. Dai, *J. Chem. Phys.* **1994**, *101*, 8554.

[22] Y.-P. Lee, P.-S. Yeh, G.-H. Leu, W.-C. Hung, S.-C. Huang, I. C. Chen., *J. Chinese Chem. Soc. (Taipei)* **1995**, *42*, 205.

[23] G. V. Hartland, D. Qin, H.-L. Dai, *J. Chem. Phys.* **1993**, *98*, 2469.

[24] D. Qin, G. V. Hartland, H.-L. Dai, *J. Mol. Spectr.* **1994**, *168*, 333.

[25] G. V. Hartland, W. Xie, D. Qin, H.-L. Dai, *J. Chem. Phys.* **1992**, *97*, 7010.

[26] G. V. Hartland, D. Qin, H.-L. Dai, *J. Chem. Phys.* **1995**, *102*, 6641.

[27] A. H. Laufer, *Rev. Chem. Intermed.* **1980**, *4*, 225.

[28] M. N. R. Ashfold, M. A. Fullstone, G. Hancock, G. W. Ketley, *Chem. Phys.* **1981**, *55*, 245.

[29] A. O. Langford, H. Petek, C. B. Moore, *J. Chem. Phys.* **1983**, *78*, 6650.

[30] M. Heymann, H. Hippler, D. Nahr, H. J. Plach, J. Troe, *J. Phys. Chem.* **1988**, *92*, 5507.

[31] N. Nakashima, K. Yoshihara, *J. Chem. Phys.* **1981**, *77*, 6040.

[32] T. J. Bevilacqua, R. B. Weisman, *J. Chem. Phys.* **1993**, *98*, 6316.

[33] J. R. Barker, B. M. Toselli, *Int. Rev. Phys. Chem.* **1993**, *12*, 305.

[34] J. D. Brenner, J. P. Erinjeri, J. R. Barker, *Chem. Phys.* **1993**, *175*, 99.

[35] G. V. Hartland, D. Qin, H. -L. Dai, *J. Chem. Phys.* **1995**, *102*, 8677.

Mikrochim. Acta [Suppl.] 14, 157 163 (1997)
© Springer-Verlag 1997

Photoacoustic Depth Profiling, Dynamic Rheo-optics, and Spectroscopic Imaging Microscopy of Polymers by Step-Scanning FT-IR Spectrometry

Curtis Marcott*, **Gloria M. Story**, **Anthony E. Dowrey**, **Robert C. Reeder**, and **Isao Noda**

The Procter and Gamble Company, Miami Valley Laboratories, P. O. Box 538707, Cincinnati, OH 45253-8707, U.S.A.

Abstract. Three different applications of step-scanning Fourier transform infrared (FT-IR) spectrometry to polymeric samples (photoacoustic depth profiling, dynamic rheo-optics, and spectroscopic imaging microscopy) are described. Each approach is used to study a commercial polymer laminate film consisting of poly(ethylene terephthalate), ethylene vinyl alcohol copolymer, and low-density polyethylene layers. In the near-infrared region, where saturation problems are less significant, the phase delay in the photoacoustic response to sinusoidally modulated incident source radiation is shown to be linearly related to the depth in the sample from which the photoacoustic signal originates. Phase-resolved near-IR spectroscopy performed with a step-scanning spectrometer elucidates the time-dependent rheo-optical response of polymer film samples to an externally applied small-amplitude sinusoidal strain perturbation. A step-scanning FT-IR spectrometer coupled to a refractive microscope with a liquid-nitrogen cooled indium antimonide (InSb) focal-plane array detector is used for IR spectroscopic imaging microscopy.

Key words: step-scanning FT-IR, photoacoustic, rheo-optics, imaging, polymers.

Infrared (IR) spectroscopy has long been widely used for characterizing the chemical functionality and molecular orientation of polymers. More recently, additional structural and dynamic information about polymeric systems has been provided by time-resolved IR studies. Photoacoustic spectroscopy (PAS), for example, can be used to depth-profile layered samples [1–3]. The time-resolved photoacoustic response to sinusoidally modulated incident source radiation has a characteristic phase delay which depends on the depth at which the photoacoustic signal originates in the sample. We refer to a measurement of the phase delay as a function of wavenumber as *phase-resolved* IR spectroscopy. Phase-resolved IR spectroscopy has also been successfully used to study the dynamic rheo-

optical response of polymer films to an externally applied small-amplitude sinusoidal strain [4–7]. Both of these powerful phase-resolved IR techniques are difficult to perform on conventional rapid-scanning Fourier transform infrared (FT-IR) spectrometers, where each wavelength of light is modulated at a different frequency. Use of a step-scanning interferometer circumvents this difficulty because a single frequency is used to label all wavelengths. A staring indium antimonide (InSb) focal-plane array (FPA) detector attached to a step-scanning FT-IR microscope is also used to spectroscopically image polymeric samples. This paper presents three examples where use of a step-scanning FT-IR spectrometer has provided insights into polymer systems that could not have been achieved by use of a conventional rapid-scanning instrument.

Background

Photoacoustic Spectroscopy

In a rapid-scanning FT-IR measurement, the signal at each wavenumber in the spectrum is modulated at a different acoustic range frequency by the interferometer. Sampling depth is a function of the modulation frequency of the incident radiation in PAS, with faster modulation frequencies probing shallower sample depths than do slower modulation frequencies. With a typical rapid-scanning FT-IR spectrometer, the photoacoustic spectrum generated at a given optical-path-difference velocity contains a gradient of depth information. Low-wavenumber spectral signals modulated at lower frequencies originate from deeper parts of the sample than higher wavenumber signals

---

* To whom correspondence should be addressed

do. It is, however, possible to depth-profile layered samples uniformly when a quadrature (or two-phase) lock-in amplifier is used in conjuction with a step-scanning interferometer. With a step-scanning FT-IR spectrometer, each wavelength of light can be modulated at the same frequency, and the photoacoustic spectrum obtained will originate from the same range of depths within the sample for all wavelengths. This feature is especially useful when the compositional and structural nature of the samples is totally unknown.

Proper implementation of FT-IR PAS requires some careful considerations. For example, an interferogram contains asymmetric distortions due to optical, electronic, and sampling effects. These distortions are typically removed by applying an instrumental phase-correction function. It is important to preserve the thermal phase information needed for PAS depth-profiling when correcting for the instrument response function. A sample with a strong surface absorbance, such as carbon black, is typically used to create the instrument phase-correction function. Such a sample produces an entirely positive photoacoustic spectrum which is totally in phase with the modulated incident radiation after normal phase correction. This is because the thermal phase is constant across the entire spectrum of carbon black. The phase-correction function (i.e. phase array) produced during the computation of the in-phase spectrum of carbon black can then be stored for future use during the Fourier transformation of actual sample interferograms. Other samples, where the photoacoustic signal originates over a finite range of depths in the sample, often produce spectra with both positive and negative signals as a function of wavenumber. Performing normal Mertz or Foreman phase corrections on these interferograms will force all but the sharpest spectral features to be positive [8–10]. By using the stored phase-correction function generated during the computation of the carbon black spectrum, the instrument response phase can be effectively removed from the spectrum without distorting the thermal phase information.

It was previously demonstrated that the phase delay in the photoacoustic response is the same for all non-saturating bands originating from the same layer in a film [3]. In this work, a series of polystyrene films of known thickness is used to demonstrate there is a strict linear relationship between the photoacoustic phase angle and layer thickness. A commercial packaging material consisting of laminated polymer layers is then depth-profiled by PAS. Both of these examples require

the use of near-IR excitation in order to avoid highly absorbing fundamental mid-IR bands which often exhibit photoacoustic saturation.

Dynamic Rheo-Optical Infrared Spectroscopy

Dynamic rheo-optical IR spectroscopy is a polymer characterization technique based on the measurement of a time-resolved IR response to a small-amplitude oscillatory strain applied to a polymer film [4, 7]. The resulting dynamic signals are typically very small, with an absorbance of the order of 10^{-4} or less. Dynamic IR spectra collected in this manner can provide new and useful information about polymers, such as morphology and deformation mechanisms [5]. Furthermore, a two-dimensional IR spectrum defined by two independent wavenumbers can be generated by applying a correlation analysis to the dynamic fluctuation of IR signals induced by the external perturbation [11]. A dynamic two-dimensional IR spectrum may provide significantly more information than the normal IR absorbance spectrum if different wavenumbers in the spectrum respond to the perturbation out of phase with each other.

As in FT-IR PAS, there are some fundamental difficulties associated with making this type of dynamic rheo-optical IR measurement with conventional rapid-scan Fourier transform (FT) spectrometers. Many polymer systems have a selective range of perturbation frequencies which produce characteristic time responses for different wavenumbers in the spectrum. Such frequencies often fall within the acoustic to ultrasonic range. This frequency range may not be easy to probe with conventional rapid-scan FT-IR measurements. The signal at each wavenumber in the spectrum is modulated at the same acoustic range frequency by a step-scanning interferometer, thereby allowing the time-dependent response of the sample to be retrieved more conveniently [12, 13]. In this paper, dynamic IR spectroscopy is extended into the near-infrared region of the spectrum for the first time. This enables us to study thicker, real-world samples, such as packaging materials.

Spectroscopic Imaging Microscopy

Infrared spectroscopic images recorded by means of an FT-IR microscope attachment have typically been constructed by translating a mapping stage a single pixel at a time through the sample area of interest. This can be a very tedious and time-consuming procedure.

Recently, a technique for rapidly performing high-fidelity FT-IR imaging spectroscopy by using an indium antimonide (InSb) focal-plane array (FPA) detector coupled to an IR microscope and a step-scanning FT-IR interferometer has been developed [14, 15]. These multichannel IR detectors were originally developed for thermal-imaging applications (mainly by the military), but they have tremendous potential as chemical imaging detectors when used as part of a spectrometer. The multiple detector elements enable images from all pixels to be collected simultaneously for each mirror retardation position of the interferometer. Use of an interferometer allows the entire IR spectrum over some wavelength range to be measured. The combination of a step-scanning FT-IR microscope and an InSb FPA detector provides unprecedented speed and image quality, which is limited only by the diffraction limit and/or the number of detector elements on the array. For example, we have recorded infrared chemical image data sets containing 16,384 spatially resolved FT-IR spectra at $16 \, cm^{-1}$ resolution with data acquisition times of less than five minutes. Since the spectral resolution for each data set depends on the total distance traveled by the moving mirror of the interferometer during the measurement, higher spectral-resolution images can be achieved by simply extending this distance and collecting more data points.

Experimental

Spectrometer

All of the measurements described were made with a Bio-Rad FTS-60A FT-IR spectrometer operating in step-scan mode. The standard glowbar source and KBr beam-splitter were used for the mid-IR measurements, and a tungsten–halogen source and a CaF_2 beam-splitter were in place for the near-IR measurements. A $7900 \, cm^{-1}$ long-pass optical filter was placed in the beam near the aperture of the interferometer for all the near-IR measurements. The precise position of the moving mirror of the interferometer is monitored through the use of a small-amplitude 16-kHz dither signal applied to the fixed mirror of the interferometer. A triangular arrangement of three photodiodes is used to detect the signal from the expanded He-Ne alignment/calibration laser beam. A feedback loop from these photodiodes is used to adjust the piezoelectric transducers which control the position of the fixed mirror of the interferometer. This dynamic alignment system keeps the interferometer in alignment during the scan and compensates in real time for any drift in the position of the fixed mirror of the interferometer while the moving mirror is stopped in step-scan mode. The piezoelectric transducers on the fixed mirror can also be used to impart a phase-modulation chopping signal of up to 2.0 He-Ne laser wavelengths ($1.25 \, \mu m$) to the optical path difference. A phase modulation frequency of 400 Hz with an amplitude of 1.0 He-Ne laser

wavelengths was used for all of the near-IR photoacoustic and dynamic infrared measurements described. The phase modulation was turned off for the spectroscopic imaging experiments with the staring InSb FPA detector. The 400-Hz phase modulation signal was analyzed by using the demodulator (lock-in amplifier) supplied by Bio-Rad with the FTS-60A spectrometer. The use of phase modulation can dramatically improve the signal-to-noise ratio (S/N) of step-scanning FT-IR measurements by: (1) encoding light originating from the source to distinguish it from background black body radiation; (2) enabling the experiment to be performed at a higher frequency in a range where detectors are often more sensitive and away from $1/f$ noise; (3) narrowing the bandwidth of the experiment through the use of single-frequency lock-in amplifier detection; (4) providing a first derivative of the interferogram, which is much less sensitive to baseline drift. Long-term instrument stability is a much more significant concern in a step-scanning experiment where it takes substantially more time to measure a single interferogram than it does in conventional rapid-scan mode.

Materials

Eight atactic polystyrene (PS) films of increasing thicknesses on NaCl disks (Wilmad $9 \times 5 \, mm$) were created by the following methods. Thin films (3, 8, and 11 μm) were cast on NaCl disks from dilute solutions of PS in toluene ($<0.1\%$). Thicker films (18, 27, 29, and 38 μm) were made by melting beads of PS between Teflon™ sheets and mounting circular cut-outs of the films onto NaCl disks. A thick film (100 μm) was created by melting one small PS bead onto an NaCl disk in a $100 \, ^\circ C$ oven. Film thicknesses were determined by transmittance measurements. The absorbance of the $1181\text{-}cm^{-1}$ band in each of the eight films was calibrated against an atactic polystyrene standard film of known thickness (as determined by the frequency of the interference fringes) [16].

The commercial packaging material film (Paramount) studied by all three techniques consists of a 12-μm thick surface layer of poly(ethylene terephthalate) (PET), followed by a 7.5-μm thick adhesive tie layer (possibly ethylene vinyl acetate copolymer), a 15-μm thick layer of ethylene vinyl alcohol copolymer (EVOH), another 7.5 μm thick adhesive tie layer, and a 75-μm thick layer of low-density polyethylene (LDPE). A mid-IR transmittance spectrum of this nominally 117-μm thick sample film is not particularly useful, because the fundamental IR bands absorb too strongly. An attenuated total reflectance (ATR) mid-IR spectrum of the PET side of the film is a good match with an IR reference spectrum of KODAPAK PET 5302 (Eastman Chemical Products).

Photoacoustic Spectroscopy

The photoacoustic spectroscopy (PAS) measurements were carried out by aligning an MTEC 200 PAS cell in the sample compartment of the spectrometer. The instrument response phase was determined by using a standard carbon back sample equipped with an optical screen supplied by MTEC. The Bio-Rad lock-in amplifier was tuned for the largest possible in-phase signal with the carbon black standard. The PAS cell was purged with helium gas. The interferometer was stepped at a rate of one step every two seconds. Approximately 64,000 data points were sampled following a 750 ms settling time after each step. Full double-sided interferograms were collected at a spectral resolution of $8 \, cm^{-1}$. Interferograms were undersampled by a factor of two (every other He-Ne laser zero crossing) and zero-filled 2X. The data points were acquired in three channels sequentially at each mirror retardation position. The approximately 16,000 data points in each channel were integrated, so only the average value of the signal was stored. The first and thrid channel had identical data to verify that the settling time was sufficient (the interferograms should be virtually identical). Sample interferograms were computed by using the carbon black in-phase stored phase array which contains only instrument response function contributions. The computed in-phase and quadrature sample spectra [S(I) and S(Q)] can be rotated in Cartesian space to calculate spectra at any desired phase angle S(θ) by using the equation S(θ) = S(I) sin(θ) + S(Q) cos(θ).

Dynamic Infrared Spectroscopy

The dynamic infrared spectroscopy experiment has been described in detail elsewhere [4, 12]. The same PET/EVOH/LDPE laminate sample film used for the FT-IR PAS studies was dynamically strained at 13 Hz with an amplitude of 50 μm by a dynamic micro-rheometer (designed and built by Carl C. Gryte, Columbia University) which fitted into the sample compartment of the step-scanning FT-IR spectrometer. A liquid-nitrogen cooled InSb detector and an additional 5500-cm^{-1} long-pass optical filter were used. A wire-grid linear polarizer immediately preceding the sample was oriented such that it only transmitted light polarized along the dynamic strain axis of the sample film. Two lock-in amplifiers were used to demodulate the phase modulation and dynamic strain signals. The in-phase and quadrature (90° out-of-phase) filtered outputs (25 Hz) of the Bio-Rad demodulator device (tuned to the 400-Hz phase modulation signal) were sampled sequentially by the analog-to-digital converter (ADC) after a settling time of 300 ms. Dynamic IR responses to the applied small-amplitude oscillatory strain perturbation were analyzed by using a two-phase lock-in amplifier (SR530, Stanford Research Systems) tuned to the dynamic strain frequency (13 Hz). The input to this lock-in amplifier originated from the filtered output of the in-phase signal of the Bio-Rad demodulator. An output time constant of 100 ms was used for the second lock-in amplifier. Full double-sided interferograms were collected at every other He-Ne laser-fringe zero crossing from both the in-phase and quadrature outputs of each lock-in amplifier. The spectral resolution was 16 cm^{-1}. A single scan was collected at a stepping rate of 0.25 Hz. Approximately 16,000 digital samples were collected in each of the four signal channels of interest. The total measurement time was two hours and fifteen minutes.

Spectroscopic Imaging Microscopy

An IR microscope (Bio-Rad UMA 300A) mounted on an external bench of the spectrometer was fitted with a 320 × 256 (81,920 detector element) InSb FPA detector (ImagIR, Santa Barbara Focal-plane). The microscope optics were coupled efficiently to the InSb camera by addition of a CaF$_2$ 50-mm image formation lens placed between the microscope objective and the infrared camera [15]. This lens replaces the Cassegrainian mirror normally used with this microscope for final focusing on a single-element detector. Stepping pulses from the interferometer were used to trigger the acquisition of up to 16 image frames by the infrared camera at a 30-Hz frame rate. All of the 12-bit image frames captured at each interferometer mirror retardation position were summed, averaged, and written as a 32-bit floating point file. The final data set, therefore, consisted of up to 320 × 256 floating point images collected at several hundred different interferometer retardations. The optical bandpass for the InSb FPA used in the these experiments was limited to 3–5 μm (3300–2000 cm^{-1}) by an antireflection coating on the FPA. An additional optical filter, which cut off all wavelengths longer than 4 μm, was placed in the beam near the source aperture of the interferometer to prevent the large source emission at between 4 and 5 μm (a region of little spectroscopic interest) from saturating the FPA. This reduction of spectral bandwidth enabled us to use a larger interferometer step size (undersampling) between image frames. Undersampling without optical filtering can cause spectral artifacts as a consequence of frequency aliasing. The larger step size for the experiment also provided higher spectral resolution with fewer data points, therefore minimizing the number of images to be collected. This, in turn, reduced the storage and processing requirements for the resulting data set.

The total data acquisition or "staring" time for the camera for a data set consisting of 16,384 spectra (a 128 × 128 subset of the total number of pixels available on the FPA) was less than 8 seconds. To collect each image, a frame was integrated for 2.8 ms, 10 frames being averaged for each interferometer mirror position. The moving mirror of the interferometer was stopped every eighth zero crossing of the He-Ne reference laser (an undersampling ratio of eight). A total of 280 mirror positions was recorded at a rate of one position per second. Thus, a complete interferogram image data set took less than five minutes to acquire. This image interferogram was collected entirely within the memory of a 90-MHz Pentium PC. Data collection and processing are similar to those performed for conventional FT-IR studies with the exception that much larger data sets are handled. Analysis involves first collecting a step-scan image sequence data of background, typically air. Then, using identical sampling parameters, a step-scan sequence is collected with a sample placed under the microscope. The dynamic range of the camera's analog-to-digital converter is optimized by collecting a short alignment scan consisting of several data points on either side of the center-burst, taking care that all the pixels are on scale for both the most positive point and most negative point in the interferogram. The images collected at these two extreme light levels are also used to flat-field correct the camera for non-uniform pixel response. The spatial/spectral interferometric data are organized in a 32-bit/pixel spectral image file format (SPIFF) [17]. The interferograms at each pixel in the data sets are apodized, zero-filled to a total of 512 data points, and Fourier-transformed with commercial software (Spifft 1.0, Chemlcon) to yield a total of 16,384 FT-IR spectra (for a 128 × 128 pixel subset of the image field) for both the background and sample data sets. The two data sets are then ratioed to yield infrared transmittance images of the sample over the spectral range 3950–1975 cm^{-1}. Finally, the IR image data set spectra are converted into absorbance spectra. This 16 cm^{-1} resolution 128 × 128 SPIFF absorbance file takes 32 Mbytes of disk space to store. The total amount of 90-MHz Pentium computation time needed to convert this size of SPIFF file from interferograms to absorbance spectra is about 15 minutes. A 36X Cassegrainian microscope objective was used to image a nominally 10-μm thick cross-section of the PET/EVOH/LDPE polymer laminate sample film. A standard USAF 1951 optical resolution target (Newport) was used to calibrate the spatial resolution of the images.

Results and Discussion

The phase-resolved spectrum of a sample at any phase angle can be calculated by taking the appropriate linear combination of the raw in-phase and quadrature spectra recorded during the experiment. A properly phase-corrected FT-IR photoacoustic spectrum should yield a sinusoidal plot of signal *vs.* phase angle (centered about zero signal amplitude) for each

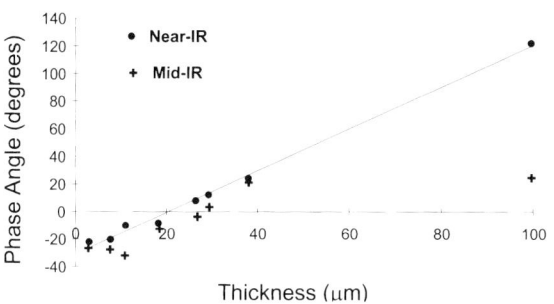

Fig. 1. Phase angle of the maximum PAS signal of PS plotted *vs.* film thickness, using both near- and mid-IR bands. The line shown is a linear best fit to the near-IR data points

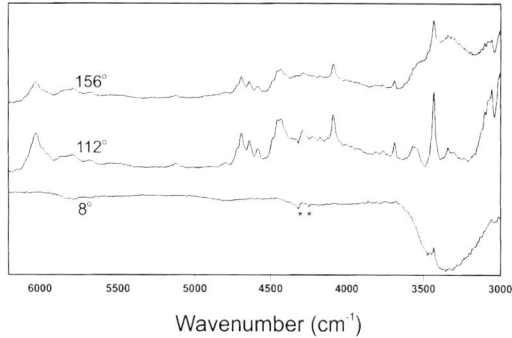

Fig. 2. Near-IR PAS spectrum of the PET/EVOH/LDPE laminate film displayed at phase angle of 8°, 112°, and 156°

wavenumber in the spectrum. The phase angle of the PAS signal is defined by subtracting 90° from the lock-in amplifier phase setting where the maximum positive signal is observed. Figure 1 shows the phase angle of the PAS signal *vs.* thickness for the series of PS films cast on NaCl plates. Data are shown for both mid- and near-IR spectral regions (both using 400-Hz phase modulation). The photoacoustic saturation causing the mid-IR result to become nonlinear at sample film thicknesses around 40 μm is no longer a problem when near-IR radiation is used. This is because the absorbances of overtone and combination bands in the near-IR region are 10 —100 times weaker than the mid-IR fundamentals. Thus, at wavelengths where the sample absorbs, much more near-IR than mid-IR radiation is available to penetrate into deeper layers of the film. Even though the thermal diffusion length of PS is only about 9 μm at 400 Hz, the observed PAS signal still responds linearly up to a film thickness of about 100 μm.

Figure 2 shows the PAS spectrum of the PET/EVOH/LDPE laminate film displayed at three different phase angles. Spectral features in Fig. 2 that are negative are from deeper in the sample than spectral features that are positive. At 8°, all of the overtone and combination bands between 4000 and 5000 cm^{-1} due to the PET are absent. The remaining negative bands in this spectrum are due to deeper layers of PE and EVOH. The 112° phase angle data were selected because the signal at 3450 cm^{-1} due to the OH-stretching fundamental of EVOH is nulled. Notice that the spectral features for PET are positive, indicating they are from the surface layer, and the small bands due to PE (4322 and 4251 cm^{-1}) are still negative. Finally, the weak LDPE bands (labeled with *) are nulled in the

spectrum displayed at 156°. Notice that the deeper into the sample the component is, the larger the phase angle. The actual phase angle determined for a given layer depends on the thickness of that layer as well as of the layer(s) above it.

Figure 3 shows the in-phase and quadrature dynamic spectra, as well as the normal near-IR absorbance spectrum of the PET/EVOH/LDPE laminate film. Since these spectra are recorded by transmittance, the signals due to the 75-μm LDPE layer dominate the responses. This is in contrast to the PAS study above, where the PET bands were the strongest, even though this layer was only 12 μm thick. This is because the PET was the surface layer of the film and the stronger PAS signals originate from the surface. The only dynamic IR signal due to the PET layer that appears in Fig. 3 is the derivative-shaped feature at 4099 cm^{-1} in the in-phase spectrum. The CH bend–stretch combi-

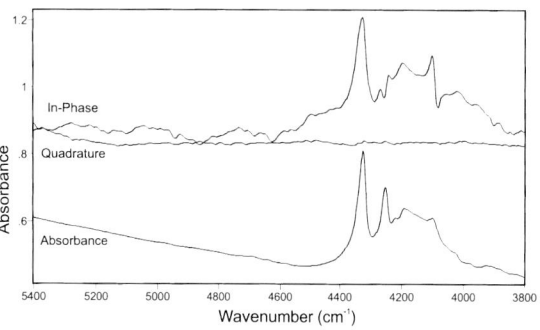

Fig. 3. In-phase and quadrature dynamic spectra, as well as the normal near-IR absorbance spectrum of the PET/EVOH/LDPE laminate film. The absorbance scale shown is for the normal IR absorbance spectrum only. The in-phase and quadrature dynamic IR spectra (plotted on the same scale as each other) are three or four orders of magnitude less intense than the absorbance spectrum

nation bands of LDPE between 4400 and 4000 cm^{-1} show many interesting dynamic responses which could eventually be used to characterize the submolecular reorientation mechanism in this layer. The absorbance of the OH-stretching fundamental in the EVOH layer is too intense (> 1.0) in this sample film. As a result, an unreliable (artifact) signal is produced in the dynamic IR spectrum near 3450 cm^{-1} (not shown). Because the level of dynamic IR signals is so low, it is difficult to achieve the dynamic range necessary to simultaneously measure strong and weak signals on the same sample film. Had we been able to do so on this sample, we probably would have observed different orientation rates (phase angles) for each component layer. Because the individual polymer layers are not mixed and have different mechanical properties, the dynamic IR response should indicate that they reorient independently of one another. It would take about one more order-of-magnitude in sensitivity improvement for us to verify this hypothesis experimentally with this particular sample film. This level of improvement in instrumentation could conceivably be achieved in the near future.

This work demonstrates for the first time that the near-IR region of the spectrum can be used to perform dynamic rheo-optical experiments. Extending this technique into the near-IR makes it possible to study real-world materials, such as packaging films, the mid-IR fundamental bands of which are typically too strongly absorbing. The assignment of these near-IR bands could be aided by correlating their dynamic responses with those of the mid-IR fundamentals obtained on a thinner sample. A two-dimensional correlation analysis of the dynamic responses in these two spectral regions should accentuate similarities and differences between the responses of individual bands. Near- and mid-IR bands originating from the same functional group should show strong synchronous cross-peaks in the two-dimensional IR map. Thus, if the assignment of a particular mid-IR band is known, a corresponding near-IR band having a response that correlates strongly with this band probably originates from the reorientation of the same functional group.

Figure 4 shows three image planes at 3200, 2972, and 2891 cm^{-1} from a spectroscopic image collected on a cross-section of the PET/EVOH/LDPE laminate sample film. The dimensions of each image plane are 184×184 µm. A spectroscopic image can be thought of as a three-dimensional cube of absorbance values with

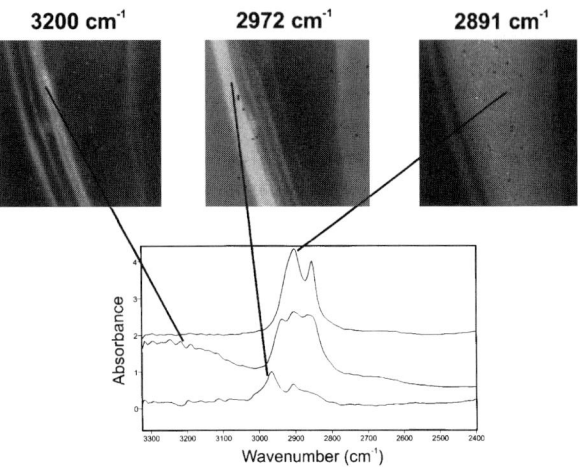

Fig. 4. Three image planes and three IR absorbance spectra from the IR spectroscopic image of a nominally 10-µm thick cross-section of the PET/EVOH/LDPE laminate film. Brighter areas represent regions of higher absorbance at the wavenumber of the image plane. The IR spectra shown are the average of 16 spectra taken from a 4-pixel x 4-pixel area indicated on the image

x–y image planes for each wavenumber of the spectrum. A line drawn in the z direction through the same x, y co-ordinates in every image plane represents the infrared absorbance spectrum of that individual pixel in the x–y plane. The brighter areas on each image represent regions of higher IR absorbance at the corresponding wavenumber of the image. The brightest stripe in the 3200-cm^{-1} image is due to the 15-µm thick layer of EVOH. The middle spectrum on the bottom of the figure is the co-added average of sixteen pixels in a 4-pixel x 4-pixel area in the middle of this bright stripe. The IR spectrum of every pixel in this stripe shows a broad OH-stretch absorbance which would be present only in the EVOH layer. The middle image highlights the PET portion of the film. The bottom IR absorbance spectrum is from a 4-pixel x 4-pixel area on the left side of the film that shows up as brighter in the 2972-cm^{-1} image plane. The bands shown are consistent with PET. The top spectrum is taken from a 4-pixel x 4-pixel area on the right side of the film which shows up more brightly in the 2891-cm^{-1} image plane. The spectrum is consistent with that of LDPE.

These results illustrate the powerful ability of spectroscopic imaging to uniquely bring out information related to the chemical nature of a sample. This approach, along with the other step-scanning FT-IR techniques of photoacoustic depth-profiling and dynamic infrared rheo-optical studies, promise to provide unique insights into polymeric systems.

Acknowledgments. The authors wish to thank Drs. Neil Lewis and Ira Levin of the National Institutes of Health, Professor Patrick Treado of the University of Pittsburgh, Drs. Richard Crocombe and Raul Curbelo of Bio-Rad, and Dr. Arn Adams of Santa Barbara Focalplane for many helpful discussions and for technical assistance in interfacing the FPA detector with the interferometer. They thank Dr. Matthew Chestnut of Procter and Gamble for preparing the cross-sectioned sample of the polymer laminate film.

References

[1] C. J. Manning, R. M. Dittmar, R. A. Palmer, J. L. Chao, *Infrared Phys.* **1992**, *33*, 53.

[2] R. M. Dittmar, J. L. Chao, R. A. Palmer, in: *Photoacoustic and Photothermal Phenomena III* (O. Bicanic, ed.) Springer, Berlin Heidelberg New York Tokyo, 1992, pp 492–496.

[3] G. M. Story, C. Marcott, I. Noda, *Proc. SPIE* **1993**, *2089*, 242.

[4] I. Noda, A. E. Dowrey, C. Marcott, *Appl. Spectrosc.* **1988**, *42*, 203.

[5] S. D. Smith, I. Noda, C. Marcott, A. E. Dowrey, in: *Polymer Solutions. Blends and Interfaces* (I. Noda, D. N. Rubingh, eds.), Elsevier, Amsterdam, 1992, pp. 43–64.

[6] I. Noda, A. E. Dowrey, C. Marcott, *J. Molec. Struct.* **1990**, *224*, 265.

[7] C. Marcott. A. E. Dowrey, I. Noda, *Anal. Chem.* **1994**, *66*, 1065A.

[8] L. Mertz, *Transformations in Optics*, Wiley, New York, 1965.

[9] M. L. Forman, W. H. Steele, G. A. Vanasse, *J. Opt. Soc. Am.* **1966**, *56*, 59.

[10] P. R. Griffiths, J. A. de Haseth, *Fourier Transform Infrared Spectrometry*, Wiley–Interscience, New York, 1986, pp. 93–97.

[11] I. Noda, *Appl. Spectrosc.* **1990**, *44*, 550.

[12] R. A. Palmer, C. J. Manning, J. L. Chao, I. Noda, A. E. Dowrey, C. Marcott, *Appl. Spectrosc.* **1991**, *45*, 12.

[13] C. Marcott, A. E. Dowrey, I. Noda, *Appl. Spectrosc.* **1993**, *47*, 1324.

[14] E. N. Lewis, I. W. Levin, P. J. Treado, *US Patent* 5,377,003, December 1994.

[15] E. N. Lewis, P. J. Treado, R. C. Reeder, G. M. Story, A. E. Dowrey, C. Marcott, I. W. Levin, *Anal. Chem.* **1995**, *67*, 3377.

[16] N. B. Colthup, L. H. Daly, S. E. Wilberley, *Introduction to Infrared and Raman Spectroscopy*, Academic Press, New York, 1990, p. 87.

[17] *ChemImag Technical Reference Manual*, ChemIcon, Pittsburgh, 1995.

Mikrochim. Acta [Suppl.] 14, 165–167 (1997)
© Springer-Verlag 1997

FT-Raman (NIR) and FT-IR Study of the Aromaticity Changes in Complexes of *o*-, *m*-, and *p*-Iodobenzoic Acids with Chosen Metals

Piotr Koczoń[1], **Halina Barańska**[2], **Krzysztof Bajdor**[2], and **Włodzimierz Lewandowski**[1,*]

[1] Institute of General Chemistry, Agricultural University of Warsaw, SGGW, Rakowiecka 26/30, 02-528 Warsaw, Poland
[2] Department of Technology Fundamentals, Industrial Chemistry Research Institute, Rydygiera 8, 01-793 Warsaw, Poland

Abstract. Vibrational spectra (IR and Raman) of *o*-, *m*- and *p*-iodobenzoic acid complexes with Na^+, K^+, Hg(II), Pb(II), Nd(III) and Dy(III) were recorded and studied. Characteristic spectral changes of $\nu(C \cdots C)_{ar}$, and $\gamma(CH)$ have been observed. The correlation between the observed changes of frequency and intensity of these bands and the ionic potential of the central ion was investigated.

Key words: metal complexes, iodobenzoic acids, FT-NIR Raman, FT-IR.

The main topic of our vibrational spectroscopy studies has been the aromaticity change in complexes of different derivatives of benzoic, salicylic and nicotinic acids with selected metals [1–4]. We have treated these compounds as models to study enzymes and other biologically important materials. Without using FT-NIR Raman spectrometers it was not possible to get Raman spectra of many such materials because of their fluorescence and, for iodide derivatives, because of their photodecomposition, taking place very often in measurements with laser excitation in the visible region. We characterize the aromaticity by comparison of the vibrational spectrum of each studied benzene derivative with that of benzene. As a molecular standard of an aromatic system we take some normal vibrations of the benzene ring, after Versanyi [5], mainly in the ranges: $\sim$1650–1350 cm^{-1} [$\nu(C \cdots C)$ stretching], $\sim$1000 cm^{-1} [ring vibration], $\sim$1000–700 cm^{-1} [$\gamma(CH)$ deformation out-of-plane], $\sim$700–550 cm^{-1} [$\varphi(CC)$ deformation out-of-plane]. Our intention is to find any correlation between changes of the molecular reactivity and the electronic and topological structure of the molecule considered.

Hence, one of the first steps is to learn the basic features of the molecular vibrations, which is not possible without detailed knowledge of the vibrational spectra. The aim of this work has been the preparation, measurement and assignment of vibrational spectra of iodo derivatives of benzoic acids in complexes with selected metals: Na^+, K^+, Hg(II), Pb(II), Nd(III) and Dy(III).

Experimental

Sodium and potassium *o*-, *m*- and *p*-iodobenzoates were prepared by dissolving the acids (Sigma, analytical reagent grade) in an aqueous solution of sodium or potassium hydroxide (Marck, analytical reagent) in stoichiometric ratio. Then the water was evaporated at 90 °C in a drier. Other salts were obtained by mixing an aqueous solution of the sodium salt of the respective iodobenzoic acid with a water-soluble salt of a given metal in stoichiometric ratio. Then the precipitate of iodobenzoate was collected, washed with redistilled water and dried as for the sodium salt. Pressed KBr pellets were used for IR measurements within the range 400–4000 cm^{-1}, with a Perkin–Elmer System 2000 FTIR spectrometer. Raman spectra of solid samples within the range 0–4000 cm^{-1} were recorded with a Perkin–Elmer FT-NIR Raman 2000 R System.

Results and Discussion

In Tables 1 and 2 are presented examples of vibrational spectra (frequencies and relative intensities) of the iodobenzoic acids studied and their complexes. The influence of the iodine position in the benzene ring on the frequency and intensity of selected vibrational bands that are characteristic for an aromatic system differs markedly. In the series of *o*-, *m*- and *p*-iodobenzoic acids (Table 1) the frequency of the $\varphi(CC)$ vibration near 550 cm^{-1} is decreased, whereas that of $\gamma(CH)$

* To whom correspondence should be addressed

Table 1. Infrared wavenumbers of selected bands of *o*-, *m*- and *p*-iodobenzoic acids[a]

o-	555 m	739 vs	1016 s	1467 m	1562 w	1583 m	1683 m
m-	550 m	747 m	997 m	1477 m	1564 m	1589 m	1683 m
p-	542 m	755 vs	1008 s	1483 m	1586 s	1565 m	1676 s
Assign.	$\varphi(CC)$	$\gamma(CH)$	ring vibr.	$v(CC)_{ar}$	8a	8b	$v(C=O)_{dimer}$

[a] s = strong, m = medium, w = weak, sh = shoulder, v = very.

Table 2. Infrared and Raman wavenumbers in spectra of studied complexes

		Na	K	Hg	Pb	Nd	Dy	Assignment
o-	IR	744 s	748 s 754 s		741 s		743 m	
	R	–	–					
m-	IR	753 vs	763 s	749 s 755 s	754 vs	755 s	756 vs	$\gamma(CH)$
	R	–	–		–	–	–	
p-	IR	763 s	767 s	767 s	762 m	766 s	766 s	
	R	–		–		–	–	
o-	IR	1013 m	1013 m		1014 m		1016 m	
	R	1014 m	1013 m	–	–	–	–	ring
m-	IR	997 m	995 m	997 m	997 m	997 w	997 w	vibr.
	R	999 vs	996 vs	–	997 vs	997 vs	997 vs	
p-	IR	1015 m	1009 m	1009 m	1009 m	1011 m	1009 m	
	R	1016 m	–	1009 w	–	1011 w	1011 w	
o-	IR	1460 w	1461 w		1461 w		1468 m	
	R	1462 vw	1463 vw					
m-	IR	1469 w	1462 vw	1472 w	1470 w	1473 m	1474 m	$v(C{\cdots}C)_{ar}$
	R	1470 vw	1465 vw		1469 w	1476 w	1476 vw	
p-	IR	1461 vw	1459 vw	1459 w	1459 vw	1458 w	1469 vw	
	R	–	–	–	–	–	–	
o-	IR	1569 vs	1569 vs		1567 s		1569 s	
	R		1572 w					
m-	IR	–	–	–	1578 s	1579 m	1579 m	$v(C{\cdots}C)_{ar}$ 8a
	R	1587 m			1573 w			
p-	IR	1585 vs	1586 vs	1586 vs	1584 vs	1584 vs	1590 vs	
	R			1584 vs				
o-	IR	1586 m	1584 m		1582 m		1584 m	
	R	1583 s	1581 vs	–	–	–	–	
m-	IR	1600 vs	1593 vs	1595 s	1587 m	1590 s	1590 m	$v(C{\cdots}C)_{ar}$ 8b
	R	1608 vw	1588 w	–	1586 s	1588 s	1588 m	
p-	IR	–	–	–	1579 sh	–	–	
	R	1590 s	–	1608 m	–	1584 vs	1585 vs	

near 750 cm^{-1} and that of ν(C$\cdots$C) near 1470 cm^{-1} are increased in frequency, but the "ring vibration" near 1000 cm^{-1} and 8a, 8b near 1580 cm^{-1} do not change systematically.

The influence of the metal on the frequencies of the selected bands is weak (Table 2). It is easily seen that the frequencies of particular modes along a series of the metals studied are either changed only slightly or not at all. On the other hand spectral changes occur when the position of the ring substituent is altered. In the series of *o*-, *m*- and *p*-iodobenzoic acid complexes the frequencies of γ(CH) near 750 cm^{-1} and 8a near 1580 cm^{-1} are increased and the others are scattered (Table 2). Therefore it seems that the effect of the substituent in the benzene ring (iodine) on the vibrational frequencies dominates the effect of the metal. This conclusion is supported by the small values of the correlation coefficients calculated by statistical analysis for several bands and different central ions. The correlation between the atomic mass or electronegativity or ionic potential of the central ions and the frequency of the same type of vibration gives correlation coefficients down to 0.7 for the *m*- and *p*-isomers and higher, between 0.8 and 0.9, for the *o*-isomers and the only ionic potential. Such weak correlation confirms the small effect of the metals.

Conclusions

The influence of metals on the frequencies of selected bands is weak and seems to be determined mainly by the position of iodine in the aromatic ring. However, varying the position of a substituent produces different changes in the aromatic ring of the acids and complexes. The fact that only very few vibrational bands change their frequencies systematically, and most of the others erratically, suggests that a very heavy substituent in the aromatic ring affects the composition of the normal modes (in terms of linear combinations of internal coordinates).

Acknowledgement. This work was supported by the State Committee for Scientific Research (Grant KBN - 3T09A 054 08).

References

[1] W. Lewandowski, H. Barańska, *Vib. Spectrosc.* **1991**, *2*, 211.

[2] P. Mościbroda, H. Barańska, T. Drapata, W. Lewandowski, *J. Mol. Struct.* **1992**, *267*, 255.

[3] W. Lewandowski, H. Barańska, P. Mościbroda, *J. Raman Spectrosc.* **1993**, *24*, 819.

[4] P. Mościbroda, H. Barańska, A. Tinti, W. Lewandowski, *Vib. Spectrosc.* **1995**, *9*, 69.

[5] G. Versanyi, *Assignments for Vibrational Spectra of 700 Benzene Derivatives,* Akadémiai Kiadó, Budapest, 1973.

Mikrochim. Acta [Suppl.] 14, 169–172 (1997)
© Springer-Verlag 1997

FT-IR Study on Triple Complexes Between Vesicles, Polynucleotides and Metal Ions

Paolo Bruni, Liberato Cardellini, Elisabetta Giorgini, Marco Iacussi, Eziana Maurelli, and **Giorgio Tosi***

Dipartimento di Scienze dei Materiali e della Terra, Università degli Studi di Ancona, Via Brecce Bianche, 60131 Ancona, Italy

Abstract. The FT-IR spectra of complexes between polynucleotides (polyadenilic and polyuridilic monophosphoric acids, calf thymus DNA) with metal cations and phosphatidylcholine are investigated. By changes in the main vibrational modes, the influence of the metal as well as of the vesicle on the spectral features of the complex can be elucidated.

Key words. infrared spectra, polyadenilic monophosphoric acid, polyuridilic monophosphoric acid, calf thymus DNA, phosphatidylcholine.

Studies on triple complexes of polynucleotides and phosphatidylcholine liposomes in the presence of some cations have already been reported and they suggest that the structure of nucleic acids may be influenced by the lipid part of membranes [1, 2]. The interaction of cations with polynucleotides and liposomes has been studied with different techniques such as EPR, thermal denaturation, turbidimetry, FT-IR, etc. [3, 4]. We report an FT-IR study of the interactions of some metal ions (K^+, Ca^{2+}, Mg^{2+}, Mn^{2+}) on complexes of egg phosphatidylcholine unilamellar vesicles as model membrane, with polyadenylic, polyuridilic acids and calf thymus DNA as polynucleotides.

Experimental

The spectra were obtained with a Nicolet 20-SX FT-IR spectrometer equipped with a Spectra Tech. diffuse reflectance accessory ("collector"). Samples were prepared by spreading and evaporating a few drops of the aqueous solution on a metal mirror [5, 6]. Resolution was 4 cm^{-1} and 250 scans were made. A Galactic software package was used for data treatment.

---

* To whom correspondence should be addressed

Results and Discussion

Main Bands in Starting Compounds (cm^{-1})

Phosphatidylcholine, PC: C=O, 1739; CH_2 scissoring mode, 1466; asymmetric phosphate stretch, $vaPO_2^-$, 1261; CO–O–C asymmetric stretch, 1180; symmetric phosphate stretch, $vsPO_2^-$, 1093; CO–O–C symmetric stretch, 1072 [7].

Polyadenilic monophosphoric acid, PA: contributions from NH_2 bending modes, 1184 and 1645; ring vibrational bands, 1604, 1576, 1476; $vaPO_2^-$, 1247; C-NH_2 stretch (convoluted), 1213; $vs PO_2^-$, 1086 (maximum of a highly convoluted band).

Polyuridilic monophosphoric acid, PU: C(2)=O, 1702; C(6)=O, 1650; NH bending mode, 1687, 1679; ring vibrational bands, 1620; $vaPO_2^-$, 1260, 1248; $vs PO_2^-$, 1105, 1065.

Calf thymus DNA: C=O, secondary structure (B form), 1709, (Z form), 1688, 1650; $vaPO_2^-$, 1246; $vsPO_2^-$, 1095, 1071.

An abundant literature deals with the assignment of these and several other bands (e.g. those arising from the saccharide backbone) [8]. Considering that [5, 6] different changes in the polynucleotide are induced when one or two moles of the metal are added, the ratio of the components in the mixtures was polynucleotide, PC, metal cation 1, 1, 1 or 1, 1, 2.

Complexes between Double Stranded PA/PU, PC and Metal Ions (all wavenumbers in cm^{-1})

(a) PA/PU and metal ions. The spectrum of double stranded PA/PU is the sum of the spectra of the

components. Addition of K^+ does not introduce noticeable changes. In the presence of Mg^{2+}, strong bands at 1663 (PU, 1682) and 1642 (PA, 1639), an increase of PO_2^- modes and a new profile for the $vaPO_2^-$, are found. With a 1/2 ratio a consistent drop of the bands at 1707, 1682 and 1663 is found, the band at 1642 is shifted to 1646 and the $vsPO_2^-$ is stronger than $vaPO_2^-$ with broadening. With Ca^{2+} the band of medium intensity at 1705 is almost undetectable (due to the weak interaction of CO with Ca^{2+} appreciably linked to the NH_2 of PA); a marked drop of the bands at 1600 and 1578 occurs; small changes occur in the region of PO_2^- modes (more in the sym mode than in the asym). A decrease in intensity and broadening are found in bands lower than $900\ cm^{-1}$. A 1/2 ratio does not introduce further changes, except for a red shift of $5\ cm^{-1}$ of the 1248 $vaPO_2^-$ and a large intensity increase of the $vsPO_2^-$ modes. Small changes occur in the presence of Mn^{2+}, mainly in the 1250–1000 region.

(b) Complexes of PA/PU, PC and metal ions. The addition of PC causes the following changes: (*i*) a little red shift (3–4 cm^{-1}) of the ester band of PA is observed with all the ions; (*ii*) the NH_2 band is not affected by the metal; (*iii*) the $vaPO_2^-$ mode is more intense than the $vsPO_2^-$ mode; (*iv*) a double amount of the metal does not introduce further changes with respect to the 1/1 molar ratio.

Complexes of DNA, PC and Metal Ions

(a) Complexes of DNA and metal ions. When K^+ is the metal a decrease of the band at 1648 (NH_2) and a change in band profile is found around 1600; the band at 1701 can be attributed to the secondary structure. Shifts of $vaPO_2^-$ (from 1243 to 1240) and of $vsPO_2^-$ (1096 to 1090, 1056 to 1064), with the second mode more intense than the former, appear. The intensity ratios between bands around 1700–1660 and those of the PO_2^- modes remain unchanged. When the amount of K^+ is doubled the secondary structure (1700, 1712) is more evident with respect to the free nucleotide and the ratio $vaPO_2^-/vsPO_2^-$ is in favour of the former. With Mg^{2+}, the band at 1695 is absent while the one of free DNA at 1640 is stronger and shifted to 1635 and, with respect to free DNA, the intensity ratio $vaPO_2^-/vsPO_2^-$ is higher. In the region 700–500, changes in band profiles are registered. A higher content of Mg^{2+} in the complex introduces a noticeable change of band profile in the region 1700–1600: the bands at 1647 and 1637

are stronger than the one at 1600 while the intensity ratio between the bands at 1710 and 1647 is unaffected; instead the ratio between the intensities of the bands at 1700–1600 and $vsPO_2^-$ is increased. Large changes are found in the profiles of PO_2^- modes and more peaks appear in the region 700–500 (Fig. 1). With Ca^{2+} a slight decrease of bands at 1700–1600 is observed: the band at 1647 is unchanged. A slight prevalence of $vsPO_2^-$ vs $vaPO_2^-$ modes also occurs. With a 1/2 ratio an increase of the band at 1646 with respect to those at 1600 and 1418 is observed together with an enhancement of the intensity ratio $vsPO_2^-/vaPO_2^-$; the bands in the region 700–600 are more pronounced. When Mn^{2+} is the cation and a 1/1 ratio is used, an increase in intensity of the band at 1648 and a modification in the profile of the composite band at 1594 take place. The most evident change in the PO_2^- region is a different profile of the $vsPO_2^-$ mode, with the maximum $5\ cm^{-1}$ blue shift. More evident changes occur in the presence of 2/1 ratio of the metal: the 1707 band is

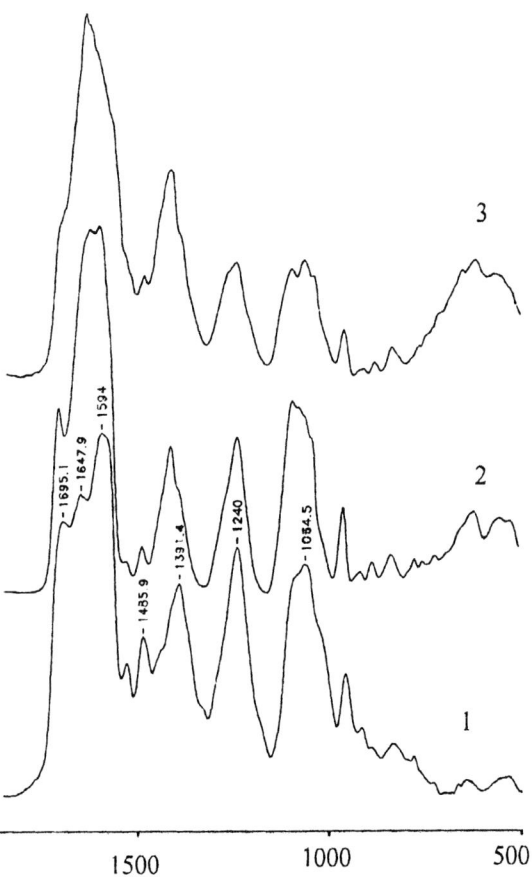

Fig. 1. FT-IR spectra (Kubelka–Munk), dry films, of DNA (*1*), DNA + Mg^{2+} (*2*) and DNA + $2Mg^{2+}$ (*3*)

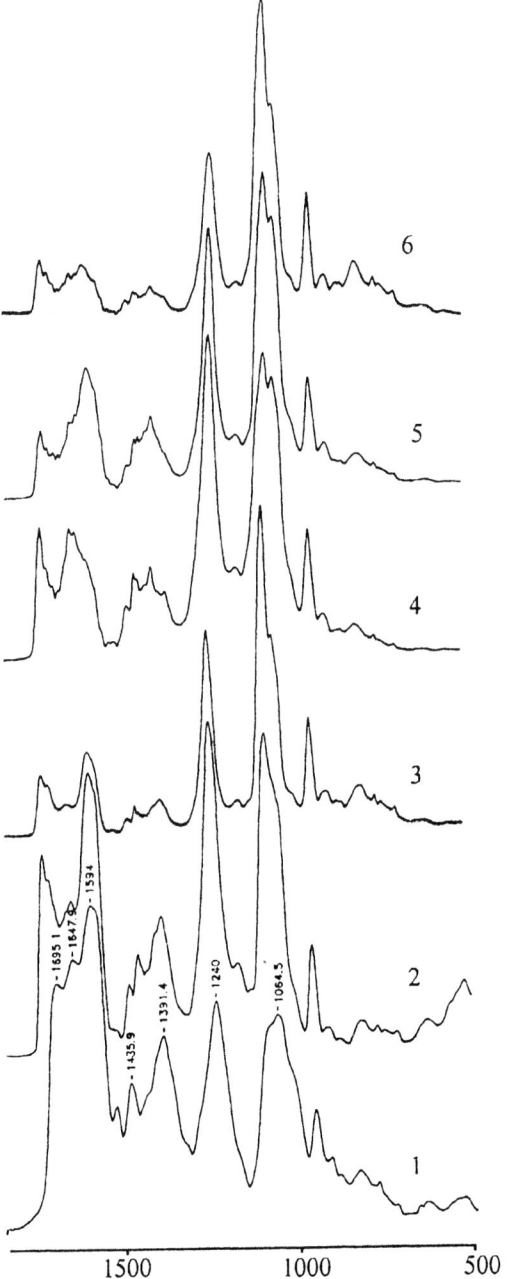

Fig. 2. FT-IR spectra (Kubelka–Munk), dry films, of DNA (*1*), PC (*2*) and of a 1/1 complex with K$^+$ (*3*), Mg^{2+} (*4*), Ca^{2+} (*5*) and Mn^{2+} (*6*)

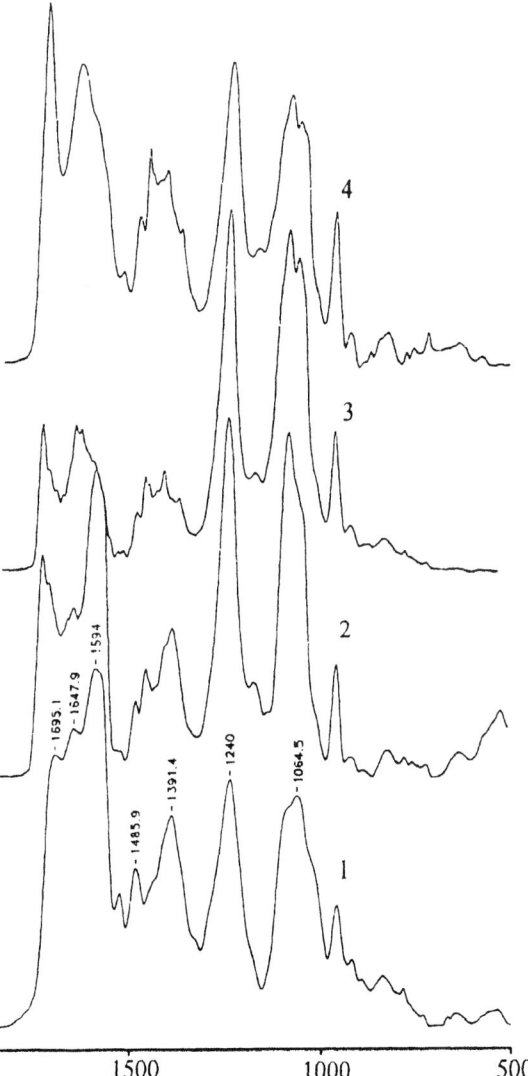

Fig. 3. FT-IR spectra (Kubelka–Munk), dry films of DNA (*1*), DNA + PC (*2*), DNA + PC + Mg^{2+} (*3*) and DNA + PC + 2Mg^{2+} (*4*)

a well resolved shoulder; the vaPO$_2^-$ mode is 5 cm^{-1} red shifted and the profile of the vsPO$_2^-$ mode is modified; changes are also found in the region 900–500.

(b) Complexes of DNA, PC and metal ions. In the presence of PC and 1/1 ratio of metal ion, the following

modifications are observed (Fig. 2): (*i*) the ester band is increased in intensity in the presence of K$^+$ while it retains its intensity with the other ions; (*ii*) small changes in the region 1720–1600 occur, mainly when Mg^{2+} is the cation; (*iii*) except with Ca^{2+}, changes are induced in the profile of the convoluted band between 1700 and 1600; (*iv*) the intensity ratio vaPO$_2^-$/vsPO$_2^-$ is higher than unity. With 2/1 ratio of metal we observe: (*i*) an evident increase in intensity of the ester band at 1734; (*ii*) with K$^+$ the band at 1645 is more intense and shifted to 1665; (*iii*) with Mg^{2+} the bands between 1700 and 1600 are highly convoluted (Fig. 3 shows the dependence of DNA and PC modes on the concentration of Mg^{2+}); (*iv*) changes occur in the region 1500–1400;

(v) the intensity ratio $vaPO_2^-/vaPO_2^-$ depends on the metal ion but the profiles are unaffected; (vi) the intensities of the bands between 700 and 500 are increased in the presence of bivalent cations.

Some final considerations can be deduced: the nature of the metal ion and its concentration strongly influence the interactions; in the presence of PC the interactions become more evident when the concentration of the ion is doubled; more apparent changes occur in the region 1750–1600 even if PO_2 and CO–O–C, as well as modes below 900 cm^{-1}, also appear important for a study of the interaction.

References

[1] V. G. Budker, A. A. Godovikov, L. P. Naumorva, J. A. Slepneva, *Nucleic Acids Res.* **1980**, *8*, 2499.

[2] V. V. Kuvichkin, A. G. Sukhomudrenko, *Biofizika* **1987**, *32*, 628.

[3] B. Bonora, G. Palka, R. Jovine, S. Miscia, E. Caramelli, F. A. Manzoli, *Physize. Chem. Phys.* **1981**, *13*, 19.

[4] V. V. Kuvichkin, L. A. Volkova, E. P. Naryshkina, F. Sh. Isangalin, *Biofizika* **1989**, *34*, 405.

[5] A. J. P. Alix, I. Bernard, M. Manfait, *Spectroscopy of Biological Molecules*, Wiley, Chichester, 1985.

[6] Yu. Bao-Zhu, W. N. Hansen, *Proc. SPIE* **1989**, *1145*, 409.

[7] M. Jackson, H. H. Mantsch, *Spectrochim. Acta Rev.* **1993**, *53*, 15.

[8] B. Schrader (ed.), *Infrared and Raman Spectroscopy*, VCH, Weinheim, 1995, p. 345 (and references therein)

Mikrochim. Acta [Suppl.] 14, 173–174 (1997)

Photothermal Absorption Spectroscopy of Mica

James A. Burt[1,*], **Kirk H. Michaelian**[2], and **Shuliang Zhang**[2,**]

[1] Department of Physics, York University, 4700 Keele Street, North York, Ontario, M3J 1P3, Canada
[2] Natural Resources Canada, CANMET, Western Research Centre, P.O. Bag 1280, Devon, Alberta, T0C 1E0, Canada

Abstract. Infrared absorption by mica produces an elastic wave that is detected by lead zirconate titanate (PZT) in intimate contact with the sample. The interferogram recorded with an FT-IR spectrometer at a mirror speed of 0.070 cm/s does not contain a centre-burst at zero path difference.

Key words: PZT, lead zirconate titanate, photoacoustic, mica, photothermal absorption.

One of the main attractions of photoacoustic and photothermal spectroscopy is the simplicity of the required experimental apparatus. For example, lead zirconate titanate (PZT), in direct contact with a sample, can detect the elastic wave resulting from absorption of modulated light. This technique was first employed by Hordvik and Schlossberg [1] to measure weak absorption coefficients of solids at Ar^+, CO_2 and CO laser wavelengths. Subsequently, Eyring and co-workers [2–4] used dispersive or Fourier-transform spectrometers and PZT detection to obtain absorption spectra of powdered samples in the visible region. Burt et al. [5] measured the transmission of laser radiation in optical fibres in intimate contact with PZT.

The use of PZT to measure absorption of mid-infrared radiation by mica is described in this paper. Data were acquired with a commercial FT-IR spectrometer, in which modulation is effected by the interferometer moving mirror. This modulation makes conventional photoacoustic infrared spectroscopy possible; in fact, the frequencies involved are similar to those in some of the published experiments mentioned

above. The objective of the current work is to use this principle to obtain photoacoustic FT-IR spectra of mica in direct contact with PZT.

Experimental

A PZT disc 7 mm in diameter and 2 mm thick was glued to a strip of mica of approximate dimensions $35 \times 10 \times 0.1$ mm and placed in the focal plane of the infrared beam in the sample compartment of a Bruker IFS 88 FT-IR spectrometer. The unamplified voltage from the PZT was input directly to the main amplifier board, yielding signals more than two orders of magnitude smaller than those obtained with conventional infrared detectors. In other words, the PZT output was about 10–100 times less than that from a commercial gas-microphone photoacoustic cell.

Interferograms were obtained by co-adding 100 scans at mirror velocities of 0.070 or 0.32 cm/s without checking for correlation of centre-burst locations from scan to scan. The lower velocity is normally used for photoacoustic spectroscopy, while the higher one is appropriate for a standard DTGS detector. For the hydroxyl stretching region near 3600 cm^{-1}, these speeds correspond to modulation frequencies of about 500 and 2300 Hz, respectively.

Results and Discussion

Qualitatively, the interferogram recorded at low mirror velocity (Fig. 1a) does not seem unusual; however, closer examination reveals that there is no centre-burst at zero path difference, and that the strongest peak occurs 6 points ($\sim 3.8\,\mu$m) away from this location. Interferograms that lack a centre-burst can occur in a differential spectrum containing both positive and negative bands [6]. Special attention should be given to phase correction of such data, since common algorithms assume positive intensities at all frequencies. The spectrum in Fig. 1b, calculated with the standard Mertz phase correction algorithm in the Bruker software, is thus not accurate; nevertheless the prominent ~ 3600 cm^{-1} band and several other peaks are visible.

---

 * Present address: Department of Chemistry, University of Alberta, Edmonton, Alberta, T6G 2G2, Canada
 ** To whom correspondence should be addressed

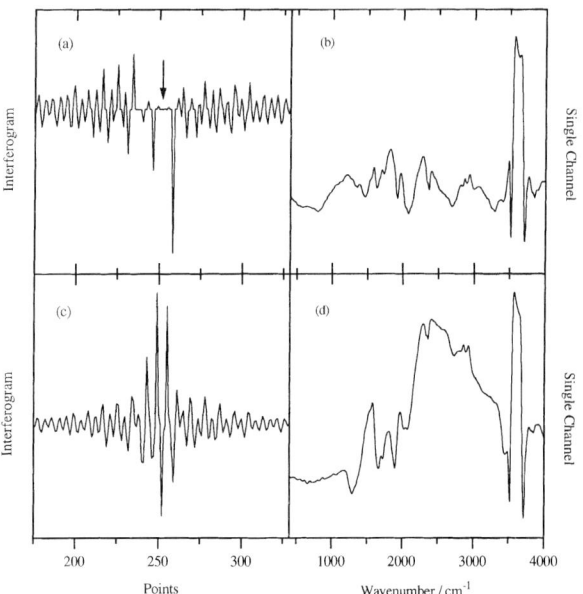

Fig. 1. Results for mica/PZT at interferometer mirror velocity 0.070 cm/s (upper boxes) and 0.32 cm/s (lower boxes): *a* and *c* show averages of 10 interferogram files (central region); the arrow shows the zero-path-difference point; *b* and *d* are spectra calculated from these average interferograms. The modulation frequencies are 56–560 Hz for *b* and 250–2500 Hz for *d*

The lower boxes in Fig. 1 show similar data, obtained with the higher mirror speed. In this case, the central region of the interferogram (Fig. 1c) displays a familiar well-defined centre-burst. Moreover, the spectrum derived from this interferogram (Fig. 1d) is noticeably improved with respect to that in Fig. 1b.

Several features are superimposed on the expected sloping background: the previously mentioned mica band; a negative-going peak near 2300 cm^{-1}, due to atmospheric CO_2; and positive bands near 2850 and 2920 cm^{-1}, attributable to CH_2 and CH_3 groups in the glue. The positive 1630 cm^{-1} band probably arises from water in the mica.

The negative lobes on either side of the 3600 cm^{-1} band are artifacts, and occur because of imperfect phase correction. The phase spectrum calculated for this region resembles a rectangle function, changing by approximately π radians in the wings of the band; this result leads to inaccurate calculation of the spectrum, since multiplicative phase correction is based on the assumption that the phase angle varies slowly with frequency. This phenomenon also affects parts of the spectrum between 1250 and 2000 cm^{-1}. These artifacts are eliminated by restricting the phase to just two quadrants instead of the usual calculation from 0 to 2π radians [7].

References

[1] A. Hordvik, H. Schlossberg, *Appl. Opt.* **1977**, *16*, 101.
[2] M. M. Farrow, R. K. Burnham, M. Auzanneau, S. L. Olsen, N. Purdie, E. M. Eyring, *Appl. Opt.* **1978**, *17*, 1093.
[3] L. B. Lloyd, S. M. Riseman, R. K. Burnham, E. M. Eyring, M. M. Farrow, *Rev. Sci. Instrum.* **1980**, *51*, 1488.
[4] L. B. Lloyd, R. K. Burnham, W. L. Chandler, E. M. Eyring, M. M. Farrow, *Anal. Chem.* **1980**, *52*, 1595.
[5] J. A. Burt, K. J. Ebeling, D. Efthimiades, *Opt. Commun.* **1980**, *32*, 59.
[6] C. A. McCoy, J. A. de Haseth, *Appl. Spectrosc.* **1988**, *42*, 336.
[7] S. L. Zhang, K. H. Michaelian, J. A. Burt, *Opt. Eng.* **1997**, in press.

Mikrochim. Acta [Suppl.] 14, 175–177 (1997)

Radiometric Applications of Fourier-Transform Spectrometers

Christopher J. Chunnilall*, Nigel P. Fox, and **Evangelos Theocharous**

Division of Quantum Metrology, National Physical Laboratory, Queen's Road, Teddington, TW11 0LW, U.K.

Abstract. The relative spectral response of a filter radiometer, centred at 800 nm, has been measured with a Fourier-transform (FT) spectrometer, using a trap detector as the reference detector. Comparison with high accuracy laser spectrophotometric measurements gives agreement at the 1–2% level, and has demonstrated the viability of the technique for the measurement of filter radiometers, as well as the principle of characterizing FT instruments with a standard detector. FT spectrometers have therefore been shown to have potential uses in the characterization of radiometric detector (and source) transfer standards.

Key words: characterization, filter radiometer, Fourier-transform spectrometer, radiometry.

Fourier-transform spectrometers have many applications, each making use of different features of such instruments. This paper describes the possibility of a new radiometric application, the characterization of a filter radiometer, which makes use of the following features: high wavenumber accuracy, high throughput, low stray light, spectrally invariant polarization and a relatively fast measurement time.

In the last decade there has been a rapid growth in the requirement for high accuracy radiometric measurements in a wide variety of applications ranging from remote sensing of the Earth through to the measurement of temperature in steel furnaces, and from the output of household light bulbs to changes in the UV radiation caused by atmospheric ozone depletion. Each of these applications and of many others, requires the measurement of radiation in a narrow spectral band ranging from a few nm to a few hundred nm in width. These measurements are usually performed with filter radiometers although these instruments are often called by names specific to their application, e.g. pyrometers, photometers, radio-meters, black-light meters etc. The most important radiometric properties of their design are essentially generic, consisting of a detector, filter (coloured glass or interference) and aperture.

The characterization of such instruments requires the measurement of their response to radiation at high spectral resolution across their spectrally sensitive band. The radiation used for this purpose must also be spatially uniform and be non-polarized. This characterization is usually performed by measuring with a grating spectrometer the transmittance or responsivity of the individual components making up the filter radiometer, and then combining the results for the finished instrument. This technique can be time-consuming and does not always meet the required accuracy because of the compromises required in operating the grating spectrometer, i.e. spectral resolution, radiant power, spatial uniformity.

The NPL has recently developed an alternative technique based on the use of a novel laser spectrometer [1]. While achieving high accuracy, approaching 0.05% uncertainty, the measurement time for a nominal 20 nm bandpass filter can take several weeks. This paper compares the measurement of the response of a 20 nm bandpass filter radiometer centred around 800 nm obtained by the high-accuracy laser-based technique and by a Fourier-transform spectrometer.

Experimental

The experimental set-up is shown in Fig. 1. The output of the spectrometer, a Bomem DA8.1, was directed into a $BaSO_4$ coated integrating sphere which had three ports. The sphere diameter was 3 cm, the input port was 1 cm in diameter and the other two ports were 3 mm in diameter. One of the smaller ports was used as the output port, the remaining port, shown dashed in Fig. 1, being used as a monitor channel using a trap detector [2]. The uniform output from the integrating sphere was then imaged on to either the

---

* To whom correspondence should be addressed

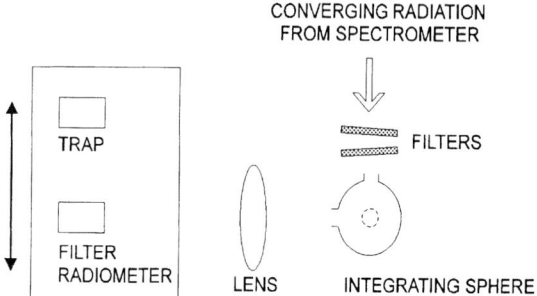

CONVERGING RADIATION
FROM SPECTROMETER

TRAP

FILTER
RADIOMETER   LENS   INTEGRATING SPHERE

FILTERS

Fig. 1. Experimental set-up

Light entering the trap detector needed to have a divergence not exceeding 4° and the lens was used in a magnifying configuration, giving a convergence angle of ~2.5°. The wavenumber (wavelength) accuracy of the instrument had been previously checked against molecular absorption and ion emission lines. A check was also made against the 869.35 nm line of the McCrone wavelength standard by inserting the YAG crystal into the beam between the lens and the trap detector. The measurable accuracy was found to be within 0.05 cm^{-1} at the resolution (4 cm^{-1}) used. The filter radiometer was enclosed in a water jacket for temperature stabilization at 20 °C, since such instruments can have significant temperature sensitivity. The measurement time was approximately 2 hours [45 minutes for each device (2700 accumulations), plus ~20 minutes changeover time and re-stabilization].

Results and Discussion

In this experiment $G(\lambda)$ was taken to be a constant G, and the relative spectral response of the filter radiometer was measured. G was chosen such that the relative response obtained was normalized at the 799.5 nm value, the laser-based measurements having been normalized at the 799.575 nm value. Figure 2 compares the high accuracy calibration (0.05% uncertainty) of the filter radiometer with our measurements (left-hand ordinate), and also shows the difference (right-hand ordinate). Agreement between the curves is good, the difference being less than 2% over most of the range and the integrated areas agreeing to within 0.4%. There is some evidence that our measurements are low on the long-wavelength edge of the filter response curve. Whether this is due to a shift in the characteristics of the filter radiometer, which can change with time, or due to a systematic error in the experimental technique, will need to be decided by further measurements.

reference detector or the filter radiometer. The beam non-uniformity was measured by scanning a 0.3 mm aperture trap detector over the beam and was found not to exceed 1%. One of the limitations of this method is the reliance on the source remaining stable over the time of the experiment. The reference detector was a trap detector [2], which is a novel design constructed from 3 Hammamatsu S1337 silicon photodiodes. This type of detector has good linearity and uniformity, a high S/N ratio, excellent long and short term stability, and is relatively insensitive to temperature. This detector is "quantum flat" to within 0.1% in its spectral range of operation in this experiment, allowing its absolute response $R_T(\lambda)$ to be given by

$$R_T(\lambda) = \frac{\varepsilon\lambda}{1239.45} \tag{1}$$

where λ is measured in nm and ε, the external quantum efficiency (~99.8%), takes account both of reflection losses and departure from 100% internal quantum efficiency.

Background spectra $B_T(\lambda)$ and $B_F(\lambda)$ were obtained with the trap and filter radiometer respectively. Both spectra were obtained at 4 cm^{-1} resolution (~0.25 nm at 800 nm) with weak Norton–Beer apodization, with the same configuration and settings of the FT spectrometer. These settings were chosen to be appropriate for both devices. The absolute response of the filter radiometer $R_F(\lambda)$ is then given by

$$R_F(\lambda) = K(\lambda)B_F(\lambda) \tag{2}$$

where

$$K(\lambda) = G(\lambda)[R_T(\lambda)/B_T(\lambda)] \tag{3}$$

The $G(\lambda)$ factor accounts for any difference in the trap and filter radiometer aperture areas, and also any non-equivalence in placing the filter radiometer at exactly the same position in the output beam as the trap detector. If the spectral distribution of the light falling on the detectors does not vary with position, $G(\lambda)$ reduces to a constant.

Equations (2) and (3) show that the use of a standard detector allows an absolute characterization of the output of the FT spectrometer to be made. Furthermore, even if $G(\lambda)$ is not known, but can be taken to be a constant, a relative characterization can be made. This use of a standard detector to characterize an FT spectrometer complements the use of standard sources such as black bodies [3, 4] to characterize FT spectrometers for measuring sources.

The spectral content of the radiation incident on the detectors was selected by using two filters placed in front of the integrating sphere. The filters, a 900 nm edge short-wave pass filter and a 750 nm edge long-wave pass filter, selected the range between 750 and 900 nm, thereby blocking any radiation which lay outside the region where Eq. (1) holds. The total spectral power was at the 1 μW level, which is within the range over which both devices exhibit a linear response.

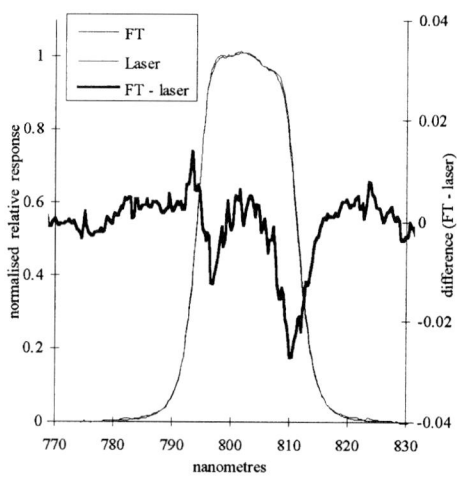

Fig. 2. Comparison of laser and FT measurements

Our measurements represent a first attempt at using this technique, and given the promising agreement with the high-accuracy laser measurement, the technique appears worthy of further refinement.

Further improvement may be achieved by optimizing the optics for this particular application. The Bomem 55 W tungsten–halogen lamp was used as the light source, and was found to be stable to around 0.5% over a half-minute period during the experiment, by using the monitor channel of the integrating sphere. However, it was also observed that occasionally the reading of the monitor trap would show an "excursion" for 1–5 minutes, and any data collected in this period would have to be discarded. The short-term noise was at the 0.1% level. An external source could be used that would enable short-term noise and the uncertainty due to source drift to be reduced to potentially 0.01%, thereby also enabling longer data acquisition times. Some gain in noise rejection may be obtained by setting the scan conditions to match the pass-band of the filter radiometer more closely, but in these measurements the same settings were used for both measurements and therefore the larger spectral region covered by the trap was sampled. However, a change of filters in front of the integrating sphere to narrow the effective range of the trap is a possibility.

Relative measurement, such as obtained here, can be made absolute by calibration at a fixed wavelength. An absolute measurement can also be made since the response of the reference trap detector is absolutely known. Higher resolution is also available from the FT spectrometer, at the cost of longer length interferograms and reduced signal due to reduced source aperture.

It follows from this work that the absolute or relative spectral throughput of an FT spectrometer can be obtained by the use of an absolute or reference standard detector together with a standard source, thereby allowing an FT spectrometer to be employed for the characterization of both detector and source transfer standards.

Conclusion

The relative spectral response of a filter radiometer measured by a Fourier-transform spectrometer, using a trap detector as the reference detector, gives agreement at the 1–2% level when compared with high-accuracy laser spectrophotometric measurements, and has demonstrated the principle of characterizing FT spectrometers with a standard detector. FT spectrometers have been shown to have potential uses in the characterization of radiometric detector (and source) transfer standards.

Acknowledgements. We thank J. E. Martin for providing the data from his accurate laser-based calibration of the filter radiometer and we also acknowledge helpful discussions with W. S. Hartree and S. N. Mekhontsev.

References

[1] V. E. Anderson, N. P. Fox, D. H. Nettleton, *Appl. Optics* **1992**, *31*, 536.
[2] N. P. Fox, *Metrologia* **1991**, *28*, 197.
[3] D. D. LaPorte, R. Howitt, *Proc. SPIE* **1982**, *364*, 2.
[4] H. E. Revercomb, H. Buijs, H. B. Howell, D. D. LaPorte, W. L. Smith, L. A. Sromovsky, *Appl. Optics.* **1988**, *27*, 3210.

Mikrochim. Acta [Suppl.] 14, 179 180 (1997)

Depth Profiling of Wool Fibres by FT-IR ATR Spectroscopy

Jeffrey S. Church* and Andrea L. Woodhead

Commonwealth Scientific and Industrial Research Organisation, Division of Wool Technology, PO Box 21, Belmont, Victoria, 3216, Australia

Abstract. The use of Fourier-transform infrared attenuated total reflectance (ATR) spectroscopy to depth profile Merino wool fibres has been investigated. The results obtained from untreated fibres are compared to those obtained from polyamide fibres. Second order derivatives were used to assist in data analysis. The use of ATR depth profiling to study the near surface chemistry of oxidation reactions is also demonstrated.

Key words: FT-IR ATR spectroscopy, depth profile, wool fibres, polyamide fibres.

Wool fibres belong to a group of proteins known as α-keratins. They have two main morphological components, a central cortex consisting of closely packed cells arranged parallel to the fibre axis and an outer sheath of cuticle cells or scales [1]. Fine Merino wool fibres have diameters of less than 22 µm and the cuticle layer is normally one cell thick (~ 0.5 µm) except where overlap occurs. The bulk of the wool fibre is made up of cortical cells. The main components of the wool fibre differ not only in their amino-acid composition but also in the secondary structure associated with their protein chains. The cortical cells have a significantly higher α-helical content compared to the cuticle cells. The modification of the surface of the wool fibre is an important aspect of wool processing; for example, surface oxidation is a common step in most shrink-proofing treatments [1]. The ability to monitor structural and chemical changes to the near surface of the wool fibre is therefore important. In this study we have investigated the use of ATR spectroscopy to depth profile Merino wool fibres.

Experimental

Merino wool (20.2 µm) and polyamide (20.0 µm) fibres were cleaned prior to analysis by three extractions with petroleum spirit and three

with ethanol, followed by a rinse in reverse-osmosis water. Oxidized wool fabric samples were prepared by treatment with aqueous solutions of 5% on the weight of wool (oww) $KMnO_4$ followed by 10% oww acidic $NaHSO_3$. Spectra were recorded with a Perkin–Elmer System 2000 Fourier-transform infrared spectrometer equipped with a SPECAC model 11900 variable angle ATR accessory and a narrow-band MCT detector. Both ZnSe and Ge internal reflective elements (IREs) with 45° bevels were used. Data acquisition and manipulation were done with PE-Grams/2000 version 2.0. Data were collected at a resolution of $4 \, cm^{-1}$, and 512 scans were co-added.

Results and Discussion

It is well known that for ATR spectroscopy, the depth of penetration (d_p) of the incident radiation into the sample is a function of wavelength, angle of incidence and the refractive indices of the sample and the IRE [2]. For samples such as those used in this work it is important to realise that the d_p values calculated can only be used as rough guides. The amide I region of a series of spectra obtained from polyamide fibres is shown in Fig. 1 (left). The spectra obtained from the shallowest depths are very similar. The amide I peak obtained with the Ge IRE at 45° is sharp and exhibits a slight high-wavenumber tail. The spectrum obtained with the ZnSe IRE at 60° is broader but very symmetrical. In contrast, the spectrum obtained from the deepest d_p (ZnSe 45°) is quite different. The second-derivative analysis of the amide I peak of this latter spectrum reveals a broad main peak with three minor components on the low wavenumber side. These data support a fibre structure consisting of a crystalline outer sheath and a less ordered inner region, as expected for extruded polyamide fibres.

The corresponding series of spectra obtained from Merino wool fibres, shown in Fig. 1 (right), is significantly more complex. The results of the second-derivative analysis of the amide I peak of these spectra are summarized in Table 1. These results suggest that as

---

* To whom correspondence should be addressed

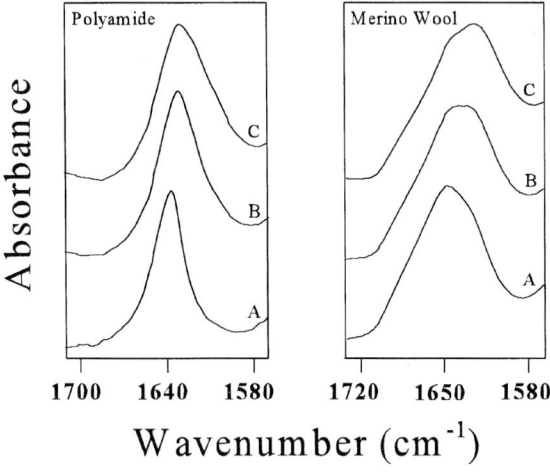

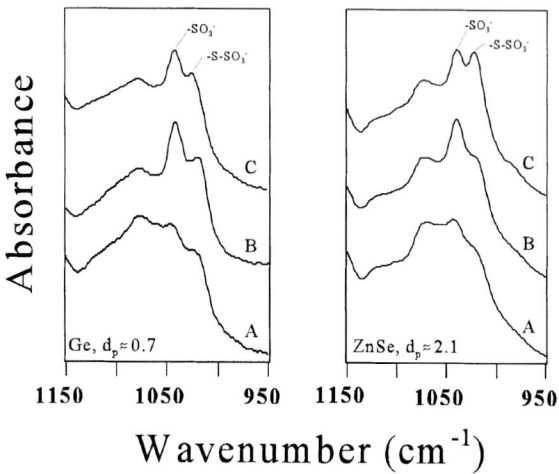

Fig. 1. The amide I region of a series of spectra obtained from different depths of penetration from polyamide (left) and Merino wool (right) samples. Approximate d_p values are $A \approx 0.3$, $B \approx 0.9$ and $C \approx 1.4 \, \mu m$ at $1600 \, cm^{-1}$

Fig. 2. The $1150–950 \, cm^{-1}$ region of ATR spectra obtained at two different depths of penetration from A untreated, B KMnO$_4$ treated and C KMnO$_4$/NaHSO$_3$ treated wool fabric samples

Table 1. Second-derivative analysis of the amide I peak for the Merino wool spectra shown in Fig. 1

Spectrum	Components (cm^{-1} and relative intensity)				
A	1654 w	1649 s	1642 w	1625 m	–
B	1655 w	1649 s	1645 w	1622 s	1613 w
C	–	1649 m	1645 w	1622 s	1613 w

s = strong, m = medium, w = weak.

the ATR signal arises from deeper within the fibre the amide I band component at lower wavenumber becomes more dominant. As Jurdana et al. [3] found the amide I frequency of decuticled wool to be at higher wavenumber than that of whole wool, it is possible that the spectral changes that we are detecting are due to the large refractive index changes associated with strong absorbances.

Oxidative treatments applied to the wool fibre result in the cleavage of the disulphide cross-links and the formation of oxidized sulphur species [1]. In general it is desired to restrict the effect of these treatments to the fibre surface. The reaction of KMnO$_4$ with wool creates a layer of MnO$_2$ on the fibre surface. The NaHSO$_3$ after-treatment removes this layer. The S–O stretching region of two series of ATR spectra obtained at 45°

from wool fabric samples taken during a KMnO$_4$/ NaHSO$_3$ treatment is shown in Fig. 2. By comparison, the spectra obtained from the untreated fabric with the ZnSe ($d_p \approx 2.1$) and Ge ($d_p \approx 0.7$) IREs are very similar. After KMnO$_4$ oxidation, similar amounts of the sulphonate, –SO$_3$, and S-sulphonate, –S–SO$_3$, species are detected at both depths. In contrast, after the NaHSO$_3$ clearing treatment the S-sulphonate concentration deeper within the fibre is significantly increased compared to that nearer the surface. These results indicate that the clearing treatment is attacking disulphide bonds internal to the fibre as well as removing the MnO$_2$ from the fibre surface.

Acknowledgement. The authors recognize the support of the Australian wool growers and the Australian Government, who fund research and development through the IWS.

References

[1] J. A. Maclaren and B. Milligan, in: *Wool Science: The Chemical Reactivity of the Wool Fibre*, Science, Marrickville, Australia, 1981, pp. 1–17, 53–57.
[2] N. J. Harrick, *Internal Reflection Spectroscopy*, Wiley, New York, 1967, p. 30.
[3] L. E. Jurdana, K. P. Ghiggino, I. H. Leaver, C. G. Barraclough, P. Cole-Clarke, *Appl. Spectrosc.* **1994**, *48*, 44.

Mikrochim. Acta [Suppl.] 14, 181–182 (1997)
© Springer-Verlag 1997

Mid-Infrared as a New Tool for Detecting Adulteration in Fruit Products

Marianne Defernez*, **Evelyn K. Kemsley,** and **Reginald H. Wilson**

Institute of Food Research, Norwich Laboratory, Norwich Research Park, Colney, Norwich NR4 7UA, U.K.

Abstract. This paper illustrates the use of attenuated total reflectance (ATR) spectra of fruit purees or jams, and diffuse reflectance (DRIFT) spectra of jams, to provide the chemical information necessary to check the composition and authenticity of these products. The fingerprint region contains information reflecting the carbohydrate, acid, and pectin content. Although differences between products from different fruit types, or between genuine and adulterated samples, can be seen in the spectra by eye, the samples cannot be identified unambiguously by visual inspection. Statistical methods can aid this identification, however. An example of the use of principal component analysis is presented, enabling genuine raspberry purees to be distinguished from raspberry purees doctored by the addition of sucrose solution to a level of about 6% sucrose.

Key words: adulteration, ATR, DRIFT, fruit, PCA.

The issue of food adulteration is highlighted by cases where the consumption of adulterated products has endangered the consumer. However, the full extent of the problem is unknown. Detecting adulteration involves the identification of often tiny differences in the composition of raw materials and finished products. Methods commonly directed at the problem require lengthy and cumbersome wet chemical processes. Fourier-transform infrared spectroscopy (FT-IR) provides a quick alternative to traditional procedures, and newly developed techniques mean that the method is particularly applicable to the determination of food authenticity. In the work presented here, FT-IR was applied to gather information on the composition of fruit purees and jams.

Experimental

Spectroscopy

ATR [1] was used to study fruit purees and both ATR and DRIFT [1] were employed for jams. All spectra were collected on a Spectra-Tech (Applied Systems Inc.) MONITIR FT-IR spectrometer system. The jams or the purees were spread directly onto the ATR plate. For the DRIFT experiments the insoluble solids of the jams were collected on a filter paper after dissolution of the sample, and from these a spectrum was recorded.

Chemometric Analysis

The spectral information was analysed by data reduction techniques based on PCA [2] with Win-Discrim, a specialized package for discriminant analysis.

Results and Discussion

The sampling techniques used produced good quality spectra of high signal-to-noise ratio. ATR spectra of both purees and jams were very reproducible. The reproducibility of jam DRIFT spectra was not so good, owing to the inhomogeneity of the material deposited on the filter paper. The fingerprint region between 800 and 2000 cm^{-1} mainly contained absorptions from the coupled C–O modes of the carbohydrates (900–1200 cm^{-1}), and the C=O stretch of the pectins and acids, superimposed on the water band (1500–1700 cm^{-1}). Jams provided two different types of DRIFT spectra in terms of relative intensities, which corresponded to "normal" and "reduced" sugar content jams [3]. ATR spectra of purees or jams produced from different fruit species were quite similar by visual inspection because they contained the same basic constituents [3, 4].

---

* To whom correspondence should be addressed

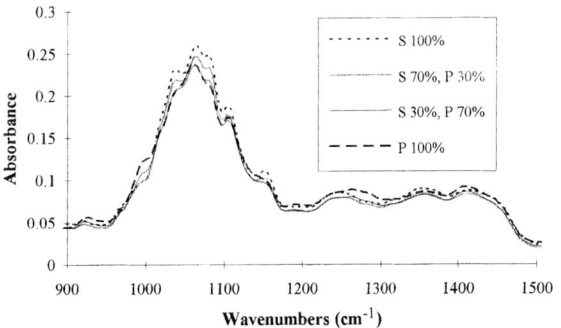

Fig. 1. Series of spectra of mixtures of strawberry (S) and plum (P) purees

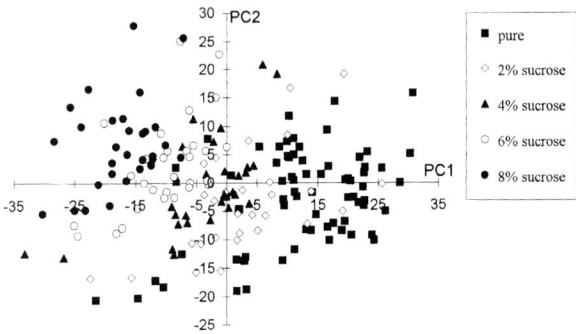

Fig. 2. PCA score plot for a set of raspberry purees, some being pure, and some adulterated with sucrose

Different chemometric methods allowed purees or jams from different fruit species to be distinguished [3, 4]. Detecting adulterated samples is a more difficult problem, but a simple approach is presented here to demonstrate the potential of chemometric methods. Figure 1 shows an example of the alteration undergone by the spectrum of a fruit puree doctored by increasing percentages of adulterant. Determination of authenticity by eye was not possible, as the differences were very subtle, and also because of the natural variability of the spectra from one fruit to another. Figure 2 shows the

scores on the first and second axes yielded by a principal component analysis (PCA) performed with a set of puree spectra, some being from genuine raspberry, and some from raspberry adulterated by sucrose at different levels. PCA discriminated the spectra according to their sucrose level on the first axis, and although very low levels of adulteration could not be distinguished from genuine purees on the first axis, samples with 6% sucrose and above formed a separate cluster.

Conclusion

This paper shows that good quality mid-infrared spectra were obtained from opaque samples such as fruit purees and jams by ATR and DRIFT. Although adulterated samples could not be detected by visual inspection of the spectra, an analysis by PCA was shown to offer a solution. The performance and limits of other chemometric methods are currently being examined. The relatively low cost and the rapidity of this procedure make it particularly suited to use in the industrial environment.

Acknowledgements. The authors are grateful to the United Kingdom Preserves Manufactures' Association Authenticity Committee for its support and provision of samples. MD thanks the AIR programme of the European Union for a Travel and Mobility Grant. KK and RHW acknowledge the funding by MAFF and the BBSRC.

References

[1] P. S. Belton, R. H. Wilson, in: *Perspective in Modern Chemical Spectroscopy* (D. Andrews, ed.), Springer, Berlin Heidelberg New York Tokyo, 1990, pp. 67–86.
[2] D. L. Massart, L. Kaufman, in: *The Interpretation of Analytical Chemical Data by the Use of Cluster Analysis* (P. J. Elving, J. D. Winefordner, I. M. Kolthoff, eds.) Wiley, New York, 1983, pp. 39–62.
[3] M. Defernez, R. H. Wilson, *J. Sci. Food Agric.* **1995**, *67*, 461.
[4] M. Defernez, E. K. Kemsley, R. H. Wilson, *J. Agric. Food Chem.* **1995**, *43*, 109.

Mikrochim. Acta [Suppl.] 14, 183–184 (1997)

An FT-IR and Raman Spectroscopy Study of Electrochemically Modified Woollen Fabric

Victoria Fredline, Serge Kokot*, and **Christina Gilbert**

Centre for Instrumental and Developmental Chemistry, School of Chemistry, QUT, Brisbane, Australia

Abstract. A woollen fabric sample has been oxidized by electro-generated oxygen species produced at a platinum electrode. This treatment produced three visually discoloured or bleached regions. The areas of the fabric that were located close to the electrode surface showed dark brown discoloration. On either side of this region, bleaching of the wool was apparent to various degrees. Fabric well removed from the electrode appeared unaffected by the treatment, and was used as a reference. Such samples were analysed by the complementary techniques of FT-IR and FT-Raman spectroscopy, and the pattern recognition method of principal components analysis (PCA). The samples were successfully separated into three groups on the basis of the levels of known oxidation intermediates from cystine and other readily oxidizable amino-acid residues present in wool-keratin.

Key words: wool, electro-generated oxidation, keratin, cystine, amino-acid residues.

When wool is oxidized, the most susceptible amino-acid residues are tyrosine, methionine, tryptophan, cysteine and cystine. Of these cystine is the most abundant, constituting $\sim 10\%$ of the total fibre mass, and oxidation of the disulphide bonds proceeds through a series of oxidative intermediates, to the final product of cysteic acid [1–3].

$-SO-S-$	$-SO_2-S-$	$-S-SO_3$	$-SO_3$
cystine-S-monoxide	cystine-S-dioxide	cystine-S-sulphonate	cysteic acid
$1078\,cm^{-1}$	$1121\,cm^{-1}$	$1024\,cm^{-1}$	$1040/1171\,cm^{-1}$

In this paper, we discuss a comparison of FT-IR and FT-Raman spectra taken from different regions of wool samples treated with electro-generated oxygen species. The disulphide intermediates absorb in the infrared region, while the cystine disulphide bond, tyrosine, tryptophan and backbone skeletal informa-

tion can be monitored by using Raman spectroscopy [4, 5] (see wavenumbers above).

Experimental

The electrochemical treatment was similar to that used by Meyer et al. [6]. The fabric sample is wrapped around a platinum electrode, and subjected to constant current electrolysis ($80\,°C$, $135\,mA$, 7 hours). The electrolyte was a 1:1 sodium acetate/acetic acid buffer ($0.1\,M$; pH 4.74). FT-IR spectrometer: Perkin–Elmer 1600; sampling: KBr pressed disc, 64 scans, $4\,cm^{-1}$ resolution, nitrogen purge $15\,l/min$; background: blank KBr disc. FT-Raman spectrometer: Perkin–Elmer 2000/NIR FT-Raman, interfaced with a 386 computer, IRDM software; sampling: 250 scans, $400\,mW$ laser power, $4\,cm^{-1}$ resolution.

Results and Discussion

Both the IR and Raman spectra generally resemble those of untreated wool featured in the literature [1–3] with the common wool assignments agreeing within $5\,cm^{-1}$.

The infrared spectra show changes in the amide I and II bands at 1630 and $1530\,cm^{-1}$ respectively, and these can be attributed to conformational changes in the secondary protein structure (i.e. from α-helix or β-sheet to more random forms).

There are also changes evident in the region 1350–$1000\,cm^{-1}$, which includes both the disulphide oxidative intermediates and the amide III band, indicating that oxidation of the cystine bond has occurred.

Changes in the Raman spectra are much more difficult to detect as the spectra appear visually similar. However, there are some noticeable changes in the $515\,cm^{-1}$ region corresponding to the cystine ($-S-S-$) disulphide stretch. This band can be seen to reduce in intensity

---

* To whom correspondence should be addressed

with the electro-generated oxygen treatment, suggesting that the disulphide bond is cleaved by oxidation.

PCA [7] of the infrared data was performed over the region 1350–1000 cm^{-1}, to include the amide III band and the disulphide oxidation products. The plot of PC2 vs. PC3 effectively separated the samples into three groups, corresponding to the brown, bleached and reference regions. PC2 separated the spectra, with the reference scores positive, the bleached scores centring around zero, and the brown scores negative. The PC2 loadings vs. variables plot showed that the amide III bands, the oxidative intermediates and cysteic acid were responsible for the discrimination. Thus, IR spectra of the three different regions – brown, bleached and reference – have been separated on their levels of oxidation of the disulphide bond, as well as on conformational changes in the secondary structure of the protein.

PCA analysis was performed on the Raman data over the region 800–400 cm^{-1}, to include the disulphide stretch, as well as tyrosine and tryptophan residues (two other amino-acids susceptible to oxidation). The PC1 vs. PC2 plot showed a clear separation of the samples on PC1. The loadings vs. variables plot for PC1 showed the basis for this separation as the 515 cm^{-1} band of the cystine disulphide stretch, as well as bands corresponding to C–S stretch and the tyrosine residue (665 cm^{-1}) and tryptophan (750 cm^{-1}).

Thus, FT-Raman spectroscopy provides complementary information about the treatment of wool-keratin with electro-generated oxygen species, and confirms that the spectra distinguish the treated samples on the basis of different levels of oxidation of the fibrous protein.

References

[1] U. Schumacher-Hamedat, C. Laurini, V. Schneider, *Proc. 8th Intern. Wool Text Res. Conf.* **1990**, *4*, 451.

[2] F. J. Douthwaite, D. M. Lewis, U. Schumacher-Hamedat, *Textile Res. J.* **1993**, *63*, 177.

[3] L. Coderch, R. Pons, P. Erra, *J. Soc. Dyers Colour.* **1991**, *107*, 410.

[4] L. J. Hogg, H. G. M. Edwards, D. W. Farwell, A. T. Peters, *J. Soc. Dyers Colour.* **1994**, *110*, 196.

[5] E. Carter, P. M. Fredericks, J. Church, R. J. Denning, *Spectrochim. Acta* **1994**, *50A*, 1927.

[6] U. Meyer, S. Kokot, R. Weber, J. Zürcher-Vogt, *Textilveredlung* **1987**, *22*, 185.

[7] A. Thielemans, D. L. Massart, *Chimia* **1985**, *39*, 236.

Mikrochim. Acta [Suppl.] 14, 185–186 (1997)

FT-IR and FT-Raman Spectroscopy of Processed Cotton Fabrics – A Chemometric Study

Christina Gilbert and **Serge Kokot***

Centre for Instrumental and Developmental Chemistry, Queensland University of Technology, Brisbane, Australia

Abstract. Two different sets of processed cotton fabrics were investigated by FT-IR and FT-Raman spectroscopy. The chemometric technique of principal component analysis, was applied to the IR and Raman spectra from each fabric series in an attempt to discriminate between the various processing stages. The results indicate that a combination of the two techniques and chemometrics adequately discriminates between cotton samples from each processing stage.

Key words: cotton fabrics, fabric processing, FT-IR spectroscopy, FT-Raman spectroscopy.

Previous work [1, 2] has shown that DRIFT spectroscopy may be used to distinguish between differently processed cotton fabrics sampled sequentially during their normal processing stages. The differentiation between the processing stages is partially due to molecular changes as well as to physical differences between the processed fibres. In general, FT-Raman spectroscopy provides complementary information to IR. Therefore, in this study, the same two sets of differently processed cotton fabrics were studied by both vibrational spectroscopic techniques for comparison. Principal component analysis (PCA) was applied to the information contained in the spectra from each fabric series.

Experimental

FT-IR

Circular cloth samples (ca. 6 mm diam.) were placed in the centre of a DRIFT sampling cup filled with dried, preground KBr.

Sampling

PE 1600 spectrometer, DRIFT accessory, 128 scans, resolution $4 \, cm^{-1}$, N_2 purge, $\log (1/R)$ display.

FT-Raman

Each fabric ($1.5 \times 6 \, cm$) was folded upon itself several times to produce a thick sample and was tightly clamped into the sample holder.

Sampling

PE Series 2000 NIR/FT-Raman spectrometer, 200 scans, resolution $4 \, cm^{-1}$, laser power 400 mW.

Spectral Data Treatment

Spectra, converted into ASCII format (SPECTRA CALC [3]), were entered into an Excel 4.0 spreadsheet. SIRIUS [4] software was used for chemometrics.

Results and Discussion

Poplin Fabric (Source – Müller AG, Seon, Switzerland)

The fabric types were raw, singed, mercerized then boiled, bleached and finished with optical brighter. The FT-Raman and DRIFT spectra of the processed poplin fabric are quite different but provide essentially complementary information. The IR spectral region shows many bands from 1800 to $800 \, cm^{-1}$ in comparison to the Raman spectra, which show some sharp peaks in the range $1500–900 \, cm^{-1}$, and also below $550 \, cm^{-1}$. The suspected IR carbonyl band (ca. $1710 \, cm^{-1}$) from samples containing a fluorescent brightener is apparently Raman-inactive. Similarly, the IR band at $1646 \, cm^{-1}$, often attributed to adsorbed H_2O, is also Raman-inactive. On the other hand, the

* To whom correspondence should be addressed

IR spectrum below 800 cm^{-1} does not show any recognizable bands, while the Raman spectrum below 550 cm^{-1} contains many bands attributed to different skeletal modes of cotton cellulose and the C_1–O–C_4 bridge structure.

Thus, the chemometric analysis of the two sets of spectra concentrated on different spectral ranges. The range chosen for PCA of the DRIFT spectra was from 1800 to 1500 cm^{-1}, as most of the visual differences between the spectra were observed in this region; for the Raman spectra two regions from 1550 to 850 and from 550 to 300 cm^{-1} were combined for PCA, as most of the spectral bands, for cotton-cellulose appeared in these ranges. The PC1 vs. PC2 scores plot of the IR data produced three clusters, consisting of the raw and singed spectral group, the bleached and mercerized group and the fluorescently brightened samples. In contrast, the PC1 vs. PC2 scores plot of the Raman data produced, initially, two clusters – a raw and singed group and a large group with all the other samples. The IR and Raman spectra of the raw and singed samples were differentiated from the others by PCA. Examination of the PC2 loadings plot suggested that the IR spectra are probably differentiated by the broad shoulder at ca. 1708 cm^{-1} present only in spectra of these samples. It is possible that this band arises from a naturally-occurring impurity or remnants of a sizing agent on the fabric surface. The differentiation in the Raman spectra is apparently influenced by C–H, C–C and skeletal vibrations that in principle, are more likely to be present in spectra of non-whitened samples (PC1 loadings plot).

The IR spectra of the mercerized and bleached samples were differentiated only when that group was analysed independently [1]. However, PCA applied to all of the Raman spectra showed a clear separation of these two classes. For the IR region, the separation of bleached and mercerized samples was influenced by subtle differences in the areas attributed to OH stretching and bending, possible carbonyl or carboxylate formation from oxidation of the OH groups in bleaching, and ring stretching or C–O–C bridge frequencies (PC2 loadings plot). The IR spectra of the fluorescently whitened fabric were clearly separated by the strong carbonyl absorption of the whitening agent, present only in spectra of these samples. In comparison, PCA of the Raman spectra did not show any clear distinction between the bleached and fluorescently whitened fabric samples because the strong carbonyl peak observed in the DRIFT spectra was absent.

Muslin Fabric (Source: Raduner AG, Horn, Switzerland)

Fabric types were raw, desized, boiled, bleached and mercerized. The DRIFT spectra of this fabric appeared very similar and, unlike the poplin spectra, did not show many significant differences. Only the spectra of the raw and desized samples showed the presence of a shoulder at ca. 1710 cm^{-1}. The Raman spectra were also very similar to each other; the minor differences observed were difficult to assign because of their weak intensities. PCA was performed for the IR analysis in the 1800–800 cm^{-1} range while for the Raman analysis the combined regions of 1550–850 cm^{-1} and 550–300 cm^{-1} were used.

PC1 vs. PC3 scores plot of the DRIFT spectra of the processed muslin fabric produced four distinct clusters – the raw samples group, the desized and mercerized groups, and the boiled and bleached cluster. Similarly, the PC1 vs. PC2 scores plot of the Raman spectra produced four distinct clusters, but the separation observed was slightly different. The raw and desized samples clustered together, but the boiled, bleached and mercerized samples formed separate groups. The differentiation of the raw and desized samples in the IR study was not observed in the corresponding Raman analysis. Also it is interesting to note that the differentiation of the boiled and bleached samples observed in the Raman analysis was not possible in the IR analysis.

Conclusions

The IR spectra from each processing stage appeared very similar, with only some minor variations present from spectrum to spectrum, and differences in the Raman spectra of these samples were barely discernible. The chemometric investigations showed that the sample discrimination observed for the Raman spectra was slightly different from that for the IR study, but the chemical basis for the observed separations was essentially the same. The use of the two techniques successfully discriminated between all fabric processes.

References

[1] C. Gilbert, S. Kokot, U. Meyer, *Appl. Spectrosc.* **1993**, *47*, 741.
[2] C. Gilbert, S. Kokot, *Vib. Spectrosc.* **1995**, *9*, 161.
[3] Spectra Calc, *Users Guide*, Galactic Industries, Salem USA, 1990.
[4] O. M. Kvalheim, *Sirius (Version 2.3) Software*, Department of Chemistry, University of Bergen, Norway.

Mikrochim. Acta [Suppl.] 14, 187–189 (1997)
© Springer-Verlag 1997

FT-IR Investigations and Modelling of Anisotropic Materials: Application to Carbon Fibre Composites

Wulf Grählert* and **Volkmar Hopfe**

Fraunhofer Institute for Materials Physics and Surface Engineering (IWS), P.O. Box 16, D-01171 Dresden, Federal Republic of Germany

Abstract. The application of specular reflectance FT-IR spectroscopy for the investigation of carbon fibre reinforced epoxy composites is discussed. The use of the general 4×4 matrix algorithm allows determination of the dielectric functions of the sample, parallel and perpendicular to the optical axis. The parallel component shows a Drude-like spectral behaviour of the carbon fibre, and the perpendicular component is mainly determined by the polymer.

Key words: specular reflectance, FT-IR, carbon fibre epoxy composites, dielectric function.

The characterization of carbon fibre reinforced composites in the IR region is usually done by the diffuse reflectance FT-IR (DRIFT) [1, 2] or photoacoustic FT-IR (PAS) techniques [3]. However, these methods do not take into account the optical anisotropy of the material.

The off-normal specular reflectance measured with polarized light is strongly influenced by the fibre orientation causing the optical anisotropy of the sample. On the one hand the usual treatment of the measured spectra by the isotropic Kramers–Kronig Transformation (KKT) has to be modified and on the other hand the back calculation of the reflectance spectra based on the dielectric functions has to be extended from the 2×2 to the 4×4 matrix algorithm [4, 5].

Experimental

A Bruker IFS-66 FT-IR spectrometer equipped with a DTGS detector was used in conjunction with a Seagull specular reflectance attachment (Harrick) complemented by a wire-grid polarizer (Au on AgCl) to collect the data. The spectra were recorded in a frequency range of 400–$6000\,\mathrm{cm}^{-1}$ at a resolution of $4\,\mathrm{cm}^{-1}$, with angles of incidence of $5°$ and $70°$. The sample consisted of highly ordered carbon fibres and an epoxy resin matrix (bisphenol A – diglycidyl ether); the surface was polished. The calculations were done with the FSOS-program package [6].

---

* To whom correspondence should be addressed

Results and Discussion

The anisotropic optical behaviour of multilayered systems is conveniently described by the general 4×4 matrix algorithm [4, 5]. This method takes into account the adjustable measuring parameters such as the type of optical spectrum, the angle of incidence α, and the state of polarization of the incident and detected waves (TM, TE), and the intrinsic properties of the system such as the dielectric functions in any space direction for all media ($\hat{\varepsilon}_j = \varepsilon'_j + i\varepsilon''_j$; $j = x, y, z$), and the orientation of each optical axis relative to the sample coordinate system, described by the Euler angles. The rotation angle β was introduced in this algorithm to describe a rotation of the sample by the user (see the optical model in Fig. 1e). The case of the plane of incidence being parallel to the optical axis has been chosen as $\beta = 0$. The 4×4 matrix algorithm based on the calculation of the Characteristical matrix C of the system resulting from the Propagation matrix P(m), the Dynamical matrix D(m) and the inverted one $\mathrm{D}^{-1}(m)$ of each medium m ($\mathrm{C} = \mathrm{D}_0^{-1}\mathrm{D}_1\mathrm{P}_1\mathrm{D}_1^{-1}\mathrm{D}_2\mathrm{P}_2 \cdots \mathrm{D}_{m-1}^{-1}\mathrm{D}_m\mathrm{P}_m$). Problems arose with layer stacks containing both isotropic and anisotropic media. To overcome this problem, special matrices the components of which depend on the angle of rotation β, were substituted by D(m)isotropic and $\mathrm{D}^{-1}(m)$isotropic in the case of isotropic media.

The fibre reinforced composites represent an uniaxial material ($\hat{\varepsilon}_y = \hat{\varepsilon}_z$; $\hat{\varepsilon}_x = c$-axis). Those reflectance spectra were measured which depend on the dielectric function either only perpendicular (Fig. 1a) or only parallel (Fig. 1b) to the optical axis. The highest intensity is observed with spectra measured under polarization in the fibre direction. From the measured spectra the dielectric functions of both components were calculated by the isotropic Kramers–Kronig Transformation. There is no significant difference in

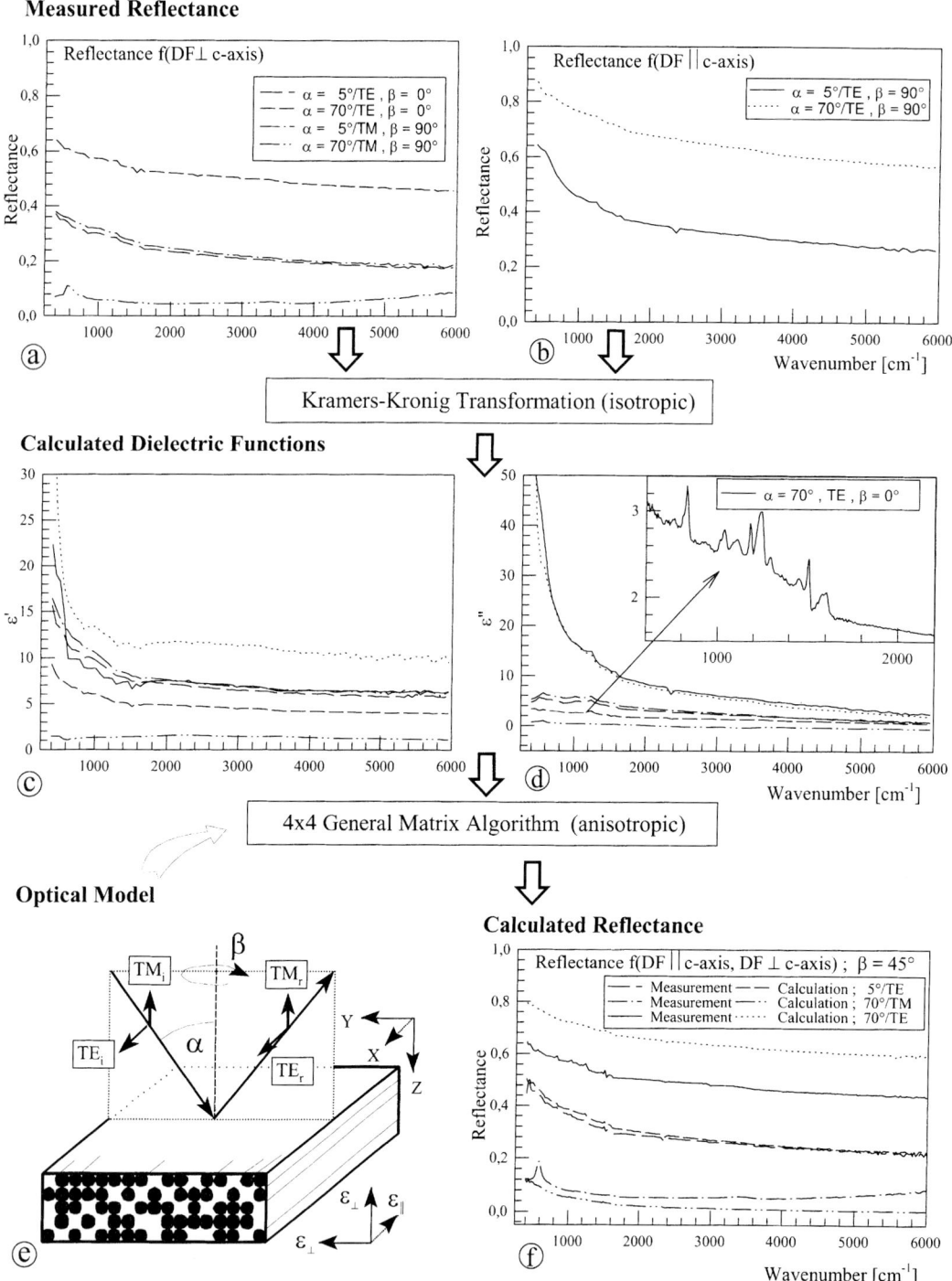

Fig. 1. Scheme for the determination and verfication of the dielectric functions of a carbon fibre reinforced epoxy composite, parallel and perpendicular to the fibre orientation

$$
D(m)_{isotropic} =
\begin{bmatrix}
-\sin\beta & -\sin\beta & -\dfrac{\sqrt{\varepsilon-\sin^2\alpha}\,\cos\beta}{\sqrt{\varepsilon}} & -\dfrac{\sqrt{\varepsilon-\sin^2\alpha}\,\cos\beta}{\sqrt{\varepsilon}} \\[3mm]
-\sqrt{\varepsilon-\sin^2\alpha}\,\sin\beta & \sqrt{\varepsilon-\sin^2\alpha}\,\sin\beta & -\sqrt{\varepsilon}\,\cos\beta & \sqrt{\varepsilon}\,\cos\beta \\[3mm]
\cos\beta & \cos\beta & -\dfrac{\sqrt{\varepsilon-\sin^2\alpha}\,\sin\beta}{\sqrt{\varepsilon}} & -\dfrac{\sqrt{\varepsilon-\sin^2\alpha}\,\sin\beta}{\sqrt{\varepsilon}} \\[3mm]
-\sqrt{\varepsilon-\sin^2\alpha}\,\cos\beta & \sqrt{\varepsilon-\sin^2\alpha}\,\cos\beta & \sqrt{\varepsilon}\,\sin\beta & -\sqrt{\varepsilon}\,\sin\beta
\end{bmatrix}
$$

$$
D(m)_{isotropic}^{-1} =
\begin{bmatrix}
-\dfrac{\sin\beta}{2} & -\dfrac{\sin\beta}{2\sqrt{\varepsilon-\sin^2\alpha}} & \dfrac{\cos\beta}{2} & -\dfrac{\cos\beta}{2\sqrt{\varepsilon-\sin^2\alpha}} \\[3mm]
-\dfrac{\sin\beta}{2} & \dfrac{\sin\beta}{2\sqrt{\varepsilon-\sin^2\alpha}} & \dfrac{\cos\beta}{2} & \dfrac{\cos\beta}{2\sqrt{\varepsilon-\sin^2\alpha}} \\[3mm]
-\dfrac{\cos\beta\,\sqrt{\varepsilon}}{2\sqrt{\varepsilon-\sin^2\alpha}} & -\dfrac{\cos\beta}{2\sqrt{\varepsilon}} & -\dfrac{\sin\beta\,\sqrt{\varepsilon}}{2\sqrt{\varepsilon-\sin^2\alpha}} & \dfrac{\sin\beta}{2\sqrt{\varepsilon}} \\[3mm]
-\dfrac{\cos\beta\,\sqrt{\varepsilon}}{2\sqrt{\varepsilon-\sin^2\alpha}} & \dfrac{\cos\beta}{2\sqrt{\varepsilon}} & -\dfrac{\sin\beta\,\sqrt{\varepsilon}}{2\sqrt{\varepsilon-\sin^2\alpha}} & -\dfrac{\sin\beta}{2\sqrt{\varepsilon}}
\end{bmatrix}
$$

the spectral behaviour of the real parts (Fig. 1c). Compared with ε' the imaginary parts ε'' show a considerable influence of the fibre alignment. The axial component is mainly influenced by the carbon fibre, obvious from the Drude-like behaviour (Fig. 1d). The perpendicular component is characterized by the polymer, and a detailed band assignment is possible. However, the dielectric functions obtained do not represent those of the pure materials but are a mixture of them.

The dielectric functions determined were verified by comparing a set of measured and calculated spectra after rotation of the sample by $\beta = 45°$. In accordance with theory both the perpendicular and parallel components influence the reflectance. The anisotropic 4×4 matrix algorithm was used for the calculations of the spectra. It was shown that there is good agreement between the corresponding spectra. There is a difference in the intensity of the $70°/\mathrm{TE}$ spectra but the spectral behaviour is correctly described.

In conclusion, carbon fibre reinforced composites can be considered as an optically anisotropic system and therefore can be described by the 4×4 matrix algorithm to a good approximation. The required dielectric functions are obtained by using the isotropic KKT for selected measuring conditions.

References

[1] K. C. Cole, C. Lehto, M. Yuhasz, *Proc. SPIE* **1986**, *665*, 243.
[2] P. R. Young, B. A. Stein, A. C. Chang, *28th National SAMPE Symposium* 1983, 824.
[3] J. M. Chalmers, J. Wilson, *Mikrochim. Acta* **1988**, *II*, 109.
[4] P. Yeh, *Surf. Sci.* **1980**, *96*, 41.
[5] A. N. Parikh, D. L. Allara, *J. Chem. Phys.* **1992**, *96*, 927.
[6] V. Hopfe, *FSOS: Software Package for IR–VIS–UV Spectroscopy of Solids, Interfaces and Layered Systems*, TU Chemnitz, Reprint 122.

Mikrochim. Acta [Suppl.] 14, 191 192 (1997)

Infrared and Raman Spectroscopic Studies of Asbestos, Transite and Concrete

Peter R. Griffiths*, **Ian R. Lewis**, and **Nathan C. Chaffin**

Department of Chemistry, University of Idaho, Moscow, ID 83844, U.S.A.

Abstract. Methods of determining the presence of asbestos and transite (an asbestos/concrete blend) for decommissioning and decontamination of hazardous waste sites have been developed, based on use of infrared diffuse reflectance (DR) spectrometry [in both the mid- (MIR) and near-infrared (NIR) spectral regions] and Raman spectroscopy. A comparison of the use of 632.8, 785 and 1064 nm excitation wavelengths for Raman spectroscopy was made, to determine which wavelength was most appropriate for this study.

Key words: asbestos, transite, concrete, diffuse reflactance spectrometry, hazardous waste sites, Raman spectroscopy.

It was found that the highest success rate for all the materials tested was achieved by using NIR DR spectroscopy, because of the minimal need for sample preparation, short measurement times, and transmission efficiency of fibre-optic material in the NIR. MIR DR spectrometry was limited by specular reflectance interference or the need for sample preparation, while the use of Raman spectrometry was severely curtailed with both 632.8 and 1064 nm excitation by fluorescence and laser absorption (heating). Nevertheless, both MIR, DR, after grinding and dilution of the materials, and Raman spectroscopy at 785 nm produced spectra significantly enhanced in comparison with previous work [1–3].

The feasibility of obtaining NIR and Raman spectra by use of fiber-optic interfaces has been investigated with reference to available fibers, wavelength coverage, and suitability for use in a hostile environment. Both silica ($v = 12000$–$4500\,cm^{-1}$) and heavy metal fluoride {HMF} ($v = 8000$–$2700\,cm^{-1}$) fibers have been compared for use in studying the NIR and extended NIR wavenumber regions. It was found that both fibers produced spectra with excellent signal-to-noise ratios in under 60 seconds and could be used for remote

measurements with a sample to spectrometer distance of 5 m. It was found that extended NIR HMF fibers have one major advantage over silica fibers in that they may be used to obtain spectra of the CH, OH, and NH fundamentals and therefore identify materials at much lower concentration than would be possible by observing only the first overtones of these modes. Disadvantages include cost, low bend-radius tolerances, lower transmission characteristics, and difficulties in cutting and polishing the fibers.

Figure 1 shows the NIR DR spectra of four common asbestos minerals, chrysotile, crocidolite, amosite, and tremolite, acquired in less than 45 seconds by using a fiber-optic probe based on one hundred 200-μm core silica fibers. The observed vibrations are the first overtones of the OH stretching modes, and the relative intensity of the vibrations reflects the occupancy of

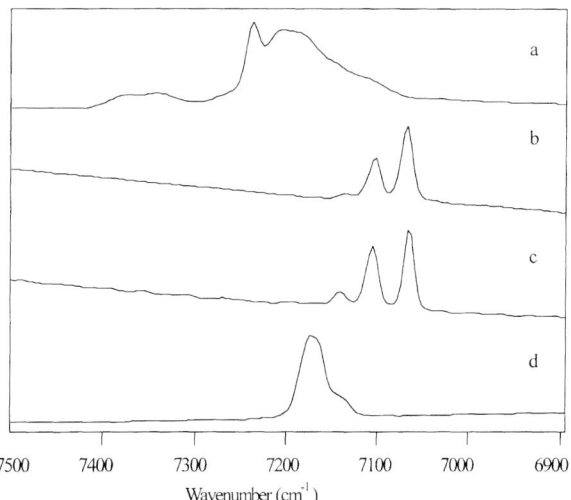

Fig. 1. NIR DR spectra of common asbestos minerals: *a* chrysotile, *b* crocidolite, *c* amosite, and *d* tremolite, recorded by using a silica fiber-optic probe

---

* To whom correspondence should be addressed

cation sites as well as differences in the anharmonicity of the fundamental modes [4]. Chrysotile is a serpentine mineral which possesses a different crystal packing arrangement from the three other asbestos minerals, which are all amphiboles.

When studying concrete, it was found that the spectrum possesses OH fundamentals and overtones, originating from the Si–OH base structure, at lower wavenumber positions than for the asbestos minerals. For example the concrete OH overtone's band envelope extends from approximately 7150 to $6200\,cm^{-1}$, while from Fig. 1 it can be seen that for all the asbestos minerals the band envelope is limited to the region between 7450 and $7000\,cm^{-1}$. The samples of transite analyzed in this study possessed a band envelope from 7400 to $6200\,cm^{-1}$, indicating that the transite sample contained both chrysotile and concrete. A similar band envelope could be produced by mixing appropriate amounts of the pure concrete spectrum and the spectrum of the mineral chrysotile.

Acknowledgements. The authors would like to thank the U. S. Department of Energy for funding this project through a contract from the Coleman Research Corporation.

References

[1] V. C. Farmer (ed.), *The Infrared Spectra of Minerals*, Mineralogical Society, London, 1971.

[2] A. N. Lazarev (ed.), *Vibrational Spectra and Structure of Silicates*, Consultants Bureau, New York, 1972.

[3] J. J. Blaha, G. J. Rosasco, *Anal. Chem.* **1978**, *50*, 892.

[4] I. R. Lewis, N. C. Chaffin, M. E. Gunter, P. R. Griffiths, *Spectrochim. Acta A* **1996**, *52*, 315.

Mikrochim. Acta [Suppl.] 14, 193–195 (1997)
© Springer-Verlag 1997

Investigation of Electrically Stressed SF$_6$ Gas by Using FT-IR Spectroscopy

H. M. Heise[1,*], **P. R. Janissek**[2], **R. Kurte**[1], **P. Fischer**[1], **S. M. A. Segundo**[2], and **D. Klockow**[1]

[1] Institut für Spektrochemie und Angewandte Spektroskopie, Bunsen-Kirchhoff-Str. 11, D-44139 Dortmund, Federal Republic of Germany
[2] Laboratòrio Central de Electrotécnica e Eletrōnica, COPEL, Rua Coronel Dulcidio 800, 80420 Curitiba, Brasil

Abstract. Compressed gaseous sulphur hexafluoride is commonly used as an insulating medium in high-voltage equipment. In the presence of traces of oxygen and moisture, SF$_6$ can undergo decomposition under electrical stress. FT-IR-spectroscopic methods provide the possibility for a multicomponent trace gas analysis in the SF$_6$-matrix with the option of on-line measurements. Decomposition compounds, produced by partial discharges, can be determined by multivariate spectral data analysis. Time-dependent concentration changes of the SF$_6$ impurities in different infrared gas cells have been evaluated.

Key words: FT-IR spectroscopy, trace gas analysis, sulphur hexafluoride decomposition, partial discharge.

SF$_6$ is used as an insulator gas in electrical high-voltage equipment and most often in gas-insulated switchgear, since it possesses a number of favourable properties such as having a high dielectric strength, the ability to quench electrical arcs and, in addition, shows chemical inertness and low toxicity. However, it undergoes some decomposition under electrical stress such as in arcs, sparks or partial discharges, particularly in the presence of oxygen and moisture. The decomposition products can be toxic and corrosive. A method for multicomponent analysis of SF$_6$ and its breakdown products is provided by FT-IR spectrometry [1]. For this purpose we coupled an electrical discharge chamber to an FT-IR spectrometer. Studies were made into contaminant concentrations in stressed SF$_6$ and into apparent changes as a function of time, with the gas resident in a two-cell system. Multivariate spectral data analysis was applied within limited spectral intervals because of the many strong absorption bands of SF$_6$ in the mid-infrared range studied. Results from different discharge experiments showed the importance of on-line measurements.

---

* To whom correspondence should be addressed

Experimental

Our discharge chamber, with an inner volume of 300 ml, was constructed from stainless steel and Teflon™, housing a needle–plate configuration with a gap distance of 10 mm. The needle tip usually had a radius of 30 μm. In a typical experiment, an alternating voltage of 29 kV was applied, with the virtual charge measured to be 50 pC per discharge event at the beginning and a higher value at the end of the experiment (see also Table 1). The SF$_6$ gas was from Messer Griesheim (Krefeld, Germany), grade 3.0, equivalent to a purity of 99.9%. The compounds SF$_4$ and SOF$_2$ were purchased from Fluorochem Ltd. (Old Glossop, U.K.), both of purity 97%. A mixture containing known concentrations of SOF$_4$ and SO$_2$F$_2$ was supplied by Prof. Minkwitz of Dortmund University (Germany) so that infrared standard spectra of all these species could be measured. The concentrations of S$_2$F$_{10}$ were estimated by using literature data from Wilmshurst et al. [2].

Infrared spectra were recorded by using a Bruker IFS 113v FT-IR spectrometer (Karlsruhe, Germany) equipped with a KBr beamsplitter and an MCT detector (cut-off at 530 cm^{-1}). One hundred interferograms were accumulated, yielding a spectral resolution of 0.5 cm^{-1}. The interferograms were Fourier-transformed after correction for detector non-linearity [3] and triangular apodization. The stressed SF$_6$ gas from the discharge chamber was directly expanded into an evacuated two-cell system, containing a heatable transmittance cell of 10 cm optical path-length (SPECAC, Orpington, U.K.) and a multiple reflection cell with a path-length set to 3 m (Long Path Mini Cell from Infrared Analysis, Anaheim, U.S.A.) allowing the simultaneous analysis for medium-concentrated and trace components. The experimental conditions are summarized in Table 1.

Results and Discussion

Absorption bands were assigned to decomposition compounds over the accessible spectral range and quantitative concentration values were determined. The following spectral intervals were considered for least-squares analyses by fitting standard spectra to the discharged SF$_6$ spectra: SO$_2$F$_2$ 1530–1410 cm^{-1}, SOF$_4$ and S$_2$F$_{10}$ 840–765 cm^{-1}, each with the 10 cm cell; SOF$_2$ and SF$_4$ 760–710 cm^{-1} with the 3 m cell. In Fig. 1 the SF$_6$ spectra recorded by using the 10 cm cell

Table 1. Details of the partial discharge experiments with SF_6 as the insulator gas

Experiment no.	Exp. time (hours)	Partial discharges[a] initial	final	Voltage (kV)	Pressure[b] (kPa)
1	4.7	15	35	20	112.8
2	39.5	50	100	29	113.2
3	68.8	50	120	29	110.7
4	140.0	50	100	30	111.0
5	3.5	20	100	14	36.7
6	68.0	50	160	30	63.3

[a] Virtual charge per discharge event between the electrodes measured in pC.

[b] Final pressure in the FT-IR cell system at the end of the experiment. The initial pressure in the discharge chamber was 300 kPa for all experiments. A chamber leakage occurred in the fifth and sixth experiments.

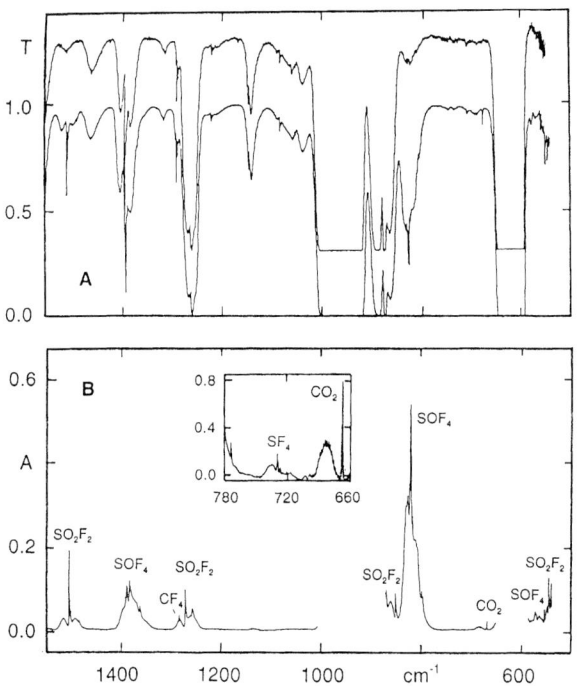

Fig. 1. A Transmittance spectrum of stressed SF_6 (path-length 10 cm, 105 kPa, discharge duration 140 hr, see also Table 1) and that of pure SF_6 (offset). **B** Difference spectrum obtained by scaled subtraction of the spectra above (inset shows the spectrum recorded with optical path-length of 3 m)

for experiment no. 4 and the difference spectrum, for which the SF_6 spectral features are compensated, to provide evidence of the decomposition products, are presented. The inset displays the spectral range for the determination of SF_4 and SOF_2 (optical path-length 3.0 m).

There are significant quantities of sulphur (VI) oxy-fluorides, i.e. SOF_4 and SO_2F_2, detected, which increase with duration of the discharge. A similar dependence exists for S_2F_{10}. The discharge conditions are rather favourable for the production of S_2F_{10}, contrary to the situation in arcs. On the other hand, sulphur(IV) compounds such as SOF_2 or SF_4, were found only at lower concentrations (see Table 2), which is opposite to the production rates given for corona discharges by Sauers et al. [4]. Experiment no. 6 still fits into the scheme, because the insulating strength was reduced, allowing much higher virtual charges per discharge, as noticed at the end of that experiment (final SF_6 pressure in the discharge chamber was 200 kPa). Experiment no. 5 is an exception, as the SOF_2 concentration after a short exposure time was above 100 ppm. Here, the leakage in the chamber was so significant that the insulating SF_6 was only at atmospheric pressure, enabling uncontrolled diffusion

Table 2. Volume fraction (ppm) of several decomposition products measured directly after the discharge experiment (for details see text)

Experiment no.	SO_2F_2	SOF_4	S_2F_{10}	SOF_2	SF_4
1	2.0 ± 0.8	3.0 ± 1	0.0	3.4 ± 0.4	1.0 ± 0.2
2	61 ± 2	501 ± 10	83.4 ± 1	39.0 ± 2	9.0 ± 0.2
3	76 ± 2	667 ± 20	136 ± 1	20.0 ± 2	8.3 ± 0.2
4	182 ± 5	1005 ± 20	149 ± 2	18.3 ± 3	9.2 ± 1
5	38 ± 1	128 ± 1	5.3 ± 0.4	126 ± 1	0.8 ± 0.1
6	273 ± 8	1614 ± 32	237 ± 3	22.2 ± 3	9.0 ± 0.3

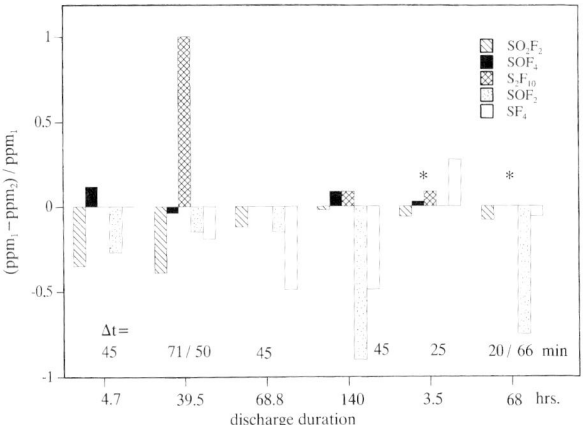

Fig. 2. Time-dependent concentration changes for different discharge experiments. SOF$_2$, SF$_4$ were measured with use of the 3 m cell, the others by means of the 10 cm cell; time difference Δt is between two measurements (if two values are given, differences refer to measurements with the 3 m and 10 cm cells, respectively; experiments marked by an asterisk are with chamber leakage

of oxygen and moisture into the chamber. In Fig. 2, the changes in the cell concentrations after an average time elapse of 45 min are presented. We see a concentration increase in the stable SO$_2$F$_2$, which is the hydrolysis product of SOF$_4$. The decay of S$_2$F$_{10}$ is monitored, which is the result of contact with the cell walls. This decomposition has already been discussed for storage containers [4]. The reaction products are SF$_6$ and SF$_4$, which can explain the increased concentration meas-

ured for the latter, although it is rather sensitive to hydrolysis, producing SOF$_2$. Further studies are needed under more closely defined conditions with regard to moisture traces adsorbed onto the cell walls, but the trends can be presented here.

Conclusion

Difficulties due to the reactivity of the compounds and limitations in the method are discussed. As seen, a satisfactory understanding of the processes leading to the final decomposition products requires the identification and quantification of primary and long-lived by-products and the study of the reaction mechanisms which lead to the final species. For this purpose, IR-techniques offer the opportunity of direct and *in-situ* detection of the SF$_6$ by-products.

Acknowledgements. Financial support by the Ministerium für Wissenschaft und Forschung des Landes Nordrhein-Westfalen and the Bundesminister für Bildung und Forschung is gratefully acknowledged. One author (P.R.J) acknowledges the travel support granted by the Volkswagen-Stiftung.

References

[1] H. D. Morrison, J. R. Robins, *Proc. SPIE* **1994**, *2089*, 326.
[2] J. K. Wilmshurst, H. J. Bernstein, *Can. J. Chem.* **1957**, *35*, 191.
[3] A. Keens, A. Simon, *Proc. SPIE* **1994**, *2089*, 222.
[4] I. Sauer, G. Harman, J. K. Olthoff, R. J. van Brunt, in: *Gaseous Dielectrics VI* (L. G. Christophorou, I. Sauers, eds.), Plenum, New York, 1991, p. 553.

Mikrochim. Acta [Suppl.] 14, 197–199 (1997)

An FT-IR Study of Alizarin Red S Adsorbed at the Fluorapatite–Water Interface

Allan Holmgren*, **Willis Forsling**, and **Liuming Wu**

Department of Inorganic Chemistry, Luleå University, S-971 87 Luleå, Sweden

Abstract. The diffuse reflectance infrared Fourier transform (DRIFT) technique was used to study sodium 1,2-dihydroxyanthraquinone-3-sulphonate (ARS) adsorbed onto the surface of fluorapatite mineral. The acidity of the water solution was varied between pH 5.4 and pH 8.4. In this pH range the infrared spectra of adsorbed ARS show that the fluorapatite surface contains precipitated calcium-ARS. The chelation of calcium strongly affects the resonance structure of the ARS molecule.

Key words: FT-IR spectroscopy, diffuse reflectance, adsorption, fluorapatite, Alizarin Red S.

The two hydroxyl groups and the sulphonate group in ARS are known to make this molecule suitable for the formation of chelates with metal ions. Recently it has been reported that ARS as a modifying agent can achieve selective separation between calcium minerals in a flotation process [1, 2]. ARS is found to react with calcium sites at the mineral/water interface. The adsorption is generally combined with complexation in solution owing to the rather high solubility of calcium minerals [3]. In the present study ARS was adsorbed onto the surface of fluorapatite and examined by the surface-sensitive DRIFT technique to explore the difference in vibrational behaviour of ARS adsorbed at the mineral surface and precipitated calcium-ARS. Precipitation may be expected to occur at the mineral surface because of the rather high solubility of calcium minerals in water. If so, the efficiency of ARS as a depressor in the flotation process would be reduced. To demonstrate the most probable position for calcium in precipitated Ca-ARS, we also recorded DRIFT spectra of solid ARS at varying degree of protonation.

Experimental

Pure crystals of Canadian fluorapatite were crushed, ground and sieved. The $-5\,\mu m$ fraction was used for adsorption studies. The

BET specific surface of this fraction was 9.3 m^2/g. ARS was used without further purification and contained about 5 wt% water. DRIFT spectra were recorded on a Perkin–Elmer PE 1760X spectrometer and plotted as the Kubelka–Munk function. Two hundred scans at 4 cm^{-1} resolution were collected for all samples. Unfortunately, ARS gave very poor NIR FT-Raman spectra, so complementary information from Raman spectra could not be obtained.

Results and Discussion

When a β-OH hydrogen atom in ARS is removed, enhanced resonance involving the quinoid group at the 10-position is possible. The change in resonance structure is clearly detected in the DRIFT spectra, as shown in Fig. 1. When less than 10 % of the beta hydrogens are removed ($pK_a = 5.9$), the intensity of the C=O stretching vibration caused by the quinoid group at the 10-position (1669 cm^{-1}) is reduced and a new band appears at about 1545 cm^{-1}. This band is probably due to the vibration of the C$\cdots$O entity formed. In addition a new absorption band appears at 1490 cm^{-1} and the intensity of the weak 1355 cm^{-1} band in ARS at lower pH is strongly enhanced. Note that the frequency of the C$\cdots$O group at the 9-position (1634 cm^{-1}) and 1355 cm^{-1} is not changed. The lower frequency of this C=O vibrator is caused by intramolecular hydrogen bonding offered by the alpha phenolic hydrogen [4, 5].

The most distinct difference between the infrared spectrum of ARS (pH 6.3) and the spectrum of precipitated Ca-ARS (pH 6.3) appears at 1545 cm^{-1}, 1490 cm^{-1} and 1355 cm^{-1} (Fig. 2). Obviously the calcium ion affects the resonance structure. However, the peak frequency at 1634 cm^{-1} is virtually unchanged. This indicates the calcium ion to be located between the beta phenolic oxygen and the sulphonate group.

ARS adsorbed at the fluorapatite surface was easily detected at pH values between 6.4 and 8.4, although the

---

* To whom correspondence should be addressed

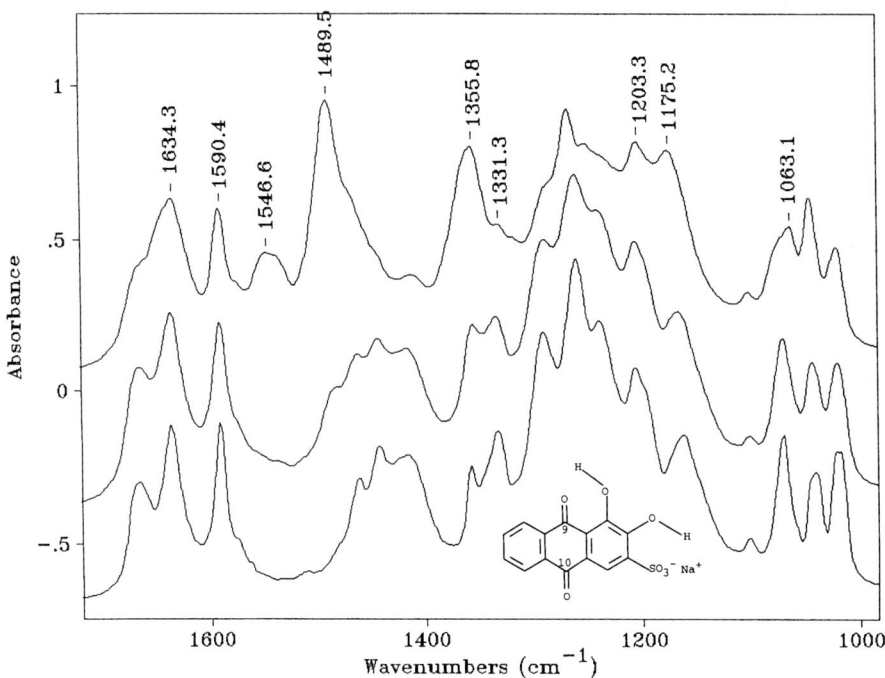

Fig. 1. DRIFT spectra of solid ARS obtained from water solutions adjusted to pH 4.0 (lowest trace), pH 5.0 (middle trace) and pH 6.0 (upper trace). The second pK_a-value for ARS is 5.9. The structure of ARS is included

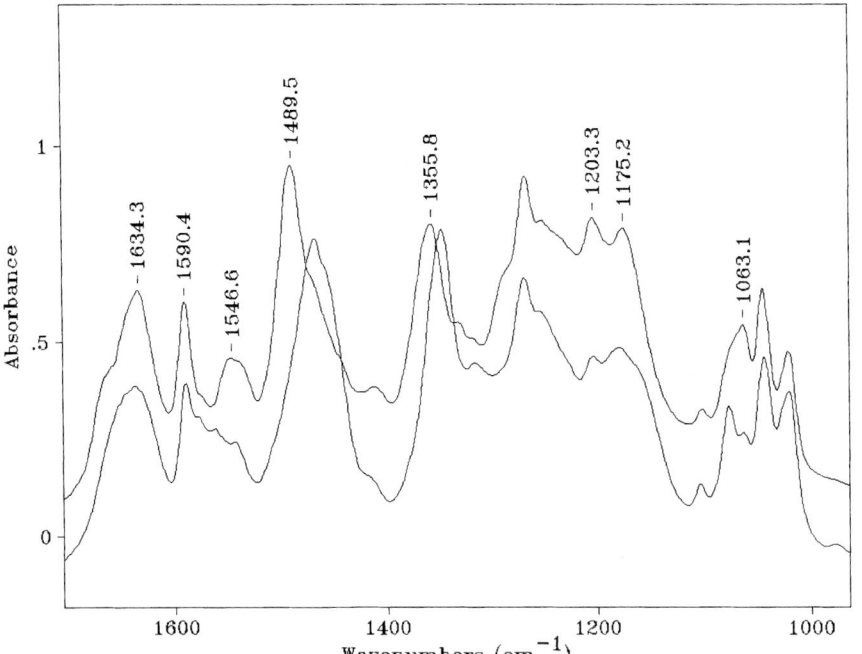

Fig. 2. Infrared spectra showing the spectral differences between ARS at pH 6.3 and precipitated calcium-ARS (pH 6.3). The spectra are shifted along the absorbance scale for clarity

intensity is lower at higher pH owing to the increased negative charge of the surface. It may be noted that the solubility of the mineral also decreases with increasing pH. According to Fig. 3, the subtraction of a spectrum of fluorapatite from a spectrum of ARS adsorbed on fluorapatite is very similar to the infrared spectrum of precipitated Ca-ARS. Therefore the surface of fluorapatite seems to be covered with precipitated Ca-ARS (pH 6.4). Also ARS adsorbed from water solutions at higher pH values, *viz.* 8.4 and 10.3, showed spectral indications of precipitated Ca-ARS. This may be due to the inability of the DRIFT technique to

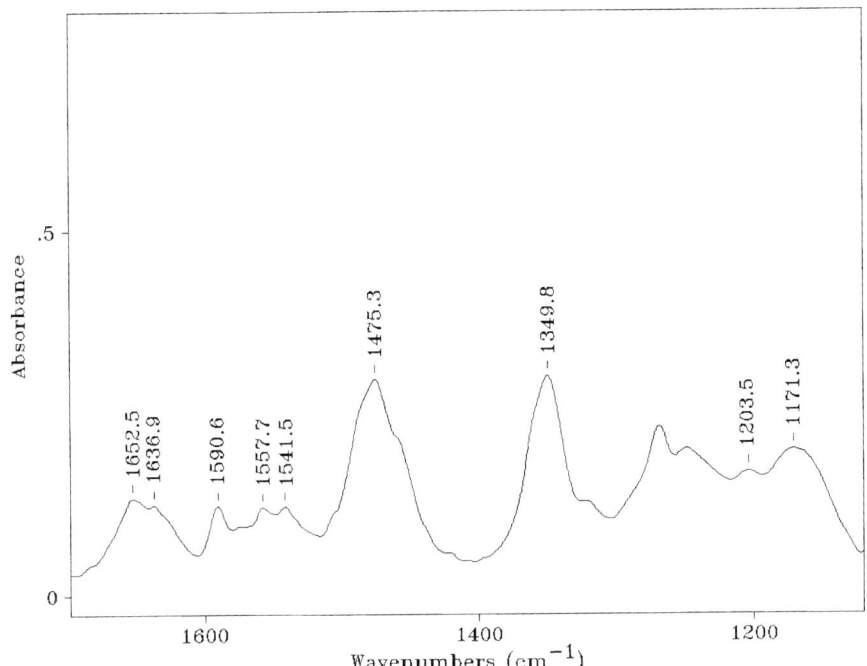

Fig. 3. DRIFT spectrum of ARS absorbed at the fluorapatite surface at pH 6.4. The spectrum of fluorapatite equilibrated at the same pH value has been subtracted

detect the low concentrations of chemically adsorbed ARS. AT pH 10.3 the fluorapatite surface shows a trace of violet colour, which means that some ARS molecules have lost all three protons. However, no infrared absorption characteristic of the C⋯O stretching vibration at 1545 cm^{-1} nor any change in the frequency of the hydrogen bonded C=O group at 1634 cm^{-1} could be detected.

Acknowledgement. This work was financially supported by the Swedish Board for Technical Development.

References

[1] L. Xiao, P. Somasundaran, *Miner. Mevall. Process.* **1989**, *6*, 100.
[2] L. Wu, W. Forsling, *J. Colloid Interface Sci.* **1995**, *174*, 178.
[3] L. Wu, W. Forsling, *Acta Chem. Scand.* **1992**, *46*, 418.
[4] D. Hadzi, N. Sheppard, *Trans. Faraday Soc.* **1954**, *50*, 911.
[5] H. Bloom, L. H. Briggs, B. Cleverley, *J. Chem. Soc.* **1959**, 178.

Mikrochim. Acta [Suppl.] 14, 201–202 (1997)

A Drift Spectroscopy Study of Oxidatively Modified Cotton-Cellulose

Serge Kokot*, **Lia Marahusin,** and **Desmond Paul Schweinsberg**

Centre for Instrumental and Developmental Chemistry, School of Chemistry, Queensland University of Technology, Brisbane 4001, Australia

Abstract. Cotton fabrics may be oxidatively bleached and damaged by electro-generated species at metal electrodes. Such fabrics were investigated by DRIFT spectroscopy. The spectra agree well with the literature, although spectra from differently damaged fabrics cannot be readily distinguished. Spectral discrimination is facilitated by principal components analysis, and the electrochemical parameters which control the fabric damage can be successfully predicted by PC regression.

Key words: modified cotton-cellulose, electro-generated oxidation, damaged fabrics, principal components analysis, chemometrics.

It has been shown [1–3] that electro-generated oxygen species can bleach and damage cotton fabric, and create cotton fibre/fabric effects consistent with those of the "catalytic damage" phenomenon observed in practice during the bleaching of cotton fabric by hydrogen peroxide. The degree of change may be varied by controlling the temperature and applied constant current of the reaction process [1].

In this paper we demonstrate that DRIFT spectra from the differently treated samples, when interpreted by chemometrics, can be used to predict some reaction parameters, such as temperature and applied current, that control the oxidation process.

Experimental

Unbleached fabric was treated by electro-generated oxygen species at a Pt anode by methods that have been described elsewhere [2]. The electrolyte was $7 \times 10^{-3} M$ Na_2SO_4 adjusted to pH 11 with NaOH. Samples were produced at five different current settings (50–100 mA) and 80 °C, and five samples at 100 mA but at different temperatures (60–80 °C). The samples were rinsed in water and dried after treatment.

---

* To whom correspondence should be addressed.

Spectroscopy

The details can be found elsewhere [4]; spectrometer: PE 1600 equipped with a DRIFT wavenumber accessory; samples: 6 mm diam. fabric discs, placed on a bed of dried, ground KBr; range: 4000–400 cm^{-1}, 90 scans, 4 cm^{-1} resolution; display: log $1/R$.

Spectral Analysis

Wavenumber range: 1800–1500 cm^{-1}; baseline corrected and zero off-set; spectra imported into QUANT + and autoscaled; submitted to principal component analysis (PCA). Three spectra from each sample were used for calibration and a fourth for validation.

Results and Discussion

When the major spectral features obtained from our oxidatively damaged samples are compared with typical band assignments reported in the literature [5], in general, close agreement is observed. No single band can be obviously associated with cellulosic species containing carbonyl and carboxylic groups, which form during the oxidative damage process. The presence of such groups can be readily demonstrated by selective staining [1].

Yang [6] has shown that C=O stretch frequencies of the carbonyl (1700–1750 cm^{-1}) and carboxylate groups (1550–1575 cm^{-1}) are sometimes important in IR analysis of modified cottons. Since carbonyl and carboxylate groups form during oxidation of cotton cellulose, we chose to use the 1800–1500 cm^{-1} spectral region for PCA analysis.

The PC1 *vs.* score–score plot from DRIFT spectra of fabric samples treated at different applied current levels and 80 °C accounted for 95.5% of the variance. PC1 discriminated between the spectra from untreated and treated fabrics, while PC2 distinguished between the spectra from cotton treated at different applied current

Table 1. Predicted applied current and temperature values of voile samples

Parameter: applied current			Temperature		
$I_{expt.}$ (mA)	$I_{pred.}$ (mA)	Difference (%)	$T_{expt.}$ (°C)	$T_{pred.}$ (°C)	Difference (%)
100	96	4.0	80	79	1.2
85	96	13	75	73	2.7
75	70	6.7	70	68	2.9
65	66	1.5	65	65	0.0
50	45	10	60	65	8.3
Average differences		7.0%	Average differences		3.0%

values. A similar PCA experiment with the spectra from fabrics treated at 100 mA and the five different temperatures also showed discrimination of the different treatments, with the first three PCs accounting for 98.9% of the data variance.

From this qualitative work, it is quite clear that DRIFT spectroscopy reflects the different degrees of oxidation of cotton-cellulose, and consequently, we attempted to predict the key experimental parameters, applied current and temperature, which control the oxidation. Principal component regression (PCR) calibration models were built for the prediction of each parameter. Each contained three significant PCs and 15 samples. On this basis, a small set of validation samples was reliably predicted as shown in Table 1.

Conclusion

The key parameters, applied current and temperature, which control the degree of oxidation of cotton fabric treated by electro-generated oxygen species produced at a Pt electrode, were successfully predicted from DRIFT spectra of the oxidized samples. Chemometrics methods were essential for prediction. Such a combination of DRIFT spectroscopy and chemometrics may lead to a development of a damage index that may be useful in practice.

References

[1] U. Meyer, S. Kokot, R. Weber, J. Zürcher-Vogt, *Textilveredlung* **1987**, *22*, 185.
[2] L. D. Marahusin, S. Kokot, D. P. Schweinsberg, *Corror. Sci.* **33**, 1281 (1992).
[3] S. Kokot, L. Marahusin, D. P. Schweinsberg, *Textile Res. J.* **1993**, *63*, 313.
[4] C. Gilbert, S. Kokot, U. Meyer, *Appl. Spectrosc.* **1993**, *47*, 741.
[5] D. Fengel, M. Ludwig, *Papier (Darmstadt)*, **1991**, *47*, 45.
[6] C. Q. Yang, *Appl. Spectrosc.* **1991**, *45*, 102.

Mikrochim. Acta [Suppl.] 14, 203–205 (1997)

FT-IR Spectroscopic Investigation of the Transformation of Allyl Cyanide in the Presence of Butyl-lithium

Zoltán Kónya[1], **István Hannus**[1,*], **Árpád Molnár**[2], and **Imre Kiricsi**[1]

[1] Department of Applied Chemistry, József Attila University, Rerrich B. tér 1, Szeged, H-6720 Hungary
[2] Department of Organic Chemistry, József Attila University, Dóm tér 8, Szeged, H-6720 Hungary

Abstract. The transformation of allyl cyanide in THF solution in the presence of butyl-lithium was investigated by FT-IR spectroscopy. The large shifts in the CN and C=C stretching vibration regions testify to the generation and existence of a carbanion intermediate in this base-catalysed double bond isomerization of allyl cyanide to crotononitrile.

Key words: FT-IR spectroscopy, allyl cyanide, crotononitrile, base catalysis, carbanion.

Base-catalysed reactions occurring via carbanionic intermediates play an important role in organic synthesis. However, few spectroscopic investigations of the nature of anionic intermediates have been described [1]. In this paper, we report the spectroscopic results obtained in the course of investigation of the transformation of allyl cyanide in tetrahydrofuran (THF) in the presence of butyl-lithium.

Experimental

The experiments were done under dry nitrogen in a glove box in order to exclude contamination by air and moisture. The reactants, products and the solvent THF were rigorously purified. Fresh butyl-lithium was used. To a $0.25\,M$ solution of allyl cyanide in THF a 50% excess of butyl-lithium was added and the mixture was introduced into the IR cell operating as the reactor under a dry nitrogen atmosphere. In each case a blank experiment, in which the solution did not contain BuLi, was also done. Spectra were run on a Mattson Genesis FT-IR spectrometer.

Results and Discussion

Unsaturated nitriles exhibit some unique IR spectral features which can be easily used for their identification. The CN group in nitriles gives a sharp absorption

close to the $2249\,cm^{-1}$ found for allyl cyanide (Figs. 1a and 2a). Conjugation between the double bond and the CN group shifts this adsorption to lower wavenumber and gives rise to a stronger absorption band ($2221\,cm^{-1}$ in the case of crotononitrile, Fig. 2b). The C=C stretching band appears at $1645\,cm^{-1}$ for allyl cyanide, which is usual for non-conjugated compounds. Conjugation simplifies the multiple bond/single bond character, and thereby lowers the frequency of vibration for crotononitrile. In our case a mixture of *cis* and *trans* compounds was used, resulting in a doublet in the spectrum at 1637 and $1627\,cm^{-1}$.

The C–H bond can vibrate by bending both in-plane and out-of-plane in these compounds. The scissoring in-plane vibration for the terminal C=C double bond is stronger at $1419\,cm^{-1}$ in allyl cyanide than at $1442\,cm^{-1}$ in crotononitrile. The C–H symmetric in-plane bendings of CH_3 characteristic of crotononitrile appear near $1380\,cm^{-1}$. The C–H out-of-plane bending vibration in allyl cyanide containing a monosubstituted double bond produces two strong bands at 991 and $934\,cm^{-1}$ (see Fig. 1a). In crotonitrile (containing a 1,2 disubstituted double bond) absorptions at $729\,cm^{-1}$ (in *cis*-crotononitrile) and a $956\,cm^{-1}$ (in *trans*-crotonitrile) were found (Fig. 1c).

Non-conjugated unsaturated compounds can be changed into conjugated isomeric derivatives through base catalysis:

$$CH_2=CH-CH_2-CN \underset{-H^+}{\rightleftharpoons} [\overset{\cdots\cdots\cdots\cdots}{CH_2-CH-CH-CN}]^-$$

$$\underset{+H^+}{\rightleftharpoons} CH_3-CH=CH-CN$$

In protonic or not absolutely dried solvent the base-

* To whom correspondence should be addressed

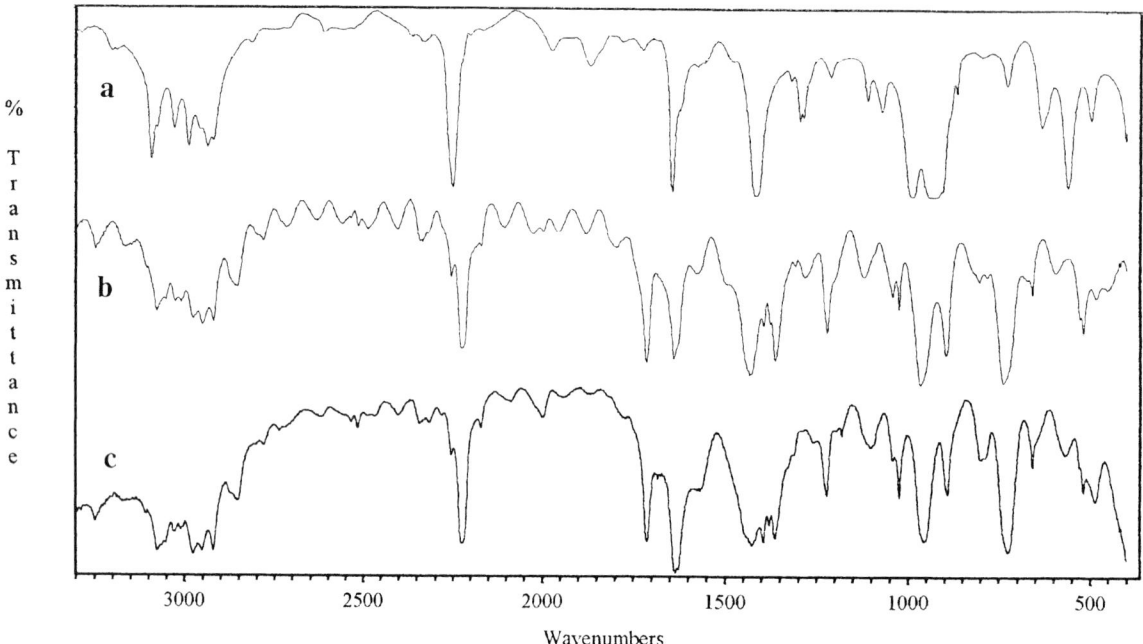

Fig. 1. Spectra of nitriles in methanol solution; *a* allyl cyanide, *b* allyl cyanide with butyl-lithium, *c* crotononitrile

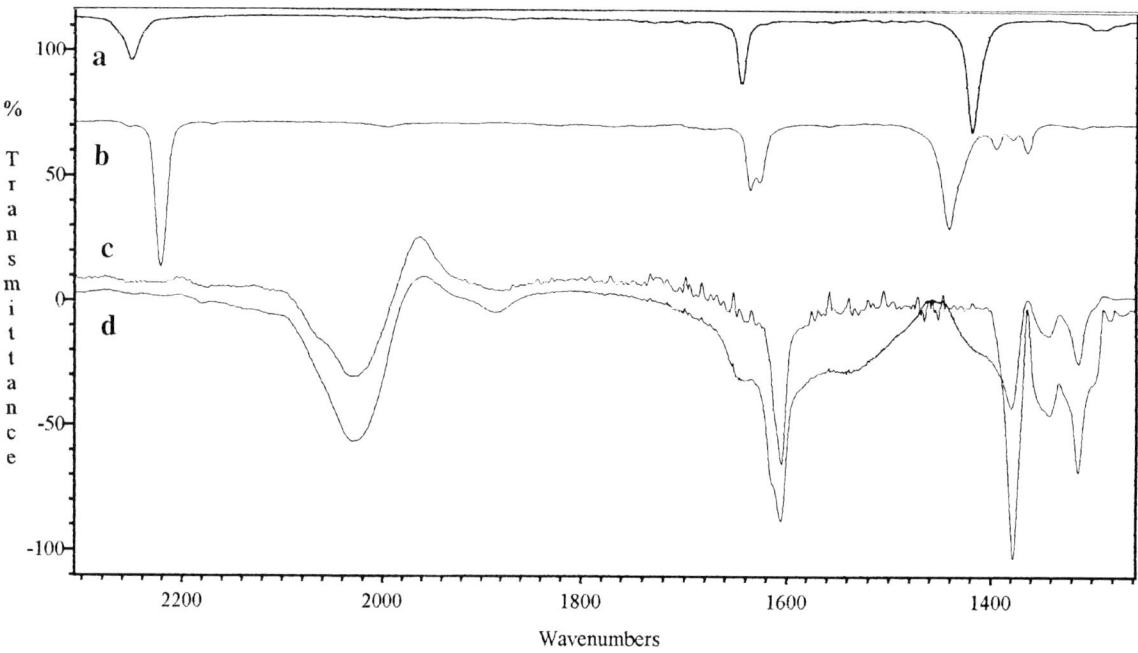

Fig. 2. Spectra of nitriles; *a* liquid allyl cyanide, *b* liquid crotononitrile, *c* allyl cyanide in dry THF with butyl-lithium, *d* crotononitrile under the same conditions as for *c*

catalysed isomerization of allyl cyanide to crotononitrile is fast, not allowing the observation of the intermediate carbanion. The transformaton, however, can be easily followed by IR spectroscopy, measuring the changes in the $v(CN)$ vibration region (from 2249 cm^{-1} to 2221 cm^{-1}) and in the $v(C=O)$ stretching vibration region (from 1645 cm^{-1} to the 1637, 1627 cm^{-1} doublet), as depicted in Fig. 1 a and b. These spectral changes clearly indicate the transformation to crotononitrile.

In contrast, in dry THF in the presence of an excess amount of BuLi a large shift (about $200\,cm^{-1}$) in the $v(CN)$ region was found. The new band at $2030\,cm^{-1}$ is probably due to the intermediate carbanion. A similar shift was observed in the case of the carbanion derived from phenylacetonitrile under identical conditions [2]. In the $v(C{=}C)$ stretching region a new band appeared at $1606\,cm^{-1}$ (see Fig. 2c). This spectrum was registered immediately after mixing the reactants and proved to be independent of time. Very similar spectral features were found for crotononitrile under the same conditions (Fig. 2d). In accordance with literature observations [1] these red shifts also indicate the appearance of an anionic intermediate.

The large shifts in the $v(CN)$ and $v(C{=}C)$ vibration regions unequivocally testify to the generation and the existence of a carbanionic intermediate in the base-catalysed double bond isomerization of allyl cyanide to crotononitrile.

Acknowledgement. Stimulating discussion with Dr. J. Corset is gratefully acknowledged.

References

[1] J. Corset, in: *Comprehensive Carbanion Chemistry* (E. Buncel, T. Durst, eds.), Elsevier, Amsterdam, 1980, Part A, Chapter 4.

[2] D. Croisat, J. Seyden-Penne, T. Strzalko, L. Wartski, J. Corset, F. Froment, *J. Org. Chem.*, **1992**, 57, 6345.

Mikrochim. Acta [Suppl.] 14, 207–209 (1997)
© Springer-Verlag 1997

FT-IR Study of the HEHEHP–n-Heptane–Water Systems

I – Variation of Spectra of Different Water Content Systems

Yan Li, Dujin Wang, Jinguang Wu*, Weijin Zhou, Zhenhua Xu, Shifu Weng, and **Guangxian Xu**

State Key Laboratory of Rare Earth Materials Chemistry and Applications, College of Chemistry and Molecular Engineering, Peking University, Beijing 100871, Peoples Republic of China

Abstract. The FT-IR method was used to investigate the properties of the microemulsion system formed by potassium-saponified HEHEHP. The subtraction spectra prove that all the polar groups of the surfactant molecule are highly hydrated. It is the strong interaction between the surfactant and water molecules that determines the structure and properties of the system. The water molecules in the microemulsion are divided into four types, namely highly bonded water, bound water, bulk-like water and isolated water, by deconvolution and curve-fitting.

Key words: FT-IR, 2-ethylhexyl phosphonic acid mono-2-ethyl-hexyl ester, water, microemulsion.

In previous work, we have reported that a microemulsion can form when water is added to saponified 2-ethylhexyl phosphonic acid mono-2-ethylhexyl ester (HEHEHP) solution in n-heptane [1]. To investigate the properties of this system, the FT-IR spectra of the system water–potassium-saponified HEHEHP (0.85 mole fraction of saponification)–n-heptane, with varied water content, were measured. Water in the microemulsion is of special interest for its special properties, similar to those of membrane water in cells [2]. The FT-IR method has been used to determine the structure of water in the microemulsion system [3,4].

Experimental

Reagents

HEHEHP, industrial product, purified by the copper salt method (content 99%), and saponified with metallic potassium; the water

used was processed by demineralization followed by distillation; all other reagents used were of analytical reagent grade or better.

Procedure

The IR spectra were taken (BaF$_2$ window) with a Nicolet 7199B FT-IR spectrometer, with 4 cm^{-1} resolution. The spectrum subtraction was done with the CH$_3$ deformation band at 1378 cm^{-1} as the reference peak. The frequencies of different structures in the v_{OH} band were determined from Fourier-deconvolution spectra. The quantitative estimation of the different water structures was achieved by curve-fitting analysis of Gaussian peaks.

Results and Discussion

The Subtraction Spectra

The subtraction spectra obtained from the spectra of the microemulsion systems with different water content and those of the heptane solution of potassium-saponified HEHEHP are shown in Fig. 1. It can be seen that the band intensities of P=O, P–O–C and P–O–H all increase when water is added. This proves that these polar groups are all highly hydrated in the microemulsion system [5].

The Changes in $v_{P=O}$, v_{P-O-C} and v_{P-O-H} with Water Content

Figure 2 shows that, with the increase of water content, the frequency of $v_{P=O}$ initially decreases, whereas that of v_{P-O-C} and v_{P-O-H} increases, but when the water content is higher than 7.5%, the frequency of $v_{P=O}$ does not change, but that of v_{P-O-C} and v_{P-O-H} decreases. This

---

* To whom correspondence should be addressed

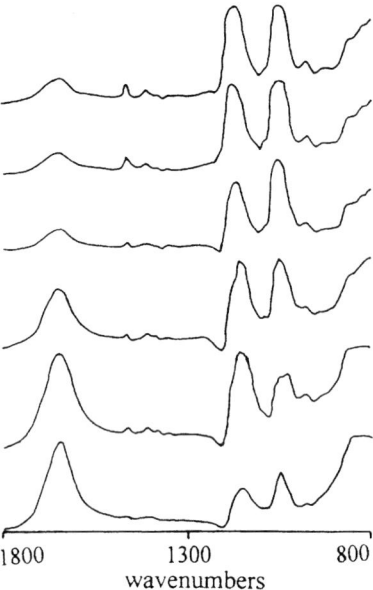

Fig. 1. The subtraction spectra (from top to bottom, the water contents are as follows: 1.0%, 4.0%, 6.0%, 10.0%, 20.0%, 68.0%)

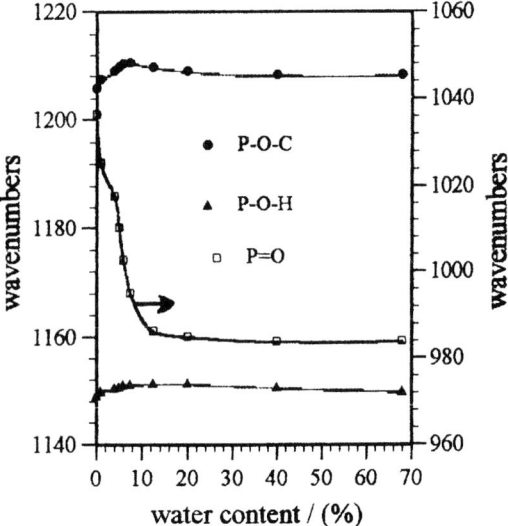

Fig. 2. Variation of $v_{P=O}$, v_{P-O-C} and v_{P-O-H} with water content

Table 1. The v_{OH} frequency of water in the microemulsion

Water content (%)	1.0	2.5	4.0	5.0	6.0	7.5	68
Frequency (cm^{-1})	3356	3362	3368	3372	3377	3387	3402

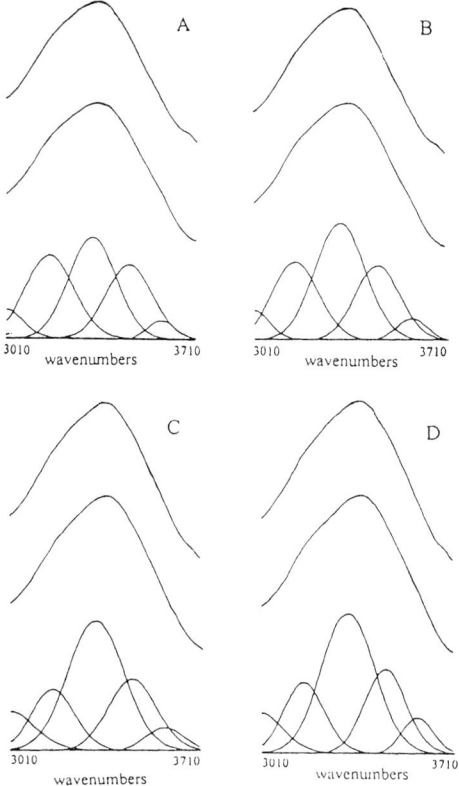

Fig. 3. The results of curve-fitting v_{OH} (above, the original band; middle, the curve-fitting results; bottom, the fitting curves). Water content: **A** 1.0%; **B** 4.0%; **C** 6.0%; **D** 12.0%

indicates that a water content of 7.5% is a turning point in the structure and properties of the system.

The Band Shift of v_{OH}

It can be seen from Table 1 that the v_{OH} band of water shifts to 3356 cm^{-1} when the water content is very low, and the absorption frequency of v_{OH} increases with the

increase of water content. It was also found that the half width of the δ_{OH} band of water in the microemulsion becomes larger than that of pure water. These facts indicate that the structure of water in the microemulsion is different from that in a bulk aqueous phase. There is a more complicated hydrogen bond structure in the microemulsion system.

The Quantification of Different Water Structures

The deconvoluted spectrum gives four peaks for v_{OH} at 3192, 3345, 3472 and 3585 ± 10 cm^{-1}. The three components at 3310 ± 30, 3480 ± 15 and 3610 ± 10 cm^{-1} of the v_{OH} band of water in AOT [Aerosol OT; sodium

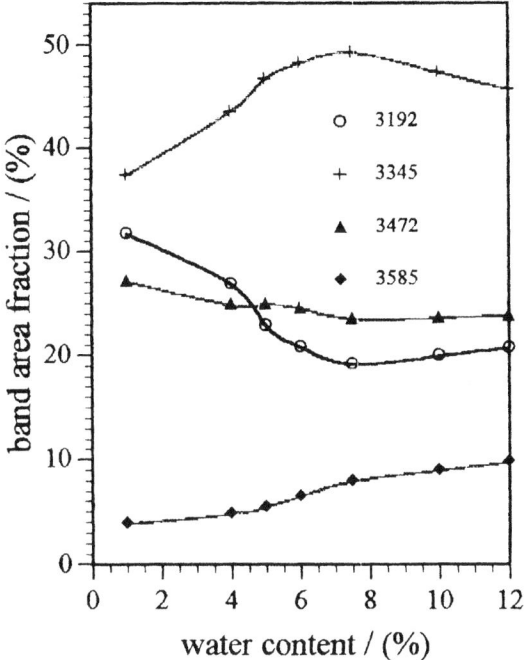

Fig. 4. Water-content dependence of band area fraction in curve-fitting results

bis(2-ethylhexyl) sulphosuccinate] microemulsion have been attributed to bulk-like water, bound water and monomeric or matrix-isolated water [3,4]. We make similar assignments as those in the AOT system for the last three components in our system, and consider the first as the contribution of highly bonded water which forms a strong hydrogen bond with the surfactant anion and a coordination bond with the surfactant cation.

These four components were fitted to the v_{OH} band with a gaussian curve-fitting program. The results are shown in Fig. 3. The variation of band area fraction with water content is shown in Fig. 4. The components at 3192 and 3472 cm^{-1}, which represent the highly bonded water and the bound water, decrease with increase of the water content, but the peaks at 3345 and 3585 cm^{-1}, which represent the bulk-like water and

isolated water, increase. The curves in Fig. 4 also have a turning point at a water content of 6–8%.

Discussion

(1) The turning points in Figs. 2 and 4 are almost the same. This indicates that interaction between the surfactant and water molecules determines their structural changes.

(2) Water molecules in the water pool of the microemulsion are often divided into three types, as described above. The component at 3192 cm^{-1} has never been reported. We assign it as an absorption band of highly bonded water. It was found in our previous work [6] that the ^{1}HMR spectra of water in such a system shifts to 5.4 ppm, compared with 4.80 ppm for pure water. This also indicates strong interaction between the surfactant and water molecules.

(3) The microemulsion system is often used in biochemical studies because its water of special structure is similar to the membrane water in cells. It was found here that the property of water in this system is different from that of the AOT system, which is the one most often used. The HEHEHP molecule has a more similar structure to lipid molecules in organisms than does AOT. This means it should be better to use the HEHEHP microemulsion system for biochemical studies.

Acknowledgement. This work was supported by The Climbing Program and National Natural Science Foundation of China.

References

[1] J. Wu, N. Shi, H. Gao, D. Chen, H. Guo, S. Weng, G. Xu, *Sci. Sin., Ser. B* (*Engl. Ed.*) **1984**, *27*, 249.
[2] L. M. Gierasch, K. F. Thompson, J. E. Lacy, H. L. Rockwell, *Reverse Micelles* (P. L. Luisi and B. E. Straub, eds.), Plenum, New York, 1986, p. 265.
[3] T. K. Jain, M. Varshney, A. Maitra, *J. Phys. Chem.* **1989**, *93*, 7409.
[4] G. Onori, A. Santucci, *J. Phys. Chem.* **1993**, *97*, 5430.
[5] W. Liu, Y. Li, N. Shi, J. Wu, X. Wang, J. Bian, L. Li, G. Xu, *Guangpuxue Yu Guangpu Fenxi* **1995**, *15*(3), 31.
[6] J. Wu, H. Gao, D. Chen, T. Jin, G. Xu, Chem. *J. Chinese Univ.* (*Engl. Ed.*) **1985**, *1*(1), 89.

Mikrochim. Acta [Suppl.] 14, 211–212 (1997)
© Springer-Verlag 1997

Far-Infrared Spectra of Kaolinite/Alkali Metal Halide Complexes

Kirk H. Michaelian[1,*], **Kelly L. Akers**[1], **Shuliang L. Zhang**[1], **Shmuel Yariv**[2], and **Isaak Lapides**[2]

[1] Natural Resources Canada, CANMET, Western Research Centre, P.O. Bag 1280, Devon, Alberta, Canada T0C 1E0
[2] Department of Inorganic and Analytical Chemistry, The Hebrew University of Jerusalem, 91904 Jerusalem, Israel

Abstract. Photoacoustic and diffuse reflectance spectra between ~ 200 and $700\,cm^{-1}$ are reported for six kaolinite/alkali metal halide complexes. Significant modifications of the kaolinite bands are caused by intercalation of the salts and water.

Key words: FT-IR, photoacoustic, diffuse reflectance, kaolinite, alkali metal halide.

The grinding of kaolinite $[Al_2Si_2O_5(OH)_4]$ with caesium chloride or caesium bromide in the presence of water results in the intercalation of the caesium salt and one or more water molecules and the formation of a clay/caesium halide/water complex. Since alkali metal halides do not absorb infrared radiation, spectroscopic evidence of their effect on the structure of kaolinite has to be derived from changes in the hydroxyl stretching or lattice vibrations of the clay. The 400–$4000\,cm^{-1}$ region of the infrared spectra of these two complexes has already been characterized [1–4].

Recently, we have expanded this study to include a wider range of kaolinite/alkali metal halide/water complexes. The current work extends this investigation to lower frequencies, presenting far-infrared (~ 200–$700\,cm^{-1}$) photoacoustic (PA) and diffuse reflectance (DRIFT) spectra of kaolinite complexes with six different caesium and potassium halides.

Experimental

Samples were prepared by grinding the kaolinite/DMSO precursor complex together with an excess of the appropriate alkali metal halide, and heating to remove DMSO. This procedure produces kaolinite/alkali metal halide/water complexes dispersed in alkali metal halide matrices.

Photoacoustic Spectra

PA spectra were obtained at $4\,cm^{-1}$ resolution with a Bruker IFS 113v FT-IR spectrometer equipped with a globar source, a $3.5\,\mu m$ Mylar beam-splitter, and a Princeton Applied Research Corporation cell. Because the mixtures were dilute, they yielded very weak

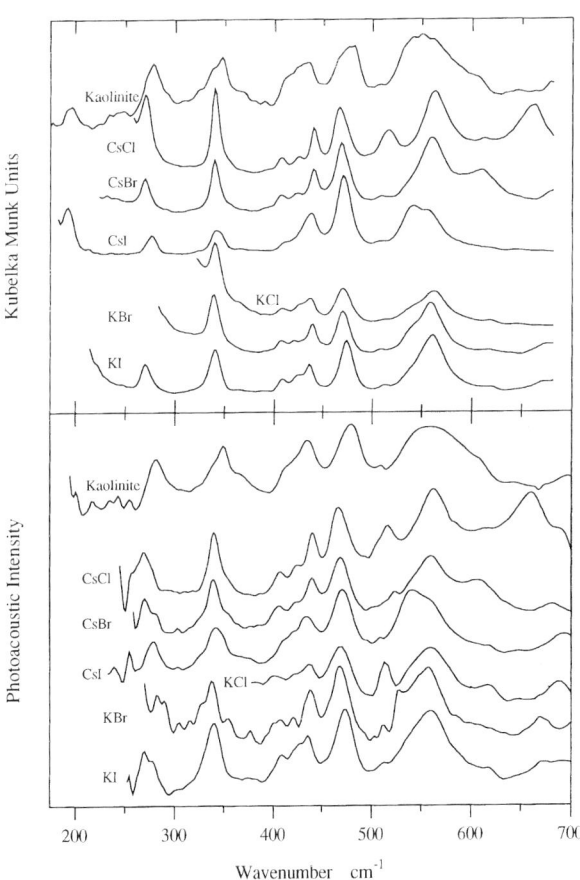

Fig. 1. Far-infrared spectra of kaolinite/alkali metal halide/water complexes. Upper box: diffuse reflectance spectra; lower box: photoacoustic spectra. Spectra have been rescaled and offset for clarity

---

* To whom correspondence should be addressed

Table 1. Far-infrared frequencies (cm^{-1}) of kaolinite/alkali metal halide/water complexes[a,b]

| | Kaolinite/alkali metal halide/water complexes | | | | | |
Kaolinite	CsCl	CsBr	CsI	KCl	KBr	KI
196 m			193 m			
278 s	270 s	270 s	276 m			270 m
347 s	340 s	340 s	343 m	340 s	339 s	340 s
411 sh	407 w	407 w	412 w, sh	408 w	407 w	408 w
421 sh	425 w	423 w		426 w, sh	420 w	425 w, sh
434 m	440 m	440 m	438 m	437 m	439 m	436 m
475 s	467 s	469 s	471 s	471 m	471 m	474 s
508 vw	516 m	510 vw	508 vw	513 vw	508 w	512 w
			541 s			
550 s	563 s	560 s	555 m, sh	562 m	559 s	561 s
	612 w	610 m		618 w	617 vw	618 w, sh
	661[c] s					
696 m	690 sh	683 w	681 w	688 m	670 w	687 m

[a] Frequencies (except last row) from diffuse reflectance spectra.
[b] s, strong; m, medium; w, weak; vw, very weak; sh, shoulder.
[c] Due to hydrated sodium ion [7], present as an impurity.

far-infrared spectra, necessitating extensive signal averaging: as many as one hundred 200-scan spectra, i.e. a total of up to 20000 scans, were averaged.

Diffuse Reflectance Spectra

To record DRIFT spectra, a DTGS detector and Harrick "praying mantis" accessory were used instead of the PA cell. Most data presented here are averages of twenty 100-scan spectra.

Results and Discussion

Figure 1 shows PA and DRIFT spectra in the ~200–700 cm^{-1} region for six kaolinite complexes; the extent of each spectrum is determined mainly by the low-frequency absorption by the alkali-metal halide. Spectra of uncomplexed kaolinite are included for comparison. Band positions are summarized in Table 1, and can be considered accurate within ± 2 cm^{-1}. The frequencies in the DRIFT spectra are reported wherever possible, because those bands are usually better defined.

It is evident that intercalation causes significant changes in the spectrum of kaolinite. Most bands become narrower with complexation; frequencies shift by as much as 13 cm^{-1}, relative intensities are modified, and significantly, identification of weak or over-lapping bands in the kaolinite spectra is facilitated by the existence of data for a series of complexes. The 411 and 421 cm^{-1} bands of kaolinite are an example of the latter observation.

Assignments of the major low-frequency infrared bands of kaolinite have been given previously [5]. In general, these bands are due to mixed SiO deformations and octahedral sheet vibrations. Translational vibrations of intercalated cations [6] are expected to occur at frequencies well below the range accessible in this study.

References

[1] K. H. Michaelian, S. Yariv, A. Nasser, *Can. J. Chem.* **1991**, *69*, 749.
[2] S. Yariv, A. Nasser, Y. Deutsch, K. H. Michaelian, *J. Thermal Anal.* **1991**, *37*, 1373.
[3] K. H. Michaelian, W. I. Friesen, S. Yariv, A. Nasser, *Can. J. Chem.* **1991**, *69*, 1786.
[4] S. Yariv, A. Nasser, K. H. Michaelian, I. Lapides, Y. Deutsch, N. Lahav, *Thermochim. Acta* **1994**, *234*, 275.
[5] V. C. Farmer, in: *Data Handbook for Clay Materials and Other Non-Metallic Minerals* (H. H. Van Olphen, J. J. Fripiat, eds.), Pergamon, Oxford, 1979.
[6] M. Ishii, T. Shimanouchi, M. Nakahira, *Inorg. Chim. Acta* **1967**, *1*, 387.
[7] S. Yariv, S. Shoval, *Appl. Spectrosc.* **1985**, *39*, 599.

Mikrochim. Acta [Suppl.] 14, 213–215 (1997)
© Springer-Verlag 1997

Vibrational Spectra of Metal Dithiocarbamates

Khaled Baghat[1,*] and **János Mink**[1,2]

[1] Institute of Isotopes of the Hungarian Academy of Sciences, P.O.B. 77, H-1525 Budapest, Hungary
[2] Department of Analytical Chemistry, Veszprém University, P.O.B. 158, H-8201 Veszprém, Hungary

Abstract. FT-IR (4000–50 cm^{-1}) and Raman spectra of the dithiocarbamates of zinc and cadmium and of their deutero-derivatives were measured. Complete vibrational assignment is proposed on the basis of a normal co-ordinate analysis and comparison with the spectra of the dithiocarbamate anion.

Key words: IR spectra, Raman spectra, dithiocarbamates.

The molecular structure and vibrational spectra of some dithiocarbamate complexes have been studied earlier [1–3]. The assignments proposed in those articles were based on the IR spectral data and normal co-ordinate analysis for the in-plane modes of the 1:1 metal–ligand model of the complexes.

We have measured the FT-IR spectra (4000–50 cm^{-1}) and Raman spectra of the M(S$_2$CNH$_2$)$_2$ and M(S$_2$CND$_2$)$_2$ (M = Cd, Zn) complexes. We have also made a complete normal co-ordinate analysis of a 1:2 model of the complexes, taking into account the CS$_2$M ring puckering. The calculated frequencies are compared with the measured spectra and the results of a normal co-ordinate analysis for dithiocarbamate and dithiocarbamate-d$_2$ anions.

Experimental

Dithiocarbamate complexes of Zn(II) and Cd(II) were prepared by addition of an aqueous solution of ammonium dithiocarbamate to the aqueous solutions of CdCl$_2 \cdot 2\frac{1}{2}$H$_2$O and ZnCl$_2 \cdot 2$H$_2$O, respectively. The precipitates obtained were washed with water, then diethyl ether and dried. Deuteration of the complexes was performed by adding D$_2$O to a dioxan solution of the metal complex and separating the precipitate. The process was repeated several times.

IR spectra were obtained on the Bio-Rad Digilab Fourier spectrometers FTS-60A and FTS-40A, with the samples in KBr disks and in nujol mull. Raman spectra were measured with the Bio-Rad SP 3200 FT-Raman spectrometer with Nd-YAG laser at 1064 nm excitation.

* To whom correspondence should be addressed

Normal Co-ordinate Analysis

Calculations were made for 1:2 metal–ligand models with suitable programs [4]. The central metal atom has tetrahedral co-ordination and the planes of the two chelate rings are perpendicular to each other. The notation of the internal co-ordinates used is given in Fig. 1. In addition, the following out-of plane co-ordinates were used: ρ_1 for the wagging movement of the NH$_2$ group, ρ_2 for the out-of plane rocking mode of the C–N bond, χ_1 for the torsion mode around the C–N bond, χ_2 for the four-member ring puckering. The structural parameters used were $d = 1.04$ Å, $R = 1.34$ Å, $r = 1.65$ Å, $D = 2.35$ Å (Zn) and 2.53 Å (Cd), $\varphi = \gamma = 120°$, $\alpha = \beta = 102°$, $\varepsilon = 80.37°$, $\delta = 79.83°$, $\pi = 126.71°$. The initial values of the force constants were taken from Durgaprasad [3] and these values were refined during the work. The final sets of the force constants for the Cd and Zn complexes, respectively, and for the anion (in parenthesis) are: $K(NH) = 6.042$, 6.025, (6.0); $K(CS) = 4.945$, 5.060, (2.78); $K(CN) = 6.890$, 6.736, (6.68); $K(MS) = 3.249$; 3.185; $F(NH,NH) = -0.14$, -0.14; $F(CS,CS) = 0.145$, 0.145; $F(MS,MS) = -0.229$, -0.222 (all in 10^2 N/m units); $H(HNH) = 0.454$, 0.450, (0.453); $H(HNC) = 0.458$, 0.208, (0.279); $H(NCS) = 0.362$, 0.362, (0.362);

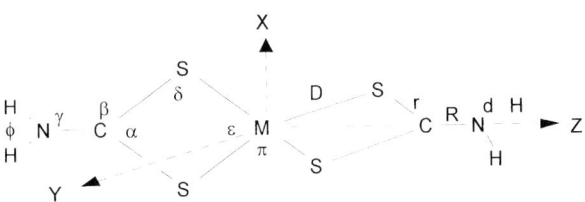

Fig. 1. Internal co-ordinates for the dithiocarbamate complexes. Additional co-ordinates are: torsions around CN bond χ_1, and χ_2 for the four-membered ring puckering, out-of-plane NH$_2$ wagging ρ_1, and out-of-plane rocking mode of CN bond ρ_2

Table 1. Calculated and observed frequencies (cm^{-1}) for the dithiocarbamate anions H$_2$NCS$_2^-$ and D$_2$NCS$_2^-$

Symmetry	H$_2$NCS$_2^-$				D$_2$NCS$_2^-$			
	Exp.	Calc.	Assign.	PED[a]	Exp.	Calc.	Assign.	PED[a]
	3290	3300	ν^sNH$_2$	99d	2420	2392	ν^sND$_2$	97d
	1594	1600	δ^{sc}NH$_2$	$69\gamma + 27R$	1080	1064	δ^{sc}ND$_2$	$82\gamma + 5\beta$
A_1	1328	1325	νCN	$54R + 30\gamma + 8\beta$	1447	1449	νCN	$71R + 14\gamma + 7\beta + 5r$
(IR,R)	503	502	ν^sCS	$56r + 27\beta + 17R$	465	484	ν^sCS	$55r + 23\beta + 18R$
	324	318	δCS$_2$	$63\beta + 36r$	325	317	δCS$_2$	$65\beta + 34r$
A_2(R)	324	323	τNH$_2$	100χ	230	230	τND$_2$	100χ
	3405	3403	ν^{as}NH$_2$	100d	2520	2524	ν^{as}ND$_2$	100d
B_1	863	860	ν^{as}CS	$77r + 18\gamma + 5\beta$	842	847	ν^{as}CS	$86r + 8\gamma + 5\beta$
(IR,R)	611	609	δ^{rock}NH$_2$	$80\gamma + 19r$	469	472	δ^{rock}ND$_2$	$88\gamma + 11r$
	165	144	δSCN	95φ	150	131	δSCN	94φ
B_2	1002	999	ρ^wNH$_2$	$99\rho_1$	780	785	ρ^wND$_2$	$99\rho_1$
(IR,R)	165	158	ρ^wCS$_2$	$99\rho_2$	150	154	ρ^wCS^2	$99\rho_2$

[a] Only co-ordinates with a contribution of more than 5% are included.

Table 2. Calculated and observed frequencies (cm^{-1}) of dithiocarbamate complexes

Symmetry	Zn(H$_2$HCS$_2$)$_2$, exp		Cd(H$_2$NCS$_2$)$_2$				Cd(D$_2$NCS$_2$)$_2$			
	h$_4$	d$_4$	Exp.	Calc.	Assign.	PED[a]	Exp.	Calc.	Assign.	PED[a]
A_1 (R)	3256	2354	3275	3271	ν^sNH$_2$	99d	2370	2371	ν^sND$_2$	97d
	1604	1089	1590	1590	δ^{sc}NH$_2$	$88\varphi + 16R$	1098	1107	δ^{sc}ND$_2$	93φ
	1409	1485	1413	1414	νCN	$61R + 15r$	1449	1452	νCN	$74R + 13r$
	619	598	609	607	νCS	$82r + 16R$	592	594	νCS	$79r + 16R$
	422	416	424	420	νMS	$64D + 32\alpha$	409	415	νMS	$67D + 33\alpha$
	199	196	186	188	δring	$64\alpha + 33D$	185	182	δring	$73\alpha + 31D$
A_2(i.a.)	690	471	686	689	τNH$_2$	$100\chi_1$	490	491	τND$_2$	$100\chi_1$
B_1(IR)	690	503	667	671	τNH$_2$	$100\chi_1$	489	478	τND$_2$	$100\chi_1$
	198	192	154	153	δMS$_2$	100π	152	153	δMS$_2$	100π
B_2 (IR, R)	3256	2354	3257	3260	ν^{as}NH$_2$	97d	2370	2361	ν^{as}ND$_2$	97d
	1604	1089	1590	1587	δ^{sc}NH$_2$	$84\varphi + 21R$	1098	1108	δ^{sc}ND$_2$	92φ
	1404	1485	1413	1408	νCN	$59R + 17r + 14\varphi$	1449	1454	νCN	$76R + 13r + 9\varphi$
	604	597	609	610	νCS	$70r + 15D + 14R$	598	598	νCS	$63r + 23D + 12R$
	555	541	554	544	νMS	$78 + 14\alpha + 10r$	530	540	νMS	$71D + 12\alpha + 12r$
	353	358	369	349	δring	$83\alpha + 7D$	362	346	δring	$83\alpha + 7D$
E (IR, R)	3357	2520	3357	3355	ν^{as}NH$_2$	100d	2485	2491	ν^{as}ND$_2$	99d
	1190	1200	1176	1183	νCS	$70r + 24\varphi$	1160?	1156	νCS	$83r + 10\varphi$
	1141	853	1092	1088	ρ^wNH$_2$	$99\rho_1$	854	863	ρ^wND$_2$	$96\rho_1$
	839	671	837	869	δ^{rock}NH$_2$	$75\varphi + 25r$	742	678	δ^{rock}ND$_2$	$86\varphi + 11r$
	538	541	404	447	νMS	$92D + 6\pi$	402	447	νMS	$92D + 6\pi$
	389	358	385	389	ρCN	$96\rho_2$	383	375	ρCN	$94\rho_2$
	247	214	218	183	δSCN	93α	173	167	δSCN	92α
	99	97	68	62	δMS$_2$	89π	77	62	δMS$_2$	88π
	30	28	19	17	χring	$97\chi_2$	18	17	χring	$97\chi_2$

[a] Only co-ordinates with a contribution of more than 5% are included.

$H(SCS) = 1.62,\ 1.42,\ (1.42);\ H(SCM) = 0.566,\ 0.564;$
$H(SMS) = 1.452, 0.1450$ (all in 10^{-18} N m rad^{-2} units).

Results and Discussion

The dithiocarbamate anion has C_{2v} symmetry. The experimental and observed frequencies are presented in Table 1. The six vibrations of the NH$_2$ group show isotopic shifts by deuteration and the calculations reproduce these shifts very well. The assignment for the NCS$_2$ skeleton is more tentative, owing to strong coupling of the vibrations. The frequency of the νCN mode increases from 1328 to 1447 cm^{-1} after deuteration. Such a "strange" behaviour is explained by the lowering of this frequency in the non-deuterated species, owing to "repulsion" with the δNH$_2$ mode of the same symmetry. The low frequency shift of δNH$_2$ after deuteration destroys this interaction. Another strong coupling takes place between the νCN and νCS modes. Potential energy distribution (PED) calculation shows the substantial contribution of CN stretching to the νCS mode. Hence the high frequency of νCN in dithiocarbamates can be explained not only by the increased bond order of this C–N bond due to the contribution of the canonical structure S$_2^{2-}$ C=N$^+$ [1–3], but also by the strong coupling with the CS vibrations. In the B_1 class mode, where such interaction is forbidden by symmetry, the νCS frequency is much higher (860 cm^{-1}) than for the A_1 class (503 cm^{-1}).

Cd dithiocarbamate belongs to the point group of symmetry D_{2h}. The experimental and calculated frequencies are presented in Table 2. The assignments for some frequencies is tentative, because of the strong vibrational coupling which is clear from the PED from the analysis of the data permits us to conclude that direct comparison of the anion and complex frequencies cannot be justified. For example, the νCN frequency in the complex (1413 cm^{-1}) is higher than in the anion (1328 cm^{-1}), but for the deuteroderivatives the frequencies are practically the same (1449 and 1447 cm^{-1}). For the lighter molecules the difference is explained by vibrational coupling with δNH$_2$.

The ν^sCS frequencies in the complex (~ 600 cm^{-1}) are higher than in the ligand (500 cm^{-1}). For asymmetrical νCS vibrations the difference is much greater (1176 and 863 cm^{-1}). Actually, there is no pure νCS vibration, in the complex, except for the stretching–bending mode of the chelate ring CS$_2$Cd. The coupling of CS and CdS vibrations leads to the increase of the νCS frequencies. The frequency 1176 cm^{-1} seems to be very high for the νCS mode, but the calculation reproduces this frequency quite well. The absence of isotopic shift confirms the proposed assignment. Metal–ligand stretching make the major band contribution to the modes at 424, 554 and 404 cm^{-1}. The rest of the assignments are shown in Table 2.

For Zn dithiocarbamate, the spectra, the results of the normal co-ordinate analysis, and the assignments are very similar to those presented for the Cd complex. Therefore the calculations only the experimental frequencies for the undeuterted Zn complex are given in Table 2.

The experimental spectra, especially in the low frequency region, are more complicated than theory predict. It is possible that in the solid state the complexes have a more complex structure, for example, forming dimers (2:4 complexes) as was shown for their N,N-diethyldithiocarbamate analogues [5].

References

[1] J. Chatt, L. A. Duncanson, L. M. Venanzi, *Suomen Kemistilehti* **1956**, *B29*, 75.
[2] K. Nakamoto, J. Fujita, R. A. Condrate, Y. Morimoto, *J. Chem. Phys.* **1963**, *39*, 423.
[3] G. Durgaprasad, D. N. Sathyanarayana, C. C. Patel, *Can. J. Chem.* **1969**, *47*, 631.
[4] J. Mink, L. M. Mink, *Computational Program System for Vibrational Analysis of Polyatomic Molecules*, Erlangen 1993.
[5] A. Domenicano, L. Torelli, A. Vaciago, L. Zambonelli, *J. Chem. Soc. (A)*, **1968**, 1351.

Mikrochim. Acta [Suppl.] 14, 217–219 (1997)

FT-IR and FT-Raman Spectroscopic Studies on Phase Transitions of a Mesogenic Long-Chain Tetraphenylporphyrin

Koji Ohta*, **Kimiko Nakao,** and **Yo Shimizu**

Osaka National Research Institute (ONRI), AIST, MITI, 1-8-31 Midorigaoka, Ikeda, Osaka 563, Japan

Abstract. Phase transitions of a mesogenic long-chain tatraphenyl-porphyrin have been investigated by FT-IR and FT-Raman spectro-scopic methods. It was revealed that the alkyl chain disorder occurs in the mesomorphic phases. Furthermore, some spectral changes related to the tautomerization of the N–H group as well as the arrangement in the mesophases were found through the mesomor-phic phase transitions.

Key words: tetraphenylporphyrin, liquid crystal, phase transition, infrared, Raman.

Macrocyclic discotic liquid crystals with a largely ext-ended π-conjugation system are quite attractive candi-dates for novel advanced materials. In particular, phthalocyanines and porphyrins are interesting mol-ecular species because of their mesomorphic nature, which is related to a variety of functional physical and chemical properties. Porphyrins, however, have not been investigated so much from the viewpoint of mesomorphic properties, whereas mesogenic phthalocyanines have been extensively studied [1]. Our recent works revealed that a series of tetraphenyl-porphyrins having a longer alkyl chain than pentyl (C_5) at the p-site of each phenyl ring exhibits a lamellar type of mesophase [2]. The general formula of these compounds is shown in Fig. 1. The homologues with the shorter alkyl chains (C_6–C_{10}) show only one type of lamellar mesophase and for the longer homologues (C_{11}–C_{16}) were found to have two types of lamellar mesophases, proposed as shown in Fig. 2. These high- and low-temperature mesophases have a common fea-ture of molecular orientation order, that is, a lamellar structure where the molecules align in a parallel man-ner to form a layered structure which is normal to the molecular planes. The high-temperature lamellar

phase (D_L phase) is considered not to have columnar structure made of molecular stacks, though the low temperature one (D_{LC} phase) has columnar arrays to form a layer [3]. However, the details of molecular arrangements within a layer for both lamellar phases have not yet been clarified. Vibrational spectroscopy could detect a change of the molecular order accom-panied with mesomorphic phase transitions and in this work a mesogenic porphyrin, 5,10,15,20-tetrakis (4-n-pentadecylphenyl)porphyrin ($C_{15}TPP$), was invest-igated by FT-IR and FT-Raman spetroscopy in rela-tion to the mesomorphic phase transitions. $C_{15}TPP$ shows a phase transition sequence as cryst ($56\,^{\circ}C$) D_{LC} ($66\,^{\circ}C$) D_L ($135\,^{\circ}C$) iso.

Experimental

The synthesis of $C_{15}TPP$ was described in a previous paper [2]. Powdered samples were sandwiched between KBr plates, and the infrared spectra were taken with a Nicolet 60SXR FT-IR spec-trometer under a Spectra-Tech IR-Plan microscope equipped with a temperature controller (Mettler FP80 and FP82); 400 scans, resolution $2\,cm^{-1}$ for IR measurements. A SPEX FT-Raman spec-trometer equipped with a specially designed temperature controller was used for measuring Raman spectra. The powdered sample was put in a glass capilllary for Raman measurements; 400 scans, resol-ution $4\,cm^{-1}$.

Results and Discussion

Figure 3 represents the temperature dependence of the FT-IR spectrum of $C_{15}TPP$ in the 3100–$2700\,cm^{-1}$ region, which shows the representative bands assigned to the C–H vibrations in the alkyl chains linked to the phenyl rings of $C_{15}TPP$. The bands at 2852, 2921, and $2955\,cm^{-1}$ in the figure correspond to CH_2 symmetric,

---

* To whom correspondence should be addressed

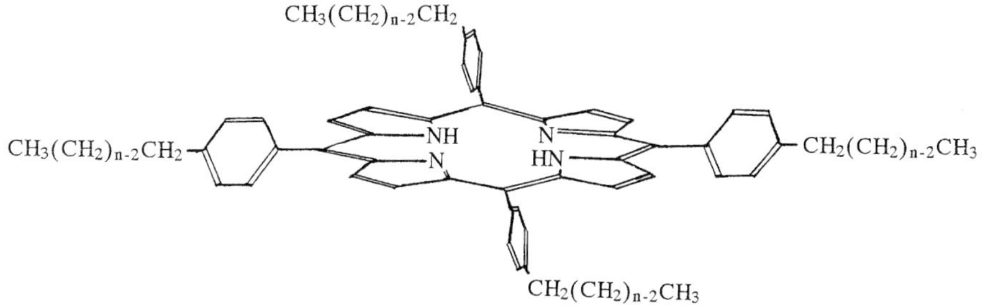

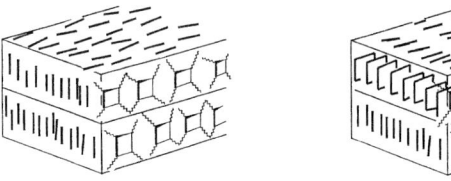

Fig. 1. General formula of the C_nTPP compounds

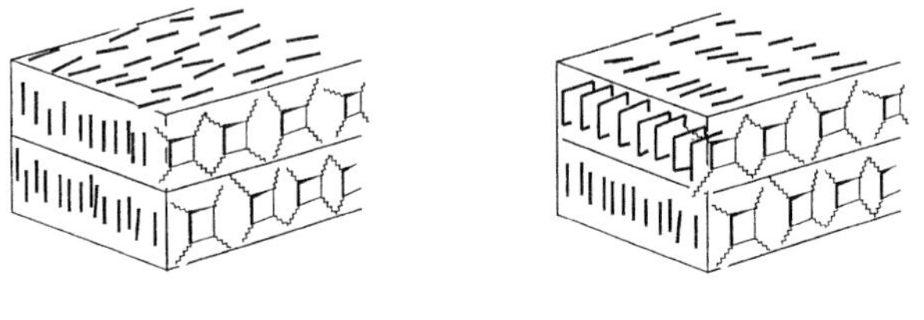

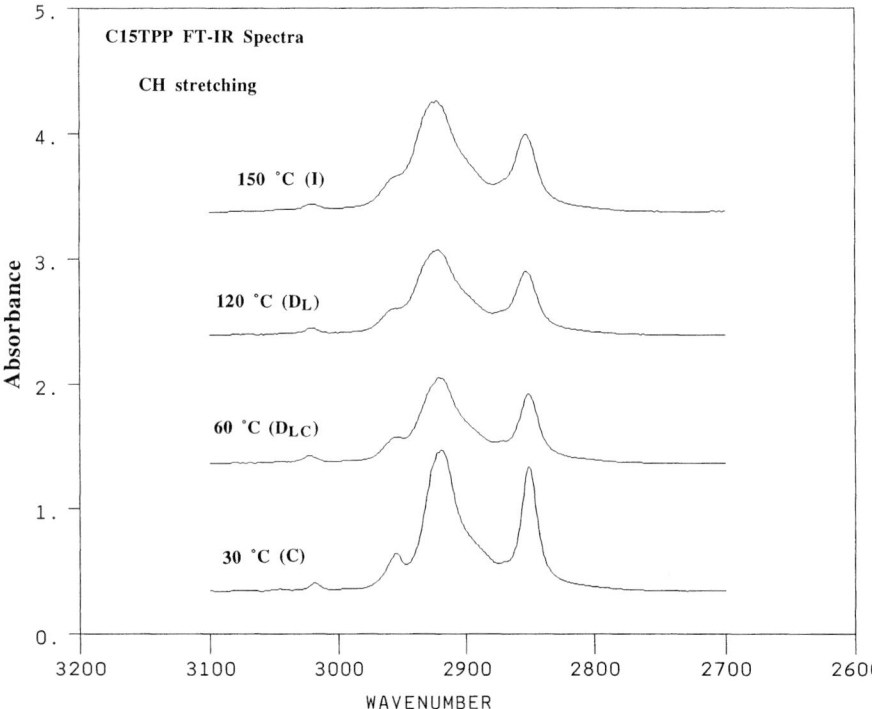

Fig. 2. Schematic representation of the proposed structures of two lamellar mesophases exhibited for C_{15}TPP. The left and right show the high temperature (D_L: dicotic lamellar) and low temperature (D_{LC}: dicotic lamellar columnar) mesophases, respectively

CH_2 antisymmetric, and CH_3 stretching vibrations, respectively. It can be seen that these three bands show broadening in their band-widths and a small shift of the peak positions at the phase change from cryst to D_{LC}. Similar band broadenings were also observed at the $1467 \, cm^{-1}$ band assigned to CH_2 scissoring and the

$796 \, cm^{-1}$ band assigned to CH_2 rocking motion at the same phase transition. All these vibrations in the band-width indicate that the peripheral chains are in a disordered state even in the lower D_{LC} phase, although in the higher D_L phase they are in a state of higher disorder (molten state). These observations in the infrared spectra are also supported by the temperature dependence of the several bands in the FT-Raman spectra measured for the same sample. The CH_2 twisting vibration band at around $1295 \, cm^{-1}$, for example, observed as a distinct band in the cryst phase, apparently disappeared above the cryst-D_{LC} transition temperature.

The most striking difference in the behaviour of the infrared bands between the D_{LC} phase and other phases is observed in the N–H stretching vibration of the porphyrin ring at around $3320 \, cm^{-1}$. As shown in

Fig. 3. FT-IR spectra of the C_{15}TPP sample in the 3100–$2700 \, cm^{-1}$ region at 30, 60, 120, and 150 °C

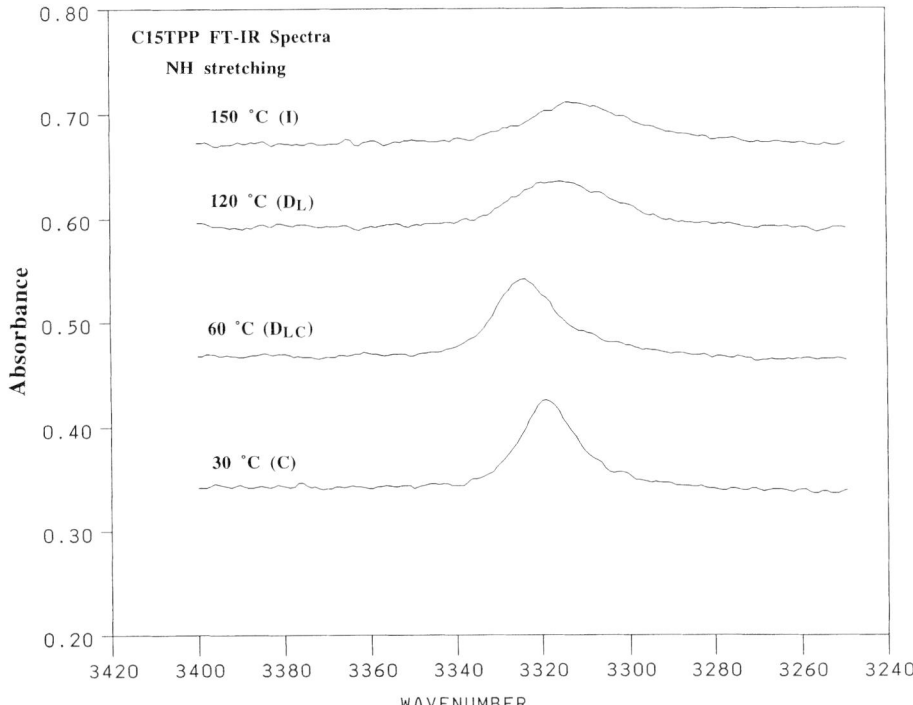

Fig. 4. FT-IR spectra of the $C_{15}TPP$ sample in the 3400–3250 cm^{-1} region at 30, 60, 120, and 150 °C

Fig. 4, this band shows a small shift by 5 cm^{-1} to higher wavenumber for the lower D_{LC} phase without any change in the bandwidth. In contrast to this, at the transition from the D_{LC} to the D_L phase, the peak position of the band shifts to the lower wavenumber near that in the cryst phase, but the band-width shows some broadening, representing a state of higher disorder.

In the porphyrin skeletal vibrations in the FT-Raman spectra, band splittings between 1208 and 1200 cm^{-1}, and between 1185 and 1179 cm^{-1} observed in the cryst phase, disappeared in the mesomorphic phases. This may indicate structural changes of the molecular arrangements.

Sarai reported that the proton migration (tautomerization) process of the N–H group is strongly coupled to the vibrational motion of the porphyrin skeleton, and the N–H motion is also important [4, 5]. It is reasonable to consider that some molecular motions could occur in the mesophase (disorder), which are related to the vibrational spectral change. Therefore, the behaviour of the band shift and FWHM for N–H stretching described above may reflect a change of the tautomerization process.

References

[1] A.-M. Giroud-Godquin, P. M. Maitlis, *Angew. Chem. Int. Ed. Engl.* **1991**, *30*, 375.
[2] Y. Shimizu, M. Miya, A. Nagata, K. Ohta, I. Yamamoto, S. Kusabayashi, *Liq. Cryst.* **1993**, *16*, 705.
[3] Y. Shimizu, M. Miya, A. Nagata, S. Kusabayashi, in preparation.
[4] A. Sarai, *J. Chem. Phys.* **1982**, *76*, 5554.
[5] A. Sarai, *Chem. Phys. Lett.* **1981**, *83*, 50.

Mikrochim. Acta [Suppl.] 14, 221–222 (1997)

FT-Raman, FT-IR and Inelastic Neutron Scattering Studies of Maleimides

Stewart F. Parker

ISIS Facility, Rutherford Appleton Laboratory, Chilton, Didcot, Oxon OX11 0QX, U.K.

Abstract. FT-Raman, infrared and inelastic neutron scattering spectra of maleimide and *N*-phenylmaleimide have been recorded. The complementarity of the three techniques is shown and has allowed almost all of the vibrations to be located.

Key words: infrared spectroscopy, FT-Raman spectroscopy, inelastic neutron scattering spectroscopy, maleimide, *N*-phenylmaleimide.

Nitrogen-substituted maleimides in the form of bis-maleimides are widely used in the aerospace industry as engineering materials that offer similar mechanical properties to metal alloys but with the advantage of lowerdensity. The commercial materials are complex, both chemically and spectroscopically, and have been little studied by spectroscopic means to date. Vibrational spectroscopy has the ability to follow all stages of the processing from the monomers to the cross-linked polymer and give information on the state of the cure process and the species present [1].

As a first step to characterizing the commercial materials, FT-Raman, FT-IR and inelastic neutron scattering (INS) spectra have been obtained for two model compounds, maleimide and *N*-phenylmaleimide (see Fig. 1 for the structures). Both molecules have been studied previously [2, 3] but there are some ambiguities in the assignments. The complementarity of the three techniques is shown and has allowed almost all of the vibrations to be located, including the out-of-plane modes of the maleimide ring and the deformations associated with the N–X linkage.

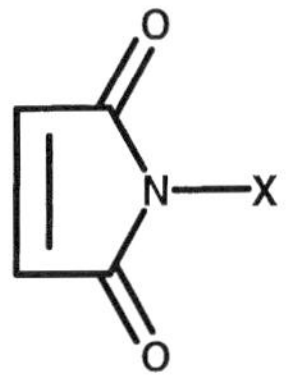

Experimental

Maleimide (Aldrich, 99%) and *N*-phenylmaleimide (Aldrich, 97%) were used as received. Infrared (KBr disk) and Raman (pure powder) spectra were recorded at 2 cm^{-1} resolution with a Digilab 896 FT-IR spectrometer and a Perkin–Elmer 1700 FT-Raman spectrometer respectively. The inelastic neutron scattering (INS) experiments were performed with the high resolution ($\sim 2\%$ $\Delta E/E$), broad-band (16–4000 cm^{-1}) spectrometer TFXA at the Rutherford Appleton Laboratory [4].

Results and Discussion

The symmetry of the maleimide ring is assumed to be C_{2v}, so $C_4H_2NO_2X$ (X = H or C_6H_5) has 24 modes associated with the maleimide ring: ($9A_1 + 3A_2 + 8B_1 + 4B_2$). Figure 2 shows the FT-Raman, in-

Fig. 1. Structure of maleimide (X = H) and *N*-phenylmaleimide (X = phenyl)

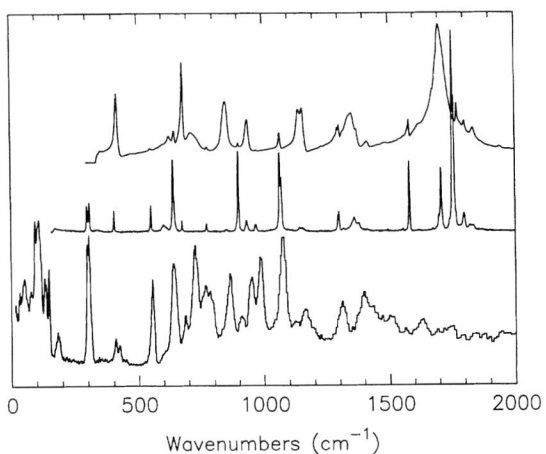

Fig. 2. Infrared (top), FT-Raman (middle) and inelastic neutron scattering (bottom) spectra of maleimide

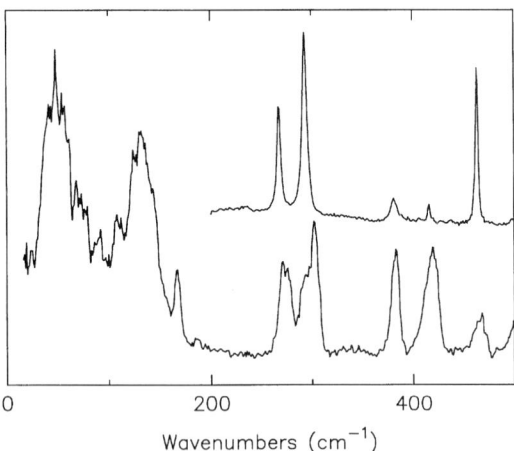

Fig. 3. FT-Raman (upper) and inelastic neutron scattering (lower) spectra of N-phenylmaleimide in the low frequency region

frared and INS spectra of maleimide in the 0–2000 cm^{-1} region. Similarities and differences are apparent between all three spectra and stress the need to have all three types of spectra for a complete analysis. The three forms of vibrational spectroscopy exploit different properties of the molecule. Infrared spectroscopy requires a change in dipole moment, so is sensitive to the polar motions of the molecule, while Raman spectroscopy requires a change in polarizability, and thus is more sensitive to the non-polar motions of the molecule. These two techniques provide information on the heavy-atom motions of the molecule, e.g. $v_{C=O}$, $v_{C=C}$. In contrast, the INS spectrum is dominated by the hydrogen motions. This arises because the intensity of an INS band is proportional to the product of the inelastic scattering cross-section and the amplitude of vibration. Since the scattering cross-section for hydrogen is at least 20 times larger than for any other type of atom present in the molecule, the INS spectrum emphasizes the modes that involve substantial hydrogen motion either directly (e.g. C–H bend) or where the hydrogen is "riding" an another atom (e.g. torsions or other ring deformations). This allows all the out-of-plane A_2 and B_2 modes to be clearly seen in the INS spectrum.

Figure 3 shows a comparison of the FT-Raman and INS spectra of N-phenylmaleimide in the low fre-

quency region. (The region above 500 cm^{-1} has been assigned previously [4]). Both maleimide and the aromatic ring have vibrations in this region and this complicates the analysis. Below 150 cm^{-1} the bands are mainly due to the lattice modes of the crystal. By comparison with maleimide (see Fig. 2) the bands at 166 (INS), 293 (Raman)/303 (INS) and 464 (Raman/INS) cm^{-1} are assigned as maleimide ring vibrations. Monosubstituted aromatic systems C_6H_5Y typically show six bands ($3A_1 + B_1 + 2B_2$) that are sensitive to the nature of Y [5]. The A_1 modes involve the C_6H_5–Y stretch; two are observed at 1208 and 711 cm^{-1} [4] and the third at 381 (Raman/INS) cm^{-1}. The out-of-plane B_2 modes are assigned as the bands at 182 (INS) and 508 (Raman/INS) cm^{-1}. The B_1 mode is assigned to the feature at 271 (Raman/INS) cm^{-1}. Also in this region a "Y insensitive" A_2 mode is expected; this is usually strong in the INS spectrum and is assigned as the 418 (Raman/INS) cm^{-1} feature. The band at 293 cm^{-1} is assigned as the in-plane N-phenyl bend, although the alternative assignment of the 271 cm^{-1} to the in-plane bend and the 293 cm^{-1} band to the B_1 mode is tenable. The out-of-plane N-phenyl bend will lie at lower energy; comparison with maleimide suggests that the B_2 in-phase out-of-plane C=O bend at 166 cm^{-1} is anomalously intense and that the bend probably overlaps this peak. The torsion about the N-phenyl bond is difficult to assign unambiguously. In biphenyl, the torsion occurs at 70 cm^{-1}, so the band at 50 cm^{-1} is provisionally assigned to this mode.

Acknowledgement. The Rutherford Appleton Laboratory is thanked for access to neutron beam facilities.

References

[1] S. F. Parker, *Vib. Spec.* **1992**, *3*, 87.
[2] T. Woldbæk, P. Klaboe, C. J. Nielsen, *J. Mol. Struct.* **1975**, *27*, 283.
[3] S. F. Parker, S. M. Mason, K. P. J. Williams, *Spectrochim. Acta* **1990**, *46A*, 315.
[4] J. Penfold, J. Tomkinson, *The ISIS Time Focused Crystal Spectrometer*, TFXA, RAL-82-038.
[5] D. Lin-Vien, N. B. Colthup, W. G. Fateley, J. G. Grasselli, *The Handbook of Infrared and Raman Characteristic Frequencies of Organic Molecules*, Academic Press, Boston, MA, 1991.

Mikrochim. Acta [Suppl.] 14, 223–225 (1997)
© Springer-Verlag 1997

Ethylenediaminetetra-acetic Acid (EDTA) Adsorption on a Modified Gamma Alumina Surface

Janusz Ryczkowski[1,*], **Dobiesław Nazimek**[1], and **Gabor Keresztury**[2]

[1] Department of Chemical Technology, Faculty of Chemistry, University of Maria Curie-Sklodowska, 3 M.Curie-Sklodowska Square, 20-031 Lublin, Poland
[2] Central Research Institute for Chemistry, Hungarian Academy of Sciences, POB 17, H-1525 Budapest, Hungary

Abstract. The adsorption of EDTA on a modified γ-alumina support has been investigated by transmission FT-IR. The results allow the conclusion that acidic and alkaline procedures used for γ-Al$_2$O$_3$ modification have an insignificant influence on the EDTA adsorption.

Key words: EDTA, γ-alumina, adsorption, FT-IR.

Infrared (IR) spectroscopic investigations of surface phenomena occurring on metal catalysts were pioneered by Eischens and Pliskin almost forty years ago [1]. The usefulness of vibrational spectroscopy (VS) in identifying surface species, determining adsorbate structures or studying surface reactions has been widely demonstrated in the literature, e.g. [2, 3].

VS has increased rapidly in importance as a tool for the study of adsorbed molecules, especially in the beginning of the 1970s with the development of the Fourier-transform infrared (FT-IR) technique. In the investigation of surfaces, IR spectroscopy is employed in two distinct and complementary roles [4]: (1) to determine the chemical nature of the adsorbed species, and (2) to characterize the substrate by observing the spectrum of some adsorbed probe molecule. EDTA is a commonly used substance in chemical analysis, pharmaceutical and chemical products as well as in preparation of dispersed nickel alumina-supported catalysts [5–8].

Preparation of Ni/Al$_2$O$_3$ catalysts with high metal dispersion involves adsorption of the EDTA compound on the alumina support surface [5, 6]. The adsorption phenomena of EDTA on γ-Al$_2$O$_3$ has been confirmed by transmission IR spectroscopy [9, 10].

The aim of this work was to investigate EDTA adsorption on the modified γ-alumina surface and to compare the results with previous findings.

Experimental

Adsorption of EDTA (disodium salt) on γ-alumina was performed with a 0.1 M aqueous solution of the salt for 30 minutes at 353 K on a pure alumina support Al-0104T (BET area = 90 m^2/g) purchased from Engelhard. Prior to adsorption of the chelating agent, the alumina support was modified as shown in Fig. 1.

As modifiers, aqueous solutions (0.1 M) of nitric acid and ammonia were used (a part of the support being washed in the appropriate solution for half an hour and dried prior to the EDTA adsorption).

FT-IR transmission spectra of the investigated samples, pressed in KBr pellets, were recorded on Perkin–Elmer (1725x) and Nicolet (Magna-IR 750) spectrometers.

Results and Discussion

The FT-IR spectra of EDTA adsorbed on an alumina surface exhibit strong bands in the region 1800–1200 cm^{-1}, the positions and assignments of which are summarized in Table 1. The carboxyl group is known to give a strong band in the range between 1735 and

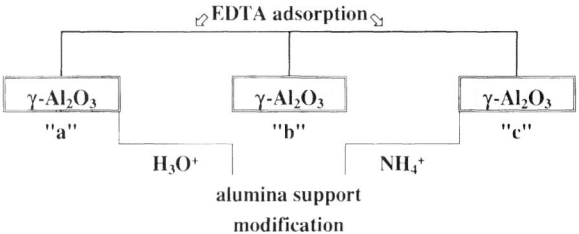

Fig. 1. Scheme of alumina support modification prior to the EDTA adsorption

* To whom correspondence should be addressed

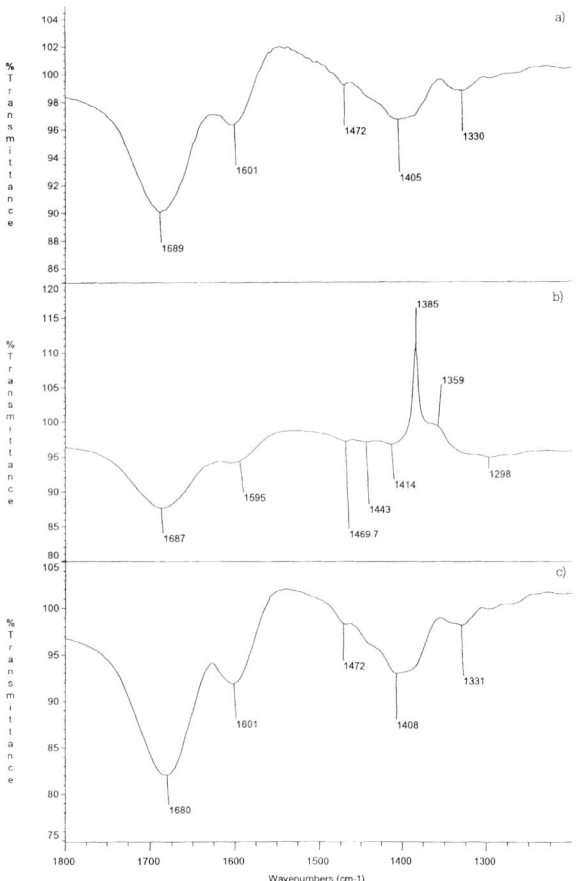

Fig. 2. FT-IR spectra of EDTA adsorbed on alumina surface: **a** unmodified, **b** modified by H_3O^+, **c** modified by NH_4^+

$1550 \, cm^{-1}$, which is due to the C=O stretching vibration [11]. A model for the adsorption of polyacetic amino-acids onto γ-alumina has been developed, with hydrogen bonding as the adsorption mechanism [12]. Based on the literature data [12] and IR results obtained [9, 10], a model of specific EDTA adsorption on an alumina surface has been proposed [10]. Because in the method used for preparation of the catalyst [5, 6] the EDTA is adsorbed from feebly acidic solution, the question is whether the pH of the chelate impregnation is of great importance or not.

From the preliminary results presented here it seems that the alumina surface modifications tested have no significant influence on the chelate adsorption (Fig. 2).

The principles of the adsorption mechanism proposed [10] were as follows in acidic solution the chelate molecule (pH < 7) interacts with hydroxyl (i.e. –OH and $-OH_2^+$) groups of the alumina support to from adsorbed species. It should be noted that, taking into account the isoelectric point of the Al_2O_3 surface, in the pH range above 7.5 new surface groups are formed ($\equiv AlO^-$). However, $\equiv AlOH$ groups exists on the support surface in the pH range from 5.8 to 10. Probably these groups are involved in the mechanism of EDTA adsorption in the neutral and feebly basic solutions.

Figure 3 shows the spectrum of $(NH_4)_2$ EDTA adsorbed on alumina, for comparison with the spectra

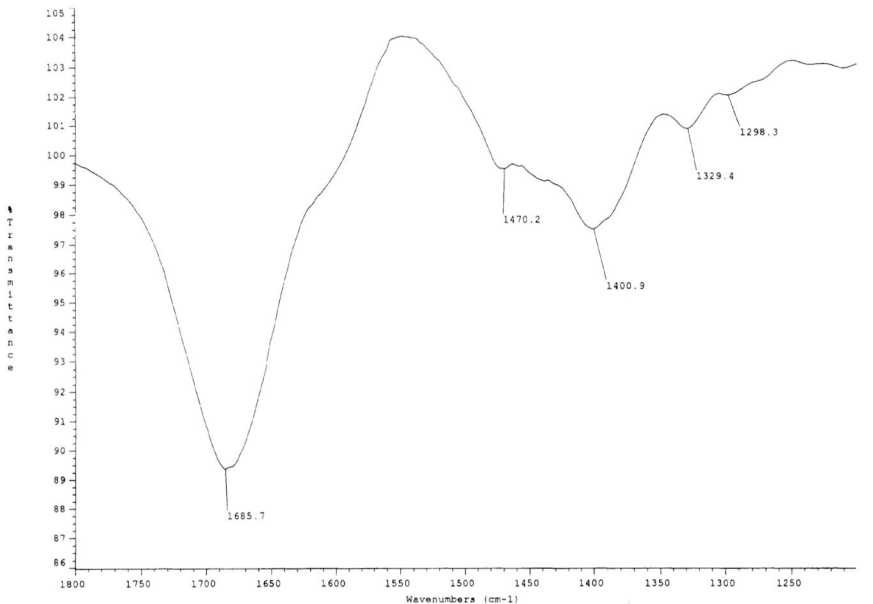

Fig. 3. Transmission spectrum of $(NH_4)_2EDTA$ adsorbed on alumina

Table 1. Assignment and band position in recorded spectra

Spectrum of Na$_2$ EDTA	C=O cm^{-1}	(O$\cdots$C$\cdots$O)$^-$	(C—H)CH$_2$	COO$^-$	COO$^-$
Unadsorbed compound	1672	1625	1475	1395	1315
Adsorbed on unmodified alumina	1689	1601	1472	1405	1330
Adsorbed on alumina modified by H$_3$O$^+$	1687	1595	$\approx$ 1470	1414	
Adsorbed on alumina modified by NH$_4^+$	1680	1601	1472	1408	1331

presented in Fig. 2. The similarity of the spectra can be an additional proof for the specific adsorption of chelate compounds on an alumina surface.

Conclusions

The results obtained allow the conclusion that acidic and alkaline procedures used for γ-Al$_2$O$_3$ modification have an insignificant influence on the adsorption of EDTA on an alumina support.

Acknowledgements. This work was supported by Grant No. 3P 40502804 of the State Committee for Scientific Research (KBN).

References

[1] R. P. Eischens, W. A. Pliskin, *Adv. Catal.* **1958**, *10*, 1.

[2] J. Mink, *Mikrochim. Acta* **1987**, *III*, 63.

[3] J. Mink, T. Szilagyi, *Croat. Chem. Acta* **1988**, *61*, 719.

[4] P. Hollins, *Surf. Sci. Rep.* **1992**, *16*, 51.

[5] J. Barcicki, D. Nazimek, W. Grzegorczyk, T. Borowiecki, R. Frak, M. Pielach, *React. Kinet. Catal. Lett.* **1981**, *17*, 169.

[6] D. Nazimek, J. Ryczkowski, *Appl. Catal.* **1986**, *26*, 47.

[7] J. Ryczkowski, *React. Kinet. Catal. Lett.* **1989**, *40*, 189.

[8] J. Ryczkowski, T. Borowiecki, *React. Kinet. Catal. Lett.* **1993**, *49*, 127.

[9] J. Ryczkowski, G. Keresztury, *Proc. SPIE* **1992**, *1575*, 540.

[10] J. Ryczkowski, *Proc. SPIE* **1993**, *2089*, 182.

[11] L. J. Bellamy, *The Infrared Spectra of Complex Molecules*, Wiley, New York, 1958, p. 162.

[12] A. R. Bowers, C. P. Huang, *J. Colloid Interface Sci.* **1985**, *105*, 197.

Mikrochim. Acta [Suppl.] 14, 227–228 (1997)

Qualitative Analyses of Iminodiacetic (IDA) and Nitrilotriacetic (NTA) Acids on Alumina by FT-IR

Janusz Ryczkowski* and **Dobiesław Nazimek**

Department of Chemical Technology, Faculty of Chemistry, University of Maria Curie-Sklodowska, 3 M.Curie-Sklodowska Square, 20–031 Lublin, Poland

Abstract. Adsorption of IDA and NTA and their sodium salts on a γ-alumina support has been investigated by transmission FT-IR. The results obtained imply, that the observed changes in the IR spectroscopic properties of the adsorbed chelates are mainly due to interaction of the chelate molecules with alumina hydroxyl groups.

Key words: IDA, NTA, γ-alumina, adsorption, FT-IR.

The catalytic and adsorption properties of aluminium oxides, the interactions of these species with the adsorbed components, their reactivity in solid-phase synthesis of complex oxides, and other properties are largely determined by the peculiarities of the local environment of Al(III) in the crystal lattice [1].

Surface spectroscopy techniques have recently provided important contributions to understanding of the influence of preparation conditions on the properties of catalysts. Several bonding schemes have been suggested to explain the adsorption of organic molecules on hydrous solids [2–5].

There are few literature data dealing with this problem that are based on IR investigations. (e.g. adsorption of EDTA-type compounds, on γ-alumina [6–8]). IDA and NTA are the simplest complexing agents in the EDTA family. They are not as popular as the main representative of the group (EDTA), but probably could be utilized in the preparation of high-dispersion metal catalysts [9–11].

The aim of this research on the interaction of IDA, NTA and their sodium salts with γ-alumina, was to compare the results with previous findings for EDTA adsorption on an Al_2O_3 surface.

Experimental

Adsorption of IDA, NTA and their sodium salts on γ-alumina was performed with their $0.1\,M$ aqueous solutions (for 30 minutes at $353\,K$) on a pure alumina support Al-0104T (BET area = $90\,m^2/g$) purchased from Engelhard. The adsorption procedure is shown schematically in Fig. 1.

The FT-IR transmission spectra of the samples investigated, pressed in KBr pellets, were recorded on Perkin–Elmer (1725x) and Nicolet (Magna-IR 750) spectrometers.

Results and Discussion

The chemical structures of the chelating agents used are shown in Scheme 1.

The carboxyl group is known to give a strong band in the range between 1735 and 1550 cm^{-1}, which is due to to the C=O stretching vibration [12]. The FT-IR spectra of the unadsorbed and adsorbed chelates on an alumina surface, in region 1800–1200 cm^{-1}, exhibit strong bands, the positions and assignments of which are summarized in Tables 1 and 2. For all the unadsorbed molecules there are characteristic vibrations of both the protonated and ionized carboxylic groups (Table 1).

The large number of bands between 1500–1300 cm^{-1} makes correct assignment difficult because

---

* To whom correspondence should be addressed

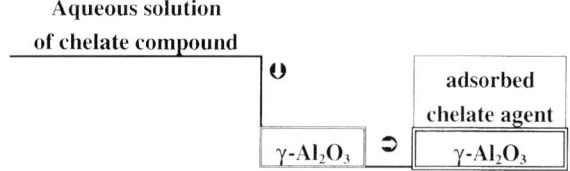

Fig. 1. Scheme of chelate compound adsorption on alumina support

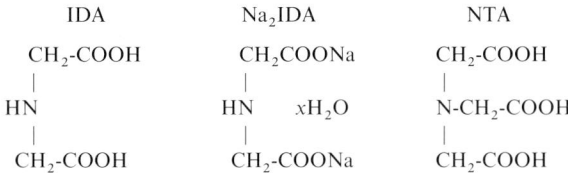

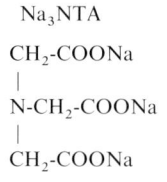

Scheme 1

Table 1. Band positions in "free" chelating agents

Spectrum	C=O	COO⁻ cm⁻¹	COO⁻
IDA	1717	1582	1434
Na₂IDA	1682	1592	1418
NTA	1734	1436	1335
Na₃NTA		1598	1407

Table 2. Band positions in chelating agents adsorbed on alumina

Spectrum	C=O	COO⁻ cm⁻¹	COO⁻
IDA	1648		1397
Na₂IDA	1647		1397
NTA	1650		1396
Na₃NTA	1648	1615	1407

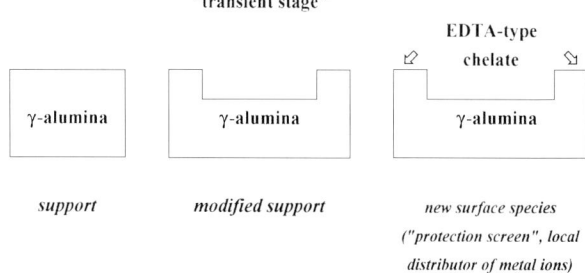

Fig. 2. Possible interactions of EDTA-type chelates with alumina support

Conclusions

The results obtained that the observed changes in the IR spectroscopic properties of the supported chelates are mainly due to interaction of the chelate molecules with alumina hydroxyl groups.

Acknowledgements. This work was supported by Grant No. 3P 40502804 of the State Committee for Scientific Research (KBN).

δNH^+, CH_2 deformation and various COO^- vibrations occur in the same region.

The IR spectra of the chelates adsorbed on alumina, in the region 1800–1200 cm⁻¹, exhibit two strong bands which could be assigned to the asymmetrical and symmetrical stretching of COO^-.

There are also several weak and medium bands between 1600 and 1400 cm⁻¹. The spectra of the chelates adsorbed on alumina indicate that chelate molecule–alumina interaction is the result of forming hydrogen bonds between protonated or neutral ($\equiv AlOH_2^+$ or $\equiv AlOH$) hydroxyls groups and carboxyl chelate groups. This is in agreement with previous findings [6, 7, 13]. From the results obtained it seems that not only EDTA but also other chelates from the EDTA group can be utilized in the preparation of high-dispersion metal catalysts [9–11]. In the case of EDTA it seems that this compound exhibits triple action activity (in catalyst preparation): as support modifier, as "protection screen" for the modified support surface and finally as local distributor for metal ions [1] (see Fig. 2).

The results obtained support this hypothesis.

References

[1] J. Ryczkowski, *React. Kinet. Catal Lett.* **1995**, in press.
[2] A. R. Bowers, C. P. Huang, *J. Colloid Interface Sci.* **1985**, *105*, 197.
[3] R. Kumert, W. Stumm, *J. Colloid Interface Sci.* **1980**, *75*, 373.
[4] W. Stumm, R. Kummert, L. Sigg, *Croat. Chem. Acta* **1980**, *53*, 291.
[5] G. Furrer, W. Stumm, *Chimia* **1983**, *37*, 338.
[6] J. Ryczkowski, G. Keresztury, *Proc. SPIE* **1992**, *1575*, 540.
[7] J. Ryczkowski, *Proc. SPIE* **1993**, *2089*, 182.
[8] J. Ryczkowski, *First European Congress on Catalysis Europa Cat-I*, Montpellier, 12–17 September 1993, *Abstracts*, II-197, p. 785 and II-198, p. 786.
[9] J. Barcicki, D. Nazimek, W. Grzegorczyk, T. Borowiecki, R. Frak, M. Pielach, *React. Kinet. Catal. Lett.* **1981**, *17*, 169.
[10] D. Nazimek, J. Ryczkowski, *Appl. Catal.* **1986**, *26*, 47.
[11] J. Ryczkowski, *React. Kinet. Catal. Lett.* **1986**, *40*, 189.
[12] L. J. Bellamy, *The Infrared Spectra of Complex Molecules*, Wiley, New York, 1958, p. 162.
[13] J. Ryczkowski, *Proc. SPIE* **1993**, *2089*, 504.

Mikrochim. Acta [Suppl.] 14, 229–231 (1997)

FT-IR Investigation of Hydroxy-Acid Adsorption on an Alumina Surface

Janusz Ryczkowski[1,*], **Tadeusz Borowiecki**[1], and **Gabor Keresztury**[2]

[1] Department of Chemical Technology, Faculty of Chemistry, University of Maria Curie-Sklodowska, 3. M Curie-Sklodowska Square, 20-031 Lublin, Poland
[2] Central Research Institute for Chemistry, Hungarian Academy of Sciences, POB 17, H-1525 Budapest, Hungary

Abstract. The adsorption of some hydroxy-acids (i.e. citric, tartaric and salicylic) on a γ-alumina support has been investigated by transmission FT-IR. The results imply that the observed changes in the IR spectroscopic properties of the supported hydroxy-acids are mainly due to interaction of the acids with alumina hydroxyl groups. The IR spectra in the C=O stretching region of the hydroxy-acids supported on alumina are quite different from those of the unsupported compounds.

Key words: citric, tartaric and salicylic acids, γ-alumina, adsorption, FT-IR.

The interaction of cations, weak acids and anions with hydrous oxide surfaces are of importance in colloid chemistry as well as in the preparation of metal-oxide supported catalysts. The hydroxylated oxide support surface can be treated like a polymeric oxo-acid (or base) and the specific adsorption of cations and anions can be interpreted in terms of co-ordination reactions at the oxide–aqueous solution interface [1–5]. In earlier investigations the influence of organic impregnants on the physicochemical properties of Ni/Al_2O_3 catalysts was studied [6]. It was concluded that in the preparation method used [7, 8], γ-alumina surface hydroxyl groups are blocked by support–organic impregnant interaction [6, 9].

This study has been undertaken in order to investigate these interactions in the case of some hydroxy acids, by the infrared (IR) method.

Experimental

Adsorption of citric, tartaric and salicylic acids on γ-alumina (Ketjen CK-300, Akzo) has been performed with their aqueous solutions

(0.050 M for citric and tartaric acids and 0.015 M for salicylic acid) as shown schematically in Fig. 1.

The FT-IR transmission spectra, obtained before and after hydroxy-acid interaction with the alumina, were recorded with Perkin–Elmer (1725x) and Nicolet (7000 series) spectrometers.

Results and Discussion

The most important hydroxycarboxylic acids used as industrial sequestering agents are gluconic and citric acids and, to lesser extent, tartaric and salicyclic acids [10]. In neutral and feebly alkaline solutions, the citrate ion is found to be the most effective. Its metal complexes are more stable than those of the other acid anions, partly because citrate carries a greater negative charge (-3) than tartrate (-2). Another reason is the stronger co-ordination power of the carboxylate group compared with that of the hydroxyl group. The co-ordination of citrate to a metal ion involves two carboxyl groups and one hydroxyl group, whereas co-ordination with tartrate occurs between one carboxyl group and two hydroxyl groups.

In Figs. 2 and 3 the spectra of the "free" acids and their adsorbed forms on γ-alumina are shown.

The spectrum of unadsorbed citric acid exhibits the presence of the acid dimer (1430 and 1392 cm^{-1}) as well as the C=O group (1754 and 1705 cm^{-1}). For the adsorbed form, the bands at 1645 and 1405 cm^{-1} indicate salt formation. It is a similar situation to the ethylenediaminetetra-acetic acid (EDTA) interaction with γ-alumina [11]. A shift to higher frequency can indicate mixed interactions, and/or can be evidence for formation of a new species between citric acid and the alumina support (there is disappearance of the C=O vibration). From this fact, it can be concluded that

* To whom correspondence should be addressed

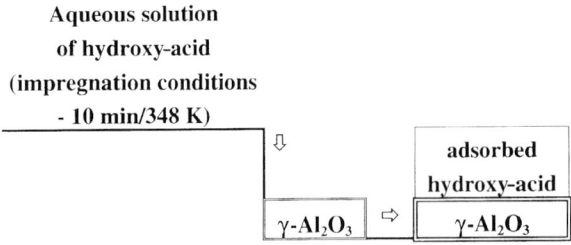

Fig. 1. Scheme of hydroxy-acid adsorption on γ-alumina

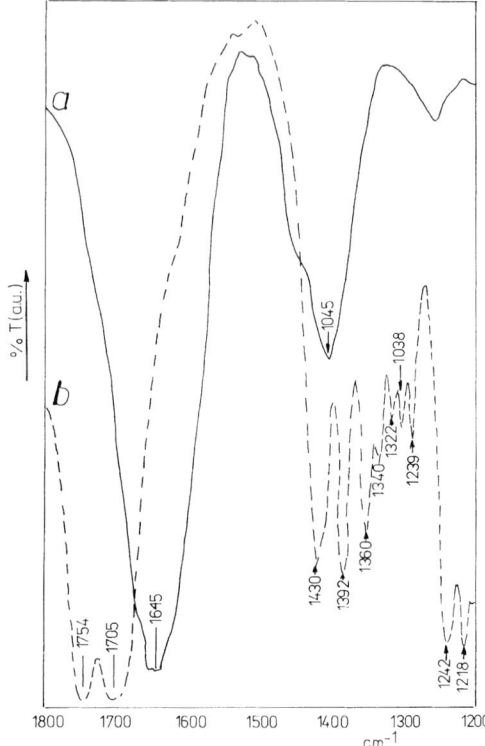

Fig. 2. IR spectra of: *a* citric acid adsorbed on alumina, *b* "free" citric acid

citric acid adsorption on alumina involves interaction of the support's surface hydroxyl groups with the acids carboxyl groups.

In the case of unadsorbed tartaric acid, the peaks at 1448 and 1398 cm^{-1} can be treated as acid dimer bands and additionally there is a C=O vibration (1739 cm^{-1}). In the spectrum of the adsorbed form, the bands in the range 1454–1382 and 1320–1211 cm^{-1} can be treated as due to deformation vibration of OH and stretching of C–O groups. The shift of the carbonyl group band (1674 cm^{-1}) and the low band for the COO$^-$ vibration suggest tartaric acid adsorption on the support.

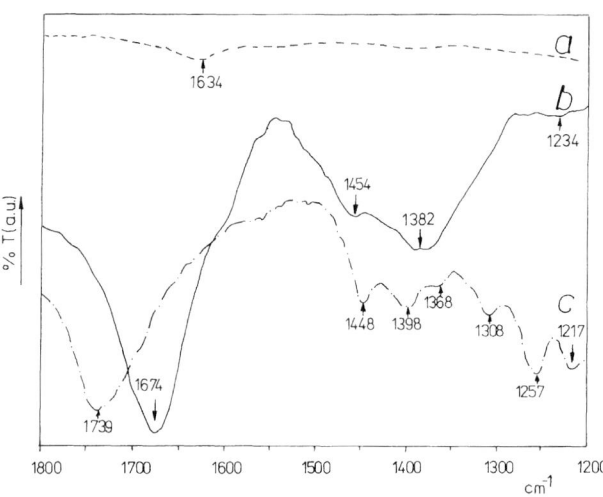

Fig. 3. IR spectra of: *a* alumina support; *b* tartaric acid adsorbed on alumina, *c* "free" tartaric acid

The higher value of the C=O frequency compared to that for citric acid, can be interpreted as due to weaker tartaric acid adsorption on γ-alumina.

From the spectrum of salicylic acid adsorbed on alumina (Fig. 4) it is obvious that the adsorbate has no longer an acid-like structure (absence of the C=O

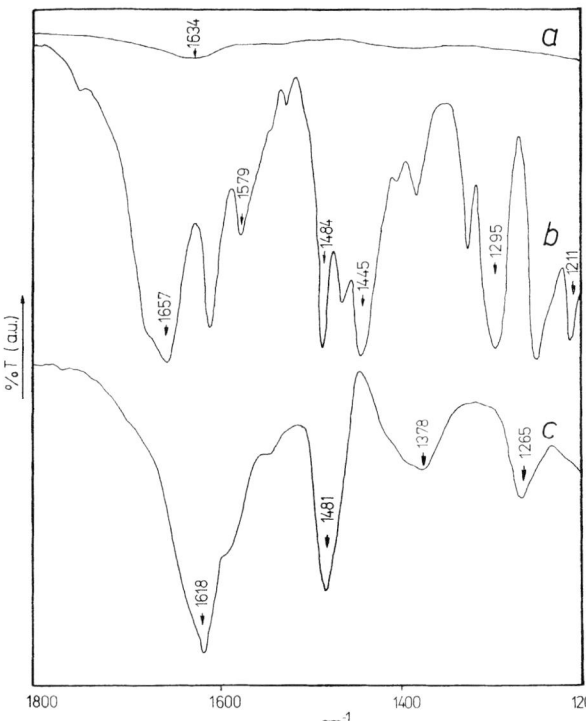

Fig. 4. IR spectra of: *a* alumina support; *b* "free" salicylic acid; *c* salicylic acid adsorbed on alumina

stretch vibration at around 1657 cm^{-1}). Also there is growth of bands at 1618, 1481 and 1378 cm^{-1}. The last two bands are ascribed to the asymmetric and symmetric stretch vibrations of the salicylate complex. Also a shift to lower wavenumbers and/or a decrease in the intensity of the bands which are ascribed to vibrations of the aromatic ring can be observed. The appearance of the vibration of the aromatic ring on γ-Al$_2$O$_3$ at 1618 cm^{-1} would lead us to conclude that the aromatic ring is stabilized, but other aromatic bands are shifted downward. The band at 1265 cm^{-1} for salicylic acid adsorbed on alumina is similar to that of phenol adsorbed on the same support (1263 cm^{-1}). It is obvious, from the similarity in position of the bands, that on γ-Al$_2$O$_3$ the salicylic acid hydroxyl group takes part in the co-ordination and the result is a phenolate-like structure.

Conclusions

The results obtained imply that changes observed in the IR spectroscopic properties of the supported hydroxy-acids are mainly due to interaction of the acids with alumina hydroxyl groups. The IR spectra in the C=O stretching region of the alumina-supported hy-droxy-acid are quite different from those of the unsupported compounds and this implies the formation of an appropriate hydroxy-acid complex with the alumina surface.

Acknowledgement. We would like to thank Mrs Barbara Sczczypa (UMCS, Lublin) for some analyses.

References

[1] R. Kumert, W. Stumm, *J. Colloid Interface Sci.* **1980**, *75*, 373.

[2] W. Stumm, R. Kummert, L. Sigg, *Croat. Chem. Acta* **1980**, *53*, 291.

[3] G. Furrer, W. Stumm, *Chimia* **1983**, *37*, 338.

[4] P. W. Schindler, W. Stumm, in: *Aquatic Surface Chemistry: Chemical Processes at the Particle-Water Interface* (W. Stumm, ed.), Wiley, New York, 1987, pp. 83–110.

[5] W. Stumm, G. Furrer, in: *Aquatic Surface Chemistry: Chemical Processes at the Particle-Water Interface* (W. Stumm, ed.), Wiley, New York, 1987, pp. 197–219.

[6] J. Ryczkowski, D. Nazimek, *React. Kinet. Catal. Lett.* **1991**, *44*, 427.

[7] J. Barcicki, D. Nazimek, W. Grzegorczyk, T. Borowiecki, R. Frak, M. Pielach, *React. Kinet. Catal. Lett.* **1981**, *17*, 169.

[8] J. Ryczkowski, *React. Kinet. Catal. Lett.* **1989**, *40*, 189.

[9] J. Ryczkowski, W. Grzegorczyk, D. Nazimek, *Appl. Catal. A* **1995**, *126*, 341.

[10] C. F. Bell, *Principles and Applications of Metal Chelation*, Clarendon Press, Oxford, 1977, pp. 128–138.

[11] J. Ryczkowski, *React. Kinet. Catal. Lett.* **1993**, *51*, 501.

Mikrochim. Acta [Suppl.] 14, 233–234 (1997)
© Springer-Verlag 1997

FT-IR and Resonance Raman Study of Photo-Induced Valence Tautomerization of 2,3-Diphenylindenone Oxide

Hiroaki Takahashi*, **Akira Inoue,** and **Noboru Kunimatsu**

Department of Chemistry, School of Science and Engineering, Waseda University, Tokyo 169, Japan

Abstract. FT-IR and resonance Raman spectra of the transient species of 2,3-diphenylindenone oxide generated by irradiation with UV light have indicated that the reversible photo-coloration reaction of this compound is well interpreted in terms of valence tautomerization involving transient species having a benzopyrylium olate structure. Frequency shifts and intensity changes in both the IR and Raman spectra of the transient species in different solvents suggest, however, that at least two transients are involved in the photo-coloration.

Key words: FT-IR, resonance Raman, valence tautomerization, photochemistry, 2,3-diphenylindenone oxide.

The reversible photo-coloration reaction of 2,3-diphenylindenone oxide (DPIO) has been considered to be attributable to the valence tautomerization between the stable colourless epoxy form and the short-lived red pyrylium olate form produced by a cleavage of the C–C bond of the epoxy ring [1] (Fig. 1). However, the identification of the transient red species as 1,3-diphenyl-2-benzopyrylium-4-olate (DBPO) has not been established spectroscopically and the mechanism of this photochemical reaction is controversial [2–4]. The purpose of this study is to establish the identity of the transient species and obtain information on the photochemical reaction mechanism of DPIO, by using FT-IR and resonance Raman spectroscopy.

Results and Discussion

Difference spectra obtained from the IR spectra measured before and after UV (308 nm) irradiation are shown in Fig. 2. The positive band at $1738\,\mathrm{cm}^{-1}$ assignable to the C=O stretch of the parent DPIO

disappears in the spectrum of the transient, and there is no band in the $1736–1604\,\mathrm{cm}^{-1}$ region. These facts indicate that the bond order of the C=O bond is considerably decreased in the transient. The negative bands at 1560, 1528, 1252 and $1227\,\mathrm{cm}^{-1}$ for the transient shift to 1549, 1513, 1265 and $1226\,\mathrm{cm}^{-1}$, respectively, on phenyl deuteration of the indenone group. The relatively small down-shifts or up-shift of these bands on phenyl deuteration and the decrease in the C=O bond order are consistent with the ben-zopyrylium olate structure of the transient.

Comparison of the spectra for the acetonitrile and benzene solutions reveals that the intensity ratio between the bands at 1560 and $1528\,\mathrm{cm}^{-1}$ and that between the bands at 1252 and $1227\,\mathrm{cm}^{-1}$ change markedly with the solvent: the intensity of the band at $1528\,\mathrm{cm}^{-1}$ relative to that of the band at $1560\,\mathrm{cm}^{-1}$ and the intensity of the band at $1227\,\mathrm{cm}^{-1}$ relative to that of the band at $1252\,\mathrm{cm}^{-1}$ increase on going from benzene solution to acetonitrile solution. It is also seen in Fig. 2 that both bands at 1560 and $1528\,\mathrm{cm}^{-1}$ exhibit high-frequency shifts of a few wavenumbers with benzene as the solvent.

Resonance Raman spectra of the transient of DPIO in different solvents are shown in Fig. 3. These spectra are also consistent with the benzopyrylium olate structure: no band is observed in the frequency range above $1609\,\mathrm{cm}^{-1}$, and bands are observed at 1563, 1528, 1257 and $1277\,\mathrm{cm}^{-1}$ for the acetonitrile solution. The frequency shifts of the bands at 1563 and $1528\,\mathrm{cm}^{-1}$ on deuteration of the indenone-phenyl group are not very clear, because of the weak intensities of these bands and overlapping by nearby bands. The bands at 1257 and $1277\,\mathrm{cm}^{-1}$ shift to 1267 and $1228\,\mathrm{cm}^{-1}$, respectively, on the indenone-phenyl deuteration. These frequency

---

* To whom correspondence should be addressed

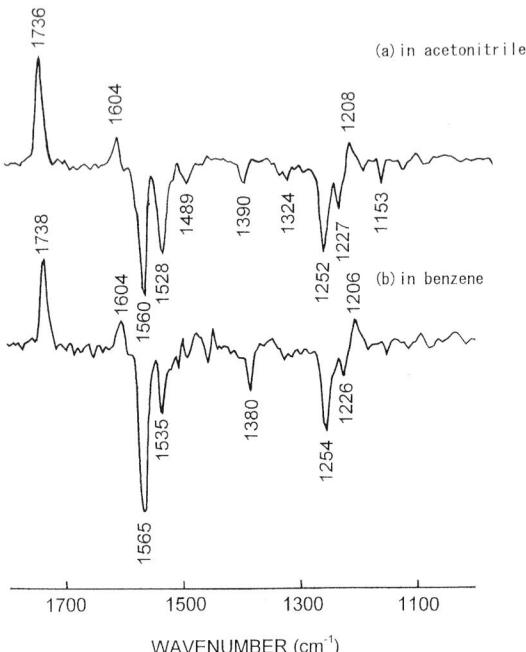

Fig. 1. Reversible photoreaction of 2,3-diphenylindenone oxide

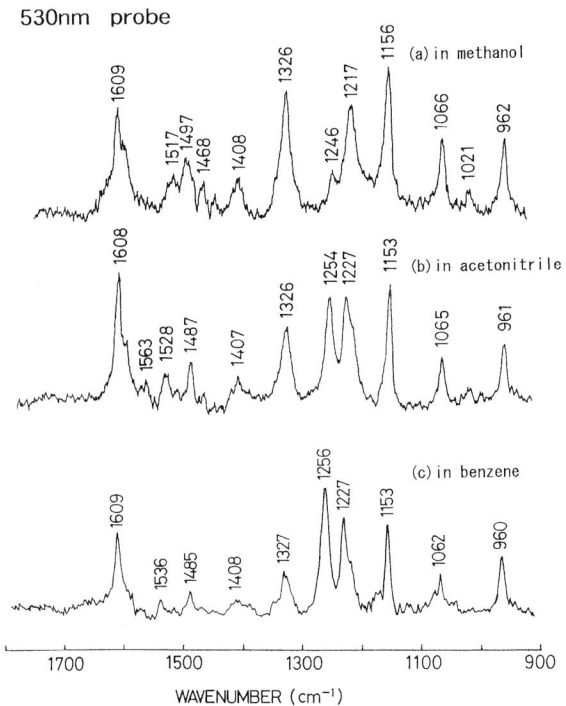

Fig. 3. Resonance Raman spectra of the transient species of DPIO in different solvents. Pump wavelength, 308 nm. Probe wavelength, 530 nm

Fig. 2. FT-IR spectra of the transient species of DPIO in different solvents. Pump wavelength, 308 nm

shifts coincide well with those of the corresponding bands in the infrared.

The intensity changes of the Raman bands at 1254 and 1227 cm^{-1} of the transient in different solvents are more drastic than those in the infrared. In non-polar benzene the band at 1256 cm^{-1} is stronger than the band at 1227 cm^{-1}. In polar acetonitrile the higher-frequency band shifts to 1254 cm^{-1} and decreases in intensity and the shoulder at around 1217 cm^{-1} becomes stronger. In hydrogen-bonding methanol the higher-frequency band shifts to the even lower frequency at 1246 cm^{-1} and reduces in intensity drastically, and the lower-frequency band is now almost completely replaced by the band at 1217 cm^{-1}.

The intensity and frequency changes of the Raman bands at 1563 and 1528 cm^{-1} are not clearly seen because of their weak intensities. It appears, however, that the band at 1536 cm^{-1} in benzene solution shifts to 1523 cm^{-1} in acetonitrile and to 1517 cm^{-1} in methanol. In methanol solution the band corresponding to the band at 1563 cm^{-1} for the acetonitrile solution is too weak to be detected.

The observed spectral changes in different solvents strongly suggest that at least two transient species are involved in the photoreaction of DPIO. The relative populations of these transients in solutions appear to be strongly influenced by the polarity and hydrogen-bonding ability of the solvent.

References

[1] E. F. Ullman, J. E. Milks, *J. Am. Chem. Soc.* **1962**, *84*, 1315; **1964**, *86*, 3814.
[2] E. F. Ullman, W. A. Henderson, Jr., *J. Am. Chem. Soc.* **1964**, *86*, 5050; **1966**, *88*, 4942.
[3] R. C. Bertelson, in: *Photochromism* (G. H. Brown, ed.), Wiley–Interscience, New York, 1971, p. 375.
[4] E. Hadjoudis, G. Bersos, *J. Photochem.* **1981**, *5*, 47.

Mikrochim. Acta [Suppl.] 14, 235–237 (1997)

The Possibilities of Identifying Photocopy Toners by Means of Infrared Spectroscopy (FT-IR) and Scanning Microscopy (SEM-EDX)

Beata M. Trzcińska* and **Zuzanna Brożek-Mucha**

Institute of Forensic Research, Westerplatte 9, 31–033 Cracow, Poland

Abstract. Nowadays, one of the most interesting problems in examination of questioned documents is the identification of toners for photocopy. Many different methods have been used for this purpose. In the present work infrared microspectroscopy (MK-FT-IR) and scanning electron microscopy with energy-dispersive X-ray analysis (SEM-EDX) were applied for differentiation among back toner powders and particles of toners taken from the printed line.

Key words: Questioned document, toner analysis, MK-FT-IR, SEM-EDX.

For the past two decades photocopy machines and consequently photocopy documents have been widely used. Therefore the toners for photocopy machines have extended the group of materials which can be used for document preparation. Photocopy machines allow many copies to be obtained in a short time period. The copies may be not identical as a result of interference by the operator and as such should be treated as questioned documents. Moreover, the toners are powders, unlike the majority of writing materials (which are liquids) and the mechanism of text creation is different. All these differences provide many analytical problems which have to be solved in the field of examination of questioned documents. In criminalistics the problem of identification may be expressed as follows: "Are two samples, the evidence and reference material, identical, i.e. do they come from the same source, or not?" The best way to find the answer is to look for significant, even though small, differences. At first sight all toner powders look very similar but they can be relatively easily classified into two main groups. Toners belonging to the first contain iron compounds, whereas those in the second group do not. This classification is not sufficient for forensic purposes and further differentiation among samples of the same class is necessary. At present a number of analytical methods can be involved in solving the problem of toner identification, e.g. HPTLC, pyrolysis gas chromatography, MS, DRIFT, SEM-EDX [1–6]. The aim of our studies was to check whether the combination of MK-FT-IR and SEM-EDX could be useful for forensic identification of toners.

Experimental

Twelve samples of black toner powders as well as particles of toner taken from the printed line were examined. The toners in bulk were chosen from the most popular photocopy machines, i.e. Rank Xerox, Canon, Ricoh, Minolta and Toshiba.

The investigations were done with an FTS 40 A spectrometer with UMA 500 microscope (Bio-Rad) and a JSM 5800 scanning electron microscope (Jeol) with an ISIS X-ray microanalysis attachment (Link, Oxford). Qualitative X-ray microanalysis was performed for the toner powders without carbon coating. The infrared transmission technique was applied both to the toner in bulk and its particles from the printed line. It should be pointed out that the toner particles can be separated from a printed line without transfer of any paper fibres and so without damaging the document.

Results

It was found that particles of the examined toners, observed under SEM at a magnification of 1000, look similar (Figs. 1 and 2). However, small differences in size and shape can be noticed from sample to sample.

More pronounced differences were observed in the IR spectra, with respect to both position and intensity of some absorption bands, usually the less intense ones (Fig. 3). Moreover, the spectra of a bulk toner and of its particles from a printed line are nearly identical (Fig. 4). Thus, it is possible to differentiate among toners even of the same make.

---

* To whom correspondence should be addressed

Fig. 1. Particles of TA 2020 toner under SEM (magnification about 1000)

Fig. 2. Particles of Canon FC 2 toner under SEM (magnification about 1000)

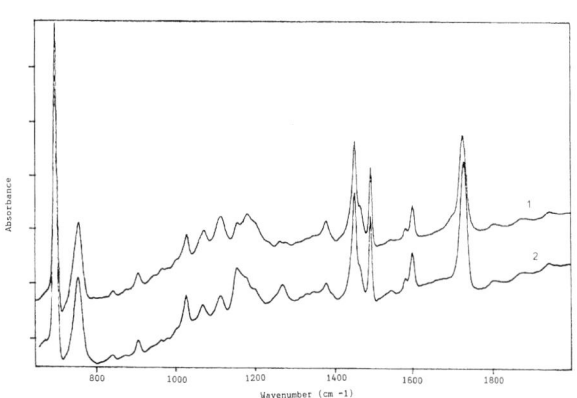

Fig. 3. IR spectra of toner powders: *1* TA 2020; *2* Canon FC 2

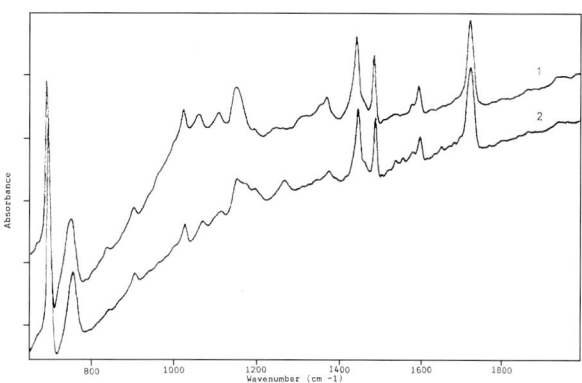

Fig. 4. IR spectra of Canon FC 2 toner: *1* from printed line; *2* in bulk

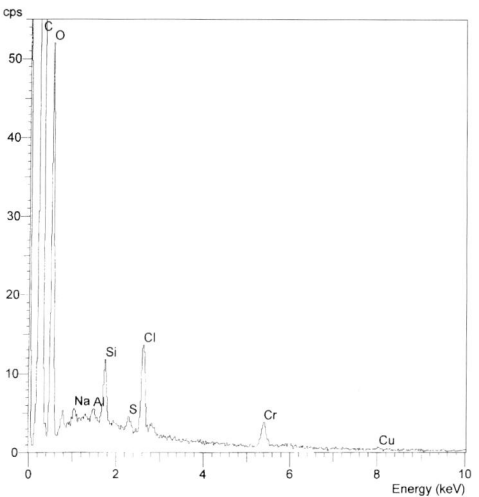

Fig. 5. EDX spectrum of TA 2020 toner powder

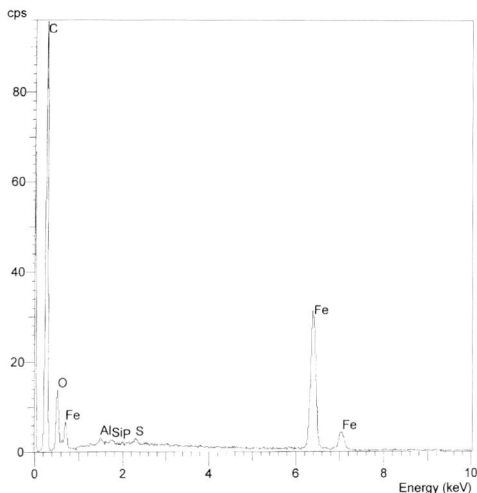

Fig. 6. EDX spectrum of Canon FC 2 toner powder

Table 1. The elemental composition of toner powders in order of decreasing amount

Toner	Chemical elements						
1 Ricoh FT 5010	C,	O,	Al				
2 Minolta EP 470z	C,	O,	Cu,	Al			
3 Ricoh 4000	C,	O,	Cl				
4 Xerox 1025	C,	O,	S,	Al,	Si		
5 TA 2020	C,	O,	Cl,	Si,	Cr		
6 Minolta 4300	C,	S,	O,	P,	Si		
7 Minolata 490z	C,	O,	S,	Si,	Na		
8 Canon NP 1550	C,	O,	S,	Na,	Si		
9 Rank 1025/1038	C,	O,	S,	Al,	Si		
10 Toshiba 1650	C,	O,	S,	Si,	Na		
11 Canon NP 1215	Fe,	C,	O,	Si,	S,	Al	
12 Canon FC 2	C,	Fe,	O,	Al,	S,	Si,	P

Table 2. Selected peak integral ratios for the most common elements in toner powders

Toner[a]	C:O	C:Na	C:Al	C:Si	C:P	C:S	C:Cl	C:Cu
1	1:0.021		1:0.004					
2	1:0.016		1:0.003					1:0.006
3	1:0.021						1:0.003	
4	1:0.030		1:0.005			1:0.009		
5	1:0.075			1:0.014			1:0.024	
6	1:0.011				1:0.010	1:0.068		
7	1:0.030			1:0.006		1:0.011		
8	1:0.048	1:0.006				1:0.009		
9	1:0.063		1:0.006			1:0.010		
10	1:0.026			1:0.005		1:0.006		
	C:O	Fe:C	Fe:O	Fe:Al	Fe:Si			
11	1:0.083	1:0.647	1:0.054		1:0.023			
12	1:0.131	1:1.439	1:0.189	1:0.033				

[a] See Table 1 for identity of toners.

The final confirmation of the individuality and forensic identification of toners of the same make was made by X-ray microanalysis. In the toners examined the following elements were found: carbon, oxygen, sulphur, silicon, phosphorus, chlorine and some metals such as chromium, iron, sodium, aluminium and copper (Figs. 5 and 6). Table 1 shows the elemental composition in order of decreasing amount of elements and Table 2 selected ratios of peak integrals. It can be seen that either the sequence of elements or the set of peak integral ratios does not repeat.

Conclusions

The results obtained in our laboratory and also by other research groups showed that it is possible to differentiate among the toners and that the differentiation is suitable for forensic purposes. In criminalistic investigations, identification can be made not only with relatively large samples but also in cases of questioned document examination when a small sample, e.g. one letter, is to be analysed. It is worth mentioning that for a reliable identification more than one analytical technique should be used. Microspectroscopy in the infrared (MK-FT-IR) and scanning electron microscopy with energy-dispersive X-ray analysis have proved to be good tools for this purpose.

References

[1] T. O. Munson, *J. Forensic Sci.* **1989**, *34*, 352.
[2] C. J. Lennard, W. D. Mazzella, *J. Forensic Sci. Soc.* **1991**, *31*, 365.
[3] W. D. Mazzella, C. J. Lennard, P. A. Margot, *J. Forensic Sci.* **1991**, *36*, 820.
[4] W. T. Chang, C. W. Huang, Y. S. Giang, *J. Forensic Sci.* **1993**, *38*, 843.
[5] J. A. Andrasko, *J. Forensic Sci.* **1994**, *39*, 226.
[6] G. Tandon, O. P. Jasuja, V. N. Sehgal, *Forensic Sci. Int.* **1995**, *73*, 149.

Mikrochim. Acta [Suppl.] 14, 239–241 (1997)

FT-IR Spectroscopic Studies on Cyclic Models of H-Bonded Folded Polypeptide Structures

Elemér Vass[1], Sándor Holly[2], and **Miklós Hollósi[1,*]**

[1] Department of Organic Chemistry, Eötvös University, 112. P.O.B. 32, H-1518 Budapest, Hungary
[2] Central Research Institute of Chemistry, P.O.B. 17, H-1525 Budapest, Hungary

Abstract. FT-IR spectroscopic studies on a variety of cyclic and linear models of β- and γ-turns show that the amide I band of acceptor amide groups in well-established $1 \leftarrow 4$ H-bonded β-turns appears between 1645 and 1630 cm^{-1}. Bands in the 1655–1641 and 1627–1617 cm^{-1} regions are due to weakly and strongly $1 \leftarrow 3$ H-bonded γ-turns, respectively. Bands appearing near 1600 cm^{-1} in protic solvents are indicative of acceptor amide groups involved in bifurcated H-bonding.

Key words: FTIR, β-turn, γ-turn, bifurcated H-bonding.

Recently there has been a growing interest in the Fourier-transform infrared (FT-IR) spectroscopic characterization of H-bonded folded polypeptide structures (turns) [1]. The amide groups of peptides give rise to several strong infrared bands of which the amide I band (1600–1700 cm^{-1}) has attracted particular attention during the last decade. The major factors which influence the position of the amide I band are H-bonding and the coupling of transition dipoles. The band shifts caused by H-bonding strongly depend on the orientation and distance apart of the interacting amide groups and thus provide information about their relative steric position (conformation). Mathematical approaches (Fourier self-deconvolution and Fourier derivation) based on Fourier transformation [2] are used to analyse the complex amide I band contour in terms of contributions of individual secondary structures (α-helix, β-sheet etc.) [1, 3].

Absorption in the 1660–1650 cm^{-1} region of polypeptide IR spectra is due to α-helix structure, while the characteristic band of β-sheet conformation appears at 1640–1620 cm^{-1} [3]. Generally, component bands at above ∼1670 cm^{-1} have been associated with β-turns

[4]. On the basis of detailed FT-IR studies on the bridged cyclic peptides **1a–c** (Scheme 1), Mantsch and co-workers have assigned a component band near 1640 cm^{-1} to the acceptor amide C=O of $1 \leftarrow 4$ (C$_{10}$) H-bonded β-turns [5]. However, the FT-IR spectra of other bridged cyclic peptides (e.g. **2**), featuring only a β- and/or a γ-turn, were found to show bands also near 1650, 1625 and 1660 cm^{-1} [6]. γ-Turns, which represent the other main type of folded polypeptide conformations, may contain a $1 \leftarrow 3$ (C$_7$) intramolecular H-bond.

This paper reports FT-IR spectroscopic studies on a selection of cyclic and linear models of β-and γ-turns. Special emphasis is given to the vibrational spectroscopic characterization of peptides containing $1 \leftarrow 3$ H-bondings.

Experimental

Details of the syntheses, circular dichroism, NMR, molecular dynamics and X-ray crystallographic characterization of the models in Scheme 1 have been published earlier [7–11]. FT-IR measurements were performed at room temperature on a Nicolet 170SX (models **2** and **4a, b**) or a Bruker IFS-55 spectrometer (models **1d, e** and **5a, b**). Trifluoroethanol (TFE) was generally used as an H-bond promoting solvent [12], and DMSO as an H-bond breaking solvent. The FT-IR spectra of models **1a–c** and **3a, b** have been reported [5] and [6].

1 **cyclo** [Gly-Pro-Xxx-Gly-NH-(CH$_2$)$_4$-CO]: Xxx = (**a**) Gly; (**b**) Ser(O^{t}Bu); (**c**) Thr(O^{t}Bu); (**d**) Asn; (**e**) Gln.
2 **cyclo** [Gly-Ser(O^{t}Bu)-Ser(O^{t}Bu)-Gly-NH-(CH$_2$)$_4$-CO]
3 **cyclo** [Pro-Ala-NH-(CH$_2$)$_5$-CO]$_n$: n = (**a**) 1; (**b**) 2.
4 **cyclo** [Gly-His-Xxx-Arg-Trp-Gly]: Xxx = (**a**) Phe; (**b**) D-Phe.
5 (**a**) Ac-Pro-NHCH$_3$; (**b**) Ac-Ala-NHCH$_3$.

Scheme 1

---

* To whom correspondence should be addressed

Results and Discussion

Results of the FT-IR studies together with data on the conformation of models **1–5** are summarized in Table 1. The amide I frequencies of turn models can be divided into six groups. Bands centred at ~ 1675 and $\sim 1665\ cm^{-1}$ can be assigned to shielded (non-solvated or weakly solvated) and solvated amide groups, respectively (see Table 1, footnote b). The other four groups of bands, presented in Table 1, are due to acceptor amide groups of turns. The acceptor amide group of well-established $1 \leftarrow 4$ H-bonded β-turns was found to absorb between 1645 and $1630\ cm^{-1}$.

In agreement with recent literature data [6] the amide I band of the acceptor amide groups in γ-turns was observed at 1655–1641 or $1627–1617\ cm^{-1}$. The appearance of high and low frequency bands of accep-

tor carbonyls can be associated with stronger and weaker $1 \leftarrow 3$ H-bonds. Based on X-ray crystallographic data, *inverse* γ-turns ($\varphi_2 \cong -80°$, $\psi_2 \cong 80°$) form stronger H-bondings than the *classic* ones ($\varphi_2 \cong 80°$, $\psi_2 \cong -80°$) [13].

The infrared spectra of some cyclic and linear models show an intense band at $\sim 1600\ cm^{-1}$. *Bifurcation* (one acceptor with two donor amides, or oriented solvation of the acceptor amide group by protic solvents) may account for the extremely low amide I stretching frequency. Our data give support to the idea that protic solvents may have a special stabilizing effect upon both $1 \leftarrow 3$ and $1 \leftarrow 4$ H-bonded structures. (Note the appearance of three low-frequency bands in TFE in the FT-IR spectrum of Ac-Pro-NHCH$_3$ (**5a**) which has only one donor amide NH group, allowing the formation of only $1 \leftarrow 3$ H-bonding. In contrast to this,

Table 1. Conformation and Fourier-transform infrared spectra of models of *β-* and *γ*-turns

Model	Type of turn[a]	Solvent	FTIR spectra[b] Frequencies (cm^{-1}) and intensities of the amide I component bands of aceptor amide groups				Ref.[c]
			$1 \leftarrow 4$	$1 \leftarrow 3$ weak	$1 \leftarrow 3$ strong	bifurcated	
1a	βII + pβ [7]	DMSO	1642s				[5]
		CHCl$_3$	1642s				
		D$_2$O	1640s		1621s		
1b	βI + pβ [7]	TFE	1635s				[5]
		CH$_3$CN	1645s				
1c	βI + pβ [7]	D$_2$O	1640s		1622s		[5]
1d	βII + pβ [8]	TFE	1640m				d
1e	βII + pβ [8]	TFE	1643w		1627m		d
2	βII [9]	TFE	1636s				d
3a	β + γ [10]	TFE	1641s			1600s	[6]
		D$_2$O	1641s			1595m	
		CH$_3$CN	1631m	1648m	1621m		
3b	β + γ [10]	D$_2$O	1636s			1602s	[6]
		CH$_3$CN	1632s				
4a	2β [11]	DMSO	1643m	1655s		1603s	d
		TFE	1630m	1649m		1599m	
4b	2β [11]	DMSO	1641m	1654s		1603s	d
		TFE	1642m			1598s	
5a	γ	DMSO		1645s	1617w		d
		TFE		1641s	1624w	1604s	
		CH$_2$Cl$_2$		1651m	1625m	1604.5w	
5b	γ	DMSO	1641w				d
		TFE		1646s	1622m		
		CH$_2$Cl$_2$[e]					

[a] Abbreviations: β β-turn; γ γ-turn; pβ pseudo β-turn (C$_{10}$ H-bonded structure formed with the participation of δ-amino valeric acid residues).

[b] Frequencies corresponding to non-acceptor amides ($> 1660\ cm^{-1}$) are not given. Abbreviations: s strong; m medium; w weak.

[c] Source of FTIR studies.

[d] This paper.

[e] Bands at 1681 and $1663\ cm^{-1}$.

Ac-Ala-NHCH$_3$ (**5b**) having two amide NH groups shows no band below 1660 cm^{-1} in the IR spectrum in CH$_2$Cl$_2$).

On the basis of the data shown in Table 1, FT-IR spectroscopy allows discrimination between β- and γ-turns and the analysis of their mixtures which present in solution.

Acknowledgement. This work was supported in part by the Hungarian National Scientific Research Foundation (OTKA grant T017432).

References

[1] M. Jackson and H. Mantsch, *Crit. Rev. Biochem. Mol. Biol.* **1995**, *30*, 95.

[2] H. H. Mantsch, D. J. Moffatt, H. Casal, *J. Mol. Struct.* **1988**, *173*, 285.

[3] W. K. Surewicz, H. H. Mantsch, D. Chapman, *Biochemistry* **1993**, *32*, 389.

[4] W. K. Surewicz, H. H. Mantsch, *Biochem. Biophys. Acta* **1988**, *952*, 115.

[5] H. H. Mantsch, A. Perczel, M. Hollósi, G. D. Fasman, *Biopolymers* **1993**, *33*, 201.

[6] R. A. Shaw, A. Perczel, H. H. Mantsch, G. D. Fasman, *J. Mol. Struct.* **1994**, *324*, 143.

[7] A. Perczel, M. Hollósi, B. M. Foxman, G. D. Fasman, *J. Am. Chem. Soc.* **1991**, *113*, 9772.

[8] E. Láng, B. Hargittai, Zs. Majer, A. Perczel, M. Mák, M. Kajtár-Peredy, L. Radics, G. D. Fasman, M. Hollósi, *Protein Peptide Lett.* **1996**, *3*, 9.

[9] M. Hollósi, K. E. Kövér, S. Holly, G. D. Fasman, *Biopolymers* **1987**, *26*, 1527.

[10] A. Perczel, G. D. Fasman, *Protein Sci.* **1992**, *1*, 378.

[11] M. S. Prachand, M. M. Dhingra, A. Saran, J. Bódi, H. Süli-Vargha, K. Medzihradszky, in preparation.

[12] M. Jackson, H. H. Mantsch, *Biochim. Biophys. Acta* **1992**, *1118*, 139.

[13] E. J. Milner-White, *J. Mol. Biol.* **1990**, *216*, 385.

Mikrochim. Acta [Suppl.] 14, 243–244 (1997)
© Springer-Verlag 1997

FT-IR Spectra of Cadmium Calcium Pyrophosphate

Danita de Waal* and **Christina Hutter**

Department of Chemistry, University of Pretoria, 0002 Pretoria, South Africa

Abstract. The FT-IR spectra of a solid solution of cadmium and calcium pyrophosphate, $Cd_{1.25}Ca_{0.75}P_2O_7$, are reported here with reference to the IR spectra of the pure compounds, which are used to obtain further information on the structure of the compound. The modes of the pyrophosphate ion can be obtained without interference from site and factor group splitting due to the C_1 site and factor groups.

Key words: pyrophosphate, solid solution, FT-IR spectra.

Experimental

Samples of β-$Ca_2P_2O_7$ and $Cd_2P_2O_7$ were obtained through the high-temperature reaction at $900\,°C$ of the appropriate carbonates and ammonium hydrogen phosphate. $Cd_{1.25}Ca_{0.75}P_2O_7$ was prepared from a similar high-temperature reaction between $Ca_2P_2O_7$ and $Cd_2P_2O_7$. All reactions were performed in platinum crucibles and products were allowed to cool down slowly to room temperature before removal from the furnace. Infrared spectra were recorded in the form of KBr and polyethylene discs, with a Bruker IFS 113v FT-IR spectrometer. A resolution of $2\,cm^{-1}$ was used throughout.

Results and Discussion

Of the vibrational modes expected for the $P_2O_7^{4-}$ ion with the eclipsed conformation (C_{2v} symmetry): $7A_1 + 4A_2 + 4B_1 + 6B_2$, six are terminal ($PO_3$) and two bridge stretching vibrations (POP), while the remaining thirteen are bending modes ($6PO_3$, $6OPO_3$, 1POP). In both $Cd_2P_2O_7$ and $Cd_{1.25}Ca_{0.75}P_2O_7$ the anion is situated on the C_1 site so that no further splitting is expected. Under the factor group C_i of $Cd_2P_2O_7$ the expected modes are: $21A_g + 21A_u$. $Cd_{1.25}Ca_{0.75}P_2O_7$ should have 21A modes under factor group C_1. The FT-IR spectra of $Cd_{1.25}Ca_{0.75}P_2O_7$ are shown in Figs. 1 and 2, and the bands are compared to those observed in the same region for $Cd_2P_2O_7$ (see Table 1). For the solid solution sixteen of the expected twenty-one infrared active modes were observed, and for the stretching modes five of the six predicted terminal modes. Both bridge stretching vibrations could be identified. Most of the bands in the bending region could be assigned to terminal bending modes. The $240\,cm^{-1}$ band in the solid solution corresponds to two bands for $Cd_2P_2O_7$ at 225 and $275\,cm^{-1}$, while the $172\,cm^{-1}$ band for $Cd_{1.25}Ca_{0.75}P_2O_7$ corresponds to two bands at 162 and $148\,cm^{-1}$ for $Cd_2P_2O_7$. This

The vibrational spectra [1–4] and crystal structures [5–8] of many pyrovanadates of the type $A_2P_2O_7$ (A = metal cation) in different phases have already been reported. The crystal structure of $Ca_2P_2O_7$ is C_4^2 (β-phase) and C_{2h}^5 (α-phase) while that of $Cd_2P_2O_7$ is C_i^1 [5–7]. The crystal chemistry of the whole $Cd_{2-x}Ca_xP_2O_7$ series, $0 \leqslant x \leqslant 2$, has also been investigated before [8]. It has also been determined [8] that solid solutions with x up to 0.40 are isostructural with $Cd_2P_2O_7$, while with values of x between 1.70 and 2.00 compounds are formed that are isostructural with β-$Ca_2P_2O_7$. For x between 0.40 and 1.00, a new and unknown phase could be observed, and in this region of x-values single crystals of $Cd_{1.25}Ca_{0.75}P_2O_7$ have been isolated and the space group has been determined as C_i^1. For x between 0.70 and 1.00 a mixture of this new phase and crystals isostructural with β-$Ca_2P_2O_7$ has been found to occur [8]. FT-IR spectra of the new phase $Cd_{1.25}Ca_{0.75}P_2O_7$ are investigated here. The vibrational spectra should, to a large extent, reflect the $P_2O_7^{4-}$ internal modes for C_{2v} symmetry, $7A_1(IR/R) + 4A_2(R) + 4B_1(IR/R) + 6B_2(IR/R)$, as the factor and site groups (both C_1) are predicted not to cause further splitting of these modes.

---

* To whom correspondence should be addressed

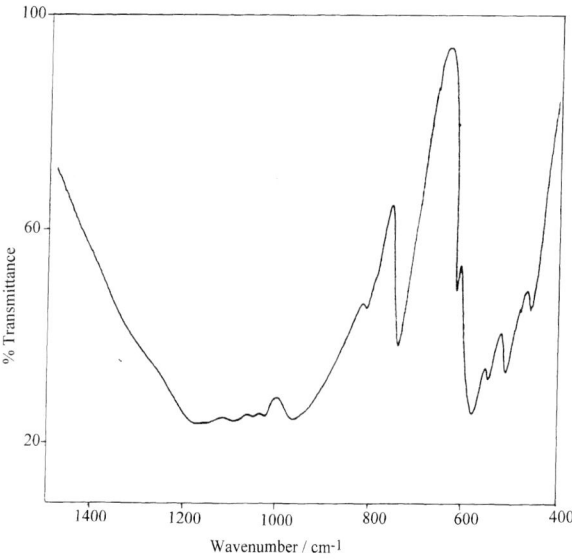

Fig. 1. The IR spectrum of $Cd_{1.25}Ca_{0.75}P_2O_7$ between 400 and 1500 cm^{-1}

Table 1. Comparison of IR modes observed for $Cd_{1.25}Ca_{0.75}P_2O_7$ and those for $Cd_2P_2O_7$

$Cd_{1.25}Ca_{0.75}P_2O_7$	$Cd_2P_2O_7$	Assignment
1150 vs	1154 s	$\nu(PO_3)$ asym
1073 vs	1094 vs	
1031 s	1030 s	$\nu(PO_3)$ sym
1007 s	1010 s	
946 s	934 s	$\nu(POP)$ asym
–	862 sh	
798 sh	–	$\nu(POP)$ sym
725 m	721 m	
609 m	607 m	$\delta(PO_3)$ asym
565 s	565 s	
–	545 sh	
533 s	528 m	$\delta(PO_3)$ sym
496 s	488 m	
447 m	435 m	
397 vw	409 m	
367 vw	–	
–	275 w	$\delta(POP)$ and lattice
240 w	225 w	modes
172 w	162 w/148 w	

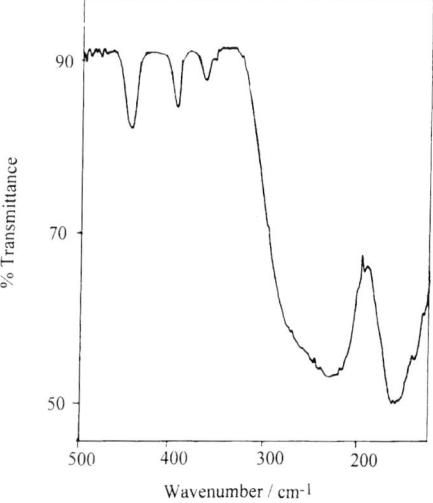

Fig. 2. The IR spectrum of $Cd_{1.25}Ca_{0.75}P_2O_7$ between 150 and 400 cm^{-1}

serves to illustrate the lack of splitting of bands under site and factor groups, which has already been predicted for the solid solution.

Conclusions

From the FT-IR spectra of $Cd_2P_2O_7$ and $Cd_{1.25}Ca_{0.75}P_2O_7$ it can be confirmed that the solid solution has a similar, although not identical, structure to that of the pure compound. The modes of the $P_2O_7^{4-}$ ion seem to remain relatively simple in $Cd_{1.25}$-$Ca_{0.75}P_2O_7$ as a result of the C_1 site and factor group symmetries.

References

[1] B. C. Cornilsen, *J. Mol. Struct.* **1984**, *117*, 1.
[2] B. C. Cornelsin, R. A. Condrate, *J. Solid State Chem.* **1978**, *23*, 375.
[3] E. Steger, B. Kässner, *Spectrochim. Acta* **1967**, *24A*, 447.
[4] A. Hezel, S. D. Ross, *Spectrochim. Acta* **1968**, *24A*, 131.
[5] B. E. Robertson, C. Calvo, *Can. J. Chem.* **1968**, *46*, 605.
[6] C. Calvo, *Inorg. Chem.* **1967**, *7*, 1345.
[7] C. Calvo, P. K. L. Au, *Can. J. Chem.* **1969**, *47*, 3409.
[8] A. Alaoui, El Belghitti, A. Boukhari, M. E. Holt, *J. Solid State Chem.* **1993**, *106*, 506.

Mikrochim. Acta [Suppl.] 14, 245–246 (1997)

Characterization of Reaction Products on Solid Supports by FT-IR

Inge E. A. Deben[1], Theo H. M. van Wijk[1], Emile M. van Doornum[1], Pedro H. H. Hermkens[2], and Edwin R. Kellenbach[1,*]

[1] Department of Analytical Chemistry for Development and [2] Department of Medicinal Chemistry II, NV ORGANON, P.O. Box 20, 5340BH OSS, The Netherlands

Abstract. Solid-phase organic chemistry (SPOC) is at present widely used to create combinatorial "libraries" of organic compounds for drug discovery and to lead finding. SPOC involves the synthesis of organic compounds on a solid support, usually a polymer resin. Since neither conventional NMR nor TLC is capable of following reactions on a solid support, we have applied FT-IR successfully to evaluate reactions on a resin. Some examples as well as advantages and shortcomings of this technique will be discussed.

Key words: FT-IR, solid-phase organic chemistry, combinatorial chemistry.

Solid-phase organic chemistry is currently being employed for the construction of "libraries" of organic compounds by using a solid support and a split/synthesize procedure (Fig. 1). This approach towards the simultaneous synthesis of a large number of compounds is called "combinatorial chemistry" [1, 2].

The recent interest in solid-phase organic chemistry for the construction of combinatorial libraries of organic compounds to lead discovery and synthesis of drugs has renewed the interest towards organic chemistry on solid supports [1, 2]. This has raised the need for analytical methods to monitor reactions during the optimization of reactions on a solid support. The organic chemist can no longer use his favourite analytical tools, NMR and TLC, to follow reactions. Structural analytical techniques such as NMR, TLC and MS usually fail to yield structural information about compounds bound an solid supports. Since FT-IR has previously been used to characterize organic compounds on a solid support [3, 4] we have used it to monitor reactions on a solid support.

Experimental

Approximately 3.5 mg of the resin, in the case of a KBr-pellet measurement, or 4.5 mg of the resin in the case of a DRIFT measurement, were used for the analyses. Spectra were recorded on an FTS-45A or an FTS-60 FT-IR spectrometer (Bio-Rad, U.S.A.). Typically, 128 scans were acquired, with a resolution of 2 cm^{-1}. The resins consisted of functionalized polystyrene.

Results and Discussion

If the resin, the starting material and the reaction product are available as a reference, comparison of the IR spectra can yield information on qualitative, and in some cases even quantitative, aspects of reactions on solid supports. FT-IR requires only small (<5 mg) amounts of resin to perform an analysis. In particular, functional group conversions and hybridization changes can easily be demonstrated by FT-IR. Problems encountered are difficulties in grinding some of the more rigid resins to obtain a KBr disc spectrum and difficulties in observing the weak IR signals of the

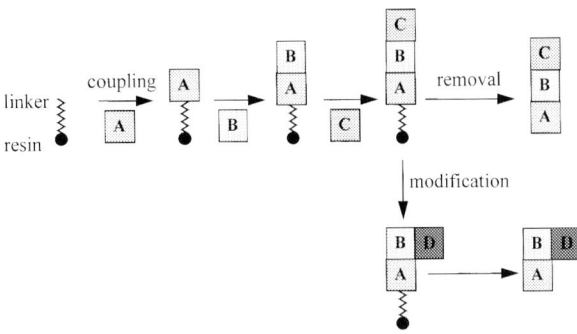

Fig. 1. Schematic representation of the creation of a library of organic compounds by using solid-phase organic chemistry

* To whom correspondence should be addressed

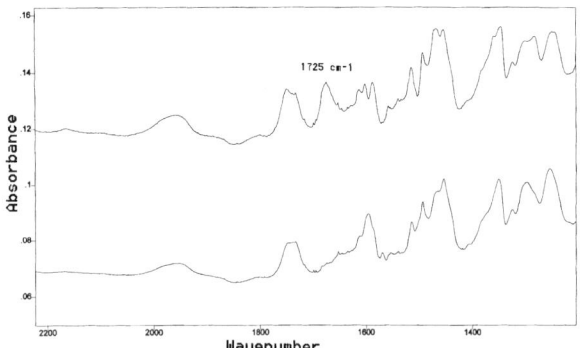

Fig. 2. IR spectra showing the conversion depicted in Scheme I, of a carbonyl C=O (bottom spectrum) into a C=S group (top spectrum). The disappearance of the carbonyl peak at 1675 cm^{-1} is indicative of the progress of the reaction. It is clear that the IR resonances of polystyrene obscure a large part of the spectrum

compound of interest amidst the strong background of the resin (see also Fig. 2). Only in cases where the functional group vibrations lie outside the spectral regions assigned to the polymer bands, can reasonable conclusions be drawn.

The resins used were mostly made from polystyrene which can be functionalized with a number of different linkers. One of the conversions studied is listed below. KBr disc spectra give the best signal-to-noise ratio. However, when information on hydroxyl groups or other exchangeable protons is required, DRIFT is the method of choice. DRIFT allows the observation of those regions in the IR spectrum which are usually hard to interpret in KBr disc spectra because of the broad water-band present in the KBr disc spectra. An example of a reaction (scheme 1) successfully monitored by FT-IR is listed below.

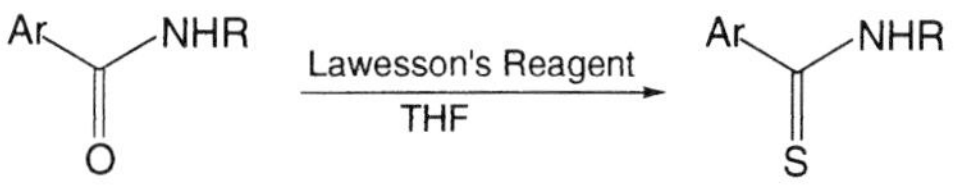

Scheme I

On a number of occasions FT-IR indicated that reactions had either not or only partly, taken place.

These results were corroborated by cleaving the compounds of the solid support and studying them by NMR methods. This validated the use of FT-IR for monitoring this type of reactions.

Recently, NMR methods have been applied to yield information on the structure of compounds attached to a solid support [5–7]. Although these methods have a number of advantages over FT-IR, notably that they do not require undisturbed spectral regions and suitable functional groups, they are more time-consuming. Moreover, they require either carbon-13 labelled material or special probes. Matrix-assisted laser desorption/ionization time of flight mass spectrometry (MALDI-TOF) has also been applied to analyse reaction products on solid supports [8], but has been reported to be applicable only when the acid-labile Rink amide linker is used to anchor the reaction product to the solid support. Therefore, FT-IR seems to be the method of choice for quickly monitoring reactions occurring on a solid support.

Note added in proof: results similar to ours were recently reported in the literature [9].

References

[1] M. A. Gallop, R. W. Barret, W. J. Dower, S. P. A. Fodor, E. M. Gordon, *J. Med. Chem.* **1994**, *37*, 1233.
[2] E. M. Gordon, R. W. Barret, W. J. Dower, S. P. A. Fodor, M. A. Gallop, *J. Med. Chem.* **1994**, *37*, 1385.
[3] C. Chen, L. A. A. Randall, R. B. Miller, A. D. Jones, M. J. Kurth, *J. Am. Chem. Soc.* **1994**, *116*, 2661.
[4] J. M. Frechet, C. Schuerch, *J. Am. Chem. Soc.* **1971**, *93*, 492.
[5] W. L. Fitch, G. Detre, C. P. Holmes, G. J. Shoodery, P. A. Keifer, *J. Org. Chem.* **1994**, *59*, 7955.
[6] C. P. Holmes, J. P. Chinn, G. C. Look, E. M. Gordon, M. A. Gallop, *J. Org. Chem.* **1995**, *60*, 7328.
[7] R. C. Anderson, M. A. Jarema, M. J. Shapiro, J. P. Stokes, M. Ziliox, *J. Org. Chem.* **1995**, *60*, 2650.
[8] B. J. Egner, G. J. Langley, M. Bradley, *J. Org. Chem.* **1995**, *60*, 2652.
[9] B. Yan, G. Kumaravel, H. Anjaria, A. Wu, R. C. Petter, C. F. Jewell, Jr., *J. Org. Chem.* **1995**, *60*, 5736.

Mikrochim. Acta [Suppl.] 14, 247–249 (1997)

FT-Raman Spectroscopy of Modified Tetraisopropyltitanate Hydrolysates

Paul A. Venz[1], **Ray L. Frost**[1,*], **John R. Bartlett**[2], **James L. Woolfrey**[2], **Lisa W. Y. Wong**[1], and **Sandra M. Dutt**[1]

[1] School of Chemistry, Queensland University of Technology, 2 George St, Brisbane, Qld. 4001, Australia
[2] Advanced Materials Program, Australian Nuclear Science and Technology Organisation, Private Mail Bag 1, Menai N. S. W. 2234, Australia

Abstract. Modification of metal alkoxides with complexing agents, such as carboxylic acids, which act both as a catalyst and a ligand, is commonly used in sol–gel processing to alter the hydrolysis and condensation rates of the alkoxide. Hydrolysates produced by the reaction of modified tetraisopropyltitanate with water were X-ray-amorphous but exhibited spectra similar to the spectrum of rutile; broad bands assigned to the E_g and A_{1g} Ti–O stretching modes of the "rutile-like" phase were observed at *ca.* 425 and 605 cm^{-1}, respectively. During peptization of the hydrolysate with nitric acid at 60 °C, the intensity of these bands decreased substantially, and new peaks appeared at 154, 405, 515 and 630 cm^{-1}, which were assigned to the δ_{O-Ti-O} (E_g), δ_{O-Ti-O} (B_{1g}), ν_{Ti-O} (A_{1g}/B_{1g}) and ν_{Ti-O} (E_g) modes, respectively, of anatase. However, the hydrolysates were also found to undergo a similar phase change when processed at the same temperature in the absence of peptizing agents, but the rate of transformation was slower. The carboxylic acid also had a significant influence on the induction time for the phase change, with the transformation occurring more slowly with increasing carboxylate chain length.

Key words: sol–gel processing, tetraisopropyltitanate, FT-Raman spectroscopy.

In recent years, sol–gel methods have become increasingly important for the production of advanced ceramics. A large body of work on polymeric sol–gel systems exists. An important technique in these sol–gel processes is the chemical modification of metal alkoxides with complexing agents such as carboxylic acids, to give new substances with new reactive properties [1–4]. Significantly fewer data are available for colloidal sol–gel systems, where hydrous oxides produced from alkoxides or salt solutions are de-aggregated (or peptized) with mineral acids, to from colloidal dispersions [5–7]. To date little or no information has been available for colloidal systems produced from chemically-modified precursors. This study investigates the effects

of heating and peptization on the structure of hydrolysates obtained from tetraisopropyltitanate (TPT) modified with carboxylic acids. FT-Raman spectroscopy is extremely useful for characterizing the materials, owing to its ability to examine samples which are weak scatterers and/or contain water, such as the hydrolysates produced here, and the colloidal dispersions produced from them.

Experimental

Acetic acid, propanoic acid or butanoic acid was mixed with TPT in equimolar ratio. The resulting solution was added rapidly, with stirring, to cold water, yielding a coarse hydrolysate which settled readily. The solid was washed several times with water, mixed with enough water to give a Ti(IV) concentration of 0.6 M and peptized at 333 K with nitric acid (0.1 mole of HNO$_3$ per mole of TiO$_2$). A sample of hydrolysate produced from TPT modified with acetic acid was peptized in an identical manner, but with no nitric acid present. The spectra of the residual solids present during peptization were recorded with a Perkin–Elmer model 2000R NIR/FT-Raman spectrometer.

Results and Discussion

The two most common forms of titania are anatase and rutile [8]. Hydrolysates produced by the reaction of modified TPT with water were X-ray-amorphous, but the Raman spectra exhibited broad poorly-defined bands at $\sim$425 and 605 cm^{-1}, Figs. 1 and 2, which were attributed to the E_g and A_{1g} Ti–O stretching modes, respectively, of a "rutile-like" phase [9–12]. An additional broad band was observed at $\sim$205 cm^{-1}, which may arise from disorder or second-order scattering from the "rutile-like" phase [12]. During heating and peptization at 60 °C, significant changes occurred in the Ti–O vibrational modes in all the systems, as follows.

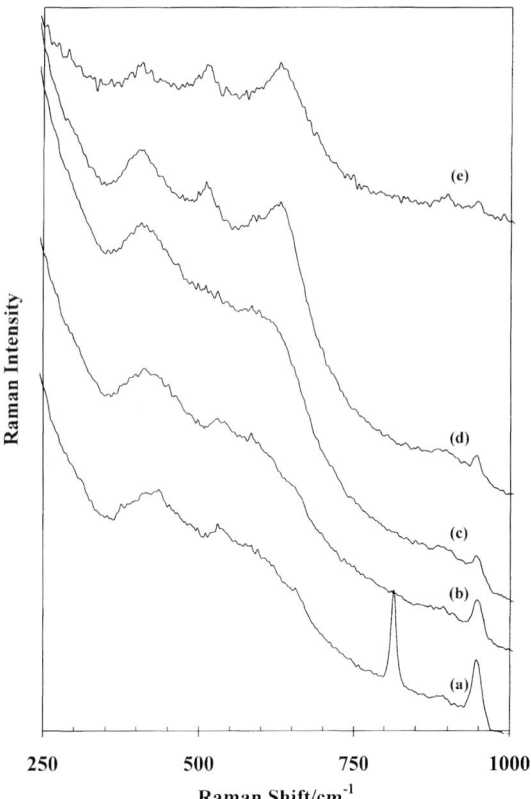

Fig. 1. The FT Raman spectra of titania hydrolysates obtained from TPT/acetic acid mixtures. As produced (*a*), washed (*b*), and peptized for 1 (*c*), 2 (*d*) and 3 hours (*e*)

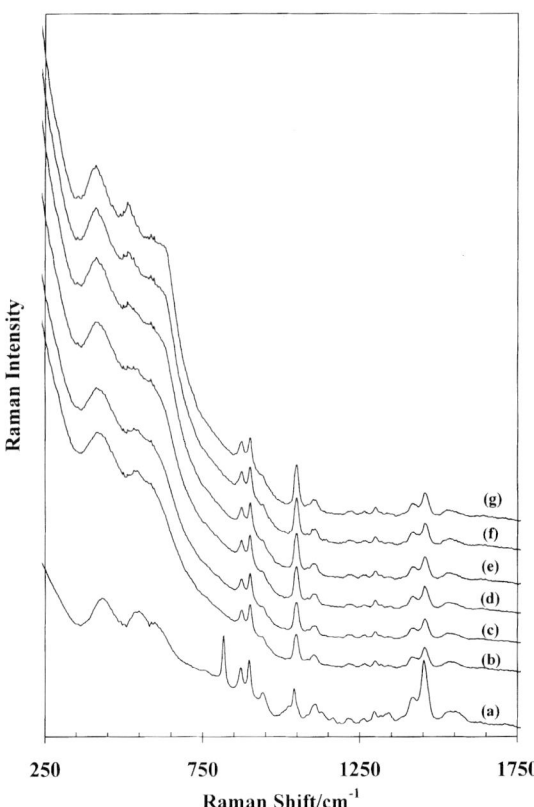

Fig. 2. The FT Raman spectra of titania hydrolysates obtained from TPT/butanoic acid mixtures. As produced (*a*), washed (*b*) and peptized for 1 (*c*), 2 (*d*), 3 (*e*), 4 (*f*) and 5 hours (*g*)

(*a*) A strong band appeared at 154 cm^{-1}, which was attributed to the $\delta_{(O-Ti-O)}$ (E_g) mode of anatase. The band narrowed and increased in relative intensity with time, suggesting an increase in the crystallinity of the hydrolysates.

(*b*) The intensity of the bands associated with the "rutile-like" phase decreased monotonically with time.

(*c*) Additional bands appeared at ∼410, 510 and 630 cm^{-1}, which were assigned to the $\delta_{(O-Ti-O)}$ (B_{1g}), $\nu_{(Ti-O)}$ (A_{1g}/B_{1g}), and $\nu_{(Ti-O)}$ (E_g) modes, respectively, of anatase.

Effects of Carboxylic Acid Chain Length

Acetic acid system. The TPT/CH$_3$COOH hydrolysate exhibited well-defined anatase modes at 154 and 410 cm^{-1} after peptization for one hour. The additional characteristic anatase modes were not clearly resolved until after two hours of peptization, because of overlap with the spectrum of the "rutile-like" phase. In contrast, the intense 154 cm^{-1} anatase mode was not

observed in the spectra of samples heated in water, without added nitric acid, until after two hours of heating. These results indicate that the crystallization of anatase in this system is mainly due to the effect of temperature, although the rate of crystallization is accelerated by the presence of nitric acid.

Propanoic and butanoic acid systems. The strong 154 cm^{-1} anatase band was observed in the TPT/C$_2$H$_5$COOH hydrolysates after peptization for two hours. The intensity of the ∼205 cm^{-1} band, associated with the "rutile-like" phase, showed a corresponding decrease in intensity during this time. In contrast, the characteristic anatase modes were not evident in the spectra of TPT/C$_3$H$_7$COOH hydrolysates until after peptization for four hours.

The spectra of TPT/CH$_3$COOH hydrolysate samples obtained during peptization reveal that the majority of acetate species are desorbed from the hydrolysate, Fig. 1. In contrast, significant quantities of butanoate species remain sorbed on the corresponding hydrolysate samples, Fig. 2, even after peptization for six hours. Partial charge model calculations [4] predict that the Ti–OOCR bond becomes more susceptible to nucleophilic attack, and hence less stable in an aqueous environment, with decreasing carboxylic acid molecular weight. These observations show that the "rutile-like" -to-anatase transformation occurs progressively more slowly with increasing carboxylate chain length, suggesting that the sorption of longer chain length species retards the crystallization of anatase.

References

[1] C. Sanchez, J. Livage, M. Henry, F. Babonneau, *J. Non-Cryst. Solids*, **1988**, *100*, 65.

[2] S. Barboux-Doeuff, S. Sanchez, *Mat. Res. Bull.* **1994**, *29*, 1.

[3] S. Doeuff, M. Henry, C. Sanchez, J. Livage, *J. Non-Cryst. Solids* **1987**, *89*, 206.

[4] J. Livage, M. Henry, C. Sanchez, *Prog. Solid State Chem* **1988**, *18*, 259.

[5] C. Guizard, N. Cygankiewicz, A. Larbot, L. Cot, *J. Non-Cryst. Solids* **1986**, *82*, 86.

[6] C. W. Turner, *Anu. Ceram. Soc. Bull.* **1991**, *70*, 1487.

[7] D. L. Segal, *J. Non-Cryst. Solids* **1984**, *63*, 183.

[8] J. Barksdale, *Titanium: Its Occurrence, Chemistry and Technology*, 2nd Ed., Ronald Press, New York, 1966.

[9] T. Ohsaka, F. Izumi, Y. Fujiki, *J. Raman Spectrosc.* **1978**, *7*, 321.

[10] S. P. S. Porto, P. A. Fleury, T. C. Damen, *Phys. Rev.* **1967**, *154*, 522.

[11] M. Ocaña, J. V. Garcia-Ramos, C. J. Serna, *J. Am. Ceram. Soc.* **1992**, *75*, 2010.

[12] M. Ocaña, V. Fornes, J. V. Garcia-Ramos, C. J. Serna, *J. Solid State Chem.* **1988**, *75*, 364.

Mikrochim. Acta [Suppl.] 14, 251–252 (1997)

Far Infrared Spectroscopic Characterization of Sugars and their Metal Complexes

Luqin Yang*, **Jinguang Wu**, **Qiao Zhou**, **Jiang Bian**, **Yanmin Yang**, **Duanfu Xu**, and **Guangxian Xu**

Department of Chemistry, State Key Laboratory of Rare Earth Materials Chemistry and Applications, Peking University, Beijing 100871, Peoples Republic of China

Abstract. This paper mainly reports the far IR spectra of mono- and disaccharides and their metal complexes. The results suggest that the far IR spectra may be used to identify different saccharides and provide definite information about metal–saccharide complex formation.

Key words: far IR, saccharide, metal complexes.

Saccharides can regulate the flow of metal ions in organisms and carry a great amount of biological messages [1, 2]. Thus interactions between metal ions and sugars are very important. Vibrational spectroscopy is an effective method to investigate these interactions, but so far only a few such IR spectra have been reported [3, 4]. In this paper, we mainly discuss the far IR spectra of some saccharides and their metal complexes. The far IR spectra give definite information about metal–saccharide complex formation.

Experimental

The solid saccharide–metal complexes were prepared by the reaction of saccharides and metal salts in alcohol or water–ethanol–diethyl ether mixed solvent. The complexes obtained were dried over P_2O_5 under vacuum. The far IR spectra were recorded with a Bio-Rad FTS-65A spectrometer, using CsI pellets or mineral oil nujol mull. The spectra (256 scans) were obtained with 8 cm^{-1} resolution.

Results and Discussion

Saccharides

Each of the crystalline saccharides has a characteristic far infrared spectrum with a special pattern of strong and sharp bands, which can be used for identification,

e.g. D-glucose: absorption maxima at 460, 420, 401, 314, 273, 249, 171 cm^{-1}; D-mannose: 403, 375, 308, 267, 198, 161 cm^{-1}; D-xylose: 432, 400, 388, 314, 262, 208, 181 cm^{-1}; D-galactose: 407, 290, 199 cm^{-1}; D-galacturonic acid: 500, 379, 265 cm^{-1}, etc. The far IR spectra of some mono- and disaccharides are shown in Figs. 1 and 2. It can easily be observed that the configuration (a,e, bonds), anomers (α, β) and glucosidic bonding (α, β) all have a strong influence on the far IR spectra of the sugars.

The far IR spectra of saccharides between 450 and 100 cm^{-1} can be assigned to the deformation of the CCC and CCO groups, and below 100 cm^{-1} to torsional CC vibrations [5]. Since the far IR is very sensitive to the saccharide structure, configuration and conformation, it may be used to identify different saccharides.

Sugar–Metal Complexes

Saccharides can form various types of metal complexes, with 1:1. 1:2, 2:1, 3:2 coordination ratios. Different metal ions tend to coordinate with a specific anomer. For example, the lanthanides Pr, Sm, Dy mainly coordinate with β-D-galactose to form 1:2 complexes, whereas Nd, Er form 1:1 complexes with α-D-galactose. These complexes have different molecular structures [6].

The evidence for coordination of metal ions by saccharides is provided by the significant differences between the far IR spectra of the sugars and of their metal complexes. In the solid sugar–metal complexes, all the absorption bands attributed to the sugars become weaker or even disappear. Strong broad M–O vibration bands appear between 400 and 100 cm^{-1}.

---

* To whom correspondence should be addressed

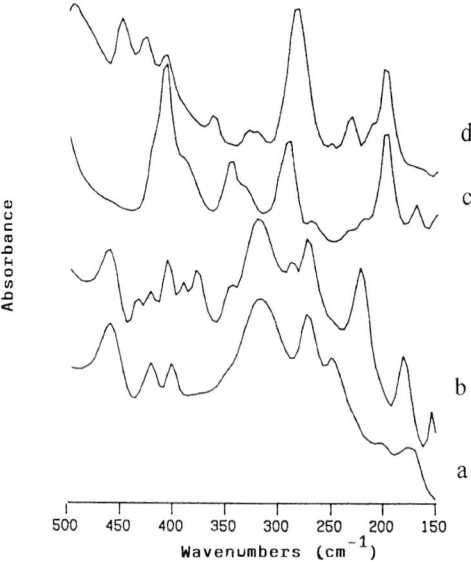

Fig. 1. Far IR spectra of some monosaccharides: *a*, β-D-glucose; *b*, α-D-glucose; *c*, D-galactose; *d*, D-mannose

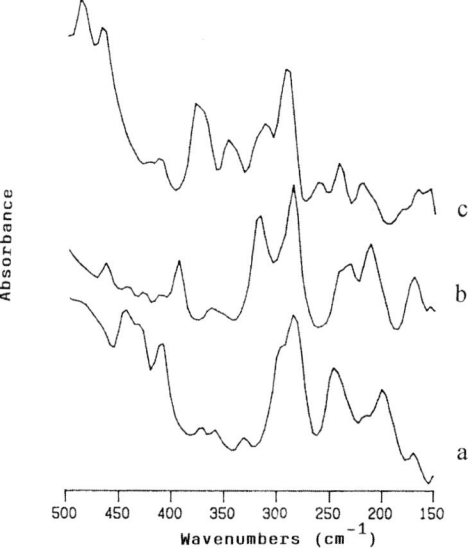

Fig. 2. Far IR spectra of some disaccharides: *a*, α, α-trehalose; *b*, maltose; *c*, cellobiose

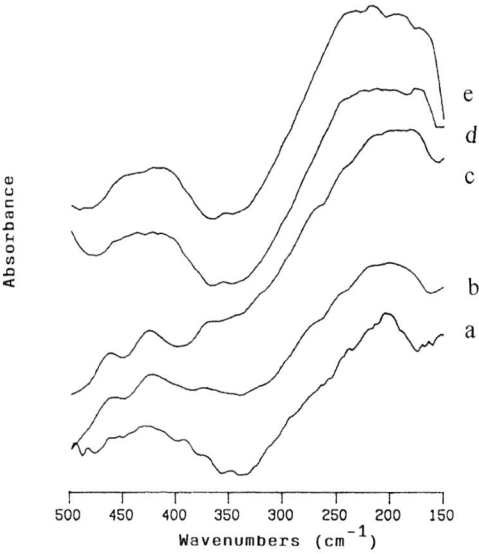

Fig. 3. Far IR spectra of some Metal complexes with saccharides: *a*, S_mCl_3–glucose; *b*, $ErCl_3$–galactose; *c*, S_mCl_3–α, α-trehalose; *d*, S_mCl_3–maltose; *e*, $NiCl_2$–maltose

identification of mono- and disaccharides and can provide direct information of interaction between metal ions and saccharides. Mid-IR and near-IR spectra also provide some evidence about these actions. The detailed results will not be discussed here.

The effect of temperature on the spectra was also studied for D-galactose, D-glucose, D-mannose and their metal complexes. The strong absorption bands at 198 and 292 cm^{-1} for galactose shift to lower wavenumbers when the temperature is increased, but that at 345 cm^{-1} is not changed. In mannose the band at 308 cm^{-1} is shifted to lower wavenumbers and that at 267 cm^{-1} does not change position.

Acknowledgements. This project was supported by the National Postdoctoral Foundation of China and the Climbing Program of China.

For lanthanide complexes with glucose, fructose, sorbose, sucrose, maltose, lactose etc., the Ln–O vibrations show up at around 200 cm^{-1} and (Ln–O) is located at 230 cm^{-1} in the lanthanide complexes with d-galacturonic acid. For calcium mucate (Ca–O) is at about 350 cm^{-1} [7]. Figure 3 shows some far IR spectra of saccharide–metal complexes.

It can easily be deduced from Figs. 1–3 that far infrared spectroscopy provides a powerful method for

References

[1] D. M. Whitfield, S. Stojkovski, B. Sarkar, *Coord. Chem. Rev.* **1993**, *122*, 171.

[2] R. L. Hudgin, W. E. Pricer Jr., G. Ashwell, R. J. Stockert, A. G. Morell, *J. Biol. Chem.* **1974**, *249*, 5536.

[3] H.-A. Tajmir-Riahi, J. T. Agbebavi, *Carbohydr. Res.* **1993**, *241*, 25.

[4] H. Liang, H. Guo, D.-F. Xu, J.-G. Wu, *Mikrochim. Acta.* **1988**, *I*, 215.

[5] M. V. Korolevich, R. G. Zhbankov, V. V. Sivchik, *J. Mol. Struct.* **1990**, *220*, 301.

[6] L. Yang, *Postdoctoral Diss.*, Peking University, 1995.

[7] J. R. Ferraro, *Anal. Chem.* **1968**, *40*, *No.4*, 24A.

Mikrochim. Acta [Suppl.] 14, 253–255 (1997)
© Springer-Verlag 1997

Local and Long-range Ordering of α-Phenylcinnamic Acid Stereoisomers – a Mid and Far FT-IR Spectroscopic Investigation

István Pálinkó* and **János T. Kiss**

Department of Organic Chemistry, József Attila University, Dóm tér 8, Szeged, H-6720 Hungary

Abstract. The local and long-range ordering of α-phenylcinnamic acid stereoisomers (with the emphasis on the Z isomer) has been studied by mid and far FT-IR spectroscopy in solution and in the solid phase. The main structural feature in solution and in the solid state was the presence of strong C=O ⋯ H–O hydrogen bonding interactions between the carboxylic groups (local order). The degree of ordering was higher in the solid state, owing to relatively strong C (aromatic) -H ⋯ O hydrogen bonds (long-range order). A similar type of bonding was not detectable in solution.

Key words: α-phenylcinnamic acid stereoisomers, ordering in solution, ordering in the solid phase, mid and far FT-IR spectroscopy.

It is well-known that carboxylic acids form dimers in the solution phase through the association of the carboxylic groups, assisted by hydrogen bonding (mainly of C=O ⋯ H–O). Analysis of the crystal structures of cinnamic acids and their derivatives, drawn from the Cambridge Crystallographic Database (CSD) [1], revealed that other hydrogen bonds, mostly of C(aromatic)–H ⋯ O type, were also prevalent and contributed to the formation of more extended structures [2]. In order to search for similar interactions and longer range ordering in the liquid phase, we started a detailed IR investigation (mid and far IR, Raman) on solutions of α-phenylcinnamic acid stereoisomers. The concentration of the acids was varied to allow the unambiguous identification of hydrogen bonds. Solid-state structures were also studied for comparison and because the single-crystal structures of these compounds are not known. Results described in the following mainly correspond to the Z isomer; for a similar study on the E isomer, see [3]. It is to be noted that the vast majority of related structures in the CSD are of the

E isomer, and therefore a study on the Z isomer fills a considerable gap.

Experimental

α-Phenylcinnamic acid isomer mixture was prepared by a modified Perkin condensation [4]. The stereoisomers were separated by selective precipitation and purified by crystallization or sublimation. FT-IR spectra were obtained in the 3600–100 cm^{-1} range. For measurements in the liquid phase, CCl_4 was the solvent and the concentration of the Z isomer was varied between 10^{-2} and 10^{-4} M. In solid-state measurements either KBr or PE (high-density polyethylene) were used as matrix materials. Spectra were recorded on Bio-Rad spectrometers [mid-IR: FTS-65A/896 with a liquid-nitrogen cooled MCT detector; far-IR: FTS-40 with DTGS detector in mixed vacuum and purge mode; optical resolution was 2 cm^{-1} in both cases; 256 (or 2560 for the most dilute solution) and 1024 scans were collected, respectively] and were analysed with the associated software package.

Results and Discussion

The spectral behaviour of the two isomers was different both in the solid state (Fig. 1) and in the solution (for the spectra of the E isomer, see [3]). A comparison of the solid-state spectra (especially in the 1800–1100 cm^{-1} region) clearly shows that the crystal structures, and thus the association characteristics, of the two isomers differ significantly.

As regards the aggregation properties of the Z isomer in the solid phase and in solution, a higher degree of ordering was found in the solid state (Fig. 2). Free carbonyl was not found in the crystalline state IR spectrum; instead, bands for C(aromatic)–H ⋯ O=C (1713 cm^{-1}) and O–H ⋯ O=C (1635 cm^{-1}) hydrogen bonds were observed.

When the molecule is dissolved, the weak C(aromatic)–H ⋯ O=C hydrogen bond breaks up and the

---

* To whom correspondence should be addressed

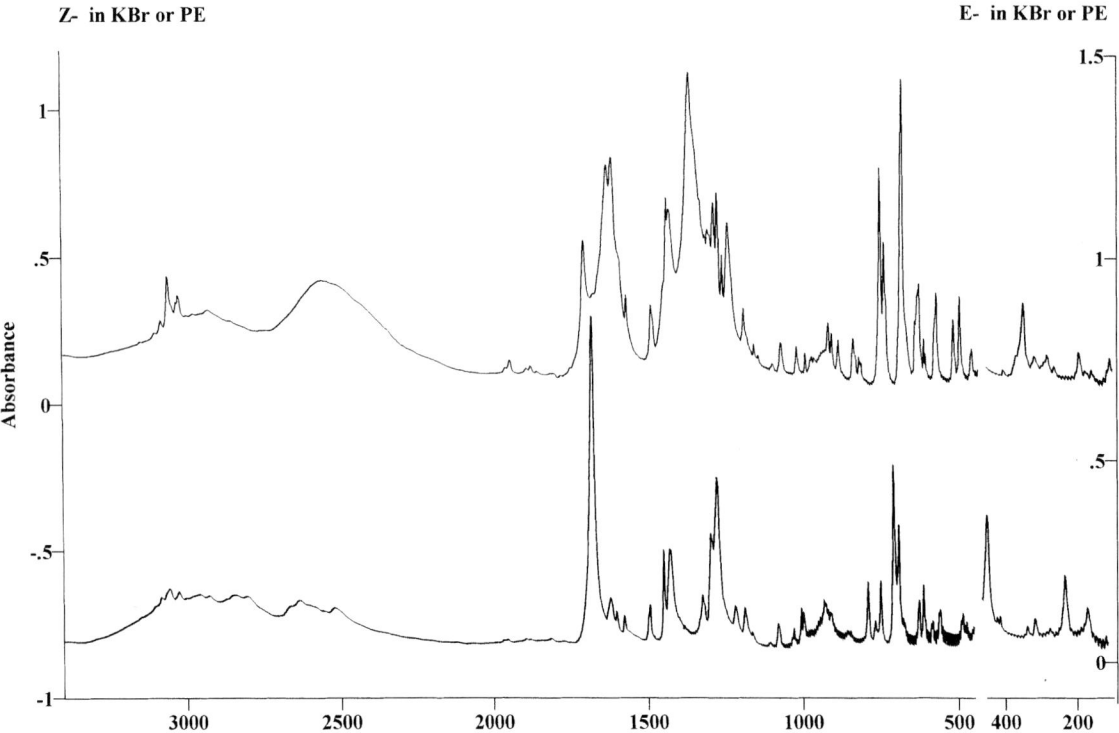

Fig. 1. Combined mid and far FT-IR spectra of *E*- and *Z*-α-phenylcinnamic acids in the solid state

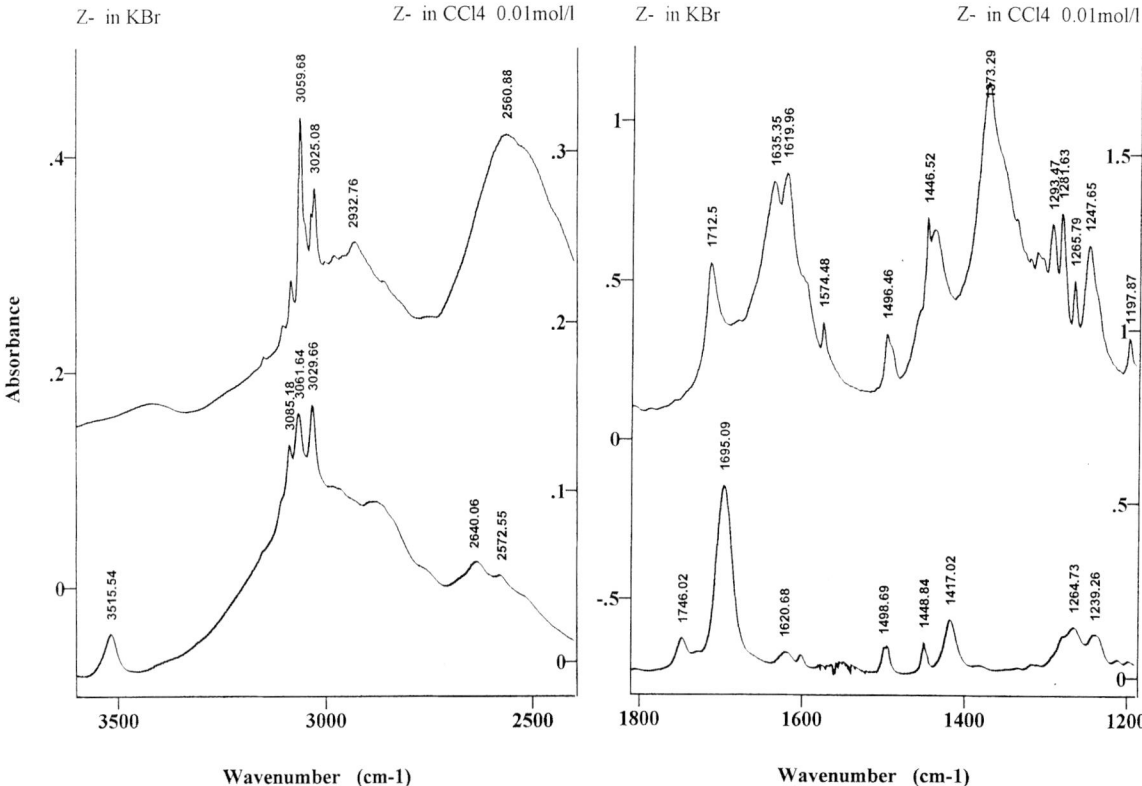

Fig. 2. The OH range and the carbonyl range (and its vicinity) for the *Z* isomer in the solid state (top) and in solution (bottom, acid concentration: 10^{-2} *M*)

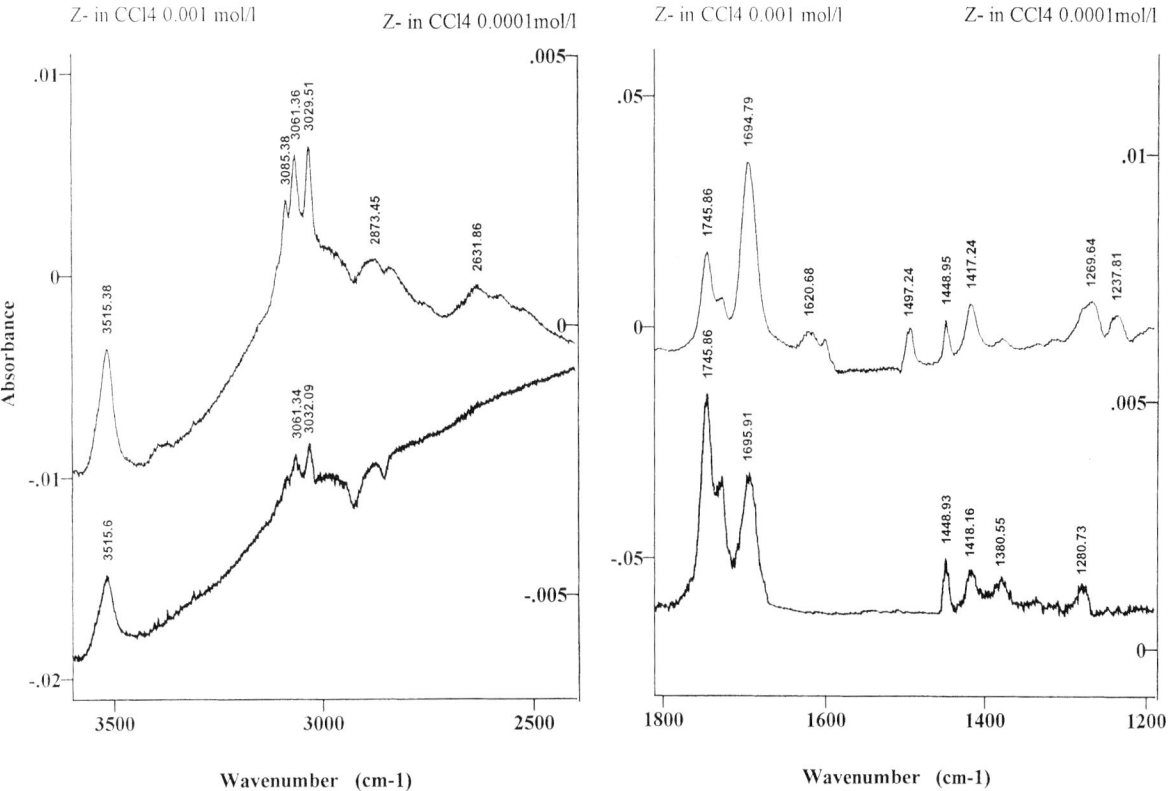

Fig. 3. Concentration dependence in the mid FT-IR spectra (OH range and the carbonyl range) of the Z isomer (top: $10^{-3}\ M$, bottom: $10^{-4}\ M$)

carboxyl–carboxyl associates loosen as well. A band for free carbonyl appears ($1746\ \mathrm{cm}^{-1}$) and becomes ever more intense as the acid concentration decreases (Fig. 3).

Similar observations can be made in the OH region ($3600–2500\ \mathrm{cm}^{-1}$). There is no free OH in the solid state, but in solution, as the acid concentration decreases, the carboxyl–carboxyl interaction weakens and an increasingly larger proportion of the carboxylic acid exists as monomer.

As far as carboxyl–carboxyl aggregates are concerned, dimers predominate over other cyclic hydrogen-bonded oligomers both in the solid state and in solution. However, the structured OH band in the mid-IR spectrum indicated significant amounts of higher oligomers. Trimers, and even tetramers, may also be present. Their accurate structure and their relative proportion are still under scrutiny.

Conclusion

Vibrational spectroscopic investigations revealed that intermolecular C(aromatic)–H $\cdots$ O=C hydrogen bonding is responsible for long-range ordering (typical for the crystalline state), while locally (both in the solid state and in solution) hydrogen-bonded carboxyl–carboxyl oligomer formation predominates.

Acknowledgements. This work was financed by the National Science Foundation of Hungary through grant F4297/1992. Spectroscopic instruments used in preparing this work were purchased with the help of the PHARE-ACCORD program no. H-9112 0152. Both forms of support are highly appreciated.

References

[1] F. H. Allen, J. E. Davies, J. J. Galoy, O. Johnson, O. Kennard, C. F. Macrae, E. M. Mitchell, G. F. Mitchell, J. M. Smith, D. G. Watson, *J. Chem. Inf. Comput. Sci.* **1991**, *31*, 187.

[2] I. Pálinkó, in: *16th European Crystallographic Meeting, Book of Abstracts*, Lund, Sweden, 1995, p. 126.

[3] I. Pálinkó, B. Török, M. Rózsa-Tarjáni, J. T. Kiss, Gy. Tasi, *J. Mol. Structure* **1995**, *348*, 57.

[4] L. Fieser, in: *Experiments in Organic Chemistry, 3rd Ed.*, Heath, Boston, 1958, p. 182.

Mikrochim. Acta [Suppl.] 14, 257 258 (1997)
© Springer-Verlag 1997

Phase Transition and Magnetic Structure in $In_2Cu_2O_5$ and $Sc_2Cu_2O_5$ by High-resolution FTS of a Rare Earth Probe

Marinw N. Popova[1,*], Yurii A. Hadjiiskii[1], Robert Troć[2], and Zbignew Bukowski[2]

[1] Institute of Spectroscopy, Russian Academy of Sciences, 142092 Troitsk, Moscow Region, Russia
[2] Institute for Low Temperature and Structure Research, Polish Academy of Sciences, 50–950 Wroclaw, Poland

Abstract. Magnetic ordering of $In_2Cu_2O_5$ at 29 ± 1 K and $Sc_2Cu_2O_5$ at 16 ± 1 K was found by high-resolution Fourier-transform spectroscopy. The most probable magnetic structure in an ordered state of $In_2Cu_2O_5$ is suggested.

Key words: high-resolution optical spectra; magnetic ordering.

The compounds $In_2Cu_2O_5$ and $Sc_2Cu_2O_5$ crystallize in the $Pna2_1$ space group and belong to the family of the $R_2Cu_2O_5$ cuprates that first received considerable attention as the phases related to high T_c superconductors. It turned out later that they have interesting magnetic properties [1, 2] and deserve study in connection with the problem of low dimensional magnetism. Our earlier spectral measurements of the $R_2Cu_2O_5$ (R = Y, Tb–Lu) compounds have shown that high-resolution Fourier-transform spectroscopy (FTS) of a rare earth probe is a powerful tool for investigating their magnetic properties. A review of this work is given in [3]. In the present work, we extend these studies to the two remaining members of the $R_2Cu_2O_5$ family, namely, $In_2Cu_2O_5$ and $Sc_2Cu_2O_5$.

Experimental

Polycrystalline X-ray single-phase samples of $R_2Cu_2O_5$ (R = In or Sc) with 1% of Er introduced as a probe have been prepared by solid-state reaction as described in [1]. Powder samples were mixed with ethanol and put on a sapphire support plate just before the InSb detector inside the helium vapour cryostat. High resolution (down to 0.1 cm^{-1}) spectra in the region of the $^4I_{15/2} \to {}^4I_{13/2}$ optical transition in the Er^{3+} ion were recorded at 4.2–110 K with a Bomem DA 3.002 FT spectrometer.

Results and Discussion

$In_2Cu_2O_5$

Figure 1 shows the most intense lowest frequency line at 6531 cm^{-1} in the $^4I_{15/2} \to {}^4I_{13/2}$ transition at different temperatures. There are two non-equivalent four-fold low symmetry (C_1) positions for the R^{3+} ions in the structure of $R_2Cu_2O_5$ cuprates. Spectral lines due to transitions in two non-equivalent positions are superimposed. It is possible to separate them in the case of sharp lines at low temperatures. The ground doublet splittings Δ at 4.2 K are for erbium ions in two non-equivalent positions Er1, Er2 respectively:

$$\Delta(1) = 19.2 \, \text{cm}^{-1}, \Delta(2) = 19.5 \, \text{cm}^{-1}.$$

These splittings are very close to the appropriate splittings of Er^{3+} probe levels in the $R_2Cu_2O_5$ cu-

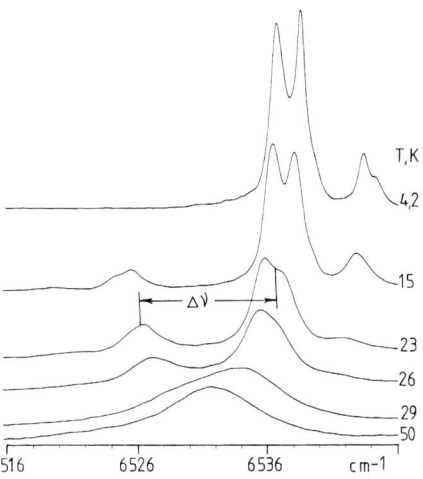

Fig. 1. Low-frequency line of the $^4I_{15/2} \to {}^4I_{13/2}$ absorption of the Er^{3+} probe in $In_2Cu_2O_5$ at different temperatures

---

* To whom correspondence should be addressed

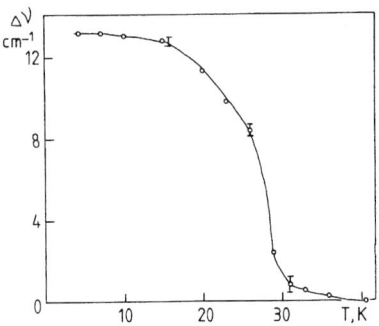

Fig. 2. $In_2Cu_2O_5$. The splitting Δv of the $6531\,cm^{-1}$ line of the $^4I_{15/2} \to {}^4I_{13/2}$ absorption of the Er^{3+} probe, as a function of temperature

prates with R = Y, Tb, Er, Lu [3]. It follows from this fact [3] that the structure of the ordered Cu magnetic moments in $In_2Cu_2O_5$ is the same as in $R_2Cu_2O_5$ with R = Y, Tb, Er, Lu, namely, quasi-two-dimensional ferromagnetic layers coupled antiferromagnetically, with Cu moments aligned along the $\vec{b}$ axis.

The splitting as a function of temperature is plotted in Fig. 2. The strong broadening of the closely spaced lines near the temperature of magnetic ordering and the low intensity of the high-frequency components of the lines make it difficult to trace the splitting of the ground state Δ; therefore Fig. 2 shows the temperature dependence of some splitting Δv (see Fig. 1) averaged for two crystallographic positions. Our spectral measurements gave a value of 29 ± 1 K for the magnetic ordering temperature. This temperature was determined as the abscissa of the point of inflection in the Δv (T) curve and is in agreement with $T_N = 30$ K found earlier from the magnetic susceptibility measurements [1].

$Sc_2Cu_2O_5$

The spectral lines of the Er^{3+} probe in $Sc_2Cu_2O_5$ are broad, in contrast to the case of $In_2Cu_2O_5$. The reason for a large inhomogeneous broadening lies, probably, in the big difference between the ionic radii of Sc^{3+} $(0.75\,\text{Å})$ and Er^{3+} $(0.89\,\text{Å})$. The temperature dependences of the half-width and central frequency of the spectral lines of the Er^{3+} probe in $Sc_2Cu_2O_5$ reveal a temperature of 16 K for a magnetic phase transition in this compound, in accordance with the results of magnetic susceptibility measurements [1]. The only conclusion concerning the magnetic structure of $Sc_2Cu_2O_5$ that can be drawn from our spectral data is that this structure is quite different from that of $R_2Cu_2O_5$ (R = Y, Tb, Er, Lu, In). This conclusion does not contradict the results of earlier neutron-scattering studies [4].

Conclusions

In conclusion, the use of high-resolution FTS enabled us to detect magnetic phase transitions in $In_2Cu_2O_5$ and $Sc_2Cu_2O_5$ and the specify the transition temperatures. The most probable magnetic structure in an ordered state of $In_2Cu_2O_5$ is suggested.

Acknowledgements. This work was made possible in part by Grant No. 95–02–03796 from the Russian Foundation for Fundamental Research and by Grant No. JEJ100 from the International Science Foundation and Russian Government.

References

[1] R. Troć, J. Klamut, Z. Bukowski, R. Horyń, J. Stępien-Damm, *Physica B* **1989**, *154*, 189.
[2] Z. A. Kazei, N. L. Kolmakova, R. Z. Levitin, B. V. Mill, V. V. Moshchalkov, V. N. Orlov, V. V. Snegirev, Ya. Zoubkova, *J. Magn. Magn. Materials* **1990**, *86*, 124.
[3] M. N. Popova, I. V. Paukov, *Opt Spektrosk.* **1994**, *76*, 285; *Opt. Spectrosc.*, **1994**, *76*, 254.
[4] A. Murasik, P. Fischer, R. Troć, Z. Bukowski, *J. Magn. Magn. Materials* **1993**, *127*, 365.

Mikrochim. Acta [Suppl.] 14, 259–261 (1997)
© Springer-Verlag 1997

FT-IR Study of the Coordination of an Organophosphonic Acid with Lanthanide Ions

Dujin Wang, Shuxin Yao, Jinguang Wu*, Nai Shi, Xu Zhang, and **Guangxian Xu**

Department of Chemistry, Peking University, Beijing 100871, People's Republic of China

Abstract. The coordination interaction between saponified 2-ethyl-hexylphosphonic acid mono-2-ethylhexyl ester (HEHEHP) with lanthanide ions [Ln(III)] in an extracted organic phase has been studied by FT-IR spectral analysis. The result indicates that with the increase of lanthanide loading in the organic phase, the microstructure of the water/oil microemulsions formed by the saponified extractant is gradually destroyed. The P=O group of the extractant was found to coordinate with the Ln(III). The coordination ability of the lanthanide series with the P=O group was found to increase from La(III) to Pr(III), but decrease from Pr(III) to Lu(III).

Key words: FT-IR, coordination, microemulsion, extraction, lanthanides.

Saponified organophosphorus acid extractants are used to separate and purify non-ferrous and lanthanide metals in processing industries. Wu and co-workers [1, 2] have proved the formation of a microemulsion in the organic phase of saponified extractants, such as naphthenic acid, HDEHP, HEHEHP etc., and investigated its effect on the extraction mechanism of bi-valent metals and lanthanides. Neuman et al. [3] have suggested a general model for aggregation of metal–extractant complexes in acidic organophosphorus solvent-extraction systems. The aggregation behaviour of the extracted complexes in a non-polar diluent has been reported [4]. Osseo-Asare and Keeney [5], and later Osseo-Asare [6] found that the microemulsion formation in the organic phase enhances the solvent extraction of transition metals. In this paper, we report our studies of the coordination interaction between saponified HEHEHP and the lanthanide series by FT-IR spectroscopy, and our attempts to find the relationship between the extraction process and the variation in aggregation in the organic phase.

Experimental

Materials and Method

HEHEHP was supplied by Beijing Chemical Works, and purified according to the method of Patridge and Jensen [7]. The purity was determined to be 99.7%. The concentration of extractant was expressed in monomer units. All the other reagents used were of analytical reagent grade.

A Nicolet Magna 750 FT-IR spectrometer was used to characterize the extracted organic phase, with a demountable BaF_2 cell to prepare IR samples. IR spectra were averaged over 32 scans in the range 800–4000 cm^{-1}, with a resolution of 4 cm^{-1}.

Sample Preparation

A 1.0 M HEHEHP solution in n-heptane was saponified with 10.4 M sodium hydroxide solution to form a microemulsion of water-in-oil type (w/o), with a composition of 0.8 M HEHEHP + 0.2 M NaEHEHP. Identical volumes of this w/o microemulsion were then added to different volumes of lanthanide chloride solutions to obtain a series of extracted organic phases with different lanthanide loadings. The initial aqueous concentration of any lanthanide chloride solution was 0.0667 M. All the samples were maintained at room temperature (20 ± 1 °C) and left standing to allow phase separation. FT-IR spectra measurements were performed on the equilibrated organic phase.

Results and Discussion

The P=O band of HEHEHP in n-heptane solution is located at 1198 cm^{-1}. For the partly saponified material, the P=O absorption frequency shifts to lower wavenumbers (1194, 1166 cm^{-1}) owing to its hydration and coordination with Na^+. The band at 1194 cm^{-1} is assigned to the absorption of the free P=O group in HEHEHP, and that at 1166 cm^{-1} corresponds to the P=O bond of the extractant anion coordinated with Na^+. The stretching vibrational frequency of H_2O at about 3400 cm^{-1} is also observed,

---

* To whom correspondence should be addressed

demonstrating that the w/o microemulsions solubilize water.

On equilibration of the extractant with the aqueous solution of lanthanide chloride, Ln(III) ions were extracted into the organic phase and Ln–HEHEHP complexes were formed, thus leading to variation of the composition of the organic phase. The FT-IR spectra (Fig. 1) show that with loading of Gd(III) in the organic phase, the absorption intensity of the P=O group (at ca. 1166 cm^{-1}) bonded with the Gd(III) becomes stronger, compared with that of the free P=O group (at ca. 1194 cm^{-1}). This is due to coordination of the extractant with Gd(III), which increases the dipole moment of the P=O bond, thus leading to the increase of the molar absorption coefficient of the extracted complex. So, with increase of the lanthanide concentration in the organic phase, the absorption intensity at 1166 cm^{-1} correspondingly increases. Through Fourier self-deconvolution, we find that at L = 100%, i.e., all the sodium ions in the organic phase were replaced by lanthanide ions, the absorption frequencies of the P=O group coordinated with Ln(III) are different for the

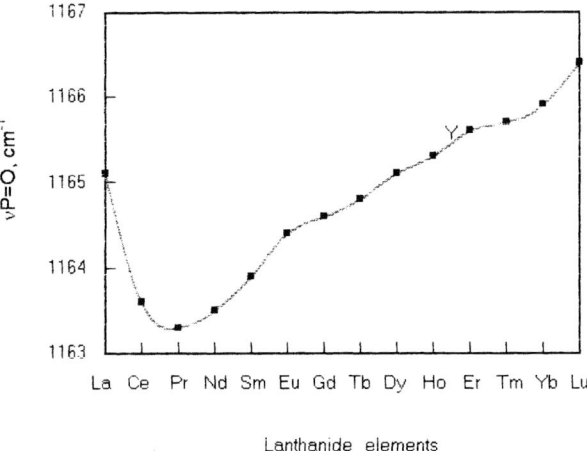

Fig. 2. Relationship between the peak positions for P=O coordinated with lanthanide ions and the atomic numbers at L = 100% in (0.8 M HEHEHP + 0.2 M NaEHEHP) in n-heptane

various lanthanides (Fig. 2). From La(III) to Pr(III), the P=O peak position frequency decreases (1165.1 → 1163.3 cm^{-1}), in accord with the coordination ability of the lanthanide series, but from Pr(III) to Lu(III), it increases (1163.3 → 1166.4 cm^{-1}). This apparent abnormality may be caused by the multiple coordination interactions in the complex system we studied. In w/o microemulsions, there is solubilized water in the polar core, which can compete with HEHEHP for coordination of the metal ion. The hydration of Ln(III) will reduce its coordination ability with HEHEHP. From La(III) to Pr(III), the coordination of Ln(III) with P=O dominates, for the mean difference between the effective radii of two consecutive Ln(III) ions is bigger (0.021 Å); from Pr(III) to Lu(III), the effect of hydration of Ln(III) becomes more important, for the mean difference between the effective radii of two ions is relative small (0.011 Å). The peak position of the P=O group bonded with Y(III) is interposed between the Ho(III) and Er(III) complexes, in accord with the radius sequence.

With the entry of Ln(III) into the organic phase and the formation of extracted complexes, the amount of free extractant anions decreases, inducing the destruction of w/o microemulsions in the organic phase. The FT-IR spectra show that with the increase of Ln(III) loading, the water contents in the extracted organic phase gradually decrease, indicating that the water solubilized in the polar core of the w/o microemulsions returns into the aqueous phase.

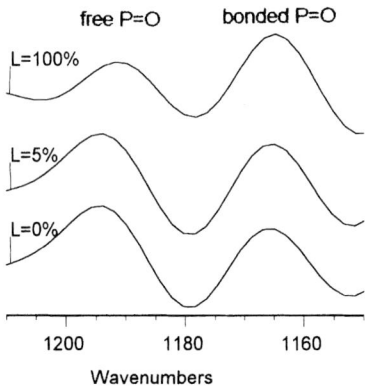

Fig. 1. IR spectra of P=O band in the extracted organic phase containing Gd(III) in (0.8 M HEHEHP + 0.2 M NaEHEHP) in n-heptane (L = loading of lanthanide in organic phase)

Acknowledgement. Financial support from the National Natural Science Foundation of China and the Climbing Program of China is gratefully acknowledged by the authors.

References

[1] C.-K. Wu, H.-C. Kao, T. Chen, S.-C. Li, T.-C. King, K.-H. Hsu, *Proc. Solvent Extr. Conf. Chemistry*, 1980, Paper No. 80–23.

[2] J. Wu, N. Shi, H. Gao, D. Chen, H. Guo, S. Weng, G. Xu, *Sci. Sin. Ser. (Engl. Ed.)* **1984**, *27*, 249.

[3] R. D. Neuman, N. Zhou, J. Wu, M. A. Jones, A. G. Gaonkar, S. J. Park, M. L. Agrawal, *Sep. Sci. Technol.* **1990**, *25*, 1655.

[4] Y. Li, H. Liao, Q. Peng, W. Zhou, H. Fu, J. Wu, G. Xu, *J. Alloys Compd.* **1994**, *216*, L21.

[5] K. Osseo-Asare, M. E. Keeney, *Sep. Sci. Technol.* **1980**, *15*, 999.

[6] K. Osseo-Asare, *Sep. Sci. Technol.* **1988**, *23*, 1269.

[7] J. A. Patridge, R. C. Jensen, *J. Inorg. Nucl. Chem.* **1969**, *31*, 2587.

Mikrochim. Acta [Suppl.] 14, 263–264 (1997)
© Springer-Verlag 1997

Ceramides/Cholesterol Mixtures as Characterized by FT Raman Spectroscopy

Siegfried Wartewig[1,*], **Reinhard Neubert**[2], **Willi Rettig**[2], and **Matthias Wegener**[2]

[1] Department of Physics, Martin-Luther-University, Hoher Weg 7, D-06099 Halle/Saale, Federal Republic of Germany
[2] Department of Pharmacy, Martin-Luther-University, D-06120 Halle/Saale, Weinbergweg 15, Federal Republic of Germany

Abstract. Ceramides IV, hydrated ceramides IV and ceramides/cholesterol mixtures were studied by FT Raman spectroscopy and differential scanning calorimetry (DSC). The temperature-dependence of the conformationally sensitive bands in the CH_2 stretching region was used to estimate the degree of order of hydrocarbon chains in terms of the relative population of *trans* and *gauche* conformers. The temperature and width of the phase transition, derived from Raman data, are similar to those from the DSC studies. The addition of both cholesterol and water lowers the phase transition temperature and broadens the transition range.

Key words: Raman spectroscopy, ceramides, ceramides/cholesterol mixture, hydrated ceramides, phase behaviour.

Increasing interest is currently focused on the molecular structure of the outermost layer of the skin, the stratum corneum. Stratum corneum lipids consist mainly of three fractions: ceramides, fatty acids and cholesterol. Mixtures of stratum corneum lipids have been used to study the structure of these lipids. However, it is necessary to characterize interactions between the main fractions of the stratum corneum lipids.

Ceramides IV can be considered a model for the fraction of ceramides. Vibrational spectroscopy of lipids uses conformationally dependent modes for elucidating the population of *trans* and *gauche* conformers and the order/disorder behaviour of alkyl chain residues [1, 2]. In this study, we focus our attention on the temperature dependence of the intensity ratio of the asymmetric to the symmetric CH_2 stretching modes as a measure of the degree of order of alkyl chains of ceramides, using Raman spectroscopy for the measurements. In addition, the phase behaviour of the systems was studied by using differential scanning calorimetry (DSC).

---

* To whom correspondence should be addressed

Experimental

Ceramides type IV and cholesterol were obtained from Sigma Chemie GmbH and used without further purification. Mixtures of ceramides IV and cholesterol were cast from chloroform solution. The solvent was evaporated under vacuum. The samples were annealed at 40 °C over 24 hours before measurements.

Raman spectra were recorded on a Bruker IFS 66 FT-IR spectrometer with the FRA 106 Raman module, and use of diode-pumped Nd:YAG laser at an operating wavelength of 1064 nm. The temperature dependence of the spectra was studied in the range from 40 to 100 °C (relative stability ± 0.2 K). The deconvolution of the very complex feature in the CH stretching region was done by applying the Bruker OPUS fit procedure. Raman intensities were determined as integrated band intensities.

Calorimetric scans were usually performed at 5 K/min on a Perkin–Elmer DSC-2 differential scanning calorimeter.

Results and Discussion

The major part of the Raman spectrum of ceramides IV and hydrated ceramides IV consists of bands ascribed to the alkyl chain residue, but there are also bands of lower intensity that belong to the head group of the ceramides [3]. Though the spectrum of the mixture of ceramides and cholesterol is very complex, the CH_2 stretching modes of the alkyl chains could be identified. Fortunately, the Raman spectrum of cholesterol exhibits only a weak temperature dependence in the range from 20 to 100 °C. Therefore, we assume that the temperature-induced changes in the spectra of the mixture are related to structural alterations of the ceramides part, shown in Fig. 1 for the 79 wt% ceramides/21 wt% cholesterol sample.

The appearance of the conformationally dependent band v_a (CH_2) in all systems studied indicates clearly the ordered structure of the hydrocarbon chain in the solid state. The intensity of the asymmetric CH_2 stretching line at about 2880 cm^{-1} decreases with in-

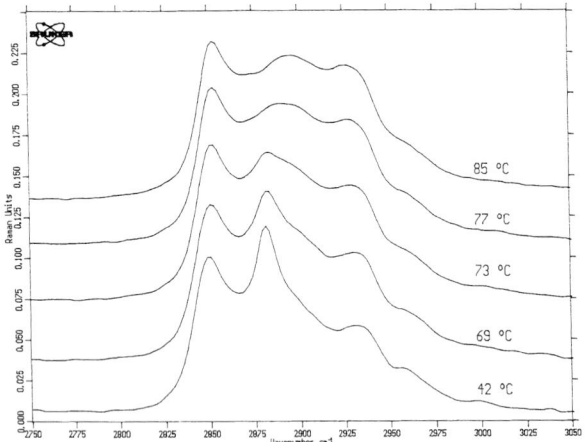

Fig. 1. Temperature dependence of the Raman spectra in the CH_2 stretching region of the 79 wt% ceramides IV/21 wt% cholesterol mixture

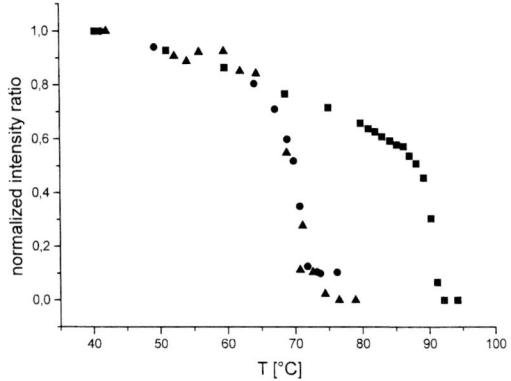

Fig. 3. Temperature dependence of the normalized intensity ratio $I[v_a(CH_2)]/I[v_s(CH_2)]$ ■ ceramides IV, ● hydrated ceramides IV, ▲ 79 wt% ceramides IV/21 wt% cholesterol mixture

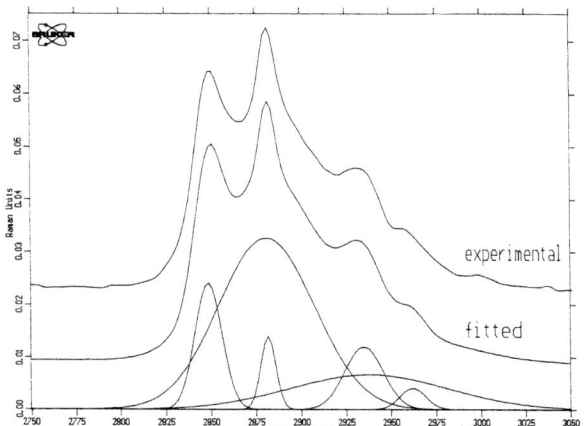

Fig. 2. Deconvolution of the overlapping CH_2 bands of the 79 wt% ceramides IV/21 wt% cholesterol mixture ($T = 42\,°C$)

effect. Figure 2 displays an example of the results for the mixture 79 wt% ceramides/21 wt% cholesterol at 42 °C. The temperature dependence of the intensity ratio $I[v_a(CH_2)]/I[v_s(CH_2)]$, as a measure of the *trans/gauche* ratio of the hydrocarbon chains of ceramides IV is shown for the three samples, in Fig. 3. Obviously, the addition of both cholesterol and water to ceramides IV induces a lowering of the phase transition point by about 20 K and a broadening of the transition range. DSC investigations confirm these results.

Further studies are in progress.

Acknowledgement. This work was supported by the Deutsche Forschungsgemeinschaft (DFG), Sonderforschungsbereich 197, Project A8 and B3.

creasing temperature and disappears in the underlying background at temperatures above the phase transition point. We have deconvoluted the overlapping bands of the CH_2 modes between 2800 and 3050 cm^{-1} into six bands with Gaussian line shape to quantify this

References

[1] D. F. H. Wallach, S. P. Verma, J. Fookson, *Biochim. Biophys. Acta* **1979**, *559*, 153.

[2] R. G. Snyder, D. G. Cameron, H. L. Casal, D. A. C. Compton, H. H. Mantsch, *Biochem. Biophys. Acta* **1982**, *684*, 111.

[3] M. Wegener, R. Neubert, W. Rettig, S. Wartewig, *Intern. J. Pharm.* **1996**, *203*, 128.

Mikrochim. Acta [Suppl.] 14, 265–266 (1997)
© Springer-Verlag 1997

FT-IR Study of Crystalline Tetrachlorometallates of Cinchonine

Aleksandra Weselucha-Birczyńska* and **Czesława Paluszkiewicz**

Regional Laboratory of Physicochemical Analyses and Structural Research, Jagiellonian University, Ingardena 3, 30-060 Kraków, Poland

Abstract. FT-IR spectra of cinchoninium tetrachlorocadmate(II) and tetrachlorozincate(II) dihydrate complexes in the 50–650 cm^{-1} region were obtained and analysed. Normal coordinate calculations of the MX_4^{2-} moieties in these isomorphous compounds and their bromide analogues were done. The force constants obtained confirm that the geometry of the tetrachlorometallates investigated is strongly influenced by the cinchoninium counter-cation.

Key words: far FT-IR, tetrachlorometallate, cinchonine, normal coordinate calculations.

Cinchonine (cin = $C_{19}H_{22}N_2O$) belongs to the group of the four most important cinchona alkaloids. Aqueous solutions of cinchonine hydrochloride, cin·$(HCl)_2$, treated with stoichiometric amounts of transition metal chlorides, MCl_2, are capable of formation of tetrachlorometallates with the general formula $(cinH_2)^{2+}(MCl_4)^{2-}·nH_2O$ [1]. The tetrahedral MCl_4^{2-} anion is linked to the doubly protonated cinchonine molecule and to water molecules by hydrogen bonding. It was established that for M = Zn(II) [2] the structure is isomorphous with that for M = Cd(II) [3]. The X-ray structure analysis and FT-IR in the 400–4000 cm^{-1} region allowed analysis of some details of their molecular structure [4]. In this contribution we describe the vibrational spectra of tetrachlorocadmate(II) and tetrachlorozincate(II) anions, which form the first metal-coordination sphere of the complexes discussed. However, these tetrachlorometallate ions adopt a significantly distorted tetrahedral geometry. In order to assign the normal modes of vibration of the MX_4^{2-} complex, where M = Cd(II) or Zn(II) and X = Cl$^-$ or Br$^-$, the analogous bromometallates of cinchonine were also synthesised and measured. Normal coordinate analysis of the vibrational data of the

MX_4^{2-} part of the molecules is also presented and discussed.

Experimental

Preparation

Cinchoninium tetrachlorocadmate(II) dihydrate and cinchoninium-tetrachlorozincate(II) dihydrate were synthesized according to the method described by Dyrek [1]. The corresponding tetrabromo derivatives were prepared analogously with cinchonine hydrobromide, cin·$(HBr)_2$, instead of cin·$(HCl)_2$.

Spectral Measurements

The Fourier-transform infrared spectra were recorded on a Bio-Rad FTS-60 v spectrometer by the CsI and polyethylene pellet techniques with 2 cm^{-1} resolution in the 50–650 cm^{-1} region.

Computational Method

In the normal coordinate calculations a GVFF approach was used [5].

Results and Discussion

Vibrational Spectra of MX_4^{2-}, where M = Cd(II), Zn(II) and X = Cl$^-$, Br$^-$, in $(cin·H_2)^{2+}MCl_4^{2-}·2H_2O$

For the MX_4^{2-} five-atom moiety, nine normal vibrations are expected [6]. Since the investigated species are distorted (the distances between the edges of the $CdCl_4^{2-}$ tetrahedron measured along the crystallographic axis are equal to: L1 = 2.67 Å, L2 = 3.02 Å and L3 = 2.78 Å) all vibrations are expected to be not only Raman active but also IR-active. Seven, six or five fundamentals are observed for the complexes studied (see Fig. 1 and Table 1).

---

* To whom correspondence should be addressed

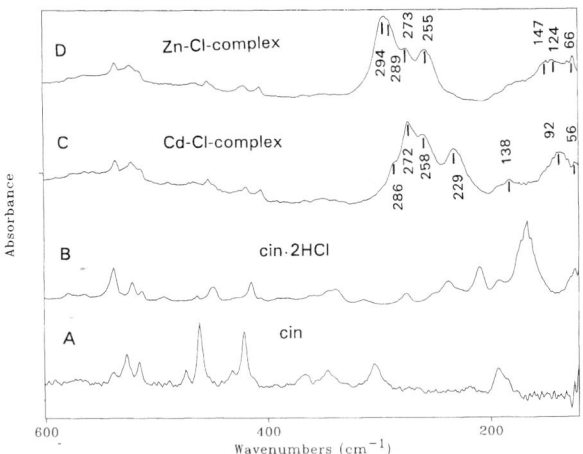

Fig. 1. FT-IR spectra (50–600 cm^{-1} region) of A cinchonine; B cin·(HCl)$_2$; C (cinH$_2$)$^{2+}$ CdCl$_4^{2-}$·2H$_2$O; D (cinH$_2$)$^{2+}$ CdCl$_4^{2-}$· 2H$_2$O

Table 1. Comparison of observed and calculated frequencies (cm^{-1}) in the 50–650 cm^{-1} region

	ZnCl$_4^{2-}$	ZnBr$_4^{2-}$	CdCl$_4^{2-}$	CdBr$_4^{2-}$
obs.	294	225	286	196
calc.	296	219	279	190
obs.	289	204	272	177
calc.	283	206	271	180
obs.	273	192	258	169
calc.	276	197	266	174
obs.	255	180	229	162
calc.	255	181	229	163
obs.	147	90	138	69
calc.	142	87	119	67
obs.	124			
calc.	130	85	113	64
obs.			92	
calc.	118	79	102	60
obs	66	55	56	
calc.	62	59	70	49
obs.				
calc.	22	58	56	46

Normal Coordinate Calculations

On the basis of the vibrational spectra we assume symmetry 1 of the MX$_4^{2-}$ species. So, to calculate the G matrix, the exact bond distances and angles found by X-ray analysis [3] were used. The redundant set of internal coordinates consists of distances r_{ij} and angles α_{ij}, where $i \neq j$ and $i,j = 1$–5. The corresponding

Table 2. The force constants (mdyn/Å) for MX$_4^{2-}$ where M = Zn(II), Cd(II) and X = Cl$^-$, Br$^-$, in (cinH$_2$)$^{2+}$ MCl$_4^{2-}$·2H$_2$O

	f_1	f_2	f_3	f_4	f_5
ZnCl$_4^{2-}$	1.106	0.105	0.595	0.259	0.047
ZnBr$_4^{2-}$	0.993	0.195	0.451	0.061	0.105
CdCl$_4^{2-}$	1.103	0.209	0.678	0.218	−0.017
CdBr$_4^{2-}$	0.803	0.157	0.355	0.072	−0.032

force constants, five in number, are: f_1 (force constant for stretching interaction along a bond), f_2 (force constant for stretching interaction along two bonds), f_3 (force constant for angle deformation), f_4 (force constant for interaction between two adjacent angle deformations), f_5 (force constant for interaction between a stretch and an adjacent angle deformation). There is a very good relation between the force constants calculated for all the isomorphous complexes (see Table 2). It is interesting to note that the interactive force constants f_3, f_4 and f_5 have quite significant values. On the contrary the ideal transition-metal tetrahalides, e.g. as present in solution, have interactive force constants close to zero [7].

Conclusions

The present study confirms that the tetrahedral geometry of the MX$_4^{2-}$ moieties, in the investigated series of isomorphous complexes, is significantly influenced by the cinchoninium counter-cation through its size and the way in which it is hydrogen bonded with the anions. A further quantitative study on monocrystals is planned.

References

[1] M. Dyrek, *Rocz. Chem.* **1976,** *50,* 2027.
[2] J. Chojnacki, B. Oleksyn, S. Hodorowicz, *Rocz. Chem.* **1975,** *49,* 429.
[3] B. J. Oleksyn, K. M. Stadnicka, S. A. Hodorowicz, *Acta Cryst.* **1978,** *B34,* 811.
[4] A. Weselucha-Birczyńska, M. Dyrek, C. Paluszkiewicz, *J. Mol. Struct.* **1990,** *219,* 73.
[5] J. H. Schachtschneider, F.S. Mortimer, *Technical Reports Nos. 231-64, 57-65,* Shell Development Co. , Emoryville, CA, 1964.
[6] G. Herzberg, *Molecular Spectra and Molecular Structure, Vol, II, Infrared and Raman Spectra of Polyatomic molecules,* Van Nostrand, Princeton, 1960.
[7] L. J. Basile, J. R. Ferraro, P. LaBonville, M. C. Wall, *Coord. Chem. Rev.* **1973,** *11,* 21.

Mikrochim. Acta [Suppl.] 14, 267–270 (1997)

Oxidation of Fullerene Materials

Michael Wohlers, Andrea Bauer, and **Robert Schlögl***

Fritz-Haber-Institut der Max-Planck-Gesellschaft, Abteilung Anorganische Chemie, Faradayweg 4–6, D-14195 Berlin,
Federal Republic of Germany

Abstract. Diffuse reflectance infrared Fourier transform spectroscopy (DRIFTS) was used to study the thermally induced oxidation of C_{60}, C_{70}, and the insoluble fullerene black with molecular oxygen at 570 K under *in-situ* conditions. The nature of the fullerene oxides formed under these conditions will be discussed. The reactivity of the investigated fullerene materials is compared to that of active carbon, which was chosen as reference system. Additional oxidation experiments using ozone as oxidizing reagent were performed at room temperature.

Key words: DRIFTS, fullerenes, oxidation.

Fullerenes are spherical molecules consisting exclusively of sp^2-hybridized carbon atoms. In contrast to graphite the non-planar π-electron system of molecules like C_{60} or C_{70} is not completely delocalized. As a consequence of this, almost localized double bonds are present in C_{60} or C_{70} between adjacent six-membered rings. These double bonds behave like electron-deficient olefins in a number of organic reactions ranging from Diels–Alder reactions [1] to epoxidations performed with ozone [2]. Reaction with molecular oxygen at room temperature under irradiation with UV/VIS light also leads to epoxides [3], whereas thermally induced oxidation of fullerenes leads to formation of degradation products [4] which were characterized as insoluble oxygen-containing polymers [5].

We have investigated the oxidation of several carbonaceous materials to decide whether this high reactivity of C_{60} and C_{70} towards oxygen is due to their molecular nature or due to their typical bent structures caused by the presence of five-membered rings. In the present study we compare the oxidation behaviour of C_{60} and C_{70} to that of insoluble fullerene black, which

is observed as the main product in the fullerene synthesis and exhibits bent structures analogous to the molecular fullerenes [6]. Oxidation reactions with molecular oxygen were performed at 570 K, which is below the temperature of a macroscopic burn-off of the fullerene materials [7], but is sufficient to overcome the diffusion barrier of C_{60} and C_{70} for the intercalation of gaseous molecules [8]. Active carbon was investigated as a reference system because of the absence of bent structures in this highly reactive carbonaceous material.

Experimental

Sublimed C_{60} and C_{70} powder samples with purities better than 99.5% were used, in both cases no traces of epoxides were present in the material, as determined by analytical HPLC. Fullerene black was prepared according to the standard conditions of the electric arc evaporation method in a helium atmosphere, leading to approximately 10% yield of soluble fullerenes [6]. Part of the fullerene black was directly used in this form for oxidation experiments ("Fullerene Black not extracted" = FBnex), the other part was Soxhlet-extracted with toluene for 48 hours to remove all soluble fullerenes ("Fullerene Black extracted" = FBex). A commercially available active carbon (Norit A, Aldrich) was used without further treatment.

DRIFT spectra were collected with a Bruker IFS 66 equipped with a Graseby-Specac P/N 19900 DRIFT accessory. All oxidation experiments with molecular oxygen were performed under *in-situ* conditions in a modified Graseby-Specac environmental chamber with a mass-flow controlled gas supply in an atmosphere of synthetic air with a flow rate of 20 ml/min. Difference spectra were calculated by subtracting the spectrum of the sample recorded at 570 K under helium from the spectrum collected after exposure of the sample to synthetic air.

Oxidations with ozone were performed at room temperature for 15 minutes in a fixed-bed reactor in a stream of oxygen containing approximately 1% of ozone. Additional experiments were performed in toluene at room temperature. The ozone-containing oxygen stream was bubbled through solutions of C_{60} and C_{70} (0.8 mM) and suspensions of FBnex, FBex, and active carbon (2 mg/ml) for 15 minutes. DRIFT spectra were collected after removal of the solvent and drying in vacuum.

---

* To whom correspondence should be addressed

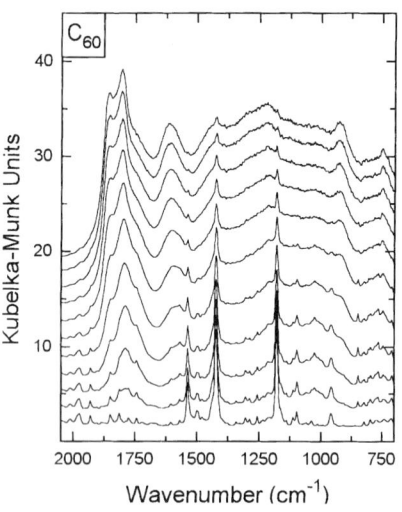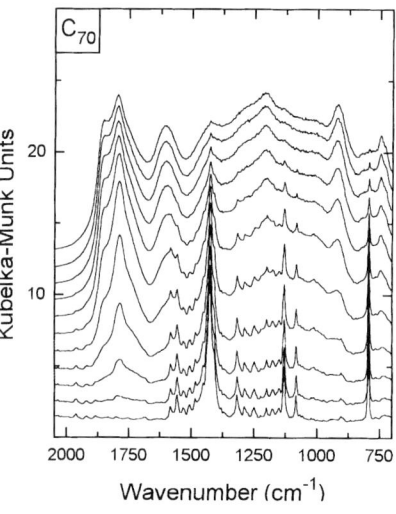

Fig. 1. DRIFT spectra of C_{60} and C_{70} recorded at 570 K in synthetic air, time of exposition from 0 hours (bottom) to 20 hours (top). Time intervals are 2 hours

Results and Discussion

Figure 1 shows a series of DRIFT spectra of C_{60} and C_{70} collected at 570 K in synthetic air. With increasing reaction time a decrease in the intensity of the absorption bands of the pristine fullerenes is observed. This is accompanied by a strong increase of broad new absorption bands due to the formation of oxidation products. The DRIFT spectra of C_{60} and C_{70} after 20 hours of oxidation are qualitatively almost identical, so they will be discussed together. Two strong bands at 1850 and 1800 cm^{-1} were assigned to $C=O$ groups of an anhydride R—CO—O—CO—R', where R and R' are unsaturated organic substitutents. The presence of such $C=O$ groups is necessarily connected to a breaking of carbon–carbon bonds and therefore an opening of fullerene cages. Absorption bands due to the presence of $C=O$ groups were already found in oxidation experiments with synthetic air at 470 K within a time period of 60 minutes, which means that the onset temperature for oxidative degradation of C_{60} and C_{70} is not higher than 470 K (data not shown). Further absorptions at 1220, 920 and 750 cm^{-1} were assigned to epoxide species; a broad underlying absorption ranging from 1300 to 1000 cm^{-1} indicates the presence of several different C—O containing species. Absorption bands at around 1610 cm^{-1} most presumably arise from carbon–carbon double bonds still being present in non-oxidized fragments of the respective opened fullerene frameworks. Another possible origin for these bands may be the presence of carboxylate groups COO^{-}, which also exhibit typical absorptions in this region.

A comparison of the changes of the different materials occurring during the oxidation process is given in Fig. 2. Difference spectra for C_{60}, C_{70}, and FBnex clearly reveal reaction of the respective molecular fullerenes by the bands pointing downwards; the formation of oxidation products is indicated by the peaks which were discussed in the previous section. Regarding only the peaks due to oxidation products, the spectra of all fullerene materials are in qualitative agreement whereas the difference spectrum of the active carbon shows almost no changes even after prolonged oxygen treatment, for 48 hours. This is indicative of a significant higher reactivity of all fullerene materials towards molecular oxygen compared to that of a commercially available active carbon.

The results of the ozone oxidation experiments at room temperature are summarized in Fig. 3. The spectra obtained after ozone treatment in the fixed-bed reactor clearly reveal oxidation of the active carbon, whereas no reaction was detected after exposition to molecular oxygen at 570 K. The spectrum of the oxidized active carbon is almost identical to the spectra of the ozone treated FBnex and FBex.

No oxidation products were observed after exposing solid C_{60} and C_{70} to ozone in the fixed-bed reactor, whereas complete degradation of the C_{60} and C_{70} units was found in the ozonization experiments in toluene solution. Taking into account the total surface areas of the investigated materials (soluble fullerenes 1–2 m^2/g, fullerene blacks 250–300 m^2/g, active carbon 550 m^2/g), it becomes obvious that for a significant oxidation of C_{60} or C_{70} in the solid state oxygen or ozone molecules have to diffuse into the lattices of their

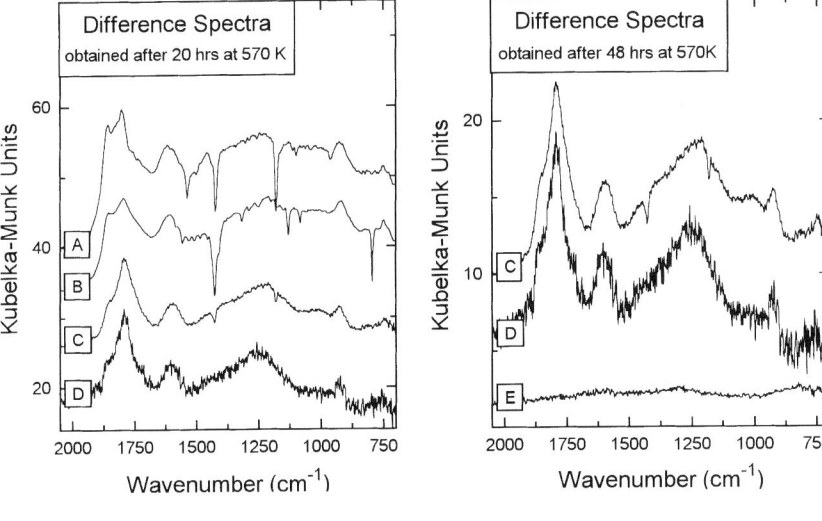

Fig. 2. Difference DRIFT spectra of A C_{60}; B C_{70}; C FBnex; D FBex; E active carbon. Time of exposition is denoted above, spectra A–D are normalized for better comparison

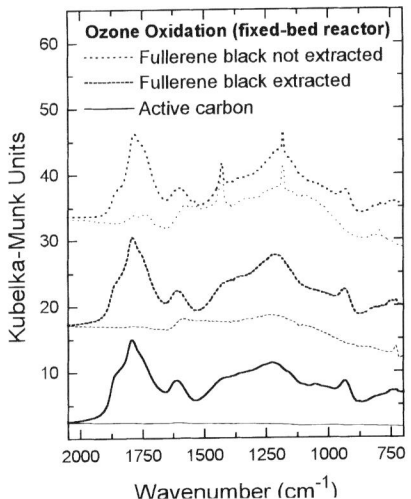

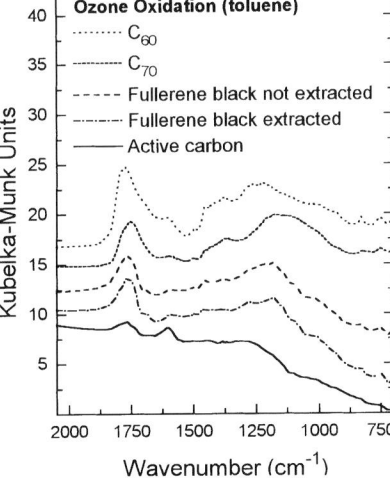

Fig. 3. DRIFT spectra obtained after oxidation with ozone at room temperature in a fixed-bed reactor (left) and in toluene (right), all spectra are normalized. Bold lines in the left plot indicate oxidized samples, the thin lines just below indicate the respective pristine materials

molecular crystals. Temperatures above 370 K are necessary for significant diffusion into these molecular crystals [8], so it has to be concluded that the reaction of solid C_{60} and C_{70} with ozone at room temperature is limited by diffusion of ozone molecules into the fullerene crystals. In toluene solution no diffusion limitation can occur, leading to complete reaction of C_{60} and C_{70} with ozone. The low intensity of the C=O absorptions found for active carbon in the analogous experiment indicates that adsorption of toluene onto the surface of the solid has occurred, which hinders its oxidation.

In conclusion, it has to be stated that the investigated fullerene materials exhibit a higher reactivity towards molecular oxygen than does active carbon, which was chosen as reference in this study. These differences in reactivity are caused by the bent structures common to all fullerene materials; possible skeletal defect structures may further be considered for the case of the fullerene blacks. The high reactivity of ozone leads to oxidation of all investigated carbonaceous materials even at room temperature, with the exception of solid C_{60} and C_{70}. The diffusion characteristics intrinsic to those soluble fullerenes prevent a significant oxidation at room temperature in the solid state.

References

[1] R. Taylor, D. R. M. Walton, *Nature* **1993**, *363*, 685.

[2] D. Heymann, L. P. F. Chibante, *Rec. Trav. Chim. Pays-Bas* **1993**, *112*, 531.

[3] R. Taylor, J. P. Parsons, A. G. Avent, S. P. Rannard, T. J. Dennis, J. P. Hare, H. W. Kroto, D. R. M. Walton, *Nature* **1991**, *351*, 277.

[4] H. Werner, Th. Schedel-Niedrig, M. Wohlers, D. Herein, B. Herzog, R. Schlögl, M. Keil, A. M. Bradshaw, J. Kirschner, *J. Chem. Soc. Faraday Trans* **1994**, *90*, 403.

[5] M. Wohlers, H. Werner, D. Herein, T. Schedel-Niedrig, A. Bauer, R. Schlögl, *Synthetic Metals* **1996**, *77*, 299.

[6] H. Werner, D. Herein, J. Blöcker, B. Henschke, U. Tegtmeyer, Th. Schedel-Niedrig, M. Keil, A. M. Bradshaw, R. Schlögl, *Chem. Phys. Lett.* **1992**, *194*, 62.

[7] I. M. K. Ismail, S. L. Rodgers, *Carbon* **1992**, *30*, 229.

[8] H. Werner, M. Wohlers, D. Bublak, T. Belz, W. Bensch, R. Schlögl, *Electronic Properties of Fullerenes*, Springer, Berlin Heidelberg New York Tokyo, 1993, p. 16ff.

Mikrochim. Acta [Suppl.] 14, 271–273 (1997)
© Springer-Verlag 1997

FT-IR and Raman Spectroscopy of $C_{58}BN$

K. Antonova[1,*], **P. Byshewski**[2], **G. Zhizhin**[3], **J. Piechota**[4], and **M. Marhevka**[5]

[1] G. Nadjakov Institute of Solid State Physics, blvd. Tzarigradsko chaussee 72, 1784 Sofia, Bulgaria
[2] Institute of Vacuum Technology, ul. Dluga 44/50, 00–241 Warsaw, Poland
[3] Institute of Spectroscopy, Russian Academy of Sciences, Troizk, Moskow region, Russia
[4] Institute of Physics, Polish Academy of Sciences, Al. Lotnikow 46, Warsaw 62–668, Poland
[5] Institute of Low Temperature and Structural Research, Polish Academy of Sciences, Wroclaw, Poland

Abstract. The standard electric arc generator was used to prepare fullerenes substitutionally doped on the surface with boron and nitrogen. The material is a unique case of a molecular crystal of high symmetry, consisting of molecules of extremely low C_s symmetry. FT-IR and FT-Raman spectra of the $C_{58}BN$ were measured in the spectral region $2000–30 \, cm^{-1}$ and over a large temperature interval: 14–300 K.

Key words: surface-doped fullerenes, $C_{58}BN$, molecular crystals.

The discovery of an efficient method [1] to prepare macroscopic quantities of the very stable fullerene molecules immediately opened the question of modifying the properties of the fulleride crystallites either by doping them in the solid form [2] or by enclosing foreign atoms inside the cage of the fullerenes [3]. Another modification of the fullerene cage involves substitution of carbon atoms by other elements e.g. nitrogen and boron, which could create one extra delocalized electron in the lowest unoccupied molecular orbital (LUMO) or a hole in the highest occupied molecular orbital (HOMO) [4]. In contrast to the case of exodoping, the effects of the substitutional doping on the structure are strongly localized around the impurity. They amount to a stretching of the impurity-carbon bond lengths by a few percent. The bond dissociation energy (DE) of various molecules which might be formed in such a plasma and the molecule bond length (BL) indicate the probability of formation of mixed fullerene during the plasma condensation and their deviation from the perfect spherical shape of C_{60}.

These parameters for the diatomic molecules show (see Table 1 that the strongest bonds are those with carbon and nitrogen, but structural deformation is more important for the case of boron substitution [5,6].

$C_{58}BN$ is a compensated system for which no localized impurity state is observed [4]. Apart from a splitting of all the states due to the lowering of symmetry, it has been observed that there is a slight decrease of the HOMO-LUMO gap of ~ 0.2 eV. The distortion of the molecules due to the substitution of carbon has to lead to modification of the molecule vibrations, which may be observed by IR and Raman spectroscopy.

Experimental

The samples were prepared in the standard electric-arc soot generator. In order to generate a high density of boron and nitrogen or boron nitride (BN) gas, one of the graphite rods was drilled and filled with high purity boron nitride. Because of the decreased conducting cross-section of the BN-filled electrode, the electric current had to be proportionatly reduced from the standard arc conditions, but other conditions of the discharge such as the He pressure and the distance between the electrodes were kept the same as in the usual process. The fullerenes were extracted from the soot with toluene in a standard Soxhlet vessel.

The FT-IR and FT-Raman spectra in the spectral region $2000–30 \, cm^{-1}$ were measured. Vacuum ampoules containing a few mg of the extract were used as samples in the Raman experiment. IR measurements were under air. One FT spectrophotometer a Bruker IFS 88, was used. The FT-Raman part of the equipment was equipped with a 1064 nm exciting laser. The spectra were measured under the following conditions: resolution $6 \, cm^{-1}$, 2000 scans and a laser power density of about $0.5 \, W/cm^{-2}$. The FT-IR part was equipped with a cryostat for low-temperature measurements down to 14 K. The spectra were recorded with a resolution of $0.5 \, cm^{-1}$, 100 scans.

---

* To whom correspondence should be addressed

Table 1. Bond dissociation energies (DE) and bond lengths (BL) for the diatomic molecules

	C–C	C–N	C–B	B–N
DE (eV)	6.5	8.2	3.9	4.0
BL (Å)	1.31	1.17	1.56	1.28

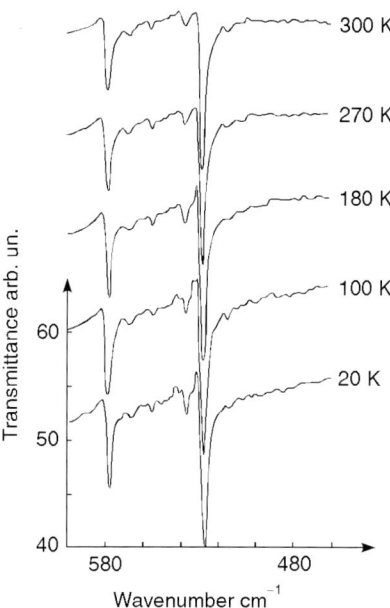

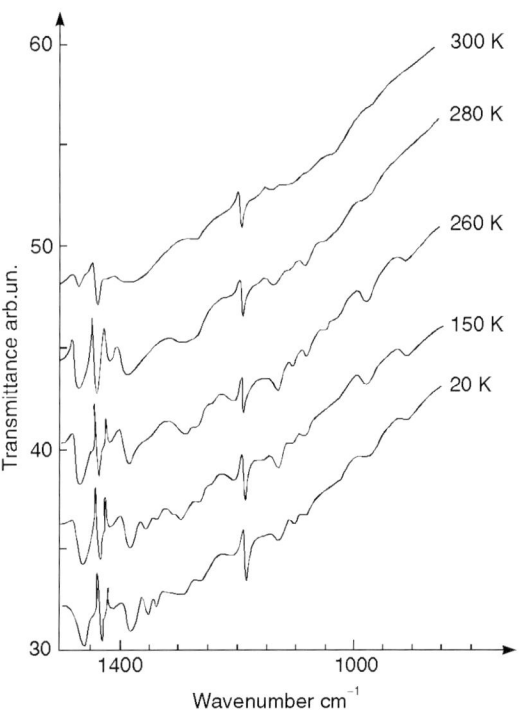

Fig. 1. FT-IR spectra of $C_{58}BN$ at different temperatures

Fig. 2. FT-IR spectra of $C_{58}BN$ at different temperatures

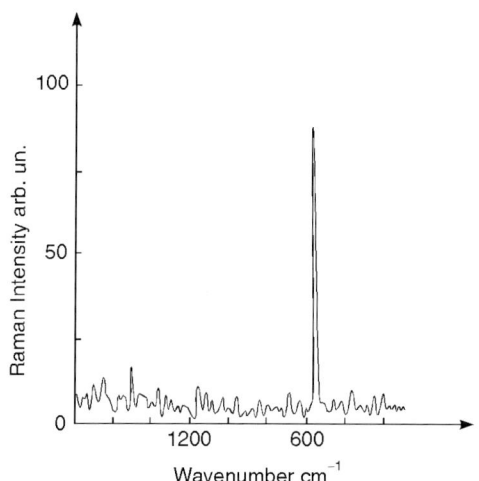

Fig. 3. FT-Raman spectra of $C_{58}BN$

Results and Discussion

The distortion of the fullerene structure by substitution of the carbon atoms by elements of different mass and bonding force, lowers the symmetry of the fullerene cage. A similar situation exists for pure crystals of C_{60} where isotopic impurities, the crystal-field and defects relax the selection rules and allow observation of silent modes [7]. It might be expected that boron and/or nitrogen built into the sphere of fullerene would not only relax the selection rules but also shift the lines from their positions for pure carbon fullerenes. The substitution of BN for two C atoms, bound to each other by the double bond in normal C_{60}, changes the molecular icosohedral symmetry to very low C_s symmetry with the symmetry plane passing through the BN-bond and the centre of the ball. Thus, some additional bands and lines appear in the IR spectrum.

Around the lines of the radial molecular vibrations at 577 cm^{-1} and 527 cm^{-1} shoulders and satellites arise, the most intense of them being at 566 and 536 cm^{-1} (see Fig. 2). We assume these bands to be due to C–B and C–N vibrations.

The significant decreasing of the molecular symmetry changes the Raman spectrum too (see Fig. 3). The pinch Raman mode appears at 1494 cm^{-1} instead of the 1469 cm^{-1} for pure C_{60} and is very weak. We observed the breathing Raman mode as a strong line at 539 cm^{-1} (530 cm^{-1} for pure C_{60}). Obviously, the

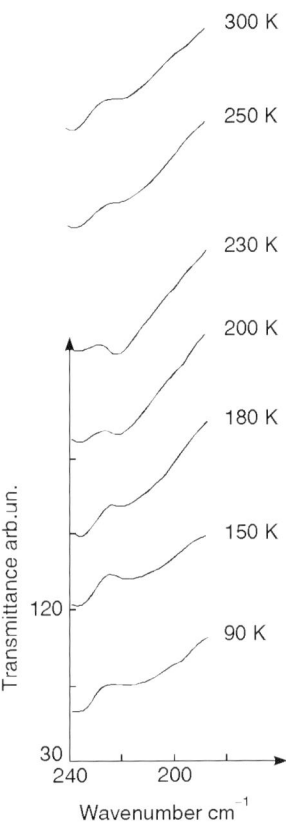

Fig. 4. FT far IR spectra of $C_{58}BN$ at different temperatures

The low-frequency spectra give information about the $C_{58}BN$ crystal lattice vibrations. Theoretically speaking the site position of the molecule can be as low as its own symmetry (the lowest possible here), but in the crystal, due to orientational mobility, the effective site symmetry can be much higher. The X-ray diffraction shows a very high face-centred cubic symmetry. The most sensitive band to temperature decrease, i.e. to the orientational ordering of the molecules, is that at 217 cm^{-1}. Its intensity approximately doubles at 240 K (see Fig. 4).

The Van der Waals character of the molecular interactions is first of all dependent on the dipole moment. The $C_{58}BN$ molecules possess a small dipole moment, of about 0.9 D [4]. This fact is important for observation of rotational modes but we did not succeed in seeing any of them.

Acknowledgements. We kindly acknowledge Prof. H. Ratajczak from the University of Wroclaw in whose laboratory the spectroscopic measurements were made. This work was partly financed by the Bulgarian National Foundation for Scientific Investigations, under No. F-529.

References

[1] W. Krätchmer, L. D. Lamb, K. Fostiropoulos, D. R. Huffman, *Nature* **1990**, *347*, 354.
[2] A. F. Hebard, M. J. Rosseinsky, R. C. Haddon, D. W. Murphy, S. H. Glarum, T. T. M. Palstra, A. P. Ramirez, A. R. Kortan, *Nature* **1991**, *350*, 600.
[3] J. R. Heath, S. C. O'Brien, Q. Zhang, Y. Liu, R. F. Curl, H. W. Kroto, F. K. Tittel, R. E. Smalley, *J. Am. Chem. Soc.* **1985**, *107*, 7779.
[4] W. Andreoni, M. Parinello, *Proc. First Italian Workshop on Fullerenes: Status and Perspectives*, Bologna, Italy, 6–7 Feb. 1992 (G. Taliani, G. Ruani, R. Zamboni (eds), World Scientific, Singapore, 1992, Vol. 2, pp.191–200.
[5] H. B. Gray, *Electrons and Chemical Bonding*. Benjamin, New York, 1979.
[6] T. Guo, C. Jin, R. E. Smalley, *J. Phys. Chem.* **1991**, *95*, 4948.
[7] M. C. Martin, X. Du, J. Kwon, L. Mihaly, *Phys. Rev.* **1994**, *B50*, 173.

surface substitution of carbon atoms influences the pinch mode more strongly.

It is seen (see Fig. 1 and Fig. 3). that some of the IR vibrations are Raman-active as well: 1357, 1122, 1078, 944, 831 cm^{-1}. The selection rules permit this for electron transitions. We suppose that they are concerned with the low-energy electron transitions to the new levels located in the region 0.13–0.17 eV of the split h_u level [4].

Mikrochim. Acta [Suppl.] 14, 275–276 (1997)
© Springer-Verlag 1997

Numerical Structure Representation and IR Spectra Prediction

Knut Baumann[1,*], **Christian Affolter**[1], **Ernö Pretsch**[2], and **Jean-Thomas Clerc**[1]

[1] Department of Pharmacy, University of Berne, Berne, Switzerland
[2] Department of Organic Chemistry, Swiss Federal Institute of Technology (ETH), Zurich, Switzerland

Abstract. The prediction of IR spectra of organic compounds containing carbon, nitrogen, oxygen and halogen atoms, from a spectroscopic database, is outlined. Structure similarity searches are performed to determine appropriate neighbours that are used for the prediction of the unknown molecule. The reliability and performance of the system was tested by "leave one out" procedure. In about 96% of all cases the system gave reliable answers.

Key words: spectra prediction, structure similarity, database.

The logical base for any attempt to predict the spectral properties of organic compounds is the relation between structural and spectral properties. Because the theory is still not fully understood, true *ab initio* approaches are mostly beyond the current state of the art. Furthermore, *ab initio* approaches are computationally very expensive and cannot be used for mid-sized to large-scale applications, such as structure elucidation tasks, where often hundreds or thousands of spectra are to be predicted. Practical applications therefore rely on approximate models, based on empirical data. The model used here to predict IR-spectra in the fingerprint region is basically a nearest neighbour search in a structure library. The predicted spectrum is constructed from the spectra of the nearest neighbours in structure space. The quality of this prediction model critically depends on the way structural similarity is established, i.e. the (numerical) representation of the problem to be solved. The numerical representation of chemical structures, however, is not yet solved in general. The optimal representation can be quite different for different molecular properties to be studied and even for the same molecular property very different approaches may give similar performance.

Structural Similarity and Structure Representation

The key assumption for the following calculations is the reasonable one that similar chemical compounds give similar IR spectra. Thus the way to determine the similarity has to take into account that molecules found to be similar in structure space should have similar IR spectra. Consequently the structure representation has to be chosen properly, so as to ascertain that this fundamental condition is fulfilled (Fig. 1). An inventive representation of the molecular structure is one of the most important prerequisites for the empirical prediction system to be successful. In the present study, we use two contrasting structure representations, that complement each other. The first code is based on a path-count statistic of the coloured molecular graph of the chemical compound to be encoded [1]. The second is an atom-centred fragment code [2]. The main characteristic of the first code is the high sensitivity to subtle changes in topology and geometry of the molecule and the lower sensitivity to different functionalities. The main characteristic of the latter code is just the opposite. Using both codes simultaneously, ensures that the similarity search in the structure library adequately considers both the constitutional properties and chemical functionalities of the query molecule.

Algorithm used for Prediction of IR Spectra

The IR spectra of the reference library are reduced to 256 data points by averaging the intensity values and only the last 128 points (fingerprint region) are processed. The structures of the reference library, as well as that of the query molecule, are encoded by using the above-mentioned codes. A nearest neighbour search is carried out for the query molecule. The similarity of the

* To whom correspondence should be addressed

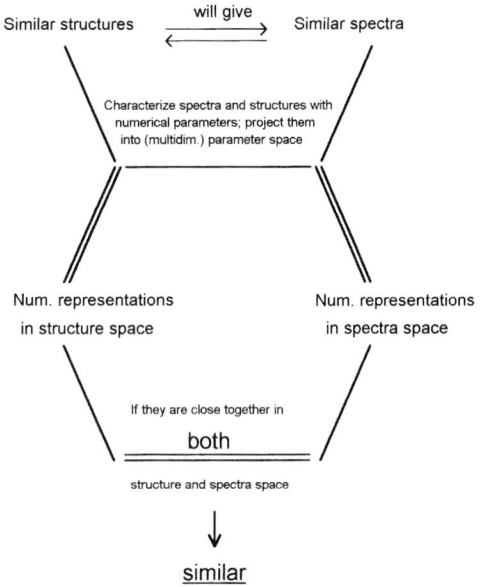

Fig. 1. Illustration of the key assumption

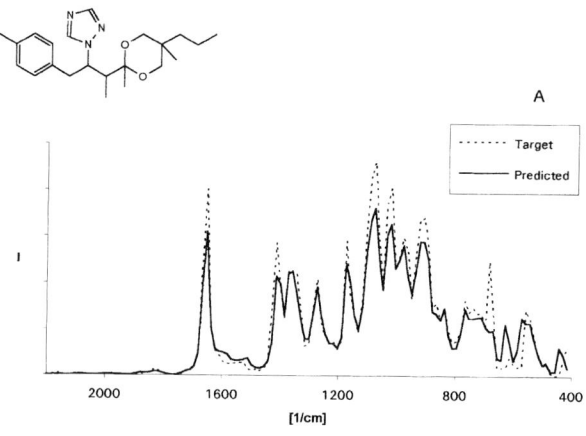

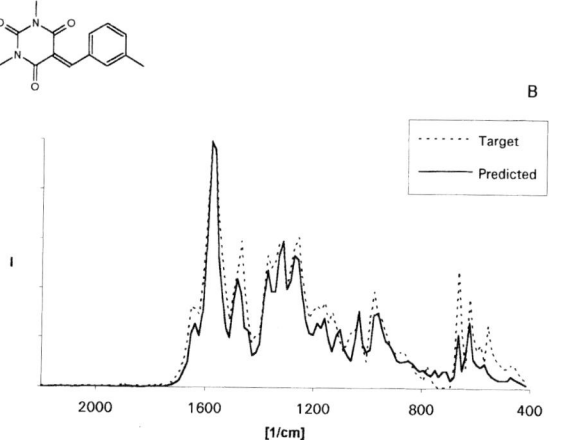

Fig. 2. A, B Examples of satisfactory prediction results

query molecule and the nearest neighbours should exceed a prespecified limit, otherwise the procedure is aborted. Common neighbours found by each structure representation are determined. If no common neighbours exist, stricter rules for the numerical similarity coefficient apply. The spectra of the (common) neighbours exceeding the limit are examined next, and are checked for spectral homogeneity by using different (parametric and non-parametric) correlation measures (coefficients). If the neighbourhood is homogeneous the spectra are brought in for prediction of the query spectrum.

Results

The performance of the empirical prediction system depends strongly on the reference library used. If no similar structures are available in the library, no reliable prediction can be made. The performance has been estimated by using a reference library with nearly 10,000 entries. An entry was selected at random, removed from the library and then used as a query. In about 96% of all cases (150 spectra were tested) the spectrum prediction was either satisfactory, or in somewhat less than half the cases, the system was not able to locate suitable reference compounds, so did not attempt a prediction, and altered the used. In 4% of all cases, however, the predicted spectrum deviated

strongly from the expected one, and the system did not alert the user.

These preliminary results depend to a large extent on the contents of the library. Furthermore, they are biased. Removal of the query entry from the library ensures that the best reference is missed. Thus, real world performance is expected to be at least as good as these test results, and presumably significantly better.

Random association of spectra with a query structure gave no reliable results throughout.

Acknowledgement. We thank Chemical Concepts, Weinheim, Germany, for the IR spectra library.

References

[1] J.-T. Clerc, A. L. Terkovics, *Anal. Chim. Acta* **1990**, *235*, 93.
[2] G. W. Adamson, M. F. Lynch, W. G. Town, *J. Chem. Soc.* **1971**, 3702.

Mikrochim. Acta [Suppl.] 14, 277–279 (1997)
© Springer-Verlag 1997

An Extension of the LOMEP Line-Narrowing Method

János Gácser and **Lajos Sztraka***

Department of Physical Chemistry, Technical University of Budapest, H-1521 Budapest, Hungary

Abstract. By introduction of an optimized deconvolution function, the applicability of the LOMEP line-narrowing method can be extended to overlapped bands having different half-bandwidths. The limit of the extended method has also been considered.

Key words: resolution enhancement, different half-bandwidths, optimization of the deconvolution function.

The recorded IR and Raman spectra of molecules in condensed phase consist generally of overlapped bands. The band shapes are determined by physical effects, so it is not possible to resolve the broad band systems by increasing the optical resolution of the spectrometer. In some cases (e.g. curve-fitting) it is important to know how many transitions are found in a band system. For obtaining this information line-narrowing methods are widely used. A new method is the LOMEP [1–3] which is a combination of the well known FT self-deconvolution and the maximum entropy method (MEM) with linear prediction. Unfortunately the correct application of the method requires that all bands in the manipulated spectral range have the same (and known) line shape (LOMEP criterion).

Extension of the LOMEP

During the test of the method, we found difficulties. Even if we apply the method to a simulated spectrum which contains Lorentzian bands of identical half-bandwidths it is difficult to find the best value of the order M of the prediction-error filter. Figure 1 shows results computed by LOMEP with different values of M. The original simulated spectrum may be seen in Fig. 3. The results show distortions, splittings and

shifts relative to the perfect line-narrowed spectrum ($M = 99$). The phenomenon can originate from two effects. The first is that the multiplication of a strongly damped interferogram by an exponential function extremely increases the round-off error of data even at a not so high order of M. The effect is more characteristic if the original spectrum is covered by noise. The second effect is that the prediction error coefficients computed by MEM converge well or not, depending on the order of M. If the end of the series of coefficients diverges, distortions and splittings occur in the resolution-enhanced spectrum.

For eliminating the problems we introduce a new deconvolution function which contains three parameters as follows:

$$\frac{\exp(2\pi\gamma_1|x|)}{r + (1-r)\ \exp\left[(\gamma_1 - \gamma_2)|x|\right]}$$

where γ_1 corresponds to the original γ_L, $\gamma_2 \leqslant \gamma_1$ and $1 \geqslant r \geqslant 0$. The optimal γ_1 value can eliminate completely the damping from the interferogram at low x values. If the $\gamma_1 - \gamma_2$ difference is not zero the remainder damping at higher x values can remove the distortion effects mentioned above. Kauppinen et al. proposed a quality factor q [2, 3]. Our idea for finding the optimal order of M is the optimization of the parameters γ_1, γ_2 and r to the maximum of q. The optimization is possible since the quality factor is a smooth function of the three parameters at a fixed M value. This requirement, however, is not true for the M order. Figure 2 shows the q factor as a function of M in the case of the problem pictured in Fig. 4. It can be seen that the factor q fluctuates wildly. Because of this each cycle of optimization contains two steps. First of all the program computes the q factors in a wide M range, using the parameters modified in the previous cycle.

* To whom correspondence should be addressed

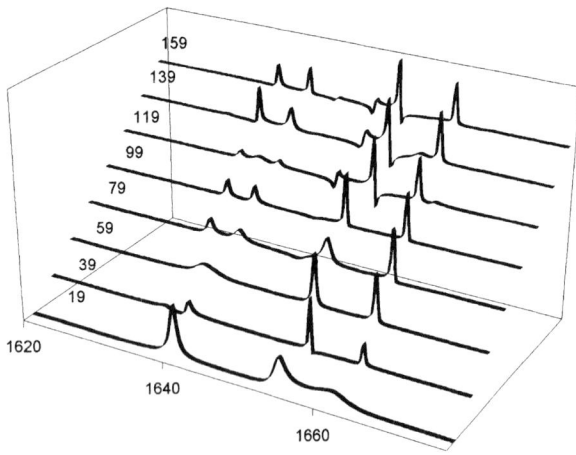

Fig. 1. Resolution enhancement of a test spectrum of four Lorentzian bands ($\gamma_L = 5 \, \text{cm}^{-1}$) as a function of the order of M

Then the program searches for the maximum of q and the corresponding M.

Results and Discussion

We have tested the extended LOMEP with simulated spectra and spectra recorded from liquid and solid samples. Figure 3 shows a resolution-enhanced spectrum obtained from four bands of identical half-bandwidths ($\gamma_L = 5 \, \text{cm}^{-1}$). Figure 4 shows, however, a spectrum obtained from four bands of different half-bandwidths ($\gamma_L = 10, 10, 5$ and $7.5 \, \text{cm}^{-1}$). The method perfectly separates the bands, but two small ghost bands appear. The aliphatic and aromatic stretching transitions of the Raman spectrum of liquid 2-methylpyrazine can be seen in Fig. 5. The number of bands is in accordance with the theory and the curve-fitting also

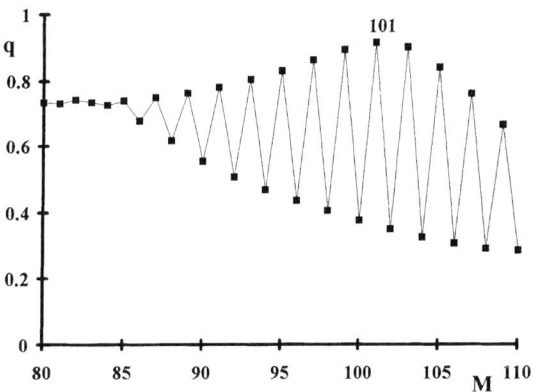

Fig. 2. The quality factor q as a function of M, calculated with the optimized parameters of the example in Fig. 4

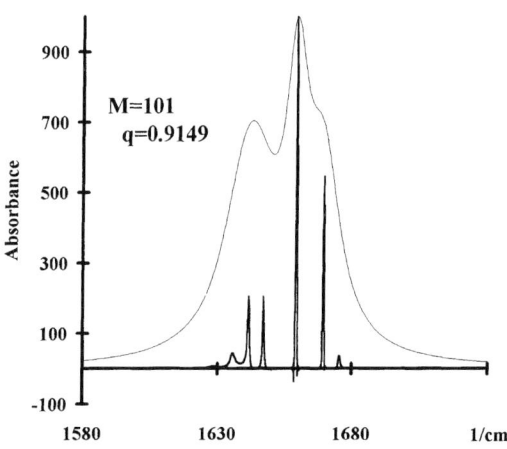

Fig. 4. Optimized resolution enhancement of a test spectrum of four Lorentzian bands ($\gamma_L = 10, 10, 5,$ and $7.5 \, \text{cm}^{-1}$)

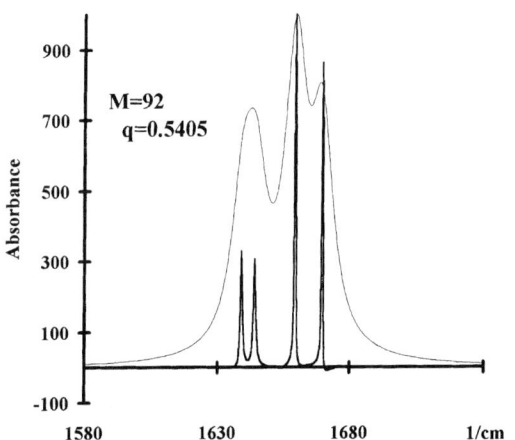

Fig. 3. Optimized resolution enhancement of a test spectrum of four Lorentzian bands ($\gamma_L = 5 \, \text{cm}^{-1}$)

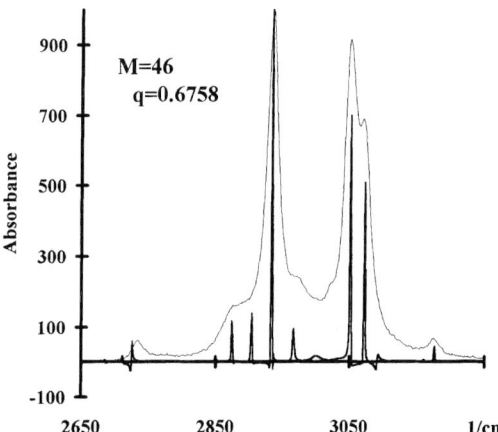

Fig. 5. Optimized resolution enhancement of a part of the Raman spectrum of 2-methylpyrazine

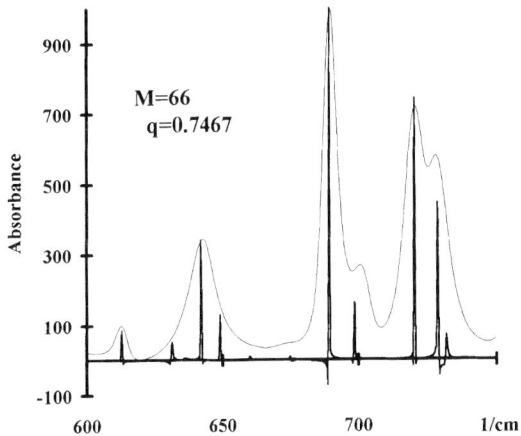

Fig. 6. Optimized resolution enhancement of a part of the IR spectrum of the cyclodextrin matrix of amygdalic acid in KBr pellet

gives an excellent result. The IR spectrum of a cyclodextrin matrix of amygdalic acid in a KBr pellet can be seen in Fig. 6. The resolution-enhanced spectrum seems to be real.

By the extension we have introduced an optimized damping function for the interferogram. The damping function corresponds to an optimized average line shape in the spectrum. It is obvious that the method works well only in a limited range of half-bandwidth ratios. We have found that a ratio of 2–3 is acceptable. Studying a higher ratio (10), we found that small ghost lines occurred. We have a similar experience for the asymmetric bands too.

The spectrum of 2-methylpyrazine was measured on a Nicolet FT-Raman 950 instrument and the spectrum of the amygdalic acid sample was recorded on a Perkin-Elmer FT-IR 2000 spectrometer.

Acknowledgements. The authors thank Dr. S. Holly (Central Research Institute for Chemistry of the Hungarian Academy of Sciences) and Dr. V. Izvekov (Department of General and Analytical Chemistry, Technical University of Budapest) for the spectra. One of the authors (J. G) thanks the József Varga Foundation for a Fellowship. The work was supported by the Hungarian Research Foundation (T-014975).

References

[1] J. K. Kauppinen, D. J. Hollberg, H. H. Mantsch, *Appl. Spectrosc.* **1991**, *45*, 411.
[2] J. K. Kauppinen, D. J. Moffatt, M. R. Hollberg, H. H. Mantsch, *Appl. Spectrosc.* **1991**, *45*, 1516.
[3] J. K. Kauppinen, D. J. Moffatt, H. H. Mantsch, *Can. J. Chem.* **1992**, *70*, 2887.

Mikrochim. Acta [Suppl.] 14, 281–282 (1997)

Supervised and Unsupervised Methods of Classification of Raman Spectra of Explosives and Non-Explosives

Nelson W. Daniel, Jr., Ian R. Lewis, and **Peter R. Griffiths***

Department of Chemistry, University of Idaho, Moscow, ID 83844, U.S.A.

Abstract. Various pattern-recognition techniques have been examined to evaluate their usefulness in automated detection and identification of explosives that constitute a threat when used in bombs in public place or airline baggage. Self-organizing map neural networks proved more useful than principal components analysis.

Key words: explosives, automated detection and identification, self-organizing map neural networks, principal components analysis.

Detection and disposal of explosives are important elements of commercial aviation security. The US Federal Aviation Administration is currently reviewing the development of automated techniques for the detection and identification of explosives, especially the highly energetic nitramine-based "plastic" explosives, which may be easily concealed in airline baggage [1]. Unfortunately, the most promising explosive detection devices (EDDs), such as thermal neutron activation, do not provide the chemical information needed to identify the specific explosive threat (i.e. type and composition of explosive) [1, 2]. A portable fiber-optic based "intelligent" Raman spectrometer could be used to automatically identify explosive materials, once the explosive threats have been detected by one or more of the very sensitive EDDs.

To determine whether Raman spectra of explosives lend themselves to automated interpretation, supervised and unsupervised pattern-recognition techniques were used for the clustering and classification of the Raman spectra of 32 explosive and 107 non-explosives. All but one of the explosives (3-nitro-1,2,4-triazol-5-one, abbreviated to NTO) were representative of 3 broad classes: nitro-aromatics ($Ar—NO_2$) [19 explosives], nitrate esters ($R—O—NO_2$) [8 explosives],

and nitramines ($>N—NO_2$) [4 explosives]. NTO is a unique 5-membered heterocyclic explosive.

Unsupervised mapping and classification were performed by principal-components analysis (PCA) and self-organizing map (SOM) neural networks. PCA is a mathematical technique useful for linear projection of large numbers of complex high-dimensional spectra into lower dimensional spaces. When PCA is used to reduce spectra into a 2-dimensional plot, clusters of spectra can be detected visually [3]. The SOM is an unsupervised computational neural network which is used to perform non-linear mapping of high-dimensional patterns into a 2-dimensional lattice of points such that the topology (i.e. spatial proximity) of the patterns is conserved. The SOM attempts to preserve both the global and the local structure of the input pattern space, while PCA preserves only the global information based on the total variance of the pattern space [4].

The back-propagation neural network (BPNN) is a supervised regression-based network similar to maximum likelihood approximation techniques. BPNNs have been frequently used for the classification of molecular vibrational spectra [4].

Experimental

All Raman spectra were collected at $4\,cm^{-1}$ resolution over a spectral range of $300–3600\,cm^{-1}$ with a Perkin–Elmer System 2000 FT-Raman spectrometer operating with 1064 nm laser excitation. The spectra were manually baseline-corrected and exported as AS-CII-Y data files by using Grams 386 (Galactic Industries, Salem, NH). Neural network analyses were performed with the Neural Works Professional II/Plus software package (NeuralWare, Pittsburg, PA). The SOM and BPNN networks were trained with default training parameters for 750 and 200 random presentations per spectrum respectively. PCA was performed by software written in Visual Basic (Microsoft, Redmond, WA), using the NIPALS

---

* To whom correspondence should be addressed

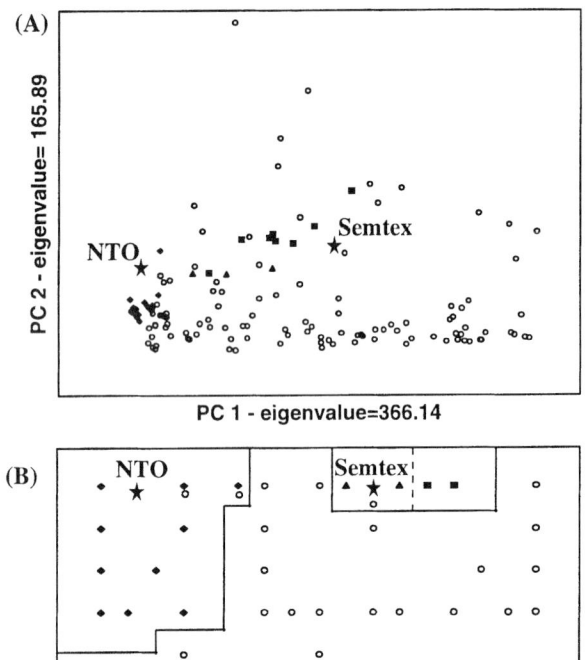

Fig. 1. A PCA and **B** SOM of explosives and non-explosives. Filled diamonds – nitro-aromatics filled squares – nitrate esters; filled triangles – nitramines; open circles – non-explosives

algorithm [5]. Some explosives required the evaporation or spectral subtraction of methanol used as solvent) before data analysis. The non-explosives used in this study were randomly selected from our laboratory, including solvents, hydrocarbons and polymers.

Results and Discussion

Two-dimensional PCA and SOM plots of the 32 explosives revealed three well separated clusters corresponding to the three classes of explosives, with some notable outliers. The most common outlier spectra are those of substituted nitro-aromatic explosives such as 1,3,5-triamino-2,4,6-trinitrobenzene (TATB) and 2,4,6-trinitrophenol (picric acid). Substituted nitro-aromatics are known to have complex IR and Raman spectra because of conjugation and mesomeric effects. The plastic explosive Semtex, a combination of the nitrate ester explosive pentaerythritol tetranitrate (PETN) and the nitramine explosive 1,3,5-trinitro-1,3,5-triazocyclohexane (RDX) is projected at a position which is between PETN and RDX, about twice as far from RDX as from PETN, which is consistent with the relative band intensities of each constituent observed in the spectrum of Semtex.

A two-dimensional PCA projection of explosives and non-explosives is shown in Fig. 1A. PCA is unable to cleanly separate the explosives from the non-explosives; however, the three classes of explosives are somewhat distinguishable. The SOM shown in Fig. 1B displays both excellent separation of explosives from non-explosives and good discrimination of the individual classes of explosives. In Fig. 1B NTO maps into the nitro-aromatic class and Semtex is intermingled with the nitramines because of the contribution of RDX.

This study also demonstrated that BPNNs perform as excellent supervised classifiers of Raman spectra of explosives but BPNNs are unable to consistently identify unusual explosives such as TATB, when they are not well represented in the training set.

Acknowledgements. The US Federal Aviation Administration and the US Federal Bureau of Investigation are thanked for support of this work.

References

[1] *Detection of Explosives for Commercial Aviation Security*, National Academy of Sciences, Washington, DC 1993.
[2] J. Yinon, S. Zitrin, *Modern Methods and Applications in Analysis of Explosives*, Wiley New York, 1993.
[3] M. J. Adams, *Chemometrics in Analytical Spectroscopy*, Royal Society of Chemistry, Cambridge, 1995.
[4] J. Zupan, J. Gasteiger, *Neural Networks for Chemists: An Introduction*, VCH, New York, 1993.
[5] B. G. M. Vandeginste, C. Sielhorst, M. Gerritsen, *TRAC* **1988**, 7, 286.

Mikrochim. Acta [Suppl.] 14, 283–285 (1997)
© Springer-Verlag 1997

Numerical Simulation of IR-Spectroscopic Experiments

M. Hiller, B. Mizaikoff[1,*], R. Kellner[1], W. Theiß[2], and P. Grosse[2]

[1] Institute for Analytical Chemistry, Vienna University of Technology, Getreidemarkt 9, A-1060 Wien, Austria
[2] Physikalisches Institut der RWTH Aachen, Sommerfeldstraße Turm 28, D-52074 Aachen, Federal Republic of Germany

Abstract. In this work we demonstrate the application of numerical simulation for the investigation of the light distribution in an optical system, such as an FT-IR microscope. Because of the complexity of simulating the optical arrangement of a complete IR-microscope we have applied a program based on a Monte Carlo type wavelength-dependent ray-tracing technique (SPRAY). The program is capable of computing spectral features in a complex three-dimensional system. Spectral features of organic layers and bodies are calculated from the real and imaginary parts of the dielectric functions determining the interaction of light with the material considered. The dielectric functions can be derived from various models. Previous work has shown that for organic substances a harmonic oscillator model can be used with sufficient accuracy. As microscopy usually exploits plane sample carriers, which do not seriously affect the path of the IR-beam, we calculated and tested cylindrical and spherical carriers, resulting in a higher energy throughput to the detector. In the simulation virtual screens could be placed before, inside and after the carrier to visualize the path of the radiation and the focusing effects, corroborating the experimental results. This work demonstrates the importance of wavelength-dependent numerical simulation of IR-spectroscopic experiments. The software developed offers a valuable tool for the verification of experimentally obtained IR-spectroscopic results and for the development and optimization of optical systems.

Key words: ray-tracing, Monte Carlo, simulation, microscope, curved sample carriers, infrared.

Ray-tracing is a commonly used procedure for computing pictures and animated sequences in the film and entertainment industry, flight and drive simulators, and also for developing, designing and testing optical set-ups.

Since three-dimensional optical set-ups are a complex subject, use of Monte Carlo techniques is implied for obtaining results in an acceptable computation time. The Monte Carlo principle is based on sum-marizing a large number of random decisions – in our case ray–interface interactions – to approximate the performance in real life. In the simulation, a light beam hitting a component of the set-up can be reflected, transmitted or absorbed, depending on the properties of the material.

Fresnel's equations determine the limits for the random decisions: for the amplitude reflection coefficients $r_\perp, r_\parallel$ and amplitude transmission coefficients $t_\perp, t_\parallel$, the equations are [1]

$$r_\perp = \left(\frac{E_{0r}}{E_{0i}}\right)_\perp = \frac{n_i \cos\theta_i - n_t \cos\theta_t}{n_i \cos\theta_i + n_t \cos\theta_t}$$

$$t_\perp = \left(\frac{E_{0t}}{E_{0i}}\right)_\perp = \frac{2n_i \cos\theta_i}{n_i \cos\theta_i + n_t \cos\theta_t}$$

$$r_\parallel = \left(\frac{E_{0r}}{E_{0i}}\right)_\parallel = \frac{n_t \cos\theta_i - n_i \cos\theta_t}{n_t \cos\theta_i + n_i \cos\theta_i}$$

$$t_\parallel = \left(\frac{E_{0t}}{E_{0i}}\right)_\parallel = \frac{2n_i \cos\theta_i}{n_i \cos\theta_t + n_t \cos\theta_i}$$

E_0 denotes the energy of the electric field vector component of the beam perpendicular $\perp$ and parallel $\parallel$ to the incident surface. The indices i, r and t designate properties assigned to the incident, reflected or transmitted beam; n represents the refractive index and θ is the angle perpendicular to the interface. The given equations are simplified providing $\mu_i = \mu_r \approx \mu_t \approx \mu_0$ which can be assumed with sufficient accuracy when dealing with linear, isotropic, homogeneous media. The refractive index n depends on the wavelength of the incident beam, and r and t are also functions of the wavelength λ. These functions can be approximated by a harmonic oscillator model [2] as shown in the litera-

---

* To whom correspondence should be addressed

ture. The ability to calculate this is incorporated in the SPRAY software developed by W. Theiß [3], in contrast to commercially available ray-tracing programs.

We have previously shown that cylindrical sample carriers such as optical fibres can have a positive effect on the signal intensity in absorption spectra obtained by IR-microscopy [4]. By a simulation procedure we will show the contribution of cylindrical and spherical shaped carriers as well as of parts of the Bruker IR-microscope towards the energy throughput.

The SPRAY Software

The SPRAY software was specially designed to simulate the interaction of IR-radiation with condensed matter such as silicon, germanium or any organic substance. The imaginary and real parts of the dielectric functions of the materials have to be supplied to the program.

The program covers use of plane, concave and convex mirrors, point-shaped, extended and transparent light sources, round, elliptical and rectangular stops, and furthermore can define plane layer-stacks and the interfaces of fibres, spheres and spherical segments, which can also consist of several layers of different materials. Virtual screens which do not affect the radiation can be placed at any point of the set-up.

Experimental

A chalcogenide fibre with a diameter of 500 μm was placed on the sample stage of the IR-microscope. The z-position of the fibre was changed in 25-μm steps and a single-beam spectrum with 100 scans and a resolution of 4 cm^{-1} was recorded with a Bruker IFS 113v spectrometer. An arbitrary intensity value was derived by integrating the single-beam spectrum in the range from 800 to 3000 cm^{-1}.

Simulations

A simulation set-up was developed which describes the IR-microscope in its orginal design. It consists of a spherically extended IR-light-source and an elliptical mirror to focus the radiation onto a fixed spot. Above the sample stage are the first cassegrain objective, consisting of a concave mirror with a hole and a convex mirror, an aperture stop with variable diameter and the second cassegrain objective focusing the light onto an MCT-detector element.

We used this set-up plus a cylindrical fibre object at the focus height and several positions above and below it to show the correlation between the simulation and the experiment described above. Two kinds of fibres made from chalcogenide and from silver halide, with a constant refractive index of $n = 2.84$ and $n = 2.21$ respectively in the considered range from 800 to 3000 cm^{-1}, were investigated. A single simulation for each fibre position had to be performed.

For the computation of energy throughput through the empty microscope we placed virtual screens before and behind each cassegrain objectives and a third between the two cassegrain mirrors. Virtual screens were also positioned before and after the aperture stop.

The simulation of the focusing effects of the cylindrical and spherical objects was done with virtual screens placed at the same position as the focusing object itself and at several positions above and below in 25-μm steps. This simulation gives an idea of the real intensity distribution in various plane cuts of the fibre or sphere, which cannot be obtained experimentally. The whole intensity distribution was retrieved from a single simulation run.

Results

As can be seen in Fig. 1 the simulation for the 500 μm fibre fits the experiment very well. This leads to the conclusion that the theoretical approach of our simulation is correct.

From the screens placed in the fibre and sphere positions we were able to calculate that the intensity achieved by focusing with the sphere is about 5 times that with the fibre and 10 times that with a plane sample carrier. The maximum intensity with 500 μm chalcogenide fibres was observed at a distance of 150 μm below the upper fibre surface. With 900 μm silver halide fibres the point of maximum intensity was at 200 μm below the upper surface and with 500 μm carriers at 300 μm below the upper surface (Fig. 2). This points out the importance of adjusting the IR-focus properly in an IR-microscopic experiment in order to achieve the maximum energy throughput and hence maximum signal at the detector.

Further evaluation of the screens placed throughout the whole microscope (Fig. 3) showed that the first cassegrain objective reduces the intensity by approximately 50%. The aperture reduces the energy by approximately 0.0013%. There was no more loss of energy through the second cassegrain objective. The total loss of energy throughout the microscope from sample illumination to the detector is 99.99935% or 1:155000.

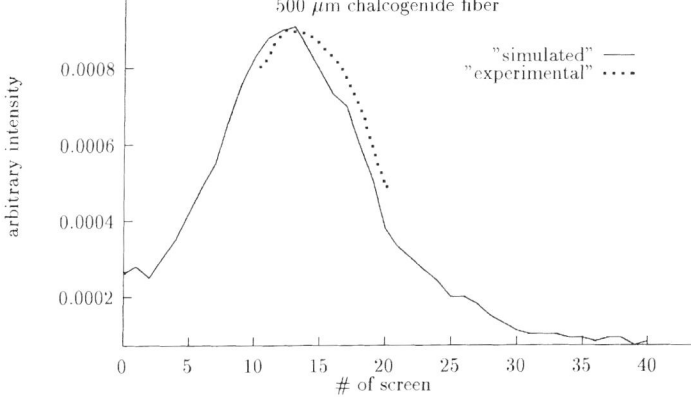

Fig. 1. Simulated and experimentally determined intensity distribution in a chalcogenide fibre

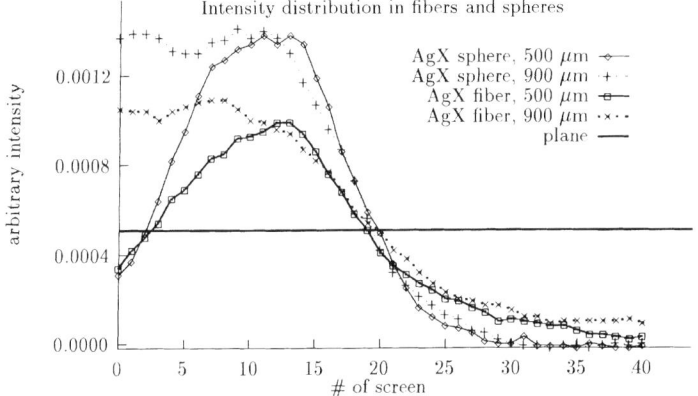

Fig. 2. Simulated intensity distribution in silver halide fibres and spheres

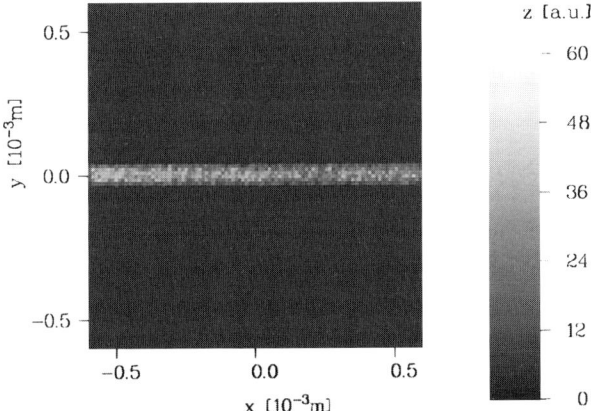

Fig. 3. Screen image at maximum intensity position in a fibre (a.u. = arbitrary absorbance units)

Conclusion

In this work we could show that simulations performed with the SPRAY software do fit our experimental conditions very well. The aperture is the only item which can be altered to increase the energy throughput significantly. The loss of lateral resolution when using larger aperture diameters may be compensated by the use of focusing sample carriers. Since the focus points when using chalcogenide and silver halide fibres and spheres were inside the carrier itself, simulations could help to calculate an optimal refractive index, which can be achieved by changing the material composition to yield a focus on top of the sample carrier surface.

This work demonstrates the importance of wavelength-dependent numerical simulation of IR-spectroscopic experiments, by use of the SPRAY software. The program proves to be a valuable tool for the simulation of optical set-ups. It offers the possibility to avoid time-consuming and expensive experiments in the development and optimization of optical systems and to verify experimentally obtained results.

References

[1] E. Hecht, *Optics, 2nd Ed.*, Addison-Wesley, New York, 1990.
[2] P. Grosse; *Mikrochim. Acta [Wien]* **1991**, *II*, 309.
[3] B. Mizaikoff, R. Kellner, K. Taga, W. Theiß, P. Grosse, *Proc. SPIE* **1993**, *2089*, 164.
[4] R. Kellner, B. Mizaikoff, K. Taga, W. Theiß, P. Grosse, *Fresenius J. Anal. Chem.* **1993**, *346*, 612.

Mikrochim. Acta [Suppl.] 14, 287–288 (1997)
© Springer-Verlag 1997

Neural Networks as a Tool for Identity Confirmation of IR Spectra

Tom Visser[1,*], **Hendrik-Jan Luinge**[2], and **John van der Maas**[2]

[1] National Institute of Public Health and Environmental Protection/LOC, PO Box 1, 3720 BA Bilthoven, The Netherlands
[2] Analytical Molecular Spectrometry, Utrecht University, Sorbonnelaan 16, 3584 CA Utrecht, The Netherlands

Abstract. A feasibility study has been made of the use of artificial neural networks for identity confirmation of IR spectra in trace analysis. Results were compared with predictions from library search and peak matching experiments. Artificial neural networks were found to be the most robust.

Key words: artificial neural networks, IR spectra, identity confirmation, trace analysis.

IR spectroscopy is widely applied for identification of analytes by comparing sample and reference spectra. Peak matching and library search algorithms are useful methods to express the spectral similarity as a value that can be used as an objective criterion [1, 2]. In forensic trace analysis, however, these methods are often insufficiently accurate, as a result of the high noise level in the spectra and the presence of interfering peaks.

Artificial neural networks (ANN) are thought to be less sensitive to artefacts, and their performance in pattern recognition is excellent. Therefore, ANN may be useful for the identification of spectra of poor quality, e.g., as obtained in residue analysis.

Methods and Materials

Three methods of spectrum identification were compared: (1) neural networks, (2) library search and (3) peak matching. IR spectra were obtained for cryotrapped gas-chromatography specimens of Clenbuterol-TMS, fluoranthene and perylene (standards and samples; amounts 0.5–50 ng). Total collection: 260 spectra (70 for Clenbuterol-TMS, 55 for fluoranthene, 40 for perylene, 95 others). Data point resolution was $4\,cm^{-1}$ over the range 2000–$700\,cm^{-1}$. Spectra were normalized between 0 and 1.5 absorbance and randomly divided into sub-sets for training/calibration and validation. For Clenbuterol-TMS: 8 sets for training and 8 for validation, 60 spectra per set (30 for Clenbuterol-TMS, 30 others). For fluoranthene 5 sets each of 45 spectra for training and validation (22 for fluoranthene, 23 others). For perylene 5 sets each of 40 spectra for training and validation (20 for perylene 20 others).

Neural Networks

Backward error-propagation with 325 data points per spectrum as input, 1 hidden layer of 10 neurons and 1 output neuron (target value 1 or 0). Learning rate: 0.1, momentum 0.9. Training was performed with synchronous prediction of the validation set to a maximum of 150 iterations. The minimum in the root mean-square error (RMSE) of the prediction value of the validation set was used to determine convergence.

Library Search

Grams/Spectracalc™ software (Galactic Industries Corp.) was used. Algorithm: first derivative absolute value. The library contained all spectra of the calibration set. For each spectrum searched, the similarity index of the first hit was used for further evaluation.

Peak Matching

The criterion for matching was agreement in absorption frequency within $4\,cm^{-1}$ and of intensity within 0.20 absorbance, for 6 bands. Reference peaks were selected on the basis of structural and spectroscopical relevance, from high quality reference spectra with resolution $1\,cm^{-1}$. Peak matching program: Sadtler IR Search software.

Similarity Values

Individual prediction values p, obtained from ANN classification of the validation spectra, were used to calculate two average correctness of prediction (ACP) values for each compound class: ACP_+ for the spectra that should have been classified as positive and ACP_- for the negatives. For the library search experiments, analogous ACP values were calculated from the 0/1 normalized similarity indices by regarding these as p. In both cases, the criterion for confirmation was set as a p-value of >0.9 in order to minimise the chance of false positives. The results of peak matching were considered as "matching" or "not-matching", resulting in the classifications correct positive (CP), correct negative (CN), false positive (FP) or false negative (FN).

---

* To whom correspondence should be addressed

Table 1. Results of identification of spectra of Clenbuterol-TMS obtained from ANN, library search and peak matching

	ANN	Library search	Peak matching
CP	240	162	227
CN	230	234	240
FP	10	6	0
FN	0	78	13
Total	480	480	480
ACP_+	0.98	0.91	–
ACP_-	0.93	0.34	–

CP = correct positive, CN = correct negative, FP = false positive, FN = false negative.

Results and Discussion

Clenbuterol TMS

The identity of all Clenbuterol-TMS spectra was confirmed by the neural networks (Table 1). The corresponding ACP_+ values were very high. Most non-Clenbuterol spectra were also classified correctly. Occasionally, spectra were identified as false positive. In all cases this concerned spectra of Bromobuterol-TMS, a closely related compound. It occurred particularly when spectra of this compound were absent in the training set, endorsing the importance of representative composition of the training set. The similarity of Clenbuterol- and Bromobuterol-TMS also appeared from the results of the library search experiments; the false positives were spectra of Bromobuterol-TMS. A large number of spectra of Clenbuterol-TMS were not confirmed by library search. These spectra exhibited a high noise level. It confirms that ANN are less sensitive to noise than is the chosen library search method. Also the ACP values of the "non-Clenbuterol" spectra were low, which also indicates that the search algorithm is less discriminating than a neural network.

The results of peak matching were good, but in a few cases, spectra of Clenbuterol-TMS were not confirmed, as either the peaks could not be distinguished from noise or the band intensities were affected too much. False positive classifications were absent, even for the spectra of Bromobuterol-TMS, but the absorption frequency of some peaks differed by as much as 5 or 6 cm^{-1} instead of the allowed 4 cm^{-1}.

Perylene and Fluoranthene

ANN classified all spectra of these polyaromatic hydrocarbons (PAHs) correctly and the ACP values were high (Tables 2 and 3). Library search confirmed

Table 2. Results of identification of spectra of fluoranthene obtained from ANN, library search and peak matching

	ANN	Library search	Peak matching
CP	110	91	98
CN	165	165	165
FP	0	0	0
FN	0	19	12
Total	275	275	275
ACP_+	0.96	0.94	–
ACP_-	0.96	0.16	–

Table 3. Results of identification of spectra of perylene obtained from ANN, library search and peak matching

	ANN	Library search	Peak matching
CP	100	71	97
CN	100	100	100
FP	0	0	0
FN	0	29	3
Total	200	200	200
ACP_+	0.97	0.91	–
ACP_-	0.97	0.23	–

a smaller number of spectra, again because of spectra with noise. False positives were absent but the ACP values of the "non-fluoranthene" and "not-perylene" spectra were very low, indicating that the similarity index was close to the critical level of 0.9. This endorses the limited usefulness of the first derivative absolute value search-algorithm for confirmational purposes. Some spectra were not confirmed by peak matching, because of the high noise level. False positives were not produced. Apparently, the peak positions and corresponding intensities differ sufficiently to prevent false positive identification.

Conclusions

Artificial neural networks are less sensitive to spectral noise than library search and peak matching methods, and hence more suitable for identification in trace analysis. Peak matching is most discriminative when the S/N ratio and resolution of the spectra are high. The first derivative absolute value algorithm is not very useful for identity confirmation of spectra of poor quality.

References

[1] W. G. de Ruig, R. W. Stephany, G. Dijkstra, *J. Chromatogr.* **1989**, *489*, 89.
[2] J. R. Hallowell, Jr., M. F. Delaney, *Anal. Chem.* **1987**, *59*, 1544.

Mikrochim. Acta [Suppl.] 14, 289–290 (1997)

Optimization of Spectral Search Results for ATR Spectra in the Mid IR Region

Michael Boruta* and **John D. Baker**

Bio-Rad, Sadtler Division, 3316 Spring Garden Street, Philadelphia, PA 19104, U.S.A.

Abstract. The ATR technique has increasingly become a common technique in the analysis of unknown materials. However, many commercial and personal spectra databases have been generated by use of transmission techniques. Because of this difference, spectral searching of these databases can sometimes fail or produce confusing results. An understanding of the differences between the transmission technique and the ATR technique coupled, with the knowledge of what effects these will have on the search algorithms, will lead to improvements in algorithm selection and data preprocessing options.

Key words: infrared, ATR, search strategies.

The path-length for an ATR experiment varies with the wavelength of light used. The path-length for a transmission experiment is constant throughout the experiment. The result, when comparing spectra from these two techniques, is that the ATR spectrum will have weaker peaks (shorter path-length) in the high wave number regions and stronger peaks (longer path-length) in the low wavenumber regions. Thus, the spectroscopist must have a means of reconciling these variations in the effective path-lengths.

For this work, two data bases from Sadtler were used; the Solvents and the ATR of Solvents. Each of 387 spectra in the ATR data base was treated as the query spectrum and searched against the Solvents database. The search strategies and their results are described below.

The first strategy used was a full range euclidean distance search. This algorithm is commonly found in most IR search software and is usually the default algorithm. Table 1, column 2 shows the distribution of the HQI (Hit Quality Index) values for this strategy. The average HQI was 873. The Standard deviation was 65. The query was found as the number one hit in 289 cases, or 75% of the trail.

The second strategy used was a full range first derivative search. This algorithm is also commonly found in most IR search software. It was chosen because it is effectively less sensitive to peak heights but more sensitive to peak positions than is the euclidean distance algorithm. This is due to its greater emphasis on the change in slope of the peak rather than its area. This relative insensitivity to peak height should make this a better choice for an ATR spectrum where it is known that the peak heights will vary from those in a transmission spectrum. Table 1, column 3 shows the HQI values for this strategy. The average HQI was 709. The standard deviation was 142. The query was found as the number one hit in 312 cases, or 80% of the total. Although the HQI values show a significant degradation both in absolute value and in standard deviation compared to those from the euclidean distance algorithm, the number of times the query was found as the number one hit improved.

The third strategy used was a limited range euclidean distance search (1800–$648 \, \text{cm}^{-1}$). Table 1, column 4 shows the HQIs for this strategy. The average HQI was 908. The standard deviation was 64. The query was found as the number one hit in 312 cases, or 80% of the total. A limited search was chosen to minimize the effects of the varying path-length of the ATR spectra. This strategy found 312 query spectra as the number one hit, the same as by the first derivative search, but substantially improved the average HQI and the standard deviation.

The fourth strategy used was another limited range euclidean distance search (1400–$700 \, \text{cm}^{-1}$). Table 1, column 5 shows the HQI values obtained. The average HQI was 992. The standard deviation was 60. The query was again found as the number one hit in 312 cases. This strategy was tried in order to find the effects of severely limiting the range. This strategy showed no improvement in the number of query spectra found as

---

* To whom correspondence should be addressed

Table 1. A breakdown of HQI (hit quality index) *vs* search strategy. Column 1 gives the HQI, broken into bins. Column 2–6 show the number of hits found in each HQI bin by the various search strategies

HQI (hit quality index)	Full range euclidean	Full range 1st derivative	1800–648 cm^{-1} euclidean	1400–700 cm^{-1} euclidean	ATR Correction euclidean
131	0	0	0	0	0
167	0	1	0	0	0
203	0	0	0	0	0
239	0	0	0	0	0
275	0	0	0	0	0
311	0	3	0	0	0
347	0	3	0	0	0
383	0	7	0	0	0
419	0	7	0	0	0
455	0	10	0	0	0
491	0	10	0	0	0
527	0	8	0	0	1
563	1	18	1	0	1
599	0	31	0	1	0
635	0	25	0	0	0
671	2	31	1	0	0
707	6	24	2	3	1
743	13	39	6	3	3
779	23	40	18	9	7
815	52	48	28	25	9
851	74	45	34	26	23
887	85	30	55	39	46
923	67	7	89	88	53
959	61	0	131	140	100
995	3	0	22	53	143

the number one hit and only a slight improvement in the average HQI and standard deviation.

The fifth strategy used was a full range euclidean distance search but first the query spectra were pre-processed with an ATR correction algorithm. Table 1, column 6 shows the HQI values obtained. The average HQI was 940. The standard deviation was 62. The query was found as the number one hit in 329 cases or 85% of the total. Processing the ATR spectrum to make it look more like a transmission spectra gave a significant improvement in the number of query spectra found as the number one hit, and in the average HQI.

Conclusion

Using an ATR correction algorithm can significantly improve the search results when searching against a database generated by use of transmission techniques. If a correction algorithm is not available, then choosing a limited range euclidean distance search or a first derivative search can also lead to better search results than the "default" full range euclidean distance search.

Mikrochim. Acta [Suppl.] 14, 291–292 (1997)
© Springer-Verlag 1997

Simple Techniques for Phase Correction in VCD Measurements

Andrew J. Turner* and **Robert A. Hoult**

Perkin–Elmer Ltd, Post Office Lane, Beaconsfield, Buckinghamshire, U.K.

Abstract. Two simple techniques are presented for the phase correction of circular vibrational dichroism FT-IR spectra which have been generated by using a photoelastic modulator.

Key words: vibration circular dichroism, FT-IR spectroscopy.

Circular vibrational dichroism (VCD) spectra (see e.g. [1, 2]) can contain both positive and negative features, and this can lead to problems in phase correction of the spectra [2]. In this paper, we present a method whereby the phase information used during the processing of the dichroism signal is derived from a previous reference scan, thus removing ambiguity in the processed spectra.

The normal instrument background provides a suitable signal for the reference. Since the phase characteristics of the demodulation circuit may be different from those of the reference channel, the signal needs to be processed through the same electronic circuits as the dichroism signal. This can be done by amplitude modulation of the detector signal at the photoelastic modulator (PEM) frequency in order to produce a signal suitable for demodulation by the lock-in amplifier.

In the simplest method of modulation, the sample is replaced by a second linear polarizer set close to 45° from horizontal. This generates a modulated component at twice the fundamental PEM frequency, so the lock-in amplifier has to be set to demodulate components at the second harmonic. The drive power to the polarization modulator may need to be reduced so that there are no nulls in the modulation efficiency in the wavelength range of interest.

The alternative, preferred, method is to take a reference scan with the signal generated by electronic amplitude modulation at the photoelastic modulator frequency of the normal background signal from the detector. The amplitude modulator circuit need only be a simple unpowered circuit involving a field effect transistor and a few resistors, and can easily be switched in and out of circuit as required.

The experimental set-up (including the additional amplitude modulator circuit) is shown in Fig. 1. The Hinds II/ZS37 polarization modulator and sample are mounted in the sample area of the Perkin–Elmer Spectrum 2000 and a KBr lens is used to image the signal onto the liquid-nitrogen cooled MCT detector, this being done to minimize the polarization effects which occur when the normal reflective optics are used. The lock-in amplifier and filters were custom made by Barman Instruments. An obey program is used to ensure that the correct phase curve is generated and used.

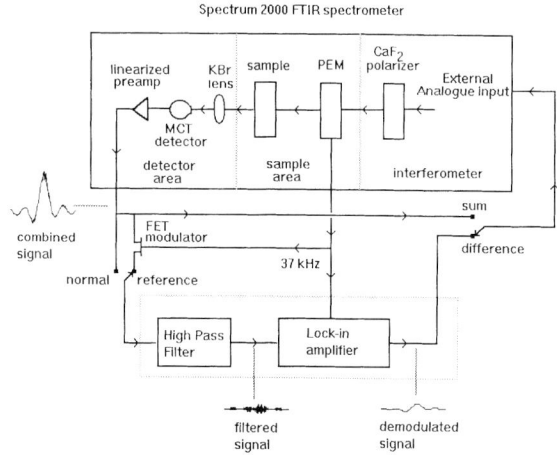

Fig. 1

---

* To whom correspondence should be addressed

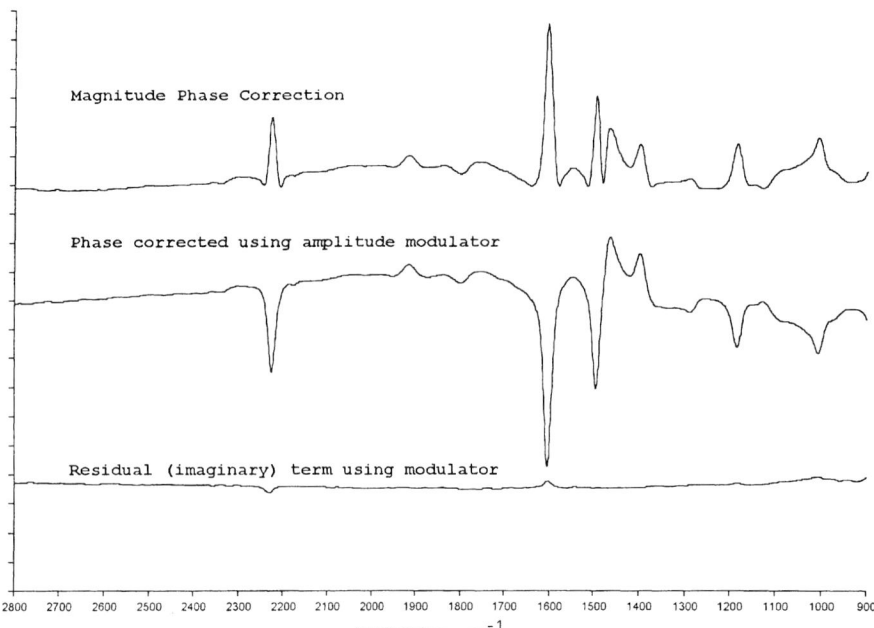

Fig. 2. 5CB liquid crystal spectra using different phase correction techniques

Figure 2 shows the spectrum of a liquid crystal sample (5CB). The windows of the cell were prepared in a manner to induce a twist in the liquid crystal orientation (similar to that used in liquid crystal displays) and therefore artificially induce circular dichroism in the material. The centre trace uses the electronic technique described, with the corresponding imaginary (error) term shown at the bottom. The low value of this term indicates the effectiveness of the technique. For comparison, the normal, magnitude corrected spectrum is shown at the top.

References

[1] P. R. Griffiths, J. A. de Haseth, *Fourier Transform Infrared Spectrometry*, Wiley, New York, 1986, Chapter 8.
[2] L. A. Nafie, in: *Advances in Applied Fourier Transform Infrared Spectroscopy*, (M. W. Mackenzie, ed.) Wiley, Chichester, 1988, Chapter 3.

Mikrochim. Acta [Suppl.] 14, 293–294 (1997)
© Springer-Verlag 1997

Absolute Infrared Intensities of Binary Mixtures of Liquid Chlorobenzene and Toluene

John E. Bertie* and **Yoram Apelblat**

Department of Chemistry, University of Alberta, Edmonton, Alberta, T6G 2G2, Canada

Abstract. Absolute infrared intensities of binary mixtures of liquid chlorobenzene and toluene were determined from transmission measurements. In general, the integrated intensities between isosbestic points of the $\tilde{v}\alpha_m''$ spectra were linear with respect to the mole fraction of the pure components. The α_m'' spectra of the pure liquids were curve-fitted with bands of classical damped harmonic oscillator shape. For most fundamentals, the experimental spectra of the mixtures and synthetic spectra obtained by mixing data from the fitted bands of the pure components agreed to within 2–4%. Larger deviations (7–15%) occurred for some of the bands, indicating molecular interaction between chlorobenzene and toluene.

Key words: absolute infrared intensities, curve-fitting, chlorobenzene, toluene, molecular interaction.

Experimental absorbance spectra of binary mixtures of liquid chlorobenzene and liquid toluene ranging from pure chlorobenzene to pure toluene, were obtained from transmission measurements between 4800 and 400 cm^{-1} at 25 °C. From these spectra, absolute intensities were determined, calculated as the molar absorption coefficient spectrum, $E_m(\tilde{v})$, the absorption index spectrum, $k(\tilde{v})$, the dielectric loss spectrum, $\varepsilon''(\tilde{v})$ and the imaginary molar polarizability spectrum $\alpha_m''(\tilde{v})$. Integrated intensities were measured as areas under the $\tilde{v}\alpha_m''$ spectra. The α_m'' spectra of the pure liquids were curve-fitted with bands of classical damped harmonic oscillator (CDHO) shape. Spectra of each mixture were synthesized from those of the pure components, according to the prediction for ideal solutions that the α_m'' spectrum of a mixture is the sum of the α_m'' spectra of the pure components, weighted by their mole fractions in the mixture. For each mixture two such synthesized spectra were created: from the fitted CDHO bands and from the experimental spectra of the pure liquids.

Results

In general, integrated intensities between isosbestic points of the $\tilde{v}\alpha_m''$ spectra are linear with respect to the mole fraction of the component. In the lower box of Fig. 1 the region between 1640 and 1550 cm^{-1} is shown. The prominent band is the CC stretch, v_{8a}, located at 1583 cm^{-1} for chlorobenzene and 1604 cm^{-1} for toluene. In the upper box of Fig. 1, the integrated intensities under the α_m'' spectra between the isosbestic points at 1634.4, 1596.3 and 1553.9 cm^{-1} are given. The areas, as well as most areas for other bands, are linear within the estimated experimental error for the pure liquids [1, 2].

The α_m'' spectra of the pure liquids were fitted with CDHO bands. The minimum number of bands was used to fit the spectra, and essentially all were evident in the experimental spectra: 193 CDHO bands were used for chlorobenzene and 166 bands for toluene. The large number of bands is due to the numerous weak but accurately measured [1, 2] features. For each mixture, spectra synthesized from the fitted bands and from the experimental spectra of the pure liquids agreed very well, reflecting the good fit of the pure liquid spectra. For most fundamentals, the experimental spectra of the mixture and the spectra synthesized from the fitted bands agree within 2–4%. An example for v_{8a} is given

---

* To whom correspondence should be addressed

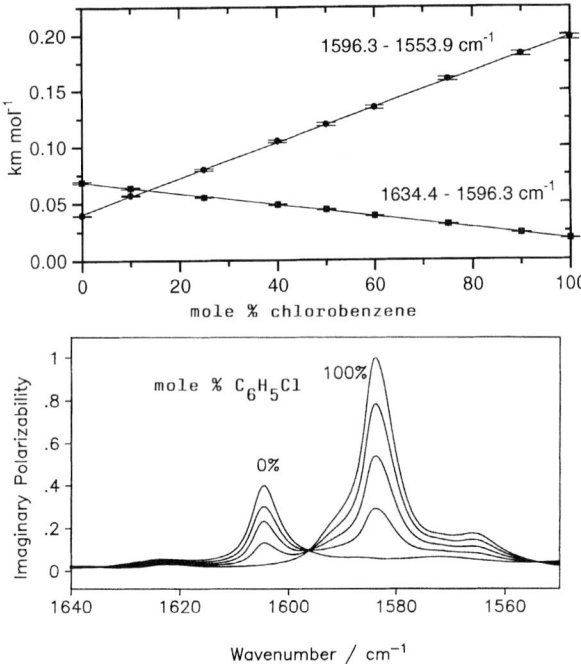

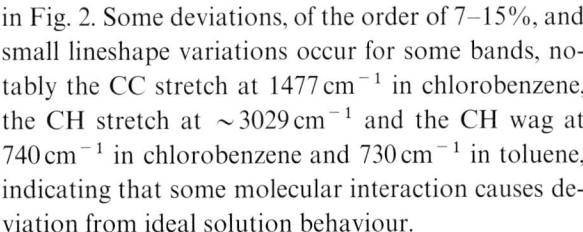

Fig. 1. The α_m'' spectra of binary mixtures of liquid chlorobenzene and toluene are given in the lower box. The integrated intensities between isosbestic points are given in the upper box

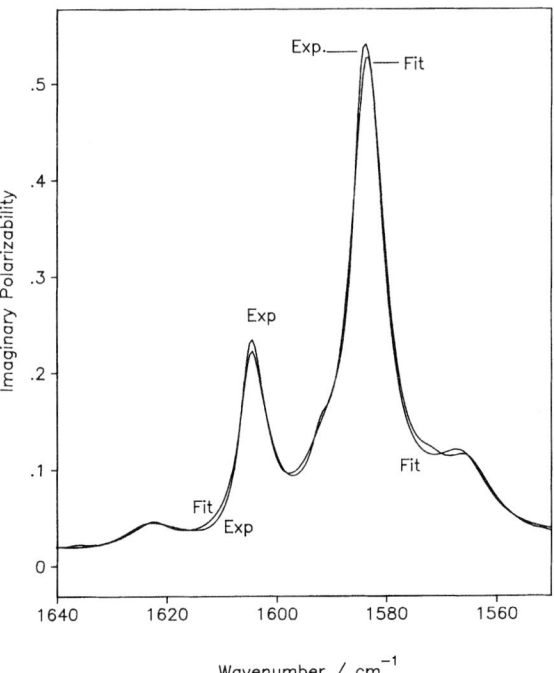

Fig. 2. The experimental α_m'' spectrum and that synthesized from the fitted CDHO bands for an equimolar mixture of liquid chlorobenzene and toluene. Similar comparisons for other compositions yield similar results

in Fig. 2. Some deviations, of the order of 7–15%, and small lineshape variations occur for some bands, notably the CC stretch at $1477\,\mathrm{cm}^{-1}$ in chlorobenzene, the CH stretch at $\sim 3029\,\mathrm{cm}^{-1}$ and the CH wag at $740\,\mathrm{cm}^{-1}$ in chlorobenzene and $730\,\mathrm{cm}^{-1}$ in toluene, indicating that some molecular interaction causes deviation from ideal solution behaviour.

Acknowledgement. This work was supported by the Natural Sciences and Engineering Research Council of Canada.

References

[1] J. E. Bertie, R. N. Jones, Y. Apelblat, C. D. Keefe, *Appl. Spectrosc.* **1994**, *48*, 127.
[2] J. E. Bertie, R. N. Jones, Y. Apelblat, *Appl. Spectrosc.* **1994**, *48*, 144.

Mikrochim. Acta [Suppl.] 14, 295–296 (1997)

An Approximate Method for the Calculation of the Infrared Molar Absorption Coefficient and Absorption Index Spectra of Liquids from Transmission Measurements

John E. Bertie* and **Yoram Apelblat**

Department of Chemistry, University of Alberta, Edmonton, Alberta, Canada T6G 2G2

Abstract. A simple and effective approximate method for the calculation of infrared molar absorption coefficient spectra, $E_m(\tilde{v})$, and absorption index spectra, $k(\tilde{v})$, of liquids from transmission measurements is presented. In the approximate method, the apparent absorbance due to reflection losses is calculated by assuming the cell is a single window. This simplifies the correction of the experimental absorbance for the contribution of reflection and eliminates the need for an iterative calculation. Although simpler, the method gives k and E_m values accurate to 1% for all but the strongest infrared absorptions. Areas under these spectra are accurate to 0.5%. The method is demonstrated on a mixture of 90 mole% toluene and 10 mole% chlorobenzene.

Key words: transmission measurements, absorption index spectra, molar absorptivity spectra.

Experimental absorbance spectra of a 9:1 binary mixture of toluene and chlorobenzene at 25 °C were obtained from transmission measurements between 4800 and 400 cm^{-1}. The experimental absorbance, EA, is given by $EA = -\log(I_t/I_0)$ and is the sum of the absorbance by the liquid, A, the apparent absorbance due to reflection from the interfaces of the cell, AA_R, and the apparent absorbance due to unexplained baseline errors, AA_B.

Previously, $k(\tilde{v})$ was obtained in an iterative procedure in which the correction of EA for the apparent absorbance due to reflection is exact for an ideal experiment [1–3]. Thus, to calculate AA_R, Fresnel's equations were applied to all interfaces of the cell, air–window–liquid–window–air. The interference fringes from multiple reflections within the thin liquid layer were calculated. In contrast, the multiple reflections within the window were averaged so that they yielded no interference fringes, as is appropriate for

~5 mm thick alkali halide windows and 1 cm^{-1} resolution. To correct for baseline errors, measurements made in long path-length cells were used to determine the linear absorption coefficient, $K(\tilde{v})$, at specific wavenumbers in the baseline, so-called *anchor points*. $K(\tilde{v}) = A/\ell$ where ℓ is the path-length through the liquid and A was taken to be $EA - AA_R$. The K values were used to remove the apparent absorbance due to baseline errors in the cells of normal length, ℓ', by moving EA to the correct place, calculated as $EA = K\ell' + AA_R$. The $k(\tilde{v})$ thus obtained is exact in the ideal case. In the approximate method the apparent absorbance due to reflection, AA_R, is calculated by assuming the cell is simply a single window. The transmittance of the window, T_w, is given by $T_w = 2n(\tilde{v})/[n^2(\tilde{v}) + 1]$ and the apparent absorbance due to reflection losses is $AA_R(\tilde{v}) = -\log\{2n(\tilde{v})/[n^2(\tilde{v}) + 1]\}$. The K values at the anchor points are calculated from long path-length cell measurements, using the same assumption. In this method, the absorbance by the liquid, A, is simply the baseline-corrected EA minus AA_R. The k and E_m spectra are then calculated from the equations $k(\tilde{v}) = 2.303 A(\tilde{v})/(4\pi\tilde{v}\ell)$ and $E_m(\tilde{v}) = A(\tilde{v})C\ell$.

This approximation to AA_R is poorest on the sides of strong bands (upper box in Fig. 1). Its effect on peak heights and areas is shown in the next section.

Results

In the lower box of Fig. 1 the E_m spectra obtained by using the exact and approximate methods are shown. The spectra are not distinguishable, indicating that the approximate method is very effective. Neither the peak heights nor the line shapes are changed by the approxi-

* To whom correspondence should be addressed

Table 1. Comparison of intensities and areas calculated under the $E_m(\tilde{v})$ spectrum by the exact and approximate methods

$v\,(\mathrm{cm}^{-1})$	Intensity $(\mathrm{l\,mole}^{-1}\,\mathrm{cm}^{-1})$		100(Approx. − Exact)/Exact %
	Exact	Approx.	
1604.7	27.82	27.87	0.18
1495.7	95.83	95.90	0.07
1477.7	38.70	38.80	0.26
1460.7	22.02	22.08	0.27
1445.7	21.72	21.78	0.28

Range (cm^{-1})	Area $(\mathrm{l\,mole}^{-1}\,\mathrm{cm}^{-2})$		100(Approx. − Exact)/Exact %
	Exact	Approx.	
1698.1–1547.1	597.1	598.4	0.22
1547.1–1398.6	1971.4	1977.3	0.30

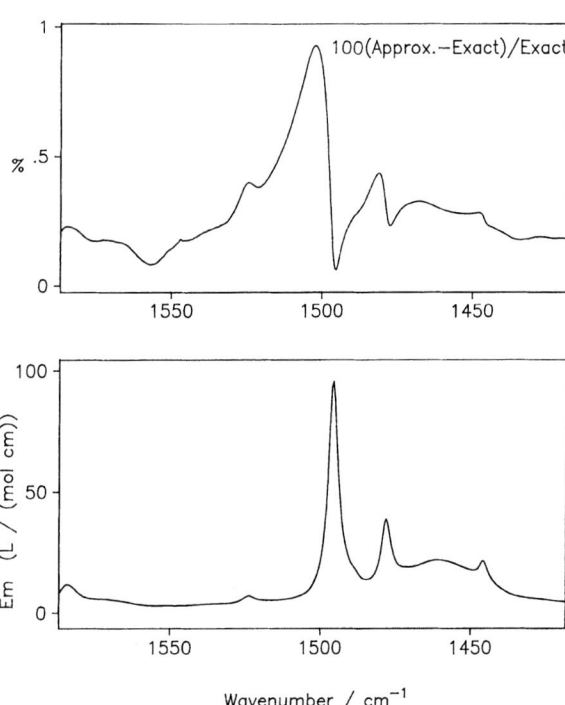

Fig. 1. The exact and approximate E_m spectra of a mixture of 90 mole% toluene and 10 mole% chlorobenzene given in the lower box, but are not distinguishable. The percentage difference between them is given in the upper box

mation. In the upper box of Fig. 1, the percentage difference between the spectra from the two methods is shown. The maximum deviation is less than 1% and occurs to high wavenumber of the peak. The deviation at the peak, located at 1495 cm^{-1}, is under 0.1%. Weaker peaks yield similar or lower percentage difference. For stronger peaks, the percentage deviation is larger, as expected. In fact, for the 3 times stronger 740 cm^{-1} band of pure chlorobenzene the deviation at the peak is − 1.5%. The areas under the k or E_m spectra are accurate to better than 0.5% for all bands. Some illustrative peak intensities and area are given in Table 1.

Acknowledgement. This work was supported by the Natural Sciences and Engineering Research Council of Canada.

References

[1] J. E. Bertie, C. D. Keefe, R. N. Jones, *Can. J. Chem.* **1991**, *69*, 1609.

[2] D. G. Cameron, J. P. Hawranek, P. Neelakantan, R. P. Young, R. N. Jones, *Computer Programs for Infrared Spectrophotometry XLII to XLVII*, National Research Council of Canada Bulletin 16, 1977.

[3] J. P. Hawranek, P. Neelakantan, R. P. Young, R. N. Jones, *Spectrochim. Acta* **1976**, *32A*, 75.

Mikrochim. Acta [Suppl.] 14, 297–299 (1997)

Quantitative Analysis of Alkylbenzenesulphonates in Water Solution by ATR/FT-IR Spectroscopy

Núria Ferrer*, Monserrat Roura, and **Montserrat Baucells**

Serveis Científico Tècnics, Universitat de Barcelona, Lluís Solé i Sabarís, 1, 08028 Barcelona, Spain

Abstract. Fourier-transform infrared spectroscopy has been used for the determination of dodecylbenzenesulphonate in commercial samples of liquid detergents. Analyses have been performed without sample treatment. Attenuated total reflectance with use of a ZnSe crystal allows measurement of the bands of the detergent together with the water spectrum. The precision, recovery and accuracy obtained by use of calibration line and standard additions have been optimized.

Key words: dodecylbenzenesulphonate, FT-IR, quantitative analysis, detergents, attenuated total reflectance.

Common substances found in detergents, and consequently in waters and sediments, are alkylbenzenesulphonates, which are used as surfactants in commercial products. Nowadays there is a demand for rapid and selective instrumental analytical techniques for determination of the chemical composition of detergents. Although various HPLC methods have been applied [1–3], the use of this technique requires the extraction of the sample into a solvent and a large dilution factor, because the concentration of these compounds is at the per cent level.

Infrared spectroscopy measurements applied to quantitative analysis of aqueous solutions allow us to obtain a spectrum rapidly without sample pretreatment, and to manipulate the data easily. The use of attenuated total reflectance (ATR) spectroscopy allows determination of these compounds in water solution.

Experimental

Apparatus

A Bomen DA. 3 Fourier-transform infrared spectrometer equipped with a potassium bromide beam-splitter, MCT (narrow range) detector and an attenuated total reflectance accessory with a ZnSe crystal was used for all the samples. FT-IR spectra were measured with a resolution of $4\,cm^{-1}$ and 150 scans were taken in order to obtain an appropriate signal to noise ratio. The spectral data were processed with the GRAMS/386 (Galactic) program.

Materials

Solid dodecylbenzenesulphonate (Carlo Erba) was used as a reference material. All dilutions were done in "Milli-Q" distilled water. The liquid samples analysed were commercially available and those of a known brand (product A). Product X is a commercially available ecological detergent not containing dodecylbenzenesulphonate.

Sample Handling

Liquid commercial samples and solid standard were treated with water. Because of the foam produced during the mixing, it is necessary to use an ultrasound bath and to let the dilution stand for about 24 hours.

Results and Discussion

The characteristic bands of dodecylbenzenesulphonate are at 1180, 1128, 1036 and $1010\,cm^{-1}$. The detection limit for dodecylbenzenesulphonate was about 0.1%.

Precision Tests

Precision of the analysis was measured for all absorptions by using 10 different samples of the same concen-

---

* To whom correspondence should be addressed

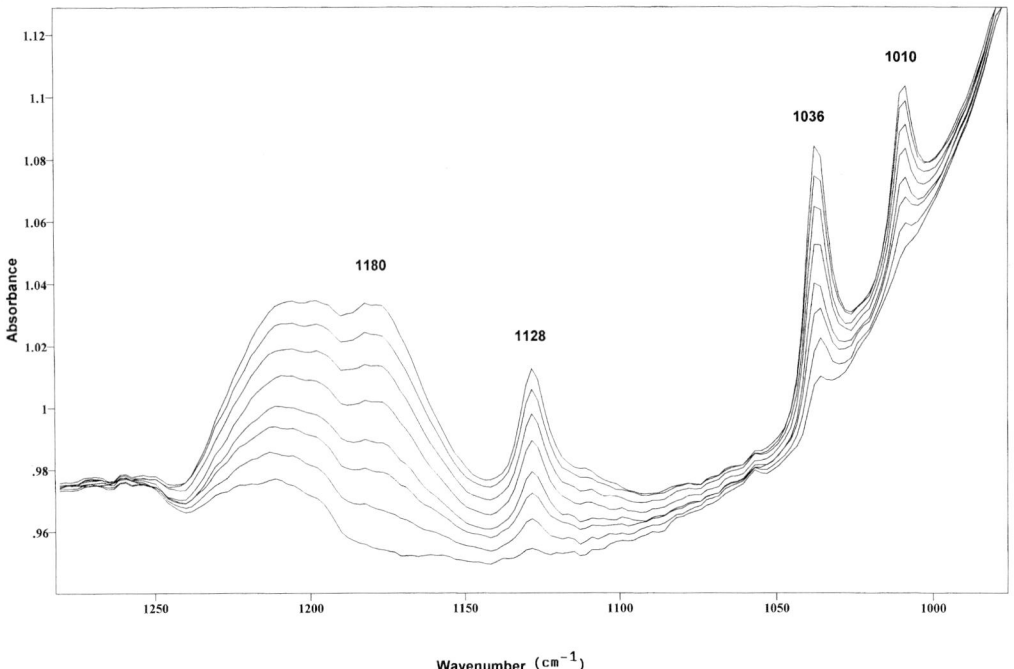

Fig. 1. Recovery of dodecylbenzensulphonate in detergent X

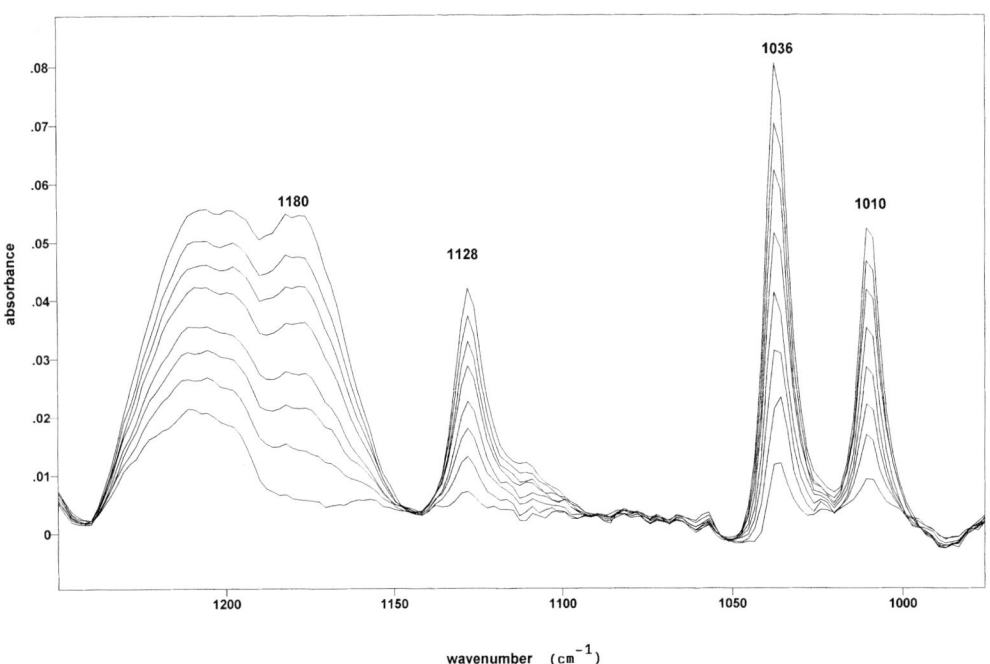

Fig. 2. Standard addition of dodecylbenzensulphonate in detergent A

tration (1% standard in water). The relative standard deviation (RSD) value includes the errors due to the instrument, sample handling and area or height measurement. The RSD for use of height measure- ments was 5, 9, 9 and 10% respectively for the four bands mentioned above. Measurements of areas were normally worse than the height measure- ments.

Recovery Tests

Recovery was measured for product X doped with 0.2% dodecylbenzenesulphonate, by the standard addition method. In order to minimize the effect of the matrix, we chose product X because its spectrum did not have the absorption peaks due to dodecylbenzenesulphonate (Fig. 1). The best results were for the peaks at 1180 and $1128 \, cm^{-1}$, with recoveries of 110 and 95% respectively. The peaks at 1036 and $1010 \, cm^{-1}$ showed recoveries of 175 and 32%.

Quantification Procedure

Calibration line and standard addition techniques were used for the quantification of dodecylbenzenesulphonate by infrared spectroscopy. In order to create a calibration line, different amounts of dodecylbenzenesulphonate were added to a commercial sample that did not contain it (product X). The standards ranged from 0.75 to 2% and correlation coefficients were better than 0.99 for all peaks. The final concentration of dodecylbenzenesulphonate found in the product A was: 32, 40, 28, or 26%, depending on the peak used.

The standard addition procedure is sometimes better to match standards with the samples when the effect of the matrix is important. Product A was diluted to 5% in water, from which eight aliquots were taken. To all but one of these aliquots, known increasing amounts of the compound were added (from 0 to 1.75%) (Fig. 2). The intercept of the straight line with the x axis corresponds to the concentration of the analyte in the product. The concentrations of dodecylbenzenesulphonate found in product A were 36, 27, 31 and 36%, depending on the peak used.

Conclusions

Fourier-transform infrared spectroscopy is an optional technique for quantifying dodecylbenzenesulphonate in commercial samples of liquid detergents. Precision and recoveries are acceptable enough for some peaks to allow their use in the determination, using either the calibration line or standard addition methods. The choice of peak to use depends on the matrix effects found.

References

[1] L. S. MacDonald, B. G. Cooksey, W. C. Campbell, *Anal. Proc.* **1986**, *23*, 448.
[2] L. Smedes, J. C. Kraak, C. F. Werkhoven-Goewie, U. A. Th. Brinkman, R. W. Frei, *J. Chromatogr.* **1982**, *247*, 123.
[3] A. Nakae, K. Tsuji, M. Yamanaka, *Anal. Chem.* **1981**, *53*, 1818.

Mikrochim. Acta [Suppl.] 14, 301–303 (1997)
© Springer-Verlag 1997

Application of Detector Non-linearity Correction for FT-IR Spectrometric Gas Analysis of SF$_6$ Samples

H. Michael Heise[1,*], Paulo R. Janissek[2], and **Peter Fischer[1]**

[1] Institut für Spektrochemie und Angewandte Spektroskopie, Bunsen-Kirchhoff-Strasse 11, D-44139 Dortmund, Federal Republic of Germany

[2] Laboratòrio Central de Electrotécnica e Eletrônica, COPEL, Rua Coronel Dulcidio 800, 80420 Curitiba, Brasil

Abstract. Fourier-transform infrared spectroscopy can achieve the low spectral noise level necessary for trace analysis. In the mid-infrared a photoconductive HgCdTe detector is routinely used, which suffers from a substantial non-linear response for high radiation flux such as found when recording the interferogram centre-burst. The photometric accuracy, close to detector cut-off, is thereby dramatically reduced for strongly absorbing samples, and the line shapes can be seriously influenced by phase errors. The effects on the spectra of gaseous SF$_6$, and the improvements gained by software correction are presented.

Key words: detector non-linearity correction, photometric accuracy, Fourier transform infrared spectroscopy.

Sensitive spectroscopic measurements in the mid-infrared routinely require the use of photoconductive mercury cadmium telluride (HgCdTe) detectors. The detector is ideal for low radiation flux conditions, but shows non-linear response to higher intensities, as experienced in FT-spectroscopy during recording of the interferogram centre-burst. The problems caused for the photometric accuracy are well described in the literature; remedies based on different hard- and software approaches for avoiding or correcting the non-linear response effects in the spectra have been suggested [1].

An estimation of the non-linear response is made, based on the method proposed by Keens and Simon [2], and a second order approximation correction is applied to the interferogram, which yields, after the usual data processing, a spectrum with improved overall photometric accuracy. The advantages for spectrometric gas analysis of SF$_6$ are demonstrated with

spectra recorded with use of different optical path-lengths. Spectral intervals for a quantitative analysis of contaminants can reliably be defined. The interesting low wavenumber region is then accessible for analysis, as is demonstrated by an example.

Experimental

Measurements were made with a Bruker IFS 113v FT-spectrometer equipped with a DTGS-detector and an MCT detector (Infrared Ass., Suffolk, England) with cut-off at 530 cm^{-1}. For the gas measurements a cell from SPECAC (Orpington, UK) with a path-length of 10 cm, and a multiple reflection cell (long path Mini-Cell, Model 6) from Infrared Analysis (Anaheim, U.S.A.) with an optical path-length variable between 1.2 and 6.0 m, were used. For each spectrum, 100 interferograms, providing a resolution of 0.5 cm^{-1}, were co-added. The computer program for detector non-linearity correction [3] was supplied by Bruker (Karlsruhe, Germany).

Results and Discussion

FT-IR spectrometry offers the possibility of multicomponent analysis within an SF$_6$ matrix. SF$_6$ is extensively used as the insulating gas in electrical high voltage equipment. The gas also constitutes the arc quenching medium in modern circuit-breakers and switches. To reach the appropriate sensitivity for trace analysis, long optical path-lengths are required. Since SF$_6$ possesses many strong absorption bands in the mid-infrared, only a limited number of spectral intervals is accessible for the analysis of trace constituents.

The data evaluation is complicated by the non-linear response of the MCT detector operated in the photoconductive mode, which affects the recording of the interferogram. The photometric accuracy is considerably reduced, particularly for strong absorption bands

---

* To whom correspondence should be addressed

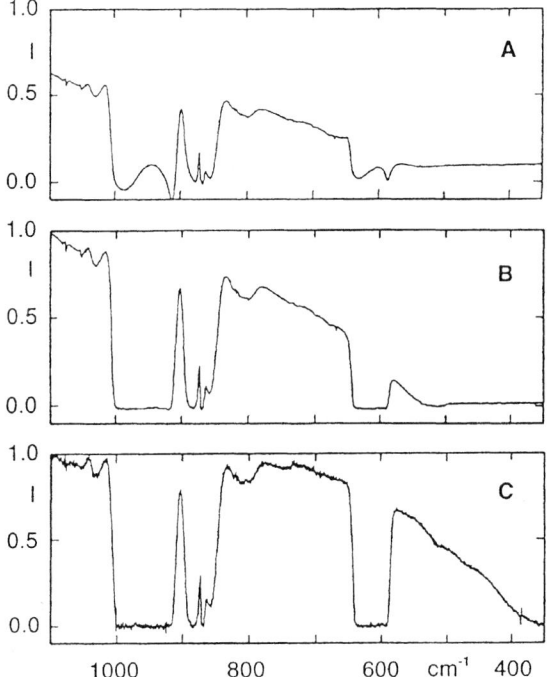

Fig. 1. Effects of non-linearity correction of the MCT detector response. **A** single beam spectrum of SF$_6$ recorded at 1005 hPa in a 10 cm cell; **B** same as A, calculated with correction of the non-linear response; **C** single beam spectrum recorded by using a DTGS detector and the same sample

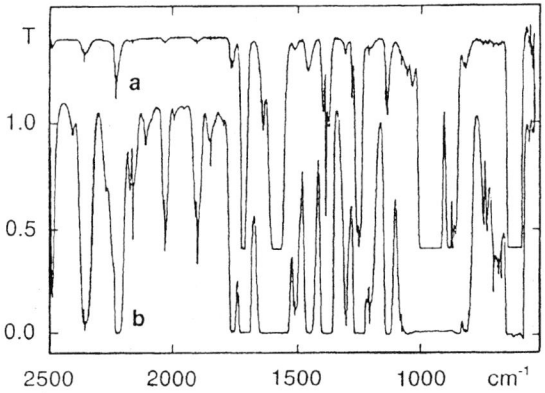

Fig. 2. Transmittance spectra of SF$_6$ at different path-lengths [offset: 10 cm (*a*) and 4.8 m (*b*)] recorded with an MCT detector and application of non-linear response correction (pressure 1005 hPa, temperature 25 °C)

Table 1. Opaque spectral ranges (in cm^{-1}) not available for the analysis of contaminants in SF$_6$. The spectral ranges are obtained by using the quotient of two consecutive single beam spectra of SF$_6$ samples recorded at 1005 hPa (100%-line)

Cell path-length		
4.8 m	3.0 m	0.1 m
2236–2209	2229–2220	–
1777–1680	1729–1688	1718–1707
1664–1531	1646–1539	1602–1563
1472–1438	–	–
1411–1355	1405–1365	—
1282–1230	1276–1233	–
1154–1117	1142–1123	–
1062–805	1058–835	1003–870
652–576	649–578	643–585

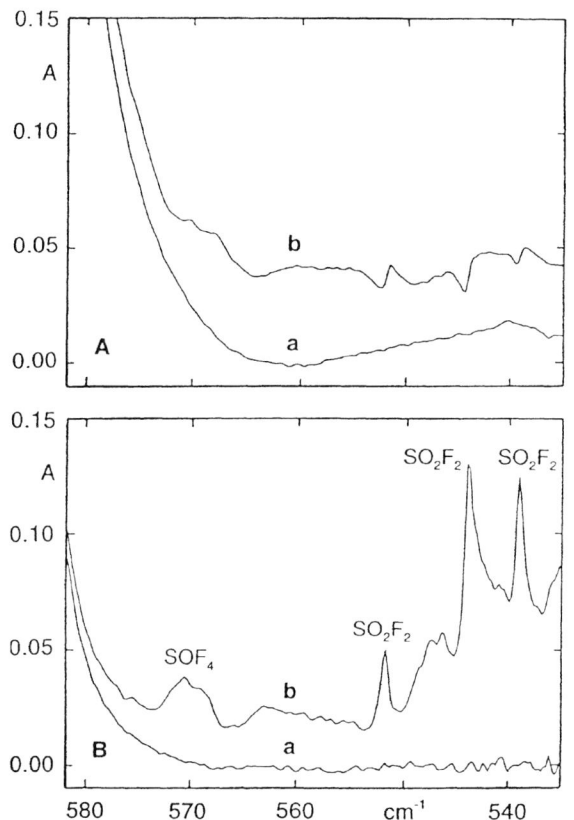

Fig. 3. Absorbance spectra of pure SF$_6$ (*a*) and SF$_6$ stressed by partial discharges (*b*); path-length 10 cm, pressure 63.3 kPa: (**A**) without correction of non-linear detector response and (**B**) with correction applied

after direct Fourier transformation and phase correction as implemented for routine spectroscopy (see Fig. 1). Negative spectral intensities can result, and the single beam spectrum at wavenumbers below cut-off shows positive values instead of zero. We applied the

non-linearity correction to the recorded interferograms and achieved a satisfactory photometric accuracy for the wavenumber region above detector cut-off. The spectra are similar to those recorded with the DTGS detector, which is known to provide spectra which have high photometric accuracy [4], but show a significantly lower noise level. Multivariate analysis requires good photometric accuracy and well defined stable spectral intervals which are exploited for quantitative analysis. For several optical path-lengths SF_6 gives opaque spectral windows which cannot be exploited for contaminant gas analysis (see Fig. 2 and Table 1). The absorbance of strong SF_6 bands can be shown to be a linear function of optical path-length (e.g. at around $820\,cm^{-1}$), so scaled absorbance subtraction can be considered for stripping the matrix absorption features.

Correction of the detector non-linearity enabled us to consider the smaller absorption bands of SO_2F_2 and SOF_4 at just above $530\,cm^{-1}$. Without correction the band shapes are seriously affected by phase errors, as demonstrated in Fig. 3. This spectral range is important for the simultaneous determination of several SF_6 decomposition products, such as SO_2F_2, SOF_4, SOF_2, SF_4 and S_2S_{10}.

Conclusions

The application of a liquid-nitrogen cooled MCT detector, with correction of its non-linearity response by computer software, enabled us to reach high signal/ noise ratios combined with satisfactory photometric accuracy, which is necessary for multivariate quantitative analysis of spectra from electrically stressed gaseous SF_6.

Acknowledgements. Financial support by the Ministerium für Wissenschaft und Forschung des Landes Nordrhein-Westfalen and the Bundesminister für Bildung und Forschung is gratefully acknowledged. One author (P.R.J) acknowledges the travel support granted by the Volkswagen-Stiftung.

References

[1] M. C. Abrams, G. C. Toon, R. A. Schindler, *Appl. Optics* **1994**, *33*, 6307.
[2] A. Keens, A. Simon, *Proc. SPIE* **1994**, *2089*, 222.
[3] *US Patent* No. 4,927,269 (22 May), **1990**.
[4] D. A. C. Compton, J. Drab, H. S. Barr, *Appl. Optics* **1990**, *29*, 2908.

Mikrochim. Acta [Suppl.] 14, 305–306 (1997)

Quantitative Infrared and Sub-millimetre Spectroscopy of Dielectric Single Crystals, Ceramics and Thin Films

Jan Petzelt*, Vladimir Železný, Stanislav Kamba, Ilja Fedorov, and **Filip Kadlec**

Institute of Physics, Czech Academy of Sciences, Na Slovance 2, 180, 40 Praha 8, Czech Republic

Abstract. Results recently obtained from IR reflectivity and far-IR transmission measurements of various ferroelectric and related crystals, bulk ceramics and thin films are summarized. In almost all cases the complex dielectric function was evaluated and compared with lower-frequency results. Transmission measurements provide a more accurate way than reflectivity for evaluating the dielectric response.

Key words: infrared reflectivity, far-infrared transmission, dielectric response, ferroelectrics, microwave ceramics.

We present a survey of quantitative IR results in the 15–3000 cm^{-1} range obtained on various solid samples within the last few years with a Bruker IFS 113v Fourier-transform spectrometer. Our interest was focused especially on low-lying polar phonon modes in ferroelectrics and related complex dielectrics, and on sub-millimetre absorption in transparent materials for microwave applications.

Experimental

In the range of strong one-phonon absorption the bulk samples were investigated by using specular normal reflection from a carefully polished flat surface. In the case of anisotropic samples, the light polarized along the principal directions of the dielectric permittivity tensor $\varepsilon_{ij}(\omega)$ was used to evaluate the spectra of each component of this tensor separately. The measured reflectivity spectra were evaluated by using various model fitting procedures (sum of classical oscillators, coupled oscillators, product of generalized 4-parameter oscillators) or Kramers–Kronig analysis, depending on the form of the spectra and the type of problem investigated.

In the range of weak absorption well below the strong polar mode eigenfrequencies, the normal transmission through a plane-parallel slab (thickness typically 0.1–1 mm) was used. The transmission spectra were evaluated either by simple model fitting simultaneously with higher-frequency reflectivity or by calculation of the complex dielectric function, independently of any model, from the interference pattern due to the multiple passage of light through the sample (possible if the absorption peaks in the spectra are much broader than the interference period). In the case of thin films the optical response was evaluated by using both normal transmission and reflection from a system of (semi-transparent) plane-parallel substrate plus film (typically 0.1–1 μm thick). The response of each layer was modelled by a sum of damped oscillators, where the fitting parameters of the substrate were determined from independent measurements of the free substrate. Our experience shows that the transmission measurements (if possible) are more accurate than the reflection measurements. In the case of polycrystalline samples (bulk ceramics, films) with non-cubic structure (anisotropic optical response) some caution is needed to evaluate the polar phonon frequencies from the measured optical response because the effective medium response shifts the measured frequencies slightly upwards from the transverse eigenfrequencies towards the corresponding longitudinal optical eigenfrequencies.

Main Results

Bulk Ferroelectrics

For the first known ferroelectric, Rochelle salt $NaKC_4H_4O_6 \cdot 4H_2O$ (RS), by combining the known microwave and millimetre-wave data with our IR and Raman data [1] we were able to discuss quantitatively for the first time the overall picture of the soft-mode dynamics. It was shown that the driving mechanism is due to an optic phonon near 75 cm^{-1}, partial softening of which causes the critical slowing down of the microwave relaxation observed earlier, through its linear coupling with another optical mode near 20 cm^{-1} and hard relaxation near 13 cm^{-1}. The phase transitions in RS can therefore be treated as a mixture of displacive and order–disorder type.

For the weak (pseudoproper) ferroelectric $Li_2Ge_7O_{15}$ the theory predicts a change of the sign of the soft-mode effective charge near 234 K ($T_c = 284$ K) [2]. We have confirmed this behaviour by using transmission IR spectroscopy on thick samples (1–4 mm)

* To whom correspondence should be addressed

[3]. This seems to be the first time that a disappearence of an IR mode was observed in a limited temperature range without any change of symmetry in the crystal lattice.

Betaine calcium chloride dihydrate $(CH_3)_3NCH_2 COO \cdot CaCl_2 \cdot 2H_2O$ (BCCD) exhibits below 164 K a unique sequence of at least 18 phase transitions into various commensurate and incommensurate phases (so-called incomplete devil's staircase) locking in a simple ferroelectric phase below ~ 46 K. Our IR studies together with Raman data combined with specific selection rules for all the phases enabled us to predict the dispersion of the 8 lowest lying phonon branches in both BCCD [4] and its deuterated analogue [5], confirmed by a direct inelastic neutron scattering experiment [5].

Dipolar Glasses

We have studied the IR dielectric response of the most popular dipolar glass system, mixed ferroelectric-antiferroelectric crystals of the $Rb_{1-x}(NH_4)_xH_2PO_4$ (RADP-100x) type and its fully deuterated analogue (DRADP-100x) for $x = 0.5$ and 0.25 [6–8]. The results are discussed together with the lower-frequency dielectric spectroscopy data in terms of a soft-mode freezing obeying overall Vogel–Fulcher behaviour between 10^{12} and 10^6 Hz with the freezing Vogel–Fulcher temperature of 10 and 34K for RAPD-50 and DRADP-50 respectively.

Ceramics

In relaxor ferroelectric ceramics the IR reflectivity can be used to detect the local order in the nanometer range. This was proven for several $A(B'B'')O_3$ complex perovskite ceramics with different degrees of B-site ordering, such as $Pb(Sc_{1/2}Ta_{1/2})O_3$, $Pb(Mg_{1/3}Nb_{2/3})O_3$, $Pb(Fe_{1/2}Nb_{1/2})O_3$, $Pb(In_{1/2}Nb_{1/2})O_3$ [9] and the well known transparent ceramic PLZT [10].

In the case of the relatively transparent microwave ceramics used for microwave resonators, a correlation was established between the microwave complex permittivity, sub-mm transmission and IR reflectivity for more than 80 different compounds with permittivity ε' varying between 10 and 100 [11, 12]. Sub-mm transmission data give more accurate data and the extrapolation to the microwave range is more reliable than that from IR reflectivity.

Ferroelectric Thin Films

We have accomplished the first IR soft-mode study of ferroelectric thin films [13]. For $PbTiO_3$ (PT) and $PbZr_{1-x}Ti_xO_3$ ($x = 0.25, 0.47$; PZT-25, PZT-47) films on a sapphire substrate transmission spectroscopy enables more accurate investigation of the soft-mode behaviour to be performed than does reflectivity spectroscopy on bulk crystals. Transmission data below the soft-mode frequency as well as comparison with lower-frequency dielectric data provide evidence for additional dispersion below the soft phonon mode frequency in our thin films.

Acknowledgement. This work was supported by the Czech Grant Agency (projects No. 202/93/0691, 202/95/1393 and A 1010507) and the Alexander von Humboldt Foundation.

References

[1] S. Kamba, G. Schaack, J. Petzelt, *Phys. Rev. B* **1995**, *51*, 14998.
[2] A. K. Tagantsev, *Ferroelectrics.* **1988**, *79*, 57.
[3] F. Kadlec, J. Petzelt, V. Železný, A. A. Volkov, *Solid State Commun.* **1995**, *94*, 725.
[4] S. Kamba, V. Dvořák, J. Petzelt, Yu. G. Goncharov, A. A. Volkov, G. V. Kozlov, *J. Phys., Condens. Matter* **1993**, *5*, 4401.
[5] S. Kamba, J. Petzelt, V. Železný, F. Smutný, V. Dvořák, J. Hlinka, M. Quilichini, A. A. Volkov, B. P. Gorshunov, G. V. Kozlov, R. Currat, J. F. Legrand, *Ferroelectrics* **1994**, *159*, 97.
[6] J. Petzelt, V. Železný, S. Kamba, A. V. Sinitski, S. P. Lebedev, J. A. Volkov, G. V. Kozlov, V. H. Schmidt, *J. Phys. Condens. Matter* **1991**, *3*, 2021.
[7] J. Petzelt, S. Kamba, A. V. Sinitski, A. G. Pimenov, A. A. Volkov, G. V. Kozlov, R. Kind, *J. Phys. Condens. Matter* **1993**, *5*, 3573.
[8] S. Kamba, J. Petzelt, V. Železný, F. Smutný, B. P. Gorshunov, G. V. Kozlov, V. V. Voitsekhovskii, A. A. Volkov, *Ferroelectrics* **1994**, *157*, 227.
[9] I. M. Reaney, J. Petzelt, V. V. Voitsekhovskii, F. Chu, N. Setter, *J. Appl. Phys.* **1994**, *76*, 2086.
[10] J. Pokorný, J. Petzelt, V. Železný, I. Gregora, V. Vorlíček, A. Pronin, *Proc. EMF 8*, 1995, *Ferroelectrics* in press.
[11] J. Petzelt, N. Setter, *Ferroelectrics* **1993**, *150*, 89.
[12] J. Petzelt, S. Kamba, G. V. Kozlov, A. A. Volkov, *Ferroelectrics* **1996**, *176*, 145.
[13] I. Fedorov, J. Petzelt, V. Železný, G. A. Komandin, A. A. Volkov, K. Brooks, Y. Huang, N. Setter, *J. Phys. Condens. Matter* **1995**, 7. 4313.

Mikrochim. Acta [Suppl.] 14, 307–309 (1997)

Systematic Errors of FT-IR Transmission Spectra

Angelika Reklat*, **Waltraud Bessau**, and **Anka Kohl**

Bundesanstalt für Materialforschung und -prüfung, Rudower Chaussee 5, D-12489 Berlin-Adlershof, Federal Republic of Germany

Abstract. The effects of systematic errors in data processing and in connection with the selection of the resolution value should be tested by changing the apodization functions, the phase correction procedure, and the extent of zero-filling. FT-IR spectra were recorded at a resolution of 1, 2, 4, and 8 cm^{-1} and calculated with different apodization functions (Boxcar, Fourpoint, Norton–Beer, Happ–Genzel, Blackman–Harris, Triangular). The influence of the number of data points and the phase correction procedure (Mertz method for single-sided interferograms and Power method for double-sided interferograms) on the band parameters is determined. We exhibit examples of FT-IR bands measured by using different spectrometers, different beam-splitters, and different detectors.

Key words: FT-IR spectroscopy, quantitative determination, systematic errors, resolution, data processing.

Quantitative determination of infrared band parameters, such as absorbance, band areas and half-widths, by using the FT-IR techniques requires the selection of experimental conditions which make possible the measurement of sufficiently accurate spectra. Systematic errors concerning optical effects within the sample cell or the sample layer can be reduced or eliminated. Random errors in FT-IR spectroscopy are small. Systematic errors in data processing and in connection with the selection of the resolution value are often reproducible.

A number of authors [1–4] have discussed the effects of some procedures on spectra, but not all factors were discussed in this connection. In this work, a systematic study of the influence of experimental conditions and mathematical treatments on band parameters of practical samples has been performed.

Experimental

The FT-IR spectra were recorded with Bruker FT-IR spectrometer models IFS66 and IFS66v with an external sample box.

The 2νCH band was recorded CHCl$_3$ in a quartz sample cell (thickness 0.5 mm) at constant position in the sample chamber during all measurements, at different resolution values and with all apodization functions. The interferograms were measured single-sided and phase-corrected by the Mertz method. The following parameters were obtained under correct conditions (2 cm^{-1} resolution and boxcar apodization).

Wavenumber position 5910 cm^{-1}; FWHH 31.25 cm^{-1}; area 33.40; peak absorbance above baseline 0.808.

The 2νCH band of CHCl$_3$ was measured with two different spectrometers, different beam-splitters (quartz, CaF$_2$, KBr) and different detectors (Ge-diode, InSb, DTGS, MCT), a CaF$_2$ sample cell ($n = 1$, 4) with thickness of 0.414 mm, at a resolution of 2 cm^{-1} and with boxcar apodization. The dyestuff (benzothiazolchalkon-O,N-crown) band was measured in a KBr pellet. Interferograms were recorded at different resolution values (1, 2 and 4 cm^{-1}). At first the spectra were calculated with boxcar and Norton–Beer medium apodization. After that, they were calculated again after zero-filling at factors 2, 4 and 8. The quartz band was measured in a KBr pellet at 2 cm^{-1} resolution and with Norton–Beer medium apodization. Spectra were calculated from double-sided interferograms by the Power method and from single-sided interferograms by the Mertz method, using different numbers of phase interferogram points (111, 222, 444, 888 and 1777). All spectra were obtained with the same KBr pellet. The precision of all these results was better than 0.4%.

Results and Discussion

The results of experimental measurements are shown in Fig. 1 and Tables 1–3. For correct measurement of a band, the resolution must be better than 10% of the half-width of the band; spectra can then be measured with boxcar apodization [1]. In the case of correct selection of resolution, the results exhibit only small deviations if such apodization functions as Fourpoint

---

* To whom correspondence should be addressed

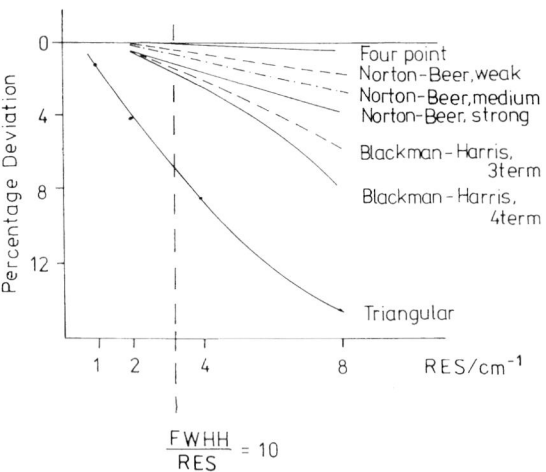

Fig. 1. Deviation of the peak absorbance obtained with different apodization functions and resolution values

Table 1. The influence of resolution, apodization function and zero-filling on the band parameters of the 1183 cm^{-1} band of a dyestuff

	Absorbance	Area	FWHH
RES = 1 Boxcar	0.482	4.674	7.7

	Deviation by use of other conditions and zero-filling to an FT size of 64K data points		
RES = 2 Boxcar	−0.4%	−0.5%	0
RES = 4 Boxcar	−1.5%	−1.0%	+1.3%
RES = 1 Norton–Beer, medium	−1.0%	−0.4%	+1.3%
RES = 2 Norton–Beer, medium	−3.4%	−1.3%	+3.9%
RES = 4 Norton–Beer, medium	−2.9%	−4.3%	+15.4%

Table 2. The effect of the number of phase interferogram points and the phase correction mode on the absorbance of a quartz band

Phase resolution	Phase interferogram points	Absorbance of the 779 cm^{-1} band	
		Power double-sided	Mertz single-sided
512	111	0.545	0.516
256	222	0.545	0.530
128	444	0.545	0.542
64	888	0.545	0.545
32	1777	0.545	0.545

and Norton–Beer are used. The apodization function which yields the best results is boxcar. For measurements made at fairly high resolution, only the triangular apodization function gives poor results (Fig. 1).

The zero-filling interpolation does not increase the spectral information. Table 1 gives an example for absorbance at a resolution value of 4 cm^{-1}. For spectra recorded with different resolution values (corresponding to 16K, 32K and 64K data points) and boxcar apodization, the results obtained showed less than 1.5% variation of the band parameters. For the same measurement of the interferogram and use of the Norton–Beer apodization, results with deviations of up to several per cent in the band parameters were obtained, Table 1.

Systematic errors of up to several per cent in the transmittance spectrum can be obtained if the Mertz method is used to calculate the spectra (Table 2). A minimization of the phase error is possible if several hundred data points are used in the phase calculation [4, 5]. The acquisition of double-sided interferograms and the subsequent calculation of the spectra by the Power method is the best method for superior reproduction of the spectra. However, if noise is present, noise contributions computed from the Power method

Table 3. The influence of different FT-IR spectrometers, beam-splitters, and detectors on the band parameters of the 2 νCH band of CHCl$_3$ at 5910 cm^{-1}

Spectrometer	Beam-splitter	Detector	FWHH	Absorbance	Area
IFS66	quartz	Ge-diode	31.3	0.665	27.92
IFS66	KBr	DTGS	31.5	0.663	27.87
IFS66v	KBr	DTGS	31.4	0.658	27.51
IFS66v	CaF$_2$	InSb	31.3	0.660	27.23
IFS66v/external sample box	KBr	MCT	32.1	0.638	28.19

are always positive and in the worst case, larger by a factor of about $\sqrt{2}$ than the noise amplitudes computed from the Mertz method.

Spectra recorded with the two different spectrometers IFS66 and IFS66v, different beam-splitters (quartz, CaF_2, KBr) and different detectors (Ge-diode, InSb, DTGS) show a very small deviation of the band parameters – less than 1%, Table 3.

References

[1] P. R. Griffiths, J. A. de Haseth, *Fourier Transform Infrared Spectrometry*, Wiley, New York, 1986.

[2] J. L. Domenech, M. V. Garcia, M. A. Raso, *J. Mol. Struct.* **1986**, *142*, 213.

[3] P. B. Tooke, S. F. Parker, *Proc. SPIE* **1989**, *1145*, 629.

[4] D. B. Chase, *Appl. Spectrosc.* **1982**, *36*, 240.

[5] Bruker Analytische Meßtechnik GmbH, *Handbook of OPUS/IR, Version 2.0*, 1994.

Mikrochim. Acta [Suppl.] 14, 311–313 (1997)

Propagation of Calibration Errors in FT-IR Emission Spectroscopy of Gases

V. Tank

Deutsche Forschungsanstalt für Luft- und Raumfahrt (DLR), D-82234 Wessling, Federal Republic of Germany

Abstract. Radiometric instrument calibration in FT-IR emission spectroscopy delivers absolute (radiance) spectra of the object of measurement. From these spectra, quantitative information such as densities of gas columns can be derived, usually through back-calculation. This may result in misleadingly low parameter errors because of a "perfect fit". Thus propagating errors introduced by the calibration procedure, imperfect calibration standards and conditions, are masked. Simulations of calibration, measurement and data evaluation based on realistical parameters have been performed. They enable tracking of error propagation. Different calibration procedures and source properties are included in the simulation. This aids in understanding the process and gives quantitative information on final errors. Means to reduce the errors are deduced.

Key words: FT-IR, emission spectroscopy, calibration, blackbody, emissivity, responsivity.

The environment-related task of this work is the determination of gas column densities from the spectra of gas emissions from smoke stacks, aircraft engines and the like, as has been published by several authors [1–3]. There is, however a lack of information on the calibration quality of such investigations. Simulations have been performed for three different calibration procedures, different calibration source emissivities, temperatures and temperature errors. The gases investigated are SO_2, CO, and HF at two different temperatures.

Calibration Procedures

When emission spectroscopy is used as a remote-sensing technique the measured spectrum can be described by:

$$U_D(\sigma) = R(\sigma)[\tau(\sigma)L_O(\sigma) + L_I(\sigma)] \qquad (1)$$

where σ is the wavenumber, $U_D(\sigma)$ is the measured "raw" spectrum, $\tau(\sigma)$ is the spectral transmission of the radiation transfer medium, $R(\sigma)$ is the spectral respon-

sivity of the spectrometer, $L_O(\sigma)$ is the spectral radiance of the emitter, and $L_I(\sigma)$ is the spectral radiance of the spectrometer's inner surfaces. To be determined are $R(\sigma)$ and $L_I(\sigma)$ through calibration with a blackbody. The blackbody radiance $L_P(\sigma)$ follows Planck's law and is given by its temperature T_P. It replaces $L_O(\sigma)$ in Eq.(1). Blackbody temperature errors cause erroneous radiance $L_P^{\bullet}(\sigma)$ and instrument data $R^{\bullet}(\sigma)$ and $L_I^{\bullet}(\sigma)$.

Three different calibration procedures are in use. Single-source calibration considers $L_I(\sigma)$ to be negligible. A single blackbody measurement at a given temperature is performed, yielding $R^{\bullet}(\sigma)$ [4]. Two-source calibration uses measurements at two different blackbody temperatures; from the resulting two spectra $R(\sigma)$ and $L_I(\sigma)$ are calculated [5, 6]. Three-source calibration is based on measurement at three different blackbody temperatures, leading to three spectra, from which $R(\sigma)$, $L_I(\sigma)$, and also the three temperatures $T_P(1, 2, 3)$ are determined. This reduces the error introduced by limited blackbody temperature-measurement accuracy [7].

Calibration Sources

Usually the spectrometer employs an extended aperture, such as a mirror telescope. Consequently, for the calibration "extended area blackbodies" are used, which do not easily permit a very high spectral emissivity $\varepsilon(\sigma)$. The spectral emissivity depends on the surface structure and coating. Commonly used are simple flat plates, plates with machined parallel v-grooves or pyramid structures. To further enhance $\varepsilon(\sigma)$ high emissivity coatings are applied to the surfaces [8, 9]. The effective spectral emissivity can be calculated as indicated in [10].

For this work an $\varepsilon(\sigma)$ resembling that of 3 M velvet black varnish has been chosen [9]. Since $\varepsilon(\sigma) < 1$, then besides the reduced emitted radiance the reflected ambient radiation $L_A(\sigma)$ must also be taken into account. Most often $\varepsilon(\sigma)$ and $L_A(\sigma)$ are not well known. Therefore in the calibration procedure $\varepsilon(\sigma)$ is set equal to 1 and erroneous instrument data $R^{\bullet}(\sigma)$ and $L_I^{\bullet}(\sigma)$ result:

$$R^{\bullet}(\sigma) = \varepsilon(\sigma) R(\sigma); \quad L_I^{\bullet}(\sigma) = \frac{L_A(\sigma) + L_I(\sigma)}{\varepsilon(\sigma)} - L_A(\sigma) \quad (2)$$

Simulations

Gas radiance spectra $L_G(\sigma)$ (free from instrument noise) were calculated by using the Fascod2 model [11] for: SO_2 with $N = 10^{18}$ mol/cm^2 from 1100 to 1200 cm^{-1}, CO with $N = 4 \times 10^{17}$ mol/cm^2 from 2050 to 2250 cm^{-1}, and HF with $N = 10^{17}$ mol/cm^2 from 3600 to 4000 cm^{-1} (N is the "gas column density" used in quantitative spectroscopy); gas temperatures $T_G = 373$ and 573 K (industrial stack and aircraft jet engine exhaust respectively); gas pressure 950 mbar; background $T_B = 298$ K blackbody radiance $L_B(\sigma)$; instrument's inner radiance $L_I(\sigma)$ blackbody at $T_I = 288.15$ K; highest allowable temperature for 3M varnish $T_P = 473$ K; calibration ambient temperature $T_A = 293.15$ K; spectral resolution $\Delta\sigma = 0.5$ cm^{-1}.

The emitter radiance shown in Fig. 1 for the 573 K gas is given by:

$$L_O(\sigma) = L_G(\sigma) + L_B(\sigma) \quad (3)$$

The erroneous object radiance is calculated according to:

$$L_O^{\bullet}(\sigma) = \frac{R(\sigma)[L_O(\sigma) + L_I(\sigma)]}{R_{(\sigma)}^{\bullet}} - L_I^{\bullet}(\sigma) \quad (4)$$

From these spectra the inversion algorithm calculates the erroneous column density $N^{\bullet}$, gas temperature $T_G^{\bullet}$, and background temperature $T_B^{\bullet}$; the gas pressure is considered to be known *a priori*.

Results and Discussion

Extensive simulations have been performed. The very few examples that can be presented here have been limited to the two-source calibration as the most widely used. They are also valid for the three-source calibration, except for the calibration temperature errors. The three-source calibration yields the temperature itself, hence erroneous temperatures cannot be introduced into the simulation procedure. The errors for the single-source calibration are in general higher.

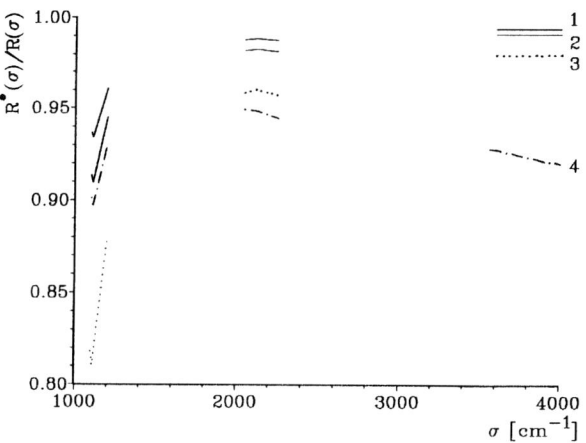

Fig. 2. Two/three-source calibration, ratio of erroneous to true responsivity. Surface structure: 1: v-grooves 50°; 2: v-grooves 60°; 3: flat; 4: v-grooves 60°, additionally temperature error of +3 K (appropriate only for two-source calibration)

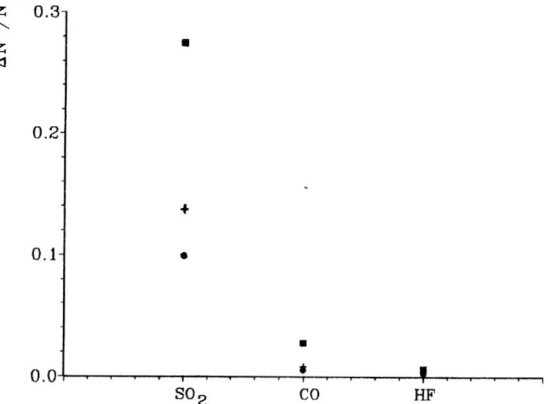

Fig. 3. Two/three-source calibration, relative error of colunm density. Calibration temperatures 473 and 373 K; gas temperature 373 K; surface structure: dots: v-grooves 50°; crosses: v-grooves 60°; squares: flat

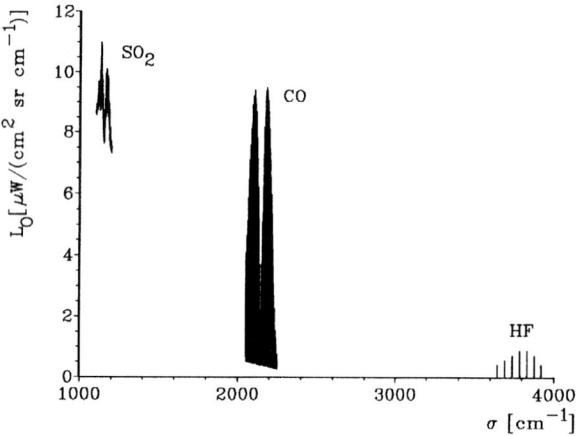

Fig. 1. Gas emission spectra, gas temperature $T = 573$ K

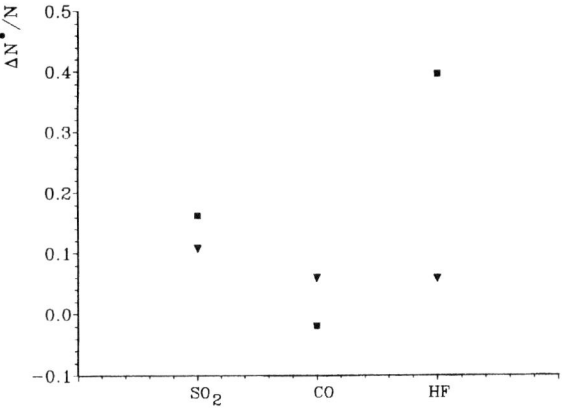

Fig. 4. Two-source calibration, relative error of colunm density. Calibration temperatures: 473 and 373 K; calibration temperature error: + 3 K; surface structure: v-grooves 60°; gas temperature: triangles: 573 K, squares: 373 K

For the case of setting the calibration source's spectral emissivity as $\varepsilon(\sigma) = 1$, in the absence of temperature errors the resulting ratio of erroneous to true spectral responsivity $R^{\bullet}(\sigma)/R(\sigma)$ is given in Fig. 2. For the two- and three-source calibrations (Fig. 2) these curves (1, 2, 3) resemble the blackbody spectral emissivity according to Eq. (2). Errors occurring in temperature measurement by the two-source calibration lead to additional deviations (curve 4). The relative error in gas column densities resulting when the calibration source's emissivity is set to $\varepsilon(\sigma) = 1$ is given in Fig. 3. As can be expected from Fig. 2 it decreases with increasing wavenumber. It becomes as high as nearly 30%. Finally the relative error in gas column densities resulting from additional calibrational temperature errors is given in Fig. 4. The calibration temperature error of 3 K is about 0.63% and 0.8% of the true temperatures (473 and 373 K, respectively). For most cases consider-

ed the column density error was in the region 6–18% and could become as high as nearly 40%.

Conclusions

Significant masked errors in the final results of radiance emission spectroscopy may be caused by calibration uncertainities. They can be reduced by high spectral-emissivity calibration-source design and low-error calibration-source temperature determination. It is recommended to use calibration sources based on honeycomb structures resembling a field of cavity sources. Ambient radiation can be shielded by a highly reflective housing. For attaining low temperature errors the three-source calibration should be applied.

References

[1] W. L. Smith, H. M. Woolf, H. E. Revercomb, *Appl. Optics* **1991**, *30*, 1117.

[2] H. Fischer, *NATO ASI Series* **1993**, *19*, 341.

[3] E. Lindermeir, *Ann. Geophysicae* **1994**, *12*, 417.

[4] R. Haus, K. Schäfer, W. Bautzer, J. Heland, H. Mosebach, H. Bittner, T. Eisenmann, *Appl. Optics* **1994**, *33*, 5682.

[5] B. J. Vastag, S. R. Horman, *Proc. SPIE* **1981**, *289*, 74.

[6] H. Oelhaf, Th.v. Clarmann, F. Fergg, H. Fischer, F. Friedl-Vallon, Ch. Fritzsche, Ch. Piesch, D. Rabus, M. Seefeldner, W. Völker, *Proc. 10th ESA ESA-Symposium on European Rocket and Ballon Programmes*, 1991.

[7] E. Lindermeir, P. Haschberger, V. Tank, H. Dietl, *Appl. Optics* **1992**, *31*, 4527.

[8] A. Leupin, H. Vetsch, F. K. Kneubühl, *Infrared Phys.* **1990**, *30*, 199.

[9] D. Dreßler, *Velvet Coating 2010 schwarz*, 3M Deutschland GmbH, private communication.

[10] G. J. Zissis (ed.), *The Infrared and Electro-Optical Systems Handbook*, Vol. 1, *Sources of Radiation ERIM*, Ann Arbor, M, SPIE, Bellingham, Wash., 1993.

[11] S. A. Clough, F. X. Kneizys, E. P. Shettle, G. P. Anderson, *Atmospheric Radiance and Transmittance* in FASCOD2, *Proc. 6th Conference on Atmospheric Radiation*, Williamsburg, 1986, 141.

Mikrochim. Acta [Suppl.] 14, 315–316 (1997)
© Springer-Verlag 1997

A Procedure for Testing the Radiometric Accuracy of Fourier-transform Infrared Spectrometers

Zhuomin M. Zhang[1,*,**], **Leonard M. Hanssen**[1], **Jack J. Hsia**[1], **Raju U. Datla**[1], **Changjiang Zhu**[2], and **Peter R. Griffiths**[2]

[1] National Institute of Standards and Technology, Gaithersburg, Maryland 20899 U.S.A.
[2] Department of Chemistry, University of Idaho, Moscow, Idaho 83844, U.S.A.

Abstract. We present a simple procedure for evaluation of FT-IR radiometric accuracy, which incorporates the use of calibrated standard reference materials. It starts with the standard practice for testing the FT-IR performance recommended by ASTM, as well as using a wavelength/wavenumber standard reference material developed at the National Institute and Standards and Technology. Further steps are proposed to test the noise level, linearity, and radiometric accuracy. A set of neutral-density filters is being developed for testing the radiometric accuracy. The effects of instrumental resolution and apodization function on the radiometric accuracy for absorption band measurements are also addressed.

Key words: Fourier-transform spectrometer, radiometric accuracy, transmittance.

In recent years, there has been an increase in the use of Fourier-transform infrared (FT-IR) spectrometers in quantitative measurements. There are potentially many sources of errors in the measurements [1, 2]. A standard practice for testing the FT-IR performance was developed by ASTM [3]. The main purpose of the tests (level zero and level one) was to check the instrumental stability and repeatability. Recently, a standard reference material (SRM) was certified by the National Institute and Standards and Technology (NIST) [4]. It is made of matte-finish polystyrene film and can be used for wavelength/wavenumber calibration. It is important, however, to develop a standard practice method and standard reference materials for evaluating the radiometric accuracy of FT-IR spectrometers. As an initial step, we propose a simple procedure for testing the FT-IR radiometric accuracy. A step-by-step approach is elaborated as follows.

---

* Present address: Department of Mechanical Engineering, University of Florida, Gainesville, FL 32611, U.S.A.
** To whom correspondence should be addressed

Procedure

Step 1. Follow ASTM standard practice [3]. The absorption lines of NIST polystyrene standard [4] (or similar standard developed at other national laboratories) or an appropriate gas should be used to check the wavelength/wavenumber accuracy, since the errors in the wavelength scale can affect the radiometric scale.

The stability and noise level must be tested (by taking 100% lines) every time the major configuration parameters (such as the aperture size and resolution) are changed. For example, a strong time variation of the spectral response has been observed using a pyroelectric detector when the aperture was changed [5].

Step 2. Obtain a zero-signal spectrum by blocking the beam between the interferometer and the detector. The ratio of the zero-signal spectrum to the reference spectrum (0%) should indicate whether there is any scattered light reaching the detector or whether there is any electronic pick-up noise. Different gain factors of the preamplifier can be tested. This is important for low-level transmittance measurements, since FT-IR spectrometers have been used for quantitative measurements of infrared filters with high attenuation [6, 7].

Step 3. Compare with broadband reference materials. Crystalline or amorphous samples (typically 0.5–2 mm thick) can be used to test the radiometric accuracy as well as non-linearity. For example, Si and Ge have near 50% transmittance in the region from 2 to 25 µm except for some absorption bands (for high-quality samples, the refractive index data are available in handbooks); CaF_2, MgO, and $SrTiO_3$ transmit below 9–10 µm; glass, quartz, and sapphire transmit below

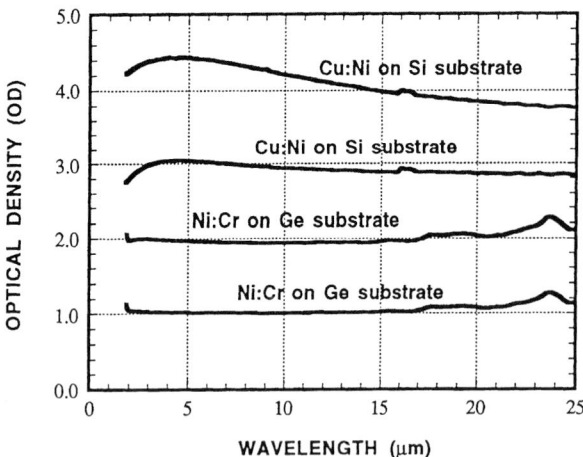

Fig. 1. Optical density of a set of neutral-density filters

3–5 μm. It is often easier to observe the non-physical energy caused by detector non-linearity by using the samples that are opaque for longer wavelengths.

Researchers at NIST are developing infrared transmittance SRMs. These are neutral-density filters having optical density (defined as $OD = -\log_{10} T$, where T is the transmittance) near 1, 2, 3, and 4 in the mid-infrared. The optical density of four filters at wavelengths between 2 and 25 μm is shown in Fig. 1. These filters are made by vacuum depositing alloy films onto dielectric substrates. The FT-IR spectrometer used to characterize these filters is calibrated against laser measurements at 10.6 μm, by using specially designed ultrathin filters [8].

Inter-reflection between the sample and the interferometer can cause errors in the transmittance measurements. This problem can be avoided by tilting the sample by up to 10° from the optical axis. At the same time, beam deviation or deflection by the sample can cause measurement error if the detector's receiving element is small or non-uniform. Decreasing the substrate thickness reduces beam deviation. The resolution should be lower than the free spectral range (the spacing between interference extrema) to average out the interference fringes [8]. For a 0.5 mm thick silicon substrate, the free spectral range is near $3\,\mathrm{cm}^{-1}$.

Step 4. Check the effect of incident power level. Detector non-linearity or non-equivalent responsivity [5] may cause significant measurement error if the incident power is too strong. This can be checked by comparing the measurements of reference samples as discussed in Step 3. The power level can be changed by changing the aperture size and/or using attenuators (metal-mesh or neutral-density filters). For low-signal measurements, however, it is suggested to use the neutral-density filters as the reference to improve the signal-to-noise ratio [7].

Step 5. Check resolution and apodization effects. As discussed by Anderson and Griffiths [9], the resolution and apodization function can affect the radiometric accuracy for absorption band structures. Zhu and Griffiths [10] developed a technique to test the radiometric accuracy of FT-IR measurements of Lorentzian bands. Several solute bands in the spectra of solutions of cyclohexane and benzonitrile are Lorentzian and have minimal intermolecular interactions leading to deviations from Beer's law. When the measured peak absorbance of these bands is plotted against the true peak absorbance, the closeness to which the experimental plots follow the theoretical behaviour allows the radiometric accuracy of the instrument to be evaluated. The effect of resolution and apodization function can be varied for a given instrument to yield the optimized performance.

Conclusion

In summary, a test procedure is presented for quantitative measurements with FT-IR spectrometers. The use of a set of neutral-density filters as the standard reference materials to calibrate an FT-IR instrument is discussed.

Acknowledgements. The authors acknowledge financial support from the U.S. Army through the Calibration Coordination Group (CCG), and thank Michael Zander for insightful comments.

References

[1] T. Hirschfeld, in: *Fourier Transform Infrared Spectroscopy, Vol. 2* (J. R. Ferraro, L. J. Basile eds.), Academic Press, New York, 1979, pp. 193–242.
[2] J. R. Birch, F. J. J. Clarke, *Spectrosc. Europe* **1995**, in press.
[3] ASTM E 1421–91, *Annual Book of ASTM Standards, Vol. 14.01*, 1991.
[4] D. Gupta, L. Wang, L. M. Hanssen, J. J. Hsia, R. U. Datla, *NIST Spec. Publ.* 260–122, U.S. Government Printing Office, Washington D.C., 1995.
[5] M. I. Flik, Z. M. Zhang, *J. Quant. Spectrosc. Radiat. Transfer* **1992**, *47*, 293.
[6] D. A. C. Compton, J. Drab, H. S. Barr, *Appl. Opt.* **1990**, *29*, 2908.
[7] Z. M. Zhang, L. M. Hanssen, R. U. Datla, *Opt. Lett.* **1995**, *20*, 1077.
[8] A. Frenkel, Z. M. Zhang, *Opt. Lett.* **1994**, *19*, 1495.
[9] R. J. Anderson, P. R. Griffiths, *Anal. Chem.* **1975**, *47*, 2339.
[10] C. Zhu, P. R. Griffiths, 1995, unpublished.

Mikrochim Acta [Suppl.] 14, 317–319 (1997)
© Springer-Verlag 1997

A Computer Simulation of the Non-linearity Effect on FT-IR Measurements

Zhuomin M. Zhang*·** and **Leonard M. Hanssen**

National Institute of Standards and Technology, Gaithersburg, MD 20899, U.S.A.

Abstract. We have developed a computer simulation algorithm for quantitative examination of the errors induced in FT-IR measurements by non-linearity. Using this program, we have investigated the effect of non-linearity on the measurement of transmittance for several types of optical filters and materials, including neutral-density filters, bandpass filters, and samples with absorption bands.

Key words: FT-IR, detector non-linearity, quantitative measurements.

It is well known that detector non-linearity can result in significant radiometric error in the measurements made with Fourier-transform infrared (FT-IR) spectrometers [1]. In many cases, quantitative prediction of the error caused by non-linearity is required but cannot be obtained easily. The effect of non-linearity depends on the shape of the spectrum, and the resulting error is a function of wavelength. We have developed a computer simulation algorithm that allows determination of the error in FT-IR measurements that is due to detector non-linearity. We have investigated the effect of non-linearity on the transmittance of different spectral shapes, including neutral-density filters, bandpass filters, cut-off filters, and absorption bands.

Simulation Procedure

Non-linearity directly affects the interferogram measured by an FT-IR instrument, because the output voltage of the detector is not a linear function of the incident radiation power. In order to predict non-linearity errors in the measurements, the phase error and phase correction must also be considered. On the basis of the Fourier-transform process described in [2], we have developed a non-linearity simulation algorithm. The key steps are summarized as follows. (1) Take an arbitrary response spectrum, i.e., the reference spectrum, $V(v)$, where $v \in [0, v_m]$ is the wavenumber (frequency). (2) Make an even function: $V(v) = V(v)$ for $v < v_m$ and $V(v) = V(2v_m - v)$ for $v_m < v < 2v_m$. This is required for the complex fast Fourier-transform (CFFT). (2) Add a pre-stored phase error computed from an actual interferogram. (4) Perform a CFFT to get an interferogram, $V(t)$. (5) Offset the interferogram by adding the center-burst value (V_{max}) to each point, so that there will be no negative points. (6) Add first-order non-linearity by multiplying the interferogram $V(t)$ by $[1 - f \cdot V(t)]$, where f is the percentage of non-linearity. (7) Offset the interferogram so that the values away from the center-burst are around zero. (8) Perform an inverse CFFT. (9) Correct the phase error by the Mertz method (or Forman method) [2]. (10) Take half of the spectrum, which is the resulting reference spectrum that includes non-linearity, $V'(v)$. (11) Multiply a transmittance spectrum, $T(v)$, by the reference spectrum, $V(v)$. This yields the sample spectrum, $V_s(v)$. (12) Repeat steps 2–10 to get the sample spectrum with non-linearity, $V'_s(v)$. (13) The transmittance with non-linearity is $T'(v) = V'_s(v)/V'(v)$. The error in transmittance caused by non-linearity can thus be determined at different wavenumbers.

Results and Discussion

The predicted non-linearity effects on the transmittance of different spectral types are shown in Fig. 1. The solid lines indicate the original spectra with 0% non-linearity error. For a 10% transmittance neutral-density filter (Fig. 1a), the non-linearity effect is the

---

  * Present address: Department of Mechanical Engineering, University of Florida, Gainesville, FL 32611, U.S.A.
 ** To whom correspondence should be addressed

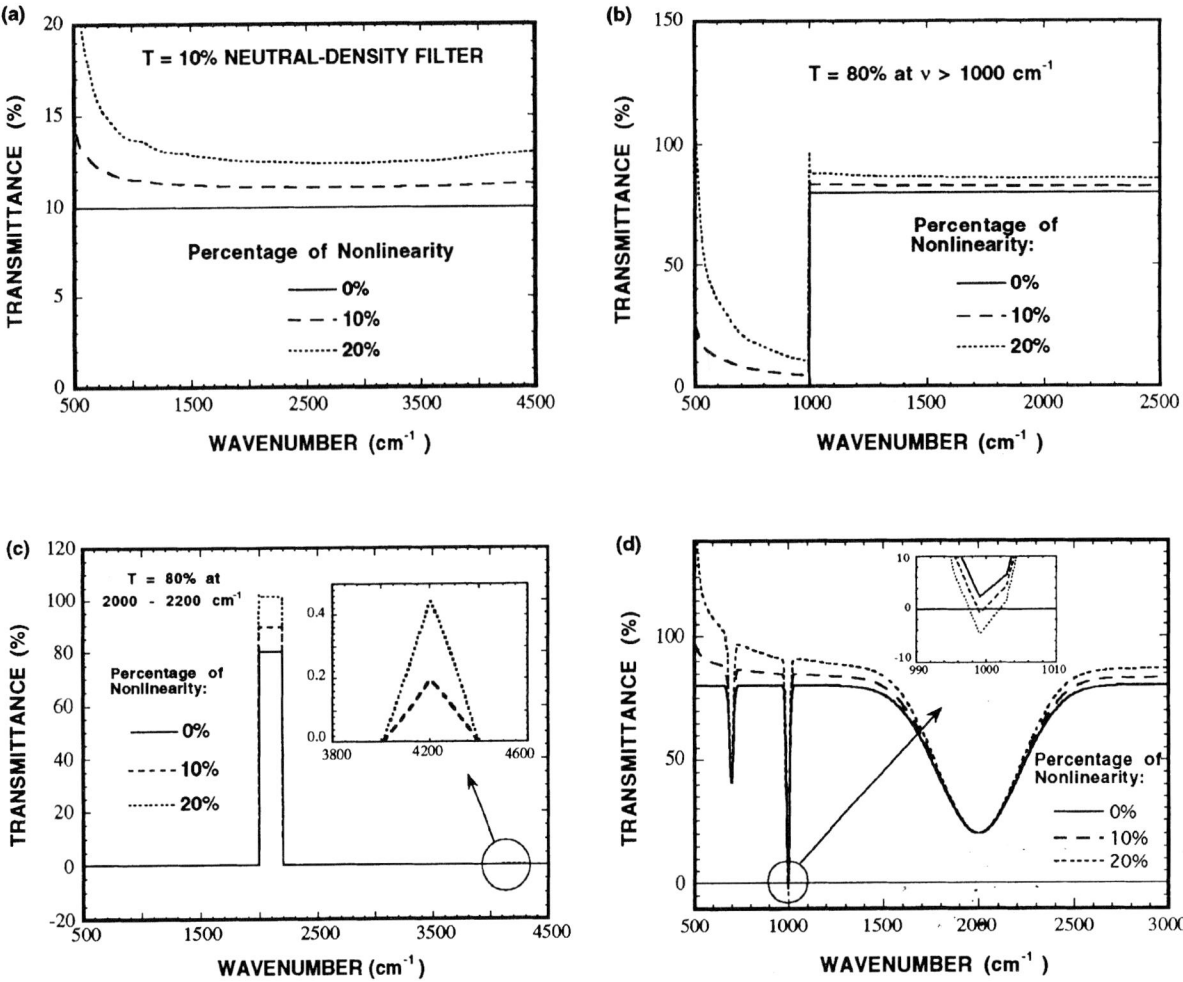

Fig. 1. Effects of non-linearity on different types of optical filters and materials: **a** neutral-density filter; **b** short-wavelength-pass filter; **c** narrow-band filter; **d** absorption bands

strongest near the long-wavelength cut-off. The minimum error is near $2500\,cm^{-1}$, which roughly corresponds to the maximum of the reference spectrum. Figure 1b shows the transmittance spectra with and without non-linearity of a cut-off filter which has 80% transmittance for $v > 1000\,cm^{-1}$ and 0% for $v < 1000\,cm^{-1}$. Many dielectric materials have a similar behaviour, e.g. MgO, CaF_2, and $SrTiO_3$. There is a strong non-physical signal beyond the cut-off wavelength. Hence, it is easier to observe non-linearity error in the measurements by using one of these samples. The simulation results for a narrow-band filter is shown in Fig. 1c. A non-physical response at twice the wavenumber is seen (as shown in the insert). Figure 1d shows the non-linearity effect on a sample with three symmetric absorption bands. Again, the

effect of non-linearity is stronger near the long-wavelength cut-off. The minimum transmittance for a narrow absorption band at $1000\,cm^{-1}$ changes from 2% to a negative value with increasing non-linearity. The depth of the narrow absorption bands at 700 and $1000\,cm^{-1}$ increases, whereas the broad band at $2000\,cm^{-1}$ is not deepened.

Both software and hardware methods are being proposed to correct for the errors caused by non-linearity [3–5]. Further research is needed for quantitative evaluation of the detectors and the correction methods.

In summary, an algorithm has been developed for quantitative prediction of the non-linearity effects on FT-IR measurements. The effect of non-linearity on different types of optical filters and materials has been studied and demonstrated by using this program. This study facilitates quantitative characterization of FT-IR spectrometers.

Acknowledgements. The authors acknowledge financial support from the U.S. Army through the Calibration Coordination Group (CCG). They thank Jack Hsia, Raju Datla and Michael Zander for insightful comments.

References

[1] D. B. Chase, *Appl. Spectrosc.* **1984**, *38*, 491.
[2] P. R. Griffiths, J. A. de Haseth, *Fourier Transform Infrared Spectrometry*, Wiley, New York, 1986, pp. 108–119.
[3] G. Eppeldauer, L. Novak, *Proc. SPIE* **1989**, *1110*, 267.
[4] R. M. Carangelo, D. G. Hamblen, C. R. Brouillette, *U. S. Patent* No. 5136154, August 1992.
[5] A. Keens, A. Simon, *Proc. SPIE* **1993**, *2089*, 222.

Mikrochim. Acta [Suppl.] 14, 321–323 (1997)

Near FT-IR Spectroscopy of Epoxy Diamine Formulations (Curing and Interactions)

Bernard Chabert, Gilbert Lachenal*, and **Isabelle Stevenson**

Laboratoire des Matériaux Plastiques, Université Claude Bernard, UMR5627, 43, Bd du 11 November 1918, 69622 Villeurbanne Cedex, France

Abstract. In recent years NIR spectroscopy has been successfully used to study the kinetic reactions of cross-linked resins. Since the role played by hydrogen bonding strongly influences the chemical, physical and mechanical properties of polymers containing hydroxyl, urethane, amine or amide functional groups, it is not surprising that the change of hydrogen bonding during the cure can modify the absorptivity of chemical groups under investigation (up to 10% for the NH_2 combination band in a solvent amine blend) and can lead to inaccurate infrared quantitative analysis. More care should be taken in the choice of the reference band to monitor the reaction cure.

Key words: near infrared spectroscopy, hydrogen bonding, epoxy, curing, Beer's law.

Infrared and Raman spectroscopy are excellent techniques for the characterization of hydrogen bonding in organic molecules or polymeric materials [1]. Mid IR spectra are widely interpretable in terms of chemical structure, but NIR spectra, although less informative, provide analysis with minimal sample preparation [2]. Usually NIR analysis is used with chemometric methods, but in the field of polymers some absorbance bands can be directly used to monitor chemical reactions or crystallinity and to study inter- and intramolecular interactions.

Epoxy–amine formulations are widely used as resins and matrices for composites. A clear understanding of these network structures has not yet been achieved but is indispensable in order to improve the performances of these products. The mechanism for curing epoxy resins with amine involves three main reactions: epoxy–primary amine addition, epoxy–secondary amine reaction and epoxy–hydroxyl reaction; this involves, for each step, the formation of hydrogen-bonded epoxy and amine species.

Owing to the strong overlapping of N–H and O–H bands the investigation of curing is not easy by mid IR but is easier by near IR, which provides valuable information concerning cure mechanisms by monitoring each chemical group [3]. Moreover, the polymer processing can be influenced by hydrogen bonds between the polymer(s), hardener and other species. Specific interactions (hydrogen bonding) have been extensively studied by mid infrared [4], but little attention has been given to studying hydrogen bonding by near IR spectroscopy [5]. Though the study of hydrogen bonding can provide valuable information for the investigation of polymer blends and the understanding of curing mechanism of epoxy formulations, the quantitative determination of functional groups can be complicated by the change of each specific absorptivity during the cure.

Hydrogen Bonding

Hydrogen bonding involves the interaction between a proton-donating (R–X–H) and a proton-acceptor group (R'–Y). The hydrogen bonding forces lower the $v(X–H)$ and $v(R'–Y)$ stretching frequencies, whereas the deformation frequencies associated with the motion of the hydrogen atom perpendicular to the X–H bond (bending) will be increased. For example in the mid IR, the water molecule has three modes of vibration: the bending (v_2, 1600–1650 cm^{-1}) and the symmetric and antisymmetric stretching (v_1, v_3, observed near 3500–3750 cm^{-1}). When hydrogen bond-

---

* To whom correspondence should be addressed

ing increases, the frequency position of the OH band decreases for the stretching mode (v_1, v_3) but increases for the bending mode. Owing to the fact that hydrogen bonding has opposite effects on the frequency shifts of the v_2 and v_3 bonds, a weak shift of the ($v_2 + v_3$) combination band is observed from 5100 to 5300 cm^{-1} whereas the first harmonic v_3 can shift from 6700 to 7400 cm^{-1}. On the other hand, the absorptivity of each type of hydroxyl is different, according to the environment [6].

Although NIR spectra and analysis of mixtures of amines were used in the 1950s [7], little information on the effect of hydrogen bonding in the amine mixtures is available.

Epoxy Resins

The fabrication of composite materials such as poly-epoxy/glass fibre has to go through a step consisting of impregnating the reinforcement fibres with the reactive epoxy–hardener mixture, sometime solubilized (dissolved) in a solvent such as methyl ethyl ketone (MEK). Adding this "non-reactive" solvent decreases the viscosity of impregnating baths and the impregnating temperature of the prepregs and increases the bath life times (pot life). This solvent is eliminated before filament winding. By thermal and mechanical analysis, it has been shown that solvent traces, which can remain on the prepreg after a period in ventilated ovens, were causing modifications of the mechanical properties of the composites [8]. The solvent seems to interact with the hardener. Near IR spectroscopy (transmission, diffuse reflectance, hot stage and microscopy) has been used for studying the amine hardeners (DDM, DDS) and epoxy (DGEBA) with and without solvent.

Solvent–amine Blend

Although a very slight shift of combination band was observed, as shown in Fig. 1, the molar absorptivity changes more than 10% when the concentration of DDM increases from 0.03 M to 1 M. This expected result for the determination of amine groups [9] and the recent paper of Mijovic et al. [10] pointing out the inaccuracy of the determination of epoxy content at 915 cm^{-1} (information that was published nearly 40 years ago by Dannenberg et al. [11], but has perhaps been forgotten, since over the years numerous researchers have used this band in their resin investigations without any correction or control) underline the

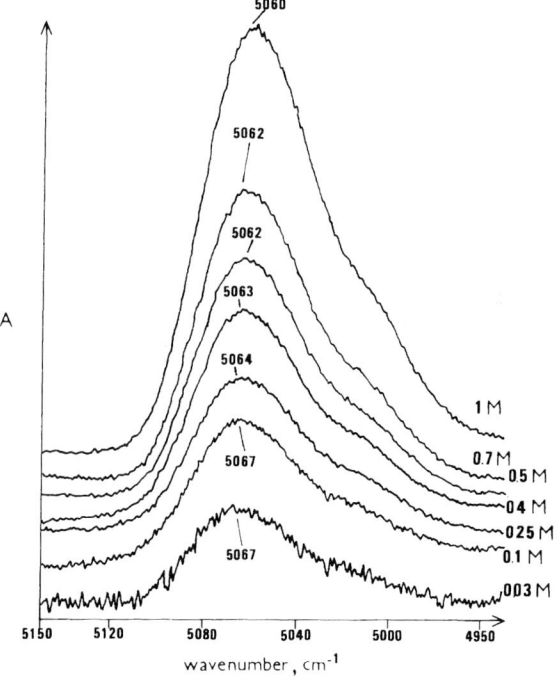

Fig. 1. NIR spectra of DDM/MEK blends

need for care in the quantitative infrared analysis (MIR or NIR) of epoxy/amine formulations.

Epoxy–amine Blend

NIR spectra of epoxy/amine mixtures show changes in the position and absorptivity of the amine combination band for the formulations used (see Fig. 2). This fact can be correlated with the observations of Kradjel

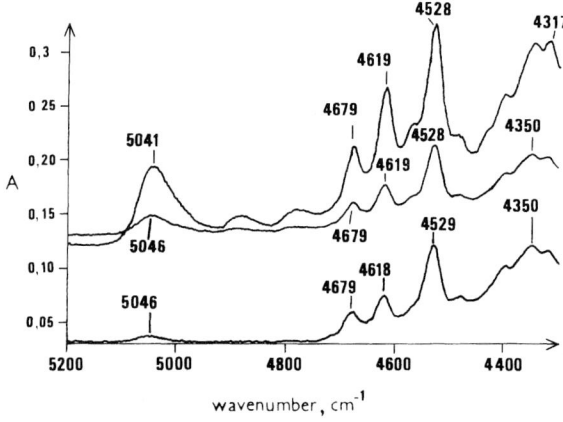

Fig. 2. NIR spectra of epoxy/DDM blends

et al. [12] or Mijovic et al. [13] regarding a curious increase in the absorption intensity of the amine overtone peak at $6700\,cm^{-1}$, during the reaction. On the other hand the epoxy band at $4530\,cm^{-1}$ is not significantly modified (peak position or molar absorptivity) by concentration changes or by the type of hardener (in the liquid state and at room temperature). This peak is probably a combination of the C–H stretching fundamental ($v\,CH = 3050\,cm^{-1}$) with the CH_2 deformation fundamental ($\delta\,CH_2 = 1460\,cm^{-1}$) of the oxyrane ring and is less sensitive to hydrogen bonding than are the amine combination (and probably overtone) bands. As underlined by Dannenberg [11, 14], determination of the epoxy content of cured epoxy by infrared includes the assumption that the absorbance is constant regardless of the degree of cure (from liquid to solid state) and that there is no deviation from Beer's law. The lack of calibration standards for cross-linked resins complicates the determination of all functional groups in the cured epoxy resins. However, the use of two-dimensional correlation can bring useful information for understanding the complicated NIR spectra [15, 16].

Conclusion

In the study of polymers and polymer blends, very useful information can be obtained from near infrared spectra. However, for the insoluble (cross-linked) polymers involving changes in hydrogen bonding during the cure, the determination of some functional groups by using the NIR (or MIR) cannot be fully accurate since hydrogen bonding causes serious deviations from Beer's law even if the peak position is not significantly shifted. Nevertheless NIR presently appears to be one of the best techniques to follow the extent of reaction, both in the laboratory and in process control, but great care must be taken in the quantitative analysis and the spectra carefully interpreted.

References

[1] K. A. Bunding Lee *Appl. Spectrosc. Rev.*, **1993**, 28, 213.
[2] G. Lachenal, in: *Advances in Practical Spectroscopy*, ESIS 93 (G. Lachenal, H. W. Siesler, eds.), UCB Lyon, 1994, p. 62.
[3] C. J. deBakker, G. A. George, N. A. St John, P. M. Fredericks, *Spectrochim. Acta* **1993**, 49A, 739.
[4] H. Zhang, D. E. Bhagwagar, J. F. Graf, P. C. Painter, M. M. Coleman, *Polymer* **1994**, 35, 5379.
[5] B. Chabert, G. Lachenal, C. V. Tung, *Macromol. Symp.* **1995**, 94, 145.
[6] S. Venz, B. Dickens, *J. Biomed. Mater. Res.* **1991**, 25, 1248.
[7] K. Whetsel, W. E. Roberson, M. W. Krell, *Anal. Chem.* **1958**, 30, 1598.
[8] B. Chabert, G. Lachenal, *ISPAC 8*, Sinabel, USA, 1995.
[9] J. H. Lady, K. B. Whetsel, *J. Phys. Chem.* **1964**, 68, 1001.
[10] J. Mijović, S. Andjelić, *Macromolecules* **1995**, 28, 2787.
[11] H. Dannenberg, W. R. Harp, *Anal. Chem.* **1956**, 28, 86.
[12] C. Kradjel, L. McDermott, *Handbook of Near Infrared Analysis* (D. A. Burns, E. W. Ciurczak eds.) Dekker, New York, 1992, p. 565.
[13] J. Mijović, S. Andjelić, C. F. W. Yee, F. Bellucci, L. Nicolais, *Macromolecules* **1995**, 28, 2797.
[14] H. Dannenberg, *SPE Trans.* **1963**, 3, 78.
[15] Y. Ozaki, Y. Liu, M. Czarnecki, I. Noda, *Macromol. Symp.* **1995**, 94, 51.
[16] I. Noda, Y. Liu, Y. Ozaki, M. A. Czarnecki, *J. Phys. Chem.* **1995**, 99, 3068.

Mikrochim. Acta [Suppl.] 14, 325–327 (1997)
© Springer-Verlag 1997

FT-NIR Comparative Quantitative Performance for Complex Samples

David L. Wetzel* and **Joe A. Sweat**

Microbeam Molecular Spectroscopy Laboratory, Kansas State University, Shellenberger Hall, Manhattan, KS 66506, U.S.A.

Abstract. Quantification in routine near IR spectroscopy is based on mA differences requiring signal conservation and limitation of instrument noise to a few μA units. An FT-NIR bench equipped with a fiber optic sampling probe was used for quantitative analysis of clear liquid chemical mixtures, emulsions, and homogeneous and heterogeneous granular solids. The combination of hardware, sampling optics, and calibration equation produced analytical standard errors comparable to those obtained with both dispersive and acousto-optic NIR instruments. The sampling area of the probe required averaging of multiple probes for heterogeneous solids. Excellent wavelength reproduction of the FT-NIR enabled the use of spectral subtraction for both qualitative and quantitative purposes.

Keywords: near infrared, quantitative FT-NIR, NIR optical fiber probe.

In the past 20–25 years, near IR spectroscopy has become established as a rapid routine quantitative technique that requires minimal sample work-up. Most of this work has been done by diffuse reflectance, enabling the direct analysis of solid samples and powders by using statistically derived expressions (chemometrics) to relate reflectance data to quantitative analysis. With quantification dependent on absorbance differences of a few thousandths, the signal must be conserved and instrument noise must be held to an absorbance of a few millionths. Greater than 90% of the instruments in use for quantification employ interference filters. When near infrared scanning instruments were proposed two decades ago, interferometry was considered unnecessary and too expensive. In rare instances, a mid IR instrument was converted to near IR use by adding a different source and detector and using an alternative beam-splitter. More recently the field of FT-Raman spectroscopy with near IR excitation has brought about the necessity for inter-

ferometers designed, optimized, and dedicated to the near IR. Progression of FT-NIR instrumentation has gone from (1) accessory modules provided for a mid IR instrument to (2) modified FT instruments intended for dual range, and currently to (3) dedicated FT-NIR instruments. The interferometer benches manufactured for FT-Raman are now competitively priced and available for routine analytical applications. There are on the market at least three dedicated FT-NIR spectrometers equipped with fiber optic probes. The commercial fiber optic FT-NIR instruments designed as quick qualitative probes are not addressed in this paper.

Instrumentation

Let us consider potential advantages and limitations of FT *vs* dispersive scanning instruments. From the theoretical perspective, the traditional mid IR multiplex advantage and throughput advantages would be expected, but for FT-NIR, the detector sensitivity and source intensity are relatively high so the IR throughput is less of an issue. Also, the limitation of the dynamic range of the A/D converter results in leveling off of the signal-to-noise ratio at a moderate throughput level, negating any increased throughput beyond that level. The expected FT-NIR advantages include resolution, sensitivity, and precision of both the wavelength and intensity axes. High resolution, band shape precision, and particularly wavelength precision may be an advantage when dealing with narrow band measurements. We previously reported quantitative comparisons for industrial chemical waste samples analyzed on a dual range commercial FT-NIR, a homemade NIR acousto-optic tunable filter spec-

---

* To whom correspondence should be addressed

trometer (TFS) and a commercial grating mono-chromator instrument. Those equipped with custom calibrations worked equally well on the same optically challenging sample set [1]. In this early work, chemicals were used that have a distinct spectroscopic signature, compared to the commodities routinely analyzed by NIR. This report deals with the performance of a late model standard bench equipped with a fiber optic probe covering the range 1100–2200 nm for quantitative analysis of liquids, emulsions slurries, and both heterogeneous and homogeneous granular solids.

Results and Discussion

On a series of emulsions (cutting oil) containing from 1 to 30% oil, the standard error of calibration (SEC) was 0.29. This was comparable to previous results obtained with our homemade acousto-optic TFS (AOTF) and a commercial grating monochromator instrument. Liquid mixtures of five organic chemicals were also run and regression results are reported in Table 1 that are comparable to those of earlier work.

Previous experiments in our laboratory with these analytes [2] in the same concentration range in a petroleum base yielded SEC and r values for benzene, cyclohexane, octanol, and butylamine (by AOTF) of 0.173/0.9999, 0.254/0.9992, 0.292/0.9993, and 0.269/0.9995 respectively. With the InfraAnalyzer (dispersive) I/A 500, the corresponding results were 0.194/0.9999, 0.244/0.9992, 0.213/0.9996, and 0.272/0.9994. Solid samples used included wheat flour milling streams (range 7.8–15.7% protein) for which an SEC of 0.26 and SEP (standard error of performance) of 0.33 resulted from a three-wavelength regression equation or 0.24 for four partial least square factors. These results were only slightly higher than those obtained with a conventional filter or grating instrument. Ground whole wheat resulted in an SEC of 0.26 when averaging was used and 0.39 by taking individual probes. Similar samples run on a conventional NIR

quantitative instrument with an integrating sphere gave an SEC of 0.22. The 2–3 mm diameter sampling area of the probe is considerably less than that used with routine diffuse reflectance instruments. To test sampling error, a single homogeneous stream was probed 20 times with stirring of the sample between each probe. The standard deviation was 0.11% for a 10.9% protein sample. A heterogeneous ground whole wheat sample of 11.4% protein similarly probed gave a standard deviation of 0.16%. This explains why averaging the results of three replicate probings for each sample analyzed reduced the sampling error.

Use of the probe for on-line monitoring was performed in a one-time clinical experiment. In this case, the probe was placed against the forehead of a cardiac rehabilitation patient on a treadmill to monitor blood glucose while exercise was used to change the glucose level from 197 to 149 (see Figs. 1 and 2). Colorimetric "finger stick" meter readings were used every 10 minutes for 60 minutes. FT-NIR scans were run every minute. Twenty-one scans were regressed, assuming that the seven color tests were applicable to three adjacent scans. The three-wavelength calibration produced an SEC of 5 units for the range of nearly 50, and interpolated color values were predicted from the NIR spectra with reasonable accuracy. Exceptions were probably due to poor probe contact. Spectral subtraction capability is exemplified in Figs. 3 and 4 where "Coffeemate" minus "Coffeemate Lite" produced an excellent lipid spectrum and where C-18 with nitrobenzene minus the "C-18 Empore" extraction disk (3M, Minneapolis MN) spectrum revealed nitrobenzene. Spectral subtraction was also used in an attempt to semi-quantify a drug concentrated from a dilute sol-

Table 1. Results of five-component mixture analysis

Component	SEC	Range	Corr. coeff. (r)
Benzene	0.119	0.0–50.4	0.9999
Cyclohexane	0.147	0.0–20.2	0.9997
Octanol	0.284	0.0–39.2	0.9995
Bytylamine	0.181	0.0–28.3	0.9998

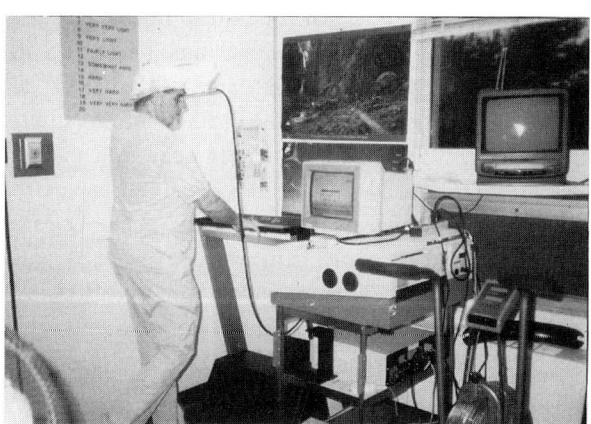

Fig. 1. Patient on treadmill

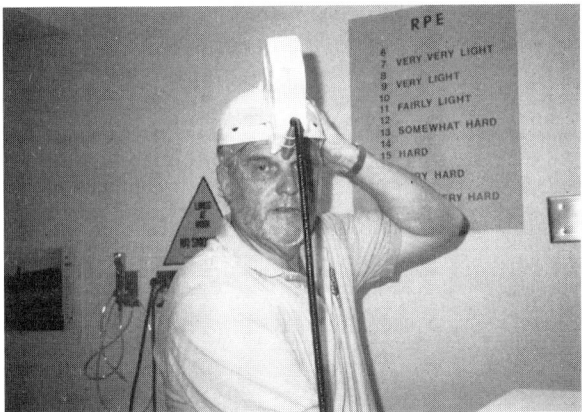

Fig. 2. NIR probe for glucose

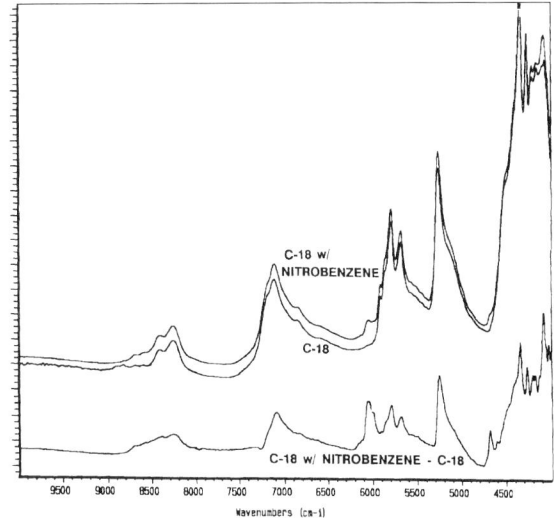

Fig. 4. Nitrobenzene on C-18 disk

Conclusion

A commercial FT-NIR (Nicolet Magna model) bench equipped with a fiber optic probe has wavelength reproducibility adequate for spectral substraction, and for quantitative determination performs competitively with dispersive and AOTF NIR spectrometers. Aside from sampling error introduced when using a small area probe for heterogeneous solids, the results were comparable to those with an AOTF NIR spectrometer or a commercial dispersive or filter NIR instrument for liquids, emulsions, hetrogeneous and homogeneous solids, and on-line monitoring of flowing liquids.

Acknowledgement. Contribution No. 96-77-1 Kansas Agricultural Experiment Station, Manhattan.

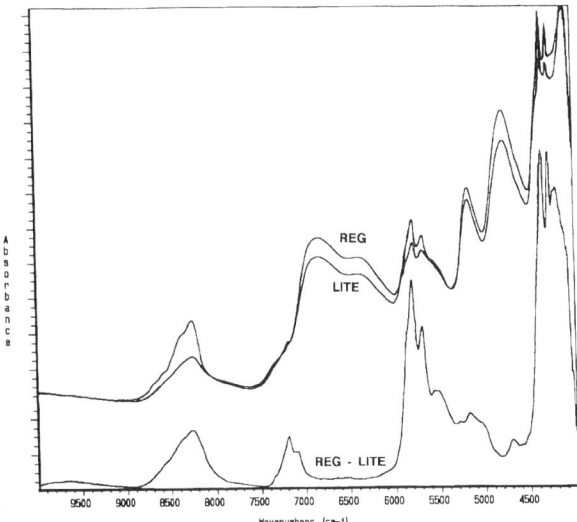

Fig. 3. "Coffeemate" minus "Coffeemate Lite"

ution onto C-8 extraction disks, with the difference regressed against the known solution concentration. This first exploration was reasonable and encouraging.

References

[1] D. L. Wetzel, A. J. Eilert, *Proc. SPIE* **1991**, *1575*, 523.
[2] A. J. Eilert, *PhD. Dissertation*, Kansas State University, 1995, p. 221.

Mikrochim. Acta [Suppl.] 14, 329–332 (1997)
© Springer-Verlag 1997

FT-IR Spectroscopy and Microscopy with a Diamond Cell Applied to Forensic and Industrial Samples

Núria Ferrer

Serveis Científico Tècnics, Universitat de Barcelona, Lluís Solé i Sabarís, 1, 08028 Barcelona, Spain

Abstract. Fourier-transform infrared spectroscopy, using the diamond cell technique, has been shown to be a useful tool for the analysis of small samples. The diamond cell can be used both in an infrared microscope and in a microbeam compartment. In our laboratory both techniques have been applied to forensic and industrial analysis and some examples are reported. Sample preparation normally consists of physical separation with the help of tungsten needles, microtoming techniques after embedding the samples in paraffin blocks, or the use of solvents to extract some compounds before evaporation on different kinds of plate. The diamond cell technique seems to provide good results for all samples analysed.

Key words: infrared microscopy, microsamples, forensic analysis.

Microscopical analysis using FT-IR spectroscopy has greatly advanced recently. Many samples from industry or forensic analysis require the use of very small amounts of material. It is often necessary to identify a compound or to compare different substances without destroying the sample.

Infrared microscopy has been applied to a variety of samples. A recent article [1] reviews some applications of vibrational microspectroscopy to analytical chemistry. Other applications are reported in some books [2, 3]. Analysis of forensic material [4], small samples of fibres [5], paints [6], polymers [7] and human body specimens [8–10] has been reported.

Experimental

Apparatus

All analyses were done with a Bomem MB-120 Fourier-transform infrared spectrometer equipped with a potassium bromide beamsplitter, DTGS detector and 4 × microbeam compartment. The microscope accessory was an IR Plan Analytical Microscope Spectra-Tech with an MCT narrow-range detector. The spectra were measured from 4000 to 400 cm^{-1} with the 4 × microbeam compart-

ment and from 4000 to 750 cm^{-1} with the microscope. Resolution was 4 cm^{-1} and 100 and 200 scans were taken with use of the microbeam compartment and the microscope respectively, in order to obtain an appropriate signal-to-noise ratio. The spectral data were processed with GRAMS/386.

Handling of Samples

Samples were pressed in a diamond cell (two diamond windows) before the analysis in order to spread them and to produce a thin film. Once the sample had been pressed, one of the windows was removed and the analysis was performed.

Results and Discussion

Crystals that appeared in a damaged lung of a child were analysed by using the infrared microscope with a diamond cell (Fig. 1). Some pieces of crystals were prepared for microtoming by embedding them in paraffin blocks. The thickness of the sample was about 15 μm. The most difficult thing was to find a section where a relatively high concentration of crystals could be observed. The crystals were about 20 μm in size before pressing and it was necessary to use a variable aperture in order to mask them and to avoid the spectral information from the rest of the sample. The crystals were identified as aragonite (orthorhombic calcium carbonate), which shows a characteristic spectrum different from that of calcite (trigonal calcium carbonate). The characteristic bands of aragonite are at 1474, 1082 and 858 cm^{-1}.

White crystals that appeared in an industrial filter were analysed with the diamond cell. These crystals were about 100 μm in size and could be handled with tungsten needles. Therefore, it was possible to use the microbeam compartment. Apart from other absorp-

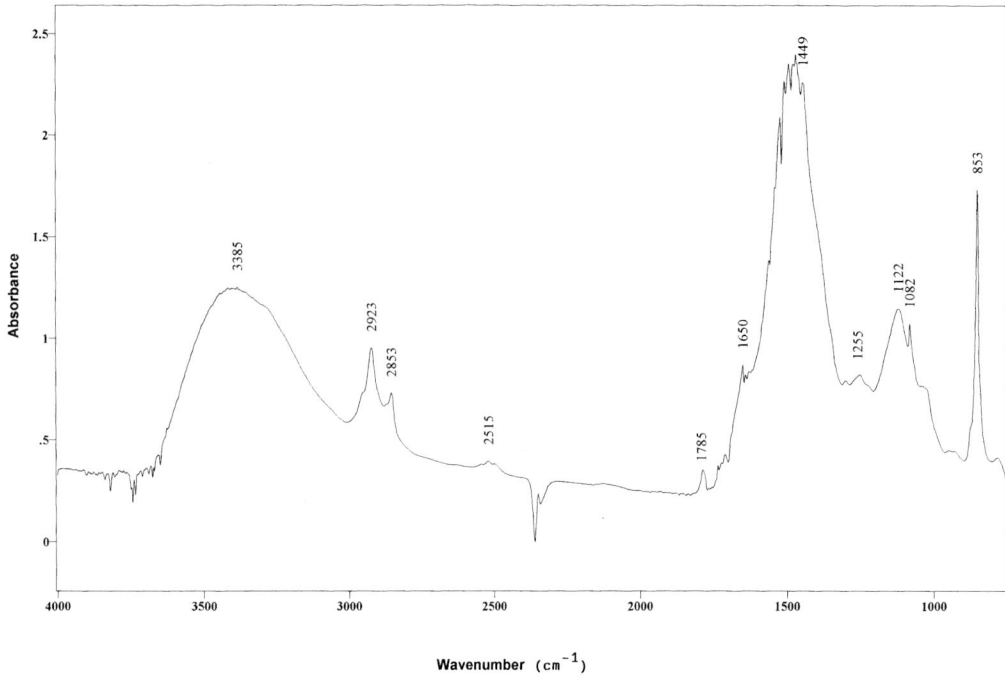

Fig. 1. Infrared spectrum of crystals from a lung

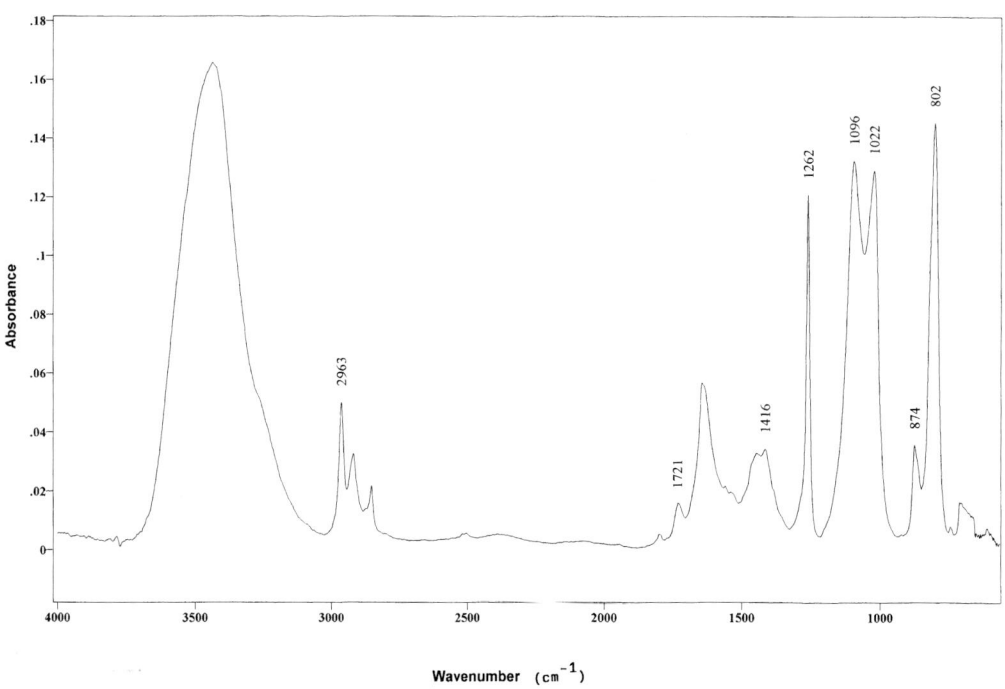

Fig. 2. Infrared spectrum of a sodium chloride crystal polluted with silicone

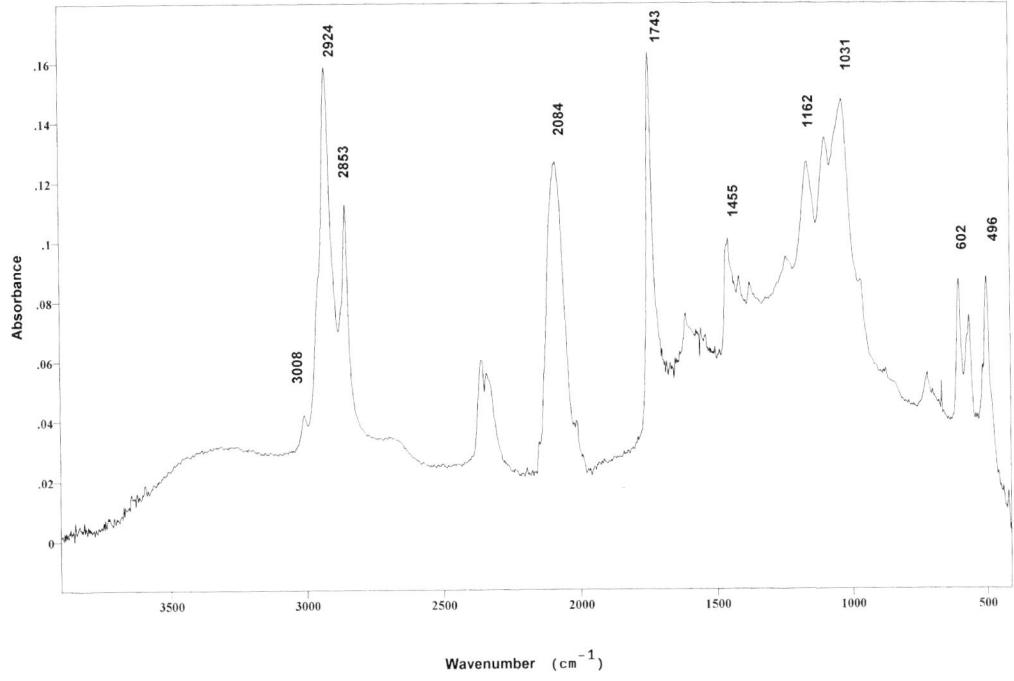

Fig. 3. Infrared spectrum of a particle of ink from an ancient copper surface

tions, some bands of a silicone were identified: 1262, 1093, 1024, 877 and 801 cm^{-1} (Fig. 2). When the microscope was used it was possible to focus the beam onto the centre of the pressed crystal and also onto the perimeter, which showed a different colour. The spectra from the centre of the crystal did not present any bands. Other techniques such as scanning electron microscopy showed that it was sodium chloride. The spectra from the perimeter showed strong bands corresponding to a silicone. To sum up, the crystals found in the industrial filter were sodium chloride covered with silicone. The origin of such contamination could be located and dealt with.

Some ancient inks on copper surfaces in a museum were analysed by two techniques: reflection with the microscope and transmission by using a diamond cell in the microbeam compartment. Although the same information could be obtained from both techniques, the transmission spectra showed much better signal-to-noise ratio. Due to the fact that some bands corresponding to inorganic compounds were seen in the spectra, we dissolved the sample in trichlorotrifluoroethane in order to separate the soluble organic fraction. This fraction was deposited onto a glass sur-

face and removed with the help of tungsten needles. A band corresponding to isothiocyanate (2084 cm^{-1}) was identified (Fig. 3).

Other applications, to animal hairs, defects in polymers, electrical components, falsified cheques or votes, and traffic accidents have been made in our laboratory by use of these techniques. Some samples that recover their original shape after dismantling of the cell, such as some kind of gums, must be analysed by using both diamond windows.

Conclusions

Using a diamond cell as a handling tool to make a sample thinner and as a window for transmittance analysis is a technique that can solve a number of forensic and industrial problems we receive in our centre.

The use of the microscope is recommended when the area of the sample or the impurity to be measured, after pressing between the two diamonds, is less than 200 µm^2. Using a beam condenser instead of the microscope is recommended when we need to measure bands below 750 cm^{-1}.

References

[1] J. E. Katon, *Vib. Spectrosc.* **1994**, *7*, 201.

[2] R. G. Messerschmidt, M. A. Harthcock (eds.), *Infrared Micro-spectroscopy: Theory and Applications*, Dekker, New York, 1988.

[3] J. R. Ferraro, K. Krishnan (eds.), *Practical Fourier Transform Infrared Spectroscopy: Industrial and Laboratory Chemical Analysis*, Academic Press, New York, 1990, Chap. 3.

[4] E. G. Bartick, M. W. Tungol, J. A. Reffner, *Anal. Chim. Acta* **1994**, *288*, 35.

[5] M. W. Tungol, E. G. Bartick, A. Montaser, *Appl. Spectrosc.* **1993**, *47*, 1655.

[6] D. J. McEwen, G. D. Cheever, *J. Coat. Technol.* **1993**, *65*, 35.

[7] W. E. Steger, S. Machill, K. Herzog, R. Gerhards, I. Jussofie, H. Schator, *Fresenius J. Anal. Chem.* **1992**, *344*, 203.

[8] K. S. Kalasinsky, J. Magluilo, T. Schaefer, *Forensic Sci. Int.* **1993**, *63*, 253.

[9] K. S. Kalasinsky, J. Magluilo, T. Schaefer, *J. Anal. Toxicol.* **1994**, *18*, 337.

[10] J. A. Centeno, F. B. Johnson, *Appl. Spectrosc.* **1993**, *47*, 341.

Mikrochim. Acta [Suppl.] 14, 333–337 (1997)

A New Infrared Microspectrometer Accessory for External- and Internal-Reflection Spectroscopy

John A. Reffner*, **Gregg Ressler**, **David W. Schiering**, and **William T. Wihlborg**

Spectra-Tech, Inc., P.O. Box 869, 2 Research Dr., Shelton, CT 06484–0869, U.S.A.

Abstract. A new FT-IR spectrometer accessory for infrared-micro-analysis has been developed that allows external and internal (ATR) spectral analysis of micro-samples. FT-IR microanalysis refers specifically to conducting quantitative or qualitative analysis on samples ranging from 0.1 to 10 mg. This FT-IR microanalysis accessory, called the InspectIR™, combines a unique, infinity-corrected optical system with digital video-imaging technology, providing a system that is simple to operate, rugged in its optical design, and versatile in its analytical applications. The InspectIR may be used to collect infrared spectra of solids or liquids by both ATR and external-reflection techniques.

Key words: infrared microanalysis, sampling instrumentation, reflection.

Microspectroscopy has shown us that external and internal reflection spectral measurements simplify FT-IR microanalysis. The ATR objective [1] is widely used to simplify or eliminate sample preparation for FT-IR microanalysis. Most samples analyzed by infrared spectroscopy are visible to the unaided eye and weigh 10 mg or more. Classically, microanalysis refers to analyzing samples ranging from 0.1 to 10 mg. FT-IR microscopes are capable of analyzing samples in the nanogram to picogram range. Conventional sampling methods generally require samples larger than 0.1 g. The InspectIR is a sampling accessory that is specifically designed for microanalysis, filling the gap between the microscope and conventional sampling accessories.

Experimental

Optical Design

The InspectIR introduces an infinity-corrected optical design that allows the nearly-collimated external beam from the spectrometer to

be directed through one-half of the aperture of a Schwarzschild reflecting optic. The reflected radiation from the sample is collected by the opposite half of the Schwarzschild optic and is directed to a dedicated MCT detector. Though micro-samples generally can be detected by the unaided eye, the ability to observe the samples at higher magnification is very useful in selecting the appropriate area of the specimen for analysis. The samples are observed at high magnification by the process of directing visible light through the Schwarzschild lens, collecting the reflected radiation by the lens, and then imaging the reflected radiation onto a dedicated CCD video camera. The visual image is seen on a video monitor or can be displayed on the computer monitor through a video input board.

Spectral Analysis Modes

The InspectIR is an accessory system that is able to collect infrared spectra of micro-samples in either an external or internal reflection mode [2]. For external reflection measurements, the radiant energy is incident on the sample through an angular range from 16 to 40°. For internal reflection measurements, an internal-reflection element (IRE) of silicon, germanium, or zinc selenide is positioned in the Schwarzschild objective lens so that the IRE can come into contact with the specimen. The focus of the Schwarzschild objective is at the sample–IRE interface. The infrared radiant energy from the objective is incident on this interface at angles that exceed the critical angle, producing total internal reflection.

External Reflection

There are three primary modes of external reflection; specular, diffuse and reflection-absorption (RAS), each having its own advantages and applications. For external reflection, the sample geometry will determine which type of reflection occurs. Specular reflection requires a level, lustrous, dielectric material. A rough or finely ground sample, such as a powder, will undergo mostly diffuse reflection. RAS requires thin samples on reflective (metallic) substrates.

For specular reflection, a reflective surface is simply brought into focus, and the reflectivity of the surface is measured relative to that of a gold reference mirror. The specular-reflection data can easily be transformed by using the Kramers–Kronig transform to produce

---

* To whom correspondence should be addressed

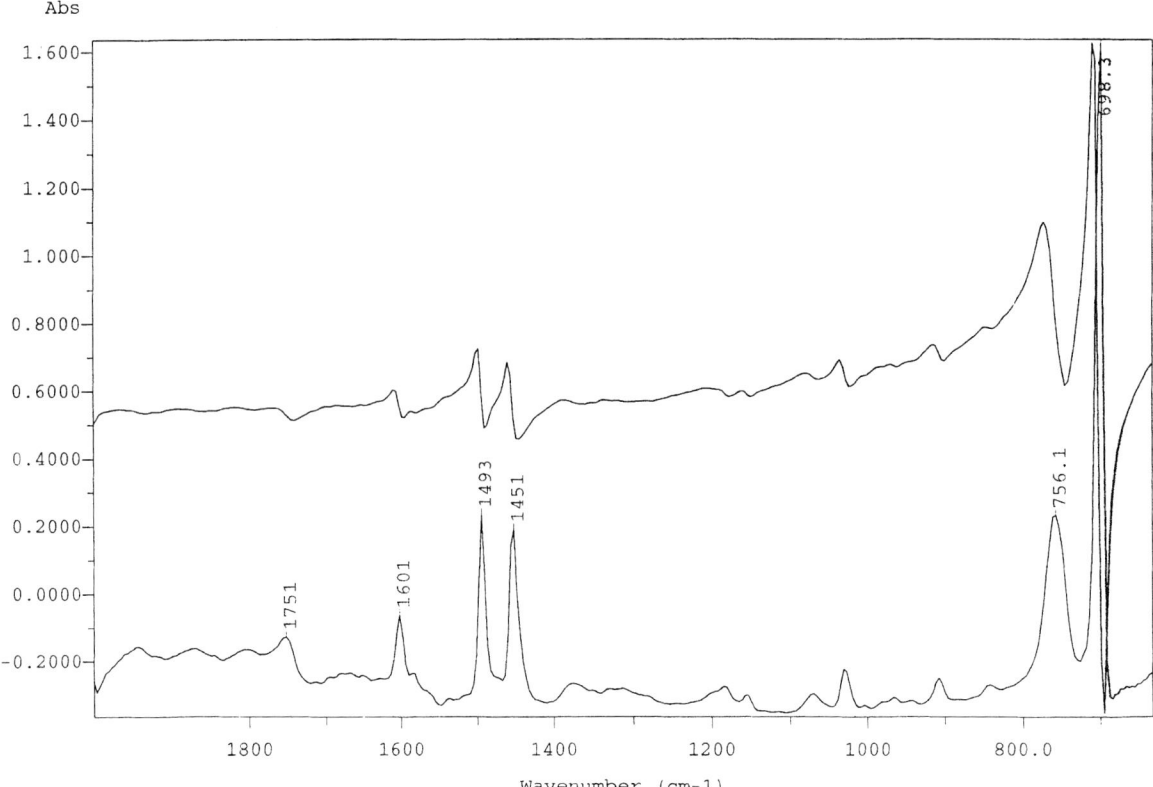

Fig. 1. Specular reflection spectrum of a plastic part; the top spectrum gives the raw reflection data, and the bottom one is the spectrum produced by applying the Kramers–Kronig (KK) integration

transmission-like data. A good example of a specular sample is a flat plastic plate. A spectrum of a plastic part is seen in Fig. 1. The top spectrum shows the raw reflection data, and the bottom spectrum is that produced by applying the Kramers–Kronig (KK) integration. The KK transformation is easily and quickly applied by most infrared data-processing software packages, making specular reflection a more useful analytical technique. The limitation of this mode is that the sample must be a good specular reflector.

Powders or roughened surfaces can be analyzed by diffuse-reflectionIR spectroscopy (DRIFTS). DRIFTS data may be collected directly with the InspectIR from some samples, but the best results are obtained when the micro-sample is dispersed in a non-absorbing matrix. For these data, a background or reference of ground KBr is recommended. Another convenient method is to rub the sample with a small piece of fine-grit silicon carbide or diamond paper. Figure 2 is the spectrum of varnish removed from a cabinet with diamond abrasive paper. The background reference spectrum was of the clean diamond abrasive paper and the sample spectrum was of the paper with the varnish dispersed on it. In diffuse reflection spectra, weak bands appear stronger and strong bands appear weaker. A Kubelka–Munk (K-M) data transformation was used to correct the intensity of the absorption bands [3].

For RAS measurements, samples are placed upon reflecting substrates (such as gold or aluminum mirrors), and spectra are collected directly, with the substrate spectrum used as the reference. Although the InspectIR is a reflection-only accessory, transmission experi-

ments may still be accomplished by using RAS. For RAS microanalysis, the sample is simply flattened on a reflective substrate. In addition, samples such as contaminants on circuit boards and coated metals are naturally RAS candidates.

The RAS spectrum obtained of a dye residue on a gold mirror is shown in Fig. 3. Since RAS is a double-pass transmission experiment (the beam is transmitted through the sample, reflected off the mirror, transmitted a second time through the sample, and collected by the objective), it is a highly efficient mode. This dye was identified as a Rubia type dye, found in the roots of plants belonging to the *Relbunium* or *Galium* genera.

Internal Reflection

Internal reflection occurs when radiation is incident on a high refractive index surface at angles exceeding the critical angle [4]. For a non-absorbing material in contact with this interface, the radiant energy is totally internally reflected. The electric field of the radiation extends beyond the interface and the intensity of reflection is diminished when the material is absorbing. Because the total reflected intensity is reduced by an absorbing material is contact with the IRE, this technique is called *attenuated total reflection* (ATR).

For the ATR measurements, an internal-reflection element (IRE) of silicon, germanium, or zinc selenide may be inserted into the light path. The surface of the IRE, when in contact with the specimen, is at

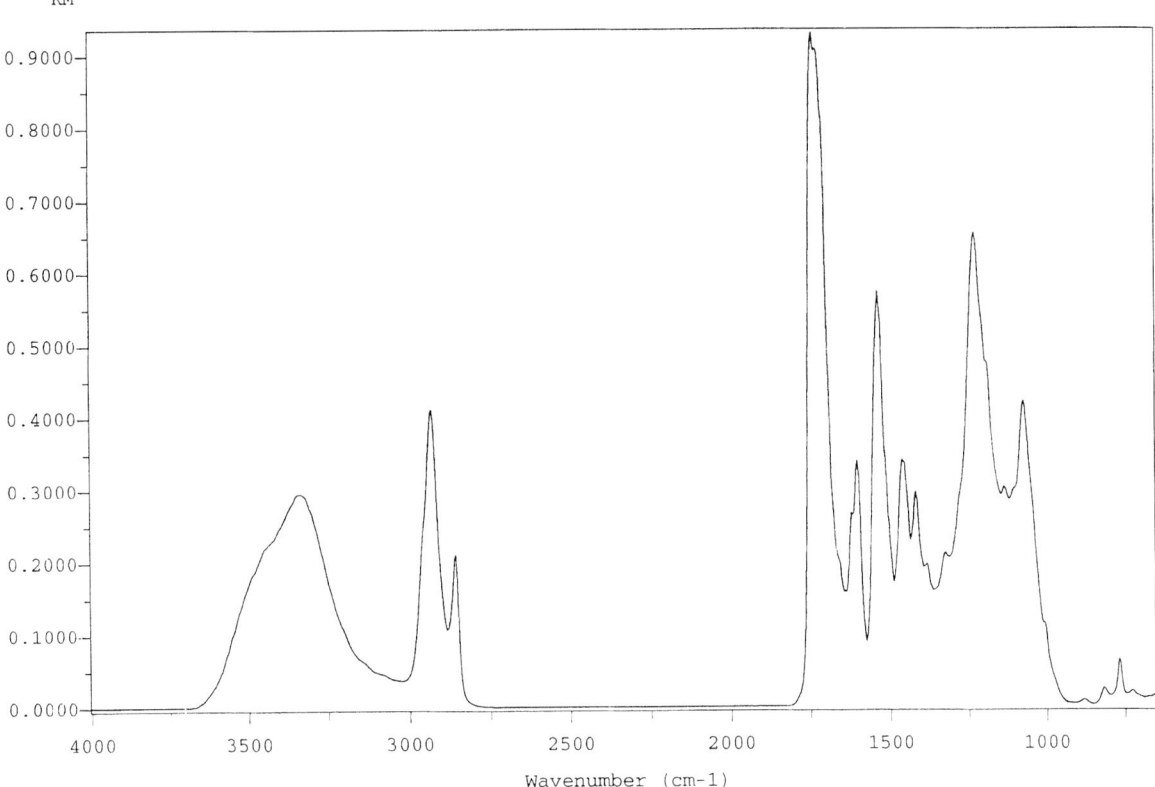

Fig. 2. DRIFTS spectrum of varnish removed from a cabinet with diamond abrasive paper

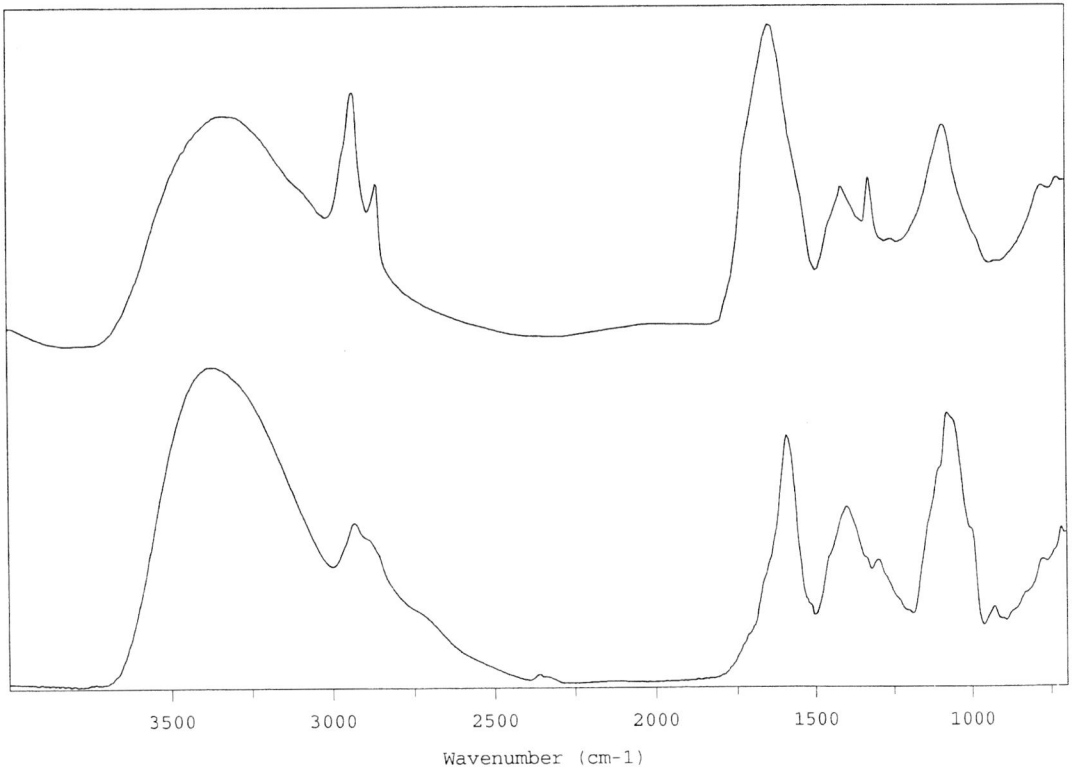

Fig. 3. The RAS spectrum of a red dye residue on a gold mirror

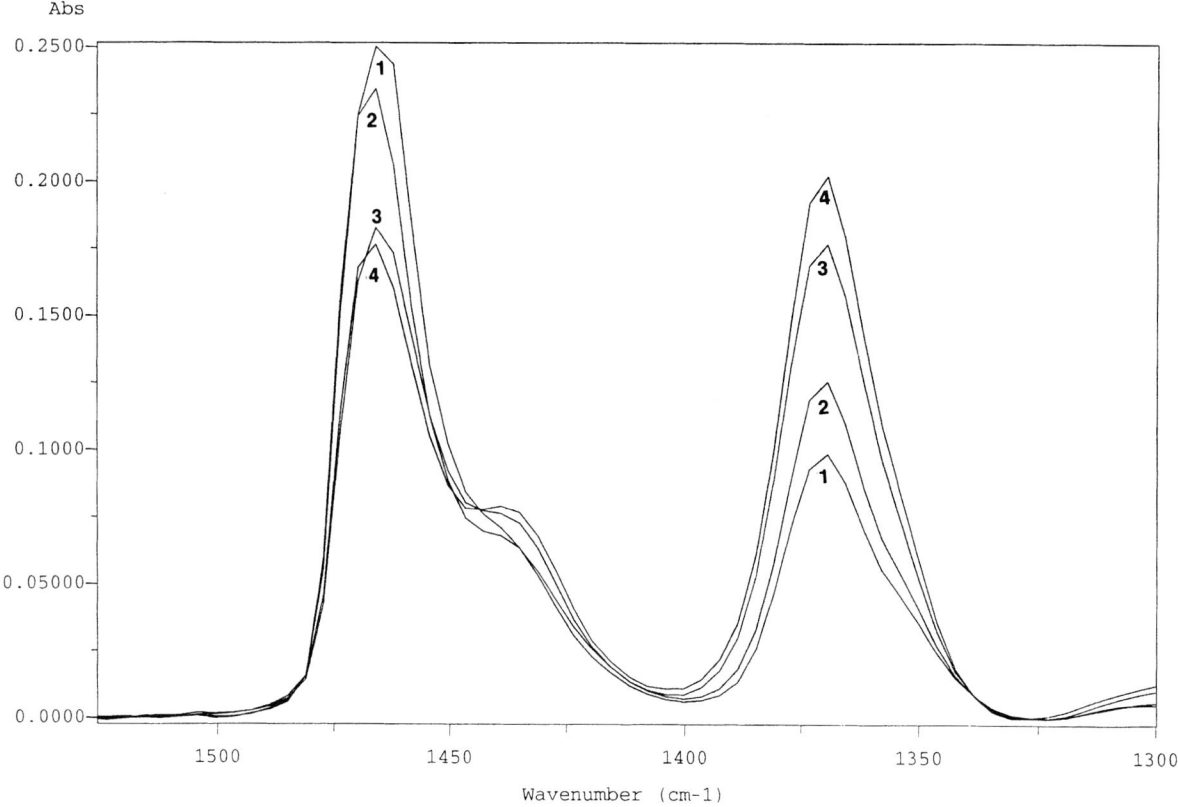

Fig. 4. ATR spectra from four different polyethylene/vinyl acetate copolymers. *1* (9%), *2* (14%), *3* (25%) and *4* (33%) vinyl acetate

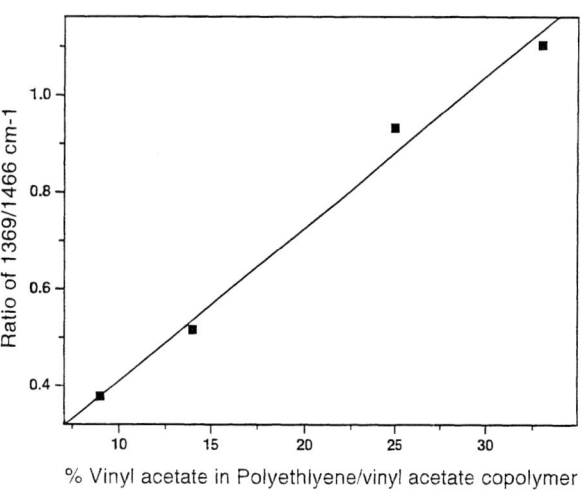

Fig. 5. Beer's law plot of concentration *vs* absorbance ratio of 1369/1466 cm⁻¹ bands

objective lens is slid into position, and the sample is raised to make contact with the IRE. Once contact is made, the ATR spectrum of the sample may be recorded. The reference for ATR spectral measurements is usually the clean IRE with nothing but air contacting its surface.

Although the InspectIR produces a spot size of 250 × 250 μm, the IRE itself reduces this by increasing the effective magnification of the optical system. When a ZnSe IRE is used, it reduces the infrared sample spot to ~100 μm. Other IREs can also act as magnifiers. A silicon IRE will yield a sampling size of ~70 μm and for a germanium IRE ~60 μm.

ATR spectra obtained with the InspectIR from four different polyethylene/vinyl acetate copolymers are shown in Fig. 4. The spectral acquisition parameters were 128 scans at a resolution of 8 cm⁻¹. The ratio between the peak heights at 1369 cm⁻¹ (due to vinyl acetate copolymer) and 1466 cm⁻¹ (due to ethylene copolymer) was calculated and plotted *vs.* the known vinyl acetate percentage in the copolymer. The Beer's law plot is seen in Fig. 5. A good linear regression fit is obtained, with $R = 0.9945$, and a standard deviation of 0.04399.

the primary focus of the Schwarzschild objective. To collect an ATR spectrum of a micro-sample, the specimen is placed on a support (such as a microscope slide), positioned on a stage, and moved to the center of the viewing field. When the area to be analyzed has been centered in the field, the IRE crystal on the

Conclusions

The InspectIR offers the easiest, most efficient, and advanced microanalysis accessory for FT-IR spectral analysis. All of the advantages of both internal and external reflection techniques are realized in this new system. The video image-capturing capabilities make it simple to include the sample images into documents and archives, resulting in extremely powerful and valuable documentation. The InspectIR makes it possible to obtain high-quality spectral data of micro-samples, faster and better than ever before.

References

[1] D. W. Sting, R. G. Messerschmidt, J. A. Reffner, *U.S. Patent* 5,019,715 (assigned to Spectra-Tech, Inc., Shelton, CT), 1991.

[2] J. A. Reffner, and P. A. Martoglio, *Practical Guide to IMS*, Dekker, New York, 1995.

[3] G. Kortum, *Reflectance Spectroscopy*, Springer, New York, 1969.

[4] H. J. Harrick, *Internal Reflection Spectroscopy*, Harrick Scientific Corp., Ossining, NY, 1979.

Mikrochim. Acta [Suppl.] 14, 339–341 (1997)
© Springer-Verlag 1997

Infrared Microspectroscopy with Synchrotron Radiation

John A. Reffner[1,**], **G. Lawrence Carr**[2], and **Gwyn P. Williams**[3]

[1] Spectra-Tech, Inc. P.O. Box 869, 2 Research Dr. Shelton, CT 06484–0869, U.S.A.
[2] Northrop Grumman Corp. 1111 Stewart Ave. Bethpage, NY 11714, U.S.A.
[3] NSLS Brookhaven National Laboratory, P.O. Box 5000, Upton, NY 11973–5000, U.S.A.*

Abstract. The brightness, spectral distribution, and low noise of synchrotron radiation are major advantages for Fourier-transform infrared microspectroscopy. The infrared radiant emission from the National Synchrotron Light Source (NSLS) is 100–1000 times brighter than that from a 1500 K thermal-emission source [1]. Using synchrotron radiation, an Irμs (Nicolet Corp.) microspectrometer system demonstrated spatial resolution unequaled by conventional sources. With this system, spectra with a signal-to-noise ratio greater than 200:1 could be collected in less than 5 seconds from a $6 \times 6 \mu m$ sample area (defined by dual confocal remote apertures). With $6 \times 6 \mu m$ confocal dual apertures, the system clearly resolved $4 \mu m$ wide layers in a multilayered laminate.

Key words: infrared microspectroscopy, synchrotron radiation, infrared, polymer.

In a synchrotron, the light intensity depends upon the number of electrons in the circulating bunches. The loss of electrons from these bunches is a slow exponential function with a half-life of several hours. This means that synchrotron-produced infrared radiation is very stable. This infrared radiation is free from the noise associated with thermal sources and is highly collimated; additionally, the source size is small, only $200 \times 400 \mu m$. All these features of synchrotron infrared emission combine to increase the signal-to-noise ratio (SNR), improve spatial resolution and extend the useful spectral range for infrared microspectroscopy.

The first installation of an FT-IR microspectrometer on a synchrotron took place at Brookhaven National Laboratory (NSLS) on 12 September 1993, a few weeks after the 9th International FT-IR Conference in Calgary, Alberta, Canada. This research was a direct result of the work reported by Gwyn Williams at that conference.

Experimental

An IRμs infrared microspectrometer system was installed on the U2b infrared beam line on the VUV ring at the NSLS at Brookhaven National Laboratory. With this system, a 40-fold increase in the SNR for a 10-μm diameter spot was immediately demonstrated. The beam of synchrotron radiation was focused onto a diffraction-limited spot, with 70% of the total incident beam energy transmitted through a 12.5-μm diameter pinhole. The experimentally measured profile of the beam intensity incident on the sample is reported in Fig. 1. With this system, spectra with a SNR greater than 200:1 could be collected in less than 5 seconds from a $6 \times 6 \mu m$ sample area (defined by dual confocal remote apertures).

To determine the spatial resolution of the IRμs system with synchrotron radiation, a series of spectra was recorded of a defined fine structure in steps of 1.0 μm. Figure 2 is a photograph of a cross-section of a sample of photographic film emulsion. Five layers can be seen, the outermost one being at the top of the picture. The remains of the backing paper are at the bottom. The layer thickness are 6, 4, 4, 8 and 32 mm. Figure 3 shows a series of infrared absorption spectra taken across the layers of the film at intervals of 1 μm, and in the range 3750–1000 cm⁻¹. The chemical signatures of the various layers are easily visible. The outermost layer is gelatin, the following three layers are different mixtures of poly(vinyl acetate) and poly(vinyl alcohol) and the thick layer is polyethylene. These data were taken at a spectral resolution of 8 cm⁻¹, 16 scans being co-added for a total acquisition time of about 5 seconds per spectrum. The interface between the outer gelatin layer and the adjoining polymer provides a test of the system's spatial resolution. Figure 4 is a set of spectra extracted from the data shown in Fig. 3, where the interface between layers 1 and 2 is traversed. The spectra shown in Fig. 4 correspond to shift of the beam in 2-μm steps, with a total of 8 μm from the first to the last spectrum. It is clear that within this 8 μm, the 6 μm spot has moved from the gelatin totally into the adjacent polymer.

Results

In these tests, we found that we could get most of the flux from the NSLS through a 10 μm aperture. This is

* Supported by the U.S. D.O.E., under contract D.E.-A.C. 02-76C.H.00016
** To whom correspondence should be addressed

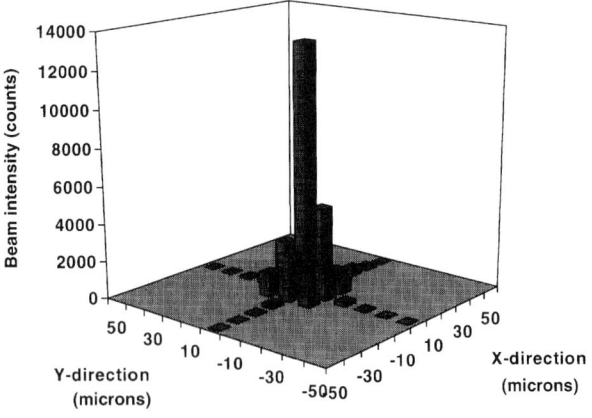

Fig. 1. The experimentally measured profile of the beam intensity incident on the sample, measured by stepping a 12.5 µm diameter pinhole in 10 µm steps along two orthogonal directions

Fig. 2. A photomicrograph of the cross-section of a multilayered photographic material

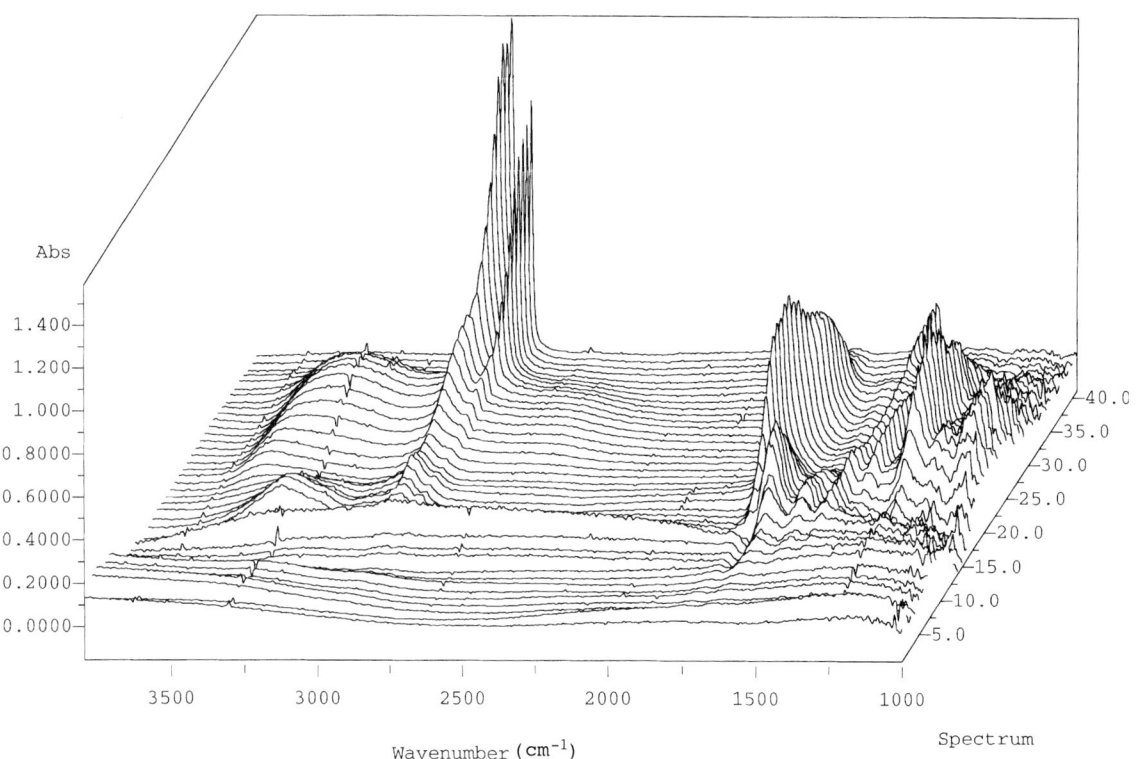

Fig. 3. A sequential collection of spectra at 1 µm intervals across the multilayered photographic material cross-section; the dual confocal apertures were set at 6 × 6 µm and each spectrum was collected at 8 cm^{-1} spectral resolution, with 16 scans co-added (5 second per spectrum)

consistent with what would be expected for a 1 mm beam size demagnified by a factor of 100 on going from $f/100$ optics to $f/1$ optics. It is also significantly better than that obtained by using the conventional blackbody source installed in the instrument, in which roughly the same signal was obtained for an aperture of 100 µm. Overall, then, the enhancement was around 100 times, although the improved stability of the synchrotron source produced spectra of lower noise for a given measuring time.

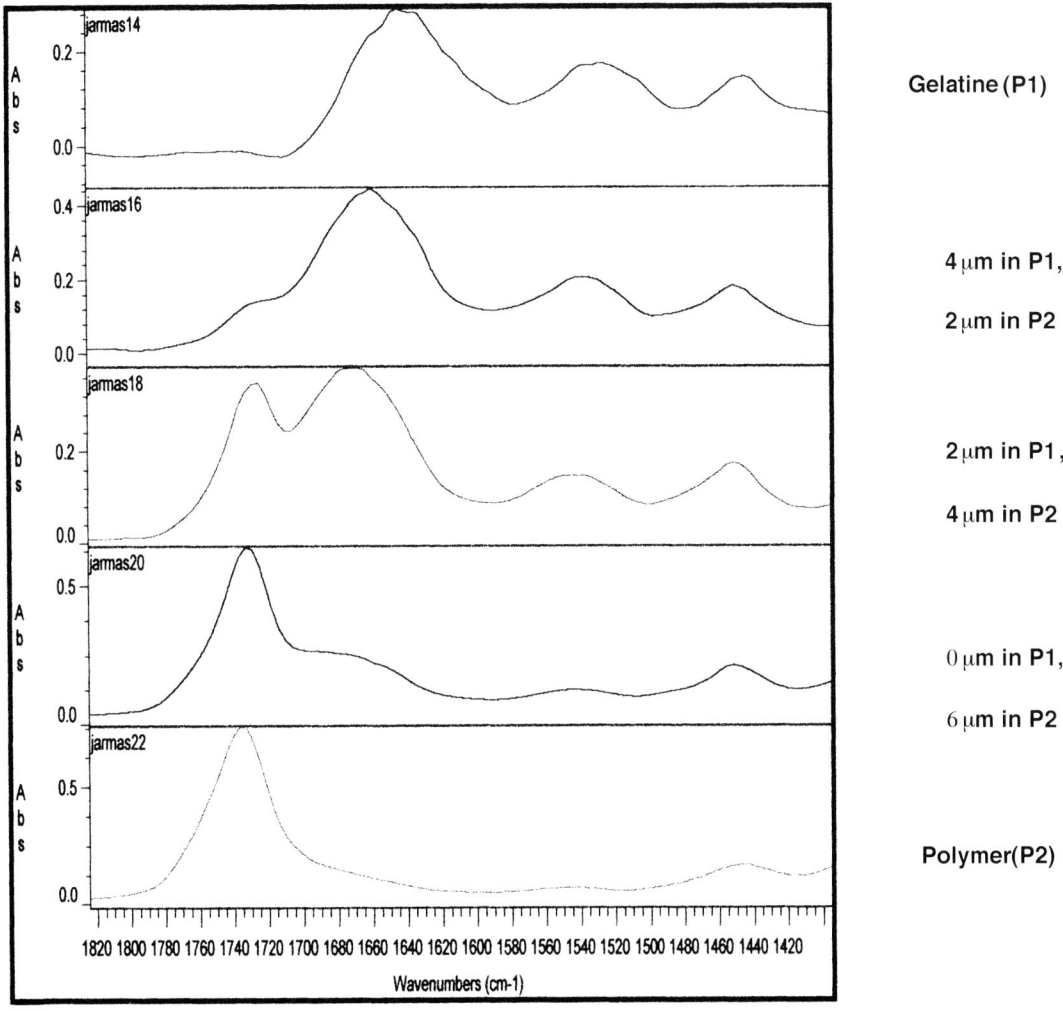

Fig. 4. A series of spectra at 2 μm intervals across the interface between layers 1 and 2; a 6 × 6 μm dual aperture was used to define the infrared beam size on the sample, and complete spectral resolution was accomplished in 8 μm

Conclusions

Infrared microspectroscopy with synchrotron radiation has been applied to the analysis of multilayered laminates, electronic devices, array detectors, geological materials, biological specimens, and forensic evidence [23]. These analyses illustrate the new levels of performance and resolution that are achievable through the use of synchrotron radiation. The confocal aperturing, diffraction limited optical system and the intensity of the synchrotron source combine to produce unequaled resolution in infrared microspectroscopy. In addition, use of a copper-doped germanium

detector in recent studies has extended the spectral range for infrared microspectroscopy to $400 \, \text{cm}^{-1}$. This extended spectral range is particularly important in the study of inorganic materials. With synchrotron radiation, new areas of research have been opened for infrared microspectroscopy.

References

[1] G. P. Williams, *Nucl. Instrum. Methods Phys. Res. Sect. A* **1990**, *A291*, 8.
[2] J. A. Reffner, P. A. Martoglio, G. P. Williams, *Rev. Sci. Instrum.* **1995**, *66*, 1298.
[3] C. Meade, J. A. Reffner, E. Ito, *Science* **1994**, *264*, 155.

Mikrochim. Acta [Suppl.] 14, 343–344 (1997)

Cross-sectional Infrared Transmission Measurements for Highly Sensitive Thin Film Characterization

Adele Sassella[1,*], **Alessandro Borghesi**[2], and **Branko Pivac**[3]

[1] Dipartimento di Fisica, Università degli Studi di Milano, Via Cetoria 16, I-20133 Milano, Italy
[2] Dipartimento di Fisica, Università degli Studi di Modena, Via Campi 213a, I-41100 Modena, Italy
[3] R. Boskovic Institute, P.O. Box 1016, HR-10000 Zagreb, Croatia

Abstract. The results of transmission measurements of SiO_2 films in silicon performed in the standard experimental configuration and in a new cross-sectional configurat†ion have been compared. The cross-sectional measurements proved to be the most suitable for characterization of very thin (few nanometres thick) films and in addition are very sensitive.

Key words: thin films, infrared characterization, thickness measurement.

Infrared (IR) spectroscopy can be successfully employed in the study of dielectric films, giving information on their stoichiometry, structure, and impurity content. It is also a very good characterization tool for use in the microelectronics industry, being non-destructive and very sensitive. One of the limits of standard IR transmission measurements is that they are not suitable for the characterization of very thin films (thinner than 50 nm), the optical response being directly related to the sample volume investigated.

Recently, it has been demonstrated [1] that a new optical arrangement for IR transmission measurements permits films that are a few nanometres thick to be successfully studied. This new technique leads to highly sensitive thickness measurements and to the determination of the frequency ω_{LO} of the longitudinal optical phonon of the material composing the film, strictly related to its main properties [2].

Here, the suitability of the new optical configuration for thin film characterization is discussed by comparing the experimental results obtained for two SiO_2 films embedded in silicon, one 59 Å thick and the other 296 Å thick, with those from standard IR transmission

measurements of the same samples. Such a comparison shows the adequacy of standard transmission measurements for thick films and of the new technique for very thin ones.

Experimental

The samples studied were four bonded wafers with the SiO_2 film of different thickness, i.e. 59 Å, 296 Å, 0.945 μm, and 1.872 μm, manufactured by Shin-Etsu Handotai. Strips 200 μm wide and a few cm long were cut from such wafers for measurement with the light beam impinging on the wafer cross-section. The IR transmission measurements were performed in the spectral range from 5000 to 500 cm^{-1}, by using a Bruker IFS 113v Fourier-transform spectrometer, with a Globar source, a KBr beam splitter, and an N_2-cooled mercury–cadmium telluride detector. Resolution was 4 cm^{-1} and 1024 scans were accumulated for each spectrum to improve the signal-to-noise ratio. Two optical configurations were used, i.e. with the IR beam impinging at near-normal incidence either on the wafer cross-section (with the beam parallel to the SiO_2 film surfaces) or on the wafer surface (with the beam perpendicular to the film, as in standard measurements). To improve spatial resolution an IR microscope was used for the cross-sectional measurements, with a ×36 objective and 13 × 100 μm rectangular spots, parallel to the film. Performing the measurements on such small spots led to spectra with a higher noise level, owing to the lower intensity of the IR beam on the sample. Spectra taken on bare silicon in the two configuration were used as reference.

Results and Discussion

Figure 1 reports the transmission spectra of the 59 Å thick film (continuous line) and of the 296 Å film (dashed line), as obtained in the cross-sectional configuration schematically illustrated in the inset. The strong absorption band at 1256 cm^{-1} in the spectrum of the thinner film is related to the interaction of surface

---

* To whom correspondence should be addressed

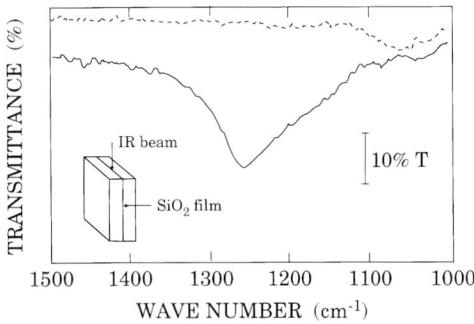

Fig. 1. Transmittance spectra of a 59 Å thick SiO$_2$ film (continuous line) and a 296 Å thick film (dashed line) embedded in silicon, taken in the experimental configuration sketched in the inset

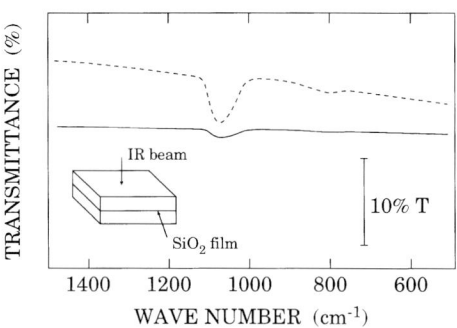

Fig. 2. Transmittance spectra of a 59 Å thick SiO$_2$ film (continuous line) and a 296 Å thick film (dashed line) embedded in silicon, taken in the standard experimental configuration sketched in the inset

modes in the film and is demonstrated to be exactly at ω_{LO} of the material composing the film (ω_{LO} of amorphous SiO$_2$ is precisely 1256 cm^{-1}). In the thicker film such an interaction is no longer effective.

In these spectra, two absorption peaks should be present, as predicted by the expression of the transmittance of the film, which, in the particular configuration used, can be regarded as a grazing angle transmittance. In the thin film limit, its approximate expression is:

$$T \simeq \frac{\cos\theta/n}{\cos\theta/n + \delta[(\cos\theta/n)^2 Im\,\tilde{\varepsilon} + n^2 \sin^2\theta \; Im \; (-1/\tilde{\varepsilon})]} \quad (1)$$

where θ is the angle of incidence on the film surface, n is the real part of the refractive index of the medium surrounding the film, δ is proportional to the film thickness, and $\tilde{\varepsilon}$ is the complex dielectric function of the material composing the film. The term proportional to $[Im\,\tilde{\varepsilon}]^{-1}$ and that proportional to $[Im\,(-1/\tilde{\varepsilon})]^{-1}$ are responsible for the absorption at ω_{TO} and at ω_{LO}, respectively. For SiO$_2$ films in silicon the latter term is demonstrated [1] to give rise to a much stronger absorption at ω_{LO} with respect to that at ω_{TO}, which is almost undetectable.

The sensitivity of this kind of measurement for very thin films can be further increased by using a polarized IR light beam with the electric field perpendicular to the film. In this case, the transmission of the 59 Å thick film at the ω_{LO} peak is as low as about 30%, while for the thicker film the improvement is very small.

To point out the advantages and the limitations of the cross-sectional configuration, in Fig. 2 the IR spectra of the same samples of Fig. 1 are reported, as obtained in the standard configuration for transmission measurements sketched in the inset. The con-

tinuous line is the spectrum of the 59 Å thick film and the dashed line is the spectrum of the 296 Å thick one. The figure clearly shows a behaviour opposite to that in the results of Fig. 1, namely, the transmittance of the 296 Å thick film shows the characteristic absorption band at about 1080 cm^{-1}, i.e. at ω_{TO} of amorphous SiO$_2$, and a slight absorption at about 810 cm^{-1}, whereas the spectrum of the 59 Å thick film is almost completely flat. In the spectra of the thicker films studied (not shown) all the three bands at 1080, 810 and 450 cm^{-1}, characteristic of SiO$_2$, are detected.

For this experimental configuration, neglecting reflectance, the transmittance of the film can be written as $T = \exp(-\alpha t)$, where α is the absorption coefficient of SiO$_2$ and t the film thickness. As α is maximum at ω_{TO}, in this spectral position the transmittance is expected to be minimum. From the literature data [3] for the optical functions of SiO$_2$, the transmittance of the 59 Å thick film at the ω_{TO} peak is about 98%, while that of the 296 Å thick film is about 90%, in agreement with the experimental findings of Fig. 2.

From the comparison of the spectra in the two figures the cross-sectional measurements turn out to be the most adequate to detect and characterize films that are a few nanometres thick, and almost undetectable by means of standard transmission measurements. On the other hand, the standard configuration leads to more sensitive measurements of films thicker than about 20–30 nm.

References

[1] A. Borghesi, A. Sassella, *Phys. Rev. B* **1994**, *50*, 17756.
[2] C. Martinet, R. A. B. Devine, *J. Appl. Phys.* **1995**, *77*, 4343.
[3] H. R. Philipp, in: *Handbook of Optical Constants of Solids* (E. D. Palik, ed.), Academic Press, Orlando, 1985, p. 749.

Mikrochim. Acta [Suppl.] 14, 345–347 (1997)
© Springer-Verlag 1997

Micro FT-IR and EXAFS Studies on Periodic and Chaotic Precipitation in Bile Salt Systems

Libo Wang[1], **Tiandou Hu**[2], **Jinguang Wu**[1,*], **Zhenhua Xu**[1], **Xu Zhang**[1], **Duanfu Xu**[3], and **Guangxian Xu**[1]

[1] Department of Chemistry, Peking University, Beijing 100871, Peoples Republic of China
[2] Synchrotron Radiation Laboratory, Institute of High Energy Physics, Beijing 100039, China
[3] Institute of Chemistry, Academia Sinica, Beijing 100080, Peoples Republic of China

Abstract. Micro FT-IR and EXAFS spectroscopy methods were utilized to characterize the periodic and chaotic precipitation occuring in some metal bile salt systems in sol–gel medium. The results indicate that the precipitates were just the corresponding metal bile salt complexes, and the agar gel medium has very weak interaction with the dispersed ions. In addition, comparison between the EXAFS spectra of the precipitate formed in the CoCl$_2$–NaDC gel system and of the complexes Co(DC)$_2$ synthesized in water–ethanol solvent suggested that the two complexes were of similar composition but the precipitate in the gel had a more ordered coordination structure.

Key words: Micro FT-IR, EXAFS, periodic precipitation, chaotic precipitation, bile salt

During gallstone growth, the precipitation of salts and metals often takes the form of periodic precipitation (Liesegang rings) or the form of chaotic precipitation [1]. Very similar phenomena have also been observed in vitro, when metal ions and bile salt ions are dispersed in sol–gel medium [2]. It is believed that these experimental results must have some correlation with the biological phenomena, so it is very useful for gallstone research to study the periodic and chaotic precipitation in bile salt system in vitro, and it is also a new way of using the non-linear scientific principle to study the etiology of gallstone formation. In our study, the precipitation phenomena in some metal–sodium deoxycholate (NaDC) systems, including CoCl$_2$–NaDC, MnCl$_2$–NaDC, CuCl$_2$–NaDC and NdCl$_3$–NaDC systems in agar gel medium were examined. Under certain conditions, periodic precipitation was found in the CuCl$_2$–NaDC and NdCl$_3$–NaDC systems (see Fig. 1), and manifold chaotic precipitation in the other two systems (see Fig. 2). In the present paper, micro FT-IR and EXAFS spectroscopic methods were used to analyse the precipitates formed in the gel.

Experimental

In a boiling-tube, 18 ml of agar gel containing NaDC at known concentration, 7 ml of agar gel, and 5ml of agar gel containing metal

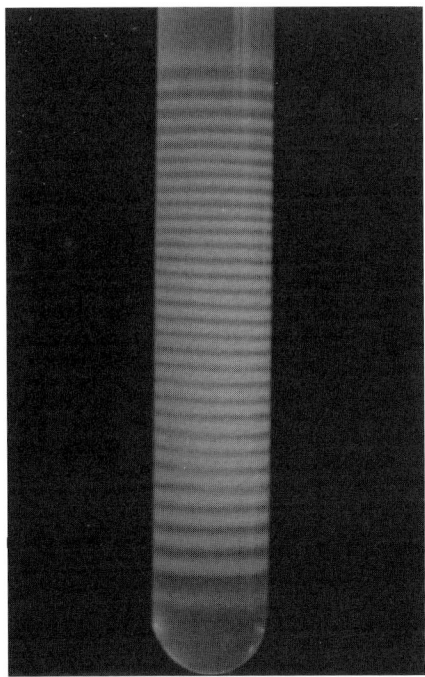

Fig. 1. Periodic precipitation in NdCl$_3$–NaDC system

---

* To whom correspondence may be addressed

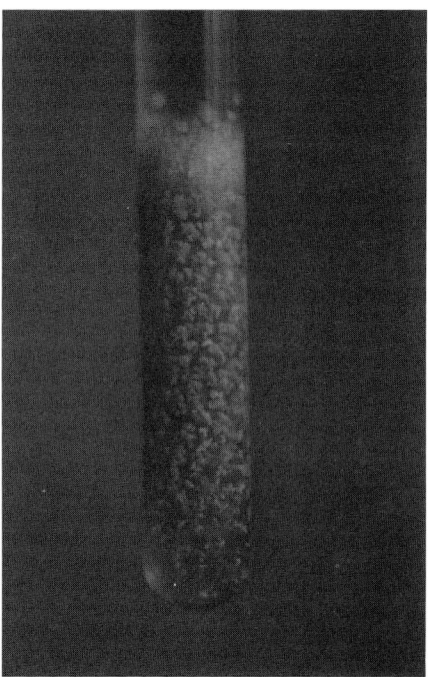

Fig. 2. Chaotic precipitation in CoCl$_2$–NaDC system

ions at known concentration were placed in that order. After 10 days or so, the periodic or chaotic precipitation took place. The precipitates were then washed out from the gel medium and dried thoroughly for micro FT-IR and EXAFS measurement.

Micro FT-IR spectra were recorded on a Nicolet 750 FT-IR spectrometer with a Nic-plan microscopy accessory, in the region 4000–650 cm^{-1}.

The EXAFS experiments were performed in the EXAFS experiment station at BSRF (Beijing Synchrotron Radiation Facility) in the 4W1B beam line in transmission mode at the Co K-edge. The storage ring was run at 2.2 GeV and the current was in the range 20–50 mA, with as Si(III) double-crystal monochromator. The energy resolution was about 2 eV at the Co K-edge. Each sample was measured three times.

The corresponding metal bile salt complexes Co(DC)$_2$, Mn(DC)$_2$, Cu(DC)$_2$ and Nd(DC)$_3$ were synthesized in water–ethanol solvent to provide references for the precipitates formed in the gel.

Results and Discussion

Table 1 shows the infrared frequency and the assignment of the main peaks of the precipitates formed in some metal bile salt systems in agar gel medium. The data for the synthesized complexes are also listed for comparison. The results show that the FT-IR spectra of the precipitates formed in the gel resembled the spectra of the synthesized complexes in both peak location and absorption intensity. This indicates that the precipitates were just the corresponding complexes and that interaction between the agar gel medium and the dispersed ions was not detected.

Furthermore, the precipitate formed in the CoCl$_2$–NaDC system and the synthesized Co(DC)$_2$ have been studied by EXAFS. The EXAFS data are listed in Table 2. We can see that the samples obtained in the two different ways were identical in coordination number N and the M–O bond length R, but some small differences can be seen clearly by comparing their EXAFS spectra (see Fig. 3), which showed that the precipitate formed in the gel have clear curves corresponding to the first and second

Table 2. EXAFS data

R (A)		N	σ^2	
CoCl$_2$–NaDC (g)		2.07	6.0	0.009
Co(DC)$_2$ (s)		2.07	5.5	0.009

g = sample formed in gel, s = synthesized sample, σ^2 = Debye–Wallar factor.

Table 1. Infrared frequency and assignment of some main peaks (cm^{-1})

Precipitate	ν_{O-H}	ν_{C-H}	ν_{as}COO–	ν_sCOO–
CoCl$_2$–NaDC (g)	3625 3367	2932 2905 2861	1553 1524	1413
Co(DC)$_2$ (s)	3624 3368	2932 2905 2861	1552 1524	1413
MnCl$_2$–NaDC (g)	3624 3346	2938 2863	1603 1572 1553	1414
Mn(DC)$_2$ (s)	3624 3351	2938 2863	1603 1572 1556	1413
CuCl$_2$–NaDC (g)	3405	2937 2865	1597 1561	1416
Cu(DC)$_2$ (s)	3402	2938 2865	1613 1564	1416
NdCl$_3$–NaDC (g)	3613 3406	2937 2864	1541 1538	1418
Nd(DC)$_3$ (s)	3671 3396 3609	2936 2865	1562 1547 1536	1422 1416

g = samples formed in gel, s = synthesized samples.

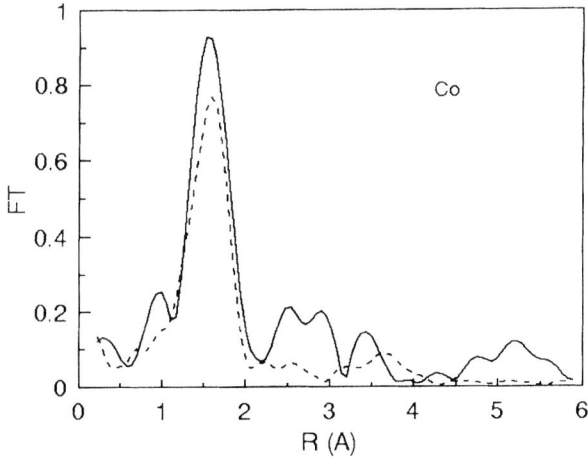

Fig. 3. Comparison of the spectra between the precipitate Formed in $CoCl_2$–NaDC System and the Synthesized $Co(DC)_2$. ———— precipitate in gel, --------- synthesized $Co(DC)_2$

coordination shells whereas the synthesized material had a clear curve for only first coordination shell; in other words, the precipitate formed in the gel had a more orderly coordination structure.

Acknowledgements. This project was supported by NSF of China and National Climbing Program of China.

References

[1] X. F. Li, R. D. Soloway, W. D. Huang, J. G. Wu, Y. Huang, W. Z. Huang, Z. X. Zhu, N. F. Zhou, D. F. Xu, *Clin Res.* **1993**, *41*, 158A., 415.

[2] L.-B. Wang, R. D. Soloway, J.-G. Wu, S. F. Weng, W.-D. Huang, D.-F. Xu, N. Shi, G.-X. Xu, *Gestroentesology* **1995**, *108*, A440.

Mikrochim. Acta [Suppl.] 14, 349–351 (1997)
© Springer-Verlag 1997

Single Fiber Characterization by Polarization FT-IR Microspectroscopy

David L. Wetzel* and **Liling Cho**

Kansas State University, Microbeam Molecular Spectroscopy Laboratory, Shellenberger Hall, Manhattan, KS 66506, U.S.A.

Abstract. Polarization FT-IR microspectroscopic data obtained from unstretched and stretched polyethylene terephthalate single fibers showed a pronounced dichroic effect for six of the fundamental vibrations and two overtones. The change in dichroic ratio, in one instance, was from 1.06 (unstretched) to 0.38 (stretched). Orientation of the macromolecules in a fiber during manufacture is a useful forensic science tool for individualization. The spatially resolving infrared microspectrometer performs this function in addition to localized chemical analysis.

Key words: infrared microspectroscopy, FT-IR, fiber analysis, forensic spectroscopy.

In forensic science, fiber individualization is necessary to narrow down the field of choice and indicate mismatch or probable match between fibers. Chemical information regarding the composition of the polymer itself and anything adhering to the fiber may potentially be contributed by infrared (IR) spectroscopy. Furthermore, orientation of the macromolecules in the fiber may be revealed from IR spectroscopic data. The techniques used in this paper demonstrate the utility of a state-of-the-art integrated microspectrometer for examining single fibers to enhance fiber individualization based on the degree of orientation or crystallinity brought about by the particular manufacturing conditions used. The utility of the dichroic ratio in fiber identification has been demonstrated by work performed at the FBI Forensic Laboratory [1].

Experimental

The instrument used was an infrared microspectrometer (IR μs, Spectra-Tech, Shelton, CT). The system consists of an IR microscope as an integral part of the interferometer bench of a Fourier-transform IR spectrometer. This integrated system, designed as a unit, is optically optimized to conserve throughput. Front surface optics (Cassegrainian objective and condenser) are used to pass infrared radiation to and from the specimen on the microscope stage. Small areas in a single fiber in the field of view are selectively examined by the use of remote (projected) apertures before and after the focusing optics. A liquid-nitrogen cooled small area MCT detector is used [2]. An infrared polarizer accessory equipped with a micro positioning dial may be placed in the incident IR beam. With minimal sample preparation, nearly all fibers can be analyzed without chemical alteration. The polyester fiber in the present study, polyethylene terephthalate (PET), was stretched on the microscope stage until a "neck" could be seen on the stretched fiber. Infrared spectra were obtained in both polarization orientations, parrallel and perpendicular to the longitudinal axis of the fiber. Differences in the orientation under polarized radiation affected the absorptivity coefficients. Ratioing the absorbance of a particular band under polarization parallel to the molecular chain axis to the absorbance in the perpendicular orientation resulted in a dichroic ratio. We compared the dichroic ratio between the undrawn, neck, and drawn transition portions of the stretched PET fiber. In the on-stage stretching the fibers were typically elongated to 140% of their original length. Spectra were obtained in the transmission mode with a $15\times$ objective, a $10\times$ condenser, 12×48 μm dual apertures, 256 co-added scans and resolution of 4 cm^{-1}.

Results and Discussion

Six fundamental and two overtone bands were selected for the present study. These bands were chosen because they were relatively free from overlapping bands and were weak enough not to be over-absorbing for thick, unflattened fibers [3, 4]. Figure 1 shows the difference between parallel and perpendicular polarization for a 3.19 draw-ratio PET fiber. The spectra showed stronger peaks at 973, 1455, 1505, 1579 cm^{-1} when the polarization was parallel to the fiber axis. When the fiber was stretched, the changes of the dichroic ratio were not uniform (Fig. 2). For the peaks at 876, 1376, 1960 and 3435 cm^{-1} the dichroic ratio decreased from the undrawn to the drawn portion (Table 1).

---

* To whom correspondence should be addressed

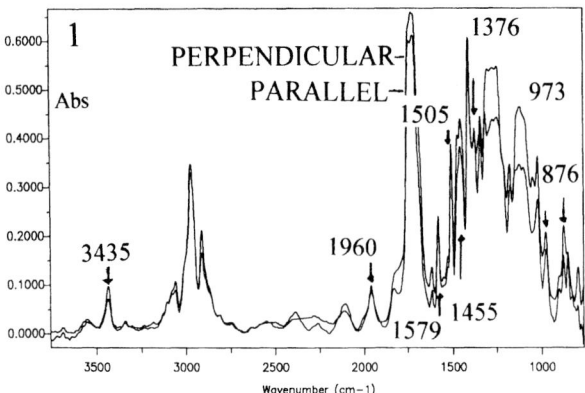

Fig. 1. Infrared spectra of 3.19 draw ratio polyester fiber

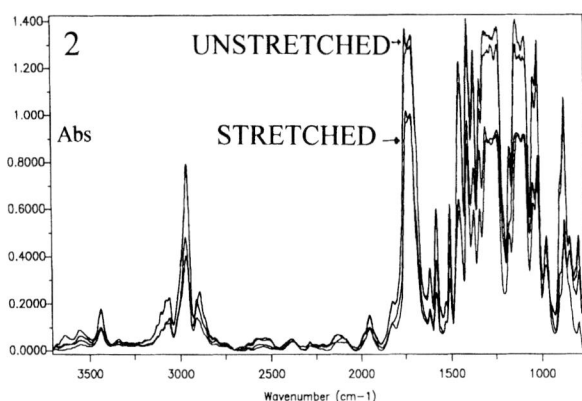

Fig. 2. Infrared spectra of stretched and unstretched fiber

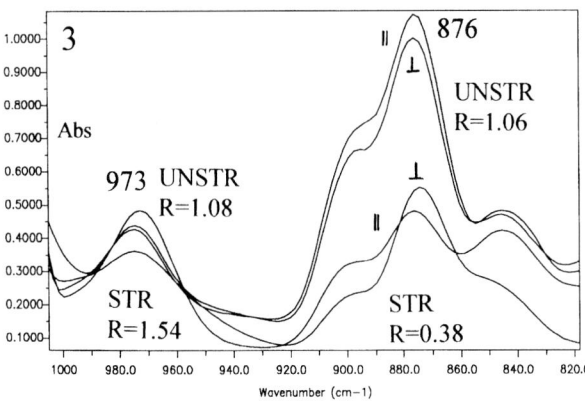

Fig. 3. Infrared spectra with bands at 876 and 973 cm⁻¹

Table 1. Dichroic ratio decreased on drawing

Frequency (cm^{-1})	Assignment	Drawn portion (R)
876	Phenyl C–H band	0.3758
1376	CH$_2$ Wagging	0.8828
1960	Phenyl overtone	0.8741
3435	Carbonyl overtone	0.8492

In the stretching process, the terephthalate groups are assumed to become oriented perpendicular to the molecular chain axis. This is consistent with the observation that for the peaks at 973, 1455, 1505 and 1579 cm^{-1}, the dichroic ratio increased from undrawn to drawn portions (Table 2), i.e. the band intensity with perpendicular polarization is increased from the undrawn to the drawn portion of the fibers examined (Figs. 3–6). The data presented are typical of those from numerous experiments.

Table 2. Dichroic ratio increased on drawing

Frequency (cm^{-1})	Assignment	Drawn portion (R)
973	C–O Stretch	1.0951
1455	CH$_2$ Bending	1.0782
1505	Phenyl C–C stretch	1.7802
1579	Phenyl C–C stretch	2.0812

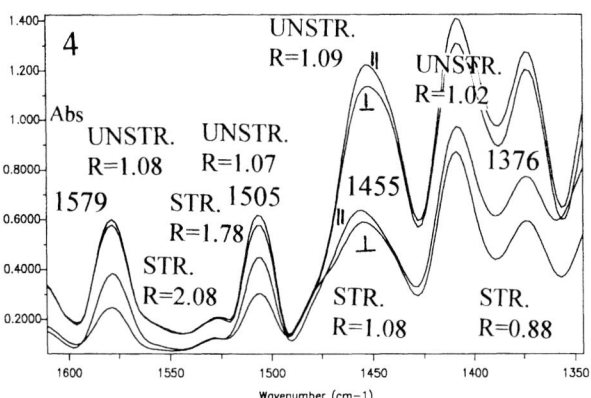

Fig. 4. Infrared spectra with bands at 1579, 1505, 1455 and 1376 cm^{-1}

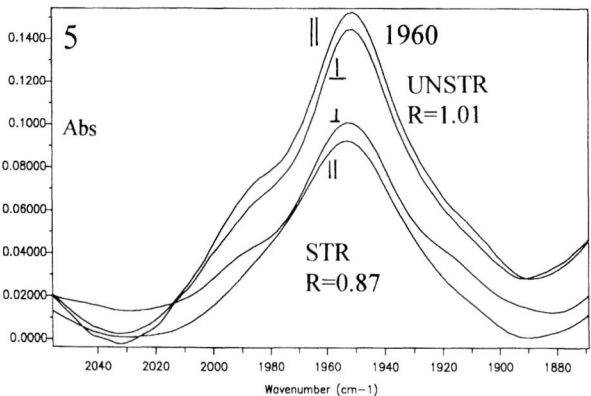

Fig. 5. Infrared spectra with band at 1960 cm^{-1}

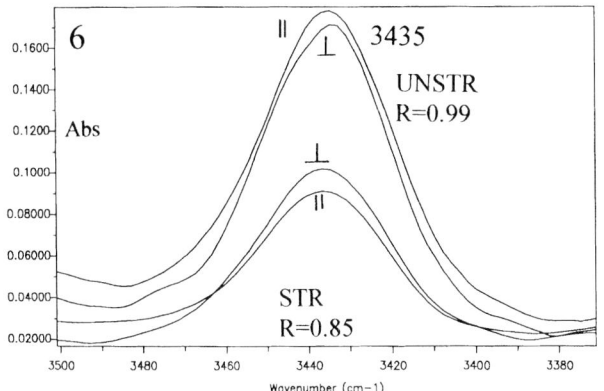

Fig. 6. Infrared spectra with band at 3435 cm^{-1}

Conclusion

In this work the polymer PET was used as a model to demonstrate that polarized IR spectroscopy has in fact become practical for single fibers. The dichroism of the unstretched fiber was contrasted with that of the stretched fiber and eight wavelengths were chosen for documentation of the dichroic characteristics. We believe that achievement of single fiber information of this type may assist in fiber individualization and forensic analysis.

Acknowledgements. We wish to thank John A. Reffner for useful discussions, and NSF EPSCOR: OSR-9255223 for support. Contribution No. 96-785 from Kansas Agricultural Experiment Station, Manhattan.

References

[1] M. W. Tungol, E. G. Bartick, A. Montaser, in: *Practical Guide to Infrared Microspectroscopy* (H.J. Humecki, ed.), Dekker, New York, 1995, pp. 245–285.
[2] D. L. Wetzel, in: *Food Flavors, Generation, Analysis and Process Influence* (G. Charalambous, ed.), Elsevier, Amsterdam, 1995, pp. 2039–2108.
[3] S. S. Sikka, H. H. Kausch, *Colloid Polymer Sci.* **1979**, *257*, 1060.
[4] S. P. Church, N. Khan, *Polymer Bull (Berlin)* **1993**, *30*, 559.

Mikrochim. Acta [Suppl.] 14, 353–355 (1997)
© Springer-Verlag 1997

Synchrotron-Powered FT-IR Microspectroscopy: Single Cell Interrogation

David L. Wetzel[1,*], **John A. Reffner**[2], and **Gwyn P. Williams**[3]

[1] Kansas State University, Microbeam Molecular Spectroscopy Laboratory, Shellenberger Hall, Manhattan, KS 66506, U.S.A.
[2] Spectra-Tech, Inc., 2 Research Drive, Shelton, CT 06484, U.S.A.
[3] Brookhaven National Laboratory, National Synchrotron Light Source, U2B, Bldg. 725B, Upton, NY 11973, U.S.A.

Abstract. Excellent spatial resolution is achieved with an integrated FT-IR microspectrometer (IRµs, Spectra-Tech) in which the globar (thermal) source is replaced by synchrotron radiation. The infrared portion of the spectrum extracted from the vacuum ultraviolet storage ring of the National Synchrotron Light Source (NSLS) of the Brookhaven National Laboratory, is not only brighter, but is free from thermal noise and concentrated into a small area. Passing the beam with low divergence through an aperture does not severely attenuate the throughput as is the case with a thermal source. Spectra obtained with 5 and 6µm diameter apertures from adjacent tissue in a mouse retina, a wheat aleurone cell and cell wall, and two cells in wheat primary root, have shown localized chemical differences. Functional group contour maps and 3D maps of mouse cerebellum, a grass vascular bundle and a cross-section of rye illustrate the utility of the system described.

Key words: infrared, microspectroscopy, FT-IR, synchrotron.

Excellent spatial resolution is required to allow spectroscopic interrogation of single cells in the presence of adjacent tissue on the microscope stage. Recent developments in instrumentation have made it possible to obtain good spectra, with small-aperture scans, in a reasonable time.

Instrumentation

IR microspectroscopy instrumentation has progressed to this optical state-of-the-art in four stages from (1) a microscope accessory on a dispersive spectrometer to (2) a microscope with remote-projected apertures to allow microscopic selection of the part of the field to be sampled, interfaced to an FT-IR instrument with a miniature detector; then (3) an integrated instrument designed and optically optimized as a single unit, which has provided good enough signal-to-noise ratios for smaller apertures to be used before and after the microscope stage. The IRµs (Spectra-Tech, Shelton CT) equipped with a programmable motorized stage has allowed functional group mapping from multiple spectra obtained in a grid pattern [1, 2]. This has been used to map brain tissue [3, 4] grain sections [5–7], and other specimens. Finally, stage (4) resulted a recent International FTS Conference (Calgary 1993) where synchrotron work in the far IR was presented and two of the authors of this paper (G.P.W. and J.A.R.) met. Within two months the synchrotron source and the slightly modified IRµs were united in a hasty shotgun marriage. IR radiation was obtained from a standard synchrotron port without modification. Since that hasty union, a synchrotron beam line equipped with a permanent IR microspectrometer has been established [8]. Synchrotron radiation is 1000 times brighter than a standard thermal (1200 K) source and thus offers considerable advantages for IR microspectroscopy. The brightness advantage of the synchrotron source [8] is shown in Fig. 1, in which the signal at the NSLS U2B beam-line (Brookhaven), through a 10 micron pinhole at $f/1$ optics, is shown and compared with that from a 1200 K thermal source with the same optics. Also shown in the typical noise level for the commonly used liquid-nitrogen cooled photo-conductive mercury cadmium telluride detector. The synchrotron S/N advantage is evident in this case. For larger samples, this source may not be an advantage. Since signal-to-noise in the synchrotron-powdered IRµS is favorable even with small double apertures, excessive

---

* To whom correspondence should be addressed

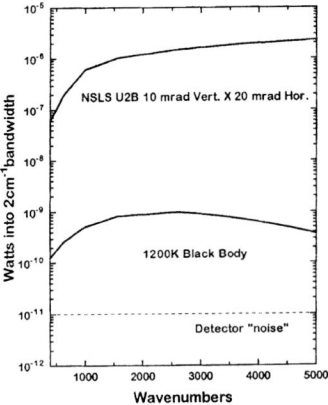

Fig. 1. Relative intensities (synchrotron and thermal)

co-adding of scans is not necessary. This permits single cell interrogation, detailed mapping with cellular dimensions, and mapping of large microscopic sections from many small spot measurements.

Results and Discussion

The single-cell spectra shown in Figs. 2 and 3 show the chemical difference between a barley aleurone cell and the adjacent cell wall (5 μm diameter round spot), and between different cells in wheat primary root (6 × 6 μm spot). Figures 4 and 5 show two different spectra from a line map of retinal tissue (mouse) that reveal in contrast the high protein and high lipid portions. Tissue appearing homogeneous, when mapped with IR may reveal chemical differences hitherto undetected without staining. Figure 6 is a wavelength-specific contour map of mouse cerebellum in which the features coincide with visual observations. The 3D map in Fig. 7 shows a grass leaf vascular bundle expressed in

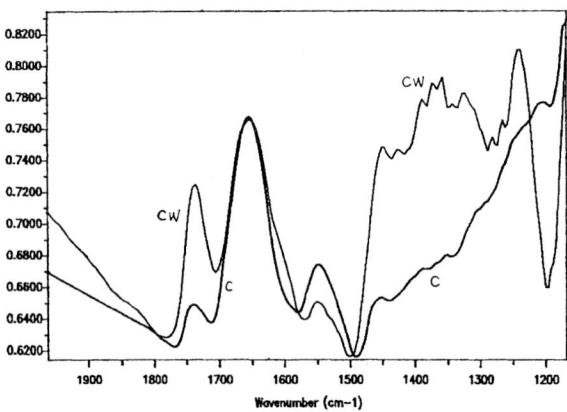

Fig. 2. Aleurone cell (*c*) and cell wall (*cw*) (5 μm)

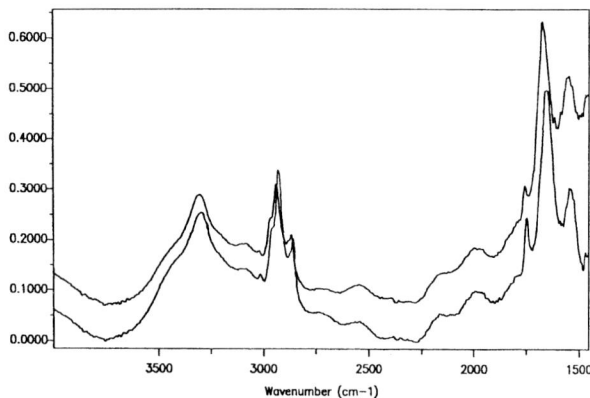

Fig. 3. Wheat primary root cells (6 μm)

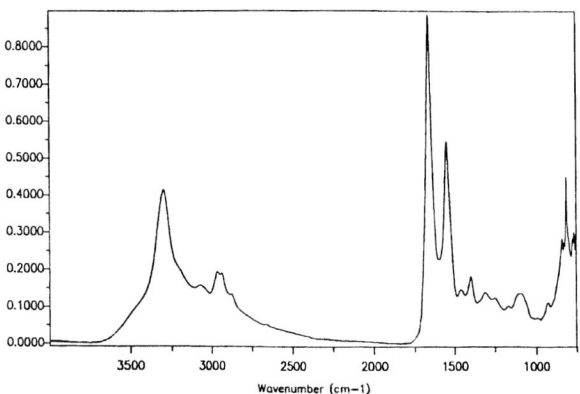

Fig. 4. Mouse retina (high protein)

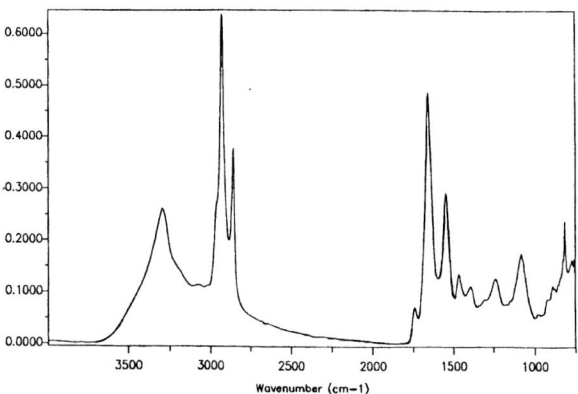

Fig. 5. Mouse retina (high lipid)

terms of the area of the 1509 cm^{-1} peak. The presence of this absorption peak represents low digestibility. The 3D map of a cross-section of rye shows the lipid localization in the germ and around the edges of the kernel. Current work involves ATR of the finish on

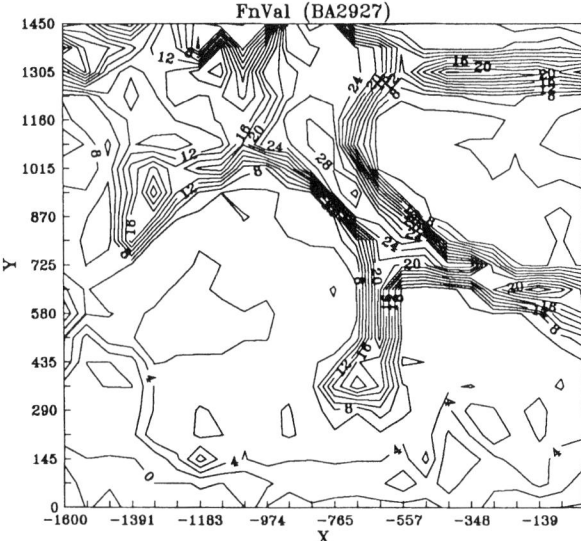

Fig. 6. Mouse cerebellum (2927 cm^{-1} peak area)

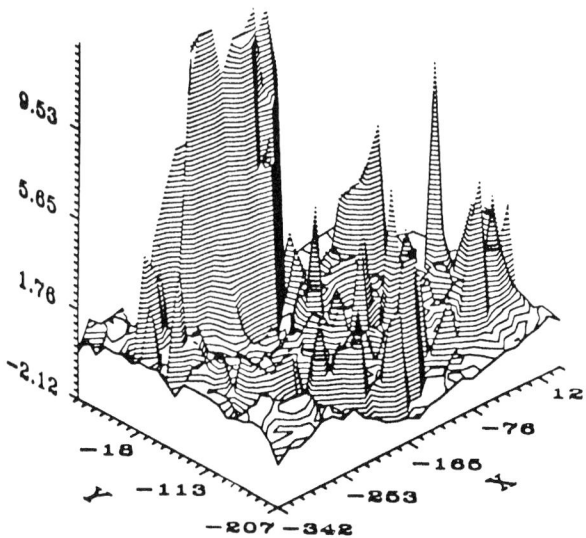

Fig. 8. Rye kernel (1740 cm^{-1} peak area)

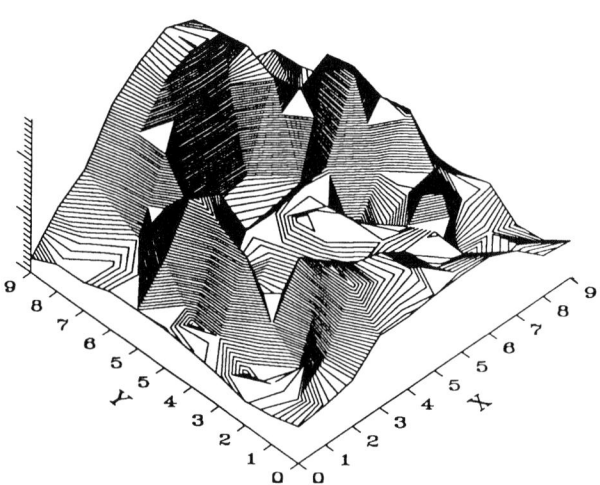

Fig. 7. Vascular bundle of grass leaf (1509 cm^{-1} peak area)

single textile fibres and efforts to localize drugs in human hair.

Conclusion

Single cells or parts of cells are conveniently examined with an IR beam 5–6 μm in diameter and a minimum of co-added scans. Chemical differences from neighboring tissue are revealed for biological material, including grain, leaf, mouse brain, and mouse retina. This application was achieved by using the high brightness, low noise, and low divergence of synchrotron radiation.

Acknowledgements. This work was supported by NFS EPSCOR: OSR-9255223 and USDOE Contract DE-AC0276CH00016. Contribution No. 96-80-J from Kansas Agricultural Experiment Station, Manhattan.

References

[1] J. A. Reffner, *Inst. Phys. Conf. Ser.* (*EMAG-MICRO 89*, Vol. 2) **1990**, *98*, 559.
[2] D. L. Wetzel, in: *Food Flavors Ingredients, and Composition* (G. Charalambous, ed.), Elsevier, Amsterdam, 1993, pp. 679–729.
[3] D. L. Wetzel, S. M. LeVine, *Spectroscopy* **1993**, *8*, 40.
[4] S. M. LeVine, D. L. Wetzel, *Appl. Spectrosc. Rev.* **1993**, *28*, 385.
[5] D. L. Wetzel, J. A. Reffner, *Cereal Foods World* **1993**, *38*, 9.
[6] D. L. Wetzel, in: *Food Flavors: Generation, Analysis and Process Influence*, (G. Charalambous, ed.), Elsevier Amsterdam, 1995, pp. 2039–2108.
[7] D. L. Wetzel, A. J. Eilert, *Proc. SPIE* **1994**, *2089*, 464.
[8] G. L. Carr, J. Reffner, G. P. Williams, *Rev. Sci. Instrum.* **1995**, *66*, 1490.

Mikrochim. Acta [Suppl.] 14, 357–359 (1997)
© Springer-Verlag 1997

The Use of Micro Fourier-transform Infrared Spectroscopy and Scanning Electron Microscopy with X-ray Microanalysis for the Identification of Automobile Paint Chips

Janina Zięba-Palus

Institute of Forensic Research, Westerplatte 9, 31-033 Cracow, Poland

Abstract. Infrared microspectroscopy (FT-IR microscopy) and electron scanning microscopy with energy-dispersive X-ray analysis (SEM-EDX) have been applied for examination of multilayer fragments of automobile paints, for criminalistic purposes.

Key words: Forensic analysis, automobile paints, FT-IR microscopy, SEM-EDX.

Criminalistic examination of automobile paints includes a comparison of a "sample" coming from the car accident site with a "control" taken from the suspected car body in order to establish whether they come from the same source. Infrared spectroscopy, besides microscopic morphological examination, is very often applied in this type of investigation [1–3]. These techniques make it possible to determine the type of paint and to differentiate between particular samples in respect of their qualitative composition. However, the interpretation of the results may be difficult if the spectra are not different enough for it to be postulated that they come from qualitatively different sources [4].

In the analysis of multilayer fragments of automobile paints, it is necessary to establish the chemical composition of consecutive layers of varnish. Infrared spectroscopy allows identification of the polymer matrix as well as the main pigments, and atomic emission spectrometry enables determination of the elemental composition. Thus, the two methods provide information about all the pigments and fillers. The application of these methods requires, however, prior separation of every layer from the paint chip to be examined. This is difficult and sometimes impossible to do because of the small amount of the sample chip to be investigated. Microscopic spectrometry in the infrared

and scanning microscopy together with X-ray microanalysis (SEM-EDX) allow the analysis without use of the separation procedure. This paper describes an attempt made to apply both methods in identification of paint chips for criminalistic purposes.

Experimental

Fragments of ten different multilayer automobile paints were examined. Samples were prepared by cutting each paint chip into 0.02 mm thick slices, perpendicular to the layers. The samples were placed on the microscope stage of the spectrometer, in the infrared beam, the field of view being limited to one layer by means of a variable aperture and the IR spectra were measured by transmission techniques. Then the samples were coated with carbon and placed on the stage of the scanning microscope.

IR spectra were obtained with a Digilab FTS 40A spectrometer equipped with a UMA 500 microscope attachment. Each spectrum represented a collection of 512 scans at a resolution of 8 cm^{-1}. The spectra were searched automatically by the computer for the maxima and peak positions.

A Jeol JSM 5800 scanning electron microscope equipped with an energy dispersive X-ray microanalysis attachment (Link ISIS, Oxford) was used for the determination of the elemental contents in each layer of a particular paint sample.

Results and Discussion

The samples examined had from 3 to 7 layers. The chemical composition of every layer was determined for all samples. A detailed discussion of the results obtained will be presented for two samples selected as an example.

Sample 1 consists of 6 layers and sample 2 of 3 layers, which differ in colour, thickness and granulation (Fig. 1) as well as in chemical composition. The differences in their constitution are first of all in the resins and some pigments. The IR spectra obtained are presented in Figs. 2–4. The elemental composition of each layer of paint is shown in Table 1. By comparison of the IR spectra obtained, with those of standard resins, and taking into consideration the elemental contents, the composition of each layer observed in a cross-section of the paint chip was determined (Table 2).

The application of both methods in analysis of multilayer paint fragments enables determination in a simple way of the kind of paint composing a particular layer and so may be useful in identification of the paint fragments as well as in differentiation between them.

Fig. 1. Multilayer automobile paint (sample 1)

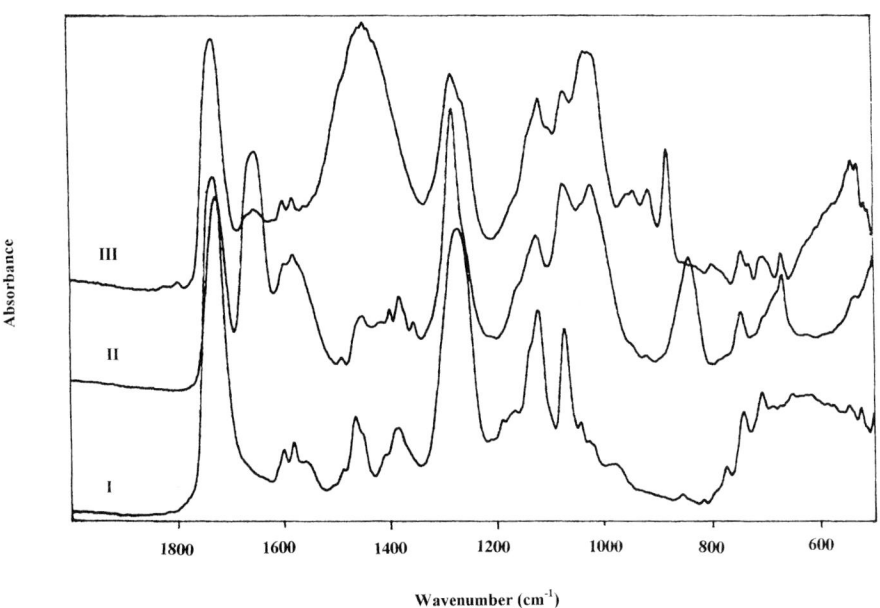

Fig. 2. Infrared spectra of layers I, II, III of sample 1

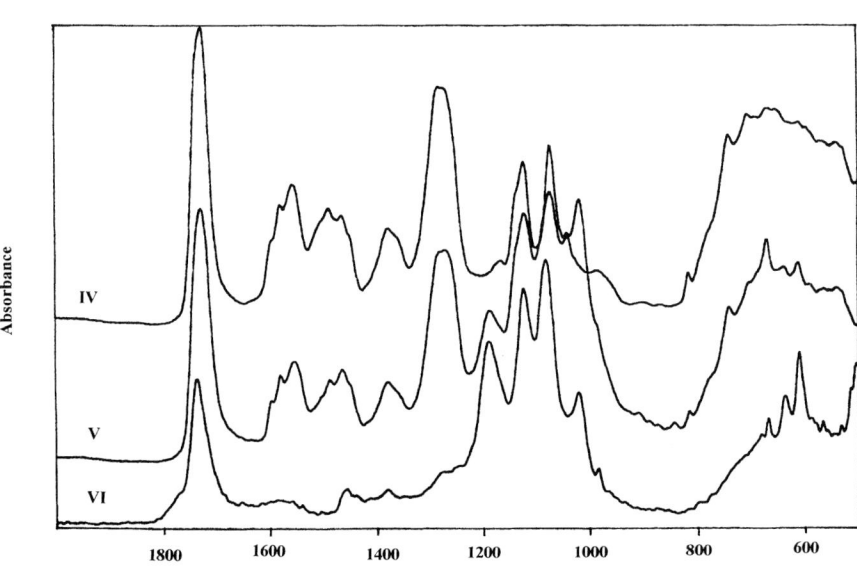

Fig. 3. Infrared spectra of layers IV, V, VI of sample 1

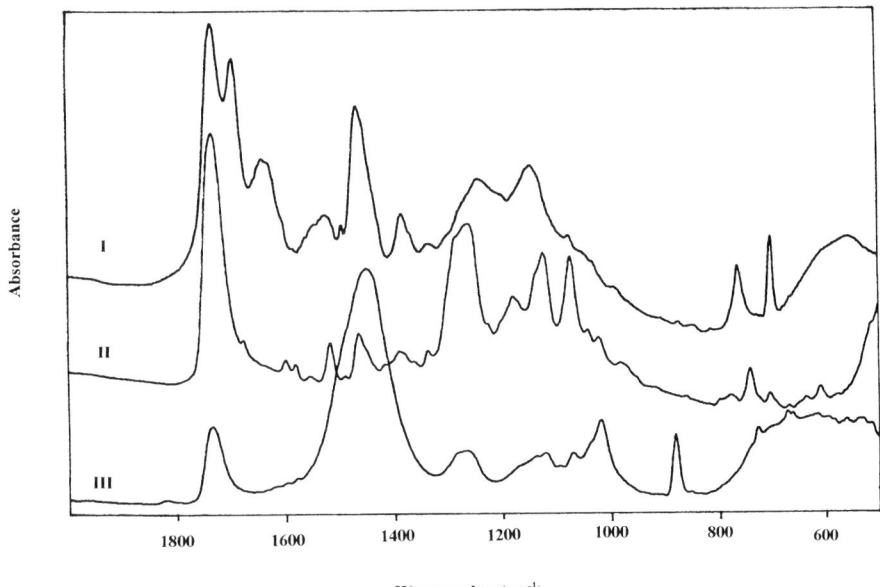

Fig. 4. Infrared spectra of particular layers of sample 2

Wavenumber (cm⁻¹)

Table 1. Elemental composition of the paint chips

Sample	Layer colour	Elements
1	I turquoise	Ti, Ba, S, $\gg$ Al, Si Mg
	II silvery	Zn, Si, Mg $\gg$ Al, Ca, Ti
	III red	Ca, Si, Fe, Mg $\gg$ Zn, Cr, Al
	IV blue	Ti $\gg$ Fe, Al, Si
	V white	Ti, Ba, S, Si, Mg
	VI grey	Zn, Ba, S, Si, Mg $\gg$ Al, Fe
2	I white	Ca, Pb, S, Zn
	II orange	Cr, Pb, S $\gg$ Ca, Mg
	III white	Ca, Ti

Table 2. Chemical composition of the particular layers in the paint chips

Sample	Layer	Composition
1	I	alkyd resins, TiO_2, $BaSO_4$
	II	alkyd and nitrocellulose resins, talc, ZnO
	III	alkyd resins, $CaCO_3$, talc, Fe_2O_3
	IV	alkyd and melamine resins, TiO_2
	V	alkyd and melamine resins, TiO_2 $BaSO_4$, talc
	VI	acrylic resin, ZnO, talc, $BaSO_4$,
2	I	styrene acrylic resins, $PbSO_4$, ZnO, $CaCO_3$
	II	alkyd resins, $PbSO_4$, $PbCrO_4$,
	III	acrylic resins, $CaCO_3$, TiO_2

Conclusions

The application of FT–IR microscopy as well as the SEM-EDX method has proved to be a good tool for forensic identification of multilayer paints. These two methods provide information on the chemical composition of all layers of a paint chip, without their prior separation.

References

[1] R. Goebel, W. Stoecklein, *Scanning Microsc.* **1987**, *1*, 1007.
[2] E. M. Suzuki, *J. Forensic. Sci.* **1989**, *34*, 164.
[3] J. M. Wilkinson, J. Locke, D. K. Laing, *Forensic Sci. Int.* **1988**, *38*, 43.
[4] J. Zięba, A. Pomianowski, *Forensic Sci. Int.* **1981**, *17*, 101.

Mikrochim. Acta [Suppl.] 14, 361–362 (1997)

Application of Transmittance and Reflectance FT-IR Microscopy to Examination of Paints Transferred onto Fabrics

Janina Zięba-Palus

Institute of Forensic Research, Westerplatte 9, 31-033 Cracow, Poland

Abstract. Infrared microspectroscopy has been applied for examination of paints transferred onto different kinds of fabrics, without their prior separation from the fabrics.

Key words: Forensic analysis, automobile paints, FT-IR-microscopy.

During examination of the clothing of a car accident victim a great many traces proving its contact with a moving car can be found. The most frequently observed traces are coloured streaks coming from paint transferred through attrition from the vehicle to the fabric. The amount of paint transferred is usually very small and in some cases it is set into the textile fibres, and it may therefore be impossible to separate it from the fabric. In forensic analysis it is necessary to compare the paint of the suspected car body and the paint transferred to the fabrics. So far, infrared spectroscopy has been successfully applied to examination of automobile paint chips and of particles of such paints separated from the fabric. FT-IR microscopy allows examination of a sample of a material on a matrix of another material. In the present work an attempt was made to check the possibility of car paint identification directly on clothing, so that the process of separation, usually difficult and time-consuming, could be omitted.

Experimental

Ten samples of alkyd car paints transferred to different kinds of fabrics (made from cotton, wool, polyamide and acrylic fibres) were examined.

For use of the transmission technique, paint microfragments separated from a particular fabric were placed on a KBr plate under the microscope.

For use of the reflection technique, no sample preparation was performed. Paint samples were examined directly on a fabric. The field of view was limited to $20 \times 20\,\mu m$.

Spectra were recorded on a Digilab FTS 40A spectrometer attached to a UMA 500 microscope, at a resolution of $8\,cm^{-1}$ in the mid-infrared spectral range; 512 scans in the transmission mode and 1024 scans in the reflection mode were collected.

The Kramers–Kronig transform procedure was applied for each reflection spectrum.

Results

The main absorption bands visible in the spectra of all the samples examined originated from the paint. The spectra obtained by both techniques were similar with regard to the number, position and intensity of the absorption bands. However, in some cases absorption bands of the fabric (Fig. 1) or deformation of the band shape of the paint (Figs. 2 and 3) could be observed. Nevertheless, the interpretation of the spectra and so the reliable identification of the paint was possible.

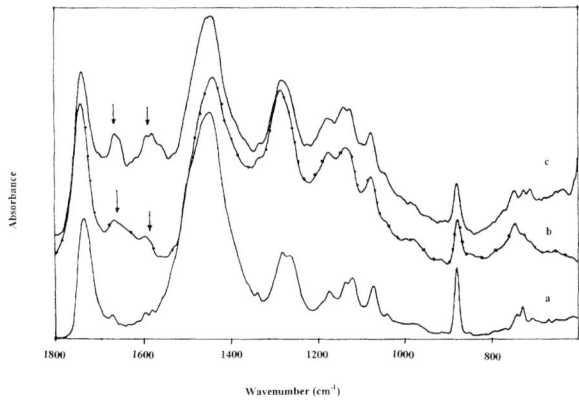

Fig. 1. Infrared spectra of green alkyd paint: *a* transmission spectrum (paint separated from fabric), *b* reflectance spectrum (paint on green nylon fabric), *c* reflectance spectrum (paint on black Perlon fabric); arrow: amide absorption bands of fabrics

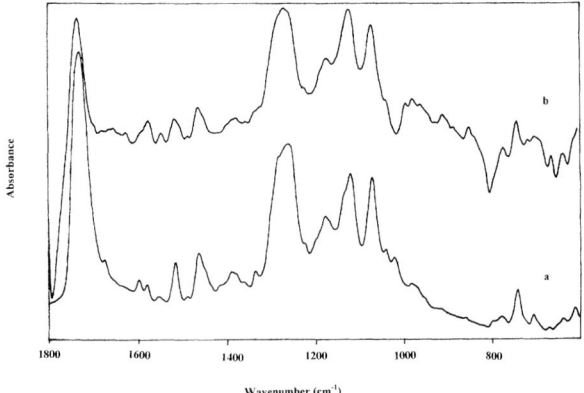

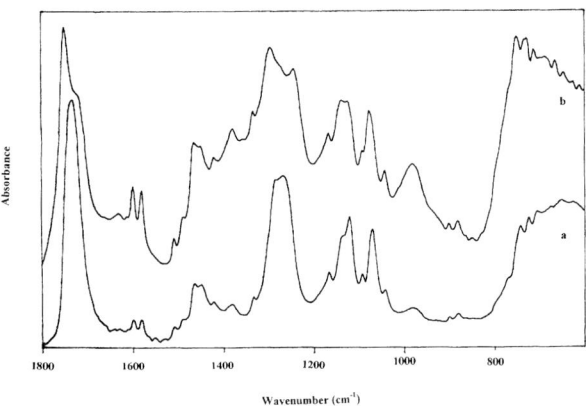

Fig. 2. Infrared spectra of blue alkyd paint: *a* transmission spectrum (paint separated from fabric), *b* reflectance spectrum (paint on green nylon fabric)

Fig. 3. Infrared spectra of yellow alkyd paint: *a* transmission spectrum (paint separated from fabric), *b* reflectance spectrum (paint on white Orlon fabric)

Conclusions

The investigation of a car paint directly on a fabric by using reflectance FT-IR-microscopy is possible and the identification of the paint can be achieved. Thus this method enables differentiation of a paint sample taken from the car body from the sample transferred to the fabrics if they have different chemical composition.

Moreover, the separation process can be avoided and the whole paint analysis shortened.

References

[1] A. Bassl, *Plaste Kautsch.* **1988**, *35*, 102.
[2] D. R. Cousins, C. R. Platoni, L. W. Russell, *Forensic Sci. Int.* **1984**, *24*, 183.
[3] J. M. Wilkinson, J. Locke, D. K. Laing, *Forensic Sci. Int.* **1988**, *38*, 43.

Mikrochim. Acta [Suppl.] 14, 363–365 (1997)

An in Situ Infrared Study of Hydrogenation of CO over Rh/ZrO$_2$

James A. Anderson** and **Mahmoud M. Khader***

Department of Chemistry, The University, Dundee DD1 4HN, Scotland, U.K.

Abstract. FT-IR has been used to follow the formation and reaction of surface species during the hydrogenation of CO over a Rh/ZrO$_2$ catalyst at high temperature and pressure. Particular attention has been paid to the relationship between carbonate and formate species. The reaction was studied at various reaction temperatures and for two catalyst reduction temperatures. Carbonate species which are probably located at the metal/support interface are hydrogenated to give formate species. Pretreatments which modify the metal/support interface influence the relationship between carbonate and formate.

Key words: in situ IR, CO hydrogenation, Rh/ZrO$_2$.

The hydrogenation of CO over Rh/ZrO$_2$ has received attention [1–4], but little mechanistic detail is available [2–4]. Carbonate, bicarbonate and formate species have been identified on zirconia following its exposure to CO/H$_2$ or CO$_2$/H$_2$ [5–7] and the role of these species in methanol synthesis has been postulated. A relationship between bicarbonate/ carbonate and formate species has been suggested for hydrogenation on ZrO$_2$ [5–7] and Cu/SiO$_2$ [8] but none of these studies was conducted under high-pressure/high-temperature conditions. FT-IR studies made *in situ* under such conditions may be used to identify which species exist during reaction, and whether they play a part in the overall reaction mechanism.

Experimental

Zirconiun hydroxide (3.5% SiO$_2$, Magnesium Electron Ltd., XZO645/1) was converted into the oxide and then impregnated with aqueous rhodium nitrate solution to give a 1% loading of rhodium metal. The product was calcined at 773 K for 2 h to give a BET surface area of 88.5 m^2/g. Samples discs of 25 mm diameter were heated at 773 K for 30 min in a flow of N$_2$ (100 ml/min) before

---

* Present address: Chemistry Department, U.A.E University, Al-Ain, United Arab Emirates

** To whom correspondence should be addressed

adjustment to reduction temperature (573 or 773 K). The nitrogen was replaced by flow (100 ml/min) of 3.5% H$_2$ in Ar and the temperature was maintained at 573 or 773 K for 2 h. The sample was then outgassed in dynamic vacuum while the cell was adjusted to reaction temperature, and the CO/H$_2$ mixture was introduced at the desired ratio and total pressure.

Results and Discussion

Rh/ZrO$_2$ reduced at 573 K and exposed to CO/H$_2$ (1:2 v/v) at 533 K and 20 bar showed bands at 1440, 1423, 1394 and 1376 cm^{-1} in addition to maxima at 1582(sh), 1576 and 1561 cm^{-1}. During the initial period of contact, bands at 1440, 1423, 1394 and 1376 cm^{-1} all increased in intensity whereas at longer reaction times, the maxima at 1440 and 1423 cm^{-1} decreased in intensity and were barely visible in spectra recorded after 150 min. During this period, bands in the region 1400–1350 cm^{-1} coninued to grow, with four bands in this region at 1394, 1386, 1380 and 1376 cm^{-1} being detected. A weaker feature at 1310 cm^{-1} became apparent after extended reaction periods. The absence of bicarbonate species on Rh/ZrO$_2$, which were detected for reaction on ZrO$_2$ above (1620 and 1230 cm^{-1}) and of a band due to formate species at 1394 cm^{-1} in addition to the bands at 1386 and 1365 cm^{-1} observed for ZrO$_2$ alone under the same conditions, indicates that the species formed are influenced by the presence of rhodium.

The intensities of the bands at 1394 and 1440 cm^{-1} as a function of time for Rh/ZrO$_2$ reduced at 773 K may be compared with those for a sample reduced at 573 K under the same reaction conditions (Fig. 1). The plots show that the band at 1440 cm^{-1} reaches a maximum after a short period of reaction, whereas the band at 1394 cm^{-1} continues to grow. The plots produced for a sample reduced at 773 K are similar to those of the sample reduced at 573 K with an initial rapid growth in

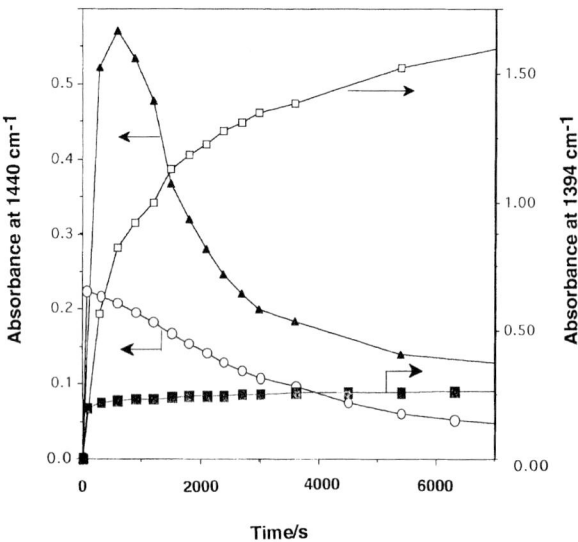

Fig. 1. Band intensities as a function of reaction time at 533 K in CO/H$_2$ (1:2), pressure 20 bar, for Rh/ZrO$_2$ reduced at 573 K (1394, □; 1440, ▲) and reduced at 773 K (1394, ■; 1440, ○)

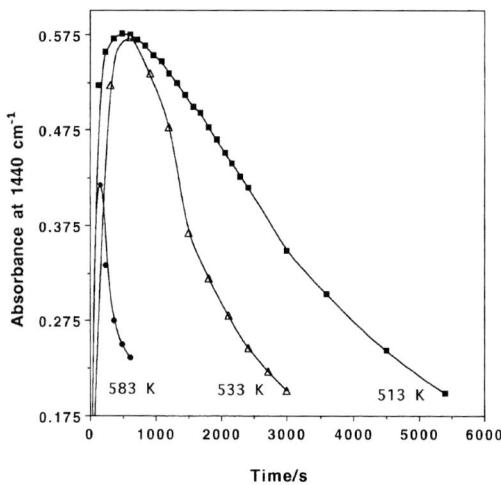

Fig. 2. Normalized band intensities at 1440 cm^{-1} as a function of reaction time in CO/H$_2$ (1:2) at 20 bar and reaction temperatures of 513 (■), 533 (△) and 583 K (●)

the band at 1394 cm^{-1} and then a slow continual increase, while for the 1440 cm^{-1} band, a maximum is followed by a decrease in intensity. However the maximum in band intensity at 1440 cm^{-1} is reached after a shorter period of reaction time and the intensities at 1394 and 1440 cm^{-1} throughout the course of the reaction are much lower than for the sample reduced at 573 K. The bands at 1440 and 1423 cm^{-1} may be assigned to carbonate species. A band at 1416 cm^{-1} has been assigned to carbonate species on Cu/SiO$_2$ [8]. Carbonate species give bands within the region 1630–1200 cm^{-1} with the positions and separation of the vibrational modes depending on the mode of coordination. Since $\Delta v =$ ca. 160 cm^{-1} for the two modes of a unidentate carbonate ion and ca. 320 cm^{-1} for a bidentate carbonate ion [9] it is unlikely that the species is ligated in these forms. Uncoordinated carbonate (D_{3h} symmetry) shows only one IR band at ca. 1470–1420 cm^{-1} [9]. Unlike the call of Cu/SiO$_2$ [8], two maxima are detected here. As the bands grow and decrease concomitantly it is likely that there is only one type of species that lies flat on the surface, but owing to the heterogeneity of the latter, a lowering of the symmetry from D_{3h} occurs, giving two bands but with a limited (17 cm^{-1}) value of Δv.

The influence of reaction temperature on the formation and reactivity of the carbonate species was followed by monitoring the 1440 cm^{-1} band during reaction at 513, 533 and 583 K. Figure 2 illustrates the

time taken for the normalized absorbance value to drop below 0.25 (i.e. the concentration of the species is halved). An influence of temperature on the reactivity of the carbonate is apparent, indicating that it is an intermediate in some process during the reaction. An approximate experimental activation energy of 85 kJ/mol was calculated for reaction of the carbonate, by using rate constants from the half-life of the species, assuming first order.

Results in Fig. 1 suggest a relationship between uncoordinated carbonate and formate species to be of consecutive reaction type

$$A \overset{k_1}{\longrightarrow} B \overset{k_2}{\longrightarrow} C \rightarrow D, \qquad (1)$$

where B and C are carbonate and formate respectively, and the rate constant k_2 is less than k_1. The maximum in carbonate concentration at shorter reaction times (Fig. 1) indicates that an increase in reduction temperature leads to an increase in k_1. Since samples were preheated at 773 K and exhibited no difference in BET area for reduction at 573 or 773 K, it is possible that k_1 was influenced by a charge in particle size that would inherently affect the morphology and the number of interface sites. A decrease in the number of interface sites for the surface reduced at 773 K would be consistent with the diminished concentrations of formate and carbonate species, if these species are adsorbed at these sites.

If carbonate species are an intermediate during CO hydrogenation, then the rate of their disappearance must be linked to the formation of some other species.

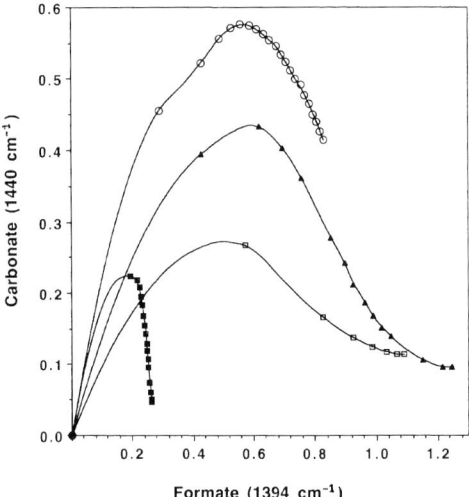

Fig. 3. Relationship between band intensities due to carbonate (1440 cm^{-1}) and formate (1394 cm^{-1}) species for Rh/ZrO$_2$ reduced at 773 K, reaction temperatures 513 K (■), and Rh/ZrO$_2$ reduced at 573 K, reaction temperatures 513 (○), 553 (▲), and 583 K (□), all in CO:H$_2$ (1:2) at 20 bar total pressure

The relationship between formate and carbonate for a range of reaction temperatures and for two reduction temperatures leads to a series of curves (Fig. 3). The maximum in carbonate concentration for the samples reduced at 573 K always exists at the same formate concentration, although the relative quantities of carbonate at its maximum concentration vary according to the reaction temperature. A change in reduction temperature shifts this maximum to correlate with lower surface formate concentrations. The results in Fig. 3 indicate that a link between carbonate and formate species exists. The non-linear correlation between the two species may be related to the fact that unlike previous studies [6, 8] the reaction here was studied under high-pressure high-temperature conditions, which would shift reaction (1) towards product, D.

Studies using Group IB metals and ZrO$_2$ [10] suggest that ZrO$_2$ is involved in the reaction sequence and that the adsorbed species is located at the metal/support interface. The relationship between the two species in reaction (1) relates to two species adsorbed at the same type of site, consistent with the lesser quantities of both species on the surface reduced at higher tempera-

ture. High-temperature reduction modifies the surface and reduces the number of adsorption sites available. Studies of Pt on an identical zirconia sample [11] indicate that at ca. 823 K, hydrogen is spilled over from the metal to the support. Some of this is adsorbed by the support while the remainder results in a partial reduction of the zirconia. These processes may be responsible for the modification of the surface leading to the depleted concentrations of adsorbed species.

Reduction at 773 K may also result in a more dehydroxylated surface, which would hinder the transfer of spill-over hydrogen to hydrogenate the carbonate species, should this be located on the support surface. However, this would lead to a build-up in carbonate species for the sample reduced at 773 K, which is not consistent with the results in Fig. 1.

Conclusions

Infrared studies of CO hydrogenation carried out under *in situ* conditions of high temperature and high pressure have allowed a relationship between adsorbed carbonate and formate species to be established. Modifications to the Rh/ZrO$_2$ by variation in the catalyst pretreatment lead to changes in the carbonate/formate relationship which may be significant in terms of modifying catalyst selectivity.

References

[1] M. Ichikawa, *Bull. Chem. Soc. Jpn.* **1978**, *51*, 2268, 2273.
[2] M. Ichikawa, K. Shikakura, *Stud. Surf. Sci. Catal.* **1981**, *7*, 925.
[3] M. Ichikawa, K. Sekizawa, K. Shikakura, M. Kawai, *J. Molec. Catal.* **1981**, *11*, 167.
[4] J. P. Hindermann, A. Kiennemann, S. Tazkritt, *Stud. Surf. Sci. Catal.* **1988**, *48*, 481.
[5] M. Y. He, J. G. Ekerdt, *J. Catal.* **1984**, *87*, 381.
[6] T. Chafik, D. Bianchi, S. J. Teichner, *Topics Catal.* **1995**, *2*, 103.
[7] T. Onishi, K. Maruya, K. Domen, H. Abe, J. Kondo, in *Proc. Int. Cong. Catal. 9th* (M. J. Phillips, M. Ternan, eds.), Chem Inst. Canada, Ottawa, 1988, p. 507.
[8] G. J. Millar, C. H. Rochester, C. Howe, K. C. Waugh, *Mol. Phys.* **1991**, *76*, 833.
[9] L. H. Little, *Infrared Spectra of Adsorbed Species*, Academic Press, London, 1966.
[10] A. Baiker, M. Kilo, M. Maciejewski, S. Menzi, A. Wokaun, in *Proc. Int. Cong. Catal. 10th.* (L. Guczi, F. Solymosi, P. Tetenyi, eds.), Elsevier, Amsterdam, 1993, p. 1257.
[11] D.-L. Hoang, H. Berndt, H. Lieske, *Catal. Lett.* **1995**, *31*, 165.

Mikrochim. Acta [Suppl.] 14, 367–368 (1997)

Novel H-bonding in Acid Soaps and Pressure-induced Proton Shift. An FT-IR Approach

Jiang Bian, Shifu Weng, Jinguang Wu*, Luqin Yang, Xu Zhang, and **Guangxian Xu**

Department of Chemistry and State Key Laboratory of Rare Earth Materials Chemistry and Applications, Peking University, Beijing 100871, Peoples Republic of China

Abstract. Acid potassium soaps (KHA_2, in which A stands for carboxylate) of myristic acid, palmitic acid, and stearic acid were studied with a Fourier-transform infrared spectrometer. The infrared spectra of the acid soaps showed only one carboxylate mode in the range $1800–1500\,cm^{-1}$. It is suggested that a new type of H-bonding, in which the proton is shared by two carboxylate groups, exists in the acid soaps. In this work, it was also observed that these acid soaps underwent structural transitions with increasing pressure. The spectral results are analysed and discussed.

Key words: FT-IR, hydrogen bond, acid salt, phase transition.

In recent years, long-chain amphiphilic molecules have received much attention, and an important problem is the properties of hydrogen bonds on interfaces. As is known, hydrogen bonds play key roles in maintaining the conformations of supermolecular assemblies. Some researchers have reported strong H-bonding in acid salts [1, 2]. However, in this work, a new style of H-bonding was found to exist in acid potassium soaps with long alkyl chains, and pressure-induced phase transitions of these acid soaps were identified by FT-IR spectroscopy. The pressure used was far lower than that needed by pressure-tuning of structural changes of fatty acids [3].

Experimental

The acid potassium soaps (KHA_2) of myristic acid (C14), palmitic acid (C16), and stearic acid (C18) were obtained by recrystallization from their aqueous ethanol solutions. The compositions of the crystalline products were determined to be 1:1 acid soaps through elemental analysis. The infrared spectra were measured with a Nicolet Magna 750 Fourier-transform infrared spectrometer with an NIC-Plan microscope. The spectral resolution was $4\,cm^{-1}$.

A pressurizer with a maximum pressure of no more than 0.70 kbar was used to process the crystalline samples.

Results and Discussion

The H-Bond of KHA_2

In the infrared spectra (Fig. 1), the v_{OH} mode of KHA_2 was found in the range $1500–800\,cm^{-1}$, instead of the broad v_{OH} band (in the range $3400–2400\,cm^{-1}$) of the fatty acid. The large shift to lower frequency indicated that strong H-bonding was present in KHA_2.

In the range $1800–1500\,cm^{-1}$, only a $1644\,cm^{-1}$ band could be observed for KHA_2, whereas both the C=O stretching mode (ca. $1700\,cm^{-1}$) of the fatty acid and the COO^- asymmetrical stretching band (ca. $1560\,cm^{-1}$) of a potassium soap were undetectable. These spectra suggest that there was only one type of carboxylate group in KHA_2. If so, the proton in the hydrogen bond was perhaps shared by two carboxylate groups.

The Pressure-Induced Phase Transition

In Fig. 2, it can be seen that the CH_2 bending (at ca. $1465\,cm^{-1}$) and rocking (at ca. $720\,cm^{-1}$) modes changed with increasing pressure. Since both vibrations are related to allkyl chain packing in lattices [2], the spectral results show that a phase transition took place. The original orthorhombic packing of the chains, which is characteristic of factor group-splitting of CH_2 rocking bands (729 and $721\,cm^{-1}$), partly evolved into a hexagonal phase which only displayed a band at $721\,cm^{-1}$.

---

* To whom correspondence should be addressed

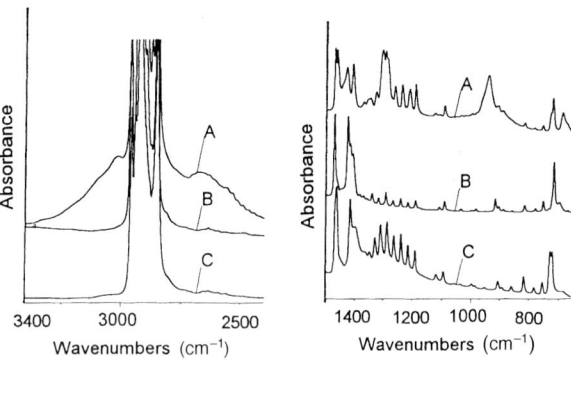

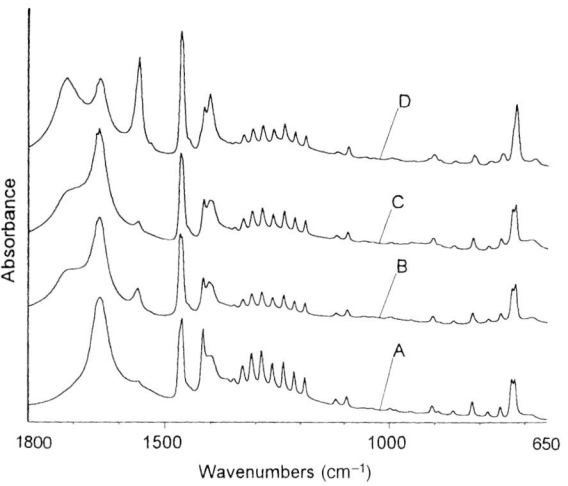

Fig. 2. Infrared spectra of potassium hydrogen dimyristate. *A* original sample; *B* 0.21; *C* 0.35; *D* 0.55 kbar

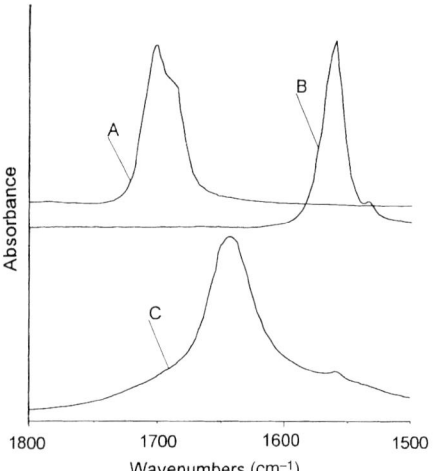

Fig. 1. Infrared spectra. *A* myristic acid; *B* potassium myristate; *C* potassium hydrogen dimyristate

Conclusions

The spectral observations above showed that a new type of H-bonding, in which the proton was shared by two carboxylate groups, existed in KHA_2. KHA_2 underwent phase transitions with increasing pressure. The new type of hydrogen bond in KHA_2, which differed from the H-bonds in short-chain analogues of acid salts (e.g. acid acetates and acid laurates) and could be tuned by pressure, might imply that the H-bonds on interfaces have some unknown properties, besides structural stabilizing roles.

Acknowledgement. This work was supported by the NSFC and Climbing Program of China.

References

[1] D. Hadzi, *Chimia* **1972**, *26*, 7.
[2] H. H. Mantsch, S. F. Weng, P. W. Yang, H. H. Eysel, *J. Mol. Struct.* **1994**, *324*, 133.
[3] W. W. Ley, H. D. Drickamer, *J. Phys. Chem.* **1990**, *94*, 7366.
[4] G. Zerbi, G. Conti, G. Minoni, S. Pison, A. Bigotto, *J. Phys. Chem.* **1987**, *91*, 2386.

In the range 1800–1500 cm⁻¹, the original acid potassium myristate, showed only one broad band at 1644 cm⁻¹; however, a shoulder near 1720 cm⁻¹ and a band at 1526 cm⁻¹ appeared and increased steadily with increasing pressure, though the pressure employed was relatively low (< 0.70 kbar). The appearance of $v_{C=O}$ (1720 cm⁻¹) and v_{COO^-} (1562 cm⁻¹) modes indicated that the proton shared by two carboxylate groups was put to one side of the H-bonded dicarboxylate, by which the acid-soap systems reached a new equilibrium.

Mikrochim. Acta [Suppl.] 14, 369–371 (1997)
© Springer-Verlag 1997

Variable Temperature and Variable Pressure Spectroscopic Studies of Some Perovskites – LnTMO$_3$

Richard Mortimer*, **James G. Powell**, and **Nagasampagi Y. Vasanthacharya**

Defence Research Agency, Fort Halstead, Sevenoaks, Kent, TN14 7BP U.K.

Abstract. The infrared spectroscopic manifestations of phase changes in lanthanum nickel ferrate (LaNi$_{0.5}$Fe$_{0.5}$O$_3$) are reported. LaNi$_{0.5}$Fe$_{0.5}$O$_3$ shows no anomalous temperature-dependence (in the range 13–300 K) at atmospheric pressure, but at 300 K there is a pressure-induced phase transition at around 14×10^8 Pa. We ascribe this transition to a redistribution of electronic charge at and above the phase transition pressure. *Ab initio* calculations utilizing spin-density function methods support and enhance our interpretation of the behaviour of LaNi$_{0.5}$Fe$_{0.5}$O$_3$ from infrared vibrational spectroscopic data.

Key words: infrared, pressure, temperature, phase-change, LaNi$_{0.5}$Fe$_{0.5}$O$_3$.

New electro-optic devices might arise if electronic phase changes in materials can be understood and manipulated. The electronic properties of perovskite-like materials such as LaNiO$_3$, LaFeO$_3$ and LaCoO$_3$ are dictated by the spin-state [1] of the transition metal, so low-spin perovskites, for instance LaNiO$_3$ [2] are metallic, whereas high-spin analogues, such as LaFeO$_3$ [2], are insulators. Where the magnitude of the crystal field gap ($10 Dq$) is of the order of kT (with T about 300 K), there is an equilibrium [3] between high-spin and low-spin species, giving rise to interesting electrical, magnetic and infrared vibrational spectroscopic [4] properties: this is the case for LaCoO$_3$.

Of interest to us in the present paper is the system lanthanum nickel ferrate, (LaNi$_{0.5}$Fe$_{0.5}$O$_3$), which is iso-electronic with LaCoO$_3$. We have investigated the temperature (13–300 K) and pressure (0–100 $\times 10^8$ Pa) dependence of manifestations for phase transitions in the v_{TM-O} (TM = transition metal) spectral region. In particular, we have demonstrated the ability of quan-

tum mechanical calculations using spin-density function techniques [5] to support and advance our interpretation that is based upon experimental data.

Experimental

Lanthanum nickel ferrate (LaNi$_x$Fe$_{1-x}$O$_3$) compounds were prepared [6] by mixing appropriate amounts of lanthanum nitrate, nickel nitrate and iron nitrate as aqueous solutions, followed by drying to form the precursor material. The precursor material was then pressed into pellets and sintered for a number of days to yield the desired material. All materials were analysed by X-ray powder crystallography to measure phase purity and crystallinity.

Ab initio quantum mechanical investigations were made by using the spin-density function approximation [5] of the effects of applied pressure on the LaNi$_{0.5}$Fe$_{0.5}$O$_3$ system. Our model for this system was the R$\bar{3}$ structure of LaCoO$_3$ [7] at 668 K, with one of the two independent cobalt sites occupied by one nickel and the other by one iron ion. Three cases were considered: $r_{Ni-O} > r_{Fe-O}$, $r_{Ni-O} = r_{Fe-O}$, and $r_{Ni-O} < r_{Fe-O}$ (where r is the bond length), simply accomplished by displacing the x co-ordinate of the oxygen atom by $+0.02$, 0 or -0.02 Å, respectively, relative to its x fractional coordinate in the R$\bar{3}$ structure of LaCoO$_3$. These three cases were then systematically subjected to an isotropic expansion or contraction of the rhombohedral unit cell axes, or neither.

Results and Discussion

Some infrared spectra of lanthanum nickel ferrate (LaNi$_{0.5}$Fe$_{0.5}$O$_3$) at 300 K in the pressure range 0–34×10^8 Pa are shown in Fig. 1. A reversible transition is found to occur at about 14×10^8 Pa. We do not find any comparable spectral changes in the temperature region 13–300 K at atmospheric pressure.

We note that the isomer shifts for iron derived from Mössbauer studies of LaNi$_x$Fe$_{1-x}$O$_3$ for $x = 0.05$, 0.1, 0.2 [2] and $x = 0.5$ [8] are more characteristic of Fe^{3+} than Fe^{4+}. Also, the infrared spectra obtained at zero

---

* To whom correspondence should be addressed

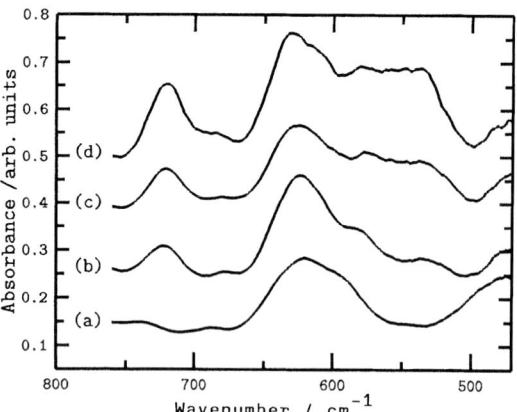

Fig. 1. Infrared spectra in the $\nu_{(Fe/Ni)-O}$ region at a 0, b 12.3 × 10⁸, c 14.3 × 10⁸, and d 34.1 × 10⁸ Pa

applied pressure also suggest that both ions are in the same oxidation state, assuming that the force constants for the Ni–O and Fe–O stretching modes are comparable. Thermally-driven changes of the spin-state in the analogous material $LaCoO_3$ appear to change the vibrational force field very little [4]. We conclude therefore that under zero applied pressure both transition metal ions are in the tetravalent state and that whatever the spin states, their vibrational force fields are quite similar. As our pressure measurements show, however, in $LaNi_{0.5}Fe_{0.5}O_3$ there is a dramatic and reversible change at around 14 × 10⁸ Pa, the origins of which we now explore.

The pressure-induced changes of the vibrational spectrum of $LaNi_{0.5}Fe_{0.5}O_3$ indicate a strong change of the vibrational force field. On going through and beyond the transition there is no significant mixing of the nickel/iron–oxygen stretching $\nu_{Ni/Fe-O}$ and the nickel/iron–oxygen deformation $\delta_{Ni/Fe-O}$ co-ordinates. Our evidence for this comes from the fact that the integrated area in the $\nu_{Ni/Fe-O}$ region remains constant before, during and after the phase transition and that the deformation modes do not show any measurable intensity changes near the phase transition. The parent peak (622 cm⁻¹) and the three peaks at 722, 583 and 529 cm⁻¹ are, then, ascribed to predominantly ν_{TM-O} modes.

For a number of reasons, we discount perturbations to the infrared spectra arising solely because of a high-spin to low-spin transition, that might be expected for tetravalent iron, which is commonly found in the high-spin state, but not for nickel, which is expected to possess a low-spin state. In the series $LaNi_xFe_{1-x}O_3$

outside $x = 0.4$–0.6, it is impossible to cause a similar perturbation to the infrared spectra with the pressures we typically achieve, about 100 × 10⁸ Pa. It appears that the presence of both transition metal ions in approximately equal concentration is required. Also, when $LaCoO_3$ is subjected to increasing applied pressure [and thus the e_g orbitals (in a pseudo-octahedral environment) are depopulated], at room temperature, there is evidence for changes of band intensity in the ν_{Co-O} region but the wavenumber shifts are either small or insignificant. Further, the relative constancy of the vibrational integrated intensity in the 760–500 cm⁻¹ region, *vide supra*, suggests that the new features originating at about 14 × 10⁸ Pa derive their intensity at the expense of a diminution of parent intensity centred around 622 cm⁻¹. What, however, would be the consequences of hetero-atomic transition metal to transition metal electron transfer?

Ar Ni–O distances around 1.9–2.15 Å there are very small differences between the energies for the three cases $r_{Ni-O} > r_{Fe-O}$ (●), $r_{Ni-O} = r_{Fe-O}$ (▲), and $r_{Ni-O} < r_{Fe-O}$ (■) (Fig. 2). Upon further isotropic compression of the unit cell there is an increase in the overlap of the electron wave functions and the case where, as a result, the oxygen atom has migrated towards the iron ion, becomes the significantly more stable configuration (Fig. 2). For the more stable situation the amount of charge on the lanthanum and oxygen ions changes very little on compression (Fig. 3), but there is a significant donation of electron density from the nickel ion and a commensurate quantity of charge accepted by the iron ion. We conclude, therefore, that there is indirect

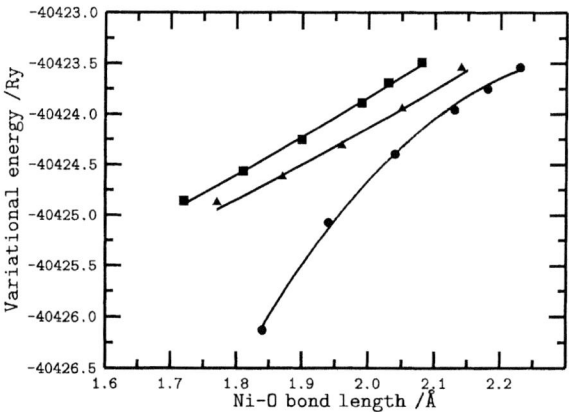

Fig. 2. Variational energy of $LaNi_{0.5}Fe_{0.5}O_3$. Zero compression (triangles), 5% expansion (squares) and 5% compression (circles) for differing Ni–O, and therefore rhombohedral unit cell, lengths

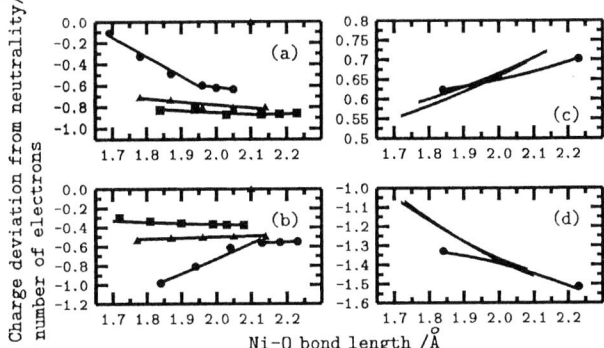

Fig. 3. Deviation from neutrality of $LaNi_{0.5}Fe_{0.5}O_3$. Zero compression (triangles), 5% expansion (squares) and 5% compression (circles) for differing Ni–O, and therefore rhombohedral unit cell, lengths for *a* iron, *b* nickel, *c* oxygen, and *d* lanthanum. For clarity, in *c* and *d* limiting data points are shown only for the most stable situation which develops under pressure

charge transfer from nickel ions to iron ions, probably mediated through the oxygen ions.

Since electron transfer is likely to cause a significant change in the ionic component of bonding, we tentatively assign the peaks at $702\,cm^{-1}$ and the $583/529\,cm^{-1}$ pair to $v_{Ni-O}^{-\delta q}$ and $v_{Fe-O}^{+\delta q}$ species, respectively.

Acknowledgements. We thank Professor C. N. R. Rao for provision of some of the experimental facilities for preparation of materials.

References

[1] C. N. R. Rao, *J. Indian Chem. Soc.* **1974**, *51*, 979.

[2] P. Ganguly, N. Y. Vasanthacharya, *J. Solid State Chem.* **1986**, *61*, 164.

[3] K. Asai, P. Ghering, H. Chou, G. Shirane, *Phys. Rev. B* **1989**, *40*, 10982.

[4] R. Mortimer, J. G. Powell, N. Y. Vasanthacharya, *Phys. Stat. Sol. B* **1995**, *188*, K47.

[5] *ESOCS* (*Electronic Structure of Close Packed Systems*), Biosym Technologies Inc., San Diego, USA.

[6] N. Y. Vasanthacharya, K. K. Singh, P. Ganguly, *Rev. Chim. Miner.* **1981**, *18*, 333.

[7] G. Thornton, B. C. Tofield, A. W. Hewat, *J. Solid State Chem.* **1986**, *61*, 301.

[8] K. Asai, H. Sekizawa, *J. Phys. Soc. Jpn.* **1980**, *49*, 90.

Mikrochim. Acta [Suppl.] 14, 373–375 (1997)
© Springer-Verlag 1997

An FT-IR Spectral Study on the Binuclear Complex of Dysprosium with a Tetraketone at High External Pressure

Zhen Hua Xu[1,*], **Lu Qin Yang**[1], **Jin Guang Wu**[1], **I. S. Butler**[2], and **Guang Xian Xu**[1]

[1] State Key Laboratory of Rare Earth Material Chemistry and Applications, Peking University, Beijing 100871, Peoples Republic of China
[2] Department of Chemistry, McGill University, Montreal, PQ, Canada, H3A 2K6

Abstract. The FT-IR and FT-Raman spectra of the binuclear complex of dysprosium with a tetraketone 1,5-bis(1′-phenyl-3′-methyl-5′-pyrazolone-4′)-1,5-pentanedione were measured at high external pressure. The results show that there is not a pressure-induced phase transition and that the complex underwent only geometrical transformation at high external pressure; the pressure-induced change of this complex is small.

Key words: FT-IR spectra, FT-Raman spectra, high pressure, complex.

Organic, inorganic (complex, organometallic etc.) and biochemical molecular crystals often undergo dramatic structure changes when subjected to very high external pressure [1–3]. These changes include pressure-induced phase transitions and/or geometrical, spin-state and conformational transformations. Vibrational spectroscopy at high external pressure is a straightforward method for investigating these changes [2]. 4-Acetyl-bispyrazolone represents a new type of β-diketone, which has two constituent β-diketones. Their rare earth complexes have strong fluorescence and herbicidal activities [4]. We have measured the FT-IR and FT-Raman spectra of $Dy_2L_3 \cdot 8H_2O$ at high external pressure ($H_2L = 1,5$-bis(1′-phenyl-3′-methyl-5′-pyrazolone-4′)-1,5-pentanedione, Fig. 1) and discussed the structure change of this complex.

Experimental

The FT-IR spectra were measured at different high pressures (below 50 kbar) with the aid of a commercial diamond-anvil cell (DAC) [2]. The sample and calibrant were placed in a 300 µm diameter hole

drilled in the centre of a 270 µm thick stainless-steel gasket, located between the parallel faces of the pair of diamonds in the DAC. The DAC was mounted under the microscope of a Bruker IFS-48 spectrometer with an MCT-detector. The pressures were calibrated with $NaNO_2$ as calibrant [5]. The spectra were recorded at 4 cm^{-1} resolution (1000 scans). The wavenumber range is 4000–600 cm^{-1}. We also measured the FT-Raman spectra of the sample at zero and 33.4 kbar pressure, using the same DAC cell. The pressure was determined by means of the high-frequency component of the diamond band (about 1333 cm^{-1}) at the diamond interface [6]. The wavenumber range is 3600–50 cm^{-1}. The spectra of the DAC were recorded at 2.6 cm^{-1} resolution (50 scans). A filter was used to cut off the range of about 1400–1200 cm^{-1} to get rid of the very strong diamond band when the FT-Raman spectra of the samples were measured. The DAC was mounted under the microscope of a Bruker IFS-88 spectrometer with an MCT-detector. The spectra of the samples were recorded at 4 cm^{-1} resolution (10000 scans).

Results and Discussion

The FT-IR spectra data are listed in Table 1. Assignments are given only for main constituent vibrational modes.

The frequency shifts and the intensity changes of bands in the FT-IR spectra are obvious. All the peaks exhibit the expected blue shift towards high wavenumber with increases pressure. The 1616.1 cm^{-1} peak of v(phenyl ring), v(C=O) and v(pyrazolone ring) shows a little higher dv/dp (0.506 cm^{-1}/kbar) below 35 kbar and then becomes lower (0.444 cm^{-1}/kbar) (Table 1, Fig. 2a). The peak for stretching of the phenyl and pyrazolone rings [7], at 1488.2 cm^{-1} also appears to yield two dv/dp regimes (Table 1). These effects may be due to each of these bands representing constituents with different vibrational modes. The 1442.7 cm^{-1} peak is mainly from overlapping of the CH_2 deformation and CH_3 asymmetric deformation vibrations; the CH_3 symmetric deformation is near 1374.9 cm^{-1}. Both

---

* To whom correspondence should be addressed

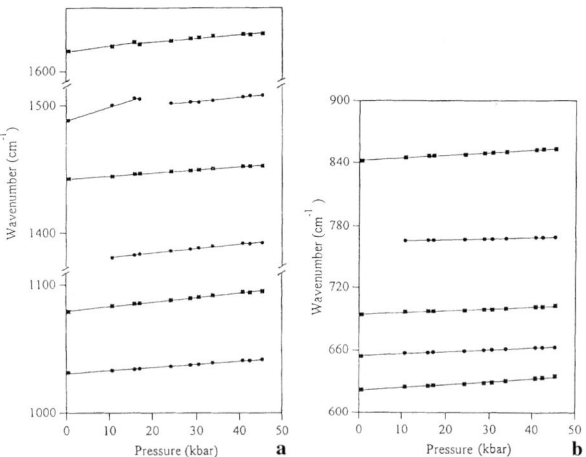

Fig. 1. The structure of H_2L

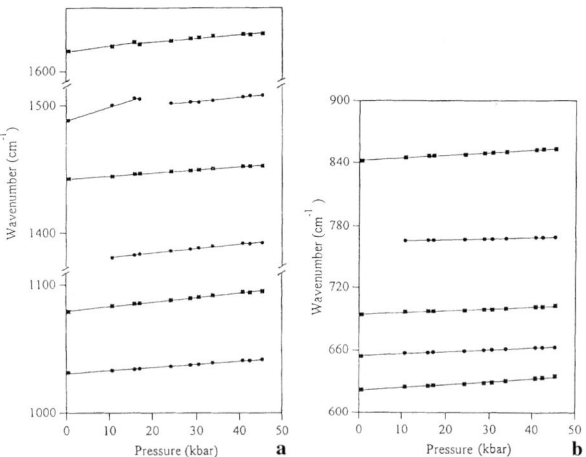

Fig. 2. Pressure dependences of some observed FT-IR bands of the complex

have similar dv/dp values, 0.270 and 0.294. The 1079.6 and 1031.7 cm^{-1} bands are mainly from in-plane CH bending of monosubstituted benzene, but their dv/dp value are different, the 1079.6 cm^{-1} band having a higher value (0.367) than the 1031.7 cm^{-1} band (0.256). The reason may be that there are some C–C stretching vibrational modes in the 1079.6 cm^{-1} band. For monosubstituted benzene, the in-plane quadrant bending appears at 841.8 (dv/dp 0.237) and 622.2 cm^{-1} (dv/dp 0.221) (Table 1, Fig. 2b). The bands at 654.2, 694.1 and 759.2 cm^{-1} have lower dv/dp values (0.203, 0.184 and 0.107). All the dv/dp values (cm^{-1}/kbar) are relatively low. Most of them are between 0.4 and 0.2 (Table 1). This means that the pressure sensitivities are low and the compressibility of the complex is small. The crystal structure of the complex $Dy_2L_3 \cdot 5DMF$ [8], which is very similar to $Dy_2L_3 \cdot 8H_2O$, shows that each L as a bridge ligand chelates two dysprosium ions (Dy^{3+}) by means of its two β-diketonates. Each Dy^{3+} coordinates with three β-diketonates, each from a different ligand. The coordination mode is more efficient than others for fixing the organic ligand. This makes

the complex a densely packed structure with lower compressibility. There is no pressure-induced phase transition for the complex, although breaks are observed in the two slopes (Fig. 2a). These breaks may result from each of these bands being constituted of different vibrational modes. Some bands overlapped and disappeared at the high pressure (Fig. 3). This indicates that some bands were broadened and that the symmetry of this crystal changed at high external pressure. The FT-Raman spectra of this complex at zero and 33.4 kbar pressure are shown in Fig. 4. The 1596.2 cm^{-1} band is from the stretching vibration of the monosubstituted benzene ring, together with

Table 1. The pressure dependences of some observed FT-IR bands of the complex

Bands (cm^{-1})	dv/dp (cm^{-1}/kbar)	Assignments
1616.1	0.506 0.444	v(phenyl ring), v(C=O), v(pyrazolone ring)
1488.2	0.410 0.278	v(phenyl and pyrazolone ring)
1422.7	0.270	deformation of CH$_2$ and asymmetric deformation of CH$_3$
1374.9	0.294	symmetric deformation of CH$_3$
1079.6	0.367	bending of CH in monosubstituted benzene, with v(C–C)
1031.7	0.256	bending of CH in monosubstituted benzene
841.8	0.237	in plane quadrant bending of monosubstituted benzene
759.2	0.107	CH wagging of monosubstituted benzene, CH deformation out-of-plane pyrazolone ring
694.1	0.184	monosubstituted benzene ring bending, out-of-plane bending of pyrazolone ring
654.2	0.203	out-of-plane bending of monosubstituted benzene ring
622.2	0.221	in-plane quadrant bending of monosubstituted benzene

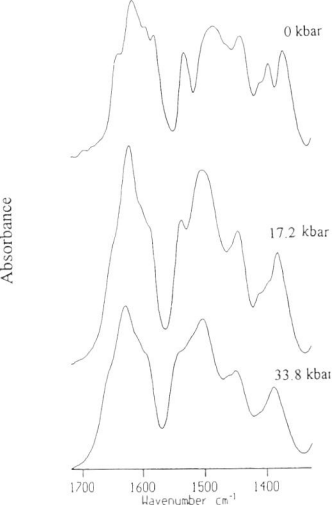

Fig. 3. FT-IR spectra of the complex at selected pressures (1720–1330 cm^{-1})

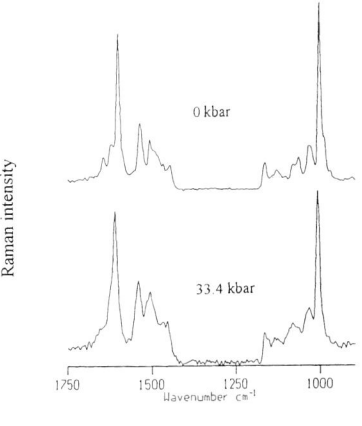

Raman schift

Fig. 4. FT-Raman spectra of the complex at selected pressures (1750–900 cm^{-1})

a little ring-stretching vibration of pyrazolone. It shifted to 1606.7 cm^{-1} at 33.4 kbar pressure. The in-plane CH deformation of monosubstituted benzene at

1001.9 cm^{-1} moved to the higher frequency of 1007.7 cm^{-1}. The dv/dp values are about 0.314 and 0.174 cm^{-1}/kbar respectively, very close to the values for the FT-IR spectra. These results suggest that this complex underwent geometrical transformation at high external pressure.

Conclusion

FT-IR spectroscopy and FT-Raman spectroscopy are powerful methods for studying structure changes in a new type of rare earth complex with 4-acetyl-bispyrazolone subjected to high external pressure. The results show that there is no pressure-induced phase transition and that the complex undergoes only geometrical transformation at pressures below 50 kbar. The dysprosium complex has a densely packed structure and the pressure-induced change is small.

Acknowledgements. Z. X. Xu and G. X. Xu (China) and I. S. Butler (Canada) gratefully acknowledge the award of travel grants from NNSFC and NSERC, respectively, as part of a bilateral exchange project between Peking and McGill Universities.

References

[1] S. M. Barnett, R. D. Markwell, S. H. R. Brienne, I. S. Butler, D.F. R. Gilson, N. T. Kawai, A. Vlcek Jr., *J. Raman Spectrosc.* **1993**, *24*, 471.

[2] J. R. Ferraro, *Vibrational Spectroscopy at High External Pressures: The Diamond Anvil Cell*, Academic Press, Orlardo, 1984, p. 12; p. 120.

[3] J. R. Howlett, A. A. Ismail, D. W. Armstrong, P. T. T. Wong, R. K. Wong, *Biochim. Biophys. Acta* **1992**, *1159*, 227.

[4] L. Yang, R. Yang, *J. Coord. Chem.* **1994**, *33*, 303.

[5] D. D. Klug, E. Whalley, *Rev. Sci. Instrum.* **1983**, *54*, 1205.

[6] S. K. Sharma, H. K. Mao, P. M. Bell, J. A. Xu, *J. Raman Spectrosc.* **1985**, *16*, 350.

[7] N. B. Colthup, L. H. Daly, S. E. Wiberley, *Introduction to Infrared and Raman Spectroscopy*, 3rd Ed., Academic Press, 1990, p. 261.

[8] L. Yang, R. Yang, *J. Mol. Struct.* **1996**, *380*, 75.

Mikrochim. Acta [Suppl.] 14, 377–379 (1997)

Sample Preparation of Illicit Drugs for FT-IR Microspectrophotometry

T. Gál*, **T. Veress,** and **I. Ambrus**

Institute for Forensic Science, 1903 Budapest P.O. Box 314/4, Hungary

Abstract. A method of sample preparation is proposed that permits identification (by FT-IR microspectrophotometry) of the physiologically active principles and inactive components of illicit drugs. It is based on the differential solubility of the components in methanol.

Key words: Drug identification, heroin, amphetamines, FT-IR, infrared microspectrophotometry.

The identification of both psychologically active principles and inactive components of illicit drug preparations is an important part of forensic drug analysis [1–3]. The majority of illicit drug preparations such as powders and tablets appear in the black market in diluted form, containing sugars (saccharose, glucose, lactose, sucrose, mannitol), starch, ascorbic acid, etc. [4, 5]. The determination of the diluents is useful for comparative analysis in order to confirm the common origin of samples.

In this work we aim to develop a simple and fast sample preparation procedure, which allows the separation of the carbohydrate diluents from the active principle in order to identify the components by the FT-IR microspectrophotometric method. We choose powders containing heroin and tablets containing different amphetamine derivatives as model materials.

Experimental

Materials and Equipment

The illicit preparations were seized in the black market by police. The methanol used for the extraction was of LiChrosolv grade (Merck, Darmstadt, Germany). The potassium bromide was of IR-spectroscopic grade (Fluka, Buchs, Switzerland) and for filtration GF/D glass fibre (Whatman) was used. The columns for separation of sample components were prepared by using plastic pipette tips.

For concentration of solutions a Reacti-Therm heating module (Pierce) was applied, by flushing with nitrogen. The IR-spectra were recorded with a Nicolet 710 FT-IR spectrophotometer equipped with a Spectra-Tech IR-PLAN II analytical infrared microscopic unit.

Sample Preparation

The powder and tablet samples were crushed and homogenized in an agate mortar. The homogenized sample (ca. 10 mg) was filled into a plastic column packed with glass fibre and potassium bromide (ca. 300 mg). The sample layered onto the potassium bromide was washed with 5 ml of methanol and the eluate was collected in a test-tube. The methanolic solutions were concentrated to 50–100 µl. A 10-µl portion of the concentrated solution was dropped onto the surface of a glossy metal plate. The solvent was left to evaporate and the absorption/reflection IR spectra were recorded with the infrared microscope. The solvent residues from the samples remaining in the column were removed by flowing nitrogen through the column. The dried material was removed by gentle tapping of the reversed column and the mixture of potassium bromide and sample residue was pressed into a disc and its spectra were recorded.

IR Spectrophotometric Conditions

The infrared microscope is equipped with a high sensitivity, narrow band, mercury-cadmium-telluride (MCT type-A) detector. The infrared radiation is focused at the sample by a 15X Cassegrainian objective (N. A. ≈ 0.58) containing only mirror elements for both visual inspection and the subsequent FT-IR analysis.

The unit was purged with dried air at 800 l/h. Bands for residual CO_2 and H_2O were removed from the spectra by subtraction, and the baselines were corrected by use of the Nicolet SX software correction routine.

The infrared spectra were recorded over the range 700–4000 cm^{-1}. A 4 cm^{-1} resolution was used, with Happ–Genzel apodization, and 1000 co-added scans were run in most cases.

For identification purposes we also used the Toronto Forensics Library and the Georgia State Crime Lab Library of illicit drugs.

Results

In our experience, the majority of the "street" heroin samples contain just one diluent component besides the heroin and its impurities (acetyl codeine, mono-acetyl morphine, papaverine, narcotine) and adulterants (procaine, caffeine, etc.). A typical "street" heroin sample is shown in Fig. 1A, the spectrum of material

---

* To whom correspondence should be addressed

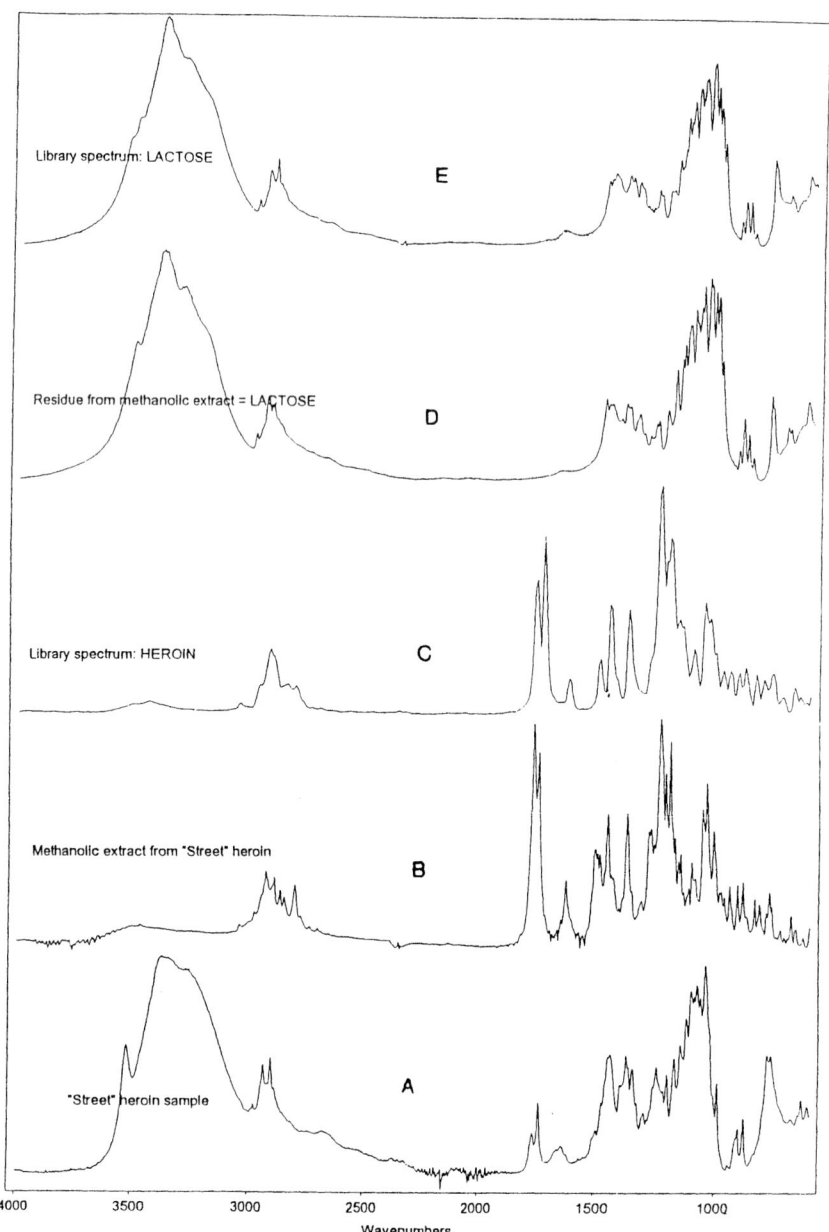

Fig. 1. Spectra of "street" heroin (*A*), methanolic extract from "street" heroin (*B*), residue of "street" heroin after extraction with methanol (*D*) and library spectra of heroin (*C*) and lactose (*E*), respectively

obtained after washing the "street" heroin sample with methanol can be seen in Fig. 1B. The spectrum is identical with that of the heroin (Fig. 1C). The spectrum obtained from the residue of methanolic extract can be seen in Fig. 1D. The spectrum is identical with that of lactose (Fig. 1E).

In Fig. 2A a spectrum of a white tablet containing some amphetamine is shown. The library search revealed that the psychologically active component of the tablet was methamphetamine. The binding component of the tablet was methamphetamine. The binding component of the tablet was lactose (Fig. 2B). The

spectrum obtained after extraction of the tablet material with methanol is shown in Fig. 2C.

Conclusion

The majority of the "street" heroin samples could be cleaned sufficiently by the extraction procedure developed, for their carbohydrate diluent content to be determined by IR spectroscopy. The procedure is applicable for samples which contain just one type of diluent, the diluent being insoluble in methanol. The heroin, its impurities and adulterants are removed with

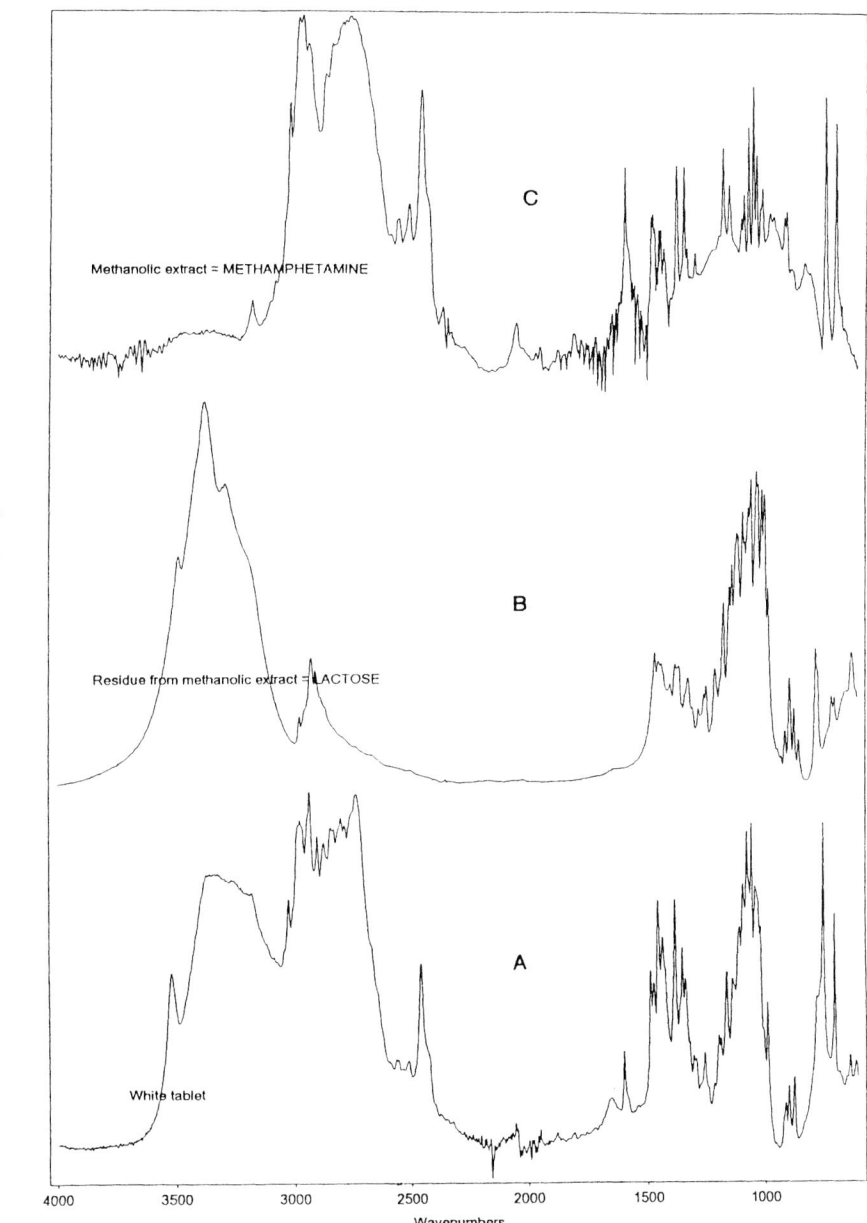

Fig. 2. Spectra of illicit tablet (*A*), residue of illicit tablet after extraction with methanol (*B*), methanolic extract from illicit tablet (*C*)

the methanol. The psychologically active principle and the binding material of illicit tablets can be separated by the proposed procedure if no other components are present in the preparation. The sample preparation procedure proposed is simple and fast and has been applied with success in the analysis of several real samples seized in the black market.

References

[1] M. Ravreby, *J. Forensic. Sci.* **1987**, *32*, 20.
[2] J. C. Shearer, D. C. Peters, T. A. Kubic, *Trends Anal. Chem.* **1985**, *4*, 246.
[3] E. M. Suzuki, *J. Forensic. Sci.* **1992**, *37*, 467.
[4] E. Kaa, *Arch. Pharm. Chem., Sci. Ed.* **1986**, *14*, 87.
[5] M. Chiarotti, N. Fucci, C. Furnani, *Forensic. Sci. Int.* **1991**, *50*, 47.

Mikrochim. Acta [Suppl.] 14, 381–382 (1997)
© Springer-Verlag 1997

FT-IR Study of the Interaction Between Sidnone Derivatives and Butyl-lithium in Cold Solutions

M. G. Ezernitskaya*, B. V. Lokshin, E. I. Kazimirchuk, V. N. Khandozhko, and **V. N. Kalinin**

A. N. Nesmeyanov Institute of Organoelement Compounds, Russian Academy of Sciences, Vavilova Str., 28, 117813 Moscow, Russia

Abstract. The reaction of 3-phenyl- and 3-pyridylsidnones with butyl-lithium and 3-phenylsidnone with MeYbI in cold THF solutions was monitored by FT-IR spectroscopy. Butyl-lithium was found to form with sidnone derivatives chelate adducts which are intermediates in the preparation of the 4-derivatives. MeYbI reacts with 3-phenylsidnone to give a complex, in which the Yb atom is bound to the exo-oxygen atom.

Key words: sidnones, IR spectroscopy, complex formation.

Sidnones belong to the group of non-benzenoid aromatic compounds which possess a sextet of electrons shared between all the atoms comprising the five-membered cycle [1]. They are termed mesoionic compounds and can be reproduced as a resonance hybrid of a number of canonical forms with different degrees of charge distribution, for example:

Sidnone derivatives are biologically active and are widely used in pharmacology. For this reason, sidnone chemistry is very important. Sydnones react with butyl-lithium to form lithium derivatives, which are thermally and oxidatively unstable; they are usually obtained at −70 °C in THF (tetrahydrofuran) solutions and are used in situ in subsequent reactions with unsaturated compounds for syntheses of various 4-substituted sidnones [2].

The reaction of 3-phenyl- and 3-pyridylsidnones with butyl-lithium and MeYbI in cold THF solutions was monitored by FT–IR spectroscopy in order to reveal the structure of the adducts, which can be intermediates in preparative reactions.

* To whom correspondence should be addressed

Experimental

All the procedures were carried out in an argon atmosphere. A THF solution of 3-substituted sidnone was cooled to −70 °C, and a THF solution of BuLi or MeYbI was added. The mixture was stirred, then the solution was transferred by the argon pressure to a Karl Zeiss low-temperature cell precooled to the same temperature. FT–IR spectra of the cold reaction mixture were recorded with Bruker IFS-113v and IFS-25 FT–IR spectrometers with a resolution of 2 cm^{-1}.

Results and Discussion

In the IR spectra of the initial solutions of 3-phenyl- and 3-pyridylsidnones, the v_{CO} modes exhibit a doublet (Fig. 1, curve a). The component positions and intensity ratio are fairly dependent on the solvent employed; in DMSO solutions, a single band is observed. Consequently, these doublets originate from Fermi resonance. In THF solutions, the v_{CO} position was determined as the centre of gravity of the resonance doublet,

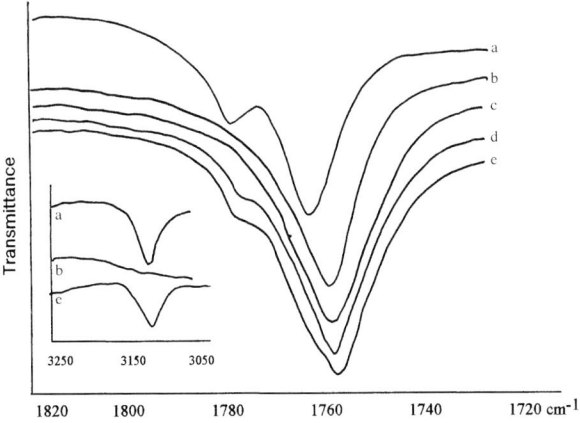

Fig. 1. Temperature-dependent FT–IR spectra of a THF solution of 3-phenylsidnone in the presence of BuLi: *a* initial compound; *b* 70 °C; *c* 20 °C; *d* 0 °C; *e* room temperature

1768 cm^{-1} for 3-phenylsidnone and 1770 cm^{-1} for 3-pyridylsidnone.

These values are unusually high, compared to those of tropones (1638 cm^{-1}) and 4-thiapirone (1574 cm^{-1}). They are closer to the carbonyl absorption of γ-lactones and so evidence for the major contribution of the canonical structure with the normal exo-cyclic carbonyl group (**I**) rather than that of the betaine structure (**II**).

Substituted sidnone molecules have various possibilities for interaction with a metal atom, namely:

Temperature-dependent FT–IR spectra of a THF solution of 3-phenylsidnone in the presence of BuLi are presented in Fig. 1. At $-70\,°$C, a band at 1760 cm^{-1} is observed instead of the resonance doublet of the ν_{CO} modes (Fig. 1, curve b). This picture is maintained until the solution is heated to $-20\,°$C; at $0\,°$C, the bands of the initial doublet appear in the spectrum along with the band at 1760 cm^{-1}; the ν_{CH} band of the sidnone ring at 3125 cm^{-1} is absent in the spectrum of the adduct; it appears again at $0\,°$C together with the ν_{CO} bands of the initial 3-phenylsidnone.

Upon the interaction with butyl-lithium, the ν_{CH} band disappears, indicating the formation of the 4-lithio-derivative. At the same time, one down-shifted band is observed in the ν_{CO} range instead of the resonance doublet of the initial compound. The exo-cyclic oxygen atom seems to be also involved in the interaction with the lithium atom, causing a decrease in the ν_{CO} wavenumber. This results in the resonance requirements being broken, and one band is present in the spectrum instead of the doublet. The interaction leads to a low-frequency shift of 8 cm^{-1}, which also indicates that sidnone derivatives contain the normal carbonyl group.

For 3-pyridylsidnone under the same conditions, the ν_{CO} bands of the initial Fermi resonance doublet do not disappear completely, thus indicating a lower basicity of the exo-cyclic CO group of 3-pyridylsidnone compared to that of 3-phenylsidnone.

3-Substituted sidnones seem to form with butyl-lithium chelate adducts of the following structure:

These adducts are intermediates in preparative reactions leading to various 4-derivatives.

Sidnone derivatives also give adducts with organo-ytterbium compounds, which cannot, however, be involved in subsequent reactions with the same reagents as lithium adducts are, to form 4-substituted sidnones. We measured temperature-dependent FT–IR spectra of THF solutions of 3-phenylsidnone in the presence of MeYbI. The interaction with MeYbI under the same conditions as for BuLi results in the new ν_{CO} band at 1710 cm^{-1} ($\Delta\nu = 58$ cm^{-1}). Such a shift is typical of complex formation with the CO group. The complex with Yb is supposed to have the following structure:

The complex obtained is stable even at room temperature. The formation of such a type of complex can explain the fact that the metallation reaction with organoytterbium compounds does not lead to 4-substituted sidnones.

Conclusions

(1) 3-Phenyl- and 3-pyridylsidnones interact with butyl-lithium at $-70\,°$C to give unstable chelate adducts. (2) The basicity of the exo-cyclic oxygen atom in 3-phenylsidnone is higher than that of 3-pyridylsidnone. (3) Lithium adducts are unstable and can enter subsequent reactions to give various 4-derivatives. (4) 3-Phenylsidnone interacts with MeYbI at $-70\,°$C to form a stable complex, where the Yb atom is bound to the exo-oxygen atom. (5) The Yb complex of 3-phenylsidnone does not enter subsequent reactions as the lithium adducts do. It cannot be considered an intermediate in syntheses of 4-substituted sidnones. (6) The possibility of using the metallation reaction for the synthesis of 4-substituted sidnones depends on the structure of the adducts formed with organometallic compounds at low temperatures.

Acknowledgement. The research was supported by the International Science Foundation (Grants MEQ000 and MEQ300).

References

[1] C. A. Ramsden, *Comprehensive Organic Chemistry* (D. H. R. Barton, W. D. Ollis, eds.), Pergamon Press, Oxford, 1979, Vol. 4, p. 1171.

[2] V. N. Kalinin, S. F. Min, *J. Organomet. Chem.* **1988**, *352*, C34.

Mikrochim. Acta [Suppl.] 14, 383–385 (1997)

The Fourier-transform Infrared Spectra of the 1:1 Electron Donor–acceptor Complexes of Boron Trifluoride with Dimethyl Ether and Dimethyl Sulphide in Cryogenic Matrices

Lawrence M. Nxumalo[1] and **Thomas A. Ford**[2,*]

Department of Chemistry, University of the Witwatersrand, Johannesburg, P. O. Wits 2050, Republic of South Africa

Abstract. The infrared spectra of boron trifluoride co-deposited with dimethyl ether and dimethyl sulphide in nitrogen and argon matrices have been recorded at cryogenic temperatures. The wavenumber shifts of the monomer bands resulting from the formation of 1:1 electron donor–acceptor complexes have been measured and rationalized on the basis of the interactions of the Lewis acid (boron trifluoride) with hard (dimethyl ether) and soft (dimethyl sulphide) bases.

Key words: infrared spectroscopy, matrix isolation, donor–acceptor complexes, boron trifluoride.

Boron trifluoride readily forms charge-transfer complexes with electron donors, particularly oxygen, nitrogen, sulphur and phosphorus bases. The mode and strength of interaction are dependent on the type of base involved, and hard and soft bases provide examples of quite different complex properties. Infrared spectroscopy is one of the most powerful tools available for the study of such interactions.

The use of *ab initio* molecular orbital theory is nowadays a routine procedure for predicting infrared spectra and its use as a valuable aid in the interpretation of matrix-isolation infrared spectra has been demonstrated [1, 2]. The structures, interaction energies and infrared spectra of the complexes of boron trifluoride with dimethyl ether and dimethyl sulphide have recently been predicted *ab initio* [3].

The spectra of BF_3 co-deposited with $(CH_3)_2O$ and with $(CH_3)_2S$ in argon and nitrogen matrices at cryogenic temperatures have been recorded, and are interpreted in terms of the interaction of BF_3 with a hard (ether) and a soft (sulphide) base, using the computed spectra as aids in the assignments. The spectrum of the $BF_3 \cdot O(CH_3)_2$ complex is in good agreement with that reported earlier by Hund and Ault [4].

Experimental

The spectra were recorded with a Bruker IFS 88 Fourier-transform infrared spectrometer; the cryostat used was an Air Products Displex CSA-202 closed-cycle helium refrigerator. Samples were deposited at ca. 17 K and routinely annealed to ca. 35 K; matrix ratios were in the range 250/1 to 1000/1 and deposition rates were approximately 2–6 mmol/h.

Results

Tables 1 and 2 list the wavenumbers of the bands observed in the spectra of the complexes that were not present in the spectra of BF_3, $(CH_3)_2O$ or $(CH_3)_2S$ alone, with assignments based on the computed spectra reported earlier [3].

Discussion

The forms of the intramolecular normal modes of the two complexes were generally calculated to be very similar to the corresponding modes of the monomers, making identification fairly easy [3]. The computed wavenumbers indicate that very large red shifts are to be expected for the antisymmetric BF_3 stretching

---

[1] Present address: Department of Chemistry, Vista University, Soweto Campus, Private Bag X09, Bertsham 2013, Republic of South Africa

[2] Present address: Department of Chemistry and Applied Chemistry, University of Natal, Durban, Private Bag X10, Dalbridge 4014, Republic of South Africa

* To whom correspondence should be addressed

Table 1. Observed infrared absorption bands of the $BF_3 \cdot O(CH_3)_2$ complexes isolated in nitrogen and argon matrices

Symmetry species	Mode[a]	Wavenumber/cm^{-1}			
		Nitrogen		Argon	
		$^{10}BF_3$	$^{11}BF_3$	$^{10}BF_3$	$^{11}BF_3$
a′	$\delta_a(CH_3)$	1454.0	1454.0	1461.3	1461.3
	$\delta_a(CH_3)$	1435.9	1435.9	1446.1	1446.1
	$\nu(BF)$	1240.1	1197.8	1251.6	1220.8
	$\rho(CH_3)$	1166.9	1166.9	1219.2	1219.2
	$\nu_s(COC)$	923.2	919.8	999.1	950.2
	$\nu_s(BF_2)$	817.1	812.0	815.1	811.6
	$\delta(BF_2)$	652.5	637.1	637.8	624.1
	$\delta_a(BF_3)$	480.9	480.9	–	–
	$\delta(COC)$	474.2	474.2	–	–
a″	$\delta_s(CH_3)$	1424.2	1424.2	–	–
	$\nu_a(BF_2)$	1247.3	1207.8	1260.1	1222.7
	$\rho(CH_3)$	1094.3	1094.3	–	–
	$\nu_a(COC)$	1031.3	1031.3	–	–

[a] See [3] for assignments.

Table 2. Observed infrared absorption bands of the $BF_3 \cdot S(CH_3)_2$ complexes isolated in nitrogen and argon matrices

Symmetry species	Mode[a]	Wavenumber/cm^{-1}			
		Nitrogen		Argon	
		$^{10}BF_3$	$^{11}BF_3$	$^{10}BF_3$	$^{11}BF_3$
a′	$\delta_a(CH_3)$	1459.1	1459.1	–	–
	$\nu(BF)$	1453.9	1393.2	1443.1	1395.1
	$\rho(CH_3)$	1096.8	1096.8	1040.1	1040.1
	$\rho(CH_3)$	1028.0	1028.0	1034.1	1034.1
	$\nu_s(BF_2)$	853.5	853.5	873.7	873.7
	$\delta(BF_2)$	690.5	670.2	688.2	668.2
	$\delta_a(BF_3)$	481.1	481.1	482.8	482.8
a″	$\delta_a(CH_3)$	1498.7	1498.7	1498.3	1498.3
	$\nu_a(BF_2)$	1478.2	1424.1	–	–
	$\delta_s(CH_3)$	1363.8	1363.8	1363.9	1363.9
	$\delta_a(BF_3)$	474.8	474.8	473.9	473.9

[a] See [3] for assignments.

vi-bration, which splits into a doublet on complexation, due to the removal of the degeneracy present in the monomer. A smaller red shift is predicted for the our-of-plane BF_3 bending mode, while the antisymmetric BF_3 bending vibration is calculated to be relatively insensitive to complexation [3].

In the case of the $BF_3 \cdot S(CH_3)_2$ complex, the wavenumber shifts on complexation are predicted to be far smaller than their counterparts in $BF_3 \cdot O(CH_3)_2$; here the major perturbation is expected to be

the red shift of the out-of-plane BF_3 bending mode, with a smaller wavenumber decrease predicted for the pair of antisymmetric BF_3 stretching vibrations [3].

The experimental shifts for $BF_3 \cdot O(CH_3)_2$ are indeed very large; the antisymmetric BF_3 stretching vibrations, ν_7 and ν_{25}, shift by -245.4 and -235.4 cm^{-1} respectively and the out-of-plane bending, ν_{12}, by -25.2 cm^{-1} (see Table 1). The observed BF_3 stretching wavenumber shifts of the $BF_3 \cdot S(CH_3)_2$ complex are also in the same directions as predicted, but are far smaller (-50.0 and -19.1 cm^{-1} for ν_6 and ν_{22} respectively, while that of the out-of-plane bending mode, ν_{12}, is 7.9 cm^{-1} (see Table 2).

The mutual perturbation of the spectra of the BF_3 and the base monomers on complexation is evidently far greater for $BF_3 \cdot O(CH_3)_2$ than for $BF_3 \cdot S(CH_3)_2$. This is due to the greater hardness of $(CH_3)_2O$, on Pearson's scale [5], than that of $(CH_3)_2S$ (8.0 *vs* 6.0 eV).

References

[1] G. A. Yeo, T. A. Ford, *Struct. Chem.* **1992**, *3*, 75.

[2] L. M. Nxumalo, T. A. Ford, *S. Afr. J. Chem.* **1995**, *48*, 30.

[3] L. M. Nxumalo, T. A. Ford, *J. Mol. Structure (Theochem.)*, in press.

[4] R. L. Hunt, B. S. Ault, *Spectroscopy, Int. J.* **1982**, *1*, 45.

[5] R. G. Pearson, *J. Org. Chem.* **1989**, *54*, 1423.

Mikrochim. Acta [Suppl.] 14, 387–390 (1997)

Co-condensation of Cu and CO in an N₂ Matrix

Stella Nunziante Cesaro[1,*], and **Sándor Dobos**[2]

[1] C.S. CNR Termodin. Chim. Alte Temp., University of Rome, P.le Aldo Moro 5, 00185 Rome, Italy
[2] Spectroscopic Department, Institute of Isotopes, Hungarian Academy of Sciences, P.O.B. 77, H-1525 Budapest, Hungary

Abstract. Copper carbonyl complexes were produced by high temperature vaporization of Cu with excess of carbon powder. FT-IR spectra of copper carbonyls suspended in nitrogen are presented for the first time and compared with the corresponding spectral features in argon medium, reported in the literature. Polycarbonyl complexes are nearly unaffected by the isolating medium whereas the monomeric species show large frequency shifts, suggesting the existence of more complex molecules, in which N₂ ligands and/or Cu_2, Cu_3 clusters are also involved.

Key words: matrix isolation, FT-IR, copper carbonyls.

The removal of nitrogen oxides from exhaust gases emitted by internal combustion engines involves the catalytic reactions of CO, NO_x, N_2, O_2, N_2O and hydrocarbons [1–3]. Copper seems to offer a new and cheap catalyst in these processes [4–6]. In order to systematically model the interaction of Cu with these ligands, as a first step we studied the reaction of Cu with CO in N_2 cold matrices, i.e. in a large excess of nitrogen. The spectral data obtained are compared with the literature data for mono-copper carbonyls, and carbonyls of copper clusters trapped in Ar [7, 8]. In order to make the comparison more reliable, we reinvestigated the interaction of Cu and CO in Ar matrices.

Experimental

The experimental apparatus and procedures have been described elsewhere [9]. In most experiments Cu (Aldrich 99.999%) was vaporized from graphite, tantalum or molybdenum Knudsen cells together with an excess of ^{12}C (Strem Chemicals) in the 1400–1850 K temperature range (5×10^{-5}–5×10^{-3} mmole/h). Only Ta or Mo crucibles were used when vaporizing Cu with ^{13}C (Cambridge Isotopes Laboratories, 99%). Separate vaporizations of Cu, ^{12}C and ^{13}C were also performed in order to check their purity spectroscopi-

cally. Therefore, the occurrence of CO resulted from the reduction of the partially oxidized Cu surface in the presence of C powder.

Traces of H_2O were detected after long-lasting depositions, and CO_2 in traces was also present in vaporizations involving C powder. Reaction products of copper with carbon dioxide were never detected [10].

Matrix gases were high-purity N_2 and Ar (Caracciolo, Rome) or their mixtures in controlled ratios. In all cases, the gas flow was regulated at about 1 mmol/h through a standardized needle valve.

Since this experimental procedure does not allow a quantitative estimate of the CO trapped, a few experiments were also performed in which copper was deposited on Ar:CO and N_2:CO mixtures at a ratio in the 300:1–1000:1 range in order to have a reasonable evaluation "a posteriori" for comparing the free CO areas [9]. Depositions lasted from 5 to 60 min. Annealing cycles were usually performed at up to 33 K and spectra were recorded after recooling to 12 K. For routine spectra at 1 cm⁻¹ resolution, 200 scans were accumulated.

Results and Discussion

Infrared spectra of Cu atoms suspended in nitrogen matrices doped with CO are presented in the following and compared with those in argon matrices. Since monocarbonyl complexes seemed to be more affected by the isolating medium, the appropriate experimental conditions were chosen for obtaining their preferential formation in the matrices. Mono- and polycarbonyl species are therefore discussed separately, and more details are given for the former.

Copper Monocarbonyls

On deposition of Cu atoms in argon doped with ^{12}CO, the spectrum reported in Fig. 1a was obtained. The excellent agreement with spectra obtained by Moskovits and co-workers [7, 8] on co-condensation of copper atoms with 1.300 CO:Ar mixtures clearly indicates that the same complex species are formed in spite of the different experimental conditions used. In accord with previous assignments [7, 8, 11], the main doublet

* To whom correspondence should be addressed

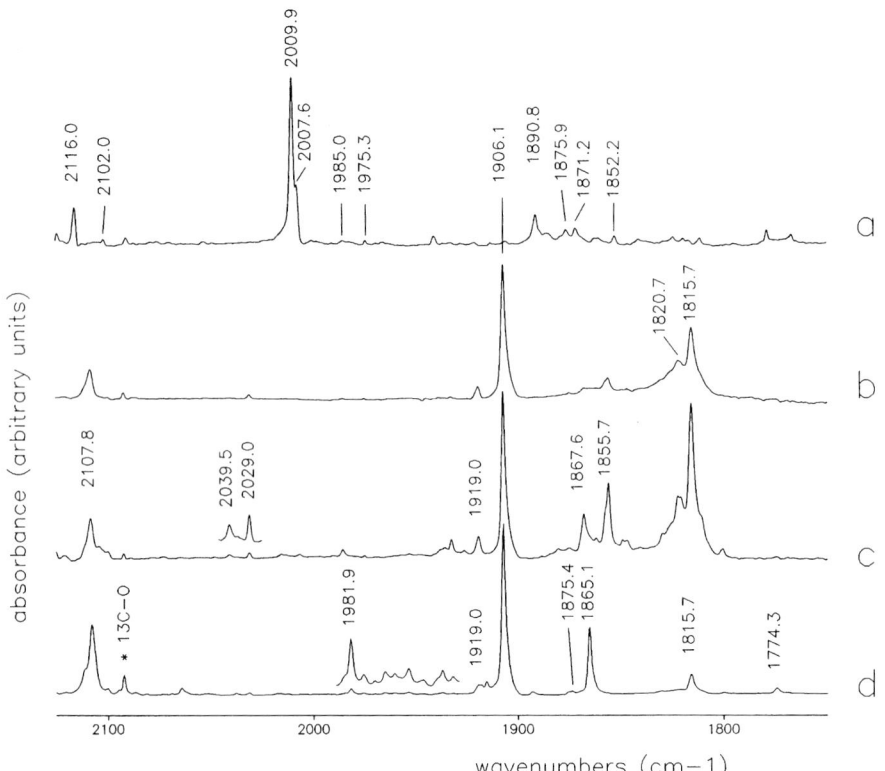

Fig. 1. FTIR spectra of copper carbonyls; *a* Cu:Ar = 1:1000; ^{12}CO:Ar = 1:300, *b* Cu:N$_2$ = 1:10,000; ^{12}CO:N$_2$ = 1:1000, *c* Cu:N$_2$ = 1:1000; ^{12}CO:N$_2$ = 1:300, *d* Cu:N$_2$ = 1:10,000; ^{12}CO:^{13}CO:N$_2$ = 3:1:1000

at 2009.9/2007.6 cm^{-1} could be attributed to CuCO at two different matrix sites. The less intense doublet at 1890.8/1875.9 cm^{-1} belongs to the asymmetric stretching mode of a linear Cu(CO)$_2$ species. The species Cu(CO)$_3$ absorbs weakly at 1985.0/1975.3 cm^{-1}, owing to the high dilution of CO. The remaining bands have been assigned to carbonyl complexes of Cu clusters: 2116.0 cm^{-1} (Cu–Cu–CO), 2102.0 cm^{-1} (Cu$_3$CO), 1871.2 cm^{-1} (Cu$_2$CO, with bridging CO) and 1852.2 cm^{-1} (Cu$_2$CO of unsettled geometry).

Figures 1b and 1c show the spectra obtained when using nitrogen as isolating gas at different copper concentrations (Cu:N$_2$ = 1:10000 and Cu:N$_2$ = 1:1000, respectively). Significant differences were observed on comparing the spectra reported in Figs. 1a and c, which had been obtained under identical experimental conditions except for the isolating gas. This finding suggests the role of the matrix gas to be more important than expected.

The copper monocarbonyl (**I**) which was the most abundant reaction product observed in argon at 2009.9/2007.6 cm^{-1} was apparently absent in the nitrogen system. In the same spectral region only a doublet was detected (Fig. 1c), consisting of a very weak component at 2039.5 cm^{-1} and a weak one at 2029.0 cm^{-1}. When copper was isolated in nitrogen doped with ^{12}CO:^{13}CO (3:1) mixture, the ^{13}CO-substituted counterparts of the doublet could be identified at 1993.7 cm^{-1} and approximately 1985 cm^{-1}, in agreement with the expected isotopic shift. Unfortunately the component at around 1985 cm^{-1} was overlapped by a more intense band at 1981.9 cm^{-1} (Fig. 1d, see below) and its wavenumber could not be precisely determined.

The prominent feature in all the nitrogen experiments consisted of a doublet at 1919.0/1906.1 cm^{-1}, (Figs. 1b and 1c), undoubtedly attributable to a complex having a single CO group shifted under ^{13}C isotopic enrichment to 1875.4/1865.1 cm^{-1} (Fig. 1d). A correspondence of this doublet with the predominant one detected in argon at 2009.9/2007.6 cm^{-1} was readily ruled out, because it would imply an exceedingly high matrix shift. Alternative hypotheses were therefore taken into account. The possibility of a nitrogen-perturbed copper monocarbonyl complex of formula $(N_2)_x$CuCO ($x = 1$–2) seemed a straightforward one. However, it was discarded because the nitrogen perturbation on copper would cause a blue shift of the CO stretching mode. In particular, for $(N_2)_2$CuCO (**II**) a blue shift of 32 cm^{-1} has been predicted for the CO mode as calculated by the DFT method [12]. In the second instance, the formation of a Cu_3CO complex (**III**) was considered. In fact, a recent theoretical treatment of mononuclear and polynuclear carbonyl complexes of Cu [13] evaluated the CO stretching mode of the Cu_3CO molecule as being at 2105 cm^{-1} and 1917 cm^{-1} in C_{2v} and C_s symmetry respectively. Both geometries are probable, the second one being only few kcal/mole higher in energy. However, the presence of Cu_3CO in C_{2v} symmetry was readily eliminated. We observe a band at 2107.8 cm^{-1} only in N_2 matrices, but the lack of isotopic effect provided evidence that the band was not due to a CO mode. The assignment of the doublet at 1919.0/1906.1 cm^{-1} to the CO fundamental mode of Cu_3CO in C_s symmetry, on grounds of the closeness of the experimental and theoretical values, seems at present uncontradicted. However, the copper vapour is greater than 99% monoatomic under the conditions employed in these experiments and this assignment would imply the additional condition that the Cu atoms are induced to react in cold matrices to produce a consistent amount of polynuclear complexes. As an alternative, the 1919.0/1906.1 cm^{-1} doublet can be attributed to a binuclear complex of formula $[(N_2)_x$Cu$]_2$CO (with $x = 1$–2), with bridging CO (**IVA**) which is based on the greater reactivity of the copper dimer than that of the atom ([13] and references therein). In addition, the frequency value of the doublet lies in the middle of the typical spectral regions where the vibrations of the terminal and bridging CO are expected. For one or more Cu atoms bound to one or more N_2 molecules a tightening of the CO bond is expected, with a consequent increase of its frequency value. In this model, the band at 2107.8 cm^{-1} is assigned to the stretching mode of a nitrogen molecule bound to a copper atom. The parallel intensity enhancement of the 2107.8 cm^{-1} band and the 1919.0/1906.1 cm^{-1} doublet in dependence on the vaporization tem-

perature and/or the time of deposition supported their assignment to the same molecular complex. For a better comprehension of the N_2 role in the origin of the complex, experiments were also performed with Cu isolated in Ar:N_2:CO mixtures of different Ar:N_2 ratios. For Ar:$N_2 = 10$:1, bands separately observed in Ar and N_2 coexisted. At higher N_2 concentration, the spectra reproduced all the features observed for the pure nitrogen medium, which suggests for the N_2 molecule the role of a reactant rather than of an isolating medium.

A third monocarbonylic species has been detected in N_2 at 1820.7/1815.7 cm^{-1} in the presence of ^{12}CO, which is shifted to 1779.0/1774.3 cm^{-1} in the case of ^{13}CO doped matrices (**IVB**). The two frequency value suggested the presence of a bridging CO group. A complex species with formula Cu_2CO of C_{2v} symmetry could be proposed on the grounds of theoretical calculations, predicting for this molecule a frequency value of 1823 cm^{-1}[13]. It is, however, worth noticing that the formation of a bridged Cu_2CO complex is consistent with the generation of the aforementioned complex in the nitrogen medium, involving the addition of one or more N_2 molecules, which would not be surprising in a nitrogen medium.

Copper Polycarbonyls

In agreement with literature data [7, 8], the CO stretching mode of $Cu(CO)_3$ isolated in argon has been found at 1985.0 cm^{-1}, accompanied by a second minor site feature at 1975.3 cm^{-1}. Under isotopic conditions the main component splits to give a quartet pattern (1985.0, 1966, 1951.5 and 1940.5 cm^{-1}) as expected for a trigonal planar structure (D_{3h}). We detected the corresponding frequency values in nitrogen at 1981.9, 1964.2, 1953.2 and 1936.9 cm^{-1}, showing a negligible matrix shift. The possible existence of multiple matrix sites could not be investigated in our experiments, because of the weakness of the bands observed in isotopic experiments.

The CO stretching mode of $Cu(^{12}CO)_2$ isolated in argon gives rise to a doublet pattern at 1890.8 and 1875.9 cm^{-1}, in accordance with the literature [7, 8]. With the nitrogen matrix a doublet at 1867.6/1855.7 cm^{-1} can be assigned to the same species. The weakness of the doublet components and their proximity to the Cu_2CO species at 1820.7/1815.8 cm^{-1} did not allow the observation of a clear isotopic pattern. As a consequence, the last assignment is only tentative.

References

[1] K. I. Choi, M. A. Vannice, *J. Catal.* **1991**, *131*, 22.

[2] M. Shelef, *Catal. Lett.* **1992**, *15*, 305.

[3] W. K. Hall, J. Valyon, *Catal. Lett.* **1992**, *15*, 311.

[4] G. J. Millar, C. H. Rochester, K. C. Waugh, *J. Chem. Soc. Faraday Trans.* **1991**, *87*, 1467.

[5] C. J. G. van der Grift, A. F. H. Wielers, B. P. J. Joghi, J. van Beijnum, M. de Boer, M. Versluijs-Helder, J. W. Geus, *J. Catal.* **1991**, *131*, 178.

[6] Y. Fu, Y. Tian, P. Lin, *J. Catal.* **1991**, *132*, 85.

[7] H. Huber, E. P. Kündig, M. Moskovits, G. A. Ozin, *J. Am. Chem. Soc.* **1975**, *97*, 2097.

[8] M. Moskovits, J. E. Hulse, *J. Phys. Chem.* **1977**, *81*, 2004.

[9] A. Feltrin, M. Guido, S. Nunziante Cesaro, *Vib. Spectrosc.* **1995**, *8*, 175.

[10] J. Mascetti, M. Tranquille, *J. Phys. Chem.* **1988**, *92*, 2177.

[11] J. H. B. Chenier, C. A. Hampson, J. A. Howard, B. Mile, *J. Phys. Chem.* **1989**, *93*, 114.

[12] I. Pápai, *Private Communication.*

[13] R. Fournier, *J. Chem. Phys.* **1995**, *102*, 5396.

Mikrochim. Acta [Suppl.] 14, 391–393 (1997)
© Springer-Verlag 1997

Correlations of Attenuated Total Reflectance Spectra to Physical Properties or Polyurethane Foams

Sanmitra A. Bhat and **James A. de Haseth***

Department of Chemistry, University of Georgia, Athens, Georgia 30602-2556, U.S.A.

Abstract. Polyurethane foams are produced from five basic components: polyol, silicone surfactant, tin and amine catalyst, water and an isocyanate. Inconsistency in the measurement of physical properties of the foam arises if there is a change in reactivity of any of the components. The change in reactivity is often the result of batch-to-batch variability of the chemical components. Production of a consistent foam in these cases is difficult and manufacturing takes time and wastes chemicals. The purpose of this study is to correlate attenuated total reflectance (ATR) infrared spectra of the physical properties of the foam. Correlations were made with the use of partial least squares (PLS) software. Both PLS-1 and PLS-2 algorithms were implemented for correlation. The correlation coefficient (R^2) was better with the PLS-1 than the PLS-2 algorithm. The values predicted by the PLS model fall within the standard deviation.

Key words: ATR, PLS, polyurethane foams.

Partial least squares is a spectral decomposition technique used for the quantitative analysis of spectroscopic data. It is a multivariate statistical technique applied for quantitative analysis of ultraviolet, near-infrared and chromatographic data. Excellent details about application of this technique to infrared data have been presented by Fuller et al. [1] and Haaland and Thomas [2].

Polyurethane foams produced in this study are referred to as slabstock foams. The foams are produced with the use of five basic components. Lot-to-lot variability of these components results in the production of inconsistent foams. So, trial and error methods are used to come up with consistent foams with good physical properties. In the present study, a calibration equation is developed with partial least squares (PLS) software, the correlates the attenuated total reflectance (ATR) spectra to silicone surfactant concentration and physical properties of the foam. This correlation would reduce the waste, both in terms of time and chemicals.

Experimental

Slabstock foams were produced in our laboratory by a hand-mix procedure. The basic components: polyol (Voranol 3137), TDI (80/20 mixture of 2,6- and 2,4-toluene di-isocyanate), silicone surfactants, tin and amine catalysts were supplied by Dow Chemical Co. A known quantity of all components, except TDI, was weighed in a plastic beaker and then mixed with a stirrer at 2500 rpm for two minutes. The TDI was poured into the mixture and stirred for another eight seconds. The viscous creamy mixture was poured immediately into a paper bucket to form the polymer. It takes about 2–3 minutes for the full foam rise. Two different silicone surfactants, DC-5160 and L-5340, were used. To study the effect of the surfactant, only the level of silicone surfactant was varied, while maintaining constant the level of all other components. Twenty foams were produced with each of the surfactants. All the foams were cured for 24 hours at room temperature and attenuated total reflectance spectra (ATR) were recorded on a Perkin-Elmer System 2000 FT-IR spectrometer. An ATR spectrum of L-5340 is shown in Fig. 1. The foams were then further cured for seven days to determine the physical properties (tensile strength and air-permeability). Physical properties were determined according to American Society of Testing and Materials (ASTM) standards. Samples were taken from the interior of the foam in the direction of foam rise. An Instron Universal Testing Instrument, Table Model 1130, with Instron Series Automated Testing System was used to determine tensile strength. The sample was elongated at a rate of 8.3 ± 0.8 mm/s, until it broke. An air-permeability tester (constructed in our laboratory according to ASTM standards) was used to determine the air flow through the samples. Samples $2'' \times 2'' \times 1''$ in size were taken from the bottom, middle and upper part of the foam. A differential pressure of 125 pascals was maintained across the thickness of the sample. An average value for the three samples was recorded. Physical properties were measured for all the foams made with each surfactant. Both PLS-1 and PLS-2 algorithms were used to generate the calibration

---

* To whom correspondence should be addressed

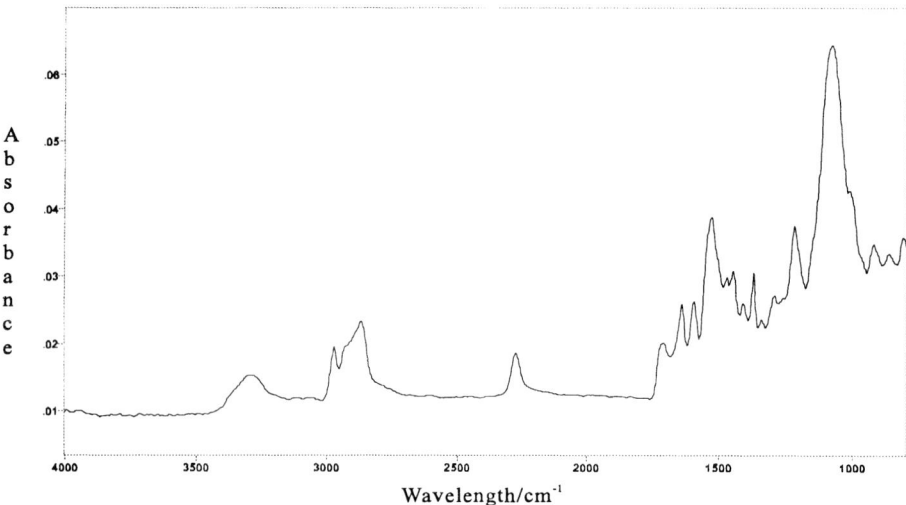

Fig. 1. ATR of L-5340 foam

equations. Based on a correlation spectrum, spectral regions of 3500–2700, 2350–2200 and 1800–800 cm^{-1} were selected. The correlation coefficient (R^2) was better with PLS-1 then PLS-2; therefore, correlation was done only by the PLS-1 model.

Results and Discussion

The average relative standard deviation (RSD) for tensile strength (TS) was around 7% and that for air-permeability (AP) around 21%. The RSD for tensile strength mainly arises from the standard method of measurement. During tensile testing, the sample is clamped at each end in metallic jaws. Uneven pressure at each jaw causes an error in the measurement. This error results in an RSD of 3–12%. So samples having an RSD greater than 12% were discarded during the calibration. Owing to the complex nature of polyurethane formation reactions, the high RSD for air-permeability was expected. During polymer formation, as the foam rises, cell opening is better towards the upper part of the foam. Thus the air-permeability value was greater for the sample taken from the upper part of the foam than for the lower part. The air-permeability tester itself had an RSD of 4–5%. Samples having an RSD greater than 28% for air-permeability were discarded from the calibration. A calibration file that contained information about the spectra, component names (TS, AP, surfactant) and measurements, was created for each of the surfactants. The PLS-1 diagnostics calculation was run to find out the PRESS (predicted residual error sum of squares) value for each of the surfactants. The number of factors at which the PRESS value is minimum is generally chosen for the calibration. The PRESS plot of surfactant, TS and AP

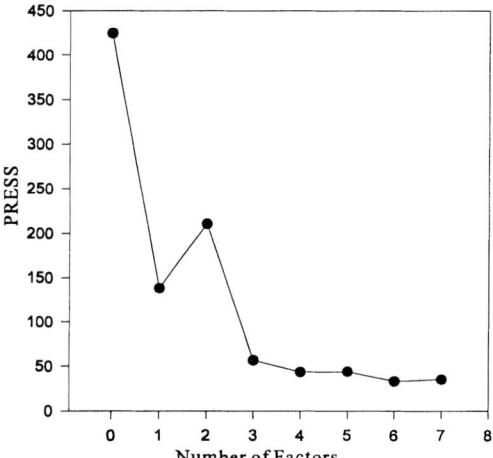

Fig. 2. PRESS value of surfactant

for L-5340 foam is shown in Fig. 2 and 3. The PRESS value was minimum at six factors for surfactant with an R^2 value of 0.93, at seven factors for average AP ($R^2 = 0.79$) and TS ($R^2 = 0.23$). For DC-5160 foam, the minimum PRESS value was at ten factors for surfactant ($R^2 = 0.85$), at eight factors for AP ($R^2 = 0.35$) and six factors for TS ($R^2 = 0.57$). The R^2 value for TS and AP suggests average to weak fits for both of the surfactants. For good fit, the R^2 value should be 0.98–1. For the DC-5160 foams, the actual RSDs were 6.03% and 19.2% for TS and AP, and predicted RSDs were 7.21% and 19.7%. A correlation plot of TS is shown in Fig. 4. The actual RSDs of TS and AP for L-5340 foams were 6.24% and 23.3% and the predicted RSD were 9% and 13.7%. A correlation plot of AP is

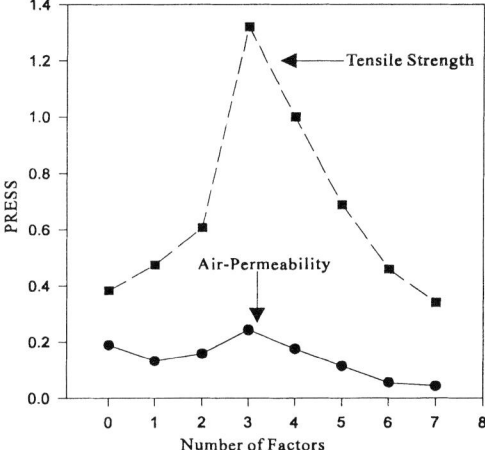

Fig. 3. PRESS plot for TS and AP for L-5340 foam

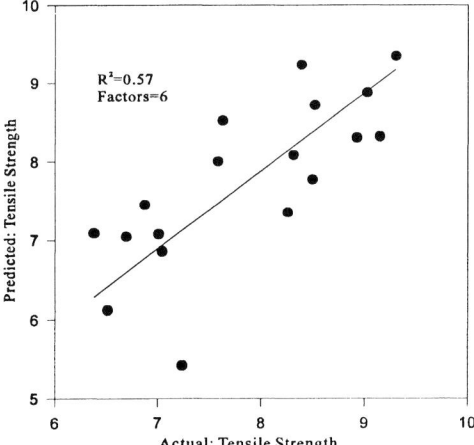

Fig. 4. Correlation plot of TS for DC-5160 foam

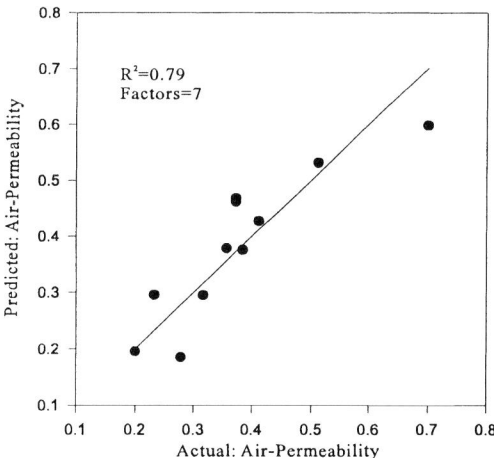

Fig. 5. Correlation plot of AP for L-5340 foam

Conclusion

Correlations between ATR and physical properties are possible. Correlation models for both of the surfactants seemed average and predicted the values within the RSD for TS and AP. The models are hindered by the RSD present in the standard methods of measurement. The models performed better than the standard methods of measurement of TS and AP. The study showed that this correlation would reduce waste involved during the production of polyurethane foams, as infrared spectra can be measured more readily than direct physical properties.

References

[1] M. P. Fuller, G. L. Ritter, C. S. Draper, *Appl. Spectrosc.* **1988**, *42*, 228.
[2] D. M. Haaland, E. V. Thomas, *Anal. Chem.* **1988**, *60*, 1202.

shown in Fig. 5. The model did as well as the ASTM standard methods of measurement in predicting the TS and AP.

Mikrochim. Acta [Suppl.] 14, 395–397 (1997)

Study of the Dynamics of Photoinduced Orientation of Azo Polymers By Infrared Spectroscopy

T. Buffeteau[1,*] and **M. Pézolet**[2]

[1] Laboratoire de Spectroscopie Moléculaire et Cristalline, Université de Bordeaux I, F-33405 Talence, France
[2] Centre de Recherche en Sciences et Ingénierie des Macromolécules, Département de Chimie, Université Laval, Québec, G1K7P4, Canada

Abstract. Infrared spectroscopy coupled with the polarization modulation technique has been used to characterize the molecular orientation in azo polymers irradiated by polarized light. The results obtained with an amorphous azo polymer containing Disperse Red-1 side-chains reveal that this technique is highly effective to determine quantitatively the time dependence of the degree of orientation of several chemical groups during the orientation (laser on) and relaxation (laser off) processes.

Key words: birefringence, optical anisotropy, linear dichroism, polarization modulation, Disperse Red-1, azo polymer.

Polymers containing azobenzene side-chains can become birefringent when they are irradiated with linearly polarized light. This birefringence results from the reorientation of the azobenzene groups through *trans–cis–trans* photoisomerization. As the optical anisotropy can be erased by using circularly polarized light, these photopolymers can be used for reversible storage of optical information and have thus aroused increasing interest in the past few years.

Infrared linear dichroism (IRLD) is a technique well suited to characterize the molecular orientation in polymers [1, 2]. In order to improve the sensitivity of IRLD and to allow the *in situ* investigation of the dynamics of orientation and relaxation processes, infrared spectroscopy has been coupled with the polarization modulation technique.

In this paper, we present the time-dependence of the orientation and relaxation processes on an amorphous azo polymer containing Disperse Red-1 side-chains (PDR1A). Bands due to several chemical groups have been analysed in order to follow simultaneously the chromophore and main-chain orientation. The effect of the film thickness on the orientation function has also been investigated.

Experimental

Thin films of poly {4′[[2-(acryloyloxy)ethyl]-ethylamino]-4-nitroazobenzene} (pDR1A) were obtained by spin-coating chloroform solutions of the polymer onto barium fluoride substrates. Films were annealed above the T_g of the polymer (1 hour at 95 °C).

IRLD spectra by polarization modulation were recorded at room temperature with a Bomem Michelson MB-100 spectrophotometer, using the optical set-up presented in Fig. 1 and the two-channel electronic processing previously described [3]. The optical anisotropy was induced in the polymer films by using a polarized frequency-doubled Nd–YAG laser at 532 nm with an irradiance of 10 mW/cm².

Results and Discussion

By use of a proper calibration procedure [3], the differential signal obtained by polarization modulation can be related quantitatively to the dichroic difference spectrum $\Delta A = A_{//} - A_{\perp}$ (where $A_{//}$ and $A_{\perp}$ are the absorbances for polarization parallel and perpendicular to the laser polarization, respectively).

Experimental dichroic difference spectra are presented in Fig. 2 for the orientation (laser on) and relaxation (laser off) kinetics. As seen in this figure, spectra with a very high signal-to-noise ratio are obtained, even for small values of ΔA, allowing the precise measurement of the time-dependence behaviour of each band.

Considering a uniaxial orientation distribution, the orientation function F_{θ} is related to the dichroic differ-

---

* To whom correspondence should be addressed

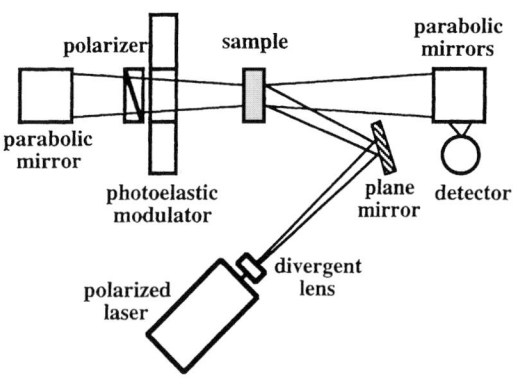

Fig. 1. Optical set-up (top view the of MB-100 interferometer)

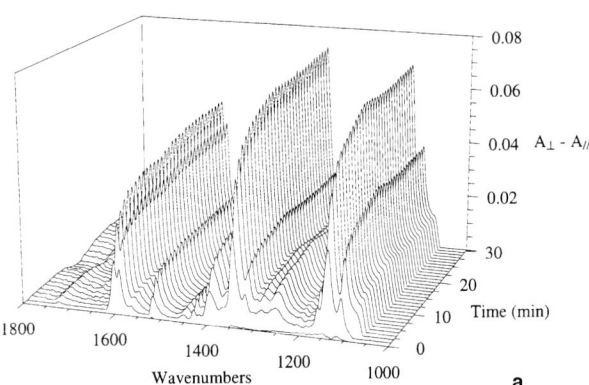

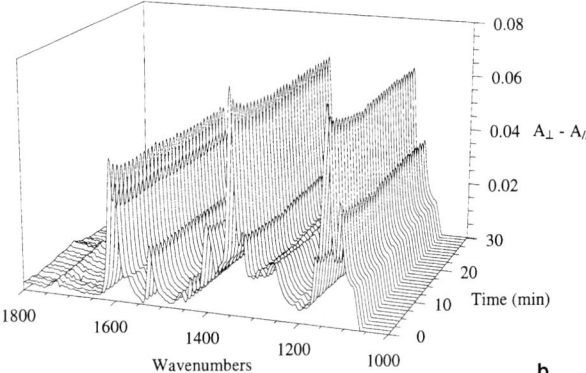

Fig. 2. Time-dependence of the dichroic difference spectra for orientation (**a**) and relaxation (**b**) processes of a pDR1A film. The experimental conditions were 4 cm^{-1} resolution and 50 scans per spectrum at a rate of 53 scans per minute

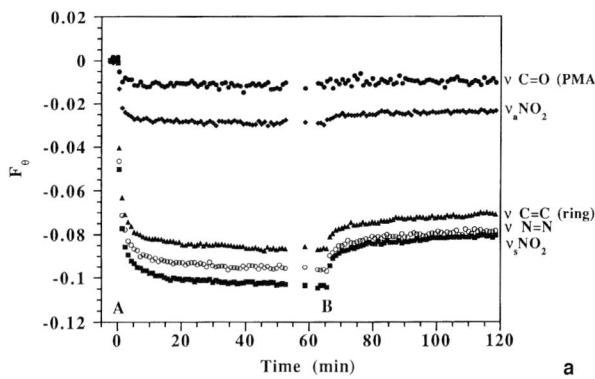

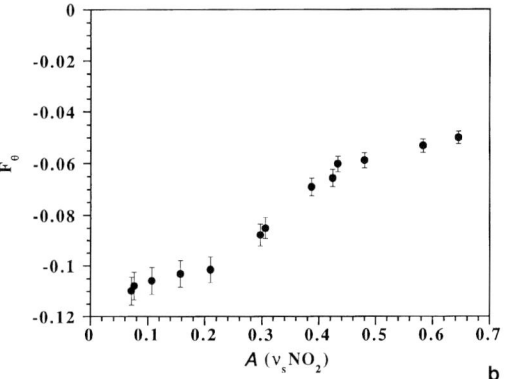

Fig. 3. Time-dependence of the orientation function for several chemical groups of pDR1A (**a**) and dependence of the orientation function of the v_sNO_2 vibration with the absorbance of this band (**b**)

ence by $F_\theta = \Delta A / 3A_0$ where A_0 is the absorbance of the sample before the irradiation.

The time-dependence of the orientation function F_θ is presented in Fig. 3a for several chemical groups of pDR1A. At point A in time, the writing (linearly polarized) laser beam is turned on, activating the *trans–cis–trans* isomerization process. Vibrations with their transition moment essentially parallel to the long axis of the azobenzene groups [v_sNO_2, v C=C (ring), v N=N] have negative values of their orientation function, indicating that the irradiation of the polymer produces a preferred orientation of these groups perpendicular to the writing polarization direction [1].

The orientation function of the v_aNO_2 vibration is also negative, even though this vibration has its transition moment perpendicular to that of the v_sNO_2 vibration. This result suggests that, for the written polymer, the NO_2 groups and most likely the azobenzene groups do not display a cylindrical symmetry of orientation. Finally, the time-dependence of the v C=O vibration reveals that the orientation of the azobenzene groups induces a slight co-operative orientation of the main chain of the polymer.

When the laser beam is turned off (point B in Fig. 3a), the orientation function of each vibration decreases by about 20%. This decrease can be associated with the thermal reorientation of some azobenzene groups.

Since it has already been shown that the light-induced birefringence in azo polymers depends on the film thickness [4], we have studied the dependence of the orientation function F_θ of the $v_s NO_2$ vibration (measured after 1 hour irradiation time) with the absorbance of this band (proportional to the film thickness). Figure 3b shows that for absorbances lower than 0.2, the value of F_θ is nearly constant, indicating that the writing light penetration depth is much larger than the film thickness. On the other hand, for absorbances higher than 0.4, complete orientation saturation is not achieved since the writing beam is completely absorbed by the sample. This result highlights the importance of

considering the film thickness when the level of orientation of different samples is compared.

Acknowledgement. We express our thanks to A. Natansohn and P. Rochon for providing pDR1A samples and for helpful discussions. T.B. is also grateful to NATO for a fellowship.

References

[1] A. Natansohn, P. Rochon, M Pézolet, P. Audet, D. Brown, S. To, *Macromol.* **1994**, *27*, 2580.

[2] U. Wiesner, N. Reynolds, C. Boeffel, H.W. Spiess, *Liquid Cryst* **1992**, *11*, 251.

[3] T. Buffeteau, B. Desbat, M. Pézolet, J.M. Turlet, *J. Chim. Phys., Phys. Chim. Biol.* **1993**, *90*, 1467.

[4] P. Rochon, D. Bissonnette, A. Natansohn, S. Xie, *Appl. Optics* **1993**, *32*, 7277.

Mikrochim. Acta [Suppl.] 14, 399–402 (1997)
© Springer-Verlag 1997

Applications of FT-Raman and FT-IR Microspectroscopy in the Near Infrared to Polymer Analysis

B. Chabert[1], **J. L. Gardette**[2], **G. Lachenal**[1,*], and **I. Stevenson**[1]

[1] Université Claude Bernard, Laboratoire des Matériaux Plastiques et des Biomatériaux, UMR 5627, 43, Bd du 11 Novembre 1918, 69622 Villeurbanne Cedex, France
[2] Université Blaise Pascal, Laboratoire de Photochimie, 63177 Aubière Cedex, France

Abstract. FT-IR microspectroscopy in the near infrared of polymers yields selective structural and conformational information as well as data on crystallinity changes and inter/intra molecular interactions and can be obtained on microscopic samples. The recent developments in FT-Raman microspectroscopy allow small areas to be analysed without interfering with their surroundings. FT-Raman microspectroscopy, which is non-destructive, is thus well adapted to both the analysis of very small samples and the analysis of heterogeneous media. A few new examples of applications of these two techniques in the polymer field will be presented and their advantages or drawbacks will be discussed.

Key words: microspectroscopy, FT-Raman, FT-NIR, polymer.

Near infrared spectroscopy is widely used for quantitative analysis of compounds containing C–H, N–H, O–H groups so can easily be applied to the analysis of polymers and composites [1]. The analysis of epoxy resin, widely used for adhesives, protective coatings and matrices for many advanced materials, has been extensively studied by FT-IR and high-resolution solid-state NMR [2]. DeBakker et al. [3] have monitored by FT-Raman and near IR spectroscopy the spectral changes in terms of mechanism of cure for a commercial epoxy resin.

Mid IR microspectroscopy is now a widely applied technique for characterization of heterogeneous materials, defects and aging [4] in a huge range of objects, and the development of new microscope objectives allows scanning not only in transmission and reflection modes but also in ATR and grazing angle. Unfortunately this technique requires samples of suitable thickness i.e. between 5 and 15 μm, which are not easy to prepare [5] and require skilled staff. Near infrared microspectroscopy, in contrast, can provide spectra on relatively thick samples (50–500 μm) either in transmission or reflectance mode [6] between 4000 and 6000 cm^{-1} and can bring much useful information, even though the identification of impurities or embedded particles seems to be excluded for the moment.

Micro FT-IR

The NIR microtransmission spectra of different areas of an amine hardener crystallized from a melt onto CaF_2 have shown some interesting features, probably due to different molecular orientation or crystallinity [6]. Samples of a heterogeneous mixture of this amine hardener and epoxy resin (200–500 μm thickness) have given NIR spectra characteristic of areas rich in non-reacted epoxy groups and others rich in amine groups [6].

Figure 1 shows different areas of crystallized DGEBA on the same sample and melted DGEBA. It

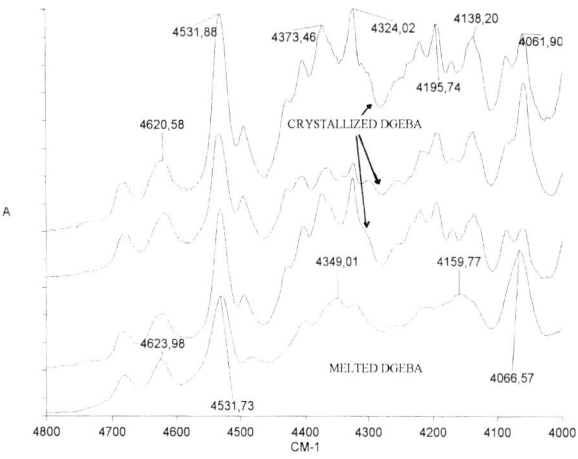

Fig. 1. Different areas of crystallized DGEBA on the same sample and melted DGEBA (400 × 400 μm aperture, 200 scans, 4 cm^{-1} resolution)

---

* To whom correspondence should be addressed

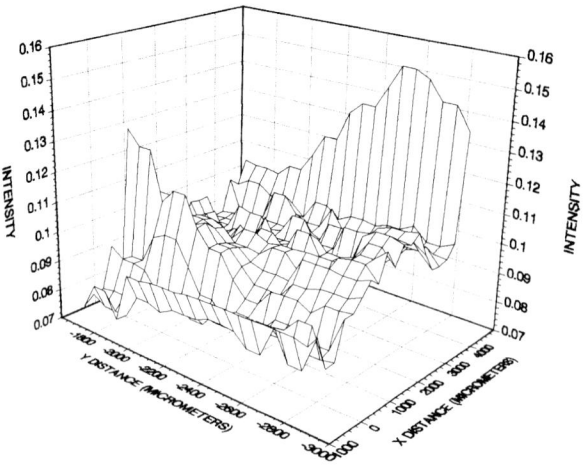

Fig. 2. Mapping of a 500 μm thick sample of PET with a Nicolet micro FT-IR, showing different crystalline areas

Micro FT-Raman

The recent developments of FT-Raman microspectroscopy permit small areas to be analysed without interfering with their surroundings. Improvements of the spatial resolution can allow analysis down to 10 μm with no problem (the spatial resolution is theoretically 2 μm with a 1.064 μm excitation). FT-Raman normally requires little or no sample preparation. The sample is not destroyed during the measurements and it can be used for further experiments.

The FT-Raman spectra of a thermoplastic polymer included in an epoxy matrix have been recorded. Several spectra were recorded by moving the analysed area along the x-axis by successive 25 μm steps. Figure 3 compares the spectra of the matrix with the spectrum of the thermoplastic. Both spectra appear rather different: the spectrum of the epoxy matrix is characterized by an intense band at 2860 cm^{-1} and a narrow band at 1613 cm^{-1}, whereas the spectrum of the thermoplastic may be characterized by a band at 3070 cm^{-1} and a band at 1780 cm^{-1}. Both these sets of bands permit us to discriminate easily between the two polymers. Figure 4 shows the spectrum recorded by focusing on the interface. The monitored spectrum appears as a convolution of the two spectra presented in the previous figures. By moving the measured zone from the interface to the bulk of the thermoplastic or in the opposite direction

shows that the degree of crystallization is heterogeneous over the whole sample. It is possible to follow nucleation zones. If the nucleant is a fibre or a powder, it is possible to follow the crystallization at the interface over a sufficiently wide area (5 μm range if a suitable instrument is available). Figure 2 shows the mapping of a PET sample 500 μm thick, showing different crystallinity areas by using the peak area ratio of the 4681 and 4360 cm^{-1} bands.

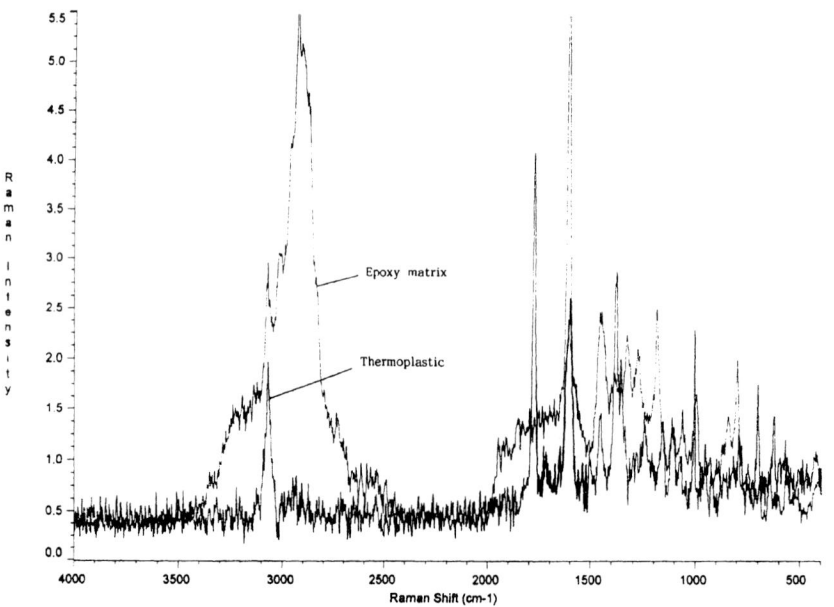

Fig. 3. Micro FT-Raman spectra of the epoxy resin and of the thermoplastic inclusion

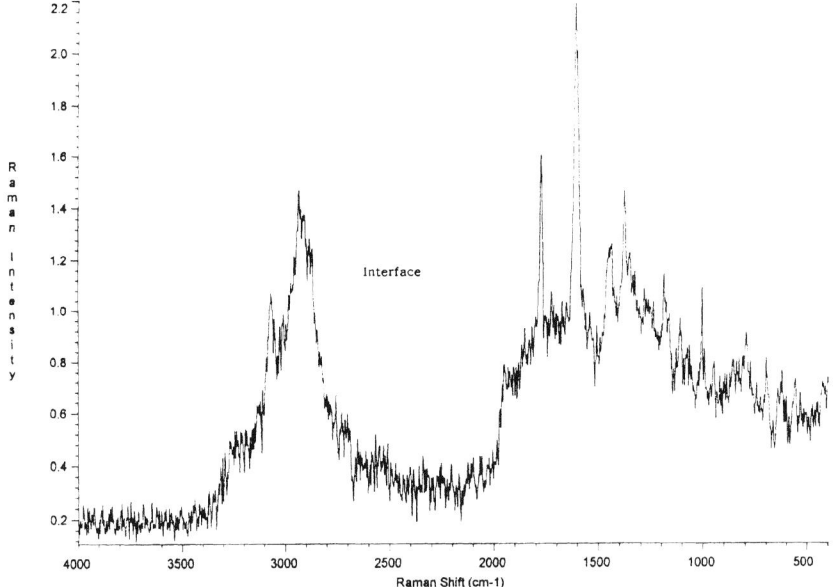

Fig. 4. Micro FT-Raman of the interface (epoxy resin/thermoplastic inclusion)

to the bulk of the epoxy resin, it is possible to record the distribution profiles.

Experimental

Micro FT-IR spectra were recorded either with a Perkin–Elmer FT-IR 1760X (ceramic source–KBr beam-splitter) coupled to a Spectra-Tech microscope (MCT detector) (Fig. 1) or with a Nicolet FT-IR 550 MAGNA (tungsten/halogen source–CaF$_2$ beam-splitter) coupled with a Nic-Plan microscope (MCT detector) (Fig. 2).

Micro FT-Raman spectra were recorded with a Nicolet FT-Raman 910 spectrometer (CaF$_2$ beam-splitter – germanium detector – excitation source Nd:YAG laser at: 1064 nm) coupled to a Nicolet Nic-Plan microscope. The size of the area analysed with the microscope was less than 300 μm^2. All the Raman spectra illustrated were measured under the same standard conditions of 4 cm^{-1} resolution, 200 scans.

Table 1 surveys the advantages and limitations of the techniques available.

Table 1. Advantages and drawbacks of microspectroscopic methods

Mid IR (25–2.5 μm)	Near IR (2.5–1 μm)	FT-Raman
sample preparation necessary and time consuming	no sample preparation	no sample preparation
sample sometimes embedded in resin	sample recoverable	non-destructive analysis
strong absorption due to water and CO$_2$	atmospheric water absorbs less	no absorption of atmospheric water
sample thickness smaller than 20 μm	better results for samples at least 50 μm thick	better results for samples at least 100 μm thick
spectra library available and identification easier	no spectra library, and assignment difficult	spectra library not available
better resolution	less good resolution weaker sensitivity due to weaker absorption coefficients	very often complementary to IR analysis
transparent materials in the mid infrared	glass slides transparent in the NIR and cheap	no need for sample holder for solid materials
expensive or hygroscopic		

Conclusion

It seems that micro-NIR spectroscopy can provide useful information on the modification of the curing of resin in the vicinity of reinforcement fibres or on the difference of crystallinity during a thermal or mechanical treatment. In some cases micro-NIR, with its sampling facilities can be favourably compared with mid IR microspectroscopy in regard to the information yielded. The micro FT-Raman technique offers possibilities of microspectroscopic imaging with minimal preparation of the sample. FT-Raman microspectroscopy is well adapted to the analysis of very small samples and the analysis of heterogeneous media.

These two techniques (near infrared and FT-Raman microscopies) allow good quality spectra to be obtained with a minimum of sampling time.

References

[1] B. Chabert, G. Lachenal, C. Vinh Tung, *Macromol. Symp.* **1995**, *94*, 145.
[2] E. Mertzel, J. L. Koenig, *Adv. Polym. Sci.* **1986**, *75*, 73.
[3] C. J. Bakker, G. A. George, N. A. St. John, P. M. Fredericks, *Spectrochim. Acta* **1993**, *49A*, 739.
[4] J. E. Katon, *Vib. Spectrosc.* **1994**, *7*, 201.
[5] G. Nichols, *Proc. ESIS 93, Lyon* (G. Lachenal, H. W. Siesler, eds.), 1994, p. 72.
[6] G. Lachenal, I. Stevenson, *Nir News,* **1995**, *6*, (*2*), 10.

Mikrochim. Acta [Suppl.] 14, 403–405 (1997)

Orientation and Conformation in PET: New Information from Specular Reflection FT-IR

Kenneth C. Cole*, Jacques Guèvremont, Abdellah Ajji, and Michel M. Dumoulin

Industrial Materials Research Institute, National Research Council Canada, 75 boul. de Mortagne, Boucherville, Québec J4B 6Y4, Canada

Abstract. Specular reflection FT-IR spectroscopy was used to study thick films of amorphous poly(ethylene terephthalate) that had been subjected to either uniaxial drawing or thermal annealing. The high quality of the spectra made it possible to perform, for the first time, a detailed analysis of the most intense absorption bands. Curve-fitting techniques were used to decompose these bands into individual component peaks, and the effect of drawing upon the intensity and dichroism of these peaks was analysed. The results provide new insight into the conformational and other structural changes that occur upon drawing and thermal annealing of PET.

Key words: poly(ethylene terephthalate), PET, conformation, orientation, FT-IR.

It has been demonstrated recently that infrared specular reflection from the front surface of thick oriented samples can be used to obtain detailed information about the molecular structure, and we have reported some initial results obtained for poly(ethylene terephthalate) (PET) by this technique [1, 2]. An important advantage of specular reflection is the fact that it makes it possible to study the more intense absorption bands in the spectrum. Although PET has been widely studied, these bands have received little attention, because in order to study them by transmission it is necessary to use very thin films, which cannot be easily drawn. Hence we undertook a detailed analysis of a series of samples in hopes of obtaining new information on the structure of PET.

Conformation changes in PET are of great importance in determining its structure [1, 2, and references therein]. Especially important is rotation about the glycol C–C bond, which gives rise to *gauche* and *trans* conformers (G and T). Similar rotation can occur about the glycol C–O bond (g and t conformers). Finally, the carbonyl groups can rotate about their bond with the benzene ring, giving rise to *cis* (C_B) and *trans* (T_B) conformers. The notation used is that of Štokr et al. [3], who attempted to determine the conformational structure of PET. In the crystalline state the molecule is entirely *trans* (100% T, t, and T_B). However, in the amorphous state it is quite different, with the structure estimated to involve only 10% T, 75% t, and 50% T_B.

Experimental

Rather thick (about 0.5 mm) films of amorphous PET were uniaxially drawn at 80 °C to various draw ratios between 1 and 5. These samples are designated IA, for "initially amorphous". In addition, two samples were prepared by annealing the amorphous sheet at 200 °C for 15 min, and one of these was drawn to a draw ratio of 2.6. These are designated TC, for "thermally crystallized". Specular reflection spectra were measured with polarization parallel and perpendicular to the draw direction, then converted into absorption index spectra by means of the Kramers–Kronig transformation. "Structural factor" spectra were calculated by combining the spectra for the two polarizations. Further details and examples of spectra are given in previous publications [1, 2]. The high quality of the spectra made it possible to perform, for the first time, a detailed analysis of the most intense absorption bands. These include the carbonyl band at 1725 cm^{-1}, the complex bands at around 1265 cm^{-1} and 1100 cm^{-1}, and the out-of-plane benzene ring C-H band at 729 cm^{-1}. With the aid of the commercial software Spectra Calc™ from Galactic Industries Corp., curve fitting was used to decompose these bands into individual Gaussian component peaks, and the effect of drawing upon the intensity and dichroism of these peaks was characterized. To follow the peak intensities, the areas were normalized with respect to the 1410 cm^{-1} peak, which has been shown to be insensitive to orientation and crystallinity. To follow the dichroic behaviour, the quantity $(D-1)/(D+2)$ was calculated, where D is the usual dichroic ratio [1]. This quantity is proportional to the orientation function f, but f can only be calculated if the angle α between the chain axis and the transition moment is known.

---

* To whom correspondence should be addressed

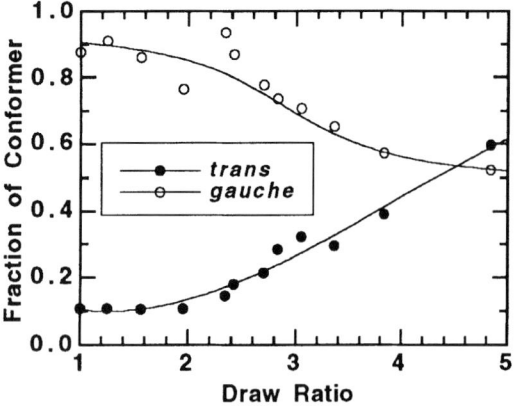

Fig. 1. Fraction of *trans* and *gauche* conformers as a function of draw ratio

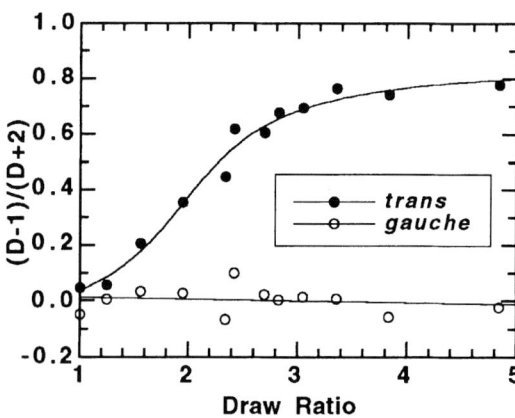

Fig. 2. Dichroism of *trans* and *gauche* conformer peaks as a function of draw ratio

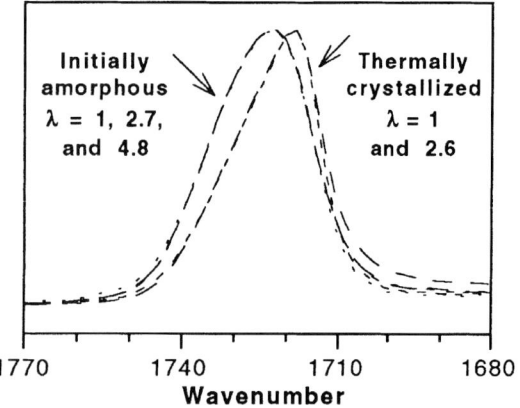

Fig. 3. The carbonyl band shape for different samples

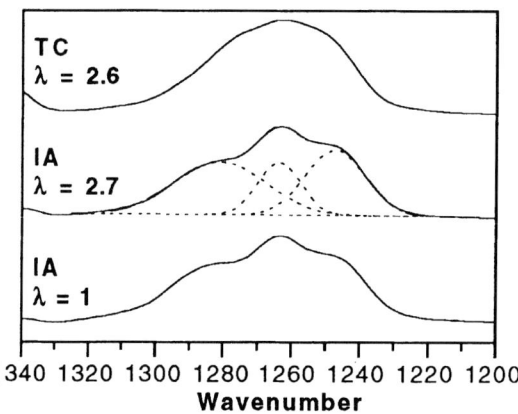

Fig. 4. The shape of the composite peak near 1263 cm^{-1}

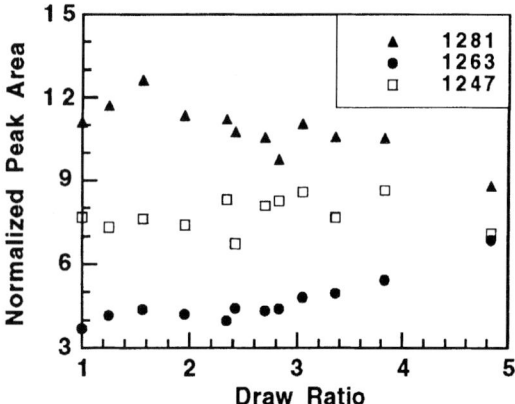

Fig. 5. Variation of component peak areas for the composite peak at 1263 cm^{-1}

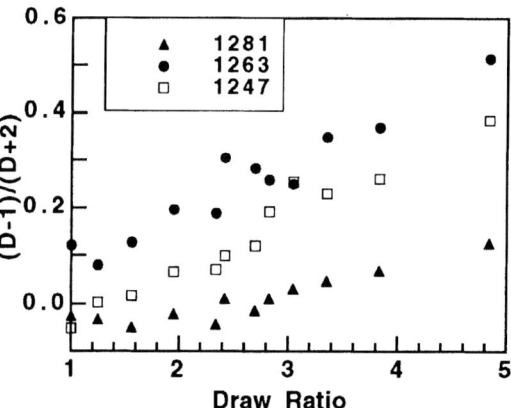

Fig. 6. Dichroism of the components of the 1263 cm^{-1} peak

Results and Discussion

The CH_2 wagging peaks at $1370\,cm^{-1}$ (*gauche*) and $1340\,cm^{-1}$ (*trans*) were used to calculate the fraction of G and T conformers, as well as their orientation, for the drawn samples (Figs. 1 and 2). The *trans* content remains steady at about 10% up to a draw ratio of 2, at which point conversion of *gauche-trans* begins and increases continuously with draw ratio. The *trans* conformers begin orienting right from the start of drawing and are highly oriented by draw ratio 3, whereas the *gauche* conformers do not show detectable orientation. These results concur with earlier studies by other workers. However, there is little information available on the conformational behaviour of the terephthalate ring moiety. Figure 3 shows the carbonyl band for some of the samples studied here, and a very interesting behaviour is observed. For the IA samples, the band shape remains exactly the same up to draw ratio of 5; the maximum occurs at $1723\,cm^{-1}$. Thermal annealing, however, produces a definite change in shape and the maximum shifts to $1719\,cm^{-1}$, but drawing of the annealed sample produces no further change. We interpret these results to imply that drawing at 80 °C easily changes the glycol conformation but is incapable of producing a change in the benzene ring conformation. This can only be achieved by heating at higher tem-

perature, and possibly involves a *cis* to *trans* conversion. The behaviour of the benzene ring out-of-plane C–H band at $729\,cm^{-1}$ is very similar to that of the carbonyl band. The complex band near $1263\,cm^{-1}$ is also interesting (Fig. 4). For the IA samples, it can be decomposed into three component peaks. The $1263\,cm^{-1}$ component increases for draw ratios >3 (Fig. 5) and shows strong orientation (Fig. 6), suggesting that it is associated with the *trans* glycol conformer. The $1281\,cm^{-1}$ peak decreases on drawing and shows low orientation, so it probably arises from the *gauche* conformer. Thermal annealing results in a more complex band shape (Fig. 4). Similar behaviour is observed for the other complex band near $1100\,cm^{-1}$. All these results suggest that the behaviour of PET can be explained in terms of two distinct conformational rearrangements with quite different kinetics. Further details and discussion of the implications of this will be given in a future publication.

References

[1] K. C. Cole, J. Guèvremont, A. Ajji, M. M. Dumoulin, *Appl. Spectrosc.* **1994**, *48*, 1513.
[2] J. Guèvremont, A. Ajji, K. C. Cole, M. M. Dumoulin, *Polymer* **1995**, *36*, 3385.
[3] J. Štokr, B. Schneider, D. Doskočilová, J. Lövy, P. Sedláček, *Polymer* **1982**, *23*, 714.

Mikrochim. Acta [Suppl.] 14, 407–409 (1997)

Fourier-transform Infrared Dichroism Investigation of Molecular Orientation in Model Networks of Poly(dimethylsiloxane)

Liliane Bokobza[1,*], **Bernard Desbat**[2], and **Thierry Buffeteau**[2]

[1] Laboratoire de Physico-Chimie Structurale et Macromoléculaire, ESPCI, 10 rue Vauquelin, F-75231 Paris Cedex 05, France
[2] Laboratoire de Spectroscopie Moléculaire et Cristalline, Université de Bordeaux I, F-33405 Talence Cedex, France

Abstract. Molecular orientation in uniaxially stretched model networks of poly(dimethylsiloxane)(PDMS) is investigated by Fourier-transform infrared dichroism. Precise determination of the dichroic effects, in the mid- as well as in the near-infrared range, are obtained by polarization modulation of the incident electromagnetic field. Introduction of a reinforcing filler such as silica in PDMS leads to an increase in orientation, attributed to a higher microscopic strain in the elastomeric part.

Key words: poly(dimethylsiloxane), orientation, infrared dichroism, near infrared spectroscopy, filled elastomers.

Mechanical properties of polymers are strongly influenced by molecular orientation occurring during stretching or during various forming processes. Measurement of this orientation is of particular importance for a better molecular understanding of the mechanisms involved in polymer deformation and also for a better understanding of rubber elasticity. The directions of chain segments in an isotropic network are randomly distributed. When the network is macroscopically distorted, the orientations of the segments become anisotropic. The state of anisotropy under strain may be accurately characterized by infrared spectroscopy which can probe the orientational behaviour of network chains at a molecular level, in contrast to the macroscopic information provided by most other characterization techniques.

Fourier-transform infrared dichroism is used, in the present work, to investigate molecular orientation in model networks of poly(dimethylsiloxane) (PDMS). These nearly ideal end-linked networks present only a small number of defects such as dangling chains, loops or double connections. Their controlled polydispersity as well as their junctions of almost constant functionality allow a quantitative comparison with theoretical predictions [1].

Experimental

Samples

Model networks of PDMS were obtained by end-linking stoichiometric mixtures of α, ω-di(hydrogeno)-poly(dimethylsiloxane) having a molecular weight of 12000 g/mole with 1,3,5,7-tetravinyl-1,3,5,7-tetramethylcyclotetrasiloxane, used as a tetrafunctional cross-linking agent. The two PDMS samples (unfilled and filled with 25 parts by weight of silica) used to investigate the reinforcement effect were kindly supplied by the Rhône-Poulenc Industries (Silicon Division). These samples have the same molecular weight between cross-links, M_c, of 17500.

Infrared Dichroism Measurements

Segmental orientation in a network submitted to uniaxial elongation may be conveniently described by the second Legendre polynomial [2]:

$$\langle P_2(\cos\theta) \rangle = (3\langle \cos^2\theta \rangle - 1)/2$$

where θ is the angle between the direction of extension and the local chain axis of the polymer.

Parameters commonly used to characterize the degree of optical anisotropy in stretched polymers are the dichroic difference $\Delta A = A_\parallel - A_\perp$ or the dichroic ratio $R = A_\parallel/A_\perp$ where $A_\parallel$ and $A_\perp$ are the absorbances of the investigated band, measured with radiation polarized parallel and perpendicular to the stretching direction, respectively.

The orientation function $\langle P_2(\cos\theta) \rangle$ is related to the dichroic ratio R by the expression:

$$\langle P_2(\cos\theta) \rangle = \frac{2}{(3\cos^2\beta - 1)} \cdot \frac{(R-1)}{(R+2)} = F(\beta)\frac{(R-1)}{(R+2)}$$

where $F(\beta)$ depends only on the angle β between the transition moment vector of the vibrational mode considered and the local chain axis of the polymer or any directional vector characteristic of a given chain segment.

---

* To whom correspondence should be addressed

$\langle P_2(\cos\theta)\rangle$ is related to the structural absorbance $A = (A_\parallel + 2A_\perp)/3$ by:

$$\langle P_2(\cos\theta)\rangle = F(\beta)\frac{A_\parallel - A_\perp}{A_\parallel + 2A_\perp} = F(\beta)\frac{\Delta A}{3A}$$

On account of the very high chain flexibility, PDMS exhibits a low level of orientation. In this case, as the magnitude of ΔA is small and the dichroic ratio R close to unity, the classical method lacks sensitivity and a polarization modulation approach can then be used to measure small dichroic effects. The polarization modulation method consists basically of a fast modulation of the polarization state of the incident field between directions parallel and perpendicular to the stretching direction of the investigated polymer. By improving considerably the signal-to-noise ratio, this method is particularly recommended for films under low deformation or for polymers exhibiting anisotropies nearly undetectable by the standard methods.

Results and Discussion

A mid-infrared spectrum of a PDMS film, about 100 μm thick, exhibits very strong absorption bands associated with fundamental modes. Hence we have investigated the dichroic behaviour of the band located at $2500\,\mathrm{cm}^{-1}$, ascribed to the overtone of the CH_3 symmetrical bending vibration located at $1260\,\mathrm{cm}^{-1}$. The transition moment associated with both vibrational modes lies along the CH_3–Si bond, which is a symmetry axis of the methyl group. We have also characterized bands between 4000 and $4500\,\mathrm{cm}^{-1}$ located in the near infrared (NIR) region. Let us point out that bands appearing in the NIR range arise from overtones and combinations of fundamental absorbances. As these bands are much weaker than the corresponding fundamental absorptions, the advantage of NIR analysis is the ability to examine thick specimens.

Figure 1 represents the strain dependence of the dichroic difference, ΔA, for the band at $2500\,\mathrm{cm}^{-1}$ of the mid-IR and for the spectral pattern between 4000 and $4500\,\mathrm{cm}^{-1}$ in the NIR region together with the proposed assignment for each band. The absolute value of ΔA, and thus the anisotropy of the sample, increases with the draw ratio α ($\alpha = l/l_0$, l_0 and l being the undeformed and deformed lengths, respectively). On account of its high sensitivity, the polarization modulation technique is able to detect precisely the onset of orientation and a dichroic effect can be measured for a draw ratio as low as 1.05, whereas the standard method leads to significant measurements only for $\alpha = 2.5$.

Values of the orientation function, $\langle P_2(\cos\theta)\rangle$ are plotted against what we call the strain function $(\alpha^2 - \alpha^{-1})$ in Fig. 2. Two different measurements were made. One is related to the band located at $2500\,\mathrm{cm}^{-1}$,

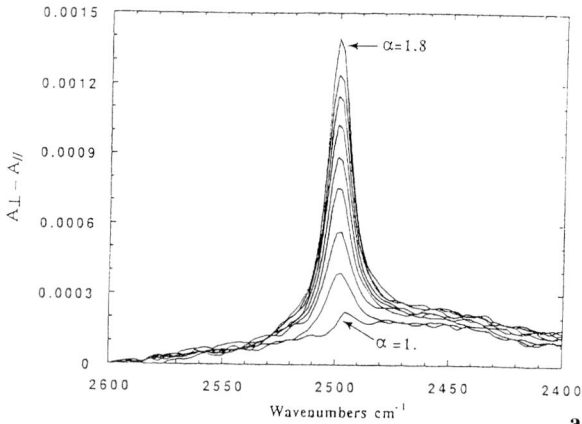

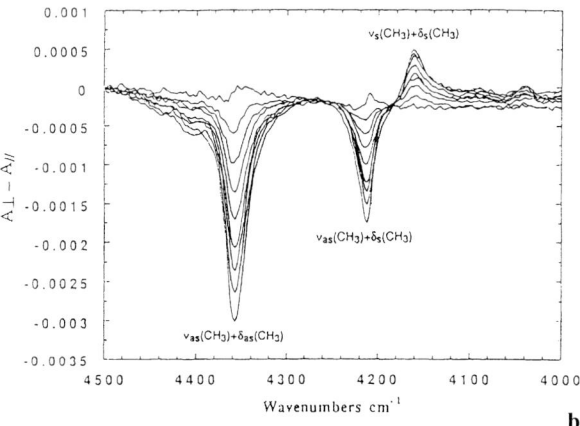

Fig. 1. Strain dependence of the dichroic difference: **a** for the band at $2500\,\mathrm{cm}^{-1}$, **b** for the bands located between 4000 and $4500\,\mathrm{cm}^{-1}$

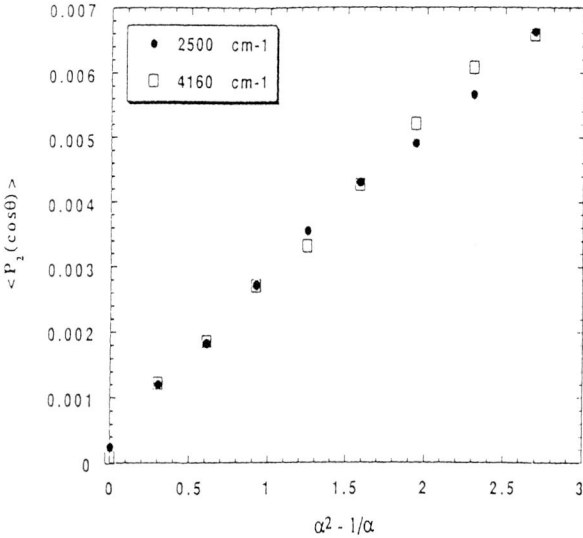

Fig. 2. Strain dependence of the orientation function derived for experiments carried out in the mid-IR ($2500\,\mathrm{cm}^{-1}$) and in the NIR ($4160\,\mathrm{cm}^{-1}$)

the other concerns the band at $4160 \, cm^{-1}$ ascribed to a combination of the symmetrical stretching mode of the methyl group at about $2905 \, cm^{-1}$ with the symmetrical bending mode of the same group at about $1260 \, cm^{-1}$. The polarization modulation technique leads to a perfect linear relationship between the experimentaly observed $\langle P_2(\cos\theta)\rangle$ values and the strain function $(\alpha^2 - \alpha^{-1})$. Such a behaviour is predicted by the various theoretical models of rubber elasticity used to estimate molecular orientation under deformation. In the simplest from, $\langle P_2(\cos\theta)\rangle$ may be expressed as the product of two contributions:

$$\langle P_2(\cos\theta)\rangle = D_0\,(\alpha^2 - \alpha^{-1}),$$

the front factor, D_0, which represents the "orientability" of the chain segments and the strain function $(\alpha^2 - \alpha^{-1})$, which reflects the effect of the macroscopic deformation on orientation.

The data plotted show that similar information can be obtained from mid- and near-infrared spectroscopy. This demonstrates that, in combination with the polarization modulation technique, NIR provides a nice way of measuring small anisotropy in thick polymer films with high precision.

Incorporation of fillers into elastomers is of significant commercial importance owing to the enhancement of the physical properties of the final materials. Analysis of the properties of filled elastomers under strain is an area of particular interest for understanding the molecular origin of the reinforcement effect.

Stress-strain measurements carried out on an unfilled PDMS and on the silica-filled network reveal clearly that the incorporation of the reinforcing particles significantly increases the modulus and the ultimate properties like the stress at rupture and the maximum extensibility. Orientational analysis, including birefringence and infrared studies, also shows that, at a given elongation, the orientation is increased by a factor of about 1.5 by the addition of the filler particles. Such a behaviour can be explained by the fact that the microscopic strain in a filled system is higher than the applied macroscopic strain since the particles are rigid and cannot be stretched. The enhancement, X, of the slope of the curves representing Δn or $\langle P_2(\cos\theta)\rangle$ $vs.$ $(\alpha^2 - \alpha^{-1})$ can be estimated by the Guth and Gold expression [3]:

$$X = (1 + 2.5\,\varphi + 14.1\,\varphi^2)$$

where φ is the volume fraction of the spherical filler particles. In our filler system, this expression predicts a value of 1.4 for X which is quite similar to that observed experimentally.

References

[1] J. E. Herz, A. Belkebir-Mrani, P. Rempp, *Eur. Polymer J.* **1973**, 9, 1165.

[2] B. Jasse, J. L. Koenig, *J. Macromol. Sci. Rev. Macromol. Chem.* **1979**, *C17*, 61.

[3] E. Guth, O. Gold, *Phys. Rev.* **1938**, 53, 322.

Mikrochim. Acta [Suppl.] 14, 411–412 (1997)
© Springer-Verlag 1997

Rheo-Optical FT-IR Spectroscopy of Polyurethane–polyolefine Blends During Cyclic Elongation and Recovery

W. B. Fischer[1,*]**, P. Pötschke**[2]**, K.-J. Eichhorn**[2]**, and H. W. Siesler**[3]

[1] Institut für Analytische Chemie, TU Dresden Zellescher Weg 19, D-01062 Dresden, Federal Republic of Germany
[2] Institut für Polymerforschung Dresden, e.V., Hohe Straße 6, D-01069-Dresden, Federal Republic of Germany
[3] Institut für Physikalische Chemie, Universität Essen, Schützenbahn 70, D-45117-Essen, Federal Republic of Germany

Abstract. FT-IR spectra have been recorded during cyclic uniaxial elongation and recovery of films of thermoplastic polyester- and polyether–polyurethane elastomers (TPU) and their blends with polyolefines (polyethylene and polypropylene). From these spectra the orientation functions of the individual segments have been calculated. The TPU hard segments are built up to 4,4'-diphenyl-methane di-isocyanate (MDI) units with 1,4-butanediol as chain extender. The soft segments used in this study consists of polyethers or polyesters. We find reduced stress and strain levels for the blends as well as weaker orientation of the segments. For polyether–polyurethane blends with polypropylene we notice extended strain.

Key words: rheo-optical FT-IR spectroscopy, polyurethane–polyolefine blends, cyclic elongation, recovery.

Rheo-optical FT-IR spectroscopy has been used to analyse the orientational behaviour of films of thermoplastic polyester- and polyether-polyurethane elastomers (TPU) and their blends with polyolefines during uniaxial elongation and recovery. The TPU hard segments are based on 4,4'-diphenylmethane di-isocyanate (MDI) units with 1,4-butanediol as chain extender. The soft segments used in this study consist of polyethers or polyesters. Blends of both TPUs with 20% w/w polyethylene or polypropylene were produced by melt mixing. The blend components are highly incompatible and the resulting blends are therefore heterogeneous.

FT-IR spectra were recorded during cyclic elongation and recovery of the polymer films. From these spectra the orientation functions of the individual segments have been calculated. These are then correlated with the mechanical behaviour. In our presentation we focus on three issues: (*i*) how does the incorporation of

polyolefine change the typical orientation behaviour of TPUs?, (*ii*) which behaviour shows the orientation functions of the polyolefine segments? and (*iii*) how do the orientations correlate?

Experimental

Commercially available polymers were used, such as Elastollan® C64D Polyester-TPU (TPU1), Elastollan® 1195A Polyether-TPU (TPU2), Polyethylene (PE) Lupolen® 4261A and Polypropylene (PP) Novolen® 2900NCX (containing a rubber phase of ethylene–propylene copolymers with ethylene-rich semi-crystalline regions [1]). Film samples (thickness ca. 8–12 μm) of pure TPU were cast from 2% (w/v) dimethylformamide solutions on a surface-roughened glass plate and dried at 323 K in vacuum overnight. Film samples of the TPU-blends were prepared by microtom cutting (thickness ca. 15–20 μm). FT-IR spectra were recorded with a Bruker IFS 88 at 303 K by the rapid-scan technique. Ten interferograms were co-added to calculate the absorbance spectra (resolution 4 cm^{-1}). The orientation function was calculated for individual absorption bands [2]. Details of the stretching device have been given elsewhere [3]. The speed of elongation was 100% per minute.

Results and Discussion

For calculating the orientation function of the hard and soft segments (hard seg., soft seg.) of the polyurethanes and their blends we used the band intensities (band area) of the $v_{\text{CNarom (urethane)}}$ (hard. seg. 1310 cm^{-1}) and v_{COC} (soft seg.; for TPU1 and blends at 1081 cm^{-1}, for TPU2 and blends at 1230 cm^{-1}). Both hard and soft segments of pure TPU reflect linear orientation under stress (Fig. 1). After addition of the polyolefinic compounds the orientation of both hard and soft segments decreases during cyclic elongation and recovery especially for the TPU1 blends [4]. Calculation of the orientation function of the polyolefinic blend

* To whom correspondence should be addressed

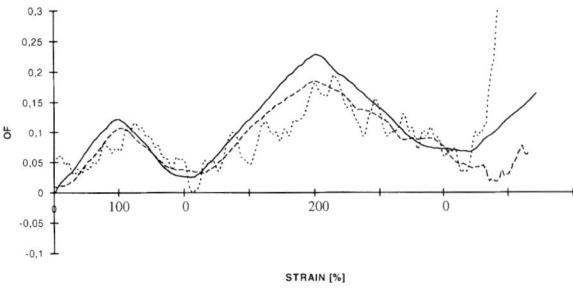

Fig. 1. Orientation function (OF) of the hard segments of TPU2 (—) and TPU2/PP (---). (.....) Orientation function of polypropylene in TPU2-matrix

components in TPU1 is hindered by the weak band intensities of the resulting CH_2 vibration at $1463 \, cm^{-1}$. Because of the non-negligible contributions of rubber in PP we are able to identify orientation in the polyolefinic phase of TPU2/PP (Fig. 1). The orientation of the PP component in the TPU2/PP blend shows similar behaviour to that of the segments in pure TPU2.

Our stress–strain diagrams monitor reduced stress level and lower elongation during the mechanical treatment of the polyolefinic blends (TPU1/PE, TPU1/PP and TPU2/PE) compared to TPU1 (~ 200–250%) and TPU2 (~ 300–350%) [4]. TPU2/PP is an exception, showing reduced stress, but can be extended to higher strains (~ 380–420%) compared to TPU2 (~ 300–350%).

Data not shown here from DSC-experiments support the picture of the 'non-miscibility' of the blend components. The curve of the heat flow for TPU1/PP overlaps with the individual curves of TPU1 and PP. However, for TPU2/PP the heat flow displays values which can only be explained as due to interaction of the two phases.

Our data can be discussed in terms of the disperse distribution of the polyolefinic blend components in the TPU phases. Especially for TPU2/PP, a fine disperse distribution of olefinic particles in the TPU matrix is observed [4]. Since TPU2 has more hydrophobic contribution (poly-tetrahydrofuran units in the soft seg.) than TPU1 (adipinic acid ester in the soft seg.), interactions between the blend components through Van der Waals forces are promoted. This explains the mechanical behaviour as well as the orientation in the polyolefine matrix. As a consequence, the higher contribution of polar groups in TPU1 weakens the compatibility of this polyurethane with the polyolefines.

Conclusion

The interaction of hydrogen-bond donors (NH in urethane linkages) and acceptors (C=O in ester and/or urethane linkages, O in ether groups) within the TPU chains in general influences the mechanical properties of the TPUs and their blends.

Incorporation of "miscible" components *improves* the flexibility of the TPU-blends. The contribution of polar groups and the reduction of the extended network of hydrogen bonds in the TPU-blend segments *decrease* the flexibility.

Acknowledgement. We would like to thank U. Hoffmann, I. Zebger, F. Pfeifer and N. Völkl (Department of Physical Chemistry, University of Essen, Germany) for technical assistance and helpful discussion. For the support by DSC data we thank G. Pompe (IPF, Dresden, Germany). We also gratefully acknowledge the supply of the polymers by BASF AG, Ludwigshafen.

References

[1] H.-G. Braun, in: *Aufbereiten von Polymerblends*, Tagung in Baden-Baden, 29–30 September 1989, VDI, Düsseldorf, 1989.
[2] M. M. Coleman, D. J. Skrovanek, J. Hu, P. C. Painter, *Macromolecules* **1988**, *21*, 59.
[3] H. W. Siesler, *Ber. Bunsenges. Phys. Chem.* **1988**, *92*, 641.
[4] W. B. Fischer, P. Pötschke, K.-J. Eichhorn, H.W. Siesler, manuscript in preparation.

Mikrochim. Acta [Suppl.] 14, 413–415 (1997)
© Springer-Verlag 1997

Vibrational Spectra and Rotational Isomerism of Polydialkoxyphosphazenes

Boris V. Lokshin*, Lidiya I. Komarova, Irina A. Garbuzova, Natalia N. Lapina, and **Dzidra R. Tur**

A. N. Nesmeyanov Institute of Organoelement Compounds, Russian Academy of Sciences, Vavilova Str. 28, 117813 Moscow, Russia

Abstract. FT-IR and Raman spectra of polydialkoxyphosphazenes (PDAP) – $[P(OC_mH_{2m+1})_2 = N]_n$ – with $m = 1$–9 in the temperature interval from -100 to $+100\,°C$ were studied. It is shown that the temperature-dependent conformational changes occur in amorphous and mesomorphic phases due to internal rotation about P–O, C–O and C–C bonds in the side chains. The main chain does not change its form (presumably helix structure).

Key words: infrared spectra, Raman spectra, polydialkoxyphosphazenes, phase transitions, rotational isomerism.

Polydialkoxyphosphazenes (PDAP) – $[P(OC_mH_{2m+1})_2 = N]_n$ – are interesting not only from the practical point of view, but also as systems for the study of the different forms of self-arrangement of the polymer molecules, and in particular of the mesomorphic states, which appear for the PDAP with $m = 3$–5 only when the content of defects in the polymer chains is less than 2% [1]. In this work the FT-IR and Raman spectra of PDAP ($m = 1$–9) with $MM > 10^6$ and defect content $<1\%$ were studied in the temperature range from -100 to $+100\,°C$. Vibrational assignment is discussed and temperature-dependent spectra are analysed from the point of view of conformational changes in the main and side chains of the polymers in connection with their phase transitions.

Experimental

The synthesis of PDAP is described in [1]. IR spectra (4000–400 cm^{-1}) were measured with the Bruker FT-IR spectrometers IFS-113v and IFS-25. Raman spectra were obtained on the Jobin Ivon Ramanor HG2S instrument with 514.5 nm excitation (Ar$^+$-laser).

Results and Discussion

Assignment of the Main Chain Vibrations

The assignment of the main chain vibrations was done by comparing the IR and Raman spectra and using the model normal coordinate analysis for the simplest PDAP – polydimethoxyphosphazene (PDMP). The results are presented in Table 1.

Normal coordinate analysis was done for *cis-trans* configuration of the main chain with the force constants from related compounds. The calculation of the frequency branches was made in the zero approximation for $n = 15$. Increase of n above 15 does not change the shape of the branches. Two strongly curved branches near 1250 and 700 cm^{-1} correspond to asymmetrical stretching and symmetrical stretching–bending vibrations of the main chain. A very strong and broad band near 1245 cm^{-1} observed only in the IR spectra and a very strong band near 600 cm^{-1} appearing only in the Raman spectra, correspond to these vibrations. Such "mutual exclusion" is indicative of high symmetry of the main chain and confirms its *helix* structure. Upon lowering of the temperature to $-100\,°C$, the band-widths decrease to some extent but the same spectral pattern is observed for all the PDAP studied. Such behaviour demonstrates that the configuration of the main chain is not changed with temperature and the existing phase transitions are connected with the changes in the side chains.

Internal Rotation about the P–O Bond

In PDAP molecules, internal rotation is possible about the signal bonds P–O, C–O and C–C in the side chains. In PDMP only the rotation about P–O bonds can take place. Normal coordinate analysis for three

---

* To whom correspondence should be addressed

Table 1. Vibrational frequencies (cm^{-1}) of PDMP and their assignments

Raman	IR	Calculated	Assignment	Raman	IR	Calculated	Assignment
3001 w					1063 sh	1073	δCCH
2994 w				1042 m	1040 vs	1059	v^{as}CO + δCH$_3$
2952 s	2950 w	2986	v^{as}CH$_3$		918 w	908	vPN + vPO
	2916 w			856 m	863 w	855	vPO + vCO + vPN
	2900 w			816 w	820 s		
	2848 w	2866	v^sCH$_3$		756 m	774	vPO
2849 s	1466 w						
	1459 w	1464	δ^{as}CH$_3$	596 vs		696	v^sPN + δPNP + vPO + δPOC
1458 m	1451 w	1442	δ^sCH$_3$			470	vPO + δOPN + δPOC + δOPN
	1329 w						
	1245 s		v^{as}PN			400	δOPN
1186 w	1185 w	1142	vPN + vPO + vCO + δCH	318 w		334	vPN + δPNP + δOPN
						223	δOPN + δPOC
1164 w	1152 sh	1116	δCCH			146	δPNP
1087 m		1075	v^sCO				

conformers with the different orientation of the methyl groups relative to the P–O bonds (*trans-trans, trans-gauche* and *gauche-gauche*) shows that the region of 1000–1100 cm^{-1} must be sensitive to the conformational changes, but the frequency shifts are small (about 10 cm^{-1}). FT-IR spectra of PDMP were studied in the temperature range from −80 to +100 °C. PDMP is amorphous, no phase transitions are observed and the configuration of the main chain is kept over all this temperature interval. The only effect in the 1000–1100 cm^{-1} region is the 10 cm^{-1} low frequency shift of the very strong band at 1045 cm^{-1} on lowering of the temperature. The isosbestic point shows, however, that this change can be connected with the conformational equilibrium. Another change is observed in the δCH$_2$ frequency region, where the doublet band with maxima at 1450 and 1462 cm^{-1} appears. The relative intensity of the 1450 cm^{-1} bond increases with rise of temperature. The spectral changes are obviously connected with the rotational isomerism and with an increase of the *gauche*-conformers at high temperatures.

Internal Rotation about the C–O Bond

In molecules of polydiethoxyphosphazene (PDEP), additional possibilities of rotation about the C–O bond appear compared to PDMP. The temperature dependence of the IR spectrum (from −70 to +95 °C) was studied. In this temperature range, the polymer is amorphous, has no phase transitions and keeps the main chain configuration. The normal coordinate

analysis predicts that a series of bands in the region of 1100–600 cm^{-1} can be sensitive to the internal rotation. It was noticed that when the temperature was raised, the relative intensities of the bands at 759, 837, 813 cm^{-1} decreased and those of the bands at 752, 1100 cm^{-1} increased. Such changes are supposed to be connected with the rotational isomerism.

Internal Rotation About C–C Bonds in PDAP (m = 3–9)

PDAP with $m = 3–5$ are mesomorphic at room temperature. The main chains are ordered, but the side chains are disordered in this phase [1]. At high temperatures (>160 °C), they can be converted into the amorphous phases, while at low temperatures ($\approx −110$ °C), they pass to the glassy state without crystallization. Temperature changes of the spectra might be detected from the rotations about the C–C, C–O and P–O bonds in the side chains. polymers with $m = 6–9$ are amorphous at room temperature but polymers with $m = 7–9$ form ordered crystal phases where all the side chains have the *all-trans* conformation as proposed in [1].

For all the polymers studied in the region of the scissoring δCH$_2$ modes the relative intensity of the 1450 cm^{-1} band increases with temperature, and at temperatures about 100 °C this band becomes predominant in the spectra. On cooling, the 1466–1470 cm^{-1} band increases, while the intensities of the other bands decrease. For polymer C$_8$, which is crystalline at temperatures below 4 °C, only one band at

$1470 \, \text{cm}^{-1}$ is observed at low temperatures. It permits assignment of this band to the methylene group vibrations of the *all-trans* form. Then the $1458 \, \text{cm}^{-1}$ band corresponds to the scissoring CH_2 modes of the *all-gauche* or *partially-gauche* conformer. Such an assignment is in agreement with the literature data for paraffins. The analysis of the spectra of the PDAP studied shows that the contribution of the *trans*-conformers in the side chains increases with lowering of the temperature in the amorphous and mesomorphic phases. Temperature changes in the IR-spectra were also observed in the regions of 1100–900 and 900–700 cm^{-1}. The frequencies, half-widths and relative intensities of the bands are changed. For PDAP ($m = 7$–9), which crystallize at -17 and $4 \, ^\circ C$ respectively, dramatic changes in some regions of the spectra take place. A series of narrow bands appears instead of the very broad spectra due to the formation of the highly ordered structures with *all-trans* structure of the aliphatic side chains.

Acknowledgements. The authors are grateful to the International Science Foundation for Financial support (Grants MEQ000 and MEQ300).

Reference

[1] D. R. Tur, N. P. Provotorova, S. V. Vinogradova, V. I. Bakhmutov, M. V. Galakhov, V. P. Zhukov, I. I. Dubovik, D. Ja. Tsvankin, V. S. Papkov, *Makromol. Chem.* **1991**, *192*, 1905.

Mikrochim. Acta [Suppl.] 14, 417–420 (1997)
© Springer-Verlag 1997

FT-IR Study of the Molecular Orientation of Rubbed Polyimide Alignment Films

Takeshi Matsunobe*, Naoto Nagai, Ritsu Kamoto, Yoshitsugu Nakagawa, and **Hideyuki Ishida**

Toray Research Center, Inc., Sonoyama 3-Chome, Otsu, Shiga 520, Japan

Abstract. The molecular orientations and surface morphology of mechanically rubbed polyimide films have been extensively studied by FT-IR, ellipsometry and atomic force microscopy (AFM). In-plane and out-of-plane molecular orientations of rubbed polyimide films have been examined by means of the angular dependences of polarized infrared transmission spectra. It has been confirmed that the polyimide chain inclination against the surface plane is induced by the rubbing.

Key words: polyimide alignment film, FT-IR, molecular orientation, tilt angle.

It is very important to obtain homogeneous controlled alignment of liquid crystal molecules in the manufacturing process of liquid crystal displays (LCD). It is well known that such homogeneous alignment of LC molecules can be achieved by rubbing the polyimide thin film on indium tin oxide (ITO) substrates with nylon or rayon cloth. However, the orientation behaviour of the polyimide molecular chain is not clear even at the present stage. In the present paper, thin polyimide films rubbed mechanically at different intensities [1] are studied mainly by FT-IR spectroscopy.

Experimental

Polyimide thin films were deposited by a spinning method on the indium tin oxide (ITO) coated onto silicon substrates. The thicknesses of the polyimide film and ITO were 800 and 470 Å, respectively. Polyimide thin films were rubbed at different intensities (strong rubbing: 225 mm, medium rubbing: 112 mm, weak rubbing: 36.8 mm). The rubbing was done mechanically [1] with the film mounted on a flat moving stage and in contact with a rotating band of cloth (Fig. 1). The IR absorption spectra were measured with a Bruker IFS-120HR spectrometer equipped with a mercury cadmium tellurium (MCT) detector. Figure 2 shows arrangements for the measurement of the angular dependence of the infrared transmission spectra. The polarized infrared transmission spectra were measured for the analysis of in-plane molecular orientation by rotating the silicon substrates with respect to the plane of polarization (Fig. 2a). For analysis of the out-of-plane molecular orientation, the silicon substrates were tilted against the direction of polarization (Fig. 2b). A 10 mm diameter mask was used in order to eliminate the effect on the measurement area caused by the tilting.

An atomic force microscope (Nano Scope 3 AFM, Digital Instruments) was used for observation of the surface morphology of the rubbed polyimide films.

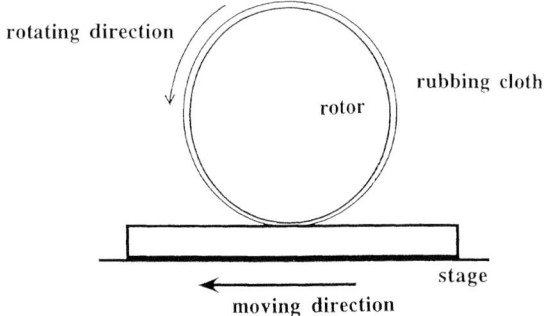

Fig. 1. Method for mechanical rubbing. The rubbing intensity (mm)= NM $[(2\pi\ r\ n/60\ V) - 1]$, where $N =$ number of rotations, $M =$ contact depth of cloth (mm), $r =$ radius of the rotor (mm), $n =$ speed of rotation (rpm), $V =$ speed of stage movement (mm/sec)

---

* To whom correspondence should be addressed

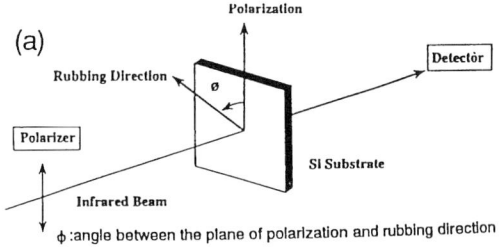

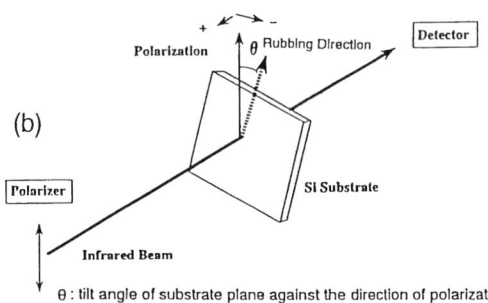

Fig. 2. Arrangements used for the measurement of the molecular orientation of rubbed polyimide film by use of the angular dependence of infrared transmission spectra: **a** analysis of in-plane orientation, **b** analysis of out-of-plane orientation

Results and Discussion

Surface Microstructure of Rubbed Polyimide Films

Figure 3 shows AFM images of the rubbed polyimide surface. Microgrooves are observed along the rubbing

direction, at intervals from 100 to 1000 Å. With increasing rubbing intensity, the depth and number of grooves increase. The surface roughness estimated from the mean squared roughness (Rq) was found to be proportional to the rubbing intensity.

Molecular Orientation of Rubbed Polyimide Films Studied by Means of the Angular Dependences of Polarized Infrared Transmission Spectroscopy

Figure 4 shows the polarized infrared transmission spectra of strongly rubbed films on ITO/Si substrates measured at different rotating angles. The stretching vibrations of C–O–C and carbonyl C=O have their transition moments parallel and perpendicular to the molecular chain, respectively. In-plane and out-of-plane molecular orientations of rubbed polyimide films have been examined by monitoring the absorption intensity of the C–O–C band.

The angular dependences of the IR absorption band (C–O–C) are shown in Fig. 5 for the films rubbed at different rubbing intensities. It is clearly shown that the molecular chains are oriented parallel along the rubbing direction. Figure 6 shows the relationship between the rubbing intensity and the IR dichroic ratios. The molecular orientation estimated from the IR dichroic ratios tends to saturate with increasing rubbing

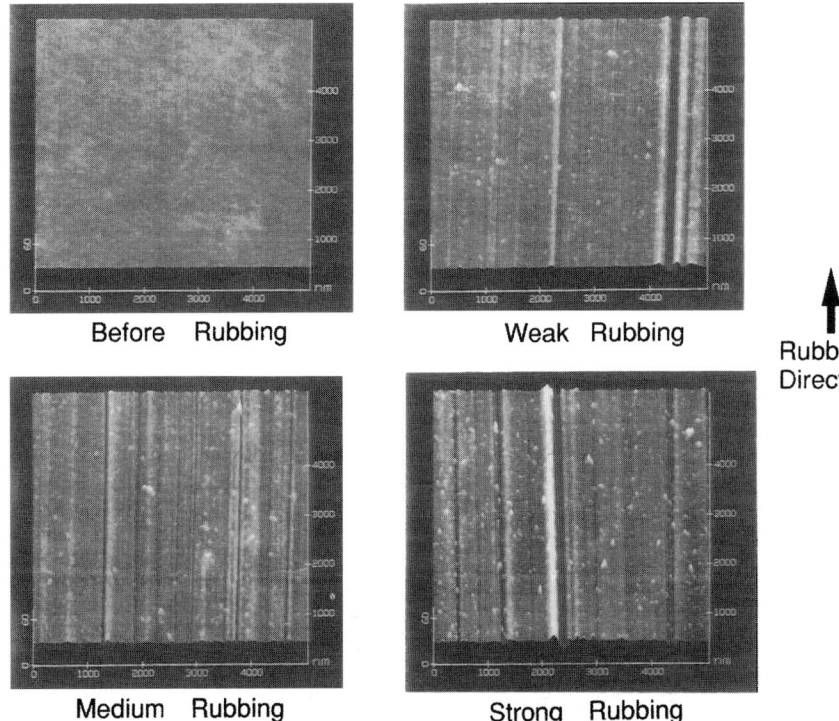

Fig. 3. AFM images of rubbed polyimide films

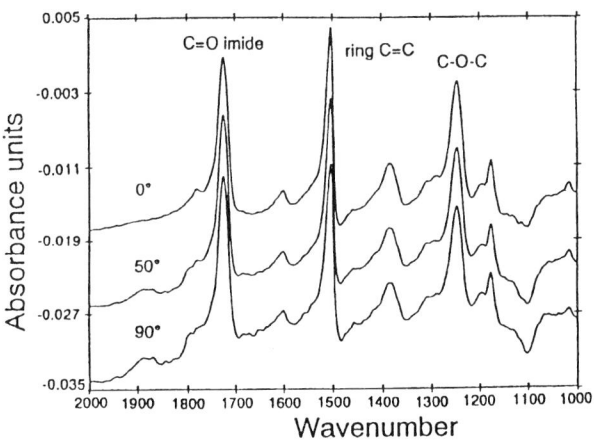

Fig. 4. Angular dependence of IR absorption spectra of rubbed polyimide film

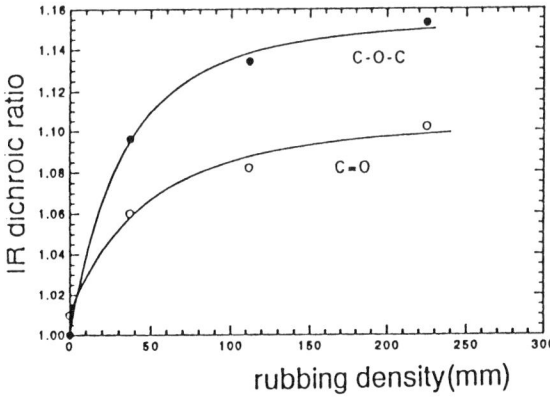

Fig. 6. Relationship between the IR dichroic ratios and rubbing intensity

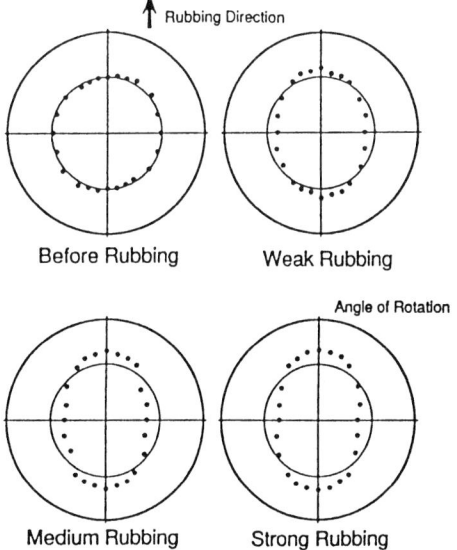

Fig. 5. Angular dependences of IR absorption band of rubbed polyimide films (C–O–C at 1245 cm⁻¹)

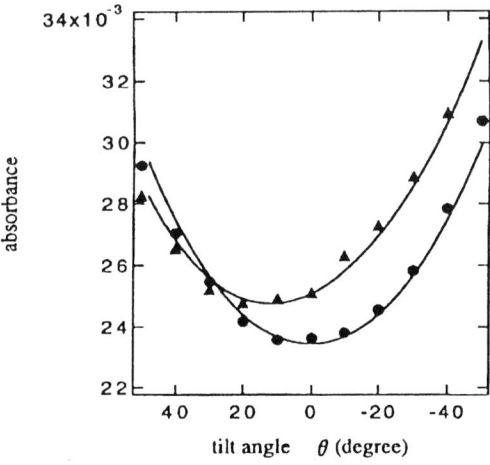

Fig. 7. Angular dependence of IR absorption band (C–O–C at 1245 cm⁻¹). The filled circles and triangles are the data points for before rubbing and after strong rubbing, respectively. The solid curves represent the best-fit calculated curves

intensity. The mechanical rubbing process can not homogeneously orient the overall polyimide film. The surface thin layer of the polyimide film is induced to orient by the stress caused by the rubbing. This is the reason why the molecular orientation tends to saturate with increasing rubbing intensity.

The out-of-plane molecular orientation can be estimated by measurement of the tilt angle of the molecular chain against the surface plane. Figure 7 shows the angular dependences of the IR absorption intensity of the C–O–C stretching band for the samples before rubbing and after strong rubbing. The IR band intensity shows symmetrical angular dependence with respect to the vertical axis for the sample before rubbing, while the sample after strong rubbing shows asymmetric angular dependence. The asymmetrical angular dependence strongly suggests the influence of the tilt angle.

Table 1 summarizes the calculated [2] results together with the pre-tilt angle of the LC molecule for the rubbed polyimide films. The tilt angle of the polyimide chain ranges from 20° (weak rubbing) to 31° (strong rubbing). The thickness of the surface oriented layer ranges from 113 to 138 Å, which is about 15% of the entire film thickness. It is shown that the tilt angle of the molecular chain in the rubbed film increases with

Table 1. Tilt angle of molecular chain, thickness of surface oriented layer and the pre-tilt angle of LC molecule for the polyimide films rubbed at different intensities

	Rubbing intensity (mm)	Pretilt angle of LC molecule	Tilt angle of molecular chain	Thickness of surface oriented layer (Å)
Weak rubbing	36.8	4.2°	20°	113
Medium rubbing	112	3.8°	28.6°	123
Strong rubbing	225	3.52°	30.5°	138

increasing rubbing intensity. On the other hand, the pre-tilt angle of the LC molecule on the rubbed polyimide film shows the opposite change with the rubbing intensity. These results suggest that the polyimide chain inclination induced by rubbing has no direct correlation with the pre-tilt angle of the LC molecule. Further detailed study of the molecular interaction of the LC molecule with the rubbed polyimide surface is now in progress in order to clarify the origin of the pre-tilt angle.

References

[1] T. Uchida, M. Hirano, H. Sakai, *Liq. Cryst.* **1989**, 5, 1127.
[2] L. M. Blinov, N. V. Dubinin, L. V. Mikhnev, S. G. Yudin, *Thin Solid Films* **1984**, 120, 161.

Mikrochim. Acta [Suppl.] 14, 421–424 (1997)

Influence of Cross-linking Processes on the Orientation in Thermotropic Copolyesters Studied by FT-IR Spectrometry

Karin Sahre*, **Klaus-Jochen Eichhorn**[1], **Uwe Hoffmann**, and **Heinz W. Siesler**[2]

[1] Institute of Polymer Research Dresden, D-01069 Dresden, Federal Republic of Germany
[2] Department of Physical Chemistry, University of Essen, D-45117 Essen, Federal Republic of Germany

Abstract. Dynamic (rheo-optical) FT-IR measurements during deformation of a cross-linkable thermotropic main chain copolyester are presented. Maximum orientation functions of 0.6 for the flexible spacers and 0.8 for the stiff mesogenic units were calculated from the dichroic ratios of specific IR bands. The characterization of thermal and photochemical cross-linking processes in the 400% stretched copolyester foils to achieve a fixation of aligned domain-structure in the liquid crystalline state was the focus of interest. A complete cross-linking is reached only by UV irradiation at 135 °C for 120 min, and is connected with a decrease of the orientation of the mesogens and spacers by about 50 and 35% respectively. The IR results were confirmed by solubility experiments and size-exclusion chromatography.

Key words: thermotropic copolyesters, cross-linking, liquid crystal, mesogens, spacers.

In earlier studies we investigated the orientation of macromolecular chains in cross-linkable liquid crystalline copolyesters based on itaconyl-bis-(4-oxybenzoic acid) **(I)**, sebacoyl-bis-(4-oxybenzoic acid) **(S)** and 2-methylhydroquinone **(M)** before and after uniaxial drawing [1, 2]. By means of a cross-linking reaction using the double bonds in the itaconyl groups a fixation of oriented chain segments should be possible. With that, the anisotropy is decreased and sufficient strength should be obtained not only in the drawing direction, which is typical for main chain polymers, but also perpendicular to this. In this paper we report the influence of different cross-linking processes on the orientation of the copolyester mentioned.

Experimental

Materials

The ternary **I/S/M** copolyester was synthesized by the high-temperature solution polycondensation of the dichlorides of the acids **(I)** and **(S)** with 2-methylhydroquinone **(M)** in 1-chloronaphthalene at 180 °C. The liquid crystalline phase of the **I/S/M** copolyester was obtained at above 120 °C; the clearing point is higher than the cross-linking temperature, at about 200 °C.

Preparation of Polymer Films

Isotropic homogeneous films, as initial samples for all the experiments, were cast from a solution of the polymers in chloroform. The film thickness was approximately 10–15 µm.

Rheo-optical FT-IR Spectrometry

This method is based on the simultaneous recording of polarization spectra and stress–strain diagrams during the deformation process [3]. In our case the **I/S/M** copolyester was drawn to 280% strain (0.98% strain/spectrum) at 60 °C, above the glass transition point. FT-IR spectra were recorded on a Bruker IFS 88 spectrometer with a resolution of 4 cm^{-1}; 20 scans were co-added.

Thermal and Photochemical Cross-linking Processes

Copolyesters stretched 400%, with the orientation functions $f = 0.72$–0.75 of mesogens and $f = 0.53$–0.60 of spacers were used for these cross-linking processes. For the thermal cross-linking process the samples were treated according to different temperature programmes. One of these, and the resultant orientation behaviour, are shown in this paper. For the photochemical cross-linking process the copolyester foils were held at 22 °C (room temperature) or heated to 80 or 135 °C and irradiated for 30, 60 and 120 min by UV radiation (spectral range 200–400 nm).

---

* To whom correspondence should be addressed

Results and Discussion

Rheo-optical FT-IR Measurements

FT-IR polarization spectra of the CH_2 stretching and ring stretching vibration regions are shown in Fig. 1a, b. The dependences of the dichroic ratios R (Fig. 2) and the resultant orientation functions f (Fig. 3) on the strain are presented. The orientation functions are calculated according to Fraser [4].

The samples drawn to approximately 280% elongation show interesting orientational behaviour. The polymer chains were aligned in the drawing direction; the stiff units (mesogens) were more rearranged in the preferred direction than were the flexible groups. Maximum orientation functions of 0.6 and 0.8 were calculated for the flexible and mesogenic units by using the characteristic bands of the CH_2 stretching vibrations at $2985–2820\,cm^{-1}$ and the aromatic ring stretching vibrations at $1630–1568\,cm^{-1}$ and $1531–1477\,cm^{-1}$. Relaxation effects were not observed.

Thermal and Photochemical Cross-linking Processes of the 400% Stretched I/S/M Copolyesters

Thermal cross-linking process. Figure 4 (a and b) shows the temperature programme and the simultaneous orientation behaviour of the copolyesters in different liquid crystalline phases. The samples were heated to the corresponding temperatures, annealed for 60 min at these temperatures and then cooled to room temperature and measured.

In the case of a step-by-step increase of temperature the orientation functions of the mesogens and spacers decrease only to about 55 and 70% respectively of the values for the initial samples. At a temperature above 200 °C the cross-linking is started and produces networks in defined orientation states. In contrast to this, a continuous linear heating from room temperature up to temperatures > 200 °C causes a complete loss of orientation. Therefore thermal cross-linking produces cross-linked structures of the previously oriented polymer chains only under defined conditions of temperature. However, in the polarized IR spectra one of the typical bands of the unsaturated itaconoyl unit (cross-linkable group) at $1637\,cm^{-1}$ does not show the cross-linking exactly, but the band at $969\,cm^{-1}$ indicates the cross-linking process. Moreover, in all cases soluble parts remain. That means that cross-linking is not complete, because fully cross-linked products are insoluble in the solvent used.

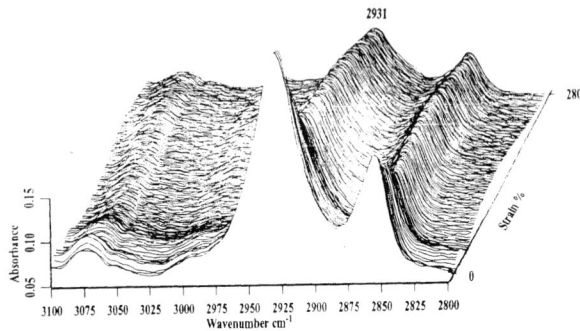
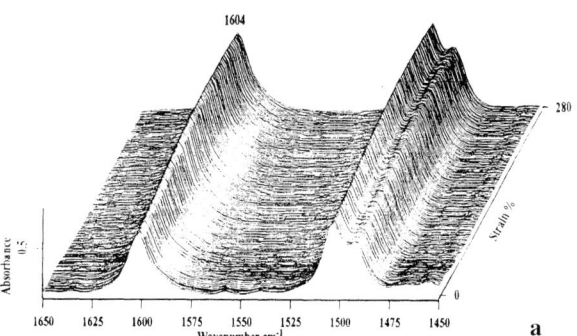
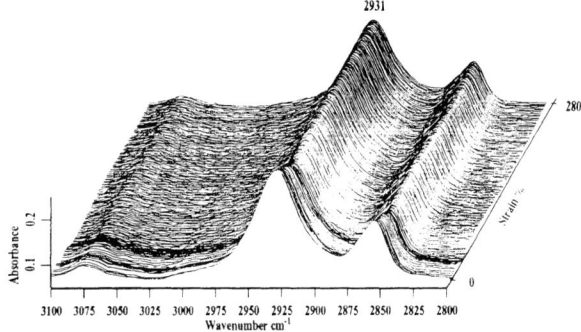
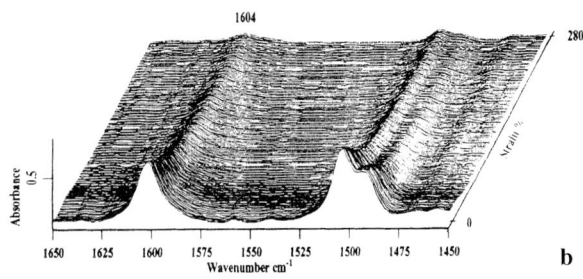

Fig. 1. IR spectra of the chosen regions for the **I/S/M** copolyester: **a** parallel to the stretching direction; **b** perpendicular to the stretching direction

Photochemical Cross-linking Process. The polarized IR spectra of the C=C stretching and ring stretching vibration regions of an **I/S/M** copolyester irradiated for 120 min at 22 °C in comparison with those for the initial state are shown in Fig. 5.

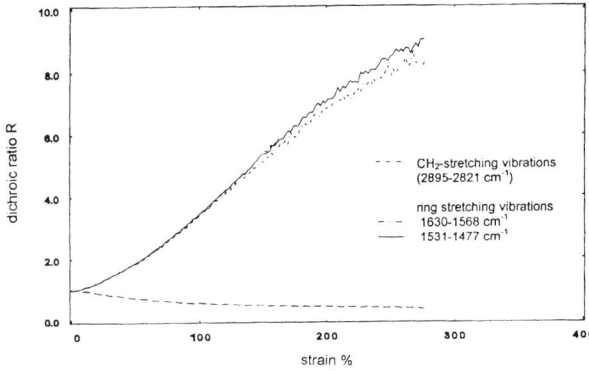

Fig. 2. Dichroic ratio *vs* strain plot

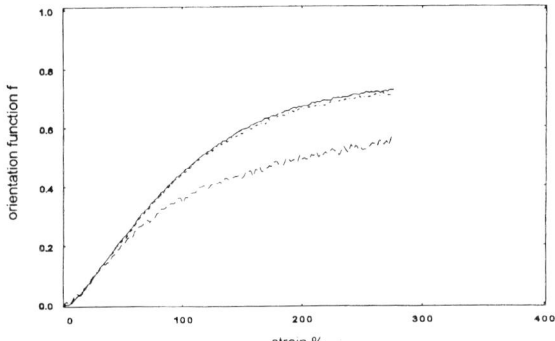

Fig. 3. Orientation function *vs* strain plot

The dependence of the solubility of the irradiated samples on the irradiation time and temperature (a) and the dependence of the weight-average molecular weights M̄w of the soluble parts (determined by size-exclusion chromatography) on the irradiation time and temperature (b) are demonstrated in Fig. 6. The orientation functions of the mesogens and spacers for

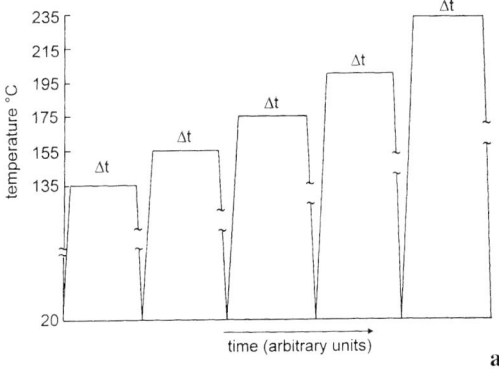

Fig. 4. a Temperature programme, $\Delta t = 60$ min (polarized IR spectra were measured every cycle at 20 °C; **b** orientation functions of the mesogens and spacers following the described temperature programme

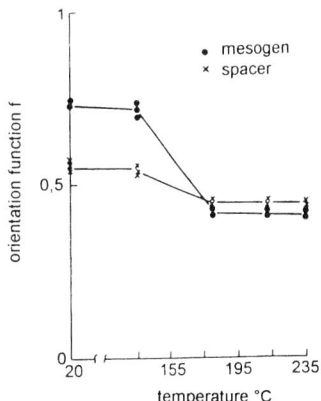

Fig. 4b.

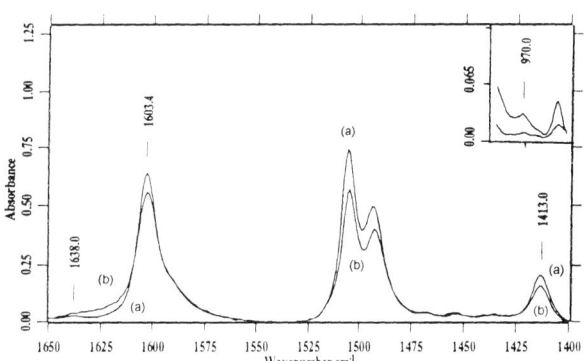

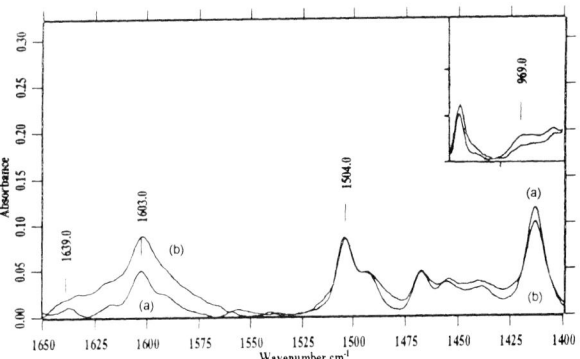

Fig. 5. Polarized IR spectra of the C=C stretching and ring stretching vibration regions for the 400% stretched **I/S/M** copolyester (initial (*a*) and irradiated (*b*) states). Upper: parallel polarization, lower: perpendicular polarization

different temperatures and times of irradiation are listed in Table 1.

The polarized IR spectra of the **I/S/M** copolyester irradiated for 120 min at 22 °C show the decrease of the intensity of the characteristic band at 970/969 cm^{-1}. Moreover, the region of the C=C stretching and ring stretching vibrations is broader than for the unir-

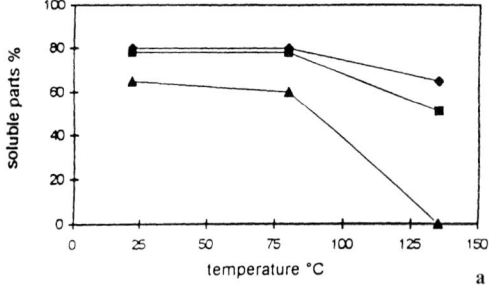

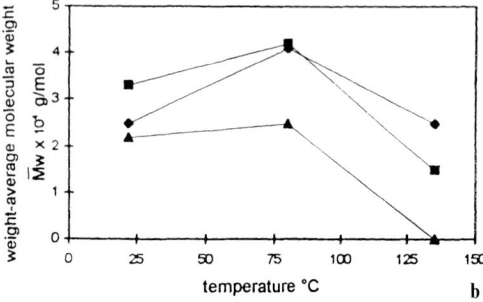

Fig. 6. The dependences of the solubility and the $\bar{M}$w of the soluble parts of the irradiated samples on irradiation time and temperature: **a** the solubility of the irradiated samples; **b** the $\bar{M}$w of the soluble parts. Rhombus: 30 min irradiation, squares: 60 min irradiation, triangles: 120 min irradiation

radiated initial state (Fig. 5). This is a typical phenomenon for all the **I/S/M** copolyester samples irradiated under the various designated conditions.

In the case of irradiation at 135 °C (for 120 min) it was not possible to determine $\bar{M}$w, and no soluble part was detected. That means that only cross-linked parts existed. It can be concluded that in contrast to the partial thermal cross-linking of the **I/S/M** copolyesters a complete cross-linking of the stretched samples is reached by UV irradiation at 135 °C for 120 min.

Table 1. The dependence of the orientation function on the irradiation time and temperature

Temperature °C	Orientation function f	
	of mesogen before/after	of spacer before/after
	irradiation	
22	0.78/0.64 $\approx 20\%$[a]	0.57/0.53 $\approx 5\%$[a]
80	0.77/0.53 $\approx 30\%$[a]	0.57/0.45 $\approx 20\%$[a]
135	0.79/0.38 $\approx 50\%$[a]	0.56/0.36 $\approx 35\%$[a]

[a] Decrease of the value of $f(\%)$.

The orientation functions of the mesogens and spacers are decreased by about 50 and 35% respectively during this irradiation at increasing temperature and time (Table 1).

Hence we can conclude that defined UV cross-linking in the liquid crystalline phase is the most attractive and effective procedure (and also from the technological point of view) for producing a network in thin films without dramatic decrease of orientation in the thermotropic copolyester.

References

[1] K. Sahre, K.-J. Eichhorn, N. Reichelt, D. O. Hummel, *Acta Polym.* **1994**, *45*, 36.
[2] K. Sahre, K.-J. Eichhorn, N. Reichelt, D. O. Hummel, *Proc. SPIE* **1993**, *2089*, 190.
[3] H. W. Siesler, *Adv. Polym. Sci.* **1984**, *65*, 1.
[4] R. D. B. Fraser, *J. Chem. Phys.* **1958**, *29*, 1428.

Mikrochim. Acta [Suppl.] 14, 425–427 (1997)
© Springer-Verlag 1997

Infrared Spectroscopic Characterization of the Crystallinity and the Hydrocarbon Chain Packing of the Polyethylene Segments in Poly(ethylene-co-dimethylaminoethyl Methacrylate)

Huiping Zhang[1], Duanfu Xu[1,*], Yizhuang Xu[2], Shifu Weng[2], and Jinguang Wu[2]

[1] Institute of Chemistry, Academia Sinica, Beijing 100080, Peoples Republic of China
[2] Department of Chemistry, Peking University, Beijing 100871, Peoples Republic of China

Abstract. The multicrystalline behaviour and the crystallinity of poly(ethylene-co-dimethylaminoethyl methacrylate) were investigated by infrared spectroscopy. The metastable monoclinic packing structure of the polyethylene segments of the copolymers usually co-exists with the stable orthorhombic form in the isotropic samples. The reason for the occurrence of this multicrystalline structure is discussed. The relative content of the monoclinic form in the crystalline regions changed greatly with the co-unit contents and the crystallization conditions. The crystallinity of the samples was also estimated by spectroscopic methods.

Key words: infrared spectroscopy, poly(ethylene-co-dimethylaminoethyl methacrylate), multicrystalline behaviour, crystallinity.

Poly(ethylene-co-dimethylaminoethyl methacrylate) (EDAM), a useful aid for polyolefin processing, is known to be a segmented copolymer. Our previous electron microscope studies on the morphology of a series of EDAM samples have shown that the polyethylene (PE) segments of the molecules generally form lamellae while dimethylaminoethyl methacrylate (DAM) co-units usually disperse homogeneously in the non-crystalline regions [1]. Various crystalline architectures of the PE segments of the copolymers, from random dispersed lamellae to spherulites, can be observed in the samples. In this paper, we investigate the multicrystalline behaviour and the crystallinity of EDAM by means of Fourier-transform infrared (FT-IR) spectroscopy.

Experimental

The EDAM films were prepared by casting $\sim 3\,\text{wt}\%$ solutions of EDAM samples in xylene onto thermostatic glycerol set at 120 °C,

* To whom correspondence should be addressed

which is above the melting temperature of EDAM, evaporating the solvent to obtain molten films, then slowly cooling (SC) the films just by switching off the heater, or quenching (Q) the films by quickly transferring them into water at ambient temperature. The films obtained were washed with water to remove residual glycerol and dried, then put into the infrared spectrometer for measurements. All infrared spectra were recorded on a Bio-Rad 65A FT-IR spectrometer at a resolution of $1\,\text{cm}^{-1}$. The methylene rocking bands, 750–$700\,\text{cm}^{-1}$, were used to characterize the packing modes of the PE segments, i.e. the hydrocarbon chain, and the crystallinity of the samples. For PE, there are at least four bands in this region, appearing near 731, 723, 720 and $718\,\text{cm}^{-1}$ respectively [2]. The first and third bands are assigned to the orthorhombic crystalline structure. The fourth band is assigned to the monoclinic crystalline structure. The second band is associated with the amorphous fraction of PE.

The EDAM samples studied are listed in Table 1.

Results and Discussion

The results of the second derivative spectra of EDAM samples (Fig. 1b) indicate that the monoclinic form $(717\,\text{cm}^{-1})$ usually co-exists with the stable orthorhombic form (730 and $720\,\text{cm}^{-1}$) in the crystalline regions of our isotropic samples. The monoclinic form is a metastable packing structure of PE chains, and was previously known to occur in some anisotropic PE samples. However, its occurrence in isotropic samples of ethylene copolymers has not been reported yet.

We define the relative content of the monoclinic form in crystalline regions, c_m, as

$$c_m = I\,(717)/I(730 + 720 + 717) \qquad (1)$$

where I represents the integrated intensities for the bands indicated by number. The infrared intensities were obtained by curve-fitting methods. The estimated c_m values of the samples are listed in Table 1. As can be seen from Table 1, the relative content of the mon-

Table 1. Relative content of monoclinic form and crystallinity of EDAM

Sample	DAM content mol %	wt %	Crystallization conditions	Relative content of monoclinic form, c_m, (%)	Crystallinity, x_c (%)
1#–03	2.9	14	slow-cooled	0	25
			quenched	23	19
2#–07	6.7	29	slow-cooled	9	15
			quenched	34	12
3#–11	10.6	40	slow-cooled	21	10
			quenched	37	8

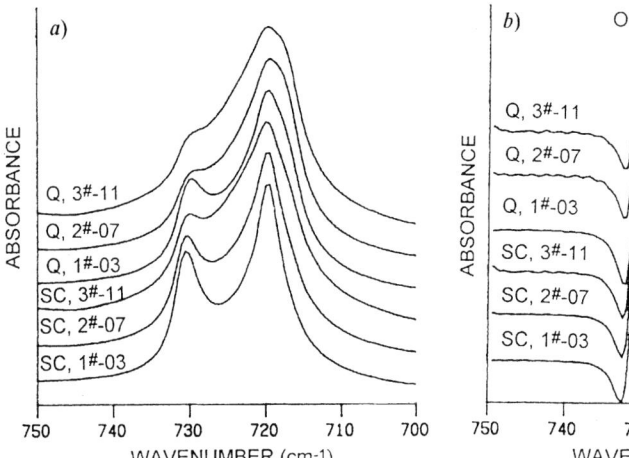

Fig. 1. a Infrared spectra and **b** second derivative spectra of EDAM samples. From the bottom are the slow-cooled (SC) samples of 1#–03, 2#–07, 3#–11, and the quenched (Q) samples of 1#–03, 2#–07, 3#–11

oclinic form is much higher in quenched samples than in slow-cooled samples, and increases dramatically with the increasing DAM content. Although the monoclinic structure is thermodynamically less stable, it can be formed in a relatively large amount in EDAM samples. We assume that the movement of PE segments is restrained, owing to the segmental insertion of DAM co-units into the chain, so the crystallization of some PE segments seems to be stopped at the metastable monoclinic state, in view of the kinetics, rather than reach the thermodynamically stable orthorhombic state. It seems that the quenching procedure and the high DAM content can strengthen such a tendency.

The curve-fitting results show that the amorphous band appears near 722 cm^{-1}. The crystallinity of EDAM, x_c, can be defined as:

$$x_c = \frac{I(730 + 720) + \beta I(717)}{I(730 + 720) + \alpha I(722) + \beta I(717)} \times (1 - x_{DAM})$$

(2)

where x_{DAM} is the weight fraction of DAM component in the samples. The coefficient α is defined as the ratio of the molar absorptivity of the orthorhombic crystalline fraction to that of the amorphous fraction in PE, and has been found to be 2.1 from our previous work [3]. The coefficient β is defined as the ratio of the molar absorptivities between the orthorhombic and monoclinic fractions in PE. We assumed the value of β to be 1, as it was not yet available. The estimated values of x_c of the samples are also listed in Table 1. These spectroscopically measured values are comparable with those obtained from calorimetic measurement and reported in the literature [4]. Compared with c_m, the crystallinity of the samples varies with the DAM contents and crystallization conditions slightly, and in quite a different way. For example, quenched samples have lower crystallinity but higher monoclinic contents in the crystalline regions.

Acknowledgements. This work was supported by the National Nature Science Foundation of China, the Key Project Foundation of Academia Sinica, and the Science Foundation of the Polymer Physics Laboratory of Academia Sinica.

References

[1] H. Zhang, F. Guo, D. Xu, *Natl. Symp. Polym. Phys. '94, Zhengzhou, China*, 1994, p. 89.

[2] H. Hagemann, H. L. Strauss, R. G. Snyder, *Macromolecules* **1987**, *20*, 2810.

[3] H. Zhang, D. Xu, Y. Xu, X. Liu, J. Yang, X. Zhang, S. Weng, J. Wu, *Preprints Inttl. Microsymp. Polym. Phys. '95, Xian, China*, 1995, p. 32.

[4] T. Ohmae, S. Hosoda, H. Tanaka, H. Kihara, B. Jiang, Q. Ying, R. Qian, T. Masuda, A. Nakajima, *Pure Appl. Chem.* **1993**, *65*, 1825.

Mikrochim. Acta [Suppl.] 14, 429–432 (1997)
© Springer-Verlag 1997

In-vivo Measurements of Skin Tissue by Near-infrared Diffuse Reflectance Spectroscopy

Andreas Bittner, Stefan Thomaßen, and **H. Michael Heise**

Institut für Spektrochemie und Angewandte Spektroskopie, Bunsen-Kirchhoff-Strasse 11, D-44139 Dortmund, Federal Republic of Germany

Abstract. Non-invasive transcutaneous sensors based on near infrared spectroscopy are feasible. For metabolite monitoring the variability of the spectra of skin tissue exceeds by far the signals expected from the analytes. Important measurement parameters such as photon penetration depth are estimated by a Monte Carlo simulation of the radiation transport in tissue. Different variance factors and their magnitude, important for the calibration stage, are studied by using two diffuse reflectance accessories.

Key words: in vivo spectroscopy, near infrared, skin tissue, variance analysis.

Demands for gentle medical diagnostic methods have caused a recent increase in biomedical applications using optical spectroscopy. To obtain quantitative information on various analytes in tissue or blood, optical techniques usually measure a broad spectral range, allowing multivariate calibration modelling. This strategy was applied for the non-invasive blood glucose monitoring of diabetic patients [1, 2]. The analytical performance, in particular for the lower and critical glucose concentrations, was not sufficient to allow miniaturization of the measuring device. The quantitative analysis is complicated by the variability of the measurement, for which differing lip position and contact pressure were suggested as the main factors. Additionally, the biological variance between individuals because of their tissue texture is much greater than experienced with measurements performed on the lip tissue of one person. In this investigation the diffuse reflectance spectra of skin tissue, as measured by using different accessories, i.e. a diffuse reflectance device based on mirrors [2] and a fibre optic probe, are compared. The reproducibility of spectral data collection is studied and different factors influencing the measurement are evaluated in detail.

---

* To whom correspondence should be addressed

Experimental

Spectral measurements were made with a Bruker IFS 66 FT-spectrometer equipped with a tungsten lamp and a CaF_2 beam-splitter. The diffuse reflectance (DR) accessory housing with transfer optics (custom-made) and InSb detector from Infrared Ass. (Suffolk, U.K.) [2] and a bifurcated, Y-shaped fibre optic accessory with common end diameter of 4 mm (the fibre bundle contained fibres of 0.2 mm diameter, total length 1.5 m) purchased from Bruker (Karlsruhe, Germany), were attached to the spectrometer for measuring the skin tissue spectra of different subjects. A total of 1200 interferograms with a spectral resolution of $32\,cm^{-1}$ were averaged for each spectrum, for which a subsequent reference spectrum using Spectralon reflectance standard from Labsphere (North Sutton, NH, U.S.A.) was recorded.

Results and Discussion

Monte Carlo simulations of the radiation transport in tissue were performed to elucidate important measurement parameters, with use of optical constants estimated for dermis at $6400\,cm^{-1}$, where absorption bands for glucose exist (for details see [2]). Results from 500,000 photons launched are shown in Fig. 1. Mean values for maximum photon penetration depth are: 0.58 mm (ideal accessory), 0.54 mm (DR) and 0.73 mm (fibre); mean values for integral photon pathlength are: 2.12 mm (ideal), 2.06 mm (DR) and 2.32 mm (fibre). The reflectance values are calculated with a white Lambertian reflection standard: 7.88% (ideal), 6.28% (DR) and 8.50% (fibre). Owing to the numerical aperture of 0.22 for the fibre bundle as stated by the distributor, only a small cone with a half opening angle of 9° can be used for the collection of diffusely reflected radiation from tissue, whereas the solid angle for the accessory with reflection optics is much larger (hemisphere above sample, except for an illumination cone with half opening angle of 35°). As a consequence, the

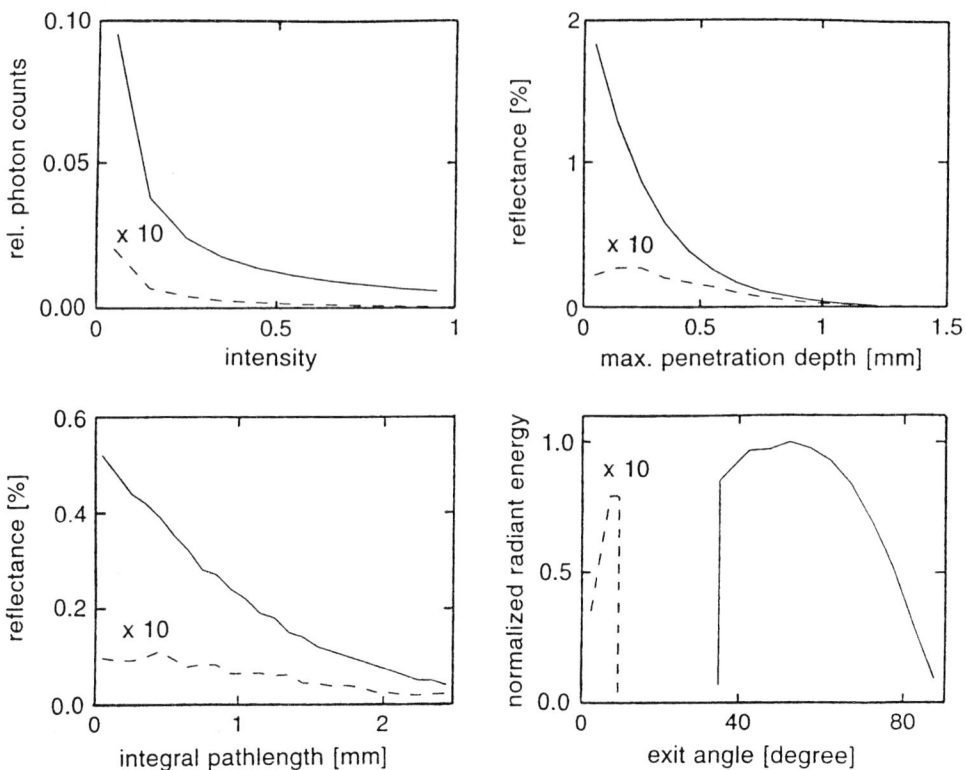

Fig. 1. Distribution functions calculated by Monte-Carlo simulation for the DR and fibre optic accessories (dashed curve) under matched boundary conditions (parameters for dermis at $6400\,cm^{-1}$: $\mu_a = 10\,cm^{-1}$, $\mu_s = 50\,cm^{-1}$, absorption and scattering coefficients, respectively; $g = 0.8$ anisotropy parameter; refractive index $n = 1.37$)

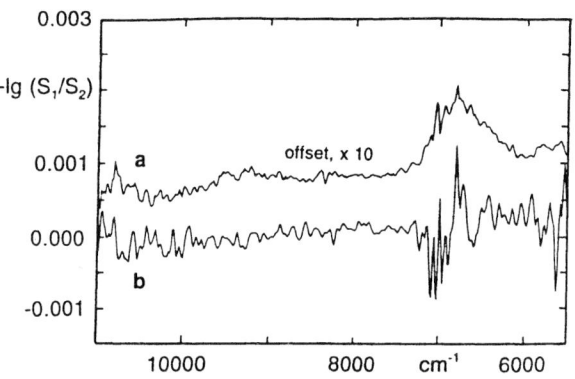

Fig. 2. Spectral noise level obtained by using different accessories with inner lip: *a* DR accessory with reflection optics (InSb detector) *b* fibre optic accessory (Ge detector)

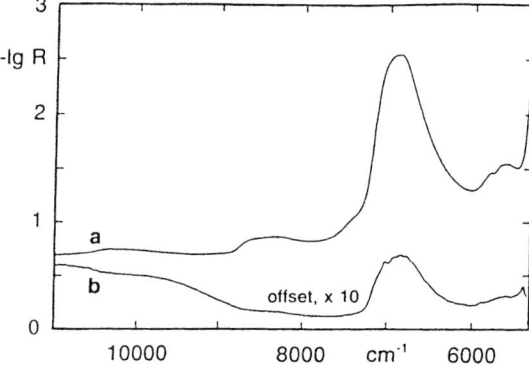

Fig. 3. Spectrum of inner lip measured by fibre optic accessory: *a* normal contact pressure, *b* difference spectrum obtained from normal *vs.* high contact-pressure

collection efficiency differs by a factor of about 50, which is also noticeable in the spectral signal/noise ratio, as shown in Fig. 2, where the logarithm of the quotient of two single-beam spectra obtained with the inner lip attached to the accessories is presented.

An important variance factor is contact pressure, which is demonstrated in Fig. 3. In particular, the low wavenumber region below $6000\,cm^{-1}$ shows noticeable variations due to absorption features of the outer skin layer, the stratum corneum (see Fig. 4). Spectral

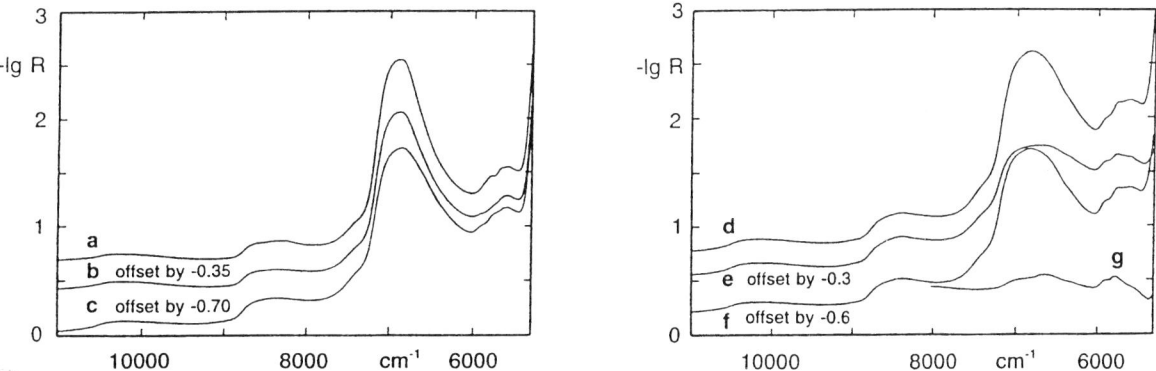

Fig. 4. Spectra of various skin tissues from a test person, measured by fibre optic accessory: *a* mucous tissue of inner lip, *b* outer lip, *c* thumb, *d* heel with a thick layer of stratum corneum, *e* as *d*, highly scattering, *f* thumb measured through nail, *g* hair

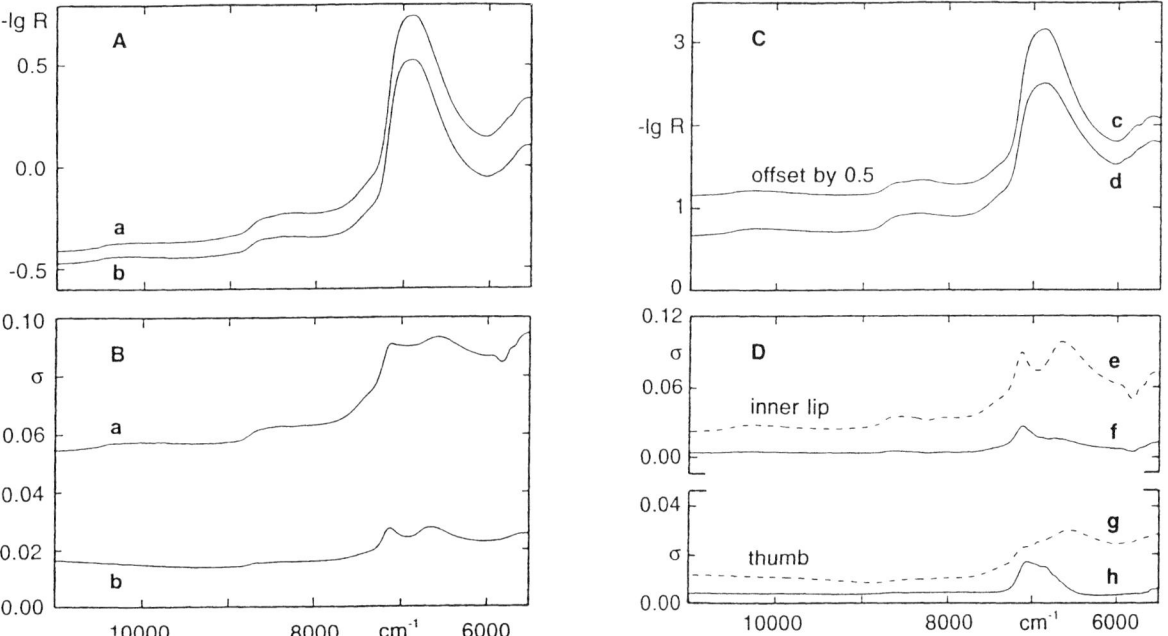

Fig. 5. Tissue spectra and their reproducibility. Average inner lip spectra (**A**) and their standard deviations (**B**) measured by DR accessory against a grey standard ($R \approx 0.1$): *a* from a multi-person experiment (133 persons, 390 spectra) and *b* from a single person experiment (219 spectra). Average lip (*c*) and thumb spectrum (*d*) measured by fibre optic accessory against a white standard (**C**) and standard deviations (**D**) with (*e*, *g*) and without repositioning of the fibre probe (*f*, *h*)

variations are much smaller for experiments with a single person than with several persons; the corresponding standard deviations are given in Fig. 5 A and B. In Fig. 5, C and D, the effect of repositioning is demonstrated for 20 measurements, compared to use of a fixed position of the measuring device. Another variance factor is skin temperature. The temperature-dependence of biotic fluids is well defined (see Fig. 6, A and B), but can easily be controlled. In an in-vivo experiment, however, other physiological factors such as microcirculation will influence the measurement (see Fig. 6C and D), and have to be investigated in the future.

Conclusions

The need for a rapid turn-around time for analytical laboratory results and the availability of technological

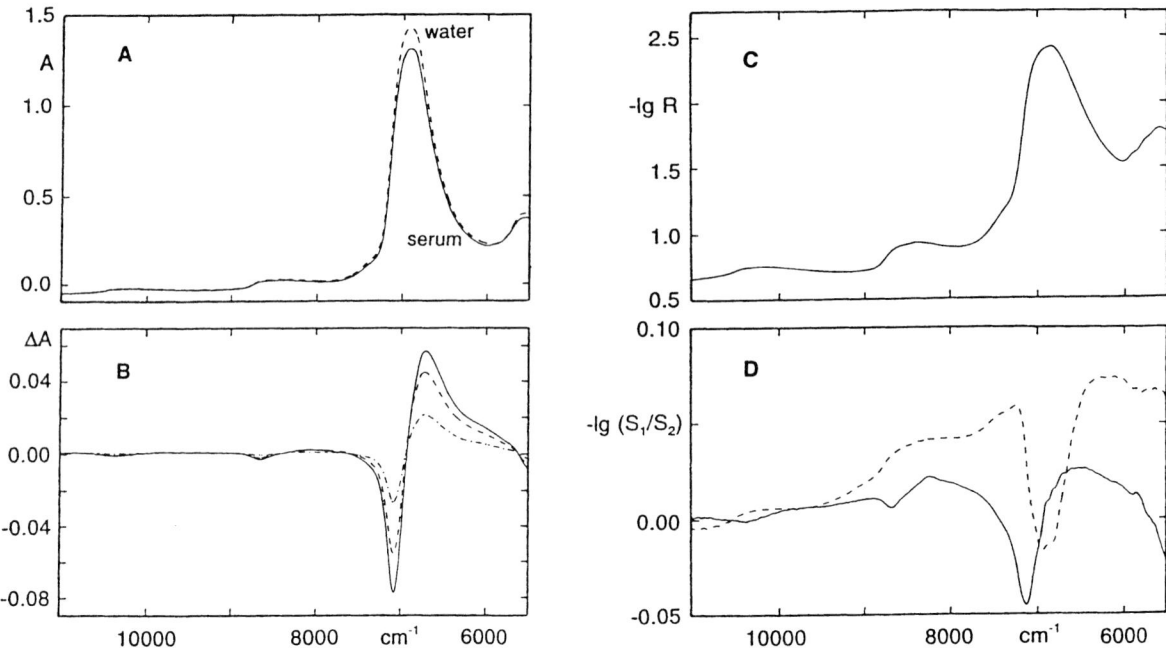

Fig. 6. Temperature-dependence of physiological samples: absorbance spectra of water and serum at 37 °C, cell path-length 1 mm (**A**); difference spectra for serum at various temperatures *vs.* 40 °C: $\Delta T = -15$ °C (—), $\Delta T = -10$ °C (---) and $\Delta T = -5$ °C (—·) (**B**); DR spectrum of thumb skin at normal body temperature, measured by fibre optic accessory (**C**), and difference spectra for cooled skin (—) and heated, hydrated skin (---) *vs.* normally conditioned skin (**D**)

innovations make it desirable to carry out many tests directly at the patient. Our findings are important for the further development of non-invasive systems and spectrometric *in vivo* sensors under development for diabetic patients and for direct monitoring in an intensive care unit or an operating room by using near infrared spectrometry of skin tissue.

Acknowledgements. Financial support by the Ministerium für Wissenschaft und Forschung des Landes Nordrhein-Westfalen, the Bundesminister für Bildung and Forschung and Boehringer Mannheim GmbH is gratefully acknowledged.

References

[1] H. M. Heise, R. Marbach, Th. Koschinsky, F. A. Gries, *Artificial Organs* **1994**, *18*, 439.
[2] R. Marbach, H. M. Heise, *Appl. Optics* **1995**, *34*, 610.

Mikrochim. Acta [Suppl.] 14, 433–435 (1997)
© Springer-Verlag 1997

Infrared Spectroscopy of Chemically Preserved Cervical Specimens on IR-transparent Disposable Filters

M. Cohenford[1,*], **P. S. Bhandare**[1], **B. Rigas**[2,3], and **K. Krishnan**[1]

[1] Bio-Rad, Digilab Division, 237 Putnam Ave, Cambridge, MA 02139, U.S.A.
[2] The New York Hospital-Cornell Medical Center, New York, NY 10021, U.S.A.
[3] Cornell University Medical College, New York, NY 10021, U.S.A.

Abstract. In this study, we focused on methods to improve the handling and the processing of cervical scrapings for FT-IR spectroscopy. More specifically, emphasis was placed on the chemical preservation of cervical scrapings, and on the acquisition of spectra with disposable IR-transparent filters. Studies showed that there were no demonstrable differences in the spectra of chemically preserved cervical scrapings, and cervical scrapings that were stored in liquid nitrogen. No differences in the spectra were also noted when BaF_2 windows were replaced with disposable IR-transparent filters. The clinical applications of FT-IR spectroscopy can be greatly enhanced with the aid of chemical preservatives, and disposable IR-transparent filters.

Key words: FT-IR spectroscopy, cervical scrapings, disposable IR-transparent filters, chemical preservative.

In 1991, Wong et al. [1] described the detection of pre-malignant and malignant cervical scrapings by FT-IR spectroscopy, and demonstrated spectral features that distinguished normal cervical scrapings from scrapings with dysplasia and cancer. Their study relied on features of the mid-IR region (3000–950 cm^{-1}) to discriminate between the samples. The spectra of normal samples exhibited a prominent peak at 1025 cm^{-1} characteristic of glycogen, and other less pronounced bands at 1047, 1082, 1155 and 1244 cm^{-1}. The spectra of specimens diagnosed with cancer exhibited significant changes in the intensity of the bands typically associated with glycogen at 1025 and 1047 cm^{-1}, and demonstrated a peak at 970 cm^{-1} which was absent in normal specimens. Samples with cancer also showed shifts in the peaks normally appearing at 1082, 1155 and 1244 cm^{-1}. The cervical specimens diagnosed cytologically as dysplasic exhibited spectra intermediate in appearance between those of spectra for normal and malignant specimens. On the basis of observations, Wong et al. concluded that FT-IR spectroscopy may provide a reliable and cost-effective alternative for screening cervical specimens.

One serious concern with studying biological specimens is maintaining the integrity of materials during transport and storage. Wong et al. relied on liquid nitrogen as a means of preserving cervical scrapings [1]. This technique, however, has several limitations that may preclude its use in a clinical setting. While the utilization of chemical preservatives may be an attractive option, care must be taken to ensure that the preservatives used do not interfere with or mask the spectra of cervical specimens. More importantly, if screening of cervical scrapings by FT-IR is to become routine, it would be ideal to acquire spectra on a disposable IR-transparent matrix, and not on BaF_2, KBr or AgCl discs as was reported earlier [1–3].

The aim of this study was to explore different sampling and preservation techniques, and to focus on disposable matrices that would facilitate the FT-IR spectroscopic screening of cervical specimens. This report compares the spectra obtained with an FT-IR spectrometer and a microscope, for chemically preserved cervical scrapings on 3 M disposable IR cards (filters) and on BaF_2 windows.

Materials and Methods

Cervical scrapings were collected by the standard brushing procedure. Exfoliated cells from each brush were harvested in separate

---

* To whom correspondence should be addressed

vials which contained either normal saline, or Preserv Cyt (Cytyc Corporation, Marlborough, MA). The cell suspensions in each vial were split into two equal portions. One portion of the cell suspension was spread on a microscope slide, fixed and stained with Papanicoleau stain, and examined by a pathologist. The other portion, which was intended for spectroscopy, was processed and stored as follows: Specimens in normal saline were centrifuged and the pellets were frozen in liquid nitrogen. Specimens in Preserv Cyt were kept for a maximum of two weeks at refrigeration temperature. Prior to spectroscopy, the specimens in Preserv Cyt were centrifuged and the pellets used for spectral acquisition. Cervical squamous carcinoma cell lines from ATCC were propagated in RPMI 16/40 supplemented with 2% fetal bovine serum. Following growth, the cultures were trypsinized, the cells were washed in saline and thereafter centrifuged, and the pellets were stored in liquid nitrogen or Preserv Cyt.

Spectroscopic Analysis

Mid-infrared spectroscopic studies were performed at room temperature on a Bio-Rad Digilab FTS 165 spectrometer equipped with a DTGS detector. Spectra were collected at a resolution of $4 \, cm^{-1}$, and 100 scans were co-added. Microspectrometric studies were conducted on BaF_2 windows, with a Bio-Rad UMA-500 microscope.

Results and Discussion

Figure 1 compares the spectra of a normal cervical scraping, and of an ATCC culture on BaF_2 windows and 3M filters (KC-0061-microporous polyethylene substrate, 3M Disposable Products Division St. Paul, MN). Spectra A and C were collected by use of BaF_2 windows, and B and D with 3M filters. A and B represent the spectra of a normal cervical scraping, and C and D the spectra of the M-180 ATCC culture, respectively. While no significant differences between the spectra obtained with the two matrices were observed, the normal cervical scrapings and the ATCC cultures exhibited distinct spectral features. These

spectral features confirmed the earlier observation that the spectra of normal cervical scrapings have a pronounced glycogen peak at $1025 \, cm^{-1}$, whereas cancer cells do not. A prominent peak at $972 \, cm^{-1}$ for cancer cells was also shown to be absent in the spectra of normal cervical scrapings. Although the spectra obtained by use of the filters were reproducible, the question still remained whether freezing by liquid nitrogen could be replaced by a simpler procedure. Consequently, attention was focused on chemical preservatives.

Most chemical preservatives, however, absorb strongly in the mid-IR. Therefore, it was imperative to seek only those chemicals that caused the least amount of interference in the spectra. The selection criteria also required that the preservative be compatible with cervical scrapings, and maintain the integrity of the specimens in a clinical setting. After screening a number of preservatives, Preserv Cyt was selected as a model system. Advantages of using Preserv Cyt included (a) knowledge that prior clinical studies supported the efficacy of the product for the storage and transport of cervical scrapings [4, 5], and (b) evaporation of the product on the filters left no detectable interference. Figure 2 shows the spectra of specimens which were stored in Preserv Cyt. The spectrum labeled A represents that of a normal cervical scraping, and the spectra in B and C were acquired for the M-180 and the SiHa ATTC cultures, respectively. Finally, our results showed (see Fig. 3) that there exists a greater diversity in the spectral patterns of cervical scrapings with dysplasia than reported earlier by Wong et al. [1]. These findings confirm previous observations [2].

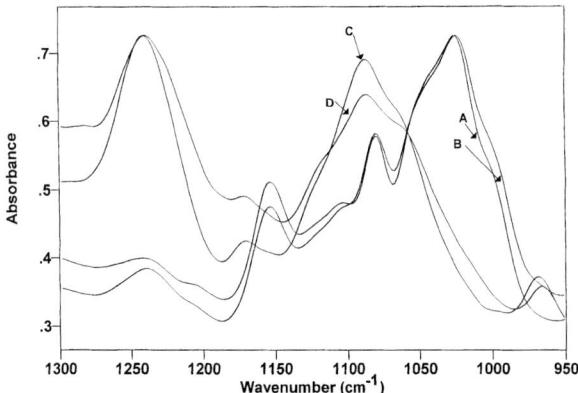

Fig. 1. Spectra of normal cervical scrapings and M-180 ATCC culture on barium fluoride windows and 3M filters

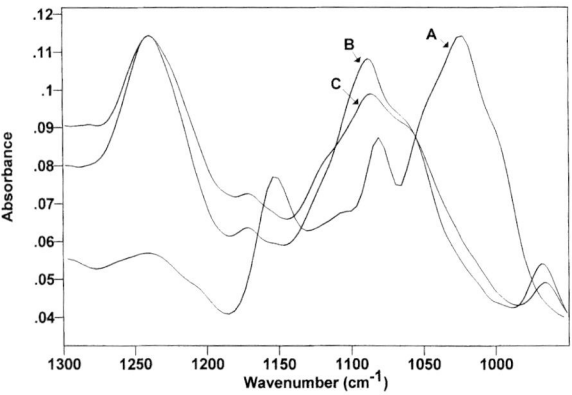

Fig. 2. Spectra of a normal cervical scraping and ATCC squamous carcinoma cell lines after storage in Preserv Cyt on 3M filters

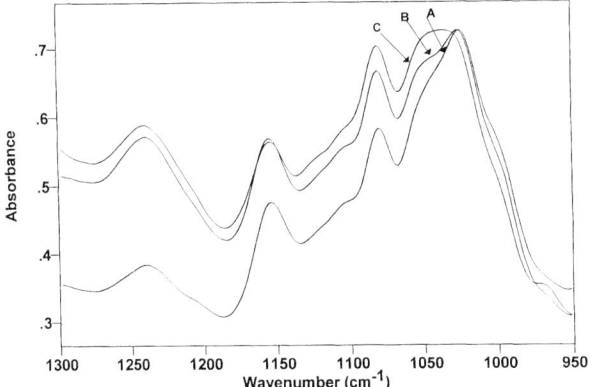

Fig. 3. Traces A, B, and C represent spectra of normal and dyplastic specimens with CIN I and CIN III diagnosis on 3M filters, respectively

Conclusion

As research in the clinical applications of FT-IR spectroscopy intensifies, a need can be expected for improved methods to facilitate the sample handling and processing of biological specimens for infrared spectroscopy. In this study, we explored the suitability of a chemical preservative for FT-IR spectroscopy. Using normal and abnormal cervical scrapings as model specimens, we demonstrated that their spectra can be acquired without any interferences from the preservative. In addition, when BaF_2 windows were replaced with disposable IR-transparent filters, no demonstrable differences in the spectra of the specimens were observed. The biological applications of FT-IR spectroscopy can be greatly enhanced with the aid of chemical preservatives, and disposable IR-transparent filters.

References

[1] P. T. T. Wong, R. K. Wong, T. A. Caputo, T. A. Godwin, B. Rigas, *Proc. Natl. Acad. Sci. U.S.A.* **1991**, *88*, 10988.

[2] B. J. Morris, C. Lee, B. N. Nightingale, E. Molodysky, L. J. Morris, R. Appio, S. Sternhell, M. Cardona, D. Mackerras, L. M. Irwig, *Gynecol. Onc.* **1995**, *56*, 245.

[3] Z. Ge, C. W. Brown, H. J. Kisner, *Appl. Spectrosc.* **1995**, *49*, 432.

[4] M. Bur, K. Knowles, P. Pekow, O. Corral, J. Donovan, *Acta Cytol.* **1995**, *39*, 631.

[5] S. L. Aponte-Cipriani, C. Teplitz, E. Rorat, A. Savino, A. J. Jacobs, *Acta Cytol.* **1995**, *39*, 623.

Mikrochim. Acta [Suppl.] 14, 437–439 (1997)

Comparison of Bacteriorhodopsin Conformational Changes with Uniaxially Elongated Aliphatic Polyamide Samples by Using FT-IR Difference Spectroscopy

W. B. Fischer[1,*], P. Wu[2], U. Hoffmann[2], K.-J. Eichhorn[3], H. W. Siesler[2], and K. Rothschild[4]

[1] Institut für Analytische Chemie, Technische Universität Dresden, D-01062 Dresden, Federal Republic of Germany
[2] Institut für Physikalische Chemie, Universität Essen, Schützenbahn 70, D-45117 Essen, Federal Republic of Germany
[3] Institut für Polymerforschung Dresden e. V., Hohe Strasse 6, D-01069 Dresden, Federal Republic of Germany
[4] Physics Department, Boston University, 590 Commonwealth Ave., Boston MA 02215, USA

Abstract. FT-IR difference spectra for the photoreactions of bacteriorhodopsin (bR) are compared to difference spectra of polyamide-6 and -11 obtained during uniaxial elongation. The purple membrane was hydrated with D_2O in order to identify bands above $3000 \, cm^{-1}$. In the bR photocycle (bR → K), the difference spectra in D_2O closely match the band contour found in the difference spectra of elongated polyamide-11. The data suggest that a structural change occurs in the backbone of the membrane protein bacteriorhodopsin, resulting in weakened and strengthened hydrogen bonds of the amide functionality of the backbone.

Key words: bacteriorhodopsin, polyamide-6, polyamide-11, FT-IR difference spectroscopy, protein backbone.

One of the best structurally characterized transmembrane ion transport systems is bacteriorhodopsin (bR) from *Halobacterium salinarium*. In the presence of light this protein catalyses proton transport from the cytoplasm to the exterior of the bacterial cell [1]. Upon absorption of light, the major event is a structural change of the retinylidene chromophore from an all-*trans* to a 13-*cis* configuration [2]. This event is accompanied by environmental and protonation changes of amino-acid residues, interstitial water and the protein backbone. The photocycle includes five major well-characterised intermediates (K, L, M, N and O). The role of several amino-acid residues (i.e. Asp-85, -96, -115, Tyr-185, Trp-86) in proton translocation is well established [3]. Recent studies detected water molecules which are active during the photocycle [4, 5]. The behaviour of the protein backbone throughout the photocycle has also been investigated [6–8]. In order to better understand the structural changes in the

backbone of membrane proteins, we started investigations on model compounds such as polyamides, under applied stress. We compare the spectral changes for the backbone structure of bR with those for uniaxially elongated films of polyamide-11 (PA-11) and polyamide-6 (PA-6) in the region above $3000 \, cm^{-1}$. The bR samples were hydrated with H_2O and D_2O. Especially during the early stage of the photocycle (bR → K) there is an overlap of positive and negative bands with those of the PA-11 sample. In both cases we find evidence for a weakening *and* strengthening of the NH-hydrogen bonds.

Experimental

bR

Purple membrane was isolated from *Halobacterium salinarium* by standard methods [9]. For a detailed description of the experimental set-up for recording and calculating the FT-IR difference spectra of bR see [5, 10].

Polyamides (PA)

Film samples of polyamide-6 (thickness ca. 10–11 μm) were cast from 1% (w/V) formic acid solution on a surface-roughened glass plate and dried in vacuum at 323 K overnight. Thin films of polyamide-11 have been obtained from injection-moulded test bars, by cutting slices 40 mm in length, 4 mm in width and 13 μm in thickness with a microtome. The samples were clamped between two polished Teflon plates, annealed for 15 min at 483 K and shock-cooled with ice-water. The PA-films were drawn at room temperature and to elongations of 60% strain. FT-IR spectra were recorded on a Bruker IFS 88 at 303 K, by the rapid scan technique. Ten interferograms were co-added to calculate the absorbance spectra (resolution $4 \, cm^{-1}$). Polarization spectra were taken with light polarized alternately parallel and perpendicular to the direction of drawing. Differ-

---

* To whom correspondence should be addressed

ence spectra were calculated by interactively subtracting the structural absorbance spectra [11] of the original film from spectra of the elongated film. For a detailed description of the stretching device see [11].

Results and Discussion

Earlier studies [12] indicate that a fairly low level of H/D exchange occurs in bR even when the sample is exposed for up to 48 h to D_2O. In the amide I region of the difference spectra of the bR → K, L and M transitions a set of bands appear which do not downshift but undergo a slight change in their intensities due to D_2O, which can be assigned to amide I vibrations (data not shown). Bands should appear in the amide A and B region (3400–3250 cm^{-1}), due to unexchanged protons of the backbone amide groups. Figure 1 shows difference spectra for the first three steps of the bR photocycle in the range 3800–2100 cm^{-1} in H_2O (up-

per panel) and D_2O (lower panel). Several bands above 3400 cm^{-1} were assigned to water [4, 5]. The difference spectrum of bR → K is dominated by a positive band at 3352 cm^{-1} along with several sub-bands. Two negative bands at 3305 and 3282 cm^{-1} can be found. In the bR → L difference spectrum a broad band pattern with two negative bands appears. This pattern is lost in the bR → M spectra in favour of several well-resolved bands. In D_2O almost the complete spectral contour of the OH-stretching region shifts down to 2750–2100 cm^{-1}. Around 3400–3250 cm^{-1} several bands are still present. Subtraction of one difference spectrum in H_2O from the difference spectrum of the same transition in D_2O results in an equivalent band pattern in the OH stretching region in the region of 2600–2100 cm^{-1} (in D_2O) except for the "frequency compression" in the latter region due to the deuterium atoms (data not shown). The bR → K difference spectrum in D_2O is characterized by an intense negative band at 3305 cm^{-1} flanked by weaker bands on both sides, at 3330 and 3282 cm^{-1} respectively. Two positive bands on the high frequency side (3374 and 3352 cm^{-1}) and a broader band at lower wavenumbers (ca. 3261 cm^{-1}) indicate a weakening and strengthening of the hydrogen bonds in the backbone at this early stage of the photocycle. For bR → L we find negative bands at 3357, 3322 and 3276 cm^{-1} and two positive bands at 3486 cm^{-1} (due to Trp-182 [13]) and 3249 cm^{-1}. In bR → M a positive/negative pattern 3370(+)/3352 (−) cm^{-1} emerges. The positive band may have a contribution from a slightly negative band at 3388 cm^{-1}. The band at 3322 cm^{-1} has lost its negative intensity.

Figure 1B shows the difference spectra of uniaxially elongated polyamide samples. We find a correspondence between bands in the difference spectra of bR → K and PA-11 (first two traces from top). All polyamide difference spectra (last three traces) share a common negative band flanked by positive bands. Whereas the PA-11 difference spectrum displays broad featureless positive bands of equal intensity at 3336 and 3261 cm^{-1}, the PA-6 difference spectrum (7% strain) has sharp positive bands at 3311 and 3303 cm^{-1} along with a weaker one at 3282 cm^{-1}. Upon increasing the strain for PA-6 to 10% the band at 3303 cm^{-1} loses intensity while at 3330 and probably 3347 cm^{-1} two weaker bands arise. In the original polyamide samples the polymer chains are randomly oriented (isotropic). The uniaxial extension of polyamides leads to (1) shear-induced highly ordered or crystalline regions with strong inter- and intramolecular interactions and

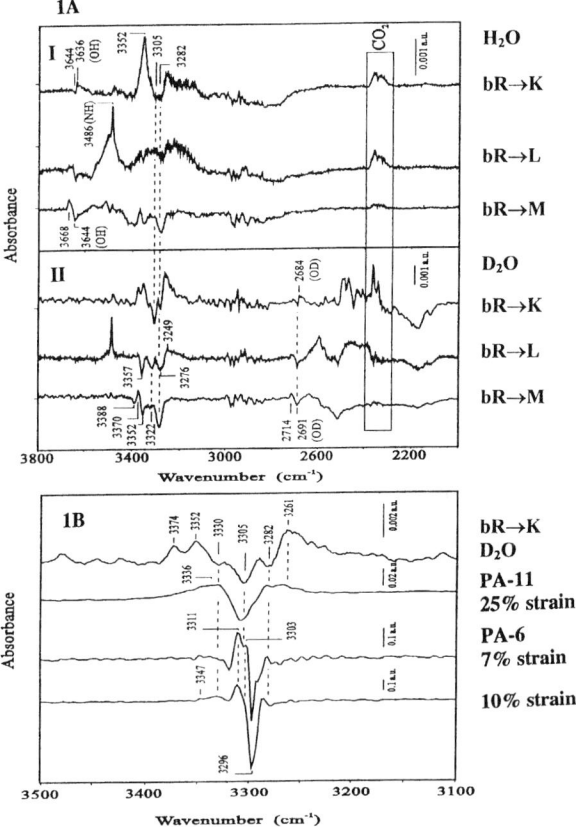

Fig. 1. A I: FT-IR difference spectra of bacteriorhodopsin in H_2O; bR → K at 80 K (upper trace), bR → L at 170 K (middle trace), bR → M at 250 K (lower trace); A II: difference spectra as for A I in D_2O. **B** Difference spectra of bR → K in D_2O (upper trace), elongated polyamide-11 (2nd trace from top) and polyamide-6 under increased strain (3rd and 4th trace)

low water absorption, which withstand further strain and H/D exchange; (2) less ordered or amorphous regions which allow the mobile macromolecules to undergo further strain and H/D exchange reactions [14,15]. The decrease of the ratio of CH_2 units to amide functionalities on going from PA-11 to -6 is related to increased structural order and thus a narrowing of the bands at 3311, 3303 and $3296\,cm^{-1}$ (Fig. 1B, last two traces) can be observed under strain.

Overall, our comparative FT-IR study of polyamides and bR in D_2O allows us to assign difference bands in the $3300\,cm^{-1}$ region to changes in hydrogen bonding of *unexchanged* backbone amide groups (e.g. groups buried in the interior of the membrane protein [12]). Furthermore, these results indicate that similar backbone structural changes may occur in polyamides upon stretching and bacteriorhodopsin upon photoexcitation, especially during the bR → K transition. One possibility is that chromophore isomerization leads to regions of compression and stretching in specific regions of the bR backbone. Recent studies employing site-directed isotope labelling (SDIL) support this picture and indicate that one region of structural changes is in helix F near Tyr185/Pro186 [8]. In order to understand the mechanical alteration during other stages of the bR photocycle, additional studies will be necessary.

Conclusions

Comparison with polyamides which undergo stress indicates that a similar process may occur in the backbone transmembrane α-helical segments of bR during the initial photoreaction. Mechanical stress due to chromophore isomerization could lead both to tilts in specific helices and to changes in hydrogen bond strength within helices and between helices. Further analysis of the structural changes of model compounds such as the polyamides should be a valuable tool not only for identifying characteristic band frequencies but also for analysing mechanical alterations of proteins in action.

Acknowledgements. We want to thank Prof. R. Salzer (TU of Dresden) for providing valuable suggestions for future experiments. Further we acknowledge the helpful discussion with Drs. S. Sonar, P. Rath and O. Bousché (BU) and their technical assistance. This work was supported by a grant from the NSF to KJR.

References

[1] W. Stoeckenius, R. A. Bogomoni., *Ann. Rev. Biochem.* **1982**, *51*, 587.

[2] S. O. Smith J. Lugtenburg, and R. A. Mathies, *J. Membr. Biol.* **1985**, *85*, 95.

[3] K. J. Rothschild, S. Sonar, in *Handbook of Organic Photochemistry and photobiology* (W. M. Horspool, P.-S. Song, eds.) CRC, London, 1995 pp. 1521–1544.

[4] A. Maeda, J. Sasabi, Y. Shichida, T. Yoshizawa, *Biochemistry* **1992**, *31*, 462.

[5] W. B. Fischer, S. Sonar, T. Marti, H. G. Khorana, K. J. Rothschild, *Biochemistry* **1994**, *33*, 12757.

[6] J. E. Draheim, J. Y. Cassim, *Biophys. J.* **1985**, *47* 497.

[7] K. J. Rothschild, T. Marti, S. Sonar, Y.-W. He, P. Rath, W. Fischer, H. G. Khorana, *J. Biol. Chem.* **1993**, *268*, 27046.

[8] C. F. C. Ludlam, S. Sonar, C.-P. Lee, M. Coleman, J. Herzfeld, U. L. Rajbhandary, K. J. Rothschild, *Biochemistry* **1995**, *34*, 2.

[9] B. Becher J. Y. Cassim, *Biophys. J.* **1976**, *16*, 1183.

[10] K. J. Rothschild, P. Roepe, J. Lugtenburg, J. A. Pardoen, *Biochemistry* **1984**, *23*, 6103.

[11] H. W. Siesler, *Macromol. Chem. Macromol. Symp.* **1992**, *53*, 89.

[12] T. N. Earnest, J. Herzfeld, K. J. Rothschild. *Biophys. J.* **1990**, *58*, 1539.

[13] Y. Yamazaki, J. Sasaki, M. Hatanaka, H. Kondori, A. Maeda, R. Needleman, T. Shinada, K. Yoshihara, L. S. Brown, J. K. Lanyi, *Biochemistry* **1995**, *34*, 577.

[14] R. Vieweg, A. Müller, *Kunststoffhandbuch Bd. VI 'Polyamide'* Hanser, München, 1966.

[15] W.B. Fischer, P. Wu, K.-J. Eichhorn, H. W. Siesler, Manuscript in preparation.

Mikrochim. Acta [Suppl.] 14, 441–443 (1997)
© Springer-Verlag 1997

Investigation of Acetylcholine Under the Influence of Different Electrostatic Fields by FT-IR Difference Spectroscopy

W. B. Fischer*, **I. Fischer,** and **R. Salzer**

Institute of Analytical Chemistry, Technical University Dresden, D-01062 Dresden, Federal Republic of Germany

Abstract. In this study we analyse structural changes of the neurotransmitter acetylcholine (ACh) under the influence of phenol and the electrolytes Li^+, Na^+, K^+, Cl^-, Br^- and I^-. By FT-IR difference spectroscopy, spectral changes, especially in the region of the $v_{C-N-C/C-O-C}$ around 1250–1240 cm^{-1}, are investigated. Our data show a shift of the band envelope in this region to lower wavenumber in the presence of phenol and I^- and to higher wavenumber in the presence of the cations. From studies with ethyl acetate we find shifts of the v_{C-O-C} to lower wavenumber under the influence of Li^+ ions. We also find major changes of the band envelope in the presence of phenol. No such changes are found if AChCl is compared to AChI.

Key words: acetylcholine, electrolytes, FT-IR difference spectroscopy.

Acetylcholine (ACh) [1] is a natural agonist activating the proteinogenic nicotinic acetylcholine receptor (n-AChR). The n-AChR is found at both neuromuscular junctions and neuronal synapses mediating the electrochemical signal transduction. In a 2:1 stoichiometry (ACh:n-AChR) [2] the agonist triggers the facilitated diffusion process of ion transport through the integral membrane protein. Amino-acids such as cysteine, tryptophan and tyrosine play a crucial role for the acetylcholine receptor binding process ([3 and references therein]).

FT-IR difference spectroscopy (for a review see [4]) is a valued tool for probing the conformational changes of integral membrane proteins such as n-AChR [5] and bacteriorhodopsin [6] during the ion transport. With this technique minimal structural changes caused by the activity of the protein can be monitored. In the study of proteins the difference spectra are covered by a enormous number of bands due to changes of individual side groups and the protein backbone. A major goal is to assign the individual bands. In this study we focused our attention on the structural changes of ACh caused by electrostatic interactions with solutes in aqueous solution. ACh is exposed to several electrolytes such as Cl^-, Br^-, I^-, Li^+, K^+ and electron-rich molecules like those found in the binding pocket of the receptor. Therefore we used phenol as a model substance for an aromatic π-system. We found changes in the shape of the bands in the 1800–1600, 1350–1150 and 980–920 cm^{-1} regions representing the $v_{C-N-C/C-O-C}$ and $v_{C-N^+-(CH_3)_3}$ of ACh.

Experimental

AChCl, -Br and -I (all 99% purity) were obtained from Sigma Chemical GmbH, Germany. All other compounds, from Sigma, Merck, Laborchemie Apolda or Berlin-Chemie, Germany, were of the highest degree of purity available. Ester spectra were recorded from the saturated aqueous solution. Additional experiments with tetramethylammonium hydroxide (TMAH) and ethyl acetate (EA) were accomplished in the same way. Calculation of the difference spectra was done by interactive subtraction of the individual absorbance spectra. Solution spectra (512 scans for each spectrum) and spectra with KBr tablets (128 scans for each spectrum) were recorded on either a Bruker IFS 88 with an MCT detector or a Nicolet 5PC with a DTGS-detector. For recording the solution spectra the ATR-technique was used.

Results and Discussion

Interaction of ACh (1 M) with the model electron-rich compound, phenol (0.5 M) in aqueous solution, causes a shift of the $v_{C-N-C/C-O-C}$ band to lower wavenumbers (see Fig. 1A). In this negative/positive feature two bands are involved ([1273 (−), 1257 (−)/1240(+), 1232(+)]. At around 950 cm^{-1} the $v_{C-N^+-(CH_3)_3}$ shows

---

* To whom correspondence should be addressed

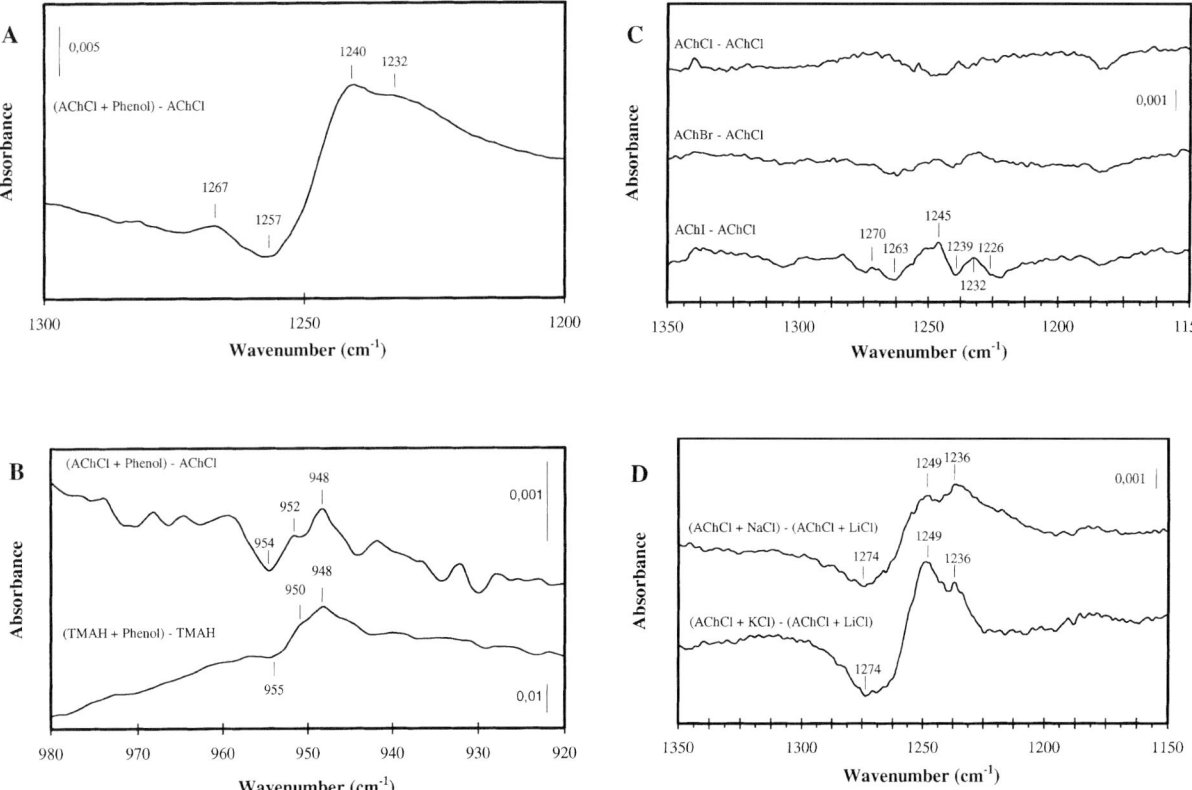

Fig. 1. A: Difference spectrum of ACh (1 M)/phenol (0.5 M) solution minus ACh solution (1 M) in the region 1300–1200 cm^{-1}. **B**: The same spectrum as in A. (upper trace), difference spectrum of tetramethylammonium hydroxide (TMAH)/phenol (same concentrations as for A); (lower trace) in the range 980–920 cm^{-1}. Difference spectra of **C**: AChCl, AChBr and AChI (each 1 M) minus ACh (1 M); of **D**: AChCl (1 M) in the presence of NaCl (1 M) and KCl (1 M) minus AChCl (1 M)/LiCl (1 M)

a shift from 954(−) to 952(+) and 948(+) cm^{-1} (Fig. 1B). At this frequency a pure phenol solution does not show any absorbance (data not shown). Also the $\nu_{C-N^+-CH_3}$ of TMAH shifts in the same direction when phenol is present (Fig. 1B, lower trace). The region of the ν_{C-N-C} (1500–1450 cm^{-1}) overlaps with the aromatic ring vibration of phenol. Here we find a negative/positive feature [1489 cm^{-1} (−)/1479 cm^{-1} (+)] without any indication of a second band being involved. If ACh is in contact with electron-rich and easy polarizable counterions, such as iodide in aqueous solution, changes in the difference spectra at around 1250 cm^{-1} are weak. Nevertheless for the difference spectra of AChI against AChCl we observe again the negative/positive "double contour" [1273(−), 1263(−)/1245(+), 1232 cm^{-1} (+)] (Fig. 1C, third trace from top). The shift of the frequencies is supported by theoretical calculations [7]. In order to assess the contribution of the ν_{C-O-C} to the changes of the overall band envelope in the difference spectra we

analysed ACh-solutions with NaCl, KCl and LiCl (see Fig. 1D). These difference spectra of AChCl + NaCl or KCl against (AChCl + LiCl) verify at least a positive "double contour" with bands at 1249 and 1236 cm^{-1}. Thus, in the presence of LiCl the broad band envelope at around 1250/1240 cm^{-1} from ACh gains intensity on the high frequency side from two band compounds. This is in contrast to the band shifts in the presence of negatively charged compounds. All difference spectra discussed so far have in common a shift of the $\nu_{C=O}$ to lower wavenumbers [8]. Our difference spectra give evidence for negligible alterations of the structure of ACh in the presence of the physiologically relevant ions Na$^+$, K$^+$, Cl$^-$ or Br$^-$.

Spectral changes are much better resolved in the difference spectra of the KBr-tablets [8]. We find different frequencies for the vibration of ACh but no change in the principle frequency shifts compared to those obtained in aqueous solution by the ATR-technique. This supports the NMR-data for ACh which indicate similar conformation of ACh on going from the "crystalline state to dilute solutions" (H_2O/salt = ca. 50) [9]. Since in KBr the water content is minimized, the spectral effects are due to the attraction of the opposite charges, which is not lowered by the dielectric properties of bulk water.

To verify the changes in frequencies of the bands due to the oxygen-rich part in ACh we studied ethylacetate (EA). Suprisingly, the addition of NaCl and LiCl to saturated solutions of EA resulted in a complete reversion of the shifts in the difference spectra (EA + LiCl) minus (EA + NaCl) compared to those in AChCl solutions [8]. Use of Li^+ in instead of Na^+ caused a shift from $1253\,cm^{-1}$ [($-$), broad] to $1230\,cm^{-1}$ (sharp). Also the $\nu_{C=O}$ shifted from $1745\,cm^{-1}$ ($-$) to $1735\,cm^{-1}$ ($+$). This we explain by a different structure formation of EA molecules in aqueous solution. Probably the alkyl tails assemble, thus allowing both oxygens to interact with the same positive charge. In conclusion, ACh is monomolecularly distributed in solution at the concentration studied here.

Our data show that electron-rich compounds such as phenols and iodide shift parts of the band envelope to lower frequencies; cations on the contrary shift the envelope to higher frequencies. The latter shift indicates preferentially interaction with the carbonyl function as the electron-rich part of the molecule (see [10] and references therein). The negative/positive "double contour", as mentioned above, may also have its origin in shift of the bands in opposite directions when the solutes are present. For example, phenol also interacts through hydrogen bonding with the negatively charged tail of ACh, while the π-system interacts with the ammonium function. At the same time the addition of cations is accompanied by the same concentration of anions. However, this correspondingly increases the anionic electrolyte concentration and should definitively affect the ammonium group of ACh.

Conclusion

The $+ - \pi$ interactions of ACh with aromatic systems result in intense spectral changes. Phenol, representing the aromatic amino-acids present in the ACh-binding site of the n-AChR, does largely affect the neurotransmitter. Point charges like I^- in its salt with ACh do not show an equivalent alteration or effect on ACh. This underlines the important character of this type of interaction in determining the binding of the neurotransmitter to the receptor.

Acknowledgement. We gratefully acknowledge the helpful discussions with G. Steiner, Ch. Kuhne and S. Hockeborn (TU Dresden).

References

[1] M. J. Michelson, E. V. Zeimal, *Acetylcholine: An Approach to the Molecular Mechanism of Action*, Pergamon Press, Oxford, 1973.
[2] J. A. Reynolds, A. Karlin, *Biochemistry* **1978**, *17*, 2035.
[3] J.-P. Changeux, *Fidia Res. Found. Neurosci. Award Lectures*, Vol. 4, Raven Press, New York, 1990, pp. 21–168.
[4] R. Salzer, *Wiss. Z. Karl-Marx-Univ. Leipzig, Math.-Naturwiss. R.* **1986**, *35*, 12.
[5] J. E. Baenziger, K. W. Miller, K. J. Rothschild, *Biochemistry* **1993**, *32*, 5448.
[6] W. B. Fischer, S. Sonar, T. Marti, H. G. Khorana, K. J. Rothschild, *Biochemistry* **1994**, *33*, 12757.
[7] W. B. Fischer, M. Mann, J. Fabian, R. Salzer, unpublished results.
[8] I. Fischer, *Diploma*, TU Dresden, Germany, 1995.
[9] K. M. Harmon, A. C. Akin, G. F. Avci, L. S. Nowos, M. B. Tierney, *J. Mol. Struct.* **1991**, *244*, 223.
[10] G. Mass, in: *Intermolecular Forces. An Introduction to Modern Methods and Results* (P. L. Huyskens, W. A. P. Luck, T. Zeegers-Huyskens, eds.) Springer, Berlin Heidelberg New York Tokyo, 1991, pp. 195–216.

Mikrochim. Acta [Suppl.] 14, 445–447 (1997)
© Springer-Verlag 1997

FT-IR Studies on T-cell Epitopic Deletion Peptides of Influenza Virus Hemagglutinin

Sándor Holly[1], Ilona Laczkó[2], Zsuzsa Majer[3], Gábor Tóth[4], and **Miklós Hollósi[3],***

[1] Central Research Institute of Chemistry, 1525 Budapest, P.O.B. 17, Hungary
[2] Institute of Biophysics, Biological Research Centre, 6701 Szeged, P.O.B. 521, Hungary
[3] Department of Organic Chemistry, Eötvös University, 1518 Budapest, P.O.B. 32, Hungary
[4] Department of Medical Chemistry, Szent-Györgyi Medical University, 6720 Szeged, Hungary

Abstract. Fourier-transform infrared spectroscopy was used to compare the conformational mobility of deletion analogues of epitopic peptides from three serotypes (HS1–HS3) of the HA1 sub-unit of the cleaved human influenza virus hemagglutinin. The peptides are present as multicomponent mixtures of conformers even in the structure-promoting solvent trifluoroethanol. Special attention is given to the question whether or not vibrational spectroscopy is capable of monitoring the conformation in solution of mid-size synthetic peptides which may have side-chain and/or counter-ion infrared contributions in the amide I region.

Key words: hemagglutinin, FT-IR spectroscopy, peptide conformation, side-chain absorption, T-cell epitope.

Our recent circular dichroism and Fourier-transform infrared (FT-IR) studies on peptides representing the C-terminal of the HA1 sub-unit of human influenza A virus have demonstrated changes of conformation in dependence on serotype-related differences of the amino-acid sequence [1]. Interestingly, the peptide showing the highest α-helix content has proved to be less potent to induce in vivo T-cell response [2].

According to X-ray crystallographic data epitopic peptides bind to the histocompatibility proteins in extended rather than helical conformation [3]. Since an amphipathic helical structure has been proposed for T-cell epitopic peptides [4], the ability of a switch from α-helix to extended conformation appears to be a prerequisite for good antigenic and immunogenic properties. This paper reports FT-IR spectroscopic studies performed on deletion analogues (Table 1) representing the C terminal epitopic region from different serotypes (HS1–HS3) of hemagglutinin.

Experimental

The peptides were synthesized by the solid-phase technique, utilizing Boc chemistry [1]. The crude peptides were purified by reversed-phase HPLC and characterized by HPLC, FAB mass spectrometry and FT-IR spectroscopy. According to the FT-IR measurements in KBr pellets, except for HS1a and HS1b all the peptides were present as trifluoroacetate (TFA) salts. Infrared measurements (at a resolution of $4\,cm^{-1}$) were performed on a Nicolet 170SX spectrometer. A KBr cell of 0.041 cm path-length was used; the concentration of the samples was $\sim 1\,mg/ml$. The FT-IR spectra were analysed by a normalized least-squares curve-fitting program using products of Gauss and Lorentz curves (Holly et al. personal communication). The position of the component bands is based on Fourier self-deconvolution [5].

Table 1. Amino acid sequence of deletion peptides from C-terminal region of the cleaved influenza virus A hemagglutinin (HA1) from different serotypes

Serotype and location	Abbreviation	Sequence
$HS1_{317-329}$	HS1	VTGLRNIPSIQSR
$HS1_{319-328}$	HS1a	GLRNIPSIQS
$HS1_{320-328}$	HS1b	LRNIPSIQS
$HS1_{320-329}$	HS1c	LRNIPSIQSR
$HS2_{317-329}$	HS2	ATGLRNVPQIESR
$HS2_{319-328}$	HS2a	GLRNVPQIES
$HS2_{320-328}$	HS2b	LRNVPQIES
$HS2_{320-329}$	HS2c	LRNVPQIESR
$HS3_{316-329}$	HS3	LATGMRNVPEKQTR
$HS3_{319-328}$	HS3a	GMRNVPEKQT
$HS3_{320-328}$	HS3b	MRNVPEKQT
$HS3_{320-329}$	HS3c	MRNVPEKQTR

---

* To whom correspondence should be addressed

Table 2. Characteristic amide I infrared frequencies of deletion peptides from the C-terminal region of cleaved hemagglutinin (HA1) serotypes HS1–HS3

Component bands (cm^{-1})

Peptides	Free (shielded) amides	Solvated amides	α-Helix or 3_{10} helix	Aperiodic	β-Turn	β-Sheet
HS1	1677 (26)	1666.5 (12)	1657.5 (18)		1642.5 (12)	1622 (5)
	1675		1660sh			1624
HS1a	1679 (18)		1662 (29)			1632 (27)
(no TFA)	1677		1663			1633
HS1b	1678 (32)		1661 (33)		1637 (6)	1621 (10)
(no TFA)	1673		1665		1634	1623
HS1c	1675.5 (48.5)		1659 (26)			1631.5 (7)
	1675		1662			1634
HS2	1680 (48)	1664 (22)	1651 (21)			1621 (7)
	1674	1662				1622
HS2a	1675 (40)		1658 (14)	1644 (11)		1622
	1676		1663	1647		1623
HS2b	1676 (39)		1659 (11)	1649 (16)		1619 (7)
	1676		1662			1622
HS2c	1676 (48)		1650.5 (11)	1645 (6)		1623 (12.5)
	1676		1660-65sh	1645-50sh		1620
HS3	1677 (22.5)		1659.5 (34)		1641 (3)	1624 (8.5)
HS3a	1676 (31)		1659 (14)		1642 (9)	1621 (9)
	1676		1658		1641	1621
HS3b	1676 (34)		1661 (22)			1626 (15)
	1676		1663			1626
HS3c	1676 (38.5)		1660 (20)			1626.5 (15)
	1676		1662sh			1625.5

In parenthesis: relative intensity in percentage. Only components with $\geqslant 5\%$ are listed. The corresponding maximum or shoulder in the FSD spectra is underlined. Weak high frequency component bands at $\sim 1690\,\mathrm{cm}^{-1}$ (shielded amide carbonyls or the satellite band of antiparallel β-sheets), $\sim 1700\,\mathrm{cm}^{-1}$ (C=O of H-bonded COOH), and $\sim 1715\,\mathrm{cm}^{-1}$ (C=O of free COOH) are not listed.

Results and Discussion

The frequencies and assignments of the component bands are listed in Table 2. The deletion peptides are present as mixtures of conformers even in trifluoroethanol (TFE) solution. The removal of the N- and C-terminal residues of the peptide HS1 leads to increased helicity, which is surprisingly high in HS1b. By contrast, truncation of HS2 and HS3 peptides decreases the relative amounts of 3_{10} or weakly H-bonded α-helical conformers. All peptides show some tendency to adopt extended or β-pleated sheet conformation. However, because of the side-chain contribution (band of the guanidinium ion at $1633\,\mathrm{cm}^{-1}$) of the Arg residues to this region, the relative amount of β-sheet is not easy to infer from the curve-fitted spectra containing one or two component bands between 1640 and $1600\,\mathrm{cm}^{-1}$. The $v_{as}\mathrm{CO}_2^-$ band of the trifluoroacetate counter-ion appears at $1673\,\mathrm{cm}^{-1}$ in overlap with the $v_{as}\mathrm{NCN}$ band of the guanidinium group. The presence of this band in the spectra of HS1a and HS1b containing no trifluoroacetate suggests that linear peptides have a uniform band in this amide I region. Thus, the component band at $\sim 1675\,\mathrm{cm}^{-1}$ in the IR spectra of epitopic peptides HS1–HS3c may also be indicative to amide carbonyls exposed to the solvent but not involved in strong oriented intramolecular or intermolecular H-bonds of ordered secondary structures. Our results suggest that an analysis of the high-frequency ($> 1660\,\mathrm{cm}^{-1}$) region of the FT-IR spectra of trifluoroacetate salts of small Arg-containing peptides requires special attention.

Acknowledgement. This work was supported by grants OTKA 1995-T017432 (to M. H.) and 1995-T016516 (to IL.).

References

[1] S. Holly, Zs. Majer, G. K. Tóth, G. Váradi, É. Rajnavölgyi, I. Laczkó, M. Hollósi, *Biochem Biophys. Res. Comm.* **1993**, *193*, 1247.

[2] É. Rajnavölgyi, *Immunology Today* **1992**, *13*, A17.

[3] D. R. Madden, J. C. Gorga, J. L. Strominger, D. C. Wiley, *Nature* **1991**, *353*, 321.

[4] C. DeLisi, J. A. Berzofsky, *Proc. Natl. Acad. Sci. USA* **1985**, *82*, 7048.

[5] H. H. Mantsch, D. J. Moffat, H. G. Casal, *J. Mol. Structure* **1988**, *173*, 285.

Mikrochim. Acta [Suppl.] 14, 449–450 (1997)

Monitoring the Stepwise Hydration of Phospholipids in Films by FT-IR Spectroscopy

Carsten Selle*, Walter Pohle, and **Hartmut Fritzsche**

Institute of Molecular Biology, Department of Biophysical Chemistry, Friedrich-Schiller University Jena, Winzerlaer Strasse 10, D-07745 Jena, Federal Republic of Germany

Abstract. In a comparative Fourier-transform infrared spectroscopic study, dialkyl- (dihexadecyl) and diacyl-(diapalmitoyl and dioleoyl) glycerophosphatidylethanolamines and -cholines have been investigated as cast films at room temperature in terms of the ambient relative humidity.

Adsorption isotherms obtained reveal that phosphatidylcholines generally imbibed much larger amounts of water than the analogous phosphatidylethanolamines. Furthermore, the presence of unsaturated chains led to a substantially higher water sorption than found for the analogues with saturated chains, i.e. dioleoylglycerophosphatidylcholine exhibited the maximal water uptake. Hydration effects can also be followed by means of the spectral behaviour of phospholipid bands assignable to sensitive groups. This allows information to be obtained about the order in which the relevant lipid binding sites are affected by hydration; in dipalmitoylphosphatidylcholine, for instance, phosphate groups bind water molecules rather than carbonyl groups.

Key words: stepwise hydration, FT-IR spectroscopy, phosphatidylcholine, phosphatidylethanolamine.

Phospholipids (PLs) are important biological molecules, not only because they provide a shape-determining frame for the membranes of cells and cell organelles, but also since they are thought to be involved in the regulation of membrane functioning.

The presence of water is a major prerequisite for realization of their intrinsic amphiphilicity which results in the particular supermolecular structural organization characterizing this class of compounds. Hence, the process of PL hydration may be considered biologically very important.

Fourier-transform infrared (FT-IR) spectroscopy has been proven to be a very useful tool in the field of lipid and membrane research, as documented in a series of recent articles (for instance [1, 2]). However, this method has been applied so far mainly to samples that are either anhydrous or dispersed in an excess of water [1, 2].

In an attempt to develop an algorithm for studying the gradual hydration of PLs by FT-IR spectroscopy, we have investigated a set of relevant compounds chosen such as to involve systematic structural variations in the headgroups as well as in the apolar tail regions. Thus, six different substances were selected, namely the dioleoyl, dipalmitoyl and dihexadecyl glycerophosphatidylcholines and -ethanolamines (DOPC, DPPC, DHPC, DOPE, DPPE, DHPE). The work was undertaken to probe the potentialities of FT-IR spectroscopy in following the effects of hydration on structural aspects of PLs.

Experimental

All PLs were commercial products, purchased from Sigma Chemicals, Munich (oleoyl and hexadecyl compounds), and Bachem Biochemica, Heidelberg (palmitoyl compounds). Films of the PLs were prepared by spreading their chloroform solutions (10 mg of lipid per ml) *in situ* over an IT-transparent ZnSe window. Spontaneous orientation of the multi-bilayers was tried to be achieved by unidirectional stroking of the lipid gels until they were dry. The films were then installed in special IR cells. The ambient relative humidities (RHs) were controlled by saturated salt solutions placed inside the cells and, in this way, varied within wide limits (0–98%). Starting from high RH, the films were made to undergo complete dehydration–rehydration cycles between 5 and 10 different RH values.

Infrared spectra were recorded for each RH step by a Bruker IFS-66 FTIR spectrometer, equipped with a shuttle device, under standard conditions and at a temperature of about 28 °C. Thirty-two scans were performed at a resolution of $2\,cm^{-1}$, with a zero-filling factor of 2. Data processing was done with the OPUS software package (Bruker).

Results and Discussion

Phospholipid hydration was monitored by FT-IR spectroscopy in two ways: (*i*) directly, by means of the water absorption bands and, (*ii*), indirectly, through the PL bands assignable to the molecular groups in-

---

* To whom correspondence should be addressed

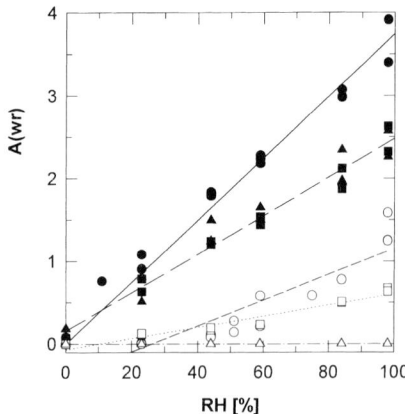

Fig. 1. Depiction of the water-sorptivity parameter A(wr) *vs.* RH, for the different PLs; full symbols are for PCs, open symbols are for the corresponding PEs; circles, squares and triangles are for the oleoyl, palmitoyl and hexadecyl analogues, respectively. The straight lines are the linear regressions for each of the PLs (one straight line for DPPC and DHPC)

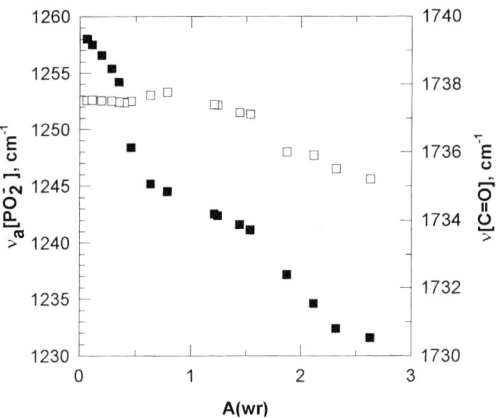

Fig. 2. Wavenumber displacements for the C=O (open-symbols) and antisymmetric PO_2^- stretching-vibration bands of DPPC as a function of the relative humidity surrounding the film

volved in (and, hence responding to) interaction with water.

(*i*) The integral absorbance of the water-OH stretching-vibration band near $3400\,\mathrm{cm}^{-1}$ relative the integral absorbance of the peak ensemble formed by the bands due to the lipid CH_2 and CH_3 stretching vibrations as reference standard, A(wr), has been taken as a quantitative measure for the water content of the lipid films. As shown in Fig. 1, water adsorption isotherms determined in this way reveal that the lipids imbibed water in the following fairly significant order

DOPC > DPCC, DHPC ≫ DOPE > DPPE > DHPE

This order allows us to draw some general conclusions. First, the water sorption by PCs surpasses that by PEs in each case and to a dramatic extent. This finding confirms the bulk of previous data concerning the capacity of lecithins and cephalins to take up water (see [3] for instance) and seems to be related to the different geometrical arrangements in the headgroup regions of PCs and PEs, where only in the latter case is the supermolecular structural organization determined by a stable network of hydrogen bonds and/or salt bridges between the PO_2^- and NH_3^+ moieties [4]. Note also the apparently complete inability of DHPE to accept water.

Secondly, the ability of oleoyl compounds to imbibe water is clearly larger than that of the saturated-chain analogues. This result may be interpreted to mean that the oleoyl analogues adopt a more "hydrophilic", *i.e* a fluid phase (*vs.* L_β phase for the latter) and/or as a consequence of the increased cross-section area of the

apolar tail of the oleoyl PLs probably resulting from a bent at the rigid C=C double bond of the hydrocarbon chain.

(*ii*) As expected, the effects of hydration are reflected only by IR bands attributable to the polar headgroups of the PLs studied. According to the difference in water sorption capacity described above, these effects are much more pronounced for PCs than PEs, *i.e* the overall frequency shifts induced by hydration are dramatically greater for the PCs.

As the plots in Fig. 2 reveal, the FT-IR spectroscopic data are in principle able to provide an order of PL sites for priority of water binding; the onsets of the significant wavenumber shifts demonstrate that, for instance in DPPC, water molecules are attached to phosphate groups rather than to carbonyl groups. In DOPC, on the other hand, both these binding sites are accessible to water molecules at nearly the same water activities in the sample (data not shown).

Acknowledgements. This work was generously supported by the Ministry for Science of Thuringia (TMWFK).

References

[1] H. H. Mantsch, R. N. McElhaney, *Chem. Phys. Lipids* **1991**, *57*, 213.
[2] R. N. A. H. Lewis, R. N. McElhaney, W. Pohle, H. H. Mantsch, *Biophys. J.* **1994**, *67*, 2367.
[3] G. L. Jendrasiak, J. C. Mendible, *Biochim. Biophys. Acta* **1976**, *424*, 149.
[4] H. Hauser, I. Pascher, R. H. Pearson, S. Sundell, *Biochim. Biophys. Acta* **1981**, *650*, 21.

Mikrochim. Acta [Suppl.] 14, 451–453 (1997)

FTIR/NIR Assessment of Ischemic Damage in the Rat Heart

Michael G. Sowa, James R. Mansfield, Michael Jackson, John C. Docherty, Roxanne Deslauriers, and **Henry H. Mantsch***

Institute for Biodiagnostics, National Research Council Canada, 435 Ellice Ave., Winnipeg, Manitoba, R3B 1Y6, Canada

Abstract. The utility of combined near- and mid-infrared spectroscopic measurements to monitor short-term changes in the heart subjected to ischemia-reperfusion and the ability to assess long-term modifications to the extracellular matrix following an ischemic insult to the heart is presented.

Key words: infrared spectroscopy, ischemia-reperfusion injury, myocardial infarction.

Disruption of the blood supply to the heart can arise when the coronary blood vessels become obstructed ("heart attack"). Prolonged ischemia can cause cell death, leading to areas of necrosis (infarct). Under conditions of limited ischemia, cellular damage may be exacerbated upon re-establishment of blood flow to the area. Reperfusion can trigger a complex series of events, including the involvement of reactive oxygen species in lipid peroxidation, which disrupts cellular membrane structures [1]. The heart can often survive the short term damage associated with a mild ischemic insult. However, over several weeks, scar tissue forms in the infarcted areas. The fibrous scar tissue, which is predominantly type I collagen, can be detrimental to the mechanical and electrical function of the heart [2]. Thus, both short term ischemia-reperfusion injury as well as long term modifications to the extracellular matrix must be considered when assessing overall damage to the heart following an ischemic insult.

The primary purpose of this study was to determine whether FTIR/NIR spectroscopy can detect changes that occur in heart tissue during ischemia and reperfusion, as well as the longer term changes to the extracellular matrix. Correlating spectral changes to the degree of tissue damage would provide a powerful means of assessing the efficacy of therapeutic interventions that

could be used to minimize the damage associated with heart attacks or with surgical procedures where the heart is temporarily deprived of oxygen.

Experimental

Perfusion Procedure

Isolated rat hearts were perfused via the aorta at a constant pressure of 80 mm Hg. Mechanical function was monitored by means of a water filled latex ballon inserted into the left ventricle and attached to a pressure transducer. Once stablized, the hearts were subjected to a 25 minute period of no-flow global ischemia, followed by 60 minutes of reperfusion. Mechanical performance and NIR spectra were monitored continuously throughout the experiment.

Myocardial Infarction

Myocardial infarction was produced in male Sprague-Dawley rats by occlusion of the left coronary artery. Under isofluorane anesthesia and positive pressure (95% O_2 and 5% CO_2) ventilation, the thorax was opened by cutting the third and fourth ribs and the heart was extruded through the intercostal space. The left coronary artery was ligated about 2–3 mm from the origin with a suture, the heart was repositioned in the chest and the wound was closed using a purse-string suture. Sham-operated animals were treated similarly, with the exception that the coronary suture was not tied.

Infrared Spectroscopy

Reflectance FT-NIR spectra of nominally $16 \, cm^{-1}$ resolution consisting of 128 co-added scans were collected in the $4000–10000 \, cm^{-1}$ ($2.5–1 \, \mu m$) range using a Bio-Rad NIR FTS-40 (Biorad, Cambridge, MA) spectrometer with a high intensity NIR tungsten-halogen source and a liquid nitrogen cooled InSb detector coupled through a bifurcated all-silica fiber optic bundle (CeramOptic, Enfield, CT). Interferograms were acquired continuously during the experimental protocol. Transmission and ATR FT-IR spectra of small tissue samples ($<1 \, mm^3$) were acquired with a Bio-Rad FTS-40 spectrometer equipped with a MCT detector. For each sample, 256 interferograms were co-added and Fourier transformed to generate a spectrum with a nominal resolution of $2 \, cm^{-1}$.

---

* To whom correspondence should be addressed

Results and Discussion

Figure 1 displays NIR reflectance spectra acquired during cardiac ischemia and reperfusion of an isolated heart. Spectra acquired prior to ischemia on healthy, beating hearts are highly reproducible indicating that the NIR reflectance spectra are either relatively unaffected by heart motion or that motional effects are adequately averaged out over the 128 scans (approximate 90 sec. acquisition time). Within the first 6 minutes of ischemia, the spectra change dramatically, especially in the region between 5300–7500 cm^{-1}. Changes in the spectra occur more slowly at the later stages of ischemia until reperfusion is initiated (25 minutes in Fig. 1). Upon reperfusion there is an abrupt change in the spectra, supporting the hypothesis that many of the effects of reperfusion on heart tissue are almost instantaneous. As reperfusion proceeds, tissue re-oxygenates and myocardial tissue spectra begin to approach the spectra aquired prior to ischemia, indicating almost full recovery of the heart following reperfusion.

Figure 2 displays a sequence of inverted second derivative NIR reflectance spectra acquired at 3 minute intervals on a beating heart. The spectral changes observed during ischemia and reperfusion are reproducible between hearts suggesting that NIR reflectance spectroscopy can detect changes in the heart tissue. In the second derivative representation, further changes are apparent. For instance, changes coincident with the contracture phase of ischemia (evident as an increase in resting pressure within the left ventricle) are clearly apparent in the spectra. These early findings suggest that the NIR reflectance measurements correlate with

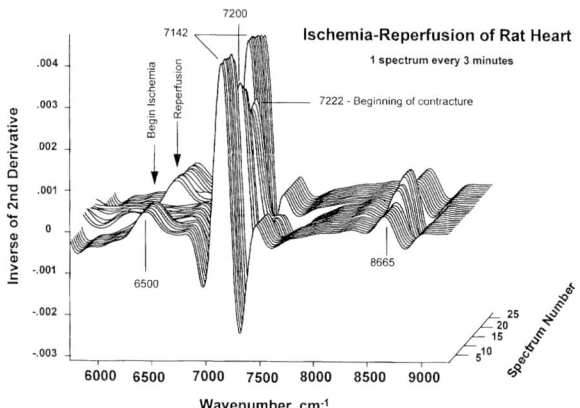

Fig. 2. Inverted second derivative of NIR reflectance spectra acquired at three minute intervals from a rat heart undergoing ischemia-reperfusion

certain mechanical and electrical parameters related to the short-term physiological changes that occur during ischemia and reperfusion.

Infrared spectroscopy can also be used to characterize the longer term modifications that occur as a result of myocardial infarction. FT-IR spectra of control and infarcted rat ventricle are shown in Fig. 3. The dominant absorptions in both tissues arise from proteins (1500–1700 cm^{-1}). In tissue from sham-operated rats, a prominent pair of absorptions is seen at 1084 and 1238 cm^{-1}, which arise from tissue DNA and RNA. In infarcted tissue, the characteristic protein bands are significantly different in shape relative to those in control tissue, suggesting protein modification. The region containing the absorptions from DNA and RNA is also significantly altered in infarcted tissue,

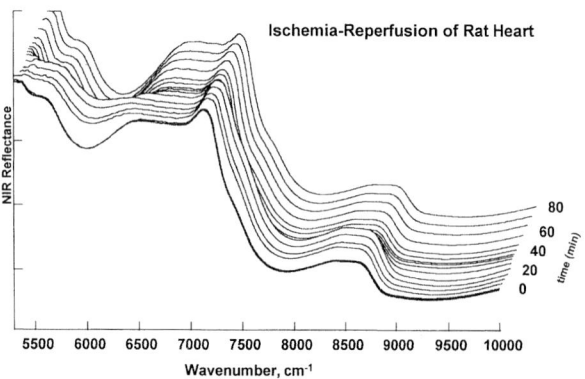

Fig. 1. NIR reflectance spectra of an isolated rat heart undergoing ischemia followed by reperfusion. Pre-ischemia (overlaid at t = 0 min.), ischemia (t = 6 to t = 25 min.), reperfusion (t = 28 to 80 min.)

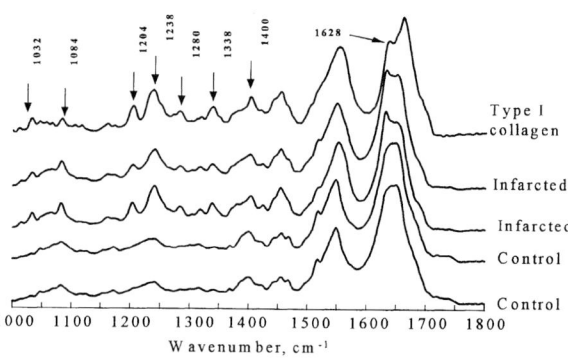

Fig. 3. Infrared spectra of type I collagen, infarcted left ventricle and sham-operated control ventricle after Fourier self-deconvolution using a band narrowing factor of k = 1.7 and a half band width of 13.5 cm^{-1}

which may suggest modifications of the nucleic acids. However, comparison of spectra from infarcted tissue and rat type I collagen show a high degree of similarity in this region, suggesting the differences represent deposition of type I collagen in infarcted tissue. Thus infrared spectroscopy can detect changes in the extracellular matrix following a myocardial infarction and potentially assess (quantify) the long term damage associated with an ischemic insult to the heart.

References

[1] K. Ytrehus, A. C. Hegstad, *Acta Physiol Scand [Suppl.]* **1991,** *599,* 81.
[2] M. A. Pfeffer, E. Braunwald, *Circulation* **1990,** *81,* 1161.

Mikrochim. Acta [Suppl.] 14, 455–457 (1997)
© Springer-Verlag 1997

Structural Characterization of Supported Planar Bilayers by Using Infrared Spectroscopy and Atomic Force Microscopy

Susan M. Stephens and **Richard A. Dluhy***

Department of Chemistry, University of Georgia, Athens, GA 30602-2556, USA

Abstract. Model membrane bilayers have been investigated by using a combination of infrared spectroscopy (IR) and atomic force microscopy (AFM). Oriented single planar bilayer membranes were prepared by direct fusion of 1,2-dipalmitoyl-*sn*-glycero-3-phosphocholine (DPPC) vesicles onto an alkane-derivatized substrate. These substrates consisted of either a self-assembled monolayer of 1-octadecanethiol (ODT) on a vapor-deposited Au-on-mica substrate, or a Langmuir–Blodgett monolayer of perdeuterated stearic acid (d-SA) transferred onto a Ge ATR element. ATR-IR spectroscopy of the DPCC/d-SA bilayer system showed that both membrane components exist in an ordered conformational state with acyl chains oriented nearly perpendicular to the bilayer plane. The absence of a crystal-field splitting in the CH_2 scissoring mode indicates a hexagonal type packing of the phospholipid monolayer. AFM images of the DPPC/ODT model system also reveal molecularly resolved phospholipid head-groups; the unit cell areas obtained from these images correspond to molecular areas which are commensurate with those obtained by X-ray diffraction from analogous phospholipid single crystals. Our results demonstrate that the combination of IR spectroscopy and scanning probe imaging is a powerful tool for structural investigation of phospholipid-supported planar bilayers.

Key words: supported planar bilayers, IR spectroscopy, atomic force microscopy, membrane biophysics.

Our laboratory has been developing methods for both in situ and ex situ study of phospholipids incorporated into supported planar bilayers for the study of model biological membranes [1–4]. For the in situ study of phospholipid monolayers at the air–water interface, we have developed an external reflectance [1] and, more recently, an internal reflectance [2] spectroscopic method which are useful for the study of Langmuir monolayers at the air–water interface. For the ex situ study of phospholipid-supported planar bilayers, our current work demonstrates that a combination of IR

spectroscopy and atomic force microscopy (AFM) provides detailed information concerning the conformation and distribution of the bilayer components.

Experimental

The surface chemistry methods used in this study have been previously described in detail [3, 4]. Monocrystalline germanium (Ge) substrates were prepared for attenuated total reflectance infrared (ATR-IR) analysis, and Langmuir–Blodgett monolayers were transferred onto these substrates, as previously described [3, 4]. Supported planar lipid bilayers were prepared by the vesicle fusion technique developed by Kalb et al. [5].

Infrared spectra of the bilayers transferred onto Ge substrates were acquired with a Digilab FTS-40 Fourier-transform infrared spectrometer equipped with $4 \times$ ATR beam condensor and a narrow-band liquid-N_2 cooled HgCdTe detector [3, 4].

Results

Ex situ IR Spectroscopy of DPPC-supported Planar Bilayers on Ge

The model membrane constructed for this study consisted of an inner monolayer of perdeuterated stearic acid deposited onto the Ge substrate by the Langmuir–Blodgett technique. A monolayer of DPPC was then deposited onto the d-SA/Ge substrate by vesicle fusion to form the outer monolayer of the bilayer. By depositing the DPPC/d-SA bilayer onto a Ge substrate, it is then possible to analyze the model system by ATR-IR spectroscopy, with the Ge substrate as the ATR element. A spectrum of the DPPC/d-SA bilayer is shown in Fig. 1. The observed band peak frequencies and their corresponding mode assignments correspond to those that have been previously reported [3].

---

* To whom correspondence should be addressed

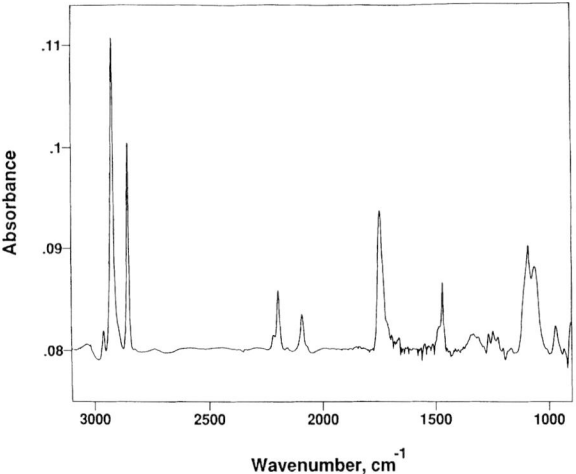

Fig. 1. Attenuated total reflectance-IR spectrum (*p*-polarized) of a DPPC/d-SA bilayer on a Ge substrate. Acyl chain peredeuterated stearic acid (d-SA) was the bottom monolayer transferred by the Langmuir–Blodgett method onto the Ge at a surface pressure of 20 mN/m. An outer monolayer of DPPC was formed by the vesicle fusion technique. The conformational state of the DPPC hydrocarbon chains was monitored by using the C–H stretching region of the spectrum (3000–2800 cm^{-1}) and the CH$_2$ scissoring vibration at ~ 1468 cm^{-1}. The d-SA component of the bilayer was monitored by using the C–D stretching region between 2300 and 2000 cm^{-1}

To characterize the orientational distribution of the hydrocarbon chains within the DPPC monolayer, the dichroic ratio (R) of the measured peak intensities of the symmetric CH$_2$ stretching vibration was used. In the same vein, the symmetric CD$_2$ stretching vibration can also be used in the dichroic analysis of d-SA. The experimental values of R obtained from this investigation revealed that the acyl chains of both DPPC and d-SA have a near unity order-parameter and are therefore oriented normal to the substrate surface, confirming previous results with phosphoglycerol lipids [3].

Another feature of the IR spectrum that can be utilized in the characterization of acyl chain packing is the presence of crystal-field splitting of the CH$_2$ scissoring band. The CH$_2$ scissoring mode will appear as a singlet when the hydrocarobon chains are oriented in a hexagonal or triclinic crystal lattice, because the interchain interactions within the crystal lattice are equivalent. In an orthorhombic structure, the interaction between non-equivalent acyl chains occurs, causing the scissoring band to split into two components. Therefore, IR spectroscopy provides a useful tool in distinguishing between lipid packing modes. The CH$_2$ scissoring band of the DPPC monolayer can be seen in Fig. 1 as a single peak at 1467.7 cm^{-1}. For hexagonal

packing, the scissoring band should be centered around 1468 cm^{-1}, which is consistent with the experimental frequency observed in our spectrum.

AFM Imaging of DPPC-supported Bilayers on Au

The supported planar bilayer model system consisting of DPPC/ODT on an Au/mica substrate was used for AFM imaging. In this configuration, the phospholipid head-groups are oriented outwards toward the probe tip and are probably components which are observed in the acquired images.

Our images of the DPPC/ODT bilayer routinely revealed well-ordered, hexagonal type packing of the phospholipid molecules. These features were regularly observed for image sizes up to 100 × 100 nm, so there appears to be a fairly uniform, homogeneous coverage of the substrate by the bilayer vesicle fusion method, indicating a lack of domain formation. The hexagonal character of the phospholipid packing is also seen in these images. This information corroborates well the IR data indicating hexagonal packing of the hydrocarbon chains within the DPPC monolayer. The surface unit cell has the dimensions $a = 5.6 \pm 0.2$ Å, $b = 8.3 \pm 0.4$ Å, $\theta = 62 \pm 6°$ corresponding to a molecular area of 41 ± 2 Å^2. The unit cell area is in excellent agreement with the molecular area of 38.9 Å^2 previously reported by Pearson and Pascher for the crystal structure of the analogous phospholipid DMPC [3].

Conclusion

We have applied the combination of infrared spectroscopy and atomic force microscopy to the study of model membrane bilayers containing DPPC as the outer monolayer. This approach of using the two techniques in tandem provides complementary information allowing us to gain greater insight into the conformation, orientation, and morphology of the bilayer components. We have shown that it is possible to form well-ordered, homogeneous supported single planar bilayers ex situ by fusion of lipid vesicles onto alkane-derivatized surfaces. Using AFM, we have observed individual DPPC head-groups in these supported planar bilayer membrane models. Therefore, the combination of IR spectroscopy and atomic force microscopy potentially provides a powerful tool for probing the structure of membranes, revealing information that would be unobtainable by either method alone.

Acknowledgement. This work was supported by the U.S. Public Health Service through National Institutes of Health grant GM40117 (RAD).

References

[1] R. A. Dluhy, *J. Phys. Chem.* **1986**, *90*, 1373.

[2] P. H. Axelsen, W. D. Braddock, H. L. Brockman, C. M. Jones, R. A. Dluhy, B. S. Kaufman, F. J. Puga, *Appl. Spectrosc.* **1995**, *49*, 526.

[3] B. W. Gregory, R. A. Dluhy, L. A. Bottomley, *J. Phys. Chem.* **1994**, *98*, 1010.

[4] F. R. Rana, S. Widayati, B. W. Gregory, R. A. Dluhy, *Appl. Spectrosc.* **1994**, *48*, 1196.

[5] E. Kalb, S. Frey, L. K. Tamm, *Biochim. Biophys. Acta* **1992**, *1103*, 307.

Mikrochim. Acta [Suppl.] 14, 459–461 (1997)

FT-IR and Resonance Raman Studies of Structural Changes of Bacteriorhodopsin under Dehydration

Evgeni L. Terpugov* and **Olga V. Degtyareva**

Institute of Cell Biophysics, Russian Academy of Sciences, 142292 Pushchino, Moscow Region, Russia

Abstract. FT-IR difference spectroscopy has been used to study the influence of water on the structure of both the retinal and protein in the dark-adapted purple membrane at low temperature. Assignment of chromophore vibration was made on the basis of FT-IR data and by comparison with resonance Raman results. We established that the BR_{506} form, generated in a dehydrated purple membrane comes from the parent BR_{548} from. The present FT-IR difference measurements in a two-channel mode showed that the retinal configuration of BR_{506} is the all-*trans*, while according to the earlier resonance Raman data the conformation of its retinal is 13-*cis*. This discrepancy is explained in the present paper.

Key words: bacteriorhodopsin, dehydration, dark-adaptation, FT-IR.

bacteriorhodopsin (BR) is a unique energy-transducting protein in the purple membrane (PM) of *Halobacterium salinarium* [1, 2]. Essential for its function is one molecule of retinal, covalently bound as a protonated Schiff base to the lysine residue of the apoprotein. In the dark-adapted state of BR, the equilibrium ratio between the 13-*cis* and all-*trans* isomers of retinal is approximately one [3]. Continuous illumination shifts this equilibrium completely to the all-*trans* from (BR_{568}). The active transport of protons is found only for the *trans* but not for the *cis* form. The structural reason for the lack of proton pumping of the 13-*cis* isomer (BR_{548}) and the physiological relevance of dark-adaptation are unknown. Earlier it was demonstrated by measurements in the visible range that an additional chromophore state with an absorption band at 506 nm is formed upon dehydration of both light- and dark-adapted PM [4]. The form is photoactive, and its photocycle has a long-lived intermediate M_{412} [4, 5]. Some experimental data point towards the

configuration of the retinal of the BR_{506} from being close to 13-*cis* [6, 7]. These facts raise a contradiction. First, it is unclear why the M_{412} intermediate, which is normally observed in the photocycle of the all-*trans* form (BR_{568}), also appears in the photocycle of the 13-*cis* form. Secondly, it is strange that the 13-*cis* isomer exists in the light-adapted PM. The present work was undertaken to clarify these questions. Using Fourier-transform infrared (FT-IR) difference and resonance Raman (RR) spectroscopy we focused on the structure of the BR_{506} form. Our observations showed that the retinal of this form is in the all-*trans* configuration, the BR_{506} form being produced from the BR_{548} form.

Experimental

The culturing of *Halobacterium salinarium* (strains ET-1001, JW-5) and preparation of PM and apomembrane were performed according to established methods [1]. The preparation of dried films of PM and apomembrane have already been described elsewhere [6]. Dark- and light-adapted thin films (OD = 1.2 at 280 nm) were investigated at 77 K. The RR spectra were recorded with a Russian DFS-24 spectrometer. The IR-spectrophotometric measurements were made with a Russian FS-02 two-channel Fourier spectrometer. To avoid the protein vibrations and obtain only the conformational changes in the retinal and amino-acid residues located inside the BR active site, we placed the films of BR without retinal (strain JW-5) in the reference channel. This was necessary because a conformation of the protein can be modified by dehydration of BR.

Results and Discussion

It has been shown that the two forms, all-*trans* BR_{568} and 13-*cis* BR_{548}, can be easily distinguished by means of the intensity distribution among the three diagnostic bands at 1185, 1215 and 1255 cm^{-1} [7]. It is seen that in the fingerprint region from 1150 to 1300 cm^{-1}, the

---

* To whom correspondence should be addressed

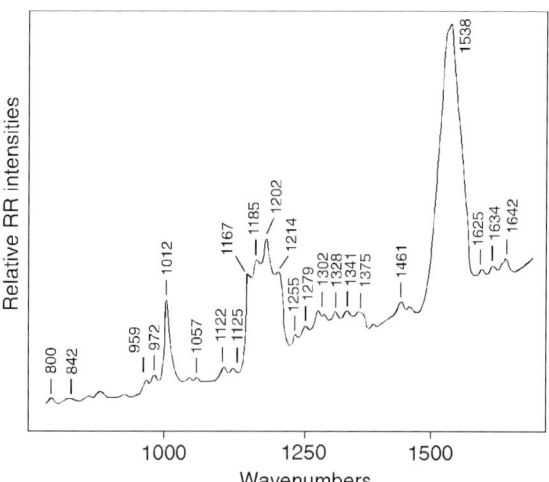

Fig. 1. RR spectrum from dried PM film; sample at 77 K. The probe beam, and additional pump beam were coaxial. The probe beam at 514.5 nm from an Ar^+ laser at a power level of 5 mW, and the pump beam at 632.8 nm from an He–Ne laser at a power level of 20 mW. Spectrum was recorded at 2 cm^{-1} resolution

RR spectra of dehydrated PM and the 13-*cis* chromophore BR_{548} closely resemble each other, while the characteristic bands of the all-*trans* form at 1215 and 1255 cm^{-1} are of reduced intensity in dehydrated PM. On the basis of these data, it has been concluded that the retinal of BR_{506} is mainly in the 13-*cis* configuration, although a fraction of the chromophores will still be in the all-*trans* form. It was inferred from these RR experiments that the BR_{506} form comes from the BR_{568} form. Indeed, the broadening of the C=C stretching band of the all-*trans* form at 1528 cm^{-1} and its shift to 1538 cm^{-1} indicated the existence of a blue-shifted photoproduct in dehydrated PM. Because there was a strong overlapping of bands in the RR spectrum recorded at room temperature, it was difficult to separate these components. In the present low-temperature experiments, we tried to obtain the separate components of the C=C streching band. An RR spectrum of the dehydrated light-adapted membrane, recorded under stationary conditions at 77 K, is shown in Fig. 1. The RR spectrum of dehydrated dark-adapted membrane (not shown) recorded under the same conditions is similar. The pattern of changes observed in our RR spectra for the hydration–dehydration transition (disappearance of the 1167 cm^{-1} and 1255 cm^{-1} lines and appearance of the 1185 cm^{-1} and 1634 cm^{-1} lines and a decrease in the intensity of the 1202 cm^{-1} and 1214 cm^{-1} lines) is in agreement with that obtained in the experiments at room temperature [6]. We assume that

the broadening of the C=C stretching band and its shift to 1538 cm^{-1} is the result of overlapping of the bands at 1528, 1532 and 1543 cm^{-1} of the BR_{568}, BR_{548} and BR_{506} forms, respectively. The appearance of the BR_{548} form in this spectrum can be explained by the reversible photoreaction $BR_{568} \rightleftharpoons BR_{548}$ [3, 8]. This reaction must lead to a photosteady state for relatively high light intensities (probe beam + pump beam). To clarify what is the isomeric state of the BR_{506} form and which parent form it comes from, we have used FT-IR difference spectroscopy, applying an approach by which we could avoid protein vibrations and obtain conformational changes in the retinal and amino-acids residues, involved in the interaction with retinal (see Experimental). We recorded the spectrum of a dark-adapted PM consisting of an equimolar mixture of the all-*trans* and 13-*cis* isomers, to obtain information about the transition of BR_{568} to BR_{506} or of BR_{548} to BR_{506}.

Figure 2 shows the FT-IR difference spectrum of dehydrated and hydrated dark-adapted PM, recorded at 77 K. The peak at 1528 cm^{-1} (BR_{568}) vanishes, but a negative peak at 1532 cm^{-1} and a positive peak at 1543 cm^{-1} appear. According to the linear correlation which has been established between the frequency and wavelength of the absorption maximum, the negative peak at 1532 cm^{-1} and the positive peak at 1543 cm^{-1} should be assigned to the C=C stretchings of BR_{548} and BR_{506}, respectively. This means that the BR_{506} form comes from BR_{548}. The BR_{568} form does not take part in this process. In order to understand what is the conformation of the retinal of BR_{506}, we have analysed

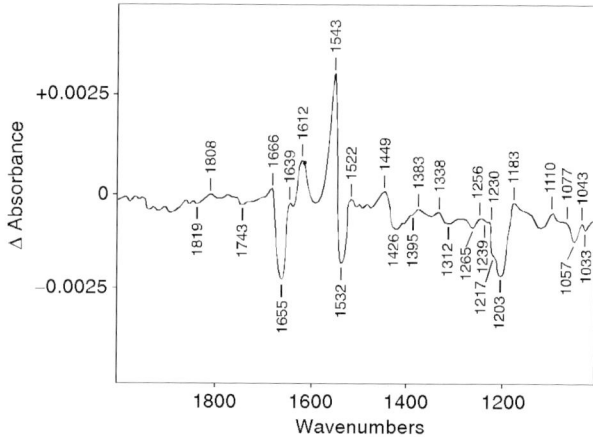

Fig. 2. FT-IR difference spectrum for the hydration–dehydration transition of dark-adapted PM at 77 K at 4 cm^{-1} resolution

all the peaks marked by wavenumbers. Assignment of the retinylidene chromophore vibration was made by comparing these data with the RR spectra obtained for BR_{568}, BR_{548}, all-*trans*, and 13-*cis* retinals [9, 10]. Appearance of all positive peaks in this spectrum is due to either (*a*) a shift of lines (for example, from -1057 cm^{-1} to $+1043$ cm^{-1}; from -1312 cm^{-1} to $+1338$ cm^{-1}; from -1395 cm^{-1} to $+1383$ cm^{-1} where $+$ and $-$ refer to positive and negative peaks, respectively) or (*b*) a change in intensity (for example the increase in intensity of the lines at 1256 cm^{-1}), which points to occurrence of the conformational change 13-*cis* $\rightarrow$ all-*trans* in the chromophore. This clearly indicates that the configuration of the retinal of the BR_{506} form is all-*trans*. Thus we can explain the occurence of intermediate M_{412} in the photocycle of BR_{506}.

We also observed a shift of C=O stretching from 1655 cm^{-1} to 1666 cm^{-1} for BR amino-acid residues. Presumbly, it is determined by breaking of hydrogen bonds between Asn or Gln and a water molecule, when the content of water is decreased. There are reasons to assume that the shift of the absorption band of BR_{548}

to 506 nm is determined by a change of electrostatic interaction between the carboxylate side-group of BR and the retinal. However, this assumption calls for further investigation.

References

[1] D. Oesterhelt, W. Stoeckenius, *Methods Enzymol.* **1974**, *31A*, 667.

[2] K. J. Rothschild, *J. Bioenerg. Biomembr.* **1992**, *24*, 147 (and references therein).

[3] W. Sperling, P. Carl, C. N. Rafferty, N. A. Dencher, *Biophys. Struct. Mech.* **1977**, *3*, 79.

[4] Yu.A. Lazarev, E.L. Terpugov, *Biochim. Biophys. Acta* **1980**, *590*, 324.

[5] R. Korenstein, B. Hess *Nature* **1977**, *270*, 184.

[6] E. L. Terpugov, L. N. Chekulaeva, Yu.A. Lazarev, *Mol. Biol.* **1982**, *16*, 814.

[7] P. Hildebrandt, M. Stockburger, *Biochemistry* **1984**, *23*, 5539.

[8] T. Kouyama, R.A. Bogomolni, W. Stoeckenius, *Biophys. J.* **1985**, *48*, 201.

[9] S. O. Smith, M. S. Braiman, A. B. Myers, J. A. Pardoen, J. M. L. Courtin, C. Winkel, J. Lughtenburg, R. A. Mathies, *J. Am. Chem. Soc.* **1987**, *109*, 3108.

[10] S. O. Smith, J. A. Pardoen, J. Lugtenburg, R. A. Mathies, *J. Phys. Chem.* **1987**, *91*, 804.

Mikrochim. Acta [Suppl.] 14, 463–466 (1997)

Microbeam FT-IR Spectroscopic Examination of Diseased Brain Tissues

Steven M. LeVine[1,*], **David L. Wetzel**[2], and **Dennis W. Dickson**[3]

[1] Department of Physiology and the Smith Mental Retardation and Human Development Center, University of Kansas Medical Center, 3901 Rainbow Blvd., Kansas City, KS 66160, U.S.A.
[2] Microbeam Molecular Spectroscopy Laboratory, Kansas State University, Shellenberger Hall, Manhattan, KS 66506, U.S.A.
[3] Departments of Pathology and Neurology, Albert Einstein College of Medicine, 1300 Morris Park Ave., Bronx, NY 10461, U.S.A.

Abstract. Fourier-transform infrared (FT-IR) microspectroscopy is a powerful technique for obtaining infrared spectra from microscopic regions of tissue sections. Individual spectra can be collected from tissue areas of interest, or several spectra can be collected along a grid pattern, e.g. across a lesion site, and transformed into maps of chemical functional groups. In either case, chemical information can be directly correlated with histological or histopathological information, in situ. In this paper we review our applications of FT-IR microspectroscopy to examination of normal and diseased brain tissues, and present data from the brains of HIV-infected individuals.

Key words: FT-IR microspectroscopy, myelin, HIV, white matter, microbeam.

Fourier-transform infrared (FT-IR) microspectroscopy has been used to examine the chemical characteristics of normal and diseased brain tissues. In normal mice, the spectral features of white matter and gray matter were compared [1–3]. The light hue of white matter is due to myelin, which is a multilamellar, lipid-rich structure. The concentrations of lipid functional groups were greater in white matter than in gray matter, and the degree of difference in absorbance varied with functional group: $2927\,cm^{-1}$ $(CH_2) > 1469\,cm^{-1}$ $(CH_2) > 1085\,cm^{-1}$ $(HO-C-H) > 1740\,cm^{-1}$ $(C=O) > 1235\,cm^{-1}$ $(P=O)$. The absorbance at $1550\,cm^{-1}$ (amide II) was similar for white and gray matter. White and gray matter were then examined in shiverer mice, which have a mutation in their myelin basic protein gene and a greatly diminished amount of myelin. Spectra from white matter of shiverer mice revealed that the lipid functional groups were significantly lower in concentration than in the

white matter of normal mice [3]. FT-IR microspectroscopy was also used to detect a specific chemical compound in white matter [4]. For this study, twitcher mice, which are an authentic animal model of Krabbe's disease in humans, were used. Twitcher mice have a gene mutation in their galactocerebroside β-galactosidase gene, which results in the accumulation of the enzyme's substrate, psychosine, in myelin-producing cells. We observed that the peak position for CH_2 in psychosine ($2919\,cm^{-1}$) was shifted from its position in white matter ($2925\,cm^{-1}$). A mathematical function $(A_{2919} - A_{3000})/(A_{2923} - A_{3000})$ where A_{2919}, A_{2923}, and A_{3000} are the absorbance at the wavenumbers indicated by the subscripts, was used to determine whether the CH_2 peak in the twitcher mice spectra was shifted towards the peak position observed in psychosine. In twitcher mice, 36% of the spectra collected from the hind brain white matter and 19% of the spectra collected from the subcortical white matter displayed such a shift relative to the peak positions in the white matter of normal mice. There was no shift observed in the cerebellar cortex (gray matter) of twitcher mice [4]. In the examination of another animal model of white matter disease, blood was injected into the subcortical white matter of the normal rat, and chemical changes at different areas of the lesion sites were examined. Spectra were collected from normal white matter, which was away from the injection site, at the penumbra region of the injection site, and at an area of white matter that was moderately infiltrated by blood products [5]. In normal white matter, high absorbance values were observed at $2927\,cm^{-1}$ (CH_2), $1740\,cm^{-1}$ $(C=O)$, and $1085\,cm^{-1}$ $(HO-C-H)$, and

---

* To whom correspondence should be addressed

a moderate value was observed at 1550 cm^{-1} (amide II). In the penumbra area, the CH$_2$, C=O, and HO–C–H absorbances were lower than in the normal tissue, and that of amide II was increased. In the moderately affected area of white matter, the CH$_2$, C=O, and HO–C–H absorbances continued to decrease and the amide II absorbance continued to increase.

In the present study, we extended these investigations of white matter diseases by examining HIV-infected brain tissues. HIV infection can result in several pathological changes in various regions of the brain, including the white matter. We collected infrared spectra from pathological sites in the white matter of these brains in order to obtain information about the chemical alterations that result from this disease.

Experimental

The brains from individuals that had died from AIDS and non-AIDS related conditions were fixed in formalin, and tissue blocks from various areas were embedded in paraffin. Paraffin sections (8 μm thick) from the cortex/subcortical white matter were mounted onto barium fluoride disks, and the paraffin was removed by treatment with xylenes and alcohol.

An IRμs molecular microscopy system (Spectra-Tech Inc., Shelton, CT), which has an FT-IR spectrometer fully integrated with microscope optics, was used for these studies.

The following conditions were typically used for the collection of infrared spectra: 24 × 24 μm aperture, 256 co-added scans, and 4 cm^{-1} resolution. Mapping experiments were also performed. These studies utilized a computer-driven motorized stage that enabled the collection of spectra along a grid pattern covering the tissue area of interest. These experiments typically used a larger aperture and fewer co-added scans. Functional group maps were generated by interpolating the individual peak areas from the spectra collected in a mapping experiment.

Results and Discussion

The pathological sites in the HIV-infected tissues were sparse and difficult to find by FT-IR microspectroscopy alone. In order to locate a lesion site, we used an adjacent section that was stained with hematoxylin and eosin (H + E). The unstained section, which was on

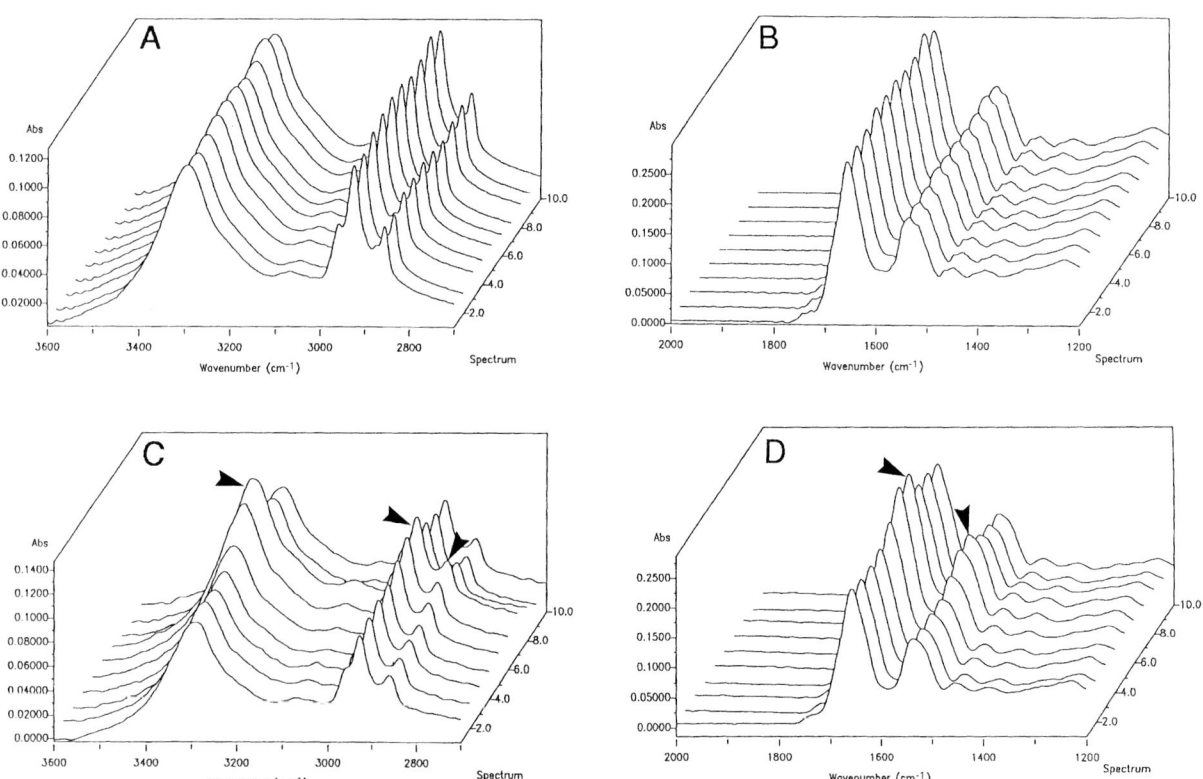

Fig. 1. Spectra were collected along a line in the white matter from normal (**A**, **B**) and HIV-infected (**C**, **D**) individuals. The spectra in normal white matter appear to be relatively uniform, but the spectra from HIV-infected white matter show spectral heterogeneity. In particular, the intensity of the peaks at ∼3300, ∼2927, ∼2850, ∼1650, and ∼1550 cm^{-1} (arrowheads) changed between spectra collected from different points along the line. Increased absorbances at these wavenumbers were observed in spectra that were collected from sites of inflammation in the HIV-infected white matter

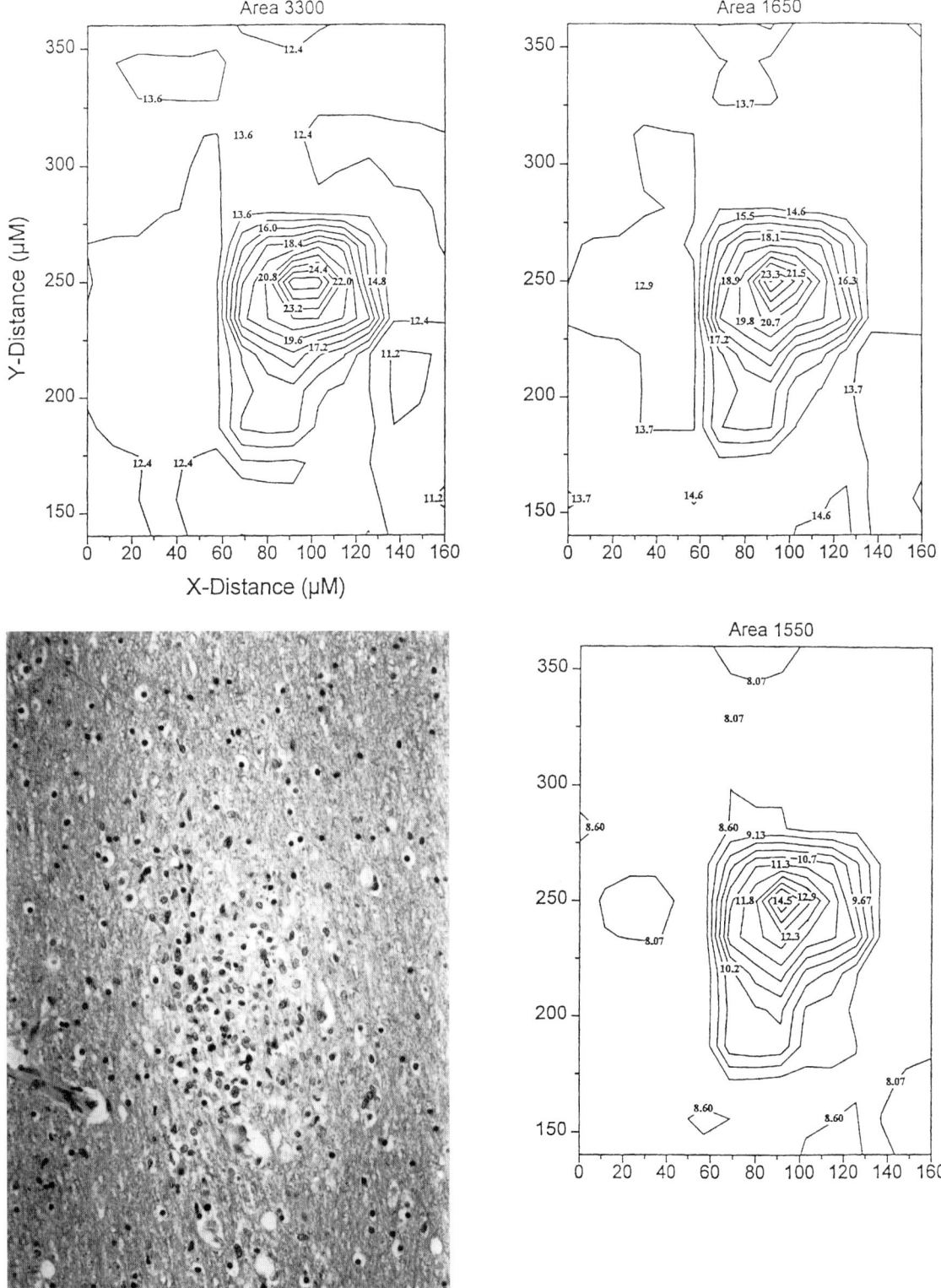

Fig. 2. Spectra were collected along a grid pattern that encompassed a lesion site in HIV-infected white matter. Contour maps were generated by interpolating the peak areas between each collection site. Maps for peak areas at 3300, 1650 and 1550 cm^{-1} all show greater absorbances in a region near the center. The photomicrograph is from an adjacent section that was stained with H + E, and illustrates the lesion site that was mapped (note the inflammatory cells are the aggregation of the nuclei in an equivalent region to that where the contour maps displayed increased absorbances)

the barium fluoride disk, was lined up to the location that matched the lesion area observed on the H + E slide. Once a lesion site was identified, spectra were collected from it and the surrounding tissue areas, and functional group maps were generated.

Owing to the potentially hazardous nature of the human tissue, all samples were fixed in formalin and then processed through paraffin, xylenes and alcohols. These processing steps changed the spectral profile from that observed for frozen, unfixed tissues. In particular, the spectra from the white matter of the processed tissues had several peaks that were diminished or shifted from previously observed locations on frozen white matter tissues [1, 2].

Besides spectral changes due to tissue processing, we observed spectral changes at lesion sites, compared to the surrounding tissues. Spectra that were collected along a line of white matter in normal brain tissues were relatively uniform, whereas spectra from HIV-infected brains showed spectral alterations which corresponded spatially to lesion sites (Fig. 1). Mapping experiments revealed that the N–H ($3300 \, \text{cm}^{-1}$), amide I ($1650 \, \text{cm}^{-1}$), amide II ($1550 \, \text{cm}^{-1}$) and CH_2 ($2927 \, \text{cm}^{-1}$) bands were concentrated at an area of inflammation in the tissue from the HIV-infected individuals (Fig. 2). These changes were quite pronounced and are likely due to the inflammatory cells located at the lesion site. These changes are probably not specific to HIV infection, because increases in the heights of

N–H and amide peaks have been observed in an experimental model of extravasated blood damaged brain tissue [5]. In that experimental model, however, there was a decrease in the CH_2 band but in the HIV-infected tissue the CH_2 band for the lesion site was increased.

In the present study, the spatial resolution never exceeded $24 \times 24 \, \mu\text{m}$. In studies that employ a light source with significantly greater intensity, it is possible to increase the resolution to close to the theoretical limit based on the wavelength [6]. Since a greater amount of spectral heterogeneity is observed within a tissue area when the resolution is increased [6], it is possible that other spectral features of HIV-infected tissue would be revealed with greater spatial resolution. The present study represents our first attempt at examining HIV-infected tissues, and we realize that there are many challenges ahead in more fully characterizing the chemical changes that occur in this disease.

References

[1] S. M. LeVine, D. L. Wetzel, *Appl. Spec. Rev.* **1993**, *28*, 385.
[2] D. L. Wetzel, S. M. LeVine, *Spectroscopy* **1993**, *8*, 40.
[3] S. M. LeVine, D. L. Wetzel, *Neuro Protocols* **1994**, *5*, 63.
[4] S. M. LeVine, D. L. Wetzel, A. J. Eilert, *Inter. J. Dev. Neurosci.* **1994**, *12*, 275.
[5] S. M. LeVine, D. L. Wetzel, *Am. J. Path.* **1994**, *145*, 1041.
[6] D. L. Wetzel, A. J. Eilert, J. A. Reffner, L. Carr, S. M. LeVine, S. S. Miller, L. Pietrzak, *Pitts. Conf. Anal. Chem. Appl. Spectrosc.* **1995**, *46*, paper 246.

Mikrochim. Acta [Suppl.] 14, 467–469 (1997)

The Effect of Implantation Conditions on the Formation of SiC

Pearl R. Berndt*, **Johannes H. Neethling,** and **Jacobus A. A. Engelbrecht**

Department of Physics, University of Port Elizabeth, P. O. Box 1600, Port Elizabeth 6000, Republic of South Africa

Abstract. Analysis of silicon implanted with 5×10^{17} C^+ cm^{-2} and 2×10^{18} C^+ cm^{-2} and annealed at 1200 °C for 30 min, revealed that the lower dose implantation produced a substantial yield of SiC, while the higher dose implantation resulted in the formation of carbon clusters which retarded the synthesis of SiC.

Key words: SiC, carbon clusters, ion implantation.

Silicon carbide is an intrinsic semiconductor, which has exhibited considerable potential as a material suitable for the fabrication of devices operating in hostile environments [1]. Fabrication of SiC by carbon ion implantation in silicon was first reported by Borders et al. [2]. Further investigations revealed that the synthesis of SiC was aided by increasing the temperature of the substrate during implantation [3]. Nussupov et al. [4] established that ion doses in excess of 5×10^{17} C^+ cm^{-2} resulted in the formation of carbon clusters within the material. These clusters not only inhibited the formation of SiC, but also required extremely high temperatures to facilitate their degradation. This investigation considers the influence of implantation dose, substrate temperature and post-implantation annealing on the degree of SiC formation.

Experimental

Si(001) substrates were implanted with 150 keV carbon ions to total fluences of 5×10^{17} C^+ cm^{-2} and 2×10^{18} C^+ cm^{-2}. Implantation was performed with the substrates maintained at temperatures of 100–500 °C. Post-implantation annealing was performed at 1200 °C for 30 minutes, in a flowing argon atmosphere. Fourier-transform infrared absorption spectroscopy (FT-IR) results were generated, with a Perkin–Elmer 1600 Series spectrometer. These

results were supported by transmission electron microscopy (TEM) studies.

Results and Discussion

Absorption spectroscopy of samples implanted with the lower ion dose, at a temperature of 100 °C, showed a substantial increase in the concentration of substitutional carbon in the silicon, evidenced by the peak situated at 610 cm^{-1}. Implantation at the same dose, but at a temperature of 500 °C, resulted in a lower concentration of substitutional carbon with further evidence of the increased precipitation of tiny SiC particles, shown by a peak at 835 cm^{-1}. These precipitates, embedded in the silicon matrix, are thought to be nucleation sites for further SiC formation [5].

As can be seen in Table 1, implantation at 350 °C with the higher carbon ion dose, resulted in an even lower concentration of substitutional carbon, with a corresponding higher concentration of SiC precipitates. These precipitates result from carbon being expelled from substitutional lattice sites [5], because of either the heat treatment or the use of large ion implantation doses.

It is evident from Fig. 1(a), that none of the absorption spectra of the implanted samples exhibited the peak at 800 cm^{-1} which is indicative of the presence of bulk SiC in the material [6].

Post-implantation annealing at 1200 °C was necessary to produce bulk SiC [Fig. 1(b)]. Although the substrates implanted at a temperature of 100 °C did not yield bulk SiC, the FT-IR absorption spectra of these samples clearly indicated an increase in the concentration of SiC precipitates. While all samples implanted at

---

* To whom correspondence should be addressed

Table 1. Normalized FT-IR absorption peak heights for implanted material, and implanted material annealed at 1200 °C for 30 min

Temperature of substrate during implantation (°C)	Implantation dose ($C^+ cm^{-2}$)	Implanted material		Annealed implanted material	
		Peak A 610 cm^{-1}	Peak C 835 cm^{-1}	Peak B 800 cm^{-1}	Peak C 835 cm^{-1}
100	5×10^{17}	1.77	0.15	-	0.29
350	2×10^{18}	1.29	0.29	0.01	0.29
500	5×10^{17}	1.35	0.27	0.02	0.25

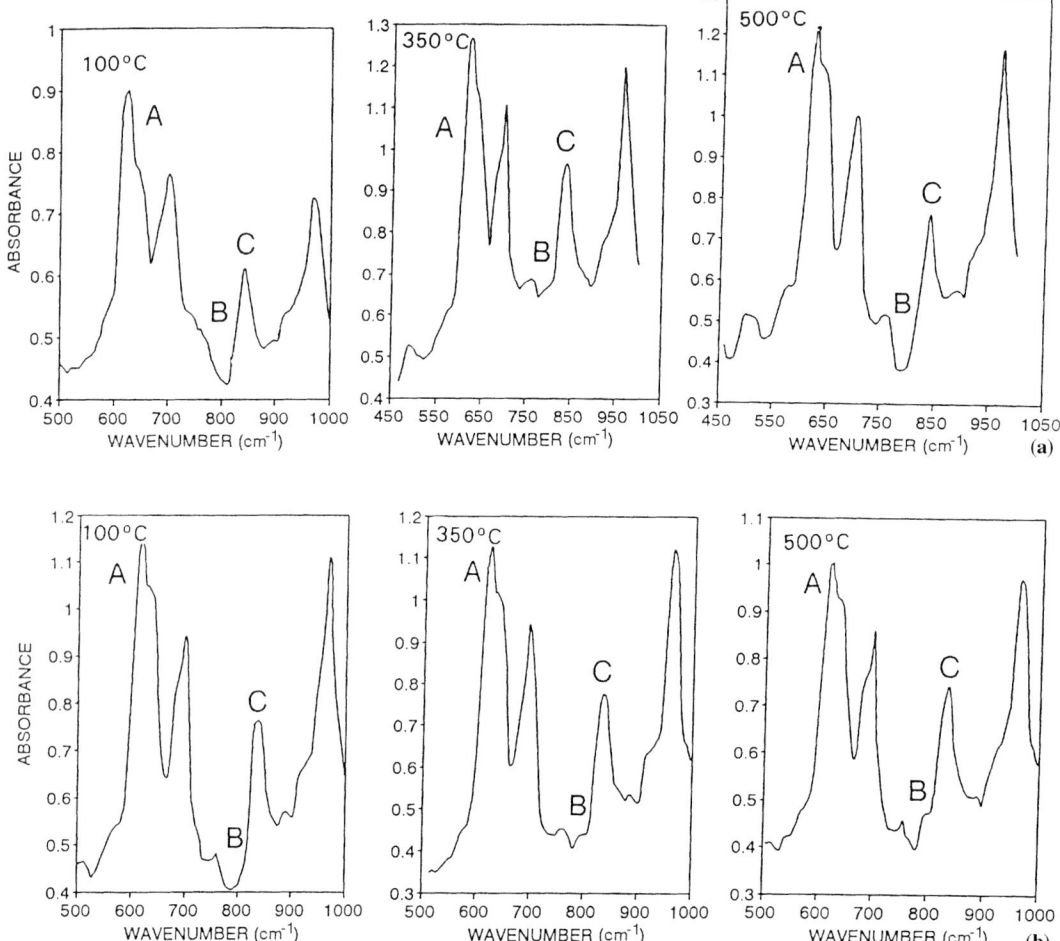

Fig. 1. (a) Typical FT-IR absorbance spectra of Si samples, implanted at various temperatures, but not annealed. **(b)** Typical FT-IR absorbance spectra of Si samples implanted at various temperatures, and annealed at 1200 °C

temperatures of 350–500 °C produced bulk SiC upon annealing, the implantation at the higher temperature yielded a greater concentration, with a corresponding decrease in the concentration of the SiC precipitates that function as SiC nucleation centres. Despite the higher implantation dose, the samples implanted at

350 °C yielded a lower dose of bulk SiC, which can be attributed to the formation of carbon clusters.

As it is not possible to detect the carbon clusters by using FT-IR absorption spectroscopy, TEM was employed. Figure 2 is a micrograph of the material implanted with the lower dose at 500 °C. The distinct

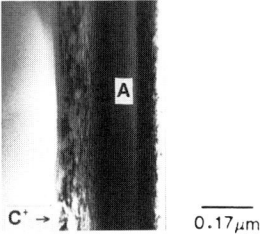

Fig. 2. Cross-sectional TEM micrograph of unannealed implanted Si (5×10^{17} C$^+$ cm^{-2} at 150 keV)

(a) (b)

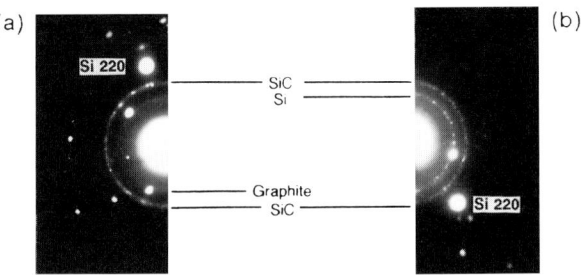

Fig. 3. **(a)** Selected area diffraction pattern of the annealed implanted region (2×10^{18} C$^+$ cm^{-2}) showing Si reflections and rings from polycrystalline SiC and graphite; **(b)** Selected area diffraction pattern of the annealed implanted region (5×10^{17} C$^+$ cm^{-2}) showing Si reflections and rings from polycrystalline Si and SiC

implanted region A evident in the micrograph was observed in similar studies of all samples, before and after annealing. Following annealing, electron diffraction examination of region A in the samples implanted at 350–500 °C revealed that the region consisted of Si and SiC (Fig. 3). The diffraction pattern of the higher-dose implanted material exhibited an additional feature in the form of a graphite diffraction ring [Fig. 3(a)], thereby confirming the presence of carbon clusters in the implanted region.

In summary, it is evident that carbon ion implantation in silicon does yield SiC, although it is in the form of tiny SiC precipitates. Implantation at temperatures of at least 350 °C with subsequent annealing at 1200 °C is necessary to produce bulk SiC. Analysis of the implanted silicon revealed that implantation doses in excess of 5×10^{17} C$^+$ cm^{-2} produced SiC with accompanying formation of carbon clusters. These carbon clusters are reported to inhibit SiC formation.

Acknowledgement. The authors wish to thank the Schonland Research Centre for Nuclear Sciences, University of Witwatersrand, South Africa, for performing the carbon implantation.

References

[1] G. Kelner, S. Binari, M. Shur, K. Sleger, J. Palmour, H. Kong *Mat. Sci. Eng.* **1992**, *B11*, 121.
[2] J. A. Borders, S. T. Picraux, W. Beezhold *Appl. Phys. Lett.* **1971**, *18*, 509.
[3] T. Kimura, S. Kagiyama, S. Yugo, *Thin Solid Films* **1982**, *94*, 191.
[4] K. K. Nussupov, V. O. Sigle, N. B. Bejsenkhanov, *Nucl. Instrum Methods. Phys. Res. Sect. B* **1993**, *82*, 69.
[5] A. R. Bean, R. C. Newman, *J. Phys. Chem. Solids* **1971**, *32*, 1211.
[6] W. G. Spitzer, D. A. Kleinman, C. J. Frosch, *Phys. Rev.* **1959**, *113*, 133.

Mikrochim. Acta [Suppl.] 14, 471–472 (1997)
© Springer-Verlag 1997

FT-IR Determination of Dopant Content in GaAs Structures

J. A. A. Engelbrecht*, **J. R. Botha**, and **E. E. van Dyk**

Physics Department, University of Port Elizabeth, P.O. Box 1600, Port Elizabeth 6000, South Africa

Abstract. Fourier-transform infrared reflectance spectroscopy is used to establish the peak height of the transverse optic mode of AlAs and InAs in GaAs structures. The peak height is found to be linearly dependent on alloy composition.

Key words: infrared spectroscopy, $Al_xGa_{1-x}As$, $In_xGa_{1-x}As$.

Quantum-well structures and superlattices of $(Al_x/In_x)Ga_{1-x}As$ have many potential uses for opto-electronic applications owing to the ability to tailor the bandgap of the material by varying the alloy composition of the structures. The assessment of the mole fraction of a substituent after growth of a particular structure is hence of great interest. Many techniques have been reported for determination of the mole fraction in such alloy systems [1]. Optical techniques are preferred for characterization purposes, owing to their non-destructive nature. However, all investigations of $Al_xGa_{1-x}As$ where IR reflectance spectroscopy was employed were done in conjunction with curve-fitting procedures.

An earlier publication [2] reported a linear relationship between the height of the absorption band at $362 \, cm^{-1}$ of substitutional Al in GaAs and the Al dopant concentration. However, the concentrations reported were in the ppb range and IR measurements were performed at liquid-nitrogen temperatures. Other references to the band in the vicinity of $360 \, cm^{-1}$ have reported the position of this band to be independent of aluminium content [3, 4]. However, visual inspection of the *height* of this band as a function of mole fraction, by Fukasawa et al. [5] seemed to indicate the band to

be linearly dependent on the mole fraction of aluminium. This observation was subsequently tested, also with samples from our laboratories. Structures that were evaluated included epilayers [3, 6], stand-alone films [4] and superlattices [5].

Experimental

$(Al_x, In_x) Ga_{1-x}As$ layers of varied Al (or In) content and thickness were grown by atmospheric pressure organometallic vapour phase epitaxy (OMVPE) on semi-insulating GaAs substrates. Both single-side polished (SSP) and double-side polished (DSP) substrates were investigated. The thickness of the substrates was 300–$350 \, \mu m$, and that of the layers was 2–$3 \, \mu m$. The composition of the layers was determined by photoluminescence (PL) measurements. PL measurements were performed at $12 \, K$, using the $5145 \, Å$ line of an argon laser. The alloy composition determined by this method is accurate to ± 0.01 (mole fraction).

Fourier-transform IR reflectivity measurements were performed on a Nicolet 20 DXB instrument at a resolution of $8 \, cm^{-1}$ and near-normal incidence. The peak height of the transverse optic (TO) mode of AlAs at around $360 \, cm^{-1}$ was measured after plotting the region of interest (250–$400 \, cm^{-1}$) using the standard procedure to obtain a suitable beseline [7]. Similar measurements were performed for the InAs TO peak, which is found at $\sim 220 \, cm^{-1}$.

Infrared spectra for $Al_xGa_{1-x}As$ and $In_xGa_{1-x}As$ structures, as found in the literature, were also evaluated to test the hypothesis.

Results and Discussion

The peak height of the Al TO absorption band, as reported for infrared reflectivity measurements by various research groups, together with data obtained by ourselves, was plotted as a function of mole fraction, and a linear relationship was obtained (Fig. 1). A similar linear dependence is observed for $In_xGa_{1-x}As$.

In summary, a fast and simple technique is proposed to measure the Al or In mole fraction in $Al_xGa_{1-x}As$ and $In_xGa_{1-x}As$ layers on GaAs substrates, by

---

* To whom correspondence should be addressed

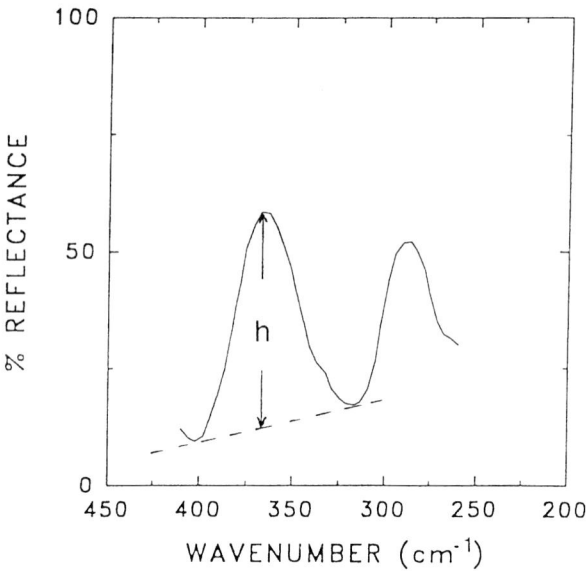

Fig. 1. An example of the reflection spectrum of an $Al_xGa_{1-x}As$ layer on GaAs, indicating baseline for measurement of peak height h

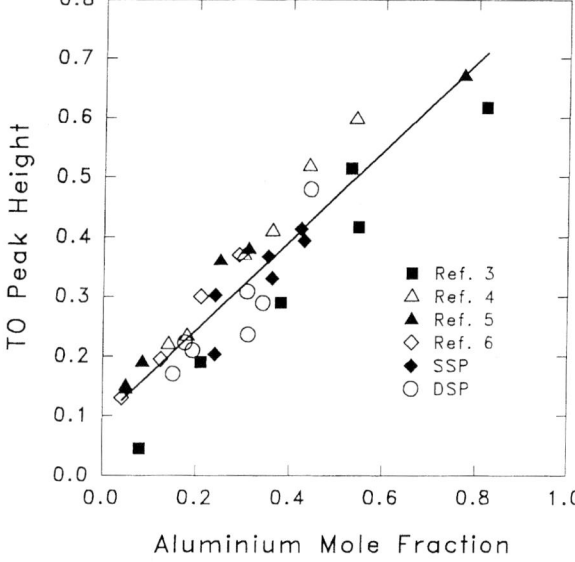

Fig. 2. Display of mole fraction as function of TO peak height as obtained by various research groups, together with the linear regression fit through the datapoints for $Al_xGa_{1-x}As$

measuring the height of the relevant dopant peak in a reflection spectrum.

Acknowledgements. The financial assistance of the Foundation for Research Development, Pretoria and the University of Port Elizabeth, and technical advice from Dr Ann Conibear are gratefully acknowledged.

References

[1] J. A. A. Engelbrecht, J. R. Botha, E. E. van Dyk, *Appl. Spectrosc.* **1997** in print.

[2] O. G. Lorimor, W. G. Spitzer, M. Waldner, *J. Appl. Phys.* **1966**, *37*, 2509.

[3] M. Ilegems, G. L. Pearson, *Phys. Rev. B* **1970**, *1*, 1576.

[4] O. K. Kim, W. G. Spitzer, *J. Appl. Phys.* **1979**, *50*, 4362.

[5] R. Fukasawa, M. Wakaki, K. Shirawachi, S. Nishizawa, K. Ohta, *Jpn. J. Appl. Phys. Part 1* **1992**, *31*, 2409.

[6] I. Lukes, J. Humlicek, V. Vorlicek, M. Zavetova, *Phys. Status Solidi A* **1989**, *111*, 655.

[7] *1980 Annual Book of ASTM Standards, Part 42*, American Society for Testing and Materials, Philadelphia, PA, 1980, E 168.

Mikrochim. Acta [Suppl.] 14, 473–474 (1997)

Application of the Fourier Self-deconvolution Technique to Infrared Spectra of Chemical Impurities in Semiconductors

L. T. Ho

Institute of Physics, Academia Sinica, Taipei, Taiwan, Republic of China

Abstract. An example is given to show that the Fourier self-deconvolution technique, i.e., spectral deconvolution using Fourier transforms and the intrinsic line-shape, can be used to better resolve the infrared absorption spectra of chemical impurities in semiconductors.

Key words: Fourier transform, deconvolution, infrared spectrum, impurity, semiconductor.

The infrared absorption spectra of the transitions from the ground-state to the excited-state of chemical impurities in semiconductors have been investigated intensively for many years. Recently, there has been a renewed interest in this area as the result of the observation of fine structure in the spectra measured by high-resolution Fourier-transform infrared spectrometers. In addition, a theory of Fourier self-deconvolution [1], i.e., spectral deconvolution based on use of Fourier transforms and the intrinsic line-shape, has also been developed. We have used this method in trying to resolve the excitation lines of chemical impurities in semiconductors and have obtained some excellent results. The purpose of this paper is to present an example to demonstrate that the Fourier self-deconvolution technique can be very useful in this area as well.

The basic Fourier self-deconvolution theory can be described as follows [1]. Any experimental spectrum, E, can be expressed as a convolution of a line-shape function, G, and a spectrum, E', i.e., $E = G \cdot E'$. The inverse Fourier transform of both sides gives $I = F^{-1}G \cdot I'$, where $I = F^{-1}E$ and $I' = F^{-1}E'$ are the interferograms of E and E', respectively. Therefore, $I' = (1/F^{-1}G) \cdot I$, i.e., the deconvolution operation requires a particular form of apodization of the interferogram. E' is then simply obtained by taking the Fourier

transform of I'. By following this procedure, it is possible to deconvolute the intrinsic line-shape function from a spectrum comprised of overlapping bands, thus narrowing considerably the lines in the deconvoluted spectrum.

Shown in Fig. 1 is the infrared absorption spectrum of beryllium impurities in silicon, measured at liquid-helium temperature with an Oxford cryostat and a Bomem DA3 Fourier-transform infrared spectrometer. Beryllium is a group II element and behaves like an acceptor when introduced into silicon by diffusion [2]. The spectrum in Fig. 1 clearly shows that there is a set of very strong absorption lines centred around 180 meV. Previous study [3] suggests that these lines are due to beryllium atoms occupying substitutional lattice sites. Some weaker lines have also

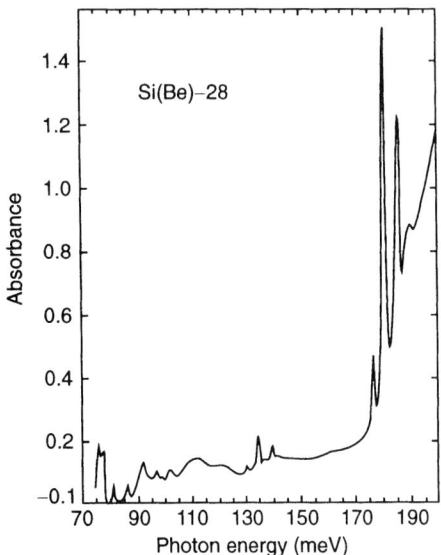

Fig. 1. Infrared spectrum of Si(Be)

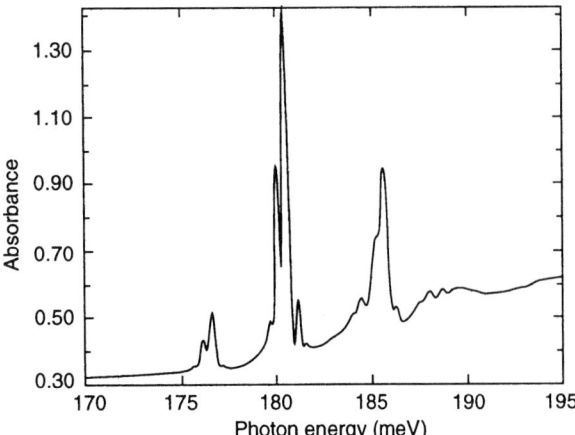

Fig. 2. Self-deconvolution of the spectrum of Si(Be)

been observed in this spectrum but will not be discussed here.

Being a group II substitutional impurity and thus a double acceptor, the energy level structure of beryllium is different from that of group III substitutional impurities, which are single acceptors in silicon. The main difference is that the ground-state splits into two levels for double acceptors but remains a single state for single acceptors (4). As a result, each absorption line due to transition from the ground-state to each excited-state for a double acceptor will show a line-splitting feature.

The absorption lines observed in Fig. 1 apparently do not show any hint of line splitting. It is interesting to find out, therefore, whether applying the Fourier self-deconvolution technique can help to resolve the spectral lines better. The result is shown in Fig. 2, and clearly shows that splitting of the lines does occur as predicted. The ground-state splitting has also been observed for beryllium [4] and mercury [5] double acceptors in germanium. Self-deconvolution of the spectrum also confirms unambiguously the splitting of the ground-state for the case of beryllium double acceptors in silicon.

In conclusion, we have demonstrated that the Fourier self-deconvolution technique can indeed be used to better resolve the infrared absorption spectra of chemical impurities in semiconductors. Such a technique is easy to apply for identifying spectral lines not evident in the unresolved spectra and to determine the positions of the spectral lines more accurately. The technique is particularly useful to facilitate the study of spectra in which relevant data are contained in overlapped band contours.

Acknowledgement. This work was partially supported by the National Science Council of the Republic of China.

References

[1] J. K. Kauppinen, D. J. Moffatt, H. H. Mantsch, D. G. Cameron, *Anal. Chem.* **1981**, *53*, 1454.
[2] J. B. Robertson, R. K. Franks, *Solid State Commun.* **1968**, *6*, 825.
[3] R. K. Crouch, J. B. Robertson, T. E. Gilmer, Jr., *Phys. Rev. B* **1972**, *5*, 3111.
[4] J. W. Cross, L. T. Ho, A. K. Ramdas, R. Sauer, E. E. Haller, *Phys. Rev. B* **1983**, *28*, 6953.
[5] R. A. Chapman, W. G. Hutchinson, *Phys. Rev.* **1967**, *157*, 615.

Mikrochim. Acta [Suppl.] 14, 475–477 (1997)

Magneto-Optical Rotation Spectrum of Semiconductors

Joseph K. McDonald[†], **Charles R. Christensen, John. A. Grisham, George A. Tanton***, and **A. J. Syllaios****

U. S. Army Missile Command, AMSMI-RD-WS-CM, Redstone Arsenal, AL 35898-5248, U.S.A.

Abstract. A Fourier-transform infrared spectrometer was used as a broad-band source to obtain magneto-optical rotation (MOR) spectra of semiconductor materials. Carrier concentrations obtained from the MOR data showed good agreement with those obtained from conventional Hall measurements. This technique has been shown to be a powerful tool in the study of impurity and defect levels in semiconductors.

Key words: magneto-optical rotation, Faraday-effect, semiconductors, FT-IR.

Magneto-optical rotation (MOR) techniques using Faraday rotation are capable of providing a topographic map of carrier concentration, mobility, and resistivity of each slice of semiconducting material. This technique does not require electrical contact to the material, such as used in the Hall bar test, and therefore can be used at different stages of semiconductor production. An objective of the MOR program is to develop an advanced capability for characterizing current and future IR focal-plane array detector materials, by high-resolution mapping of carrier concentration, mobility, and resistivity, in order to significantly reduce device manufacturing costs and improve producibility. The technique is applicable to binary semiconductors such as CdS, GaAs, InSb, and tertiary compounds such as HgCdTe (MCT), AlGaAs, and PbSnTe. The system is applicable to slices of solid state recrystallized material (bulk material), liquid phase epitaxially produced material, or metallo-organo-chemically vapor deposited material.

When first developed at the U.S. Army Missile Command, CO_2 and infrared diode lasers were used as light sources to measure MOR. This paper presents a modified MOR technique which uses a Fourier-transform infrared spectrometer (FT-IR) as a light source instead of the infrared laser used in previous designs [1]. This Fourier-transform-magneto-optical rotation (FT-MOR) technique permits recording of the Faraday rotation spectrum over a broad spectral band compared to the narrow range available from the lasers used in previous work. The enhanced analytical capability, combined with spatial mapping, produces a powerful technique for identifying and mapping the distribution of carrier concentration and mobility [2].

Experimental

The basic FT-MOR experimental set-up is shown schematically in Fig. 1. The probe beam was from a Mattson Sirius 100 or a Mattson

* Present address: Morgan Research Corporation, 2707 Artie St., Suite 17, Huntsville, AL 35805-4769, U.S.A.

** Present address: Texas Instruments, P.O. Box 655936, Dallas, TX 76203, U.S.A.

[†] To whom correspondence should be addressed

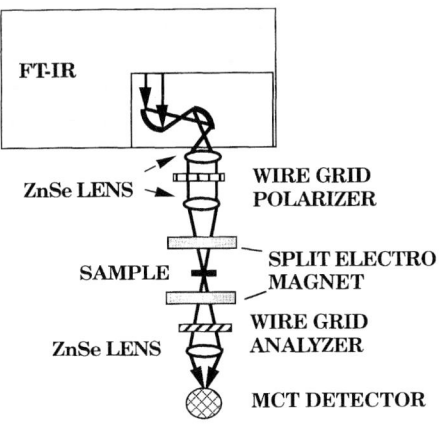

Fig. 1. Fourier-transform spectrometer and magneto-optical rotation set-up

RS2 FTIR spectrometer that covered the spectral region from 2.5 to 20 μm. The practical spectral range was limited to 7–15 μm owing to water vapor, ZnSe optics and detector response. Spectra were obtained at 4 cm^{-1} resolution, with up to 200 co-added scans. The output from the MCT detector was fed back into the electronics of the spectrometer to transform the signal into the wavelength-dependent transmission curves. The magnet was a GMW model 3470 electromagnet with field strengths up to 15 kG. Most spectra were obtained at 5 kG. An in-house fabricated variable temperature dewar was used in obtaining low-temperature data. Samples of MCT and InSb were obtained from Texas Instruments and Johnson Matthey, respectively.

Results and Discussion

In semiconductors, the amount of MOR is a function of the free carrier concentration $N = n_i + N_d - N_a$ where n_i is the intrinsic carrier concentration and $N_d - N_a$ is the net donor concentration; n_i is approximately 10^{16} cm^{-3} at 300 K but decreases to $<10^{14}$ cm^{-3} at 77 K. At 77 K, n_i drops below the detectable level and only the donor carrier concentration is detected. The MOR can then be used to map N in a semiconductor wafer, thereby determining those parts of a wafer where N is within acceptable limits for further processing into detectors and focal plane arrays. The total Faraday rotation (δ) in semiconductors is a sum from three major sources [3–6]: free-carrier plasma (δ_p), interband (δ_i), and free-carrier spin-induced (δ_s). Each of the components depends on the magnetic field strength, the thickness of the sample, and the wavelength of light. Only the plasma and spin components depend on the free carrier concentration. The magnitude of the interband rotation was found to be too small to be considered with the experimental

parameters used in these experiments. The wavelength dependence of the MOR is given by:

$$\delta(\lambda) = \delta_p \lambda^2 + \frac{\delta_s}{(1 - \chi^2)} \qquad (1)$$

where $\chi = \lambda_g/\lambda$; λ_g is the bandgap wavelength. The contribution to the Faraday rotation from the spin component is largest as λ approaches λ_g. At high magnetic fields full spin alignment of the carriers is achieved and δ_s becomes saturated and makes no further contribution to $\delta(\lambda)$. The largest contribution to the MOR from δ_s is near the bandgap, and δ_p dominates at longer wavelengths.

Using the equation, $I = I_0 \cos^2(\theta)$, for light transmitted through two polarizers, the ratio of the light transmitted with and without the applied magnetic field is given by

$$\frac{\cos^2(\theta + \delta)}{\cos^2(\theta)} = T(\lambda). \qquad (2)$$

T is the transmission intensity at λ, and θ is the angle of the analyzer relative to the polarizer. The MOR was obtained by determining the transmission at selected wavelengths and using the formula:

$$\delta(\lambda) = \arccos[\sqrt{T}\cos(\theta)] - \theta. \qquad (3)$$

θ was maintained at 40° for all the experiments.

In order to obtain the MOR transmission spectrum, a background interferogram of the sample was obtained with the magnet off and another interferogram was obtained with the magnet on. The Fourier transform of these two interferograms gave the transmission spectra (see Fig. 2). A transmission greater or less than

Table 1. Carrier concentrations from FT-MOR data for MCT and InSb samples

Sample	Carrier concentration ($\times 10^{15}$ cm^{-3})		
	Hall	δ_p	δ_s
MCT-1 (77 K)	2.81	3.19	3.56
MCT-2 (77 K)	1.06	2.27	2.10
MCT-3 (77 K)	0.629	0.657	0.899
MCT-4 (77 K)	0.716	3.05	2.85
MCT-5 (77 K)	1.50	1.02	1.02
InSb-1 (300 K)	31.0	20.8	21.8
InSb-2 (300 K)	100	76.3	74.6
InSb-3 (300 K)	100	78.3	79.4
InSb-4 (300 K)	210	150	17.7
InSb-5 (300 K)	626–866	741	12.4

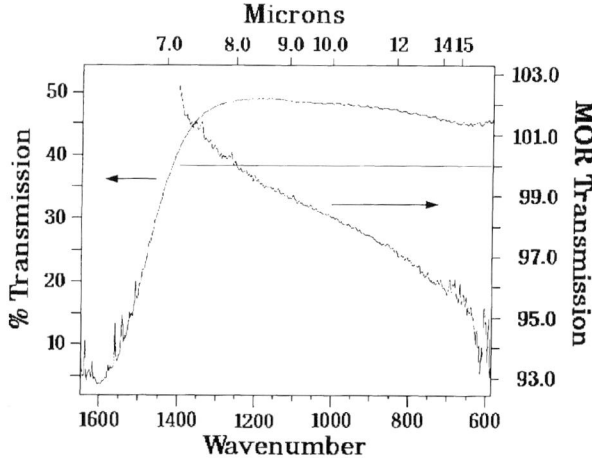

Fig. 2. Transmission and FT-MOR spectrum of a 57 μm epitaxial layer MCT sample

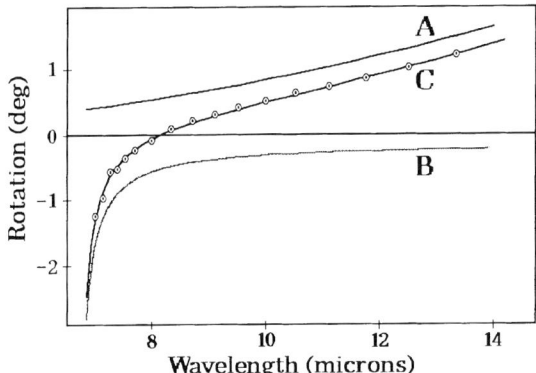

Fig. 3. MOR contributions to data in Fig. 2; *A* from free carrier plasma; *B* spin-induced; *C* sum of A and B

100% indicated a negative or a positive rotation of the probe wavelength, respectively.

The MOR spectrum was smoothed and the % transmissions at $50\,\mathrm{cm}^{-1}$ intervals were obtained. Intervals of $10\,\mathrm{cm}^{-1}$ near the bandgap were used to obtain a better fit. By Eq. (1), a non-linear fit of the rotation angles as a function of wavelength was used to determine the contribution of each source of MOR. The coefficients determined corresponded to the magnitude of the contribution from the various rotational components. An approximate bandgap wavelength (λ_g) was determined from the transmission spectrum of the sample and was adjusted to give the best fit of the data. Adding a constant to Eq. (1) resulted in better fits of the data. This constant compensated for any magnitude shifts during taking and smoothing of the data, and small contributions from the interband rotation.

The calculated contributions and fit for a typical MCT sample at $300\,\mathrm{K}$ are shown in Fig. 3. The carrier concentration was then determined from the coefficients obtained for the plasma and spin. Comparing the results in Table 1 for the two rotational components shows that in general the δ_s and δ_p carrier concentrations agree and are in reasonable agreement with the Hall test measurements. The Hall test provides an average measurement of the carrier concentration across the bulk of the sample and large differences

from the MOR measurements can be observed since the MOR measurements are at discrete positions on the samples. The discrepancy observed in InSb-4 and InSb-5 between δ_p and δ_s is due to a saturation of the Fermi energy levels as the donor concentration increases. Only δ_p can be used to obtain useful data for concentrations greater than $1 \times 10^{17}\,\mathrm{cm}^{-3}$ in the InSb samples.

At $77\,\mathrm{K}$ the bandgap can also shift and in the case of MCT this shift is to longer wavelengths. In previous work, when the bandgap wavelength shifted to overlap the IR laser wavelengths, the MOR could not be determined. The wide wavelength range of the FT-MOR allowed the rotation data to be measured and carrier concentrations determined for these samples.

Conclusions

The FT-MOR set-up has the advantage of collecting data over a broad spectral range with a single scan and co-adding of scans to yield better signal-to-noise ratio than does the infrared laser set-up. The carrier concentrations calculated from the plasma and spin induced rotation coefficients allow for an internal check of the accuracy of the measurement. Using FT-MOR measurements opens up the possibility of extending the technique as a general analytical mapping tool for other classes of materials, with potential applications in other fields.

References

[1] J. K. McDonald, G. A. Tanton, J. A. Grisham, C. R. Christensen, A. J. Syllaios, *Proc. SPIE* **1994**, *2228*, 316.
[2] S. D. Smith, T. S. Moss, K. W. Taylor, *J. Phys. Chem. Solids* **1959**, *11*, 131.
[3] S. Y. Yuen, P. A. Wolff, P. Becla, D. Nelson, *J. Vac. Sci. Technol.* **1987**, *A5*, 3040.
[4] L. M. Roth, *Phys. Rev.* **1964**, *133*, A542.
[5] A. J. Syllaios, C. L. Littler, X. N. Song, V. C. Lopes, C. R. Christensen, J. A. Grisham, F. W. Clarke, *Semicond. Sci. Technol.* **1993**, *8*, 953.
[6] H. Piller, in: *Semiconductors and Semimetals*, Vol. 8 (R. K. Willardson, A. C. Beer, eds.), Academic Press, New York, 1972, p. 103.

Mikrochim. Acta [Suppl.] 14, 479–483 (1997)
© Springer-Verlag 1997

Emission and Absorption Experiments on Strontium Titanate in Reducing Atmospheres

Fred A. Meyer*, **Roland Pohle**, **Josef Gerblinger**, and **Hans Meixner**

Siemens AG, Corporate Research and Development, Otto-Hahn-Ring 6, D-81739 München, Federal Republic of Germany

Abstract. We performed two types of experiments on polycrystalline $SrTiO_3$ thin films at temperatures between 573 and 1273 K in reducing gas mixtures containing CO or CH_4. Using an FT-IR spectrometer with a long-path gas cell, we analysed the composition of the gas mixture by means of absorption measurements before and after its exposure to the sample, and in a second step we measured the emission from the surface of the metal oxide. The absorption experiments revealed two different catalytically activated processes of CO oxidation, at 950 and 1100 K respectively. In the temperature region where catalytic conversion predominates, the oxidation of CO has an Eley–Rideal type characteristic, whereas the oxidation of CH_4 obeys a Langmuir–Hinshelwood mechanism. The experiments with CH_4 showed a pronounced dependence on the oxygen partial pressure above 1040 K, which could originate from reduction of the $SrTiO_3$ surface. We found emission from physisorbed CO_2 in CO-containing atmospheres, but no emission from adsorbed molecules was found in the experiments with CH_4.

Key words: infrared emission spectroscopy, infrared absorption spectroscopy, strontium titanate, reducing atmospheres.

The sensitivity of strontium titanate $SrTiO_3$ to oxygen is due to charged oxygen vacancies in the crystal lattice interacting with oxygen molecules in the gaseous phase [1]. When present in oxygen-containing atmospheres, this material also exhibits cross-sensitivity to reducing gases such as carbon monoxide or methane [2]. To clarify whether this is due to chemical reactions in the gaseous phase changing the oxygen partial pressure, to annihilation or generation of oxygen vacancies on the surface of the metal oxide, or to a direct interaction with the surface by chemisorption processes, we performed two types of experiments. Using an FT-IR spectrometer in absorption mode with a long-path gas cell, we analysed the composition of the gas mixture before and after its exposure to the sample, thus revealing any chemical reaction either in the gaseous phase

or catalytically activated on the surface of the sample. To gain insight into the binding mechanisms and reactions of the gas molecules adsorbed on the surface, we then measured the emission from the surface of the metal oxide.

Experimental

The experimental set-up (Fig. 1a) consists of a computer-controlled gas mixing unit, a sample chamber and a commercial FT-IR-spectrometer (Bruker IFS 66v) equipped with a KBr beam-splitter, an MCT detector and a variable path gas cell (optical path-length 0.8–8.0 m). The gas mixing unit allows preparation of mixtures

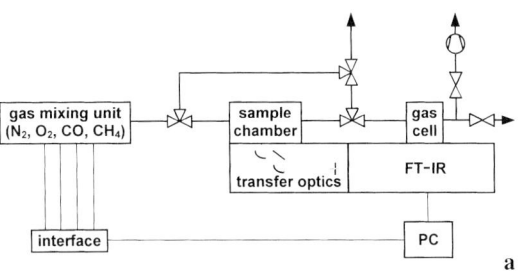

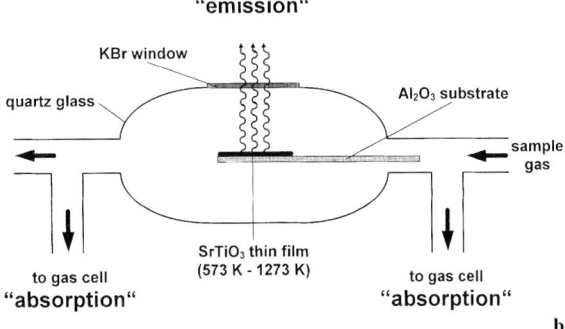

Fig. 1. a Experimental set-up. **b** Schematic diagram of sample chamber

---

* To whom correspondence should be addressed

containing 0.4–5.0% CO (purity 4.7) and 400–10000 ppm CH$_4$ (10%, remainder Ar). Oxygen is added by means of synthetic air (purity 5.0), and nitrogen (purity 5.0) is used as carrier gas. All experiments were done with a gas flow of 1000 ml/min.

The sample chamber (Fig. 1b) and the sample holder are made of quartz glass, which is said to be inert in the reducing gas mixtures used [3]. A KBr window allows IR emission from the heated surface of the sample to be passed into the spectrometer by transfer optics. This radiation was then analysed in the emission experiments, with the internal IR source off. At the maximum sample temperature of 1273 K, we estimate a maximum temperature of about 600 K at the inner surface of the quartz glass and the KBr window, so we cannot completely exclude an influence of the sample chamber on the reactions studied.

The samples consist of an Al$_2$O$_3$ substrate with electrodes (screen-printed Pt, thickness 6 µm) on one side and a heating element (screen-printed Pt) on the other side, with which temperatures up to 1273 K can be achieved. Both sides are covered with thin films (≈ 2.5 µm) of sputtered polycrystalline SrTiO$_3$ [1] in order to suppress the influence of the Pt, which is known to be highly catalytically active in the oxidation of CO or CH$_4$ in the temperature range considered. The geometrical size of the heated surface is 50 mm^2, but we cannot estimate the active surface area because we do not know the porosity of the polycrystalline films. The electrodes are used to monitor the resistance of the SrTiO$_3$ thin film, to ensure that an equilibrium state has been reached.

For the sake of reproducibility, the moisture contained in the substrates was removed by annealing the samples before every series of measurements, for at least 1 hour at 1273 K in synthetic air. This also helped to oxidize the sample surface when it had been reduced in a preceding series of measurements with oxygen deficiency.

The spectral investigations were performed over a range of sample temperatures between 573 and 1273 K, always according to the following measurement sequence. First, the desired gas mixture and sample temperature was set. After 15 min (which proved to be long enough for an equilibrium state to be reached at the surface, the gas in the gas cell to be completely exchanged, and the walls of the tubes and the gas cell to be completely saturated) a spectrum was recorded with a spectral resolution of 0.5 cm^{-1} and 2.0 cm^{-1} for the absorption and emission experiments, respectively. Co-addition of 256 scans was used for the absorption spectra and of 1000 scans for the emission spectra. Four-term Blackman–Harris apodization and a zero-filling factor of 2 were applied. The quantitative determination of the reaction products in the absorption measurements was made by using the peak method [4] with a digitized library of spectra from Infrared Analysis Inc. [5]. Only lines with a peak absorbance lower than 0.1 were taken into account, to avoid the distorting effect of non-linearity in the Lambert–Beer law [6].

Results and Discussion

Absorption Experiments

In the absorption experiments with CO in oxygen-containing atmospheres, no other reaction product than CO$_2$ was found in the temperature range of 673–1273 K and a range of oxygen concentrations of 0.8–20.0%, with the optical path-length of the gas cell set to 6.4 m. The CO concentration was always 2.0%.

To ensure reproducibility and exclude an irreversible change of the surface morphology of the sample

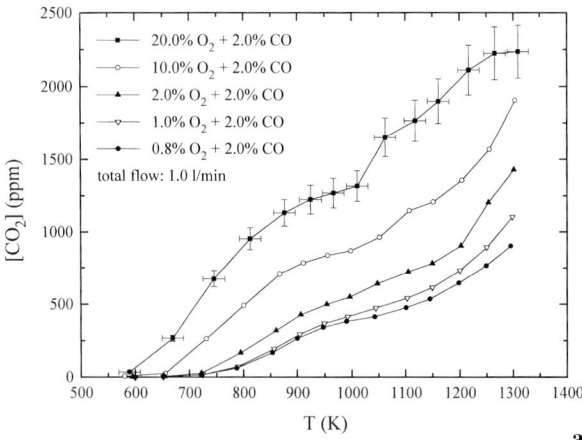

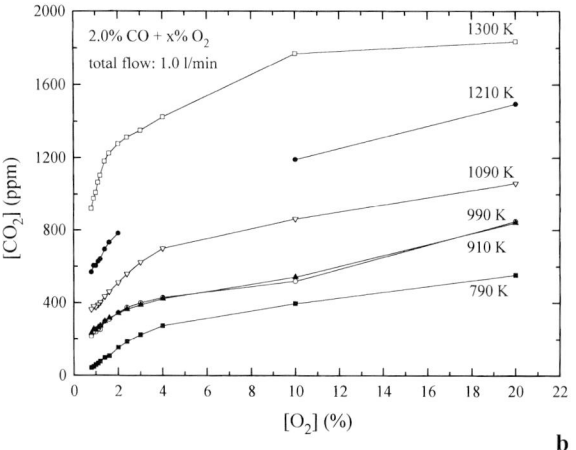

Fig. 2. Dependence of the conversion of 2.0% CO into CO$_2$, **a** on the sample temperature, **b** on the oxygen partial pressure

during the measurements, we examined the temperature dependence of the CO$_2$ production in 2.0% CO and 2.0% O$_2$ from time to time. The absolute values varied by a maximum of $\pm 8\%$, while no distinct tendency or change in the curve characteristic was observed.

Figure 2a shows the temperature-dependence of the conversion of CO into CO$_2$ at various oxygen partial pressures, with the gas cell length set to 0.8 m. Each data point represents the average concentration calculated from 8 different rotation lines of the asymmetric stretch mode v_3 of CO$_2$, centred at 2349 cm^{-1} [7]. The CO$_2$ production increases with increasing temperature, with a shoulder at 950 K for every gas composition examined. Above an oxygen concentration of 2.0% another shoulder-like characteristic arises at approximately 1100 K. At about 973 and 1123 K in 2.0% CO plus 20.0% O$_2$ mixture, the temperature at the sample surface, measured by the resistance of the heating element at constant heating voltage, was 25 and 60 K, respectively, higher than that determined during calibration in pure nitrogen, thus hinting at exothermic surface reactions. These results will be published in more detail elsewhere [8]. To further distinguish between a non-catalytic thermal reaction and a surface reaction, we then examined the CO_2 production without sample, by letting the gas mixture flow through a tube furnace (diameter equal to sample chamber, length of homogeneous zone 5 cm). Starting at 723 K, we obtained a steady rise of CO_2 production with increasing temperature without any structure. We therefore conclude that the two shoulders originate from catalytically activated CO oxidation on the surface of the $SrTiO_3$ thin film, superimposed on the gas phase reaction. At present, it is not yet clear whether these two shoulders can be assigned to two different adsorption sites for CO_2 resulting in two different desorption temperatures, to two different adsorption mechanisms (physisorption or chemisorption) or to two different reactions (CO oxidation with lattice oxygen or oxygen from the gas phase).

In another series of measurements we obtained the dependence of the CO_2 production on the oxygen concentration (Fig. 2b). Here, the shoulder at 950 K in Fig. 2a is confirmed, since the curves for 910 and 990 K are almost identical, whereas all the other curves show increasing CO_2 production with increasing temperature. At all temperatures, a steady rise of CO_2 production is visible with increasing oxygen concentration, with saturation at the highest concentrations. In the lower temperature region (790, 910, 990 K), where catalytically activated CO oxidation is predominant, this characteristic can be ascribed to an Eley-Rideal mechanism [9, 10]: one sort of gas molecules is adsorbed and reacts directly with another sort in the gas phase. At present, we cannot decide whether CO or O_2 plays the role of the adsorbed species.

In the case of CH_4 conversion, the results of the absorption experiments are much more difficult to interpret for we observe three different reaction products —CO, CO_2 and H_2O— and secondary reactions cannot be excluded (in experiments with an optical path-length of 8.0 m none of the other possible products, such as ethylene, acetylene or ethane, was identified).

The temperature-dependence of the conversion into CO_2 at various oxygen partial pressures is shown in

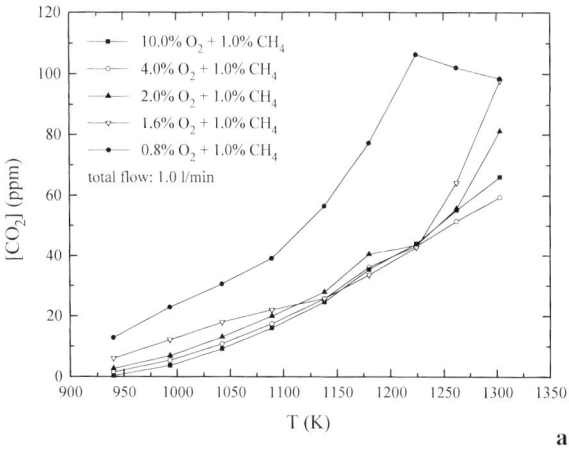

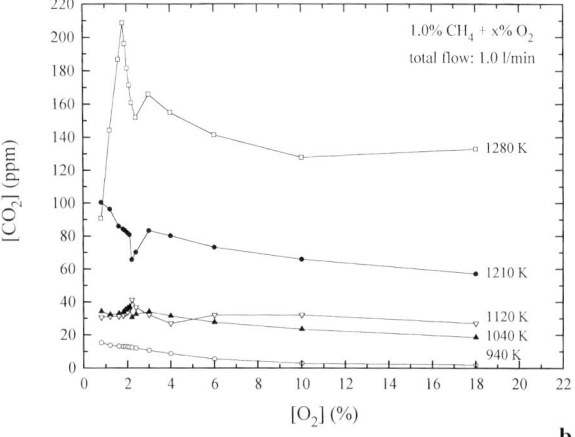

Fig. 3. Dependence of the conversion of 1.0% CH_4 into CO_2, **a** on the sample temperature, **b** on the oxygen partial pressure

Fig. 3a. Because the shape of the curves of water production is consistent with that of the CO_2 curves, only the concentration of carbon dioxide [CO_2] is displayed. CO_2 and H_2O appear at temperatures higher than 873 K. CO was observed in traces at above 1203 K, but in amounts comparable to CO_2 beyond 1123 K at the lower oxygen concentrations (0.8%, 1.6%). Similarly to the experiments with CO, the gas phase reaction is superimposed on the catalytic reactions at temperatures above 1073 K. (An experiment without sample in the furnace revealed a sudden rise of CO_2 production at 1073 K and saturation above 1150 K, consistent with the results of Seiyama [11]). Contrary to the behaviour expected from the gas phase reaction, the highest CO_2 concentrations were observed at the lowest oxygen concentration (0.8%). This leads to the conclusion that the main part of the conversion takes place on the surface.

Figure 3b shows the dependence of the conversion of methane at various sample temperatures on the oxygen partial pressure. In the temperature range between 873 and 1073 K, the conversion decreases continuously with increasing oxygen partial pressure. This behaviour is typical of a Langmuir–Hinshelwood reaction [9, 10], where both reactants have to be adsorbed for catalytic conversion. As described in [11] the catalytic oxidation is mainly operated by the lattice oxygen. Here, the thin film does not undergo a permanent reduction, because adsorbed oxygen can migrate into the lattice and replace the lattice oxygen removed by the catalytically activated CH_4 oxidation. This is enabled by the high diffusion coefficient of oxygen vacancies in $SrTiO_3$ at temperatures above 873 K [12]. At the highest temperatures (1213 and 1283 K), a strong dependence on the oxygen partial pressure appears for $[O_2]/[CH_4]$ ratios lower than 4. This could be due to reduction of the surface, assuming that the above-mentioned mechanism can no longer provide sufficient oxygen on the surface at these temperatures, which would also explain the pronounced formation of CO and the sudden decrease of the CO_2 production with 0.8% O_2 above 1223 K shown in Fig. 3a.

Emission Experiments

As the intensity scale for our emission data, we chose the relative emittance, i.e. the emission of the sample in a reducing atmosphere relative to the emission in the same gas mixture without the reducing component. We are thus eliminating the wavenumber dependence of the instrument response, but with the disadvantage of an enormously increasing noise level at higher wavenumbers, where the emission of the substrate and $SrTiO_3$ layer tends to zero.

Figure 4 shows the relative emittance in 2.0% CO and 1.0% O_2 at various temperatures. A linear slope has been removed but the spectra have not been smoothed in any way. Above 1000 K a broad emission feature clearly appears, superimposed by the adsorption of gas phase CO_2 between the sample and the KBr window (see Fig. 1b). This feature seems to be slightly red-shifted with respect to the v_3 CO_2 gas phase absorption. Hence we consider this to be only slightly distorted, physisorbed CO_2. Any attempt to subtract absorption spectra of gas phase CO and CO_2 failed, for we could not produce reference spectra which take into

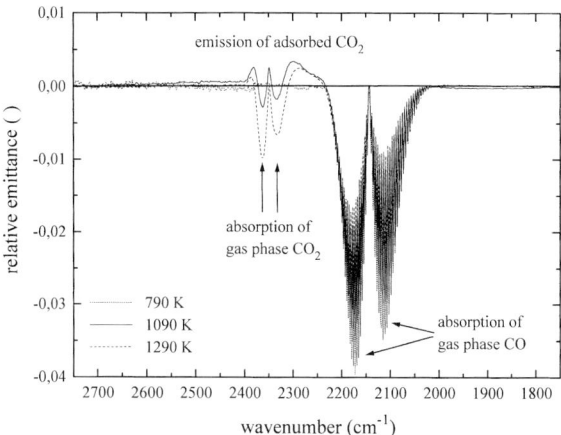

Fig. 4. Emittance of an $SrTiO_3$ sample under 2.0% CO plus 1.0% O_2, relative to 1.0% O_2. Emission of adsorbed CO_2 is superimposed on absorption of gaseous CO_2

account the temperature gradient of the gas mixture between the heated sample and the KBr window. We are therefore not able to reveal the structure of the CO_2 emission feature or any emission due to adsorbed CO, hidden behind the CO gas phase adsorption. At higher oxygen concentrations (above 2.0%), emission at around 2350 cm^{-1} is clearly visible even at a sample temperature of 773 K.

In CH_4-containing atmospheres, no IR emission from adsorbed molecules was observed, whether for CH_4 or any of the reaction products. We consider the following reasons for this: (*i*) because of a very weak bond strength, the total number of adsorbed hydrocarbon species is too small to give sufficient intensity in emission, (*ii*) the amount of CO_2 produced in the CH_4 experiments is more than one order of magnitude smaller than in the experiments with CO.

In forthcoming experiments, a modified sample geometry has to be thought of in order to reduce the background emission of the substrate relative to the emission of adsorbed molecules.

References

[1] J. Gerblinger, H. Meixner, *J. Appl. Phys.* **1990**, *67*, 7453.
[2] J. Gerblinger, U. Lampe, H. Meixner, I. V. Perczel, J. Giber, *Sens. Actuators B* **1994**, *19*, 529.
[3] J. R. Anderson, K. C. Pratt, *Introduction to Characterisation and Testing of Catalysts*, Academic Press, Sydney, 1985.
[4] H. Günzler, H. Böck, *IR-Spektroskopie: eine Einführung*, VCH, Weinheim, 1983.
[5] P. L. Hanst, S. T. Hanst, *Infrared Spectra for Quantitative Analysis of Gases*, Infrared Analysis Inc., Anaheim.

[6] P. L. Hanst, S. T. Hanst, *Gas Measurement in the Fundamental Infrared Region*, Infrared Analysis Inc., Anaheim.

[7] J. Staab, *Industrielle Gasanalyse*, Oldenbourg, Munich, 1994.

[8] F. A. Meyer, to be published.

[9] A. Clark, *The Theory of Adsorption and Catalysis*, Academic Press, New York, 1970.

[10] S. R. Morrison, *The Chemical Physics of Surfaces*, Plenum, New York, 1977.

[11] T. Seiyama, *Properties and Application of Perovskite-Type Oxides* (L. G. Tejuca, J. L. G. Fierro, eds.), Dekker, New York, 1993.

[12] A. Müller, K. H. Härdtl, *Appl. Phys. A*, **1989**, *49*, 75.

Mikrochim. Acta [Suppl.] 14, 485–487 (1997)
© Springer-Verlag 1997

Infrared Study of Oxygen Segregation at Structural Defects in Polycrystalline Silicon

B. Pivac[1,*], **A. Sassella**[2], and **A. Borghesi**[3]

[1] R. Bošković Institute, P.O. Box 1016, HR-10000 Zagreb, Croatia
[2] Dipartimento di Fisica, Università degli Studi di Milano, via Celoria 16, I-20133 Milano, Italy
[3] Dipartimento di Fisica, Università degli Studi di Modena, via Campi 213a I-41100 Modena, Italy

Abstract. We have studied oxygen- and carbon-doped multicrystalline silicon samples. It was shown that a significant part of the oxygen incorporated into the samples was accumulated at extended structural defects and therefore could not be quantified. Furthermore, we demonstrated that oxygen was accumulated in SiO_x clusters, with x close to 1. This explains the changes in the IR spectrum of the samples after thermal annealing, related to the dissolution of such clusters.

Key words: polycrystalline silicon, structural defects, oxygen-doped silicon, FT-IR spectroscopy.

Polycrystalline (or multicrystalline) silicon is a promising material for solar energy conversion owing to its low cost of production. Having grains with macroscopical dimensions (typically from a few mm to a few cm), such material keeps all the advantages and electrical characteristics of single-crystal silicon since each grain consists of one single crystal, typically grown in columnar form with crystallographic orientation close to that of the others. The major difference between this kind of material and electronic grade single-crystal silicon is that the poly-material typically contains various structural defects, such as grain boundaries, dislocations, etc., and a much higher concentration of various impurities that might be detrimental for device performance. Furthermore, the different production techniques (casting, ribbon growth, etc.) introduce different impurities in concentration typical for each specific technique of growth.

We have paid particular attention to samples grown by the edge-defined film-fed growth (EFG) technique, developed at Mobil Solar [1]. Due to the particular growth conditions (carbon container and dies), EFG introduces a significant carbon concentration in the material; in some cases, oxygen is also intentionally introduced into the growing samples by adding CO and/or CO_2 to the inert argon atmosphere of the puller around the meniscus [2]. The presence of oxygen in EFG material was found to significantly influence its photovoltaic characteristics [3], but the exact mechanism for that is not yet known.

In the silicon lattice at thermal equilibrium oxygen is incorporated in interstitial positions [4]. However, in multicrystalline silicon rich with various kinds of structural defects, oxygen already tends to segregate at such defects in the course of production and/or during the subsequent thermal processing of the material [5]. It is very important to know the exact place of oxygen in the silicon lattice in order to predict and model its behaviour during thermal annealing in the course of technological processes.

Experimental

Silicon sheets for the present study were grown at Mobil Solar Energy Corp. (now ASE Americas Inc.) in EFG ribbon system described in detail elsewhere [2]. Though the modern EFG growing system produces closed nonagon tubes [6], we used samples grown with an older system to illustrate better the oxygen segregation at structural defects.

The samples were boron-doped, with 2–$4\,\Omega\,cm$ resistivity and about $300\,\mu m$ thickness, and were grown in an inert argon atmosphere with the addition of CO_2.

Room temperature infrared (IR) spectra in the 5000–$400\,cm^{-1}$ wavenumber range were taken with a Bruker model IFS 113 v Fourier-transform IR (FT-IR) spectrometer. A globar source, a DTGS detector and a KBr beam-splitter were used. Instrumental resolution was $4\,cm^{-1}$. A differential technique was employed in the

* To whom correspondence should be addressed

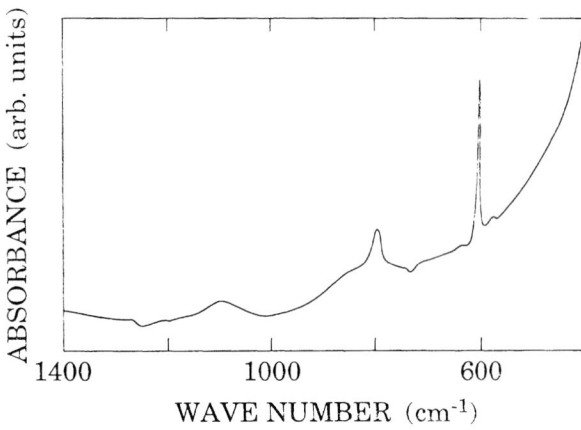

Fig. 1. Differential absorbance spectrum of as-received EFG multi-crystalline silicon sample taken with respect to float-zone, oxygen- and carbon-free single-crystal silicon

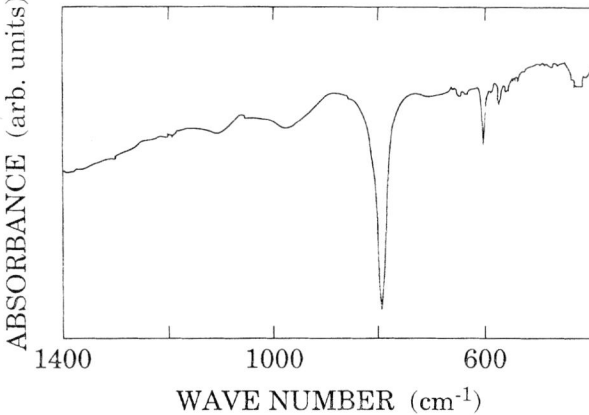

Fig. 2. Differential absorbance spectrum of multicrystalline EFG silicon sample annealed at 950 °C for 72 h in air, taken with respect to the same as-received sample

transmission measurements with a float-zone oxygen- and carbon-free monocrystalline silicon as reference.

Results and Discussion

Figure 1 shows the IR absorbance spectrum of an as-received EFG sample, normalized to the reference to remove all multiphonon contributions. The peaks observed are attributed to impurities present in the material; in particular, a significant presence of carbon up to 9×10^{17} atoms/cm^3 is demonstrated by the pronounced peak at 605 cm^{-1}. Such a high concentration of carbon is typical for this type of crystal growth. Also, the strong peak at 797 cm^{-1} is attributed to the presence of SiC nanocrystallites at the surface of the sample exposed to CO_2. The broad band beneath this one is attributed to [CO] complexes. Furthermore, the broad peak centered at about 1100 cm^{-1} is due to oxygen. Such oxygen is, however, not represented only by isolated atoms placed at interstitial positions, otherwise we would see only a sharper peak at 1107 cm^{-1}. In fact, deconvolution showed two contributions, one due to interstitial oxygen and the other at lower wavenumbers, indicating the presence of oxygen agglomerated in the suboxide-like form SiO_x. Such oxygen is very likely to be agglomerated at the structural defects, as indicated before. Moreover, the very high carbon concentration present in the material would further promote such oxygen agglomeration.

Now, an important question rises regarding the stability of these oxygen agglomerates to the thermal

treatments that the material endures during typical technological processes. We performed annealing in air for 72 h at temperatures ranging from 650 to 1150 °C. Figure 2 shows the differential absorbance spectrum of the sample annealed at 950 °C with respect to the same as-received sample. A negative peak at about 980 cm^{-1} with a shoulder at about 1100 cm^{-1} appears. Something similar was not observed after annealing at lower temperatures. The peak centred at about 980 cm^{-1} suggests that the oxygen was agglomerated in SiO_x form, with x close to 1 [7], as suggested by the deconvolution results cited above. Therefore, oxygen in this from is demonstrated to be unstable upon annealing at high temperature. Furthermore, though annealing was performed in air, any contribution from atmospheric oxygen can be ruled out, since no positive peak at 1107 cm^{-1} is observed. On the contrary, it is clearly shown in the figure that interstitial oxygen as well as substitutional carbon present in the material left their positions upon annealing, as indicated by the negative peaks at 1107 and 605 cm^{-1}. A pronounced negative peak at 797 cm^{-1} shows the dissolution of the SiC nanoprecipitates, as described earlier [8].

From the experiment performed we can draw the following conclusions. First, as received multicrystalline material (owing to the high concentration of structural defects) typically contains a significant concentration of dissolved oxygen. The oxygen atoms are placed in part at thermodynamically stable interstitial sites and in part are accumulated at structural defects (as evidenced by the broad IR band from 1080 to $920\,\mathrm{cm}^{-1}$). Oxygen present at interstitial sites can be quantified only by low temperature IR measurements, and oxygen accumulated at the structural defects can not be quantified. Moreover, oxygen accumulated at structural defects forms SiO_x clusters, with x close to 1. Such clusters, therefore, do not have the thermal stability of SiO_2 precipitates. Second, oxygen present at interstitial sites in EFG material is not affected even by a long thermal treatment at temperatures up to $950\,^{\circ}\mathrm{C}$. Furthermore, oxygen accumulated at structural defects also dissolves and enters the bulk, therefore increasing the overall content available for interaction with defects.

References

[1] J. P. Kalejs, in: *Silicon Processing for Photovoltaics II* (C. P. Khattak, K. V. Ravi, eds.), North-Holland, Amsterdam, 1987, p. 187.

[2] J. P. Kalejs, L.-Y. Chin, F. M. Carlson, *J. Cryst. Growth* **1983**, *61*, 473.

[3] B. Mackintosh, J. P. Kalejs, C. T. Ho, F. V. Wald, in: *Proceeding 3rd EC Photovoltaic Solar Energy Conference* (W. Palz, ed.), Reidel, Dordrecht, 1981, p. 553.

[4] A. Borghesi, B. Pivac, A. Sassella, A. Stella, *J. Appl. Phys.* **1995**, *77*, 4169.

[5] B. Pivac, A. Sassella, A. Borghesi, *Mater. Sci. Eng. B* **1996**, *36*, 55.

[6] A. S. Taylor, B. H. Mackintosh, L. Eriss, F. Wald, *J. Cryst. Growth* **1987**, *82*, 134.

[7] B. Pivac, A. Borghesi, M. Geddo, A. Sassella, A. Stella, *Appl. Surf. Sci.* **1993**, *63*, 245.

[8] B. Pivac, K. Furić, M. Milun, T. Valla, A. Borghesi, A. Sassella, *J. Appl. Phys.* **1994**, *75*, 3586.

Mikrochim. Acta [Suppl.] 14, 489–490 (1997)

FT-IR Spectroscopy of Tetrahedrally Coordinated Impurities in Sillenites

Enrico Bandini[1], **Paola Beneventi**[1], **Danilo Bersani**[1], **Rosanna Capelletti**[1,*], **Marin Gospodinov**[2], and **László Kovács**[3]

[1] INFM and Physics Department, University of Parma, Viale delle Scienze, I-43100 Parma, Italy
[2] Institute of Solid State Physics, Bulgarian Academy of Sciences, 72 Tzarigradsko Chaussee, 1784 Sofia, Bulgaria
[3] Research Laboratory for Crystal Physics, Hungarian Academy of Sciences, P. O. Box 132, H-1502 Budapest, Hungary

Abstract. FT-IR absorption spectroscopy in the temperature range 9–300 K was used to study the vibrational modes induced by aliovalent impurities (such as P^{5+} and V^{5+}) in $Bi_{12}SiO_{20}$(BSO), $Bi_{12}GeO_{20}$(BGO), and $Bi_{12}TiO_{20}$(BTO) and by traces of Ge^{4+} in $Bi_{12}SiO_{20}$ and of Si^{4+} in $Bi_{12}TiO_{20}$. Peaks due to the asymmetric F stretching mode, to its overtone, and to the combination with the symmetric A mode of the tetrahedral ImO_4 unit (Im being the impurity) were detected. Weak anharmonicity effects were displayed. The role played by the host matrix and by temperature on the peak positions was examined and discussed.

Key words: impurities, vibrational modes, oxides, FTIR spectroscopy, anharmonicity.

BSO, BGO, and BTO sillenites are body-centred cubic crystals with the general formula $Bi_{12}MeO_{20}$ (with Me = Si, Ge, and Ti). Some properties of these large-gap semiconductors (e.g. the photorefractive effect), relevant for advanced technological applications, are affected by impurities. Vibrational spectroscopy is a sensitive tool to detect impurities and to probe their environment. The impurity-related absorptions can be separated from the intrinsic ones, since the IR spectra of undoped BSO, BGO, and BTO are known [1–3]. These are peculiar to each sillenite in the wavenumber region above $600\,cm^{-1}$, where the stretching modes (fundamental, overtone, and combination transitions) of the tetrahedral MeO_4 unit are excited [1–3]. Additional weaker absorption lines have been observed at room temperature in doped samples and attributed to the presence of P^{5+} and V^{5+} [2,4]. In this work the more sensitive FT-IR spectroscopy is exploited to investigate (1) the presence and the coordination of quinquevalent (such as P and V) and of quadrivalent

impurities (such as Ge and Si) in BSO, BGO, and BTO single crystals, (2) possible anharmonicity effects, and (3) the temperature dependence of the line positions.

Experimental

The single crystals were grown in air by the balance-controlled Czochralski method (BSO and BGO) and from solution by the top seeding technique (BTO). The absorption spectra of single crystals and powdered samples (in KBr pellets) were measured by means of a Bomem DA8 FT-IR spectrometer in the $500–4000\,cm^{-1}$ range ($0.04\,cm^{-1}$ resolution) and by using a CTI-Cryogenics 21SC model Cryodine Cryocooler in the 9–300 K range.

Results and Discussion

The spectra of pellets (in the $600–1000\,cm^{-1}$ range) and of massive crystals (in the $1000–2000\,cm^{-1}$ range) of doped sillenites show some weaker peaks in addition to those typical of the undoped materials. Figure 1 displays, as an example, the spectra (in the $1000–2000\,cm^{-1}$ range) of pure (curve a) and doped BTO crystals (curves b, c and d): the intrinsic absorptions, peaking at 1328 and $1378\,cm^{-1}$, are very strong and common to all the spectra. The positions (at 9 K) of the additional peaks are summarized in Table 1 for each matrix and impurity Im: they are very close to those of the intrinsic peaks of $Bi_{12}MeO_{20}$ crystals (with Me = P, V, Ge, and Si) [3, 5]. The peaks detected at the lowest frequency v_F can be attributed to the asymmetric F stretching mode of the tetrahedral ImO_4 unit, resulting from the impurity Im being at the centre of an oxygen tetrahedron. The weaker peaks at v_{2F} and at v_{F+A} (see Table 1 and Fig. 1 for BTO) are ascribed to two-phonon transitions, i.e. the photons of such fre-

---

* To whom correspondence should be addressed.

Table 1. Wavenumbers v_F, v_{2F}, and v_{F+A} (cm^{-1}) of the peaks, measured at 9 K, and induced by different impurities Im in sillenite crystals

Matrix	Im	v_F	v_{2F}	v_{F+A}	matrix	Im	v_F	v_{2F}	v_{F+A}
BSO	P^{5+}	968.2	1930.0	1866.3	BTO	V^{5+}	767.8	1531.3	1560.6
BGO	P^{5+}	966.2	1927.9	1863.6	BSO	Ge^{4+}	682.2	1359.3	1394.5
BTO	P^{5+}	965.5	1923.8	1860.2	BTO	Si^{4+}	826.3	1647.4	1600.1

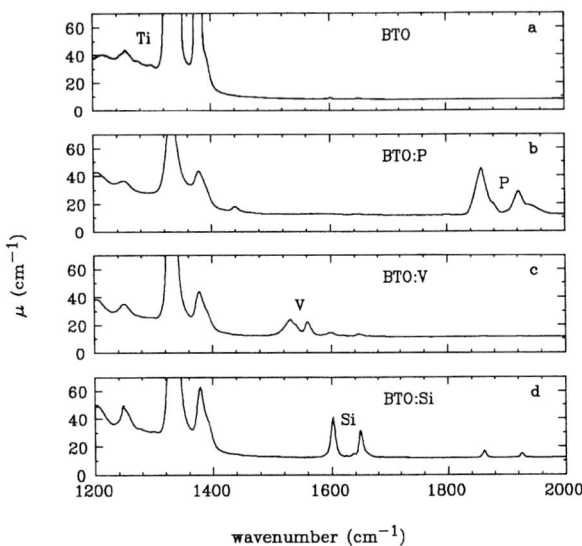

Fig. 1. Role of the impurities Im in the FT-IR absorption spectra of BTO single crystals. Spectra are measured at 9 K. **a** Undoped BTO; **b** BTO doped with P; **c** BTO doped with V (traces of P and Si are also detected, compare with curves **b** and **d**, respectively); **d** BTO doped with Si

quencies are exciting the first F-mode overtone and an F asymmetric mode + an A symmetric mode, respectively. The hypothesis is supported by the Raman measurements, which provided v_A, i.e. the frequency of the A-mode (not infrared active), and by the analogy with the multi-phonon transitions of the MeO$_4$ tetrahedron in undoped Bi$_{12}$MeO$_{20}$ [3, 5]. The two strong intrinsic lines at 1328 and 1378 cm^{-1} in Fig. 1 are just an example of the $2F$ and $F + A$ transitions of the TiO$_4$ tetrahedron. The present results suggest that both quadrivalent impurities, such as Ge and Si, and quinquevalent ones, such as P and V, are tetrahedrally coordinated, as Me^{4+} in the host matrix. The positive charge excess introduced by P^{5+} (or V^{5+}), could be easily compensated by an anti-site Bi^{3+}. The presence of an appreciable fraction of these defects [6] might

favour the substitution of an Im^{5+} for an Me^{4+}. The anharmonicity effects, responsible for the F overtone and for the $F + A$ combination-mode, are weak. In fact, the anharmonicity parameter x_e is 2–3 × 10^{-3}, as evaluated according to the Morse potential model.

For a given impurity the peak positions v_F, v_{2F} and v_{F+A} are only slightly affected by the host matrix, see Table 1. For quadrivalent impurities, such as Ge in BSO and Si in BTO, the peak positions, listed in Table 1, are shifted with respect to those of Ge and Si in undoped BGO and BSO, respectively, the former result being in nice agreement with results reported for mixed BGO–BSO crystals [3]. On increase of the temperature from 9 to 300 K the peaks weaken, broaden and shift. For Si the shift is appreciable ($\sim$ 6 cm^{-1}) and independent of its presence as a host Me^{4+} or as an impurity Im^{4+}. On the contrary, very weak or nearly negligible shifts are observed for V^{5+} and P^{5+}: the enhanced Coulomb interaction between Im^{5+} and O^{2-} and the smaller Im^{5+} radius might cause a contraction of the ImO$_4$ group and, as a consequence, a looser interaction with the surroundings.

As a concluding remark, FT-IR spectroscopy is very sensitive in monitoring the environmental symmetry of impurities present even in very weak traces in sillenites.

References

[1] W. Wojdowski, T. Łukasiewicz, W. Nazarewicz, J. Zmija, *Phys. Status Solidi* **1979**, *94b*, 649.

[2] A. V. Khomich, Yu. F. Kargin, P. I. Perov, V. M. Skorikov, *Inorg. Mater.* **1990**, *26*, 1635.

[3] P. Beneventi, R. Capelletti, L. Kovács, *Radiation Effects and Defects in Solids*, **1995**, *134*, 293.

[4] V. V. Volkov, Yu. F. Kargin, A. V. Khomich, P. I. Perov, V. M. Skorikov, *Inorg. Mater.* **1989**, *25*, 701.

[5] M. Devalette, J. Darriet, M. Couzi, C. Mazeau, P. Hagenmuller, *J. Solid State Chem.* **1982**, *43*, 45.

[6] S. C. Abrahams, P. B. Jamieson, J. L. Bernstein, *J. Chem. Phys.* **1967**, *47*, 4034.

Mikrochim. Acta [Suppl.] 14, 491–492 (1997)
© Springer-Verlag 1997

High Resolution FT-IR Spectroscopy of OH⁻ Ion Perturbed by Sr²⁺ in KCl

Paola Beneventi, Patrizia Bertoli, and **Rosanna Capelletti***

INFM and Physics Department, University of Parma, Viale delle Scienze, I-43100 Parma, Italy

Abstract. High resolution (0.04 cm⁻¹) FT-IR absorption spectroscopy was applied in the range 9–300 K to study vibrational modes of OH⁻ in KCl: Sr²⁺ mixture. The lines (very narrow at 9 K) in the range 3500–3650 cm⁻¹ were attributed to the stretching of OH⁻ embedded in Sr²⁺-complexes. The spectra (fundamentals, overtones, and isotopic replicas) were analysed in by means of the Morse anharmonic oscillator model, the parameters of which were evaluated and discussed. A single phonon coupling model, applied to the temperature-dependence of the line position, provided the frequencies of the coupled phonons.

Key words: OH stretching, FT-IR spectroscopy, alkali metal halides, impurities, anharmonicity.

The stretching frequency of the OH⁻ ion, embedded in complexes formed by bivalent cation impurities (Me²⁺) in alkali metal halides, is a very sensitive probe of even slight changes in the surroundings. In fact the absorption spectra of LiF and NaF (Me²⁺ = Mg²⁺) [1] and of NaCl (Me²⁺ = Cd²⁺ or Ca²⁺) [2] display in the range 3500–3700 cm⁻¹ a large number of narrow lines, due to OH⁻ interacting with different Me²⁺-related lattice defects. In order to obtain a more general picture of Me²⁺–OH⁻ interaction alkali metal halides, high resolution FT-IR absorption spectroscopy has been applied to study the KCl: Sr²⁺ system in the range 9–300 K. The aim of the work was (1) to gain an insight into the nature of the Sr²⁺–OH⁻ complexes, (2) to analyse the OH⁻ stretching modes on the basis of the Morse model, (3) to compare the Morse parameters with those obtained for similar systems, and (4) to study the phonon coupling of the OH⁻ stretching modes.

Experimental

KCl single crystals were grown by the Kyropoulos method in a nitrogen atmosphere. The Sr²⁺ and OH⁻ dopings were achieved by adding to the melt SrCl₂ (10^{-4}–10^{-3} mole fraction) and KOH pellets. The isotopic substitution of OH⁻ with OD⁻ was achieved by heating the samples at ∼730°C in dry N₂ flowing through D₂O. The absorption measurements were performed by means of a Bomen DA8 FT-IR spectrometer in the range 500–7500 cm⁻¹, with a resolution as good as 0.04 cm⁻¹. The spectra were measured in the range temperature 9–300 K achieved by using a CTI-Cryogenics 21SC model Cryodine Cryocooler.

Results and Discussion

Figure 1 (curves a and b) shows the absorption spectra of KCl:Sr,OH in the 3550–3650 cm⁻¹ range, i.e. in the region typical for the OH⁻ stretching mode: curve a displays the broad spectrum measured at 300 K, while curve b gives that taken at 9 K. At low temperature the lines are very narrow (< 0.06 cm⁻¹). The substitution of OD⁻ for OH⁻ induces the growth of a line at 2654.61 cm⁻¹, which is the replica of that at 3598.35 cm⁻¹. This result stresses that the two lines are due to hydroxyl (deuteroxyl) ions. The 3642.10 cm⁻¹ line, present also in KCl:OH, is due to OH⁻ not interacting with any cation impurity [3], while the other lines of Fig. 1 (a and b) appear only if KCl is co-doped with Sr²⁺ and OH⁻. By changing the Sr²⁺ concentration in the melt, the relative weight of the lines in the spectra is changed, confirming that they are due to the OH⁻–Sr²⁺ interaction. Hints on the nature of defects is obtained by submitting the samples to proper thermal treatments. In fact, quenching of a thin sample from 600°C induces an increase in the 3642.1 cm⁻¹ and 3598.35 cm⁻¹ lines, while the other lines disappear. A subsequent heat treatment at 600°C,

---

* To whom correspondence should be addressed

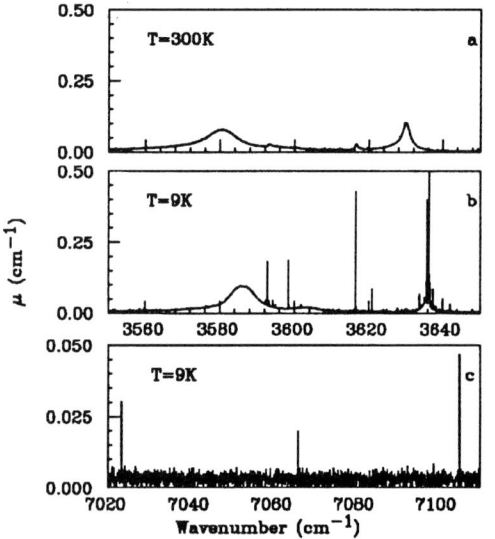

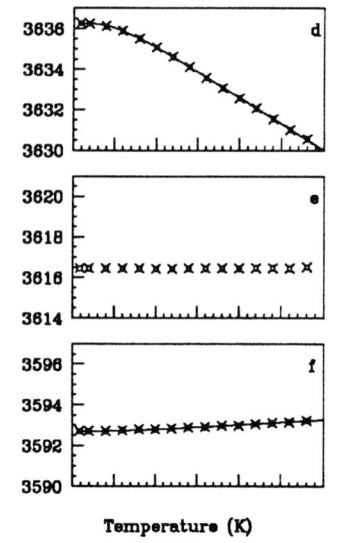

Fig. 1. High resolution (0.04 cm⁻¹) optical absorption spectra and temperature dependence of OH⁻ stretching modes in KCl:Sr²⁺ (10⁻⁴ mole fraction), OH⁻ samples: **a** optical absorption spectrum measured in the region of OH⁻-stretching modes (fundamental transitions) at 300 K (thickness $x = 22.2$ mm); **b** as **a**, but spectrum measured at 9 K; **c** spectrum measured in the region of OH⁻-stretching modes (overtone transitions) at 9 K ($x = 27.6$ mm). **d, e, f** temperature-dependence of the line position $\tilde{v}_{str}(T)$ for the lines peaking at 3636.25, 3616.44 and 3592.70 cm⁻¹ at 9 K: the crosses are the experimental data and the full line is the fitting according to [4]

followed by a slow cooling (annealing), recovers the 3616.44 and 3636.25 cm⁻¹ lines, present already in the original samples. These results can be interpreted as follows. The quenching breaks complexes formed by Sr²⁺ and OH⁻ ions (3616.44 and 3636.25 cm⁻¹ lines) leaving only isolated OH⁻ (3642.1 cm⁻¹ line) and a very simple defect, as an OH⁻ near an Sr²⁺ (3598.35 cm⁻¹ line). The subsequent annealing favours the recovery of complexes formed by OH⁻ and Sr²⁺-clusters (3616.44 and 3636.25 cm⁻¹ lines).

At wavenumbers twice (within 2%) those of the lines displayed in Fig. 1b, weak overtone absorptions were monitored, see Fig. 1c. Lines due to bending + stretching combination modes were detected in the 520–570 and 4100–4200 cm⁻¹ respectively. The stretching mode spectra (fundamentals, overtones, and isotopic replicas), analysed by means of the Morse anharmonic oscillator model, gave the following values for the parameters: the anharmonicity x_{em} and D_e, i.e. the binding energy of the OH molecule, were in the range 0.022–0.023 and 41000–43200 cm⁻¹, respectively. They were practically temperature-independent in the range 9–150 K, where the overtone could be detected. The ratio ρ between the reduced masses of OH and OD, embedded in the same complex, (ρ referring to the hydroxyl ion coupling to the lattice) was 0.5302 for the line at 3598.35 cm⁻¹, i.e. very close to the free diatomic molecule value (0.5300). The results agree with those for Me²⁺–OH defects in LiF and NaF (Me²⁺ = Mg [1]), in NaCl (Me²⁺ = Ca or Cd [2]), and for the isolated OH⁻ ion [3].

Since the spectra are heavily affected by temperature (compare curves a and b in Fig. 1), the line positions $\tilde{v}_{str}(T)$ and the widths $\Delta W(T)$ were analysed as a function of the temperature T. With increasing T the lines broaden and shift in different ways: towards lower wavenumbers (Fig. 1d) and towards higher wavenumbers (Fig. 1f). In a few cases no shift was observed (Fig. 1e). The experimental data of $\tilde{v}_{str}(T)$ fit well the single phonon coupling model [4] (full lines in Fig. 1, d and f). This analysis made it possible to evaluate the frequency $\tilde{v}_{ph,OH}$ of the coupled phonon: the values for different Sr–OH defects were in the range 60–150 cm⁻¹, i.e. in the range of the allowed phonon frequencies of the host KCl lattice. As for very narrow lines ΔW was comparable with the apparatus resolution, the lineshape could not be correctly analysed, at least at low T. In such cases $\Delta W(T)$ could not be tested against the model [4].

References

[1] R. Capelletti, P. Beneventi, E. Colombi, W. B. Fowler, *Nuovo Cimento Soc Ital. Fis.*, D **1993**, *15 D*, 415.

[2] P. Beneventi, R. Capelletti, M. Darra, R. Fieschi, W. B. Fowler, A. Gainotti, *Radiation Effects and Defects in Solids*, **1995**, *134*, 379.

[3] A. Afanasiev, C. P. An, F. Luty, in: Defects in Insulating Materials, Vol. 1 O. Kanert, J.-M. Spaeth, eds, World Scientific, Singapore, 1993, p. 551.

[4] P. Dumas, Y. J. Chabal, G. S. Higashi, *Phys. Rev. Lett.* **1990**, *65*, 1124.

Mikrochim. Acta [Suppl.] 14, 493–496 (1997)
© Springer-Verlag 1997

Vibrations of Monosubstituted Octasilasesquioxanes

Claudia Marcolli, **Roman Imhof**, and **Gion Calzaferri***

Institute for Inorganic and Physical Chemistry, University of Berne, Freiestrasse 3, CH-3000 Bern 9, Switzerland

Abstract. The vibrational spectra of monosubstituted sphero-siloxanes of the type $RH_7Si_8O_{12}$ can be understood as a superposition of the spectral features of $H_8Si_8O_{12}$ and of the substituent R. We show this for the FT-Raman spectra of $RH_7Si_8O_{12}$ with R = $-CH_2-CH_2-C_6H_5$ and $-CH=CH-C_6H_5$. The force fields of $H_8Si_8O_{12}$ and of the organic substituent, ethylbenzene and styrene, respectively, were combined and normal coordinate analysis was applied. The spectra of phenethyl–$H_7Si_8O_{12}$ and styryl–$H_7Si_8O_{12}$ were correlated with the spectrum of $H_8Si_8O_{12}$. The Si–X stretching frequencies in the spherosiloxanes are larger than in most other siloxane compounds. We have shown that this is due to the different X–Si–O–Si conformations, which are anti and syn, respectively, and can be explained by the dependence of the bond order on the X–Si–O–Si dihedral angle.

Key words: Raman spectrometry, octasilasesquioxanes, normal coordinate analysis, valence force field.

A number of mono and more highly substituted silasesquioxanes have become available in the last few years [1–15]. As the application range of substituted spherosiloxanes increases, a deeper insight into the vibrational pattern of these molecules is desirable. In [15] the IR spectrum of hexyl–$H_7Si_8O_{12}$ was analysed according to group frequencies. For this work IR and Raman spectra have been measured and are used to describe the vibrational structure of phenethyl–$H_7Si_8O_{12}$ and styryl–$H_7Si_8O_{12}$ as a superposition of the spectral features of $H_8Si_8O_{12}$ and of the organic substituents ethylbenzene and styrene, respectively. Normal coordinate analysis was applied to assign the frequencies.

Experimental

Synthesis

The synthesis of phenethyl–$H_7Si_8O_{12}$ is described in [2]. Starting from $H_8Si_8O_{12}$ and phenylacetylene styryl–$H_7Si_8O_{12}$ was synthesized by hydrosilylation, in the same way as phenethyl– and hexyl–$H_7Si_8O_{12}$ [2, 15].

Raman Spectroscopy

The FT-Raman spectra were recorded with a Bomem DA3.01 FTIR spectrometer. The interferometer was equipped with a quartz beam-splitter and a liquid-nitrogen cooled InGaAs detector. The cw Nd^{3+}: YAG laser (Quantronix Model 114) was run in the TEM_{00} mode at 9395 cm^{-1}. Rayleigh scattering was blocked by three holographic super notch filters (Kaiser Optical Systems) in 6° angle position.

Normal Coordinate Analysis

The vibrational analysis of the molecules investigated was performed by the Wilson GF matrix method [16] with the computer package QCMP067. To analyse the spectra of phenethyl–$H_7Si_8O_{12}$ and styryl–$H_7Si_8O_{12}$ the force fields of $H_8Si_8O_{12}$ and of the organic substituents were combined. For the siloxane cage a harmonic force field in terms of internal force constants was used [17]. For phenethyl–$H_7Si_8O_{12}$ a force field of ethylbenzene determined by Snyder and Painter [18] was used and for styryl–$H_7Si_8O_{12}$ a force field of styrene that was generated by combining the force field of the phenyl ring [18] with the force field of propene [19].

Correlation of the Spectra

In Fig. 1 the FT-Raman spectra of the monosubstituted compounds are divided into the lines assigned to the siloxane cage, the lines belonging to the substituent, and the vibrations involving the Si–C bond. This presentation is based on the result of the normal coordinate analysis, which we illustrate in Fig. 2. The

---

* To whom correspondence should be addressed

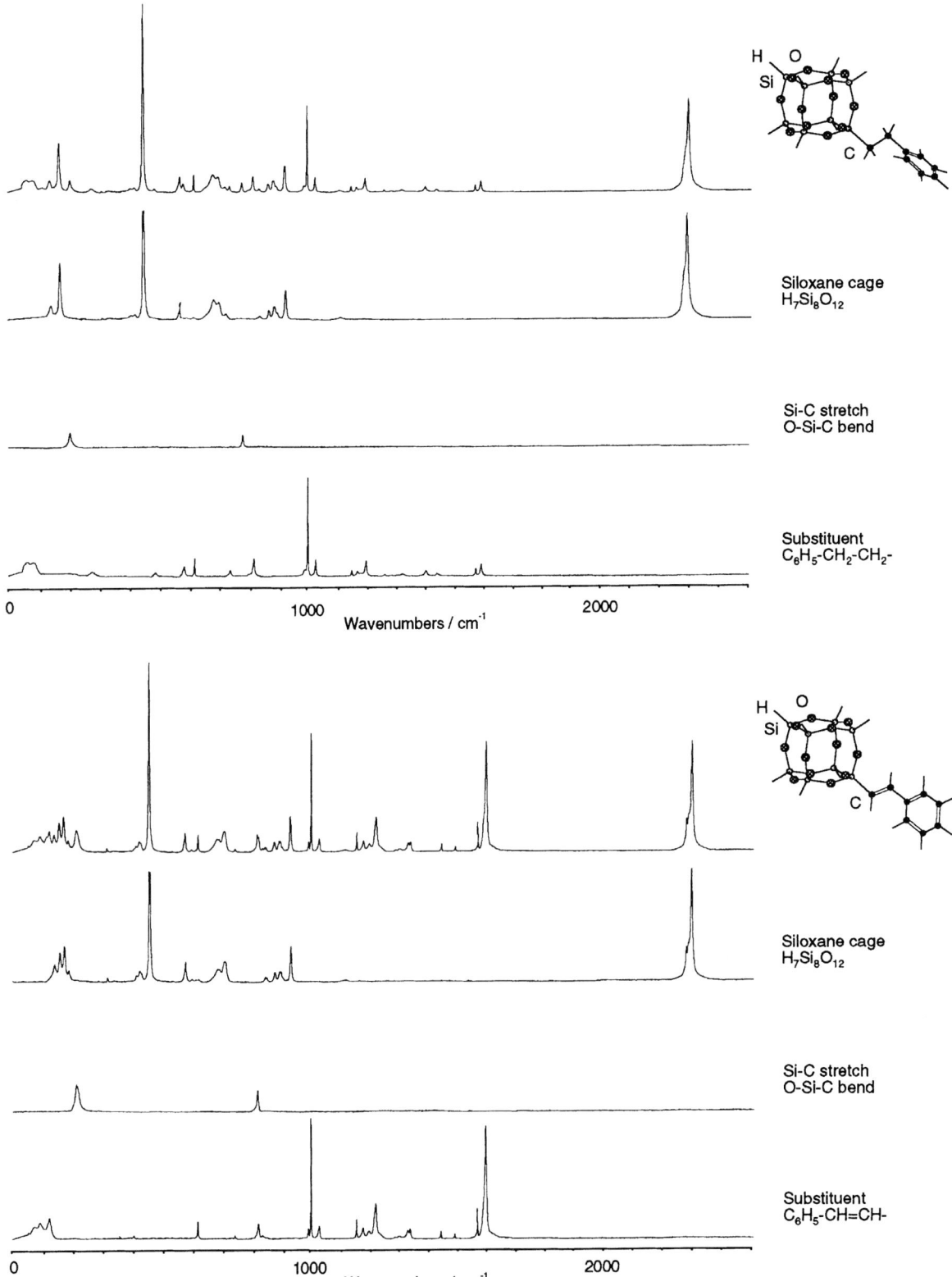

Fig. 1. FT-Raman spectra of phenethyl–$H_7Si_8O_{12}$ (top) and styryl–$H_7Si_8O_{12}$ (bottom) divided into the lines of the siloxane cage, of the substituent, and of the Si–C stretch and the O–Si–C bend. The highest peak in the spectra of $H_7Si_8O_{12}$ is cut

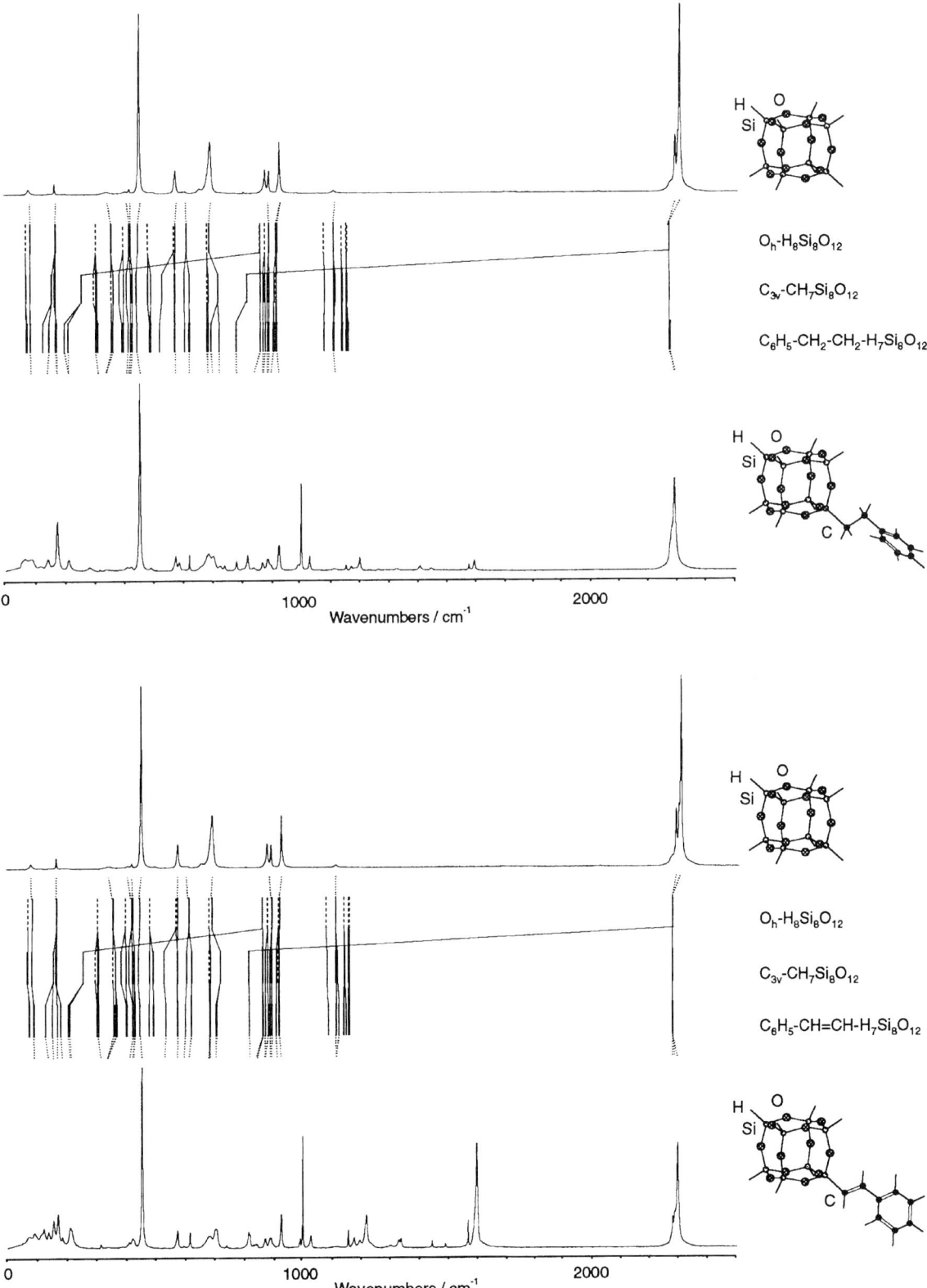

Fig. 2. Correlation diagram of the siloxane cage vibrations of phenethyl–$H_7Si_8O_{12}$ (top) and styryl–$H_7Si_8O_{12}$ (bottom) with $H_8Si_8O_{12}$. The middle part shows the calculated spectra and the correlation of the lines

FT-Raman spectra of $H_8Si_8O_{12}$ and of the substituted silasesquioxanes are correlated in two steps. In a first step the substituent is treated as a point mass with the weight of a C atom, what leads to C_{3v} symmetry. In a second step the whole substituent is included in the calculation.

The symmetry reduction of the siloxane cage has two effects: inactive modes in $H_8Si_8O_{12}$ become active in the substituted compounds and degenerate modes split. Both effects can be observed in the spectra of phenethyl–$H_7Si_8O_{12}$ and styryl–$H_7Si_8O_{12}$. In particular, the O–Si–H bending vibrations in the region from 850–930 cm^{-1} can be understood as a result of the symmetry reduction. They form a typical pattern for monosubstituted octahydrosilasesquioxanes. The only two vibrations that shift considerably compared to $H_8Si_8O_{12}$ are the Si–C stretch which correlates with the Si–H stretches of $H_8Si_8O_{12}$, and the O–Si–C bending, which correlates with the O–Si–H bendings of $H_8Si_8O_{12}$. The IR spectra of these compounds were also analysed. We found that they can be understood in the same way as the Raman spectra.

Si–C and Si–H Stretching Modes

The Si–C stretching modes appear at 784 cm^{-1} for R = –CH_2–CH_2–C_6H_5, at 790 cm^{-1} for R = –$(CH_2)_5$–CH_3 [15], and at 821 cm^{-1} for R = –CH = CH–C_6H_5. The Si–H stretches show only small frequency shifts. The totally symmetric ν(Si–H) is at 2295 cm^{-1}. In the substituted and in the unsubstituted $H_8Si_8O_{12}$ the Si–C and Si–H stretching frequencies are larger than in most other siloxane compounds (Table 1). A calculation of the EHMO-ASED type on

$XSi(OSiMe_3)_3$ (X = H, CH_3, Cl) in its syn and anti position (Fig. 2 in [20]) showed that the Si–X bond order depends on the X–Si–O–Si conformation [20]. For X = H the Si–H bond order is highest for the anti conformation ($\varphi = 0$) and decreases with increasing dihedral angle φ (Fig. 3 in [20]). As the cage structure of $X_8Si_8O_{12}$ requires the anti conformation and syn is the stable one in $XSi(OSiMe_3)_3$, the Si–X bond order for $X_8Si_8O_{12}$ is higher than for $XSi(OSiMe_3)_3$.

Acknowledgement. This work was financed by the Schweizerischer Nationalfonds zur Förderung der wissenschaftlichen Forschung (project NF 20-040598.94/1) and by the Schweizerisches Bundesamt für Energiewirtschaft (project BEW (93)034).

References

[1] J. D. Lichtenhan, *Comments Inorg. Chem.* **1995**, *17*, 115.
[2] G. Calzaferri, D. Herren, R. Imhof, *Helv. Chim. Acta* **1991**, *74*, 1278.
[3] V. W. Day, W. G. Klemperer, V. V. Mainz, D. M. Millar, *J. Am. Chem. Soc.* **1985**, *107*, 8262.
[4] H. Bürgy, G. Calzaferri, *Helv. Chim Acta* **1990**, *73*, 698.
[5] D. Herren, H. Bürgy, G. Calzaferri, *Helv. Chim. Acta* **1991**, *74*, 24.
[6] D. Hoebbel, I. Pitsch, D. Heidemann, *Z. Anorg. Allgem. Chem.* **1991**, *592*, 207.
[7] G. Calzaferri, R. Imhof, *J. Chem. Soc. Dalton Trans.* **1992**, 3391.
[8] M. Morán, C. M. Casado, I. Cuadrado, J. Losada, *Organometallics* **1993**, *12*, 4327.
[9] A. R. Bassindale, T. E. Gentle, *J. Mater. Chem.* **1993**, *3*, 1319.
[10] G. Calzaferri, R. Imhof, K. W. Törnroos, *J. Chem. Soc. Dalton Trans.* **1993**, 3741.
[11] B. J. Hendan, H. C. Marsmann, *J. Organomet. Chem.* **1994**, *483*, 33.
[12] P. Jutzi, C. Batz, A. Mutluay, *Z. Naturforsch.* **1994**, *49b*, 1689.
[13] A. Sellinger, R.M. Laine, V. Chu, C. Viney, *J. Polym. Sci. Part A* **1994**, *32*, 3069.
[14] F. J. Feher, K. J. Weller, J. J. Schwab. *Organometallics* **1995**, *14*, 2009.
[15] G. Calzaferri, R. Imhof. K. W. Törnroos, *J. Chem. Soc. Dalton Trans.* **1994**, 3123.
[16] D. F. McIntosh, M. R. Peterson, *General Vibrational Analysis System*, QCPE No. QCMP067, 1988.
[17] M. Bärtsch, P. Bornhauser, G. Calzaferri, R. Imhof, *J. Phys. Chem.* **1994**, *98*, 2817.
[18] R. W. Snyder, P. C. Painter, *Polymer* **1981**, *22*, 1629.
[19] B. Silvi, P. Labarbe, J.P. Perchard, *Spectrochim. Acta* **1973**, *29A*, 263.
[20] M. Bärtsch, G. Calzaferri, C. Marcolli, *Res. Chem. Intermed.* **1995**, *21*, 577.

Table 1. Experimental FT-Raman Si–X stretching frequencies/ cm^{-1}

Molecule	X = H	X = CH_3	X = Cl
$X_8Si_8O_{12}$[a]	2302	819	733
$XSi(OSiMe_3)_3$	2207	748	588

[a] Totally symmetric stretching vibration.

Mikrochim. Acta [Suppl.] 14, 497–499 (1997)
© Springer-Verlag 1997

Far-Infrared Spectra and Potential Energy Surface for 5,6-Dihydro-4H-thiopyran

Jaebum Choo[1], **Nils T. Meinander**[2], **John R. Villarreal**[3], and **Jaan Laane***

Department of Chemistry, Texas A&M University, College Station, TX 77843-3255, U.S.A

Abstract. The far infrared spectrum of 5,6-dihydro-4H-thiopyran shows eleven ring bending bands (near 120 cm^{-1}), four ring-twisting bands (near 275 cm^{-1}), and twelve sum and difference bands in the 383–397 cm^{-1} and 148–166 cm^{-1} regions. From these frequency data a detailed energy map was constructed for many of the excited vibrational states of the two coupled out-of-plane ring vibrations. A two-dimensional potential energy surface, which satisfactorily fits the observed data, was determined in terms of the bending and twisting coordinates respectively. The minima on the potential energy surface correspond to twisting angles of $\pm48°$ (half-chair conformation). The lowest-energy bent (boat) conformation corresponds to a saddle point 1500 cm^{-1} above the twisted conformation on the potential energy surfaces, and the barrier to planarity was estimated to be 6000 cm^{-1}.

Key words: far-infrared spectra, potential energy surface, ring-bending vibrations.

Over the past several years we have studied the low-frequency vibrational spectra of cyclohexene (**I**) [1,2], several of its oxygen-containing analogues (**II–V**) [3], and 5,6-dihydro-2H-thiopyran (**VI**) [4].

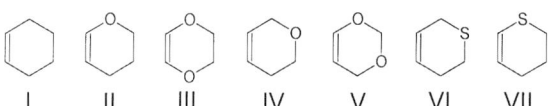

These molecules possess three out-of-plane vibrations: the ring-bending, the single-bond ring-twisting (SB twisting), the double-bond twisting (DB twisting). The first two have large amplitudes, are generally

coupled, and can be utilized to represent the primary conformational changes. For molecules **I–VI** we have determined two-dimensional potential energy surfaces in terms of the bending and SB twisting vibrations. In the present study we report on 5,6-dihydro-4H-thiopyran (**VII**) in order to provide additional insight into the conformational energetics of cyclohexene analogues and sulfur-containing ring systems.

Experimental

Far infrared spectra of the vapor-phase sample in a long-path cell were recorded on a Bomem DA3.002 Fourier-transform infrared interferometer equipped with a liquid-helium cooled germanium bolometer as a detector. More details can be found elsewhere [1–6].

Assignment of Spectra

Figure 1 shows a survey vapor-phase far infrared spectrum of 5,6-dihydro-4H-thiopyran in the 70–350 cm^{-1} region. The two principal absorption regions result from the ring-bending (100–125 cm^{-1}) and the single bond ring-twisting (270–280 cm^{-1}), respectively. Figure 2 shows the far infrared spectrum of the ring-

Permanent addresses: [1] Department of Chemistry, Hanyang University, Kyeongkido 425–791, Korea; [2] Radiocarbon Dating Laboratory, FIN-00014, University of Helsinki, Finland; [3] Department of Chemistry, University of Texas-Pan America, Edinburg, TX 78539, U.S.A
* To whom correspondence should be addressed

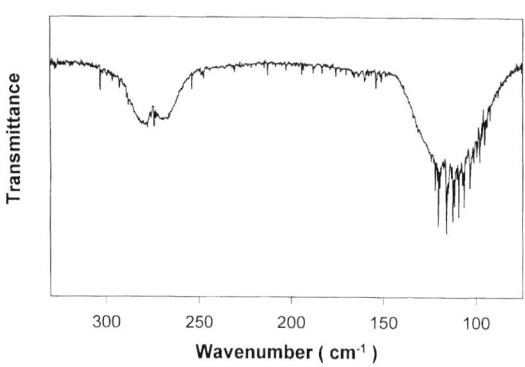

Fig. 1

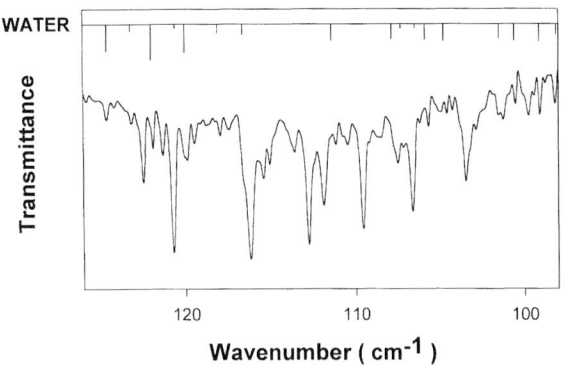

Fig. 2

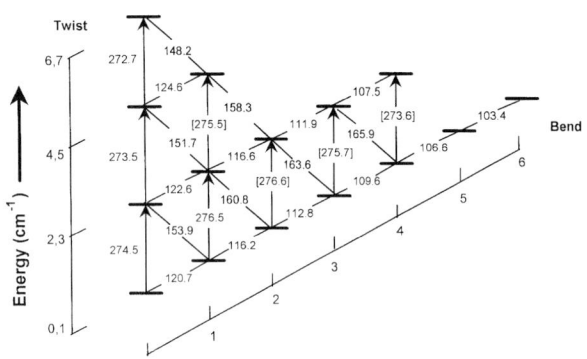

Fig. 3

bending region in greater detail. Several twist–bend difference bands (148–166 cm^{-1}) and sum bands (383–397 cm^{-1}) were also observed. Figure 3 presents the energy level diagram showing some of the observed vibrational transitions. The twist–bend difference and sum bands are extremely useful in helping to confirm many of the assignments. Table 1 presents a listing of observed frequencies and their assignments in terms of the quantum numbers V_{twist}, V_{bend}. The twisting energy levels are doubly degenerate as a result of the double-minimum potential well for this vibration.

Calculations

Kinetic Energy Expansions

The reduced masses for the large-amplitude ring-bending and ring-twisting vibrations are not constant, but depend on the vibrational coordinates. The kinetic energy functions for the ring-bending (x) and ring-twisting (τ) were calculated and are given elsewhere [5,6].

Table 1. Observed and calculated vibrational frequencies (cm^{-1}) for the ring-bending and ring twisting transitions of 5,6-dihydro-4H-thiopyran

| Transition[a] | Observed | Frequency | | Relative intensity | |
		Calculated	Δ	Observed	Calculated
Ring-twisting bands					
(0,0)-(2,0)	274.5	274.9	−0.4	(1.00)	(1.00)
(2,0)-(4,0)	273.5	273.2	0.3	0.38	0.51
(4,0)-(6,0)	272.7	272.0	0.7	0.19	0.20
(0,1)-(2,1)	276.5	278.0	−1.5	0.53	0.15
(2,1)-(4,1)	275.5	276.7	−1.2	–	–
(0,2)-(2,2)	276.6	278.8	−2.2	–	–
(0,3)-(2,3)	275.7	279.5	−3.8	–	–
Ring-bending bands					
(0,0)-(0,1)	120.7	119.9	0.8	0.95	0.99
(0,1)-(0,2)	116.2	115.5	0.7	(1.00)	(1.00)
(0,2)-(0,3)	112.8	112.3	0.5	0.73	0.81
(0,3)-(0,4)	109.6	109.4	0.2	0.59	0.60
(0,4)-(0,5)	106.6	106.6	0.0	0.54	0.43
(0,5)-(0,6)	103.4	103.9	−0.5	0.41	0.29
(2,0)-(2,1)	122.6	123.0	−0.4	0.32	0.31
(2,1)-(2,2)	116.6	116.3	0.3	0.29	0.27
(2,2)-(2,3)	111.9	113.0	−1.0	0.25	0.22
(2,3)-(2,4)	107.5	119.6	−2.1	0.19	0.16
(4,0)-(4,1)	124.6	126.5	−1.9	0.09	0.11

[a] $(V_T, V_P) - (V_T', V_P)$ where V_T and V_P are the twist and bend quantum numbers.

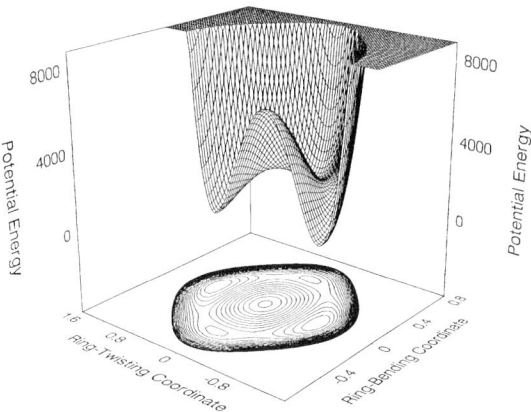

Fig. 4

Potential Energy Surface

The two-dimensional potential energy surface was calculated in a manner similar to that previously reported for five- [7–9] and six-membered [1,3,4] rings. The two-dimensional vibrational Hamiltonian for this calculation has the form

$$\mathscr{H}(x,\tau) = -\tfrac{1}{2}h^2(\partial/\partial x\, g_{44}\, \partial/\partial x + \partial/\partial\tau\, g_{55}\, \partial/\partial\tau)$$
$$+ V(x,\tau) \tag{1}$$

where x and τ are the ring-bending and ring-twisting coordinates, respectively, and where $V(x,\tau)$ is the potential energy function. The spectroscopic data for 5,6-dihydro-4H-thiopyran extend to about 800 cm^{-1} and the two-dimensional potential energy surface up to this energy is well-defined by these data. However, since the barrier to planarity and barrier to interconversion are much higher in energy, they can only be estimated by extrapolating the potential energy function to higher values. A plausible potential energy surface which fits the experimental data is

$$V(x,\tau) = 9.48 \times 10^4 x^4 - 4.13 \times 10^4 x^2 + 1.37 \times 10^4 x_2^4$$
$$- 1.82 \times 10^4 x_2^2 + 1.10 \times 10^5 x^2 x_2^2 \tag{2}$$

This function has a barrier to planarity of 6000 cm^{-1} and a bending energy of 1500 cm^{-1}. The two-dimensional surface for this function is shown in Fig. 4. The twisting minima occur at ± 0.80 radians (48°) and the bending saddle point corresponds to a bending coordinate value of 0.47 Å. Table 1 compares the observed data with the frequencies and relative intensities calculated for this surface.

Conclusion

The far-infrared spectra do a good job of defining the potential energy surface up to about 800 cm^{-1}. However, since the barrier to planarity is considerably higher than that, the surface at higher energies can only be estimated by extrapolation. We estimate that the barrier to planarity (and its uncertainty) is 6000 ± 2000 cm^{-1}, the bending energy is 1500 ± 750 cm^{-1} and the barrier to pseudorotation is 2000 ± 1000 cm^{-1}. Although our potential energy surface shows an energy minimum nearly 500 cm^{-1} below the barrier to pseudorotation, whether there even *is* a minimum for this conformation cannot be ascertained from the data. A comparison of the results from this work for 5,6-dihydro-4H-thiopyran (molecule **VII**) with values derived for the other cyclohexene analogues (**I–VI**) is given elsewhere [1–6].

Acknowledgments The authors wish to thank the National Science Foundation, Robert A. Welch Foundation, and the Texas ARP program for financial support. Support from the NIH-Biomedical program (JRV) is also gratefully acknowledged.

References

[1] V. E. Rivera-Gaines, S. J. Leibowitz, J. Laane, *J. Am. Chem. Soc.* **1991**, *43*, 9735.

[2] J. Laane, J. Choo, *J. Am. Chem. Soc.* **1994**, *116*, 3889.

[3] M. M. J. Tecklenburg, J. Laane, *J. Am. Chem. Soc.* **1989**, *111*, 6920.

[4] M. M. J. Tecklenburg, J. R. Villarreal, J. Laane, *J. Chem. Phys.* **1989**, *91*, 2771.

[5] J. Choo, N. T. Meinander, J. R. Villarreal, J. Laane, *J. Chem. Phys.* **1995**, *102*, 9506.

[6] J. Choo, *Ph. D. Thesis,* Texas A&M University, 1994.

[7] L. F. Colegrove, J. C. Wells, J. Laane, *J. Chem. Phys.* **1990**, *93*, 6291.

[8] J. Choo, J. Laane, *J. Chem. Phys.* **1994**, *101*, 2772.

[9] S. Leibowitz, J. Laane, *J. Chem. Phys.* **1994**, *101*, 2740.

Mikrochim. Acta [Suppl.] 14, 501–502 (1997)
© Springer-Verlag 1997

Conformational Stability, Vibrational Spectra and ab initio Calculations for Ethyldifluorophosphine, Methoxydifluorophosphine and Chloromethyldifluorophosphine

James R. Durig[1,*], **James B. Robb**[2], **Zhongnan Shen**[1], **Todor K. Gounev**[1], and **Thali R. Honeycutt**[2]

[1] Department of Chemistry, University of Missouri-Kansas City, Kansas City, MO 64110-2499, U.S.A.
[2] Department of Chemistry and Biochemistry, University of South Carolina, Columbia, SC 29208, U.S.A.

Abstract. The infrared spectra of methoxydifluorophosphine, CH_3OPF_2, dissolved in liquid xenon have been recorded. From temperature studies of these spectra it is shown that only one conformer exists. Relative conformational stabilities, barriers to internal rotation, force constants, fundamental vibrational frequencies, Raman activities and infrared intensities have been obtained from *ab initio* calculations at the RHF/6-31G* or MP2/6-31G* levels for ethyldifluorophosphine, methoxydifluorophosphine and chloromethyldifluorophosphine.

Key words: conformational stabilities, infrared and Raman data, ethyldifluorophosphine, methoxydifluorophosphine, chloromethyldifluorophosphine, internal rotation barriers.

We have been studying the conformational stability of several organophosphorus molecules and recently determined that the *trans* conformer of ethyldifluorophosphine, $CH_3CH_2PF_2$, is more stable than the *gauche* rotamer in all phases [1]. This is in contrast to the earlier microwave investigation [2] where the *gauche* conformer was predicted to be the more stable. For chloromethyldifluorophosphine, $ClCH_2PF_2$, van der Veken et al. [3] reported that the *gauche* conformer was more stable in the vapor phase, but the *trans* rotamer was more stable in the liquid and the only conformer present in the solid phase. As a continuation of this series of conformational studies, we have made an infrared and Raman spectroscopic study of methoxydifluorophosphine, including an infrared variable-temperature study of the sample dissolved in liquid xenon. Additionally, the conformational stabilities, barriers to internal rotation, force constants, fundamental vibrational frequencies, Raman activities and infrared intensities have been obtained from ab initio

calculations at the RHF/6-31G* or MP2/6-31G* levels. Similar calculations have been done for ethyldifluorophosphine. These theoretical values are compared with the experimental values when appropriate.

Experimental

The mid-infrared spectrum of methoxydifluorophosphine dissolved in liquid xenon were recorded for temperatures ranging from -60 to $-100\,°C$ on a Bruker model IFS-66 Fourier-transform interferometer equipped with a Globar source, Ge/KBr beam-splitter, and a TGS detector. For each temperature investigated, 100 interferograms were collected at $0.5\,cm^{-1}$ resolution, averaged, and transformed with a boxcar truncation function. The temperature studies were performed with the liquified xenon in a specially designed cryostat cell which consisted of a copper cell with a 4 cm path-length and wedged silicon windows sealed to the cell with indium gaskets. After the cell had cooled to the desired temperature, a small amount of sample was condensed into the cell. Next, the pressure manifold and the cell were pressurized with xenon, which immediately started condensing in the cell, which allowed the compound to dissolve.

Conformational Stability

In earlier vibrational [4] and microwave [5] studies methoxydifluorophosphine was reported to exist in the *trans* conformer and no evidence for a second conformer was presented. A subsequent theoretical conformational analysis within the CNDO/2 approximation by Robinet et al. [6] suggested the *gauche* conformer was the more stable conformer by 2 kcal/mol. We have utilized a temperature study of the infrared spectrum of CH_3OPF_2 dissolved in liquified xenon to attempt to detect a second conformer if one is present. Although the bands became sharper as the temperature decreased, there was no indication of any

Table 1. Heavy atom vibrations of *trans* $ClCH_2PF_2$, $CH_3CH_2PF_2$ and CH_3OPF_2 (XYPF$_2$)

Description	$ClCH_2PF_2$		$CH_3CH_2PF_2$		CH_3OPF_2	
	Exp.[a]	Calc.[b]	Exp.[c]	Calc.[d]	Exp.	Calc.[d]
X–Y stretch	776	794	1038	1018	1044	1097
Y–P stretch	632	644	651	692	786	804
PF$_2$ symmetric stretch	817	949	830	866	829	870
XYP bend	152	156	208	217	210	231
PF$_2$ deformation	357	339	349	331	351	355
PF$_2$ wag	455	462	501	506	536	536
PF$_2$ antisymmetric stretch	796	963	819	865	797	848
PF$_2$ twist	271	291	286	295	370	366
torsion	–	86	93	94	122	150

[a] Ref. [3]. [b] Calculated frequencies from the RHF/6-31G* basis set with scaling factors of 0.9 for stretches and 0.8 for bends. [c] Ref. [1]. [d] Frequencies from the MP2/6-31G* calculation.

bands due to a second conformer or changes in relative intensities of any of the bands. On the basis of these data, only one conformer, the *trans* form, was indicated by the spectra. This conclusion is supported by our ab initio results where the *trans* conformer is predicted to be more stable than the *gauche* rotamer by 4.61 kcal/mol, from the MP2/6-31G* calculations. With such a large energy difference we do not expect to detect a second conformer of CH_3OPF_2 by vibrational spectroscopy at ambient temperature.

We have carried out ab initio calculations on ethyl-difluorophosphine to determine the conformational stability and obtain the vibrational spectra. The results predict the *trans* conformer to be more stable than the *gauche* form by 452 cal/mol, from MP2/6-31G* calculations. These results support the vibrational data where the enthalpy difference between the conformers was experimentally determined to be 275 cal/mol, with the *trans* conformer more stable.

There have been no experimental evaluations of the enthalpy difference for the conformers of chloro-methyldifluorophosphine but the earlier vibrational study indicated that the *gauche* conformer was more stable for this molecule [3]. The ab initio calculations at the RHF/6-31G* level predicted the *gauche* conformer to be more stable by 1.05 kcal/mol. Although the chlorine atom is approximately the same size as the methyl group, it is clear that the electronegativity has had a pronounced effect on the conformational stability of these molecules. The two isoelectronic molecules,

$CH_3CH_2PF_2$ and CH_3OPF_2, both have the *trans* conformers as the more stable form whereas the *gauche* conformer is the more stable form for $ClCH_2PF_2$. It would be interesting to investigate the conformer stability of the isoelectronic FCH_2PF_2 molecule.

Vibrational Assignments

The vibrational assignments for the heavy atom modes of the *trans* conformer of the studied compounds are listed in Table 1. The ab initio calculated frequencies for each molecule agree quite well with the observed values. The frequencies for the normal modes are quite similar, the only notable difference being the PF$_2$ twist, which is much higher for the CH_3OPF_2 molecule. The frequencies obtained from the MP2/6-31G* calculations agree reasonably well without any scaling. These results demonstrate the utility of ab initio calculations in the interpretation of vibrational spectra.

References

[1] J. R. Durig, J. S. Church, C. M. Whang, R. D. Johnson, B. J. Streusand, *J. Phys. Chem.* **1987**, *91*, 2769.
[2] P. Groner, J. S. Church, Y. S. Li, J. R. Durig, *J. Chem. Phys.* **1985**, *82*, 3894.
[3] B. J. van der Veken, P. Coppens, R. S. Sanders, F. F. Daeyaert, J. R. Durig, *J. Mol. Struct.* **1992**, *272*, 305.
[4] J. R. Durig, B. J. Streusand, *Appl. Spectrosc.* **1980**, *34*, 65.
[5] E. G. Codding, C. E. Jones, R. H. Schwendeman, *Inorg. Chem.* **1974**, *13*, 178.
[6] G. Robinet, J. F. Labarre, C. Leibovici, *Chem. Phys. Lett.* **1974**, *29*, 449.

Mikrochim. Acta [Suppl.] 14, 503–504 (1997)

FT-IR Spectra of Xenon Solutions of c-C₃H₅CFO and CH₂CHCFO

James R. Durig*, Gamil A. Guirgis, Shiyu Shen, Yin Li, Douglas T. Durig, Lin Zhou, and **Yanping Jin**

Department of Chemistry, University of Missouri-Kansas City, Kansas City, MO 64110, U.S.A.

Abstract. The infrared (3500–400 cm⁻¹) spectra of cyclopropylcarbonyl fluoride, c-C_3H_5CFO, and propenoyl fluoride, CH_2CHCFO, in xenon solutions have been recorded at various temperatures (from −60 to −100 °C). The data show that there is an equilibrium between the *cis* (oxygen atom of the carbonyl bond *cis* with respect to the ring or carbon–carbon double bond) and *trans* conformers in the fluid phases. Analysis of the temperature-dependent spectra shows the *cis* conformer is the more stable rotamer for the cyclopropyl compound whereas the *trans* rotamer is the more stable conformer for propenoyl fluoride.

Key words: cyclopropylcarbonyl fluoride, propenoyl fluoride, *cis–trans* equilibrium.

We have been determining the potential functions governing the conformational interchange of a number of cyclopropylcarbonyl and propenoyl halides [1–4] from the frequencies of the asymmetric torsional transitions, obtained from far infrared and low-frequency Raman spectra of the gases. For these potential function determinations we need to know the enthalpy difference between the conformers, and the structural parameters from which the F series (kinetic energy) for the internal rotor is obtained. For propenoyl fluoride [1], temperature studies of the Raman spectrum of the gas gave inconclusive results. However, ab initio RHF/6–31G* (restricted Hartree-Fock) calculations gave the *trans* conformer as the more stable rotamer by 79 cm⁻¹. This result is in agreement with the microwave-determined value [5] of 31 ± 35 cm⁻¹ but the large uncertainty casts doubt as to which conformer is the more stable rotamer.

Similarly, for cyclopropylcarbonyl fluoride, the only conformational stability study of the gas has been made by a microwave investigation of the gas [6]. In this study, the *cis* conformer was found to be more stable than the *trans* rotamer by 200 ± 30 cm⁻¹, which is in contrast to the corresponding propenoyl fluoride molecule. The *ab initio* RHF/6–31G* calculations give the *cis* conformer as more stable by only 28 cm⁻¹, which is so small that the calculations cannot be relied on to give the stable conformer. Therefore, we undertook a study of the temperature-dependent FT-IR spectra of cyclopropylcarbonyl fluoride and propenoyl fluoride to determine the conformational stability of these two molecules.

Experimental

The mid infrared spectra of the sample dissolved in liquefied xenon, as a function of temperature ranging from −60 to −100 °C were recorded on a Bruker model IFS-66 Fourier-transform interferometer equipped with a Globar source, Ge/KBr beam-splitter, and a TGS detector. For each temperature investigated, 200 interferograms were collected at 0.5 cm⁻¹ resolution, averaged, and transformed with a boxcar truncation function. The temperature studies with the liquefied noble gas were performed in a specially designed cryostat cell which consisted of a copper cell with a 4 cm path-length and wedged silicon windows sealed to the cell with indium gaskets. The temperature was monitored with two Pt thermoresistors and the cell was cooled with boiling liquid nitrogen. The complete cell was connected to a pressure manifold to allow for the filling and evacuation of the cell. After the cell had cooled to the desired temperature, a small amount of sample was condensed into the cell. Next, the pressure manifold and the cell were pressurized with xenon, which immediately began condensing in the cell, allowing the compound to dissolve.

Results and Discussion

In the infrared spectra of the xenon solutions of cyclopropylcarbonyl fluoride, there are a number of well separated doublets resulting from the heavy-atom skeletal vibrations of the two conformers. These are observed at 436 (*cis*)/475 (*trans*), 593 (*trans*)/612 (*cis*), 730 (*cis*)/723 (*trans*), and 757 (*trans*).762 (*cis*) cm⁻¹. The

---

* To whom correspondence should be addressed

Table 1. Temperature and intensity ratios for the conformational study of cyclopropylcarbonyl fluoride

$T(°C)$	$1000/T(K)$	$cis\ I_{730}$	$trans\ I_{723}$	Ratio	$\ln(I_{cis}/I_{trans})$
−60	4.69	1.3992	0.5299	2.6403	0.9709
−70	4.93	1.7186	0.6378	2.6947	0.9912
−80	5.18	2.3153	0.7808	2.9656	1.0870
−90	5.46	2.6645	0.9049	2.9446	1.0799
−100	5.78	3.2110	1.0400	3.0864	1.1271

Table 2. Temperature and intensity ratios for the conformational study of propenoyl fluoride

$T(°C)$	$1000/T(K)$	$trans\ I_{1218}$	$cis\ I_{1115}$	Ratio	$\ln(I_{cis}/I_{trans})$
−60	4.69	8.9022	12.1279	1.3623	0.30917
−70	4.93	9.3750	11.7399	1.2523	0.22498
−75	5.05	9.6649	12.0479	1.2466	0.22042
−85	5.31	10.1135	12.2622	1.2124	0.19260
−90	5.46	10.2712	12.1006	1.1781	0.16390
−100	5.78	10.5991	11.9958	1.1318	0.12381

doublet near 600 cm^{-1} is not a good candidate for obtaining the enthalpy difference, since the silicon window absorbs a large amount of the radiation at approximately 620 cm^{-1}. Therefore, the choices are either the doublet in the 400 cm^{-1} region, or one of those in the 700 cm^{-1} region. We chose the 730/723 cm^{-1} doublet, which appears to have excellent band contours and is sufficiently separated for good measurements to be made on the areas underneath the curves. From these spectra, the intensity ratios were measured at five temperatures ranging from −60 to −100 °C. The experimental data are listed in Table 1. The enthalpy difference between the cis and trans conformers was calculated by using the van't Hoff isochore, $-\ln K = (\Delta H/RT) - \Delta S/R$. A plot of $-\ln K$ vs $1/T$, where K is the ratio of the intensity of the band due to the cis conformer to the one due to the trans conformer, has a slope which is proportional to the enthalpy difference. These data gave a ΔH value of 102 ± 21 cm^{-1} (292 ± 60 cal/mol). It is clear from these data that the cis conformer, where the oxygen atom is over the three-membered ring, is the more stable rotamer for cyclopropylcarbonyl fluoride. The statistical error is relatively small, and the relatively small, and the relative intensities of the two bands as the temperature is decreased clearly show the more stable conformer to be the cis form.

Similar data were obtained for propenoyl fluoride, where the following doublets were observed: 486

(cis)/527 (trans), 510 (cis)/476 (trans), 626 (cis)/602 (trans), 796 (cis)/801 (trans), 808 (cis)/825 (trans), 1115 (cis)/1218 (trans) and 1275 (cis)/1294 (trans) cm^{-1}. Additionally, there are three fundamentals between 980 and 1000 cm^{-1} which are obviously due to conformers, but they are so badly overlapped that they cannot be utilized for enthalpy measurements. There is only one doublet, the 1115 (cis).1218 (trans) cm^{-1}, which has appropriate intensity with good signal-to-noise ratio and is sufficiently well separated for excellent intensity ratios to be obtained. The data for this doublet are listed in Table 2 and an enthalpy determination gives a value of 108 ± 14 cm^{-1} (309 ± 40 cal/mol). However, it should be noted that in this case, the trans conformer is the more stable form, in contrast to the cis rotamer for the cyclopropylcarbonyl fluoride compound.

References

[1] J. R. Durig, R. J. Berry, P. Groner, *J. Chem. Phys.* **1987**, *87*, 6303.
[2] J. R. Durig, H. D. Bist, T. S. Little, *J. Chem. Phys.* **1982**, *77*, 4884.
[3] J. R. Durig, A.-Y. Wang, T. S. Little, *J. Mol. Struct.* **1991**, *244*, 117.
[4] J. R. Durig, A.-Y. Wang, T. S. Little, *J. Mol. Struct.* **1992**, *269*, 285.
[5] J. J. Keirns, R. F. Curl, Jr., *J. Chem. Phys.* **1968**, *48*, 3773.
[6] H. N. Volltrauer, R. H. Schwendeman, *J. Chem. Phys.* **1971**, *54*, 268.

Mikrochim. Acta [Suppl.] 14, 505–506 (1997)

Experimental Determination of the Dynamic Dipole Polarizability of the Cadmium Atom

Dirk Goebel and **Uwe Hohm***

Institut für Physikalische und Theoretische Chemie Technische Universität Braunschweig, Hans-Sommer-Str.10, D-38106 Braunschweig, Federal Republic of Germany

Abstract. Dispersive Fourier-transform spectroscopy in the visible (DFTS-VIS) is used to obtain the refractive index spectrum $n(\sigma, \varrho, T)$ of cadmium vapour in the wavenumber range between 10000 and 20000 cm^{-1}. We will demonstrate that use of a carefully designed Michelson interferometer is capable of yielding these spectra up to temperatures of 1100 K. In addition the dependence of $n(\sigma, \varrho, T)$ on density yields the dispersion of the dipole polarizability $\alpha(\sigma)$ of atomic cadmium.

Key words: DFTS, refractive index, polarizability, cadmium vapour.

Dispersive Fourier-transform spectroscopy (DFTS) is known to be a powerful technique for the quasi-continuous experimental determination of the dispersion of the complex refractive index $\hat{n}(\sigma, \varrho, T) - 1$ [1, 2]. However, in most cases it is applied in the infrared region of the electromagnetic spectrum, because some of the special advantages of DFTS disappear when the visible frequency region is used. Additionally, the experiments become very difficult. Despite these drawbacks we feel that DFTS-VIS is a powerful technique for the determination of refractive index spectra $n(\sigma, \varrho, T)$ at elevated temperatures, of unusual systems such as metal vapours. At present we are examining Zn, Cd and Hg, because the dispersion of the polarizability of these atoms is not precisely known. In this contribution we will concentrate on various aspects of the experiments with cadmium. The theory of the dipole polarizability $\alpha(\sigma)$ is described in detail in [3].

Theoretical

If $I_A(\delta)$ is the interferogram of the sample (apparatus + gas) and $I_B(\delta)$ the corresponding interferogram of the empty apparatus, then the complex spectra $\hat{A}(\sigma)$ and $\hat{B}(\sigma)$ are obtained by Fourier transformation via:

$$\hat{A}(\sigma) = \int_{-\infty}^{+\infty} [I_A(\delta) - I_A(\infty)] \exp(-i2\pi\sigma\delta)\mathrm{d}\delta$$
$$\equiv P_A(\sigma) - iQ_A(\sigma) \tag{1}$$

$$\hat{B}(\sigma) = \int_{-\infty}^{+\infty} [I_B(\delta) - I_B(\infty)] \exp(-i2\pi\sigma\delta)\mathrm{d}\delta$$
$$\equiv P_B(\sigma) - iQ_B(\sigma) \tag{2}$$

Here, σ is the wavenumber, δ is the geometrical difference between the length of the two arms of the interferometer, and $I(\infty)$ is the background intensity of the interferograms $I_A(\delta)$ and $I_B(\delta)$, respectively. For non-absorbing samples, only the real part of the complex refractive index $[\hat{n}(\sigma) - 1]$ is non-zero. It can be obtained via

$$n(\sigma) - 1 = \frac{\delta_M}{l} + \frac{\varphi_A(\sigma) - \varphi_B(\sigma) + 2\pi(m_A - m_B)}{2\pi\sigma l} \tag{3}$$

where l is twice the length of the sample cell. The principal value of the phase $\varphi(\sigma)$ is obtained via

$$\varphi(\sigma) = \arctan[Q(\sigma)/P(\sigma)] = \Phi(\sigma) - 2\pi m \tag{4}$$

$\Phi(\sigma)$ is the absolute value of the phase, m_A and m_B are two integers. Here, m_B can be set to zero. To get absolute values of the refractivity $n(\sigma) - 1$, δ_M and m_A must be known; δ_M is the geometrical difference between the zero orders of the sample interferogram $I_A(\delta)$ and the apparatus interferogram $I_B(\delta)$. It is normally obtained by recording two white-light interferograms with and without sample, respectively; m_A is determined by recording one absolute value of $n(\sigma_0) - 1$ at

---

* To whom correspondence should be addressed

a reference wavenumber σ_0. Alternatively, m_A and δ_M may be calculated if one additional refractivity $n(\sigma_1) - 1$ at σ_1 is measured.

Experimental

The measurements were made with an evacuated Michelson twin interferometer with parallel guided beams [4]. The interferometric set-up consisted of one measuring and one reference interferometer. Both interferometers used the same end mirrors and the same beam-splitter cube. The arm length of both interferometers was about 2500 mm in order to ensure special thermal stability. The measuring interferometer, which contained the sample cell, was illuminated with a 100 W halogen lamp with a signal-to-noise ratio of $S/N = 200$, which was found to be independent of the sample temperature. Our attempts to use a 150 W Xe arc-lamp in order to enlarge the frequency range of our measurements failed because of low frequency flickering of the Xe-arc, although it was stabilized with a photo feed-back system. The reference interferometer was illuminated with a HeNe laser of wavenumber $\sigma_0 = 15798 \text{ cm}^{-1}$. It was used to obtain the shift δ of the common end mirror of both interferometers. The interference fringe patterns of both interferometers were recorded separately with two photomultiplier tubes, each with pinhole aperture. Cadmium of mass m ($6\,\text{mg} < m < 18\,\text{mg}$) was placed in the evacuated quartz sample cell of length $l/2 \approx 500$ mm and volume $V = 26.5 \text{ cm}^3$. The cell was sealed off by fusion and placed in the active arm of the measuring interferometer. The passive arm contained an identical evacuated reference cell. The sample was heated to approximately 1050 K until the cadmium was fully vaporized. The resulting interference fringe shift of the measuring interferometer was due to the increasing refractive index $n(\sigma, \varrho, T)$ of the sample. The interference fringe shift was measured simultaneously with two HeNe lasers at σ_0 and $\sigma_1 = 18399 \text{ cm}^{-1}$, yielding $n(\sigma_0, \varrho, T) - 1$ and $n(\sigma_1, \varrho, T) - 1$. In order to minimize the statistical error, this was done several times for four different amount-of-substance densities $\varrho = m/(MV)$ ($M = 0.11241$ kg/mol is the atomic mass of Cd). These two refractivities were then used to determine δ_M and m_A in Eq. (4).

If the sample was completely evaporated, DFTS spectra $I_A(\delta)$ were recorded in the usual manner by using the halogen lamp as white-light source and changing the geometrical path difference δ. The interferogram $I_B(\delta)$ of the empty apparatus was recorded at room temperature, because we have not found any significant difference between $\hat{B}(\sigma)$ recorded at high temperatures and $\hat{B}(\sigma)$ obtained at room temperature. $I_A(\delta)$ and $I_B(\delta)$ were analysed according to Eqs. (1)–(5), giving the refractive index spectrum in the range between 10000 and 20000 cm^{-1} with a resolution of (at present) 10 cm^{-1}. We have also tried to obtain the white-light interferogram $I_A(\delta)$ of the sample and $I(\delta)$ of the reference interferometer simultaneously, in order to determine δ_M in Eq. (4) directly. However, we have observed a small temperature-dependent tilt of the mirrors, which could not be avoided in the reference interferogram during the heating period. On account of this tilt, the interference fringe pattern of the reference interferogram became extremely small and could not be resolved sufficiently by our recording system.

Results and Discussion

All experimental parameters and measured quantities can be found in detail in [3]. The measured refractivi-

ties are analysed according to the Lorentz–Lorenz formula

$$\frac{n^2(\sigma, \varrho, T) - 1}{n^2(\sigma, \varrho, T) + 2} = \frac{N_A}{3\varepsilon_0} \alpha(\sigma, T)\varrho \qquad (5)$$

N_A being the Avogadro constant and ε_0 the permittivity of free space. In our experiments no temperature-dependence of the dipole polarizability $\alpha(\sigma, T)$ of cadmium could be detected which would be due to the non-resonant Stark effect caused by black-body radiation. The resulting wavenumber-dependence of $\alpha(\sigma)$ can be described by a one-term Kramers–Heisenberg dispersion formula

$$\alpha(\sigma) = \alpha(0)\sigma_0^2/(\sigma_0^2 - \sigma^2) \qquad (6)$$

with a static polarizability of $\alpha(0) = 8.19(27) \times 10^{-40}$ $\text{C}^2\text{m}^2\text{J}^{-1}$ and an effective transition wavenumber of $\sigma_0 = 55003(50) \text{ cm}^{-1}$. This result is in a good agreement with the best relativistic *ab initio* calculations of $\alpha(0) = 8.1(2.0) \times 10^{-40} \text{C}^2\text{m}^2\text{J}^{-1}$ [5] and with the early experimental results of Cuthbertson and Metcalfe, who obtained $\alpha(0) = 8.24(46) \times 10^{-40} \text{ C}^2\text{m}^2\text{J}^{-1}$ and $\sigma_0 = 53210(10000) \text{ cm}^{-1}$ [6].

To the best of our knowledge these data represent the first quasi-continuously determined refractive index and polarizability curves known for a vapourized metal, over a large wavenumber range. With minor improvements concerning the frequency range covered by our experiments, DFTS-VIS can be used with success as a very valuable technique for the determination of the refractive index spectra of unusual systems.

Acknowledgement. Financial support of the Deutsche Forschungsgemeinschaft and Fonds der Chemischen Industrie is gratefully acknowledged.

References

[1] T. J. Parker, *Contemp. Phys.* **1990**, *31*, 335.
[2] J. R. Birch, J. Yarwood, in: *Spectroscopy and Relaxation of Molecular Liquids* (D. Steele, J. Yarwood, eds.), Elsevier, Amsterdam, 1991, p.174.
[3] D. Goebel, U. Hohm, *Phys. Rev. A.* **1995**, *52*, 3691.
[4] U. Hohm, K. Kerl, *Meas. Sci. Technol.* **1990**, *1*, 329.
[5] T. M. Miller, in: *Handbook of Chemistry and Physics, 73rd Ed.* (D. R. Lide, ed.), CRC, Boca Raton, 1992.
[6] C. Cuthbertson, E. P. Metcalfe, *Phil. Trans.* **1907**, *207*, 135.

Mikrochim. Acta [Suppl.] 14, 507–510 (1997)
© Springer-Verlag 1997

Vibrational Spectroscopy of Lithium Silicates and Aluminosilicates in Crystalline Form

Miroslaw Handke and **Marek Nocuń***

Department of Materials Science and Ceramic, University of Mining and Metallurgy, al. Mickiewicza 30, 30-059 Kraków Poland

Abstract. The results of structural research on crystalline lithium silicates and aluminosilicates in the system Li_2O–Al_2O_3–SiO_2 are presented. Spectroscopy in the mid and far infrared was used as a main research tool. The Li–O band positions were established on the basis of isotope substitution.

Key words: lithium silicates, lithium aluminosilicates.

Li_2O–Al_2O_3–SiO_2 is a basic system for many glasses and glass-ceramics that can be called "special". The most important of them are glass-ceramics with zero linear expansion coefficient, and a recently developing field, the application of spodumene for ceramic implants. It is also worth noticing the application of orthosilicates as an ionic conductor in fuel cells. Despite such wide practical application; the structures of the crystalline and amorphous compounds in this system are still not well understood. The aim of this work was to study the structure of crystalline lithium silicates and aluminosilicates by means of infrared spectroscopy as a method of structural research.

In the Li_2O–Al_2O_3–SiO_2 system there are three lithium silicate compounds: orthosilicate – $2Li_2O\cdot SiO_2$, metasilicate – $Li_2O\cdot SiO_2$, disilicate – $Li_2O\cdot 2SiO_2$ and three aluminosilicates: β-eucripite, β-spodumene and petalite. The structures of these compounds can be found in the literature [1–6].

Preparation Procedure

Silicates and aluminosilicates were synthesised by the high temperature method. The baths were prepared with pure, lithium carbonate silica and alumina. Lithium hydroxide supplied by Euroisotop was

used as a source of ^6Li. The samples were melted in a platinum crucible at appropriate temperatures. The melt was kept at the maximum temperature for about 60 min, (for homogenization) then poured into a container filled with liquid nitrogen. The crystalline compounds were obtained by devitrification of glassy samples. The temperature of crystallization was established on the basis of DTA measurements. The chemical and phase composition of all the samples was checked by chemical analysis and X-ray diffraction. Spectroscopic analyses were done with a Bio-Rad FTS-60, Fourier transform spectrometer. The absorption spectra of the samples were measured by the KBr pellet mid infrared (MIR) and polyethylene pellet for infrared (FIR) techniques.

Results

The MIR and FIR spectra of crystalline lithium silicates are shown in Fig. 1. The MIR spectra of lithium orthosilicate are characterized by two groups of bands: 1000–800 and 600–400 cm^{-1}. Although an SiO_4 tetrahedron is isolated in the structure of orthosilicate its site symmetry C_s is lower than the ideal T_d. As a consequence, instead of the two F_2 type IR-active vibrations characteristic for T_d symmetry, we observed many bands due to activation of A_1 and E type vibrations and to the lack of degeneracy of F_2 and E type vibrations. Some extra bands due to Davidov's splitting and combination bands are also observed. The bands in the range 1000–800 cm^{-1} should be due to stretching vibrations of the Si–O bond (F_2, A_1) and in the range of 600–400 cm^{-1} to bending vibrations (F_2 and E). Bands corresponding to lattice vibrations, i.e. translational and rotational vibrations of silicate tetrahedra, lie mainly below 500 cm^{-1}.

In the spectra of lithium metasilicate, an additional band at 738 cm^{-1} is observed. This band is a result of the rocking vibrations of Si–O–Si bridges and it is not

---

* To whom correspondence should be addressed

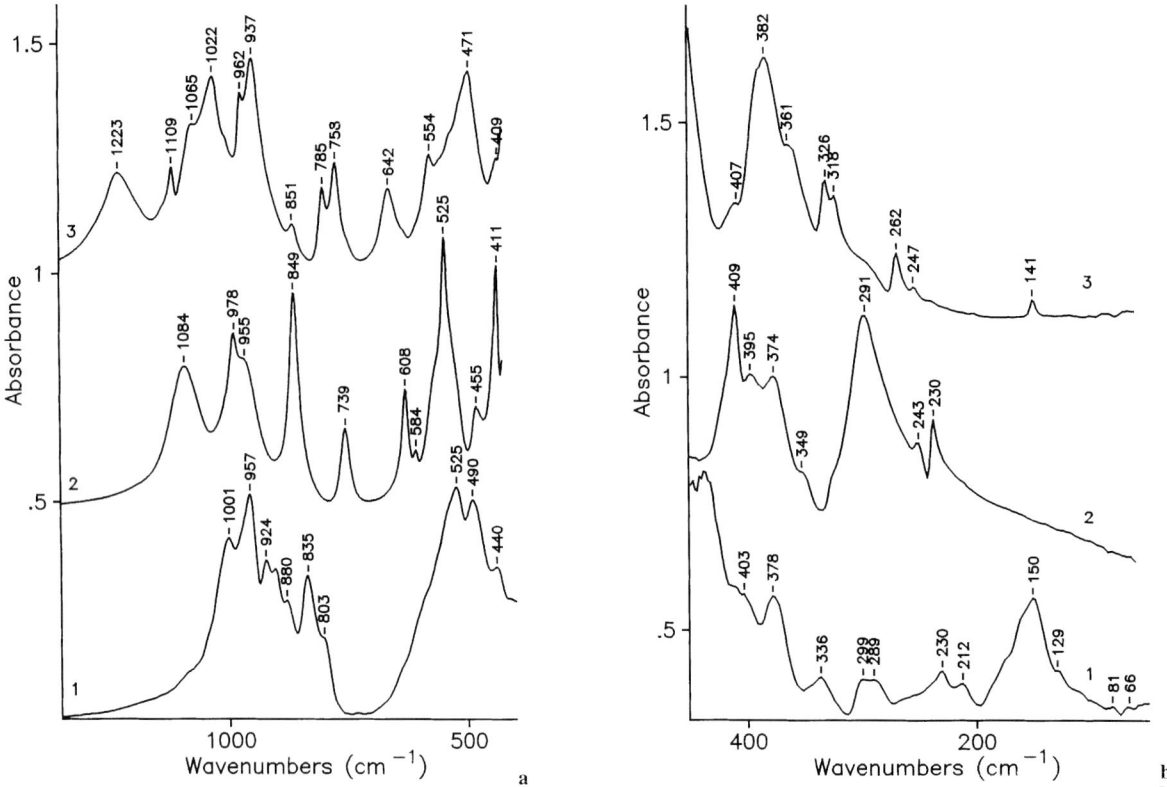

Fig. 1. IR spectra of lithium silicates; **a** MIR, **b** FIR; *1* lithium orthosilicate, *2* lithium metasilicate, *3* lithium disilicate

observed in the spectra of such compounds as orthosilicates where there are no Si–O–Si bridges. The presence of Si–O–Si results in increasing of the stretching force constants, moving the absorption bands towards higher frequencies – Fig. 1a-2. Hence the band at 1084 cm^{-1} is due to stretching vibrations of Si–O(Si) bonds and a band at ∼ 850 cm^{-1} is due to stretching vibrations of Si–O terminal bonds. The bands at 980 and 960 cm^{-1} are most probably, due to crystal field splitting of the F_2 stretching mode. The bands in the range 600–400 cm^{-1} are connected with bending vibrations of O–Si–O bonds.

In the spectra of lithium disilicate, with sheet structure, two bands in the range 780–750 cm^{-1} are observed – Fig. 1a-3. Similar bands in this range are observed in the spectrum of quartz. They can confirm the presence in the structure, of bridges with different angles. Moving from the island-structure orthosilicates to the sheet-structure disilicates, there is a shift of the bands to higher frequencies, from 1000 cm^{-1} for orthosilicates to ∼ 1200 cm^{-1} for disilicates. This

results from the rigidity of the structure, caused by the large number of bridging oxygen atoms present in the structure of disilicates. Stretching vibrations of the Si–O(Si) bonds give rise to absorption bands at higher frequencies – 1200 cm^{-1} whereas one terminal oxygen atom, weakly bonded to a silicate ion, gives rise to the band at about 950 cm^{-1}. Bands at ∼ 600 cm^{-1} are not observed in the spectra of the amorphous phase so they are characteristic of only the crystalline phase. It is assumed in the literature that these bands are due to vibrations of rings with different symmetry, and their numbers and locations depend on the ring type.

To explain the controversy connected with the location of Li–O vibration bands, the isotope substitution method was applied. This method is based on the assumption that changing the mass of vibrating atoms has no influence on the force constants and does not change the symmetry of the structure. We used the isotope ^{6}Li instead of the natural isotope ^{7}Li, so it can be expected that the bands connected with Li–O

Table 1. Wavenumber shifts in lithium silicates and β-eucriptite as a result of isotope substitution

Orthosilicate		Metasilicate		Disilicate		β-eucriptite	
ω_{Li-7} [cm^{-1}]	$\omega_{Li-6} - \omega_{Li-7}$	ω_{Li-7} [cm^{-1}]	$\omega_{Li-6} - \omega_{Li-7}$	ω_{Li-7} [cm^{-1}]	$\omega_{Li-6} - \omega_{Li-7}$	ω_{Li-7} [cm^{-1}]	$\omega_{Li-6} - \omega_{Li-7}$
150	4	228	2	140	0	81	0
175	7	244	1	247	0	184	3
213	6	291	17	262	2	247	9
231	4	348	15	318	6	297	11
285	14	371	18	326	9	360	7
300	6	394	7	360	3	407	3
338	13	411	0	382	13	438	3
357	20	455	0	408	13	457	10
409	14	525	4	470	6	471	38
438	27	552	17	554	0	654	8
488	12	584	0	642	6	675	4
523	8	608	0	758	0	708	4
565	0	739	2	785	0	744	5
600	8	849	2	850	5	758	5
799	7	954	0	937	0	880	6
837	0	978	0	962	0	997	2
870	4	1084	0	1022	0	1034	4
881	4			1065	0		
903	4			1109	0		
920	0			1223	0		
955	2						
1003	0						

vibrations will move towards higher wavenumbers. The analysis of the spectra of crystalline lithium silicate and aluminosilicate shows that these bands lie mainly below 600 cm^{-1}. All results are shown in Table 1.

The interpretation of the absorption spectra of lithium aluminosilicates in the MIR range is based on the assumption that the Si–O–Si group rather than the SiO$_4$ tetrahedron is the monomeric unit. Such a model was proposed by Bell and Dean [7] and applied by Handke and co-workers [8, 9]. The Si$_2$O monomer has the same symmetry as the H$_2$O molecule, i.e. C$_{2v}$. For such a symmetry we have three IR-active vibrations, which correspond to stretching, rocking and bending vibrations of Si–O–Si bridges. MIR spectra of crystalline lithium aluminosilicates are shown in Fig. 2a. Similarly to lithium silicates, the FIR region is dominated by lattice vibrations and, in the case of eucriptite, by vibrations of the lithium tetrahedron – Fig. 2b. The MIR spectra of the aluminosilicates reflect the C_{2v} symmetry of Si–O–(Si, Al) bridges. Three groups of bands are observed in the spectrum at 1150–850, 800–750 and 450–400 cm^{-1}, Fig. 2a. These bands are assigned to the stretching, rocking and bending vibrations of Si–O–(Si, Al), respectively. In the case of eucriptite only Si–O–Al bridges are present, so a nar-

row band at 758 cm^{-1} is observed. A sharp and narrow band at 1000 cm^{-1} reflects the high symmetry of the SiO$_4$ tetrahedron in eucriptite. This band coincides with bands due to vibration of the SiO$_4$ tetrahedron modified by the presence of an aluminate ion, and hence broadens at the bottom (Lorentz type curve) – Fig. 2a-1. The IR spectrum of spodumene is characterized by wider bands, which can mean deformation of SiO$_4$ and AlO$_4$ tetrahedra. The band at 780 cm^{-1} is connected with vibrations of the Si–O–Si bridges and its considerable width at half maximum indicates that the angles differ very much. The next band, at 750 cm^{-1}, is also very wide, and results from bending vibrations of the Si–O–Al bridges. The spectra of petalite and orthoclase confirm this opinion – Fig. 2a-3,4. The samples with the chemical composition of petalite and orthoclase are solid solutions of spodumene and SiO$_2$, which is clearly shown in the spectra – Fig. 2a. The increase in the intensity of the band due to the stretching vibrations of the Si–O bond at $\sim$1100 cm^{-1} with, at the same time, a constant ratio of other bands, provides strong evidence that there are some parts with a quartz structure besides parts with a spodumene structure. Increasing SiO$_2$ contents do not cause breaking of the Si–O–Al bonds, as was

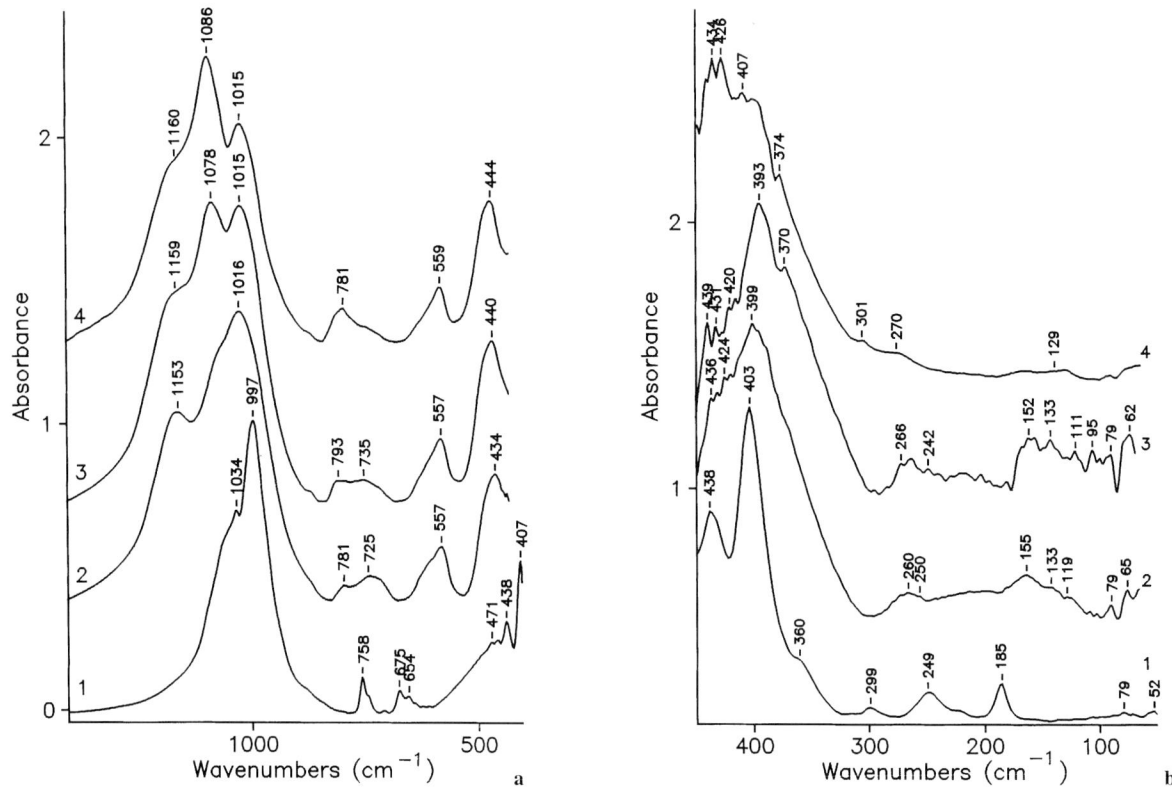

Fig. 2. IR spectra of crystalline lithium aluminosilicates; **a** MIR, **b** FIR; *1* β-eucriptite, *2* β-spodumene, *3* orthoclaze, *4* petalite

assumed earlier, but lead to a structure with a higher degree of polymerization. It can be assumed that from a structural point of view, the samples of petalite and orthoclase are heterogeneous.

References

[1] D. Tranqui, R.D. Shannon, H.-Y. Chen, *Acta Crystl.* **1979**, *B35*, 2479.

[2] H. Seemann, *Acta Crystl.* **1956**, *9*, 251.
[3] V. Devarajan, H.F. Shurvell, *Can. J. Chem.* **1977**, *55*, 2559.
[4] K.-F. Hesse, *Acta Crystl.* **1977**, *B33*, 901.
[5] C.-T. Li, D.R. Peacor, *Z. Kristall.* **1968**, *126*, 46.
[6] V. Tscherry, H. Schulz, F. Laves, *Z. Kristall.* **1972**, *135*, 161.
[7] R.J. Bell, P. Dean, *Discuss. Faraday Soc.* **1970**, *50*, 55.
[8] M. Handke, W. Mozgawa, *Vib. Spectrosc.* **1993**, *5*, 75.
[9] M. Handke, W. Mozgawa, M. Nocuń, *J. Mol. Struct.* **1994**, *325*, 129.

Mikrochim. Acta [Suppl.] 14, 511–513 (1997)

Vibrational Spectra of AlPO$_4$ as a Structure Model for Silica

Mirosław Handke*, **Włodzimierz Mozgawa,** and **Magdalena Rokita**

Department of Material Science and Ceramics, University of Mining and Metallurgy (AGH), 30–059 Kraków, al. Mickiewicza 30, Poland

Abstract. The vibrational spectra of various crystalline forms of SiO$_2$ and AlPO$_4$ are compared. Both compounds are characterized by a linked network of tetrahedra with Si–O–Si and Al–O–P oxygen-atom bridges, respectively. For the vibrational spectra band assignments, two types of model molecules, XO$_4$ and X$_2$O, are proposed. The tetrahedral model explains well the spectra of aluminium phosphate, whereas in the case of SiO$_2$ the Si$_2$O unit model is more useful. It has been shown that in geometrical terms the structures of SiO$_2$ and AlPO$_4$ are similar, whereas their spectra, which are due to the electronic structures (chemical bonds) are different.

Key words: silica, aluminium phosphate, vibrational spectra, model structural units.

Glassy and crystalline phosphosilicate materials are very promising as biomaterials. In order to design their compositions it is necessary to take into account the structure and properties of the compounds of the Al$_2$O$_3$–SiO$_2$–P$_2$O$_5$ system. AlPO$_4$ is of particular importance among these compounds.

Aluminium, silicon and phosphorus are adjacent elements in the periodic table (atomic numbers 13, 14 and 15 respectively). Their maximum oxidation levels are 3+, 4+ and 5+ respectively; thus the maximum charge of silicon is the arithmetic mean of the maximum charges exhibited by aluminium and phosphorus. Additionally the diameters of these cations are close to each other and – as a consequence – they all exhibit tetrahedral coordination with respect to oxygen. Thus the isostructural conditions for SiO$_2$ and AlPO$_4$ are fulfilled and the following isomorphous substitutions: $2Si^{2+} \leftrightarrow Al^{3+} + P^{5+}$ can be expected.

Silicon dioxide, SiO$_2$, which can be treated as silicon orthosilicate, Si[SiO$_4$], exists in various crystalline forms [1] and analogous forms of AlPO$_4$ can be found [2]. Thus low- and high-temperature quartz has a similar structure to berlinite, tridymite to phosphotridymite and cristobalite to phosphocristobalite. SiO$_2$ (considered as Si[SiO$_4$]) and AlPO$_4$ can be treated as isomorphous compounds in terms of crystalline structure, i.e. the distribution of atoms or ions in space. However, if the electronic structure (chemical bonding) is taken into account this analogy is no longer valid. Aluminium, silicon and phosphorus belong to three different main groups in the periodic table and they differ in their electronegativity. Therefore the surroundings of the oxygen atoms are totally different in the SiO$_2$ (Si–O–Si) and AlPO$_4$ (Al–O–P) structures. The former is a case of mesodesmic structure (valency of cation bonding 4/4), the latter is an example of anisodesmic structure (valency of bonding 5/4 and 3/4).

The main goal of the present work is to compare the structurally isomorphous SiO$_2$ and AlPO$_4$ (but with different electronic structure) by vibrational spectroscopy, which is sensitive to both structure (selection rules) and chemical bonding (force constants).

Experimental

AlPO$_4$ was prepared in solution. The samples were dried, then thermally treated at various temperatures. X-Ray diffraction was applied to identify the phases present in the samples. Spectroscopic studies in the mid (MIR) and far (FIR) infrared regions were performed with a Digilab FTS 60 V Fourier-transform spectrometer (Bio-Rad). Standard KBr pellet (MIR) and polyethylene pellet (FIR) techniques were used. Raman studies were made with an FT-Raman spectrometer connected to an FTS-40 near infrared spectrometer. As an excitation source an Nd-YAG laser of 1064 nm wavelength was used. Power on the sample was 0.6 W. The resolution of all measurements was 4 cm^{-1} and the number of scans was 256. X-Ray analysis showed that the AlPO$_4$ sample consisted exclusively of phosphotridymite. However, the procedure applied did not allow preparation of pure phosphocristobalite (the product contained a small amount of phosphotridymite) or berlinite (phosphotridymite and phospho-

---

* To whom correspondence should be addressed

cristobalite were also present). The spectra of the pure forms of berlinite and phosphocristobalite were obtained by spectra subtraction by the SpectraCalc™ program. The spectra of the various forms of SiO_2 were first published in [3].

Results

Structurally, the various forms of SiO_2 and $AlPO_4$ consist of a three-dimensional network which is formed by SiO_4 or PO_4 and AlO_4 tetrahedra linked by sharing corners with each other. Thus each oxygen atom belongs simultaneously to two tetrahedra. This means the existence of Si–O–Si bridges in SiO_2 and Al–O–P bridges in $AlPO_4$. However, the spectra of these type of structure should be interpreted in terms of their crystal unit cell atom-vibrations, for which we propose a much simpler model of tetrahedral XO_4 or bridge XOX unit vibrations [4].

In the 1200–400 cm⁻¹ region the spectra presented in Figs. 1 and 2 are characterized by three or four groups of bands. These could be interpreted as both XO_4 or X_2O molecular group atom vibrations such as stretching, rocking and bending modes corresponding

to the T_d symmetry of the tetrahedron and C_{2v} symmetry of the oxygen bridges.

If the isolated tetrahedron approach is considered, the analysis of spectra in terms of tetrahedron vibrations with ideal T_d symmetry should be performed. Such spectra can be obtained for $[SiO_4]^{4-}$ and $[PO_4]^{3-}$ anions in aqueous solutions [5], whereas for $[AlO_4]^{5-}$ they can be inferred from the spectra of various forms of aluminates [6]. In the case of $[PO_4]^{3-}$ and $[SiO_4]^{4-}$ the following bands are observed: at 935 and 800 cm⁻¹ corresponding to symmetrical stretching A_1 vibrations, 1080 and 1050 cm⁻¹ due to F_2 stretching vibrations, 420 and 500 cm⁻¹ corresponding to bending E vibrations and 550 and 625 cm⁻¹ due to F_2 rocking. All four bands occur in the Raman spectra, whereas in the IR spectra only the F_2 type vibrations can be observed. In crystalline structures the site symmetry of tetrahedra is usually lower and – in consequence – the selection rules change. In the case of low-temperature quartz (berlinite), tridymite (phosphotridymite) and cristobalite (phosphocristobalite) – Figs. 1 and 2 – the site symmetry of

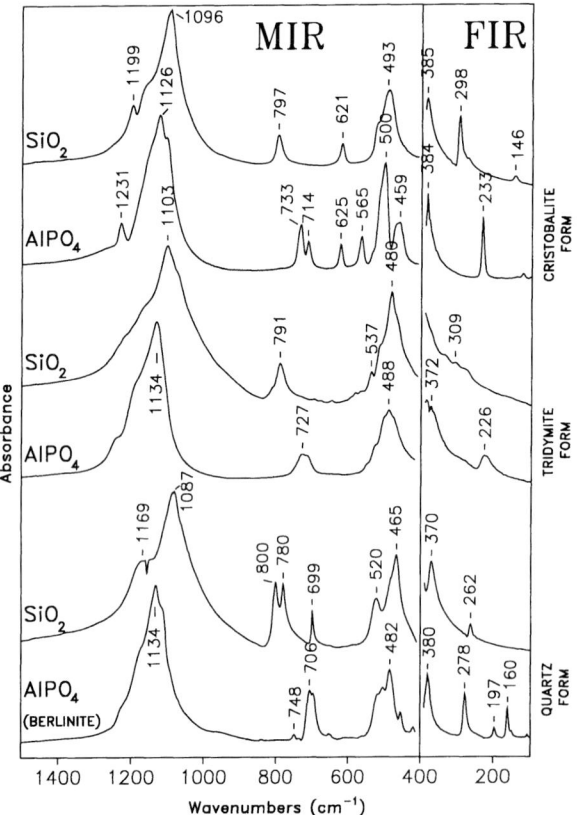

Fig. 1. MIR and FIR spectra of particular forms of SiO_2 and $AlPO_4$

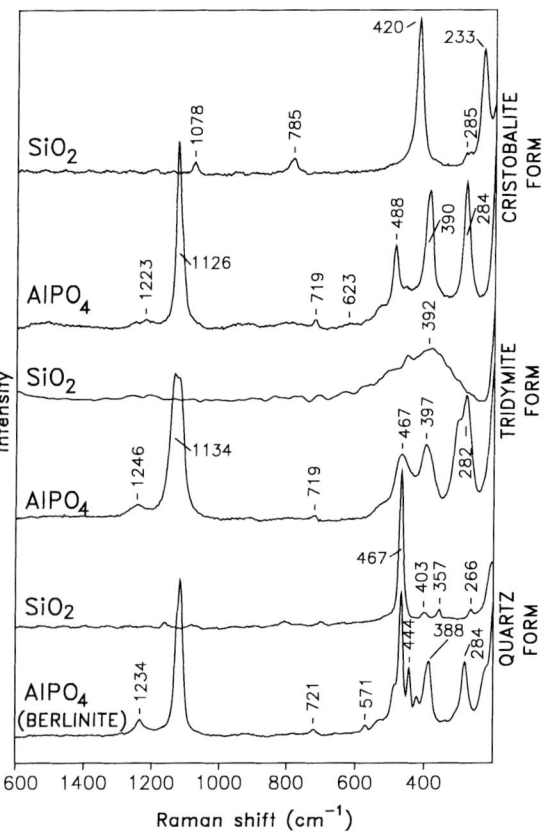

Fig. 2. Raman spectra of particular forms of SiO_2 and $AlPO_4$

Table 1. Vibrations of XO$_4$ units

	[SiO$_4$]	[PO$_4$]	[AlO$_4$]
Stretching A_1 (v_1)	~800	1120–1140	~720
Stretching F_2 (v_2)	1200–1080	1135–1100	750–700
Bending E (v_3)	~400	380–400	~300
Rocking F_2 (v_4)	625–450	560–450	400–300

Table 2. Vibrations of X$_2$O units

	Si–O(Si)	P–O(Al)	Al–O(P)
Stretching A_1	1200–1080	1235–1100	750–690
Rocking B_1	~800	~650	~400
Bending A_1	520–450	540–450	400–300

the tetrahedra is C_2 (quartz, cristobalite) or C_1 (tridymite). For the quartz factor group P3$_1$2$_1$ (D$_3^4$), tridymite C$_c$ (C$_s^2$) [or P$_1$ (C$_1$)] and P4$_3$2$_1$2 (D$_4^8$), in both the IR and Raman spectra at least one band is due to A_1-type vibrations, two bands correspond to E-type vibrations and three F_2-type vibrations can be observed. Since the samples for which spectra are shown in the present work were polycrystalline, accurate band assignment to the separate factor group vibrations is impossible. For the given anions in various structures only the wavenumber ranges which correspond to the four main vibrations of the isolated tetrahedron can be determined. This is shown in Table 1. The bands at below 300 cm^{-1} should be regarded as lattice vibrations of polyhedra.

For the second model, based on the X$_2$O unit vibrations, i.e. Si–O–Si or Al–O–P bridges, C$_{2v}$ site symmetry should be assumed and thus three types of vibrations should be observed: stretching (A_1), bending (B_1) and rocking (A_1). The assignments to the bands occurring in the spectra are given in Table 2. As shown in previous work [4] this model explains well the

spectra of SiO$_2$. In the case of AlPO$_4$ account should be taken of the asymmetry of Al–O–P bridges, in which the P–O bond is much stronger than the Al–O bond. This is expressed by the Pauling cation valances which are 5/4 and 3/4 respectively. In the case of AlPO$_4$ the PO$_4^{3-}$ anion vibrations should be considered independently. Thus the tetrahedral anion model explains well the spectra of aluminium phosphate, whereas in the case of SiO$_2$ the pseudomolecular Si$_2$O unit model is more useful.

The wavenumbers corresponding to the valence vibrations in the crystalline structures are shifted towards higher values than those for the anions in solution. This can be explained as due to the increase of the stretching force corresponding to the share of the electrons that bind oxygen in the P–O and Si–O bonds, formation of π-bonding and increase in the band order. This is corroborated by the observed shortening of these bonds [7].

To conclude this short discussion, it can be said that the IR and Raman spectra of SiO$_2$ polymorphs cannot be explained by comparison with the spectra of the corresponding AlPO$_4$ forms, but both structures are closely related.

Acknowledgement. This work was financially supported by the Polish Committee for Scientific Research (KBN) under grant 7TO8D 020 10.

References

[1] E. Gorlich, *Ceram. Int.*, **1982**, 8, 3.
[2] K. Kosten, H. Arnold, *Z. Kristall.*, **1980**, 152, 119.
[3] M. Handke, W. Mozgawa, *Vib. Spectrosc.*, **1993**, 5, 75.
[4] M. Handke, W. Mozgawa, *J. Mol. Struct.*, **1995**, 348, 341.
[5] N. Nakamoto, *Infrared and Raman Spectra of Inorganic and Coordination Compounds, 4th Ed.*, Wiley-Interscience, New York, 1986.
[6] P. Tarte, *Spectrochim. Acta*, **1962**, 18, 467.
[7] S. G. Kosinski, D. M. Krol, T. M. Duncan, D. C. Douglass, J. B. MacChesney, J. R. Simpson, *J. Non-Cryst. Solids*, **1988**, 105, 45.

Mikrochim. Acta [Suppl.] 14, 515–517 (1997)

Vibrational Spectroscopy of Borogermanate Glasses as Models for Aluminosilicate Glasses

Mirosław Handke*, Włodzimierz Mozgawa, and **Magdalena Rokita**

Department of Material Science and Ceramics, University of Mining and Metallurgy (AGH), 30-059 Kraków, al. Mickiewicza 30, Poland

Abstract. In the work presented IR spectroscopy studies of borosilicate and borogermanate glasses were made. These glasses can be regarded as a structural model for aluminosilicate glasses. Their compositions were selected in such a way that they corresponded to the framework of postassium aluminosilicates in which glass-forming components, i.e. SiO_2 and Al_2O_3, were replaced by GeO_2 and B_2O_3 respectively. Comparison of the starting aluminosilicate glasses showed that aluminogermanate glasses are good analogues, whereas borosilicate glasses are not exact models.

Key words: aluminosilicate glasses, aluminogermanete glasses, borosilicate glasses.

Difficulties encountered in the preparation of silicate and aluminosilicate glasses are due to their high melting points (often above 1500 °C). This means that in practice most of the glasses belonging to this group cannot be prepared by the conventional procedure, i.e. melting of the starting materials. Therefore in order to study the structure of these glasses, model glasses are quite often considered. In the case of aluminosilicate glasses, SiO_2 and Al_2O_3 (components that determine the high melting points) are replaced by isotypical substituents. Most frequently GeO_2 is used instead of SiO_2 and B_2O_3 instead of Al_2O_3 (i.e. the oxides of the elements adjacent to Si and Al in the main groups of the periodic table). Thus glasses of a new system are obtained according to the following scheme:

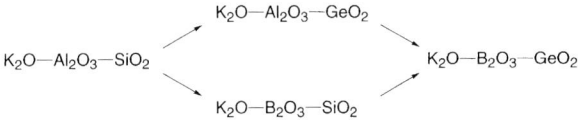

The glasses formed in this way exhibit similar structure to the corresponding silicate or aluminosilicate structures, so interpretation of the IR spectra may be facilitated by comparing them with the spectra of glasses belonging to the systems mentioned in the scheme above.

Experimental

Glasses were prepared by conventional melting of starting materials (except for potassium aluminosilicate glasses, the synthesis of which was described in [1]). Potassium carbonate, aluminium nitrate, germanium oxide and boric acid were used as starting compounds. The preparational procedure involved standard operations, such as grinding, drying, roasting of the starting materials and melting of the sinters formed, in the electric furnace. The melting points depend on the composition and lie in the range 900–1300 °C. The glasses prepared were clear and transparent. Compositions of all glasses corresponded to that of potassium aluminosilicate, and their oxide formulae are given in Table 1. Spectroscopic measurements in the mid (MIR) and far (FIR) infrared regions were made with a Bio-Rad Digilab FTS 60V Fourier-transform spectrometer. Standard KBr pellet (MIR) and polyethylene pellet (FIR) techniques were used. The resolution of all measurements was 4 cm⁻¹ and the number of scans was 256.

Results and Discussion

The ionic radius of Ge^{4+} is larger than that of Si^{4+} and the radius ratio of Ge^{4+}/O^{2-} is equal to 0.31–0.43 according to various authors [2, 3]. Thus it can be concluded that germanium is coordinated by 4 oxygen atoms, though a coordination number of 6 is also possible. In glassy GeO_2 the germanium coordination number is 4 [3], similar to that in most multicomponent glasses [4]. Octahedral coordination was confirmed only in the case of alkali metal germanate glasses. However in this case such coordination is caused by alkali metal cations which break oxygen bridges [5].

---

* To whom correspondence should be addressed

Table 1. Compositions of the synthesised glasses

System $K_2O–Al_2O_3–SiO_2$	System $K_2O–Al_2O_3–GeO_2$	System $K_2O–B_2O_3–SiO_2$	System $K_2O–B_2O_3GeO_2$
$K_2O·Al_2O_3·6SiO_2$	$K_2O·Al_2O_3·6GeO_2$	$K_2O·B_2O_3·6SiO_2$	$K_2O·B_2O_3·6GeO_2$
$K_2O·Al_2O_3·4SiO_2$	$K_2O·Al_2O_3·4GeO_2$	$K_2O·B_2O_3·4SiO_2$	$K_2O·B_2O_3·4GeO_2$
$K_2O·Al_2O_3·2SiO_2$	$K_2O·Al_2O_3·2GeO_2$	$K_2O·B_2O_3·2SiO_2$	$K_2O·B_2O_3·2GeO_2$

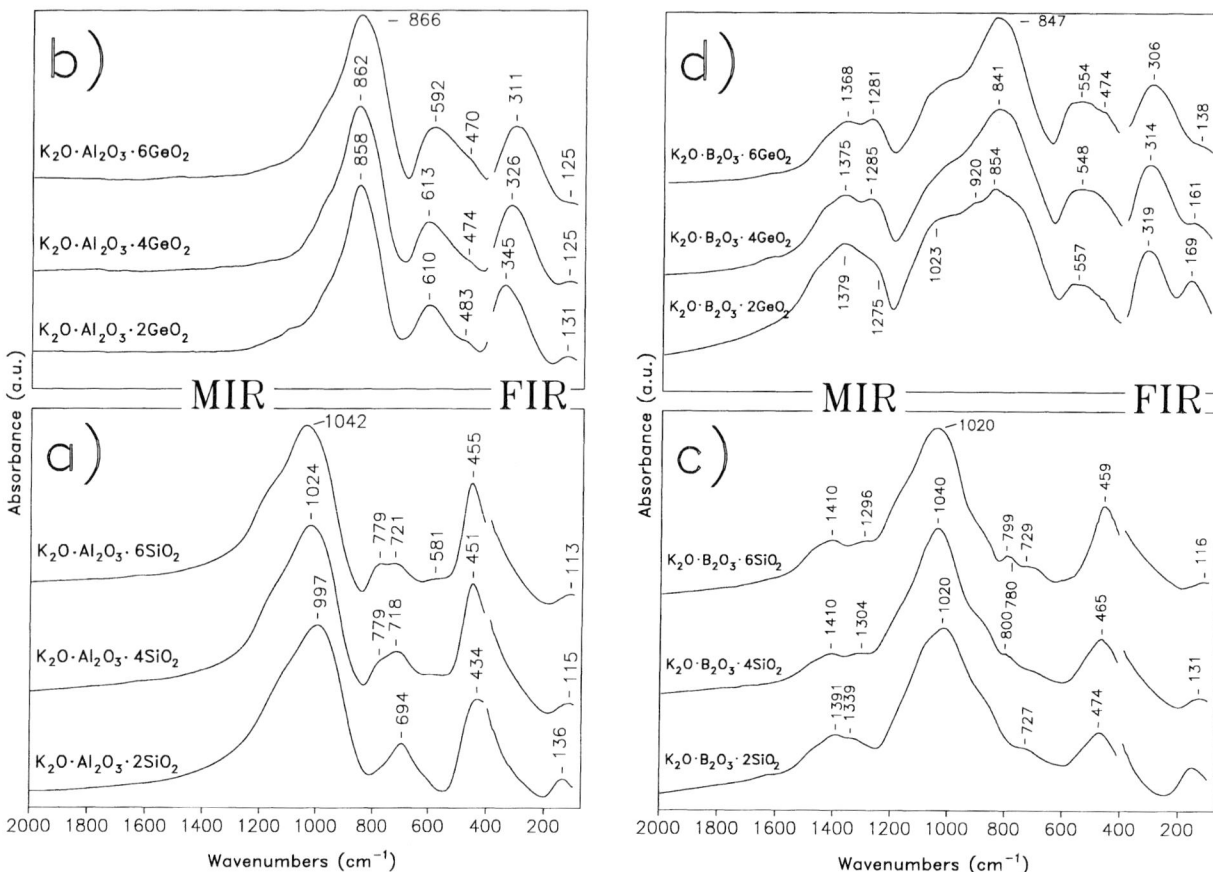

Fig. 1. MIR and FIR spectra of glasses of the systems: $K_2O–Al_2O_3–SiO_2$(**a**), $K_2O–Al_2O_3–GeO_2$(**b**), $K_2O–B_2O_3–SiO_2$(**c**), $K_2O–B_2O_3–GeO_2$(**d**)

Figure 1a shows MIR and FIR spectra of potassium aluminosilicate glasses. They were discussed in our previous papers [1, 6]. Comparing these spectra with those of the aluminogermanate glasses (Fig. 1b) reveals that all the bands are shifted towards lower wavenumbers. This fact may be explained by both the smaller force constant of the Ge–O bond and the higher mass of germanium [7]. Three main complex bands can be observed in the spectral regions 800–1000, 400–650 and 300–350 cm^{-1}. According to the network structure of these glasses, the bands occurring in their spectra can be assigned to internal vibrations of

Ge–O–Al and Ge–O–Ge bridges. Geometrically such bridges with C_{2v} symmetry are analogous to the Si–O–Al and Si–O–Si bridges occurring in aluminosilicates, but the angle in Ge–O–Ge bonds is smaller, viz. 133° [8]. The most intensive band at around 860 cm^{-1} is due to stretching vibrations of Ge–O. The second band group is associated with the rocking vibrations originating from the oxygen bridges, but in $K_2O·Al_2O_3·2GeO_2$ glass only one type of bridges is present, viz. Al–O–Ge, whereas in two other glasses ($K_2O·Al_2O_3·4GeO_2$ and $K_2O·Al_2O_3·6GeO_2$) both types of bridges exist. The third band characteristic is due to the deformational vibrations of O–Ge–O being shifted to the region 300–350 cm^{-1}. In the spectra there are no bands that could prove octahedral germanium coordination characteristic of the tetragonal (quartz-like) GeO_2 form. The band characteristic of this form should occur at 720 cm^{-1} [3]. Thus it can be concluded that judging by IR spectroscopy, aluminogermanate glasses are good models for their aluminosilicate analogues.

In potassium borosilicate glasses (Fig. 1c) the second of the glass-forming components, i.e. Al_2O_3, is replaced by B_2O_3. However, the structural similarity of these glasses is not as close as in the case of GeO_2 and SiO_2. Boron atoms can be both tetrahedrally and triangularly coordinated. In the IR spectra this is revealed by the appearance of the bands in the regions of 850–1200 cm^{-1} (stretching vibrations of B–O in BO_4 units) and 1200–1500 cm^{-1} (stretching vibrations of B–O in BO_3 units) [9]. The bands in this second spectral region are visible, which means that some boron atoms exhibit coordination number 3. Apart from that a very strong band in the region of 1020–1050 cm^{-1} appears which is due to both the stretching Si–O and B–O vibrations mentioned above (both cations 4-coordinated). The bands due to Si–O–Si and Si–O–B bridge vibrations occur in the range 700–800 cm^{-1}. The glass with composition described by the formula $K_2O \cdot B_2O_3 \cdot 2SiO_2$ contains only Si–O–B bridges and therefore a single band can be observed in its spectrum. In the spectra of $K_2O \cdot B_2O_3 \cdot 4SiO_2$ and $K_2O \cdot B_2O_3 \cdot 6SiO_2$ glasses a very weak doublet at 780–800 cm^{-1} characteristic of quartz-like structure is visible. This may result from a heterogeneous glass structure in which two types of areas co-exist: ones of aluminosilicate framework type (tetrahedral coordination of B and Si) and ones of different structure (with triangular coordination of B and some quartz is also possible). In the FIR spectra only a broad band in the range of 150–110 cm^{-1} is visible, which is due to potassium ion translational vibrations [10]; its intensity grows as the amount of K_2O in the glass increases. It should be noted that bands in this range appear in the FIR spectra of all the glasses discussed, and their evolution is similar (Fig. 1a–d). Since triangular coordination around Al atoms in the starting aluminosilicate glasses is impossible it may be concluded that borosilicate glasses are not exact structural analogues of aluminosilicate glasses. However, many features in the spectra of both types of glasses are consistent.

Figure 1d shows MIR and FIR spectra of glasses of K_2O–B_2O_3–GeO_2 system. The spectra are close to the spectra of both potassium aluminogermanate glasses and potassium borosilicate glasses. However, they cannot be regarded as simple superposition of these spectra. The most intensive broad band in the region 850–840 cm^{-1} with a shoulder at about 1020 cm^{-1} results from coincidence of stretching Ge–O(Ge) and Ge–O(B) vibrations. The bands in the region 1380–1250 cm^{-1} originate from the stretching B–O vibrations when boron is 3-coordinated [9]. Their intensity is, however, much lower than in the spectra of borosilicate glasses, which may indicate that substitution of Si atoms by Ge favours triangular coordination of boron. The other bands can be assigned to the same vibrations as in the two previous cases. From the spectra obtained it can be concluded that simultaneous substitution of both glass-forming components leads to significant differences from the structure of aluminosilicate glasses.

Acknowledgement. The present work was financially supported by the University of Mining and Metallurgy grant number 11.160.189.

References

[1] M. Handke, W. Mozgawa, *Vib. Spectrosc.* **1993**, *5*, 75.

[2] V. A. Blinov, *J. Mater. Sci.* **1969**, *4*, 461.

[3] M. K. Murthy, E. M. Kirby, *Phys. Chem. Glasses* **1964**, *5*, 146.

[4] E. Dowty, *Phys. Chem. Minerals* **1987**, *14*, 122.

[5] G. S. Henderson, M.E. Fleet, *J. Non-Cryst. Solids* **1991**, *134*, 259.

[6] M. Handke, W. Mozgawa, *J. Mol. Struct.* **1995**, *348*, 15.

[7] T. Furukawa, W. B. White, *J. Mat. Sci.* **1980**, *15*, 1648.

[8] S. K. Sharma, D. W. Matson, J. A. Matson, J. A. Philpotts, T. L. Roush, *J. Non.-Cryst. Solids* **1984**, *68*, 99.

[9] E. I. Kamitsos, A. P. Patsis, M. A. Karakassides, G. D. Chryssikos, *J. Non.Cryst. Solids* **1990**, *126*, 52.

[10] C. I. Merzbacher, W. B. White, *Am. Miner.* **1988**, *73*, 1089.

Mikrochim. Acta [Suppl.] 14, 519–521 (1997)

FT-IR Spectra of Enol-esters of 2-Aryl-2,3-dihydrophenalene-1,3-diones

Ilyana R. Karamancheva[1,*], **Tzwetanka S. Philipova**[1], **Neyko M. Stoyanov**[2], **Iwanka S. Losseva**[2], and **Boris V. Aleksiev**[2]

[1] University of Chemical Technology and Metallurgy, BG-1756 Sofia, Bulgaria
[2] Institute of Chemical Technology and Biotechnology, BG-7200 Razgrad, Bulgaria

Abstract. The FT-IR spectra of new enol-esters of 2-aryl-2,3-dihydrophenalene 1,3-diones have been measured. The influence of the structure of the compounds on the position of the C=O and OC=O stretching bands is discussed. The available experimental data from the FT-IR spectra suggest the presence of bipolar structures due to the ester resonance.

Key words: FT-IR spectra, enol-esters of 1,3-diones.

The presence of substituents substantially affects the biological activity and functionality of β-dioxo-derivatives. In this work the influence of substituents on the characteristic frequencies in new enol-acetates and enol-benzoates of dihydrophenalenediones was investigated, by using deconvoluted and derivative spectra.

Experimental

The compounds studied were of general formula I. Compounds 2, 8, and 9 were prepared by standard methods [1]. The other compounds were prepared for the purpose of this research and their syntheses will be published later. All compounds investigated were purified by recrystallization or chromatography. IR spectra (KBr) were recorded with a Perkin–Elmer 1600 FTIR spectrophotometer, in the range 4000–650 cm^{-1}, resolution 4 cm^{-1}, sample thickness 1 mm. Derivative calculation and deconvolution are incorporated functions of the apparatus. Deconvoluted spectra were computed by using the following parameters: width – 9.51 cm^{-1} and smooth – 5.66 [2].

Results and Discussion

Characteristic frequencies of the compounds studied are given in Table 1. In order to determine the influence

No.	R	X	Y	Z
1	CH$_3$	H	H	H
2	CH$_3$	Cl	H	H
3	CH$_3$	H	Cl	H
4	CH$_3$	H	H	Cl
5	CH$_3$	F	H	H
6	CH$_3$	H	F	H
7	CH$_3$	H	H	F
8	CH$_3$	CH$_3$	H	H
9	CH$_3$	OCH$_3$	H	H
10	CH$_3$	NHCOCH$_3$	H	H
11	C$_6$H$_5$	H	H	H
12	C$_6$H$_5$	OCH$_3$	H	H
13	C$_6$H$_5$	Cl	H	H

Formula 1

of substituents in the benzene ring the results are compared with those for compound 1. On the other hand, the influence of the acidilating agent was studied.

The band for the ketonic carbonyl group appears at around 1640 cm^{-1}. In this case the conjugation with the naphthalene ring is dominant and other substituents have no essential influence.

The absorption of the ester carbonyl group was of great interest. Because of interruption of the conjugated system, due to the ether oxygen atom, $\nu_{OC=O}$

Table 1. Characteristic frequencies of compounds 1–13

No.	v_{CH} (arom)	v_{CH} (aliph)	$v_{OC=O}$	$v_{OC=O}$[a]		Δv[a]	$v_{C=O}$	$v_{C=C}$
1	3055	2936	1764				1638	1624
2	3056	2930	1758	1765	1752	13	1639	1628
3	3059	2927	1753	1765	1753	12	1641	1624[a]
4	3059	2927	1762	1771	1752	19	1639	1624[a]
5	3059	2926	1761	1766	1752	14	1637	1626[a]
6	3053	2938	1765	1770	1758	12	1639	1626
7	3063	2934	1786	1790	1784	6	1640	1624[a]
8	3051	2919	1758	1766	1752	12	1638	1624[a]
9	3051	2932	1771	1771	1758	13	1636	1624[a]
10	3063	2933	1777 +1722, 1698 (NHCO)	1784	1768	16	1634	1625[a]
11	3054	2923	1744	1735			1635	1624[a]
12	3060	2953	1742	1735			1635	1623[a]
13	3055	2933	1744	1735			1645	1626[a]

[a] Computed fourth derivative.

should not be influenced by substituents in the benzene ring. However, in some cases a frequency shift of up to 36 cm^{-1} was observed. Difference spectra showed that the substituents caused the appearance of a new peak.

The essential increase in $v_{OC=O}$ for *ortho*-substituted compounds, such as 7, may be interpreted only in terms of the predominant field effect [3].

Table 1 shows that regardless of the absence of conjugation between the substituents X, Y and the ester group, an essential frequency shift is observed. A linear correlation between $v_{OC=O}$ and the σ^+ constants of the substituents was found, calculated according to [4]:

$$\sigma_p^+ = 0.38F + 0.62R,$$

$$\sigma_m^+ = 0.80F + 0.20R$$

where F is the field effect and R the resonance effect.

This means that there is a strong resonance interaction within the molecule. The negative sign of ρ (-0.17) shows that the interaction is realized through an electron-deficient centre [4]. This fact can be explained as due to the existence of bipolar reasonance structures, as a result of ester resonance [5]:

In the valence structure **III** the extension of the conjugated system compensates for the energy loss from

Fig. 1. (—) Absorption band of ester carbonyl group for compound 7; (—·—) fourth derivative spectrum of the same band; (----) deconvoluted spectrum of the same band

the appearance of separated charges. The positive charge may be delocalized in the conjugated system similarly to the case of meso-ionic compounds [4]. Structure **III** explains the higher $v_{OC=O}$ since: (**i**) substituents X, Y and the ester group are in a conjugated system and influence each other; (**ii**) these resonance structures suggest the presence of two ester carbonyl peaks. Deconvolution revealed that the peaks shown in column 4 of Table 1 were coupled vibrations, obtained as a result of two overlapping peaks. In order to avoid over-deconvolution, the data were confirmed by means of derivative spectra (Fig. 1). For compound 1 only one peak was observed, which means that the resonance interaction was determined by substituents X, Y and Z. Further, (**iii**) the ketonic carbonyl group remains out of this conjugated system and that is why it is not affected by substituents.

A comparison between compounds containing R= CH_3 and R=C_6H_5, shows lowering of $\nu_{OC=O}$ by 20 cm^{-1} for R=C_6H_5, without further influence of X, because of the already existing conjugation with the benzene ring.

In compound 10 there is another carbonyl group – the amide, which appears as a doublet (1721, 1699) because of keto–enol tautomerism.

The absorption band arising from the double bond of the cycle appears at about 1625 cm^{-1}. In most cases it is masked and observed as a shoulder of the ketonic carbonyl group band. Its proper position is determined by evaluation of the fourth derivative. The absorption frequency of the double bond is lowered and practically not influenced by the substituents, owing to the conjugation with the benzene ring.

Acknowledgement. The authors thank Prof. B. Jordanov for helpful discussion and greatly appreciate the financial support given by Perkin–Elmer BG.

References

[1] E. Gudriniece (ed.), *Structure and Tautomerism of β-Dicarbonyl Compounds*, Zinatne, Riga, 1977, pp. 381–392.

[2] J. K. Kauppinen, D. J. Moffat, H. H. Mantsch, D. G. Cameron, *Anal. Chem.*, **1981**, *53*, 1454.

[3] W. Kemp, *Organic Spectroscopy*, Macmillan, London, 1987, p. 30.

[4] J. March, *Advanced Organic Chemistry*, Wiley, New York, 1985, Vol. 1, pp. 91, 372.

[5] J. B. Lambert, H. F. Shurvell, L. Verbit, R. G. Cooks, G. H. Stout, *Organic Structural Analysis*, Macmillan, New York, 1976, pp. 121, 231.

Mikrochim. Acta [Suppl.] 14, 523–524 (1997)

FT-IR Spectroscopy of OH⁻ Ions in Borate Single Crystals

László Kovács[1,*], **Katalin Polgár**[1], **Ágnes Péter**[1], and **Rosanna Capelletti**[2]

[1] Research Laboratory for Crystal Physics, Hungarian Academy of Sciences, H-1112 Budapest, Budaörsi út 45., Hungary
[2] INFM and Physics Department, University of Parma, Viale delle Scienze, I-43100 Parma, Italy

Abstract. Infrared absorption bands were detected and attributed to the stretching vibrational transition of hydroxyl ions (OH⁻) in lithium triborate, LiB_3O_5 (LBO) and lithium tetraborate, $Li_2B_4O_7$ (LTB) single crystals. The anharmonicity of the OH⁻ vibration and the binding energy were determined from the first overtone spectrum in LBO. The model of a weakly coupled phonon was used to analyse the temperature dependence of the band parameters.

Key words: lithium triborate, lithium tetraborate, hydroxyl ions, weakly coupled phonon model.

Owing to the occurrence of boron in three or four coordination a variety of borates can form, combining the trigonal BO_3 and the tetrahedral BO_4 molecular units. Some of them, e.g. lithium triborate LiB_3O_5 (LBO), reveal superior non-linear optical properties, and lithium tetraborate, $Li_2B_4O_7$ (LTB) shows excellent piezoelectric properties. LTB doped with a proper impurity (e.g. Cu) is a good candidate for use in thermoluminescence dosimetry.

Successful growth of high quality borate single crystals was first reported in the mid 1980's [1]. Shortly afterwards borates were also prepared at the Research Laboratory for Crystal Physics. The incongruently melting LBO (m.p. 834 °C) was grown by the high-temperature top-seeded solution growth technique, whereas the congruently melting LTB (m.p. 917 °C) was prepared by the Czochralski method. LBO is an orthorhombic biaxial crystal with space group $Pna2_1$, while LTB is a tetragonal uniaxial one with space group $I4_1cd$.

Very little is known about defects induced by impurities in these crystals. Oxides melting at around 1000 °C and grown in air usually contain hydroxyl ions. In this work high-resolution FT-IR absorption spectroscopy was utilized to study the vibrational modes of hydroxyl ions incorporated in LBO and LTB crystals during the growth process. Spectra were taken by use of a Bomem DA8 Spectrometer with an apodized resolution as good as 0.05 cm⁻¹ in the 9–300 K temperature range (CTI Cryogenics Model 21 Refrigerator). Typically several hundred scans were recorded at each temperature.

Results and Discussion

A narrow (0.54 cm⁻¹) absorption band at about $v_{01} = 3460.97$ cm⁻¹ was found at 9K for LiB_3O_5 (Fig. 1a). From the typical position of the band it may be attributed to the 0→1 transition related to the stretching vibrational mode of the hydroxyl ion. A very weak signal at about $v_{02} = 6731.25$ cm⁻¹, corresponding to the first overtone transition, was also detected. The anharmonicity parameter $x_e = (2v_{01} - v_{02}) (6v_{01} - 2v_{02})^{-1} = 0.0261$ and the binding energy $D_e = (3v_{01} - v_{02})^2 (4v_{01} - 2v_{02})^{-1} = 4.34$ eV, calculated by using the Morse potential model, are in good agreement with those of OH-defects in other oxide crystals [2].

A much broader (6.4 cm⁻¹) absorption band at 3263.2 cm⁻¹ was found for both pure and doped (Cu, Mn or Fe) $Li_2B_4O_7$ at 9 K, indicating the presence of OH⁻ impurity in the crystals (Fig. 1b). No overtone of the OH⁻ band was detected, probably because of the weak and broad fudamental signal. Preliminary polarization measurements show that the dipole moment of the OH⁻ ion is at an angle of about 45° with respect to the tetragonal c axis of the crystal.

---

* To whom correspondence should be addressed

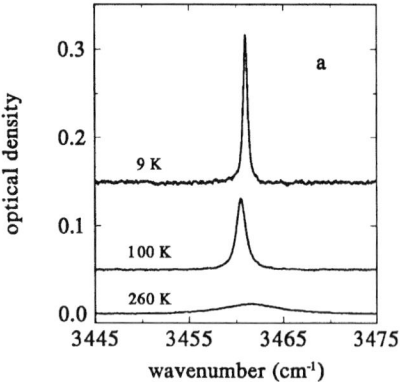

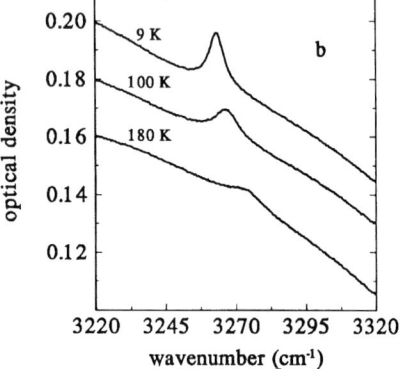

Fig. 1. The absorption bands measured at various temperatures in borate crystals: **a** LiB_3O_5, **b** $Li_2B_4O_7$

Table 1. The parameters of the coupled phonons determined in a given temperature range; $\delta\omega$ is the coupling constant, γ and ω_0 are the width and the frequency of the coupled phonon, respectively, ω_R is a phonon frequency measured by the Raman scattering method

	$\delta\omega$ (cm^{-1})	γ (cm^{-1})	ω_0 (cm^{-1})	ω_R (cm^{-1})	T (K)
$Li_2B_4O_7$	11.6	28.8	102.7	100	9–100
LiB_3O_5			160	163	9–300[a]
	−0.6		47		9–100[b]
	13.7		342	342	100–300[b]

[a] Calculated from the temperature dependence of the halfwidth
[b] Calculated from the temperature dependence of the band position

The OH⁻ band parameters are changed drastically by increasing the temperature. The intesity decreases and the halfwidth increases for both crystals (see Fig. 1). In LBO the frequency of the OH⁻ vibration shows a red-shift at temperatures up to 100 K and a blue-shift between 100 and 300 K. In LTB, however, a blue-shift of the vibrational frequency was found over the whole temperature range, which is widely believed to be characteristic for hydrogen bonding. The relatively low vibrational frequency compared to that of free OH⁻ (3555.59 cm^{-1} [3]) and the broad halfwidth even at 9 K support the hypothesis above.

The weakly coupled phonon model was successfully applied to analyse the temperature dependence of the OH⁻ band parameters in sillenite crystals [2]. The frequency ω_0, the bandwidth γ of the coupled phonon, and the coupling constant $\delta\omega$ can be determined simultaneously from the temperature dependence of the position and width of the OH⁻ band. In borates, however, the model was found to be valid only with restrictions. In LTB the experimental data fit the weakly coupled phonon model only in the 9–100 K temperature range. It should be stressed that the fitting of the data to the model proposed by Wójcik and Falk [4] for hydrogen bonded molecules also failed. In LBO, owing to the peculiar behaviour of v_{01} vs. T, the halfwidth of the band can only be fitted over the whole temperature range studied (9–300 K) by assuming a single phonon coupling. From v_{01} (T) two different phonon frequencies can be calculated, at temperatures below and above 100 K, respectively. Nevertheless the coupled phonon frequencies obtained (see Table 1) are in good agreement with those measured by the Raman scattering method [5, 6]

Acknowledgement Financial support from the National Science Fund of Hungary (OTKA T-4420 and T-014026) is acknowledged.

References

[1] C. Chen, B. Wu, A. Jiang, G. You, *Sci. Sin., Ser. B* **1985**, *28*, 235.
[2] P. Beneventi, R. Capelletti, L. Kovács, Á. Péter, A.M. Lanfredi Manotti, F. Ugozzoli, *J. Phys., Condens. Matter* **1994**, *6*, 6329.
[3] J. C. Owrutsky, N. H. Rosenbaum, L. M. Tack, R. J. Saykally, *J. Chem. Phys.* **1985**, *83*, 5338.
[4] M. J. Wójcik, M. Falk, *Chem. Phys. Lett.* **1987**, *56*, 450.
[5] G. L. Paul, W. Taylor, *J. Phys. C: Solid State Phys.* **1982**, *15*, 1753.
[6] Q.-Y. Shang, B. S. Hudson, C. Huang, *Spectrochim. Acta* **1991**, *47A*, 291.

Mikrochim. Acta [Suppl.] 14, 525–528 (1997)
© Springer-Verlag 1997

Infrared and Raman Spectra of Hydroquinone Crystalline Modifications

Miklós J. Kubinyi[1,*] and **Gábor Keresztury**[2]

[1] Department of Physical Chemistry, Technical University of Budapest, 1521 Budapest, Hungary
[2] Central Research Institute for Chemistry, Hungarian Academy of Sciences, P.O. Box 17, 1525 Budapest, Hungary

Abstract. The infrared and Raman spectra of the three crystalline modifications of hydroquinone (HQ), known as α-HQ, β-HQ and γ-HQ, have been measured and compared with the spectra of HQ solutions. The spectra have been assigned on the basis of a normal coordinate analysis of the molecule and a factor group analysis of the crystals. The hydrogen bonds are relatively the strongest in β-HQ, the weakest in γ-HQ.

Key words: hydroquinone, infrared spectrum, Raman spectrum, crystalline modifications.

Hydroquinone (HQ, 1,4-dihydroxybenzene) forms three crystalline modifications, designated α, β and γ. Of these α is the stable form at room temperature. The crystal structures of all three forms are known from X-ray diffraction studies. In α-HQ there are three types of molecules [1]. Two of these molecules are involved in forming open hydrogen-bonded cage structures, the third forming double helices of hydrogen-bonded chains of HQ molecules. β-HQ is an open cage structure in which the hydrogen bonds connect the molecules into 6-membered rings [2]. These two modifications are capable of containing small molecules in clathrate structures. In γ-HQ the hydrogen bonds link the molecules into infinite chains [3].

The structures of the HQ-clathrates have been widely studied by vibrational spectroscopy. In the present work FT-IR and Raman spectra of HQ in solution and in the three crystalline states have been recorded and the spectra have been assigned to normal vibrations on the basis of normal coordinate calculations. It is believed that these results will be of significant value in later studies on HQ clathrates and other complexes.

Experimental

Commercially available HQ was recrystallized from water three times and the material obtained was used as α-HQ and for the preparation of the other two modifications. β-HQ was prepared by dissolving α-HQ in hot n-propanol and placing a crystallite of an H_2S–HQ clathrate in the cooling solution. (The H_2S–HQ clathrate was obtained by bubbling H_2S through an aqueous solution of HQ.) γ-HQ was made by subliming α-HQ in a nitrogen atmosphere.

HQ dissolves poorly in apolar solvents. In non-polar solvents the highest band intensities were observed from the spectra of a saturated solution in benzene, the temperature of which was close to the boiling point. Many more bands could be observed in the spectra of acetonitrile and ethanol solutions at room temperature, conditions which allowed Raman polarization measurements to be made. The infrared spectra of the crystal were taken in Nujol and in poly(chlorotrifluoroethylene) mulls. The Raman spectra of α-HQ and β-HQ were taken in glass capillaries, whereas that of γ-HQ was measured on the material kept is the sublimation vessel under nitrogen, since atmospheric moisture catalyses the phase change into the stable modification. Infrared spectra in the 4000–400 cm^{-1} region were recorded on a Nicolet 170 sx FT spectrometer and in the 400–50 cm^{-1} range on a Grubb Parsons IS3 FT spectrometer. A Cary 82 Raman spectrometer was used for Raman measurements in the 4000–20 cm^{-1} region.

Results and Discussion

The vibrational frequencies observed in the experimental spectra, together with their assignments, are given in Table 1. The assignments are based on the results of normal coordinate calculations by the CNDO force method, the details of which can be found in [4]. Additional calculations with *ab initio* methods using 4–31 G* basis sets are underway.

The strengths of the hydrogen bonds in the three crystalline forms of HQ are primarily characterized by the frequencies of the νOH and γOH bonds. These indicate that the hydrogen bonds would be classified as weak in all three modifications [5]. The stretching frequencies change in the order νOH(γ-HQ) >

---

* To whom correspondence should be addressed

Table 1. Observed bands (in cm^{-1} units) in the vibrational spectra of hydroquinone

IR					
Solution		Assignment	Crystal		
Benzene	CH$_3$CN		α	β	γ
3568 s	3421 s	b$_u$, νOH	3256 vs,b	3154 vs,b	3327 vs,b
		b$_u$, νCH	3030 mm	3030 m	3041 m
		b$_u$, νCH			
		a$_g$, νCC + βH	1610 w		
	1520 w				
1512 vs	1514 vs	b$_u$, νCC + βH	1517 vs	1518 vs	1515 vs
			⌜1478 s		
1447 w		b$_u$, δOH + νCC	⊢1470 sh	⌜1469 vs	1466 vs
			⌞1460 sh	⌞1450 sh	
			⌜1366 w	⌜1363 m	1365 s
1331 m	1350 s	b$_u$, νCC + βH + δOH	⊢1355 s	⊢1345 w	
			⌞1333 w	⌞1330 w,b	1322 w
			1259 s		
1246 s	1249 s	b$_u$, νCO + βH	1243 s	1246 s	1240 s
	1231 s		1222 m		
1168 s	1215 s	b$_u$, νCC + δOH	⌜1209 s	⌜1210 s	1207 s
	1201 s		⊢1192 s	⌞1197 s	
			⊢1180 w		
		a$_g$, βH	⊢1164 w		
			⊢1150 w		
			⌞1141 sh		
			1118 w	1120 w	1123 w
			1102 w		
1091 w	1095 m	b$_u$, νCC + βCH	1097 s	1096 s	1100 s
			1081 w		
		b$_u$, α + νCC	1009 m	1010 m	1011 m
			972 w		953 w
		a$_u$, γH	940 w	942 vw	947 w
			920 w	920 vw	
			890 w,b	898 w	
			852 w		
824 s	834 s	a$_u$, γH + γO	827 s	833 s	⌜830 s
					⌞826 s
	805 w		808 w		
756 s	760 s	b$_u$, νCO + α	⌜765 sh		
			⌞758 s		
		b$_g$, Φ	702 w		
	605 m	a$_u$, τOH	613 m,b	684 m,b	557 w,b
513 s	521 m	a$_u$, γO + Φ	525 m	525 s	⌜525 w
			517 s		⌞510 m
		a$_u$, Φ			
		b$_u$, βO	390 m	388 m	388 m
					354 w
		a$_u$, Φ + γO	⌜213 s	200 s	193 s
			⌞192 s		
			180 sh		
			151 m	154 w	154 w
		lattice modes	125 s	133 w	
			100 m	111 s	102 vs
			86 w		
			76 w	60 m	

Table 1. *Continued*

			Raman		
Solution		Assignment	Crystal		
Benzene	CH$_3$CN		α	β	γ
	3437 m,b	a$_g$, νOH	3216 w,b	3203 w,b	3324 m,b
	3070 s	a$_g$, νCH	⎡ 3080 m ⊢ 3067 vs ⎣ 3049 m	3067 vs 3050 s	3068 vs
		a$_g$, νCH	3022 m	3020 m	3026 m
	1618 m,p	a$_g$, νCC	1625 m	1622 m	1626 w
	1604 m,p	a$_g$, νCC	⎡ 1608 m ⎣ 1601 m	⎡ 1612 w ⎣ 1600 m	1608 m
			1409 w		
		a$_g$, βH + δOH	1370 w,b	1370 w,b	
1262	1267 s,p	a$_g$, νCO + νCC	⎡ 1274 m ⊢ 1257 s ⎣ 1252 s	1261 s	⎡ 1265 m ⊢ 1258 m ⎣ 1240 w
	1227 s,p	a$_g$, δOH + βH		1228 m	1225 m
	1162 m,p	a$_g$, βCH	1169 m	1162 s	1166 m
			1163 m		
			1163 m		
			1104 vw		
			1036 w,b		
		b$_u$, α + νCC	1010 w		
		b$_g$, γH	922 w	932 w	
	854 vs, p	a$_g$ νCC	855 vs	855 vs	854 s
	835 s, p	b$_g$, γH	⎡ 834 s ⎣ 828 m	835 s	834 m
	806 w, dp		⎡ 813 w ⎣ 802 w	815 m	813 w
703	707 w,dp	b$_g$, Φ	704 m	704 w,b	704 w
648?	648 s, dp	a$_g$, α	650 s	⎡ 649 m ⎣ 643 w	⎡ 649 m ⎣ 644 m
295?		b$_g$, τOH			
		a$_u$, γO + Φ	520 vw		
464	490 vw	a$_g$, α + βCO	⎡ 492 m ⎣ 478 m	⎡ 488 m ⎣ 480 m	478 m
	468 s, p	a$_g$, βO + α	464 m	460 m	464 w
363?	370^e s	b$_g$, γOν + Φ	377 s	385 m 379 m	376 m
		lattice modes	230 w,b	220 m,b	
			220 w		
			186 w		
			178 w		177 w,b
			151 w		
			142 w		140 sh
			129 m		128 s
			108 w		
			99 vs	95 vs	97 vs
			90 sh		
			82 vs	82 vs	80 s
			72 w	70 vs	
			64 vs		62 m
			54 vs		53 m
			50 sh		
			45 vs		44 vs
			38 w		
			31 vs		30 m

Notations of the internal coordinates: νXY, XY stretching; α, CCC bending; βX, CCX in-plane bending; δOH OH in-plane bending; Φ, ring torsion; γX, out-of-plane deformation of C-X bond; τOH OH torsion. Notations for band intensities and polarizations: vs = very strong, s = strong, m = medium, w = weak, vw = very weak, b = broad, p = polarized, dp = depolarized.

Table 2. Correlation diagram for crystalline modifications of hydroquinone

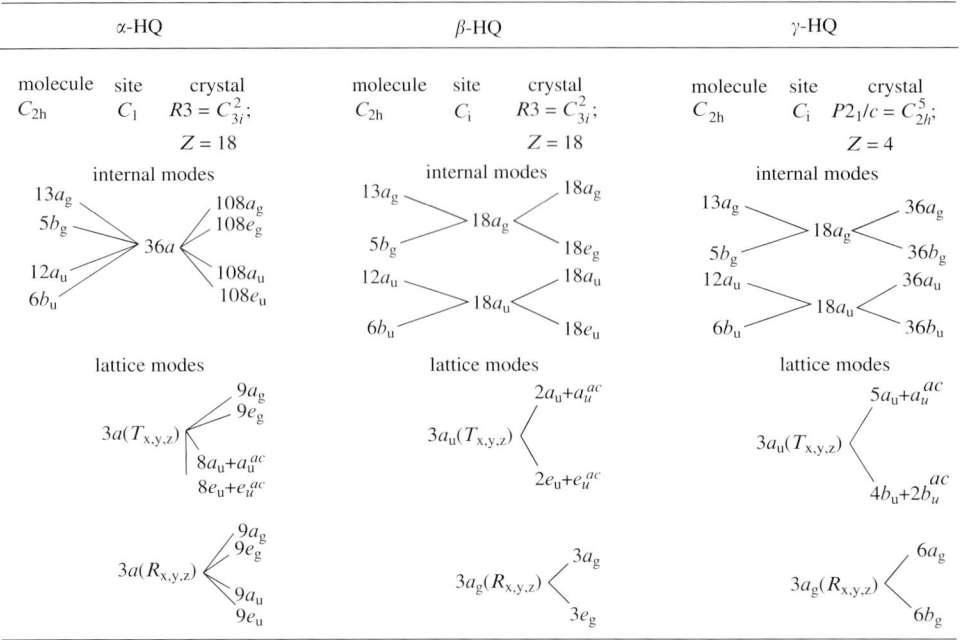

$\nu OH(\alpha\text{-HQ}) > \nu OH(\beta\text{-HQ})$, whereas the wavenumbers of the γOH bending vibrations are in the reverse order. This suggests that the hydrogen bonds are relatively the strongest in the β form and weakest in the γ form. The comparison of the distances between the oxygen atoms connected by the hydrogen bonds leads to the same conclusion, namely, this distance is 2.84 Å in γ-HQ [3], 2.678 Å in β-HQ [2] and varies between 2.662 and 2.779 Å in α-HQ with an average value of 2.719 Å [1].

The vibrational selection rules for the HQ modifications have been determined by factor group analysis, using the correlation method [6]. The correlation diagrams are shown in Table 2.

In contrast to the other two modifications, in α-HQ the molecules are not located on centrally symmetric sites of the unit cell, which means that infrared-active transitions of the molecule may be observed in coincidence with Raman transitions, and Raman-active transitions may be found to coincide with bands in the infrared spectrum. It is probable that the weak bands at 1610, 1164 and 702 cm^{-1} in the infrared spectrum of α-HQ can be assigned to Raman-active vibrations of the molecule, and the weak bands at 1010 and 520 cm^{-1} in its Raman spectrum to infrared-active modes of the molecule. Complicated splitting patterns of some molecular vibrations in the spectra of α-HQ are due to the fact that each molecular mode gives rise

to six infrared-active and six Raman-active crystal modes.

As can be seen in the correlation table a great number of lattice vibrations can be expected in vibrational spectra of α-HQ, far fewer in the spectra of γ-HQ and only a few in the spectra of β-HQ. The number of bands in the low-frequency range of the spectra is in accordance with this prediction. The available information is not yet enough for the assignment of the lattice modes. So far only Kolesov et al. [7] have attempted the assignment of the Raman-active lattice modes of β-HQ, their assignment relying on the shifts of these modes in the spectra of various clathrates.

References

[1] S. C. Wallwork, H. M. Powell, *J. Chem. Soc. Perkin Trans.* **1980,** *II*, 641.

[2] S. V. Lindeman, V. E. Shklover, Yu T. Struchkov, *Cryst. Struct. Commun.* **1981,** *10,* 1173.

[3] K. Maartmann-Moe, *Acta Cryst.* **1966,** *21,* 979.

[4] M. Kubinyi, F. Billes, A. Grofcsik, G. Keresztury, *J. Mol. Struct.* **1992,** *266,* 339.

[5] W. G. Fateley, F. R. Dollish, N. T. McDevitt, F. F. Bentley, *Infrared and Raman Selection Rules for Molecular and Lattice Vibrations: The Correlation Method,* Wiley–Interscience, New York, 1972.

[6] S. Bratos, J. Lascombe, A. Novak, in: *Molecular Interactions, Vol. 1* (H. Ratajczak, W. J. Orville-Thomas, M. Redshaw, eds.), Wiley–Interscience, Chichester, 1980, p. 301.

[7] B. A. Kolesov, G. N. Chekhova, Yu. A. Dyadin, *Zh. Strukt. Khim.* **1986,** *27,* 46.

Mikrochim. Acta [Suppl.] 14, 529–530 (1997)

Towards Interpreting the FT-IR and VCD Spectra of Selected Octadeoxynucleotides

Vanitha Maharaj[1], Hans van de Sande[2], Dimiter Tsankov[1], Arvi Rauk[1], and Hal Wieser[1,*]

Department of Chemistry[1] and Department of Medical Biochemistry[2], The University of Calgary, Calgary, AB, T2N 1N4, Canada

Abstract. Infrared absorption spectra of the octadeoxynucleotide d(CGCGCGCG)$_2$ in the 1750–900 cm^{-1} range are simulated by addition of the computed absorptions of the separate components, i.e. the appropriate number of nucleotide bases, deoxyribose and phosphate residues, calculated by *ab initio* theory at the 6-31G*$^{(0.3)}$ level. The observed absorption bands of the octadeoxynucleotide are reproduced in remarkable detail, making it possible to assign all prominent features as the necessary prelude to interpreting the VCD spectra of a series of octadeoxynucleotides containing varying numbers of G, C, A and T.

Key words: infrared spectra, VCD spectra, *ab initio* calculation, deoxynucleotides, cytosine, guanine.

The objective of this project is to examine the utility of vibrational circular dichroism (VCD) spectroscopy for characterizing DNA–drug interactions. As a first step we measured the absorption and VCD spectra by FT interferometry of six octadeoxynucleotides consisting of systematically varying sequences of d(CGNNNNCG)$_2$. where N = adenine (A), cytosine (C), guanine (G), and thymine (T), in the region of 1750–900 cm^{-1} [1], in order to extract the desired structural information, the observed VCD spectra must be compared with those theoretically simulated by suitable calculations. VCD simulations, however, require a reasonably reliable force field and a sound description of the vibrational displacements. Both the force field and VCD computations seem initially an overwhelming task for these large molecular systems. We therefore explored in the first instance the feasibility of computing *ab initio* force fields and assigning the absorptions to generate the absorption spectra by adding the computed spectra of the separate components, i.e. the amino-oxo tautomers of C, G, A

and T as appropriate, 16 deoxyribose residues, and 14 phosphate residues. In the report below we demonstrate the success of this approximate method, using d(CGCGCGCG)$_2$ as an example.

Experimental and Computational Details

The absorption and VCD spectra of d(CGCGCGCG)$_2$, d(CGCGTGCG)$_2$, d(CGCATGCG)$_2$, d(CGCTAGCG)$_2$, d(CGAATTCG)$_2$, and d(CGTATACG)$_2$ at about 17 mM duplex in D$_2$O/cacodylate buffer were measured with an FT interferometer at 4 cm^{-1} resolution in a 0.015 mm BaF$_2$ cell kept at about 5 °C, accumulating 7500 a.c. and 500 d.c. scans [1, 2]. The *ab initio* calculations were performed at the RHF/6-31G*$^{(0.3)}$ level [3, 4] for optimized geometries, force fields, and atomic polar tensors. The wavenumbers were scaled by 0.9 and Lorentzian band shapes with 10 cm^{-1} width were assumed.

Results and Discussion

Ab initio predictions of the infrared absorptions of C and G have been published at the 6–31G** level of theory [5, 6]. For phosphate, vibrational assignments were performed on dimethyl phosphate by standard empirical means [7]. Computed absorption spectra for deoxyribose have not been reported previously. For C, G and phosphate our results compare favourably with those published earlier [5–7]. Since the nucleotide bases are paired, we also computed the optimized geometries and absorption spectra of the G–C base pair. Selected observed and calculated band positions of d(CGCGCGCG)$_2$ and their assignments are listed in Table 1. In Fig. 1 the simulated spectrum is displayed as the sum of the independent components, including C and G either as single units or as base pairs.

While the two prominent absorptions at 1681 and 1649 cm^{-1} in polynucleotides such as poly (dG–dC)·

---

* To whom correspondence should be addressed

Table 1. Selected calculated and observed band positions (in units of cm^{-1}) and calculated intensities (in parentheses, in units of km/mol) of C and G in d(CGCGCGCG)$_2$

Calculated (hydrogenated)			Observed (in D$_2$O)		Mode description
Cytosine	Guanine	C–G	Infrared	VCD	
	1718 (893)	1685 (1561)	1682	1689/1678 ⎫	
1701 (821)		1661 (77) ⎫			C=O
			1655	~1665/1654 ⎭	
1642 (590)		1648 (1068) ⎭			C=C, N–C
	1616 (631)				NH$_2$ scissor
1599 (146)		1598 (489)			NH$_2$ scissor
	1578 (343)	1578 (129)	1578	1578	C=C, N–C
	1574 (239)		1560	~1565/1554	

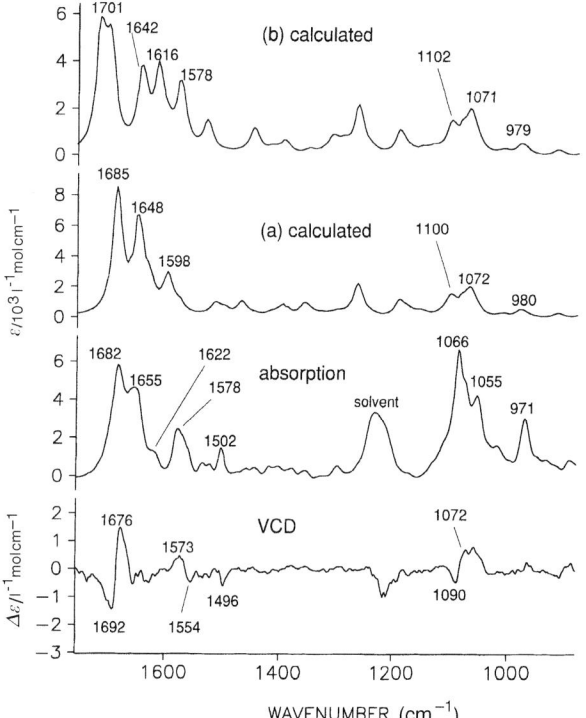

Fig. 1. Comparison of experimental VCD and absorption spectra with calculated absorption spectra of d(CGCGCGCG)$_2$. Calculated spectra consist of the sum of 16 deoxyribose residues, 14 phosphate residues, and (a) 8 C-G base pairs or (b) 8C and 8G nucleotide bases

poly(dG–dC) were recognized as arising primarily from the C=O stretching modes of C and G, respectively [8], they could not be correlated with the corresponding vibrations of the monomers. For d(CGCGCGCG)$_2$ the corresponding bands calculated for the hydrogenated species at 1685 and 1648 cm^{-1} are designated as in-phase and out-of-phase coupling of the two separate modes. They are observed (in D$_2$O)

at 1682 and 1655 cm^{-1}, respectively. The broad band at 1655 cm^{-1} can be explained as consisting of two overlapped absorptions. Other bands of interest, not included in Table 1, are the absorptions of phosphate predicted at 1082, 1067 and 1061 cm^{-1} as arising from symmetric and asymmetric O–P–O stretching modes, and deoxyribose sugar bands predicted at 1102, 1009 and 981 cm^{-1} for ring stretching modes. All these vibrations have plausible counterparts in the observed spectrum.

Although the vibrational spectra of deoxynucleotides and the constituent nucleotide bases have been the subject of many experimental and theoretical investigations, the method described above constitutes, to our knowledge, the first successful attempt to characterize the molecular vibrations of an entire deoxynucleotide as a complete unit. The method will be further explored for describing the molecular vibrations of all the octadeoxynucleotides mentioned in the introduction, with the objective of assisting with the interpretation of their VCD spectra.

References

[1] V. Maharaj, D. Tsankov, H. J. van de Sande, H. Wieser, *J. Mol. Struct.* **1995**, *349*, 25.
[2] D. Tsankov, T. Eggimann, H. Wieser, *Appl. Spec.* **1995**, *49*, 132.
[3] A. Rauk, D. Yang, *J. Phys. Chem.* **1992**, *96*, 437.
[4] A. Rauk, D. Yang, *J. Chem. Phys.* **1992**, *97*, 6517.
[5] I. R. Gould, M. A. Vincent, I. H. Hillier, L. Lapinski, M. J. Nowak, *Spectrochim. Acta* **1992**, *48A*, 811.
[6] I. R. Gould, M. A. Vincent, I. H. Hillier, *Spectrochim. Acta* **1993**, *49A*, 1727.
[7] Y. Guan, C. J. Wurrey, G. J. Thomas, Jr., *Biophysical J.* **1994**, *66*, 225.
[8] E. Taillandier, J. Liquier, J. A. Taboury, in: *Adv. Infrared Raman Spectrosc.* **1985**, *12*, 65.

Mikrochim. Acta [Suppl.] 14, 531–533 (1997)

Effects of Hydrogen Bonding and Solvation in Dielectric Media on the Amide I Frequencies: ab initio Molecular Orbital Study

Hajime Torii*, **Tomoaki Tatsumi**, and **Mitsuo Tasumi**

Department of Chemistry, School of Science, The University of Tokyo, Bunkyo-ku, Tokyo 113, Japan

Abstract. Effects of hydrogen bonding and solvation in dielectric media on the amide I frequency of N-methylacetamide are studied by ab initio molecular orbital calculations. The calculated results clarify quantitatively the factors inducing the low-frequency shifts of the amide I band observed in various solvents.

Key words: hydrogen bonding, solvation, dielectric media, amide I, ab initio molecular orbital calculation.

The amide I band is known to be sensitive to the secondary structures of proteins. The amide I bands characteristic of the secondary structures arise primarily from interactions between peptide groups in a protein molecule (the off-diagonal terms) by the transition dipole coupling mechanism [1]. However, changes in the intrinsic (uncoupled) amide I frequency due to local perturbation on each peptide group (the diagonal terms), which originates from intramolecular and intermolecular hydrogen bonding with other peptide groups and/or solvent molecules as well as from effects of the environment as a dielectric medium, may also be significantly large.

In the present study, effects of hydrogen bonding and solvation in dielectric media on the amide I frequency of the peptide group are studied by ab initio molecular orbital (MO) calculations. N-Methylacetamide (NMA) is used as a representative peptide. Only the *trans* isomer is considered in this study, because most peptide groups in proteins are in the *trans* form. The self-consistent reaction-field (SCRF) method [2] is employed to take into account the effects

_____
* To whom correspondence should be addressed

of the dielectric medium surrounding the NMA molecule. The effects of hydrogen bonding are studied by calculating the vibrational frequencies of NMA dimers and complexes of NMA with from one to three water molecules.

Computational Procedure

Ab initio MO calculations were performed at the Hartree–Fock level with the 6-31 + + G** basis set, by using the Gaussian 92 program [3]. In the calculations for hydrated NMA, up to two water molecules are hydrogen bonded to the C=O group, and one water molecule or none to the N–H group. Two different conformations are considered for the NMA dimer, which correspond to hydrogen bonded peptide groups in parallel and antiparallel β-sheets, respectively. In the SCRF calculations, the cavity radii are determined by using the VOLUME keyword in the Gaussian 92 program, so that they are consistent with the molecular volumes of the optimized structures. The calculated frequencies are scaled by 0.89 so that the amide I frequency of isolated NMA coincides with that observed in N_2 matrices (1707 cm^{-1}) [4].

Results and Discussion

Effects of Solvents as Dielectric Media

SCRF calculations of the amide I frequencies are performed for NMA in triethylamine (TEA, $\varepsilon = 2.42$), tetrahydrofuran (THF, $\varepsilon = 7.58$), dichloromethane (DCM, $\varepsilon = 7.77$), dimethylsulphoxide (DMSO, $\varepsilon = 46.68$), methanol ($\varepsilon = 32.70$) and water ($\varepsilon = 78.5$).

In the four aprotic polar solvents (TEA, THF, DCM and DMSO), the calculated amide I frequencies are lower by 10–30 cm^{-1} than that of an isolated NMA. These calculated frequencies are in good agreement with the experimental results obtained by Eaton et al. [5]. This result indicates that the low-frequency shifts of the amide I band observed in these solvents are explained by the effects of these solvents as dielectric media. Some of these aprotic polar solvents may form a hydrogen bond to the N–H group of NMA. However, the calculated results exhibit no correlation between possible ability of hydrogen-bond formation and the amide I frequency. In contrast to the cases above, the low-frequency shifts of the amide I band observed in methanol ($70 \ cm^{-1}$) and water ($79 \ cm^{-1}$) are significantly larger than the frequency shifts calculated by the SCRF method ($\sim 30 \ cm^{-1}$). Consequently, the amide I frequencies in these protic solvents are not explained solely by the effects of the solvents as dielectric media.

Effects of Hydrogen Bonding with Solvent Water Molecules

The results in the preceding section show that the effects of hydrogen bonding between a peptide group and solvent molecules are highly significant for determining the amide I frequencies. In order to examine these effects quantitatively, ab initio MO calculations have been performed for hydrated NMA. Two hydrogen-bonding sites (A and C) were assumed for the C=O group, whereas one (B) was assumed for the N–H group. As a result, frequency calculations were carried out for three forms of $NMA + H_2O$ (A, B, C), three forms of $NMA + 2H_2O$ (A&B, A&C, B&C), and one for $NMA + 3H_2O$ (A&B&C).

The calculated low-frequency shifts of the amide I band induced by hydrogen bonding at sites A, B, and C are 22, 10 and $25 \ cm^{-1}$, respectively. The hydrogen bonding to the N–H group as well as to the C=O group gives rise to low-frequency shifts of the amide I band. These frequency shifts are additive; for example, the low-frequency shift for NMA with two H_2O molecules at sites A and B is $32 \ cm^{-1}$, which is the sum of the shifts for NMA with a single hydrogen bond at A ($22 \ cm^{-1}$) and one at B ($10 \ cm^{-1}$). It is also noted that these frequency shifts are strongly correlated with the calculated C=O bond lengths. Hydrated species with lower amide I frequencies have longer C=O bonds.

The calculated amide I frequency of $NMA + 3H_2O$ (in vacuum) is $1651 \ cm^{-1}$, which is higher than the observed frequency of $1628 \ cm^{-1}$ for NMA in water solution [5]. The results in the previous section suggests that this frequency difference is explained by the

effects of surrounding water molecules (not hydrogen bonded with NMA) as dielectric media, which may be as large as $30 \ cm^{-1}$. To confirm this interpretation, vibrational frequencies are calculated by using the SCRF method for $NMA + 3H_2O$ solvated in water. Since the calculated amide I mode is coupled with the HOH bending modes of hydrogen-bonded water molecules, the average of the frequencies of the coupled modes should be taken. The resulting calculated amide I frequency is $1619 \ cm^{-1}$, in reasonable agreement with the observed frequency of $1628 \ cm^{-1}$.

Effects of Hydrogen Bonding between Peptide Groups

For both forms of the NMA dimer (parallel and antiparallel arrangements of the NMA molecules), the calculated amide I modes are delocalized over the two molecules. The strongly infrared active bands appear at 1687 and $1686 \ cm^{-1}$ for dimer 1 (parallel) and 2 (antiparallel), respectively. The shifts ($\sim 20 \ cm^{-1}$) from the amide I frequency of an isolated NMA arise from changes in both the intrinsic (uncoupled) amide I frequencies (diagonal terms) and intermolecular coupling of the amide I vibrations between the two molecules (off-diagonal terms).

In order to examine the effects of hydrogen bonding on the intrinsic amide I frequency of each molecule in the dimer, the C=O bond of one of the molecules in the dimer is isotopically substituted by $^{13}C=^{18}O$, so that the amide I vibrations of the two molecules are decoupled. Here, the dimer is supposed to consist of molecules X and Y, with a hydrogen bond between N–H of molecule X and C=O of molecule Y. When molecule X is isotopically substituted, the calculated amide I frequency of molecule Y is 1691 (dimer 1) and $1689 \ cm^{-1}$ (dimer 2). The shift from the amide I frequency of an isolated NMA is therefore $16–18 \ cm^{-1}$, which is slightly smaller than the frequency shift calculated for an NMA with its C=O bond hydrogen-bonded with a solvent water molecule. By contrast, when molecule Y is isotopically substituted, the calculated amide I frequency of molecule X is $1695 \ cm^{-1}$ (dimers 1 and 2). In other words, hydrogen bonding between the N–H bond of a peptide group (molecule X) and the C=O bond of another peptide group (molecule Y) lowers the amide I frequency of the former by about $12 \ cm^{-1}$, although the amide I vibration mainly consists of the stretch of the C=O bond, which is free in this case.

References

[1] H. Torii, M. Tasumi, *J. Chem. Phys.* **1992**, *96*, 3379.

[2] M. W. Wong, M. J. Frisch, K. B. Wiberg, *J. Am. Chem. Soc.* **1991**, *113*, 4776.

[3] M. J. Frisch, G. W. Trucks, M. Head-Gordon, P. M. W. Gill, M. W. Wong, J. B. Foresman, B. G. Johnson, H. B. Schlegel, M. A. Robb, E. S. Replogle, R. Gomperts, J. L. Andres, K. Raghavachari, J. S. Binkley, C. Gonzalez, R. L. Martin, D. J. Fox, D. J. Defrees, J. Baker, J. J. P. Stewart, J. A. Pople, *Gaussian 92*, Gaussian, Pittsburgh, PA, 1992.

[4] S. Ataka, H. Takeuchi, M. Tasumi, *J. Mol. Struct.* **1984**, *113*, 147.

[5] G. Eaton, M. C. R. Symons, P. P. Rastogi, *J. Chem. Soc., Faraday Trans. I* **1989**, *85*, 3257.

Mikrochim. Acta [Suppl.] 14, 535–537 (1997)
© Springer-Verlag 1997

FT-Vibrational Circular Dichroism (VCD) Spectra and ab initio 6-31G*(0.3) Intensities for Camphor and 2-Vinyl-exo-borneol

Dimiter Tsankov[1,*], **Vladimir Dimitrov**[1], and **Hal Wieser**[2]

[1] Institute of Organic Chemistry, Bulgarian Academy of Sciences, BG-1113 Sofia, Bulgaria
[2] Department of Chemistry, University of Calgary, Calgary, Alberta, T2N 1N4 Canada

Abstract. The FT-IR absorption and FT-IR VCD spectra in the region 1600–800 cm^{-1} are reported for 2-vinyl-exo-borneol and compared to those of camphor. Optimized geometries, harmonic force fields, and absorption spectra were calculated by ab initio means at the 6-31G*(0.3) level of RHF/SCF for both molecules. The VCD spectrum of camphor was generated via the vibronic coupling theory (VCT) with atomic axial tensors determined at the 6-31G level. Predicted and experimental absorption and VCD intensities show very good agreement.

Key words: infrared spectra, VCD spectra, camphor, 2-vinyl-exo-borneol.

Successful attempts were recently made to extend simulations of VCD spectra to larger molecules by using medium-sized basis sets at MP2 [1], DFT [2], and RHF/SCF [3] levels of theory. These achievements became a reality for MP2 after the introduction of semi-direct techniques, and a new class of density functionals (hybrid functionals) for DFT. MP2 calculations are still tedious to perform for larger systems and supercomputers are required. DFT-based methods yield results that in most cases can compete with the quality of SCF, for geometry, force fields and electric dipole transition moments. Magnetic dipole transition moments were not available in these approximations but were substituted from suitable SCF calculations. For the molecules described below we used the approach in which optimized geometries, force fields and atomic polar tensors are obtained with the 6-31G*(0.3) basis set [4] in SCF theory, and the atomic axial tensors needed for VCD are derived by the vibronic coupling theory at the 6-31G level [3]. We report here for the first time the observed and calculated absorption spectra as well as the observed VCD spectrum of

2-vinyl-exo-borneol. To verify the suitability of the computational approach by comparison with previously published observations, we also calculated the absorption and VCD spectra of camphor in the same manner.

Experimental and Calculational Details

(+)-Camphor, obtained from Aldrich with a chemical purity of 99%, was used without further purification. (+)-2-Vinyl-exo-borneol was made from camphor by initial complexation with $CeCl_3$ and subsequent addition of an appropriate Grignard reagent [5]. The IR absorption and VCD single enantiomer spectra were recorded with a Bomem MB 100 FT-IR spectrometer equipped with VCD optics as described previously [6], with 4 cm^{-1} resolution, for CCl_4 solutions in a 0.15 mm NaCl cell, with accumulation of 5000 a.c. and 100 d.c. scans taken before the a.c. collection. The base line was corrected by subtracting the solvent "VCD" spectrum. The ab initio calculations were performed by using Gaussian 92, and the VCD spectrum generated with the VCT90 program package at the distributed origin gauge [4]. The calculated wavenumbers were uniformly scaled by 0.9, and Lorentzian band shapes of 3 cm^{-1} half-width were assumed.

Results and Discussion

Figures 1 and 2 display the spectra of camphor and 2-vinyl-exo-borneol, respectively. The peak numbers are identical to the labels used by Stephens and co-workers [1, 2]. The absorption spectrum of camphor is essentially completely reproduced. Every observed band can be accounted for with comparable computed intensity. Only one minor difference in intensity occurs, for band 49 (1390 cm^{-1} observed vs. 1388 cm^{-1} calculated). The agreement is qualitatively and quantitatively at least equivalent to that of the MP2 and DFT methods. The VCD spectrum is equally well simulated. Several bands appear different in the two spectra only

---

* To whom correspondence should be addressed

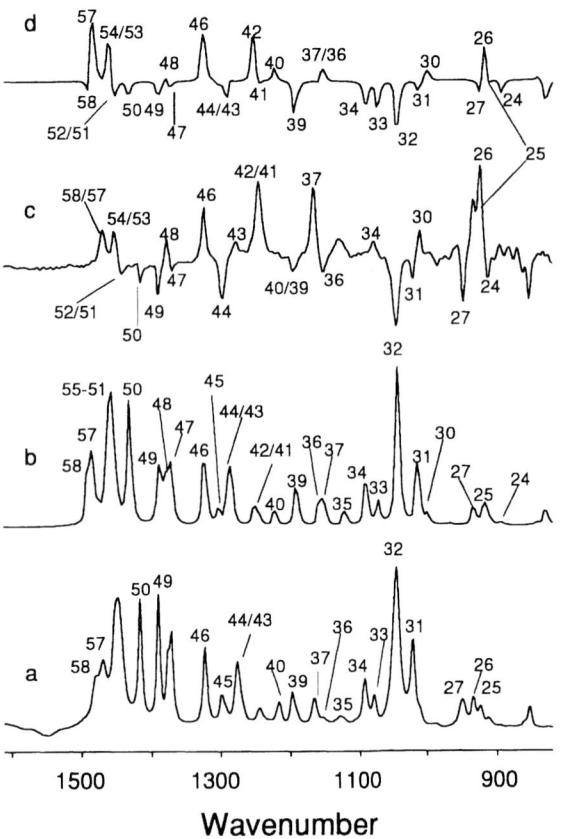

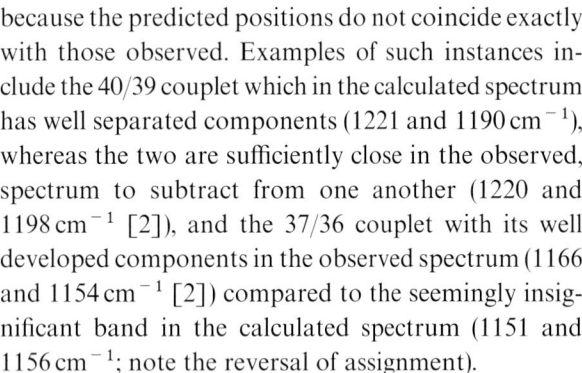

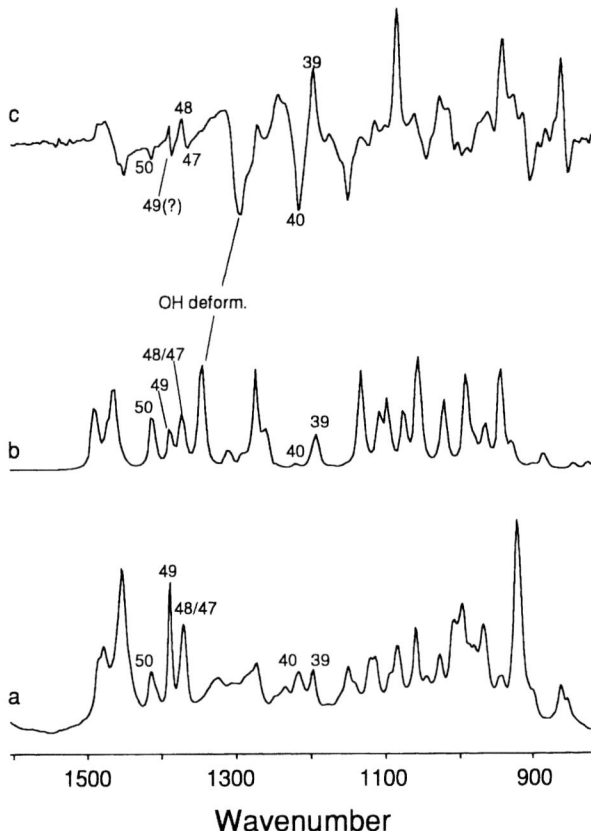

Fig. 1. Absorption and VCD spectra of camphor: **a** and **c** observed, **b** and **d** calculated

Fig. 2. Absorption and VCD spectra of 2-vinyl-*exo*-borneol: **a** and **c** observed, **b** calculated

because the predicted positions do not coincide exactly with those observed. Examples of such instances include the 40/39 couplet which in the calculated spectrum has well separated components (1221 and 1190 cm⁻¹), whereas the two are sufficiently close in the observed, spectrum to subtract from one another (1220 and 1198 cm⁻¹ [2]), and the 37/36 couplet with its well developed components in the observed spectrum (1166 and 1154 cm⁻¹ [2]) compared to the seemingly insignificant band in the calculated spectrum (1151 and 1156 cm⁻¹; note the reversal of assignment).

The absorption spectrum of 2-vinyl-*exo*-borneol appears not to be reproduced as successfully in all details. This is most likely due to intermolecular hydrogen-bonding at the concentration used for recording the spectra, as indeed the disappearance of the strong OH deformation band at 1346 cm⁻¹ would indicate. Nonetheless, the absorption bands from 47 to 58 are essentially the same in both compounds, and the similarity is retained to a degree in the VCD spectrum of the borneol. Peaks 39 and 40 seem to have corresponding

counterparts in both molecules. Thereafter, visual comparisons become difficult, and a more complete interpretation must be postponed until the calculated VCD spectrum is available. It is also too early to comment on the extent to which similar VCD features might characterize the common bicyclic ring structure, one of the objectives which motivated this study.

The results obtained to date clearly indicate that the approach to use the 6-31G*(0.3) basis set in the RHF/SCF method and in conjunction with the vibronic coupling theory, is capable of accurately reproducing the absorption and VCD spectra of camphor. We therefore expect that we can apply the same approach to simulate and interpret the spectra of a series of borneol analogues substituted at the 2-position, of which 2-vinyl-*exo*-borneol included here is but one example.

Acknowledgement. Financial support by an NSERC Operating Grant (to H. W.) is gratefully acknowledged. We are also indebted to Drs. A. Rauk and D. Yang of the University of Calgary for making available the VCT90 program package and assisting in its use.

References

[1] F. J. Devlin, P. J. Stephens, *J. Am. Chem. Soc.* **1994**, *116*, 5003.

[2] P. J. Stephens, F. J. Devlin, C. S. Ashvar, C. F. Chabalowski, M. J. Frisch, *Faraday Discussions* **1995**, *99*, 103.

[3] J. A. Nieman, B. A. Keay, M. Kubicki, D. Yang, A. Rauk, D. Tsankov, H. Wieser, *J. Org. Chem.* **1995**, *60*, 1918.

[4] D. Yang, A. Rauk, *J. Chem. Phys.* **1992**, *97*, 6517.

[5] V. Dimitrov, S. Bratovanov, S. Simova, K. Kostova, *Tetrahedron Lett.* **1994**, *35*, 6713.

[6] D. Tsankov, T. Eggimann, H. Wieser, *Appl. Spectrosc.* **1995**, *49*, 132.

Mikrochim. Acta [Suppl.] 14, 539–541 (1997)
© Springer-Verlag 1997

Badger and Herschbach–Laurie Constants for the Compounds of Xenon and Fluorine

Svetozar Milićev

"Jozef Stefan" Institute, Ljubljana, Slovenia

Abstract. Relationships between XeF bond lengths and the force constants of XeF stretching vibrations of xenon fluorides (XeF_2, XeF_4, XeF_6, $XeOF_4$), their ions (XeF^+, $Xe_2F_3^+$, XeF_5^+) and their molecular adducts are investigated. The parameters for the modified Badger relation and the Herschbach–Laurie relation are determined for 38 experimental data. The Badger values of $a_{ij} = 2.33$ and $d_{ij} = 1.00$, and the Herschbach–Laurie values of $a_{ij} = 2.25$ and $b_{ij} = 0.68$ reproduce the frequencies with an average deviation of 3.8% for the Badger rule and 3.4% for the Herschbach–Laurie correlation. Sample XeO bonds ($XeOF_4$, XeO_3, XeO_4) and XeN bonds ($FXeN(SO_2F)_2$, $FXeN(SO_2F)_2·Sb_3F_{16}$) satisfactorily pack along the curves determined for the Xe–F bonds, suggesting common parameters.

Key words: xenon fluorides, Badger's rule, Herschbach–Laurie equation, bond lengths, force constants.

Estimating various parameters of chemical bonding by empirical correlations is purposeful in cases in which experimental determination is not feasible. This may be the case for a number of xenon compounds and complexes. Among a host of equations correlating bond length and the relevant force constant (for a review see, for instance, [1] or [2]) the most widely used are Badger's rule, which we apply in the modified form, $r_e = d_{ij} + (a_{ij} - d_{ij})/F_2^{1/3}$ and the relation of Herschbach and Laurie, $r_e = a_{ij} - b_{ij} \log(F_2)$, where r_e is the equilibrium bond length in Å and F_2 is the quadratic force constant in 10^2 N/m (mdyn/Å) [3]. The parameters are defined by the positions of the bonded atoms in the rows of the periodic table, with the transition elements regarded as separate rows. The noble gases were not included in the correlations. Scarcity of data might have been one of the reasons for this in the past, but now there is an accumulation of crystallographic and spectroscopic data for xenon fluorides and their complexes in the ground state, which merits an attempt to determine the parameters.

Table 1 presents the available literature data for xenon fluorides and their compounds. For comparison some examples of XeO and XeN bond data are added. In the case of equivalents bonds the arithmetic mean is taken. This includes the basal Xe–F bonds in XeF_5^+ and the bonds in undistorted or slightly distorted molecules such as the fluorides proper and $XeOF_4$, bonds in the XeF_2 moiety of $Ag(XeF_2)_2AsF_6$ and $XeF_2(XeF_5AsF_6)_2$, and bonds in the XeF_4 moiety of $(XeF_5CrF_5)_4XeF_4$. If more equivalent bonds are present, and more vibrational bands appear due to coupling, the root mean square of the frequencies is taken to be a good approximation of the characteristic Xe–F stretching frequency [4, 5]. On the other hand, strong intermolecular interactions in some complexes of XeF_2 (including $Xe_2F_3AsF_6$) cause large differences in bond lengths and the corresponding frequencies are taken to be pure. The same is valid for the vibration of the axial F in XeF_5^+, the contribution of which to the potential energy distribution has been shown to be higher than 90% [6]). From these data, force constants are calculated by using the approximate two-atom harmonic model.

According to Fig. 1, the two-atom force constants fit very well to the simple Badger and Herschbach–Laurie correlations. The parameters determined are for Badger's rule $a_{ij} = 2.33$ and $d_{ij} = 1.00$, and for the Herschbach–Laurie equation $a_{ij} = 2.25$ and $b_{ij} = 0.68$. The average deviations of the force constants calculated from the bond lengths are 7.7% for Badger and 6.8% for Herschbach–Laurie and for the frequencies obtained by the two-atom model from the calculated force constants, 3.8% for Badger and 3.4% for Herschbach–Laurie. This is excellent agreement as no XeF data were omitted. Two-atom force constants and the

Table 1. XeF bonds and examples of XeO and XeN bonds: bond lengths in Å, stretching frequencies in cm^{-1} and force constants in 10^2 N/m[a]

XeF bond in XeF₂	Bond lengths	Frequencies	$f^b_{n.c.}$	$f^c_{2\text{-atom}}$	XeF bond in XeF₅⁺	Bond lengths[d]	Frequencies[d]	$f^b_{n.c.}$	$f^c_{2\text{-atom}}$
XeF₂	1.979[11]	538.8[11]	2.914[11]	2.84	XeF₅.GeF₅	r_{ax}1.813[25]	v_{ax}669[25]		4.38
						r_{bas}1.829[25]	v_{bas}644[25]		4.06
XeF₂.AsF₅	1.873[12]	610[12]	3.63[13]	3.63	XeF₅.AsF₆	r_{ax}1.76[15]	v_{ax}670[13]		4.39
	2.212[12]	346[12]	1.19[13]	1.17		r_{bas}1.83[15]	v_{bas}643[13]		4.04
2XeF₂.XeF₅.AsF₆	1.95[13]	550[13]		2.96	XeF₄(XeF₅.CrF₅)₄	r_{bas}1.837[18]	v_{bas}636[19]		3.96
	2.04[13]	479[13]		2.24	(XeF₅)₂NiF₆	r_{ax}1.82[26]	v_{ax}650[26]		4.13
XeF₂.XeF₅.AsF₆	1.91[13]	559[13]	3.02[13]	3.06		r_{bas}1.84[26]	v_{bas}630[26]		3.88
	2.05[13]	433[13]	1.89[13]	1.83	XeF₅.RuF₆	r_{ax}1.793[27]	v_{ax}670[28]	4.25[28]	4.39
XeF₂(XeF₅.AsF₆)₂	2.01[13]	522[13]	2.68[13]	2.67		r_{bas}1.845[27]	v_{bas}641[28]	3.95[28]	4.02
Ag(XeF₂)₂AsF₆	1.979[14]	504.5[14]		2.49	(XeF₅)₂PdF₆	r_{ax}1.813[29]	v_{ax}653[28]		4.17
						r_{bas}1.843[29]	v_{bas}638[28]		3.98
Xe₂F₃.AsF₆	1.90[15]	600[13]	3.43[13]	3.52	XeF₅.PtF₆	r_{ax}1.81[30]	v_{ax}661[28]		4.27
	2.14[15]	401[16]	1.67[13]	1.57		r_{bas}1.875[30]	v_{bas}630[28]		3.88
XeF₂.Sb₂F₁₀	1.82[17]	620[16]	3.76[13]	3.76	XeF₅.AgF₄	r_{ax}1.852[31]	v_{ax}652[31]		4.16
	2.34[17]	269[16]	0.72[13]	0.71		r_{bas}1.826[31]	v_{bas}619[31]		3.75
XeF₂.CrF₄	1.93[18]	574[19]		3.22	Xe₂F₁₁.AuF₆[d]	r_{ax}1.82[32]	v_{ax}661[32]		4.27
	2.13[18]	431[19]		1.82		r_{bas}1.84[32]	v_{bas}626[32]		3.83
					XeF₅⁺ ion[d]	r_{ax}1.76[10]	v_{ax}679[10]	4.43[10]	4.51
XeF bond in XeF₄ and XeOF₄						r_{bas}1.82[10]	v_{bas}635[10]	3.94[10]	3.94
XeF₄	1.92[20]	545[20]	3.0[20]	2.90	XeF₆[d]	1.9[33]	603[33]		3.55
XeF₄(XeF₅.CrF₅)₄	1.94[18]	550[19]		2.95					
XeOF₄	1.9[21]	568[21]	3.26[21]	3.16	**XeO bond**				
XeN bond					XeOF₄	1.71[21]	923[21]	7.08[21]	7.16
FXeN(SO₂F)₂	2.2[22]	422[23]		1.33	XeO₃	1.76[34]	804[34]	5.66[34]	5.43
FXeN(SO₂F)₂.Sb₃F₁₆	2.02[24]	632[24]		2.97	XeO₄	1.736[35]	852[36]		6.1

[a] References are given as superscripts.

[b] Stretching force constants from the normal coordinate calculations.

[c] Stretching force constants from the simple two-atom harmonic approximation, this paper.

[d] For XeF₅⁺ compounds: r_{ax} – axial bond length, r_{bas} – arithematic mean of the basal bond lengths, v_{ax} – frequency of the stretching of the Xe–F axial bond, v_{bas} – root mean square average of the basal Xe–F stretching vibrations. For Xe₂F₁₁.AuF₆ compound: XeF₅⁺ unit. For XeF₅⁺ ion: the frequency set which is in accord with the assignments of other XeF₅⁺ compounds is choosen. For XeF₆: the arithmetic mean of stretching vibrations. There are no complete data on XeF₅⁺ vibrations in mixed complexes with XeF₂.

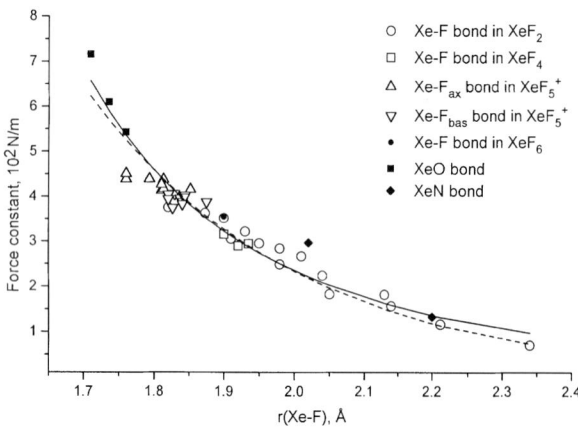

Fig. 1. Force constant–bond length correlation for XeF bonds. Full line – correlation by Badger ($a_{ij} = 2.33$, $d_{ij} = 1.00$). Dashed line – correlation by Herschbach–Laurie ($a_{ij} = 2.25$, $b_{ij} = 0.68$). XeO and XeN data are only for comparison.

constants calculated by normal coordinate analysis, where available, tally remarkably well (Table 1).

The simple two-atom relations are not successful when applied to the polyatomic molecules in which appreciable mixing of internal coordinates occurs [3, 7, 8] and the best results are obtained when dealing with molecules in which the bond character is similar, or at least no abrupt changes along the series occur [3, 9]. Both of these limitations are respected in the present case. Mixing of vibrations is relatively small, due to the heavy central atom. Where calculated, the contribution of the XeF bond stretching coordinate to the potential energy distribution of normal vibrations is high. In the XeF₅⁺ ion it is higher than 88% [6, 10]. Even in the case of the linear XeF₂ molecule it is approximately 80% [11]. The Xe–F bond monotonically changes between a proper chemical bond and a weak chemical interaction.

Inclusion of a few bonds of other type such as Xe=O or Xe–N bonds, into Fig. 1, but not into the correlations, demonstrates that even so different kinds of bonds satisfactorily cluster along the regularity determined for the Xe–F bonds. It is implied that, with larger sets of data for other bonds with Xe, a common pair of parameters may be found, introducing a new group into the Badger and Herschbach–Laurie tables. This ought to be a new group, since the parameters for the 1st and 4th period of both tables [3] give either too high (old Badger 2.33, 1.12; Herschbach–Laurie 2.36, 0.76) or too low values of the XeF force constants (new Badger parameters 2.33, 0.68).

Acknowledgement. The financial support of the Ministry of Science and Technology of Slovenia is gratefully acknowledged.

References

[1] Y. P. Varshni, *J. Chem. Phys.* **1958**, *28*, 1081.

[2] Yu. Ya Kharitonov, G. V. Kravtsova, *Koord. Khim* **1980**, *6*, 1315.

[3] D. R. Herschbach, V. W. Laurie, *J. Chem. Phys.* **1961**, *35*, 458.

[4] H. Siebert, *Anwendungen der Schwingungsspektroskopie in der Anorganischen Chemie*, Springer, Berlin. Göttingen Heidelberg New York, 1966, p. 33.

[5] W. J. Lehmann, *J. Mol. Spectrosc.* **1961**, 7, 261.

[6] S. Milićev, *Croat. Chem. Acta,* **1992**, *65*, 125.

[7] R. M. Badger, *J. Chem. Phys.* **1934**, *2*, 128; **1935**,*3*,710

[8] W. Gordy, *J. Chem. Phys.* **1946**, *14*, 305.

[9] H. O. Pritchard, H. A. Skinner, *J. Chem. Soc.,* **1951**,945

[10] K. O. Christe, E. C. Curtis, R. D. Wilson in: *Inorg. Nucl. Chem Mem Vol.* (J. J. Katz, I. Sheft, eds.) Pergamon, Oxford, 1976, p. 159

[11] N. J. Brassington, H. G. M. Edwards *J. Mol. Struct.* **1987**, *162*, 69.

[12] A. Zalkin, D. L. Ward, R. N. Biagioni, D. H. Templeton, N. Bartlett, *Inorg. Chem.* **1978**, *17*, 1318.

[13] B. Žemva, A. Jesih, D. H. Templeton, A. Zalkin, A. K. Cheetham, N. Bartlett, *J. Am. Chem. Soc.* **1987**, *109*, 7420.

[14] R. Hagiwara, F. Hollander, C. Maines, N. Bartlett, *Eur. J. Solid State Inorg. Chem.* **1991**, *28*, 855.

[15] N. Bartlett, B. G. DeBoer, F. J. Hollander, F. O. Sladky, D. H. Templeton, A. Zalkin, *Inorg. Chem.* **1974**, *13*, 780.

[16] R. J. Gillespie, B. Landa, *Inorg. Chem.* **1973**, *12*, 1383.

[17] J. Burgess, C. J. W. Fraser, V. M. McRea, R. D. Peacock, D. R. Russell in: *Inorg. Nucl. Chem.-H. Hyman Mem* (J. J. Katz, I. Sheft, eds.), Pergamon, Oxford 1976, p. 183.

[18] K. Lutar, I. Leban, T. Ogrin, B. Žemva, *Eur. J. Solid State Inorg. Chem.* **1992**, *29*, 713.

[19] S. Milićev, K. Lutar, B. Žemva, T. Ogrin, *J. Mol. Struct.* **1994**, *323*, 1.

[20] H. H. Claassen, C. L. Chernick, J. G. Malm, *J. Am. Chem. Soc.* **1963**, *85*, 1927.

[21] G. M. Begun, W. H. Fletcher, D. F. Smith, *J. Chem. Phys.* **1965**, *42*, 2236.

[22] J. F. Sawyer, G. J. Schrobilgen, S. J. Sutherland, *Inorg. Chem.* **1982**, *21*, 4064.

[23] G. A. Schumacher, G. J. Schrobilgen, *Inorg. Chem.* **1983**, *22*, 2178.

[24] R. Faggiani, D. K. Kennepohl, C. J. L. Lock, G. J Schrobilgen, *Inorg. Chem,* **1986**, *25*, 563.

[25] T. E. Mallouk, B. Desbat, N. Barlett, *Inorg. Chem.* **1984**, *23*, 3160.

[26] A. Jesih, K. Lutar, I. Leban, B. Žemva, *Eur. J. Solid State Inorg. Chem.* **1991**, *28*, 829.

[27] N. Barlett, M. Gennis, D. D. Gibler, B. K. Morrell, A. Zalkin, *Inorg. Chem.* **1973**, *12*, 1717.

[28] C. J. Adams, N. Bartlett, *Israel J. Chem.* **1978**, *17*, 114.

[29] K. Leary, D. H. Templeton, A. Zalkin, N. Bartlett, *Inorg. Chem.* **1973**, *12*, 1726.

[30] N. Bartlett, F. Einstein, D. F. Stewart, J. Trotter, *J. Chem. Soc.(A)* **1967**, 1190.

[31] K. Lutar, A. Jesih, I. Leban, B. Žemva, N. Barlett, *Inorg. Chem.,* **1989**, *28*, 3467.

[32] K. Leary, A. Zalkin, N. Bartlett, *Inorg. Chem.,* **1974**, *13*, 775.

[33] J. H. Holloway, *Noble-Gas Chemistry*, Methuen, London, 1968, p. 124.

[34] R. Hoppe, *Fortschr. Chem. Forsch.* **1965**, *5*, 213.

[35] J. S. Ogden, J. J. Turner, *Chem. Commun.* **1966**, 693.

[36] J. L. Huston, H. H. Claassen, *J. Chem. Phys.* **1970**, *50*, 5646.

Mikrochim. Acta [Suppl.] 14, 543–546 (1997)

FT-IR Spectroscopy of the CFC Replacements HFC-152a and HFC-227ea in a Supersonic Jet Expansion

D. McNaughton*, C. Evans, and **E. G. Robertson**

Department of Chemistry, Monash University, Wellington Road, Clayton, Vic. 3168, Australia

Abstract. Infrared spectra of the CFC and Halon replacements HFC-152a (1,1-difluoroethane) and HFC-227ea (1,1,1,2,3,3,3-heptafluoropropane) have been recorded in a supersonic jet expansion. The resultant rotationally cold spectra have been used in combination with *ab initio* calculations to reassign the spectra of HFC-152a and begin an assignment of HFC-227ea. High-resolution spectra have been recorded and for HFC-152a two fundamentals assigned. A least-squares fit of combination differences from these bands, together with the available microwave transitions, has led to an improved set of ground state rotational constants.

Key words: CFC replacements, supersonic jet expansion, FT-IR spectroscopy.

In response to environmental concerns, hydro-fluorocarbons (HFCs) have been proposed as substitutes for the ozone-depleting CFCs because they contain no chlorine or bromine atoms and are more readily broken down in the troposphere. HFC-134a (1,1,1,2-tetrafluoroethane) is increasingly being used to replace CFC Freon-12 (CCl_2F_2) as a refrigerant, blowing agent and propellant, HFC-152a (1,1-difluoroethane) has been proposed as a replacement refrigerant, and HFC-227ea (1,1,1,2,3,3,3,-heptafluoropropane) is in use as a fire-extinguishing gas and is proposed as a delivery system for asthma sprays. A full knowledge of the vibrational modes of these molecules is desirable in order to calculate thermodynamic quantities such as heat capacity and also to detect and monitor the gas in the atmosphere by infrared spectroscopy. Molecules as large as CF_3CH_2F, CF_2HCH_3 and CF_3CHFCF_3, however, have extremely congested rovibrational spectra at room temperature owing to the large number of

heavily populated rotational states. This congestion together with further complications from hot band spectra arising from the low frequency torsional modes makes assignment of low-resolution spectra difficult and assignment of high-resolution spectra impossible. The rotational cooling achieved in a supersonic jet expansion results in resolved and assignable high-resolution spectra and vastly simplified low-resolution spectra. We have recently made a study of HFC-134a [1] and we report here our initial results on HFC-152a and HFC-227ea.

Experimental

Our apparatus for jet–FT-IR spectroscopy consists of an external sample chamber and transfer optics box coupled to a Bruker IFS 120HR interferometer. The sample gas flows through a pyrex tube and out of a circular pinhole nozzle at the end of the tube to expand into the chamber, which is evacuated by a Varian HV12 cryopump. Mirrors within the chamber allow the infrared beam to make 11 passes of the supersonic expansion region before passing out again to be detected by a Judson Infrared Inc. HgCdTe–InSb detector. The details of this apparatus have been more fully described in a previous paper on Freon-12 [2].

The HFC samples, obtained commercially from CIG Australia (HFC-152a) and Kidde–Graviner Australia (HFC-227ea) were allowed to flow directly from the cylinder into the pyrex tubing so that the stagnation pressure was equal to the vapour pressure of the gas. A 240-µm diameter nozzle was used to deliver as great a flow-rate as possible in order to maximize absorption by the sample. Under these conditions the cryopump was overloaded, so at high-resolution only 5 scans could be taken before the pressure in the lower region of the chamber began to rise unacceptably, exceeding 0.4 Pa. A large number of scans (typically 100) were obtained by isolating the cryopump between sets of scans to allow it to recover. High-resolution jet-cooled spectra were recorded at 0.0035 cm^{-1} unapodized resolution, by using a 700–1300 cm^{-1} filter and a KBr beam-splitter. Low resolution jet-cooled spectra in the region 650–4500 cm^{-1} were also recorded at 0.1 cm^{-1} resolution by co-adding 100 scans. Room temperature spectra were recorded in a multiple

---

* To whom correspondence should be addressed

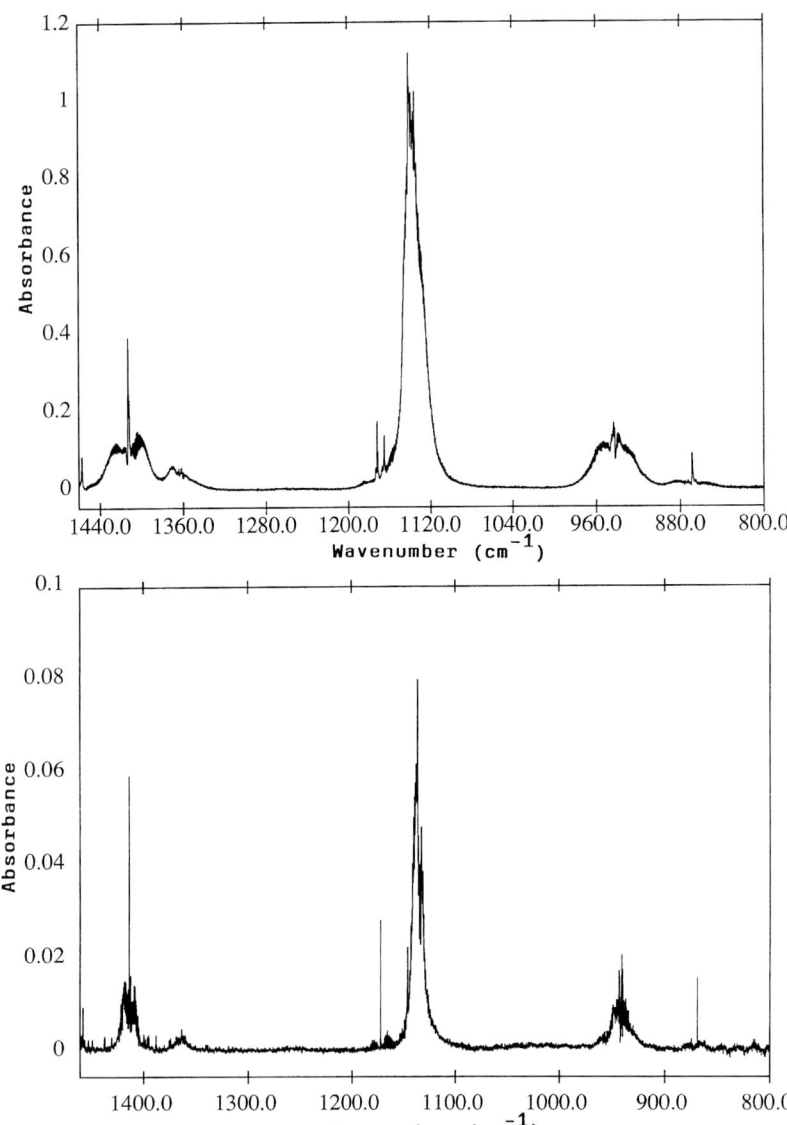

Fig. 1. Infrared spectrum of HFC-152a (1,1-difluoroethane) at room temperature (top) and in the jet nozzle (below)

traverse gas cell (set at 4 metres path-length) at pressures of ca 0.05 torr.

Ab initio calculations were done with the Gaussian 94 suite of programs on a Silicon Graphics Inc. IRIX Challenge computer system.

Results

The low-resolution spectra of HFC-152a and HFC-227ea at room temperature and in the jet nozzle apparatus are shown in Figs. 1 and 2. The jet nozzle spectra show a considerable degree of cooling, leading to spectral simplification. The major simplification is in the reduction of vibration–rotation structure in the wings of each band, although some vibrational cooling is apparent with the disappearance of bands emanating from vibrationally excited states. In order to facilitate assignments, *ab initio* calculations were made at the MP2/6–311G** level for HFC-152a and at the HF/6-311G** level for HFC-227ea. The predicted frequencies were then multiplied by a scaling factor derived from the appropriate calculations and experimental values of HFC-134a [1]. The band assignments for the HFC-152a are given in Table 1. Preliminary assignments have also been made for HFC-227ea. HFC-152a is an oblate asymmetric top ($\kappa = 0.75$) which has C_s symmetry, so that it's 18 vibrational modes belong to

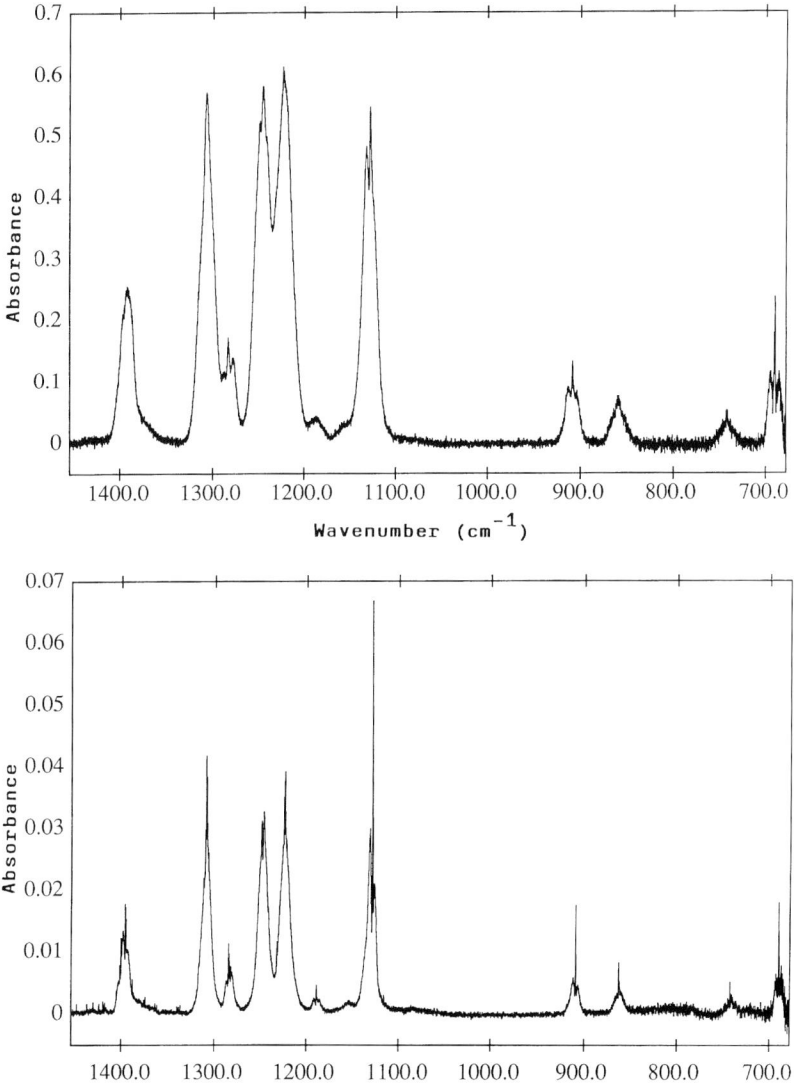

Fig. 2. Infrared spectrum of HFC-227ea (1,1,1,2,3,3,3-heptafluoropropane) at room temperature (top) and in the jet nozzle (below)

symmetry species A' (A/C-type bands) or A'' (B-type bands), thus making assignment relatively easy. As a result of jet cooling we are able to resolve the bands 1145.1/1134.9 cm^{-1} which were previously assigned by Guirgis and Crowder [3] at 1142/1149 cm^{-1}; furthermore the band at 1164 cm^{-1} which they assigned as v_{14} disappeared on cooling, which implies it is a vibrational hot band and not a fundamental. Rotational cooling also gives sharper Q-branches, which allowed better determination of the band centres and of their assignments. Further assignments will be made when we have recorded a far infrared spectrum.

Vibration-rotation Analysis

Peak-lists from the high resolution spectra of HFC-152a were examined with Macloomis, an interactive program for Macintosh which displays peaks in Loomis–Wood (Fortrat) format. This program aids in the process of assigning transitions by enabling the regularly spaced series of transitions possessing the same K_a value but differing in J to be selected. In this way assignments were obtained for the C-type band v_5 and the B-type band v_{16}. For the v_{15} (1134 cm^{-1}) band there are considerable as yet unexplained perturba-

Table 1. Experimental and *ab initio* vibrational wavenumber values for CH_3CHF_2

Mode	IR[1]	IR[2]	This work	MP2/6-311G**	Mode description
$A'\,\nu_1$	3016	3018	3016.0	3111	CH_3 asym. stretch
ν_2	2975	3001	2975.2	3055	CH stretch
ν_3	2959	2963	2958.5	3009	CH_3 sym. stretch
ν_4	1466[a]	1460	1456.9	1462	CH_3 asym. deform.
ν_5	1413	1414	1413.2	1426	CH_3 sym. deform.
ν_6	1362	1372	1359.7	1368	CCH bend
ν_7	1171	1143	1171.1	1149	CCH bend, CF stretch
ν_8	1142	1129	1145.1	1142	C–C stretch
ν_9	868	868	868.7	865	CH_3 rock
ν_{10}	569	571		559	CF_2 bend
ν_{11}	469	470		461	CF_2 wag
$A''\nu_{12}$	3016	3001		3108	CH_3 asym.stretch
ν_{13}	1457	1460		1463	CH_3 asym. deform.
ν_{14}	1164	1360	1364.1	1397	CCH bend
ν_{15}	1149[a]	1171	1134.9	1153	CCH bend, CF stretch
ν_{16}	942	930	941.7	949	CH_3 rock
ν_{17}	383	383		377	CF_2 twist
ν_{18}	222			249	CH_3 torsion

[a] Raman values, [1] Reference [3], [2] Reference [6].

Table 2. Molecular constants for the ground state of HFC-152a-A-reduction representation I^r

	Constant/cm^{-1}	Correlation matrix						
A	0.3166176 (5)	100.0						
B	0.2989653 (4)	53.3	100.0					
C	0.1724684 (4)	56.5	79.4	100.0				
Δ_J	1.627 (21) 10^{-7}	22.9	42.8	18.7	100.0			
Δ_{JK}	7.877 (1575) 10^{-8}	48.0	−7.3	6.7	4.8	100.0		
Δ_K	7.079 (2042) 10^{-8}	−38.2	16.1	10.4	6.6	91.7	100.0	
δ_J	6.404 (177) 10^{-8}	25.4	29.5	19.1	36.0	61.4	30.9	100.0
δ_K	1.451 (59) 10^{-7}	23.7	−35.3	−0.9	21.2	13.3	40.4	55.8	100.0
		A	B	C	Δ_J	Δ_{JK}	Δ_K	δ_J	δ_K

184 "lines" (151 from IR combination differences, 33 from microwave data [4]).
Standard deviation of fit = 1.2.
Numbers in parentheses are one standard deviation from the least-squares fit, in units of the least significant figures.

tions. These assignments have been confirmed by using ground-state combination difference values (GSCDs) calculated from the rotational constants of the microwave study [4]. The GSCDs from this work, combined with the previous microwave transitions have been fitted to Watson's A-reduced Hamiltonian, leading to the improved set of ground-state rotational constants presented in Table 2. Recently Villamañan et al. [5] have made a millimetre wave study that produces rotational constants comparable with ours but with more accurate and better determined centrifugal distortion constants.

Acknowledgement. The financial assistance of the Australian Research Council and the Sir James McNeill Foundation is gratefully acknowledged.

References

[1] D. McNaughton, C. Evans, E. G. Robertson, *J. Chem. Soc. Faraday Trans.* **1995**, *91*, 1723.
[2] D. McNaughton, D. McGilvery, E. G. Robertson, *J. Chem. Soc. Faraday Trans.* **1994**, *90*, 1055.
[3] G. A. Guirgis, G. A. Crowder, *J. Fluorine Chem.* **1984**, *25*, 405.
[4] N. Solimene, B. P. Dailey, *J. Chem. Phys.* **1954**, *22*, 2042.
[5] R. M. Villamañan, W. D. Chen, G. Wlodarczak, J. Demaison, A. G. Lesarri, J. C. López, J. L. Alonso, *J. Mol. Spectrosc.* **1995**, *171*, 223.
[6] D. C. Smith, R. A. Saunders, J. R. Nielsen, E. E. Ferguson, *J. Chem. Phys.* **1952**, *20*, 847.

Mikrochim. Acta [Suppl.] 14, 547–549 (1997)

Analysis of Distortions in High-resolution Fourier-transform Spectra

Wolfgang Otto, Jörg Lichtenthäler*, and **Heinz O. Tittel****

Department of Physics, University of Siegen, D-57068 Siegen, Federal Republic of Germany

Abstract. A Michelson interferometer with plane mirrors and dynamic alignment has been built for use in emission and dispersive Fourier-transform spectroscopy. The spectral range is 0.2–1.5 μm with a resolving limit of 0.03 cm^{-1}. In spite of automatic adjustment control, some disturbances in the interferogram always remain and lead to distortions in the spectra. There are deviations from linearity in the mirror drive, linear phase errors, tilt of the moving mirror, and vibrations of the mechanical set-up. To identify and quantify these errors, the phase and amplitude of a laser interferogram are analysed by means of a special algorithm. Results are presented.

Key words: Fourier transform spectroscopy, distortions, high resolution, visible, ultraviolet.

In the past two decades high-resolution Fourier-transform spectrometers (FTS) have been built for use in the UV–visible region. Examples can be found in [1–4], and show that also in this spectral range FT-spectroscopy is a powerful method mainly because of the throughput advantage, high wavenumber precision and the complete spectral record. To exploit these advantages for emission and dispersive spectroscopy we have built and tested at Siegen University a high-resolution FTS specifically for the visible and UV down to 200 nm. Plane mirrors with dynamic alignment as tilt-compensating system are used to provide the accuracy of adjustment for the mirror travel system necessary for high-resolution FT-spectroscopy [5]. In spite of automatic adjustment control, some disturbances always remain and lead to distortions in the spectra [6–9]. In this paper we describe a very precise method to identify and quantify errors, thus offering a possibility to eliminate them at their origin. Moreover, the method enables us to control the adjustment of the moving mirror over a complete scan.

---

* Present address: MPI für Radioastronomie, Bonn, Germany
** To whom correspondence should be addressed

Instrumentation

The schematic lay-out of the interferometer is illustrated in Fig. 1. The beam diameter is 45 mm and SM 1 has a focal length of 750 mm, giving an f/17 relative aperture. The maximum optical path difference is 33 cm. When double-sided interferograms are run this corresponds to a resolution limit of 0.03 cm^{-1}. For the pre-adjustment of the instrument the visibility

$$V = \frac{I_{max} - I_{min}}{I_{max} + I_{min}} \tag{1}$$

of an expanded He-Ne laser beam interferogram is measured while the voltages of the actuators PB and PC are modified. The pair of voltage giving the highest visibility is finally applied. Three parallel laser rays, which are separated from the same laser, generate three interferograms at points A, B and C (for details see [5]). During the mirror run the phases of these interferograms relative to each other must be kept at those values which were found from the highest visibility at the moment when the mirror started. This is done by an electronic circuit controlling the actuators at B and C. Signal A is also used for the internal fringe-reference sampling system. To fulfil the Nyquist theorem the period of the laser fringes is electronically subdivided by a factor of four. The origin of the interferogram is determined by an additional simultaneously recorded white light interferogram in the near IR provided by an auxiliary light source. A Transputer-based multiprocessor network is used for data read-out, adjustment control and driving control. The acquisition of an interferogram in highest resolution mode takes about 4 minutes; the same time is necessary to perform the 4 million-point FFT by a T800.

Method

To interpret the effects observed in the spectrum that are due to features occurring in the interferogram, we developed a method to measure the variations of phase and amplitude of the expanded laser beam signal for a complete scan. The method is illustrated in Fig. 2. The recorded discrete intensity values I_i of the laser interferogram can be written as

$$I_i = A_i \cos \varphi_i + A_0 \tag{2}$$

With the assumption that over a laser period the amplitude A_i is constant and the phase difference between two sample points $\varphi_{i+1} - \varphi_i = 45°$, we can calculate from (2) the local amplitude and phase of the interference signal with knowledge of three

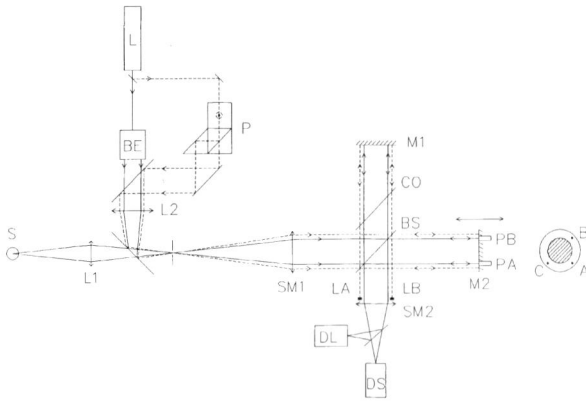

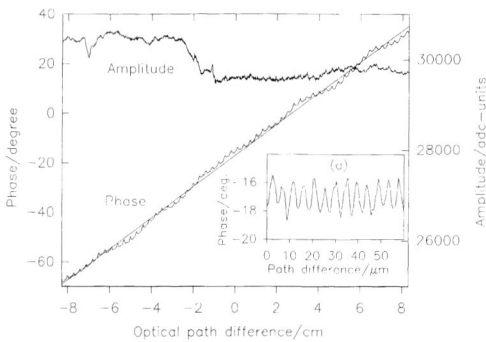

Fig. 1. Schematic diagram of the interferometer. *L* He-Ne laser; *BE* beam expander; *P* prisms; *S* light source; *L1* and *L2* lenses; *SM1* and *SM2* concave mirrors; *BS* beam splitter; *CO* compensator; *M1* fixed mirror; *M2* movable mirror; *PB* and *PA* piezo actuators; *LA* and *LB* silicon diodes; *DL* and *DS* photomultipliers. All transmitting optical components are made of fused silica

Fig. 3. Phase and amplitude of the expanded laser beam as a function of path difference. The rippled curve is the difference between the phases of LA and expanded beam signals. To eliminate random sampling position errors we calculated from (3) and (4) the average over 128 laser periods to get one data point. The straight line shows a first-order regression based on these phase values. In the inset figure (*a*) we plotted the variation of phase without any averaging. Now the period of distortion is identical with the resonance frequency of the interferometer ground plate (see text)

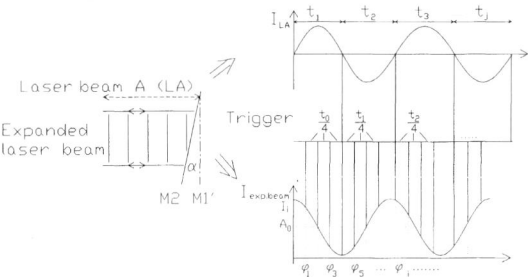

Fig. 2. Principal scheme of the method. The reference signal LA is used to generate the trigger signals. An electronic circuit measures the duration of every half period t_j. Then t_j is divided by a factor of four and on this base three trigger pulses are set in the next half-period. In general there is a phase difference between the signals of LA and the expanded beam because of a small till α between M1 and M2 and the fact that LA is reflected at the periphery of the interferometric mirrors. Disturbances changing α can be detected by our method as phase or amplitude modulation in the expanded laser beam signal

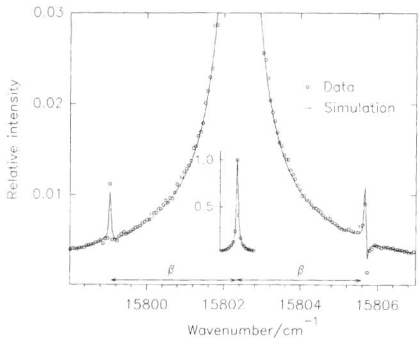

Fig. 4. Spectrum and its simulation near the centre of the laser line

measured intensities:

$$\varphi_i = \arctan\left(2\frac{I_i - I_{i+2}}{I_i - I_{i+4}} - 1\right), \qquad (3)$$

$$A_i = \frac{1}{\sqrt{2}}\sqrt{(I_i - I_{i+2})^2 + (I_{i+2} - I_{i+4})^2}. \qquad (4)$$

Results and Discussion

Figure 3 shows an example of a measurement and Fig. 4 the corresponding spectrum of the laser line. We can distinguish between four different effects.

Linear Phase Error

A misalignment between the reference beam LA and the expanded beam introduces a linear phase error

$$\Delta\Phi = 2\pi\sigma x(1 - \cos\gamma),$$

where γ is the angle enclosed by the two beams and x is the optical path difference. Here $\Delta\Phi$ is equal to $100°$ which corresponds to a misalignment angle of $\gamma = 1.4\,\mathrm{mrad}$. In the spectrum this leads to a displacement $d\sigma = 0.016\,\mathrm{cm}^{-1}$ and a broadening (leakage effect) of the laser line.

Drive Non-linearities

Even when a high-precision micrometer screw with a DC motor is used as a mirror drive a slight periodic thread error is unavoidable. The period of the ripple on the phase in Fig. 3, namely $\beta = 3.3\,\mathrm{cm}^{-1}$, corresponds exactly to the period of rotation of the micrometer. The amplitude of this distortion is about $1°$. Consequently two ghosts at $\sigma_L + \beta$ and $\sigma_L - \beta$ appear in the spectrum (Fig. 4).

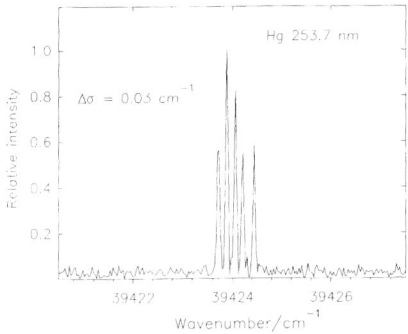

Fig. 5. Spectrum of the 253.7 nm line from a uncooled mercury hollow-cathode lamp. All five strong components of the HFS are fully resolved. One scan, without filter and apodization

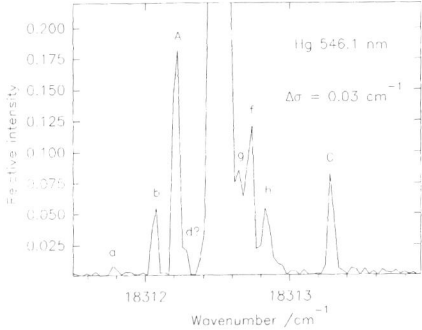

Fig. 6. Spectrum of the 546.1 nm line from a mercury germicidal lamp. Seven components (*A*, *C*, *a*, *b*, *f*, *g*, *h*) are resolved. The components *g* and *f* are separated by 0.07 cm^{-1}. One scan, without filter and apodization

Vibrations of the Mechanical Setup

In spite of the effective vibration isolation, mechanical and acoustical disturbances stimulate the resonance frequency $f_R = 180\,$Hz of the granite ground plate (Fig. 3, inserted plot).

Tilt of the Moving Mirror

The maximum phase deviation from the first order regression is 6°, which in our case means a parallelism of the interferometric mirrors to better than one seventeenth of the shortest wavelength of interest during the scan.

To verify our results we simulated the laser interferogram on the basis of the measured values. If we introduce a phase and amplitude modulation with the same frequency β, the simulation reproduces the ghosts (see Fig. 4). To show the performance of the instrument, Hg hyperfine structure measurements are presented in Figs. 5 and 6.

References

[1] P. Luc, S. Gerstenkorn, *Appl. Opt.* **1978**, *17*, 1327.
[2] A. P. Thorne, C. J. Harris, I. Wynne-Jones, R. C. M. Learner, G. Cox, *J. Phys. E. Sci. Instrum.* **1987**, *20*, 54.
[3] M. L. Parsons, B. A. Palmer, *Spectrochim. Acta B* **1988**, *43*, 75.
[4] A. Simon, L. Wunsch, G. Zachmann, *Mikrochim. Acta* **1988**, *11*, 311.
[5] R. Stenzel, W. F. Blum, H. O. Tittel, F. Romero-Borja, *Proc. SPIE* **1992**, *1575*, 236.
[6] H. O. Tittel, D. Wissmann, *Proc. SPIE* **1992**, *1575*, 244.
[7] G. Guelachvili, *Distortions in Fourier Spectra and Diagnosis*, in: *Spectrometric Techniques*, Vol. 2, (G. Vanasse, ed.), Academic Press, New York, 1981, p. 1.
[8] A. S. Zachor, I. Coleman, W. G. Mankin, in: *Spectrometric Techniques*, Vol. 2 (G. Vanasse ed.), Academic Press, New York, 1981, p. 127.
[9] E. E. Bell, R. B. Sanderson, *Appl. Opt.* **1972**, *11*, 688.

Mikrochim. Acta [Suppl.] 14, 551–553 (1997)

Vehicle Exhaust Gas Monitoring by Remote FT-IR Spectroscopy

Nicholas M. Davies*, Moira Hilton, and **Alan H. Lettington**

J.J. Thomson Physical Laboratory, University of Reading, Whiteknights, P.O. Box 220, Reading RG6 6AF, U.K.

Abstract. An optically modified Unicam 0.25 cm^{-1} resolution FT-IR spectrometer has been used for remote monitoring of absorption and emission spectra of vehicle exhaust gases. In the active mode, a beam of modulated infrared radiation was emitted by the spectrometer and the absorption by the exhaust gas was measured along the beam path to a retroreflector. The return signal was monitored by a liquid-N$_2$ cooled mercury cadmium telluride (MCT) detector external to the instrument. The concentrations of CO and other gaseous molecular species in the exhaust were compared, with good agreement, with prepared laboratory samples and models based on the AFGL atmospheric transmission data-base HITRAN. In the passive mode, the emission spectra of the warm exhaust gases have been recorded. The use of a liquid-N$_2$ background gave the spectrometer greater sensitivity to CO present in the exhaust gas. Laboratory studies at moderate (0.25 cm^{-1}) spectral resolution have been conducted to determine the spectrometer sensitivity to trace concentrations of nitrogen and sulphur dioxide.

Key words: vehicle exhaust gas, remote monitoring, carbon monoxide, sulphur dioxide.

Fourier–transform infrared spectroscopy (FTS) has been used to measure gas concentrations in vehicle exhausts [1–3]. Typically, samples of gas are collected from the exhaust pipe, diluted with air and passed into a multiple-pass White cell. The absorption of interferometer-modulated infrared radiation by the gas is then detected.

The University of Reading Physics Department has been developing FTS techniques for passive remote monitoring of stacks and aero-engine exhaust emissions [4]. This paper describes the application of our remote monitoring FTS system in both active and passive modes to vehicle exhaust emissions. This system is completely non-intrusive, removing the need for extractive sampling. In the more sensitive active mode, the absorption of the modulated infrared probe beam by the gases emitted from the vehicle exhaust pipe is

measured. In the passive mode, the spectrometer detects thermal emission from the warm gases close to the vehicle exhaust pipe. Since the radiance of the warm gases is close to that of the surrounding air, the signal-to-noise ratio is poor. The use of a liquid-nitrogen reference to improve the spectrometer's sensitivity to the warm emission gases was investigated.

Experimental

Spectra were taken with a Unicam 0.25 cm^{-1} resolution Research Series spectrometer in the mid infrared region (700–4000 cm^{-1}). In the active mode, the spectrometer was optically modified to emit a beam of modulated infrared radiation. This was achieved by using the spectrometer's internal globar source and by positioning a plane mirror immediately after the Michelson interferometer (KBr beam-splitter) to take the beam outside the spectrometer. The beam was retroreflected, by two plane mirrors set at 45° to the beam and 66 cm from the spectrometer, to the liquid-nitrogen cooled MCT detector. The detector was moved from its original position to the sample compartment of the spectrometer to make it easier to focus the infrared beam onto it. Exhaust gases were introduced, unrestricted, between the spectrometer and retroreflector, thus absorbing the infrared radiation, by using a 1.5 m length of flexible pipe attached to the vehicle's exhaust pipe. Five cars were tested in the car park outside the laboratory with their engines running at idle. Each spectrum consisted of 16 scans at 0.25 cm^{-1} resolution, with a 20 s collection time. In the passive mode, emission spectra of the warm exhaust gases were taken. The spectrometer was placed on the ground and the cars were reversed up to it; the spectrometer input window was perpendicular to the axis of the exhaust pipe and 10 cm away from it. A liquid-N$_2$ background was used to increase the spectrometer's sensitivity when the temperature of the exhaust gas was close to ambient temperature. To create the background a plane mirror was placed at 45° to reflect the image of a dish of liquid N$_2$ onto the spectrometer input window. Each spectrum consisted of 200 scans at 0.25 cm^{-1} resolution, with a 240 s collection time. A large number of scans was taken to increase the signal-to-noise ratio (SNR). At high SNR, low-intensity emission features are made more visible. The SNR achieved for the CO band was ~ 6, compared to ~ 90 in the active mode.

Laboratory studies have been conducted to determine the spectrometer's sensitivity to low concentrations of nitrogen and sulphur dioxide. Absorption measurements were carried out with a cylindri-

---

* To whom correspondence should be addressed

cal gas sample cell (12 cm in length) and a hot-plate at a temperature of 276 °C. The sample cell was evacuated and filled with nitrogen or sulphur dioxide at known pressure, then brought up to atmospheric pressure by addition of air. The hot-plate was placed behind the sample cell and absorption spectra were taken. Emission measurements were made with the same gas sample cell, but in its heated mode. Emission spectra were taken only with sulphur dioxide samples, at a temperature of 250 °C, measured inside the cell.

Results and Discussion

Absorption Measurements

When monitored in the active mode, the vehicle which produced the greatest number of molecular species was a Vauxhall Cavalier (4-star leaded petrol, 1600 cc, 1984). Therefore this section will concentrate on the spectrum from this car. Figure 1 shows the transmission spectrum of the Vauxhall Cavalier exhaust, with all the relevant features marked. The gases detected included CO, centred at 2143 cm^{-1}, CH$_4$, centred at 3017 cm^{-1} and C$_2$H$_2$, centred at 730 and 3284 cm^{-1}.

The AFGL atmospheric transmission data-base HITRAN has been used to model the CO, CH$_4$ and C$_2$H$_2$ absorption bands. The parameters entered into HITRAN are the path-length, temperature and partial pressure of the exhaust gas, the last of which is directly related to the gas concentration. The path-length has been estimated and is therefore the largest source of error. Future experiments will be conducted with the exhaust gas restricted to a known path-length.

By matching the experimental and modelled spectra the concentrations of CO, CH$_4$ and C$_2$H$_2$ in the exhaust emission were found to be 530, 70, 100 ppmv, respectively.

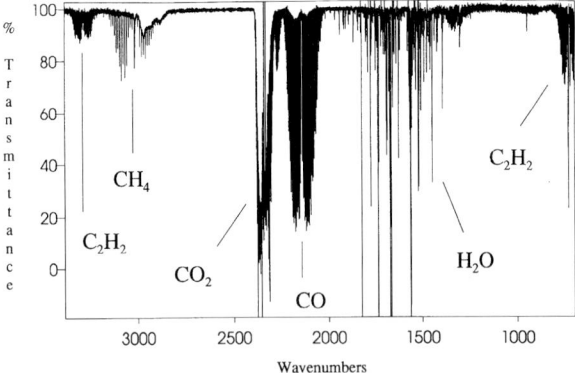

Fig. 1. Transmission spectrum of Vauxhall Cavalier (4-star petrol, 1600 cc, 1984) exhaust gas

Air samples containing different known concentrations of CO were prepared in the laboratory, in 12 cm long gas sample cell. The Vauxhall Cavalier spectrum was matched with one of the samples and a good agreement between the concentrations was found. The CO concentration in the exhaust, found by sample matching, was 0.017 ± 0.006 mol/m^3 and the concentration by using the HITRAN model was 0.022 mol/m^3.

The most interesting results from the other cars tested came from a Citroen ZX (diesel, 1900 cc, 1993) and a Jaguar XJ6 (unleaded petrol with catalytic converter, 3200 cc, 1993). There were no unburnt fuels observed and a much reduced amount of CO$_2$ and CO. This could be due to the fact that the vehicles were only built in 1993 and still had very efficient engines, or, in the case of the Jaguar XJ6, that the catalytic converter was removing CO$_2$, CO, CH$_4$ and C$_2$H$_2$.

Emission Measurements

Before the vehicles were tested they had to be driven for approximately 10 minutes; this allowed the engine to warm up and increased the temperature of the exhaust gases from 25 °C to 55 °C (ambient temperature was 25 °C). Spectra were then taken directly from the end of the exhaust pipe. Emission features due to CO$_2$ and CO could be seen in the resulting spectra.

Measurements were repeated on one car when the engine had cooled down and the exhaust gas temperature was down to 39 °C. Two spectra were taken, one with a liquid-N$_2$ background and one without. No emission features due to CO could be seen in the spectrum taken without using liquid N$_2$, but CO features were present in the spectrum taken by using liquid N$_2$. Therefore, using a liquid-N$_2$ background increases the spectrometer sensitivity to exhaust gases which are 14 °C higher than ambient temperature.

NO$_2$ and SO$_2$ Measurements

An absorption spectrum was taken with an NO$_2$ concentration of 0.144 mol/m^3 (3400 ppmv) which clearly showed the presence of an NO$_2$ absorption band centred at 1260 cm^{-1}. It was also possible to observe SO$_2$ absorption and emission features centred at 1150 and 1370 cm^{-1} with an SO$_2$ concentration of 0.041 mol/m^3 (950 ppmv).

The path length through the gas was 12 cm, but use of a longer path length would mean a lower concentration could be detected.

Conclusion

The overall conclusion from these experiments is that using a laboratory FTS spectrometer to monitor the gaseous exhaust emissions from vehicles, without dilution and sampling, is feasible. No sampling is needed, so this method is very well suited to roadside testing.

In the active mode, low concentrations of CO, CH_4 and C_2H_2 could be detected. Comparison with modelled and laboratory samples gave the concentrations of these gases.

In the passive mode, CO and CO_2 could be detected and the use of a liquid-nitrogen background gave the spectrometer greater sensitivity to CO emission.

Acknowledgements. This work is supported by the Engineering and Physical Sciences Research Council and the National Physical Laboratory, U.K.

References

[1] W. F. Herget, S. R. Lowry, *Proc. SPIE* **1991**, *1433*, 275.

[2] C. A. Gierczakr J. M. Andino, J. W. Butler, G. A. Heiser, G. Jesion, T. J. Korniski, *Proc. SPIE* **1991**, *1433*, 315.

[3] M. W. Sigrist, *Air Monitoring by Spectroscopic Techniques, 1st Ed.*, Willey–Interscience, New York, 1994, p. 439.

[4] M. Hilton, A. H. Lettington, I. M. Mills, *J. Meas. Sci. Technol.* **1995**, *6*, 1236.

Mikrochim. Acta [Suppl.] 14, 555–557 (1997)
© Springer-Verlag 1997

Infrared Gas Phase Spectrometry for Oxygen and Hydrogen Sulfide by UV Photolysis

Elisabeth S. Larsen and **Martin L. Spartz***

Department of Chemistry, Central Michigan University, Mount Pleasant, MI 48859, U.S.A.

Abstract. Oxygen, hydrogen sulfide and other weakly IR-absorbing compounds have been analyzed indirectly by photochemical conversion into other IR-active substances. Oxygen, when irradiated with intense UV light, is converted into ozone, which gives an absorption band centered at 1043 cm^{-1}. Hydrogen sulfide, a weak IR absorber, is converted into sulfur dioxide, with a band centered at 1361 cm^{-1}. Both ozone and sulfur dioxide have intense infrared spectral absorptions free from water or carbon dioxide interference. However, if this indirect method is to be employed for chemical determinations other factors must be considered. Methyl ethyl ketone and benzene have been investigated for their effect on conversion rates and percent conversion of oxygen into ozone. H_2S is converted into sulfur dioxide, but the SO_2 is subsequently dissociated. Results for oxygen and hydrogen sulfide determinations by use of various parameters will be discussed.

Key words: oxygen, hydrogen sulfide, infrared analysis, UV photolysis conversion.

Many important gas analytes absorb infrared radiation weakly and are difficult to detect by infrared methods at low to moderate concentration. Consequently, in order to obtain a measurable absorbance, a longer path-length with analytes evenly dispersed across the path is required. Open-path FT-IR spectrometry incorporates path-lengths from a few m to more than a km, depending on the application. This method, however, is limited by carbon dioxide and water, which can have intense absorptions at longer path-lengths or high concentration. Many trace gases absorb at frequencies which overlap those of water or carbon dioxide. Qualitative and quantitative analyses are difficult or impossible when water and/or carbon dioxide are present at high levels. Also, longer path-lengths may not be possible, owing to landscape and/or development problems.

Ultraviolet-assisted infrared spectroscopy has been evaluated as a potential method for the detection and analysis of trace gases [1]. UV irradiation provides an indirect detection method for non-absorbing infrared compounds such as oxygen, and difficult-to-measure compounds such as hydrogen sulfide. The indirect determination is made possible through conversion of the analyte into a strongly infrared absorbing compound. The unirradiated gas sample serves as the reference spectrum for the UV-irradiated sample. Since neither water nor carbon dioxide is lost in large quantities in UV photolysis, their concentrations remain relatively constant and they are not major contributors to the absorbance spectrum. In this study, UV-assisted infared spectrometry was evaluated for H_2S and O_2 determinations.

Equipment

A multi-pass infrared White cell [2] equipped with an internal mercury UV lamp was interfaced to a Bomem Hartmann & Braun (Quebec, Canada) MB series Fourier-transform infrared (FT-IR) spectrometer with a liquid-nitrogen cooled mercury-cadmium-telluride (MCT) detector. The White cell used was constructed by Infrared Analysis, Inc. (Irvine, CA). The cell is approximately 1 m tall and has a 46.3 L volume and 26 m path-length. This cell size is certainly not required for many applications and in some instances is not desirable. The UV lamp was a high-ozone germicidal mercury arc lamp with a quartz envelope. The primary wavelengths emitted by this lamp are 254 nm (90%) and 185 nm (2%). Cell pressure was determined by a 1000 torr capacitance manometer (MKS Instruments, Andover, MA, type 960A13TRB) connected to a digital read-out (MKS Instruments, model #270C-4). H_2S (Scott Specialty Gases, CP Grade 99.5%) and other gases, (ambient air, oxygen,

---

* To whom correspondence should be addressed

nitrogen, etc.) were introduced into the White cell through the vacuum manifold.

Experimental

Before data for the oxygen or hydrogen sulfide conversion were collected, the White cell was purged with nitrogen. The cell was then evacuated with a liquid-nitrogen trapped vacuum pump to approximately 0.5 torr. All gases were added through a vacuum manifold, with monitoring of the added partial pressure. Liquid materials (methyl ethyl ketone, benzene) were added to the manifold by freezing them with liquid nitrogen, evacuating the head space, then allowing the sample to warm to room temperature to release any dissolved air, and the process was repeated as required. To produce parts-per-million levels, the gas was added to the manifold and expanded into the total volume, including that of the multiple-pass cell.

Once a sample had been prepared in the IR cell a spectrum was obtained and used as the reference for any further data collected for that sample. The sample was then irradiated and data collected from time zero to as long as 10 minutes. Spectra were collected every 15 seconds early in the UV irradiation, with later spectra acquired at one minute intervals. The spectra of the generated ozone were collected at 4 cm^{-1} resolution and measured at 995.33 cm^{-1}, on the p-branch of ozone. The spectra for the generated SO$_2$ were collected at 1 cm^{-1} resolution and were measured at the Q-branch peak (1360.1 cm^{-1}).

Results and Discussion

Conversion of Oxygen into Ozone

The initial oxygen into ozone conversion study was performed on samples containing only controlled levels of oxygen, the balance being nitrogen. Four oxygen concentrations from approximately 20% (148.1 torr) to 2.5% (18.4 torr) were irradiated; spectra were collected and analyzed, and are displayed in Fig. 1. These spectra were obtained to ascertain whether oxygen at various levels could be determined by indirect ozone measurement. After a few minutes irradiation the ozone absorption was too large for its absorbance to be determined at a peak maximum, so 995.33 cm^{-1} was chosen for the measurement. Monitoring ozone at its peak by this arrangement could allow for low parts-per-million (v/v) (ppm) oxygen determinations. As observed in Fig. 1, all the ozone absorbances approach equilibrium. When ozone absorbances at 15 s UV irradiation were plotted vs, the initial oxygen partial pressure, a correlation coefficient (r^2) of 0.996 was obtained through least-squares analysis. Although these samples are not close to equilibrium there is a linear relationship to the initial oxygen concentration. At 600 s, when all samples except those with the highest concentration have reached equilibrium (Fig. 1), there is also a linear relationship

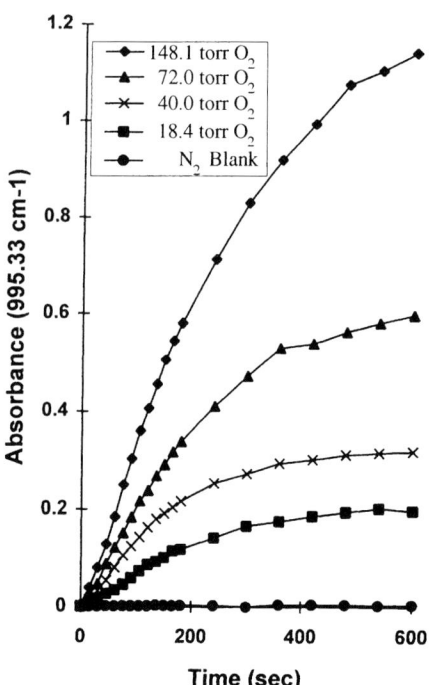

Fig. 1. Conversion of oxygen into ozone in nitrogen

($r^2 = 0.997$) for the four oxygen concentrations. If smaller sampling geometries are used, equilibrium times may be reduced.

Effect of Organics on Ozone Production

Methyl ethyl ketone (MEK, 2-butanone) and benzene were added to the oxygen–nitrogen samples to determine whether their presence influences the oxygen into-ozone conversion rate and/or equilibrium. When MEK was added to the sample and irradiated for ten minutes there was little change in the ozone production rate or equilibrium (Fig. 2). MEK is lost during UV irradiation, but even with this apparent reactivity there is no effect on the ozone generated. When benzene was added to the oxygen–nitrogen samples there was a small variation in the oxygen conversion rate and a slight decrease in total conversion. A significant absorption was noted on the high-frequency side of the ozone band. This absorption is due to the generation of formic acid.

Conversion of H$_2$S into SO$_2$

Water is involved in the conversion of H$_2$S into SO$_2$ and was expected to govern this change. In ambient air

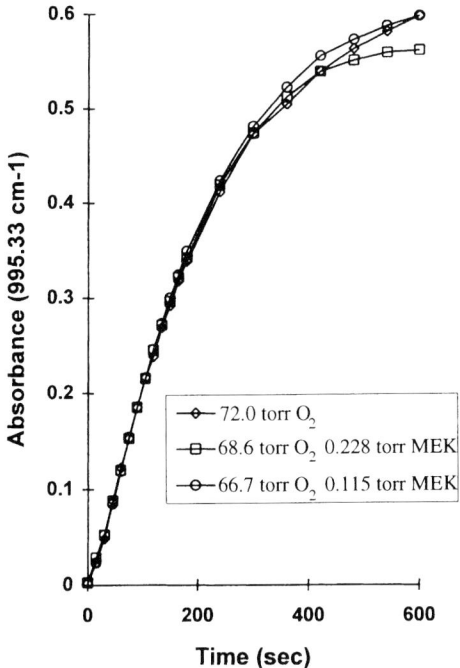

Fig. 2. Effect of MEK on ozone production

oxygen and hydrogen sulfide. Oxygen is readily converted into ozone in nitrogen by UV irradiation. The ozone concentration produced is shown to be linearly related to the initial oxygen concentration. This should provide a method for indirect oxygen quantification. The addition of MEK has little to no influence on the conversion of oxygen into ozone even though the MEK is lost during irradiation. Benzene is also lost during this irradiation, but it has an effect on the conversion rate and percent conversion. Any variations in the conversion rate or percent conversion can be modelled and the oxygen concentration determined.

Hydrogen sulfide, a weakly absorbing compound, is readily converted into sulfur dioxide, which is easily detected by infrared spectroscopy. The conversion efficiency of H_2S into SO_2 was calculated to be approximately 45% in both ambient and dry air. The conversion efficiency is limited, owing to the loss of SO_2 simultaneously with its generation. The rate of conversion and loss of SO_2 were found to be strongly dependent on sample moisture levels. However, at or below normal ambient moisture levels, the conversion efficiency remained at approximately 45%.

this conversion is rapid and peaks at between 30 and 40 s of UV irradiation. However, the SO_2 generated is also destroyed rapidly and is no longer detectable after 90 s of irradiation. The conversion efficiencies in ambient air peaked at between 41% and 47%. When the irradiation was done in dried air, it was noted that the H_2S into SO_2 peak conversion remained at 46%. However, the peak concentration was reached significantly later, after 75–105 s of irradiation.

Conclusion

This study demonstrates the potential of ultraviolet-assisted infrared spectroscopy as a method to monitor

Acknowledgements. The authors wish to acknowledge the 3M Corporation, specifically the 3M Environmental Laboratory of St. Paul, MN, which provided some instrumentation used in this study. They would also like to recognize the Dow Chemical Co. for partial student financial support. The authors also wish to thank Philip Hanst of Infrared Analysis for help and direction on this project, and Eric R. Larsen and Barbara J. Spartz for manuscript preparation, editorial advice and stimulating discussions.

References

[1] P. L. Hanst, S. T. Hanst, *Gas Measurement in the Fundamental Infrared Region: An Applications Manual*, Infrared Analysis, Anaheim, C.A.

[2] J. U. White, *J. Opt. Soc. Amer.* **1942**, *32*, 285.

Mikrochim. Acta [Suppl.] 14, 559–561 (1997)
© Springer-Verlag 1997

Emission Mode Fourier Spectroscopy of Smokestack Exhaust Gas

E. Lindermeir

Deutsche Forschungsanstalt für Luft- und Raumfahrt e.V. (DLR), Institut für Optoelektronik, Postfach 1116, D-82230 Weßling, Federal Republic of Germany

Abstract. A Fourier-transform infrared spectrometer (FTS) was used to determine the constituents of the exhaust gas of a 450 MW coal-fired power plant. For all measurements the spectrometer was operated in emission mode. Thus, the radiation of the hot exhaust gas was observed by the instrument. Two types of measurements were performed. In the first, the spectrometer was placed directly in front of an aperture in the smokestack wall. After good results had been obtained with this configuration, interferograms of the radiation of the exhaust gas leaving the smokestack at a height of 200 m were collected. By applying special evaluation algorithms the concentrations of H_2O, CO_2, CO, NO, SO_2 and the exhaust gas temperature could be derived from the infrared spectra.

Key words: spectroscopy, Fourier spectrometer, remote measurements, exhaust gas.

At present the state of the art in exhaust gas monitoring is the application of extractive methods. This is especially true as far as industrial smokestacks are concerned. Samples of the gas are transmitted through special lines to stations where they have to be properly prepared for the gas-analysers. Often these lines must be heated and the necessary temperatures depend on the species in the exhaust and the species to be determined. The exhaust gas preparation (conditioning) may include filtering, cooling (to remove water vapour) or – in the case of high concentrations – mixing, for example with nitrogen, to adapt the concentration to the analysers ranges. For example, to determine nitrogen monoxide concentrations with an NDIR (nondispersive infrared) instrument it is necessary to remove water vapour by cooling the gas, but in the resulting water some species such as NH_3 may be dissolved and thus vanish from the analyte. As a whole the gas conditioning step requires quite a lot of effort and is therefore expensive. Moreover, each gaseous species to be determined requires its own analyser.

The aim of the investigations presented in this paper was therefore to develop a method which completely eliminates exhaust gas sampling and conditioning. The basic idea is to use a Fourier transform infrared spectrometer which receives the radiation emitted by the hot gas in the smokestack or at its exit. According to its molecular structure each gaseous species emits radiation at different and well known wavelengths. The spectral radiance (the intensity of the radiation) due to one species depends on its concentration, the temperature and the pressure of the exhaust gas and the geometry of the measurement set-up. Therefore it is possible to determine the concentrations of interest from the measured spectra.

Calculation of Concentrations

The evaluation starts with the measured interferogram of the exhaust gas. After applying a Fourier-transform with Mertz's phase correction (the spectrometer measures single-sided interferograms) a "raw" spectrum is obtained. It is "raw" in the sense that it is in instrument-dependent units. For correct radiometric calibration (i.e. transformation from instrument units into units of spectral radiance) a high-quality calibration procedure must be applied. Therefore a method was chosen which uses three black bodies as reference sources. This procedure (described in [1]) automatically determines their surface temperatures together with the functions which describe the instrument radiometrically (sensitivity and offset) within a least-squares fit.

After radiometric calibration, concentrations and temperatures can be determined from the spectra. The measured (and calibrated) spectrum is compared with a theoretical one which is obtained by radiometrically modelling the measurement set-up. Basically the model is made up of three layers, the emitted radiances and transmissions of which interact: an atmospheric foreground (i.e. the atmosphere between the spectrometer and the exhaust gas), the exhaust gas itself and a background. In general it may be necessary to subdivide each layer into sub-layers. For the calculation of the model spectra the program Fascode is used. It supports simulations with up to 60 layers and contains the molecular parameters of 28 species from the Hitran database [2]. For each gaseous component in each layer the program calculates the emission lines. Finally the results for all the layers are combined to yield one overall spectrum. For a correct model the finite resolution of the spectrometer must also be included in the calculation. Therefore the Fascode-spectrum is convoluted with the instrumental line-shape function of the interferometer.

By use of the program FitFas, which was developed by the author, the parameters of the simulated spectrum are varied systematically until the squares of the differences between measurement and calculation are minimized. Then the parameters of the model are – within some uncertainties – equal to those of the measurement. FitFas also produces estimates of these uncertainties.

Experimental

In order to test this principle of measurement, an FTS was taken to a 450 MW coal-fired power plant located in the north of Munich, Germany. The smokestack at this site is 200 m high and has a diameter of 6 m. A Block Eng. spectrometer employed provides a maximum spectral resolution of $0.5 \, cm^{-1}$ and its InSb detector allows for measurements in the 2–5 μm region. In a first step "direct" measurements were performed: the spectrometer was placed on a platform at a height of 50 m in front of a hole in the smokestack. On the same platform conventional NDIR-analysers for monitoring NO, CO and SO_2 concentrations were installed, so reference data were easily obtained. On the opposite side of the FTS at the smokestack, there was another hole at which a black body at ambient temperature was placed. This black body was necessary to ensure a defined background with a brightness-temperature different from the temperature of the exhaust gas. Otherwise no emission lines could be observed, because the radiance emitted by the exhaust gas would just match the background-radiance absorbed by it.

In a second step remote measurements were performed. The spectrometer was positioned so that it received radiation from the smokestack's exit at a height of 200 m. The distance between the spectrometer's aperture and the exhaust gas in this configuration was 265 m. Thus the elevation of the receiving telescope was approximately 50°.

Results

For the evaluation the "direct" measurements were modelled with only one homogeneous layer, and

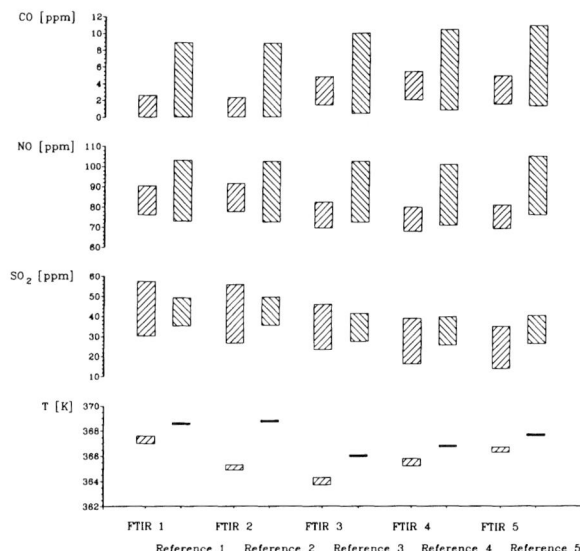

Fig. 1. Comparison of the results (with error-bars) obtained by evaluation of the spectra measured with an FTS and the reference NDIR-instruments

a black body in the background. This simple model can be applied because it is known from measurements that the temperature variation across the smokestack is negligible (within ±1 K). Furthermore before being emitted into the smokestack the exhaust gas passed through several cleaning facilities, which ensured proper mixing. A comparison of the results obtained from the spectra with the values provided by the reference instruments is displayed in Fig. 1. As can be seen, all error-bars are overlapping.

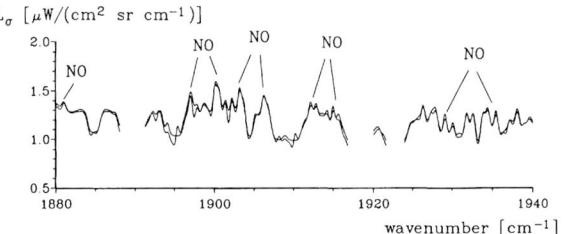

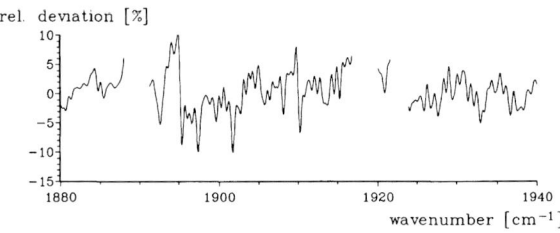

Fig. 2. Spectral region of NO emission lines. Top: measured and fitted spectrum. Bottom: relative differences of measurement and calculation

This is also the case for the remote measurements. Here the influence of the atmosphere on the radiation received had to be included in the model: one foreground layer and 8 layers for the background were used. Their initial parameters for the least-squares fit were taken from an atmospheric model included in the Fascode distribution. Figure 2 shows the part of one of the evaluated spectra in which emission lines of water vapour and nitrogen monoxide occur. At the emission lines of the gases the deviations between measurement and fitted spectrum were generally smaller than $\pm 5\%$.

As a whole, the aim of supplying a new non-contact method to measure the concentrations of exhaust gases has been achieved. At least, in the case discussed above the need for extracting gas from the smokestack has been eliminated. Moreover it is possible to detect the gases of interest, e.g. NO, CO and SO_2, with only one instrument. The remote measurements show that this kind of measurement also leads to trustworthy results.

References

[1] E. Lindermeir, P. Haschberger, V. Tank, H. Dietl, *Appl. Opt.* **1992**, *31*, 4527.

[2] L. S. Rothman, R. R. Gamache, R. H. Tipping, C. P. Rinsland, M. A. H. Smith, D. C. Benner, V. M. Devi, J. M. Flaud, C. Carny-Peyret, *J. Quant. Spectrosc. Radiat. Transfer* **1992**, *48*, 469.

[3] E. Lindermeir, *Ann. Geophysicae* **1994**, *12*, 417.

Mikrochim. Acta [Suppl.] 14, 563–564 (1997)

Monitoring Indoor Air Quality by Extractive FT-IR Spectrometry

Virginia D. Makepeace[1], **Christopher W. Chase**[1], **Charles T. Chaffin**[2], **Timothy L. Marshall**[2], **Petri T. Jaakkola**[3], **Roseann M. Hoffman**[1], **Robert M. Hammaker**[1,*], and **William G. Fateley**[1]

[1] Department of Chemistry, Kansas State University, Manhattan, Kansas 66506, U.S.A.
[2] Aero Survey, Inc. P.O. Box 1163, Manhattan, Kansas 66505, U.S.A.
[3] Temet Instruments Oy, Helsinki, Finland

Abstract. Extractive Fourier-transform infrared (FT-IR) spectrometry has been employed to monitor the indoor air quality of a commercial print shop and a commercial dry cleaning establishment. Monitoring over extended time periods demonstrates that trends in contaminant concentration may be obtained and related to activities taking place at the facility.

Key words: extractive FT-IR monitoring, indoor air.

Open-path Fourier-transform infrared (FT-IR) spectrometry has become a reliable and versatile means of monitoring several volatile organic compounds (VOCs) simultaneously in the outdoor atmosphere [1, 2]. Interest in indoor air quality has increased significantly in recent years since one important aspect of a safe working environment is good air quality. Extractive FT-IR spectrometry, originally developed for use as a stack gas analyzer, has been shown to be applicable to indoor air analysis [3]. This communication will present results obtained at a commercial print shop and at a commercial dry cleaning establishment. The general goal is to monitor for an extended time period to determine whether trends in contaminant concentration can be obtained and related to activities taking place at the facility.

Experimental

The Gasmet™ FT-IR Gas Analyzer (Temet Instruments Oy, Helsinki, Finland) used in this work consists of a complete Fourier transform infrared (FT-IR) spectrometer system, a gas cell with a 9-m folded pathlength, that is heatable to 50 °C, a 486 computer for instrument control, self-diagnostics, data collection, and automated spectral analysis, and an on-board pump and value system control-led by system software (Calcment™). A specially-designed mercury cadmium telluride (MCT) detector, which is Peltier-cooled to -30 °C, covers the spectral region from 850 to 4000 cm^{-1} with optimized performance at 8 cm^{-1} spectral resolution but with available spectral resolution from 4 to 64 cm^{-1}. In a typical sampling procedure, the Gasmet™ is allowed to warm up on-site until the gas cell temperature is stabilized with a pure nitrogen flow rate of 2 L/min. A single-beam spectrum of pure nitrogen serves as the background spectrum to create absorbance spectra from single-beam spectra for samples collected at a flow rate of 2 L/min. Collection times for background and sample spectra are 15 minutes. Reference library spectra are collected by using a custom-built reference gas generator. The spectral windows for analysis are chosen and the Calcmet™ software performs a qualitative and quantitative analysis by a classical least-squares algorithm.

Results and Discussion

Samples of the 21 solvents used in the commercial print shop were obtained and the reference gas generator was used to collect spectra from 900 to 4000 cm^{-1} for all these solvents for use in a reference library. The commercial print shop was monitored continuously from 1:00 pm on the first day to 5:15 pm on the fourth day. The spectral windows for analysis of the data collected in the commercial print shop were 3100–2400 cm^{-1}, 2200–1900 cm^{-1} and 1300–900 cm^{-1}. Analysis of the sample spectra for the monitoring period revealed the presence of five of the 21 solvents (2-propanol, Barsol D-1161-2, Barsol C-712, Varn Blanket Wash, Glaze-Phree cleaner and polish). Samples of the 20 compounds contained in these 5 solvents [determined from material safety data sheets (MSDS) and Aldrich Technical Services] were obtained and the reference gas generator was used in production of spectra from 900 to 4000 cm^{-1} for these 20 compounds for use as a reference library. Analysis of

---

* To whom correspondence should be addressed

the sample spectra for the monitoring period in terms of pure compounds rather than commercial print shop solvents revealed the presence of five of the pure compounds in addition to carbon monoxide and methane. At 4:55 pm (following the cleaning of the printing presses at the end of the work day) on the second day, the concentrations, errors and OHSA permissible exposure limits as 8-hour time-weighted averages, respectively, all in parts per million v/v (ppm) were the following: dichloromethane 2.2, 0.3, 500; dipropylene glycol methyl ether 18.5, 0.4, 100; n-hexane, 18.5, 0.1, 50; 2-propanol 147.4, 0.2, 400; p-xylene 4.5, 0.2, 100; carbon monoxide 20.7, 0.4, 35; and methane 6.8, 0.2, not listed. The variation in concentration observed over the monitoring period was consistent with the activities taking place.

Samples of four solvents (perchloroethylene and three secondary cleaning compounds) used in the commercial dry cleaning establishment were obtained and the reference gas generator was used to collect spectra from 900 to 4000 cm^{-1} for these solvents and methane, for use in a reference library. The commercial dry cleaning establishment was monitored continuously from 4:00 pm on the first day until midnight of the following day. The spectral windows for analysis of the

data collected in the commercial dry cleaning establishment were 3120–2660 cm^{-1} and 1100–900 cm^{-1}. The species detected were perchloroethylene, three secondary cleaning compounds (Pyrotex, Moth Proofing, and "Exit" Deodorizer), and methane. The perchloroethylene concentration exhibited its four highest peaks in concentration during busy working hours between 7:00 am and 11:00 am. The secondary cleaning compounds and methane also had peaks in concentration during this time period.

In both the investigations reported here as well as in others done by our group [3], trends in contaminant concentrations may be obtained and related to activities taking place at the facility.

References

[1] T. L. Marshall, C. T. Chaffin, R. M. Hammaker, W. G. Fateley, *Environ. Sci. Technol.* **1994**, *28*, 224A.

[2] W. G. Fateley, R. M. Hammaker, M. D. Tucker, M. R. Witkowski, C. T. Chaffin, T. L. Marshall, M. Davis, M. J. Thomas, J. Arello, J. L. Hudson, B. J. Fairless, *J. Mol. Struct.* **1995**, *347*, 153.

[3] C. T. Chaffin, T. L. Marshall, P. T. Jaakkola, J. K. Kauppinen, W. G. Fateley, R. M. Hammaker, in: *Proceedings of Optical Sensing for Environmental and Process Monitoring* (O. A. Simpson, ed.), A&WMA Vol. VIP-37 (SPIE Vol. 2365), 1995, pp. 140–150.

Mikrochim. Acta [Suppl.] 14, 565–566 (1997)
© Springer-Verlag 1997

Determination of Hydrocarbons in the Gases from Vehicle Exhausts by FT-IR Spectroscopy

Czesława Paluszkiewicz[1,*], **Mirosław Handke**[2], and **Jerzy Mąka**[2]

[1] Regional Laboratory, Jagiellonian University, 30–060 Kraków, ul. Ingardena 3, Poland
[2] Department of Materials Science and Ceramics, University of Mining and Metallurgy, 30–059 Kraków, Al Mickiewicza 30, Poland

Abstract. The infrared technique for measuring trace gases in the air has been evolving for many years, with rapid progress in the last ten years. The aim of our work was to determine the components of the emissions of vehicles. The gases from vehicle exhaust contains not only CO, CO_2, NO, HNO_3, NH_3, but also different hydrocarbons. Some authors suggest that ethene (C_2H_4) constitutes an important risk factor causing human cancer in urban areas. The gas samples have been taken from the exhausts of different cars. The FT-IR spectra of these gas samples were recorded in the range 4000–500 cm^{-1}, with a variable path-length cell. The data collected suggest that FT-IR spectroscopy is a valuable tool for detection of air pollutants evolved from the exhausts of cars.

Key words: FTIR, air pollution.

Toxic volatile organic air pollution has become a source of major health concern. Chemical spills, industrial sites and car exhausts are usual sources of volatile organic compounds (VOCs). Persson and Almen reported in 1990 that the cancer risk during one year of exposure to one ppb of ethene was expected to be about 3 cases in a population of one million [1]. Analysis of the emissions from vehicles is therefore very important. Conventional methods of analysing for hydrocarbons in automotive exhausts are usually indirect techniques in which the gases are preconcentrated in a cryotrap and then subjected to gas chromatographic analysis. This paper describes direct determination of hydrocarbons in the gases from different car exhausts.

Experimental

The gas samples (600 ml) were collected in busy streets at a distance of 10 cm from the exhaust pipes of different cars by a membrane pump into an analytical gas sample bag produced by NCP Analytical Instruments Incorporation. A 500 ml volume of the gas was taken from the sample bag and placed in a Harrick variable path-length cell, which was mounted in the spectrometer. The path-length used was 2 m. In the measurements KRS-5 windows, 25 mm in diameter were used in the gas cell.

FT-IR spectra were recorded in the range 4000–500 cm^{-1} on a Bio-Rad FTS-60v spectrometer with 1 cm^{-1} resolution. The spectra were identified by band assignments and confirmed with reference standards. Absorbance bands of water vapour and/or carbon dioxide in the ambient air were eliminated by spectra subtraction in the cases when they complicated the spectral data in some wavelength regions.

Results and Discussions

Examples of the spectra of the gas taken from the exhaust pipes of different cars are shown in Figs. 1–3. In the spectra the following analytical IR frequencies can be distinguished: methane (CH_4) at 3018 and 1305 cm^{-1} (F_2); ethyne ($CH{\equiv}CH$) at 3287 cm^{-1} (A_u) and 729 cm^{-1} (E_u); ethene ($CH_2{=}CH_2$) at 2988 cm^{-1} (B_{3u}), 1445 cm^{-1} (B_{3u}), 949 cm^{-1} (B_{1u}); propene ($CH_3CH{=}CH_2$) at 2954, 2932, 1443, 912 cm^{-1} (A"); propane ($CH_3CH_2CH_3$) at 2968, 2887, 1464, 1376 cm^{-1} (A_1,B_1) [2]. We can also notice the bands which are due to traces of H_2O, CO_2, CO and of aromatic hydrocarbons (aryl C—H out-of-plane bending), such as benzene at 673 cm^{-1} and ethylbenzene at 697 cm^{-1} [3, 4].

In Fig. 1 is shown a typical spectrum of exhaust gas sample from a relatively new car produced in 1994, powered by Eurosuper 95 gasoline, without catalytic converter. Figure 2 compares the spectra of gas samples from two cars produced in 1991 (gasoline powered), but one of them having a catalytic converter (Fig. 2b). We can see a significant decrease in the intensity of the bands due to CO and hydrocarbon molecules in the gas from the exhaust of the car equip-

---

* To whom correspondence should be addressed

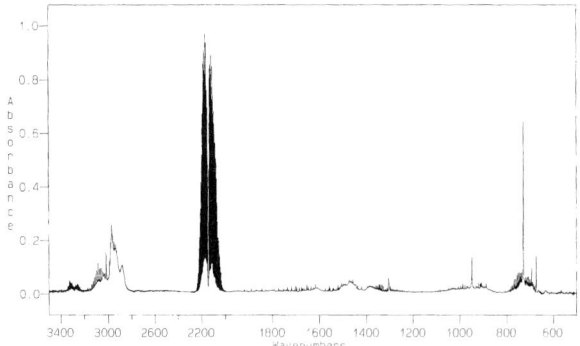

Fig. 1. FT-IR spectrum of exhaust gas sample from car produced in 1994

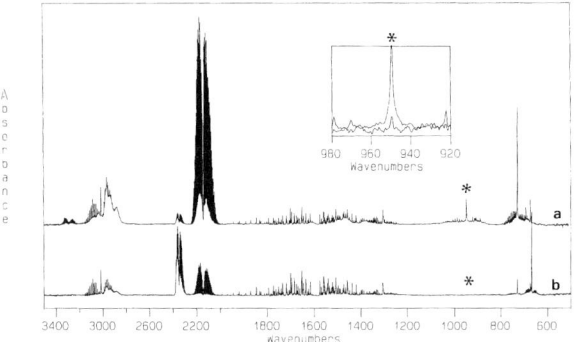

Fig. 2. FT-IR spectra of exhaust gas samples from cars produced in 1991, *a* without catalytic converter, *b* with catalytic converter. Inset: spectrum of "ideal" car exhaust

ped with a catalyst. However, looking very closely we can find in the spectra a very small amount of ethene (950 cm^{-1}) in the gas sample from an "ideal" car (Fig. 2 inset).

The spectra of the exhaust gases from a 1992 model of vehicle equipped with the catalyst are shown in Fig. 3. The concentrations of different pollutant species were simultaneously measured at intervals corresponding to two stages of fuel combustion, before and after adjustment of the car engine. After regulation

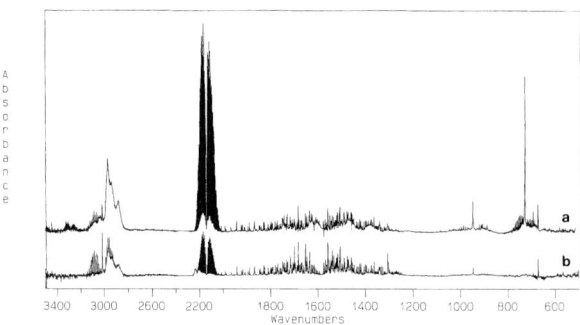

Fig. 3. FT-IR spectra of exhaust gas samples from car produced in 1992, equipped with the catalyst, *a* before engine repair, *b* after engine repair

(Fig. 3b) the analytical band due to ethyne (729 cm^{-1}) disappeared, and the intensity of bands corresponding to CO (2145 cm^{-1}) decreased. Unfortunately the bands due to ethene (at 949 cm^{-1}) and to benzene (at 673 cm^{-1}) were still present in the spectra, though at lower intensity.

The exhaust gases monitored by FT-IR spectroscopy displayed strong differences in concentrations of pollutants, including CO, CO_2 and individual hydrocarbons. These variations depended on the type of car, its engine state, and whether it was equipped or not with the catalyst converter.

The results obtained suggest that FT-IR spectroscopy is an useful analytical method for the measurement of hydrocarbons (HC) in exhaust gases (as anthropogenic HC-sources).

References

[1] K. A. Persson, J. Almen, *Swedish Env. Prot. Agency Report* **1990**, *3820*, 10.
[2] P. L. Hanst, *Fresenius Z. Anal. Chem.* **1986**, *324*, 579.
[3] L. M. Sverdlov, M. A. Kovner, E. P. Krainov, *Vibrational Spectra of Polyatomic Molecules*, Halsred, New York, 1973.
[4] J. W. Diehl, J. W. Finkbeiner, F. P. DiSanzo, *Anal. Chem.* **1993**, *65*, 2493.

Mikrochim. Acta [Suppl.] 14, 567–569 (1997)

On-line Measurements of Exhaust Gas with an FT-IR-Spectrometer

Kai Wülbern* and **Roman Windpassinger**

Lehrstuhl für Elektrische Meßtechnik, Technische Universität München, Arcisstrasse 21, D-80290 München, Federal Republic of Germany

Abstract. Gas analysis in the infrared range of radiation is quite difficult, if water vapour is present in the gas to be analysed. In this case, the spectral bands of interest are often overlapped or overlaid by the strong water absorption bands. In exhaust gas analysis, this problem gets worse, because here water vapour always occurs in large amounts. Therefore, the exhaust gas is normally diluted or dried before being analysed. It is known, however, that this conditioning can change the composition of the exhaust gas, thus causing errors. Recent investigations at the Technische Universität München have shown, that it is possible to analyse exhaust gas with an FT-IR spectrometer without any conditioning, the interference by water vapour being compensated by suitable computational methods. This paper gives an overview of the measuring system, the evaluation method and the measurements made.

Key words: exhaust gas analysis, FT-IR-spectrometry, on-line measurement.

Extractive Measurements

Measuring System

Based on a Bruker IFS66 FT-IR-spectrometer an experimental measuring system for extractive measurements has been built. The gas to be analysed is continuously extracted from the exhaust pipe and pumped through a White cell, which is installed in the sample compartment of the spectrometer (path-length 8 m). To prevent condensation of water vapour, the White cell and all gas-transmission parts of the system are heated to $130\,°C$. A gas filter with a mesh size of $0.1\,\mu m$ is used to protect the cell from dust particles. The spectrometer has a spectral resolution of $0.1\,cm^{-1}$. For the infrared source a Globar is used. With regard to the requirements in industrial sites, the spectrometer has been equipped with a pyroelectric DLATGS detector.

Evaluation Method

The strong absorption bands of water vapour and CO_2 appear in many regions of the exhaust gas spectra. Since in these regions the measured absorbance reaches high values, Beer's law is not obeyed. To allow for non-linearity in the relation between component concentration c and absorbance a, the NLS (non-linear least squares) method has been developed. In contrast to other methods known, NLS uses a non-linear model function: $a = k_0 + k_1 c + k_2 c^2$.

In the *calibration step*, a linear least-squares calculation is performed to predict the coefficients k_i of the model function by using reference spectra of known composition. The determination of the unknown concentration c in the *prediction step* leads to a non-linear minimization problem, which is solved with the iterative Levenberg–Marquardt procedure. The performance of NLS can be compared to that of the partial least squares method but the explict model makes NLS less sensitive to spectral variations. Besides that, the implementation of a reliable procedure for baseline corrections is much easier.

Laboratory Measurements

The investigations show that in practice cross-sensitivities to H_2O and CO_2 cannot be fully eliminated. However, with a sufficient spectral resolution, a careful choice of the spectral values included in the computational analysis, and a suitable composition of the set of reference spectra, it is possible to reduce these cross-sensitivities to a level which cannot be reached with standard gas analysers. For the measuring system described a resolution of $0.25\,cm^{-1}$ produced the best results. The measuring accuracy for several gaseous

---

* To whom correspondence should be addressed

pollutants (CO, NO, NO_2, N_2O, SO_2, NH_3) was determined for an integration time of 1 min. The detection limit was less than 1 ppm for all substances. Linearity error and cross-sensitivity were less than 2 ppm. The thermal drift of the measuring system was about $\pm 0.4\%$ in 24 hours.

Measurements in a Power Plant

On-line flue gas measurements have been made at a coalfired heating power plant. The flue gas components NO, SO_2, NO_2, CO, NH_3 and N_2O (concentrations below 100 ppm) were recorded simultaneously in the presence of about 12%v/v CO_2 and 10%v/v H_2O. Comparative measurements with standard gas analysers produced results consistent with those of the on-line measurements. The measuring time could be reduced to 6 seconds, i.e. the time needed to measure one interferogram (no co-addition) [1].

Gas Turbine Measurements

The gaseous emissions from a kerosene-fuelled helicopter gas turbine have been measured on-line as a function of the load moment. The measurements were performed at a gas turbine test-bench at the Technische Universität München. Besides the already mentioned substances, some hydrocarbons which have been found in the exhaust gas (methane, ethylene, formaldehyde, n-octane) were included in the analysis [2].

Conclusions

The results show that an FT-IR-spectrometer can be used to perform exhaust gas measurements without gas conditioning. With a new spectrum evaluation method outstanding accuracies can be reached even with a low sensitivity DLATGS detector. Thanks to its capability of measuring substances simultaneously, one FT-IR-spectrometer can replace several standard gas analysers. Since rugged spectrometers suitable for measurements in industrial sites are already available, it can be expected that FT-IR-spectrometers will play an important part in future industrial gas analysis.

Cross-stack Measurements

Extractive measurements of gas generally have the drawback that because of chemical transformation processes and sorptive effects the composition of the gas changes on its way from the extraction point to the gas cell of the spectrometer where it is analysed. Therefore, the only way to be sure to determine the true composition of the gas is to measure it over an open path.

To investigate the properties of open-path transmission measurements based on the common method of division by a background spectrum, on-line measurements across the smoke stack of a coal-fired power plant (see above) have been made.

Set-up and Equipment

Two openings of about 15×20 cm, at opposite sides of the stack wall, normally covered with a lid, provided an open path at the full stack diameter of 6.40 m. The permanent under-pressure of about 2 mbar in the stack allowed working with uncovered openings.

A Bruker IFS-66 spectrometer equipped with a receiving telescope (aperture 150 mm, f/4) was installed at one opening. At the other end of the optical path, a globar mounted at a second identical telescope provided a collimated IR-source.

Two detector were used. Initially, a DTGS-detector was intended to be used exclusively, but it turned out that because of its 1/f-behaviour, the detector was very susceptible to low-frequency stochastic disturbances of the measurement signal which were obviously introduced by variations of the refractive index of the flue gas. Therefore an N_2-cooled MCT-detector was also used.

Measurements and Evaluation

Series of measurements covering up to 8 hours were made. The number of co-additions for each measurement ranged from 10 to 60, the corresponding integration times ranging from 8 seconds to about 4 minutes, depending on the detector used. Because of the disturbances mentioned above, the cross-stack spectra show an extremely low signal-to-noise ratio. Before and after every measurement series, background spectra were taken across an open path in ambient air outside the stack. Spectra with the same configurations but with the IR-source removed were also taken to account for the respective background radiations.

The spectra are evaluated with the non-linear algorithm described in the first part of this paper. Reference spectra for calibration are generated by means of a gas cell (optical path-length 6.40 m) placed in the sample compartment of the spectrometer. Because of the non-linearity of the MCT-detector the corresponding spectra show strong distortions and hence have to be corrected by software before evaluation.

Results

Because of the high concentrations of about 8–12% v/v H_2O and CO_2 in the flue gas, evaluation for SO_2 and NO is difficult. The difficulties are aggravated by the stochastic disturbances, and the use of different measurement configurations for the sample measurements and the measurements of the reference spectra which results in qualitative differences between the spectra, particularly in spectral areas of high absorption. In contrast, CO shows comparatively little interference with H_2O and CO_2 and hence does not pose bigger problems. The evaluations have not yet been completed.

References

[1] K. Wülbern, *VGB Kraftwerkstechnik* **1992**, *72(11)*, 891.
[2] K. Wülbern, *Prozeßgekoppelte Messung von Rauchgasen mit einem Fourier-Spektrometer*, Fortschrittsberichte VDI, Reihe 15: Umwelttechnik, Nr. 154.

Mikrochim. Acta [Suppl.] 14, 571–574 (1997)

Optical Design and Performance of the Planetary Fourier Spectrometer (PFS)

Helmut Hirsch

Institute for Planetary Exploration, German Aerospace Research Establishment (DLR), Rudower Chaussee 5, D-12489 Berlin, Federal Republic of Germany

Abstract. The Planetary Fourier Spectrometer (PFS) is a two-channel Michelson interferometer operating in the IR from 1.25 to 45 μm. Two retroreflectors are mounted on brackets that are at an angle of 90° and fixed on an axle driven by a torque motor. The angular movement of the retroreflectors relative to the beam-splitter generates the optical path difference. The two primary degrees and the one secondary degree of freedom for misalignment are derived. The pattern of interference fringes, typical for the different types of misalignment, will be simulated, and their use for alignment will be demonstrated. This interferometer design is very robust towards slight misalignment in harsh environments, compared to the classical Michelson-type interferometer. This favours the design for applications in space. The PFS is intended for the mission Mars '96. It is optimised for the study of the Martian atmosphere. The spectra will also be used for investigations of the Martian soil.

Key words: infrared, Fourier-transform spectrometer, optical design, Mars.

The angular movement of the reflector system generates the optical path difference in the PFS [1–4] (Fig. 1). The interference fringes move typically in a slightly misaligned interferometer when the optical path difference changes, whereas in a well-adjusted interferometer, the centre of the ring system remains stationary. The monitoring of this movement yields a powerful tool for the alignment. The analysis of the parameters influencing the misalignment also shows the required degrees of adjustment. This misalignment characteristic is valid in general for all Michelson-type and related interferometers, irrespective of how the optical path difference is generated.

Misalignment of the Interferometer

The path difference Δ of a ray passing through the interferometer at an angle of incidence γ is displayed in Fig. 2. This path difference is generated by a lateral displacement q of one retroreflector and causes a misalignment:

$$\Delta = 2q \sin \gamma \qquad (1)$$

In a perfectly aligned interferometer, the retroreflectors are laterally equally displaced along the y-axes: $q_1 = q_2 = \delta/2$ (Fig. 3). The net displacement is compensated:

$$q_1 = R(1 - \cos \varphi); \quad q_2 = -R(1 - \cos \varphi);$$
$$q = q_1 + q_2 = 0 \qquad (2)$$

The corner points of the retroreflectors move along the same circular trajectory, and the partial rays move in parallel, when the reflector system is rotated by φ.

The distances between the corner point and the element of fixation may be different in real retroreflectors, owing to mechanical tolerances. Figure 4 gives an example where the right retroreflector is displaced axially by ε. This generates an uncompensated lateral displacement:

$$q = \varepsilon \sin \varphi \qquad (3)$$

There is antiparallel movement of the partial rays. This effect is of second order, and has to be aligned out when the loss of signal ΔS exceeds a certain tolerance limit:

$$\Delta S = 1 - m; \; m = 2 \left| \frac{J_1(\pi \rho)}{\pi \rho} \right|; \; \rho = 4\sigma_M \, \varepsilon \sin \varphi_M \sin \gamma_M \qquad (4)$$

The modulation factor [1,2,5,6] is indicated by m, the Bessel function of first order by J_1. The maximum wavenumber in the range of measurement is given by σ_M. The maximum angle of rotation corresponding to the spectral resolution is φ_M, and γ_M is the maximum angle of incidence determining the field of view (FOV).

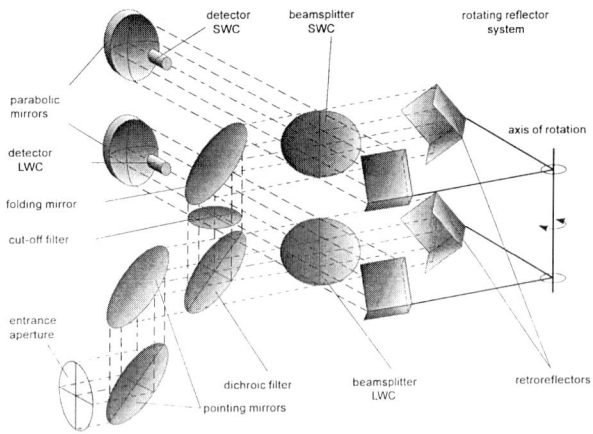

Fig. 1. The optical design of the PFS

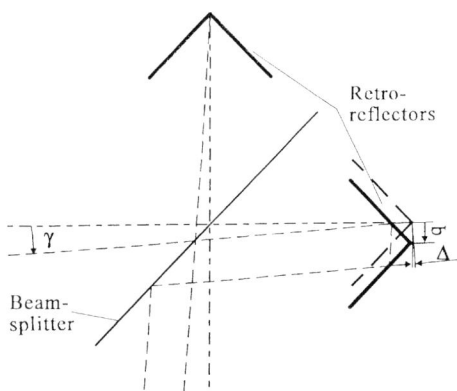

Fig. 2. The path difference Δ of a ray passing through the interferometer with an angle of incidence γ, due to the lateral displacement q of one retroreflector

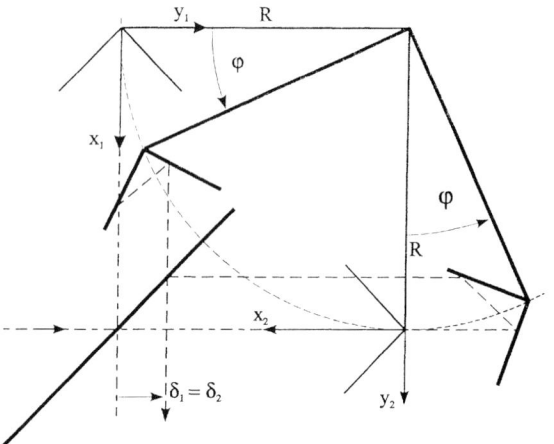

Fig. 3. The ideally aligned interferometer. The arm length is indicated by R, the angle of rotation by φ, the displacement of the partial beams by δ_1, δ_2, the co-ordinates of the corner points in the position of ZOPD by x_1, y_1 and x_2, y_2

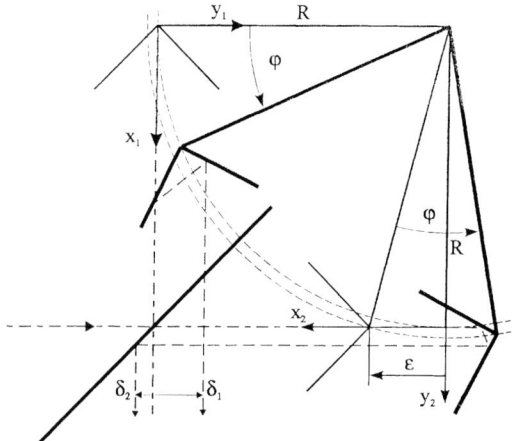

Fig. 4. The interferometer misaligned by an axial displacement of one retroreflector

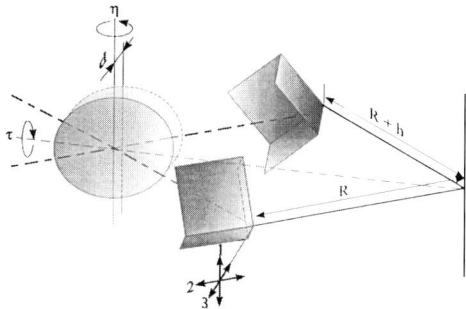

Fig. 5. The degrees of freedom of optical elements that generate a misalignment. The difference of the arm length is indicated by h and η, τ and d mean the degrees of rotation and translation of the beam-splitter. The numbered arrows denote the degrees of adjustment needed to compensate the misalignment

Figure 5 shows all the degrees of rotation and translation of the optical elements that generate a misalignment. They are related according to Eq. (5), provided that h/R, d/R, η and τ are all $\ll 1$:

$$q = \sqrt{q_i^2 + q_p^2};\ q_i = h\cos\varphi + 2R\eta + \sqrt{2}\,d;\ q_p = \sqrt{2}\,R\tau \tag{5}$$

Here, h refers to the difference in the arm length R, η and τ refer to the two angles of rotation, and d is the displacement of the beam-splitter. The numbered arrows indicate the needed degrees of freedom of alignment. The lateral displacements q_i and q_p have to be compensated by "1" and "2" in fine alignment, and "3" compensates the axial displacement ε.

Robustness of the PFS

Assuming the same loss of signal, the PFS is thirty times more robust against misalignment than is a classical Michelson-type interferometer with flat mirrors, for the same dimensions of x_M, γ_M, and σ_M. A 9% loss of signal results from $q = 5\,\mu m$ in the PFS and $0.15\,\mu m$ in a classical Michelson interferometer, for a maximum path difference $x_M = 0.5\,cm$, $\gamma_M = 17\,mrad$ and $\sigma_M = 8000\,cm^{-1}$, corresponding to Eqs. (4, 5) and Fig. 5.

The Interference Fringes

The interference fringes can be observed as the intensity distribution in the focal plane of the parabolic mirror (Fig. 6):

$$I(\sigma, x) = 1 + \cos(2\pi\sigma x);\ x = x_a + \Delta;\ x_a = 4R\sin\varphi \quad (6)$$

Here, x means the total optical path difference as the sum of the axial path difference x_a and the additional path difference Δ given in Eq. (1). A beam passing through the interferometer with an angle of incidence γ and a polar angle θ will be imaged into a point of the focal plane $(r, \theta) = (f\tan\gamma, \theta)$, where f is the focal length of the parabolic mirror. The axial path difference x_a generates the well-known coaxial ring system. The lateral displacement q_i and axial displacement ε contribute to a sagittal component Δ_s, while q_p introduces a tangential component Δ_t:

$$I = 1 + \cos(x_a - \Delta_s - \Delta_t);\ x_a = 4R\sin\varphi\cos\left(\frac{r}{f}\right);$$

$$\Delta_s = 2(q_i - \varepsilon\sin\varphi)\frac{y}{f};\ \Delta_t = 2q_p\frac{z}{f};\ r = \sqrt{y^2 + z^2} \quad (7)$$

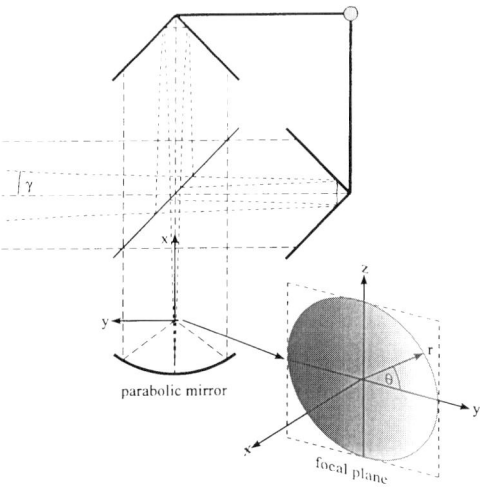

Fig. 6. The co-ordinate system in the focal plane of the parabolic mirror

The effect of q_i and q_p can be observed in the position of the zero optical path difference (ZOPD) with $\varphi = 0$, like a stripe system with the number of stripes ρ and their angle of inclination θ given by

$$\rho = 4\sigma q\gamma_M,\ \theta = \arctan\left(\frac{q_i}{q_p}\right), \quad (8)$$

provided that q and ε are $\ll R$ and φ_M and $\gamma_M \ll 1$.

All these effects together generate a stripe system, transforming into a ring system with an increasing number of rings and a retrograde moving centre when the angle of rotation is increased. Figure 7 shows

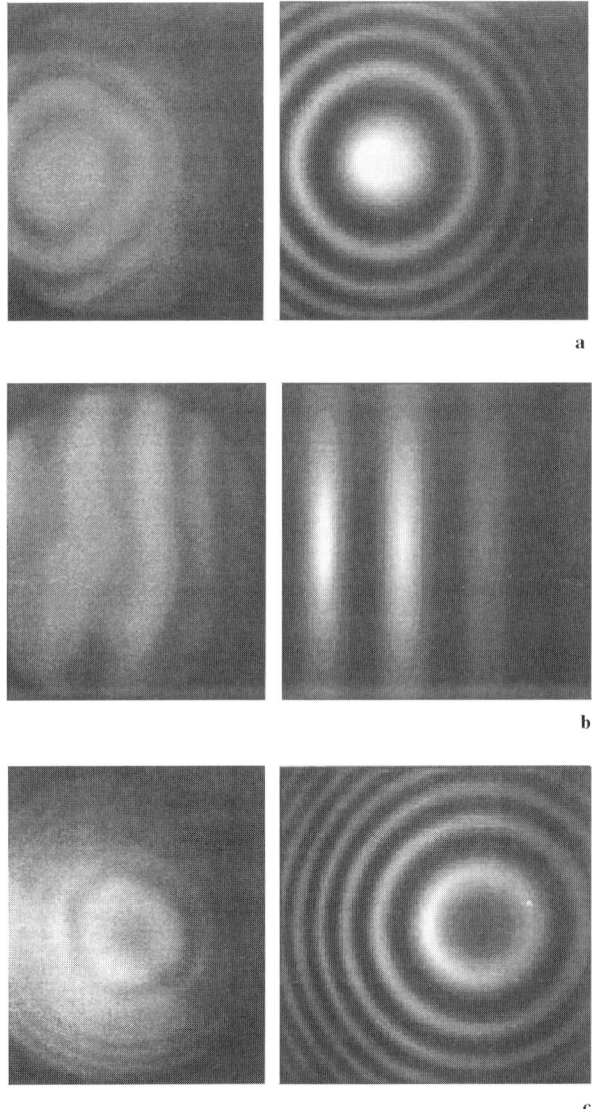

Fig. 7. Photographs and simulated images for angles of rotation **a** $\varphi < 0$, **b** $\varphi \approx 0$, and **c** $\varphi > 0$. An uncompensated sagittae displacement $q_i > 0$ is taken into account in the simulation

photographs of fringes together with images simulated according to Eq. (7) for the cases (a) $\varphi < 0$, (b) $\varphi \approx 0$, and (c) $\varphi > 0$. The intensity of the laser source is simulated by a Gaussian distribution.

Conclusions

The analysis of misalignment allows determination of all the contributing parameters as well as the needed degrees of freedom for adjustment. A comparison of an interferometer like the PFS with a classical Michelson-type interferometer demonstrates the higher robustness of the PFS towards misalignment. The two types of interferometer differ in the use of retroreflectors in full aperture instead of flat mirrors.

Monitoring of the moving fringes provides a sensitive tool for adjustment.

References

[1] H. Hirsch, A. Adriani, F. Angrilli, F. Capaccioni, S. Fonti, V. Formisano, A. Matteuzzi, G. Michel, V. Moroz, *Proc. SPIE* **1992**, *1780/2*, 677.

[2] H. Hirsch, G. Arnold, V. Formisano, V. Moroz, G. Piccioni, *Proc. SPIE* **1994**, *2268*, 331.

[3] G. Arnold, H. Hirsch, V. Formisano, V. Moroz, *45th Intern. Astronautical Congress*, Jerusalem, Israel, 9–14 October, 1994, IAF-94-Q.3.334.

[4] H. Hirsch, V. Formisano, V. I. Moroz, G. Arnold, A. Jurewicz, G. Michel, J. J. Lopez-Moreno, G. Piccioni, N. Cafaro, *Planet. Space Sci.* in press.

[5] W. H. Steel, *Interferometry*, Cambridge University Press, Cambridge, 1983, p. 185.

[6] H. Hirsch, G. Arnold, *Vib. Spectrosc.* **1993**, *5*, 119.

Mikrochim. Acta [Suppl.] 14, 575–577 (1997)
© Springer-Verlag 1997

Absolute FT-IR Line Positions with a Relative Error of 10^{-10}

Georg Ch. Mellau*, Brenda P. Winnewisser, and **Manfred Winnewisser**

Physikalisch-Chemisches Institut, Justus-Liebig-Universität, Heinrich-Buff-Ring 58, D-35392 Gießen, Federal Republic of Germany

Abstract. With very accurate calibration standards the accuracy of HR-FT spectra can be as high as 10^{-10}. We present the program AUTOCALIB, designed for the automatic calibration of spectra. Using this program we found some problems related to very accurate spectra. Finally we present a CO measurement with a relative error of 4×10^{-10}.

Key words: FTS, high resolution, peak-finder, line positions, HR-FT measurement accuracy.

In high-resolution Fourier-transform (HR-FT) spectroscopy, measurements in the mid-IR with linewidths in the range of $0.005\,\mathrm{cm}^{-1}$ with an S/N of 5000 can be achieved. Theoretically, from such spectra line positions should be obtainable with an accuracy of $\sim 0.005/5000 = 0.000001\,\mathrm{cm}^{-1}$. To achieve this limit a very accurate calibration and peak-finding procedure is needed.

Until now it was impossible to give absolute wavenumber accuracy as good as the precision for HR-FT spectra. The accuracy of the line positions used to calibrate the spectra was worse or at best the same as that of the line positions obtained from the sample spectra.

In 1995 several papers appeared with line positions from saturated absorption heterodyne measurements [1–4]. We have realized the extreme importance of these standards for the field of high-resolution spectroscopy: using these line positions allows accurate calibrations to be made, but the table of the differences between the measured (and calibrated) and the standard peak positions can also be used to find instrumental errors such as misalignment of the instrument, phase errors, peak-finder errors, or errors of the software with which the interferograms are transformed.

Automatic Calibration Program AUTOCALIB

For quick and accurate calibration we have developed the program AUTOCALIB which carries out an automatic calibration of the spectra by using standards from a database and makes a plot of what may be called the individual calibration factor for each line evaluated for the calibration. This plot is a superb diagnostic tool, which reveals systematic problems. With this program the steps of the calibration procedure are as follows.

1. Measure the calibration molecule spectrum along with the sample (internal calibration) or just after the sample measurement (external calibration). The most accurate and precise method is measuring the calibration molecule spectrum with the sample spectrum in a separate cell.
2. Create a peak-list for the calibration molecule lines with the same peak-finder that is to be used for the sample spectrum.
3. Start the program by typing: AUTOCALIB [filename] [calibration molecule] and the program will
   (a) look in the database for the standard peak-list for the calibration molecule used, (b) search the peak-list for the lines coinciding with those of the calibration gas, (c) calculate the calibration factor and the deviations for the selected lines.
4. Eliminate interactively any lines with large deviations. The program recalculates after each loop the overall calibration factor and searches for the peak with the greatest error.
5. Verify from the error plot whether there are regions with non-linearities or systematic peak-finder effects.

---

* To whom correspondence should be addressed

Previously, we chose 10–15 lines from the spectrum, determined the peak positions, and typed the values together with the standard peak positions into a small program which then calculated the calibration factor and the standard deviation. Using AUTOCALIB we can make the calibration with many lines (up to 1000) and so we get a statistically better calibration factor in only a few minutes.

Manual calibration allows selection of some "good" lines from the spectra: well resolved, not too strong and not too weak lines. The automatic calibration procedure should search also for "good" lines but this is in general not necessary. Lines that are not perfect in the sense discussed above have line positions shifted from the true line positions. This makes possible a very elegant line selection procedure. Take all the standard lines that can be found in the spectrum, and calculate the calibration factor. In this list there may be many "bad" lines shifting the calibration factor somehow from the true position. The line having the individual calibration factor that lies the farthest from the calibration factor is the "worst" line from the initial peak-list. This line can be thrown away the calibration factor recalculated, and the new "worst" line looked for. Repeating the procedure will shift the calibration factor towards its true value, and after all the "bad" lines

have been thrown away the calibration factor begins to be independent of the number of peaks used for calibration (the individual calibration factors of each line should be normally distributed about the calibration factor of the measurement). The only problem is that the number of "good" lines must be greater than that of the "bad" lines.

We have developed AUTOCALIB as an easy to use program. It accepts any form of ASCII tables as input files (standard and sample), provides for setting up a database with standard line positions, makes an interactive elimination of "bad" lines, and creates automatically an error plot of the individual calibration factors.

Problem Related to very Accurate HR-FT Measurements

Using the standards in references [1]–[4] can give a very good estimate of the absolute error for the sample measurement. Using AUTOCALIB we learned that

(a) the calibration factor depends somewhat on the peak-finder used (slight asymmetries are treated in different ways), (b) we have to use, if possible, standards from saturated absorption heterodyne measurements,

Table 1. Protocol of calibration fit of measured (and calibrated) lines $\tilde{v}_{meas}$ of CO to standard line position $\tilde{v}_{st}$ with a standard deviation $\sigma = 9 \times 10^{-7}\,\mathrm{cm}^{-1}$. The (1-0) band of $^{12}\mathrm{C}^{16}\mathrm{O}$ was measured with the Bruker IFS 120HR spectrometer on 11.11.1994

	$\tilde{v}_{st}$ cm^{-1}	Error (3σ) 10^{-8} cm^{-1}	$\tilde{v}_{meas} - \tilde{v}_{st}$ 10^{-8} cm^{-1}		$\tilde{v}_{st}$ cm^{-1}	Error (3σ) 10^{-8} cm^{-1}	$\tilde{v}_{meas} - \tilde{v}_{st}$ 10^{-8} cm^{-1}
P (20)	2059.914 677 50	583	40	R (01)	2150.856 008 19	193	80
P (19)	2064.396 920 01	423	82	R (02)	2154.595 583 28	180	204
P (18)	2068.846 945 30	300	128	R (03)	2158.299 711 86	163	44
P (17)	2073.264 606 97	206	97	R (04)	2161.968 247 11	146	79
P (16)	2077.649 758 55	140	216	R (05)	2165.601 042 23	126	125
P (15)	2082.002 253 57	93	20	R (06)	2169.197 950 53	106	43
P (14)	2086.321 945 36	70	70	R (07)	2172.758 825 20	86	57
P (13)	2090.608 687 36	60	150	R (08)	2176.283 519 68	73	106
P (12)	2094.862 332 82	56	16	R (09)	2179.771 887 22	66	52
P (11)	2099.082 735 06	63	115	R (10)	2183.223 781 36	63	4
P (10)	2103.269 747 23	73	60	R (11)	2186.639 055 47	63	101
P (09)	2107.423 222 56	86	2	R (12)	2190.017 563 18	86	48
P (08)	2111.543 014 16	103	127	R (13)	2193.359 158 02	96	94
P (07)	2115.628 975 22	123	65	R (14)	2196.663 693 62	140	0
P (06)	2119.680 958 78	143	11	R (15)	2199.931 023 74	210	6
P (05)	2123.698 817 96	163	147	R (16)	2203.161 002 13	210	44
P (04)	2127.682 405 87	180	14	R (17)	2206.353 482 61	300	116
P (03)	2131.631 575 53	193	35	R (18)	2209.508 319 11	426	167
P (02)	2135.546 180 05	203	50	R (19)	2212.625 365 64	583	127
P (01)	2139.426 072 48	200	53	R (20)	2215.704 476 22	1010	3
R (00)	2147.081 133 47	203	29				

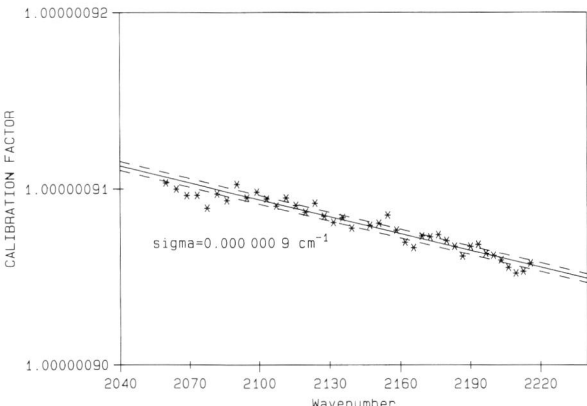

Fig. 1. Plot of the individual calibration factors for each line of CO listed in Table 1

with $d\tilde{v} \leqslant 10^{-6}\,\text{cm}^{-1}$, (c) spectra with line positions more accurate than $5 \times 10^{-6}\,\text{cm}^{-1}$ need a calibration of the type $a + bx$, (d) the best peak positions are produced by line-profile fitting procedures; other peak-finders can produce peak positions with only limited accuracy.

The first problem identified was related to the peak-finders. With the help of these standards it was for the first time possible to really test the precision of the peak-finding programs on real spectra. We tested the various programs used in our laboratory in the last few years to create peak-lists for the FT-spectra. These tests show that only by using line profile-fitting procedures, such as used in the program INTBAT [5], is it possible to achieve precision in the peak positions of $1 \times 10^{-6}\,\text{cm}^{-1}$. Because it is very time consuming, we do not use INTBAT as a peak-finder for spectra with thousands of lines.

CO Test Measurement

We tried to give an answer to a question that almost every guest in our laboratory put to us: how accurately can we measure with our Bruker IFS 120 HR? So we measured CO at $2100\,\text{cm}^{-1}$ for 4 days, obtaining a spectrum with very high S/N (7500), and could achieve an accuracy of the peak-positions of $9 \times 10^{-7}\,\text{cm}^{-1}$ over a wavenumber range of $200\,\text{cm}^{-1}$ –but only by using INTBAT as a peak-finder (see Table 1 and Fig. 1). Other peak-finders gave us 3–100 times poorer peak positions. With line positions obtained using the Bruker 2nd derivative peak-finder and standards from [1]–[4], we found a standard deviation of $5 \times 10^{-6}\,\text{cm}^{-1}$ for most of the spectra measured in the recent months in the range 1900–3500 cm^{-1}.

This works shows that with careful calibration and peak-finding line positions with a precision and accuracy of 10^{-10} can be achieved with a Fourier-transform spectrometer.

References

[1] T. George, W. Urban, A. Le Floch, *J. Mol. Spectrosc.* **1994**, *165*, 500.
[2] T. George, M. H. Wappelhorst, S. Saupe, M. Mürtz, W. Urban, A. G. Maki, *J. Mol. Spectrosc.* **1994**, *167*, 419.
[3] A. Dax, J. S. Wells, L. Hollberg, A. G. Maki, W. Urban, *J. Mol. Spectrosc.* **1994**, *168*, 416.
[4] J. S. Wells, A. Dax, L. Hollberg, A. G. Maki, *J. Mol. Spectrosc.* **1995**, *170*, 75.
[5] J. W. C. Johns, *Mikrochim. Acta* **1987**, *III*, 171.

Mikrochim. Acta [Suppl.] 14, 579–580 (1997)

High-Resolution FTS Study of Isotope Effects in LiLuF$_4$·Ho^{3+}

Marina N. Popova*, **Serguey A. Klimin,** and **Guerman N. Zhizhin**

Institute of Spectroscopy, Russian Academy of Sciences, 142092 Troitsk, Moscow Region, Russia

Abstract. Isotope structure in the spectra of the LiLuF$_4$·Ho^{3+} single crystal was found by high-resolution Fourier transform spectroscopy. This structure is shown to be connected with the isotopic disorder in the lithium sublattice.

Key words: isotopic structure, isotope effects.

Isotope effects are widely studied, both for analytical reasons and because they provide a good test of physical theories. Recently, we observed isotope shifts in the optical spectra of Ho^{3+} ions in an LiYF$_4$ crystal, associated with the isotope disorder in the lithium sublattice, and discussed physical mechanisms that might contribute to these shifts [1]. A quantitative theory was developed later; the calculated shifts were qualitatively consistent with the measured ones [2]. However, the theory in [2] did not take into account the difference of the masses of Y and Ho, and the direct comparison of the calculated and measured shifts was incorrect. LiLuF$_4$·Ho^{3+} suits the assumptions of the theory much better, so it was interesting to study this crystal experimentally. In this presentation, we report the first results of such a study.

Experimental

Absorption spectra of the LiLuF$_4$·Ho^{3+} (0.1 at.%) single crystal with the natural abundance of the ^{7}Li and ^{6}Li isotopes were measured at 5.0 K with a resolution of 0.005 cm^{-1} on a Bomem DA3.002 Fourier-transform spectrometer. The spectral regions of the $^5I_8 \rightarrow {}^5I_7$ and $^5I_8 \rightarrow {}^5I_6$ transitions near 5200 and 8600 cm^{-1} respectively were investigated.

Results and Discussion

Figure 1 presents an example of the spectrum. The hyperfine structure is clearly resolved.

In the regular ^{7}LiLuF$_4$ lattice Ho^{3+} ions substitute for Lu^{3+} ions at sites with point symmetry S$_4$. The energy spectrum of Ho^{3+} in a crystal field of S$_4$ symmetry consists of Γ_1 and Γ_2 singlets and Γ_{34} doublets. The ground state in the lower multiplet $^5I_8$ is a Γ_{34} doublet, and the nearest excited state (with energy 7 cm^{-1}) is a Γ_2 singlet.

Magnetodipole hyperfine interaction

$$A_J(\vec{J}\cdot\vec{I})$$

gives the main contribution to the energy of the interaction between the Ho^{3+} ion nucleus, with spin $I = 7/2$, and the electron shell. The diagonal matrix elements of the operator $\mathbf{J}$ are non-zero only for the doublet wave functions

$$|\Gamma_{34}^{\pm}\rangle = a_1^{\pm}|J, \pm 1\rangle + a_3^{\pm}|J, \pm 3\rangle + a_5^{\pm}|J, \pm 5\rangle + \dots$$

(where $|J, \pm M\rangle$ are the Ho^{3+} free-ion wave-functions defined by the electronic angular momentum J and its z-component M, $a_M^+ = (a_M^-)^*$, and $\langle +|J_z|+\rangle = -\langle -|J_z|-\rangle$), and in the first approximation the hyperfine sublevels of the doublet form a structure with equally spaced energies $\pm A_J\langle +|J_z|+\rangle m$, where $m = \pm 1/2, \dots, \pm 7/2$ are the z-components of the nuclear angular momentum.

In the "pure" form, eight hyperfine components of the optical transition are observed only in a crystal with homogeneous isotopic composition, e.g. in ^{7}LiYF$_4$ [1]. In the samples of mixed composition, containing isotopes ^{7}Li and ^{6}Li, the hyperfine components are split into individual narrow lines with half-width down to 0.015 cm^{-1} (see Fig. 2). This isotope structure of the hyperfine components has equal spacing in most cases; its period amounts to 0.01–0.04 cm^{-1} for LiLuF$_4$·Ho^{3+}.

---

* To whom correspondence should be addressed

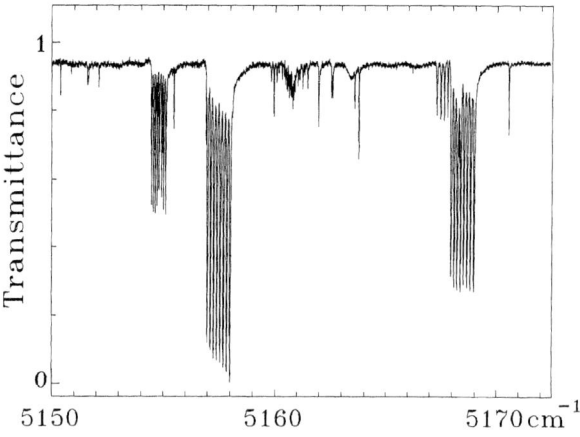

Fig. 1. The low-frequency part of the transition $^5I_8 \rightarrow ^5I_7$ of Ho^{3+} in LiLuF$_4$·Ho^{3+} (0.1 at.%) at 5.0 K

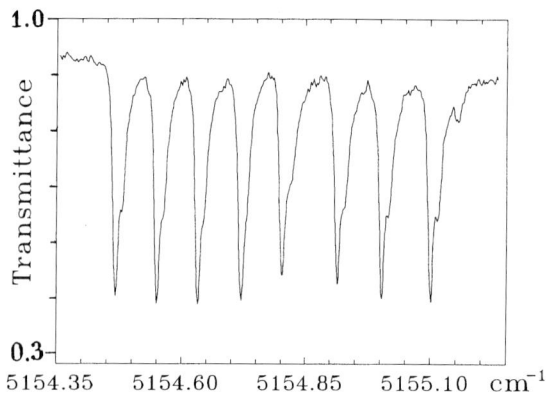

Fig. 2. The transition $\Gamma_2^{(1)}(^5I_8) \rightarrow \Gamma_{34}^{(1)}(I_7)$ in a ^{7}Li$_{0.926}$ ^{6}Li$_{0.074}$LuF$_4$·Ho^{3+} (0.1 at.%) crystal (with natural abundance of lithium isotopes)

We ascribe the strongest line of the isotope structure to the transition within Ho^{3+} ions having no ^{6}Li isotopes in their closest surroundings. In fact, the Ho^{7}Li$_8$F$_8$ molecular clusters have to be considered. The next line comes from the Ho^{7}Li$_7$^{6}LiF$_8$ clusters with one ^{6}Li isotope, and so on. Such an interpretation is supported by the relative intensities within the isotope structure, consistent with a binomial distribution.

What is the reason for the isotope shifts in the case of LiLuF$_4$·Ho^{3+}? We demonstrate the experimental evidence that the static crystal field for the Ho^{3+} ion is sensitive to lithium isotopes in the neighbourhood of this ion. We first note that the possible change in the crystal field for the Ho^{3+} ion on introduction of one foreign isotope into its surroundings, includes a lowering of the symmetry. A small low-symmetry perturbation lifts the degeneracy of the Γ_{34} electronic state and, as a result, its hyperfine components move apart. The

5155 cm^{-1} line in Fig. 2 demonstrates an experimentally observed example of such a behaviour: the middle interval between hyperfine components is larger for the cluster with one foreign isotope than for the cluster with uniform isotope composition.

Precise measurements of the observed isotope shifts will be published elsewhere.

Acknowledgements. This work was made possible in part by Grant No. 95-02-03796 from the Russian Foundation for Fundamental Research and by Grant No. JEJ100 from the International Science Foundation and Russian Government.

References

[1] N. I. Agladze, M. N. Popova, G. N. Zhizhin, V. J. Egorov, M. A. Petrova, *Phys. Rev. Lett.* **1991**, *66*, 477.

[2] N. I. Agladze, M. N. Popova, M. A. Koreiba, B. Z. Malkin, V. R. Pekurovskii, *Zh. Exp. Teor. Fiz.* **1993**, *104*, 4171 [*JETP* **1993**, *77*, 1021].

Mikrochim. Acta [Suppl.] 14, 581–583 (1997)

Electric-field-induced Reorientation of a Ferroelectric Liquid Crystal with a Tolane Ring, (S)-4-Methylhexyl-4-[4-(decyloxy) phenylethynyl]-2-fluorobenzoate Studied by Time-resolved FT-IR

Koshiro Taniike[1], Norihisa Katayama[2], Takashi Sato[1], Yukihiro Ozaki[1,*], Miroslaw A. Czarnecki[3], Masahiro Satoh[4], Tetsuya Watanabe[5], and Akio Yasuda[6]

[1] School of Science, Kwansei-Gakuin University, Nishinomiya 662, Japan
[2] School of Science, Kitasato University, Sagamihara 228, Japan
[3] Institute of Chemistry, University of Wroclaw, Wroclaw, Poland
[4] Advanced Materials Lab., Kansai Research Institute, Kyoto 600, Japan
[5] Sanyo Chemical Industries Ltd., Kyoto 605, Japan
[6] Sony Co. Research Center, Hodogaya-ku, Yokohama 240, Japan

Abstract. The electric-field-induced reorientation of (s)-4-methyl-hexyl-4-[4-(decyloxy)phenylethynyl]-2-fluorobenzoate in the chiral smectic C (Sc*) phase under various conditions have been studied by time-resolved FT-IR spectroscopy. The time-resolved spectra reveal that the reorientation occurs immediately after the electric field is applied. Of particular note is that the C=O stretching band, strong in the IR spectra, is very weak in the time resolved difference spectra.

Key words: liquid crystals, time-resolved spectroscopy, FT-IR.

Investigations of the detailed mechanism of electric-field-induced reorientation of liquid crystals (LCs) are very important in development of new display systems consisting of LCs [1]. Recently, there has been great progress in this field owing to the development of time-resolved infrared (IR) spectroscopy, which provides information about the motion of the whole LC molecule as well as the detailed time-course of the reorientation of particular fragments of the molecule [2–7].

We recently studied the electric-field-induced reorientation of (S)-4-methylhexyl4-[4-(decyloxy) phenylethynyl]-2-fluorobenzoate (MDOPEFB; see Fig. 1) in the Sc* phase by using asynchronous time-resolved FT-IR. The purpose of the present paper is to restudy the same system by using newly prepared LC cells. We have been able to obtain the spectra with

much higher signal-to-noise ratio than before, so more reliable results for the time-dependent spectral changes have been obtained.

Experimental

The sample of MDOPEFB was synthesized by the method reported in the literature [8]. We changed the method for preparing the LC cells. Previously, we employed BaF_2 windows covered with a thin layer of transparent conductive material (indium tin oxide) to construct the LC cells [5]. The windows, on which a poly(vinyl alcohol) film was deposited, were separated by poly(ethylene terephthalate) spacers (cell thickness 3.2 μm). However, since the cells were so thick that it was rather difficult to find a homogeneous well-oriented mono-domain. In the present study, we used CaF_2 windows with a thickness of 1.7 μm (the details for the preparing the cells will be reported soon [9]). By using the new cells we could easily get good domains with larger size.

The instrumentation and experimental conditions for measuring the time-resolved FT-IR spectra were the same as those described in our previous paper [5].

Results and Discussion

Figure 1a shows the raw time-resolved FT-IR spectra of MDOPEFB for delay times ranging from 0 to 500 μs and Fig. 1b exhibits the difference spectra, which represent the absorbance change from the spectrum at zero time delay (the orientation of the sample was induced by rectangular electric pulses of ±28 V and 1-kHz frequency; 27 °C). Note that the signal-to-noise ratio of the spectra in Fig. 1b is much higher than that of the

---

* To whom correspondence should be addressed

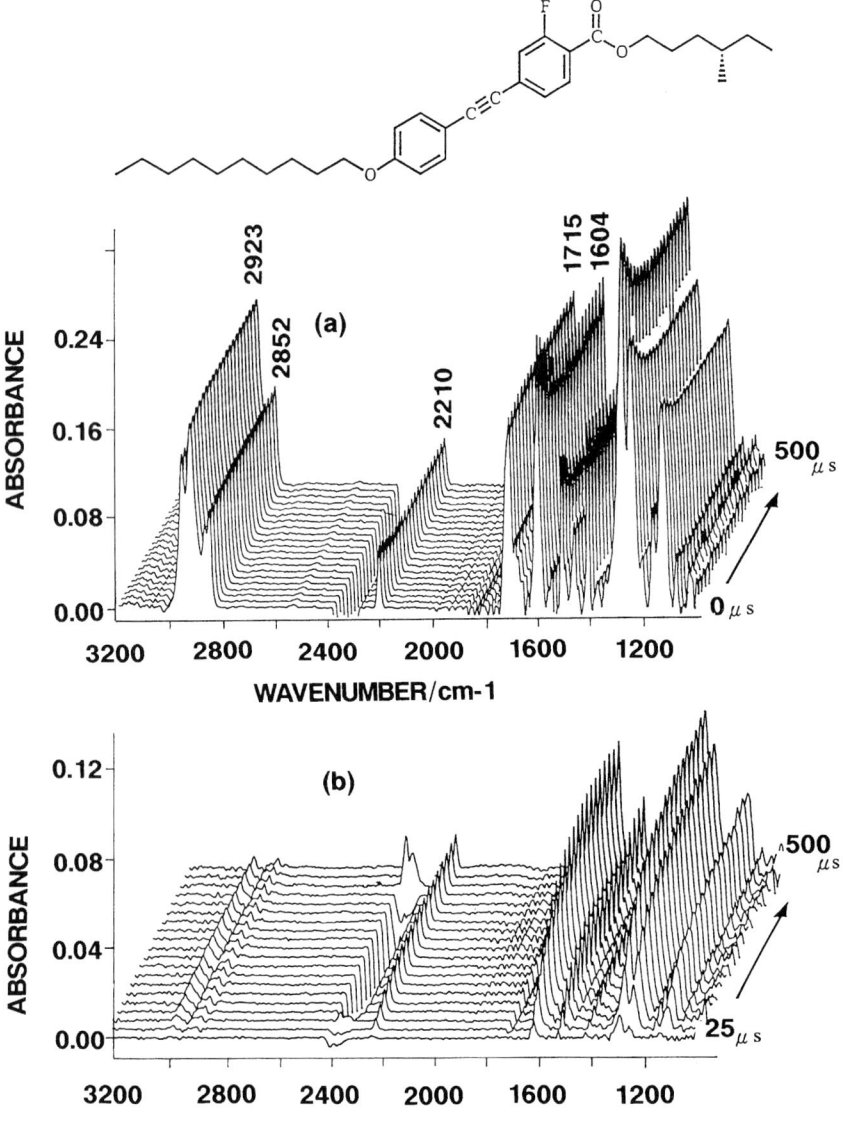

Fig. 1. Structure of MDOPEFB, and its time-resolved FT-IR spectra for times of delay ranging from 0 to 500 μs: **a** raw spectra; **b** difference spectra

corresponding spectra previously reported [5]. The band assignments for the 3200–1000 cm⁻¹ region of the spectra have been established before [5]. Briefly, the bands at 2923 and 2852 cm⁻¹ are due to CH_2 antisymmetric and symmetric stretching modes of the hydrocarbon chains, respectively, while those at 2210, 1715 and 1604 cm⁻¹ are assigned to the C≡C, C=O, and phenyl ring stretching modes, respectively.

The spectra in Fig. 1b reveal that the reorientation occurs immediately after the electric field is applied. Since the signal-to-noise ratio is much better than before we are able to monitor more precisely the movements of particular functional groups of the

MDOPEFB molecule by plotting the time-dependence of the intensity changes of bands arising from the functional groups. The time-dependences of the intensity changes of the bands at 2923, 2852, 2210, 1715 and 1604 cm⁻¹ suggest that the molecule reorients as a unit. The spectra obtained under the various conditions suggest that the temperature and applied voltage affect the tilt angle and angular velocity of reorientation of the liquid crystal, respectively.

Interestingly enough, the time-dependent intensity change of the carbonyl band at 1715 cm⁻¹ is very small. The reason for this is now under consideration and will be reported in more detail soon [9].

References

[1] J. W. Goodby, R. Blinc, N. A. Clark, S. T. Lagerwall, M. A. Osipov, S. A. Pikin, T. Sakurai, K. Yoshino, B. Zeks, *Ferroelectric Crystals, Principles, Properties and Applications*, Gordon and Breach, Philadelphia, 1991.

[2] H. Toriumi, H. Sugisawa, H. Watanabe, *Jpn. J. Appl. Phys.* **1988**, *27*, L935.

[3] T. I. Urano, H. Hamaguchi, *Chem. Phys. Lett.* **1992**, *195*, 287.

[4] K. Masutani, H. Sugisawa, A. Yokota, Y. Furukawa, M. Tasumi, *Appl. Spectrosc.* **1992**, *46*, 560.

[5] M. A. Czarnecki, N. Katayama, Y. Ozaki, M. Satoh, K. Yoshio, T. Watanabe, T. Yanagi, *Appl. Spectrosc.* **1993**, *47*, 1382.

[6] N. Katayama, T. Sato, Y. Ozaki, K. Murashiro, M. Kikuchi, S. Saito, D. Demus, T. Yuzawa, H. Hamaguchi, *Appl. Spectrosc.* **1995**, *49*, 977.

[7] M. A. Czarnecki, N. Katayama, M. Satoh, T. Watanabe, Y. Ozaki, *J. Phys. Chem.* **1995**, *99*, 14101.

[8] *Japanese Patent Applications*, Nos. 221 and 351, 1989.

[9] K. Taniike, N. Katayama, A. Yasuda, M. Sato, M. A. Czarnecki, Y. Ozaki, to be published.

Mikrochim. Acta [Suppl.] 14, 585–588 (1997)
© Springer-Verlag 1997

Advances in Photoacoustic Step-scan FT-IR Spectroscopy

David L. Drapcho[1,*], **Richard A. Crocombe**[1], and **Jürgen Seebode**[2]

[1] Bio-Rad Digilab Division, 237 Putnam Avenue, Cambridge, MA 02139, U.S.A.
[2] Bio-Rad Laboratories GmbH Bischofstrasse 86, 47809 Krefeld, Federal Republic of Germany

Abstract. An integrated system for FT-IR photoacoustic depth-profiling experiments is described, including data-visualization techniques. An example using a multilayer polymer is presented. Extensions of the current method to other spectral regions and using digital signal-processing are discussed.

Key words: FT-IR, photoacoustic, depth profiling, step-scan, digital signal processing.

It is well known that in rapid-scan FT-IR photoacoustic spectroscopy (PAS) the "sampling depth", determined by the thermal diffusion length, is a function of infrared frequency, because each IR frequency is chopped by the interferometer at a different audio-frequency [1]. However, in step-scan FT-IR PAS, employing phase modulation, the "sampling depth" is constant across the spectrum [2]. For a given sample, it is a function of the phase-modulation frequency, which is a spectrometer parameter under software control [3].

In step-scan FT-IR the spectrometer phase is distinct from the sample phase. By using a lock-in amplifier, or other demodulator, the signal can be recovered as a function of phase, giving access to depth information in the sample. In this way step-scan FT-IR PAS can be thought of as a "variable path-length solid cell" technique. This paper shows some examples of this technique and examines extensions of the current method.

Experimental

A Bio-Rad (Cambridge, MA, U.S.A.) FTS 6000 FT-IR spectrometer was used, controlled by a PC data system using Bio-Rad Win-IR Pro

software, running under Windows NT (Microsoft, Redmond, WA, U.S.A.). The photoacoustic accessory was an MTEC 300 (Ames, IA, U.S.A.), powered directly from the optical bench. It was purged with He gas.

The FTS 6000 permits step rates from 800 steps/second to one step every 250 seconds; typically a step rate of 10 Hz was used. Phase modulation (PM) amplitudes could be varied between 0.5, 1.0, 1.5 and 2.0 He–Ne fringes under digital control, and phase modulation frequencies from 5 Hz to 1 kHz were available. Typically, amplitudes of 1 fringe at 800 Hz frequency and 2 fringes at 100 Hz frequency were used. The precise conditions for each spectrum are stated in the figure captions. A ceramic source and KBr beam-splitter combination and a quartz-halogen source and quartz beam-splitter combination were employed for mid IR and near IR studies respectively.

A high dynamic range digital demodulator (Bio-Rad) was used, capable of handling the large ($\pm$10V) signals generated. In-phase and in-quadrature interferograms were collected simultaneously by using the FTS 6000's multichannel spectroscopy (MCS) capability, and stored in a single Win-IR Pro document for further processing. Win-IR Pro was used to calculate and display geometrically interpolated spectra and 2D-IR plots [4]. The layered and background samples (a high carbon-black rubber) were obtained from Dr. J. F. McClelland. The layered sample consisted of a 10-micron thick layer of polyethylene, 10 microns of polypropylene and 6 microns of polyethylene terephthalate (PET) on a thick polycarbonate substrate.

Results and Discussion

The immediate output of the experiment consists of the in-phase and in-quadrature spectra. Although all the information is present in these spectra it can be easier to visualize and analyze the data when they are presented in different forms. First, a spectral surface is generated, with wavenumber as x-axis, photoacoustic intensity as y-axis and phase angle as z-axis by geometric interpolation of the spectra. This can be rendered as a surface, or as contour maps of various types. A slice through this surface parallel to the x-axis is a spectrum at a particular phase angle, and a slice parallel to the z-axis displays the intensity of a band at a particular frequency vs phase angle. Using these slices

---

* To whom correspondence should be addressed

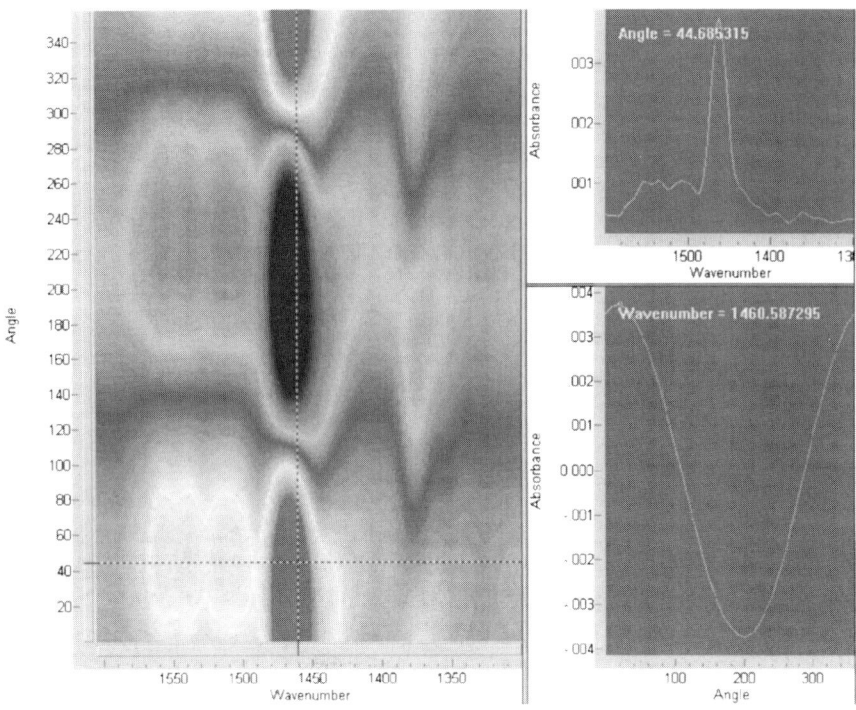

Fig. 1. 3D representation of FT-IR PAS spectra for a polymer laminate at 800 Hz phase modulation. The spectrum was obtained in 1 scan, 10 Hz step rate (0.1 s dwell time per step), 8 cm^{-1} resolution, double-sided interferogram. The total collection time was 7 minutes

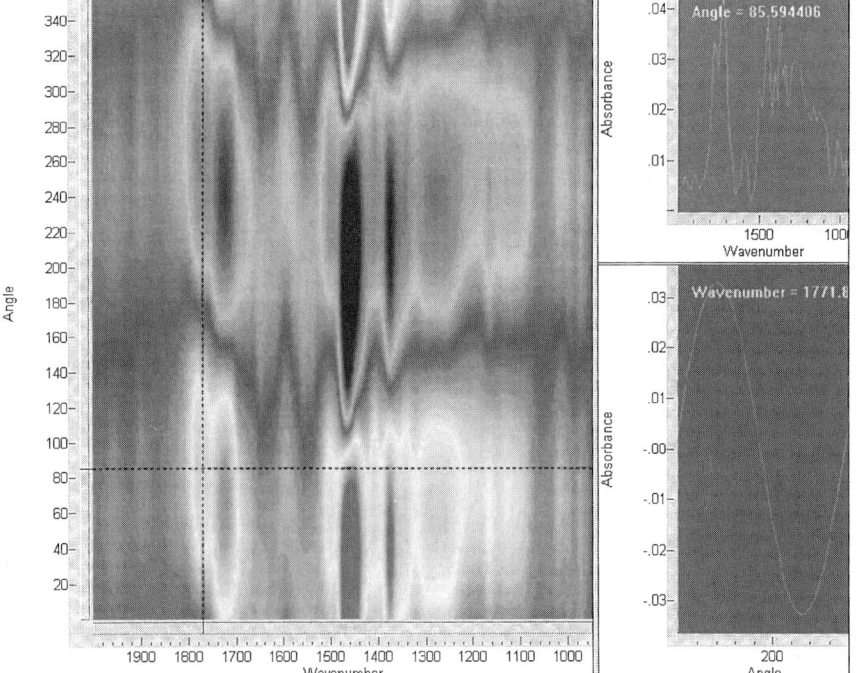

Fig. 2. 3D representation of FT-IR PAS spectra for a polymer laminate at 100 Hz phase modulation; conditions as for Fig. 1

it is easy to locate the angles at which a band reaches it maximum intensity and zero intensity, and to extract spectra at those angles. The spectral surface (or contour) can itself be used to determine the sequence of bands (as a function of phase angle), and this can also be done by comparing the intensity *vs* phase angle plots [3, 5].

Secondly, the 2D-IR formalism [4] can be used. This has been described in detail elsewhere [6], and enables the distinction of bands the changes of which as a func-

Table 1. The relationship between thermal diffusion length and phase modulation frequency for a typical polymer.

PM frequency (Hz)	Thermal diffusion length (microns)
1000	5.6
800	6.3
400	8.9
200	12.6
100	17.8
50	25.2
25	35.7
10	56.4
5	79.8

tion of phase angle are correlated, as well as determining the ordering of the layers in the sample.

Figures 1 and 2 show 3D representations of the spectra obtained for the sample at 800 and 100 Hz phase modulation frequencies. Table 1 shows the relationship between phase modulation frequency and thermal diffusion length for polymers of this type. As expected, only polyethylene features are visible in the 800 Hz PM spectra, but all of the components can be detected in the 100 Hz PM spectra. FT-IR PAS can be expanded from these mid IR applications in a number of ways: into the near IR and UV–visible regions of the spectrum, and by the use of digital signal processing. In the mid IR a strong band, even in a thin layer, will absorb all the energy in the infrared beam at that frequency. This makes "holes" in the spectra of species in lower layers; it is impossible to determine whether subsequent layers have bands at that frequency or not. Also, strong bands display photoacoustic saturation, and band contrast is lost. By obtaining the near-IR spectra of the sample these effects are minimized, as the absorption coefficients are much smaller.

Suitable step-scan spectrometers can also be equipped for use in the UV–visible region by using an Xe source and a UV-visible quartz beam-splitter. Work in this region has many applications, including the bahavior of UV additives in materials, aging and weathering processes, etc.

There are also potential advantages in using digital signal processing in place of the demodulator described above, and one such method has been described in the literature [7]. We have devised a software-based scheme [8] which involves collecting data as a function of time after each interferometer step, storing the resulting array and processing it in the data system. This

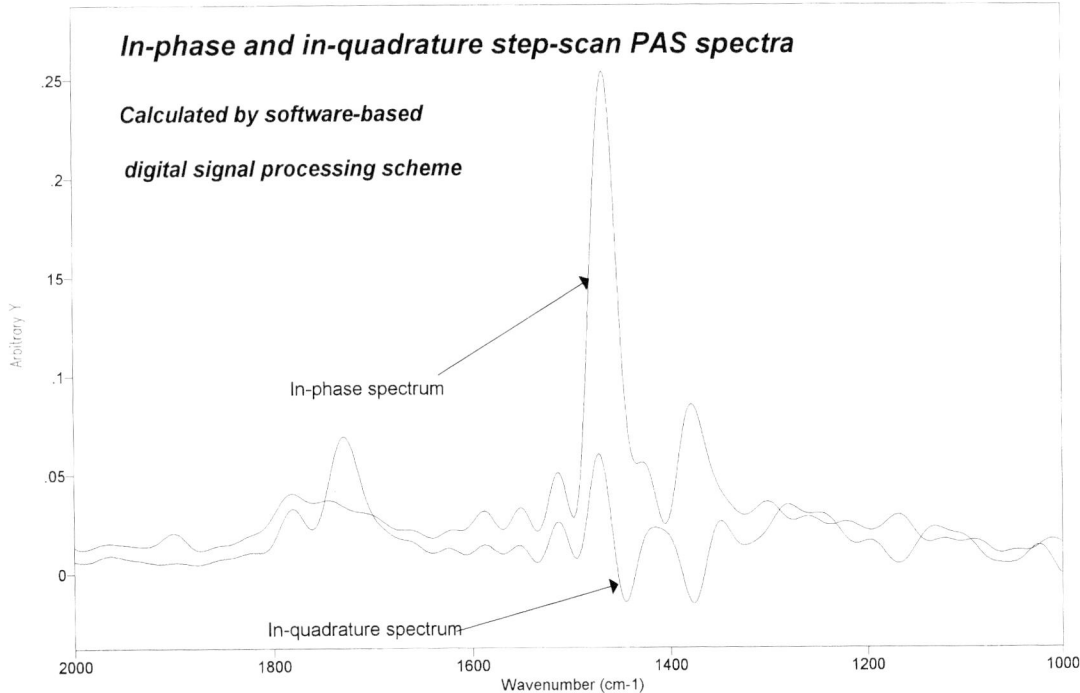

Fig. 3. Spectra obtained by a software-based digital signal-processing scheme. The spectrum was obtained in 1 scan, 5 Hz step rate (0.2 s dwell time per step), 32 cm^{-1} resolution, single-sided interferogram, zero-filling factor of four. The total collection time was 2 minutes

has the advantage over hardware-based approaches that, because the original data are saved, they can be reprocessed at will by using different software parameters, without having to be re-collected. An example of step-scan PAS data collected and computed in this fashion is shown in Fig. 3. This scheme is being extended to multiple modulation experiments [8], including those using a photoelastic modulator.

Acknowledgements. The authors thank R. Curbelo (Bio-Rad) for the use of the unpublished DSP routines and Dr. J. F. McClelland (Ames Laboratory) for the polymer sample.

References

[1] See e.g. J. F. McClelland, *Anal. Chem.* **1983**, *55*, 89A; J. A. Graham, W. M. Grimm III, W. G. Fateley, in: *Fourier Transform Infrared Spectroscopy* (J. R. Ferraro, L. J. Basile, eds.), Academic Press, New York, 1985.

[2] See e.g. R. M. Dittmar, J. L. Chao, R. A. Palmer, *Appl. Spectrosc.* **1991**, *25*, 1104.

[3] P. J. Stout, R. A. Crocombe, *Proc. SPIE* **1993**, *2809*, 300.

[4] I. Noda, *J. Am. Chem. Soc.* **1989**, *111*, 8116.

[5] J. F. McClelland, R. W. Jones, S. Ochiai, *Proc. SPIE* **1993**, *2809*, 302.

[6] I. Noda, *Appl. Spectrosc.* **1990**, *44*, 550.

[7] C. J. Manning, P. R. Griffiths, *Appl. Spectrosc.* **1993**, *47*, 1345.

[8] R. Curbelo, *Unpublished work*.

Mikrochim. Acta [Suppl.] 14, 589–590 (1997)
© Springer-Verlag 1997

Applications of Infrared Imaging with a Focal Plane Array Detector and a Step-scan FT-IR Spectrometer

E. Neil Lewis[1], Ira W. Levin[1], and **Richard A. Crocombe[2,*]**

[1] Laboratory of Chemical Physics, National Institute of Diabetes and Digestive and Kidney Diseases, National Institutes of Health, Bethesda, MD 20892, U.S.A

[2] Bio-Rad Digilab Division, 237 Putnam Avenue, Cambridge, MA 02139, U.S.A.

Abstract. A powerful new mid-infrared spectroscopic chemical imaging technique combining step-scan Fourier transform (FT) Michelson interferometry with indium antimonide (InSb) focal-plane array (FPA) image detection is described. The coupling of an infrared focal-plane array detector to an interferometer provides an instrumental multiplex/multichannel advantage. Specifically, the multiple detector elements enable spectra at all pixels to be collected simultaneously, while the interferometer allows all the spectral frequencies to be measured concurrently.

Key words: FT-IR, step-scan, focal-plane array, imaging, microscopy.

A technique for interfacing a focal-plane array detector to a step-scan FT-IR spectrometer has recently been described[1]. This paper outlines many of the instrumental aspects of this technique and presents some preliminary results. This experimental arrangement rapidly produces an array of spectra for a sample. After fast Fourier transformation (FFT) the data consist of a set of images at particular infrared frequencies for the sample. *Images* are the primary data and spectra at any spatial location which has been imaged by the spectrometer can be derived from these images.

Experimental

The instrumental set-up is shown schematically in Fig. 1. The spectrometer used was an FTS 6000 (Bio-Rad, Cambridge, MA), interfaced to a UMA 500 microscope accessory (Bio–Rad). The FTS 6000 was controlled by a PC data system running Win-IR Pro (Bio-Rad) under Windows NT. The MCT detector assembly on the microscope was replaced with an ImagIR InSb focal-plane array detector (FPA, Santa Barbara Focalplane), and a 50 mm focal length CaF_2 lens was substituted for the Cassegrainian condenser to image a sample located on the microscope stage onto the FPA. The FPA had 128×128 active elements in an array, giving a total of 16, 384 pixels. The FPA was controlled by a second PC, containing a Matrox frame grabber with a 12 bit A/D converter, and a timing card interfaced to the FTS 6000 optical bench. The optical bench was configured with a mid-IR ceramic source, KBr beamsplitter and a 3950–1975 cm^{-1} bandpass optical filter installed in the source assembly. In addition the FPA had a 2–4 µ cold filter installed in the dewar. Typical data acquisition parameters were: step-scan operation, 2 s dwell time per step, undersampling ratio of 8, and a spectral resolution of 16 cm^{-1} optical. This resulted in about 260 steps and a scan time of approximately eight minutes. The nominal spectral range acquired was 1975–3950 cm^{-1}. Thirty-two frames were co-added at each step, and at the end of a scan 16,384 interferograms had been acquired. The duty cycle (ratio of actual data acquisition time to total measurement time) was low, about 3%, because of the time required to store individual images.

The interferogram array was transformed by *Spifft* software (ChemIcon), and arrays of spectra were ratioed to give absorbance or transmittance data, and imaged by use of *ChemImag* (ChemIcon). A typical array, collected under these conditions, occupies about 16 MB of storage. Images can be rendered on the basis of intensity at particular frequencies, or derived by using simple spectral arithmetic (e.g. band ratios), and individual spectra can be extracted from the array. Images were processed for presentation by *CorelDraw* (Corel) software.

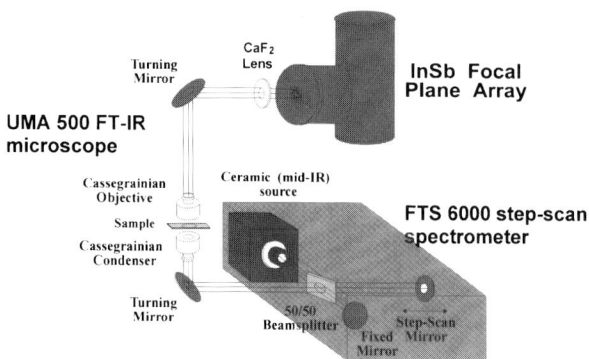

Fig. 1. Schematic lay-out of the system

* To whom correspondence should be addressed

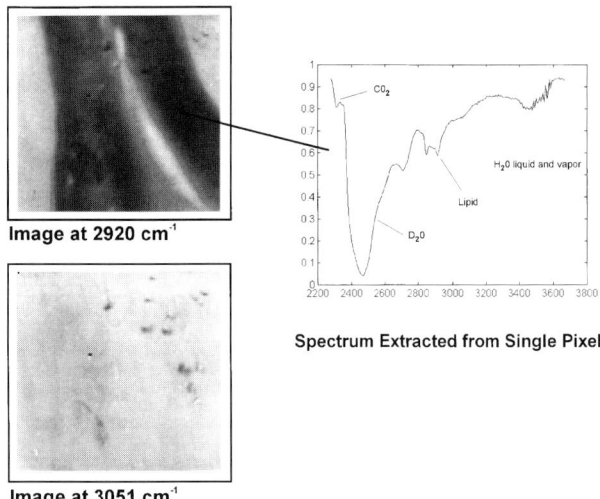

Image at 2920 cm⁻¹

Image at 3051 cm⁻¹

Spectrum Extracted from Single Pixel

Fig. 2. Images and a spectrum from a saturated C_{18} and unsaturated C_{22} lipid bilayer in D_2O. Recorded with Bio-Rad FTS 6000 and InSb focal plane array detector

The sample was a mixture of saturated C_{18} and unsaturated C_{22} lipids, in D_2O. The C_{22} lipid had six non-conjugated double bonds, and these are thought to form a helix, shortening the length of the chain, making it similar to that of the C_{18} species.

Results and Discussion

The sample forms a complex structure in D_2O solution, involving filaments and strands. Figure 2 shows the images obtained at 2920 cm⁻¹ (C–H stretch) and 3051 cm⁻¹ (a region of low overall absorbance). In the image at 2920 cm⁻¹ an inverted "Y" indicates two strands coming together. The line indicates the pixel from which the spectrum was obtained, and major features due to D_2O, H_2O and lipid can be seen, as well as minor features due to CO_2 and water vapour. An interesting point is that in visible light the refractive indices of all the materials present are so similar that there is very little contrast in the visible image. Infrared imaging, in addition to providing all the spectral information, also gives intrinsic contrast enchancement.

This technique permits the rapid chemical imaging of samples, using a non-invasive, non-destructive method. Spectra can be obtained in both transmittance and reflectance mode. Because this is an FT-IR-based experiment it benefits from the normal improvements and extensions typically associated with this method: the spectral signal-to-noise ratio can be improved by using a better duty cycle and more co-addition of data; there is no inherent limit to the spectroscopic resolution; and the technique is applicable across the spectrum, subject to availability of suitable FPAs.

The potential application areas are many, because any real sample is heterogeneous, and the spatial variation of its composition can be of interest. Conventional mapping studies have already focused on impurities and contaminants; pollutants, corrosion, weathering and aging; forensics; biological, biomedical, and pharmaceutical applications; solid-state and semiconductor samples. Applications are not limited to microscopic samples, since it is possible to image 'normal sized' samples (i.e. those several mm in diameter), and extended samples (by using a telescope). As well as transmission, reflectance and emission, FT-Raman and FT-photoluminescence techniques are possible.

The future will see fully integrated systems, both hardware and software, with a focus on working with images, and image-manipulation software in "image space", rather than spectra and spectral manipulations, regarding images as the primary data and spectra as being derived from those images. This novel, high-definition technique represents the future of infrared chemical imaging analysis, a new discipline within the chemical and material sciences, which combines the capability of spectroscopy for molecular analysis with the power of visualization.

References

[1] E. N. Lewis, P. J. Treado, C. Reeder, G. M. Story, A. E. Dowrey, C. Marcott, I. W. Levin, *Anal Chem.* **1995**, *67*, 3377.
[2] E. N. Lewis, I. W. Levin, P. J. Treado, US Patent Nr 5,377,003, December 1994.
[3] See e.g., D.R. Kodali, D. M. Small, J. Powell, K. Krishnan, *Appl. Spectrosc.* **1991**, *45*, 1310.

Mikrochim. Acta [Suppl.] 14, 591–594 (1997)

Step-Scan FT-IR Photoacoustic (S² FT-IR PA) Spectral Depth Profiling of Layered Materials

Richard A. Palmer[1,*], **Eric Y. Jiang**[1], and **J. L. Chao**[2]

[1] Department of Chemistry, Duke University, Durham, NC 29908-0346, U.S.A.
[2] IBM Corporation, Research Triangle Park, NC 27709, U.S.A.

Abstract. We have demonstrated that step-scan FT-IR photoacoustic spectroscopy (S² FT-IR PAS) has the advantages of easy extraction of signal phase and of constant probing depth applied to the entire spectrum range. High depth resolution and distinctive discrimination have been achieved by using PA phase information. The PA phase theory for multilayer materials elucidates the qualitative phase analysis rules established earlier with the experimental data and provides very useful instrument-independent phase-difference models. Quantitative determinations of thermal, optical, and geometric parameters of multilayer materials are made possible by applying the phase-difference models to the experimental data of S² FT-IR PAS.

Key words: step-scan, FT-IR, photoacoustic spectroscopy, depth profiling, leyered materials.

In step-scan Fourier-transform infrared photoacoustic spectroscopy (S²FT-IR PAS), a uniform thermal sampling depth is achieved in the absence of saturation, by applying a constant modulation frequency to the entire spectral range. A S² FT-IR PA spectral depth profile can be obtained by changing the modulation frequency for different scans. Alternatively, because the step-scan mode of data collection permits easier access to the PA phase than does the countinuous-scan mode, depth information can be obtained from the phase of the PA signal from a single (step) scan.

Two qualitative methods have been developed for using the PA phase, the quantity directly related to the spatial origin of the signal, to resolve signals from different depths more effectively and successfully in S² FT-IR PAS, namely, the continuous phase rotation method and the phase spectrum method [1], with the latter being superior in that it is significantly time-saving and provides even higher phase (depth) resol-

---

* To whom correspondence should be addressed

ution [2]. A quantitative photoacoustic phase theory for multilayer materials [3] developed in our laboratory relates the total PA signal phase response of any layer j of the sample to the modulation frequency of the incident light and to the sample's thermal diffusivity (α), optical absorption coefficient (β) and thickness (d), through Eqs. (1) and (2) for the corresponding cases: (1) layer j is *thermally thin* ($d_j \ll \mu_j$) and *optically "transparent"* ($d_j \ll 1/\beta_j$); and (2) layer j is *thermally thick* ($d_j \gg \mu_j$) and/or *optically "opaque"* ($d_j \gg 1/\beta_j$), where μ_j is the thermal diffusion depth of layer j.

$$\Phi_{j,\,\text{total}}(\sigma_j) = \Phi_0 + \frac{\pi}{2} + \left(\sum_{n=1}^{j-1} d_n \alpha_n^{-1/2} \right) (\pi f)^{1/2} \quad (1)$$

$$\Phi_{j,\,\text{total}}(\sigma_j) = \Phi_0 + \pi + \left(\sum_{n=1}^{j-1} d_n \alpha_n^{-1/2} \right) (\pi f)^{1/2}$$
$$- \tan^{-1} [\beta_j \alpha_j^{1/2} (\pi f)^{-1/2} + 1] \quad (2)$$

In the equations above, Φ_0 is the relative phase shift introduced by the instrument, and f is the modulation frequency of the incident light. The thermal and optical properties of any layer must belong to one of these two cases. From Eqs. (1) and (2), four PA phase difference models (which have the advantage of being independent of the instrument phase) and subsequently two general qualitative phase analysis rules have been established [3]. These two rules can be briefly stated as: (1) PA signals from deeper layers have greater phase lags than signals from shallower layers, owing to the thermal transport within the overlayers (*thickness effect of the overlayers*); and (2) stronger PA signals have smaller phase lags than do weaker signals from the same layer (*absorptivity effect of the same layer*). In this paper we will discuss the application of two of the PA phase difference models and the phase analysis rules in depth profiling analysis of layered polymers by use of S² FT-IR PAS.

Experimental

An IBM IR44 bench-top FT-IR spectrometer modified for optional step-scan operation and coupled with an MTEC 100 PA cell was used to collect all experimental data at room temperature. The two orthogonal single-sided interferograms (in-phase I, and in-quadrature Q) of 2048 points plus 256 phase correction points were obtained simultaneously by collecting data at intervals of one wavelength of the reference He-Ne laser to give $8 \, cm^{-1}$ resolution. The I and Q interferograms were then Fourier-transformed to obtain I and Q spectra after data collection was completed. From the I and Q spectra, the magnitude, $M = (I^2 + Q^2)^{1/2}$, and phase, $\Phi = \tan^{-1} (Q/I)$, spectra were calculated. A 60% carbon-filled polymer disk was used as both the surface phase reference and PA intensity normalization reference.

Results and Discussion

The S² FT-IR PAS experimental results for ethylene-vinyl acetate copolymer (EVAc, 12 μm thick) on polypropylene (PP, 60 μm thick), and for hexamethyl disilazane (HMDSiN)-based plasma polymer (0.7 μm thick) on *meso*-tetraphenylporphyrinatocobalt(II) (CoTPP)-based plasma polymer (1.0 μm thick) on PP (60 μm thick) were used to illustrate the applications of the theory. Figure 1 shows the PA magnitude and phase spectra of EVAc on PP collected at a modulation frequency of 400 Hz. At this modulation frequency, both layers are thermally thick $(12 \, \mu m > 10.3 \, \mu m = \mu_{EVAc}; \, 60 \, \mu m > 7.9 \, \mu m = \mu_{pp})$. Thus the PA phase response of each layer depends on its thermal and optical properties and spatial location, as suggested by the two rules stated above. The experimental results listed in Table 1 can be well explained with these two rules. The characteristic peaks of EVAc, A_1 $(1741 \, cm^{-1})$ and A_2 $(1242 \, cm^{-1})$ have similar absorptivity (relative peak heights 1.15, 1.12) and thus their phase lags are close to each other (8.3°, 8.7°). The

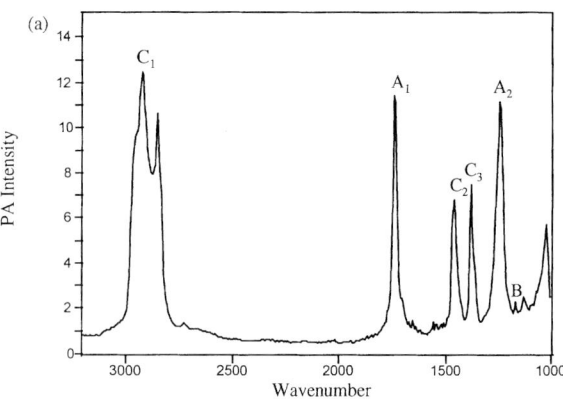

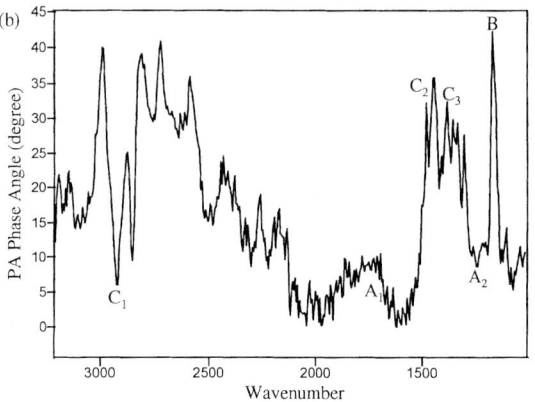

Fig. 1. PA magnitude (**a**) and phase (**b**) spectra of EVAc (12 μm thick) on PP (60 μm thick) collected at a modulation frequency of 400 Hz. The distinctive bands for EVAc and PP are A_1 $(1741 \, cm^{-1}$, C=O stretch), A_2 $(1242 \, cm^{-1}$, C–O–C asymmetric stretch) and B $(1169 \, cm^{-1}$, C–C stretch), respectively. The overlapped bands for EVAc and PP are C_1 $(2921 \, cm^{-1}$, C–H stretch), C_2 $(1459 \, cm^{-1}$, C–C stretch) and C_3 $(1378 \, cm^{-1}$, C–H bend)

characteristic peak of PP, B $(1169 \, cm^{-1})$ obviously has the greatest phase lag (40.5°) because PP is below the EVAc layer. The phase data of the overlapped bands C_1 $(2921 \, cm^{-1})$, C_2 $(1459 \, cm^{-1})$, and C_3 $(1378 \, cm^{-1})$, though not discussed by the theory, can still be quali-

Table 1. PA magnitude and phase data of EVAc (12 μm thick) on PP (60 μm thick)

Peak (cm⁻¹)	Assignment	Absorber(s)	PA mag.	Φ spec.
A_1 (1741)	C=O stretch	EVAc	1.15	8.3°
A_2 (1242)	C–O–C asymmetric stretch	EVAc	1.12	8.7°
B (1169)	C–C stretch	PP	0.22	40.0°
C_1 (2921)	C–H stretch	EVAc/PP	1.25	6.1°
C_2 (1459)	C–H bend	EVAc/PP	0.65	30.4°
C_3 (1378)	C–H bend	EVAc/PP	0.75	28.4°

Table 2. PA magnitude and phase data of characteristic peaks of HMDSiN (0.7 μm thick) on CoTPP (1 μm thick) on PP (60 μm thick)

Peak (cm⁻¹)	Assignment	Absorber(s)	PA mag.	Φ spec.
A (1257)	C–N/C–Si stretches	HMDSiN	0.68	4.3°
B (1699)	–	CoTPP	0.57	7.3°
C (1169)	C–C stretch	PP	0.52	11.3°

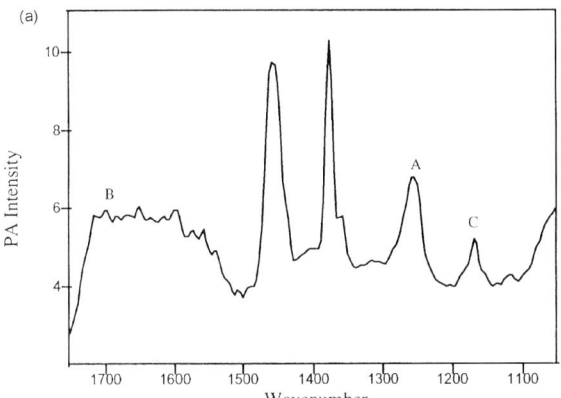

(a)

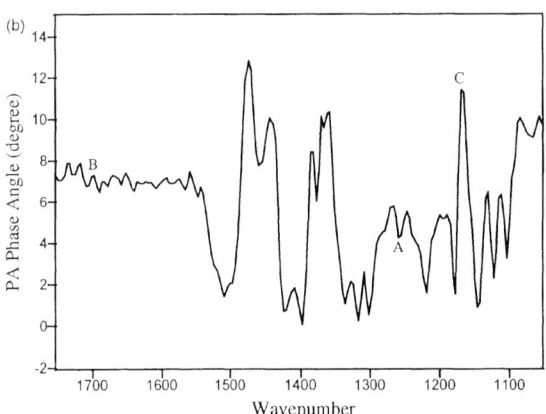

(b)

Fig. 2. PA magnitude (**a**) and phase (**b**) spectra of HMDSiN based plasma polymer (0.7 μm thick) on CoTPP based plasma polymer (1.0 μm thick) on PP (60 μm thick) collected at a modulation frequency of 200 Hz. The distinctive band of each layer from HMDSiN (top) to PP (bottom) are A (1259 cm⁻¹, C–N/C–Si stretches), B (1699 cm⁻¹) and C (1169 cm⁻¹, C–C stretch), respectively

from which the peak intensity and corresponding phase angle data are obtained as shown in Table 2. The thermal diffusivity (and thus the thermal diffusion depth) of CoTPP should be of the same order as that of HMDSiN [4]. At a modulation frequency of 200 Hz, both HMDSiN and CoTPP are thermally thin ($\mu_{HMDSIN} \approx \mu_{pp} = 12.6\,\mu m \gg 1.0\,\mu m > 0.7\,\mu m$) and optically transparent with respect to their non-saturated distinctive bands. The substrate, PP, however, is thermally thick ($\mu_{pp} = 11.1\,\mu m \ll 60\,\mu m$) at this frequency. The phase data of the distinctive peaks, given in Table 2, also support Rule (1). In addition, the phase difference between the distinctive peaks of HMDSiN and CoTPP is purely determined by d_1/μ_1, as described by Eq. (27) of reference [3]. So either the thickness (d_1) or the thermal diffusivity ($\alpha_1 = \pi f \mu_1^2$) of HMDSiN can be determined by use of the phase data. For example, presuming the thermal diffusion length of HMDSiN at 200 Hz to be 12.6 μm, and using the observed phase difference between the peaks at 1257 and 1699 cm⁻¹, 3.0° (i.e. 7.3° − 4.3°), the thickness of HMDSiN is calculated to be 0.67 μm [= 3.0° × (π/180°) × 12.6 μm], which is very close to the thickness of 0.70 μm determined by use of the mass-density method [5].

As described by the two rules stated above and by the phase difference models [3], the PA phase theory for multilayer materials provides a useful guide for qualitative depth-profiling by using the PA phase information. The surface reference phase determination is critical to relate the experimental phase data to the actual phase lags of the signals. The PA phase difference models for multilayer samples can be used to measure the thickness, the thermal diffusivities, or the optical absorption coefficients of different layers by use of the experimental phase data. The determination of the thickness or the thermal diffusivities of different layers is simplified if the layers are thermally thin at the experimental modulation frequency and optically transparent (i.e. far from saturation) with respect to their distinctive absorption bands. The optical absorption coefficient of a layer at its distinctive absorption

tatively explained by rule (2), i.e. the stronger the PA signal (1.25 > 0.75 > 0.65), the smaller is the phase lag (6.1° < 28.4° < 30.4°), because in this case the PA signal intensities are mostly contributed from EVAc, the top layer, and thus the phases of these peaks dominate.

The PA magnitude and phase spectra of HMDSiN (0.7 μm) on CoTPP (1.0 μm) on PP (60 μm) collected at a modulation frequency of 200 Hz are shown in Fig. 2,

frequency can only be determined by using this technique when this layer is prepared *thermally thick*
and/or *optical opaque* with respect to its distinctive
band at a reasonable modulation frequency for the
instrument. Though it is possible to experimentally
change the modulation frequency to fit a sample into
the thermally thick or thermally thin cases, there will
always be limits on the available modulation frequencies for any particular instrument.

References

[1] R. A. Palmer, E. Y. Jiang, *J. Physique* **1994**, [*Suppl. III (4)*] *IV,*
    C7-337.

[2] R. A. Palmer, E. Y. Jiang, J. L. Chao, *Proc. SPIE*, **1993**, *2089,*
    250.

[3] E. Y. Jiang, R. A. Palmer, J. L. Chao, *J. Appl. Phys.* **1995**, *78,*
    460.

[4] Y. S. Touloukian, R. W. Powell, C. Y. Ho, M. C. Nicolaou,
    Thermal Diffusivity, Vol. 10, FI/Plenum, New York, 1973, p.
    593–619.

[5] A. M. Wróbel, M. R. Wertheimer, in: *Plasma Deposition,
    Treatment, and Etching of Polymers* (R. d'Agosrino, ed.), Aca-
    demic Press, New York, 1990, Chapter 3.

Mikrochim. Acta [Suppl.] 14, 595–597 (1997)

Excited State Structure and Relaxation Dynamics of Polypyridyl Complexes of Low Spin d^6 Metal Ions by Means of Step-Scan FTIR Time-Resolved Spectroscopy (S^{2}FT-IR TRS)

Richard A. Palmer[1,*], **Pingyun Chen**[1], **Susan E. Plunkett**[1], and **James L. Chao**[2]

[1] Department of Chemistry, Duke University, Box 90346, Durham, NC 27708–0346, U.S.A.
[2] IBM Corporation, Research Triangle Park, NC 27709, U.S.A.

Time-resolved infrared spectroscopy has been applied to study the excited state dynamics of many biological molecules and transition metal complexes following laser-flash excitation [1–6]. The short lifetimes of the excited states involved in these processes require very high time resolution. Until recently, fast time-resolved IR measurements have typically utilized tunable IR diode lasers as the probe source. The limited tunability of these IR lasers limited the studies to systems that contain CO or CN groups as a "reporter" ligand.

Recent advances in time-resolved infrared spectroscopy using both dispersive and Fourier-transform spectrometers allow measurements to be made with nanosecond time resolution [7, 8]. The broad-band IR source employed in these systems extends measurements over the entire mid IR region. This has opened a new avenue for study of fast chemical processes by monitoring vibrational bands other than CO or CN groups.

The instrumentation used in this work comprised a step-scan Bruker IFS 88 FT-IR spectrometer, a Quanta Ray DCR-1 Nd:YAG laser and an SRS-535 pulse generator [1,7]. This set-up has been used to achieve time resolution down to 10 ns [9]. In this peper we report "snapshot" ΔA spectra of excited states of fac-[Re(dppz)(CO)$_3$(pph$_3$)](PF$_6$) (dppz = dipyrido[3,2-a,2′,3′-c]phenazine, PPh$_3$ = triphenylphosphine) and cis-[Os(bipy)$_2$(CO)(py)](PF$_6$)$_2$ (bipy = 2,2′-bipyridine, py = pyridine). These types of complexes are of great interest because of their potential applications as photochemical sensitizers in artificial photosynthesis and other solar-energy conversion schemes. Their broad absorption bands in the near UV and visible regions and the great tunability of their excited state properties through synthetic design, make them very attractive candidates for such applications. Transient absorption, emission, and resonance Raman measurements are among the mostly widely used techniques to characterize the excited state properties of such systems. The high sensitivity and high time resolution that can be achieved with these techniques make them ideal for kinetic measurements. However, the very broad, and often overlapping, bands in the absorption and emission spectra contain very little structural information. Recent developments in time-resolved IR spectroscopy will allow detailed, structure-specific vibrational information of short-lived excited states to be obtained.

The experimental set-up used for fac-[Re(dppz)-(CO)$_3$(PPh$_3$)](PF$_6$) was the same as described earlier [1] except that the excitation was achieved by using the third harmonic of the Nd:YAG laser (355 nm, 1 mJ/pulse, 10 ns FWHM). For cis-[Os(bipy)$_2$-(CO)(py)](PF$_6$)$_2$, a 200 MHz, 8-bit external digitizer (PAD board) was used for data collection on the Bruker IFS 88. Because of the limited dynamic range of the PAD board, the detector was AC-coupled to the preamplifier. This eliminates the background signal and allows only the AC component of the signal that is due to changes between the excited and ground states) to pass. This signal was further amplified by a Hewlett Packard 8447D amplifier and then digitized by the PAD board. The phase spectrum used for Mertz-

* To whom correspondence should be addressed

stored phase-correction of the AC interferogram was generated from a step-scan experiment using the same digitizer and MCT detector in DC-coupled mode.

The ground-state absorption and difference spectra (ΔA; difference in absorbance before and after the laser-flash excitation, or more exactly, $\log[I_{(t)}/I_R]$) of fac-[Re(dppz)(CO)$_3$(PPh$_3$)] (PF$_6$) in CH$_3$CN at room temperature are shown in Fig. 1A. The observed signal at 2044 cm^{-1} in the difference spectrum is consistent with the conclusion that the excited state is a $\pi\pi^*$ state localized on the dppz ligand [10]. The charge density at the metal ion is not affected by the transition, the $d\pi(\text{Re})$–$\pi^*(\text{CO})$ backbonding remains nearly the same in the excited state, and there is only a slight shift in $v(\text{CO})$ relative to that in the ground state. The ΔA spectrum shows only a weak, derivative-shaped signal.

The ground-state absorption and difference spectra following laser flash excitation of cis-[Os(bipy)$_2$(CO)(py)](PF$_6$)$_2$ are shown in Fig. 1B. In the difference spectrum there is a negative peak at 1967 cm^{-1} due to the bleaching of the ground state after laser excitation. A positive peak at 2041 cm^{-1} can be assigned to the CO stretching vibration in the excited state. The large shift in $v(\text{CO})$ to higher energy after the laser excitation is consistent with the formation of a metal to ligand charge-transfer excited state which is also indicated by emission and transient absorption studies [11]. In the excited state, osmium is effectively oxidized to Os(III) and the electron density on the metal is thus greatly reduced. This, in turn, reduces the $d\pi(\text{Os})$–$\pi^*(\text{CO})$ backbonding and increases the bond strength of CO and thus $v(\text{CO})$.

The results presented here clearly demonstrate the usefulness of the S^{2}FT-IR TRS measurement as a powerful probe to study excited state processes. With the improved sensitivity of this technique, we can easily extend the measurements to other regions such as the weaker C–C stretching modes of polypyridyl rings.

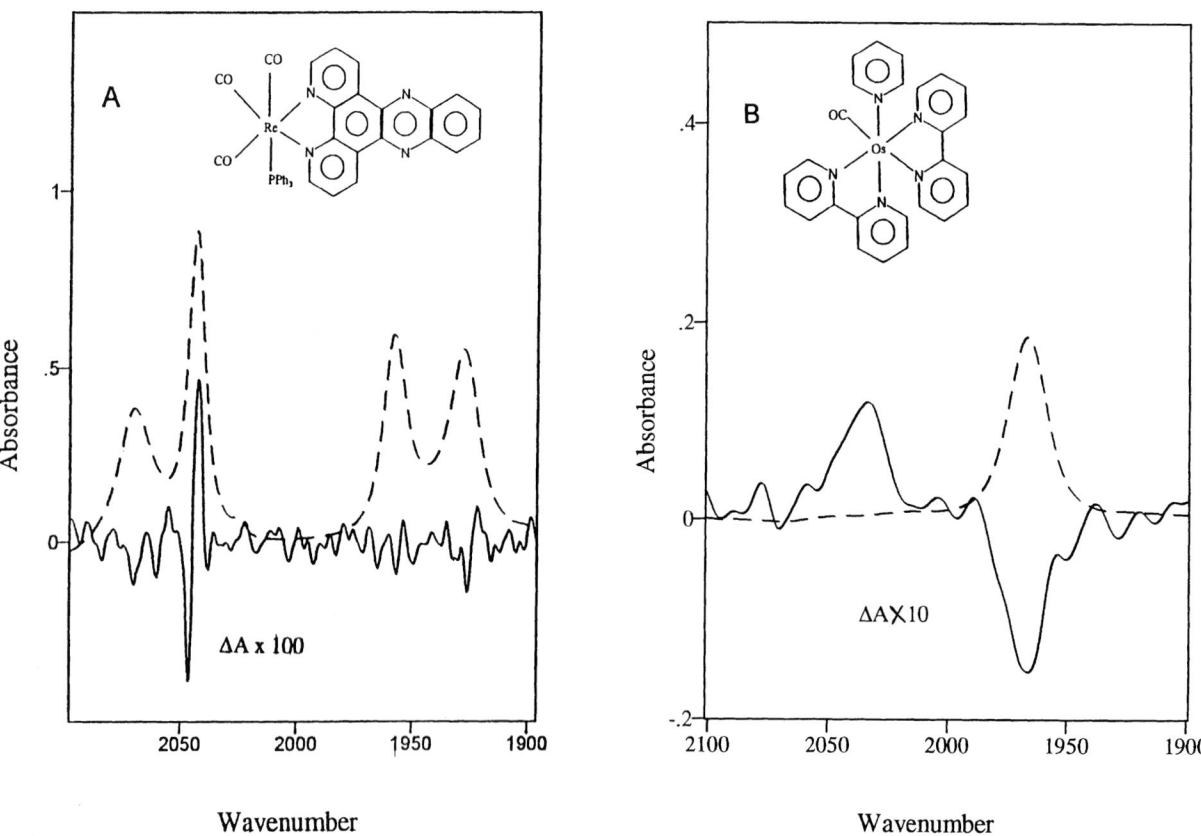

Fig. 1. Ground-state absorption (dashed line) and time-resolved difference spectra (solid line, $\Delta A \times 100$) following laser flash excitation of CH$_3$CN solutions containing **A** fac-[Re(dppz)(CO)$_3$(PPh$_3$)]$^+$ and **B** cis-[Os(bpy)$_2$(CO)(py)]$^{2+}$. The difference spectra were calculated from ln $(I(t)/I_R)$ where I_R and $I(t)$ are the light intensities before and at time t after the laser excitation. The difference spectrum for fac-[Re(dppz)(CO)$_3$(PPh$_3$)]$^+$ is a "snapshot" average of data from 5 to 50 μs after the laser flash and that for cis-[Os(bpy)$_2$(CO)(py)]$^{2+}$ is an average of data from 0 to 200 ns

Acknowledgements. We thank Bruker Instruments, Inc., Las Alamos National Laboratory, and Duke University for support of this work, and Prof. T. J. Meyer for providing the samples used in this study.

References

[1] S. E. Plunkett, J. L. Chao, T. J. Tague, R. A. Palmer, *Appl. Spectrosc.* **1995**, *49*, 702.

[2] K. Fahmy, F. Siebert, T. P. Sakmar *Biochemistry* **1994**, *33*, 13700.

[3] J. J. Turner, M. W. George, F. P. A. Johnson, J. R. Westwell, *Coord. Chem. Rev* **1993**, *125*, 101.

[4] P. O. Stoutland, R. B. Dyer, W. H. Woodruff, *Science* **1992**, *257*, 1913.

[5] J. R. Schoonover, K. C. Gordon, R. Argazzi, W. H. Woodruff, K. A. Peterson, C. A. Bignozzi, R. B. Dyer, T. J. Meyer, *J. Am. Chem. Soc.* **1993**, *115*, 10996.

[6] E. Weitz, *J. Phys. Chem.* **1994**, *98*, 11256.

[7] R. A. Palmer, J. L. Chao, R. M. Dittmar, V. G. Gregoriou, S. E. Plunkett, *Appl. Spectrosc.* **1993**, *47*, 1297.

[8] T. Yuzawa, C. Kato, M. W. George, H. O. Hamaguchi, *Appl. Spectrosc.* **1994**, *48*, 684.

[9] R. A. Palmer, S. E. Plunkett, P. Chen, J. L. Chao, T. J. Tague, Mikrochim. Acta **1997**, [*Suppl. 14*], 603.

[10] D. Bates, *Ph. D. Dissertation,* University of North Carolina, Chapel Hill, 1994.

[11] K. S. Schanze, G. A. Neyhart, T. J. Meyer, *J. Phys. Chem.* **1986**, *90*, 2182.

Mikrochim. Acta [Suppl.] 14, 599–602 (1997)

Step-Scan FT-IR Time-resolved Spectroscopy (S²FT-IR TRS) of the Electro-Reorientation of Liquid Crystalline Materials

A. Fuji*, R. A. Palmer[1],**, P. Chen[1], E. Y. Jiang[1], and J. L. Chao[2]

[1] Department of Chemistry, Duke University, Box 90346, Durham, NC 27708–0346, U.S.A.
[2] IBM Corporation, Research Triangle Park, NC 27709, U.S.A.

Abstract. Frequency domain and time domain step-scan FT-IR time-resolved spectra (S²FTIR TRS) of the homogeneous/homeotropic transition in several nematic liquid crystals are presented and discussed. These include the pure mesophases 5CB and 5PCH and the eutectic mesophase 7A. In agreement with previous studies, both frequency domain and time domain data indicate a difference in response of the rigid and "floppy" segments of the 5CB and 5PCH molecules. The time domain data show clearly that it is in the recovery (voltage off) phase that the pentyl "tails" of the molecules respond more quickly than the rigid cores. For 7A, frequency domain data collected at twice the AC field frequency and with zero DC offset, indicate a slight difference in phase of response of the two major components.

Key words: step-scan FT-IR, time-resolved spectroscopy, electro-reorientation, liquid crystals.

The success of the application of time-resolved infrared spectroscopy for monitoring the segmental dynamics of the liquid crystal switching, or electro-reorientation, of both nematic and smectic-c mesophases has been well documented during the past several years [1–12]. While some investigators have devised dispersive techniques for IR probing of mesophase dynamics [10], most have chosen to use one or more of the recently developed dynamic FT-IR methods. These have included both time domain and frequency domain measurements and continuous-scan as well as step-scan modes of data collection. Although either time- or frequency-domain measurements could, in principle, be made in either mode, the most versatile method in either case for these ns to ms processes appears to be step-scan.

In this paper the use of both time domain and frequency domain step-scan FT-IR time-resolved spectroscopy (S²FTIR TRS) to investigate the segmental dynamics of several pure and blended low molecular weight liquid crystals will be illustrated and discussed.

Experimental

The FT-IR instrumentation originally developed in our laboratories (IBM–Bruker IR 44 modified for optional step-scan operation) [13, 14] was used for the frequency domain experiments. In the results reported here frequency domain experiments both with and without a DC offset are described. For the time domain experiments the commercial step-scan modification of the Bruker Instruments IFS 88 spectrometer was employed. The general design and operational specifications of this instrument have been described [15], as have the data collection procedures for liquid crystal electro-reorientation experiments [12].

Results and Discussion

Early results of dynamic infrared studies of the electro-reorientation of 5CB (4-pentyl-4′cyanobiphenyl) [1–5 and 5PCH ([1-(4-cyanophenyl)-4′-pentylcyclohexane]) were interpreted as indicating a temporal difference in reorientational response between the rigid cores and floppy tails of the mesogen molecules. These studies have included both frequency domain and time domain measurements. More recently reports from other laboratories have indicated a failure to detect this difference [11]. Later frequency domain experiments in our laboratory [12] (Fig. 1a and b) support the earlier results for both 5CB and 5PCH when the interpretative rules for 2D frequency correlation maps outlined by Noda [16–18] are followed. Nevertheless, the apparent existence of a phase lag between rigid and floppy

* Present address: Kao Corp., 2-1-3 Bunka Sumida-ku, Tokyo 131, Japan
** To whom correspondence should be addressed

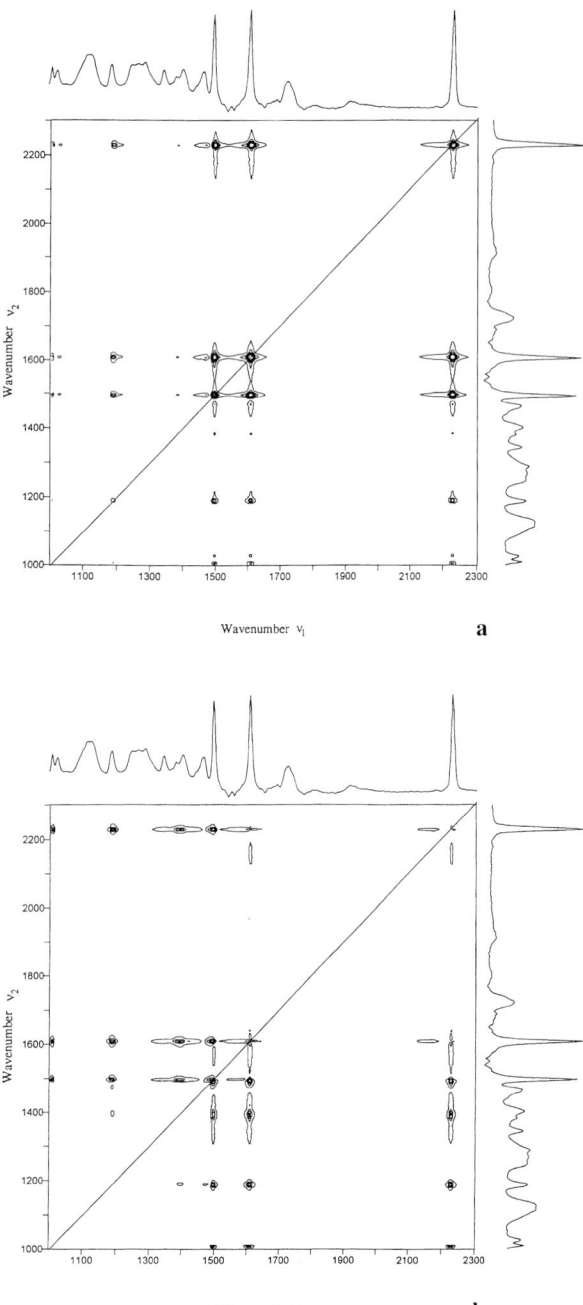

Fig. 1. Step-scan 2D FT-IR spectra of nematic liquid crystal 5CB response to sinusoidally modulated (AC) electric field. **a** Synchronous correlation; **b** a synchronous correlation. Reference spectra along the top and side of each correlation map represent the static absorbance

segments is only qualitatively indicated by the off-diagonal peaks between the cyano(bi)phenyl bands and those of the pentyl tail in the asynchronous correlation map. The 2D results do not allow any conclusion as to whether the flexible "tail" responds first as the

voltage is *increased*, or as it is *decreased*, during the sinusoidal modulation. However, in the time domain experiment, not only is it possible to separate the "rise" dynamics from the "decay" dynamics, but induction times can be measured, quantitative response rates determined and there is the possibility of detecting complex responses involving intermediate forms. Preliminary time domain measurements on 5PCH by use of the stroboscopic continuous scan technique allowed an inference that the differential response occurs in the voltage-off part of the response [19]. However, the constraints of the stroboscopic method as regards repetition rate and pulse structure make the step-scan mode more flexible for investigating the complete cycle of voltage response.

The results of time domain step-scan FTIR investigation of electro-reorientation dynamics of 5CB and 5PCH are summarized in Table 1. In Fig. 2 the voltage pulse profile for the experiment is compared with the intensity response curves for the various bands in 5CB. In the figure, each data point represents the averaged value of $\Delta A(v, t)$ for 10 consecutive 1 ms time slices. It should be noted that all of the data for each sample were collected in a single scan, that is, with *simultaneous temporal and spectral multiplexing*. The total data collection time was approximately 6.5 hr.

As seen in Table 1, both 5CB and 5PCH respond with $t_{1/2}$ of <2 ms when the voltage pulse is applied. On the other hand, the recovery (voltage-off) process, *after* the 30 ms, 11 V pulse, is much slower, requiring almost 400 ms for 5CB. However, the most interesting result is that the $t_{1/2}$ for the response of the pentyl CH_2 deformation mode at 1467 cm^{-1} in the voltage-off, homeotropic-to-homogeneous, transition appears to be approximately one third the response times for the modes associated with the rigid parts of the molecules ($t_{1/2} = 24.4$ ms *vs.* $\geqslant 70$ ms for 5CB and 6.3 *vs.* $\geqslant 15$ ms for 5PCH). At the time resolution and voltage of these experiments it does not appear that a comparable differential response occurs in the voltage-on process (nor, incidentally, does there appear to be any significant induction time after the pulse is applied before the reorientation begins) for either compound.

Thus the ambiguity of the frequency domain measurements, that is, in what part of the electro-reorientation process does the pentyl "tail" get out of phase with the rigid part of the molecule, is clearly resolved by the time domain results. Although both parts of these low molecular weight liquid crystal molecules respond ($t_{1/2} < 2$ ms) when the voltage is

Table 1. Step-scan time domain electro-reorientation half-lives for IR bands in 5CB and 5PCH

Wavenumber (cm^{-1})/Assignment	$t_{1/2}$, voltage-on (ms)		$t_{1/2}$, voltage-off (ms)	
	5CB	5PCH	5CB	5PCH
2227/CN stretch	1.6	1.5	69.2	14.9
1607/phenyl CC stretch	1.3	1.4	92.9	18.4
1503/phenyl CC stretch		1.4		15.7
1467/phenyl CH$_2$ deformation	1.9	1.9	24.4	6.3
1445/cyclohexyl CH$_2$ deformation		1.5		10.3
832/phenyl CH deformation	1.5	1.5	76.2	18.1

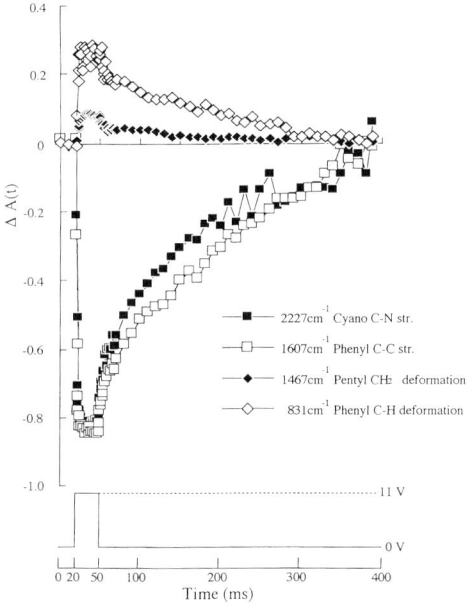

Fig. 2. Step-scan time domain data for selected IR bands in 5CB. Each point represents an average of 10 1-ms $\Delta A (v, t)$ data, each from 50 co-added pulse sequences (see text). Corresponding voltage profile for each pulse sequence is given at top of the figure

applied, the flexible "tails" of the molecules reorient faster than the rigid part as the mesophase returns (relatively slowly) to its homogeneous state after the electric field is removed. That the more flexible segments of these mesogenic molecules should return to their equilibrium (homogeneous) state more rapidly than the rigid parts correlates with the relative degrees of conformational freedom available to the different segments, which promotes a more rapid thermal relaxation of the pentyl "tail" compared to the rigid core.

Despite the power of the time domain experiment to distinguish rise and decay dynamics, as well as the fact that it explores the full range of the homogeneous/ homeotropic transition, the application of a DC volt-

age does not produce useful results in all samples. This is apparently due to the deliberate or accidental presence of ionic impurities and/or water. Thus, either the time domain experiment described above or the classical frequency domain experiment (AC modulation with a DC offset) produces inconsistent, irreproducible results. In these cases the frequency domain experiment without the DC offset may still provide useful data. In such experiments the fact that the nematic phase responds in effect to the square of the electric field leads to the production of a strong IR intensity modulated signal at the first harmonic of the AC modulation frequency. In-phase and in-quadrature data are obtained and can be used in the same ways as previously described for data collected at the fundamental frequency, with a DC offset > the AC amplitude. The frequency domain data reported earlier for 5PCH were obtained in this mode because of the inconsistent results obtained with a DC offset. A more dramatic example is the nematic eutectic 7A, which consists of 90% of a mixture of two 4-methoxy-4'-alkylazoxybenzenes, 10% nematic ester and 0.1% organic acid salt. Apparently because of the presence of the salt, no signal is obtained by use of DC voltage, either in the time or frequency domain experiment. However, the frequency domain experiment without a DC offset and with data collected at the first harmonic produces strong in-phase and in-quadrature signals. The preliminary analysis of these data by the 2D frequency correlation method indicates from the splitting of the major peaks in the asynchronous correlation that there is a small phase lag between the responses of the two major components.

Note added in proof. Recent results suggest that the apparent difference in response for the CH$_2$ deformation mode may be very dependent on experimental conditions and related to more complex response dynamics and/or differences in association with the electrode surfaces, not considered in this paper. Work is continuing, to explore these variables and factors.

Acknowledgments. The authors acknowledge helpful discussions with I. Noda and E. T. Samulski, as well as contribution of equipment from the Bruker Instrument Company and the IBM Corporation. Financial support from the Bruker Instrument Company, from Kao Corporation and from Duke University is also gratefully acknowledged.

References

[1] A. Hatta, *Mol. Cryst. Liquid Cryst.* **1981**, *74*, 195.

[2] H. Toriumi, H. Sugisana, H. Watanabe, *Japan. J. Appl. Phys.* **1988**, *27*, L279.

[3] V. G. Gregoriou, J. L. Chao, H. Toriumi, R. A. Palmer, *Chem. Phys. Lett.* **1991**, *179*, 491.

[4] R. A. Palmer, J. L. Chao, R. M. Diltmar, V. G. Gregoriou, S. E. Plunkett, *Appl. Spectrosc.* **1993**, *47*, 1297.

[5] T. Nakano, T. Yokoyama, H. Toriumi, *Appl. Spectrosc.* **1993**, *47*, 1354.

[6] S. Kohri, J. Kobayashi, S. Tahata, S. Kita, I. Karino, T. Yokoyama, *Appl. Spectrosc.* **1993**, *47*, 1367.

[7] K. Masutani, A. Yakota, Y. Furukawa, M. Tasumi, A. Yoshizawa, *Appl. Spectrosc.* **1993**, *47*, 1370.

[8] M. A. Czarnecki, N. Katayama, Y. Ozaki, M. Satoh, K. Yoshio, T. Watanabe, T. Yanagi, *Appl. Spectrosc.* **1993**, *47*, 1382.

[9] R. Hasegawa, M. Sakamoto, H. Sasaki, *Appl. Spectrosc.* **1983**, *47*, 1386.

[10] T. I. Urano, H. Hamaguchi, *Appl. Spectrosc.* **1993**, *47*, 2108.

[11] S. Okretic, U. Hoffmann, F. Pfeifer, H. W. Siesler, S. Shilov, *Abstracts Pittsburgh Conf.* **1995**, No. 606.

[12] R. A. Palmer, V. G. Gregoriou, A. Fugi, E. Y. Jiang, S. E. Plunkett, L. M. Connors, S. Boccara, J. L. Chao, in: *Multidimensional Spectroscopy of Polymers, Symposium Series 598* (M. W. Urban, T. Provder, eds.), American Chemical Society, Washington, DC, 1995, p. 99.

[13] M. J. Smith, C. J. Manning, R. A. Palmer, J. L. Chao, *Appl. Spectrosc.* **1988**, *42*, 546.

[14] C. J. Manning, R. A. Palmer, J. L. Chao, *Rev. Sci. Instrum.* **1991**, *62*, 1219.

[15] G. V. Hartland, W. Xie, H.-L. Dai, A. Simon, M. J. Anderson, *Rev. Sci. Instrum.* **1992**, *63*, 3261.

[16] I. Noda, *J. Am. Chem. Soc.* **1989**, *111*, 8116.

[17] I. Noda, *Appl. Spectrosc.* **1990**, *44*, 450.

[18] I. Noda, *Appl. Spectrosc.* **1993**, *47*, 1329.

[19] V. G. Gregoriou, *Ph.D. Dissertation*, Duke University, 1993.

Mikrochim Acta [Suppl.] 14, 603–605 (1997)
© Springer-Verlag 1997

Step-Scan FT-IR Time-Resolved Spectroscopy (S² FT-IR ΔA TRS) of the Photodynamics of Carbonmonoxy-myoglobin

Richard A. Palmer[1,*], **Susan E. Plunkett**[1], **Pingyun Chen**[1], **James L. Chao**[2], and **Thomas J. Tague**[3]

[1] Department of Chemistry, Duke University, Box 90346, Durham, North Carolina, 27708-0346, U.S.A.
[2] IBM Corporation, Research Triangle Park, North Carolina, 27709, U.S.A.
[3] Bruker Instruments, Inc., 19 Fortune Drive, Manning Park, Billerica, MA 01821, U.S.A.

Abstract. The kinetics of protein response and of CO recombination after photolysis of the Fe–CO bond in carbonmonoxy-myoglobin have been monitored by step-scan FT-IR absorption difference time-resolved spectroscopy (S² FT-IR ΔA TRS) in D$_2$O solution at ambient temperature. From simultaneous measurement of changes in vCO and in the amide I band it has been possible to correlate the CO recombination kinetics with protein secondary structural changes with µs time resolution. The spectral and kinetic data corroborate and confirm previously published single frequency infrared studies indicating that the rebinding process is accomplished with only minimal change in the protein secondary structure. Data of higher temporal resolution (to 20 ns) have also been obtained for the CO recombination process.

Key words: amide I; carbonmonoxy-myoglobin; myoglobin; step-scan FT-IR spectroscopy; time-resolved.

conformational changes during the formation of MbCO in D$_2$O solution at room temperature, by use of S² FT-IR ΔA TRS with µs temporal resolution, is reported. This is the first application of time-resolved step-scan FT-IR spectroscopy in the absorbance difference mode to study the photodynamics of an aqueous protein solution at room temperature. This work also demonstrates the potential of the technique for the sub-µs kinetic analysis of other biological molecules of interest. Recent results at 20 ns temporal resolution for the CO recombination process are illustrated.

Although the photodissociation/reassociation process in carbonmonoxy-myoglobin (MbCO) has been studied by a wide variety of techniques [1–6], until recently it has not been possible to apply the multiplex advantage of FT-IR to this problem. The development of step-scan FT-IR time-resolved spectroscopy (S² FT-IR TRS) allows sub-µs TRS data to be obtained simultaneously for both the CO ligand itself and those infrared chromophores, such as the amide I, which are associated with the conformation of the protein backbone [7–9]. Although the static absorption difference spectra indicate only a small difference in the conformations of the carbonmonoxy and deoxy forms, the nature of the conformational path between the two states is revealed explicitly by the dynamic spectra.

In this work the simultaneous monitoring of the kinetics of CO binding and of the associated protein

Experimental

The sample preparation techniques and the instrumentation and data collection procedures used for the lower temporal resolution experiments (> 1 µs), have been detailed in the preliminary report of this work [10]. In these experiments a 2.5 mM solution was found to be ideal for simultaneous monitoring of the changes in both vCO and the amide I band for the 50 µm path length used. While more concentrated solutions (7 mM) can be used to follow the rebinding kinetics of the CO, the high absorption of the amide I band in these solutions has so far precluded their use for simultaneously investigating the question of protein secondary change. The more concentrated solution was also used for the higher temporal resolution (≤ 5 µs) measurements.

A Bruker IFS 88 spectrometer was used in the step-scan mode for all experiments. For the lower temporal resolution experiments (≥ 5 µs) a photoconductive MCT detector was used in the DC-coupled mode with the 16-bit internal ADC. For these experiments, as reported earlier [10], ΔA is given by ratioing data from after and before the laser flash: $\Delta A = -\log[I(t)/I_0]$.

In the more recent higher temporal resolution experiments (≤ 1 µs) a photovoltaic MCT (Kolmar Technologies) was used in the AC-coupled mode with the external PAD fast transient digitizer board. For these experiments the ΔI spectrum is detected directly and can be converted into a ΔA spectrum by ratioing to the single-beam back-

---

* To whom correspondence should be addressed

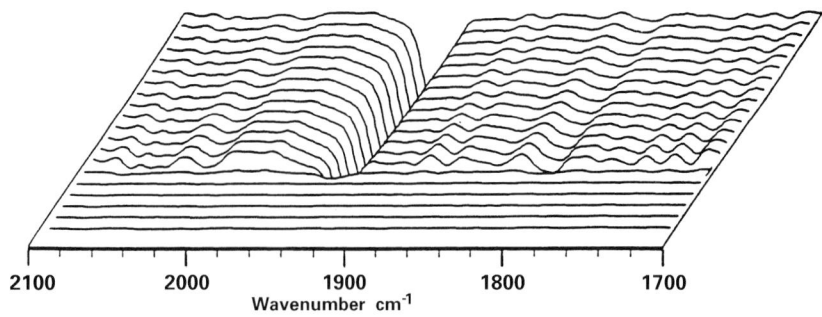

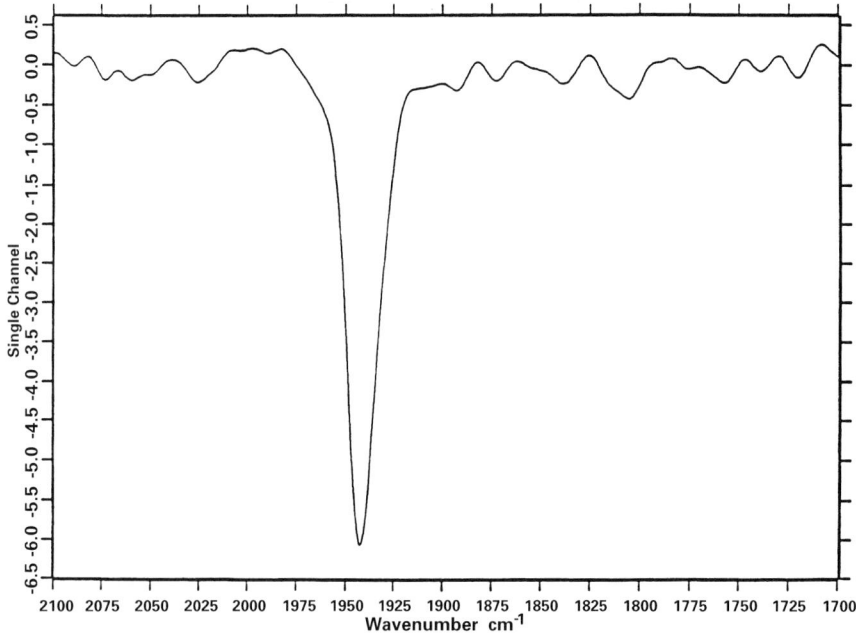

Fig. 1. S^2 FT-IR ΔA TRS of 7 mM MbCO/D$_2$O; 1 μs time resolution (slices from -5 μs to $+14$ μs and single slice at $+5$ μs shown)

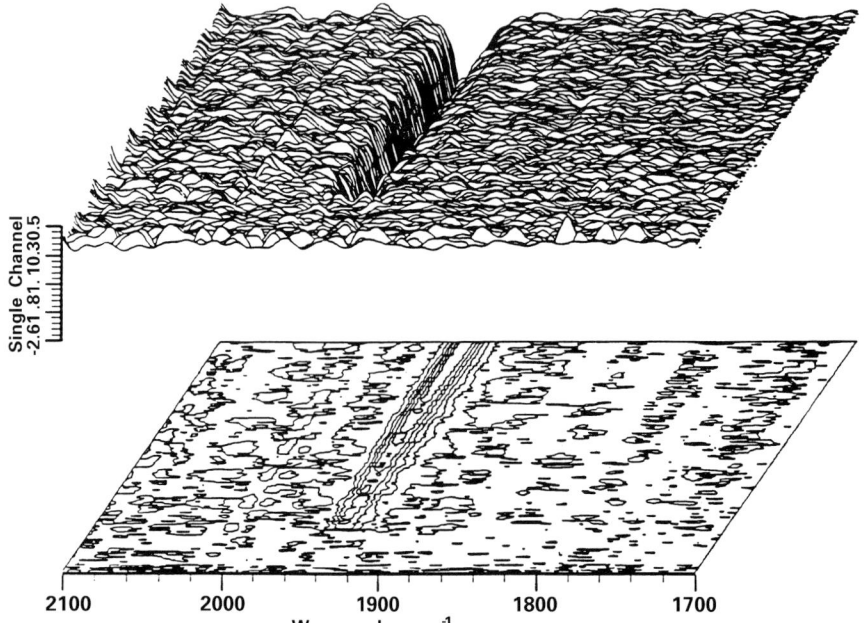

Fig. 2. S^2 FT-IR ΔA TRS of 7 mM MbCO/D$_2$O; 20 ns time resolution (slices from -700 to $+3200$ ns shown)

ground spectrum I_R, also collected in the step-scan mode, but with the laser blocked: $\Delta A = -\log[1 + \Delta I(t)/I_R]$. (Note, however, that the $\leqslant 1\,\mu s$ data presented here are unconverted $\Delta I(t)$ stack plots and contours.)

Results and Discussion

The non-geminate recombination of CO with myoglobin after photodissociation in aqueous solution at room temperature is only complete after ca. 8 ms, although the photodissociation process itself occurs in less than 1 ps [11]. Thus in the investigation of the non-geminate recombination kinetics, the DC-coupled photoconductive MCT with the internal ADC, which provides 5 μs time resolution, appears to be more than adequate to determine the kinetics of CO rebinding and the associated protein conformational changes. In the earlier report the second order A + B model produced rate constants for ΔCO of $6.7 + 3 \times 10^5$ l.-mol^{-1}s^{-1} and $9.8 + 3 \times 10^5$ l. mol^{-1}s^{-1} for Δ amide I [9], essentially confirming the previous conclusion [12] that the small but distinct conformational change is correlated with the CO recombination.

However, in other photophysical processes the ability to achieve higher temporal resolution is crucial. Therefore, it is of interest to see that in a typical heme protein such as myoglobin, sub-μs resolution can be achieved by the S^2 FT-IR ΔA TRS techniques. Figures 1 and 2 illustrate how such data can be obtained in the AC-coupled mode down to 20 ns resolution for the CO rebinding process in a fraction of the time required for the DC-coupled experiment, with comparable signal to noise ratio. The data in Figs. 1 and 2 were collected at 8 cm^{-1} spectral resolution in a single 10 coadd-per-

step scan; undersampling ratio of 7; AC-coupled PV-MCT detector with PAD digitizer; total data collection time of ca. 5 min. In principle, the PAD board should allow 5 ns resolution, although the practical limit in these measurements is 10 ns owing to the structure of the laser pulse. Achievement of such results in the amide I band region for the CO-myoglobin binding process is more difficult because of the small $\Delta I/I$ ratio as compared to that of the CO band itself.

Acknowledgements. The authors acknowledge helpful discussions with R. B. Dyer and W. H. Woodruff, as well as contribution of equipment from Los Alamos National Laboratory and from Bruker Instruments. Financial assistance from Duke University, Bruker Instruments, and Los Alamos National Laboratory through NIH Grant GM 45807 is also gratefully acknowledged.

References

[1] T. Takano, *J. Mol. Biol.* **1977**, *110*, 569.
[2] A. Ansari, C. M. Jones, E. R. Henry, J. Hofrichter, W. A. Eaton, *Science* **1992**, *256*, 1796.
[3] E. R. Henry, J. H. Sommer, J. Hofrichter, W. A. Eaton, *J. Mol. Biol.* **1983**, *166*, 443.
[4] J.-L. Martin, A. Migus, C. Poyart, Y. Lecarpentier, R. Astier, A. Antonetti, *Proc. Natl. Acad. Sci. U.S.A.* **1983**, *80*, 173.
[5] X. Xie, J. D. Simon, *Biochemistry* **1991**, *30*, 3682.
[6] E. W. Findsen, J. M. Friedman, M. R. Ondrias, S. R. Simon, *Science* **1985**, *229*, 661.
[7] R. A. Palmer, *Spectroscopy* **1993**, *8(2)*, 26.
[8] R. A. Palmer, J. L. Chao, R. M. Dittmar, V. G. Gregoriou, S. E. Plunkett, *Appl. Spectrosc.* **1993**, *47*, 1297.
[9] C. J. Manning, R. A. Palmer, J. L. Chao, *Rev. Sci. Instrum.* **1991**, *62*, 1219.
[10] S. E. Plunkett, J. L. Chao, T. J. Tague, R. A. Palmer, *Appl. Spectrosc.* **1995**, *49*, 702.
[11] J. N. Moore, P. A. Hansen, R. M. Hochstrasser, *Chem. Phys. Lett.* **1987**, *138*, 110.
[12] T. P. Causgrove, R. B. Dyer, *Biochemistry* **1993**, *32*, 11985.

Mikrochim. Acta [Suppl.] 14, 607–608 (1997)

2DIR Spectra of Polymers Based on DIRLD Spectra

Roger A. Ingemey, **Geertje Strohe**, and **Wiebren S. Veeman***

Department of Physical Chemistry, Gerhard-Mercator-Universität Duisburg, Lotharstr. 1, 47057 Duisburg, Federal Republic of Germany

Abstract. DIRLD spectra of i-PP and HDPE have been recorded and the origin of all spectral features has been pointed out. Furthermore, DIRLD spectra and the corresponding asynchronous 2DIR plot of a stretching-frequency dependent phenomenon of i-PP has been presented. A model to interpret the spectra has been proposed.

Key words: 2DIR spectra, DIRLD spectra, dynamic infrared spectroscopy, isotatic polypropylene, high-density polyethylene.

DIRLD (dynamic infrared linear dichroic) spectra [1] of three different prestretched isotactic polypropylene samples (i-pp) and a high-density polyethylene sample (hdpe) have been recorded with a Bio-Rad FTS-60A spectrometer over a wide range of stretching frequencies and tensions. The i-pp sample #1 and the pe sample were prestretched to four times their original length, the others less than that. All samples were measured after aligning the direction of the prestretching with the direction of the applied pertubation. By convention, bands with parallel dichroism show negative peaks in the dichroic spectra. Bands with perpendicular dichroism have positive signs. The relative amounts of the dipole transition moments oriented parallel or perpendicular to the stretching direction of the three different i-pp samples can be obtained from the ratio of the intensities of their dichroic spectra. They are found to be in the ratio sample #1 : sample #2 : sample #3 = 6 : 2 : 1.

Figure 1 shows typical spectra of sample #1, recorded at stretching frequencies ranging from 5 up to 20 Hz. The features of the in-phase spectrum can mainly be described in terms of frequency shifts during the stretching cycle [2]. Details such as dichroism of the bands, direction of the frequency shift and changes of the absorption intensities must be considered to ex-

plain the signs and shapes of the DIRLD bands [3]. For instance, the bands at 841 and $1103\,cm^{-1}$ in the in-phase spectrum of isotactic polypropylene show a frequency shift towards smaller wavenumbers with increasing stress (negative frequency shift), but the corresponding bipolar bands are opposite in symmetry of sign, owing to the different dichroism of the bands.

These results have been used to interpret the in-phase spectra of prestretched high density polyethylene (Fig. 2). The line-shape of the rocking modes of polyethylene in the in-phase spectrum can be described as a double bipolar band. Each of the two crystal-field split rocking modes ($720\,cm^{-1}$, $730\,cm^{-1}$) forms a bipolar band, owing to a small negative frequency shift during the stretching cycle. The symmetry in sign of these two bipolar bands is opposite to that of the bipolar band of i-pp at $841\,cm^{-1}$ (Fig. 1), but the same as that of the $1103\,cm^{-1}$ band of i-pp. This is a result of the fact that both the rocking modes of polyethylene and the $1103\,cm^{-1}$ absorption of i-pp have in common

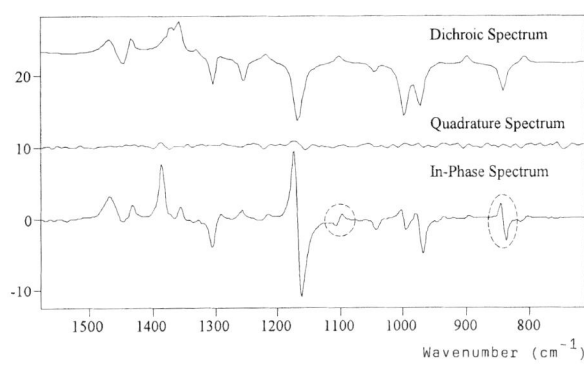

Fig. 1. Dichroic, in-phase and quadrature spectra of prestretched i-pp sample #1 at a stretching frequency of 5 Hz; spectra taken at frequencies up to 20 Hz have the same line shapes and are therefore not shown

---

* To whom correspondence should be addressed

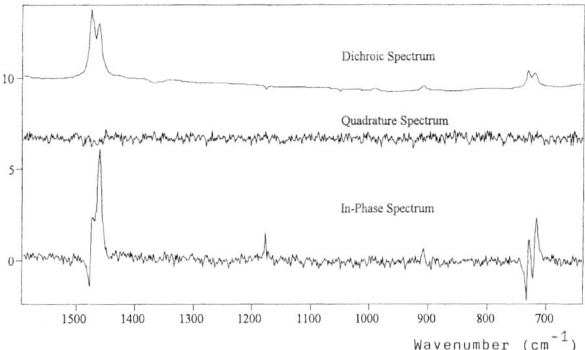

Fig. 2. Dichroic, in-phase and quadrature spectra of prestretched polyethylene at 5 Hz stretching frequency

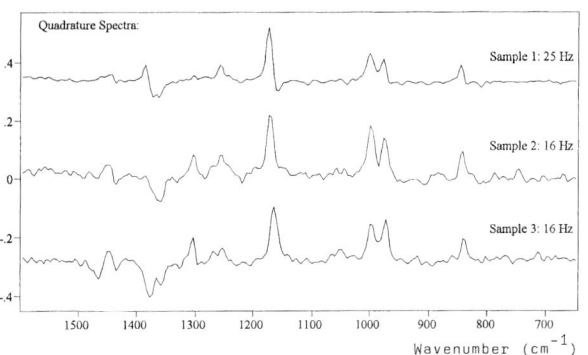

Fig. 3. Quadrature spectra of different prestretched i-pp samples; monopolar bands appear at different stretching frequencies

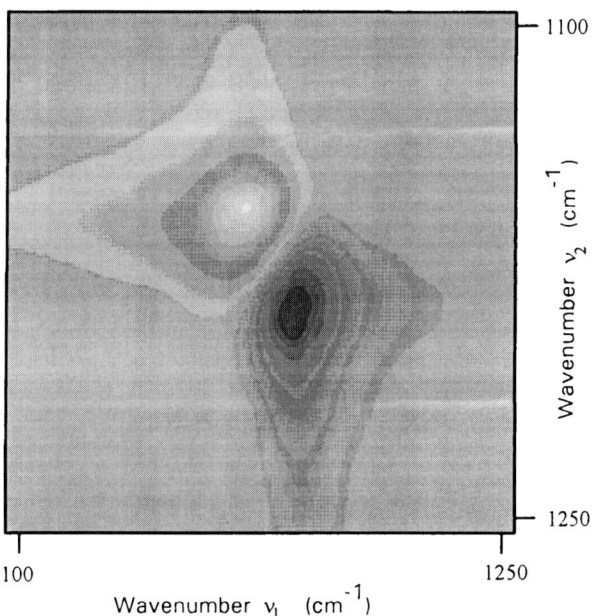

Fig. 4. Asynchronous correlation of i-pp sample #1 in the region between 1100 and 1250 cm^{-1}

a perpendicular dichroism and a small negative frequency shift with increasing stress.

Quadrature spectra with non-zero intensity were found for the prestretched i-pp samples only above a certain threshold frequency (Fig. 3). For sample #1 this threshold frequency was 25 Hz, for the other samples 16 Hz. At 25 Hz stretching frequency all three samples showed small bipolar bands, overlapping with monopolar bands. At 16 Hz sample #1 had no quadrature intensity (Fig. 1), whereas the other two samples showed monopolar bands. The appearance of these monopolar bands in the quadrature spectra is caused by a decrease of the absorption intensity of the dichroic signal during the stretching cycle, while the bipolar bands must be due to phase-lagged, stress-induced frequency shifts.

An asynchronous two-dimensional plot, created by cross-correlating the in-phase and quadrature spectra of i-pp (sample #1), taken at 25 Hz stretching frequency, is presented in Fig. 4. This correlation clearly points out that the change in absorption intensity (leading to monopolar bands) and the frequency shift (leading to bipolar bands), as reactions to the applied stress, respond differently in time. This implies that there may be two different "domains" (A and B) in the prestretched polymer: while chains in domain A are deformed and bipolar bands are caused, chains in domain B exhibit only a change in spectral intensity, probably originating from a reorientation movement of the chains in this domain away from the stretching direction. It seems that both domains consist of amorphous and crystalline regions.

References

[1] I. Noda, *J. Am. Chem. Soc.* **1989**, *111*, 8116.
[2] B. O. Budveska, C. J. Manning, P. R. Griffith, R. T. Roginski, *Appl. Spectrosc.* **1993**, *47*, 1843.
[3] R. A. Ingemey, G. Strohe, W. S. Veeman, *Appl. Spectrosc.* in press.

Mikrochim. Acta [Suppl.] 14, 609–610 (1997)
© Springer-Verlag 1997

Two-Dimensional FT-NIR Study of Dissociation of Hydrogen-Bonded N-Methylacetamide in the Pure Liquid State

Isao Noda[1,*], **Yongliang Liu**[2], and **Yukihiro Ozaki**[2]

[1] The Procter and Gamble Company, Miami Valley Laboratories, P.O. Box 398707, Cincinnati, Ohio 45239-8707, U.S.A.
[2] Department of Chemistry, School of Science, Kwansei Gakuin University, Nishinomiya 662, Japan

Abstract. Generalized two-dimensional (2D) correlation spectroscopy has been applied to the near-infrared (NIR) region to study temperature-dependent dissociation of hydrogen-bonded N-methylacetamide (NMA) in the pure liquid state. Cross peaks in the synchronous and asynchronous FT-NIR correlations spectra suggest that polymeric NMA dissociates into intermediate species such as the dimer first, and then the latter breaks down to the monomer.

Key words: two-dimensional correlation spectroscopy, near-infrared spectroscopy, FT-NIR, amide, dissociation.

The structure and hydrogen-bonding of N-methylacetamide (NMA) have long been a matter of keen interest because NMA is the simplest model for the amide groups of peptides, proteins, and polyamides. Various physicochemical techniques have been applied to the studies of the hydrogen-bonding of NMA. Recently, we employed FT-NIR spectroscopy to investigate the structure of NMA as the pure liquid and in CCl_4 [1]. The FT-NIR study provided new insight into the dissociation and thermodynamic properties of NMA [1]. The purpose of the present paper is to deepen the analysis of FT-NIR spectra of NMA by using 2D correlation spectroscopy [2, 3].

The basic concept of generating 2D IR spectra from perturbation-induced time-dependent fluctuation of IR signals was introduced by Noda in 1986 [2]. In 1993, he proposed more generalized 2D correlation spectroscopy to produce 2D correlation spectra from systematic variations of spectral variations of spectra having an arbitrary and complex wave form [3]. Now, extension to other areas of spectroscopy, such as Raman, and ultrafast time-resolved spectroscopy, has become quite straightforward. The present paper reports the application of the generalized 2D correlation approach to the analysis of a set of FT-NIR spectra of NMA as the pure liquid under variations in temperature.

Experimental

The sample of NMA of high purity (greater than 99.0%) was purchased from Aldrich Chemical Company and used without further purification. The instrumentation and experimental conditions for the FT-NIR measurements [1] and the methods for 2D analysis [4] were the same as those described previously.

Results and Discussion

Figure 1 shows FT-NIR spectra in the 11500–6000 cm^{-1} region of NMA as the pure liquid measured over

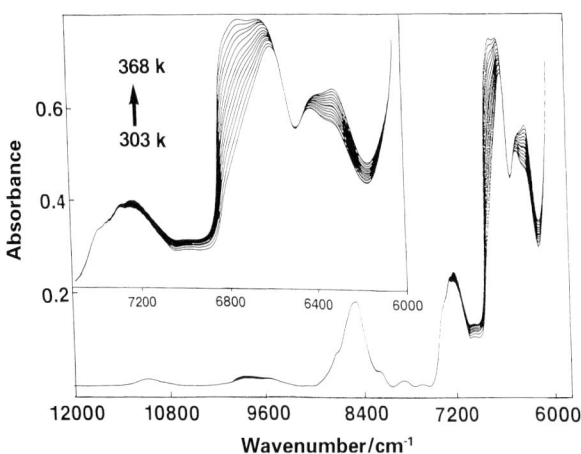

Fig. 1. FT-NIR spectra of NMA as the pure liquid, measured over a temperature range from 30 to 90 °C (after density correction). Inset: Expansion of 7500–6000 cm^{-1} region (Reproduced from ref. [1] with permission. Copyright (1994) Society for Applied Spectroscopy)

---

* To whom correspondence should be addressed

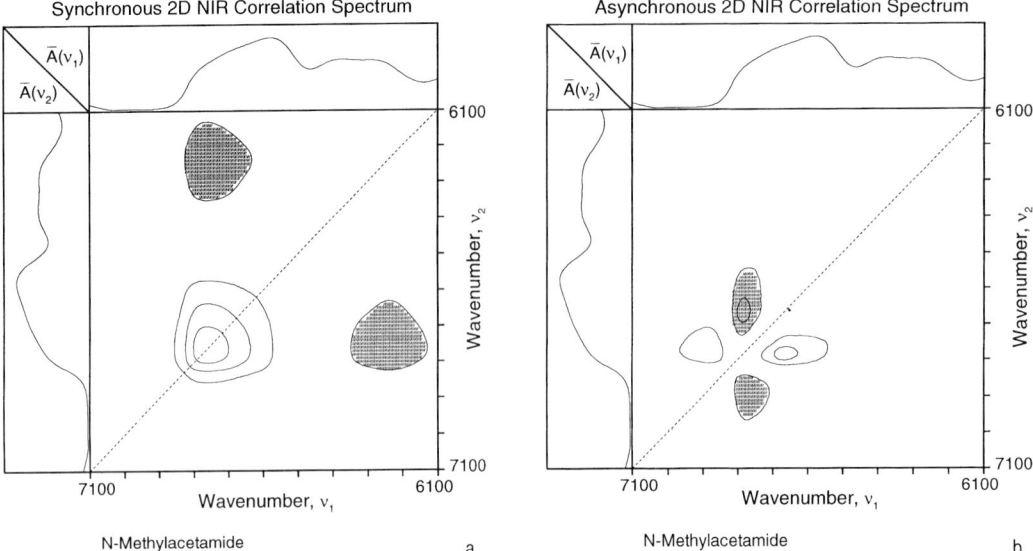

Fig. 2. a synchronous 2D FT-NIR correlation spectrum of NMA in the 7100–6100 cm^{-1} region. **b** corresponding asynchronous 2D FT-NIR correlation spectrum

a temperature range of 30–95 °C (after density correction). Band assignments have been made in our previous paper [1]. Temperature-independent bands at 10 900, 8 500, and 7 200 cm^{-1} are ascribed to the third and second overtones of the CH stretching modes of the CH$_3$ group and the combination of CH vibrations, respectively [1]. Temperature-dependent bands in the 9 900–9 600 cm^{-1} and 6 900–6 200 cm^{-1} regions are due to the second and first overtones of the NH stretching modes of free and hydrogen-bonded NMA, respectively. The temperature-dependent intensity changes in these regions suggest that the hydrogen-bonded NMA dissociates into monomeric species with increase in temperature.

Figure 2a shows a synchronous 2D FT-NIR correlation spectrum in the 7 100–6 100 cm^{-1} region of NMA in the pure liquid. The spectrum represents the temperature-dependent spectral intensity variations of NMA between 30 and 95 °C. The major autopeak at the diagonal position near 6 790 cm^{-1} dominating the synchronous spectrum corresponds to the temperature-induced peak-intensity variation of the first overtone of the NH stretching mode of monomeric NMA. There is one more autopeak near 6 300 cm^{-1}, which is too weak to be observed here. This peak arises from the first overtone of the NH stretching mode of hydrogen-bonded NMA. The essentially coincidental behaviour

of the temperature-dependent intensity change of NIR peaks around 6 790 and 6 300 cm^{-1} for NMA is indicated by the appearance of cross peaks at the appropriate spectral coordinate between the autopeaks.

Figure 2b shows the corresponding asynchronous 2D FT-NIR correlation spectrum in the same spectral region. A cross peak near 6 650 cm^{-1} in the asynchronous 2D NIR correlation spectrum probably corresponds to the first overtone of NH stretching modes of free terminal NH groups of the dimer and oligomers. The signs of the cross peaks indicate that the increase in the peak intensity at 6 790 cm^{-1} takes place at higher temperature than that in the intensities near 6 650 cm^{-1}. It seems, therefore, that the disappearance of the polymeric from does not simultaneously result in the formation of the monomeric from. The polymeric species dissociate into the intermediate species such as the dimer first, and then the latter breaks down to the monomer.

References

[1] Y. Liu, M. A. Czarnecki, Y. Ozaki, *Appl. Spectrosc.* **1994**, *48*, 1095.
[2] I. Noda, *Bull. Am. Phys. Soc.* **1986**, *31*, 520.
[3] I. Noda, *Appl. Spectrosc.* **1993**, *47*, 1329.
[4] I. Noda, Y. Liu, Y. Ozaki, M. A. Czarnecki, *J. Phys. Chem.* **1995**, *99*, 3068.

Mikrochim. Acta [Suppl.] 14, 611–612 (1997)
© Springer-Verlag 1997

Utilizing Fourier-Transform Infrared Photoacoustic Spectroscopy to Analyze Underground Storage Tank Waste Material

Stanley J. Bajic[1,*], **John F. McClelland**[1,2], and **Roger W. Jones**[1]

[1] Ames Laboratory-USDOE, Iowa State University, Ames, IA, 50011–3020, U.S.A.
[2] MTEC Photoacoustics Inc., P.O. Box 1095, Ames, IA, 50014, U.S.A.

Abstract. Hazardous underground storage tank waste from nuclear fuel processing activities poses a very challenging analytical problem. The waste is very heterogeneous and radioactive. Radiation hazards require that sample size and handling be kept to a minimum. A method is presented to determine the chemical composition of the waste tank sludge by using FT-IR-photoacoustic spectroscopy with submilligram sized samples and minimal sample handling. Preparation of samples for analysis as well as data on samples obtained from cores taken from underground storage waste tanks at the Westinghouse Hanford site in Richland, WA, is presented.

Key words: FT-IR-photoacoustic spectroscopy; waste and sludge analysis.

To safely treat and properly dispose of nuclear waste material, the chemical composition must first be determined. The waste material stored in underground storage tanks compositionally varies widely and has a consistency that ranges from peanut butter to salt-cake. Analysis by conventional methods is difficult without extensive, costly and time-consuming sample preparation. Furthermore, the radioactivity of the waste requires that samples be handled in elaborate containment cells (i.e. hot cells) and in small amounts. Several analytical spectroscopic methods have been examined by other workers [1–3]. A method is presented to determine the chemical composition of waste tank sludge by using Fourier transform infrared photoacoustic spectroscopy (FT-IR-PAS) with submilligram sized samples and minimal sample preparation. The method provides accurate compositional analysis of surrogate tank waste and actual tank waste samples supplied by the Westinghouse Hanford Company in Washington state.

* To whom correspondence should be addressed

Experimental

The method was developed with surrogate tank waste supplied by Westinghouse Hanford. Surrogate waste data were collected on a Bio-Rad Digilab FTS-60A FT-IR spectrometer with an MTEC model 200 photoacoustic cell. Tank waste data were collected on a Perkin-Elmer Paragon 1000 FT-IR spectrometer equipped with an MTEC Model 300 photoacoustic cell. Sixty-four scans were accumulated at a mirror velocity of 0.1 cm/s. All spectra were normalized to carbon black to account for spectral variations due to the source and spectrometer. Step-scan phase modulation experiments were also performed on the FTS-60A.

Results and Discussion

The developed methodology involves removing a small amount of underground storage tank sludge ($< 500 \mu g$) from a radiation containment cell, placing the sample on a specially designed sample disk and then freeze drying the sample to remove free water prior to analysis by FT-IR-PAS. This simple sample preparation method produces reproducible homogeneous samples which yield sharp photoacoustic spectra for both qualitative and quantitative analysis. Removing free water by conventional methods (e.g. air or oven drying) leads to the formation of crystals near the sample surface where evaporation occurs. The formation of these crystals is a result of water-soluble species in the sample migrating toward the surface during the drying process.

The extent of migration was studied by FT-IR step-scan phase modulation techniques. Step-scan phase modulation allows determination of whether signal generation is from the surface or the bulk of the sample [4]. Phase modulation data showed that the maximum signal for soluble species from samples that were conventionally dried, occurred at an earlier phase angle

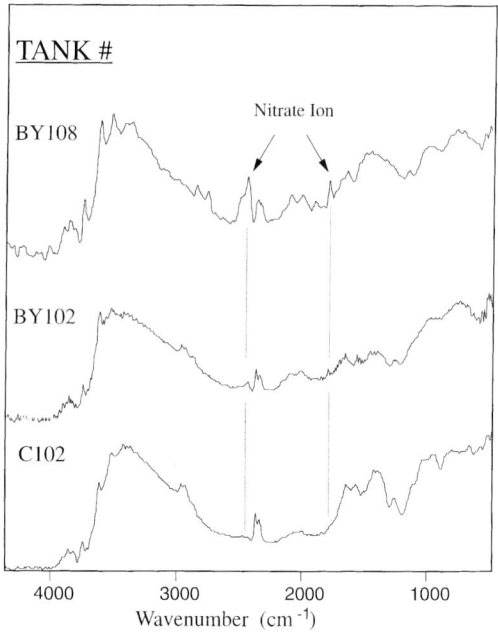

Fig. 1

surrogates were analyzed by standard addition procedures. Linear correlation plots were generated for compounds of interest: disodium nickel ferrocyanide; sodium nitrate; sodium sulfate and sodium phosphate.

The methodology developed was applied to actual radioactive underground storate tank waste at the Westinghouse Hanford Site. The spectrometer and freeze drier were set up in a hood adjacent to a "hot cell". The tank waste samples were transferred to the hood where they were freeze dried and analyzed. The figure shows typical data that were obtained from several archived tank samples. This figure demonstrates how the methodology can be used to monitor the concentration of sodium nitrate from storage tank to storage tank.

Acknowledgements. Ames Laboratory is operated for the U. S. Department of Energy by Iowa State University under contract No. W-7405-ENG-82. This work is supported by the U. S. Department of Energy, Office of Technology Development.

than that for a freeze dried sample. This indicates that during conventional drying water-soluble species migrate, causing artificial layering and variations in the thermal properties of the sample, making quantitative analysis difficult. By comparison, the phase of the maximum signal for insoluble species did not change, for all drying methods.

To demonstrate that the photoacoustic response is linear and can be applied to quantitative analysis, the

References

[1] S. A. Bryan, K. H. Pool, J. D. Matheson, in *Waste Management '93*, (R. G. Post, M. E. Wacks, eds.), University of Arizona, Tucson, 1993, p. 1093.

[2] T. V. Rebagay, D. A. Dodd, D. W. Jeppson, *Proc. SPIE* **1993**, *2089*, 320.

[3] D. R. Lombardi, C. Wang, B. Sun, A. W. Fountain, III, T. J. Vickers, C. K. Mann, F. R. Reich, J. G. Douglas, F. L. Kohlasch, *Appl. Spectrosc.* **1994**, *48*, 875.

[4] For more detail on step-scan phase modulation see R. M. Dittmar, J. L. Chao, R. A. Palmer, *Appl. Spectrosc.* **1991**, *45*, 1104 and R. A. Palmer, J. L. Chao, R. M. Dittmar, V. G. Gregoriou, S. E. Plunkett, *Appl. Spectrosc.* **1993**, *47*, 1297 and reference therein.

Mikrochim. Acta [Suppl.] 14, 613–614 (1997)
© Springer-Verlag 1997

Depth Profiling by FT-IR Photoacoustic Spectroscopy

John F. McClelland[1,2,*], **Roger W. Jones**[1], **Stanley J. Bajic**[1], and **Joan F. Power**[3]

[1] Ames Laboratory-USDOE, Iowa State University, Ames, IA 50011 U.S.A.
[2] MTEC Photoacoustics, Inc., P.O. Box 1095, Ames, IA 50014 U.S.A.
[3] Department of Chemistry, McGill University, 801 Sherbrooke St. W, Montreal, Quebec H3A 2K6, Canada

Abstract. Adding photoacoustic phase data to photoacoustic magnitude data allows for more complete and more quantitative depth profiling of samples with depth dependent composition. Phase modulation is used here to quantitatively measure layer depths in discretely layered samples. The expectation minimum principle is used to derive complete depth profiles of samples with smooth concentration gradients from rapid scan phase and magnitude data.

Key words: depth profiling, photoacoustic spectroscopy, photoacoustic phase.

Much more depth profiling information can be derived when FT-IR-photoacoustic phase and magnitude data are combined than when only magnitude data are used. Phase modulation gives easier access to phase information, and we demonstrate the use of phase modulation for quantitatively determining the depth of layers in samples. The rapid scan approach supplies data at more modulation frequencies over a wider frequency range than phase modulation, which permits determining true depth profiles, but phase data are much more difficult to extract. In addition, the problem of calculating backwards from observed photoacoustic signals to the depth profile that gave rise to them is ill posed; that is, moderate changes in the depth profile produce only small changes in the resulting signals. Here we apply to this problem the minimum expectation principle recently developed by Power and Prystay [1]. This gives a much more robust solution than other methods do. It repeatedly finds the best fit profile after seeding well characterized noise into the thermal diffusion model. The ensemble average of these stochastically distributed solutions is the expectation minimum, which is a good approximation to the global solution (best fit) of the unseeded model for general cases.

* To whom correspondence should be addressed

Experimental

All data were taken on a Bio-Rad FTS 60A FT-IR spectrometer. A 65% carbon-black filled rubber sample was used as a normalization reference, and for the rapid scan data as a phase reference as well. In all phase-modulation experiments, perpendicular interferograms with 400 Hz, 2 laser-fringe-amplitude phase modulation were acquired simultaneously. Spectra for specific phase angles were then calculated from these interferograms [2]. Sets of these spectra were then rotated in the phase angle–wavenumber plane to generate peak height *vs.* phase angle plots.

The term "rapid scan" is used here to mean any conditions under which the interferometer scanning provides the only modulation of the infrared beam (as opposed to phase modulation), whether retardation changes smoothly or stepwise. The rapid scan interferograms of samples were synchronized with those of the phase-reference rubber, and Fourier-transformed by using the phase array generated for the phase reference to give real and imaginary component spectra, from which the phase and magnitude spectra were calculated. A background signal was acquired at each scanning speed, and it was subtracted geometrically from the spectra (that is, the observed spectra were taken to be the vector sum of the background signal and the true spectra). The correction to the phase spectra required to remove instrumental phase shifts was calculated by assuming that the photoacoustic phase of the phase reference was 45°. These corrected phase and magnitude spectra were then subjected to expectation minimum analysis. Samples for the rapid scan study were made by thermally diffusing an additive into thin films or a thick sheet of polypropylene to produce smooth concentration gradients. The sheet was used free-standing and the thin films were backed by a thick polyethylene substrate.

Results and Discussion

Phase Modulation

The much simpler access to photoacoustic phase data provided by phase modulation in comparison to rapid scan has caused a substantial increase in studies involving phase data. Several groups have been working on qualitative depth profiling applications [3, 4], but very little has been done on quantitative depth profiling. For discretely layered samples, however, the relation

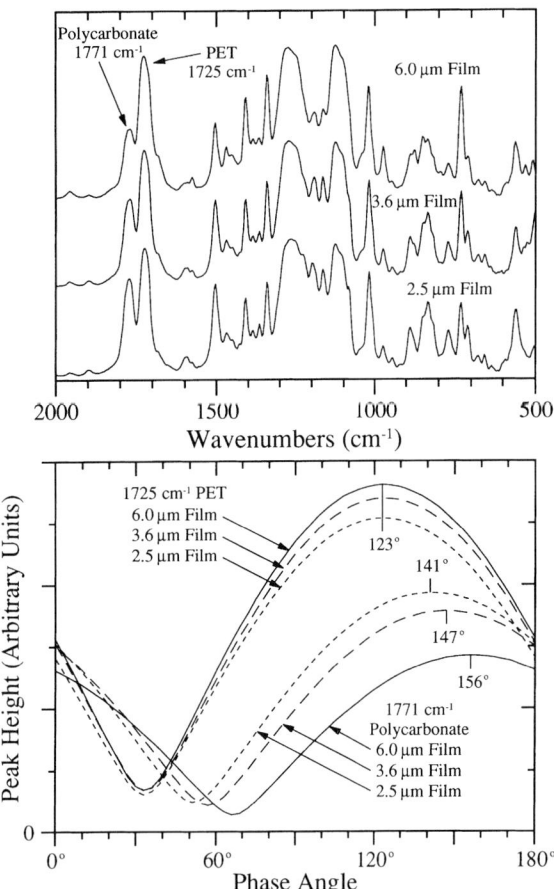

Fig. 1. Phase modulation data from three samples consisting of 2.5, 3.6 and 6.0 μm films of PET atop thick polycarbonate substrates. The top panel shows their magnitude spectra, and the bottom panel shows the phase angle dependence of the carbonyl bands of the PET and polycarbonate layers

between layer depth and the phase of a spectrum band arising from that layer is quite simple. A peak from a layer at depth d in the sample has its phase delayed by an amount $\Delta\theta$ from the value it would have had if the layer were at the sample surface, with $\Delta\theta$ given by $\Delta\theta = d/\mu = d(\pi f/D)^{1/2}$ [5], where μ is the thermal diffusion length, f is the phase-modulation frequency, and D is the thermal diffusivity.

Figure 1 shows spectra of three samples consisting of thin films of polyethylene terephthalate (PET) on thick polycarbonate substrates. The top panel shows that as the film thickness increases, the PET peaks do not change much, but the polycarbonate bands shrink. The

lower panel shows the phase angle dependences of the 1725 cm^{-1} PET carbonyl band and the 1771 cm^{-1} polycarbonate carbonyl band. The PET band phase does not change with film thickness; 2.5 μm of PET is nearly opaque at 1725 cm^{-1}, so the thicker films add little to the PET signal. The polycarbonate band phase, on the other hand, falls back as the thickness increases. With 0.0010 cm^2/s for D [6], the equation predicts $\Delta\theta$ values of 16°, 23° and 39° for the 2.5, 3.6 and 6.0 μm samples. If 123° is the phase of a surface band, then the observed phase delays are 18°, 24° and 33° for the PET films, in good agreement with theory.

Rapid Scan

Three polypropylene samples were prepared having additive concentrations that rose smoothly to a peak at a certain point in the sample; a 0.52 mm thick sample with the peak at the front surface, an 89 μm sample with the peak at the back surface, and a 99 μm sample with the peak at the mid-point (45 μm). Data at ten scanning speeds from 25 Hz to 20 kHz laser fringe frequency (retardation velocities of 0.00158–1.27 cm/s) were taken and analyzed by use of the minimum expectation principle. The analysis placed the front surface, mid-point, and back surface concentration peaks at depths of 8 (±8), 52 (±13) and 78 (±13) μm, in agreement with the actual positions, where the precisions given for the peak positions are the resolutions of the reconstructions. The complete depth profiles showed slightly elevated surface concentration features, but the subsurface concentration gradients were in agreement with the known gradients.

Acknowledgements. Ames Laboratory is operated for the U.S. Department of Energy by Iowa State University under Contract No. W-7405-ENG-82. This work was supported in part by the U.S. Department of Energy, Office of Science and Technology.

References

[1] J. F. Power, M. C. Prystay, *Appl. Spectrosc.* **1995**, *49*, 709.
[2] R. W. Jones, J. F. McClelland, *Appl. Spectrosc.*, in press.
[3] R. M. Dittmar, J. L. Chao, R. A. Palmer, *Appl. Spectrosc.* **1991**, *45*, 1104.
[4] M. G. Sowa, H. H. Mantsch, *Appl. Spectrosc.* **1994**, *48*, 316.
[5] M. J. Adams, G. F. Kirkbright, *Analyst* **1977**, *102*, 281.
[6] H. F. Mark, N. G. Gaylord, N. M. Bikales (eds.), *Encyclopedia of Polymer Science and Technology, Vol. 13*, Wiley, New York, 1970, p. 781.

Mikrochim. Acta [Suppl.] 14, 615–616 (1997)

Photoacoustic Spectroscopy vs. Chemical Categorization of Human Gallstones

E. Wentrup-Byrne[1,*], **L. Rintoul**[1], **J. M. Gentner**[1], **J. L. Smith**[2], and **P. M. Fredericks**[1]

[1] Centre for Instrumental and Developmental Chemistry, Queensland University of Technology, P.O. Box 2434, Brisbane, Q-4001, Australia
[2] Lipid Metabolism Laboratory, Department of Surgery, The University of Queensland, Royal Brisbane Hospital, Herston, Q-4029, Australia

Abstract. The application of Fourier-transform infrared photo-acoustic spectroscopy to categorization of gallstones is demonstrated.

Key words: photoacoustic spectroscopy, gallstone categorization, FT-IR/PAS.

Depending on their chemical constituents, human gallstones can be grouped into four major types: cholesterol, mixed cholesterol and bile pigments, black, and brown stones. The formation mechanism varies greatly between gallstone types. Hence rapid categorization of gallstones is a priority in many areas of modern gallstone research. Present methods of identification involve crushing the stone, extraction, and wet chemical techniques which are tedious, time-consuming and of necessity destructive [1, 2]. In this report we show how FT-IR/PAS (photoacoustic spectroscopy), in conjunction with multivariate analysis techniques, can provide semi-quantitative cholesterol concentration measurements in addition to a broad classification according to gallstone type.

Experimental

Thirty-two gallstones were obtained, with permission, from patients ndergoing elective surgery. FT-IR/PAS spectra were recorded on a Perkin-Elmer System 2000 FT-IR spectrometer, with an 8-mm sample cup in an MTEC Model 200 photoacoustic accessory, from a freshly exposed interior surface of a sectioned gallstone. A representative piece of each gallstone was finely ground prior to solvent extraction and cholesterol determination by standard enzymatic procedures. Multivariate analysis was performed with the Perkin–Elmer QUANT+ program.

Results and Discussion

In cases where the microstructural features are of interest, IR microscopy and Raman are the preferred techniques [3]. However, for categorization purposes FT-IR/PAS offers several advantages. The spectrum is obtained from a relatively large area of the gallstone surface, corresponding to the size of the IR beam. Therefore, a single spectrum contains information on the chemical composition of the whole irradiated surface. Typical spectra are shown in Fig. 1. PAS measurements are relatively insensitive to sample morphology, which means that in most cases little or no sample preparation is necessary.

Partial least-squares analysis [4] was used to model cholesterol concentration with the PAS spectra. Latent variable (LV) scores plots can be used to show groupings of the data. Figure 2 shows a scores plot of LV 1 *vs.*

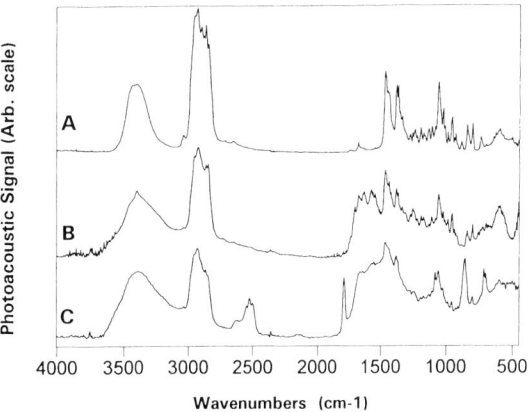

Fig. 1. PAS spectra of three representative gallstones *A* cholesterol, *B* mixed cholesterol and bilirubinate salts, *C* a mixed stone with calcium carbonate. Conventional cholesterol assay of these stones gave 100%, 67% and 89%, respectively

---

* To whom correspondence should be addressed

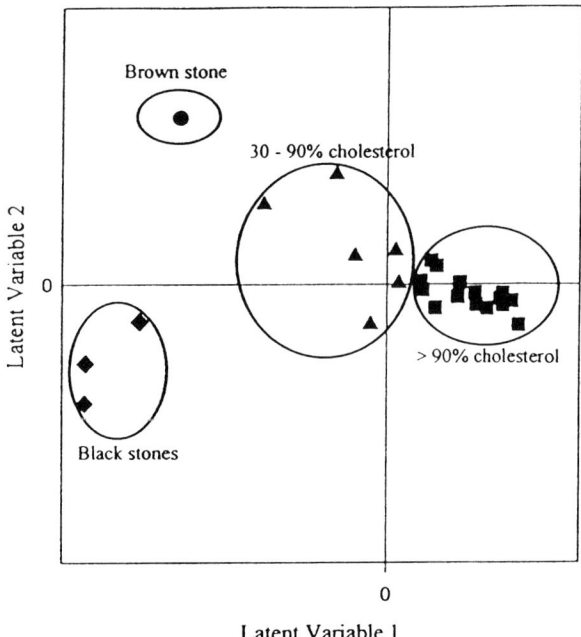

Fig. 2. Latent variable scores plot from PLS analysis

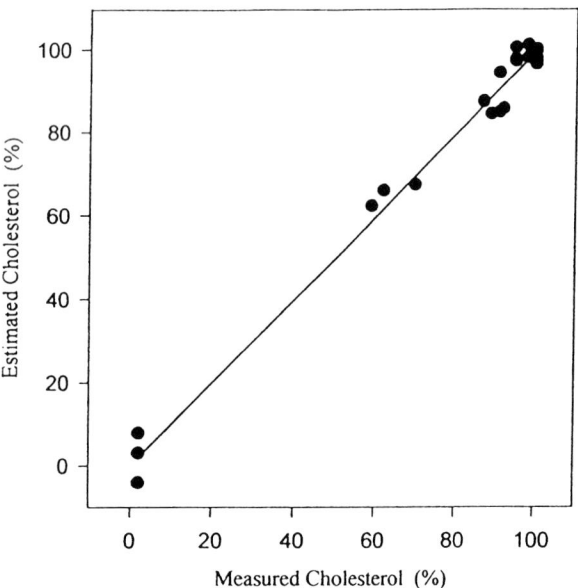

Fig. 3. Estimated verses measured cholesterol (%)

along the LV 1 axis which, in this case, largely corresponds to the cholesterol concentration. Interestingly, the brown stone is located separately from other mixed stones, despite having similar cholesterol concentration.

In QUANT+ the PLS model is developed by using only those latent variables which contribute significantly to the regression equation, to avoid over-fitting of the data. Cross validation was also used so that a more robust model was developed. The model predictions *vs.* measured cholesterol concentration are shown in Fig. 3. Reasonable agreement between predicted and measured cholesterol was obtained in all but a few examples (where the stones possessed a high degree of inhomogeneity). Despite these samples, a standard error of prediction (SEP) of 7% was obtained. Greater accuracy could be achieved, at the expense of structural integrity, if the whole stone was crushed prior to analysis.

Conclusions

PAS may be used as a rapid, relatively non-destructive method of categorizing stones into four recognized groups. Semiquantitative determination of cholesterol may be obtained, though the accuracy is limited in cases where the chemical composition of the interior is not truly representative of the whole stone. The opportunity for human error is reduced compared to conventional techniques. Grinding the gallstone achieves greater accuracy but negates the main advantage of the FT-IR/PAS technique.

References

[1] J. S. Wei, H. M. Huang, W. C. Shyu, C. S. Wu, *Clin. Chem.* **1989**, *35*, 2247.
[2] F. O. Salè, S. Marchesini, P. H. Fishman, B. Berra, *Anal. Biochem.* **1984**, *142*, 347.
[3] E. Wentrup-Byrne, L. Rintoul, J. L. Smith, P. M. Fredericks, *Appl. Spectrosc.* **1995**, *49*, 1028.
[4] *Quant+ Users Manual*, Perkin–Elmer Ltd., Beaconsfield, 1991.

LV 2. Four groupings may be identified. These are: >90% cholesterol, 30–90% cholesterol, brown, and black gallstones. Only one brown stone occurred in the group of 32. As expected, separation mainly occurs

Mikrochim. Acta [Suppl.] 14, 617–619 (1997)
© Springer-Verlag 1997

Study of ZrO_2 Coatings by Thermoanalytical and Fourier Spectroscopic Methods

J. Mihály[1], **J. Kristóf**[1,*], **J. Mink**[1], **L. Nanni**[2], **D. Patracchini**[2], and **A. De Battisti**[2]

[1] Department of Analytical Chemistry, University of Veszprém, P. O. Box 158, H-8201 Veszprém, Hungary
[2] Dipartimento di Chimica, Università di Ferrara, via L. Borsari 46, I-44100 Ferrara, Italy

Abstract. The formation mechanism of ZrO_2 – used for the stabilization of electrocatalytic thin films – from $ZrOCl_2 \cdot 8H_2O$ precursor was followed as the function of the firing temperature for the powder form as well as for the form of a thin film deposited on a titanium support. It was found that an intramolecular hydrolysis takes place upon heating, resulting in the liberation of hydrogen chloride. XRD, FT-IR and FT-Raman investigations revealed that polymorphism occurs in the oxide film. The upper surface of the film is of tetragonal structure, while the bulk is of cubic form stabilized by TiO_2 formed by oxidation of the support.

Key words: ZrO_2, $ZrOCl_2 \cdot 8H_2O$, intramolecular hydrolysis, electrocatalytic thin film, FT-IR spectrometry, FT-Raman spectrometry.

The increasing demand for efficient electrodes in industrial electrochemistry for chlorine and oxygen evolution, waste water treatment, organic electrosyntheses, etc. has prompted fundamental studies on the modifications of electrode materials. More recently, research has been extended to systems where the catalytically active component of the electrode film (IrO_2, RuO_2) is stabilized by ZrO_2 [1]. The application of Fourier transform (emission) IR spectroscopic methods for the study of the formation of TiO_2-stabilized (thermally prepared) mixed oxide systems proved to be very efficient [2, 3]. The complexity of formation of the ZrO_2-containing systems, however, requires understanding of the behaviour of the individual components as a function of temperature. Knowing the chemical path for the individual components, investigation of the formation mechanism of the two-component systems can be attempted.

In the present study the thermal behaviour of $ZrOCl_2 \cdot 8H_2O$ precursor salt is studied. The structure of the resulting ZrO_2 is investigated by applying IR

and Raman spectroscopy to the powder form as well as the form of a coating film on a titanium support.

Experimental

Thermoanalytical investigation of $ZrOCl_2 \cdot 8H_2O$ powder in a dry oxygen atmosphere was performed with Derivatograph-C type thermoanalytical equipment (Hungarian Optical Works, Budapest) at a heating rate of 5 °C/min. A thermogas titrimetric (TGT) unit was used for the simultaneous determination of HCl evolved from the heated sample. Coatings of the precursor salt were made on titanium sheets (size 4×12 mm, thickness 0.1 mm) from a 0.05 M alcoholic stock solution of the salt. The solution was deposited drop-by-drop onto the metal support and the solvent was removed by hot air (60 °C). This procedure was continued until a coating thickness of 400–800 nm was attained. Firing of the coating was done in the thermobalance under the same conditions as for the powder. Investigation of the fired coating was carried out in CsI pellets (after removal of the film) by means of a Bomen MB-102 type FT-IR spectrometer. FT-Raman investigation of the oxide coating was performed *in situ* with a Bio-Rad dedicated FT-Raman spectrometer. XRD investigations were done with a Philips PW-1050/25 powder diffractometer.

Results and Discussion

The thermoanalytical curves of $ZrOCl_2 \cdot 8H_2O$ recorded for a single sample of 101.69 mg are shown in Fig. 1. The curves indicating the total mass loss (TG), the rate of mass loss (DTG), the enthalpy change (DTA), the evolution of hydrogen chloride (HCl) and the rate of HCl evolution (DHCl) were recorded directly. The curve indicating water release (H_2O) was obtained as the difference of the TG and HCl curves. Since above 400 °C no water is formed, this difference gives the amount of elemental chlorine (curve Cl_2) released from the sample. [The formation of a small amount of chlorine (1.25%) was also confirmed by iodometric titration]. Curves designated by DH_2O and DCl_2

---

* To whom correspondence should be addressed

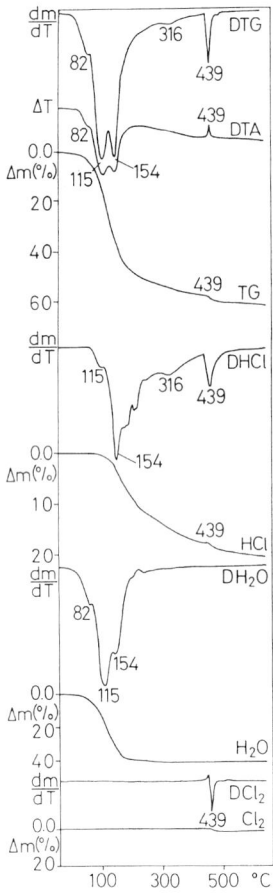

Fig. 1. Thermoanalytical curves of $ZrOCl_2 \cdot 8H_2O$

indicate the release rate of water and chlorine, respectively. Comparing the curves, it can be concluded that an intramolecular hydrolysis takes place in the heated sample, resulting in the formation of hydrogen chloride. The decomposition step at 82 °C belongs exclusively to the removal of approx. 1 mole of water. At higher temperatures, however, the simultaneous liberation of H_2O and HCl can be observed. In the last decomposition step (above 400 °C) only HCl and Cl_2 are liberated. Thus, the decomposition processes can be described by the following equations:

$$ZrOCl_2 \cdot 8H_2O = ZrOCl_2 \cdot (8-n)H_2O + nH_2O \quad (1)$$

$$ZrOCl_2 + H_2O = ZrO_2 + 2HCl \quad (2)$$

$$ZrOCl_2 + \frac{1}{2}O_2 = ZrO_2 + Cl_2 \quad (3)$$

According to the measurement results, about 99% of the chloride content is liberated in the form of HCl, and the rest is released as Cl_2. The sharp exothermic peak in the DTA curve at 439 °C indicates a structural change. X-ray diffraction investigations revealed that at this temperature an amorphous ZrO_2 structure (containing a small amount of only the cubic form) is converted into a well-structured tetragonal form. FT-Raman measurements (curve A in Fig. 3) are in harmony with this observation.

FT-IR investigations of the fired coating, however, showed that under the conditions of electrode preparation cubic ZrO_2 is formed (curve B in Fig. 2). The formation of the cubic form was also confirmed by XRD investigations. *In situ* FT-Raman investigations, however, showed the formation of a tetragonal struc-

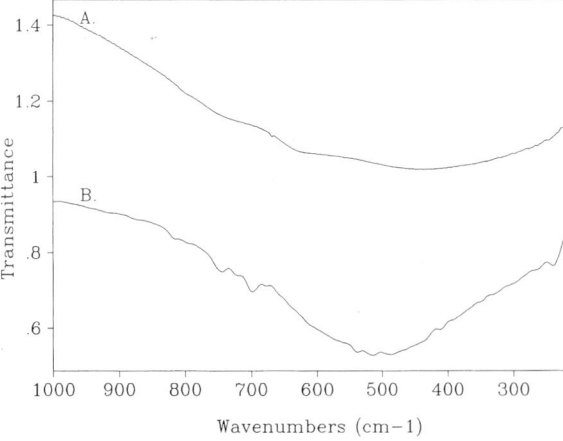

Fig. 2. FT-IR spectra of ZrO_2 (*A* in powder form; *B* as a coating)

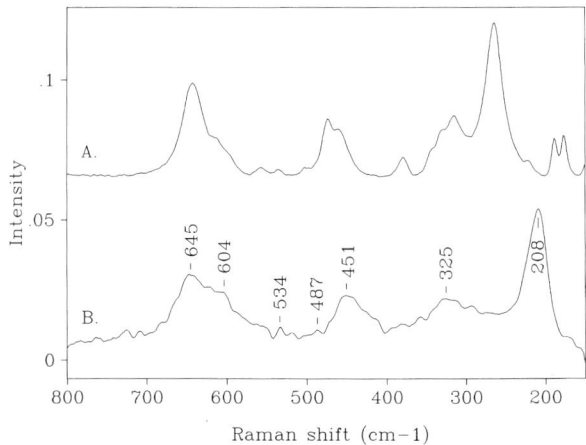

Fig. 3. FT-Raman spectra of ZrO_2 (*A* in powder form; *B* as a coating)

ture (curve B in Fig. 3). [Although the spectra of tetragonal ZrO$_2$ powder (curve A) and the ZrO$_2$ film (curve B) do not exactly match, it is obvious that a ZrO$_2$ structure close to the tetragonal form is obtained]. The phase transition pattern of ZrO$_2$ as a function of temperature follows the cubic → tetragonal → monoclinic sequence [4]. Since at 500 °C oxidation of the support material can also take place (TiO$_2$ in rutile form was identified in the film by XRD), it is reasonable to suppose that TiO$_2$ can stabilize the cubic form of ZrO$_2$. At the surface, however, no rutile is present, thus the formation of tetragonal zirconia is not hindered.

Conclusion

As a conclusion it can be said that segregation of the ZrO$_2$ film occurs under the conditions of electrode preparation. The upper layer of the film has a tetragonal structure, while the lower one is cubic. These results are extremely useful in attempts made to characterize ZrO$_2$-stabilized electrocatalysts with particular respect to microstructure and surface morphology. The joint use of thermoanalytical, XRD, FT-IR and FT-Raman techniques to follow the mechanism of film formation is very promising and deserves the attention of surface scientists.

Acknowledgement. This research was supported by the European Community in the frame of the COST Project D5 "Surfaces and Interfaces" (Contract No. ERBCIPECT 926104).

References

[1] A. De Battisti, G. Battaglin, A. Benedetti, J. Kristóf, J. Liszi, *Chimia* **1995**, *49*, 17.
[2] J. Kristóf, J. Mink, A. De Battisti, J. Liszi, *Electrochim. Acta* **1994**, *39*, 1531.
[3] J. Mink, J. Kristóf, A. De Battisti, S. Daolio, Cs. Németh, *Surf. Sci.* **1995**, *335*, 252.
[4] C. M. Phillippi, K. S. Mazdiyasni, *J. Am. Ceram. Soc.* **1971**, *54*, 254.

Mikrochim. Acta [Suppl.] 14, 621–623 (1997)

Order and Disorder in Self-Assembled Monolayers Probed by FT-RAIRS

F. Bensebaa*, Ch. Bakoyannis, and T. H. Ellis

Départment de Chimie, Université de Montréal, Montréal, Québec, H2C 3J7, Canada

Abstract. Grazing-incidence Fourier-transform reflection–absorption infrared spectroscopy (FT-RAIRS) was used to characterize self-assembled monolayers (SAMs) of alkyl thiols (HS–$(CH_2)_{n-1}$–R) tightly bonded to noble metals. The spectral features were found to be sensitive to the structural and particularly the conformational details of the monolayer film. Information on the annealing and adsorption processes of the films could be obtained with this technique.

Key words: monolayer, thiol, order, spectroscopy and kinetics.

Early published work showed that alkyl thiols adsorbed on gold [the most studied self-assembled monolayers (SAMs)] are highly ordered, packed and relatively defect-free [1]. Recent investigations have revealed more about the different types of disorder that can occur [2, 3]. Disorder due to defect chains (*gauche* conformations and surface pits), impurities (solvent retention) and incomplete packing will affect the mechanical, chemical and thermal stability of these films. These factors are critical in most technological applications such as non-linear optics and electrochemistry. This article will present an overview of our recent systematic study of the consequences of varying the substrate nature, chain composition, incubation time, temperature and annealing procedure on the quality of the films.

Experimental

RAIRS measurements were performed with a research series FTIR spectrometer (Mattson Instruments RS-1) equipped with a mid-range liquid-nitrogen cooled MCT detector. The substrate of evaporated gold on titanium-precoated glass was chemically cleaned before immersion in the thiol solution. Details of the experimental set-up and sample preparation are given elsewhere [3, 4].

Results and Discussion

Displayed in Fig. 1 are the temperature dependences of the peak position and the integrated intensity of the antisymmetric methylene stretch mode (d^-) of a C_{22} ($HS(CH_2)_{21}CH_3$) monolayer. In both cases, two regimes could be distinguished, below and above 350 K. The abrupt increase of the peak position frequency was ascribed to the appearance of bulk *gauche* defects, i.e. defects throughout the chain [1, 3, 4]. We have used the generally accepted rule that values of the d^- peak position below 2918 cm^{-1} are characteristic of a crystalline state of the SAM with an all-*trans* conformation, whereas with increasing *gauche* defects, $v(d^-)$ increases to above 2918 cm^{-1}.

The integrated intensity goes through a minimum at around 350 K. We ascribed this minimum to two competing phenomena: chain untilting and *gauche* defects. Owing to a combination of the surface IR selection rule and the initial alignment of the chains, chain untilting

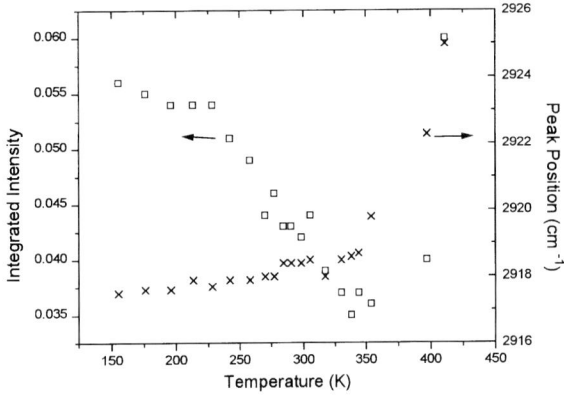

Fig. 1. Temperature dependence of the integrated intensity (□) and peak position (×) of the antisymmetric methylene stretching mode of a C_{22} thiol film

---

* To whom correspondence should be addressed

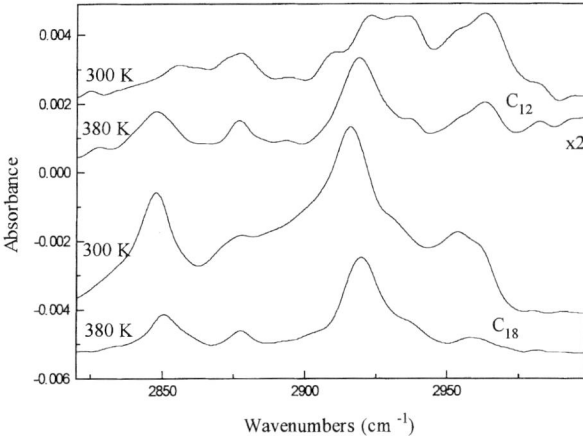

Fig. 2. RAIR spectra of C_{12} and C_{18} thiol films before and after annealing to 380 K; the two top curves are enlarged by a factor of 2

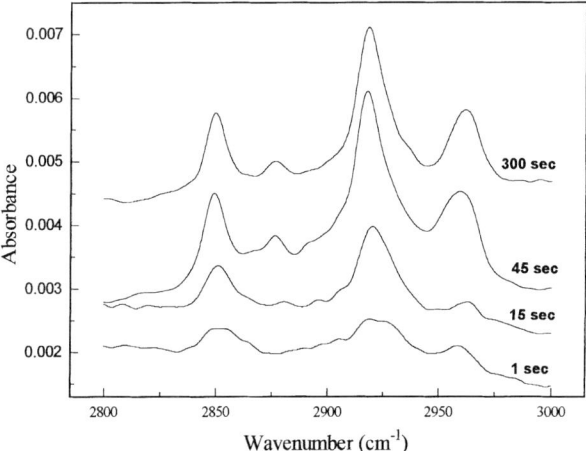

Fig. 3. Incubation time dependence of the RAIR spectra of a C_{22} thiol film

would decrease the intensity since the CH_2 dipole moments become parallel to the metal surface when the chains are aligned along the surface normal. Recent M.D. calculations [5] showed that the tilt angle relative to the normal decreases from 30° at ~100 K to 10° at ~400 K. In contrast, *gauche* defects induce an increase in the signal intensity by randomization of the dipole orientations. Similar features were observed with other long-chain films ($n = 15, 18$ and 20) [4] and end groups (R = COOH) [3].

It is tempting to ascribe the critical temperature ($T_c \sim 350$ K) to a transition from a crystalline phase to a liquid-like phase, as proposed by Fenter et al. [6] on the basis of X-ray investigations. However, the RAIR spectral features, particularly the line width and the

peak position at just above 350 K, are not characteristic of the liquid state [3, 4]. It is not until about 400 K that the integrated intensity is at maximum, as expected for an isotropic state. At even higher temperatures, there is an abrupt irreversible decrease in all measured intensities, which is assigned to film desorption.

Contradictory results have been reported in the literature, concerning the annealing effect on the degree of order of SAMs [2, 3, 6]. We have recently found that C_{12} films become ordered upon annealing to about 380 K, whereas the C_{18} films (as for C_{15}, C_{20}, and C_{22}) are irreversibly disordered when heated above 350 K [3, 4], as evidenced by an increase in the d^+ and d^- frequencies (Fig. 2). Therefore, the shortened chains that we (and others by using diffraction techniques) have found to become better ordered upon annealing, behave similarly to other chemisorption systems involving small molecules. In contrast, the disordering of longer chains is complicated by chain entanglement. At the temperature that would be required to make the long chains mobile enough to anneal, the chains themselves become conformationally disordered, and the subsequent entanglement becomes "frozen in" when the surface is cooled.

RAIR spectra of four C_{22} SAMs prepared at room temperature with different incubation times in $5\,\mu M$ ethanol solution and measured at 140 K are displayed in Fig. 3. Whereas at the lowest incubation time (1 s) the spectrum is typical of a disordered thiol film, the spectra taken for at least 45 s of incubation time were characteristic of ordered SAMs [1, 3, 4]. With increasing incubation time the peak position of the d^- mode decreases quickly from 2920 to 2917 cm^{-1}, suggesting that no appreciable improvement in conformational order occurs after 45 s. Since the conformational order depends strongly on the film density, this indicates that most of the film has been deposited and ordered after a short time, even at this low solution concentration. We also note that for all the disordered films, a peak at ~2935 cm^{-1} was observed, which indicates that this mode does not originate solely from Fermi resonance and may in fact be a specific methylene vibrational mode that is present in disordered films.

Conclusion

The present study showed the possibilities offered by FT-RAIRS for investigating the dynamic and static properties of SAMs. It is unfortunate that data from the Au–S vibration mode obtained by using synchrotron radiation and from a phase diagram obtained by using DSC (differential scanning calorimetry) are lacking. Such data would allow a better comprehension of the microscopic structure of the SAM and also the molecular mechanism of the film formation and annealing.

References

[1] L. H. Dubois, R. G. Nuzzo, *Ann. Rev. Phys. Chem.* **1992**, *43*, 437 (and references therein).

[2] C. Schönenberger, J. Jorritsma, J. A. M. Sondag-Huethorst, L. G. J. Fokkink, *J. Phys. Chem.* **1995**, *99*, 3259.

[3] F. Bensebaa, T. H. Ellis, A. Badia, R. B. Lennox, *J. Vac. Sci. Technol.* **1995**, *13*, 1331.

[4] F. Bensebaa, T. H. Ellis, A. Badia, R. B. Lennox, *J. Phys. Chem.* submitted.

[5] W. Mar, M. L. Klein, *Langmuir* **1994**, *10*, 188.

[6] P. Fenter, P. Eisenberger, K. S. Liang, *Phys. Rev. Lett.* **1993**, *70*, 2447.

Mikrochim. Acta [Suppl.] 14, 625–626 (1997)

Monolayer Formation of Organosilane Compounds on Hydroxylic Surfaces

Helmut Brunner, Ceri A. Gibson, Ulrich Mayer, and **Helmuth Hoffmann***

Department of Inorganic Chemistry, Technical University of Vienna, Getreidemarkt 9, A-1060 Wien, Austria

Abstract. In this study external reflection infrared spectroscopy and ellipsometry were used to investigate the formation and structure of monolayers of octadecyltrichlorosilane on native silicon surfaces (Si/SiO_2). It is shown that during the initial stage of the adsorption process liquid-like films with randomly oriented hydrocarbon chains are formed. With increasing surface coverage this isotropic sub-monolayer structure changes into a highly ordered, crystalline monolayer phase. The overall time required for complete monolayer formation depends strongly on the water content of the adsorbate solution.

Key words: self-assembled monolayers, external reflection infrared spectroscopy, ellipsometry.

Alkyltrichlorosilanes ($RSiCl_3$) form densely packed monolayers on a variety of non-metallic substrates containing surface hydroxyl groups, and these represent an interesting class of self-assembled monolayers from both fundamental and practical points of view [1]. It is generally believed today that the adsorption process proceeds through hydrolysis of the Si–Cl bonds and formation of silanols as intermediates, which assemble on the substrate surface initially as hydrogen-bonded aggregates. Subsequently, a condensation reaction takes place, anchoring the adsorbate molecules to the substrate through Si–O surface bonds and cross-linking the adsorbate molecules by Si–O–Si bridges, which results in the formation of highly stable alkylsiloxane surface layers [2]. However, the details of this fairly complex adsorption process and the influence of various experimental parameters on the monolayer properties are still unknown. In this study we investigate the structure of octadecyltrichlorosilane (OTS) films on Si/SiO_2 surfaces and the role of water in the adsorbate solutions, as a function of time, by preparing a series of sub-monolayer OTS films on Si wafers, using varied immersion times and varied water

concentrations of the adsorbate solutions, and characterizing these samples by means of external reflection infrared spectroscopy and ellipsometry.

Methods

P-doped, (100) oriented Si wafers (Wacker Chemitronic) were used as substrates and were cleaned by sonication in H_2SO_4/H_2O_2 solution. Film adsorption was performed in a glove box filled with dry nitrogen by immersing the substrates in $1\,mM$ solutions of OTS (Aldrich, 95%) in toluene for specified adsorption times. Afterwards the samples were thoroughly rinsed with pure solvent and dried in a stream of dry nitrogen. Specified water contents of the adsorbate solutions were obtained by mixing appropriate amounts of water-saturated and absolute toluene and determining the water concentrations of the resultant mixtures by Karl Fischer titration.

External reflection IR spectra were measured on a Mattson RS 1 FT-IR spectrometer coupled to a custom-made reflection optical system described in detail elsewhere [3, 4]. P-polarized light at an incidence angle of $80°$ was used and 1024 scans at $4\,cm^{-1}$ spectral resolution were co-added. Ellipsometric film thickness measurements were made with a PLASMOS SD 2300 ellipsometer with rotating analyser, using an He–Ne laser at $70°$ incidence as the light source. The thickness of the oxide layer was measured separately for each substrate before adsorption. An isotropic four-phase model (Si/SiO_2/adsorbate/air) was used to determine the adsorbate thickness, assuming a refractive index of 1.45 for the adsorbate.

Results and Discussion

Figure 1 shows the ellipsometrically measured thicknesses of OTS films on Si substrates prepared in four different adsorbate solutions with water concentrations between 20 and $<0.5\,mM$, as a function of adsorption time. In general, the film thicknesses approach a constant final value of 27 ± 1 Å in these experiments. Only in solutions with water levels significantly lower than the OTS concentration ($1\,mM$) no monolayers were formed (even after a week of adsorption). A monolayer thickness of 27 Å equals the calculated length of an all-*trans* octadecylsiloxane

---

* To whom correspondence should be addressed

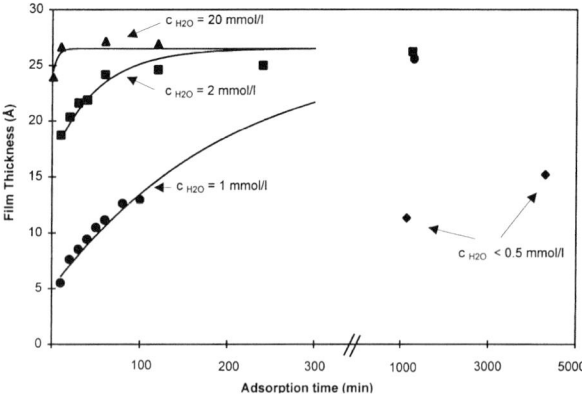

Fig. 1. Ellipsometrically determined thicknesses of adsorbate films of OTS on Si/SiO$_2$ prepared from solutions of OTS (c = 1 mM) in toluene/H$_2$O mixtures with different H$_2$O concentrations, as a function of immersion time in the adsorbate solution. The solid lines represent the best fit of the experimental data, based on a Langmuirian adsorption model (see text for details)

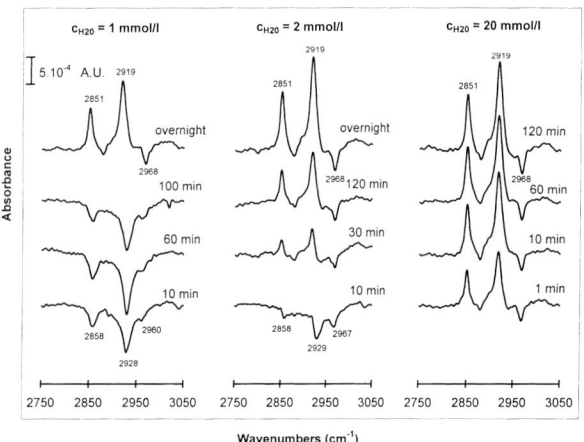

Fig. 2. External reflection IR spectra in the CH stretching region of adsorbate films of OTS on Si/SiO$_2$ prepared from solutions of OTS (c = 1 mM) in toluene/H$_2$O mixtures for different immersion times in the adsorbate solutions. Each spectrum corresponds to a specific film thickness (see Fig. 1)

molecule [1] and has been found in previous studies to be the characteristic thickness of a complete, highly ordered OTS monolayer with uniform, close-to- perpendicular orientation of the hydrocarbon chains on the substrate surface [1, 2]. The film growth curves in Fig. 1 follow to a first approximation a Langmuirian adsorption model, i.e. the film thickness increases proportionally to $(1-e^{-kt})$ (t = adsorption time, k = adsorption rate constant). The adsorption rates increase significantly with increasing water content of the adsorbate solutions. Notably, none of the curves in Fig. 1 pass through the origin. This indicates that some fast adsorption process occurs immediately after immer-

sion of the substrate in the adsorbate solution and leads to a partial coverage of the surface, after which the remaining empty surface sites are filled up comparatively slowly. A similar two-step process has been postulated recently, based on AFM images of OTS adsorbate films [5]. Further insights into the structural changes in the course of monolayer formation are obtainable from the external reflection IR spectra shown in Fig. 2. At low water concentration ([H$_2$O] = 1 mM), three absorptions at around 2858 cm^{-1} [v_s(CH$_2$)], 2928 cm^{-1} [v_{as}(CH$_2$)] and 2960 cm^{-1} [v_{as}(CH$_3$)] initially grow in the negative direction. The positions of these bands and, most significantly, their negative direction under the current experimental conditions (p-polarized radiation at 80° incidence) indicate the formation of a liquid-like, isotropic film with randomly oriented hydrocarbon chains on the surface [4, 6]. After about one hour, approximately 40% of the surface is covered (judged from the measured film thickness of $\approx$ 11 Å, Fig. 1) and the v(CH$_2$) absorptions start to decrease and convert into sharp, positive bands at lower wavenumbers (2851 cm^{-1} and 2919 cm^{-1}) characteristic of a close-to-vertical hydrocarbon chain orientation on the Si surface [4]. Monolayer formation is complete in solutions with 1 mM water after overnight adsorption. Doubling the water concentration (2 mM) shifts this disorder – order transition to the 10–30 minutes time range, while in water-saturated toluene [H$_2$O] = 20 mM) an almost complete, oriented monolayer forms after only one minute of adsorption.

A more detailed kinetic and mechanistic analysis of this adsorption process is currently in preparation and will be presented elsewhere.

Acknowledgement. This work was supported by the Fonds zur Förderung der Wissenschaftlichen Forschung (Proj. No. P 09749) and the Hochschuljubiläumsstiftung der Stadt Wien.

References

[1] A. Ulman, *An Introduction to Ultrathin Organic Films: From Langmuir–Blodgett to Self-Assembly*, Academic Press, New York, 1991.
[2] A. N. Parikh, D. L. Allara, I. B. Azouz, F. Rondelez, *J. Phys. Chem.* **1994**, *98*, 7577.
[3] H. Hoffmann, U. Mayer, H. Brunner, A. Krischanitz, *Vib. Spectrosc.* **1995**, *8*, 151.
[4] H. Hoffmann, U. Mayer, A. Krischanitz, *Langmuir* **1995**, *11*, 1304.
[5] D. K. Schwartz, S. Steinberg, J. Israelachvili, J. A. N. Zasadzinski, *Phys. Rev. Lett.* **1992**, *69*, 3354.
[6] H. Hoffmann, U. Mayer, H. Brunner, A. Krischanitz, *J. Mol. Struct.* **1995**, *349*, 305.

Mikrochim. Acta [Suppl.] 14, 627–629 (1997)

Double Beam FT-IR Reflection Spectroscopy on Monolayers

Thierry Buffeteau[1], Bernard Desbat[1,*], Eve Péré[1], and Jean Marie Turlet[2]

[1] Laboratoire de Spectroscopie Moléculaire et Cristalline (CNRS URA 124), F-33405 Talence Cedex, France
[2] Centre de Physique Moléculaire Optique et Herztienne (CNRS URA 283), Université Bordeaux I, 351 cours de la libération, F-33405 Talence Cedex, France

Abstract. For *in situ* infrared studies of ultrathin films deposited on solid substrates or spread at an air/liquid interface, we have developed a new FT-IR double beam approach based on differential reflectivity by polarization modulation. Because only the s-component (perpendicular to the plane of incidence) of the electromagnetic field is incident on the studied surface, the spectral features are easy to interpret in terms of molecular orientation. Spectra of a cadmium arachidate Langmuir–Blodgett film deposited on a gold mirror and a DMPC Langmuir film spread at the air/water interface are reported, to illustrate the advantages of this method.

Key words: double beam FT-IR, reflection spectroscopy, differential reflectivity, polarization modulation.

We have demonstrated that polarization modulation infrared reflection absorption spectroscopy (PM-IR-RAS) is a very sensitive technique which permits *in situ* obtainment of FT-IR spectra of ultrathin films deposited on metallic or dielectric substrates [1, 2]. However, as the observed differential spectrum implies simultaneously the two polarized reflectance compounds R_s and R_p, its interpretation in terms of molecular orientation is not straight forward. Indeed, to get information on vibrational modes with transition moment along the surface, it appears interesting to work only with the R_s component. For this reason we have developed an alternative approach to PM-IRRAS, which keeps the basic principle and advantages of differential reflectivity by polarization modulation but with only s-polarized light incident on the sample.

Experimental

This method combines a ZnSe photoelastic modulator (Hinds type III) and two ZnSe plates working at Brewster incident angle (Fig. 1). At the output of the FT-IR interferometer (Nicolet 740), the parallel infrared beam has an intensity modulation at frequencies ω_i situated in the 0.1–1 kHz range for the mid-infrared spectral domain. The beam passes through a first linear polarizer and the photoelastic modulator, which superimposes a polarization modulation at a fixed frequency $2\omega_m = 62$ kHz. Then, the first ZnSe plate splits the modulated beam into two parts having linear crossed polarizations. The reflected s-polarized beam undergoes another reflection on the sample before the second ZnSe plate recombines it with the transmitted beam, which after passing through the two ZnSe plates has an almost pure p-polarization. Finally, a ZnSe lens focuses the recombined light on an HgCdTe photovoltaic detector (SAT). The dynamics of detection can be optimized by adjusting the intensities of the transmitted and reflected beams to comparable levels.

If R_s and τ_p are respectively the s-polarized reflectance of the sample and the p-polarized transmittance of the two ZnSe plates, the detected intensity I_d at the output of the detector preamplifier can be written for each intensity modulation frequency ω_i:

$$I_d = C \frac{I_0(\omega_i)}{2} \{(\rho_s R_s + \tau_p) + 2(\rho_s R_s - \tau_p) J_2(\varphi_0) \cos 2\omega_m t\}$$

where J_2 is the second order Bessel function of the maximum dephasing φ_0 introduced by the photoelastic modulator, and ρ_s represent the s-polarized reflectance of the two ZnSe plates. C is a constant representative of the overall gain of the optical set-up.

Following a procedure similar to PM-IRRAS [1], band-pass filtering, demodulation by a lock-in amplifier, electronic processing and Fourier transform of the two detected interferograms give two spectra: one is proportional to the sum $(\rho_s R_s + \tau_p)$ and the other is proportional to the difference $(\rho_s R_s - \tau_p)$. Their ratio gives the differential signal:

$$S = \frac{\rho_s R_s - \tau_p}{\rho_s R_s + \tau_p} J_2(\varphi_0)$$

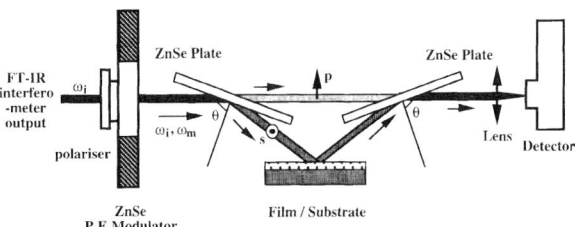

Fig. 1. Optical set-up

* To whom correspondence should be addressed

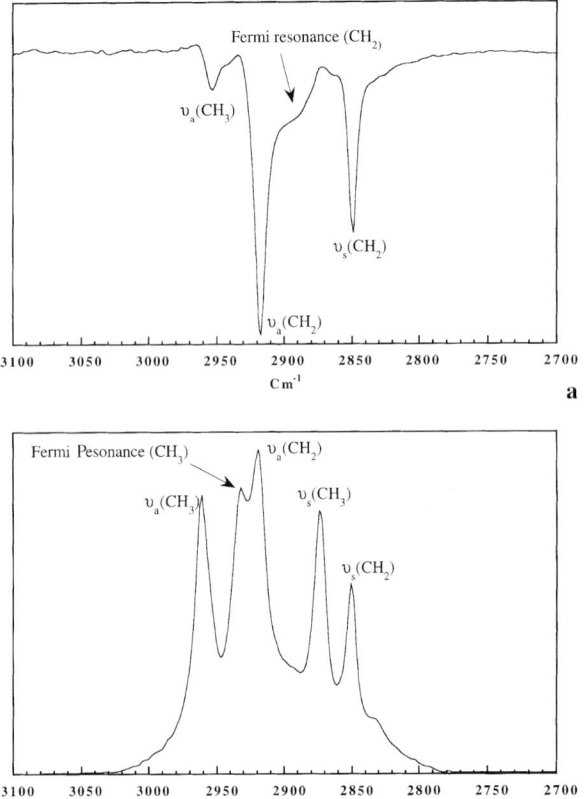

Fig. 2. a Spectrum of 10 Langmuir–Blodgett cadmium arachidate monolayers on gold. **b** PM-IRRAS spectrum of 8 Langmuir–Blodgett cadmium arachidate monolayers on gold

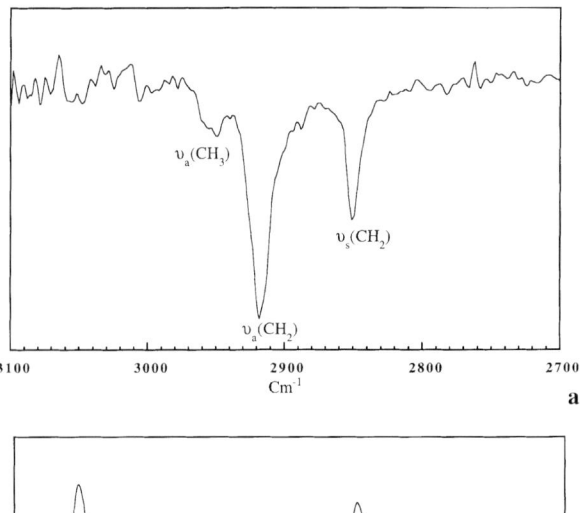

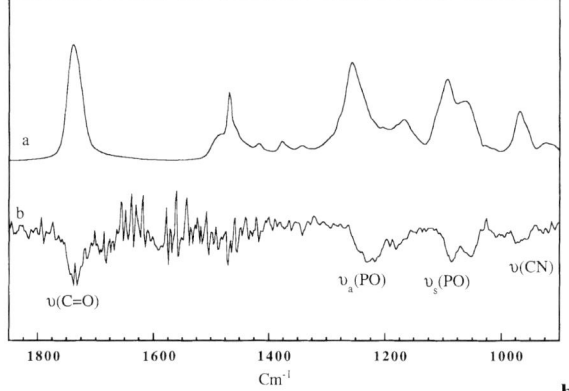

Fig. 3. a DMPC monolayer spread on water. **b** Bulk DMPC (*a*) and DMPC monolayer spread on water (*b*)

Then the J_2 Bessel function dependence may be cancelled out by mathematical modelling or by experimental normalization *vs.* a reference spectrum given by the bare substrate. As $\tau_p = 1$ at Brewster incidence, we finally obtain the normalized differential signal:

$$S_{\text{norm}} = \frac{\rho_s R_s - 1}{\rho_s R_s + 1}$$

Because S_{norm} is sensitive to only the s-polarized reflectance component of the sample, only the absorption bands with non-zero projection of their transition moment in the plane surface appear on the spectrum.

Results

In order to illustrate the potentialities of this method, we present spectra obtained in a few minutes from ultrathin films deposited on metallic and dielectric substrates. These spectra have been recorded at $4\,\text{cm}^{-1}$ resolution by co-addition of 200 interferograms (~ 7 minutes acquisition times).

Figure 2a shows, in the region of the C–H stretching modes, the normalized spectrum of 10 Langmuir–Blodgett monolayers of cadmium arachidate (CdAr)

deposited on a gold substrate (film thickness about $250\,\text{Å}$). Despite the intrinsic weak but non-zero value of the s-polarized component of the electric field near the metallic surface, the S/N ratio of this spectrum is very satisfactory. The band intensity depends on the orientation of the corresponding transition moment with respect to the surface. It is worth noting that this effect is complementary to the one observed with the same sample by PM-IRRAS (Fig. 2b): the more the moment is oriented along the surface, the more intense is the absorption band on the normalized spectrum. Indeed, the $v_a(\text{CH}_2)$ and $v_s(\text{CH}_2)$ bands with transition moment quasi-parallel to the surface are very intense, whereas the $v_s(\text{CH}_3)$ mode with a transition moment preferentially perpendicular to the surface is very weak. This is clear evidence of the quasi-perpendicular orientation of the alkyl chains relative to the gold surface.

In Fig. 3 are presented the normalized spectra of a sample more difficult to analyse: a single monolayer of dimyristoylphosphatidylcholine (DMPC) spread at the air/water interface in a solid-like phase. In the region of the C–H stretching modes (Fig. 3a), the high S/N ratio demonstrates the sensitivity of our approach. As explained above, the intense v_a (CH$_2$) and v_s (CH$_2$) bands appearing in this spectrum correspond to preferential in-plane orientation of transition moments, which is consistent with the DMPC alkyl chains being quasi-perpendicular at the interface. Figure 3b shows a comparison between bulk and monolayer spectra in the region of the absorptions associated with the $v(CO)_{ester}$ at 1736 cm^{-1}, v_a (PO) at 1178 cm^{-1}, v_s (PO) at 1093 cm^{-1} and v_a (CN) at 971 cm^{-1} stretching modes. Variations in the relative intensities of the bands observed for the bulk and monolayer samples allow an estimation of the average orientation of transition moments with respect to the interface according to the following surface selection rule: band intensities are proportional to the square of the projection along the surface of the associated transition moment. On the other hand, the band intensity will be zero if its transition moments is perpendicular to the surface.

In addition, we note that all the observed bands have the same downward orientation relative to the baseline which facilitates the analysis of the spectra, in contrast to conventional PM-IRRAS on a dielectric substrate [2], where the shape of the spectra is more complex because the bands downward or upward relative to the baseline, depending on the orientation of the transition moments.

Conclusion

This method of differential infrared spectroscopy by polarization modulation allows *in situ* study of ultrathin films deposited on metallic or dielectric substrates. With respect to conventional PM-IRRAS, this approach has the advantage of implying only the s-polarized component of the electromagnetic field at the surface. This provides complementary information and allows observation of in-plane vibrational modes on metallic surfaces as well as on dielectric surfaces.

References

[1] T. Buffeteau, B. Desbat, J. M. Turlet, *Appl. Spectrosc.* **1991**, *45*, 380.
[2] D. Blaudez, T. Buffeteau, J. C. Cornut, B. Desbat, N. Escafre, M. Pezolet, J. M. Turlet, *Appl. Spectrosc.* **1993**, *47*, 869.

Mikrochim. Acta [Suppl.] 14, 631–633 (1997)
© Springer-Verlag 1997

Determination of the Optical Constants of a Uniaxial Ultrathin Film from FT-IR Spectroscopy

Thierry Buffeteau[1], **Bernard Desbat**[1], **Eve Péré**[1,*], and **Jean Marie Turlet**[2]

[1] Laboratoire de Spectroscopie Moléculaire et Cristalline (CNRS URA 124), Université Bordeaux I, 351 cours de la libération, 33405 Talence Cedex, France
[2] Centre de Physique Moléculaire Optique et Herztienne (CNRS URA 283), Université Bordeaux I, 351 cours de la libération, 33405 Talence Cedex, France

Abstract. To evaluate the optical anisotropy of ultrathin films ($d < 400$ Å), we have developed two calculation procedures which permit us to obtain the complex refractive indexes $\tilde{n}_o$ and $\tilde{n}_e$ directly from experimental FT-IR spectra. In the first procedure, two transmission spectra recorded at different angles of incidence are used. The second one requires a transmission spectrum at normal incidence and a normalized differential reflection spectrum (PM-IRRAS) at grazing incidence. In the present work, these methods have been applied to Langmuir–Blodgett cadmium arachidate monolayers in order to illustrate their possibilities.

Key words: optical anisotropy, ultrathin films, cadmium arachidate, optical constants.

Generally, in the infrared, ultrathin films – thickness $d < 400$ Å – have a uniaxial optical response characterized by ordinary ($\tilde{n}_o = n_o + ik_o$) and extraordinary ($\tilde{n}_e = n_e + ik_e$) complex refractive indexes or by the corresponding complex dielectric functions $\tilde{\varepsilon}_o = \tilde{n}_o^2$ and $\tilde{\varepsilon}_e = \tilde{n}_e^2$. The ordinary refractive index corresponds to the electric vector perpendicular to the optical axis, and the extraordinary refractive index corresponds to the electric vector along the optical axis.

Until now, little work has been done concerning determination of the infrared anisotropic optical constants of ultrathin films. Ishino and Ishida [1] have developed a method which determines the ordinary complex refractive index from the transmittance spectrum, and the extraordinary complex refractive index is obtained by metal overlayer attenuated total reflectance (ATR) measurements. However, values found by this method do not agree with experimental data that

we have obtained on the same systems. As an alternative, we have developed a new method to determine the $n_o(\bar{v})$ and $k_o(\bar{v})$ from the transmittance spectrum at normal incidence and the $n_e(\bar{v})$ and $k_e(\bar{v})$ by using a transmittance spectrum at non-zero incidence or a PM-IRRAS spectrum.

Theory

The general dependence of the polarized reflectivity of a three-phase layered system (environment, film, substrate) has been studied by McIntyre and Aspnes [2]. Within the frame of the thin film approximation ($d \ll \lambda$), the IRRAS signal of monolayers deposited on metallic substrate can be written, for a highly reflecting metal in the infrared range, as:

$$\frac{R_p(d)}{R_p(0)} = 1 - \frac{8\pi d\bar{v}\sin^2\theta}{\cos\theta} \, \mathrm{Im}\!\left(\frac{-1}{\tilde{\varepsilon}_e}\right) \qquad (1)$$

where d is the film thickness, θ is the angle of incidence and $\bar{v}$ is the wavenumber.

Morever, as demonstrated previously [3], this signal is directly connected to the normalized PM-IRRAS signal S_{norm}, by

$$S_{\text{norm}} = 1 + \alpha\!\left(1 - \frac{R_p(d)}{R_p(0)}\right) \qquad (2)$$

The α term depends on the difference between the p- and s-polarized reflectance of the bare substrate and also between the optical set-up efficiencies for p- and s-polarizations. Relations (1) and (2) show that the normalized PM-IRRAS signal S_{norm} behaves linearly to

---

* To whom correspondence should be addressed

the imaginary part of the inverse of the extraordinary dielectric function [Im $(-1/\varepsilon_e)$].

With the same thin film approximation, the Fresnel equations lead to a very simple expression for the transmittance T of an ultrathin film deposited on a dielectric substrate. Indeed, for p-polarized incident radiation, T_p is given by:

$$T_p = T_p^{sub} \left\{ 1 - \frac{4\pi d\bar{v}}{\varepsilon_b \cos\theta + (\varepsilon_b - \sin^2\theta)^{1/2}} \right.$$
$$\left. \cdot \left[\varepsilon_b \sin^2\theta \, \text{Im}\left(\frac{-1}{\tilde{\varepsilon}_e}\right) + (\varepsilon_b - \sin^2\theta)^{1/2} \cos\theta \, \text{Im}(\tilde{\varepsilon}_o) \right] \right\} \quad (3)$$

where T_p^{sub} represents the transmittance of the bare substrate and ε_b its real dielectric function. At normal incidence ($\theta = 0°$), Eq. (3) becomes

$$T = T^{sub}\left[1 - \frac{4\pi d\bar{v}}{1 + n_b} \, \text{Im}(\tilde{\varepsilon}_o) \right] \quad (4)$$

For a non-zero angle of incidence, the p-polarized transmittance depends on both $\text{Im}(\tilde{\varepsilon}_o)$ and $\text{Im}(-1/\tilde{\varepsilon}_e)$ whereas it is proportional to only $\text{Im}(\tilde{\varepsilon}_o)$ at normal incidence.

Therefore, Eqs. (1)–(4) show that the anisotropic optical constants of an ultrathin film can be determined from PM-IRRAS or a p-polarized transmittance spectrum.

Principle of the Method and Application

Before begining the process, the film thickness which appears in all the preceding equations is evaluated by an experimental method such as ellipsometry. From the experimental transmittance spectrum at normal incidence, normalized by substrate spectrum and baseline correction, $\text{Im}(\tilde{\varepsilon}_o)$ is determined for each wavenumber $\bar{v}$ by using Eq. (4). Then, a Kramers–Krönig transformation (KKT) permits us to access the real part of the ordinary dielectric function $\text{Re}(\tilde{\varepsilon}_o)$. Knowing these two quantities, a simple arithmetical operation gives a first set of ordinary refractive index and extinction coefficient values $[n_o^1(\bar{v}), k_o^1(\bar{v})]$. To determine $[n_o(\bar{v}), k_o(\bar{v})]$ more precisely, the initial input values $[n_o^1(\bar{v}), k_o^1(\bar{v})]$ of the optical constants are successively perturbed by a Newton–Raphson (NR) method to solve the following system:

$$T^{cal}[n_o^i(\bar{v}), k_o^i(\bar{v})] - T^{exp} = 0$$

where T^{cal} and T^{exp} are the simulated and experimental transmittances, respectively.

The ability and precision of this method have been estimated by computer simulation. In the infrared range, a simulated normalized transmittance spectrum of an ultrathin film ($d = 25$ Å) deposited on a dielectric substrate ($n_b = 1.5$) is achieved from infrared complex refractive index data ($n_o^{th}, k_o^{th}, n_e^{th}, k_e^{th}$) calculated by using a Lorentz dispersion model. The calculation procedure previously described and applied to these data leads to a new ordinary complex refractive index (n_o^{sim}, k_o^{sim}) that differs from the theoretical (n_o^{th}, k_o^{th}) by less than 1%. This result shows that the proposed method leads to accurate results for ordinary optical constants of an ultrathin film.

The next step of calculation is determination of the extraordinary optical constants $[n_e(\bar{v}), k_e(\bar{v})]$ by two different procedures. In the first, knowing $n_o(\bar{v})$ and $k_o(\bar{v})$, we use a p-polarized transmittance spectrum at $\theta = 60°$ normalized by substrate spectrum and baseline correction, and Eq. (3) to determine $\text{Im}(-1/\tilde{\varepsilon}_e)$. By using the same successive operations (KKT, arithmetical operation, iterative NR method) we obtain the extraordinary optical constants $n_e(\bar{v})$ and $k_e(\bar{v})$. The calculated relative difference between these values and the theoretical ones is close to 10%.

To improve the precision of $n_e(\bar{v})$ and $k_e(\bar{v})$, we have developed another calculation procedure, using a normalized PM-IRRAS spectrum. As α can be determined from IRRAS and PM-IRRAS experiments, Eq. (1) gives $\text{Im}(-1/\tilde{\varepsilon}_e)$. Then, $n_e(\bar{v})$ and $k_e(\bar{v})$ are calculated for each wavenumber within the same frame. In that case the relative errors decrease to 1%. We conclude that the second method improves significantly the accuracy of the determination of the extraordinary optical constants.

Results

To illustrate our method, we have studied two samples of four Langmuir–Blodgett cadmium arachidate monolayers deposited on CaF_2 and gold substrates. The choice of these samples was directly motivated by the reproducibility of their preparation and by the accurate knowledge of their thickness, which is determined by the number of monolayers. The transmittance spectrum at normal incidence was recorded on a Fourier-transform infrared spectrometer (Nicolet 740) and the normalized PM-IRRAS and IRRAS spectra by the method developed in our laboratory [3]. These spectra were recorded from 3200 to 1200 cm^{-1}. The ordinary and extraordinary refractive indexes that we have obtained in this range are reported in Fig. 1,

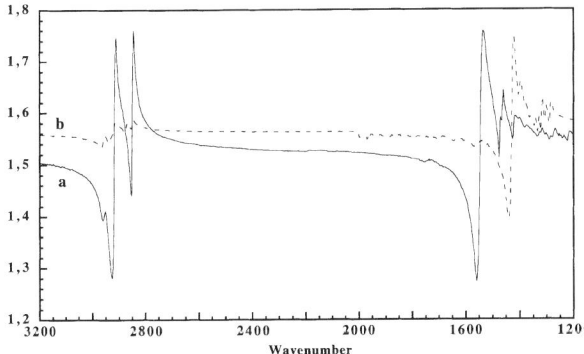

Fig. 1. Ordinary (*a*) and extraordinary (*b*) refractive index of cadmium arachidate

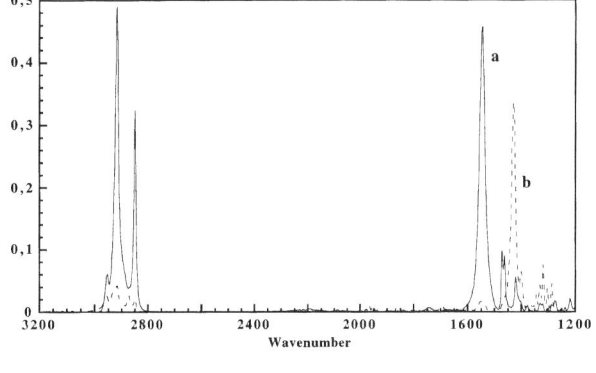

Fig. 2. Ordinary (*a*) and extraordinary (*b*) extinction coefficients of cadmium arachidate

and the ordinary and extraordinary extinction coefficients are reported in Fig. 2. The important differences between the ordinary and extraordinary optical constants is significant of the good organization of the Langmuir–Blodgett arachidate film. The set of the values k_o and k_e can give access to the orientation of chemical groups by means of the second Legendre polynomial P_2 determined from the dichroic ratio [4].

Conclusion

In this paper, we have demonstrated the possibility of obtaining the anisotropic indexes of an ultrathin film by applying a very simple calculation procedure to its transmittance and reflectance spectra. This method is limited only by the quality of the experimental spectra. The optical data obtained by the methods described allow determination of the orientation of molecular groups and comparison of the optical properties of the thin film with those of the bulk.

References

[1] Y. Ishino, H. Ishida, *Langmuir* **1988**, *4*, 1341.
[2] J. D. E. McIntyre, D. E. Aspnes, *Surf. Sci.* **1971**, *24*, 417.
[3] T. Buffeteau, B. Desbat, J. M. Turlet, *Appl. Spectrosc.* **1991**, *45*, 380.
[4] T. Buffeteau, B. Desbat, S. Besbas, M. Nafati, L. Bokobza, *Polymer* **1994**, *35*, 2538.

Mikrochim. Acta [Suppl.] 14, 635–637 (1997)
© Springer-Verlag 1997

FT-IR Studies Accompanying the Development of a Plasmaless Etching Process on the Basis of the Elemental Halogens Bromine and Fluorine

Andreas Erwin Guber*, **Uwe Köhler,** and **Wilhelm Bier**

Institut für Mikrostrukturtechnik, Forschungszentrum Karlsruhe GmbH, P.O. Box 3640, D-76021 Karlsruhe, Federal Republic of Germany

Abstract. Plasmaless etching with a Br_2/F_2 mixture allows extremely rapid etching of silicon. By selection of the gas mixture the silicon etching rate can be controlled precisely. The microstructures obtained possess very smooth surfaces and an isotropic etching profile. They can be applied to generate moulding tools made of silicon with spherical depressions, for batchwise production of small plastic microlenses.

Key words: plasmaless etching, bromine, fluorine, FT-IR, microlenses.

Within the framework of its R&D activities related to the microsystems technology project, the Institut für Mikrostrukturtechnik of Forschungszentrum Karlsruhe is also working in the field of silicon micromechanics for the production of silicon microstructures. Attention above all is focussed on dry chemical etching of monocrystalline silicon with suitable halogen-containing gases [1]. As etching techniques, reactive ion etching (RIE), laser-induced etching and plasmaless etching are applied. Owing to their high chemical reactivity, molecular fluorine and some interhalogen compounds (ClF_3, BrF_3, BrF_5, IF_5) react with silicon even at room temperature [1, 2]. Besides Cl_2, Br_2 and I_2, gaseous SiF_4 is produced as the main reaction product. As some of the initial gases and reaction products are IR-active, FT-IR spectroscopy is well suited for the accompanying analysis of plasmaless etching processes [1]. The highest silicon etching rate is attained with BrF_3. This gas, however, has a vapour pressure of only 8 mbar and lacks purity for practical application. These drawbacks have been bypassed by using a mixture of elemental bromine and fluorine. The use of Br_2/F_2 mixtures results in a constant production of intermediate BrF_3 which then reacts with silicon [3]. The surfaces of the microstructures etched into the silicon have an extremely small roughness, making it possible to produce inverse copies in plastics.

Experimental

The studies on plasmaless etching of silicon with halogen-containing gases were performed in a stainless steel reactor which also served as IR measuring cell. IR-transparent windows of AgCl were used. During the studies the gas cuvette was fixed in the specimen chamber of an FT-IR spectrometer (type: Bruker IFS 88). This allowed constant recording of the IR spectra. Usually, IR spectra between 4500 and 400 cm^{-1} are measured at a resolution of 0.5 cm^{-1}; in exceptional cases, the resolution amounts to 0.1 cm^{-1}. The etching gases studied included molecular fluorine (supplier: Solvay, Hannover), the interhalogens BrF_3 and BrF_5 (supplier: ABCR, Karlsruhe) as well as mixtures of bromine (supplier: Merck, Darmstadt) and fluorine. As pure BrF was practically not available, it was generated on the laboratory scale by the introduction of about 10% F_2 into a bromine vapour atmosphere [4]. The chemistry of the silicon etching processes was investigated by using silicon specimens with relatively large surfaces without a mask structure. To generate microstructures in silicon, adequate masks had first to be realized [5]. By means of buffered hydrofluoric acid, arrays of holes about 50 μm in diameter were etched into the SiO_2 masks generated by thermal oxidation. To obtain handy specimens for the investigations, the silicon wafer was broken into pieces 15×35 mm in size, which were then installed in the IR gas cuvette. Some of the microstructures etched with Br_2/F_2 were copied to plastic such that inverse structures of the initial microstructure were obtained.

Results and Discussion

In a first experiment, a mixture of Br_2 and F_2 was brought into contact with monocrystalline silicon in the reactor. In addition to trace amounts of HF (4250–3750 cm^{-1}) and BrF_5 (647 cm^{-1}), the reaction products SiF_4 (1029 cm^{-1}) and BrF_3 (619 cm^{-1}) were recorded in the IR spectrum. A section of this IR spectrum is represented in Fig. 1 for a gas mixture of 5 mbar Br_2 and 15 mbar F_2. At a resolution of 0.1 cm^{-1}, intermediate BrF (690–635 cm^{-1}) can be

---

* To whom correspondence should be addressed

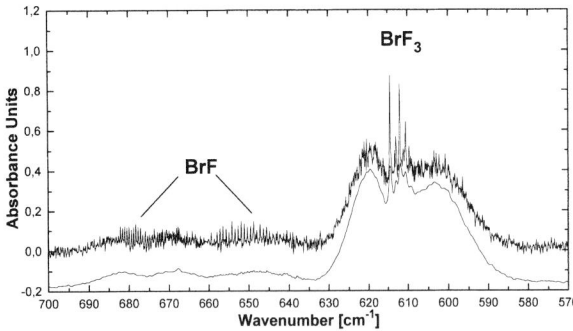

Fig. 1. Section of an IR spectrum recorded during etching of a silicon specimen with a mixture of 5 mbar Br_2 and 15 mbar F_2. The fundamental of the intermediate BrF can only be noticed at a resolution of 0.1 cm^{-1}

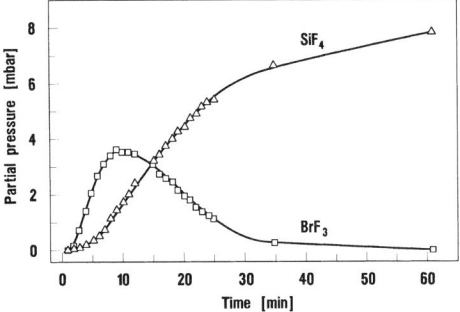

Fig. 2. Behaviour of the BrF_3 and SiF_4 partial pressures determined by IR spectroscopy. In a Br_2/F_2 mixture, immediate formation of BrF_3 takes place, which then reacts with solid silicon to give SiF_4. Owing to the developing lack of fluorine, BrF_3 partial pressure is reduced and only SiF_4 and free Br_2 are available at the end of the reaction

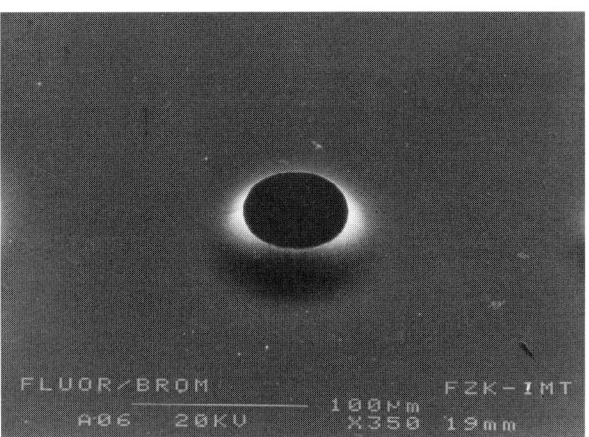

Fig. 3. SEM of a silicon surface etched with a $Br_2/F_2/N_2$ mixture. Surface roughness ranges between 5 and 10 nm

observed in the IR spectrum besides BrF_3 [4]. Figure 2 shows the SiF_4 and BrF_3 partial pressures determined by IR spectroscopy, as a function of time. Immediately after the gas inlet, the reaction of Br_2 and F_2 results in the formation of BrF_3. Following a certain delay, the latter reacts with solid silicon and SiF_4 is produced. At first, BrF_3 formation predominates. After about 10 minutes, it passes through a maximum. With increasing contact time the BrF_3 reactivity is reduced, as no additional BrF_3 can be formed, owing to the lack of fluorine. After a reaction time of 60 minutes, about 8 mbar SiF_4 has been produced from the initially supplied amount of fluorine. The initial amount of 5 mbar bromine is reconverted completely. Hence, the etching reaction comes to a standstill. The etching process can be started again by further addition of fluorine. Thus, an easily controllable silicon etching process is available with bromine being the catalyst of fluorine conversion. Etching with a Br_2/F_2 mixture can be described

by the following reaction equations:

$$Br_2 + F_2 \rightarrow 2BrF$$
$$BrF + F_2 \rightarrow BrF_3$$
$$3Si + 4BrF_3 \rightarrow 3SiF_4 + 2Br_2$$

For interhalogen bromine and fluorine compounds the following reactivity sequence has been determined with regard to monocrystalline silicon: $BrF_3 \gg BrF \gg F_2 \gg BrF_5$. Surface roughness of a silicon surface etched with F_2 amounts to about 200 nm. Addition of 25% bromine already brings about a considerable improvement of surface quality. Though the etching reaction is slowede down significantly by addition of buffer gases (N_2 or Xe) to a Br_2/F_2 mixture, the surface roughness can be reduced to 5–10 nm [6]. As in all plasmaless etching processes based on fluorine-containing gases, the microstructures generated always have an isotropic etching profile (cf. Fig. 3).

Silicon microstructures with extremely smooth surfaces can be copied to plastic by means of a moulding process. This is illustrated by the schematic representation in Fig. 4. Opening the SiO_2 mask located on the wafer allows the plasmaless etching attack of a $Br_2/F_2/N_2$ mixture to take place. With increasing contact time, isotropic under-etching of the mask is achieved such that a semispherical depression is generated. The top SEM shows the etched microstructure after the removal of the SiO_2 mask. In the subsequent moulding process, the spherical depression is filled with polymethyl methacrylate (PMMA). For this purpose, the PMMA is heated and flows into the etched depression under a vacuum. After hardening and cooling of the plastic, demoulding takes place and the

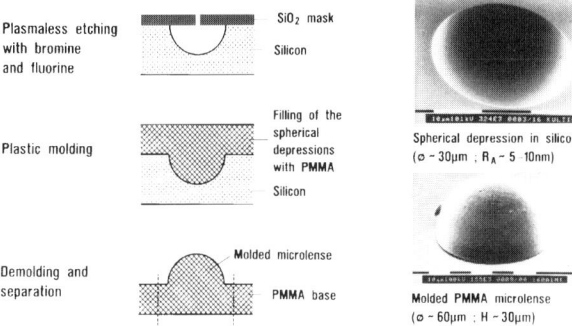

Fig. 4. Schematic representation of the main process steps in the production of plastic microlenses. By means of plasmaless etching, spherical depressions are generated in silicon. Subsequently, these depressions are filled with PMMA. After demoulding, the inverse plastic microstructures are obtained. They can be used e.g. as microlenses

inverse PMMA microstructure is obtained. Following separation, this spherical product can be used e.g. as a microlens. Such a PMMA microlens is shown in the bottom SEM of Fig. 4. As the microstructured silicon tool is not damaged during moulding, it can be used again for microlens production. Within the framework of our studies, up to 100 microlenses were produced simultaneously by means of a tool containing 100 spherical depressions. In principle, it is also possible to generate microlenses from other materials.

Conclusion

The plasmaless silicon etching process developed on the basis of a Br_2/F_2 mixture is suitable for the production of silicon microstructures. Owing to the isotropic etching profile and the extremely smooth surface, the inverse structures can be obtained easily by copying the microstructures to plastic. Thus, moulding tools are made available for the batchwise production of e.g. microlenses in the field of plastics technology based on silicon substrates.

Acknowledgements. We would like to thank Prof. Dr. W. Menz for his useful discussions and constant support of our work. Moreover, we would like to express our thanks to Messrs. P. Abaffy, G. Born and U. Merkle for the performance and evaluation of numerous experiments. Part of these activities was performed within the framework of the MINOP (Microsystems technology for use in minimally invasive neurosurgical operation techniques) joint project funded by the BMBF (grant no. 13 MV 0323).

References

[1] A. Guber, U. Köhler, W. Bier, *Proc. SPIE* **1994**, *2089*, 368.

[2] D. E. Ibbotson, J. A. Mucha, D. L. Flamm, J. M. Cook, *J. Appl. Phys.* **1984**, *56*, 2939.

[3] U. Köhler, A. E. Guber, W. Bier, *Wissenschaftliche Berichte FZKA* **1995**, 5574.

[4] H. Bürger, P. Schulz, E. Jacob, M. Fähnle, *Z. Naturforschung* **1986**, *41a*, 1015.

[5] W. Menz, P. Bley, *Mikrosystemtechnik für Ingenieure*, VCH, Weinheim, 1993.

[6] U. Köhler, A. E. Gubber, W. Bier, *Proc. 8th Intern. Conf. Solid-State Sensors and Actuators – Eurosensors IX*, p. 521.

Mikrochim. Acta [Suppl.] 14, 639–641 (1997)

The Use of FT-IR Spectroscopy in Microtechnologies

Andreas Erwin Guber

Institut für Mikrostrukturtechnik, Forschungszentrum Karlsruhe GmbH, P.O. Box 3640, D-76021 Karlsruhe, Federal Republic of Germany

Abstract. Analytical studies accompanying the development of microtechnological production methods can be carried out by means of e.g. FT-IR spectroscopy. The chemical reactions taking place during various dry chemical etching processes of silicon can be observed precisely. Microtechnological components such as micro heat exchangers or IR filters can be checked for their functionality.

Key words: microtechnologies, FT-IR, etching, micro heat exchanger, IR filters.

Within the framework of the microsystems technology project, various microtechnological processes are developed for the production of microstructures and microsystems at Forschungszentrum Karlsruhe. The best known process is the so-called LIGA technique. Here, X-ray deep-etch lithography, electroforming and plastic moulding are combined for the generation of microstructures [1]. Similar to the LIGA technique, mechanical microengineering, a high-precision method tailored to the needs of microsystems technology, is also suited for the production of microstructures of various plastics, metals, metal alloys and ceramic materials [2]. For a number of applications in microsystems technology, it is advantageous to combine the LIGA technique with silicon micromechanics. For this purpose, special dry chemical etching processes have been developed at the Institute for Microstructure Technology (IMT). These processes allow the microstructurization of the silicon substrate prior to or after the use of the LIGA technique [3, 4].

At the IMT, FT-IR spectroscopy is routinely applied in silicon micromechanics for studies accompanying the development of various dry chemical etching processes [3–5]. With microtechnologically fabricated products such as micro heat exchangers [2], FT-IR spectroscopy can be employed to determine the oil permeation rate at increased temperatures on the basis of the oil vapour IR absorption bands. IR filters based on self-contained microstructures produced by mechanical microengineering or the LIGA technique are tested by FT-IR spectroscopy after production and prior to the delivery to potential customers [6–8].

Experimental

At the IMT, two FT-IR spectrometers (Bruker types IFS 48 and IFS 88) are available for the spectroscopic studies. For both devices, various accessories can be obtained.

Usually, the studies on dry chemical etching are performed in chemical reactors which have been developed for this purpose and also serve as IR measuring cells. In general, gas cuvettes of various sizes are applied. During the experiments, they are fixed in the specimen chambers of both spectrometers [4]. Silicon specimen pieces are put into these cells and brought into contact with various etching gases or etching gas mixtures. The etching attack takes place either by plasmaless etching or by laser support and an externally applied plasma [3, 9]. If necessary, the etching gases can also flow through the measuring cells. When the IFS 48 is used, an externally adapted long-path measuring cell of 20 m optical length may be applied optionally for the exhaust gas analysis of RIE and PECVD facilities [5].

Measurement of the oil vapour permeation rate through the narrow passages of a micro heat exchanger was also accomplished in one of the measuring cells described above. For this purpose, the micro heat exchanger was flanged to an IR measuring cell. The thermo-oil in the storage tank was heated to about 100 °C and fed into the heat exchanger for an extended period of time. The oil vapour diffused through the individual passages could then be determined on the basis of the C–H stretching vibration.

Test measurements of various IR filters were performed by using the IFS 88 with the wavenumber ranging from 4500 to 20 cm^{-1}. Up to 7 IR filters were fixed simultaneously on a specimen wheel. The measurements were carried out successively. A special programme for acquisition of the measured values has been developed for complete measurement of the specimens, including automatic beam splitting variation in all five wavenumber ranges as well as for the determination of the total spectra from the individual ones.

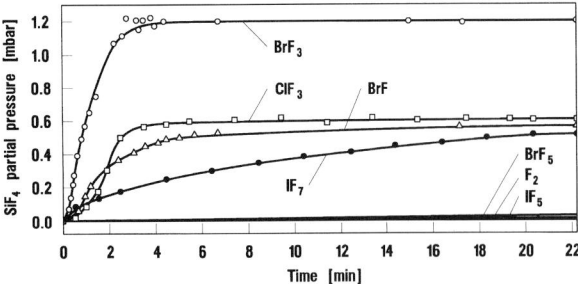

Fig. 1. SiF$_4$ partial pressures determined by IR spectroscopy for various halogen-containing etching gases. BrF$_3$ has the highest silicon etching rate

Results and Discussion

Besides molecular fluorine, several interhalogen compounds such as ClF$_3$, BrF, BrF$_3$, BrF$_5$, IF$_5$ and IF$_7$ can be used for plasmaless etching of monocrystalline silicon, owing to their high chemical reactivity. Typical results obtained for the etching gases mentioned are represented in Fig. 1. For this purpose, defined pieces of a silicon wafer were brought into contact with the gases in the IR measuring cell over an extended period of time. SiF$_4$ formation during the reaction was observed by means of IR spectroscopy. The total amount of SiF$_4$ produced per unit time represents a simple measure of the silicon etching rate. The following reactivity sequence has been determined for the etching gases studied: BrF$_3$ $\gg$ ClF$_3$ > BrF > IF$_7$ $\gg$ IF$_5$ $\sim$ F$_2$ $\sim$ BrF$_5$. By photodissociation with an excimer laser, F$_2$ and Cl$_2$ can be decomposed into fluorine or chlorine atoms at 248 and 308 nm, respectively. These atoms preferably react with silicon. During laser-induced etching of silicon with pure halogens (e.g. F$_2$, Cl$_2$) or halogen mixtures (F$_2$/Cl$_2$ or F$_2$/Br$_2$), the etching rate can be determined by IR spectroscopy [4, 9]. As is known, halogen-containing etching and cleaning gases (CF$_4$, SF$_6$, CHF$_3$, CBrF$_3$, NF$_3$, ClF$_3$) are applied in RIE (reactive ion etching) facilities as well as in plasma-supported cleaning of PECVD (plasma enhanced chemical vapour deposition) systems. On-line FT-IR spectroscopy for routine analysis and evaluation at the IMT allows conclusions to be drawn with regard to the exhaust gas composition and the chemistry of almost all etching processes [5]. To detect the smallest amounts of pollutants, the long-path cuvette is applied. Its detection limit is e.g. 13 ppb SF$_6$ [10]. By using molecular F$_2$ it was demonstrated for the first time that the pollutants produced by plasma-supported etching can be reduced considerably [11].

Fig. 2. Photography of the micro heat exchanger tested. It can be used as a compact high-performance cooler in a number of technical fields

The micro heat exchangers tested within the framework of the oil permeation experiments are characterized by a relatively small and very compact design [2]. A typical example is shown in Fig. 2. This micro heat exchanger has 4000 micro channels, a heat exchange

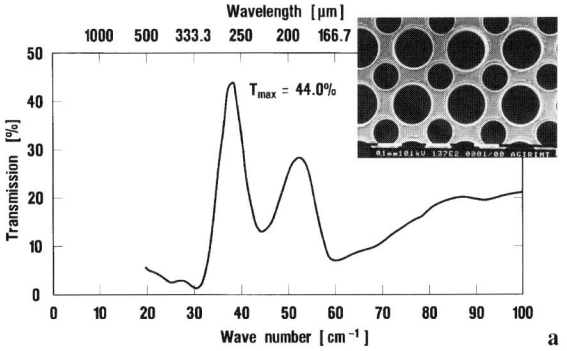

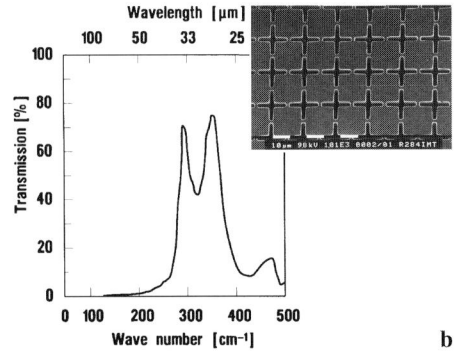

Fig. 3a, b. Spectral behaviour and typical SEMs of the IR filters studied, which have been made by microtechnological production methods

surface of 150 cm^2 and a volume of only 1 cm^3. Usually, a very high leakage rate of 1.5×10^{-8} mbar l/s is attained. To be able to determine an oil permeation rate, a heat exchanger with a much worse leakage rate was selected for the experiments. The thermo-oil diffusing through the passages at a temperature of about 100 °C was monitored by IR spectroscopy over a period of 1200 h on the basis of the oil vapour C–H stretching vibrations at 2850 and 2950 cm^{-1}. The oil permeation rate was found to be about 0.8 µg/h.

The spectral behaviour and the corresponding SEMs of the IR filters produced at the IMT and measured by means of FT-IR spectroscopy are shown in Fig. 3. In Fig. 3a, the result obtained for the combination of two microhole structures (diameters 100 and 150 µm, respectively) at a uniform filter thickness of 200 µm is represented. Maximum transmission is 44%, the cut-off is at 32 cm^{-1}. The result measured for a band-pass filter based on crosswise slot apertures is illustrated in Fig. 3b. The slot length is 18.5 µm, and the slot width only to 3 µm. The filter thickness is 20 µm [12]. In the spectral range investigated, the band-pass filter has sufficient transmission and very good cut-off behaviour.

Conclusion

The examples presented here clearly demonstrate that FT-IR spectroscopy is suited to the analytical work accompanying the development of microtechnological production processes. For chemical etching reactions in particular, major findings can be obtained by the use of FT-IR spectroscopy. Numerous microtechnological components and elements can be measured and, hence, checked for their functionality.

Acknowledgements. I would like to thank Prof. Dr. W. Menz for his constant support of this work. Moreover, I would like to express my thanks to Dr. W. Bier for the provision of the micro heat exchanger and Dr. R. Ruprecht for the supply of the IR filters produced by means of the LIGA technique. In addition, I gratefully acknowledge the recording and evaluation of a number of IR spectra by Messrs. P. Abaffy and G. Born.

References

[1] W. Menz, P. Bley, *Mikrosystemtechnik fur Ingenieure*, VCH, Weinheim, 1993.
[2] W. Bier, A. Guber, G. Linder, T. Schaller, K. Schubert, *KfK Ber.* **1993**, *5238*, 132.
[3] A. Guber, U. Köhler, W. Bier, *Proc. SPIE* **1994**, *2089*, 368.
[4] U. Köhler, A. E. Guber, W. Bier, *Wissenschaftliche Berichte FZKA* **1995**, 5574.
[5] A. E. Guber, U. Köhler, *J. Mol. Struct.* **1995**, *348*, 209.
[6] W. Bier, A. Guber, *Proc. SPIE* **1992**, *1575*, 534.
[7] A. Guber, W. Bier, G. Linder, K. Schubert, *Proc. SPIE* **1994**, *2089*, 388.
[8] R. Ruprecht, W. Bacher, P. Bley, M. Harmening, W. K. Schomburg, *KfK-Nachrichten* **1991**, *23(2–3)*, 118.
[9] U. Köhler, A. Guber, W. Bier, in: *Lasers in Engineering* (W. Waidelich, ed.), Springer, Berlin Heidelberg New York Tokyo, 1994, p. 820.
[10] Ergebnisbericht IMT 1994, *Wissenschaftliche Berichte FZKA* **1995**, *5532*, 8.
[11] A. Guber, U. Köhler, W. Bier, *VDI-Ber.* **1993**, *1060*, 295.
[12] R. Ruprecht, W. Bacher, *KfK-Ber.* **1991**, 4825.

Mikrochim. Acta [Suppl.] 14, 643–645 (1997)
© Springer-Verlag 1997

High-resolution FT-IR Spectroscopy of H_2 and D_2 Adsorbed on NaCl(Film)/NaCl(100)

Joachim Heidberg*, Natalia Y. Gushanskaya, Olaf Schönekäs, and **Richard Schwarte**[†]

Institut für Physikalische Chemie und Elektrochemie Universität Hannover, Callinstr. 3-3A, D-30167 Hannover, Federal Republic of Germany

Abstract. The adsorption of D_2 and H_2 on NaCl(film)/NaCl(100) at low gas pressures and temperatures has interesting features. We are able to discriminate between ortho- and and para-D_2 in the adsorbate as well as terrace and defect sites in the spectral region of the fundamental and coupled external and internal motion of the adsorbed molecules.

Key words: hydrogen adsorption, alkali halide, induced infrared absorption.

The system hydrogen–NaCl forms one of the simplest adsorbates. It shows a fascinating low-temperature behaviour, including for example the possibility of separating ortho- from para-species [1, 2]. Physisorption and isobaric desorption of D_2 and H_2 on NaCl (film)/NaCl(100) surfaces was studied at gas pressures of $10^{-10} < p(\text{mbar}) < 5 \times 10^{-4}$ and temperatures of $10 \leqslant T(\text{K}) \leqslant 33$. In this work the induced infrared absorption has been studied and analysed in the spectral range of the Q_1-transition ($\Delta v = 1, \Delta J = 0$) particularly the Q_{R1}-transition (combination of internal and external motions of the adsorbed molecules) and the S_1-transition ($\Delta v = 0, \Delta J = 2$).

Experimental

Our experimental set-up contained three main coupled components: the interferometer (Bruker IFS 113v) with special optics for surface measurement according to Heidberg, the UHV chamber and the separate detector compartment. Details about the UHV chamber and the sample preparation are given elsewhere [1, 2]. The FT-IR spectra were recorded in transmission at an angle of incidence of $(50 \pm 3)°$. We used an instrumental resolution of $1\ \text{cm}^{-1}$. All spectra were calculated from the interferogram with the boxcar apodization function and a zero-filling factor of 4.

* To whom correspondence should be addressed

Results and Discussion

Spectral Region of Q_1

A typical absorption spectrum of D_2 adsorbed at $T = 10$ K under a D_2-pressure of $p = 4 \times 10^{-5}$ mbar is displayed in Fig. 1. The multiplet absorption denoted with numbers in Fig. 1 is attributed to Q_1 transitions of D_2-molecules adsorbed on NaCl(100) sites. The frequencies of the eight components were determined by means of spectrum deconvolution. The splitting of the main components 2 and 5 results from the spectral

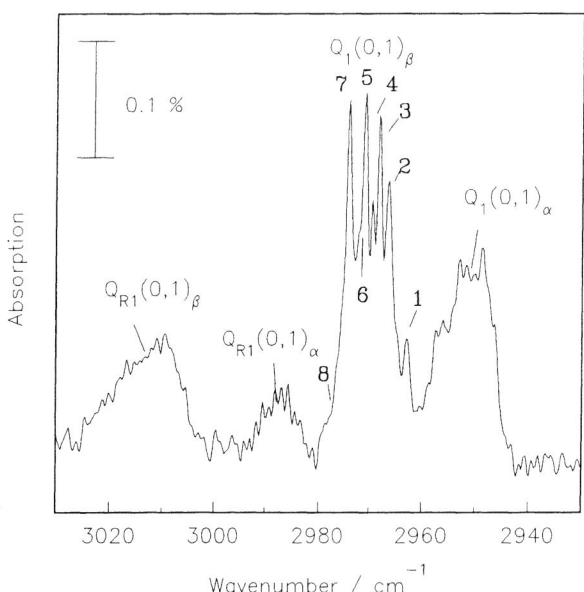

Fig. 1. Induced infrared absorption of D_2 adsorbed on NaCl(film)/ NaCl(100) at $T = 10$ K, $p(D_2) = 4 \times 10^{-5}$ mbar. Site α is atttributed to defects at the surface, whereas site β denotes adsorbate on NaCl(100) terrace sites

separation between ortho- and para-D_2. The magnitude of the splitting is comparable both to the theoretical value [3] and the frequencies of gaseous ortho- and para-D_2 under the influence of strong electric fields [4]. The splitting is primarily caused by the J-dependence of the anharmonic coupling between radial and angular variables of H_2- and D_2-molecules, but the M-dependence should not be neglected.

The behaviour of all species numbered in Fig. 1 was studied upon isobaric desorption. According to theory [5], ortho-D_2, being more weakly adsorbed than para-D_2, desorbs first. The desorption temperatures of the main species 2 and 5 differ by (2 ± 1) K. These absorptions are attributed to para-D_2 and ortho-D_2. In the adsorbate spectrum they are separated by about 4 cm^{-1}, the corresponding gas-phase separation being 2 cm^{-1} [3]. Species 7 is observable only at high pressures ($p > 10^{-5}$ mbar) and lowest temperatures ($10 \leqslant T(\text{K}) \leqslant 12$). Similarly to the adsorption of HD [6] this may be due to molecules adsorbed in a second layer located above Cl^- [6] or the centre of the NaCl unit cell [7]. Species 1 is detectable even at 29 K. Tentatively it is attributed to an adsorbate at defect sites. Species 3 and 4 disappear together with ortho-D_2 upon desorption at around 16 K. This behaviour may be due to ortho-D_2 adsorbed at different sites on NaCl(100). Another possible explanation for species 3 and 4 is transition between different rotational states, especially from $J = 0$ $(M_J = 0)$ to $J = 2$ $(M_J = 0, \pm 1, \pm 2)$. This implies a remarkable contraction of the separation of these states under the influence of a potential which extremely hinders the rotation in the surface plane [9]. All observed frequencies for ortho- and para-D_2 are given in Table 1 together with the corresponding assignments [3]. In comparison to the gas-phase values the frequencies are red-shifted for

Table 1. Observed vibrational transitions of D_2 adsorbed at site β on NaCl(film)/NaCl(100)

Species no.	Frequency [cm^{-1}]	Assignment
1	2961.6–2963.1	$Q_1(0, 1)_\alpha$
2	2966–2966.8	$Q_1(1)_\beta$
3	2968.1–2968.3	$Q_1(0)_\beta$
4	2969.3–2969.7	$Q_1(0)_\beta$
5	2970.8–2971	$Q_1(0)_\beta$
6	2972.1–2972.7	$Q_1(0)_\beta$ or $Q_1(0\ 1)_\beta$ $(\theta")$
7	2974	$Q_1(0, 1)_\beta$ $(\theta")$
8	2975.8–2976.5	$Q_1(0, 1)_\beta$ $(\theta")$

species with $J = 0$ by 27 cm^{-1} (para-H_2) and 23 cm^{-1} (ortho-D_2) and for species with $J = 1$ by 30 cm^{-1} (ortho-H_2) and 25 cm^{-1} (para-D_2). The enhanced frequency shift for both species with odd J-values (ortho-H_2, para-D_2) should be noted. It may indicate the stronger binding to the surface, in quantitative agreement with theoretical predictions [5], especially for a "helicoptering" molecule with $J = 1$, $M_J = 1$ [10].

Spectral Region of Q_{RI}

Analysis of the desorption experiments enables us to discriminate between ortho- and para-D_2 as well as terrace and defect sites, in the spectral region of coupled external and internal motions of the adsorbed molecules. The corresponding frequencies attributed to molecules adsorbed at NaCl(100) terrace sites are 3009 cm^{-1} for para-D_2 and 3018 cm^{-1} for ortho-D_2, respectively. An additional weak absorption at around 3027 cm^{-1} may be due to adsorbate in a second layer in analogy to species 7 discussed above. The frequencies of the external motions participating in the coupling can be calculated, taking into account the corresponding Q_1 species, Table 1. We obtain $v_{x,y} = 43 \text{ cm}^{-1}$ for para-D_2 and $v_{x,y} = 47 \text{ cm}^{-1}$ for ortho-D_2. These values agree quite well with the calculated value $v_{x,y} = 39$ cm^{-1} for D_2 [7], without taking into account the rotation of the molecule. The frequencies will depend upon the corrugation, which may be affected by J and M_J. However, further experiments are necessary to strengthen these assignments and to determine the influence of M_J, i.e. "helicoptering" and "cartwheeling". The site symmetry of the adsorbed D_2 molecule in the energetically favoured orientation implies that the corresponding motion has an appreciable component parallel to the surface. The hindered translation perpendicular to the surface was calculated in the static approximation for the isolated molecule [6] to be $v_Z = 137 \text{ cm}^{-1}$ [7]. We have no hint of this motion in our Q_{RI} spectra. It is observable only with a significant component of the induced dipole moment of the molecule perpendicular to the surface.

In additional experiments, special attention was focused on the possible detection of the pure rotational transition S_1 ($\Delta J = 2$) of H_2 at around 600 cm^{-1} upon adsorption on KBr(film)/KBr(100). We do not see any corresponding absorptions, whereas signals in the spectral region of the Q_1 ($\Delta v = 1$, $\Delta J = 0$) are clearly observable.

Acknowledgements Financial support of Land Niedersachsen, Fonds der Chemischen Industrie and Deutsche Forschungsgemeinchaft (DFG) is gratefully acknowledged. We thank Prof. V. Rozenbaum for fruitful discussions.

References

[1] J. Heidberg, W. Dierkes, O. Schönekäs, R. Schwarte, *Ber. Bunsenges. Phys. Chem.* **1994**, *98*, 131.

[2] J. Heidberg, N. Gushanskaya, O. Schönekäs, R. Schwarte, *Surf. Sci.* **1995**, *331–333*, 1473.

[3] C. Schwartz, R. J. Le Roy, *J. Mol. Spec.* **1987**, *121*, 420.

[4] R. W. Terhune, C. W. Peters, *J. Mol. Spec.* **1959**, *3*, 138.

[5] D. White, E. N. Lassettre, *J. Chem. Phys.* **1960**, *32*, 72.

[6] D. J. Dai, G. E. Ewing, *J. Chem. Phys.* **1993**, *98*, 5050.

[7] M. Folman, Y. Kozirovski, *J. Colloid Interface Sci.* **1972**, *38*, 51.

[8] D. J. Dai, G. E. Ewing, *J. Chem. Phys* **1994**, *100*, 8432.

[9] V. Rozenbaum, *Private Communication.*

[10] G. J. Kroes, R. C. Mowrey, *Chem. Phys. Lett.* **1995**, *232*, 258.

Mikrochim. Acta [Suppl.] 14, 647–648 (1997)

Studies of Adsorption on Mineral Surfaces by FT Spectroscopy

Ursula M. Johansson*, E. Katarina Tano, and **Willis Forsling**

Division of Inorganic Chemistry, Department of Chemical and Metallurgical Engineering, Luleå University of Technology, S-971 87 Luleå, Sweden

Abstract. Specific surface reactions have been studied by adsorbing different type of substances on the surfaces of kaolinite and γ-alumina. Analyses were performed by means of FT-Raman and diffuse reflectance mid-IR. The DRIFT spectra indicate isotopic exchange of hydroxyl groups on kaolinite and that it is possible to adsorb silanes at the surface. DRIFT and FT-Raman spectra indicate that the solvent may react with the γ-alumina surface and that phosphate adsorption occurs at the surface.

Key words: γ-alumina, kaolinite, adsorption, FT-spectroscopy.

Previous studies of isotopic exchange on kaolinite have been performed with an intercalated kaolinite to favour the exchange [1, 2]. In this study experiments were performed without intercalated kaolinite. An earlier study of silane adsorption on kaolinite has been reported [3] but not with the silanes used in this study.

Few studies of phosphate adsorption on γ-alumina have been made with FT-Raman spectroscopy [4, 5]. Here, various organic phosphates have been adsorbed on the γ-alumina surface, from different solvents.

Experimental

Materials and Sample Preparation

Kaolinite was supplied by Assi kraftliner, Piteå, Sweden, and the γ-alumina by Mandoval Ltd, England. Suspensions of kaolinite (50 mg per ml of solvent) and γ-alumina (20 mg per ml of solvent) were used.

Surface exchange reactions took place during 5–6 weeks at 60 °C and in various solvents; water, deuterium oxide, ethanol and γ-butyrolactone ($C_4H_6O_2$).

The adsorption of silanes [$(NH_2)(C_2H_4)(NH)(C_3H_6)Si(OCH_3)_3$, $(C_2H_3)(C_6H_4)(CH_2)(NH)(C_2H_4)(NH)(C_3H_6)Si(OCH_3)_3 \cdot HCl$ and $(CH_2OCH)(CH_2)O(C_3H_6)Si(OCH_3)_3$] onto kaolinite, and of phos-

phates [$(C_4H_9OPO(OH)_2 + (C_4H_9O)_2PO(OH))$, $(C_6H_5)_2P(O)OH$ and $(C_6H_5O)_3P(O)$] onto γ-alumina took place during 24–48 hours at room temperature. Phosphates were added to γ-alumina suspensions at various pH values, adjusted with 0.1 M hydrochloric acid and 0.1 M sodium hydroxide.

Instrumentation

In this study a Perkin-Elmer Fourier-transform infrared (FT-IR) 2000X spectrometer was used. IR spectra were collected by DRIFT accessory. A Perkin-Elmer PE1760X FT-IR spectrometer equipped with a near-infrared (NIR) Raman bench was used for Raman measurements. The Raman bench utilizes a Spectron SL 301 neodymium-doped yttrium aluminium garnet (Nd^{3+}: YAG) laser system providing intensity-stabilized continuous wave emission at 1064 nm. A triglycinesulphate (TGS) detector was used in the IR and Raman measurements.

Results and Discussion

Surface Exchange Reactions

In this study the experiments show that it takes a long time and that an increase in temperature is needed to deuterate kaolinite (Fig. 1). Since the OD-bands are small and the absorbance of the hydroxyl (OH) bands does not decrease during deuteration it is supposed that the deuteration occurs on only very few of the OH-groups. It is obvious that it is very difficult to make an isotopic exchange of the OH-groups to OD-groups at the surface of kaolinite.

It is apparent that the solvent used with γ-alumina is of great importance for further reactions on the surface (Fig. 2). With a solvent containing free OH-groups (water but not ethanol) a surface reaction occurs after some time and a new phase may appear [6]. A different kind of solvent, such as γ-butyrolactone, seems to be adsorbed on the surface and occupy some of the

---

* To whom correspondence should be addressed

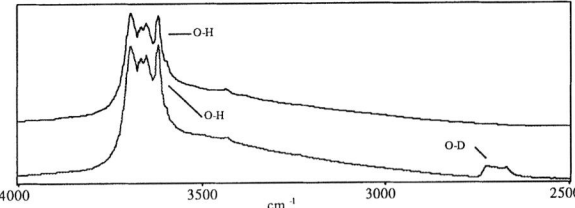

Fig. 1. DRIFT spectra of kaolinite (upper spectrum) and deuterated kaolinite (lower spectrum). The bands in the region 3700–3600 cm^{-1} are due to O–H stretching vibrations and the bands in the region 2750–2650 cm^{-1} are due to O–D stretching vibrations

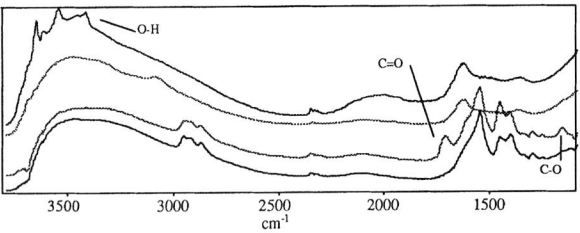

Fig. 2. DRIFT spectra showing how various solvents react with the γ-alumina surface. Changes of the OH-bands above 3200 cm^{-1} are shown for γ-alumina in water but not in ethanol (uppermost spectra). The two lower spectra show γ-alumina in γ-butyrolactone. Peaks at 1730 and 1175 cm^{-1}, which appear after some time, are due to C=O and C–O stretchings in 5-ring lactones

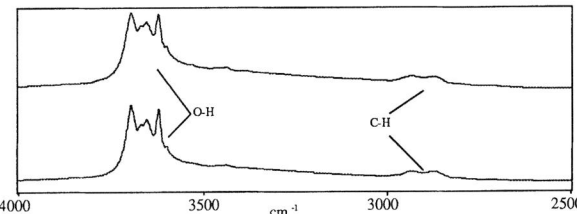

Fig. 3. DRIFT spectra showing the comparison between the adsorption of 3-glycidoxy-propyltrimethoxy silane [(CH$_2$OCH)(CH$_2$)O(C$_3$H$_6$)Si(OCH$_3$)$_3$] on kaolinite in distilled water (upper spectrum) and in ethanol (lower spectrum). The bands in the region 3000–2700 cm^{-1} are due to aliphatic C–H stretching vibrations

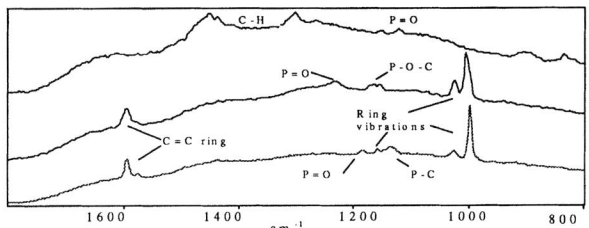

Fig. 4. FT-Raman spectra showing butyl phosphate, triphenyl phosphate and diphenyl phosphonic acid respectively, adsorbed on γ-alumina. Peaks corresponding to P=O stretchings are in the region 1300–1100 cm^{-1}. The peaks at 1450 and 1300 cm^{-1} correspond to aliphatic stretchings and the peaks around 1000, 1030, 1155 and 1595 cm^{-1} are due to ring vibrations/stretchings of phenyl groups

surface sites. This will decrease the possibility of other surface reactions to occur.

Surface Adsorption

The results show that silanes interact with OH-groups on the surface of kaolinite, forming chemical bonds at the interface. The adsorption of silanes onto kaolinite is independent of the solvent used (Fig. 3).

The FT-Raman spectra indicate that the adsorption of arenephosphates on γ-alumina occurs independent of pH although the highest adsorption occurs at pH 8.7, which should be when the γ-alumina surface is neutral.

Since adsorption of butyl phosphate on γ-alumina only takes place at low pH, when the surface will be positively charged, the reactive part of the molecule needs to be negatively charged. As a single-bonded oxygen has a stronger negative charge than a double-bonded oxygen the adsorption takes place via the free P–O oxygen. The presence of P=O stretching bands confirms this (Fig. 4).

References

[1] R. L. Ledoux, J. L. White, *Science* **1964**, *145*, 47.
[2] K. Wada, *Clay Min.* **1967**, 7, 51.
[3] T. J. Porro, S. C. Pattacini, *Appl. Spectr.* **1990**, *44*, 1170.
[4] J. M. Lewis, R. A. Kydd, *J. Catal.* **1991**, *132*, 465.
[5] E. Gantner, D. Steinert, J. Reinhardt, *Anal. Chem.* **1985**, *57*, 1658.
[6] C. Dyer, P. J. Hendra, W. Forsling, M. Ranheimer, *Spectrochim. Acta* **1993**, *49A*, 691.

Mikrochim. Acta [Suppl.] 14, 649–651 (1997)

A Novel Method for Investigation of Dibasic Aromatic Acids on Oxidized Aluminium Surfaces by NIR-FT-SERS

Ottó Klug[1],*, Ildikó Száraz[1], Willis Forsling[2], and **Maine Ranheimer[2]**

[1] RIFA Electrolytics AB, Box 98, S-56322 Gränna, Sweden
[2] Department of Inorganic Chemistry, University of Luleå, S-95187 Luleå, Sweden

Abstract. A new technique is presented for detecting dibasic aromatic acids on an SERS-inactive oxidized aluminium foil by utilizing near-infrared Fourier-transform surface-enhanced Raman spectroscopy (NIR-FT-SERS). Identification of adsorbed *ortho-*, *meta-* and *para*-phthalic acids on a typical anodic foil of electrolytic capacitors from electrolytes containing 1 ppm substrate was successfully accomplished.

Key words: NIR, SERS, phthalic acid, oxidized aluminium foil, capacitor.

Aluminium oxide is widely applied in the chemical industry. Among its well-known applications γ-alumina is also used as a dielectric in electrolytic capacitors. As part of an ongoing research project in capacitor chemistry a study has been made of the use of FT-Raman spectroscopy in the identification of adsorbed electrolyte ingredients and by-products on oxidized aluminium foils.

The adsorbates on the alumina surface layer of the foil normally showed very poor Raman intensities compared to the activity of adsorbates on γ-alumina powder. The low intensity is probably due to the much smaller surface area. However, by use of a specially adopted technique of surface-enhanced Raman spectroscopy (SERS), some adsorbed species showed much enhanced spectral activity on the alumina surface of the oxidized foils.

In the original SERS technique, silver constitutes the SERS-active layer for the adsorbates. SERS spectra have also been observed for an oxide-free aluminium surface [1], but when surrounded by air aluminium is never SERS-active, owing to its natural oxide layer. Therefore a special technique developed at Strathclyde University [2] was adopted, in which a concentrated,

citrate-reduced silver sol was prepared with a fixed particle size of around 20 nm. The silver protected by citrate would "sense" some type of molecules already at the inactive oxide surface. This method was first used on a zinc metal surface [2], then was extended to an oxidized aluminium surface by us [3]. In this work we present the SERS determination of *ortho-*, *meta-* and *para*-phthalic acids on oxidized aluminium foil.

Experimental

The oxidized aluminium foil used was provided by RIFA Electrolytics AB and was made by Becromal S. P. A. with forming voltage 690 V. Raman spectra were obtained by using a Perkin–Elmer 1760X FTIR spectrometer equipped with an NIR Raman bench, which utilizes an Nd:YAG laser operating at 1064 nm. Raman light was collected by using 180° back-scattering geometry. An indium gallium arsenide detector was applied. A silver sol was prepared and concentrated according to the method described in detail in earlier publications [2–4].

Samples were treated by immersing a piece of foil with approximately 6 cm^2 geometrical surface in 2 mM solution, of *ortho-*, *meta-* or *para*-phthalic acid in ethylene glycol and in γ-butyrolactone, solvents which are commonly used in capacitor electrolytes. The foil pieces were then directly put into an oven at 60 °C for about four hours, until they were completely dried. Later a drop of the concentrated sol was layered on the surface of the foils and dried at the same temperature. The Raman spectra were then recorded, using 800 mW power. A similar procedure was applied for treating standard glass slides.

As an SERS spectrum can be very different from the normal Raman spectrum of a compound, it was also essential to record the solid-state Raman spectra of the investigated compounds.

Results and Discussion

No Raman scattering could be obtained from any of the acids adsorbed on either the oxidized foil or the glass slides, without the silver sol. However, after sol treatment, the same Raman measurement gave rise to

---

* To whom correspondence should be addressed

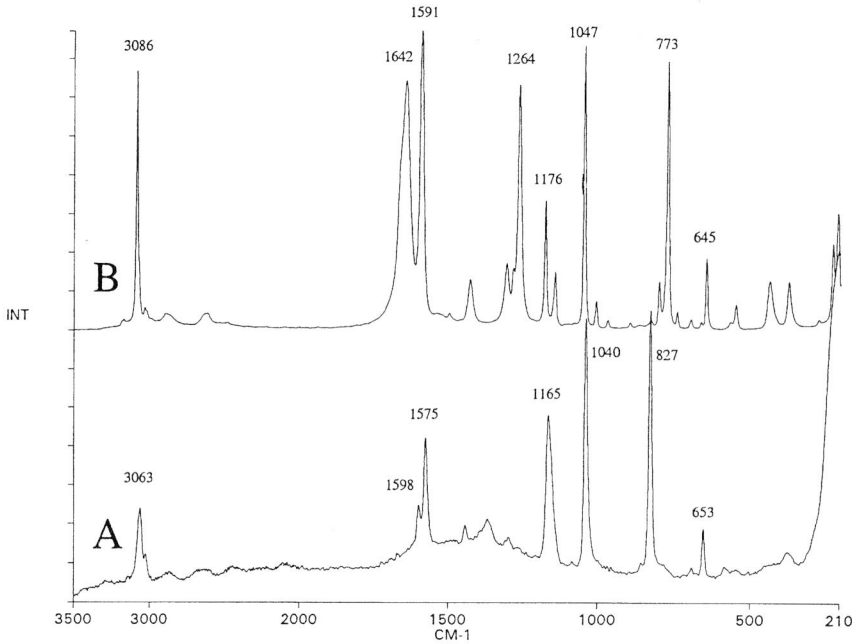

Fig. 1. *ortho*-Phthalic acid on oxidized Al-foil (*A*) Solid-state Raman spectrum (*B*)

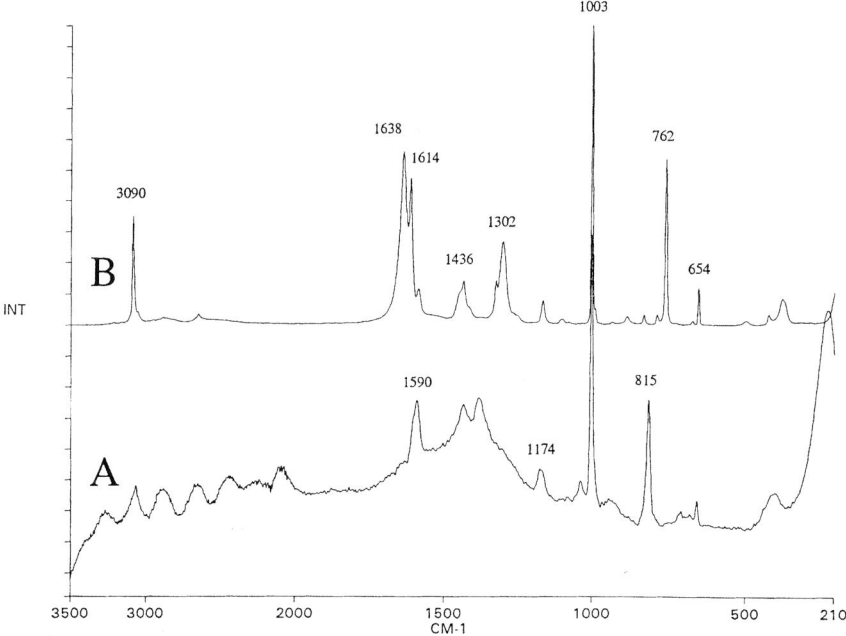

Fig. 2. *meta*-Phthalic acid on oxidized Al-foil (*A*) Solid-state Raman spectrum (*B*)

similar spectra as shown in Figs. 1–3, curves A, for the *ortho*, *meta* and *para* isomers respectively. The enhanced spectral bands were significant for each type of acid but were identical in both place and magnitude, whether the adsorption occurred from ethylene glycol or γ-butyrolactone. The reproducibility of the SERS

intensities at different parts of the sol-treated foil was found to be ± 15% or better.

A mixture of 1 cm³ of ethylene glycol solution of *ortho*- or *meta*-phthalic acid with a few drops of silver sol did not give rise to any bands except the ones from the solvent; however, a mixture contaning the *para*

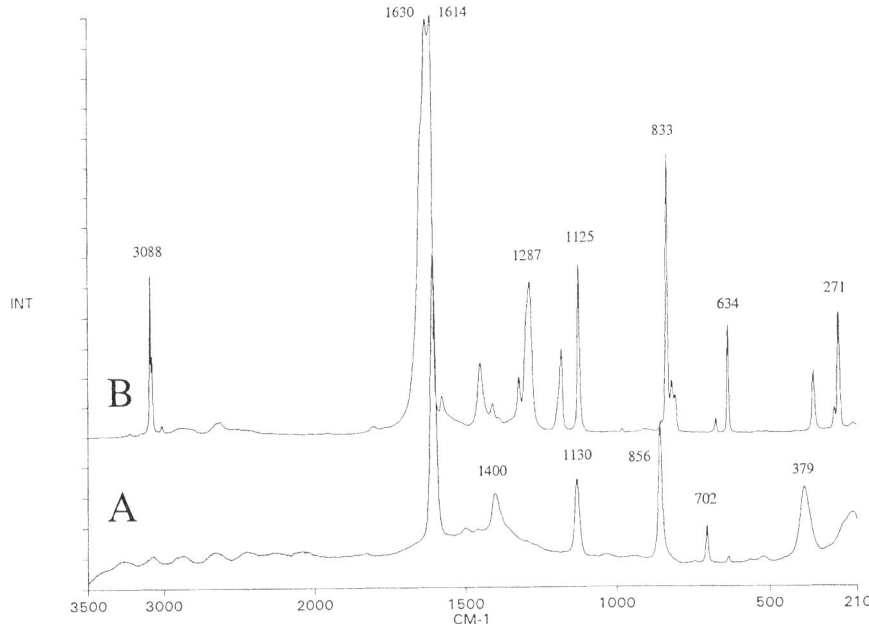

Fig. 3. *para*-Phthalic acid on oxidized Al-foil (*A*) Solid-state Raman spectrum (*B*)

isomer exhibited identical SERS to that shown in Fig. 3 (A), but at lower intensity.

When adsorbed on a non-formed, non-etched plain aluminium foil the acids produced SERS with enhanced band-intensities of the same order of magnitude as those obtained with etched capacitor foils as support, but on glass slides the acids gave intensities that were several times lower than those found with the aluminium foils, indicating the great influence of the oxidized aluminium surface. Comparing the SERS and the solid-phase Raman spectra, there are both similarities and differences. Generally either both carbon–carbon ring stretching (breathing) modes at around $1600 \, cm^{-1}$ are strongly enhanced, or only one of them is, but always there is a shift to lower wavenumbers. Another ring stretching vibration, at around $1000 \, cm^{-1}$, which is very weak for the *ortho* and *para* compounds, is significantly enhanced in the case of the *meta* isomer. There is always one of the four out-of-plane C–H bending vibrations, which gives a prominent band in the solid-phase Raman spectra, and perhaps the same one is highly enhanced in SERS, but shifted to higher wavenumbers. The aromatic C–H in-plane stretching at over $3000 \, cm^{-1}$ is only enhanced in the case of the *ortho* isomer.

In many cases it is possible to propose the orientation of the adsorbed molecule at the surface, from the vibrational modes assigned to the enhanced bands. However, in this case it is very complex, because of the two different surfaces used and also because bands from vibrations that are parallel and perpendicular to the plane of the benzene rings are enhanced at the same time. The detailed description of the mechanism needs further studies and is an object of ongoing research.

The good signal to noise ratio and the high intensity of SERS of all three phthalic acid isomers make the adsorbed species detectable on oxidized Al-foils from solutions with concentrations as low as 1 ppm, and the method may be applied with profit in the capacitor industry.

Acknowledgements. The authors acknowledge the financial support given by RIFA Electrolytics AB.

References

[1] T. Lopez-Rios, C. Pettenkofer, I. Pockrand, A. Otto, *Surf. Sci.* **1982**, *121*, L541.
[2] H. M. M. Wilson, W. E. Smith, *Proc. XIIIth Int. Conf. on Raman Spectroscopy* (W. Kiefer, M. Cardona, G. Schaack, F. W. Schneider, H. W. Schrötter, eds.), Wiley, New York, 1992, pp. 698–699.
[3] J. A. Haigh, P. J. Hendra, W. Forsling, *Spectrochim. Acta* **1994**, *50A*, 2027.
[4] P. C. Lee, D. Meisel, *J. Phys. Chem.* **1982**, *86*, 3391.

Mikrochim. Acta [Suppl.] 14, 653–654 (1997)
© Springer-Verlag 1997

Enhanced Infrared Absorption of Thiourea Adsorbed on a Rough Silver Surface

Giorgio Mattei[†], **Yujun Mo**[*], **Mario Pagannone**, and **Vladimir A. Yakovlev**[**]

Istituto di Metodologie Avanzate Inorganiche, CNR, Area della Ricerca di Roma, C. P. 10, I-00016 Monterotondo Scalo, Rome, Italy

Abstract. Thin films of different thicknesses, made by thiourea molecules deposited on a flat silver surface roughened by treatment with chemical etching solutions, have been studied by infrared reflection-absorption spectroscopy. A significant enhancement (of the order of several hundreds) of the infrared absorption cross-section of the thiourea molecules has been found. In agreement with the surface-enhanced Raman spectroscopy results for the same films, it has been found that the main enhancement is associated with the first layer of thiourea molecules chemisorbed on the silver surface.

Key words: infrared, reflection-absorption, enhancement, thiourea, SERS.

Surface-enhanced Raman scattering (SERS) is widely used for detecting extremely small amounts of adsorbed substances and for studying their interaction with the substrate. Infrared reflection-absorption (IRRA) spectroscopy is also very sensitive to surface species. It follows that the application of both techniques to the same samples gives more information about the surface of the sample [1]. On the other hand, enhancement of IR spectra had been so far demonstrated for attenuated total reflection (ATR) [2, 3] and for transmittance [4] spectra of adsorbates in the presence of island metal films. Actually, such systems are often also used for SERS experiments. Another system showing SERS is an adsorbate on the rough metal surface. It is then interesting to study IRRA spectra in the conditions under which SERS takes place. We have chosen thiourea (TU) molecules adsorbed on a roughened silver surface for such a study. Indeed, the TU molecule, well characterized by Raman and infrared spectroscopy, has been subject of SERS studies performed

in an electrochemical cell (see e.g. [5] and references therein) and recently some preliminary results of a SERS study of TU adsorbed on a chemically roughened silver surface have been reported [6].

In this paper we present the results of our study by IRRA and SER spectroscopy of TU films of different thicknesses deposited on a silver plate previously roughened by a suitable etching procedure [7]. It will be shown that for TU/Ag systems, characterized by a clear SERS effect, a significant enhancement of the infrared absorption cross-section is also present and that such an enhancement is mainly confined to the first layer of the chemisorbed TU molecules.

Experimental

A silver plate (22×6 mm in area and 0.5 mm thick) was used as a sample. One surface was carefully polished with diamond paste, then the cleaned sample was subject to an etching procedure [7] performed in two steps: (i) 30 s in a 1:1 volume/volume mixture of H_2O_2 (35%) and NH_3OH (30%), (ii) 200 s in a 0.1 M KCl solution in the same mixture. After etching, the Ag surface was wiped by spinning for a few minutes at about 12000 rpm. A dilute solution of TU in acetone was dropped on the silver surface and dried. The amount of adsorbed TU (expressed in monolayers, ML) was calculated from the known solution concentration, the amount of solution dropped and the sample area.

For IRRA measurements a Fourier-transform IR spectrometer FTS-40A (Bio-Rad) was used. A unit specially made for this experiment allowed us to obtain IRRA spectra at an average angle of incidence on the sample of $82°$. Raman measurements were made with a Spex triple spectrometer and a ccol EG&G PARC OMA detector, with excitation by the 514.5 nm Ar^+ laser line (50 mW at the sample).

Results and Discussion

The results of our measurements show that the IR and Raman bands have shifted frequencies (with respect to those of the non-adsorbed or "bulk" TU molecule)

---

* Permanent address: Institute of Physics, Chinese Academy of Sciences, P.O. Box 603, Beijing 100080, Peoples Republic of China

** Permanent address: Institute of Spectroscopy, Russian Academy of Sciences, Troitzk, Moscow reg. 142092, Russia

† To whom correspondence should be addressed

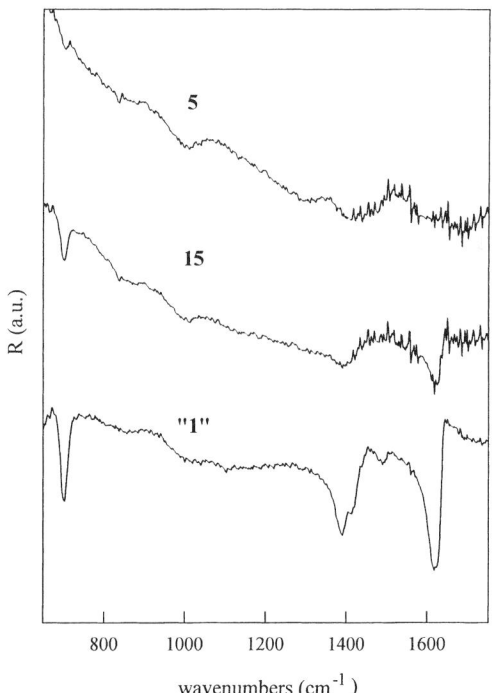

Fig. 1. IRRA spectra of three TU films; the numbers represent their average thicknesses expressed in ML. The spectrum labelled "1" is for the sample of 50 ML after washing (see text) and does not differ too much from that before washing

acetone (curve labelled as "1" ML in Fig. 1) we did not see significant changes in the spectrum. Since TU is well soluble in acetone, the several times washed sample can contain only one monolayer, or less, of chemisorbed molecules of TU. Indeed, the time stability of the "1" ML sample is much lower than that of the other samples. Once the physisorbed protective TU layer is removed by washing in acetone, the single chemisorbed layer can interact with air and in a rather short time (less than 2 hours) it is either removed or changed so that no spectra can any longer be observed.

To compare the IRRA results with the bulk spectra of TU, we have calculated the bulk dielectric function from the transmittance spectrum of the 150 ML nominal thickness film deposited by the same technique on a KBr plate. After that, we have calculated the IRRA spectrum that could be expected on a smooth silver mirror for bulk TU. From the ratio between the peak areas of the experimental and the calculated IRRA bands we obtained an enhancement factor of about 400 (or higher if the first layer is not completely covered) for the IR cross-section of the adsorbed TU molecule, and that is mainly attributable to a first layer effect. A higher enhancement of the Raman cross-section (10^4) was obtained from the SERS measurements.

References

[1] W. Knoll, M. R. Philpott, W. G. Golden, *J. Chem. Phys.* **1982**, *77*, 219.
[2] A. Hartstein, J. R. Kirtley, J. C. Tsang, *Phys. Rev. Lett.* **1980**, *45*, 201.
[3] A. Hatta, Y. Chiba, W. Suëtaka, *Appl. Surf. Sci.* **1986**, *25*, 327.
[4] M. Osawa, K.-I. Ataka, K. Yoshii, Y. Nishikawa, *Appl. Spectrosc.* **1993**, *47*, 1497.
[5] H. Kim, J.-J. Kim, *J. Raman Spectrosc.* **1993**, *24*, 77.
[6] G. Mattei, Y. Mo, M. Pagannone, in: *Proc. XIVth Intern. Conf. Raman Spectrosc.* (Y. N.-T. Yu, X.-Y. Li, eds.), Wiley, New York, 1994, p. 646.
[7] G. Mattei, L. G. Quagliano, M. Pagannone, *Europhys. Lett.* **1990**, *11*, 373.

which are typical of TU adsorbed on silver [5, 6] and their intensities are increased by adding more TU layers. The band at about 730 cm^{-1} for "bulk TU" (mainly C–S stretching) is red-shifted to around 700–707 cm^{-1}, while the band at around 1086–1090 cm^{-1} in "bulk TU" (mainly C–N stretching), very weak in the infrared but quite strong in the Raman spectrum, is blue-shifted to 1110 cm^{-1}.

As an example, in Fig. 1 the IRRA spectra of three TU films of different thickness are presented. When the film of 50 ML average amount of TU was washed in

Mikrochim. Acta [Suppl.] 14, 655–656 (1997)

Oil Analysis on Metal Surfaces by Using DR-FT-IR Mapping Techniques

George L. Powell**, **Tye E. Barber**[1], **Mariza Marrero-Rivera**[2], **David M. Williams**, **Norman R. Smyrl**, and **John T. Neu**

Oak Ridge Centers for Manufacturing Technologies*, Oak Ridge, TN 37831-8096, U.S.A.
Surface Optics Corporation, San Diego, CA 92131, U.S.A.

Abstract. The use of remote-sensing diffuse-reflectance optics to obtain Fourier-transform mid-infrared spectra from sand-blasted gold, aluminum, and steel surfaces as a function of position in order to monitor the spatial distribution of oil over the surface is described, and used to map the spread of an 0.85 µg safflower oil droplet with time, to calibrate the spectra for film thickness near 1 µm, and to monitor the oil oxidation that limited the extent of the stain.

Key words: infrared, reflectance, diffuse, surface analysis.

Remote-sensing diffuse-reflectance optics (Harrick Scientific Barrel Ellipse™) has been used to obtain Fourier-transform mid-infrared spectra from metal surfaces for the nondestructive inspection of these surfaces for films and contamination [1–4]. The practical success of this technique, using research spectrometers (Bio-Rad FTS-60), has led to the development of small portable surface inspection systems, based on the MIDAC Illuminator™ [3], used for the present work and, more recently, the Surface Optics Corporation SOC 400™, an Illuminator™/Barrel Ellipse™ reconfigured for minimum size (14 L) and weight (8 kg) [4]. Translation stages [3, 4] (both manual and automatic positioners instructed by serial commands from the spectrometer data collection system) have been implemented that position known coordinates of surfaces at the focal point of the diffuse reflectance optics. With the

SOC 400™, the spectrometer may be manipulated rather than the specimen, an important consideration since the specimen size has no limits except for a minimum concave radius of curvature of ~0.25 m. Spectra obtained from these operations may be reduced to surface maps by using the host of qualitative and quantitative spectral analysis techniques available to address vital surface film questions, not the least of which relates to the surface cleanliness of sand-blasted metals. The surfaces investigated were sand-blasted gold, aluminum, and stainless steel.

This paper describes initial attempts to establish a quantitative relationship between the substrate metal, the oil film thickness, and the spectral response. Safflower oil was chosen for its low viscosity, which allows the oil to spread to a sub-micron thick stain, its low vapor pressure to prevent evaporative losses, and its spectral variety resulting from saturated and unsaturated hydrocarbons, and carbonyls.

Experimental

The MID AC Illuminator™ system was used with a dry air purge to collect 8 cm^{-1} spectra (2 × zero-filling, 64 scans, 256-scan reference from a clean position on the same surface) at 90 s intervals, by driving a Velmex Unislide™ XY positioner (translation accuracy 0.7 mm/m) through serial commands generated in the main menu program of MIDAC/GRAMS/386™ operating in the kinetic mode. The surfaces investigated were gold, aluminum, and stainless steel, sand-blasted and cleaned as described in [1]. Spectral analysis was done with a "T" average over a 25 cm^{-1} range about the 2925 cm^{-1} maximum of the –CH$_2$– stretching band baseline corrected to the average values of the ranges over 3050–3075 cm^{-1} and 2700–2725 cm^{-1}. A drop of Wesson safflower oil was applied to each surface within a 120-s time interval by a µL-syringe, and spectra were obtained at 2-mm intervals along a 50-mm line through the initial drop position, followed by mapping a 50- by 50-mm area including the stain, with

[1] Present address: Department of Chemistry, San Houstan State University, Huntsville, TX 77341-2117, U.S.A.
[2] Present address: Department of Chemistry, Texas Agricultural and Mechanical University, Austin, TX, U.S.A.
* Managed for the U. S. Department of Energy by Martin Marietta Energy Systems, Inc. under Contract No. DE-AC05-84OR21400.
** To whom correspondence should be addressed

a 1-mm step-size. Subsequently, the aluminum and the steel specimens were area mapped in a similar manner. The calibration factor was determined by dividing the weight of oil by the numerical integral over the surface map.

Results and Discussion

Initially the oil drop spread over the surface as a visible stain with circular symmetry, slowing over time to form a stain that was 32 ± 1 mm in diameter on gold and aluminum and 36 ± 1 mm in diameter on steel. The stain appeared to change little after the first day, but spectral line mapping over time (based on the 2925 cm^{-1} band) indicated a slight growth in the diameter and slight decreases in the absorbance and baseline reflectance over several days. In this time frame, the 3004 cm^{-1} unsaturated hydrocarbon band decayed from a clearly defined band to a barely perceptible shoulder. Similar stains on gold could not be removed by ordinary cleaning methods, further supporting the conclusion that the safflower oil polymerizes through oxidation of the unsaturated hydrocarbon bonds, fixing the stain at a discrete diameter. The calibration factor for the gold surface was found to be $6.9 \text{ g(a.u.)}^{-1} \text{m}^{-2}$ or 7.5 μm/a.u. (a.u. = absorbance unit), and yielded 1.3 μm maximum film thickness compared to 1.1 μm calculated from the visible stain diameter and the 0.85 mg of oil. The calibration factors for aluminum and stainless steel were 15 and 14 μm/a.u. respectively, representing less sensitivity for these metals. Mapped regions outside the stain could be defined as clean within ± 0.0004 a.u. or, by using area spectral averaging, to ± 0.0001 a.u. For the research spectrometer (Bio-Rad FTS-60) this level is lower by a order of magnitude. There is a significant substrate dependence in the quantitative determination of oil on sandblasted surfaces. The oxidative polymerization of vegetable oils offers a method for making durable organic stain standards since they can be allowed to spread to a desired thickness in an inert atmosphere and fixed by exposure to air.

Acknowledgements. The authors gratefully acknowledge B. H. Nerren, National Aeronautics and Space Administration, Marshall Space Flight Center, Alabama, United States for his tireless efforts to implement this technology as a practical inspection technique.

This report was prepared as an account of work sponsored by an agency of the United States Government. Neither the United States Government nor any agency thereof, nor any of their employees, makes any warranby, express or implied, or assumes any legal liability or responsibility for the accuracy, completeness, or usefulness of any information, apparatus, product, or process disclosed, or represents that's its use would not infringe privately owned rights. Reference herein to any specific commercial product, process, or service by tradename, trademark, manufacturer, or otherwise does not necessarily constitute or imply its endorsement, recommendation, or favoring by the United States Government or any agency thereof. The views and opinions of authors expressed herein do not necessarily state or reflect those of the United States Government or any agency thereof.

References

[1] G. L. Powell, M. Milosevic, J. Lucania, N. J. Harrick, *Appl. Spectrosc.* **1992**, *46*, 111.

[2] G. L. Powell, N. R. Smyrl, C. J. Janke, E. A. Wachter, W. G. Fisher, J. Lucania, M. Milosevic, G. Auth, in: *The Proceedings of the Conference on Characterization and NDE of Heat Damage in Graphite Epoxy Composites, NTIAC*, Austin, Texas, 1993, pp. 97–111.

[3] G. L. Powell, N. R. Smyrl, D. M. Williams, H. M. Meyers, III, T. E. Barber, M. Marerro-Rivera, in: *Aerospace Environmental Technology Conference, NASA Conference Publication 3298* (A. F. Whitaker, ed.), MFSC, Alabama, 1995, pp. 563–571.

[4] G. L. Powell, T. E. Barber, J. T. Neu, *Proc. SPIE* **1995**, *2541*, 142.

Mikrochim. Acta [Suppl.] 14, 657–659 (1997)

Characterizing Langmuir–Blodgett Layers by Infrared Ellipsometry

A. Röseler[1,*], **R. Dietel**[2], and **E. H. Korte**[1]

[1] Institut für Spektrochemie und angewandte Spektroskopie, Laboratorium für spektroskopische Methoden der Umweltanalytik, D-12484 Berlin, Federal Republic of Germany
[2] Forschungsgruppe Dünne Organische Schichten, Institut für Festkörperphysik, Universität Potsdam, Kantstr. 55, D-14513 Teltow, Federal Republic of Germany

Abstract. Langmuir–Blodgett layers consisting of a few monolayers on a metal substrate were studied by infrared ellipsometry. The geometrical thickness and the background level of the refractive index can be precisely derived from phase shifts where absorption is negligible. With a subsequent oscillator fit, the spectra of the optical constants are determined.

Key words: infrared, ellipsometry, optical constants, Langmuir–Blodgett layers.

Different from normal reflection experiments, in ellipsometry not only the intensity and thus the reflectance of a medium is measured but also the phase shift caused by the reflection. As described elsewhere [1], from two or three ratios of spectra taken with different orientations of one polarizer, the so-called ellipsometric parameters are derived: these are the angle Ψ where $\tan^2\Psi$ is the ratio of the reflectances for parallel and perpendicular polarization, respectively, and the angle Δ which is the difference between the phase shifts introduced into the two components. The parameters are related by the Fresnel equations to the optical constants of the sample, i.e. the refractive index n and the absorption index k (or the absorptivity ε). Except for a homogeneous solid, the optical properties of which are characterized completely by two numbers per spectral interval, optical modelling and spectra simulation are necessary to reveal further parameters such as the thickness of a layer on a substrate. Layers a few nanometres thick – which are extraordinarily thin when compared to the wavelength used – can be studied in this way: examples are SiO_2 and Si_3N_4 on

silicon [2]. Their strong absorption is advantageous, but similarly thin layers of materials as weakly absorbing as organic compounds can be studied ellipsometrically when deposited on a highly reflecting metal substrate. In this contribution we report such measurements on Langmuir–Blodgett (LB) films by means of a photometric ellipsometer (experimental details can be found in [1]) coupled to a Bruker IFS55 infrared interferometer, and we comment on the advantages of this method.

Optical Modelling of the Sample

For the infrared range the optical behaviour of any medium can be simulated by a set of oscillators with appropriate parameters: from the individual complex dielectric functions (permittivity) we derive the optical constants $\hat{n} = n + ik$ of the medium. For a layer of this medium (thickness d_1) on a known substrate. Fresnel's equations lead to the reflection coefficients r_{01} and r_{12} related to the interfaces of the medium (optical constants $\hat{n}_1$) against air (refractive index n_0) and the substrate (optical constants $\hat{n}_2$), respectively; the coefficients in turn are applied in the Airy equation to calculate the (overall) reflection coefficient r of such a sample

$$r = [r_{01} + r_{12}\exp(i\delta)]/[1 + r_{01}r_{12}\exp(i\delta)] \quad (1)$$

with $\delta = 4\pi\tilde{v}d_1(\hat{n}^2 - n_0^2 \sin^2\varphi_0)^{1/2}$ where φ_0 denotes the angle of incidence and $\tilde{v}$ the wavenumber. The ratio of these coefficients for parallel and perpendicular polarization is directly related to the above-mentioned ellipsometric parameters Ψ and Δ. The difference between their simulated spectra and measured ones,

---
* To whom correspondence should be addressed

provides a criterion to optimize the parameters of the oscillator model.

In LB layers, molecules with long aliphatic chains are aligned in parallel. As a consequence of such an order, the layers are anisotropic, i.e. different dielectric functions apply for different directions. Furthermore, modified Fresnel equations and a modified Airy equation must be employed [3]. However, when the substrate is of highly reflecting metal and when the layer thickness is small compared to the wavelength, only the component of the electric vector normal to the surface is effective: this reduces the problem to virtually the isotropic situation.

Independent Evaluation of the Thickness and the Optical Constants

The layer thickness has a stronger influence on the phase shift than on the amplitude ratio and thus an evaluation based on spectra of Δ is more sensitive. However, even in a spectral range free from absorption, such a spectrum depends on the refractive index. Since Δ changes considerably with the angle of incidence, the comparison of two such spectra taken at different angles of incidence reveals the precise layer thickness along with information on the background level of the refractive index. As an example, in Fig. 1 two Δ spectra of an arachidic acid LB film are shown. The film was transferred with a uranyl subphase from water at pH 3.0 onto a gold substrate (sample #1). The evaluation in the absorption-free ranges indicates a thickness of

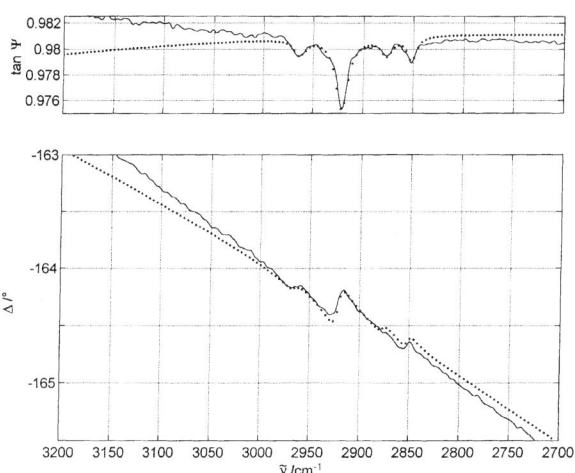

Fig. 2. Experimental spectra of tan Ψ and Δ (solid lines) for sample #2 along with simulated ones (dotted lines). The absorption pattern in the range of CH vibrations was modelled with 6 oscillators (further parameters: $d = 7$ nm, $n = 1.30$, $\varphi_0 = 70°$)

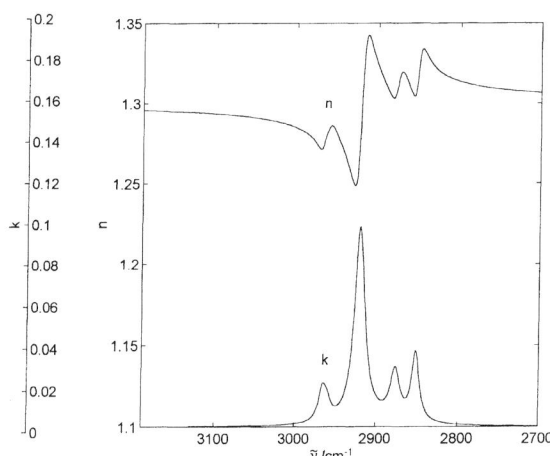

Fig. 3. Spectra of n and k derived from the oscillator set used for calculating tan Ψ and Δ in Fig. 2

11 nm for the sample, consisting of six monolayers, and the background level of the refractive index turned out to be 1.36. As pointed out before, this value refers to the z direction only (normal to the surface) and therefore depends on the orientation of the molecules with respect to the surface. The fixed values of both parameters are used in the subsequent evaluation within the absorption range.

With another sample of six monolayers, but deposited at pH 4.2 (sample #2) we measured a thickness of 7 nm and a refractive index of 1.30 when applying the previously described procedure. These considerably different values should reflect major changes in the molecular order. In Fig. 2 the Δ and Ψ spectra of the

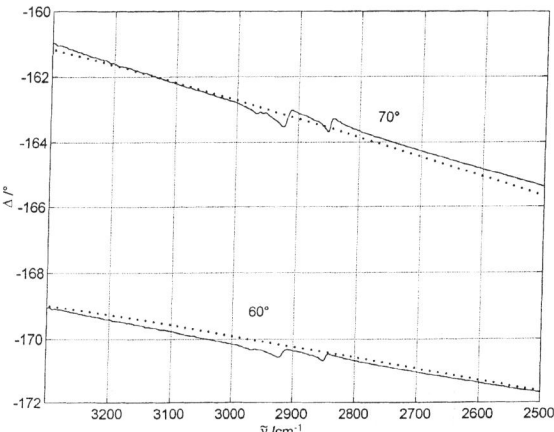

Fig. 1. Δ spectra of six monolayers of arachidic acid on gold (sample #1) for angles of incidence $\varphi_0 = 60°$ and $70°$ (solid lines); neglecting the absorption features, the simulated spectra (dotted lines) were obtained with $d = 11$ nm and a refractive index $n = 1.36$

latter sample within the range of CH vibrations are shown. The simulations (dotted lines) are based on fitting a six-oscillator model; in order to assess the agreement the expanded scale should be noticed. The related n and k spectra are given in Fig. 3.

Conclusions

Very generally an ellipsometric procedure provides us with two independent results per spectrally resolved interval. Where absorption is negligible, the geometrical thickness of a layer on a substrate can be determined with high sensitivity from the phase shift, i.e. the Δ spectra. These are not so easily accessible with other methods. The precise knowledge of the thickness contributes to the reliable simulation of the optical constants within the range of the absorption bands. Provided the layer is deposited on a highly reflecting metal substrate, only the components of n_1 and k_1

normal to the surface are observed, and these depend on the orientation of the molecules. It may be concluded that spectroscopic ellipsometry in the infrared results in a detailed and precise account of LB structures of a few monolayers. In particular the separation of thickness and absorptivity simplifies a quantitative interpretation of this state of order, for instance with respect to a tilt angle.

Acknowledgement. The financial support by the Senatsverwaltung für Wissenschaft und Forschung des Landes Berlin and the Bundesministerium für Bildung, Wissenschaft, Forschung und Technologie is gratefully acknowledged.

References

[1] A. Röseler, *Infrared Spectroscopic Ellipsometry*, Akademie-Verlag, Berlin, 1990.
[2] M. Weidner, A. Röseler, M. Eichler, *Thin Solid Films* **1993**, *234*, 337.
[3] R. M. A. Azzam, N. M. Bashara, *Ellipsometry and Polarized Light*, North-Holland, Amsterdam, 1977.

Mikrochim. Acta [Suppl.] 14, 661–663 (1997)
© Springer-Verlag 1997

Quantitative Aspects of FT-SERS of Polyaromatic Hydrocarbons on a Silver Surface

Shona D. Stewart and **Peter M. Fredericks***

Centre for Instrumental and Developmental Chemistry, Queensland University of Technology, P.O. Box 2434, Brisbane, Q4001, Australia

Abstract. A recently developed SER-active surface, prepared by electrochemically roughening a polished silver surface, has enabled the FT-SER spectra of many organic compounds, whether water-soluble or not, to be obtained. This paper reports the quantitative aspects of this surface, using a range of environmentally significant polyaromatic hydrocarbons (PAHs).

Key words: SERS, FT-Raman, silver surface, polyaromatic hydrocarbons.

Surface-enhanced Raman (SER) scattering has been studied thoroughly since it was first reported by Fleischmann et al. in 1972 [1], and yet there has been little application of the technique to analytical chemistry. This has been in spite of its known advantages of high sensitivity and selectivity. One of the main reasons for this is that it was generally necessary for samples to be water-soluble in order to be analysed by conventional SERS techniques. The recent development of a surface preparation technique [2] has enabled us to obtain reproducible SER spectra of most organic compounds with an FT-Raman spectrometer. This substrate is prepared by electrochemical roughening of a polished silver surface. It is then removed from the cell, rinsed and allowed to dry, and the analyte is applied directly as a solution in any volatile solvent, including water. The SER spectrum is obtained once the solvent has evaporated. Analysis for PAHs has been an important problem because some members of this family exhibit carcinogenic or mutagenic properties. A recent FT-SERS study [3] of 18 polycrystalline PAHs showed that each had a unique spectrum which could be readily used to distinguish the different compounds at detection limits of around 10^{-9} mol.

This paper reports the quantitative aspects of this SER-active surface, including the effect of analyte con-centration, and mixtures, on the FT-SER spectra of PAHs.

Experimental

PAH samples were obtained from Aldrich (WI, U.S.A.) and dis-solved, as received, in suitable solvents. To develop the SER-active surface, a silver working electrode was subjected to 5 minutes of prepolarization at $-1000\,\text{mV}$, then four oxidation–reduction cycles (ORCs) in an electrochemical cell in nitrogen-purged KCl solution. Each ORC consisted of 30 s of oxidation at $+400\,\text{mV}$ followed by 30 s of reduction at $-400\,\text{mV}$, after which the electrode was rinsed and then allowed to dry. Solutions of analyte were applied to the surface, and the solvent was permitted to evaporate before spectra collection. Spectra were collected on a Perkin–Elmer System 2000 FT-Raman spectrometer equipped with a Nd:YAG laser lasing at 1064 nm.

Results and Discussion

Effect of Analyte Concentration

The SER spectra of increasing concentrations of several PAH solutions were compared. Figure 1 shows typical FT-SER spectra of two different concentrations of benzo[a]pyrene. It was found that the general trend of resulting plots of band intensity $vs.$ solution concen-tration was an initial linear response, followed by a decrease in intensity. This was suspected to be a con-sequence of the thickness of the sample layer placed on the surface. Initially, as the analyte concentration was increased, more molecules were in contact with the surface, and thus the SER signal became larger. How-ever, at a certain concentration, the molecular layer became sufficiently thick for self-absorption to become significant. This self-absorption effect occurred at a dif-ferent concentration for each analyte. Figure 2 shows the concentration curve for benzo[a]pyrene, which is typical of the curves obtained for PAHs except that it

---

* To whom correspondence should be addressed

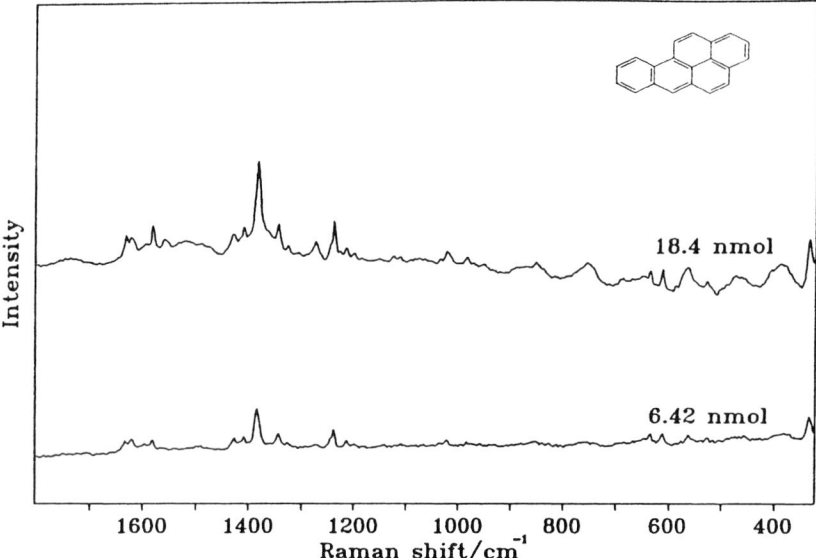

Fig. 1. FT-SER spectra of two different amounts of benzo[*a*]pyrene

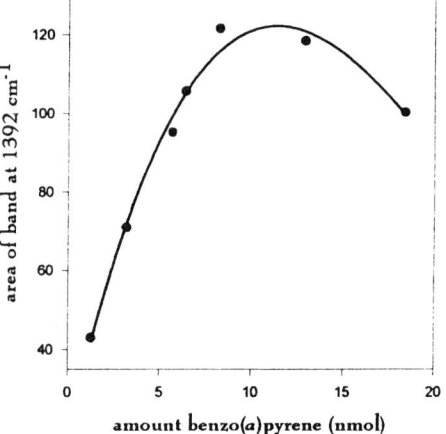

Fig. 2. Concentration curve for benzo[*a*]pyrene

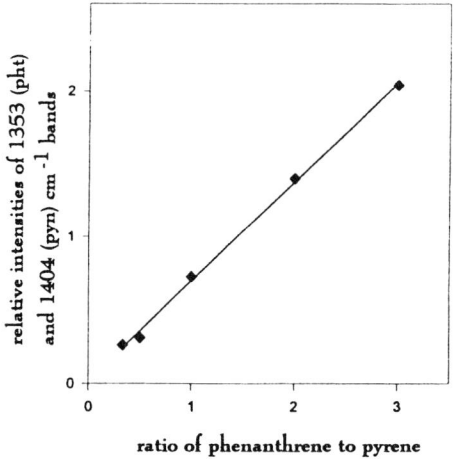

Fig. 3. Mixture containing phenanthrene and pyrene, $r^2 = 0.998$

has a relatively small linear range compared with other PAHs.

Effect of Multiple Components

The determination of a substance rarely involves the analysis of a sample containing one single component, and so the potential of this surface preparation technique for quantification was demonstrated by analysing the SER spectra of binary and ternary PAH mixtures. Angel and Myrick [4] commented that competition between components of a mixture for active sites has made quantitative analysis of solution mixtures by SER spectroscopy difficult. However, our analysis of binary mixtures of PAHs containing differing amounts of components shows that the SER spectra reflect the relative amounts of each component in the mixture, and that competition for active sites does not appear to occur. The linear plot (Fig. 3) of the ratio of band intensities of phenanthrene to those of pyrene against the proportion of each component in the mixture demonstrates this. Pyrene has already been found to have a lower detection limit on this surface [3], but nevertheless its SER spectrum does not suppress that of the phenanthrene. More complex ternary mixtures have also been studied, and in each case an approximately linear relationship was found to exist between the spectrum and component concentration over certain concentration ranges.

Conclusions

The results demonstrate that this SER technique has the potential to be used in analysis. The silver surface is simple to prepare and is reproducible. It was found that for each PAH there is a range of concentrations within which there is a linear relationship between concentra-tion and spectral intensity. Analysis of the SER spectra of binary and ternary mixtures of these compounds shows a linear relationship and demonstrates the potential for quantifying complex mixtures of organic compounds.

References

[1] M. Fleischmann, P. J. Hendra, A. J. McQuillan, *Chem. Phys. Lett.* **1974**, *26*, 163.
[2] E. Roth, G. A. Hope, D. P. Schweinsberg, W. Kiefer, P. M. Fredericks, *Appl. Spectrosc.* **1993**, *47*, 1794.
[3] S. Stewart, P. M. Fredericks, *J. Raman Spectrosc.* **1995**, *26*, 629.
[4] S. M. Angel, M. L. Myrick, *Pract. Spectrosc.* **1992**, *13*, 225.

Mikrochim. Acta [Suppl.] 14, 665–667 (1997)
© Springer-Verlag 1997

Surface-enhanced Infrared Absorption Spectroscopy (SEIRA) Using Multireflection ATR-elements

Heinz D. Wanzenböck[1,*], Boris Edl-Mizaikoff[1], Gernot Friedbacher[1], Manfred Grasserbauer[1], Robert Kellner[1], Markus Arntzen[2], Thomas Luyven[2], Wolfgang Theiss[2], and Peter Grosse[2]

[1] Institute for Analytical Chemistry, Vienna University of Technology, Getreidemarkt 9, A-1060 Wien, Austria
[2] I. Physikalisches Institut, RWTH Aachen, Sommerfeldstraße 28, D-52074 Aachen, Federal Republic of Germany

Abstract. This work shows the enhancement of spectral absorptions by noble metals (Ag, Au) deposited in nm-thick layers on the surface of ATR-crystals. By the surface-enhanced infrared absorption (SEIRA) technique, the absorption bands of p-nitrobenzoic acid can be enhanced by a factor of up to 40. This enables the detection of substance concentrations in the mM range. The structure of the metal deposit has been investigated by atomic force microscopy and correlated with the enhancement factor.

Key words: surface-enhanced infrared absorption, SEIRA, attenuated total reflection, FT-IR.

The adsorption of molecules on or near rough metal surfaces changes the light-absorption properties of the adsorbed species. This effect can be attributed to high electromagnetic fields at metal clusters, and the excitation of surface plasmons of the metal and charge-transfer complexes during the adsorption of molecules. As in surface-enhanced Raman spectroscopy (SERS), spectral enhancements have been observed in infrared spectroscopy when the analyte is deposited on nano-layers of noble metals.

It has been shown in the literature that island films with cluster sizes in the range 2–20 nm increase the spectral absorption of adsorbed analytes [1, 2]. Moreover, spectral shifts of the adsorbed species have been reported [3]. Surface-enhanced infrared absorption (SEIRA) in combination with attenuated total reflection (ATR) lowers the detection limit of ATR-spectroscopy significantly [4]. Furthermore, SEIRA is suitable for the characterization of thin organic films [5]. So far SEIRA-ATR-techniques have been restricted either to single reflections, using Kretschman-configurations

[2] or to the wavelength range confined by silicon-ATR-elements (8300–1500 cm^{-1}) [6].

The emphasis of this work is to extend the usable wavelength range for SEIRA-ATR-spectroscopy to 800 cm^{-1} by using conventional Ge and ZnSe-ATR-crystals. As a prerequisite, the deposition of chemically and mechanically stable metal clusters must be optimized. During this work the noble metals (silver, gold) were deposited on the crystal surface by various techniques such as sputtering or vapour deposition. The surface obtained was investigated by atomic force microscopy (AFM). The SEIRA-effect of the systems was tested with p-nitrobenzoic acid (PNBA) as model analyte. It could be shown that a Ge-ATR-crystal sputtered with a layer of silver with 7.0 nm nominal thickness thick results in a surface enhancement factor of 40 with respect to the ATR-measurement of the analyte in absence of the metal layer.

Experimental

For the ATR measurements crystals made of polycrystalline germanium and zinc selenide with trapezoidal shape ($50 \times 20 \times 2$ mm /$45°$) were used. The crystal surface was first polished with 0.3 μm Al_2O_3-suspension and rinsed with water, hexane and methanol (solvent quality, *pro analysis*). Silver was deposited by sputtering under ambient conditions at a base pressure of 1×10^{-4} mbar with a target current of 120 mA and an argon pressure of 4.0×10^{-3} mbar. Repeated rotation of the sample over a 5 mm aperture allowed sufficiently low sputter rates, ensuring exact control of the layer thickness. A deposition rate of 0.58 ± 0.07 nm per rotation was derived from X-ray reflection measurements. Gold was deposited from the vapour phase after evaporation of the metal by electrical resistance heating (base pressure $< 7 \times 10^{-7}$ mbar) at a deposition rate of 0.06 nm/s. The mass deposited was determined with an oscillating quartz sensor.

---

* To whom correspondence should be addressed

A thin film of *p*-nitrobenzoic acid dissolved in methanol was applied onto the ATR-crystal by depositing 5 µl on the surface. After evaporation of the solvent the measurements were performed with a Bruker IFS 66 FTIR spectrometer equipped with an optical ATR-bench and an MCT detector. All spectra were obtained by co-adding 1024 scans at a spectral resolution of $2\,cm^{-1}$. Residual water and carbon dioxide signals were eliminated by spectral subtraction. The sample spectra were ratioed against the spectra of the metal-coated ATR-crystals.

Results and Discussion

5 µl of a 1.0 m*M* PNBA solution (corresponding to approximately $188\,ng/cm^2$ PNBA) were deposited on the surface of the ATR-crystal. Analogous measurements of PNBA on silver layers with increasing nominal metal thickness from 1.2 to 8.2 nm were performed. A significant enhancement of the spectral absorptions of PNBA was found (Fig. 1). This enhancement was observed mainly for the carbonyl band at $1600\,cm^{-1}$ and the N–O vibrations at 1522 and $1345\,cm^{-1}$, although to a different extent. The detailed evaluation of the strongest band at $1345\,cm^{-1}$ shows an enhanced infrared absorption that can be correlated to the increasing layer thickness with the transmission limit at approximately 8 nm (Fig. 2). Additionally, the substrates were analysed by AFM in order to determine the metal layer structure. Figure 3 shows as an example a 3.5 nm thick silver layer with typical particle diameters from 10 to 20 nm and particle heights from 2 to 3 nm. This layer caused an absorption enhancement

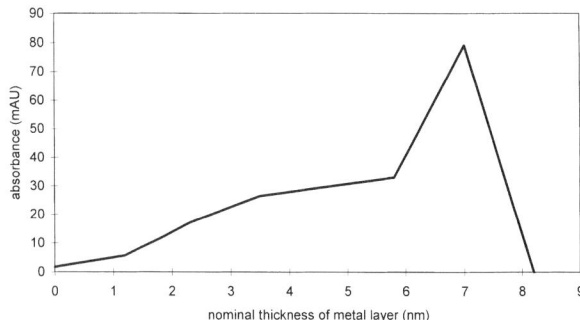

Fig. 2. Absorbance of *p*-nitrobenzoic acid (N–O vibration at $1345\,cm^{-1}$) *vs.* nominal silver layer thickness (calculated from deposition rate of 0.58 nm per rotation)

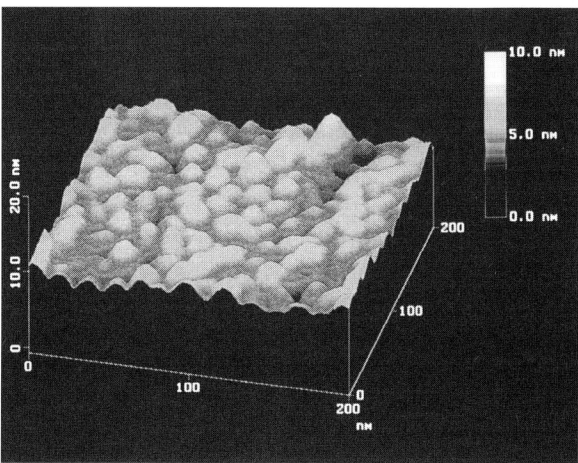

Fig. 3. AFM image of the 3.5 nm silver layer on a Ge-ATR-crystal surface (RMS: 1.0 nm; particle diameter 10–20 nm; particle height 2–3 nm)

factor of 15. Nominally thicker silver layers showed correspondingly larger clusters.

Similar results, though with lower enhancement factors, were observed with zinc selenide crystals coated with silver layers (sputtered) or gold layers (evaporated). For samples showing surface enhancement, AFM investigations showing the presence of island structures 15–25 nm in diameter and 1–2 nm in height.

Conclusions

This work proves that a combination of SEIRA and ATR with multiple ATR reflection elements is possible. With a germanium ATR-crystal coated with a 7.0 nm thick silver layer a spectral enhancement by up to a factor of 40 could be achieved for *p*-nitrobenzoic acid.

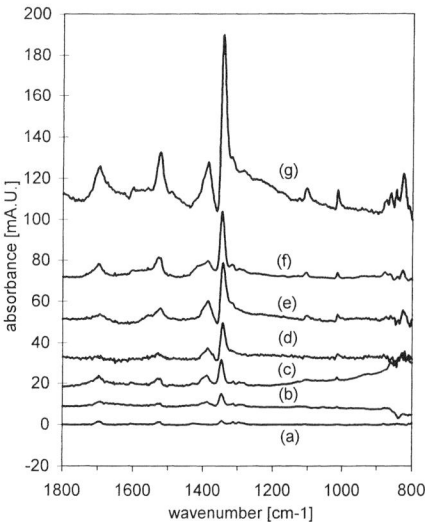

Fig. 1. 1.0 m*M* *p*-nitrobenzoic acid deposited on a Ge-ATR-crystal with *a* no silver layer, and layers of *b* 1.2 nm, *c* 1.8 nm, *d* 2.3 nm, *e* 3.5 nm, *f* 5.8 nm, and *g* 7.0 nm nominal thickness

The metal layer structure could be determined by AFM and showed that inhomogeneous metal layers with clusters in the nm-range result in a *significant spectral enhancement*.

Acknowledgements. Grateful thanks are due to C. Gmachl (Institut für Festkörperelektronik, Vienna University of Technology) for the metal layer deposition and to the "Fonds zur Förderung der wissenschaftlichen Forschung" for financial support under project "FWF 10386 CHE".

References

[1] M. Osawa, K. Ataka, *Surf. Sci.* **1992**, *262*, L118.

[2] A. Hatta, Y. Suzuki, W. Suetaka, *Appl. Phys. A* **1984**, *A35*, 135.

[3] K. P. Ishida, P. R. Griffiths, *Anal. Chem.* **1994**, 66, 522.

[4] N. J. Harrick, *Internal Reflection Spectroscopy*, Interscience, New York, 1967.

[5] Y. Nishikawa, K. Fujiwara, K. Ataka, M. Osawa, *Anal. Chem.* **1993**, *65*, 556.

[6] A. Hartstein, J. R. Kirtley, J. C. Tsang, *Phys. Rev. Lett.* **1980**, *45*, 201.

Mikrochim. Acta [Suppl.] 14, 669–670 (1997)

FT Infrared SEW Spectroscopy of Thin Films on Metal Surfaces by Using a Single Coupling Element or Composite SEW Waveguides

Guerman N. Zhizhin[1,*], **Andrei A. Sigarev**[2], and **Vladimir A. Yakovlev**[1]

[1] Institute of Spectroscopy, Russian Academy of Sciences, 142092 Moscow Region, Troitzk, Russia
[2] Zelenograd Research Institute of Physical Problems, 103460 Moscow, Russia

Abstract. The example of 4-n-octadecylphenol Langmuir–Blodgett (LB) films on a copper surface is used to show the possibility of measuring IR absorption spectra of thin dielectric films on a metal grating surface by means of surface electromagnetic wave (SEW) spectroscopy. Some methodological aspects of the use of composite surface electromagnetic wave waveguides for thin-film FT-IR spectroscopy are discussed.

Key words: infrared, spectroscopy, films, metal surface, surface electromagnetic waves.

The decay of surface electromagnetic waves (SEW) on a metal surface with a thin dielectric film between the input and output grating elements for coupling of the SEW and bulk radiation can be used to obtain the broad-band IR absorption spectrum of the film [1–3]. In the present work the possibility of measuring vibrational spectra of thin dielectric films on the metal surface of the SEW input grating element, and the versatility of the FT-IR SEW spectroscopy method in investigating the absorption spectra of thin films on smooth metal substrates by using composite SEW waveguides, are shown.

Experimental

Spectra in the wavenumber (v) range 650–2800 cm^{-1} were measured on a Digilab FTS-20B spectrometer with resolution 4 cm^{-1}, using a liquid-nitrogen cooled HgCdTe-detector and a special SEW attachment based on that described in [1]. Arrows in Fig. 1 show the path of the bulk radiation beams and the SEW propagation direction between gratings 7 and 8 in a two coupling-element configuration. In a single coupling-element configuration the role of the output coupling element was played by the interface between the smooth metal surface and the input grating element, where a part of the SEW beam was transformed into a bulk radiation beam which was propagated

at small angles to the metal surface and was directed by SEW attachment mirrors to the spectrometer detector. The scheme of the SEW attachment using two SEW input and output gratings is described in [1].

Composite SEW waveguide configurations are shown in Fig. 1b,c. A metal substrate (6) without a film or with the film (10) being investigated is brought up to the surface of a metal sample (2) between the SEW input and output gratings. The gap H between the plates (2) and (6) was set equal to 20 or 50 μm by means of nylon spacers. The input and output SEW coupling elements and the investigated LB film in an open composite waveguide were formed on three different substrates (3, 4, 5) the surfaces of which were fixed in one plane in the SEW attachment by a special sample holder. The lengths of substrates (5) and (6) were about 5 mm. Diffraction gratings of 11.4 μm period and about 0.8 μm ruling depth were used as coupling elements.

The conditions for preparation of the copper layers on glass and silicon substrates with and without surface diffraction gratings, the deposition conditions of 4-n-octadecylphenol (ODP) Langmuir–Blodgett (LB) films on the copper surface and the general procedure of measuring the optical density [D(v)] of a film are analogous to those described in [1, 2]. ODP films of 12.5 nm (5 monolayers) and 37.5 nm (15 monolayers) thickness were investigated.

Results and Discussion

The measurement of SEW transmittance spectra of a copper substrate both without and with an LB film of 5 monomolecular layers of ODP deposited on the copper surface, including the region of the coupling element, enabled us in a single coupling-element configuration to obtain an absorption spectrum of the ODP film (see Fig. 2). This technique gives similar positions and half-widths of the respective absorption bands in the ODF film spectrum in comparison with the two SEW input and output gratings technique [1], but the intensities of the absorption bands at 1260 and 1518 cm^{-1} are lower by factors of approximately 20 and 26, respectively, than those reported in [1]. This

---

* To whom correspondence should be addressed

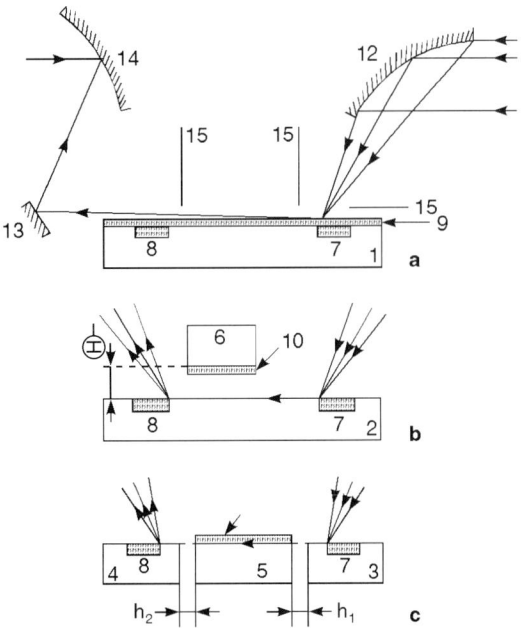

Fig. 1. Optical scheme of an SEW attachment for using a single coupling-element (**a**), and schemes of a composite SEW waveguide with two walls (**b**) and an open composite SEW waveguide (**c**): *1,2,3,4* – metal substrates with surface SEW input *(7)* and output *(8)* gratings; *5,6* – metal substrates; *9,10,11* – dielectric films; *12,13,14* – mirrors; *H* – air gap between plates 2 and 6; h_1, h_2 – air gaps between plates 3,4 and 5. A system of sceens used to cut scattering radiation is shown partially

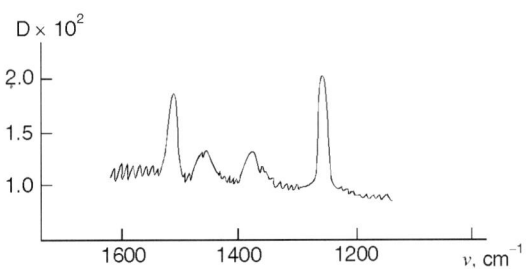

Fig. 2. The absorption spectrum of a 5 monolayer LB film of ODP on a copper grating surface (a single coupling-element method)

The application of a SEW waveguide with two walls, shown in Fig. 1b, to investigate the IR spectra of thin dielectric films on arbitrary metal substrates (6) [2] broadens the versatility of SEW spectroscopy. From the numerical calculation of the SEW dispersion for a system "copper – air gap H–ODP film 37.5 nm thick–copper" it was found that for the gaps of from 20 to 50 μm the intensity of the investigated absorption band of the film depends only weakly on the magnitude of the air gap. This agrees with the experimental results [2]. However, for the SEW waveguide with two walls the SEW transmittance decreases appreciably compared to the case of an open SEW waveguide of the same length owing to an increase in dielectric losses in the metal, which deteriorates the signal-to-noise ratio in the spectra of films.

To overcome this disadvantage the open composite SEW waveguide (see Fig. 1c) was used. Estimates based on the results obtained with CO_2-laser radiation [5] show that the SEW decrease due to the presence of about 0.1 mm gaps between the substrates is negligible. In our experiments with an open waveguide and ODP films of 5 and 15 monolayers, the signal-to-noise ratios for the peaks at 1260 and 1518 cm^{-1} were approximately as high as those in the case of a monolithic SEW waveguide [1, 2] after calculation for the same substrate length.

Thus, the use of a composite SEW waveguide broadens the applicability of the FT-SEW spectroscopy in investigating the vibrational spectra of thin films on meal substrates by the use of gratings and other kinds of coupling elements.

References

[1] G. N. Zhizhin, M. A. Moskalova, A. A. Sigarev, V. A. Yakovlev, *Opt. Commun.* **1982**, *43*, 31.
[2] G. N. Zhizhin, A. A. Sigarev, V. A. Yakovlev, *J. Molec. Liquids* **1992**, *53*, 1.
[3] G. N. Zhizhin, V. S. Bannikov, M. A. Moskalova, A. A. Sigarev, V. A. Yakovlev, *Phys. Chem. Mech. Surfaces* **1987**, *4*, 2905.
[4] K. O. Boltar, G. N. Zhizhin, A. A. Sigarev, R. A. Suris, V. A. Fedirko, *Pis'ma Zh. Tekh. Fiz*, **1983**, *9*, 1502 (in Russian).
[5] G. N. Zhizhin, M. A. Moskalova, E. V. Shomina, V. A. Yakovlev, *JETP Lett.* **1979**, *24*, 196.

discrepancy may be caused by the spectral dependence of the SEW decay coefficient on the grating [4], and the limited dimensions of the exciting IR radiation beam, but it enables estimation of the effective SEW propagating distances on the grating.

Mikrochim. Acta [Suppl.] 14, 671–674 (1997)

ATR-FT-IR Spectroscopy of Proteins Adsorbed on Biocompatible Cellulose Films

Martin Müller*, **Carsten Werner, Karina Grundke, Klaus J. Eichhorn,** and **Hans. J. Jacobasch**

Institut für Polymerforschung Dresden e.V., Hohe Strasse 6, D-01069 Dresden, Federal Republic of Germany

Abstract. The sorption of the plasma proteins human serum albumin (HSA) and human fibrinogen (FIB) onto hemodialytic cellulosic substrates was investigated by the surface-sensitive ATR-FT-IR spectroscopy. By means of this method we monitored quantitatively the kinetics of protein sorption onto acetylated and unmodified cellulose (AKZO-NOBEL), which are used as hemodialytic blood filter material. Furthermore, secondary structure alterations of the adsorbed proteins were detected and assigned. The spectroscopic findings were compared with the results of zeta potential and contact angle measurements in corresponding sorption experiments and were qualitatively related to blood compatibility.

Key words: ATR-FT-IR, adsorption, cellulose, biocompatible, plasma protein.

Cellulose is at present the most common material for hemodialytic filter membranes. However, since coagulation and clotting of blood in the filter pores cannot be prevented exclusively, there is need for deeper understanding of the polymer–blood interaction. One approach for investigation is use of medicinal tests (C5A, TAT, cell counting) with real blood. Another more basically oriented approach focuses on single or competitive plasma protein sorption experiments [1–3], since protein adorption is proven to be the first step of the complex biochemical cascade when blood interacts with the foreign surface of medicinally relevant polymers.

Here we report studies of the sorption of the plasma proteins human serum albumin (HSA) and fibrinogen (FIB) from aqueous solutions onto cellulose films, investigated by the surface-sensitive ATR-FT-IR spectroscopy. ATR-FT-IR enables quantitative *in situ* monitoring of phenomena taking place at the surface of thin polymer layers. Furthermore, FT-IR is sensitive to protein secondary structure. For the thermodynamic characterization of the polymer surface properties both contact angle and streaming potential measure-

ments were made on cellulose films equally prepared for the ATR-FT-IR experiment. To elucidate these complex processes single protein and competitive protein sorption processes on both hydrophilic unmodified and hydrophobic acetylated cellulose were studied.

Experimental

Spectroscopic Measurements

Generally, ATR-FT-IR measurements were performed on a Bruker IFS 28 spectrometer, equipped with a globar source and a nitrogen-cooled MCT detector. The ATR attachment consisted of a Perkin–Elmer four-mirror optical system and a steel flow-through cell (Prof. Fringeli, Zürich) with a $50 \times 20 \times 2\,mm$ trapezoidal silicon (refractive index $n_1 = 3.5$) internal reflection element (IRE). The angle of incidence θ was $45°$, resulting in 11 active reflections. The IRE was coated by spreading three drops of a solution of unmodified cellulose (0.4% in DMAC/LiCl) or of acetylated cellulose (0.4% in acetone; ds = 2.5), respectively. The film thickness (2000 Å) was checked by ellipsometry. For the single and the competitive protein sorption experiments protein solutions of 5 mg/ml for HSA and 1 mg/ml for FIB were injected onto the polymer-coated IRE at time zero. For the sequential protein sorption experiments, the solution of the first adsorbed protein was removed by rinsing with D_2O (5 ml) and 1 ml of the second protein solution was injected at time zero. Typically, a set of about 30–50 time-dependent. ATR-FT-IR single-channel intensity spectra (I) of 200 scans were recorded *in situ* during the sorption experiment. As references either the single-beam intensity spectra (I_0) of the blank Si-IRE or of the cellulose coated Si-IRE in contact with D_2O were used. Accordingly, computation of $A = -\log(I/I_0)$ revealed absorbance spectra due to sorbed protein and hydrated cellulose, as well as absorbance spectra due to sorbed protein. The evaluation of the amount of protein sorbed, Γ (mol/cm^2), was based on a modified Lambert–Beer law [4, 5]:

$$A = N \cdot \varepsilon \cdot \frac{\Gamma}{d} d_e$$

where A is the integrated area of the amide I' band (1700–1600 cm^{-1}), the integrated adsorption coefficient of which, ε, was taken from the literature [6], N is the number of active reflections and d_e is the so-called effective thickness, which is dependent, on n_1, n_2, n_3 and the sample thickness d.

---

* To whom correspondence should be addressed

Contact-angle Measurements

A water-in-air technique with the dry films upon the ATR-plate was used to study the contact angles at swollen cellulose films by axisymmetric drop-shape analysis (ADSA). More details of this measuring technique will be described in [7].

Zeta Potential Measurements

Streaming potential measurements [8] were applied to estimate the zeta potential at interfaces of cellulose films with aqueous solutions. The sensitivity of electrokinetic measurements with respect to changes of the interfacial composition enables monitoring of the adsorption of proteins from a laminarly flowing buffer solutions onto the substrate films *in situ*.

Results and Discussion

Initial *in situ* ATR-FT-IR measurements on the adsorption of blood components onto polymers are well documented [9–11]. Additionally, ATR-FT-IR data offer quantitative determination of the surface concentrations of adsorbed proteins [5, 6]. To elucidate protein/polymer adsorption systematically we used three basic experiments with the two typical plasma proteins HSA and FIB, *viz.* adsorption from single protein solutions (HSA, FIB), and competitive (HSA + FIB) and sequential (1 HSA, 2 FIB; 1 FIB, 2 HSA) adsorption onto cellulose.

ATR-FT-IR Spectra

In Fig. 1 a typical set of time-dependent *in situ* ATR-FT-IR absorbance spectra is shown, which were recorded during the adsorption of FIB from D_2O solution onto an acetylated cellulose film covering an

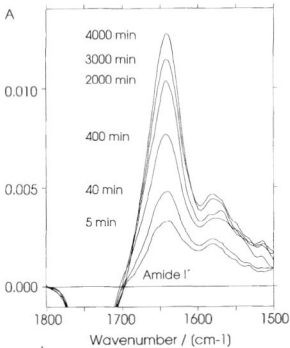

Fig. 1. Time dependent *in-situ* ATR-FT-IR absorbance spectra for adsorption of fibrinogen on acetylated cellulose. The intensity spectrum of the cellulose-coated Si-IRE was used as reference (angle of incidence: $\theta = 45°$, number of active reflections: $N = 11$

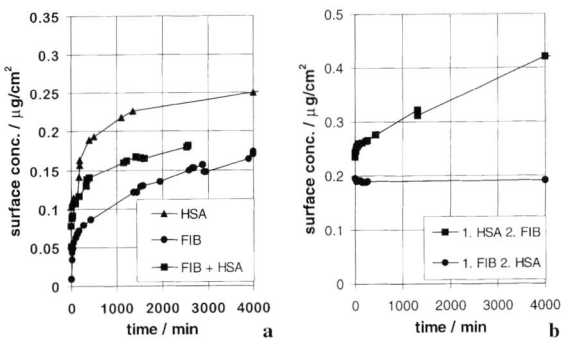

Fig. 2. a Adsorption kinetics of HSA and FIB on acetylated cellulose. Single protein adsorption of HSA (5 mg/ml) and FIB (1 mg/ml) from solution onto acetylated cellulose, and competitive adsorption between HSA (5 mg/ml) and FIB (1 mg/ml) in a mixture [FIB + HSA]. **b** Sequential adsorption kinetics of HSA (5 mg/ml) and FIB (1 mg/ml) on acetylated cellulose in the orders: 1, HSA; 2, FIB: 1, FIB; 2, HSA

Si-IRE. The significant time-dependent increase of the amide I' band, which is related to the (CO) of the protein peptide groups, was caused by the formation of surface-adsorbed protein layer.

Adsorbed Protein Amount

The kinetic curves of the single protein adsorption as well as of the competitive adsorption onto acetylated cellulose (HSA + FIB) are shown in Fig. 2a. Obviously, HSA shows greater adsorption affinity than FIB, whereas adsorption from a mixture of both proteins resulted in an adsorbed amount lying between the values for the single proteins. For both single protein layers the amount of adsorbed protein was in the range of an ideally packed monolayer, taking into account the geometrical dimensions of HSA and FIB [12] for a "side on" arrangement, i.e. the rod-like proteins are attached to the polymer surface by their biggest side.

Furthermore, sequential adsorption was performed by presorbing the acetylated cellulose with a single protein (HSA or FIB) and then changing the solution to the other protein. From Fig. 2b it is evident that for the order 1, HSA; 2, FIB the adsorption continued, whereas for the opposite order presorbing with FIB prevented further increase in the sorbed protein layer. This can be explained by further adsorption of FIB onto the HSA layer and an obviously hindered adsorption of HSA onto the compete FIB layer, which gives an idea of the different topologies of these blood proteins.

Secondary Structure

Secondary structure changes are due to irreversible adsorption of proteins onto polymeric surfaces and may therefore be a measure for biocompatibility. It is well known that FT-IR spectroscopy is sensitive to protein conformation [13–19]. Probing secondary structure is basically related to the position and the shape of the amide I band, which is often strongly overlapped by partial components representing typical secondary structures. For the analysis and significant separation of the amide I' band into partial components we applied curve-fitting of the ATR-FT-IR spectra for cellulose-adsorbed single proteins. The underlying components of the amide I' band were identified and fractionally quantified by the curve-fitting method without further manipulations of the FT-IR data [20, 21] and were assigned to protein secondary structure according to Table 1.

In Fig. 3 curve-fitting results for the amide I' region of ATR-FT-IR spectra monitoring FIB adsorption on unmodified cellulose are presented. According to Table 1, human fibrinogen adopted α-helical and β-sheet structures in the beginning of the sorption process, whereas after 2000 min these ordered structures disappeared and had changed nearly completely to random coil. The results of the conformational analysis of the proteins HSA and FIB sorbed on acetylated and unmodified cellulose are summarized in Table 2. HSA changed in secondary structure on acetylated cellulose, whereas FIB remained unaltered. However, since from

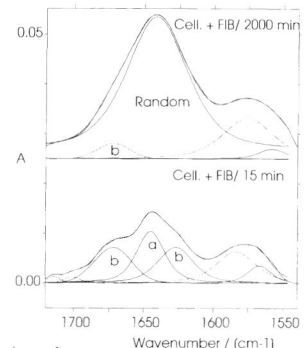

Fig. 3. Secondary structure of sorbed plasma proteins. Curve-fitting results for the secondary-structure sensitive amide I′ band (50/50% Gauss/Lorentz) of ATR-FT-IR spectra of the initial (S) and the end (E) state of fibrinogen sorption on unmodified cellulose (a α-helix, b β-sheet)

Fig. 2b we observe a further adsorption of FIB onto the presorbed HSA layer, we conclude that the irreversible adsorption of HSA could not prevent additional adsorption of FIB molecules, which have undesired thrombogenic activity (fibrinogen → fibrin) [12]. For the unmodified cellulose only FIB changed in secondary structure, whereas HSA remained stable. In that case it must be studied further whether the irreversibly adsorbed FIB may exhibit thrombogenic activity, since it may not be exchanged by further proteins with higher molecular masses. On the other hand, for blood-compatible purpose FIB may no longer be thrombogenic, since it was induced to change its native

Table 1. Assignment of amide I′ components to protein secondary structures (D$_2$O) [13–19]

Secondary structure	β-Sheet	β-Turn	Random	α-Helix	β-Sheet
Amide I′ frequency [cm^{-1}]	1690–1670	1700–1660	1660–1640	1650–1640	1630–1620

Table 2. Secondary structure fractions of HSA and FIB sorbed on cellulose. The values given in brackets refer to the fractions of the native proteins, taken from the literature [12]

	HSA (S: start → E: end)	FIB (S: start → E: end)
Acetylated cellulose	α-helix: 45 (48)% − >27% β-sheet: 21 (15)% − >39% random: 34 (37)% − >34%	-no change-
Unmodified cellulose	-no change-	α-helix: 35 (33)% − >2% β-sheet: 65 (--)% − >6% random: 0 (--)% − >92%

Table 3. Thermodynamic and spectroscopic characterization of proteins on acetylated cellulose (ac. cell)

	Θ (Final state/dry film) [°]	Zeta potential (final state) [mV]	Γ FTIR-ATR (final state) [$\mu g/cm^2$]
Acetylated cellulose	$69° \pm 2$	-15.5 mV	(0.00)
HSA + ac. cell.	$37° \pm 2$	-7.4 mV	0.25
FIB + ac. cell.	$72° \pm 13$	-11.0 mV	0.17
HSA + FIB + ac. cell.	$83° \pm 5$	-9.8 mV	0.20
1, HSA; 2, FIB + ac. cell.	$84° \pm 5$	-10.8 mV	0.42
1, FIB; 2, HSA + ac. cell.	–	-8.5 mV	0.19

conformation by the hydrophilic surface of the unmodified cellulose.

Thermodynamic Measurements

Contact angle measurements as well as zeta potential measurements were performed on uncovered and protein-adsorbed cellulose films, with the samples for the ATR-measurements. The results of both methods are summarized in Table 3.

The contact angle is an integral measure of surface energetics, including hydrophilic/hydrophobic properties. Obviously, the protein-covered acetylated cellulose films, after drying, had contact angle values different from that for the uncovered film. Moreover, adsorbed films of both FIB and HSA differed in their contact angle. Therefore it can be concluded, that (1) after protein adsorption the cellulose surface is mainly governed by the properties of the adsorbed protein film and (2) sorbed HSA films are more hydrophilic than films of adsorbed FIB. Additionally, since in the end state of the competitive and sequential sorption experiment the contact angle on the dry film is closer to the FIB value, complete coverage or exchange of the HSA layer can be deduced.

Streaming potential measurements with protein solutions allowed monitoring of the zeta potential at the water/polymer interface. For closer understanding of the zeta potential methodology applied on flat polymer films the reader is referred to the literature [8]. The negative sign of the zeta potential is due to deprotonated species or/and the preferential adsorption of hydroxide/halide anions onto the surface. Like the contact angles, the zeta potential values measured at the adsorbed protein layer differed from the values for the uncovered cellulose film. Since they revealed characteristic values for HSA and FIB, respectively, we conclude that the electrokinetic properties of biomaterial polymer surfaces are sensitively related to the character of the adsorbed protein layer.

References

[1] J. D. Andrade, *Surfaces and Interfacial Aspects of Biomedical Polymers, Vol. 1, 2*, Plenum, New York, 1985.

[2] L. Vroman, *The Vroman Effect* (C. H. Bamford, S. L. Cooper, T. Tsurutta, eds.), VSP, Utrecht, 1992.

[3] W. Norde, J. Lyklema, in: *The Vroman Effect* (C. H. Bamford, S. L. Cooper, T. Tsurutta, eds.), VSP, Utrecht, 1992, pp. 1–20.

[4] N. J. Harrick, *Internal Reflection Spectroscopy*, Harrick Sci. Corp., Ossining, New York, 1979.

[5] U. P. Fringeli, in: *Internal Reflection Spectroscopy: Theory and Application* (F. M. Mirabella, ed.) Dekker, New York, 1992.

[6] U. P. Fringeli, H-J, Apell, M. Fringeli, P. Läuger, *Biochem. Biophys. Acta* **1989**, *984*, 301.

[7] K. Grundke, T. Bogumil, T. Gietzelt, H.-J. Jacobasch, D. Y. Kwok, A. W. Neumann, *Progress Colloid and Polymer Science*, (1996, in press).

[8] H.-J. Jacobasch, J. Schurz, *Österr. Chem.-Z.* 7–8, 164 (1987) **1987**, *88*, 164.

[9] J. S. Mattson, T. T. Jones, *Anal. Chem.* **1976**, *48*, 2164.

[10] R. M. Gendreau, R. J. Jakobsen, *Appl. Spectrosc.* **1978**, *32*, 326.

[11] R. Kellner, G. Götzinger, *Mikrochim. Acta* **1984**, *II*, 61.

[12] *The Plasma Proteins, Vol. 1* (F. W. Putnam, ed.), Academic Press, New York, 1981.

[13] T. Miyazawa, *J. Chem. Phys.* **1960**, *32*, 1647.

[14] S. Krimm, J. Bandekar, in: *Advances in Protein Chemistry, Vol. 38* (C. B. Anfinsen, J. T. Edsall, M. F. Richards, eds.), Academic Press, New York, 1986, pp. 181–364.

[15] A. Elliott, E. J. Ambrose, *Nature* **1950**, *165*, 921.

[16] M. Byler, H. Susi, *Biopolymers* **1986**, *25*, 469.

[17] J. L. R. Arrondo, A. Muga, J. Castresana, F. M. Goni, *Progr. Biophys. Molec. Biol.* **1993**, *59*, 23.

[18] W. K. Surewics, H. H. Mantsch, *Biochem. Biophys. Acta*, **1988**, *952*, 115.

[19] D. C. Lee, P. I. Haris, D. Chapman, R. C. Mitchell, *Biochem.* **1990**, *29*, 9185.

[20] M. Rüegg, V. Metzger, H. Susi, *Biopolymers* **1975**, *14*, 1465.

[21] M. Müller, *Thesis ETH*, No. 10422, Zürich, 1993.

Mikrochim. Acta [Suppl.] 14, 675–676 (1997)

FT-IR Study of Langmuir–Blodgett Films of Chlorophyll *a*

Hidetoshi Sato[1], **Yukihiro Ozaki**[1,*], **Kaku Uehara**[2], **Toshinari Araki**[3], and **Keiji Iriyama**[4]

[1] School of Science, Kwansei-Gakuin University, Nishinomiya 662, Japan
[2] Research Institute for Advanced Science and Technology, University of Osaka Prefecture, Sakai 593, Japan
[3] Tokyo Gas Co. Ltd., Tsurumi-ku, Yokohama 230, Japan
[4] Institute of Medical Science, The Jikei University School of Medicine, Minato-ku, Tokyo 105, Japan

Abstract. The structure of monolayer Langmuir–Blodgett (LB) films of chlorophyll *a* (Chl-*a*) has been studied by FT-IR spectroscopy. The 1750–1450 cm^{-1} region of the FT-IR spectra shows that Chl-*a* occurs as a five-coordinate monomer in the LB films irrespective of the number of layers. The FT-IR spectra of the multilayer LB films are very close to the spectrum of a monolayer LB film, indicating that the interactions between each layer are very weak.

Key words: chlorophyll a, Langmuir – Blodgett film, FT-IR.

The structure functions of thin films of chlorophylls (Chls) have been investigated extensively, partly to understand characteristic features that Chl-*a* shows in photosynthetic systems and partly to aim at constructing functional molecular devices such as photoelectric cells based upon the Chl-*a* films [1–3]. Among various kinds of thin films Langmuir–Blodgett (LB) films have recently received keen interest because they provide the possibility of constructing artificial supramolecular organizations with designated structure and properties. Several research groups have been investigating the structure of LB films of Chl-*a* by using FT-IR spectroscopy [4, 5]. We recently demonstrated that the attenuated total reflection (ATR) technique can be used to obtain IR spectra of a monolayer LB film of Chl-*a* with a high signal-to-noise ratio and discussed the structure of Chl-*a* in the LB films [6]. The purpose of the present paper is to investigate the structure of multilayer LB films of Chl-*a* and compare it with that of the monolayer LB film.

Experimental

Methods for the preparation of Chl-*a* and its LB films were described in our previous paper [6]. The monolayer LB film of Chl-*a* was deposited on a germanium ATR prism, and the multilayer LB films were prepared on CaF$_2$ plates on which a monolayer LB film of barium arachidate had been deposited. The FT-IR spectra of the LB films were recorded at 4 cm^{-1} resolution on a Nicolet Magna 550 FT-IR spectrometer equipped with an MCT detector.

Results and Discussion

Figure 1a shows the ATR/IR spectrum of a monolayer LB film of Chl-*a*. The spectrum is almost identical with

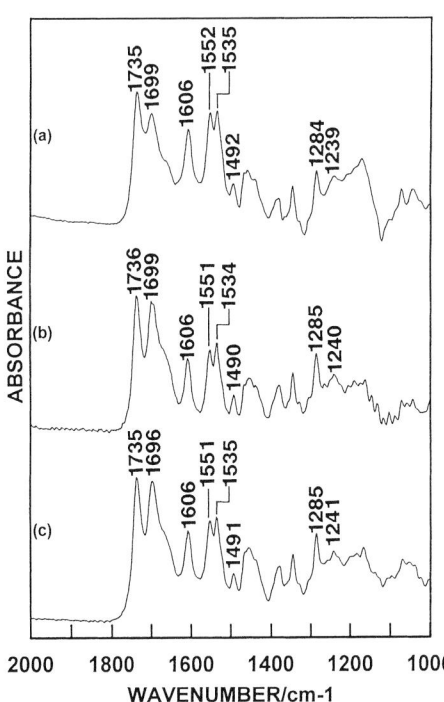

Fig. 1. ATR/FT-IR spectrum of a monolayer LB film of Chl-*a* (*a*) and transmission FT-IR spectra (*b*) six- and (*c*) twelve-layer LB films of Chl-*a*

* To whom correspondence should be addressed

the spectra previously reported for similar films [6]. Figures 1b and 1c depict IR transmission spectra of six- and twelve-layer LB films of Chl-*a*. Bands at 1736 and 1698 cm^{-1} in the three spectra are due to C=O stretching modes of the two ester groups and the free 9-keto group of Chl-*a*, respectively. Four bands at 1606, 1551, 1535, and 1491 cm^{-1} are all attributed to modes having largely methine-bridge stretching characteristics (hereafter, the four bands are referred to as IR1, IR4, IR5 and IR6, respectively) [7]. It is well known that the frequencies of the IR1, IR4, IR5, and IR6 bands are sensitive to the coordination number of the Mg atom [7].

In our previous paper [6] it was concluded from the dominant appearance of the band due to the free keto group and the frequencies of the four marker bands that Chl-*a* occurs as a five-coordinate monomer in the LB film. In the IR spectra of the multilayer LB films, the band due to the free keto group is again predominant in the 1700–1650 cm^{-1} region and the frequencies of the four marker bands are very close to those typical for the five-coordinate Chl-*a*. These observations lead us to the conclusion that Chl-*a* a five-coordinate monomer not only in the monolayer film but also in the multilayer films.

Chl-*a* always forms a dimer or oligomers in cast films and the five-coordinate monomer exists only in the LB films. Therefore, it seems that Chl-*a* molecules are highly oriented and packed tightly in the LB films. We have considered that water plays an important role in combining neighbouring Chl-*a* molecules in the LB films [6].

Comparison of the three spectra in Fig. 1 gives one more important conclusion as to the structure of the LB films of Chl-*a*. The three spectra are very close to each other in terms of both the vibrational frequencies and relative intensities of the bands, except for the slight decrease in the relative intensity of the band at 1699 cm^{-1} in the spectrum of the monolayer LB film. The close similarity between the IR spectra of the single-layer and multilayer films suggests that the longitudinal interactions between each monolayer are weak.

References

[1] C. W. Tang, A. C. Albrecht, *J. Chem. Phys.* **1975**, *62*, 2139; **1976**, *63*, 953.

[2] M. F. Lawrence, J. P. Dodelet, L. H. Dao, *J. Phys. Chem.* **1984**, *88*, 950.

[3] A. Diarra, S. Hotchandani, J-J. Max, R. M. Leblanc, *J. Chem. Soc., Faraday Trans. 2.* **1986**, *82*, 2247.

[4] C. Chapados, R. M. Leblanc, *Biophys. Chem.* **1983**, *17*, 211.

[5] J. A. Bardwell, M. J. Dignam, *J. Colloid Interface Sci.* **1987**, *116*, 1.

[6] H. Sato, Y. Ozaki, K. Uehara, T. Araki, K. Iriyama, *Appl. Spectrosc.* **1993**, *47*, 1509.

[7] M. Tasumi, M. Fujiwara, in: *Advances in Spectroscopy, Vol. 14* (R. J. H. Clark, R. E. Hester, eds.), Wiley, Chichester, 1987, pp. 407–428.

Mikrochim. Acta [Suppl.] 14, 677–678 (1997)
© Springer-Verlag 1997

Crystal Effect on Penetration Depth in Attenuated Total Reflectance Fourier-transform Infrared Study of Human Skin

Marchel Snieder* and **Wei G. Hansen**

Unichema International, P. O. Box 2, 2800 AA Gouda, The Netherlands

Abstract. A non-invasive ATR-FTIR technique has been used to directly measure skin conditions in vivo. From outside to inside, stratum corneum (SC) consists of fatty layer, one cell layer, a thick layer of lipid and another 10–20 layers of cells. Previous publications [1, 2] have reported the combination of ATR-FTIR with a skin-stripping technique to study human skin in vivo. In this paper, the subtle difference of penetration depth obtained in human skin by using different ATR crystals is discussed. With a Ge crystal (45°), only an α-helix protein band was observed in the ATR-FTIR spectrum, which indicated the penetration of light into the one-cell layer of the stratum corneum, because proteins in the cells are in α-helix form. In contrast, with a ZnSe crystal (45°), both α- and β-helix protein bands were observed in the FTIR spectrum, which indicated that IR light penetrates into the lipid layer, because the floating lipid contains β-helix form proteins.

Key words: ATR-FT-IR, skin analysis, penetration depth, crystal.

In order to study the effects of some personal care products on human skin, the ATR-FTIR technique has been utilized to analyse the skin conditions. The stratum corneum (SC) is the external part of human skin. The SC consists of layers of flattened dead cells, between which are the intercellular lipids. The thickness of the SC is 10–20 μm, or 10–20 cell layers, depending on the part of the skin and the person. On the outside of the SC is a thin fat layer (sebum) followed by one cell layer (~ 0.5–1.0 μm). A large lipid layer follows this first cell layer [1]. Skin lipids contain [3] glycerides, sterols, terpenes and esters, and between the lipids are proteins [4]. The depth profile of the SC structure can be obtained by tape stripping combined with the ATR-FTIR technique, because each strip can remove about one cell layer of the skin [1].

It has been proved from our studies that using different type of ATR crystals enables us to observe the

skin conditions at a certain few layers of the SC down to the large lipid layer, even without stripping, because different crystals give different penetration depth into human skin. Most of the previous ATR-FTIR applications to in vivo skin analysis reported in the literature [1, 2] used human forearm skin in the studies. However, our instrumental conditions are unsuitable for use with forearm skin. In this paper, all the test results were obtained by measuring the skin of human hands.

Experimental

FT-IR-ATR Instrumentation

Bio-Rad FTS40; deuterated triglycerine sulphate; Digilab ATR included ZnSe 45° and Ge 45° crystals, 32 scans; scanning range 4000–750 cm^{-1}; spectra were scaled on the amide I band.

Sample Preparation

No eating, drinking or any activity was allowed for two hours before the measurement, then after a rest for 20 min in a room conditioned at 21°C the left hand was placed on the crystal, with good contact between the crystal and the hand. The measurement was started directly, and included a delay time of 1 min. The hand was kept still during measurement, until the last spectrum had been measured.

Results and Discussion

The subtle difference of the IR radiation penetration depths in human skin that are obtainable by using different crystals was explored. The skin protein can be in the form of α- or β-helix. In the α-helix form, the NH group forms a hydrogen bond with the CO group inside the residue (n-4) of the same molecule in a coiled structure. In contrast, the β-helix is in a pleated sheet form, because the β-helix is formed by hydrogen bonds

---

* To whom correspondence should be addressed

between the NH and the CO groups from different polypeptides. Disordered bridging is also possible. Obviously, owing to the differences in hydrogen bonding energy in the α- and β-helix forms, the amide absorption bands in the FT-IR spectrum can be separated and identified as α and β peaks [5–7], with the C=O band (amide I) and N–H band (amide II). The formation energy of the α-helix is a little higher than that of the β-helix, as shown in the IR spectra, the β-helix absorption bands (~ 1632 cm^{-1}) for amide I and ~ 15375 cm^{-1} for amide II) are on the smaller wavenumber side (lower energy) than those of the α-helix (~ 1644 cm^{-1} for amide I and ~ 1547 cm^{-1} for amide II).

Depending upon the penetration depth, in the IR spectrum there are either α-helix bands for a shallower penetration, or both α-helix bands and β-helix bands for a deeper penetration. The sebum does not contain any protein [3], the skin cells contain mostly α-helix proteins [4], and in the lipid there are α- or β-helix species present [8]. In other words, the presence of α/β or α bands in the ATR-FT IR spectrum could be an indication for the penetration depth. The comparison of two spectra taken by using Ge and ZnSe crystals as ATR surfaces is shown in Fig. 1A. Figure 1B is an enlarged part of Fig. 1A. The third spectrum corresponds to the sebum, i.e. it is the spectrum of the residue on the ATR crystal after the hand has been removed.

Theoretically, IR radiation penetration depth in human skin depends on many factors, as shown by the equation for the depth calculation

$$dp = \frac{\lambda}{2\Pi n_c \sqrt{(\sin\Theta)^2 - \left(\frac{n_s}{n_c}\right)^2}}$$

In the equation dp is the penetration depth, λ is the wavelength, Θ is the incidence angle, n_s is the refractive index of the sample and n_c that of the ATR crystal (ZnSe 2.4 and Ge 4.0). The refractive index of skin in vitro is 1.6 [9]. With a ZnSe (45°) crystal, the calculated penetration depth is about 0.9 μm and with an angle of 60° is about 0.7 μm. With a Ge (45°) crystal, the penetration depth is about 0.30 μm. Thus changing the crystal type results in a larger penetration difference than does changing the angle for the same crystal.

In our studies, with a Ge crystal (45°), it is obvious that the IR radiation penetrates only into the first layer of cells in the SC tissue, because only α-proteins are detected. On the other hand, with a ZnSe crystal both

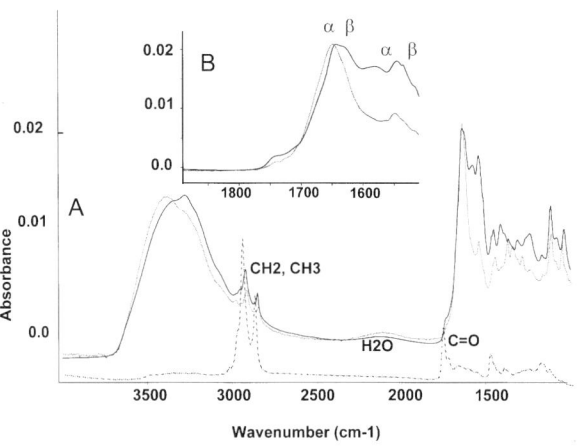

Fig. 1. A ATR-FT-IR spectrum of skin: (—) ZnSe crystal; (···) Ge crystal; (_.._..) sebum. **B** Enlargement of part of (**A**), for ZnSe and Ge crystals

α- and β-helix bands are observed in the FT-IR spectra. This indicates that the penetration reaches the structure of the lipid layer, because the β-helix absorption bands are typical absorption bands from the proteins floating in lipid layer. The IR radiation in this case has passed through the first layer of cells in SC tissue, then penetrated into the lipid.

Conclusion

As a conclusion from the discussion above, when a Ge crystal (45°) is used as the ATR surface in FT-IR to measure skin in vivo, the IR light penetration depth is ~ 0.3 μm, whereas with a ZnSe crystal (45°), the penetration depth is ~ 0.9 μm. Our observations agree well with the penetration depth estimated from the refractive index based on in vitro skin analysis.

References

[1] D. Bommannan, R. O. Potts, R. H. Guy, *J. Investigative Dermatology* **1990**, *95*, 403.
[2] H. E. Boddé, L. A. R. M. Pechtold, M. T. A. Subnel, F. H. N. de Haan, *Liposome Dermatics, Griesbach Cont.* **1992**, 137.
[3] R. Greene, D. T. Dowing, P. E. Pochi, J. S. Strauss, *J. Investigative Dermatology* **1970**, *54*, 240.
[4] B. W. Barry, H. G. M. Edwards, A. C. Williams, *J. Raman Spectvosc.* **1992**, *23*, 641.
[5] A. M. A. Pistorius *Spectrosc. Eur* **1995**, *7*, 8.
[6] K. Rahmelow, W. Hübner, Th. Ackermann, *5th Int. Conf. Spectrosc. Biol. Mol.*, 1993, 139.
[7] R. J. Cushley, P. Y. K. Chin, J. Y. N. Yang, W. D. Treleaven, *Life Sci. Adv.: Biochem.* **1993**, *12*, 89.
[8] L. Stryer, *Biochemistry*, Freeman, San Francisco, 1981, p. 222.
[9] R. J. Scheuplein, *J. Soc. Cosmet. Chem.* **1964**, *15*, 111.

Mikrochim. Acta [Suppl.] 14, 679–681 (1997)
© Springer-Verlag 1997

Internal Reflectance IR Spectroscopy of Langmuir Monolayer Films at the Air–Water Interface

Suci Widayati, Susan M. Stephens, and **Richard A. Dluhy***

Department of Chemistry, University of Georgia, Athens, GA 30602-2556, U.S.A.

Abstract. We report here the vibrational spectroscopic study of monomolecular films at the air–water interface, by using internal reflection infrared spectroscopy. Unlike previous external reflectance IR methods developed in our laboratory, this method uses attenuated total reflectance (ATR) spectroscopy to study monolayer films. The internal reflectance approach offers several advantages for the study of Langmuir monolayers at the air–water interface, including (i) improved S/N (signal-to-noise) ratio resulting from multiple reflections, and (ii) the ability to form supported planar bilayer films of defined chemical structure and orientation. One method under design in our laboratory uses a Ge internal reflection element derivatized with octadecyltrichlorosilane. Using this design, the study of spread Langmuir phospholipid monolayer films in this ATR geometry results in a supported, oriented, planar membrane bilayer with hydrated polar head groups. Preliminary studies reveal that this method of analyzing phospholipid Langmuir films at the air–water interface produces results consistent with previous ATR-IR studies of transferred Langmuir–Blodgett phospholipid monolayers.

Key words: monomolecular films, air–water interface, attenuated total reflectance.

Infrared external reflection spectroscopy has been successfully applied to the study of organic monomolecular films on reflective substrates for many years. Up until just a few years ago, the published literature on external reflection IR spectroscopy dealt exclusively with thin films on reflective metals, owing to the low reflectivity of non-metallic substrates, which leads to a low optical throughput and a degraded S/N ratio in the resulting spectrum. However, the advent of Fourier-transform infrared instrumentation along with new experimental designs specifically adapted for thin films has changed this situation; it is now possible to acquire external reflectance spectra for a variety of dielectric substrates [1]. As a consequence, researchers are able to study the structure of organic thin films on the most relevant substrates needed for their particular application [2].

This last advantage has been exploited in the study of so-called Langmuir monolayers, which are monomolecular films spread at the air–water (A/W) interface. Using the IR reflectance properties of the water subphase, our research group was the first to develop an IR reflectance method for the study of structure, conformation, and orientation in Langmuir monolayer films [1]. The development of this methodology has had an especially great impact on the use of external reflection IR spectroscopy in biophysical research, since some of the most relevant Langmuir monolayers include phospholipids, proteins, steroids, and other compounds of a biochemical nature. These compounds may now be studied directly at an aqueous interface.

While this external reflectance IR method has proven to be very successful in studying biomembrane models, there are several inherent spectroscopic properties which currently limit its routine applicability. These properties include: (i) very weak absolute reflectances which limit the possible S/N ratio in the final spectra, (ii) interferences from H_2O vapour, and (iii) baseline dispersion in regions of liquid H_2O absorbance bands, owing to the anomalous dispersion of the H_2O reflectance. The last two points especially affect the spectral region from 1400 to 1900 cm^{-1}, which encompasses the conformationally-sensitive amide I region of the IR spectrum. These are the major reasons why this method has not been routinely used to

---

* To whom correspondence should be addressed

study proteins and peptides on surfaces – the baseline dispersion due to the underlying H₂O absorbances make this a very difficult region in which to acquire external reflectance IR spectra.

Recent work has described an alternative approach to acquiring the IR spectra of insoluble biophysical monolayer films at the A/W interface, which has the potential to overcome these problems and lead to enhanced monolayer IR spectra. This method uses the principle of internal rather than external reflectance IR spectroscopy in order to increase the number of possible reflections at the monolayer interface. This increased reflectance should directly lead to higher absorbances and S/N ratios in the final spectra. Also, since this is an internal reflectance, (i.e. attenuated total reflectance [ATR]) type experiment, the problem of the underlying baseline dispersion is eliminated and the compensation of H₂O absorbances should be much easier than in the external reflectance experiment.

Experimental

The surface chemistry methods used in this study have been previously described in detail [4,5]. Monocrystalline germanium (Ge) substrates were prepared for attenuated total reflectance infrared (ATR-IR) analysis, and Langmuir–Blodgett monolayers were transferred onto these substrates, as previously described [4,5].

Infrared spectra of the bilayers transferred onto Ge substrates were acquired with a Digilab FTS-60 Fourier-transform infrared spectrometer equipped with a custom-designed optical interface and a narrow band, liquid-N₂ cooled HgCdTe detector. Specific details of the optical interface have been published elsewhere [3,6].

Results

Recently, an internal reflectance method has been developed which is specifically designed to obtain polarized IR-ATR spectra of supported monolayer films deposited onto functionalized, hydrophobic germanium crystals directly at the air–water interface [3]. To test the efficiency of this method, we have used it to obtain the IR spectra of phospholipid monolayers at various surface pressures.

In our internal reflectance design, an IR spectrometer is optically interfaced with a specially-designed Langmuir trough containing a trapezoidal germanium ATR crystal immersed in the subphase. Specific design features of this instrument include: (i) the geometry of the crystal provides for multiple reflections, thereby increasing the S/N ratio in the resulting spectra, (ii) the Ge crystal may be surface-treated, thereby allowing for the formation of supported bilayer films, in addition to the study of the monolayer film with its polar groups oriented away from the

crystal surface and towards the aqueous phase, and (iii) the IR absorptions of the underlying water subphase may be compensated by using standard spectral subtraction methods.

Figure 1 illustrates the increasing quality of the monomolecular film IR spectra collected by using the internal reflectance approach, compared to external reflectance IR spectroscopy. The top spectrum shows the IR spectrum in the range 3050–2750 cm⁻¹ of a DPPC monolayer acquired at low surface pressures (~ 10 mN/m) in the external reflection mode. Though the main CH₂ stretching vibrations are readily apparent, significant noise exists in the spectrum at such low surface pressures. In comparison, the bottom spectrum illustrates the same spectral range for the same DPPC monolayer acquired in the internal reflection mode. The spectral quality has clearly significantly improved, and with it our ability to make determinations of

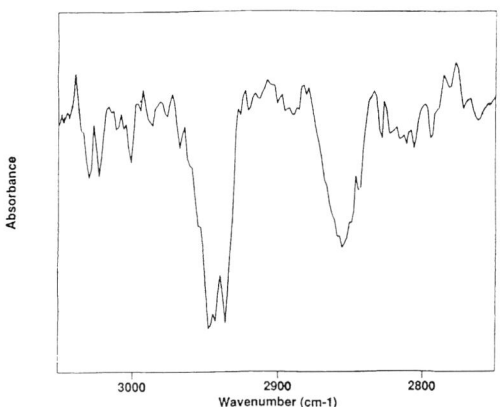

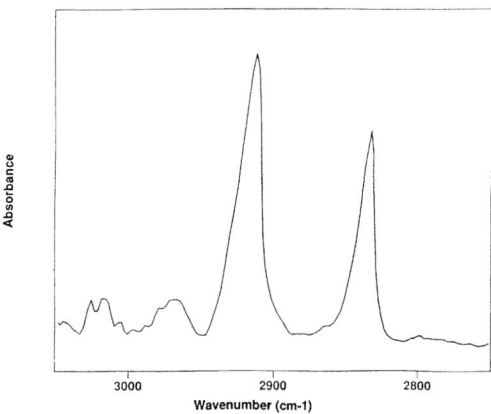

Fig. 1. A comparison of the IR spectra collected for a DPPC monolayer at the air–water interface by the external reflection and internal reflection methods. (Top) IR spectrum of a monomolecular film of DPPC acquired by external reflection IR spectroscopy at a surface pressure of ~ 10 mN/m. (Bottom) IR spectrum of a monomolecular film of DPPC acquired by internal reflection IR spectroscopy at a surface pressure of ~ 10 mN/m

molecular structure based on the IR spectrum of the monomolecular film.

Conclusion

A new method of internal reflection IR spectroscopy for the study of supported planar bilayer membranes in situ at the air–water interface is described. We have been able to demonstrate an increased spectral S/N and quality by use of this method, compared to the previously developed external reflection IR spectroscopic method. The increased sensitivity of the internal reflection method will enable us to study in detail the interaction of peptides with membrane surfaces, in situ.

Acknowledgement. This work was supported by the U.S. Public Health Service through National Institutes of Health grant GM 40117 (RAD).

References

[1] R. A. Dluhy, *J. Phys. Chem.* **1986**, *90*, 1373.
[2] R. A. Dluhy, S. M. Stephens, S. Widayati, A. D. Williams, *Spectrochim. Acta A* **1995**, *51*, 1413.
[3] P. H. Axelsen, W. D. Braddock, H. L. Brockman, C. M. Jones, R. A. Dluhy, B. S. Kaufman, F. J. Puga, II, *Appl. Spectrosc.* **1995**, *49*, 526
[4] B. W. Gregory, R. A. Dluhy, L. A. Bottomley, *J. Phys. Chem.* **1994**, *98*, 1010.
[5] F. R. Rana, S. Widayati, B. W. Gregory, R. A. Dluhy, *Appl. Spectrosc.* **1994**, *48*, 1196.
[6] S. Widayati, R. A. Dluhy, *Thin Solid Films* **1995**, in press.

Mikrochim. Acta [Suppl.] 14, 683–685 (1997)

Determination of Microdomain Size in Langmuir–Blodgett Films of Binary Mixtures by IR Spectroscopy and Atomic Force Microscopy

Suci Widayati and **Richard A. Dluhy***

Department of Chemistry, University of Georgia, Athens, GA 30602–2556, U.S.A.

Abstract. Quantitative determination of individual microdomain sizes Langmuir–Blodgett films of fatty acid binary mixtures has been accomplished by using infrared (IR) spectroscopy, and the visualization of these domain structures has been accomplished by atomic force microscopy (AFM). Ordered bilayers of binary mixtures of long-chain fatty acids were transferred onto Ge attenuated total reflectance IR elements and examined by ATR-IR spectroscopy. Samples were composed of C_{18}, C_{20} and C_{24} fatty acids mixed together with a (deuterated C_{18}) fatty acid in various mole fractions. Analysis of the crystal-field splitting in the CH_2 and CD_2 scissoring vibrations allowed us to calculate the number of hydrocarbon chains in each domain. In addition, we have used AFM to demonstrate lateral phase segregation in these same binary fatty acid mixtures on both the macroscale and at molecular level. A mixture of rectangular and hexagonal packing was observed that directly correlated with the hydrocarbon organization predicted by the IR results. These results are the first reports of the use of IR spectroscopy and AFM imaging to identify microdomain phase segregation in these two-dimensional, ultrathin films of binary alkane mixtures.

Key words: Langmuir–Blodgett films, binary mixtures, microdomains, ATR-IR, atomic force microscopy.

The study of the lateral segregation of amphiphilic molecules into microdomains, including molecular structure, domain size, shape, and formation, and the study of possible monolayer defect structures, are subjects of intense current interest in monolayer research, from both an experimental and a theoretical standpoint. From a biophysical perspective, this is due to the role that local aggregation of molecules of similar structure may have in such processes as self-organization, membrane fludity, premeability, membrane fusion, and membrane protein structure and reactivity. Unfortunately, characterization of domain structures within molecular monolayer and/or thin films is a difficult experimental process.

Historically, vibrational spectroscopy has been a widely-used method of determining structure and conformation in alkane assemblies. With advances in surface reflection techniques, it is also possible to determine structure and orientation in amphiphilic monolayers at liquid and solid surfaces. Recently, Snyder et al. have shown that IR spectroscopy can be applied to study the kinetics of microdomain segregation in solid-phase, three-dimensional bulk solutions of binary n-alkane mixtures [1]. The segregation or demixing process for the n-alkanes results in domain formation in the bulk solid phase that can be quantitatively analyzed from the splitting of the methylene scissoring vibration. Rabolt et al. have used the splitting of these same methylene scissoring bands to identify the interlayer diffusion of cadmium arachidate through polymers [2]. Our own work in this area has used the splitting observed in the C–H and C–D scissoring vibrations in mixed monolayer films of fatty acid binary mixtures for the quantitative calculation of individual microdomain sizes in transferred Langmuir–Blodgett films [3]. It is the first use of IR spectroscopy to identify microdomain phase segregation in two-dimensional, ultrathin films of binary alkane mixtures. Our recent work has used atomic force microscopy imaging techniques to visualize the domain structures first calculated by IR spectroscopy.

Experimental

The surface chemistry methods used in this study have been previously described in detail [4, 5]. Monocrystalline germanium (Ge)

---

* To whom correspondence should be addressed

substrates were prepared for attenuated total reflectance infrared (ATR-IR) analysis, and Langmuir–Blodgett monolayers were transferred onto these substrates, as previously described [4, 5].

Infrared spectra of the bilayers transferred onto Ge substrates were acquired by using a Digliab FTS-40 Fourier-transform infrared spectrometer equipped with a 4 × beam condenser and a narrow-band liquid-N_2 cooled HgCdTe detector. AFM images were acquired with a Digital Instrument Nanoscope III microscope equipped with a 0.7 µm head and a silicon nitride pyramidal tip with a cantilever spring constant of 0.12 N/m.

Results

In the work described here, we have applied the IR analysis methods described by Snyder et al. [1] for bulk phase alkanes to ultrathin model membranes formed by Langmuir–Blodgett (L–B) methods on solid substrates. In this report, we show that IR spectroscopy can be used to quantitatively determine microdomain sizes in amphiphilic hydrocarbon chain assemblies at interfaces. Our model systems consisted of ordered bilayers of binary mixtures of long-chain fatty acids transferred onto germanium attenuated total reflection crystals. We found that microdomain phase segregation can be detected in bilayers of two fatty acids (4 monomolecular layers). Experimentally, ordered bilayers of binary mixtures of long-chain fatty acids were transferred onto germanium attenuated total reflection (ATR) cyrstals and examined by IR spectroscopy. The fatty acids used were C_{18} (stearic), C_{20} (arachidic), and C_{24} (lignoceric). These fatty acids were mixed with acyl chain perdeuterated C_{18} (d_{35}-stearic) fatty acid in H:D mole ratios of 1:4, 1:1, and 4:1. Analysis of the crystal-field splitting in the CH_2 and CD_2 scissoring vibrations allowed us to calculate the number of hydrocarbon chains in each domain. It was found that microdomain phase segregation can be detected in thin films consisting of two fatty acid bilayers (four monomolecular layers). Calculations based on the magnitude of the CH_2 and CD_2 scissoring band splittings showed that the non-deuterated fatty acid domains consist of 80–100 chains for C_{24}:C_{18}-d_{35} (4:1 mol:mol) binary mixture. The domain size decreases as the chain length difference of the components decreases. The largest domains occur when the non-deuterated fatty acid is the major component and the chain length difference is the greatest. A binary mixture with equal chain lengths gives a random mixture of the two components, with no preferential microdomain phase segregation seen. This is the first use of IR spectroscopic

methods to identify microdomain phase segregation in two-dimensional, ultrathin films of binary alkane mixtures.

In a companion series of experiments [6], we have used atomic force microscopy (AFM) to visualize the phase segregation of these fatty acid binary mixtures at the molecular level. As with the IR experiments, stearic acid was used as the short chain component in binary mixtures with (i) stearic acid, (ii) arachidic acid, or (iii) lignoceric acid. We found that in binary mixtures of acids with different chain lengths, in equal mole ratios, resulted in a random distribution of the components, whereas phase separation was observed in mixtures where the short chain component was present as the major mole fraction. The molecular packing arrangements deduced from the 2-D fast Fourier transform patterns of the AFM images correlate well with the molecular packing deduced from the CH_2 and CD_2 scissoring band frequency splitting observed in the IR data taken from the same film. For the C_{18}:C_{20} mixed film, we measure a vertical distance of 2.7 Å between the domains, in good agreement with a two-methylene unit difference. In the C_{18}:C_{24} mixed film, we measure a vertical distance of 8.37 Å between the domains, in good agreement with a six-methylene unit difference. The AFM images of the phase-separated domains formed from the binary mixtures of unequal chain length and unequal mole ratio clearly show the presence of irregular-ribbon-like domains in these thin films. As predicted from the splitting of the C–H and C–D scissoring vibrations seen in the IR spectra, the AFM images showed a mixture of hexagonal and rectangular packing in all mixtures.

Conclusions

The IR spectra in conjunction with the AFM images demonstrate the ability to probe lateral separations in binary fatty acid mixtures on both the macroscale and at a molecular level. Both sets of data show that the tendency of these mixtures to lateral separation is higher in the mixtures of different chain lengths when the perdeuterated component dominates. Both rectangular (orthorhombic) and hexagonal packing arrangements are predicted from the IR spectra, and both are observed in the AFM images. This is the first report of the combined use of IR and AFM for the study of lateral segregation in multicomponent monomolecular films.

Acknowledgement. This work supported by the U.S. Public Health Service through National Institutes of Health grant GM40117 (RAD).

References

[1] R. G. Snyder, M. C. Goh, V. J. P. Srivatsavoy, H. L. Strauss, D. L. Dorset, *J. Phys. Chem.* **1992**, *96*, 10008.

[2] M. Shimomura, K. Song, J. F. Rabolt, *Langmuir* **1992**, *8*, 887.

[3] S. Widayati, R. A. Dluhy, submitted.

[4] B. W. Gregory, R. A. Dluhy, L. A. Bottomley, *J. Phys. Chem.* **1994**, *98*, 1010.

[5] F. R. Rana, S. Widayati, B. W. Gregory, R. A. Dluhy, *Appl. Spectrosc.* **1994**, *48*, 1196.

[6] S. Widayati, R. A. Dluhy, in preparation.

Mikrochim. Acta [Suppl.] 14, 687–689 (1997)
© Springer-Verlag 1997

The Phase Behavior of Binary Lipid Monolayers as Determined by Surface Chemistry and IR Spectroscopy

Amy D. Williams, Jennifer M. Wilkin, and **Richard A. Dluhy***

Department of Chemistry, University of Georgia, Athens, GA 30602-2556, U.S.A.

Abstract. We have applied both a classical surface thermodynamic analysis and quantitative IR spectroscopy to mixed monolayer films of phosphocholine (PC) and phosphoglycerol (PG) lipids in order to quantitatively describe the phase miscibility of these physiologically important species. The particular PC:PG binary mixtures were chosen to model the major components of mammalian pulmonary lung surfactant. The surface chemistry analysis revealed that saturated PC:PG mixtures were nearly ideal at all mole ratios studied, while the mixtures of saturated PC with unsaturated PG lipids were found to have positive deviations from ideality at all mole fractions studied, indicating phase-separated binary mixtures. We have also utilized Langmuir–Blodgett transfer methods, ATR-IR, and ^{31}P NMR spectroscopy as a means of determining the surface composition of the phospholipids at varying surface pressures. Using these IR methods, we have found no evidence for the selective exclusion of any component from the interface, or the so-called "squeeze-out" of monolayer components.

Key words: ATR-IR, binary mixtures, phase miscibility, surface chemistry.

Recent biophysical approaches to the study of the functional mechanisms of pulmonary surfactant have begun to yield insight into the molecular-level interactions involved. However, one of the central issues in surfactant physiology that is still under active investigation is the so-called "squeezing-out" hypothesis, which predicts the reorganization of surface monolayer films upon compression. The evidence for the enrichment of DPPC at the (A/W) interface during isotherm compression is primarily based on the $\pi-A$ behavior of mixed monolayers of DPPC with other unsaturated components.

Previous work from this laboratory has used a combination of ATR-IR spectroscopy in conjunction with

^{31}P-NMR and classical surface chemistry techniques to characterize mixed monolayers composed of acyl chain perdeuterated 1,2-dipalmitoyl-*sn*-glycero-3-phosphocholine (i.e. DPPC-d$_{62}$) and 1,2-dipalmitoyl-*sn*-glycero-3-phosphoglycerol (DPPG), a phosphoglycerol lipid containing fully saturated hydrocarbon chains, to directly test the "squeeze-out" hypothesis [1,2]. This current work involves utilizing the previously described methods to investigate the binary mixture composed of DPPC-d$_{62}$ with 1,2-dioleoyl-*sn*-glycero-3-phosphoglycerol (i.e. DOPG). The DOPG lipid is an unsaturated phospholipid that contains a double bond in each of its acyl chains between carbons 9 and 10. Unsaturation of the phospholipid acyl chains tends to fluidize monolayer films of DPPC and promote spreading [3].

Experimental

The surface chemistry methods used in this study have been previously described in detail [1–3]. Monocrystalline germanium (Ge) substrates were prepared for attenuated total reflectance infrared (ATR-IR) analysis, and Langmuir–Blodgett monolayers were transferred onto these substrates, as previously described [1,2].

Attenuated total reflectance infrared spectra of the monolayers transferred onto Ge substrates were acquired by using a Digilab FTS-40 Fourier-transform infrared spectrometer equipped with a $4 \times$ beam condenser and a narrow-band, liquid-N$_2$ cooled HgCdTe detector. Details of the ^{31}P-NMR spectroscopy and its correlation to the ATR-IR results have been previously described [2].

---

* To whom correspondence should be addressed

Results

While conventional monolayer π–A isotherms are important, the information obtained from their use alone provides only an indirect means of understanding surface film miscibility is mixed monolayers. We have used a standard thermodynamic method to analyze the π–A isotherms obtained from two binary mixtures commonly used as models of the surfactant interface: DPPC + DPPG and DPPC + DOPG [3]. By this method, detailed information concerning the thermodynamic miscibility properties of these lipids can be obtained [3].

For the binary mixture of DPPC with DPPG, the excess free energy results obtained by this analysis suggest that, on an NaCl subphase, this binary combination approximates an ideal mixture. This is consistent with other published reports that show that the DPPC:DPPG binary mixture does not form laterally segregated domains upon compression.

The interaction of the acyl chain unsaturated lipid DOPG with DPPC on an NaCl subphase is very different from that of its saturated DPPG analog. Positive ΔG_{xs}^{π} values are found for the DPPC:DOPG binary mixture at all mole fractions studied. This indicates a phase-separated system with positive deviations from ideal mixing behavior. At all mole fractions studied with DPPC:DOPG binary mixtures, increasing the surface pressure by compressing the monolayer film results in larger ΔG_{xs}^{π} values, which signifies an even greater phase separation at higher surface pressures.

We have complemented this surface chemistry isotherm work with Langmuir–Blodgett transfers, attenuated total reflectance IR spectroscopy, and ^{31}P NMR spectroscopy to characterize the mixed monolayer films composed of perdeuterated DPPC and DOPG, to directly test the "squeeze-out" hypothesis. Deuteration of the acyl chains of one component of the mixture results in a vibrational frequency shift which allows for the monitoring of the conformation and orientation of each component in the mixture.

In these experiments, a series of DPPC-d$_{62}$:DOPG binary mixture standard solutions were made at different mole ratios. The IR spectra as well as the ^{31}P-NMR spectra of these standards were recorded. The integrated intensities of the C–D and C–H stretching bands obtained from the IR spectra of the standard solutions were compared to the corresponding integrated intensities of the DPPC-d$_{62}$ and DOPG peaks in the ^{31}P-NMR spectra, as previously described [1, 2]. A calibra-

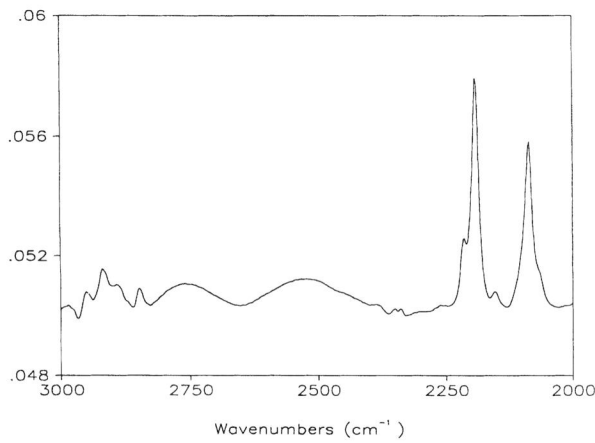

Fig. 1. Attenuated total reflectance-IR spectrum 7:1 (mol:mol) binary mixture of a DPPC-d$_{62}$:DOPG monomolecular film on a Ge substrate. The monolayer film was transferred by the Langmuir–Blodgett method onto the Ge at a surface pressure of ~ 70 mN/m. The mole fraction of the DPPC-d$_{62}$ hydrocarbon chains was monitored by using the C–D stretching region of the spectrum (2300–2000 cm^{-1}), while the DOPG mole fraction component of the monolayer was monitored by using the C–H stretching region between 3000 and 2800 cm^{-1}

tion curve was established, based on this relationship. Langmuir–Blodgett films of 7:1 DPPC-d$_{62}$:DOPG binary mixtures were transferred onto GE ATR crystals and the ATR-IR spectra of these monolayer films were obtained. Figure 1 is an example of the ATR-IR spectrum of a monomolecular film of a 7:1 DPPC-d$_{62}$:DOPG binary mixutre transferred onto a Ge ATR crystal. The mole fraction of each component present at the air–water interface was then determined by using the relative integrated intensities of the C–H and C–D vibrational bands, as previously described for the DPPC-d$_{62}$:DPPG binary system [1, 2].

Conclusion

A combination of spectroscopy and surface chemistry was applied to films composed of perdeuterated DPPC and DPPG and perdeuterated DPPC with DOPG at initial mole fractions of 7:1 (PC:PG). No evidence was found for the "squeezing-out" of either the DPPG or DOPG component at any surface pressure. For both mixtures, no evidence of enrichment of the PC component, and hence, no evidence of depletion of the PG component, was found. The films of the binary mixture appear to exist in an ordered and oriented conformation regardless of surface pressure [4]. The combination of the surface chemistry and the IR spectroscopy seem to indicate that the DPPC and DOPG components in the binary mixture are phase-separated into domains, but that each domain is highly ordered at high surface pressures [4].

Acknowledgement. This work was supported by the U.S. Public Health Service through National Institutes of Health grant GM40117 (RAD).

References

[1] F. R. Rana, A. J. Mautone, R. A. Dluhy, *Biochemistry* **1993**, *32*, 3169.

[2] F. R. Rana, A. J. Mautone, R. A. Dluhy, *Appl. Spectrosc.* **1993**, *47*, 1015.

[3] A. D. Williams, J. Wilkin, R. A. Dluhy, *Colloids and Surfaces* **1995**, *99*, in press.

[4] A. D. Williams, R. A. Dluhy, in preparation.

Mikrochim. Acta [Suppl.] 14, 691–692 (1997)

Drift Spectroscopy for Determination of Organic Acids on Montmorillonite

Ray L. Frost[1] and **Robert W. Parker**[2,*,**]

[1] Centre for Instrumental and Developmental Chemistry, Queensland University of Technology, GPO Box 2434, Brisbane, Q4001, Australia
[2] Queensland Department of Lands, Robert Wicks Research Station, P.O. Box 178, Inglewood, Q4387, Australia

Abstract. This paper reports the application of diffuse reflectance Fourier-transform-infrared spectroscopy for the analysis of mixtures of organic acids adsorbed on montmorillonitic clays. Mixtures of several straight-chain homologues: ethanoic, propanoic, butanoic, pentanoic, hexanoic and octanoic acid and 3 iso-alkyl homologues: 2-methylpropanoic, 3-methylbutanoic, and 4-methylpentanoic acid were used. Acid concentrations were determined by using principal-components regression, with recoveries ranging from 89 to 123%.

Key words: DRIFT, organic Acid, montmorillonite, chemometrics.

Diffuse reflectance Fourier-transform-infrared (DRIFT) spectroscopy is easy to use, gives quality spectra, and provides information on organic–surface interactions that is not obtainable by other infrared techniques [1]. By using chemometrics it is possible to determine concentrations of organic acids on montmorillonitic clays from their DRIFT spectra.

Experimental

A Wyoming type sodium montmorillonite, dried to constant weight at 130 °C, was used for this work. Acid–montmorillonite complexes were prepared by blending mixtures of the organic acids with the clay. Ethanoic, propanoic, butanoic, pentanoic, hexanoic octanoic, 2-methylpropanoic, 3-methylbutanoic, and 4-methyl- pentanoic acids in liquid form were used to prepare the mixtures. DRIFT spectra were collected by using a Diffuse Reflectance Accessory in a Perkin-Elmer 1600 series FT-IR, with nitrogen flushing. Spectra were recorded between 440 and 400 cm^{-1} over 16 scans at 4 cm^{-1} resolution. For each sample 5 spectra were collected. Principal components regression (PCR) analysis was used to determine component concentrations. A training set of 100 spectra, representing

* Present address: Alan Fletcher Research Station, Queensland Department of Natural Resources, P.O. Box 36, Sherwood, Q4075, Australia
 ** To whom correspondence should be addressed

known mixtures of acids adsorbed on montmorillonite, was used for calibration. Selection of an appropriate training set covering the variation of the unknowns is essential for valid results. The spectral region between 4000 and 460 cm^{-1}, but excluding the bands at 2390–2280 and 674–660 cm^{-1} corresponding to atmospheric carbon dioxide, was used for PCR analysis. Spectra were normalizsed on the O–H stretch vibration of the clay at 3630 cm^{-1}, with baseline correction by mean centring.

Results and Discussion

Figure 1 shows typical DRIFT spectra obtained for the montmorillonite and a complex formed with a mixture of the acids. Vibrations for the structural O–H stretch of the clay [2], and alkyl C–H and C=O [3] stretch of the acids produced the bands at around 3630, 2900 and 1700 cm^{-1} respectively. Because of the similarity of the infrared spectra of organic acids, there is extensive overlapping of vibrational bands in the spectra of their mixtures, and thus no distinctive bands are available for quantification of the individual acids.

However, it was found possible to determine individual concentrations by using PCR and an appropriate spectral training set. Because only 100 spectra were used in the training set, an optimal experimental design needed to be used in their selection. Table 1 contains recovery data obtained for several samples. Of the 9 acids in the mixture, 5 had recoveries within 10% of the true value, with octanoic acid showing the widest variation. Because these acids all have similar infrared spectra their determination in mixtures by using DRIFT is difficult. Analysing mixtures containing fewer acids may give better individual recovery levels by reducing the degree of spectral overlap to be dealt with. However, the use of DRIFT spectra is simpler

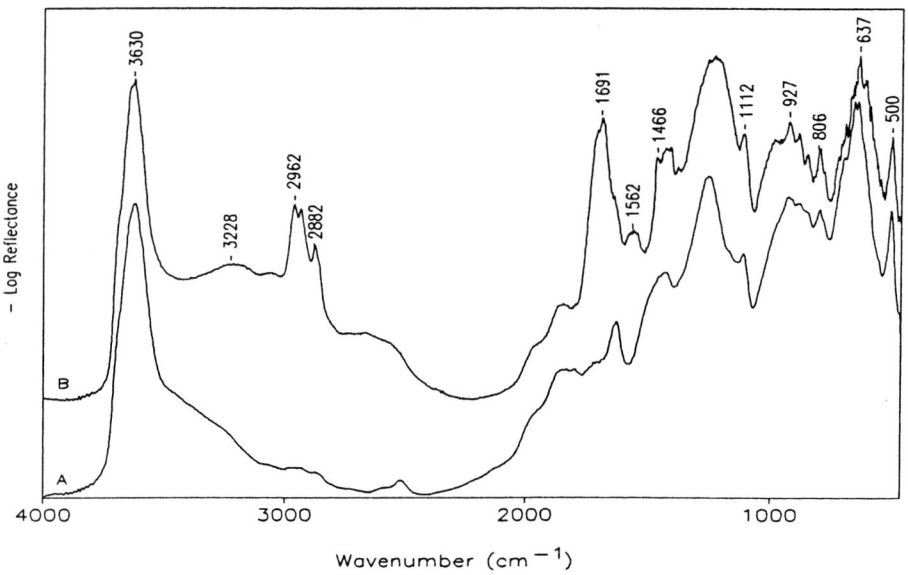

Fig. 1. DRIFT spectra for montmorillonite (*A*) and a mixture of organic acids on montmorillonite (*B*)

Table 1. Mean percentage recoveries obtained for 5 samples containing mixtures of organic acids adsorbed on montmorillonite. Individual concentrations range between 0.02 and 0.5 mol/kg for a total acid concentration range of 0.8–1.0 mol/kg

Acid	Sample 1	Sample 2	Sample 3	Sample 4	Sample 5	Overall mean
Ethanoic	89	86.5	86.5	98.8	101	92.4
Propanoic	76.4	83.2	83.4	104	100	89.4
2-Methylpropanoic	118	114	115	112	110	114
Butanoic	86.6	105	108	96.7	83.6	96
3-Methylbutanoic	114	116	115	120	124	118
Pentanoic	94.8	102	99.3	105	115	103
4-Methylpentanoic	101	102	101	115	118	107
Hexanoic	103	99.3	98	106	104	102
Octanoic	142	136	141	118	76.3	123
Total organic acid	87.5	90.1	90.3	93.7	104	93.1

than using combined solvent extraction and chromatographic techniques. DRIFT spectra could also be used to investigate molecular interactions. In conclusion, chemometric analysis of DRIFT spectra proved an adequate technique for the determination of individual organic acids present in mixtures on montmorillonitic clays.

References

[1] J. M. Bowen, S. V. Compton, M. S. Blanche, *Anal. Chem.* **1989**, *61*, 2047.

[2] V. C. Farmer, in: *Data Handbook for Clay Materials and Other Non-metallic Minerals* (H. van Olphen, J. J. Fripiat, eds). Pergamon Oxford, 1979, pp. 285–337.

[3] W. Kemp, *Organic Spectroscopy, 3rd Ed.*, Macmillan, London, 1991, pp 58–88.

Mikrochim. Acta [Suppl.] 14, 693–694 (1997)

Fourier-transform Spectroscopy of Silicon Nitride Layers: Hydrogen Concentration and Thickness Measurements

Ingrid Jonak-Auer*, **Ronald Meisels,** and **Friedemar Kuchar**

Institut für Physik, Montanuniversität, A-8700 Leoben, Austria

Abstract. The accuracy for determining both H concentration and thickness of SiN layers by FTS can be enhanced by using a polynomial fit of the spectral background outside the absorption lines rather than a simple linear baseline. With this method absorption lines which correspond to changes in transmission of down to 0.01% can be resolved and evaluated. SiN layers produced by Plasma-CVD and thermally grown layers with thicknesses between 1.5 μm and 10 nm were studied. The overall H concentration of thermally grown layers is considerably smaller than that of Plasma-CVD layers. Whereas in thermally grown layers hydrogen is mainly bound to nitrogen, in Plasma-CVD layers most of the hydrogen atoms are bound to silicon. Annealing procedures using nitrogen and oxygen cut down the H concentration to about 20% and 10% respectively compared to the untreated layers. Thickness measurements by FTS show good agreement with measurements by ellipsometry for SiN layers of thicknesses down to 30 nm.

Key words: silicon nitride layers, hydrogen concentration, thickness measurements.

We report on FT-IR measurements of the hydrogen content and the thickness of silicon nitride (SiN) layers, which rely on the fact that vibrations of impurities in semiconductor lattices yield absorption in the far IR spectral region. Those measurements are important for the semiconductor industry because the quality of SiN layers is highly dependent on the hydrogen content.

SiN has been gaining popularity in the semiconductor industry as an encapsulant for silicon integrated circuits and is now not only used as a passivation film but also as a dielectric.

Hydrogen in SiN layers is chemically bound to Si and N. If the H content is considerable, it will affect such properties as film structure, etch rate, stress and moisture permeability. On the other hand, a certain amount of H in the films is desirable because it reduces the trap density by passivating silicon dangling bonds, which improves the device characteristics. The sources of H in SiN layers are the deposition reactants NH_3, SiH_4 and SiH_2Cl_2.

Experimental

The experiments were performed with a Bruker Fourier-transform Spectrometer IFS 113 v with a Globar source, a KBr beam-splitter and a DTGS detector.

In contrast to commercially available FT-IR software and recommended procedures in the literature, which in principle use linear baselines to evaluate the areas of the N–H, Si–H and Si–N absorbance peaks at $3350\,cm^{-1}$, $2160\,cm^{-1}$ and $875\,cm^{-1}$ respectively, we use a polynomial fit of the spectral background outside the absorption lines. This yields a much higher accuracy in determining the areas of the absorbance peaks and therefore the hydrogen concentration and layer thickness.

Figure 1a shows the absorbance of a Si–H line and the appropriate polynomial fit of the spectral background. In Fig. 1b the polynomial fit is subtracted from the absorbance curve. The resultant line-shape can very well be fitted by a Gaussian curve, a fact which proves the high quality of the background fit.

Calibration factors and equations to evaluate the experimental data are taken from the literature [1, 2]:

$$\frac{N-H}{cm^2} = \frac{Area_{N-H}}{5.3 \times 10^{-18}} \quad \text{where } N-H \text{ is the number of } N-H \text{ bonds}$$

$$\frac{Si-H}{cm^2} = \frac{Area_{Si-H}}{7.4 \times 10^{-18}} \quad \text{where } Si-H \text{ is the number of } Si-H \text{ bonds}$$

$$d = \frac{Area_{Si-N}}{243 - 8.2 \times \dfrac{Si-H}{N-H}}$$

SiN layers prepared both by Plasma-CVD and low pressure thermal deposition with thicknesses between 1.5 μm and 10 nm were studied. For the thermally deposited SiN layers we also studied the influence of oxygen and nitrogen annealing procedures on the H concentration. Especially for the thinner thermal layers high intensity-resolution and stability of the measurement are essential since typical

---

* To whom correspondence should be addressed

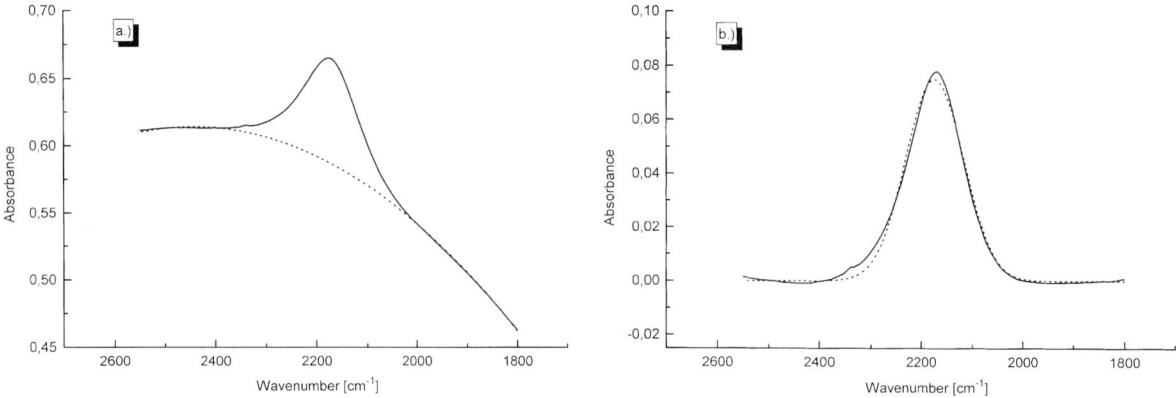

Fig. 1. a Si–H absorption line of a Plasma-CVD SiN layer and polynomial fit (dashed line) of the spectral background. **b** Polynomial fit subtracted from the Si–H line from (**a**) and Gaussian fit (dashed line) of the resultant line-shape

absorbance values correspond to changes in transmission of down to only 0.01%.

Results and Conclusions

For a high accuracy in determining the areas of absorbance peaks it is necessary to use a polynomial fit of the spectral background rather than a linear baseline. Differences in area of up to 30% are possible when comparing a simple baseline fit to an appropriate polynomial fit.

The total H concentration in Plasma-CVD layers varies between 1.3×10^{22} and $2.0 \times 10^{22} \, cm^{-3}$, the H concentration in thermal layers between 4.5×10^{20} and $5.5 \times 10^{21} \, cm^{-3}$. In Plasma-CVD layers the number of N–H bonds per cm^3 varies between 5.9×10^{21} and 7.6×10^{21}, while the number of Si–H

bonds per cm^3 is higher, namely between 5.8×10^{21} and 1.4×10^{22}. Thermal layers on the other hand contain more N–H bonds than Si–H bonds: the number of N–H bonds per cm^3 varies between 4.5×10^{20} and 5.8×10^{21}, the number of Si–H bonds is approximately an order of magnitude smaller than that of N–H bonds, but the Si–H lines are too weak to allow a reliable quantitative evaluation.

The effects of different annealing procedures of thermal layers are shown in Fig. 2. Oxygen annealing reduces the H concentration more than nitrogen annealing does; the H concentration decreases from $5.9 \times 10^{21} \, cm^{-3}$ for the untreated layer to $1.0 \times 10^{21} \, cm^{-3}$ for the nitrogen-annealed layer to $4.5 \times 10^{20} \, cm^{-3}$ for the oxygen-annealed layer.

A comparison between the thickness measurements obtained by ellipsometry and by FTS shows that apart from the very thin layers ($< 30 \, nm$) the values lie within 8% of each other. For thinner layers the thickness measured by FTS tends to be smaller than that measured by ellipsometry by approximately 20%.

Acknowledgement. Financial support for this work came from the Bundesministerium für Wissenschaft, Forschung und Kunst, GZ 601.555 and from the Steierm. Wissenschafts- und Forschungslandesfonds, GZ AAW-12Ku24-93. All samples were prepared by Austria Mikro Systeme International.

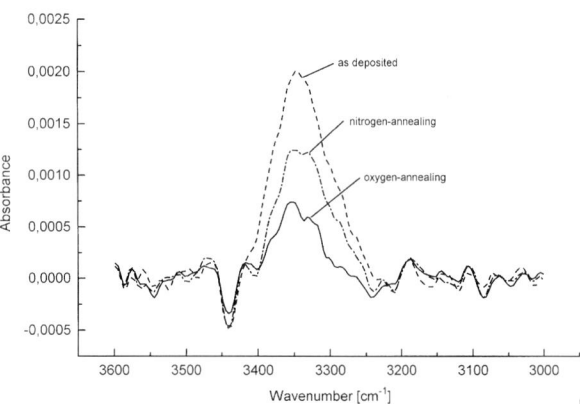

Fig. 2. Influence of nitrogen- and oxygen-annealing procedures on the N–H absorption-lines of a thermally deposited SiN layer

References

[1] W. A. Lanford, M. J. Rand, *J. Appl. Phys.* **1978**, *49*, 2473.
[2] W. R. Knolle, *Report AT & T Bell Laboratories*, Reading, PA 19607.

Mikrochim. Acta [Suppl.] 14, 695–696 (1997)

Amplitude-phase FIR and IR Spectroscopy of a Langmuir Film on Metal

Sergey A. Klimin[1,*], **Lubov A. Kuzik**[1], **Vladimir A. Yakovlev**[1], **Guerman N. Zhizhin**[1], **Nadezhda K. Matveeva**[2], and **Andrey A. Sigarev**[2]

[1] Institute of Spectroscopy, Russian Academy of Sciences, 142092 Troitsk, Moscow Region, Russia
[2] Zelenograd Research Institute of Physical Problems, 103460 Zelenograd, Moscow, Russia

Abstract. Amplitude–phase surface electromagnetic wave spectroscopy is proposed as a new method for investigation of the surface of solid states and thin films on them.

Key words: surface electromagnetic wave, surface, thin films.

Surface electromagnetic waves (SEW) are successfully applied for research on surfaces of solid states and thin films on them [1]. In general laser sources are used for SEW excitation, so the field of SEW applications is restricted. The use of the Fourier-transform spectrometer thermal radiation source and high-sensitivity detectors for weak light-flux registration allows extension of the scope of application of SEW to vibrational spectroscopy problems from the far IR to 3000 cm^{-1}.

SEW–Fourier-transform spectroscopy was used for research on metals [2] and on Langmuir–Blodgett films on metals [3], but only SEW energy characteristics were investigated. Amplitude–phase SEW spectroscopy (APSEWS) opens new possibilities. It was used for the first time in [2,4]. We used APSEWS for investigation of a thick copper film and a Langmuir-film surface of polyorganosiloxane (POS) on it.

Experimental

A DA3.002 Bomem transform spectrometer with MCT and SiB detectors and a Ge bolometer was used in the spectral range 30–3000 cm^{-1}. An SEW unit was made for amplitude–phase interferometric measurements. The scheme of the unit is shown in Fig.1. Aperture excitation of SEW was used in this unit. The globar radiation passes through the interferometer and falls on the gap between screen (2) and the sample surface, which is placed at the intermediate focus of the sample compartment, and excites the SEW on the sample. The

SEW is propagated to the end of the sample and transformed into bulk radiation at the sample edge. This bulk wave interferes with the bulk radiation from the input gap. At a fixed distance (a) between the input gap and the end of the sample the far field interference pattern depends on the frequency v_m and a coordinate z (see Fig. 1). We have measured the frequency distribution at the different fixed coordinates z. These coordinates were given by the positions of a slit (slit width ~ 0.5 mm) in screen (3). This slit was parallel to the input gap. The spectrometer detector registered the radiation transmitted through the output slit. The condition for the interference extrema at a fixed coordinate z and frequency v_m is :

$$Re k_x(v_m) = \frac{m+\Delta}{2 a v_m} + \sqrt{c^2 + z^2} - \sqrt{(c-a)^2 + z^2} \qquad (1)$$

where k_x is the effective SEW refractive index, m the doubled number of the extremum, c the distance between screens (2) and (3) and Δ the additional shift between the bulk and surface waves.

To change distance a we moved the sample, the distance c was kept constant. This provided the input slit position at the focus of the spectrometer sample compartment and the same efficiency of SEW excitation.

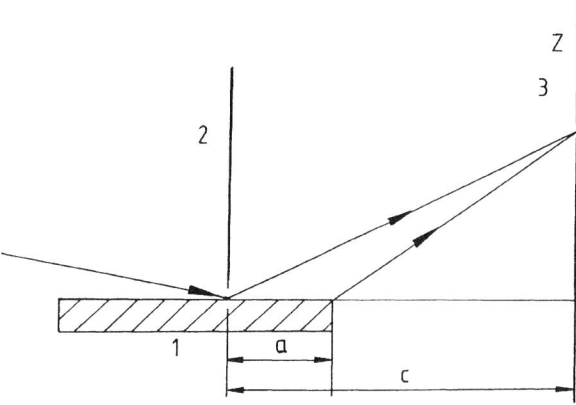

Fig. 1. Scheme of the APSEWS unit: *1* – sample, *2* – input screen, *3* – output screen

* To whom correspondence should be addressed

Results

Some of interference dependencies in the mid IR for a thick copper film on the optically polished glass substrate for different z and a are shown in Fig. 2. All SEW interferometric spectrum intensities I were normalised to the transmission spectrum intensity I_0 without screen (3). In the spectra we can clearly see that the distance between the extrema decreases with increasing a and z, in correspondence with Eq. (1).

The SEW parameters and the optical constants of copper were evaluated from these interferometric patterns. For copper the plasma frequency $v_p = 60000$ cm^{-1} and $v_\tau = 2000$ cm^{-1}. We have made some interferometric measurements in the far IR too. The interferogram patterns are in accordance with the copper optical constants evaluated.

In this work we have studied the IR absorption spectra of a POS film with a thickness of 6 nm (5 monomolecular layers) which was deposited by the Langmuir–Shaefer horizontal lift method on the surface of a thick copper film. This film has no absorption bands in the far IR, but has a strong band at

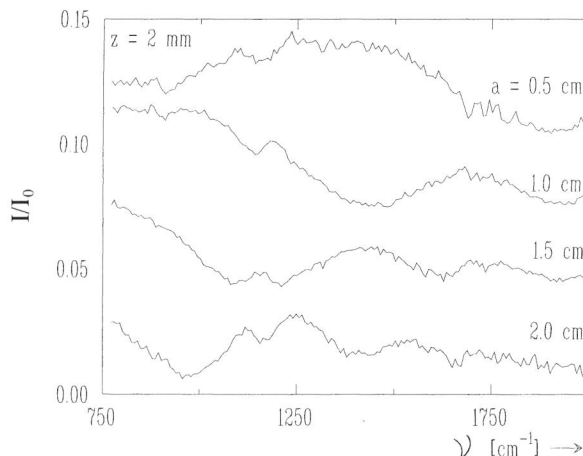

Fig. 3. Interference SEW spectra for a 5-monolayer Langmuir film of a polyorganosiloxane

1160 cm^{-1}. This band is well seen in the interferometric SEW spectra of the POS film on copper, in Fig. 3.

Conclusions

The APSEWS can be considered as a promising method for research on solid state surfaces and thin films on them, and for direct determination of optical constants without using any model. This method can be used in a wide spectral region, when coupled with a Fourier-transform spectrometer.

Acknowledgements. This work was made possible in part by the support of the Russian Fund for Fundamental Research and the International Science Foundation.

References

[1] V. M. Agranovich, D. L. Mills (eds.), *Surface Polaritons*, North-Holland, Amsterdam, 1982, p. 717.
[2] M. A. Chesters, S. F. Parker, V. A. Yakovlev, *Opt. Commun.* **1985**, *55*, 17.
[3] G. N. Zhizhin, A. A. Sigarev, V. A. Yakovlev, *J. Molec. Liq.* **1992**, *53*, 10.
[4] G. N. Zhizhin, V. A. Yakovlev, *Phys. Rep.* **1990**, *194*, 281.

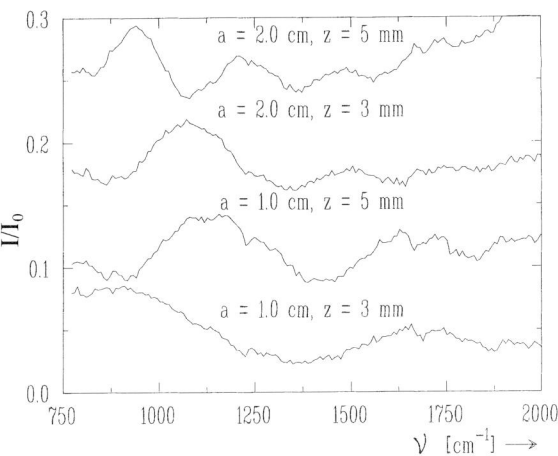

Fig. 2. Interference SEW spectra for copper at different z and a

Mikrochim. Acta [Suppl.] 14, 697–699 (1997)

In situ FT-IR Study of Ceria Reduction by CO

Ahmed Badri, Claude Binet, Jacques Saussey, and **Jean-Claude Lavalley***

Laboratoire Catalyse et Spectrochimie, URA CNRS 414, ISMRA, Université de Caen, 6, Boulevard du Maréchal Juin, 14050 Caen Cedex, France

Abstract. Use of FT-IR spectrometry coupled to gas chromatography evidenced two temperature stages for the reduction of ceria by CO: (i) below 300 °C, surface reduction occurred and mainly involved formate species as intermediate, (ii) above 300 °C, subsurface reduction involved terdentate carbonates.

Key words: ceria reduction, carbon monoxide, FT-IR spectroscopy.

Ceria is a component of the so-called "three-way catalysts" used for automotive gas exhaust treatment, owing to its ability to release and store oxygen. Thus, much work has been devoted to study of the reducibility of ceria, using H_2 as reducing agent (for instance, [1]). Only a few studies have been devoted to CO-reduction [2–4]. The aim of this paper is to correlate the formation of surface-adsorbed species with the CO_2 evolution rate in order to get an insight into the CO-reduction process. Measurements at various temperatures were hence performed, using coupled gas chromatography and in situ FT-IR techniques.

Experimental

The in situ FT-IR technique coupled with gas chromatography that was used is described elsewhere [5]. The ceria (Rhône-Poulenc) had a BET surface area of $120 \, m^2/g$. It was treated in the IR cell. Its oxidation was first performed by flowing ($15 \, cm^3/min$) an N_2, O_2 mixture (synthetic air) over it at 375 °C. After cooling to 45 °C under N_2, reduction was performed with a flow of CO ($15 \, cm^3/min$) as the temperature was raised to 375 °C at a rate of 5 °C/min. During the entire process, infrared spectra were recorded with a Nicolet FT-IR spectrometer, and the CO_2 formation rate was simultaneously measured by gas chromatography.

* To whom correspondence should be addressed

Results and Discussion

The variation of the $\nu(OH)$ bands due to hydroxy groups present on the ceria, as a function of temperature, under a flow of CO, is shown in Fig. 1 (part A). The $3550 \, cm^{-1}$ band will not be further discussed, as it is due to residual cerium hydroxide as impurity [6]. Two more or less resolved $\nu(OH)$ bands are located at $3635 \, cm^{-1}$ (denoted A) and $3655 \, cm^{-1}$ (B). Both are due to bidentate OH species differing from the presence (B) or absence (A) of an oxygen vacancy at the cation adsorption sites [6]. As the reduction temperature increases from 110 to 280 °C, band B progressively vanishes, showing that band A is less involved in the reduction process (Fig. 1). The sample reoxidation performed at 375 °C leads to an increase of the band A intensity. This result shows that, during the reduction process, hydrogen (originating from the disappearance of the initial hydroxy species during the process) was stored in the ceria, leading to hydroxy species regeneration upon reoxidation.

Concomitantly, formation of formate (denoted by F) and carbonate (C) species was observed (Fig. 1, part B). The F species are characterized by bands at $2845 \, cm^{-1}$ $[\nu(CH)]$, $2945 \, cm^{-1}$ $[\delta(CH) + \nu_a(OCO)]$, $1580 \, [\nu_a(OCO)]$ and $1375 \, cm^{-1}$ bands $[\nu_s(OCO)]$ [3, 7]. Other bands, at 1570 and $1300 \, cm^{-1}$, characterize bidentate carbonates (C_2), those at 1485 and $1385 \, cm^{-1}$ terdentate carbonates (C_3).

It is clear that formate formation involves interaction of CO with surface OH species, mainly of type B as previously shown from the study of the $\nu(OH)$ range. The variation of the number of formate species with temperature is clearly deduced from the $\nu(CH)$ band intensity. It is reported in Fig. 2, curve b: F species

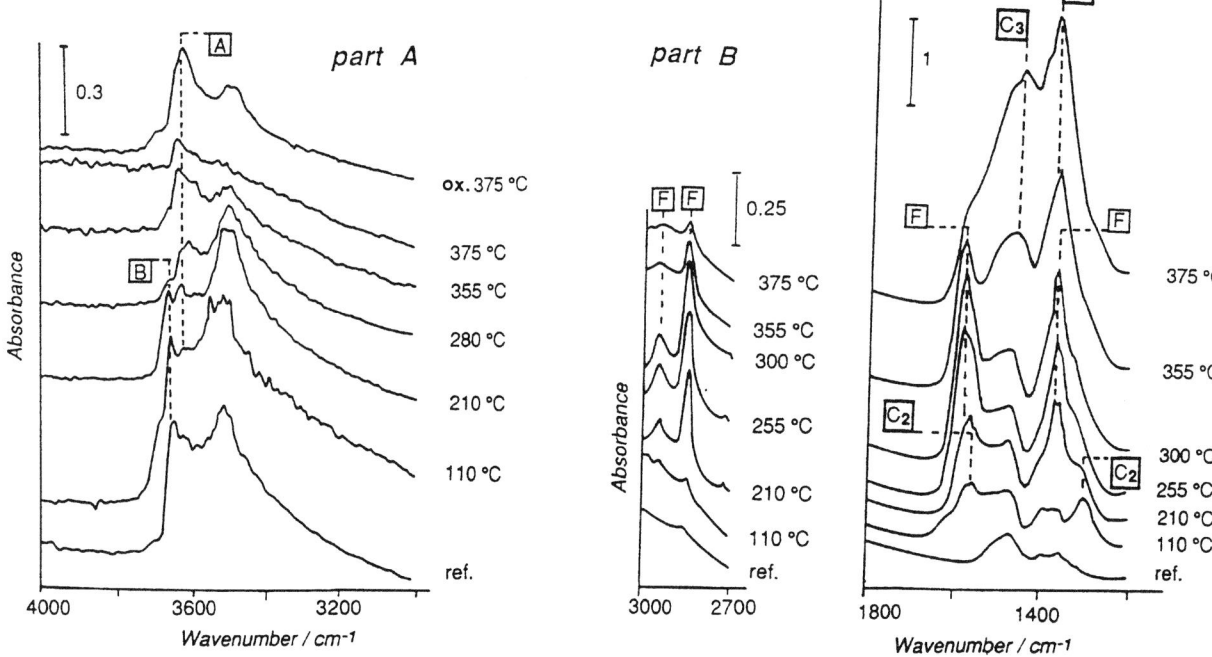

Fig. 1. Variations with the CO-flow temperature of: $\nu(OH)$ bands due to OH surface species (part A) and $\nu(CH)$ and $\nu(OC)$ formate bands (part B); bidentate (C_2) and terdentate (C_3) carbonate species (part B)

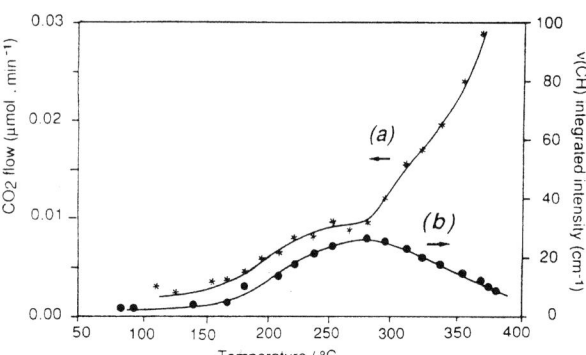

Fig. 2. Variations with the CO-flow temperature of: a CO_2 flow, b formate $\nu(CH)$ band intensity

begin appearing at $110\,°C$; their number passes through a maximum at $280\,°C$, and they vanish at $375\,°C$.

Residual terdentate carbonates are present as core-carbonates in ceria before reduction (Fig. 1). Upon reduction at $110\,°C$, bidentate carbonates (C_2) are already formed but they disappear as the reduction temperature increases (Fig. 1). C_2 species arise from CO interaction with capping O^{2-} surface species. Terdentate carbonates (C_3), possibly produced from CO

interaction with subsurface O^{2-} species, are mainly formed at above $300\,°C$.

The rate of CO_2 formation as a function of the reduction temperature is reported in Fig. 2, curve a. The extremum at $280\,°C$ fairly correlates with maximum formation of the formate species (Fig. 2, curve b). The sharp increase of the CO_2 formation rate above $300\,°C$ correlates well with the formation of the terdentate carbonates.

Conclusion

Surface reduction at below $300\,°C$ mainly proceeds through the formation of formate species (s = surface):

$$CO + OH_s^- + Ce_s^{4+} \rightarrow Ce_s^{4+} + HCOO_s^-$$
$$\rightarrow CO_2 + (Ce^{3+}, H)_s$$

with H storage by ceria.

Subsurface reduction, at above $300\,°C$, proceeds through the formation of terdentate carbonate species (b = bulk):

$$CO + 2\,O_b^{2-} + 2\,Ce_b^{4+} \rightarrow CO_3^{2-} + 2\,Ce_b^{3+}$$
$$\rightarrow CO_2 + O_b^{2-} + 2\,Ce_b^{3+}$$

The last reaction, involving the O^{2-} ion in carbonate formation, may also contribute to CO_2 production from O^{2-} surface species, intermediate bidentate carbonate surface species then being formed. However, this last route is thought to be much less important that the previous one involving formate species as intermediates.

References

[1] A. Laachir, V. Perrichon, A. Badri, J. Lamotte, E. Catherine, J. C. Lavalley, J. El Fallah, L. Hilaire, F. le Normand, E. Quéméré, G. N. Sauvion, O. Touret, *J. Chem. Soc. Faraday Trans.* **1991**, *87*, 1601.

[2] M. Guénin, *Ann. Chim. (Paris)* **1973**, *8*, 147.

[3] C. Li, Y. Sakata, T. Arai, K. Domen, K. Maruya, T. Onishi, *J. Chem. Soc. Faraday Trans. I* **1989**, *85*, 929.

[4] A. Badri, J. Lamotte, J. C. Lavalley, A. Laachir, V. Perrichon, O. Touret, G. N. Sauvion, E. Quéméré, *Eur. J. Solid State Inorg. Chem.* **1991**, *28* [Suppl.], 445.

[5] J. F. Joly, N. Zanier-Szydlowski, S. Colin, F. Raatz, J. Saussey, J.-C. Lavalley, *Catal. Today* **1991**, *9*, 31.

[6] A. Badri, *Thesis*, Caen, France, 1994.

[7] C. Binet, A. Jadi, J. C. Lavalley, *J. Chim. Phys., Phys. Chim. Biol.* **1992**, *89*, 1779.

Mikrochim. Acta [Suppl.] 14, 701–702 (1997)
© Springer-Verlag 1997

Fourier-transform Mid and Far IR and ^{23}Na MAS NMR Analysis of Framework Vibrations and Distribution of Cations in Alkali-Exchanged Y-type Zeolites

István Hannus[1,*], **István Pálinkó**[2], and **Imre Kiricsi**[1]

[1] Applied Chemistry Department, József Attila University, Rerrich B. tér 1, H-6720 Szeged, Hungary
[2] Department of Organic Chemistry, József Attila University, Dóm tér 8, H-6720 Szeged, Hungary

Abstract. The effects of alkali metal ion (Li^+, K^+, Rb^+, Cs^+) exchange on framework vibrations have been studied in Y-type zeolites, with NaY as the basis of comparison. Vibrations were monitored in the 1300–250 cm^{-1} region. It was found that the position of the bands corresponding to the internal tetrahedral structure were insensitive, while those corresponding to the external linkages were sensitive to the size of the ion introduced. ^{23}Na MAS NMR measurements revealed that Na^+ ions in the supercage can be exchanged most easily, than those in the sodalite units and in the the hexagonal prism.

Key words: alkali matal ion (Li^+, Na^+, K^+, Cs^+, Rb^+) Y-zeolites, mid and far IR measurements, ^{23}Na MAS NMR measurements, framework vibrations, cationic positions.

The preparation of alkali metal ion-exchange zeolites are of interest because the acidity and of course the basicity of zeolites, among other things, can be fine-tuned by introducing ions with various charge density into the framework structure (for recent works, see [1–3]). These materials then can be used in catalytic reactions ranging from those needing strong acid sites to those requiring basic centres. Obviously, rationalization, let alone prediction of catalytic performance, requires the knowledge of changes in the zeolite framework resulting from ion-exchange and the positions of the exchanged cations. Framework vibrations can be conveniently studied by mid [4–6] and far [5, 7–10] IR spectroscopy, and cationic positions are easily accessible to solid-state NMR measurements. In the work leading to this contribution, LiY, NaY, KY, CsY and RbY zeolites were studied by ^{23}Na MAS NMR and IR spectroscopy, the latter in the mid and far infrared region.

Experimental

LiY, KY, CsY and RbY zeolites were prepared from NaY by the wet ion-exchange method. Aqueous solutions of the appropriate metal salts were applied and the ion-exchange was repeated twice. Then the samples were air-dried and stored. Before measurements, they were heated at 573 K for two hours under evacuation. The degree of ion-exchange and the cationic positions were measured by atomic absorption spectrometry and ^{23}Na MAS NMR measurements (Bruker MSL-400 spectrometer), respectively. FT-IR spectroscopic measurements were performed in the mid and far infrared regions. For mid and far infrared measurements KBr and HDPE (high density polyethylene) were the matrix materials, respectively, and 128 scans were collected for one spectrum. Spectra were corrected with the water spectrum. Measurements were performed in the 3500–250 cm^{-1} range (but studied in the 1300–250 cm^{-1} region) on Bio-Rad spectrometers (mid: FTS-65A/896 with liquid-nitrogen cooled MCT detector; far: FTS-40 with DTGS detector in mixed vacuum and purge mode; optical resolution was 2 cm^{-1} in both cases) and the spectra were analysed by using the associated software package.

Results and Discussion

Framework IR measurements showed excellent crystallinity. Framework vibration bands were recorded and analysed in the 1300–250 cm^{-1} region, with NaY as the basis of comparison. The bands, relative to which shifts were monitored, were at 1147, 1019, 792, 734, 712, 580, 507, 459, 379 and 317 cm^{-1}. Spectra for the ion-exchanged zeolites are depicted in Fig. 1 and the frequency values are to be seen in Table 1.

The bands at 1019, 792, 580 and 507 cm^{-1} showed systematic changes upon ion-exchange, namely, their position decreased in frequency with increasing size of the cation. These bands should have been associated with the positions of the cations, being sensitive to their size (external linkages) and the others were insensitive (internal tetrahedral structure).

---

* To whom correspondence should be addressed

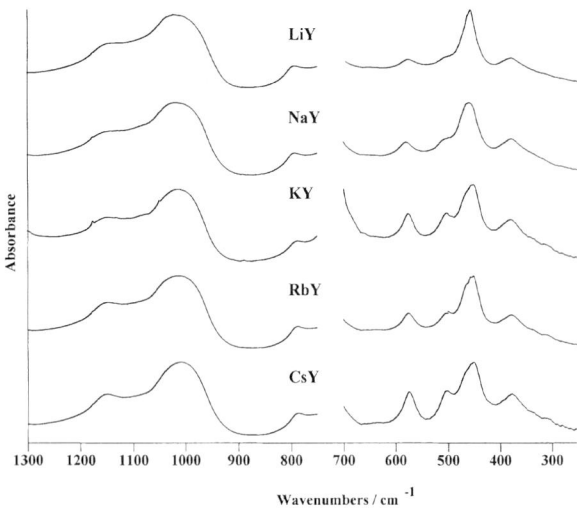

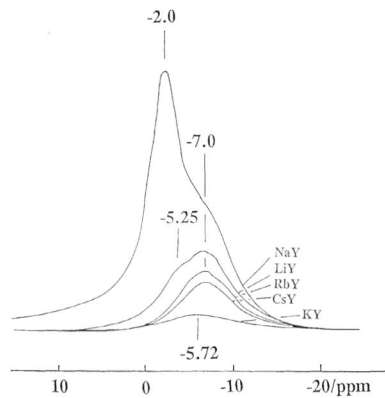

Fig. 2. ²³Na MAS NMR spectra of alkali-metal exchanged Y zeolites

Fig. 1. IR spectra of alkali-metal exchanged Y zeolites in the 1300 – 250 cm⁻¹ region

Table 1. Positions of IR-active framework vibrations on dehydrated alkali-metal exchange Y zeolites

Sample	Composition	Positions of framework vibrations/cm⁻¹									
		Asym. stretch		Sym. stretch			Dbl. ring	T–O band		Pore opening	
LiY	$Li_{38}Na_{20}$	1147^{sh}	1022^{bs}	792^{w}	733^{m}	712^{m}	580^{w}	509^{sh}	458^{s}	380^{m}	316^{vw}
NaY	Na_{58}	*1147^{sh}*	*1019^{bs}*	*792^{w}*	*734^{m}*	*712^{m}*	*580^{w}*	*507^{sh}*	*459^{s}*	*379^{m}*	*317^{vw}*
KY	$K_{54}Na_{4}$	1148^{sh}	1016^{bs}	787^{w}	735^{m}	712^{s}	576^{m}	503^{m}	451^{s}	380^{m}	316^{vw}
RbY	$Rb_{42}Na_{16}$	1148^{sh}	1013^{bs}	787^{w}	733^{m}	712^{m}	575^{m}	502^{m}	452^{s}	379^{m}	314^{vw}
CsY	$Cs_{50}Na_{8}$	1149^{sh}	1008^{bs}	785^{w}	733^{m}	710^{m}	574^{m}	501^{m}	452^{s}	377^{m}	314^{vw}

sh = shoulder, bs = broad strong, s = strong, m = medium, w = weak, vw = very weak

²³Na MAS NMR measurements on the parent and the ion-exchanged samples (Fig. 2) revealed that ion-exchange is most facile in the supercage ($\delta = -2.0$), then in the sodalite units (δ from -5.25 to -7.0), while Na⁺ ions in the hexagonal prism cannot be exchanged (for assignments, see [11]).

Thus, Na⁺ ions other than those in the hexagonal units do not remain in KY, RbY, and CsY in significant amounts, but LiY contains Na⁺ ions in the sodalite units as well.

Acknowledgements. This work was financed by the National Science Foundation of Hungary through grant T007577/1992. IR spectroscopic instruments used in preparing this work were purchased with the help of the PHARE-ACCORD program nr. H-9112 0152. Both forms of support are highly appreciated.

References

[1] Z.-H. Fu, Y. Ono, *J. Catal.* **1994**, *145*, 166.

[2] S.-B. Liu, B. M. Fung, T.-C. Yang, E.-C. Hong, C.-T. Chang, P.-C. Shih, F.-H. Tong, T.-L. Chen, *J. Phys. Chem.* **1994**, *98*, 4393.

[3] B. L. Su, D. Barthomeuf, *Appl. Cat. A: General* **1995**, *124*, 73.

[4] E. M. Flanigen, H. Khatami, H. A. Szymanski, *Adv. Chem. Ser.* (*Molecular Sieve Zeolites, Vol. I*) **1971**, *101*, 201.

[5] W. P. J. H. Jacobs, J. H. M. C. van Wolput, R. A. van Santen, *Zeolites* **1993**, *13*, 170.

[6] D. Das, M. Duttagupta, S. K. Palit, *Bull. Chem. Soc. Jpn.* **1994**, *67*, 2906.

[7] M. D. Baker, J. Godber, G. A. Ozin, *ACS Symp. Ser.* (*Perspectives in Molecular Sieve Science*) **1988**, *368*, 136.

[8] H. A. M. Verhulst, W. J. J. Welters, G. Vorbeck, L. J. M. van de Ven, V. H. J. de Beer, R. A. van Santen, J. W. de Haan, *J. Phys. Chem.* **1994**, *98*, 7056.

[9] H. Esemann, H. Förster, *J. Chem. Soc. Chem. Commun.* **1994**, 1319.

[10] H. Esemann, H. Förster, *Z. Phys. Chem.* **1995**, *189*, 263.

[11] L. B. Welsh, S. L. Lambert, *ACS Symp. Ser.* (*Perspectives in Moleculer Sieve Science*) **1988**, *368*, 33.

Mikrochim. Acta [Suppl.] 14, 703–704 (1997)

An in situ DRIFTS Study of the Active Phase of the Heteropoly Acid Catalyst $H_4[PVMo_{11}O_{40}]$ in Oxidation Reactions

B. Herzog, M. Wohlers, and **R. Schlögl***

Fritz-Haber-Institut der Max-Planck-Gesellschaft, Faradayweg 4–6,D-14195 Berlin, Federal Republic of Germany

Abstract. Heteropoly acid (HPA) catalysts based on the title compound exhibit the well-known Keggin-structure at temperatures relevant for catalytic oxidation processes (520–570 K). In situ DRIFTS (diffuse reflectance infrared Fourier-transform spectroscopy) data show that a variety of solid-state reactions occur with the dehydrogenation of isobutyric acid (IBA) leading to structural rearrangements and partial decomposition of the HPA anions. These changes involve the formation of lacunary structures and vanadyl species. Re-oxidation with O_2/H_2O at reaction temperature re-establishes the highly symmetric structure of the Keggin-anions.

Key words: heteropoly acid, dehydrogenation, isobutyric acid, in situ DRIFTS, in-situ XRD.

During the last decade, heteropoly acids (HPA) and their salts have gathered a lot of interest for their use in heterogeneous catalysis. HPAs of the Keggin- structure type, have been extensively studied, especially the title compound $H_4[PVMo_{11}O_{40}]$, because of its particular acidic and redox properties. However, the nature of this compound is still not completely understood [1–5]. Phases stable under ex-situ conditions may completely differ from those which are formed during catalytic reactions. Since the knowledge of the "active" phase is of major concern, it is mandatory to examine the properties of the catalyst under working conditions. In this contribution we will discuss the structure of the active phase in the dehydrogenation reaction of isobutyric acid (IBA) to yield methacrylic acid (MAA).

Experimental

The HPA samples were synthesized hydrothermally from mixtures of the metal oxides in dilute phosphoric acid. The quality of the freshly prepared samples was checked by in-situ UV/VIS spectroscopy, TGA, XRPD (X-ray powder diffraction), SEM/EDAX, and ^{31}P NMR. FT-IR measurements were performed on a Bruker IFS 66

---

* To whom correspondence should be addressed

spectrometer, equipped with a Graseby-Specac DRIFTS attachment. In-situ experiments were done by using the environmental high temperature chamber from Graseby-Specac connected to a home made gas-supply and evaporation system. Process control was possible by analysing the condensed product by FT-IR spectroscopy. For Raman investigations a Dilor Omars 89 retro-Raman spectrometer was used. XRPD experiments were performed with a Stoe Stadi P X-ray diffraction system [1, 6].

Results and Discussion

HPA compounds contain a great amount of water of crystallization at low temperatures. With combination of in-situ XRPD, thermoanalytical methods, Raman-, and FT-IR spectroscopy it could be shown that in the temperature range of technical applications (520–570 K) the Keggin-structure is preserved, even though all water of crystallisation has been removed. The corresponding data are displayed in Fig. 1.

In the course of the catalytic reaction in situ DRIFTS measurements reveal distinct changes within the structure of the HPA anions. Figure 2a shows a spectrum taken at 530 K under reductive load. The formation of lacunary structures reduces the symmetry of the HPA anion cages, which drastically affects the P–O valence vibration. For the intact framework this vibration is observed as a strong band at 1075 cm^{-1}, but with partial decomposition the vibration band becomes broader and splits into several new bands. Simultaneously the frequencies of the M–O–M bonds shift to lower values as a consequence of strong reduction. The M=O$_t$ band remains unaffected. A further consequence of the catalytic reaction is the separation of vanadium from the Keggin structure. This process was postulated in recent in-situ XRPD studies [1] and is strongly reflected in the DRIFT spectrum in Fig. 2a as the band at 1035 cm^{-1} can be assigned to a V–O

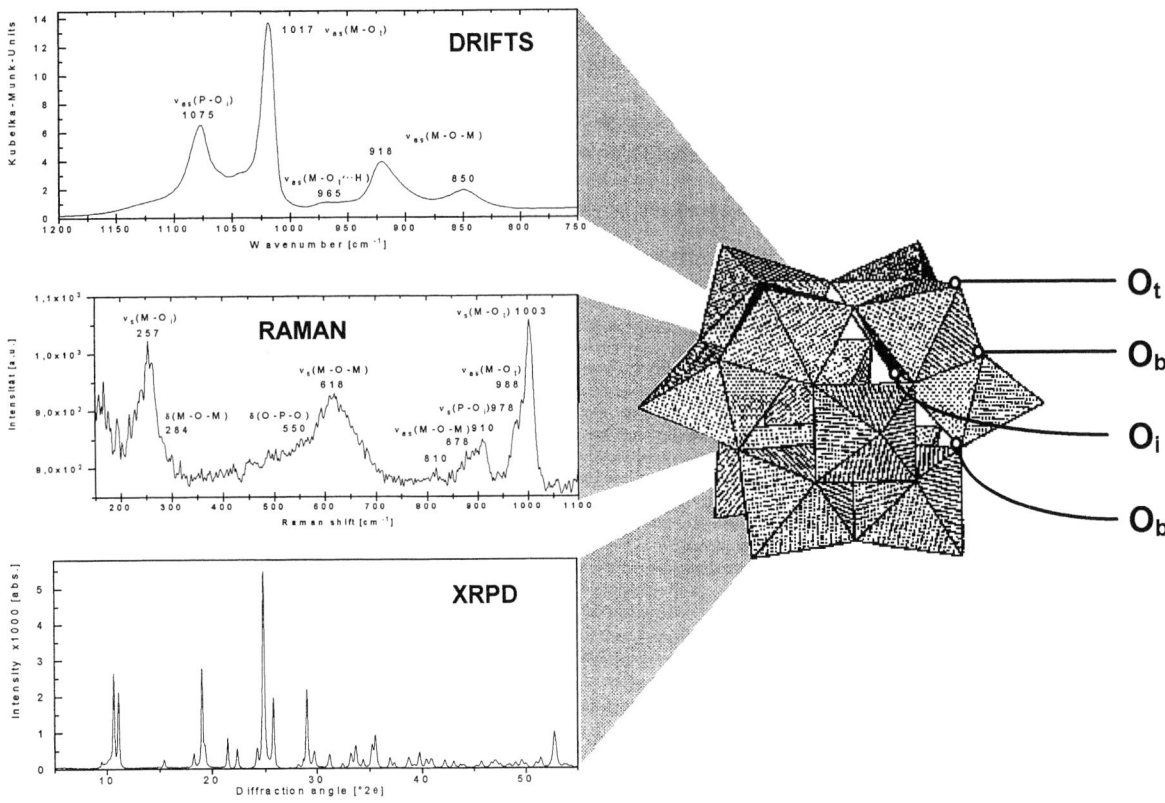

Fig. 1. Structural characterization of Keggin-type compounds at 470 K

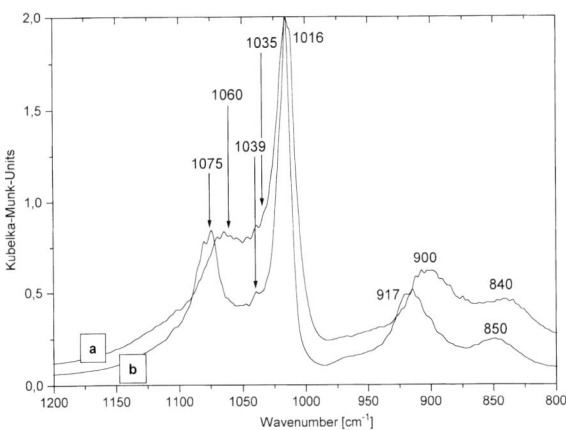

Fig. 2. DRIFT spectra of HPA recorded at 530 K under in-situ conditions. Spectrum *a* was recorded under reductive load, spectrum *b* was recorded after reoxidation

vibration [1, 7]. In [1] also a regeneration process was proposed, in the case that oxygen and water were supplied to the reactor feed. This regeneration process was also found in the DRIFT spectrum in Fig. 2b. After

a partial decomposition caused by the catalytic reaction of IBA to from MAA, exposure to an O_2/H_2O mixture rebuilt the typical high temperature spectrum of the Keggin anion. In particular the re-established high intensity of the P–O band is strong evidence for an intact polyoxo framework, but an additional band at $1039\,cm^{-1}$ also shows that some V–O groups outside the anions are still present, which might act as counter-ions.

References

[1] T. Ilkenhans, B. Herzog, Th. Braun, R. Schlögl, *J. Catal.* **1995**, *153*, 275.

[2] B. Herzog, W. Bensch, T. Ilkenhans, R. Schlögl, N. Deutsch, *Catal. Lett.* **1993**, *20*, 203.

[3] J. B. Black, J. D. Scott, E. M. Serwicka, J. B. Goodenough, *J. Catal.* **1987**, *106*, 16.

[4] C. Rocchiccioli-Deltcheff, M. Fournier, *J. Chem. Soc. Faraday Trans.* **1991**, *87*, 3913.

[5] M. Misono, *Catal. Lett.* **1992**, *12*, 63.

[6] B. Herzog, T. Ilkenhans, R. Schlögl, *Fresenius J. Anal. Chem.*, **1994**, *349*, 247.

[7] B. Herzog, *Thesis*, Univ. Frankfurt/Main, 1995.

Mikrochim. Acta [Suppl.] 14, 705–706 (1997)
© Springer-Verlag 1997

Comparison of Transmission Infrared Spectroscopy and Diffuse Reflectance Infrared Spectroscopy of a Commercial Hydrotreating Catalyst

Stewart F. Parker*,**, **Angelo Amorelli**, **Yvonne D. Amos**, **Catherine Hughes**, and **Jacquiline R. Walton**

British Petroleum International, Group Research and Engineering, Chertsey Road, Sunbury-on-Thames, Middlesex TW 16 7LN, U.K.

Abstract. Diffuse reflectance and transmission infrared spectra of CO and NO adsorbed on a Co-Mo/Al$_2$O$_3$ catalyst are compared. It is shown that the two techniques give somewhat different results and some possible reasons for the differences between the transmission and diffuse reflectance spectra are presented.

Key words: infrared spectroscopy, diffuse reflectance, hydrotreating catalyst.

Hydrotreatment of crude oil fractions to remove sulphur, nitrogen, metals and to partially hydrogenate unsaturated species is a major process in oil refining. Hydrotreating (HYT) catalysts are typically sulphided CoMo or NiMo on γ-alumina supports. Infrared spectroscopy of chemisorbed NO probe molecules has been extensively used to characterize the metal sites on the catalyst surface. The integrated intensity has been shown to broadly correlate with catalyst activity [1]. The goal of the present work was to investigate whether infrared spectroscopy could be used to characterize regenerated HYT catalysts. As an initial investigation, we have studied the chemisorption of CO and NO on the same catalyst, using both diffuse reflectance and transmission infrared spectroscopy (CO chemisorption has also been shown to provide additional information on the catalyst system [2]). Surprisingly, it was found that there were significant differences between the spectra recorded by the two techniques. The spectra are presented here and possible reasons are discussed.

Experimental

A fresh commercial Co (2.4%)-Mo (8.9%) on γ-alumina hydrotreating catalyst was used for this study. The experiments were done by *in situ* diffuse reflectance (Spectra-Tech Collector with an environment-

al chamber fitted with zinc selenide windows) and by transmission (Specac, model 5750) infrared spectroscopy. Infrared spectra were recorded on a Digilab FTS-60A Fourier-transform infrared (FT-IR) spectrometer (diffuse reflectance) and a Nicolet 7199B FT-IR spectrometer (transmission).

After loading of the sample, the cell (transmission or diffuse reflectance) was evacuated to 10^{-4} mbar. The sample was activated by (*i*) heating to 400 °C under vacuum, (*ii*) reduction/sulphidation at 400 °C with 15% H$_2$S/H$_2$ for 2 hours, (*iii*) evacuation for 2 hours at 400 °C. The sample was then allowed to cool room temperature. CO was then adsorbed and the cell evacuated. Adsorption of NO was examined similarly. Spectra were recorded in the presence of the gas phase and during evacuation for both gases.

Results and Discussion

The top trace in Fig. 1 shows the diffuse reflectance spectrum of the catalyst in the presence of CO after subtraction of that of the gas phase CO. Three bands at 2182, 2102 and 2059 cm^{-1} are clearly present. On evacuation (middle and lower traces of Fig. 1), the band at 2182 cm^{-1} is immediately lost and the bands at 2102 and 2059 cm^{-1} are greatly diminished. The band at

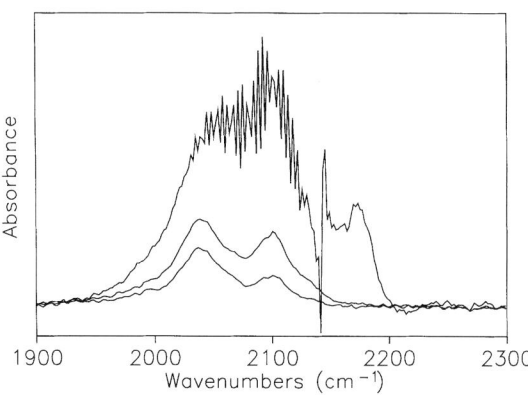

Fig. 1. The adsorption of CO on a reduced sulphided catalyst by diffuse reflectance spectroscopy: (top) the catalyst under CO (gas phase subtracted), (middle) immediately on evacuation (bottom) after 30 minutes evacuation

* Present address: ISIS Facility, Rutherford Appleton Laboratory, Chilton, Didcot, Oxfordshire OX11 OQX, U.K.

** To whom correspondence should be addressed

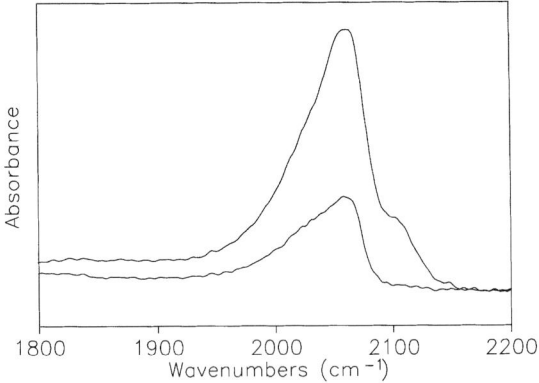

Fig. 2. The adsorption of CO on a reduced sulphided catalyst by transmission infrared spectroscopy: (upper) the catalyst under CO (gas phase subtracted), (lower) after 40 minutes evacuation

$2182\,\mathrm{cm}^{-1}$ is assigned to CO physisorbed on coordinatively unsaturated Al^{3+} sites [3]. The bands at $2105\,\mathrm{cm}^{-1}$ and $2059\,\mathrm{cm}^{-1}$ are assigned to CO chemisorbed on sulphided Mo and Co respectively [2].

Figure 2 shows the transmission spectrum obtained in the presence of CO with the gas phase contribution subtracted (upper trace) and after 10 minutes evacuation (lower trace). In this case, the $2182\,\mathrm{cm}^{-1}$ band is not observed and the $2105\,\mathrm{cm}^{-1}$ band rapidly disappears. However, the species responsible for the $2059\,\mathrm{cm}^{-1}$ band is retained on the catalyst.

The major difference observed between our diffuse reflectance and transmission measurements is the presence of the $2182\,\mathrm{cm}^{-1}$ band. The difference can be rationalized by the fact that the path-length for diffuse reflectance is not defined (as it is for transmission measurements, by the thickness of the sample) so in regions of low absorption the path-length can be very long (by multiple scattering in the solid) with a consequent increase in sensitivity. The effect is similar to that for diffuse reflectance spectra of pure materials that show anomalously strong overtones and combinations.

Figure 3 shows a comparison of the spectra obtained after exposure to NO followed by evacuation, by both transmission, 3a, and diffuse reflectance, 3b, infrared spectroscopy. It is apparent that there are both similarities and differences. The frequencies of the three strong bands at 1866, 1794 and $1696\,\mathrm{cm}^{-1}$ are in good agreement with each other and are typical of NO chemisorbed on a reduced, sulphided $Co\text{-}Mo/Al_2O_3$ hydrodesulphurization catalyst [1, 2]. However, the transmission measurements show that the CO was not immediately removed, but was slowly desorbed in vacuum. This is in contrast to the diffuse reflec-

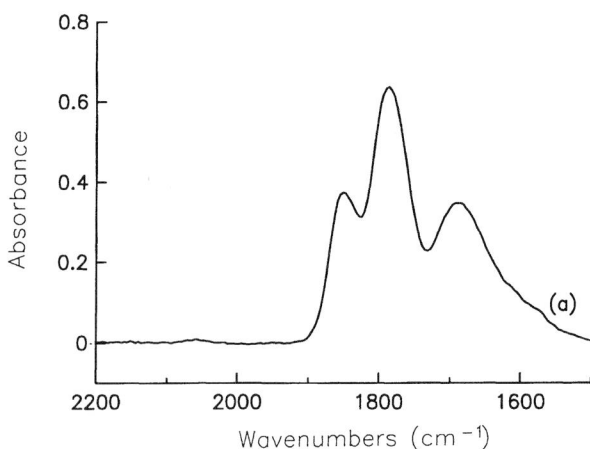

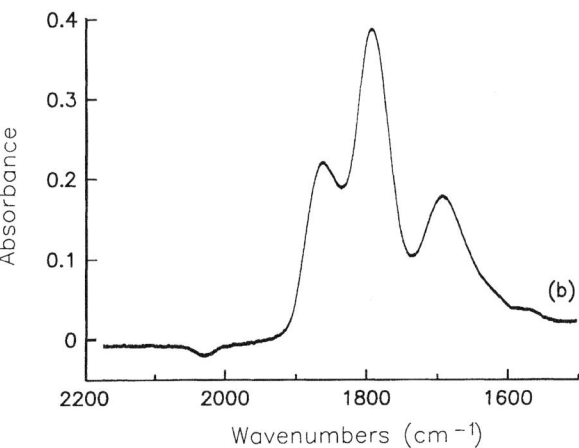

Fig. 3. The adsorption of NO on a reduced sulphided catalyst after exposure to CO. Spectrum recorded by **a** transmission infrared spectroscopy, **b** diffuse reflectance infrared spectroscopy

tance measurements, where complete displacement was seen.

NO was found to be rapidly desorbed from the catalyst surface when studied by diffuse reflectance measurements but at a lower rate when studied by transmission spectroscopy. The difference in rates is possibly because the diffuse reflectance spectra are from the top few microns of the sample where the pumping speed is greatest. In contrast the transmission spectra are from a pressed disc where the effective pumping speed in the bulk is considerably lower than at the surface.

References

[1] N.-Y. Topsøe, H. Topsøe, *J. Catal.* **1983**, *84*, 386.
[2] Q. Xin, X. Guo, R. P. Silvy, P. Grange, B. Delmon, *Proc. Int. Cong. Catal. 9th* **1988**, *1*, 66.
[3] T. H. Ballinger, J. T. Yates, Jr., *Langmuir* **1991**, *7*, 3041.

Mikrochim. Acta [Suppl.] 14, 707–709 (1997)
© Springer-Verlag 1997

IR Study of H$_2$S and SO$_2$ Adsorption on Ferric Oxide

Anne Pieplu, Odette Saur, and **Jean-Claude Lavalley***

Laboratoire Catalyse et Spectrochimie, URA CNRS 414, ISMRA, Université de Caen, 6, Boulevard du Maréchal Juin, 14050 Caen Cedex, France

Abstract. The adsorption and oxidative adsorption of SO$_2$ and H$_2$S on Fe$_2$O$_3$ have been investigated by FT-IR at various temperatures. Sulphite and sulphate adsorbed species are formed. Their IR spectra allow them to be differentiated.

Key words: Fe$_2$O$_3$, SO$_2$, H$_2$S, sulphate, FT-IR.

Adsorption of SO$_2$ on metal oxides leads to various species and most of them can be characterized by IR spectroscopy. As an example, the sulphite species give rise to IR absorption bands in the 1100–900 cm^{-1} range [1], in particular at 1060 cm^{-1} in the case of alumina [2]. Their formation probably involves O^{2-} surface basic sites. SO$_2$ also interacts with basic OH groups, leading to bisulphite species MSO$_3$H [1, 3].

Two types of H$_2$S adsorption modes have been observed on metal oxides.

(1) A dissociative adsorption giving rise to a new OH band and possibly a weak ν(SH) band according to:

$$2H_2S + O^{2-}_{surf} \rightarrow H_2O + 2SH^-_{surf}$$

or

$$H_2S + O^{2-}_{surf} \rightarrow H_2O + S^{2-}_{surf}$$

The OH and SH bands were observed near 3690 and 2585 cm^{-1}, respectively, for H$_2$S adsorption on highly activated alumina [4].

(2) A non-dissociative adsorption by coordination on coordinatively unsaturated cations or by interaction with surface OH groups; the intensity of the ν(H$_2$S) band was also weak and located at 2570 cm^{-1} for adsorption on Al$_2$O$_3$ [4].

The aim of the present paper is to extend the study to Fe$_2$O$_3$, a metal oxide used for the selective oxidation of H$_2$S to sulphur [5].

---

* To whom correspondence should be addressed

Experimental

α-Fe$_2$O$_3$ was prepared from ferric nitrate by precipitation using citric acid at 70 °C. After drying and calcination at 500 °C, its surface area is about 22 m^2/g. For IR study, the powder was pressed at 80 MPa into self-supporting disks of about 15 mg/cm^2. The wafers were activated for 2 hours by heating at 400 °C under vacuum before introduction of gases to be adsorbed. IR spectra were recorded at room temperature with a Nicolet MX-1 spectrometer.

Results and Discussion

SO$_2$ Adsorption

The spectra obtained after introduction of SO$_2$ at 800 Pa at different temperatures are shown in Fig. 1. For room temperature we notice a broad and intense IR band at about 950 cm^{-1} and two other, less intense,

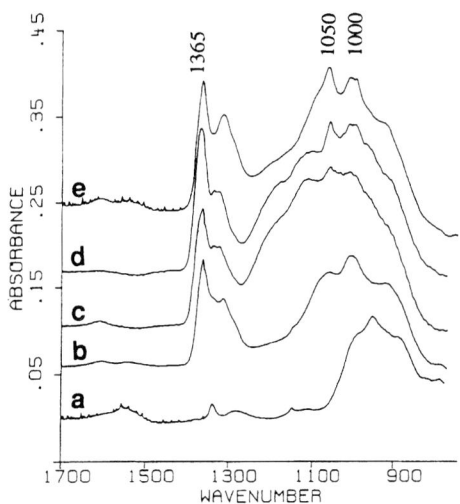

Fig. 1. IR spectrum of species formed after SO$_2$ addition to Fe$_2$O$_3$ at: *a* room temperature, *b* 250 °C, *c* 350 °C, *d* 450 °C, *e* after evacuation at 450 °C

bands at 1337 and 1140 cm^{-1}. These last two bands are close to the v_a and v_s SO$_2$ gas vibrations and are assigned to species weakly bonded to OH groups or Lewis acid sites. The weaker wavenumber band near 950 cm^{-1} is due to sulphite species bonded to basic sites.

On increase of the wafer temperature in presence of SO$_2$, new bands appear at 1365, 1312, 1050 and 1000 cm^{-1}. These bands persist after evacuation at 450 °C (Fig. 1e). They are close to bands assigned to sulphate species on ferric oxide [6] or on other oxides [7]. The bands at 1365 and 1312 cm^{-1} could be due to the v(S=O) vibration of two different surface sulphate species and the others to v(S–O) vibrations. As previously observed [7] the spectrum depends on the surface hydration state.

The spectrum of the species obtained on Fe$_2$O$_3$ heated in presence of SO$_2$ + O$_2$ (Fig. 2) confirms this assignment since it is close to that observed in Fig. 1 with SO$_2$ alone at high temperature. However, the v(S=O) band is shifted towards higher wavenumbers and is more intense. The shoulder observed near 1320 cm^{-1} is less clear because it is partially overlapped by the 1386 cm^{-1} band. These modifications are due to the larger amount of sulphate species obtained with O$_2$ present.

We can also notice shoulders near 1175 and 1110 cm^{-1}, which are more intense in Fig. 2 than in Fig. 1 c, d. They are assigned to more ionic sulphate species close to bulk ferric sulphate. These species are more numerous when a larger amount of SO$_2$ is heated with an excess of O$_2$. We have compared the thermal stability of these two types of sulphate. The bulk-type

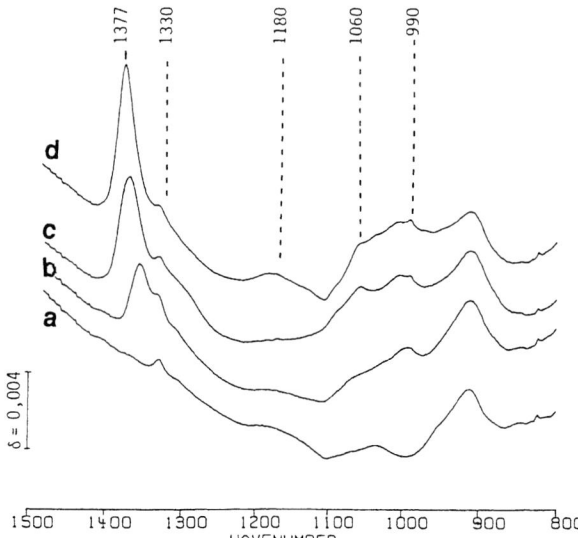

Fig. 3. IR spectrum of species formed after H$_2$S adsorption then evacuation, at *a* room temperature, *b* 200 °C, *c* 300 °C, *d* 400 °C

species disappear above 350 °C and the surface sulphate ones are still stable at 450 °C. Temperature-programmed reduction experiments under H$_2$ show that the sulphate species are reduced at 400 °C.

H$_2$S Adsorption

After adsorption of H$_2$S, the Fe$_2$O$_3$ sample shows only a broad band between 3550 and 2650 cm^{-1} and a weak one at about 1600 cm^{-1} in its spectrum, which suggests dissociative H$_2$S adsorption with formation of water. The presence of S^{2-} surface ions is evidenced by further treatment of the sample with O$_2$: a broad and intense band due to bulk sulphate species appears at around 1170 cm^{-1}. Heating the sample after H$_2$S adsorption and evacuation also leads to oxidation of sulphide ions. At 200 °C, surface sulphate species appear, and bulk sulphate species begin to be formed at 400 °C (Fig. 3).

Conclusion

SO$_2$ is easily chemisorbed on O^{2-} sites, giving rise to sulphite species owing to the basic character of Fe$_2$O$_3$. H$_2$S adsorption also involves these sites, leading to H$_2$O and S^{2-} species. The sulphite and sulphide species are oxidized to surface sulphate species by lattice oxygen atoms when the sample temperature is increased, underlining the oxidative properties of Fe$_2$O$_3$.

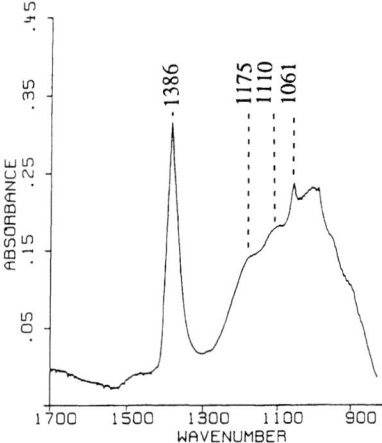

Fig. 2. IR spectrum of species formed after SO$_2$ + O$_2$ addition to Fe$_2$O$_3$, heating and then evacuation at 450 °C

References

[1] M. Waqif, A. M. Saad, M. Bensitel, J. Bachelier, O. Saur, J.-C. Lavalley, *J. Chem. Soc. Faraday Trans.* **1992**, *88*, 2931.

[2] H. G. Karge, I. G. Dalla-Lana, S. Terevizan de Suarez, Y. Zhang, *Proc. 8th Int. Congr. Catal., Berlin*, **1984**, *3*, 453.

[3] J.-C. Lavalley, A. Janin, J. Preud'homme, *React. Kinet. Catal. Lett.* **1981**, *18*, 85.

[4] O. Saur, T. Chevreau, J. Lamotte, J. Travert, J. C. Lavalley, *J. Chem. Soc. Faraday Trans. I.* **1981**, *77*, 427.

[5] P. J. van den Brink, *Thesis*, Utrecht, 1992.

[6] T. Jin, M. Machida, T. Yamaguchi, K. Tanabe, *Inorg. Chem.* **1984**, *23*, 4396.

[7] O. Saur, M. Bensitel, A. B. M. Saad, J.-C. Lavalley, C. P. Tripp, B. A. Morrow, *J. Catal.* 1986, **99**, 104.

Mikrochim. Acta [Suppl.] 14, 711–713 (1997)
© Springer-Verlag 1997

Surface Chemistry of Tungstate- and Carbonate-Treated Aluminas Studied by FT-IR Spectroscopy

Bruno Amram*, **Hervé Ponceblanc,** and **Anne-Marie Le Govic**

Rhône-Poulenc Recherches, Centre de Recherches d'Aubervilliers, 52, rue de la Haie Coq, 93308 Aubervilliers Cedex, France

Abstract. The purpose of our work was to study the surface chemistry of tungstate- and carbonate-impregnated aluminas by Fourier-transform infrared (FT-IR) spectroscopy. Direct and pyridine adsorption/desorption investigations allowed us to study the influence of the impregnation and of the catalytic test on the surface chemistry and acidity. Methanol adsorption/desorption studies allowed us to perform a partial simulation of the reaction (hydrosulphurization of methanol) and to compare the spectra with ex situ examinations of the samples before and after the catalytic test.

Key words: alumina, FT-IR, pyridine, methanol, methanethiol.

The surface chemistry of alumina impregnated with tungstate (Na_2WO_4 or K_2WO_4) or carbonate (K_2CO_3) was studied by FT-IR spectroscopy. These materials are used as catalysts for the synthesis of methylmercaptan (CH_3SH) from methanol (CH_3OH) and hydrogen sulphide (H_2S).

Direct surface studies (after simple activation) and adsorption/desorption studies of pyridine (for probing the surface acidic sites) and methanol (partial simulation of the reaction) allowed information to be obtained about the surface chemistry (influence of the treatment and the transformations induced by the catalytic test).

Experimental

Samples

The samples were prepared by impregnation of a transition alumina (Spheralite 505 from Procatalyse, 247 m^2/g), to incipient wetness with an aqueous solution of Na_2WO_4, K_2WO_4 or K_2CO_3 followed by calcination in air at 470 °C.

For the carbonate-impregnated samples, the K_2CO_3 contents were 2, 6.4 and 10 wt%. For the tungstate-impregnated samples, the Na_2WO_4 content was 13.6 wt% for the first, and the K_2WO_4 content was 14 wt% for the second. The non-impregnated alumina support (same calcination) was studied as the reference.

FT-IR Measurements

The samples were studied at various temperatures as self-supported pellets (10–15 mg) by transmission in a vacuum cell (10^{-7} mbar) placed in the sample compartment of a Nicolet 7199 FT-IR spectrometer.

After activation at 300 °C for 17 h, the samples were put in contact, at 25 °C, with pyridine or methanol for 5 min. The probe molecules were then progressively desorbed in increasing temperature steps. Infrared spectra were acquired at each step.

Results and Discussion

Influence of the Impregnation

The impregnation treatment modifies the hydroxyl populations and the surface acidic sites. Whereas, on the alumina support, three types of free hydroxyls can be clearly detected at 3760, 3730 and 3680 cm^{-1} (respectively associated with types I, II and III of the Knözinger classification [1]), the tungstate and carbonate treatments induce a drastic decrease of the OH absorbing at 3680 cm^{-1} (type III). This observation indicates that impregnation induces specific chemisorption on these OH groups of type III.

On the alumina support, medium (L2) and strong (L3) Lewis acid sites were observed by means of pyridine adsorption (chemisorbed pyridine on L2 and L3 was detected at $v8a = 1614$ and 1620 cm^{-1} and at $v19b = 1449$ and 1454 cm^{-1} respectively), and the

---

* To whom correspondence should be addressed

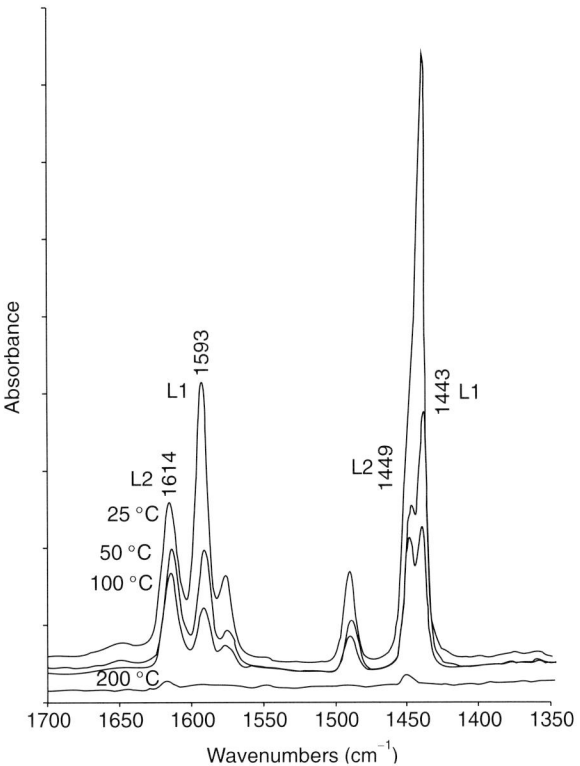

Fig. 1. FT-IR spectra of pyridine on 13.6 wt% Na_2WO_4/Al_2O_3 at 25, 50, 100, 200 °C

with the catalysis-induced transformations). The chemisorption of methanol on these catalysts is of very high importance; the formation of methoxy groups from methanol has been shown to be a crucial intermediate step for the reaction with H_2S [2, 3].

By comparing the catalyst before and after the synthesis, a strong decrease of the OH absorbing at 3730 cm^{-1} (type II) and of the remaining Lewis acid sites (medium strength, L2) was found.

Methanol adsorption showed the same strong decrease of the OH absorbing at 3730 cm^{-1} (thus showing that the same decrease induced in the catalysed synthesis could be related to the methanol adsorption) and a very strong chemisorption (methoxylation [4]) on three different sites, associated with the $v(C–O)$ modes at 1185 (sites A), 1103 (sites B) and 1011 cm^{-1} (sites C) respectively (see Fig. 2). Sites A and B are due to the alumina support and were also observed for the reference alumina, whereas sites C are due to the

tungstate treatment was found to replace the strong Lewis acid sites L3 of the alumina support by new weak Lewis sites (L1, $v8a = 1593$ cm^{-1}, $v19b = 1443$ cm^{-1}, probably located on the W atoms). This feature, shown in Fig. 1, was observed for both tungstate-impregnated samples (the chemisorbed pyridine on the weak Lewis acid sites due to W atoms cannot be interpreted as physisorbed pyridine, since it persists at above 100 °C). The carbonate treatment neutralizes progressively, as a function of the loading, the two types of Lewis acid sites on the alumina support.

Influence of the Methylmercaptan Synthesis

Surface chemistry transformations induced in the catalysed methylmercaptan synthesis were also studied (ex situ) for the carbonate-treated sample (6.4 wt% K_2CO_3) and methanol–surface interactions were investigated (partial simulation of the reaction, analogies

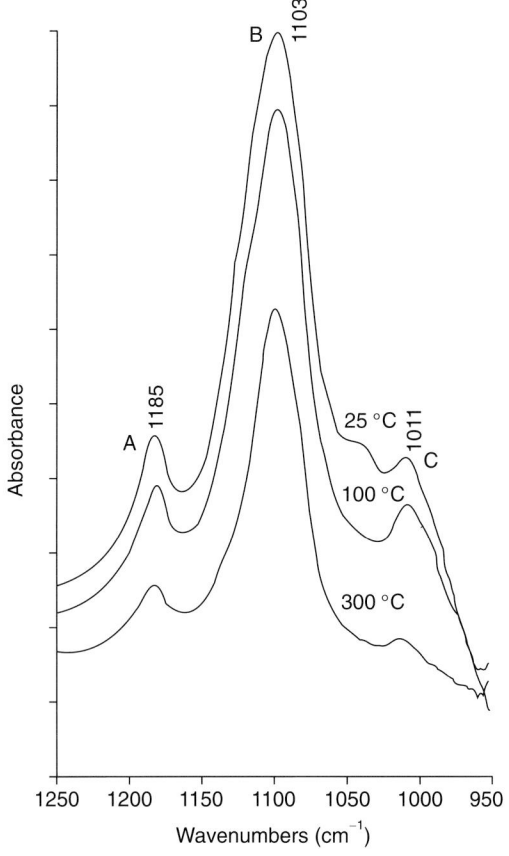

Fig. 2. FT-IR spectra of methanol on 6.4 wt% K_2CO_3/Al_2O_3 at 25, 100, 300 °C

carbonate impregnation treatment (undetected for the reference alumina).

Conclusions

Direct and probe-molecule FT-IR investigations of alumina catalyst surfaces impregnated with carbonate and tungstate allowed us to get valuable information on the modifications of the hydroxyl populations and of the surface acidity by the impregnation treatment. The induction of surface modifications by the catalysed synthesis (ex situ investigations) was also interesting to compare with methanol chemisorption experiments.

FT-IR spectroscopy was then shown to be a very informative tool (from the structural information obtained and the test-induced transformations found) for obtaining knowledge of surface chemistry and for the design of improved catalysts.

Acknowledgements. The authors wish to thank M. Arama and M. Lahaye for their skilful technical assistance.

References

[1] H. Knözinger, in: *Elementary Reaction Steps in Heterogeneous Catalysis* (R. W. Joyner, R. A. van Santen, eds.), Kluwer, Dordrecht, 1993, p. 267.

[2] M. Ziolek, J. Kujawa, O. Saur, J.-C. Lavalley, *J. Phys. Chem.* **1993**, *97*, 9761.

[3] A. V. Mashkina, V. M. Kudenkov, E. A. Paukshtis, V. Yu. Mashkin, *Kinet. Katal.* **1992**, *33*, 1128.

[4] A. A. Davydov, in: *Infrared Spectroscopy of Adsorbed Species on the Surface of Transition Metal Oxides* (C. H. Rochester, ed.), Wiley, New York, 1990, Chap. 4.

Mikrochim. Acta [Suppl.] 14, 715–717 (1997)
© Springer-Verlag 1997

FT-MIR and FT-FIR Studies of Adsorbed Molecules on Silica and Alumina

János T. Kiss[1,*] and **János Mink**[2,3]

[1] Department of Organic Chemistry, University of Attila József, H-6720 Szeged, Dóm tér 8, Hungary
[2] Department of Analytical Chemistry, University of Veszprém, H-8201 Veszprém, P.O. Box 158, Hungary
[3] Institute of Isotopes of the Hungarian Academy of Sciences, H-1525 Budapest, P.O. Box 77, Hungary

Abstract. The interaction between silica or alumina and a series of substituted aromatic compounds (toluene, benzonitrile, bromobenzene) was studied by IR spectroscopy. The weak adsorption of the organic molecules on the pure supports caused well defined spectral differences. The surface hydroxyl groups took part in the interaction with the organic molecules and gave a new broad band at a frequency lower than the original band position characteristic of free OH groups. At the same time, the intensity of this latter band turned to the opposite direction. The far IR region of the adsorbed aromatic species was also studied and the band due to the 10b deformation mode was recorded in all cases. The observed frequencies in the entire range showed small shifts relative to those measured for the pure liquid phase.

Key words: silica, alumina, FT-IR, adsorption.

High surface area ($>100\,\mathrm{m^2/g}$) oxides, as supports, play an important role in metal catalysis [1]. It is also known that they are involved even in the catalytic processes [2–4]. Revealing the properties of the supported metal catalysts would be easier if more data about the pure supports were available.

Previously, the far IR region received relatively less attention, owing to the high absorbance of the catalysts and the lower performance of the available instruments. However, Primet et al. [5] have found that the IR emission method is capable of providing information in the region of catalytic support cut-off. Mink et al. [6] have shown that most oxide type catalytic supports have a narrow transparency window in the region below $400\,\mathrm{cm^{-1}}$, which could permit measurements of adsorbed species by the transmission technique.

In this paper, some results are reported in connection with the weak adsorption of substituted aromatic compounds on silica and alumina, studied by transmission IR spectroscopy, in both the mid and far IR regions.

Experimental

Transmission spectra of the adsorbed (or physisorbed) species were recorded by means of a Bio-Rad FTS-60A spectrometer equipped with KBr or Mylar (6 μm) beam-splitters and two room-temperature DTGS detectors supplied with CsI and polyethylene windows for the mid and far IR regions, respectively.

Silica or alumina powder was pressed into 13 mm diameter self-supporting wafers which were pretreated at 723 K for 1 hour and then pumped continuously in a conventional vacuum cell equipped with silicon windows.

Organic compounds (toluene, benzonitrile and bromobenzene) were evaporated into the cell after several freeze–thaw cycles. In all cases spectra were recorded at three different vapour pressures: for tolulene, 3333, 2000, 1000 Pa; for benzonitrile 120, 80, 40 Pa; for bromobenzene 600, 400, 200 Pa.

Results and Discussion

The interaction between silica or alumina and the organic compounds causes changes in the OH region of the supports and slight shifts in the CH region of the aromatics. The most conspicuous difference is the transformation of the free OH bands into broad ones which are characteristic of H-bonded OH groups. The strength of the newly formed H-bonds depends on the type of the aromatic ring substituent. When toluene or bromobenzene interacts with OH groups, similar shifts are observed ($100–120\,\mathrm{cm^{-1}}$). For benzonitrile, however, the shift is nearly $200\,\mathrm{cm^{-1}}$, which means that this compound forms a stronger H-bond than the two

---

* To whom correspondence should be addressed

others do. Figure 1 shows the spectra of the three aromatic compounds on alumina, including the bands observed in the far region.

The spectra of the same compounds on silica show some differences. The shifts differ slightly, but the OH bands originating from the interaction are more characteristic than the equivalent bands on alumina, owing to the change in the properties of the free surface OH groups. Figure 2 shows these series of spectra and the bands below $200 \, cm^{-1}$.

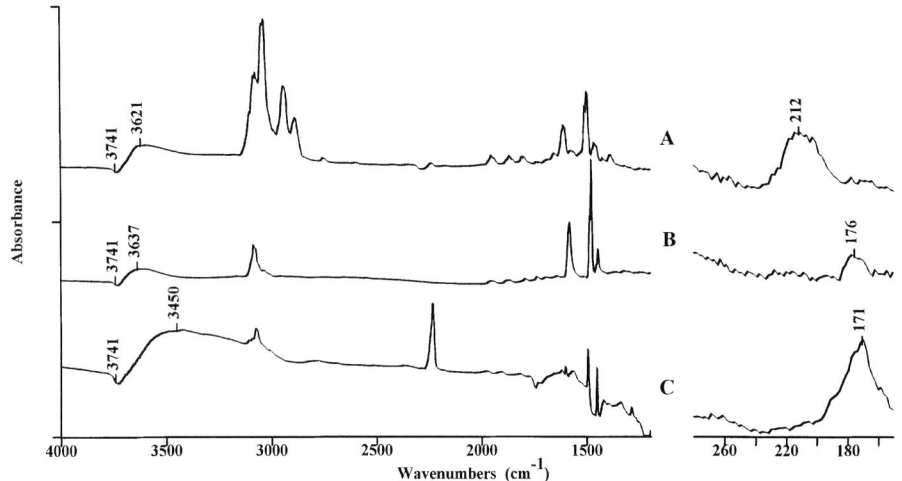

Fig. 1. *A* 2000 Pa toluene on alumina; *B* 600 Pa bromobenzene on alumina; *C* 120 Pa benzonitrile on alumina

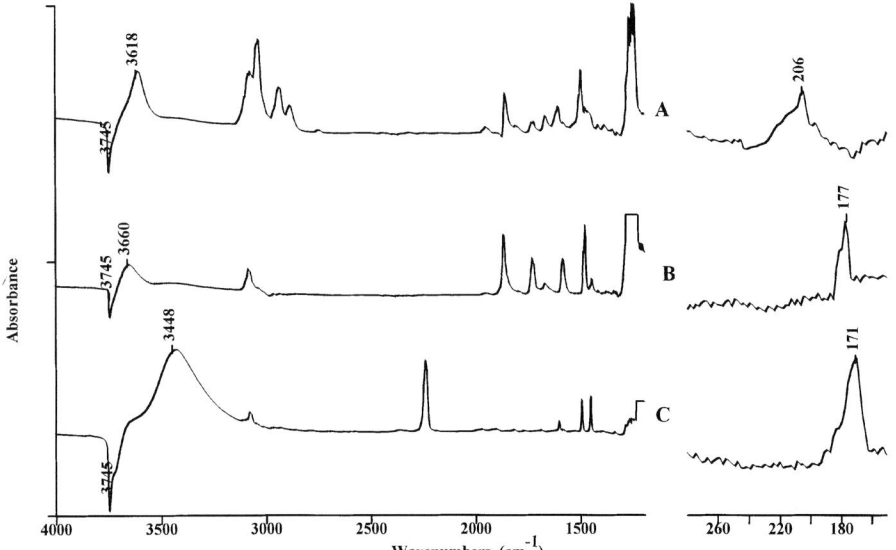

Fig. 2. *A* 2000 Pa toluene on silica; *B* 600 Pa bromobenzene on silica; *C* 120 Pa benzonitrile on silica

Table 1. For IR bands (cm^{-1}) of absorbed mono substituted benzene on pure supports

Toluene			Benzonitrile			Bromobenzene			Assignment
liq.	Al_2O_3	SiO_2	lig.	Al_2O_3	SiO_2	liq.	Al_2O_3	SiO_2	
460	–	462	–	–	–	454	–	–	16b
345	–	–	380	–	–	314	–	314	15
215	212	206	172	171	171	178	176	177	10b

In the liquid phase the aromatic compounds used give two or three bands between 500 and 50 cm^{-1}, which are assigned to OMO deformation modes. The transparent window of the pure supports gives the possibility of recording some of these bands even for the adsorbed form. The bands due to adsorbed (or physisorbed) species on alumina and on silica and observed in the far IR region are summarized in Table 1.

The positions of these bands are very close to those for the liquid phase. Some of them show a little broadening.

Conclusion

As a result of weak adsorption of organic molecules on silica and alumina supports, the free hydroxyl groups of the oxides, absorbing at 3740 cm^{-1}, form a broad band centred around 3450 or 3620 cm^{-1}, depending on the substituent of the aromatic ring. In the far IR region the 10b vibrational mode of the adsorbed organic molecules was observed as a slightly broadened band shifted towards lower frequencies compared to the liquid phase ones.

Acknowledgement. The financial support of the European Community under the COST Project D5 "Surfaces and Interfaces" (Contract No. ERBCIPECT 926104) is gratefully acknowledged.

References

[1] B. C. Gates, *Chem. Rev.* **1995**, *95*, 511.
[2] P. Basu, D. Panayotov, J. T. Yates, Jr., *J. Am. Chem. Soc.* **1988**, *110*, 2074.
[3] E. Baumgarten, C. Lentes-Wagner, R. Wagner, *J. Mol. Catal.* **1989**, *50*, 153.
[4] M. Arai, S.-L. Guo, Y. Nishiyama, *J. Catal.* **1992**, *135*, 638.
[5] M. Primet, F. Fouilloux, B. Imelik, *Surf. Sci.* **1979**, *85*, 457.
[6] J. Mink, G. Keresztury, P. L. Goggin, N. I. Agladze, G. N. Zhizhin, *J. Electron Spectrosc. Relat. Phenom.* **1990**, *54/55*, 823.

Mikrochim. Acta [Suppl.] 14, 719–720 (1997)
© Springer-Verlag 1997

Trace Analysis GC-IR by Injection of Large Sample Volumes

Thomas Hankemeier[1], Udo Brinkman[1], Marjo Vredenbregt[2], and Tom Visser[2],*

[1] Free University, Department of Analytical Chemistry, De Boelelaan 1083, 1081 HV Amsterdam, The Netherlands
[2] National Institute of Public Health and Environment/LOC, P.O. Box 1, 3720 BA Bilthoven, The Netherlands

Abstract. The analyte-detection power of gas chromatography combined with cryotrapping infrared spectrometry (GC-IR) is improved by a factor of 30–100 by injection of large sample volumes by use of a loop-type interface with concurrent solvent evaporation. The usefulness of this of the analysis of real-life samples is illustrated by the identification of polyaromatic hydrocarbons (PAHs) in river Rhine water at a level of 0.5 µg/l.

Key words: gas chromatography, infrared spectrometry, trace analysis.

The limit of detection of cryotrapping GC-IR is approximately 1 mg/l when using conventional 1 µl split/splitless injection [1]. This is insufficient for many applications in trace analysis. Injection of large sample volumes via a loop-type interface (LTI) designed for on-line LG-GC [2, 3] is a way to improve the detectability. The object of this study was to gain insight into the possibilities and limitations of this interface when used in conjunction with cryotrapping GC-IR.

Methods and Materials

The instrumental set-up is shown in Fig. 1. The transfer parameters of the LTI (transfer temperature and closure time of early solvent vapour exit) were optimized to reduce the amount of solvent reaching the IR detector and to minimize loss of analytes during transfer. Optimization was done with solutions of n-alkanes (C11–C25) and PAHs (SRM 2260) in hexane at concentrations of 20–0.05 mg/l and 60–0.06 mg/l, respectively. The volumes injected were 1 µl (split 1:10/splitless) and 100, 200, 400 µl (loop-type). Practical usefulness was tested on hexane extracts of water samples from the river Rhine spiked with PAHs (0.5 µg/l).

Gas Chromatography

A Carlo Erba MEGA 5160 GC with split-splitless injector and FID was used. The retention gap was 5 m × 0.32 mm i.d., DPTMDS-deactivated. The retaining precolumn and analytical column were 0.25 mm i.d. DB-17, 0.15 µm film thickness, 1 m and 15 m long, respectively.

IR Spectrometry

A Digilab FTS-40 FT-IR (Bio-Rad) with Tracer GC interface was used, with 2 scans/sec at a data-point resolution of 4 cm^{-1}, with 4 and 256 scans co-added.

Results and Discussion

The optimum transfer temperature and closure time of the solvent vapour exit for the solvent n-hexane were determined as 95 °C and 30 sec, respectively, with C19 as the first quantitatively recovered n-alkane. For the n-alkanes > C19, the absolute detection limit obtained from 100 µl injections with the LTI, was the same as for 1 µl split/splitless injections of solutions with a 100-fold higher concentration. This implies that the analyte detectability has been improved by a factor of 100. For

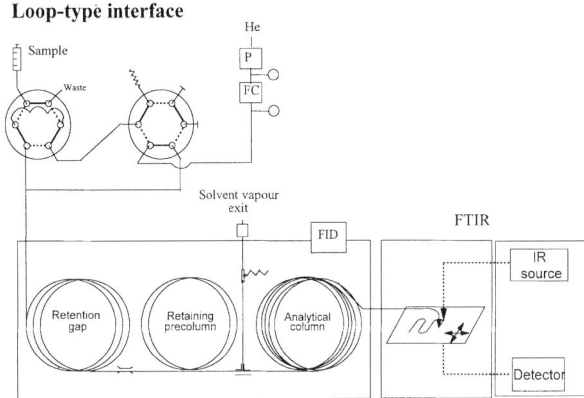

Loop-type interface

Fig. 1. Schematical set-up of the large-volume injection cryotrapping GC-IR system

---

* To whom correspondence should be addressed

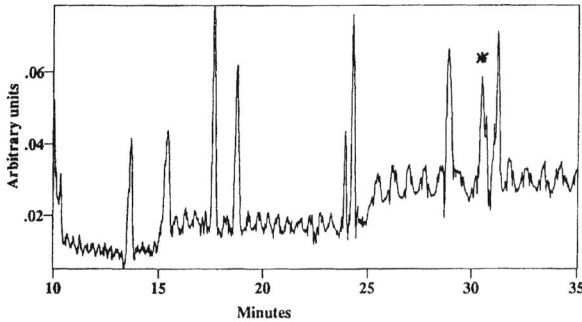

Fig. 2. Functional-group GC-IR chromatogram (950–750 cm^{-1}) of a 100 µl injection of a hexane extract of river Rhine water spiked at the 0.5 µg/l level

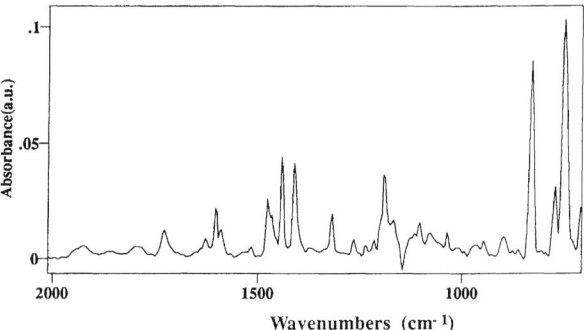

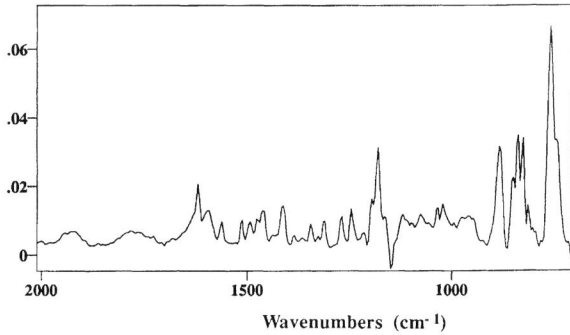

Fig. 3. Cryotrapped GC-IR spectra of benzo[e]pyrene (left) and benzo[a]pyrene (right) as derived from the peak indicated by * in the chromatogram of Fig. 2

the PAHs, however, the gain in detectability with the LTI was a factor of 30. The poorer performance for the PAHs is attributed to band broadening in the loop-type interface and a high background level of ice in the IR detector during the experiments with the PAHs. Injections of 100 µl were the most appropriate, since for larger volumes solvent impurities and solvent crystallization in the sample storage interface became apparent. The repeatability was good (relative standard deviation of 6 injections was better than 15%).

The GC-IR chromatogram of the 100 µl injection of the spiked water extract is shown in Fig. 2. All PAHs were recovered and identified by means of the IR-spectra derived from the chromatographic peaks, even when the compounds were not fully separated.

Figure 3 shows the spectra of the isomers benzo[a]pyrene and benzo[e]pyrene as obtained from the wings of the virtually single peak indicated by * in the chromatogram of Fig. 2. The recoveries for fully separated compounds were between 80 and 90% when related to the absorbance of the strongest peak in the spectrum.

From these results, the limit of detection for the water samples was estimated as 0.1 µg/l. In view of the sensitivity of cryotrapping GC-IR, better limits of detection should be attainable. However, two drawbacks of the current system should be solved to achieve this. First, the high background absorption of ice, originating from traces of water that enter the system through diffusion in the LC-values, the restrictors and the large number of (press-fit) connectors, should be reduced. Replacement of the loop-type interface by an on-column injector might be a way to achieve this. Secondly, interfering solvent crystallization in the IR detector hampers high sensitivity. A post-column detector-switching module to split off undesired chromato-

graphic (solvent) peaks should reduce this problem. A study of both modifications is currently in progress.

Conclusions

The detectability of analytes with cryotrapping GC-IR can be improved by a factor of 30–100 relative to split/splitless injections, by use of a loop-type interface with a 100 µl injection loop. Larger sample volumes are attended by interfering absorptions of solvent and solvent contaminants. In extracts, PAHs can be determined with high repeatability at concentration levels corresponding to the "alert" levels for river and drinking water.

Optimization of the interface is simple, but the system is sensitive to traces of water. Small leaks are highly detrimental.

References

[1] P. Jackson, G. Dent, D. Carter, D. J. Schofield, J. M. Chalmers, T. Visser, M. J. Vredenbregt, *J. High Resolut. Chromatogr.* **1993**, *16*, 515.
[2] K. Grob, Jr., D. Fröhlich, B. Schilling, H. P. Neukom, P. Nägeli, *J. Chromatogr.* **1984**, *295*, 55.
[3] K. Grob, J.-M. Stoll, *J. High Res. Chromatogr. Chromatog. Commun.* **1986**, *9*, 518.

Mikrochim. Acta [Suppl.] 14, 721–724 (1997)
© Springer-Verlag 1997

Conformational Effects of Reversed-Phase HPLC on Ribonuclease A and α-Chymotrypsin by Particle Beam LC/FT-IR Spectrometry

Randall T. Bishop and **James A. de Haseth***

Department of Chemistry, University of Georgia, Athens, Georgia 30602-2556, U.S.A.

Abstract. The reversed-phase liquid chromatographic behavior of ribonuclease A (Rnase) and α-chymotrypsin (α-Chy) have been examined by particle beam LC/FT-IR spectrometry. Interaction of many proteins with a non-polar stationary phase is known to induce reversible/irreversible denaturation of the protein. Reverse-phase HPLC of such proteins often produces a series of chromatographic peaks representing different solution conformations. The chromatographic behaviors of Rnase and α-Chy were examined through selective sampling of the major peaks observed in the chromatographic profiles of each protein by particle beam LC/FT-IR spectrometry. These conformations were produced and separated in this study by the use of a C-4 column with an acetonitrile gradient. These bands were examined post-column with a diode array detector at 220 nm, followed by isolation of the different solution conformers by the particle beam technique. The aerosol-based particle beam apparatus has previously been shown to be successful in the isolation of intermediate conformations of proteins. The resulting protein deposits are examined off-line by FT-IR microscopy. Additionally, Rnase was examined in various concentrations of acetonitrile to assess the degree to which the organic modifier affects the secondary structural content of the protein. Differences in secondary structure elements were evident among the chromatographic peaks of both Rnase and α-Chy. Secondary structure differences were also evident in Rnase in different concentrations of acetonitrile.

Key words: infrared, protein, conformation.

Many globular proteins are known to be denatured to various degrees under reversed-phase HPLC conditions. Denaturation is due to both the mobile phase and the adsorption of a protein onto the non-polar stationary phase [1]. This can result in several chromatographic peaks, the separation of which is based on the degree to which the different conformational states interact with the stationary phase. As protein separations are almost exclusively performed with an organic modifier gradient, the presence of organic modifiers is also a factor in inducing changes in the solution conformations of proteins [2]. It is of interest to learn about the interaction of globular proteins with reversed-phase column packings, and the factors affecting it, so that protein separations can be optimized to yield the desired results.

Fourier transform infrared (FT-IR) spectroscopy is a powerful technique for the analysis of secondary structure of globular proteins in aqueous solution. The amide I mode is the most useful of the backbone peptide vibrational modes. The amide I band contains information on the secondary structure content of proteins: α-helices, β-sheets, turns, and unordered structures [3].

The particle beam apparatus functions as an aerosol-based LC/FT-IR interface allowing for the isolation of protein solution conformers [4]. The column effluent is nebulized and introduced into a low-pressure desolvation chamber. Rapid solvent evaporation immediately traps the protein solution conformers, which are subsequently deposited on an infrared-transparent substrate for off-line analysis.

Experimental

Ribonuclease A and α-chymotrypsin were obtained from the Sigma Chemical Company and used as received. All samples were prepared in 0.1% trifluroacetric acid solution in 18-MΩ demineralized water and were 1% (w/v) protein. Instrumentation consisted of a Perkin Elmer 200 series LC pump, a Perkin Elmer 235C Diode Array detector, a PE Nelson Model 1022 Personal Integrator, and a particle beam apparatus. The design and operation of the particle beam apparatus has been described elsewhere [4]. A Vydac C-4 Protein

---

* To whom correspondence should be addressed

column was used in this study, with an increasing gradient of acetonitrile to produce the conformational separations; specific conditions are indicated in the respective figures.

Protein deposits were made onto a calcium fluoride window (25 × 2 mm) and examined off-line with a Spectra Tech IR-Plan™ Infrared Microscope interfaced to a Perkin Elmer 1760X FT-IR Spectrometer. All spectra were obtained from 1000 co-added scans at 8 cm⁻¹ resolution. All spectral manipulations, including second derivatization and deconvolution, were performed with GRAMS 386™ software (Galactic Industries Corp., Salem, NH).

Results and Discussion

The chromatograms in Fig. 1 illustrate the fractions of column effluent collected by the particle beam for Rnase and α-Chy, respectively. Figures 2 and 3 illustrate the amide I and II regions and corresponding second derivative spectra of the indicated fractions of Rnase and α-Chy. Figures 4 and 5 illustrate reference spectra of Rnase and α-Chy obtained from particle beam deposits without the use of a chromatographic column and evaporated protein films. Figure 6 illustrates the amide I and II regions and second derivative spectra of Rnase in different concentrations of acetonitrile.

The amide modes and second derivative spectra of Rnase (Fig. 2) demonstrate significant differences in secondary structure among themselves, and in reference to the spectra of unchromatographed Rnase (Fig. 4). Deposits 1–3 show the development of a strong, broad band at 1661 cm⁻¹, assigned to unordered structure, indicating the predominance of this secondary structure element in the solution conformation of deposits 2 and 3. This is consistent with the increasing retention times at which the deposits were collected, and indicates a greater degree of interaction with the stationary phase.

The amide modes and second derivative spectra of α-Chy (Figs. 3 and 5) again demonstrate significant differences in secondary structure. The spectra of deposits 1–3 show a relative reduction of the 1651 cm⁻¹ band with a concomitant increase in the relative intensity of the 1695, 1662, and 1632 cm⁻¹ bands. The bands are assigned as follows: 1695 cm⁻¹, turns; 1662 cm⁻¹, unordered; 1651 cm⁻¹, α-helix; 1632 cm⁻¹, β-sheet.

Acetonitrile is demonstrated (Fig. 6) to have a definite but not drastic effect on the solution structure of Rnase. The second derivative spectra show a parallel decrease in bands at 1695, 1660, and 1641 cm⁻¹.

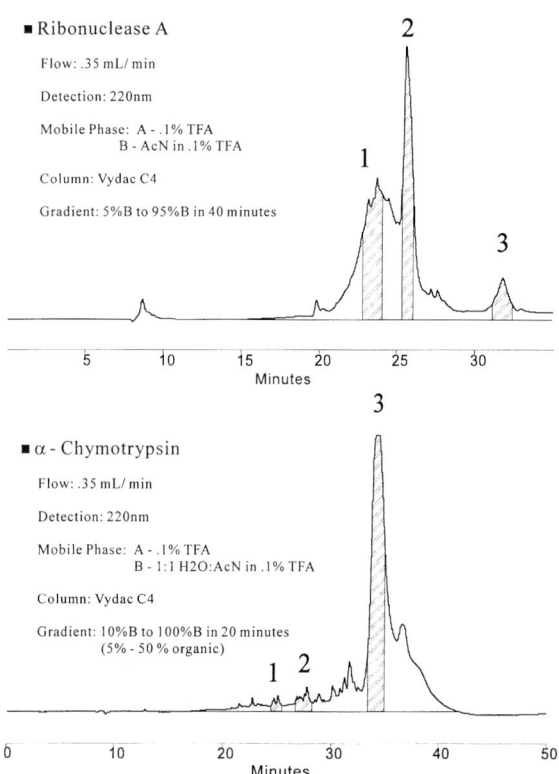

Fig. 1. RPC of ribonuclease A and α-chymotrypsin: fractions collected with particle beam

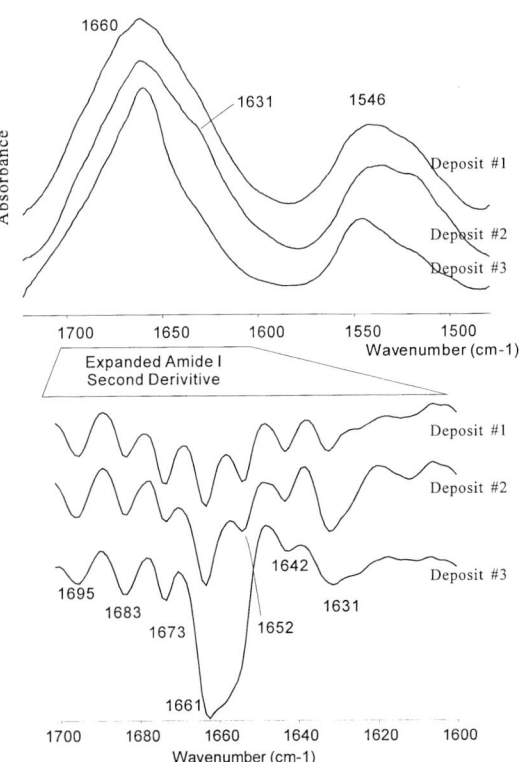

Fig. 2. Amide I and II bands and second-derivatives of particle beam LC/FT-IR collections of ribonuclease A

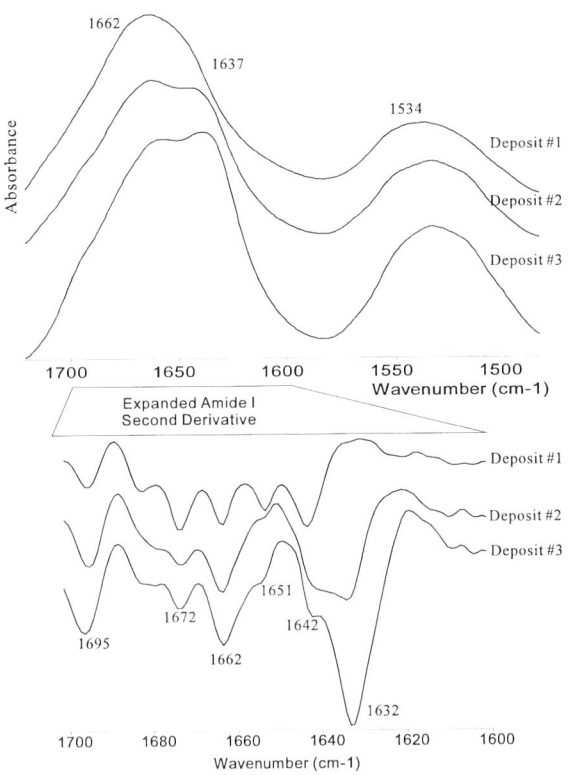

Fig. 3. Amide I and II bands and second-derivatives of particle beam LC/FT-IR collections of α-chymotrypsin

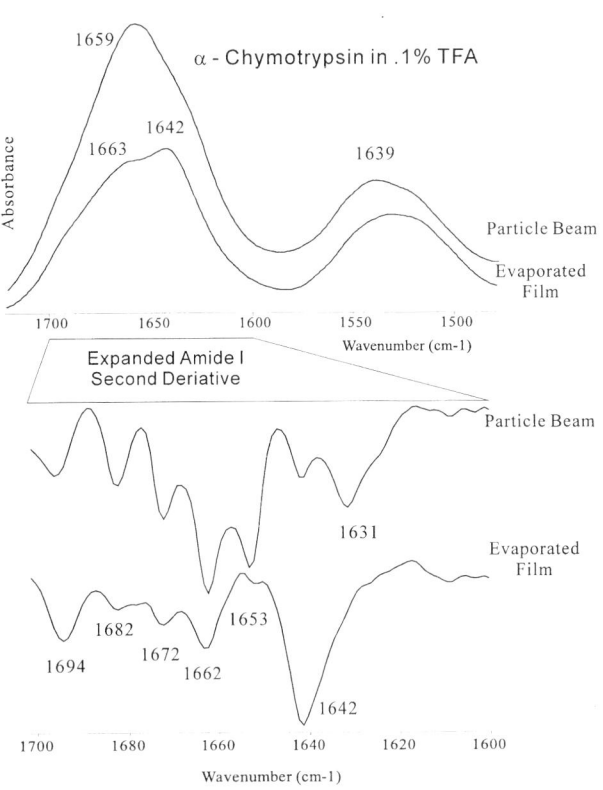

Fig. 5. Reference spectra: particle beam and evaporated film

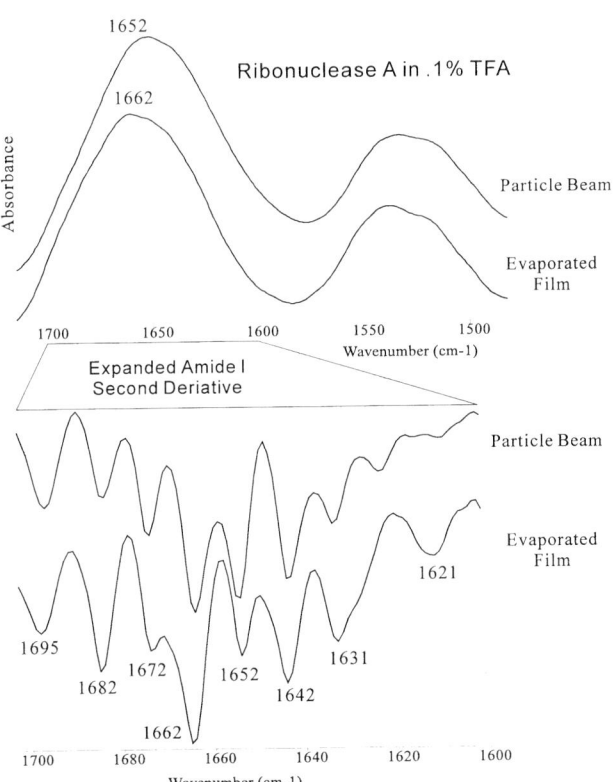

Fig. 4. Reference spectra: particle beam and evaporated film

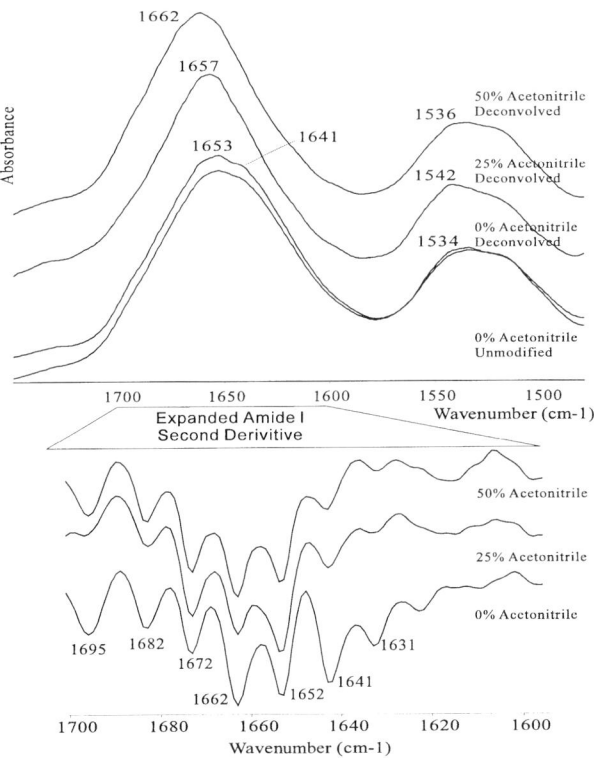

Fig. 6. Deconvolved amide I and II regions and second-derivatives of ribonuclease A with increasing acetonitrile concentration

Conclusion

Particle beam LC/FT-IR spectrometry offers a new and valuable method for assessing the conformational state of proteins in reversed-phase separations. The degree of protein-reversed phase interaction, as characterized by retention time, was shown to have a definite effect on secondary structure content, usually leading to a more denatured state. The effect of increasing acetonitrile concentration was also shown to lead to secondary structure denaturation.

References

[1] J. D. Andrade, V. Hlady, *Adv. Poly. Sci.* **1986,** *79,* 1.
[2] K. M. Gooding, F. E. Regneir (eds.), *HPLC of Biological Macromolecules: Methods and Applications, Vol. 51,* Dekker New York, 1990.
[3] S. N. Timasheff, G. D. Fasman (eds.), *Structure and Stability of Biological Molecules,* Dekker, New York, 1969.
[4] V. E. Turula, J. A. Haseth, *Appl. Spectrosc.* **1994,** *48,* 1255.

Mikrochim. Acta [Suppl.] 14, 725–727 (1997)

Flexible Off-line LC-IR Interfacing by Micro-column Switching and Addition of a Make-up Liquid

Marjo Vredenbregt[1], **Jan ten Hove**[1], **Ad P. J. M. de Jong**[1], **Tom Visser**[1,*], and **Govert W. Somsen**[2]

[1] National Institute of Public Health and Environment/LOC, P.O. Box 1, 3720 BA Bilthoven, The Netherlands
[2] Department of General and Analytical Chemistry, Free University, De Boelelaan 1083, 1081 HV Amsterdam, The Netherlands

Abstract. The versatility of a spray-jet LC-IR interface has been improved by application of capillary liquid chromatography combined with post-column addition of methanol as a make-up liquid. Fluctuations in the evaporation profile are minimized, which facilitates gradient elution including high percentages of water. Micro-column switching has been applied to allow the introduction of large sample volumes, thus improving the limit of detection. The current performance of the interface is demonstrated by the results for the separation and structure elucidation of metabolites of pyrene in an *in-vitro* microsomally treated sample.

Key words: infrared spectrometry, liquid chromatography, hyphenated techniques.

In previous papers we have reported on a spray-jet interface that combines narrow-bore liquid chromatography and infrared spectrometry [1, 2]. The principle of the interface is based on evaporation of the mobile phase prior to IR detection and immobilization of the chromatogram on an IR-transparent substrate. It has proven to be a useful tool for several applications but the performance is sensitive to the composition of the mobile phase. Also, large flow-rates and high water percentages cannot be handled. Gradient elution with high water percentages, however, is quite common in reversed-phase LC applications. In order to cope with this and to accomplish a more versatile interface, the experimental set-up has been modified. A capillary analytical column is used to reduce the analytical flow-rate, and excess of a volatile solvent is added as make-up liquid to the LC effluent to mask fluctuations in the composition of the eluent. In addition, a trap column is incorporated to preconcentrate the analytes. The usefulness of the modified interface to solve real-life problems is investigated by the analysis of extracts of microsomes that were incubated with pyrene.

Methods and Materials

The experimental set-up is shown in Fig. 1. Samples were introduced via a 20 µl loop injector and transferred to a home-made trap column for preconcentration, with water as a mobile phase. Next, the analytes were back-flush eluted to the analytical column with a gradient of water and acetonitrile. After LC separation, a volatile solvent was added as a make-up liquid in a 5:1 ratio via a T-piece connector (zero dead-volume). The spray-jet interface was used for elimination of the mobile phase and deposition of the separated compounds on a moving ZnSe window. Spectra of the immobilized LC-chromatogram were obtained by IR-microscopy. Optimization was done with acetone and methanol as a make-up liquid.

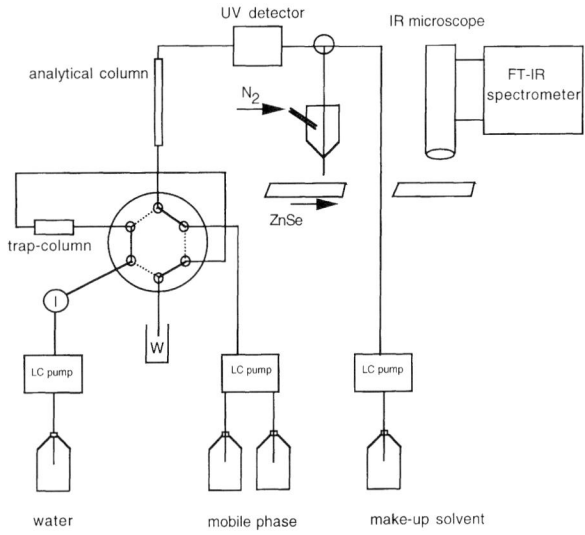

Fig. 1. Schematical set-up of the LC-IR system

---

* To whom correspondence should be addressed

Samples

Extracts in water of microsomes that were incubated with pyrene.

Chromatography

Home-made trap column, Bond-Elut C18, particle size 40 μm, 20 × 0.5 mm; mobile phase 100% water, flow rate 30 μl/min, injection volume 20 μl. Analytical column Hypersil C18 BDS, 3 μm, 150 × 0.3 mm; mobile phase water for 15 min to transfer sample from trap column then water–acetonitrile gradient changing from 70% to 0% water 100% acetonitrile for 25 min, flow rate 4 μl/min. Make-up liquid methanol, flow rate 20 μl/min.

Interface

Evaporation gas N$_2$, temp. 130 °C, pressure 5.5 bar, substrate ZnSe, table speed 350 μm/min (X/Y stepper motor).

Spectroscopy

UV-detector Applied Biosystems 785A with micro-flow cell, wavelength 254 nm, IR-spectrometer Bruker IFS55 with A590 microscope and MCT detector; 512 scans/spectrum, resolution 8 cm^{-1}. Aperture size 100 × 200 μm.

Results and Discussion

Comparison of the LC-UV chromatogram obtained after sample preconcentration, including back-flush elution, with the chromatogram acquired from loop injection, revealed that extra peak broadening caused by the trap-column is negligible.

Optimization of all parameters of the LC-IR interface for both acetone and methanol as a make-up liquid, showed that the performance of the interface is less susceptible to eluent composition changes when methanol is used. Therefore, the latter was used in further experiments.

Analysis of the extracts showed four major peaks in the LC-UV chromatogram (Fig. 2). Visual inspection of the chromatogram immobilized on the ZnSe window, revealed that peaks I and II were slightly overlapping, and III and IV were separated. The IR spectra of components III and IV were recorded by scanning the virtual centre of the corresponding spot. Spectra of I and II were obtained from positions at the beginning and the end of the overlapping spots. Components III and IV were identified by library search as 1-hydroxy-pyrene and the mother compound pyrene, respectively. For I and II matching reference spectra were not found. However, interpretation of the IR-spectra (Fig. 3) pointed to a symmetrical polyaromatic structure with conjugated carbonyl

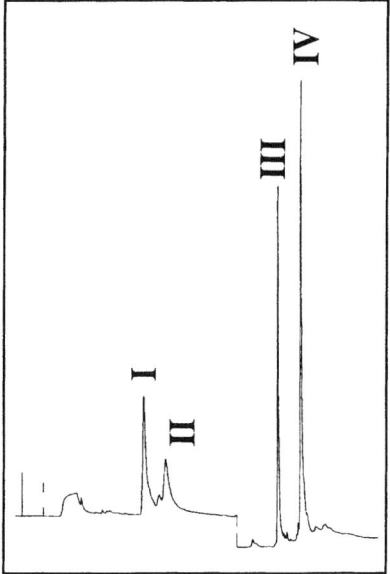

Fig. 2. LC-UV chromatogram of the extract of microsomes incubated with pyrene

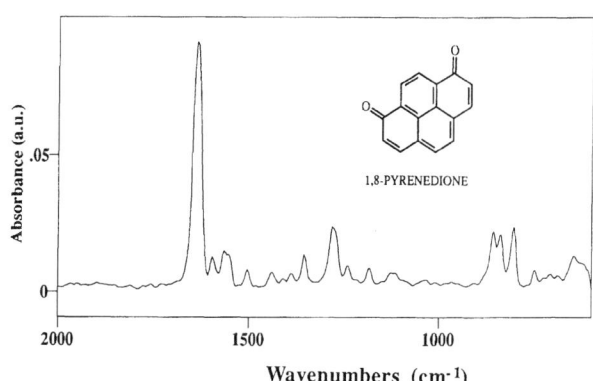

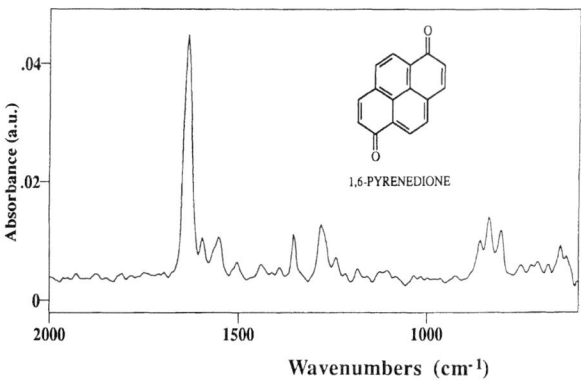

Fig. 3. IR spectra of 1,6-pyrenedione (I) and 1,8-pyrenedione (II)

groups of very low vibrational frequency (1635 and 1638 cm^{-1}, respectively). Literature study resulted in the identification of two isomers, component I as 1,6-

pyrenedione and component II most likely to be 1,8-pyrenedione [3].

Conclusions

Preconcentration of sample solutions on a trap-column combined with back-flush elution to the analytical column by micro-column switching is an elegant way to inject large volumes into a capillary LC. It enhances the overall sensitivity of the combined LC-IR system without significant peak-broadening.

By post-column addition of excess of methanol as make-up liquid, an almost uniform effluent is obtained, so that the performance of the LC-IR interface becomes independent of the composition of the mobile phase. This enables the application of both gradient elution and high water percentages up to 80%.

References

[1] G. W. Somsen, R. J. van de Nesse, C. Gooijer, U. A. Th. Brinkman, N. H. Velthorst, T. Visser, P. R. Kootstra, A. P. J. M. de Jong, *J. Chromatogr.* **1991**, *552*, 635.

[2] G. W. Somsen, L. P. P. van Stee, C. Gooijer, U. A. Th. Brinkman, N. H. Velthorst, T. Visser, *Anal. Chim. Acta* **1994**, *290*, 269.

[3] A. J. Fatiadi, *J. Chromatogr.* **1965**, *20*, 319.

Mikrochim. Acta [Suppl.] 14, 729–731 (1997)

Applications of TGA/FT-IR Analysis to Polymer Systems

Prashant S. Bhandare* and **K. Krishnan**

Bio-Rad Laboratories, Digilab Division, 237 Putnam Avenue, Cambridge, MA 02139, U.S.A.

Abstract. FT-IR spectrometry was used to perform evolved gas analysis from samples used in thermogravimetric analysis (TGA) and differential thermal analysis (DTA) techniques. In this study, we demonstrate the utility of such combined FT-IR/TGA/DTA techniques for identification of evolved gases from their infrared spectra and also for providing useful chemical insight regarding the sample.

Key words: thermogravimetric analysis (TGA), differential thermal analysis (DTA), evolved gas analysis (EGA), polymers.

In thermogravimetric analysis (TGA), the sample is placed on a sensitive balance and heated in a controlled manner, under a selected gas flow. The weight of the sample is continuously monitored as the temperature is changed. The profile of weight change plotted against temperature can provide valuable information regarding the thermal behaviour of the sample. Differential thermal analysis (DTA) provides information about energy changes in the sample during heating. However, neither TGA nor DTA can provide information regarding chemical changes occurring in the sample as a result of temperature change. Such changes can often be evidenced by evolution of gaseous products accompanying the weight loss [1, 2]. By use of FT-IR spectrometry the evolved gases can be analyzed, and their identities can often be determined [3, 4]. The objective of this study was to demonstrate the application of combined TGA/FT-IR and TGA/DTA/FT-IR techniques for analyzing different types of samples, such as polymers and organic materials.

Experimental

A Bio-Rad FTS 165 FT-IR spectrometer was interfaced with a TA Instruments model 2050 TGA or model 2960 TGA/DTA analyzer by using a specially designed TGA/IR accessory. The gases evolved from the samples were led into the accessory through an inert transfer line. The accessory consisted of a heated stainless steel gas cell having two IR-transparent windows at each end. Both the cell and the transfer line were maintained at 250 °C in order to eliminate condensation. The time resolution between spectra was set at 6 s. Absorbance spectra were collected at a resolution of 4 cm^{-1}.

Results and Discussion

Figure 1 shows the weight loss and derivatized weight loss profile for a packaging polymer. There are 3 primary weight losses during thermal decomposition at 145, 180, 250 °C (24, 34 and 37 minutes) of this sample. Figure 2 shows selected spectra gases evolved at various times during polymer decomposition. Although several gases were evolved during each of the weight losses, many of the gases could be readily identified from their characteristic spectral patterns. During the first weight loss, spectra of carbon dioxide, water vapor and nitrous oxide were clearly seen along with some spectral bands in the 3000–2800 cm^{-1} region. These bands were identified as due to dimethyl-

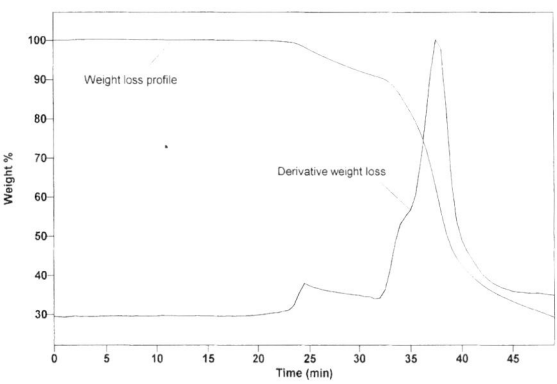

Fig. 1. Weight loss and derivatized weight loss for polymer sample

---

* To whom correspondence should be addressed

Bio-Rad Win-IR

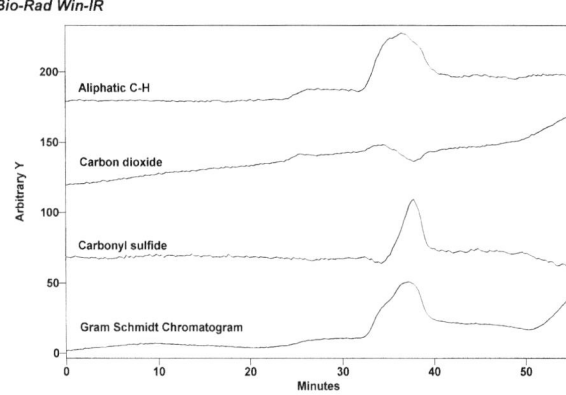

Fig. 2. Evolved gas analysis

Bio-Rad Win-IR

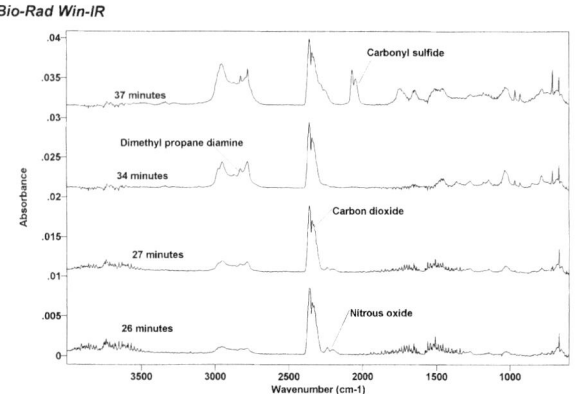

Fig. 3. Absorbance spectra of evolved gases

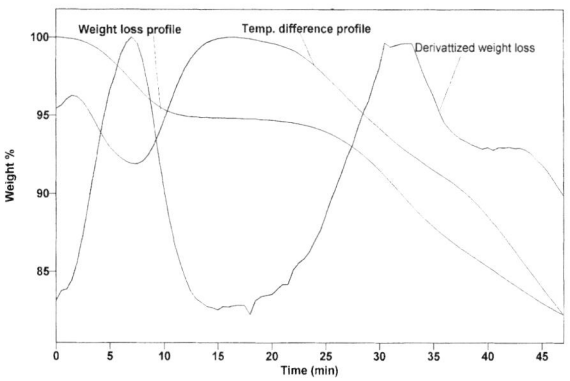

Fig. 4. Weight loss, derivatized weight loss and temperature difference for organic polymer

Bio-Rad Win-IR

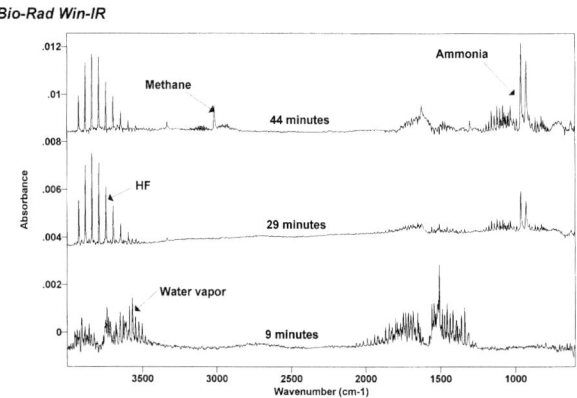

Fig. 5. Absorbance spectra of evolved gases

propanediamine by using a computerized spectral library search. During the second weight loss, more dimethylpropanediamine evolved and the amount of water vapor decreased. During the third weight loss carbonyl sulfide was also evolved. Figure 3 shows the Gram Schmidt chromatogram (total evolved gas profile, which resembles a derivatized weight loss profile) as well as functional group chromatograms (specific gas profiles). These profiles allow inspection of variation in the amount of gases evolved over a period of time. Knowing the gases evolved during the specific weight losses helps in understanding the mechanism for thermal decomposition of the polymer.

An application of combined TGA/DTA/FT-IR analysis to an organic polymer is summarized below. Figure 4 shows the weight loss and derivatized weight loss, as well as the heat flow profile for the sample. Figure 5 shows the absorbance spectra of the gases evolved during specific weight losses. It is evident that the first weight loss was due to evolution of water vapor

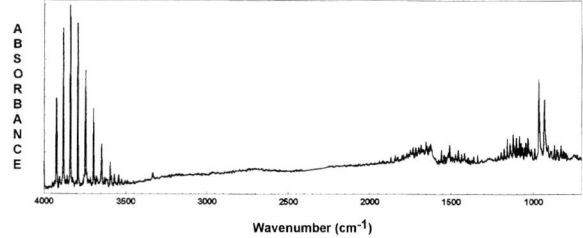

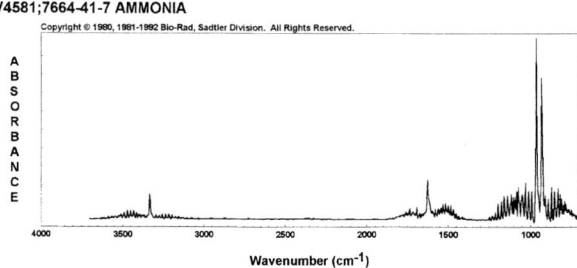

Fig. 6. Results from computerized spectral library search

only (water of hydration). The second weight loss resulted from evolution of hydrogen fluoride as well as ammonia. Since the polymer molecule did not contain fluorine, the hydrogen fluoride was regarded as a leftover from hydrofluoric acid used as a solvent in the polymer synthesis. During the third weight loss, ammonia and hydrogen fluoride were evolved along with methane. Figure 6 shows an example of results from a computerized spectral library search of the spectrum of the gases evolved at 330 °C (29 minutes).

References

[1] D. A. C. Compton, D. J. Johnson, M. L. Mittleman, *Integrated TGA/FT-IR System to Study Polymeric Materials*, FTS/IR Note No. 70, Bio-Rad Digilab Division, 1989.

[2] D. A. C. Compton, *On-line FT-IR Analysis of the Gaseous Effluent from a Thermogravimetric Analyzer: Applications of an Integrated TGA/FT-IR System*, Bio-Rad Digilab Division, FTS/IR Note No. 47, 1987.

[3] H. G. Schild, *J. Polymer Sci., Part A, Polymer Chem.* **1993**, *31*, 1629.

[4] H. G. Schild, *J. Polymer Sci. Part A, Polymer Chem.* **1993**, *31*, 2403.

Mikrochim. Acta [Suppl.] 14, 733–738 (1997)
© Springer-Verlag 1997

Fourier-transform Raman Spectroscopic Study of the Association Between Co(II) Ions and Acrylonitrile in Aqueous Solution

J. M. Alia[2,*], **H. G. M. Edwards**[1], and **J. Moore**[1]

[1] Chemistry and Chemical Technology, University of Bradford Bradford, West Yorkshire, DB7 1DP, U.K.
[2] Dpto. de Quimica Física, E.U.I.T.A., Universidad de Castilla-La Mancha, Ronda de Calatrava, s/n, E-13071 Ciudad Real, Spain

Abstract. The association between acrylo nitrile (propenenitrile) and cobalt ions in aqueous solution was investigated by Fourier-transform Raman spectroscopy. The new $v(C\equiv N)$ band at $2268\,cm^{-1}$ attributable to cation-associated acrylonitrile, displaced by $+32\,cm^{-1}$ from the wavenumber corresponding to the solvent (water/acrylonitrile mixture), was quantitatively studied in a range of concentrations of 2.452–$4.001\,M$ for cobalt [as cobalt(II) nitrate] and 2.241–$2.584\,M$ for acrylonitrile. The proposed equilibrium involves the interchange of one molecule of water and one molecule of acrylonitrile in the first solvation sphere of the cation, namely: $[Co(H_2O)_6]^{2+} + ACN \Leftrightarrow [Co(H_2O)_5ACN]^{2+} + H_2O$. The association constant calculated at *infinite dilution* was $0.315 \pm 0.024\,l/mol$ at $25\,°C$, greater than that corresponding to the same equilibrium with nickel ions ($0.221 \pm 0.014\,l/mol$). This difference is explained by the higher hydration energy of the nickel ion.

Key words: acrylonitrile, cobalt(II), solvation sphere, exchange reaction.

The interaction between the cobalt ion and acrylonitrile has recently received considerable attention. This ion is known as a superior inductor of the enzyme acrylonitrile hydratase de *Arthrobacter sp.* IPCB-3 [1, 2], which can also be induced by the presence of iron(II) ions. The formal kinetics of the copolymerization of acrylonitrile with cobalt, nickel and copper acrylates has been investigated [3] and compared with the reactivity of different cations [4]. Cobalt acetylacetonate has been proved an effective initiator of acrylonitrile–styrene copolymerization [5, 6]. The catalytic activity of cobalt ions in the hydro-esterification of acrylonitrile [7, 8] is also well known. In organometallic chemistry, several acrylonitrile coupling reactions have been found to be cobalt-promoted [9, 10] and dicyano-cobaltacyclopentanes have been obtained from a co-

balt(acrylonitrile) complex [11]. In a comprehensive review [12] of the coordination chemistry of acrylonitrile, in a compilation of the literature before 1982, cobalt ion is the most-cited coordinating agent.

A spectroscopic study of the preferential solvation of cobalt ions in water–acetonitrile mixtures has been published very recently [13]. The presence of mixed species $[Co(H_2O)_n(AN)_{6-n}]^{2+}$ (AN = acetonitrile) was detected by means of visible absorption spectra. The ionic concentration studied was $0.05\,M$.

We have studied the association between nickel ions and acrylonitrile in aqueous solution at different temperatures by means of Raman spectroscopy [14], thereby calculating the enthalpy of the association equilibrium. Although the solubility of acrylonitrile in pure water is only about $75\,g/l$, it readily dissolves in higher concentrations in the presence of inorganic salts, because of the salt/acrylonitrile association.

The very deep colour of acrylonitrile solutions concentrated by presence of cobalt precludes their Raman spectroscopic study by use of visible laser excitation. In the present study, the acrylonitrile–cobalt ion solvation process has been investigated by FT-Raman spectroscopy.

Experimental

Solutions

Cobalt nitrate hexahydrate (P. A., Aldrich) was dissolved in distilled water to give solutions in the concentration range 60–85% w/w. In each case, 1 g of acrylonitrile (BDH, Spectroscopic grade) was added to 10 g of aqueous solution. The solutions were prepared by weighing and their densities measured with a conventional pyconometer at $25.0 \pm 0.2\,°C$. Solutions were filtered through $0.65\,\mu m$ Millipore

---

* To whom correspondence should be addressed

Table 1. Solution characteristics

Co(NO$_3$)$_2$·6H$_2$O %, w/w	Density g/cm^3	Acrylonitrile M	Co^{2+} M	Water M
60	1.3079	2.241	2.452	41.14
65	1.3436	2.302	2.729	40.13
70	1.3814	2.367	3.022	39.06
75	1.4213	2.435	3.331	37.94
80	1.4636	2.508	3.659	36.74
85	1.5084	2.584	4.007	35.47

filters prior to recording of the FT-Raman spectra. Table 1 gives the composition of each solution.

FT-Raman Spectra

FT-Raman spectra were excited at 1064 nm by an Nd:YAG laser used with a Bruker IFS66 optical bench and an FRA 106 Raman accessory. The laser power was set at 80 mW and 2000 scans were accumulated with a resolution of 2 cm^{-1} (spectral density: 1 experimental point for each wavenumber).

Data Processing

OPUS$^\copyright$ software (Bruker) was used to evaluate the spectral characteristics of the bands (peak wavenumber position and integrated intensity). The original OPUS$^\copyright$ format was transferred to JCamp format, then to standard ASCII format, and finally processed to perform the band deconvolution and component fitting. Specially constructed software [15] was used to improve the resolution and to calculate the relative areas of overlapped bands. In no case were smoothing procedures or baseline correction techniques used.

The C–H rocking band (v_7) of acrylonitrile at 1288 cm^{-1} was used as an internal standard in order to normalize the spectra for band intensities. This procedure has proved to be satisfactory in previous investigations [16] because of the lack of sensitivity of this band to the presence of water and/or solutes.

Results and Discussion

In Fig. 1, the FT-Raman spectra of a saturated solution of acrylonitrile in water and of the 85% solution are shown.

The fundamental v_1(A$_1$) of the nitrate anion at 1048 cm^{-1} has a symmetric shape, unlike the bands obtained in aqueous solutions of associated nitrates [17, 18]. In addition, the presence of a new band was observed, at 2268 cm^{-1}, which originates from acrylonitrile molecules in the cation's first solvation sphere. The observed shift from the band of the bulk acrylonit-

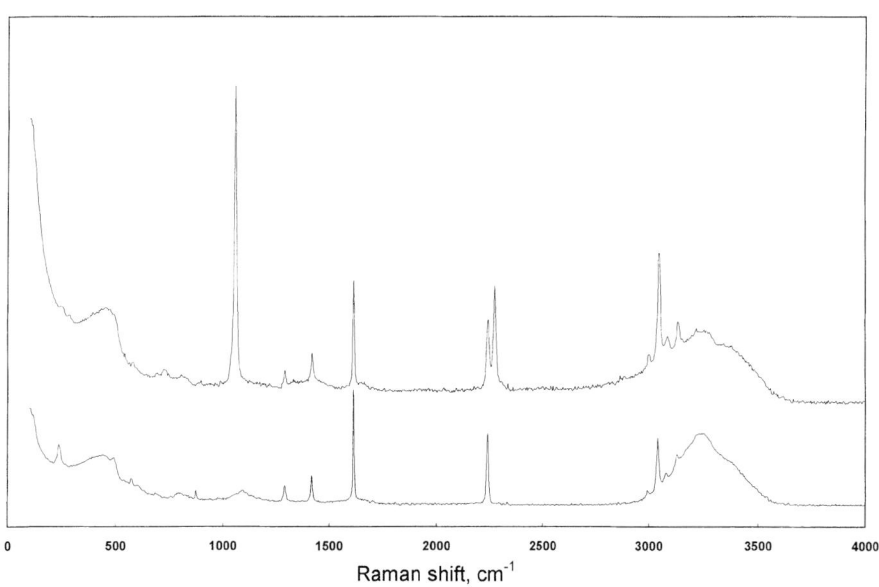

Fig. 1. FT-Raman spectra of acrylonitrile in aqueous solution (saturated) [bottom] and in w/w 85% aqueous solution of cobalt nitrate [top]

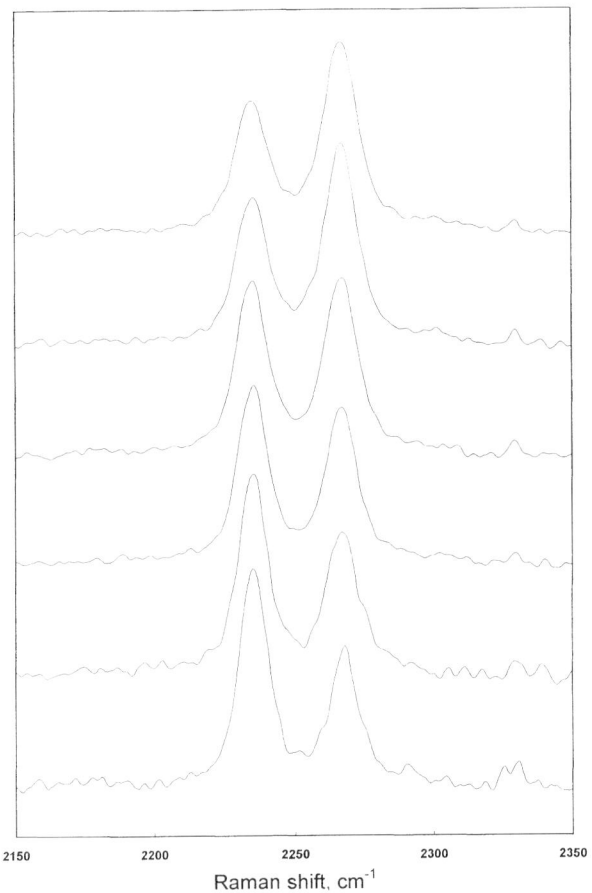

Fig. 2. The $\nu(C\equiv N)$ region in the FT-Raman spectra of acrylonitrile in several cobalt nitrate aqueous solutions; from bottom to top, w/w concentrations of cobalt nitrate are 60, 65, 70, 75, 80 and 85%

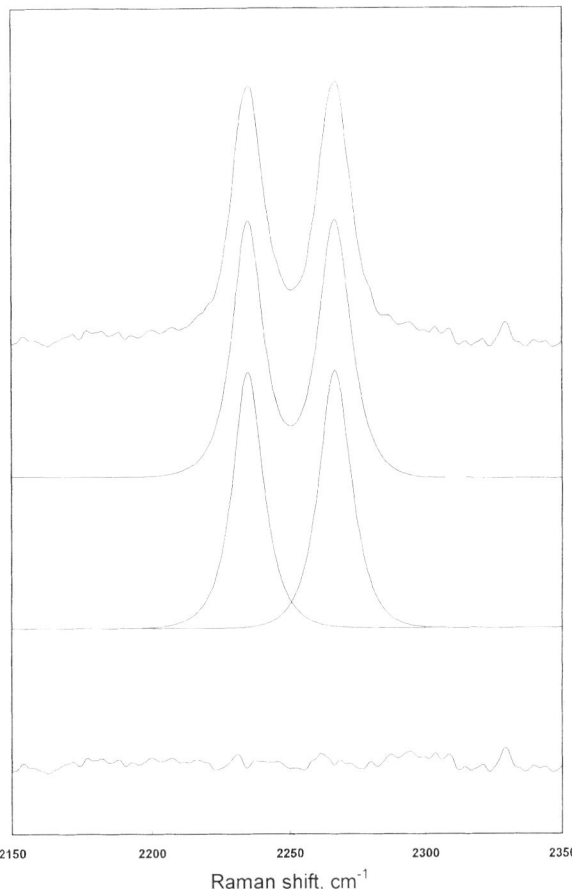

Fig. 3. Deconvoluted FT-Raman spectrum of acrylonitrile in 75% (w/w) cobalt nitrate aqueous solution; from top to bottom: experimental spectrum, calculated spectrum, resolved spectrum and residuals

rile is $+32\,cm^{-1}$, lower than the value corresponding to similar solutions with nickel ions ($\Delta\tilde{v}=40\,cm^{-1}$) [14]. The band at $2236\,cm^{-1}$ is likewise shifted by $+6\,cm^{-1}$ from the value obtained in pure acrylonitrile. This fact has already been reported [14, 19–21] and is ascribed to the breaking of the stabilizing dimers which nitriles form in their liquid state, the association of which would be disrupted by their dissolution in water or other solvents.

Figure 2 shows the FT-Raman spectra of the six solutions studied in the $\nu(C\equiv N)$ region, at $25\,°C$. It is evident that an increase in the intensity of the component at higher wavenumber occurs when the salt concentration is increased. As in other cases [14, 22, 23] the band positions remain unchanged with changes in the salt concentration. However, this does not mean, necessarily, that the stoichiometry of the association complex is retained over all the concentration range, as we have shown results elsewhere [16].

In order to quantify the solvation of the cobalt ions in the acrylonitrile/water mixture, a spectral deconvolution and subsequent component fitting of the bands was performed (Fig. 3). The numerical results are given in Table 2.

The fraction of area that corresponds to the associated acrylonitrile increases when the cobalt ion/acrylonitrile molar ratio is increased (Fig. 4), and this may be noted on inspection of the spectra (Fig. 2). For quantitative calculations of the association equilibrium, these area fractions must be transformed into molar concentrations, which requires information about the molar scattering factors of the free and associated acrylonitrile species. Although in the case of the system Ni(II)/acrylonitrile/water, it was implicitly assumed that these scattering factors were equal [14], we have shown that they are usually different [16, 24]. To approach the problem, the evolution of the band's integrated intensity with the salt concentration must be

Table 2. Band-fitting results for the $v(C\equiv N)$ bands at several concentrations of $Co(NO_3)_2$ in acrylonitrile/water mixtures; positions and half-widths in cm^{-1}

% w/w	Position		Half-width		Gauss (%)		Area (%)	
	Free	Bound	Free	Bound	Free	Bound	Free	Bound
60	2235.4	2267.7	5.9	7.0	7.7	13.4	56.9	43.1
65	2235.4	2267.2	6.0	7.1	8.2	13.2	52.2	47.8
70	2235.3	2267.1	6.3	6.9	7.9	13.6	49.1	50.9
75	2235.1	2267.1	6.7	7.6	8.1	13.0	45.7	54.3
80	2235.2	2267.1	6.9	7.2	8.0	13.8	40.5	59.5
85	2235.2	2267.1	7.0	7.6	7.8	13.7	37.4	62.6

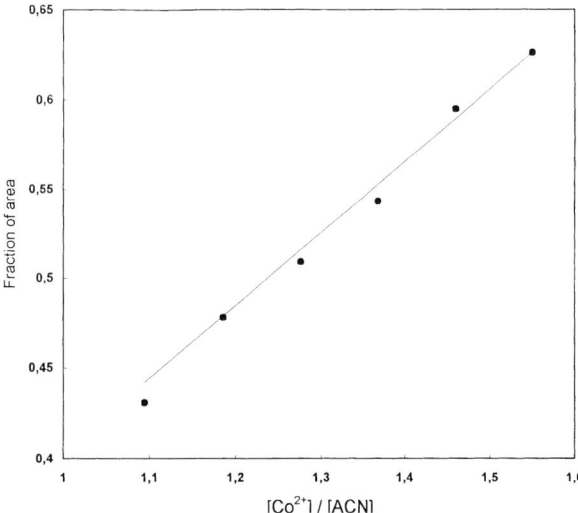

Fig. 4. Fraction of area corresponding to associated species in the $v(C\equiv N)$ envelope vs the molar ratio $[Co^{2+}]/$acrylonitrile

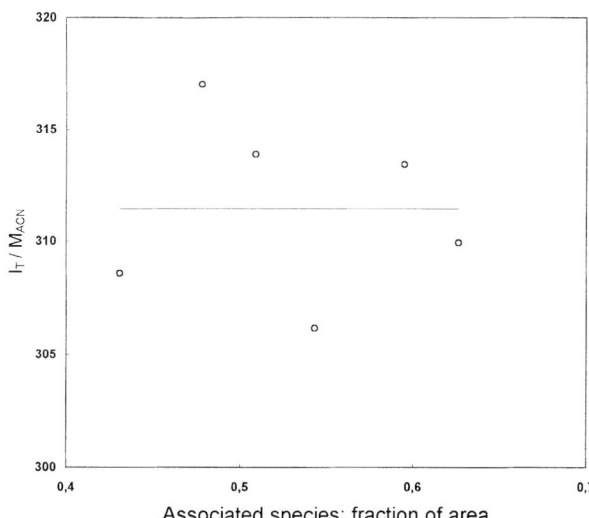

Fig. 5. Plot of the quotient between the total intensity of the envelope and the molar concentration of acrylonitrile vs the fraction of area of the associated species; the fitted line has a zero slope

studied. In the present case, the solvent (*i.e.* the major component) is the water and not the acrylonitrile; moreover, its molar concentration is not constant (see Table 1), which precludes the treatment proposed by Irish [25, 26]. However, simple data-processing provides some useful information. For the free species, if χ_f is the fraction of spectral area (integrated intensity), C_f the molar concentration and J_f the molar scattering factor; similarly if χ_b, C_b and J_b represent the same parameters for the bound species, then the total integrated intensity of the spectral band envelope, I_T, can be written as:

$$I_T = C_b J_b + C_f J_f$$

but

$$C_b + C_f = M$$

where M is the total (molar) concentration of acrylonitrile, and

$$\chi_f = 1 - \chi_b$$

Then

$$I_T = M\chi_b J_b + M(1 - \chi_b)J_f$$

which can also be written as:

$$I_T = M[J_f + \chi_b(J_b - J_f)]$$

The plot of I_T/M against χ_b will give different representations depending on the relation between J_b and J_f. If the scattering factors are equal, the plot will be a line parallel to the x-axis (slope = 0). Otherwise, if the slope of the line is greater than 0 it means that $J_b > J_f$; a negative slope will be obtained if $J_b < J_f$. Such a plot is given in Fig. 5. The constancy of the ratio I_T/M is evident, which means that the molar scattering factors of the free and associated species are equal. The fractions of the spectral band area can then be interpreted as molar fractions.

Table 3. Concentrations and quotients in the equilibrium; complex $Co(H_2O)_5ACN^{2+}$; $[Co^{2+}]_{hyd}$ $Co(H_2O)_6^{2+}$; ACN: acrylonitrile

Solution % w/w	χ_b	Complex M	Free ACN M	$[Co^{2+}]_{hyd}$ M	Quotient l/mol
60	0.431	0.966	1.275	1.486	0.510
65	0.478	1.100	1.202	1.629	0.562
70	0.509	1.205	1.162	1.817	0.571
75	0.543	1.322	1.113	2.009	0.592
80	0.595	1.492	1.016	2.167	0.678
85	0.626	1.618	0.967	2.389	0.701

As in the case of the association between acrylonitrile and Ni(II) ions in aqueous solution [14], the following equilibrium is proposed (where ACN = acrylonitrile):

$$[Co(H_2O)_6]^{2+} + ACN \Leftrightarrow [Co(H_2O)_5ACN]^{2+} + H_2O$$

from which the corresponding concentration quotient, assuming the water molar concentration is constant, is

$$Q_M = \frac{[Co(H_2O)_5ACN^{2+}]}{[ACN][Co(H_2O)_6^{2+}]}$$

where the different concentrations are calculated as follows:

$$[Co(H_2O)_5 ACN^{2+}] = \chi_b M_{ACN}(C_{comp})$$
$$[Co(H_2O)_6^{2+}] = M_{Co} - C_{comp}$$
$$[ACN] = \chi_f M_{ACN}$$

M_{ACN} and M_{Co} are the analytical (initial) molar concen-trations of acrylonitrile and cobalt nitrate, respectively (see Table 1). Table 3 gives the concentrations of the different species and the concentration quotients.

These quotients are not true equilibrium constants, because the activity coefficients of the acrylonitrile species (free and solvating) are unknown. Nevertheless, an estimation of the association constant can be obtained by extrapolation to *infinite dilution*. Figure 6 shows the plot of the concentration quotient's natural logarithm *vs* the molar concentration of the cobalt ion. With a good linear fit ($r^2 = 0.9673$), the ordinate at $M = 0$ results in an equilibrium constant value of $0.315 \pm 0.024 \text{ l/mol}$; this equilibrium constant can be compared with that corresponding to the analogous equilibrium system with nickel instead of cobalt. By using the data reported in [14], an equilibrium constant of $0.221 \pm 0.014 \text{ l/mol}$ can be calculated. This significant difference can be explained in terms of the hydration energy; literature values of ΔH_{hydr} are -503.3 kcal/mol for nickel ions and -491.1 kcal/mol for cobalt ions [27]. Hence, nickel ions are more strongly hydrated than cobalt ions, and the replacement of one water molecule in a nickel complex is more difficult, even if the acrylonitrile-complex formation is an exothermic process. It would be of interest to study the thermodynamics of the equilibrium process; in view of the present results, a greater enthalpy of association in the case of cobalt ion compared with nickel would be expected.

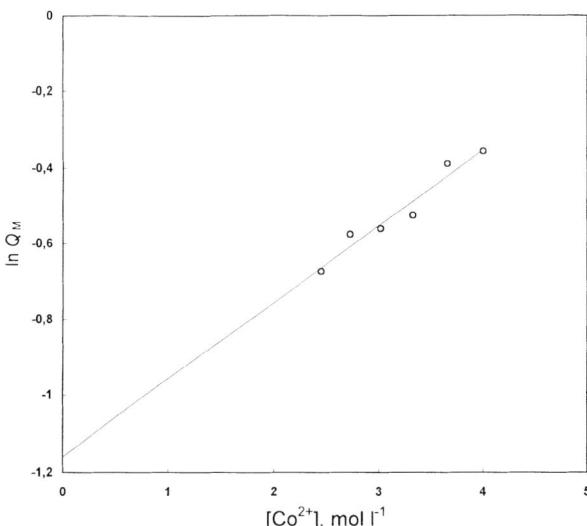

Fig. 6. Natural logarithm of the concentration quotient *vs* the molar concentration of the cobalt ion

References

[1] C. Ramakrishna, J. D. Desai, *Biotechnol. Lett.* **1993**, *15*, 267.
[2] C. Ramakrishna, J. D. Desai, *Biotechnol. Lett.* **1992**, *14*, 827.
[3] T. Czerniawski, Z. Wojtczak, *Acta Polym.* **1991**, *42*, 277.
[4] T. Czerniawski, Z. Wojtczak, *Acta Polym.* **1984**, *35*, 443.
[5] N. Kalyanam, V. G. Gandhi, S. Sivaram, *Polymer Bull. (Berlin)* **1984**, *11*, 105.
[6] V. G. Gandhi, S. Sivaram, I. S. Bhardwaj, *J. Macromol. Sci. Chem.* **1983**, *A19*, 147.

[7] F. Pesa, T. Haase, *J. Molec. Catal.* **1983**, *18*, 237.

[8] R. A. Dubois, P. E. Garrou, *J. Organometal. Chem.* **1983**, *241*, 69.

[9] J. D. Druliner, R. C. Blackstone, *J. Organometal. Chem.* **1982**, *240*, 277.

[10] J. D. Druliner, R. C. Blackstone, *Abst. Papers Am. Chem. Soc.* **1980**, *180*, 251-INOR.

[11] Y. Wakatsuki, H. Yamazaki, *J. Chem. Soc. Chem. Commun.* **1980**, *24*, 1270.

[12] S. J. Bryan, P. G. Huggett, K. Wade, J. A. Daniels, J. R. Jennings, *Coord. Chem. Rev.* **1982**, *44*, 149.

[13] E. Kamienska-Piotrowicz, J. Stangret, *J. Molec. Struct.* **1995**, *344*, 69.

[14] J. M. Alía, H. G. M. Edwards, *J. Molec. Struct.* **1995**, *354*, 97.

[15] F. Rull, F. Sobrón, *J. Raman Spectrosc.* **1994**, *25*, 693.

[16] J. M. Alía, H. G. M. Edwards, J. Moore, *J. Raman Spectrosc.* **1995**, *26*, 715.

[17] D. E. Irish, in: *Ionic Interactions. From Dilute Solutions to Fused Salts* (S. Petrucci, ed.), Academic Press, New York, 1971, p. 230.

[18] D. W. James, M. T. Carrick, R. L. Frost, *J. Raman Spectrosc.* **1982**, *13*, 115.

[19] B. H. Thomas, W. J. Orville-Thomas, *J. Molec. Struct.* **1969**, *3*, 191.

[20] S. Singh, P. J. Krueger, *J. Raman Spectrosc.* **1982**, *13*, 178.

[21] D. Jamroz, J. Stangret, J. Lindgren, *J. Am. Chem. Soc.* **1993**, *115*, 6165.

[22] A. R. Hoskins, H. G. M. Edwards, A. F. Johnson, *J. Molec. Struct.* **1991**, *763*, 1.

[23] H. G. M. Edwards, A. R. Hoskins, A. F. Johnson, I. R. Lewis, *J. Raman Spectrosc.* **1991**, *22*, 651.

[24] J. M. Alía, H. G. M. Edwards, J. Moore, *Spectrochim. Acta* **1995**, *41A*, 2039.

[25] Z. Deng, D. E. Irish, *J. Chem. Soc., Faraday Trans.* **1992**, *88*, 2891.

[26] Z. Deng, D. E. Irish, *Can. J. Chem.* **1991**, *69*, 1766.

[27] J. Burgess, *Metal Ions in Solution*, Ellis Horwood, Chichester, 1990, p. 183.

Mikrochim. Acta [Suppl.] 14, 739–740 (1997)

Biomolecular Investigations Using FT-Raman Spectroscopy

Prashant S. Bhandare[1,*], **Robert Jakobsen**[2], and **K. Krishnan**[1]

[1] Bio-Rad Laboratories, Digilab Division, 237 Putnam Avenue, Cambridge, MA 02139, U.S.A.
[2] IR-ACTS, 4892 N. High Street, Columbus, OH, U.S.A.

Abstract. This paper discusses the application of FT-Raman spectrometry for investigating different biologically important molecules such as amino acids and proteins in both the solid and native states. The utility of FT-Raman spectroscopy for examining structural changes in biomolecules, resulting from environmental changes such as pH, temperature and chemical reactions is demonstrated.

Key words: FT-Raman spectrometry, biomolecules, structural changes.

In comparison with dispersive Raman spectroscopy, FT-Raman spectroscopy has significant advantages such as increased throughput, wavelength accuracy, and higher speed of analysis [1]. Therefore, FT-Raman spectroscopy is becoming a general purpose analytical tool. Compared with IR spectra, Raman spectra are less affected by the presence of water in samples. This is particularly advantageous for biological specimens since it allows studies of biological molecules in their native (aqueous) state [2, 3]. The main objectives of this study were to (i) demonstrate the feasibility of obtaining quality FT-Raman spectra of biological molecules in the solid as well as aqueous states and (ii) investigate spectral change resulting from their structural variations.

Experimental

A Bio-Rad FT-Raman spectrometer equipped with an Nd-YAG laser (excitation wavelength 1064 nm), super holographic notch filters for Rayleigh line rejection and a germanium detector was used for these studies. Capillary tubes were used for sampling solids whereas for liquids half-silvered bulbs were used. Typically, laser excitation power at the sample ranged between 500 and 900 mW. The spectra were obtained at a resolution of 4 cm^{-1} and 500–1000 scans were co-added.

* To whom correspondence should be addressed

Results and Discussion

L-Amino acids from the basic building blocks of proteins. Differences in the side-chains of different amino acids are reflected in differences in their FT-Raman spectra. Figure 1 shows spectra of glycine, phenylalanine, and tryptophan. The strong peaks in the phenylalanine (e.g. 1004 cm^{-1}) and tryptophan (e.g. 1363, 863 cm^{-1}) spectra are also observed in the spectra of proteins. Figure 2 shows spectra of glycine in aqueous solutions at different pH. Addition of glycine to water results in a solution having a pH of 6, with the glycine in the form of the zwitterion. Decreasing the pH of the solution causes protonation of the glycine, resulting in glycine cations. Increasing the pH, on the other hand, results in formation of glycinate anions. These molecular changes cause shifts in the C–C stretching band at around 895 cm^{-1}. Figure 3 shows the spectra of solid bovine serum albumin and chymotrypsinogen. Albumin (55% α-helical structure, the remainder random coil) and chymotrypsinogen (49% β-pleated sheet, 12% α-helical and remainder random

Bio-Rad Win-IR

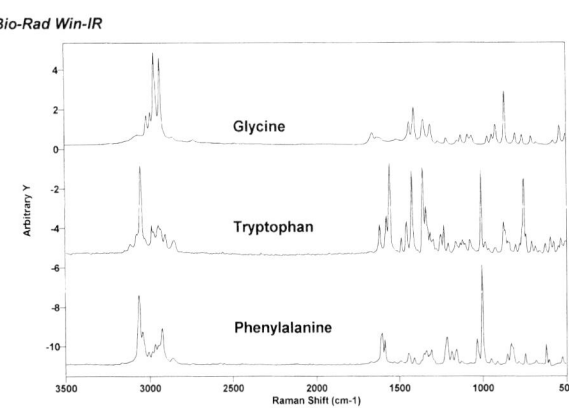

Fig. 1. FT-Raman spectra of some amino acids

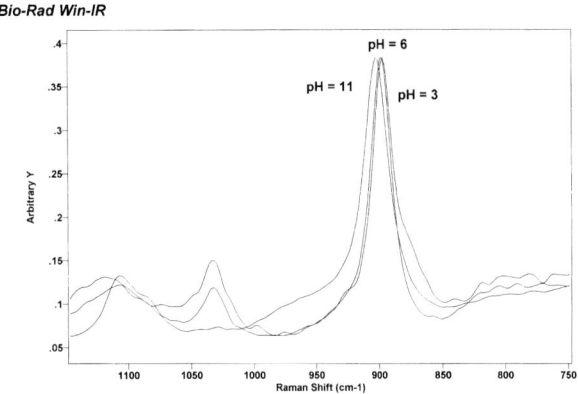

Bio-Rad Win-IR

Fig. 2. FT-Raman spectra of aqueous solutions of glycine

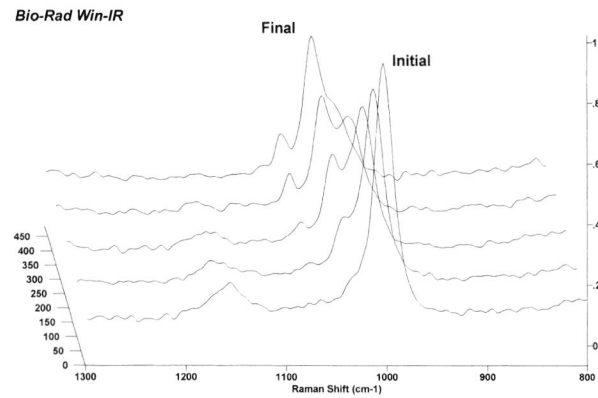

Bio-Rad Win-IR

Fig. 5. Kinetic monitoring of reaction between urea and urease

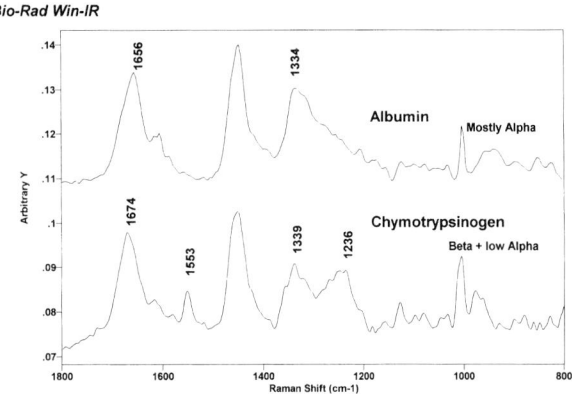

Bio-Rad Win-IR

Fig. 3. Spectra of proteins with different secondary structures

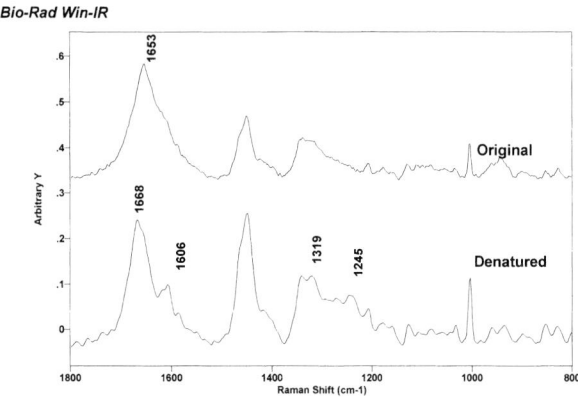

Bio-Rad Win-IR

Fig. 4. Heat denaturation of aqueous albumin solution

coil structure) show significant differences in their FT-Raman spectra in the Amide I and Amide III regions [4]. Similar differences have been observed in the FT-Raman spectra of other proteins differing in peptide backbone conformation.

Figure 4 shows spectra of an aqueous albumin solution before and after heating to 90 °C. Significant conformational changes can be seen, as the result of temperature differences. Heat denaturation of albumin appears to transform the protein from an α-helical structure to random coil and β-pleated sheet. This is evidenced by the shift of Amide I band from $1653 \, \text{cm}^{-1}$ to $1668 \, \text{cm}^{-1}$.

Figure 5 shows the feasibility of monitoring enzymatic reactions using by FT-Raman spectrometry. As the reaction between urea and urease progresses, C–N bonds in the urea molecules are broken, as demonstrated by weakening of the $1004 \, \text{cm}^{-1}$ band. Side-reactions of the products with urease probably result in bands at 1034 and $1066 \, \text{cm}^{-1}$.

References

[1] T. Hirschfeld, B. Chase, *Appl. Spectrosc.* **1986**, *40*, 133.
[2] I. W. Levin, E. N. Lewis, *Anal. Chem.* **1990**, *62*, 1101A.
[3] Y. Ozaki, A. Mizuno, H. Sato, K. Kawauchi, S. Muraishi, *Appl. Spectrosc.* **1992**, *46*, 533.
[4] P. R. Carey, *Biochemical Applications of Raman and Resonance Raman Spectroscopies*, Academic Press, New York, 1982.

Mikrochim. Acta [Suppl.] 14, 741–743 (1997)
© Springer-Verlag 1997

FT-Raman Spectroscopy of Silvered Wool Fibres: A SERS Effect?

Elizabeth A. Carter and **Peter M. Fredericks***

Centre for Instrumental and Developmental Chemistry, Queensland University of Technology, P.O. Box 2434, Brisbane, Q4001, Australia

Abstract. FT-Raman spectra have been obtained of wool and nylon fibres coated with a thin layer of silver. The spectra appear to be SER spectra and, in the case of wool, are consistent with a surface having increased lipids and increased β-sheet character for the proteins, compared with the bulk of the fibre.

Key words: FT-Raman, wool, surface-enhanced Raman scattering.

Woollen fibres and fabrics undergo a commercial process known as shrinkproofing to prevent felting shrinkage, which results when the scale-shaped cuticle cells on the fibre surface progressively entangle during laundering or milling [1, 2]. Shrinkproofing is a surface-specific process consisting of several stages: chlorination, neutralization, and polymer coating. ATR spectroscopy has been the technique of most use for studying the shrinkproofing process [3].

Recently, FT-Raman spectroscopy has been applied to the characterization of wool [4, 5]. High-quality spectra are produced with little fluorescence background, but it has been found that FT-Raman is a bulk technique and cannot distinguish samples from different stages of the shrinkproofing process.

This paper describes work towards a surface-enhanced Raman (SER) technique with the potential to increase the surface specificity of FT-Raman spectroscopy of wool.

Experimental

Wool Samples

Samples were obtained from a commercial continuous treatment plant utilizing Australian Merino wool top of 22 μm average fibre diameter and 71–73 mm mean fibre length. Samples were obtained before and after the chlorination step, and after neutralization.

---

* To whom correspondence should be addressed

Silvered Wool and Nylon

Wool or nylon-6,6 samples were mounted on glass slides and thin silver films (2.5–20 nm mass equivalent) were deposited on the fibre at room temperature by evaporation under vacuum from pure silver wire (99.99%, Johnston–Matthey).

Raman Spectra

These were collected on a Perkin–Elmer System 2000 FT-Raman spectrometer equipped with a Spectron Laser Systems SL301 Nd:YAG laser emitting at a wavelength of 1064 nm. 180° sampling geometry was employed and spectra were recorded at a resolution of 8 cm^{-1}, a mirror velocity of 0.1 cm/s, and a laser power of 200 mW.

Results and Discussion

Previous studies by Lippert et al. [6] have shown that Raman spectra of PET samples coated with silver island films give rise to a SER spectrum of the surface in contact with the silver. This SER spectrum is quite different from the spectrum of the bulk PET. This experiment seemed to have application to Raman studies of wool fibre surfaces.

Nylon Fibres

Given the complexity of wool, we decided to use nylon as a model system. The FT-Raman spectrum of silver-coated (20 nm mass equivalent) nylon fibres is shown in Fig. 1, together with the spectrum of the uncoated nylon. The spectra are broadly similar but do show some enhanced bands and shifts of around 2 cm^{-1} for several of the most intense bands. One band near 1460 cm^{-1} is much less intense for the silver-coated sample. On the basis of the PET studies mentioned above [6], the spectrum of the silver-coated nylon should represent a SER spectrum of the nylon surface

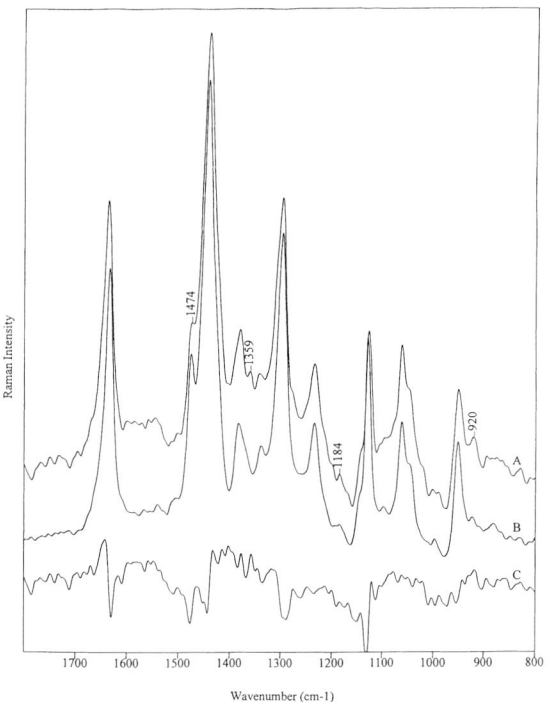

Fig. 1. FT-Raman spectra of *A*: nylon, silvered, *B*: nylon, unsilvered, *C*: difference spectrum (*A* minus *B*)

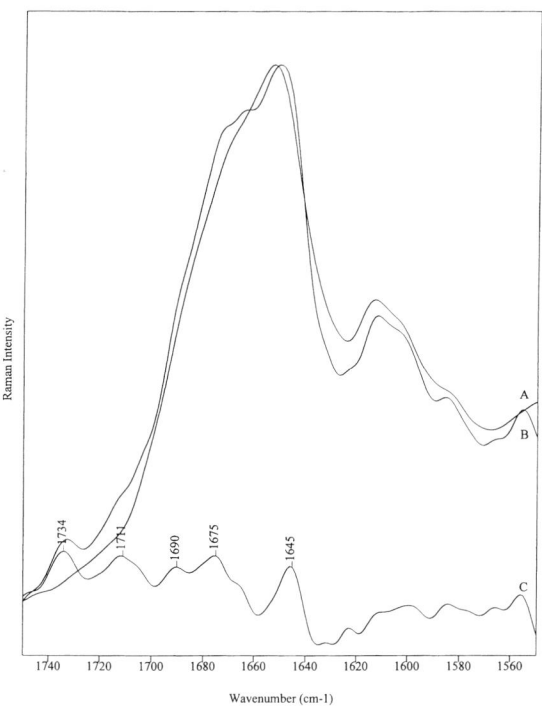

Fig. 2. FT-Raman spectra of *A*: chlorinated, silvered wool, *B*: chlorinated, unsilvered wool, *C*: difference spectrum (*a* minus *b*)

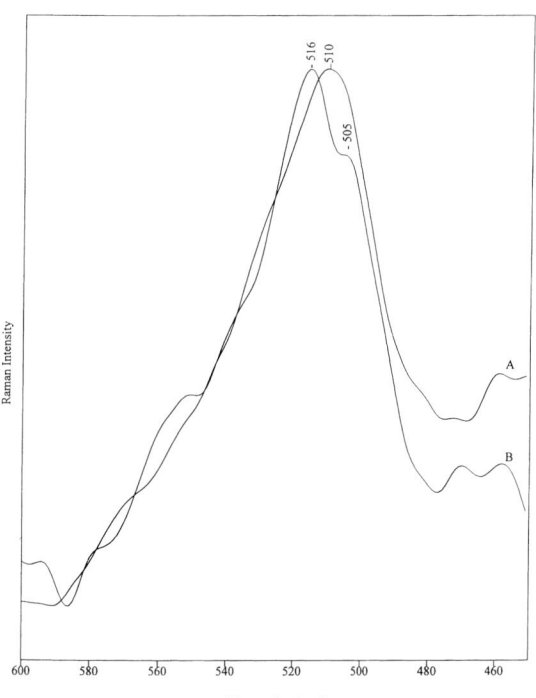

Fig. 3. FT-Raman spectra of *A*: chlorinated, silvered wool, *B*: chlorinated, unsilvered wool

(which would be expected to have similar molecular structure to the bulk nylon).

Wool Fibres

Spectra of the Amide I region of the wool spectrum for silver-coated chlorinated wool (10 nm mass equivalent) and uncoated chlorinated wool are shown in Fig. 2. It can be seen that for the silver-coated sample there has been a shift of part of the Amide I band to higher wavenumber, and a weak band at $1734 \, \text{cm}^{-1}$ has appeared. This spectral change is typical of all the silver-coated wool samples studied. If this is an SER spectrum of the surface, then it is consistent with what is known about the wool surface: that it contains a lipid layer, which might give rise to the C=O band at $1734 \, \text{cm}^{-1}$, and that the surface proteins have more β-sheet character than the bulk proteins, which have more α-helix character [7, 8].

Spectral differences can also be observed in other regions of the spectrum, especially at $700–400 \, \text{cm}^{-1}$, where vibrations due to S–S bonds occur in the FT-Raman spectrum of wool. For silvered wool, the position of the band at $510 \, \text{cm}^{-1}$ has shifted to $516 \, \text{cm}^{-1}$ and a shoulder has appeared near $505 \, \text{cm}^{-1}$ (Fig. 3).

These spectra are difficult to interpret because, if they are due to a SER effect, then spectral differences may be assigned either to a difference in the nature of the surface, or differences caused by the SER effect itself, or both. SER spectra are most different where there is strong interaction between the metal and the analyte compound. As this is possible with protein samples, the two effects of surface variation and SER spectral variation may prove difficult to separate.

Acknowledgements. A research grant from the Australian Wool Research and Promotion Organisation is gratefully acknowledged. Thanks are also due to J. Church and R. Denning, CSIRO, Division of Wool Technology, Geelong.

References

[1] J. A. Maclaren, B. Milligan, *Wool. Science: The Chemical Reactivity of the Wool Fibre*, Science Press, Sydney, 1981.

[2] D. M. Lewis (ed.), *Wool Dyeing*, Society of Dyers and Colourists, Bradford, 1992.

[3] U. Schumacher-Hamedat, C. Laurini, V. Schneider, in: *Proc. 7th Wool Textile Research Conf., Tokyo* **1985**, *VI*, 120.

[4] L. J. Hogg, H. G. M. Edwards, D. W. Farwell, A. T. Peters, *J. Soc. Dyers Colour.* **1994**, *110*, 196.

[5] E. A. Carter, P. M. Fredericks, J. S. Church, R. J. Denning, *Spectrochim. Acta* **1994**, *50A*, 1927.

[6] Th. Lippert, F. Zimmermann, A. Wokaun, *Appl. Spectrosc.* **1993**, *47*, 1931.

[7] H. Zahn, H. Messinger, H. Höcker, *Textile Res. J.* **1994**, *64*, 554.

[8] A. P. Negri, H. J. Cornell, D. E. Rivett, *Textile Res. J.* **1993**, *63*, 109.

Mikrochim. Acta [Suppl.] 14, 745–746 (1997)

Surface-enhanced Raman Spectroscopy as a Technique for Drug Analysis

E. Horváth[1,*], **J. Mink**[2], and **J. Kristóf**[2]

[1] Research Group for Analytical Chemistry, Hungarian Academy of Sciences, P.O. Box 158, H-8201 Veszprém, Hungary
[2] Department of Analytical Chemistry, University of Veszprém, P.O. Box 158, H-8201 Veszprém, Hungary

Abstract. *In situ* surface-enhanced FT-Raman (SERS) detection of heroin, codeine and cocaine samples after separation on a Kieselgel 60 type chromatographic thin layer has been performed. With a silver coating applied onto the TLC spots by vacuum evaporation a detection limit of about $0.2\,\mu g/mm^2$ could be attained. Due to the low matrix effect of the silicate-based stationary phase, SERS detection of drugs can be used as a quick and reliable identification method.

Key words: surface-enhanced Raman spectroscopy, thin layer chromatography, drug analysis.

Separation methods based solely on relative retention and traditional detection techniques often face the problem of similar chromatographic behaviour. As a consequence, attention has been paid to the application of so-called hyphenated techniques.

Recently attempts have been successfully made for the *in situ* detection of thin layer chromatographic spots by the FT-IR [1–3] and surface-enhanced Raman spectroscopic (SERS) techniques [4–7]. According to our experience [8] FT-IR analysis of TLC spots of real drugs can be done quickly and reliably in spite of the fact that the matrix effect of the chromatographic stationary phase often spoils qualitative analysis. In our judgement, surface-enhanced Raman spectroscopic detection after chromatographic separation offers reliable analysis where the eluent and/or chromatographic stationary phase will give a strong matrix effect.

In spite of the fact that the use of FT-Raman spectroscopy in TLC analysis is rather limited, owing to the spectral properties of the analyte and to the relatively high amount of sample, the *in situ* SERS analysis of the TLC spots of heroin, codeine and cocaine can be done with certainty.

---

* To whom correspondence should be addressed

Experimental

Thin Layer Chromatographic Sample Preparation

Portions (15 μl) of ethanolic heroin, codeine and cocaine solutions of different concentrations were applied onto chromatographic thin layers of the type Kieselgel 60 (Merck, 0.2 mm thickness). The chromatograms were developed with an eluent mixture of ethyl acetate–methanol–ammonium hydroxide (25%) in the ratio 10:45:45 (v/v). The developed plates were then air-dried and stored in a desiccator.

Surface-enhanced FT-Raman Measurements

After separation the chromatographic thin layers were coated with silver in a vacuum evaporator (type SCD 020, Leica Inc.). It was found that the optimum thickness of the coating can be obtained in 1.5 sec. Treating the layer with a silver sol gave a spectrum of lower quality. The SERS effect develops in 5–7 min, and lasts for about 35 min. During this period 600–800 interferograms were co-added at a resolution of $4\,cm^{-1}$ at low mirror velocity (2.5 kHz). Measurements were performed by means of a Bio-Rad Digilab FT-Raman Spectrometer (Nd:YAG laser, 560 mW laser power).

Results and Discussion

Surface-enhanced Raman spectra of heroin, codeine and cocaine are shown in Figs. 1 and 2. The normal spectra are denoted by B, D (Fig. 1) and C (Fig. 2). The detection limits for heroin, codeine and cocaine were 19.8, 22.9 and 27.3 μg on a $1\,cm^2$ TLC spot, respectively. It means that under in situ conditions a detection limit (on the spot) of $0.2\,\mu g/mm^2$ can be obtained. This quantity corresponds to the normal load of thin layers used for analytical purposes. As a comparison, a similar sensitivity requires the dissolution of 100 μg of analyte followed by its deposition on a gold mirror. Thus, with the ex situ technique a sample quantity of $10\,\mu g/mm^2$ is required to obtain a spectrum of similar quality. The quality of SERS spectra depends on the

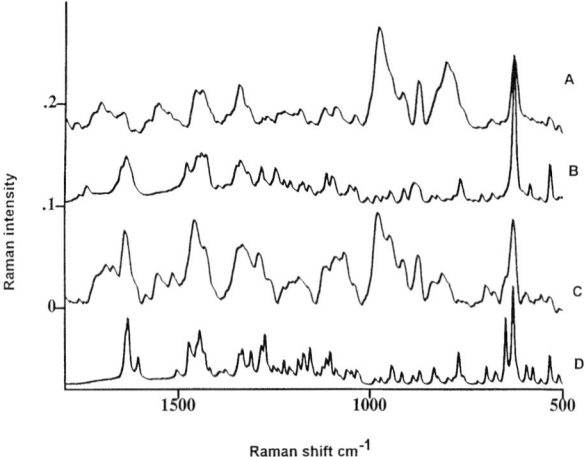

Fig. 1. FT-Raman spectra of heroin and codeine: *A* SERS spectrum of 0.2 μg/mm² heroin; *B* standard heroin spectrum (powder on gold mirror); *C* SERS spectrum of 0.2 μg/mm² codeine; *D* standard codeine spectrum (powder on gold mirror)

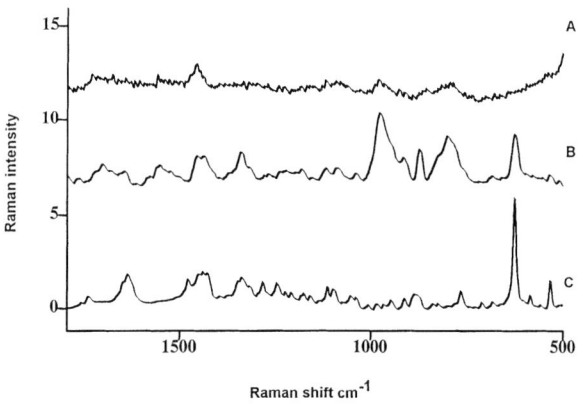

Fig. 2. FT-Raman spectra of cocaine: *A* silver-coated Kieselgel 60 reference spectrum; *B* SERS spectrum of 0.2 μg/mm² cocaine; *C* standard cocaine spectrum (powder on gold mirror)

method of sample preparation. It should be noted that the quality of the chromatographic thin layer and the composition of the eluent used can also influence the spectrum quality. If SERS detection is used after separation on a reversed phase, the strong matrix effect arising from the apolarity of the phase should be taken into account. The use of thin layers with cumulation zones as well as volatile and aqueous eluents is more advantageous, since the local concentration can be increased and disturbing effects originating from the possible adsorption of the eluent on the layer can be eliminated. It is interesting to note that hydrogen bonds play a role in the phase–analyte interaction. It is indicated by the shift of the C=O stretching frequency from 1736 to 1706 cm^{-1} and from 1721 to 1712 cm^{-1} for heroin and cocaine, respectively.

Conclusion

The results obtained for different drug samples show that the SERS technique – as a quick, reliable and *in situ* detection method – can be advantageously used for the analysis of real drugs.

References

[1] M. P. Fuller, P. R. Griffiths, *Anal. Chem.* **1978**, *50*, 1906.
[2] M. P. Fuller, P. R. Griffiths, *Appl. Spectrosc.* **1980**, *34*, 533.
[3] K. H. Shafer, P. R. Griffiths, Wang Shu-Qin, *Anal. Chem.* **1986**, *58*, 2708.
[4] C. D. Tran, *Anal. Chem.* **1984**, *56*, 824.
[5] C. D. Tran, *J. Chromatogr.* **1984**, *292*, 432.
[6] J.-M. L. Séquaris, E. Koglin, *Anal. Chem.* **1987**, *59*, 525.
[7] E. Koglin, *J. Planar Chromatogr. Mod. TLC* **1990**, *3*, 117.
[8] J. Mink, E. Horváth, J. Kristóf, T. Gál, T. Veress, *Mikrochim. Acta* **1995**, *119*, 129.

Mikrochim. Acta [Suppl.] 14, 747–749 (1997)
© Springer-Verlag 1997

Fourier-Transform Raman Spectroscopy of Kaolinite

Ray L. Frost*, **Thu Ha Tran**, and **Tri Le**

Centre for Instrumental and Developmental Chemistry, Queensland University of Technology, 2 George St., Brisbane, GPO Box 2434, Q4001, Australia

Abstract. The vibrational modes of clay minerals are uniquely accessible to FT Raman spectroscopy. Raman spectra in the 50–3800 cm^{-1} region were obtained for a suite of kaolinite clay minerals. The kaolinite clay minerals are characterized by intense bands centred at 142.7 cm^{-1} with a prominent shoulder at 127 cm^{-1} and are attributed to the O–Al–O and O–Si–O symmetric bends. Differences in the lattice modes for the kaolinite clay minerals in the 200–1200 cm^{-1} region were obtained. Five hydroxyl bands were obtained for kaolinite at 3619, 3650, 3667, 3686, and 3695 cm^{-1}; four OH bands were found for dickites at 3621, 3639, 3652 and 3703 cm^{-1}.

Key words: FT Raman spectroscopy, kaolinite, halloysite, dickite.

The Raman spectroscopy of clay minerals has received little attention [1–3] More recently, the advent of FT Raman spectroscopy [4, 5] which uses near infrared radiation as the excitation source and a Fourier-transform spectrometer enables the study of clay mineral structure to be further investigated. The use of FT Raman spectroscopy offers the advantages of reduced fluorescence, improved signal to noise ratio by co-adding of scans, and reduced sample degradation owing to the longer wavelength of light used.

Experimental

Raman spectra were obtained with a sample of the raw mineral directly in the incident beam by use of a Bio-Rad series 2 FT-IR spectrometer equipped with a Raman accessory comprising a Spectraphysics T10-1064C Nd-YAG diode laser operating at 1064 nm. Measurement times of between 0.5 and 2 hours were used to collect the Raman spectra, with a signal to noise ratio of better than 100 at a resolution of 4 cm^{-1}.

Results and Discussion

The kaolinite clay minerals are characterized by very intense bands centred at 142.7 cm^{-1} for kaolinite,

143 cm^{-1} for halloysite and 131.2 cm^{-1} for dickite. FT Raman bands were also found at 129, 127, and 120 cm^{-1} respectively for the three clays. San Juanito dickite shows complexity with bands at 143, 130 and 120 cm^{-1}. One possible assignment of the bands in this region is that the higher frequency band (143 cm^{-1}) is for the O–Al–O symmetric bend and the lower frequency band (127 cm^{-1}) is that of the O–Si–O symmetric bend. In any event the very large intensity of these bands can only occur as a result of a vibrational mode which induces a very large change in polarizability. It is likely that the variation in the peak position of a particular clay is related to the stress on the kaolinite crystal structure and as a consequence is sample-dependent. The high frequency of KCa-2 at 143 cm^{-1} is indicative of this tension compared with that of the other kaolinites and this may be related to layer stacking.

The 200–1000 cm^{-1} spectral region can be used to fingerprint clay minerals [6]. Halloysite is characterized by the intense 459 cm^{-1} band. This band is split into two bands for dickite (432 and 459 cm^{-1}) and consists of four bands for kaolinite (395.4, 426, 469 and 509 cm^{-1}) and has been attributed to the Si–O bending vibration. The band positions are in good agreement with the spectra obtained by dispersive Raman spectroscopy [1, 3]. The multiplicity of peaks for kaolinite can be explained in terms of the reduction in symmetry from C_{6v} to C_s arising from the substantial deviations from hexagonal symmetry. After the 142 cm^{-1} band, the next most intense band for kaolinite is the 636.5 cm^{-1} band. This band has been attributed to an Si–O–Si stretch. The bands at 704, 752, and 785 for kaolinite, 703, 744.5 and 792.7 for halloysite and 744.5 and 794.7 cm^{-1} for dickite are attributed to the Al–OH vibrations of surface hydroxyl groups. The peak posi-

---

* To whom correspondence should be addressed

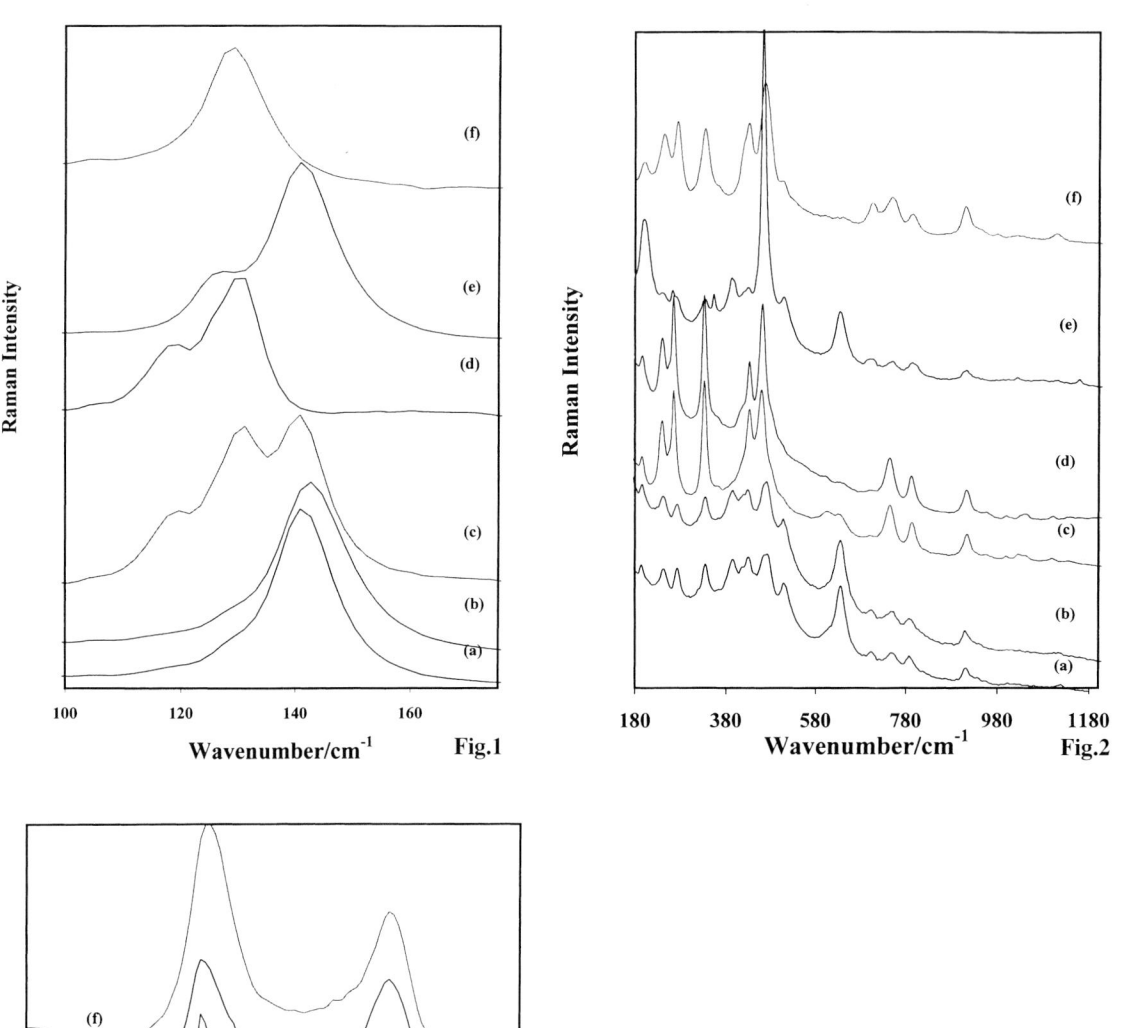

Fig. 1. FT-Raman spectra of the kaolinite clay minerals in the 100–200 cm^{-1} region. (*a*) Kaolinite (KGa-1), (*b*) kaolinite (KGa-2), (*c*) dickite (SJ), (*d*) dickite (St CL), (*e*) halloysite (Eureka), (*f*) halloysite (N. Z.)

Fig. 2. FT-Raman spectra of the kaolinite clays in the 200–1200 cm^{-1} region

Fig. 3. FT-Raman spectra of kaolinite clay mineral hydroxyl stretching region

tions at 914, 910 and 912 cm^{-1} for dickite, halloysite and kaolinite are attributed to OH librations. The bands observed at 1020, 1043 and 1110 cm^{-1} are attributed to the stretching Si–O vibrations.

The FT Raman spectra of kaolinite hydroxyl groups show better resolution, with the band profiles reaching baseline between the 3650 and 3700 cm^{-1} peaks. Low temperature studies of kaolinite show the presence of six bands, with the 3652 cm^{-1} band split into two components at 3650 and 3657 cm^{-1} and the 3621 cm^{-1} band exhibiting splitting at 3618 and 3628 cm^{-1} [3]. The FT-Raman spectra of kaolinites show five Raman peaks at room temperature and six at low temperature. The FT Raman peak is at 3620 cm^{-1} for the well crystalline kaolinites but at 3624 cm^{-1} for the less crystalline kaolinites. There is a similar frequency for dickites and ordered halloysites. Less crystalline halloysites show this peak at 3627 cm^{-1}. The dickite (Saint Claire) shows a complex spectrum in the 3620 cm^{-1} region. Similarly halloysite (Eureka) shows a similar although less well resolved pattern for this hydroxyl band. The two hydroxyl frequencies at 3652 and 3668 are in good agreement with published data [2]. The second Raman peak for dickite was at 3642 cm^{-1}, which corresponds well with the infrared data [6]. The band profile centred at 3695 cm^{-1} is complex and is structure-dependent. The Raman spectrum of kaolinite shows two bands at 3688 and 3695 cm^{-1} for KGa-1 and two bands at 3685 and 3692 cm^{-1} for KGa-2. This particular set of bands is at higher frequencies for dickite and halloysite and is centred at 3705 cm^{-1} and 3702 cm^{-1}. The 3705 cm^{-1} band is in good agreement with the infrared spectra.

The 3702 cm^{-1} band is at higher frequency in the Raman than in the infrared (3695 cm^{-1}) [6].

Conclusion

FT Raman spectroscopy with near infrared excitation has been shown to be very useful for the study of the vibrational spectrum of kaolinite clays. The technique has several unique advantages: (a) no sample treatment, as the spectra can be obtained from the solid or powdered samples; (b) low wavenumber vibrational bands which are inaccessible to mid-IR spectrometers are easily measured; and (c) vibrational bands which are not infrared-active are observable.

References

[1] A. Wiewiora, T. Wieckowski, A. Sokolowska, *Arch. Mineral.* **1979**, *35*, 5.

[2] C. T. Johnston, G. Sposito, R. R. Birge, *Clays Clay Min.* **1985**, *33*, 483.

[3] V. Pajcini, P. Dhamelincourt, *Appl. Spectrosc.* **1994**, *48*, 638.

[4] D. B. Chase, *J. Am. Chem. Soc.* **1986**, *108*, 7485.

[5] D. J. Cutler, *Spectrochim. Acta* **1990**, *166A*, 131.

[6] R. L. Frost, J. R. Bartlett, P. M. Fredericks, *Spectrochim. Acta* **1993**, *49A*, 667.

Mikrochim. Acta [Suppl.] 14, 751–754 (1997)

NIR FT-Raman Spectroscopy for Monitoring the Synthesis of Azo Dyes

David M. Lewis, Xiaoming Shen, and **Kelvin N. Tapley***

Department of Colour Chemistry, University of Leeds, Leeds, LS2 9JT, U.K.

Abstract. A circulating sampling arrangement was developed and employed for the transfer and return of an aqueous reaction mixture from a reaction vessel to an NIR FT-Raman spectrometer for on-line analysis. Spectroscopic studies from each stage of the synthesis of a reactive dye (C. I. Reactive Red 1) have been investigated and a selection of results is presented in this paper with particular interest focused on the formation of the diazonium ion ($-N_2^+$) and the azo linkage ($-N = N-$) which are very important chemical groups in dye chemistry. The 1, 3, 5-triazine ring system has been studied in some detail but only limited results are included in this paper.

Key words: FT-Raman, monitoring synthetic reactions, on-line analysis, azo dyes.

Many industrial processes, which are often potentially hazardous, involve a complex mixture of reactants, and/or require multiple steps, and whilst the major components are often amenable to analysis it is desirable to be able to monitor and preferably control the quantity of each constituent in a process. Ideally the analysis should be done on-line and in real time rather than some time later at some remote site where it may take various lengths of time to do the analysis, depending on the analytical technique employed. The technique of NIR FT-Raman spectroscopy offers the potential for on-line continuous monitoring of many chemical reactions, including those in aqueous solutions [1]. The analysis is done in real time (although the need for multiple scans, to increase the signal to noise ratio, may result in an analysis time of 2–3 minutes) and the Raman spectra obtained contain information regarding the chemical composition of the mixture analysed.

The status of a number of peaks, used for the identification of different compounds, such as products and starting materials, have been monitored during the multi-step synthesis of a reactive dye (C. I. Reactive Red 1).

Experimental

Instrumentation and Measurement

FT-Raman spectra were collected with a System 2000R FT-IR Spectrometer (Perkin–Elmer Limited) configured with an NIR FT-Raman accessory. Generally, each spectrum was obtained by averaging 100 scans (3 minutes) and was recorded by using an InGaAs detector at $8\,cm^{-1}$ resolution. The Raman laser was a vertically polarized Nd:YAG laser operating at 1064 nm and was employed at 150–350 mW laser power, depending on the nature and concentration of the sample. The system was utilized in a 180° back-scattering geometry.

The on-line analysis system was installed on a batch process reactor (see Fig. 1). The basic design for the cyclic system involved circulating the reaction mixture from the reactor to the sampling cell, contained in the spectrophotometer, and back again via a length (1.20 m) of insulated (to keep cool) tubing (internal diameter = 5 mm) by means of a peristaltic pump.

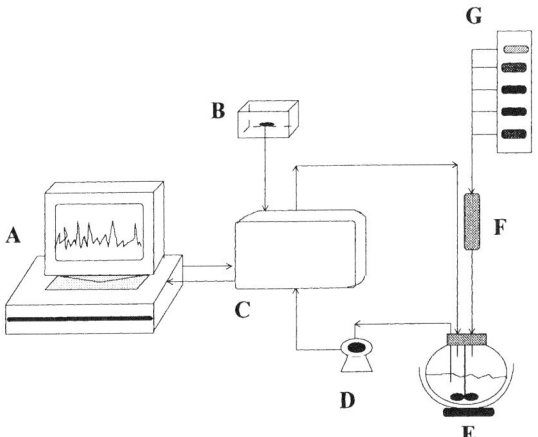

Fig. 1. Schematic diagram of the NIR-FT-Raman cyclic reactor sampling arrangement. *A* computer, *B* Nd:YAG laser, *C* FT-Raman spectrometer, *D* pump, *E* reactor, *F* reagent addition system, *G* reagent cabinet

---

* To whom correspondence should be addressed

1. Condensation Reaction

2. Diazotization Reaction

3. Coupling Reaction

All chemicals used were of laboratory grade, supplied by Aldrich Chemical Company (industrial grade chemicals have also been investigated). The synthetic steps investigated are outlined above.

Results and Discussion

Properties of the Spectrometer in the Circulating Sampling Arrangement

Short-term reproducibility. With NIR FT-Raman, wavenumber accuracy and ordinate precision are intimately affected by the alignment of the source and the collection optics [2]. It was found, in a series of experiments, that for good reproducibility focusing was required at regular intervals (this resulted in intensity deviations of about 3.1% or less). The main reason for the deviations was believed to be vibrations caused by

the peristaltic pump employed to circulate the reaction mixture.

Subtractive property of FT-Raman spectroscopy. Spectral subtraction tests were performed for a mixture of three aromatic amines, the individual spectra of which were also measured, with use of the cyclic sampling arrangement. The subtractions, to produce difference spectra, gave excellent results (not shown).

Measurements and Results

The temperature of the liquor in the cycling system during measurement increased slightly (to a maximum of 12 °C) because of the laser beam emission energy and

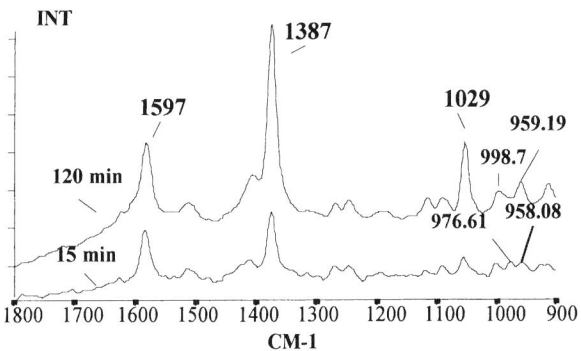

Fig. 2. FT-Raman spectra of the condensation reaction between H-acid and cyanuric chloride

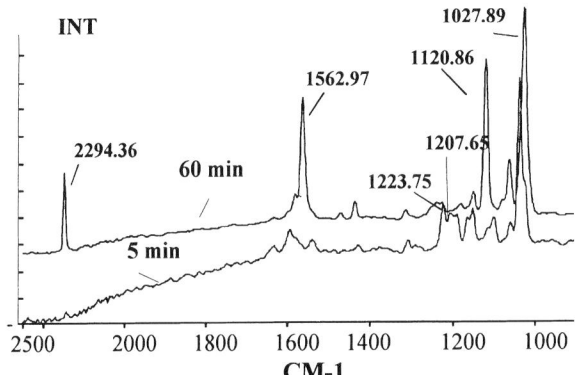

Fig. 3. FT-Raman spectral results as the diazotization reaction proceeded

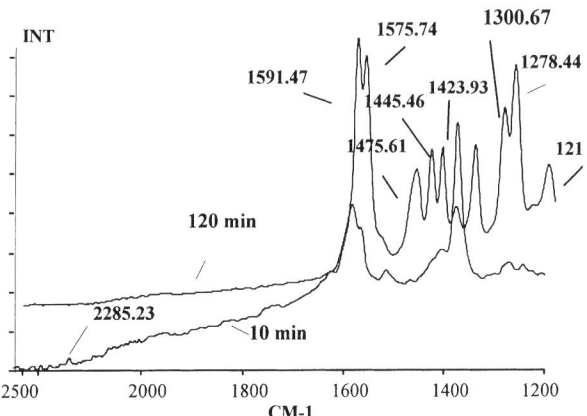

Fig. 4. FT-Raman spectral measurements from the coupling reaction

friction of the pumping system, which was subsequently replaced by an air pump for more efficient circulation of the liquor.

Condensation Reaction

Figure 2 shows the FT-Raman spectra (in the region 1800–900 cm^{-1}) obtained during the condensation reaction. The band at 977 cm^{-1}, representing a characteristic peak for cyanuric chloride, gradually became weaker as the reaction proceeded, and after 120 minutes the peak had completely disappeared.

Diazotization Reaction

Figure 3 shows FT-Raman spectral results from continuous measurement during the course of a diazotization reaction. A strong Raman band appeared at about 2300–2250 cm^{-1} due to the $-^{+}N\equiv N$ group stretching vibration. Hence, the 2300–2250 cm^{-1} re-

gion may be considered to be useful for the quantitative analysis of diazo compounds. The spectral properies of the diazo compound are different at various pH values, which can affect the quantitative analysis of the reaction process unless this is taken into account.

Coupling Reaction

Part of the FT-Raman spectral results from the coupling reaction are shown in Fig. 4. A band representing the azo group (-N = N-) appeared in the 1440–1400 cm^{-1} region of the spectrum. The diazonium species was consumed as the reaction proceeded, its characteristic band at 2285 cm^{-1} gradually decreasing in intensity.

Conclusions

The reproducibility of NIR FT-Raman spectroscopy under the developed cyclic-sampling system conditions was best when measurement was made following focusing for each spectrum. In such controlled conditions the subtraction of spectra appeared to give almost perfect results. For the condensation reaction of 1-amino-8-naphthol-3, 6-disulphonic acid (H-acid) and cyanuric chloride the disappearance of the Raman band of cyanuric chloride at 977–970 cm^{-1} can be considered as an indication of the end-point of the reaction. There was a strong Raman band at 2300–2250 cm^{-1} for the diazonium compound in the FT-Raman spectra. A characteristic Raman peak for the azo group in the product, in the range 1440–1400 cm^{-1} may be suitable for quantitative as well as qualitative analysis in this process.

The laboratory constructed circulating sampling arrangement has been shown to produce satisfactory FT-Raman spectra. Further modifications have been, and continue to be, made to improve the quality of the results, especially for quantitative analysis. The potential for this type of analysis of aqueous medium reactions, as they occur, is clearly large. In monitoring reactions such as those investigated in this study, the amount of quantitative information that can be obtained will be influenced by factors such as the quality/purity of the starting materials, the pH at which measurements are made, the control of temperature, precipitation of any components, volume changes (e.g. by the addition of reagents) and by the instrumentation (including software for data processing) plus sampling accessories employed.

References

[1] F. T. Walder, M. J. Smith, *Spectrochim. Acta* **1991**, *47A*, 1201.
[2] Perkin–Elmer Limited, Specification of System 2000R FT-IR, Company literature, 1995.

Mikrochim. Acta [Suppl.] 14, 755–756 (1997)

Comparative Raman Studies of Hydrogenated Amorphous Carbon Films by Using Infrared and Visible Laser Excitations

István Pócsik[1,*], **Margit Koós**[1], **Said H. Moustafa**[1], **József A. Andor**[2], **Ottó Berkesi**[2], and **Martin Hundhausen**[3]

[1] Research Institute for Solid State Physics, P.O. Box 49, H-1525 Budapest, Hungary
[2] Institute of Physical Chemistry, Attila József University, P.O. Box 105, H-6701 Szeged, Hungary
[3] Institut für Technische Physik, Friedrich-Alexander Universität, Erlangen-Nürnberg, Erwin Rommel Strasse 1, D-91058 Erlangen, Federal Republic of Germany

Abstract. Raman spectra of hydrogenated amorphous carbon (a-C:H) samples were recorded by traditional visible light excitation and infrared (IR) excitation, with an FT-Raman spectrometer. Highly different spectra were detected, with better resolution in the IR case, due to the resonance enhancement of scattered intensity. This improved resolution allowed the detection of fine structure not observed earlier in Raman spectra of a-C:H. The spectrum was deconvoluted in the 900–1800 cm^{-1} range into four Gaussian lines, with intensive new peaks at ~ 1218 and ~ 1462 cm^{-1}.

Key words: amorphous carbon, FT-Raman spectra, infrared excitation.

Hydrogenated amorphous carbon (a-C:H) is a very promising new substance in contemporary materials science. A number of the properties of diamond can be realized in the homogeneous film form of this amorphous matter; these properties have suggested two names synonymously used for these materials, viz. "amorphous diamond" or "diamond-like amorphous carbon" [1, 2].

Raman spectroscopy is the most frequently used non-destructive method to investigate the structural properties of amorphous carbon and its bonding features, because the principal vibration modes of diamond and graphite are Raman active. The published Raman spectra were measured by visible light excitation. The good quality Fourier-transform Raman spectrometers – as is well known – use infrared light for excitation, and have improved the measuring possibilities in many fields, but amorphous carbon spectra measured by this sort of equipment are unknown to us.

In this paper we report the first results of FT-Raman investigations of systematic sample series prepared from methane. The importance of these measurements for understanding the correlation between the structural and electronic properties of a-C:H film is unquestionable, as infrared photons preferentially excite the $\pi - \pi^*$ transitions of the lowest energy states, which at the same time determine the forbidden gap of this material.

Experimental

The samples were deposited on a silicon wafer by rf glow discharge decomposition of methane under a number of different self-bias conditions in our laboratories. The samples were prepared at 80 mtorr pressure by varying the self-bias voltage ($U_{s,b}$) between 0 and -1200 V. In Erlangen, Raman spectra were excited by a TiSa laser working at 445.3 nm and were detected through a SPEX 1404 double monocromator with a photon counting system. The FT-Raman spectra were measured in Szeged, on a Bio-Red Digilab dedicated FT-Raman spectrometer, with excitation by a Spectra Physics Topaz ND:YVO$_4$ laser, working at 1.06 µm. Specimens for FT-Raman investigations were deposited onto gold-covered wafers (~ 1 µm thick), to decrease luminescence.

Results and Discussion

The visible light excited Raman spectra in the 800–2000 cm^{-1} range for a series of samples prepared under different self-bias conditions can be seen in Fig. 1. The shape of these spectra is similar to that published earlier [3], and its variation with self-bias voltages is in excellent agreement with the results of Xu et al. [4]. These spectra are usually deconvoluted into two Gaussian components, which are related to the graphitic region (~ 1580 cm^{-1}) and to the disordered region

---

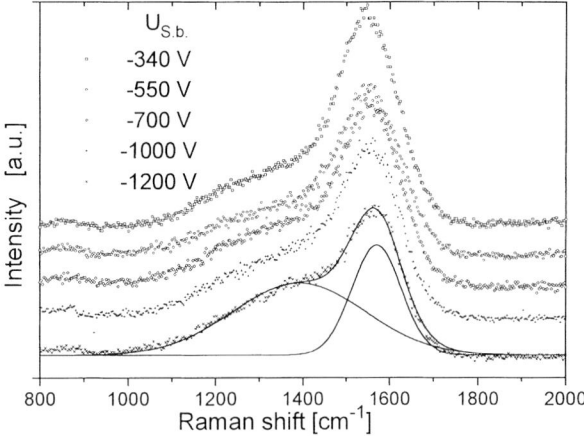

Fig. 1. Visible excitation Raman spectra

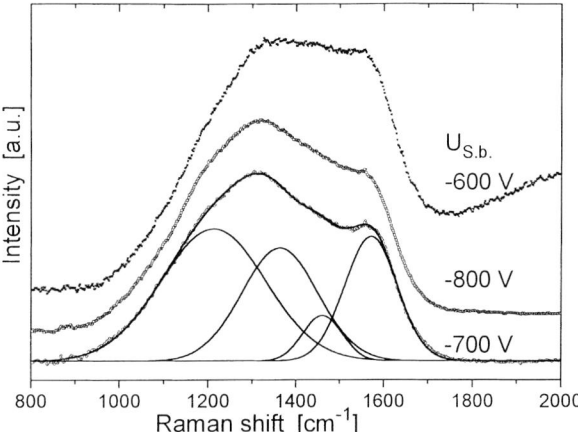

Fig. 2. IR excitation Raman spectra

($\sim 1350 \, cm^{-1}$), usually called the G peak and D peak, respectively. The actual position and intensities of these peaks depend on the preparation conditions and excitation wavelength [5].

The FT-Raman spectra of some samples are displayed in Fig. 2. A narrower bias voltage range is displayed because of the strong luminescence effect in the other samples; however, the bias-voltage dependence can be seen, similarly to Fig. 1.

From comparison of the Raman spectra obtained with visible and IR excitation the following differences can be established: (1) a strong increase in the scattered intensity of the D peak centred at $\sim 1350 \, cm^{-1}$, (ii) an unresolved shoulder appeared on the lower energy side of the spectra and (iii) a change of the peak half-width. A more detailed analysis of the ($-700 \, V$ self-bias)

spectrum shows that this D peak spectral region should be decomposed into four Gaussian lines, which are located at 1218, 1366, 1462 and $1571 \, cm^{-1}$. The bands at 1218 and $1462 \, cm^{-1}$ was not observed earlier for amorphous carbon films.

Our results can be interpreted on the basis of the structural model developed for a-C:H [1] in which the C atoms in sp^2 configuration are clustered in graphic islands of different sizes, and embedded in sp^3 configured carbon matrix. The proportion of different bonding configurations changes with self-bias voltage. The largest sp^2 bonded clusters determine the forbidden gap of this material. By decreasing the exciting photon energy from the visible to the infrared range, preferential excitation takes place for the lowest energy $\pi - \pi^*$ transition, together with resonance enhancement of the Raman scattered light on the lower energy side. The down-shift in the G peak by increasing excitation wavelength was explained in a similar way [3], but our FT-Raman spectra give clear experimental verification for resonant enhancement of the Raman intensity. Further investigations are in progress for the proper assignment of the two bands previously not detected and to separate the Raman spectrum for samples giving intense luminescence.

Conclusions

Raman spectra of hydrogenated amorphous carbon (a-C:H) samples were recorded, with visible light and infrared excitation. Strong differences were found between the two sets of spectra, and can be explained as due to strong resonance enhancement in the D peak region. The improved resolution allows the separation of two new peaks in this D region, which needs further investigation.

Acknowledgement. This work was supported by the Hungarian Science Foundation contract numbers OTKA-T-4223 and OTKA-T-017371.

References

[1] J. Robertson, *Adv. Phys.* **1986**, *35*, 317.
[2] V. S. Veerasamy, G. A. J. Amaratunga, W. I. Milne, P. Hewitt, P. J. Fallon, D. R. McKenzie, C. A. Davis, *Diam. Rel. Mat.* **1993**, *2*, 782.
[3] M. Ramsteiner, J. Wagner, *Appl. Phys. Lett.* **1987**, *51*, 1355.
[4] S. Xu, M. Hundhausen, J. Ristein, B. Yan, L. Ley, *J. Non-Cryst. Solids* **1993**, *164–166*, 1127.
[5] M. A. Tamor, J. A. Haire, C. H. Wu, K. C. Hass, *Appl. Phys. Lett.* **1989**, *54*, 123.

Mikrochim. Acta [Suppl.] 14, 757–758 (1997)
© Springer-Verlag 1997

Interpretation of Spinning-sample Bands in the FT Raman Spectra of Rotating Samples

U. Roland[*,**] and **R. Salzer**

Technische Universität Dresden, Institut für Analytische Chemie, D-01062 Dresden, Federal Republic of Germany

Abstract. The FT Raman spectra of rotating samples were studied to interpret the occurrence of additional bands that often led to a noisy and non-evaluable spectrum. It could be shown that these equidistant spinning-sample bands were due to modulation of the scattered light by the angular concentration function on the surface of the rotating sample.

Key words: FT Raman spectroscopy, sample rotation, spinning-sample.

The application of sample rotation to FT Raman spectroscopy leads to a double modulation of the detector signal. The Raman spectra obtained after Fourier transformation seem to be extremely noisy and are difficult to evaluate. In particular, thermally sensitive solid samples with a rough surface cannot usually be measured by FT Raman spectroscopy. In looking for a solution of this problem, it is essential to investigate the origin and the characteristics of these "spinning-sample bands (SSB), appearing as noise in the Raman spectrum. In particular, the dependence of their position and intensity on the rotation frequency, the spectroscopic parameters and the properties of the sample has to be clarified.

To allow the interpretation of the FT Raman spectra of rotating samples it was found necessary to use a very high rotation frequency (up to 500 Hz) and samples having a simple Raman spectrum (e.g. diamond with a single line at 1332 cm^{-1}). Under these conditions well resolved equidistant additional bands appeared in the FT Raman spectrum. The spectroscopic parameters were varied over a wide range and a clear influence on

these spinning-sample bands was found. Additionally, it could be shown that these bands represent the properties of the sample. In particular, the angular concentration function at the sample can be monitored by FT Raman measurement of the rotating sample.

Experimental

The measurements were done with a Bruker RFS 100 FT Raman spectrometer with NIR excitation by an Nd-YAG laser at 1064 nm, and a special rotor originally used in dentistry (Schick Dental, rotation frequencies ω could be varied from 0 to 500 Hz). The spectroscopic parameters were varied as follows: number of scans from 1 to 10000, resolution from 1 to 16 cm^{-1}, scanner velocity v_{scan} from 1 to 32 kHz. Additional experiments were made to observe the influence of external light sources and their modulation with different frequencies below 50 Hz. The occurrence of mechanical coupling between the rotor and the spectrometer was carefully excluded. It could be shown that the rotor did not produce additional bands in the Raman spectrum even when it was installed on the same bench in a test experiment. Scattered light from another source as well as the AC frequency (50 Hz) was also excluded as the origin of the additional bands from rotating samples. As expected, comparative measurements with a dispersive Raman spectrometer (Dilor XY 800) under the same conditions did not lead to the additional bands.

Results

Pattern of the Additional Bands of Rotating Samples

A typical FT Raman spectrum of a rotating sample (diamond, rotation frequency 500 Hz, 1 scan) is represented in Fig. 1. The pattern of the spinning-sample bands can be characterized as a number of bands with equal (wavenumber) difference (main spinning-sample bands – MSSB) and smaller side-bands (spinning-simple side-bands – SSSB) of each of these main bands. Under the applied experimental conditions the MSSBs could be detected up to the tenth order. It should be mentioned that the positions of these additional bands

---

* Present address: Umweltforschungszentrum Leipzig-Halle, Sektion Sanierungsforschung, Permoserstr. 15, D-04318 Leipzig, Federal Republic of Germany
** To whom correspondence should be addressed

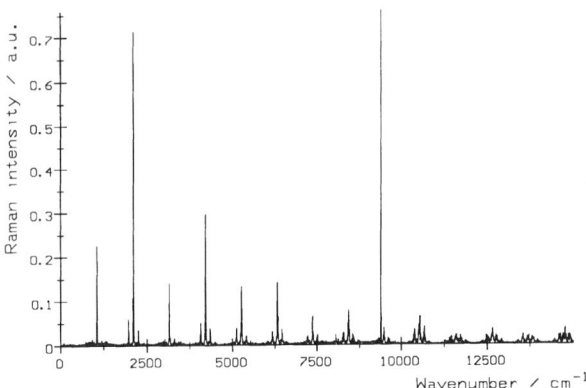

Fig. 1. FT Raman spectrum of a rotating diamond sample ($\omega = 333\,\mathrm{Hz}$, scan velocity $5\,\mathrm{kHz}$, 1 scan, resolution $4\,\mathrm{cm}^{-1}$)

$\tilde{v}_{\mathrm{MSSB}}$ and $\tilde{v}_{\mathrm{SSSB}}$ will be discussed as absolute values, i.e. as the difference between the wavenumber of the Nd-YAG excitation laser ($\tilde{v}_{\mathrm{exc}} = 9394\,\mathrm{cm}^{-1}$) and the value obtained from the Raman spectrum $\Delta\tilde{v}_{\mathrm{meas}}$:

$$\tilde{v}_{\mathrm{MSSB/SSSB}} = \tilde{v}_{\mathrm{exc}} - \Delta\tilde{v}_{\mathrm{meas}}. \tag{1}$$

Position of the MSSBs

The equidistant MSSBs were observed at the positions $\tilde{v}_{\mathrm{MSSB},n}$:

$$\tilde{v}_{\mathrm{MSSB},n} = n\tilde{v}_{\mathrm{MSSB},1} \qquad n = 1, 2, 3, \dots. \tag{2}$$

The values $\tilde{v}_{\mathrm{MSSB},1}$ and consequently the values $\tilde{v}_{\mathrm{MSSN},n}$ were proportional to the rotation frequency of the sample ω and inversely proportional to the scanner velocity v_{scan}. Their positions can be calculated from

$$\tilde{v}_{\mathrm{MSSB},n} = n\tilde{v}_{\mathrm{MSSB},1} = \frac{n\omega}{\lambda_{\mathrm{He-Ne}}\,v_{\mathrm{scan}}} \tag{3}$$

The positions of the MSSBs could be reproduced very exactly for the same experimental conditions. The line-width of these bands was less than $20\,\mathrm{cm}^{-1}$.

Intensity of the MSSBs

The pattern of the MSSBs after 1 scan, i.e. the relative intensity of these bands, was found to be exactly reproducible and characteristic for each position of the sample in the laser beam (distance of the focus from the rotation centre). The pattern did not change when the rotation frequency and the scanner velocity were varied. On the other hand, significant differences were observed for different sample positions. Therefore, the pattern of the MSSBs is correlated with the concentra-tion profile and the roughness of the sample as a function of the rotation angle, at constant distance from the rotation centre.

Position of the SSSBs

In addition to the MSSBs a number of SSSBs as satellites of the MSSBs were observed. In addition to the first-order SSSBs, higher orders of these bands were found in some cases. The distance of these bands from the MSSBs did not depend on the rotation frequency ω. Their existence can be interpreted as due to a certain non-linearity of the motion of the mirror as a function of time.

Discussion

The occurrence of the additional bands in the FT Raman spectra of rotating samples was shown to be due to the rotation of the sample. Mechanical reasons, scattered external light and the AC frequency could be excluded as the origin of the bands. The position of the equidistant spinning-sample bands was correlated with the rotation frequency when the scanner velocity was taken as the time scale [Eq. (3)].

The origin of the MSSBs can be attributed to the varying local concentration at the rotating sample, as a function of the angle. The result is a periodic modulation of the intensity of the scattered light. The spectrometer filter system for the exciting laser radiation leads to the occurrence of the bands which are due to the Fourier transformation of the modulation function, i.e. the intensity profile. In practice, the response time characteristics of the detector have a remarkable influence on the transformation.

Conclusion

The investigation of additional bands in the FT Raman spectrum of rotating samples showed that these bands are due to the inhomogeneous concentration of the scattering substance on the surface of the sample. The position of these bands can be satisfactorily explained by considering the rotation frequency in relation to the scanner velocity.

Acknowledgements. The authors thank Mr. T. K. Müller and the Dentallab Andreas for providing the rotor. Financial support from the Leopoldina (LPD 95) and the Fonds der Chemischen Industrie is gratefully acknowledged.

Mikrochim. Acta [Suppl.] 14, 759–761 (1997)

Hadamard- and Fourier-transform Infrared Imaging and Spectrometry

Michael K. Bellamy[1], **A. Norman Mortensen**[2], **Edward A. Orr**[1], **Timothy L. Marshall**[3], **Joseph V. Paukstelis**[1], **Robert M. Hammaker**[1,*], and **William G. Fateley**[1]

[1] Department of Chemistry, Kansas State University, Manhattan, Kansas 66506, U.S.A.
[2] Department of Electrical and Computer Engineering, Kansas State University, Manhattan, Kansas 66506, U.S.A.
[3] AeroSurvey, Inc., P.O. Box 1163, Manhattan, Kansas 66505, USA

Abstract. A moving (mechanical) two-dimensional Hadamard encoding mask has been employed with a Fourier-transform infrared (FT-IR) spectrometer to combine imaging in two spatial dimensions with the spectral dimension as a form of three-dimensional spectrometry in the mid-infrared and near-infrared spectral regions. Chemical images at chosen wavenumbers and mid-infrared spectra of individual pixel positions in the image have been obtained with acceptable signal-to-noise ratios and reasonable data-acquisition times.

Key words: Hadamard-transform, infrared, imaging.

For more than a decade we have utilized stationary Hadamard encoding masks based on liquid crystal technology in our application of Hadamard-transform techniques in spectrometry, depth profiling and imaging, conveniently classified as one, two, three, or four dimensional (1D, 2D, 3D, or 4D) spectrometry by the total number of spatial and spectral dimensions involved ([1–3] and references cited therein). Consequently, investigations were restricted to the visible and near-infrared spectral regions. Recently, we decided to see whether advances in the technology of motional devices and mask fabrication since the 1970s would overcome the problems that impeded the initial development of HT techniques with moving (mechanical) Hadamard encoding masks (see reference 2 in [1]). The use of a moving 2D Hadamard encoding mask with all mask elements either completely open or completely closed makes imaging and spectrometry possible in any spectral region for which an appropriate spectral separator and detector are available. Thus,

we chose to use a Fourier-transform infrared (FT-IR) spectrometer to combine the frequency precision, spectral multiplex, and throughput advantages of an FT-IR spectrometer with the spatial multiplex advantage of the moving 2D Hadamard encoding mask, hoping to obtain images in the mid-infrared (IR) spectral region and mid-IR spectra of individual pixels of the image with usable signal-to-noise ratios in a reasonable data-acquisition time. Our final goal is to provide an economical means of chemically mapping heterogeneous samples.

Experimental

The moving 2D Hadamard encoding mask, which is based on the cyclic simplex S_{255} matrix (see reference 2 in [1]) as a 15 × 17 array and was etched out of a metal substrate, was mounted vertically and was moved from one encodement pattern to the next by using a translation stage. All mounts and apertures were machined in our department. The source radiation (SiC globar powered by a 12 volt power supply) was collimated with a 19.1 mm focal length off-axis paraboloidal reflector and passed through the encoding mask, through the sample, and into a Midac MD 2400 series FT-IR spectrometer. The sample was an L-shaped piece of polystyrene with a square piece of acetate film (from a transparency for an overhead projector) placed in the open area of the L. Each spectrum taken at each of the 255 encodement patterns was for 32 co-additions at $4 \, cm^{-1}$ spectral resolution for a total data-acquisition time of approximately 40 minutes. The software that was used to perform the fast Hadamard-transform to generate a chemical image at a given wavenumber and to construct a spectrum at each of the 255 pixels of the image was written by one of us (ANM).

Results and Discussion

A data point corresponding to a given wavenumber was extracted from each file for the 255 encodements

---

* To whom correspondence should be addressed

and a fast Hadamard-transformation was performed on this collection of 255 points to generate a chemical image of the sample at that given wavenumber. The mid-IR spectra of the two components of the sample (polystyrene and acetate) have numerous absorption bands on which to image and we were able to obtain images showing the two components at a number of different wavenumbers. The images can be displayed as

either arbitrary intensity units or absorbance units when data from a spectrum of open air serves as background in the absorbance calculation. The chemical images of the two components are clearly evident and we have detected no noise spikes or artifacts in these images (see Figs. 1 and 2).

It is possible to perform a fast Hadamard-transform on all the data points of the spectra (i.e. all the data points in each file for the 255 encodements) and place the results end to end to generate a spectrum of the sample at each of the 255 pixel positions of the image of the sample. Since the polystyrene and acetate pieces are

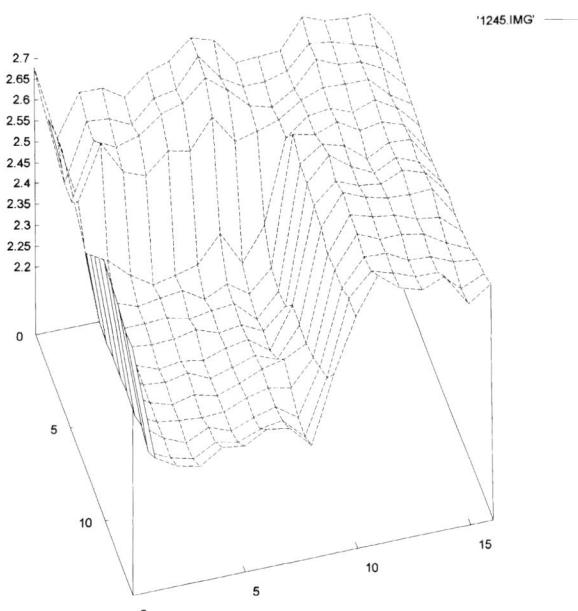

Fig. 1. An absorbance versus pixel plot of the Hadamard-transform image of polystyrene (L shape) at 2848 cm^{-1}

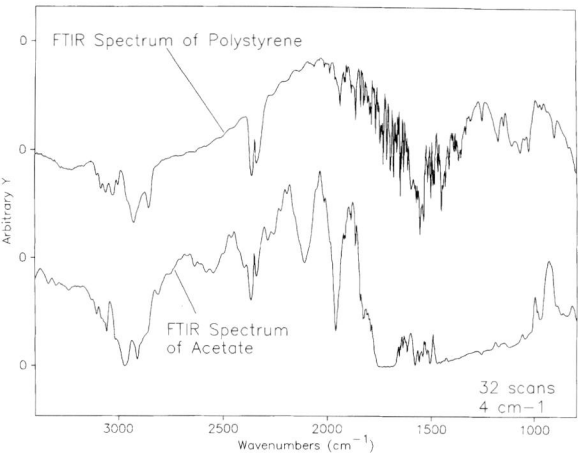

Fig. 3. Conventional Fourier-transform infrared (FT-IR) spectra of large pieces of polystyrene film (top) and acetate film (bottom)

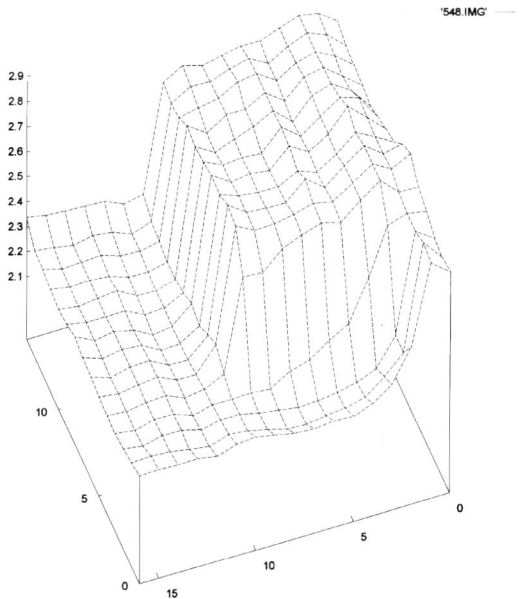

Fig. 2. An absorbance versus pixel plot of the Hadamard-transform image of acetate (square shape) at 1504 cm^{-1}

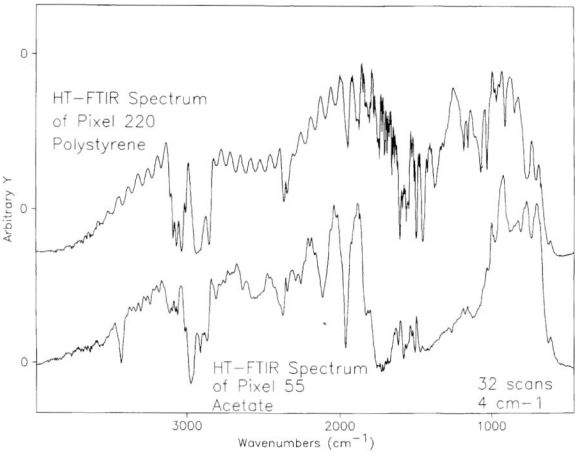

Fig. 4. Hadamard- and Fourier-transform infrared (HT-FTIR) spectra of a single pixel containing polystyrene only (top) and acetate only (bottom). The Hadamard-transform applies to the spatial dimensions and the Fourier-transform applies to the spectral dimension

spatially separated, it should be possible to obtain a spectrum of only polystyrene at some pixels and of only acetate at some other pixels. This procedure was performed and the spectra of individual pixels generated appear to be free of artifacts with no apparent loss of spectral information (see Figs. 3 and 4).

Our images and spectra have acceptable signal-to-noise ratios (SNR). Efforts to modify and improve our optical design to improve SNR and/or decrease data-acquisition time are under way. Preliminary experiments to repeat this work in the near-infrared spectral region with appropriate changes in source, beam splitter, and detector have been encouraging. Our final goal remains to provide an economical means of chemically mapping heterogeneous samples in the UV-Visible, near-IR, and mid-IR spectral regions.

Acknowledgement. This paper was prepared with the support of the U.S. Department of Energy, Chemical Sciences Division, Grant number DE-FG02-85ER13347. However, any opinions, findings, conclusions, or recommendations expressed herein are those of the authors and do not necessarily reflect the views of the D.O.E. Additional support form DOM Associates, Inc., Manhattan, Kansas, USA is gratefully acknowledged.

References

[1] W. G. Fateley, R. M. Hammaker, J. V. Paukstelis, S. L. Wright, E. A. Orr, A. N. Mortensen, K. J. Latas, *Appl. Spectrosc.* **1993**, *47*, 1464.

[2] W. G. Fateley, R. Sobczynski, J. V. Paukstelis, A. N. Mortensen, E. A. Orr, R. M. Hammaker, in: *Spectrophotometry, Luminescence and Colour; Science and Compliance*, (C. Burgess, D. G. Jones, eds.), Elsevier, Amsterdam, 1995, 315.

[3] R. M. Hammaker, A. N. Mortensen, E. A. Orr, M. K. Bellamy, J. V. Paukstelis, W. G. Fateley, *J. Mol. Struct.* **1995**, *346*, 135.

Mikrochim. Acta [Suppl.] 14, 763–765 (1997)
© Springer-Verlag 1997

Spectroscopic Ellipsometry in the Mid Infrared with a New Accessory Fitting into Standard Fourier-transform Spectrometers

Georg Dittmar[1,*], Arnulf Röseler[2], Uwe Richter[1], and Uwe Wielsch[1]

[1] Sentech Instruments GmbH, Rudower Chaussee 6, D-12489 Berlin, Federal Republic of Germany
[2] Institut für Spektrochemie und angewandte Spektroskopie, Rudower Chaussee 5, D-12489 Berlin, Federal Republic of Germany

Abstract. Spectroscopic ellipsometry in the mid and far infrared is a developing method for the characterization of bulk materials, thin-film and multilayer samples. A new and easy to use accessory is presented which combines the advantages of spectroscopic ellipsometry and Fourier-transform spectroscopy. A prism retarder is included in the set-up to improve the quality of the measurements. The function is demonstrated with a typical multilayer sample.

Key words: spectroscopic ellipsometry, infrared, instrument, Fourier-transform spectroscopy.

Spectroscopic ellipsometry in the mid and far infrared is becoming a more and more important method for the characterization of bulk materials, thin-film and multilayer samples consisting of materials which have structured dielectric properties in this spectral range. It is advantageous to cover the infrared spectral range by combining ellipsometry with Fourier-transform spectroscopy [1, 2]. Owing to the high light throughput a high spectral resolution can be easily achieved within a short time of measurement.

The sample is illuminated with polarized light of known state at a high angle of incidence. Usually a fixed polarizer is used at the azimuth $\alpha_1 = 45°$. After reflection the state of polarization is measured by taking single beam spectra at various settings α_2 of the analyzer. The detected intensity depends on α_2, the wavenumber $\tilde{v} = \omega/2\pi c$ and the ellipsometric angles of the sample Ψ and Δ, given by

$$I(\alpha_2, \Psi, \Delta, \tilde{v}, \alpha_1) \propto s_0(\Psi, \Delta, \tilde{v}) + s_1(\Psi, \Delta, \tilde{v}, \alpha_1)\cos 2\alpha_2$$
$$+ s_2(\Psi, \Delta, \tilde{v}, \alpha_1)\sin 2\alpha_2.$$

From these spectra the Stokes parameters s_1/s_0 and s_2/s_0 can be determined by using a Fourier analysis at

each measured spectral point. Furthermore the ellipsometric angles Ψ and Δ can be determined, which give amplitude and phase information from $\tan\Psi e^{i\Delta} = r_p/r_s$. Here r_p and r_s are the complex amplitude reflection coefficients of p- and s-polarized light. A retarder with a phase shift near to 90° allows an additional measurement and significantly improves the results in spectral regions with Δ near 0° or 180° and allows the unambiguous determination of Δ within the range 0–360°.

By a modelling and fitting procedure the properties of the sample (dielectric function and film thickness) can be determined from Ψ and Δ.

Instrument Description

We present a patented [3] new and easy to use accessory for spectroscopic ellipsometry. As shown in Fig. 1 it has the advantage of fitting completely into the

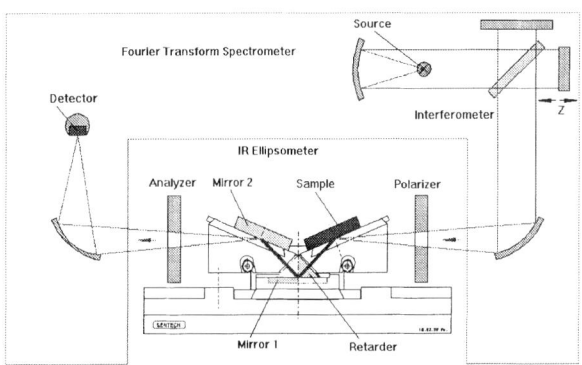

Fig. 1. Accessory for spectroscopic ellipsometry in the mid and far infrared designed for use in the sample chambers of standard Fourier transform spectrometers

---

* To whom correspondence should be addressed

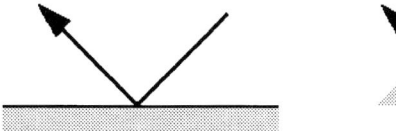

Fig. 2. Mirror (left) and retarder prism (right) can be exchanged, leading to phase shifts near -180° and near -90° respectively. In contrast to a quarter-wavelength plate the prism ensures the retardation in a broad spectral range

sample compartments with centre focus of standard Fourier-transform spectrometers from the main manufacturers. Thus any external beam coupling can be avoided and the vacuum or air purge of the sample compartment can still be used. This is known to be necessary for reliable results, without artifacts from atmospheric influences.

The light from the source is modulated in the interferometer and passes through the polarizer directly onto the sample at a high angle of incidence, usually 67.5°, which is suitable for many ellipsometric applications. Other angles are possible, but the holder then has to be exchanged. From the sample the light is reflected towards mirror 1 and then folded back into the main optical axis of the spectrometer by mirror 2. The standard detector of the spectrometer is used. Additional apertures can be inserted into the light path to decrease the size of the illuminated area of the sample.

The retarder can be automatically inserted instead of mirror 1, as shown in Fig. 2. A high-precision dovetail bearing is used to ensure proper optical alignment. The phase shift of the mirror, which is near to 180°, is then replaced by the phase shift of the internal reflection prism, which is designed to be near to 90°. The advantage of a prism, compared to quarter-wavelength plates, is the broad spectral range of nearly equal retardation.

The polarizer and the analyzer are positioned on rotation stages driven by stepper motors and providing a high angular resolution of better than 0.001°.

Data Collection

The accessory is fully remote-controlled by a comprehensive software package for ellipsometers, which includes the control of the spectrometer, calculation of measured data as well as extended modelling, an efficient fit procedure and simulation of multilayer samples.

A measurement cycle consists of the collection of a number of single-beam spectra at certain settings α_2

of the analyzer, which turns stepwise. The following spectra have to be measured and calculated once during the calibration of the accessory: the degree of polarization of the polarizers, the ellipsometric angles of the mirrors, the ellipsometric angles of the retarder. They are included in the calculation of the measured sample data and improve the results significantly.

Example

As an example we present a measurement of the multilayer sample shown in Fig. 3 which was prepared for experiments with photosensitive layers [4]. The measurement was performed with the new accessory in a Bio-Rad FTS-40A interferometer. In Fig. 4 the measured and fitted data are shown. Some small ranges, especially around 3000 cm^{-1}, are excluded from measurement, owing to absorption bands in the polarizers.

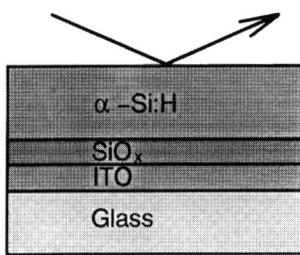

Fig. 3. Structure of the measured multilayer sample

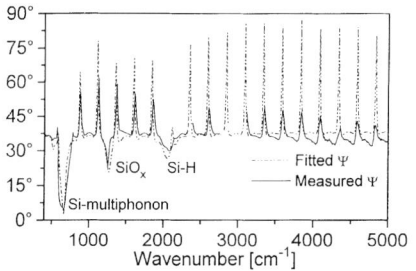

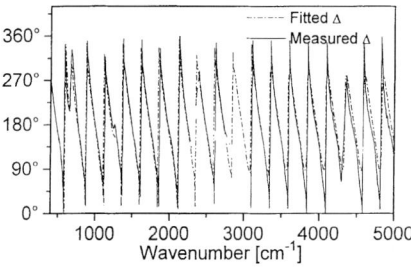

Fig. 4. Measured and fitted Ψ and Δ spectra, showing strong thickness interferences

For the fit, the dielectric functions of the materials were described by the Drude–Lorentz oscillator model [5]. The strong interference structures result from the thick α-Si:H layer on top of the sample. The indium–tin–oxide (ITO) layer is conductive, and thus has a high reflectance in the infrared for p- and s-polarized light, which leads to the lower level of the Ψ-spectrum at about 35°. Typical structures in the spectrum result from Si–H vibrations, from SiO_2 and from multi-phonon absorption in Si, as marked in Fig. 4. The Δ spectrum was measured in the whole range 0–360° because the retarder was used and mainly shows the interferences and structures from the material properties mentioned above. The following layer thicknesses were found: α-Si:H 6.59 μm; SiO_x 89 nm; ITO 100 nm.

Conclusion

The new accessory presented here makes it possible to perform spectroscopic ellipsometry in the mid infrared with standard Fourier-transform spectrometers. It is designed for easy handling and high mechanical accuracy, has all necessary features, includes necessary correction formulae and is fully supported by SpectraRay software for data collection and evaluation. Its correct functioning has been demonstrated with a typical multilayer sample.

References

[1] A. Röseler, *Infrared Spectroscopic Ellipsometry*, Akademie Verlag, Berlin, 1990.
[2] G. Dittmar, V. Offermann, M. Pohlen, P. Grosse, *Thin Solid Films* **1993** *234* 346.
[3] German patent DD 299 138.
[4] W. Senske, N. Marschall, R. Herkert, K.-H. Greeb, M. Frank, *J. Non-Cryst. Solids* **1983**, 59/60, 1239.
[5] P. Grosse, *Freie Elektronen in Festkörpern*, Springer, Berlin Heidelberg New York, 1979.

Mikrochim. Acta [Suppl.] 14, 767–768 (1997)

Portable "Double Cat's Eye" FT-IR Spectrometer

Tatiana B. Egevskaya

Institute of Semiconductor Physics, Siberian Branch of Russian Academy of Sciences, 630090, Novosibirsk, Avenue Lavrentjeva 13, Russia

Abstract. In the present paper the portable FT-IR spectrometer based on the "double cat's eye" interferometer (circular version, patented) is described. The advantages of the "double cat's eye" (DCE) interferometer, which incorporates two spherical mirrors and a movable beam-splitter, are the low sensitivity to misalignment and the doubled resolution for the unit of motion. The unique construction of the spectrometer gives the opportunity of lowering polarization effects on the beam-splitter. Compact size, low weight, and robust construction are additional features of the instrument. Because of the above-mentioned features no additional adjustment is need during the whole lifetime of the equipment, including production. The circular version of the spectrometer described here has a throughput about 20 times that of the previous version, owing to more effective use of the light.

Key words: FT-IR spectrometer, interferometer, double cat's eye.

The basic arrangement of the FT-IR spectrometer was proposed in our previous paper [1]. In the present paper the main technical specification of an FT-IR spectrometer is described, based on the circular version of the interferometer patented by the author in 1993 [2]. The construction of the DCE interferometer is the result of combining two "cat's eye" reflectors into one. The optical device incorporates two spherical mirrors and a movable beam-splitter. The interferometer has permissible large-angle adjustment of the optical elements and twice as much resolution for the unit of motion of the beam-splitter. In addition the DCE interferometer (circular version) has much higher throughput.

Description of the Equipment

The optical arrangement of the equipment is depicted in Fig. 1. The main functional part of this device is the DCE (circular version). This version is characterised by the shift of the mirrors towards each other from the confocal position and by the coincidence of the input light axis and the optical system axis. This unique arrangement is resulting in a decreasing of the throughput. These features are based on the more effective use of the light flow. As is depicted in Fig. 1 the incident light covers the whole surface of mirror (5). This mirror produces a parallel beam. The circular beam-splitter (6) accepts a spatial angle of 0.3 sr from the field of view. The input hole area is more than 3 mm^2. In this way the DCE throughput is more than 1 mm^2 × sr.. This parameter is more than 20 times the DCE throughput described in [1]. (where the entering radiation is at an angle to the mirror axis).

To combine the interferometer with photodetector (14) without loss in throughput, the double "focon" (8) was used. It transforms the circular output beam into a parallel beam and removes the central shadow zone.

The instrument has two optical channels: with sample (15) and without it (16). The He–Ne laser is used for control of the beam-splitter motion and simultaneously as a reference source. The beam-splitter motion speed is 0.36 mm/sec. The circular DCE (diameter of the mirrors and the beam-splitter 60 mm) can tolerate both high angular (± 5) and linear (± 0.3 mm) misalignment of the optical elements, calculated from the conditions for conservation of the intensity field at $\lambda = 1\mu$m. The small/large mirror distance is absolutely equal for the two CE reflectors of the DCE interferometer. Therefore the DCE does not need alignment either in the near IR range during the assembly, or in operation of the interferometer.

Application

The PC-based DCE FT-IR spectrometer is shown in Fig. 2. It contains the sample compartment with the typical infrared sampling accessories such as gas and

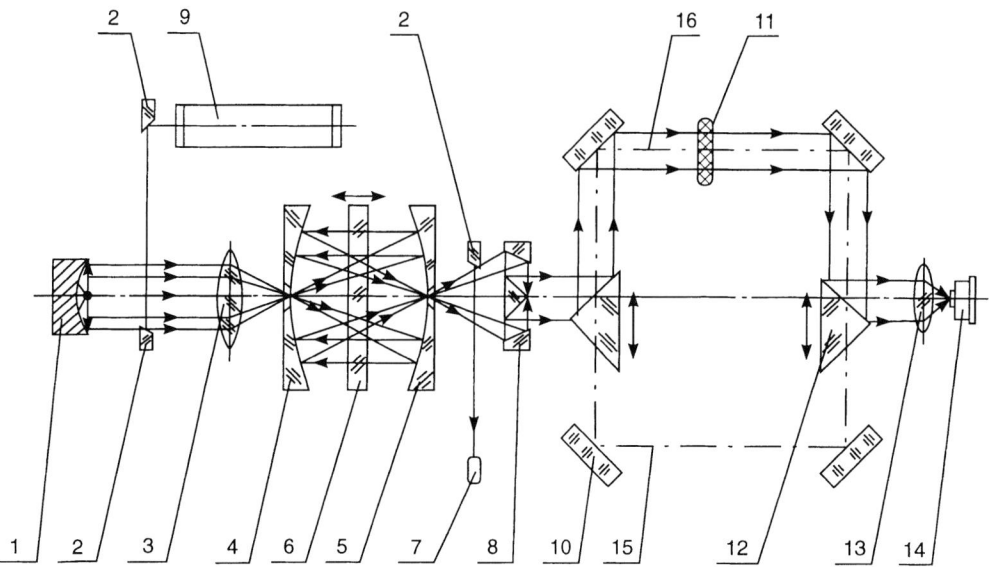

Fig. 1. The optical arrangement of the DCE FT-IR spectrometer: (1) – light source (NiCr heater); (2), (10), (12) – flat mirror; (3), (13) – KBr lens; (4), (5) – spherical mirrors of the DCE interferometer; (6) – beam-splitter, Ge on KBr; (7) – Si photodiode; (8) double "focon"; (9) – He–Ne laser; (11) – sample; (14) – pyroelectric detector MG32; (15) – reference channel; (16) – sample channel

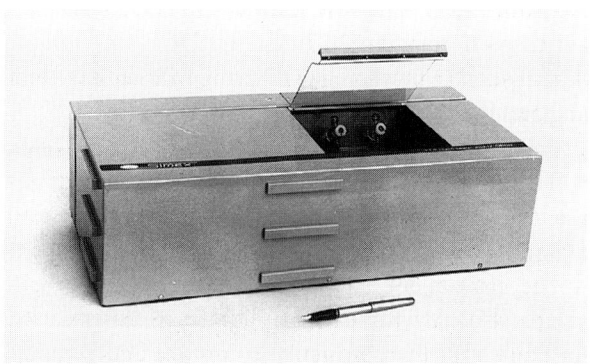

Fig. 2. Portable "Double Cat's Eye" FT-IR spectrometer

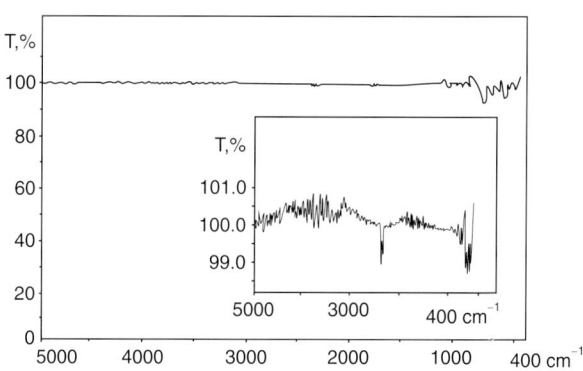

Fig. 3. 100% transmission line of the DCE FT-IR spectrometer

liquid cells, and film holders and may be used in chemistry, geology, the food industry, pharmacology and criminalistics.

Performance: resolution $1\,cm^{-1}$, spectral range $5800–400\,cm^{-1}$, wavenumber precision $0.01\,cm^{-1}$ (a stabilized He–Ne laser is used). Photometric accuracy and the flatness of the 100% T line for the optical unit without sealing and desiccating, for 15 scans at $4\,cm^{-1}$

resolution is shown in Fig. 3. The dimensions of the portable "Double Cat's Eye" FT-IR spectrometer are $572 \times 266 \times 155$ mm, weight 17 kg.

References

[1] T. B. Egevskaya, *Infrared Phys.* **1984**, *24*, 329.
[2] T. B. Egevskaya, *Patent No. 1552791*, Russia, 25 June 1993.

Mikrochim. Acta [Suppl.] 14, 769–771 (1997)

Design and Applications of a Near-Infrared Polarization Interferometer for Industrial Use

Jochen Knecht

Bühler ANATEC, Am Porscheplatz 5, D-45127 Essen, Federal Republic of Germany

Abstract. The design of the polarization interferometer is described and the resulting differences from the Michelson interferometer for operation in the near infrared region are discussed. Typical industrial applications are presented.

Key words: FT-NIR spectroscopy, polarization interferometer.

FT-NIR spectroscopy is gaining rapid acceptance for industrial applications. The fundamental advantages of measurements in the near infrared spectral region make FT-NIR especially well suited for the demands and applications of a production environment [1, 2]. These advantages are: low absorptivity, no sample preparation, use of SiO_2 fibres, at-line and on-line sampling, simple explosion-proof installations, use of chemometrics, correlation of chemical and physical parameters for closed loop process control.

However, an important prerequisite for successful implementation of FT-NIR spectroscopy within a "real world" industrial environment is the robustness and optical stability of the spectrometer.

The Polarization Interferometer

In contrast to the Michelson interferometer, the polarization interferometer [3] is a single-path interferometer, utilizing the anisotropy (birefringence) of crystalline quartz. The instrument is shown schematically in Fig. 1.

Polarized radiation entering into an anisotropic crystal is separated into an ordinary and an extraordinary beam, which differ in phase, depending on the thickness of the anisotropic crystal. By means of static

and a movable quartz wedge the thickness of a quartz plate can be varied continuously, resulting in a continuously increasing phase shift. After passing the second polarizer the beams form an interference pattern. A compensator plate is used to define the zero-position of the interferometer.

The key features of the polarization interferometer can be summarized as follows.

Both interfering beams share the same optical path. Thus, the interferometer is highly stable towards disalignment. This is specifically useful in the near infrared region where the shorter wavelengths reduce the alignment tolerances.

The small difference in optical velocity between the ordinary and extraordinary beams requires a comparatively long mechanical movement of the wedge (100 times the optical retardation). This reduces the

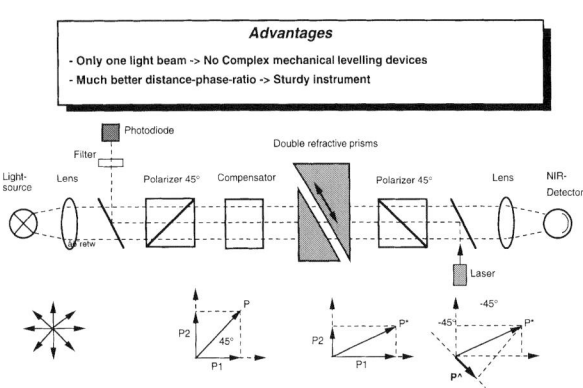

Fig. 1. Measurement principle of NIRVIS (patent pending)

precision requirements for the mechanical components and thus results in a very robust construction.

In a conventional Michelson interferometer the inverse relationship between mechanical movement and resulting optical retardation will cause a high sensitivity towards environmental disturbances such as vibrations and changing temperatures.

During every return movement of the quartz wedge to the starting position the spectrometer will record the interferogram of an internal reference channel. For each scan, the resulting spectrum is then ratioed against the actual external sample channel. This "ratio" feature effectively provides very high stability of the optical system.

The apparent drawbacks of the polarization interferometer are its limited resolution of 12.5 cm^{-1} and the fact that it can only operate in the near IR. In practice, however, these drawbacks are overcome by the broad spectral bandwidths in the near infrared region as well as the intended use of the spectrometer.

The polarization interferometer is implemented in the NIRVIS FT-NIR spectrometer. The system comprises a 10 W tungsten halogen source and a thermostatted PbS detector. This spectrometer is equipped either with a fibre-bundle for diffuse reflectance measurements or with single-fibre transmittance probes.

Typical Applications

Identifying and analysing raw materials is a classical application for FT-NIR spectroscopy [4]. In pharmaceutical production, regulatory requirements call for identification of the content of every single raw material container (Fig. 2). In general, raw materials identification helps to avoid sub-specification production batches and to maintain production safety.

The determination of octane numbers of gasoline is another major area of application for in-process FT-NIR [5]. Since the conventional test engine runs are rather time-consuming, a substantial reduction of the

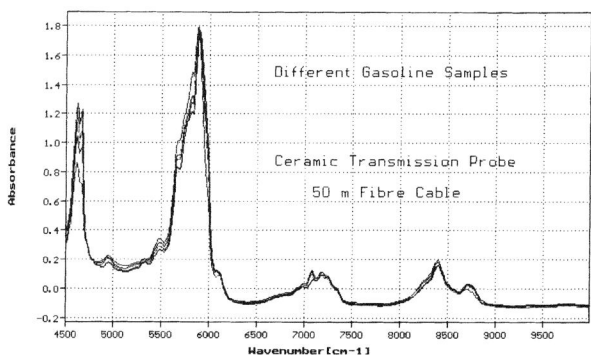

Fig. 3. FT-NIR spectra of gasoline samples

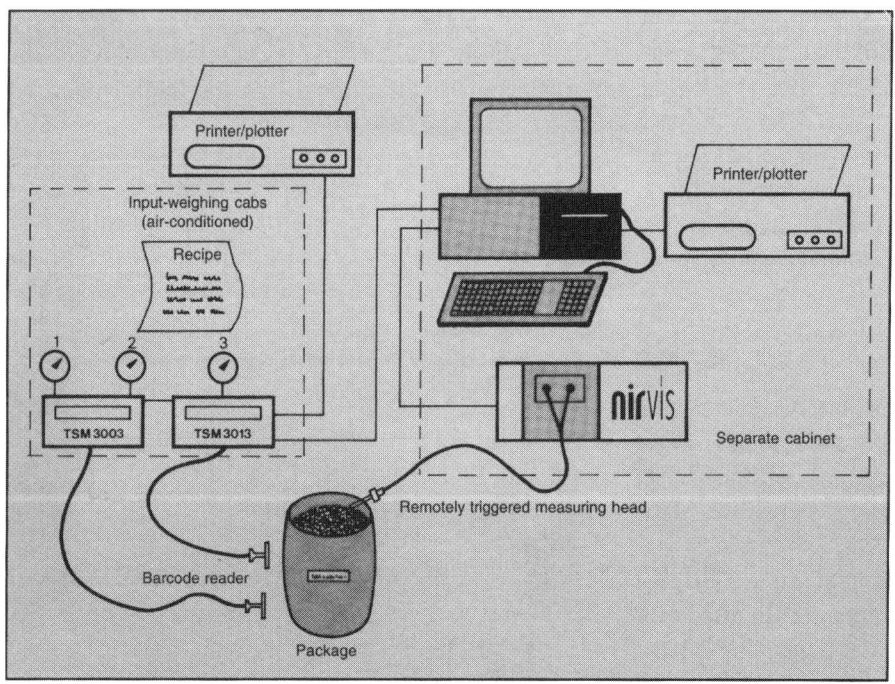

Fig. 2. Schematic of raw material identification within a computer-based manufacturing environment

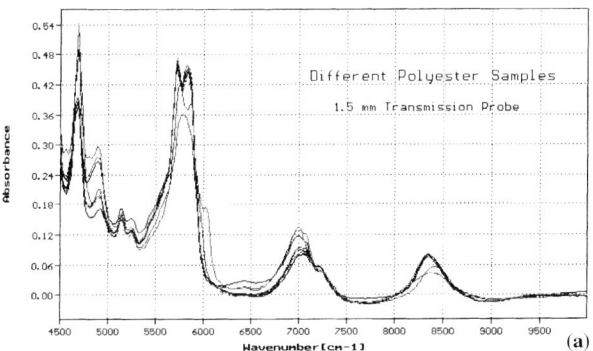
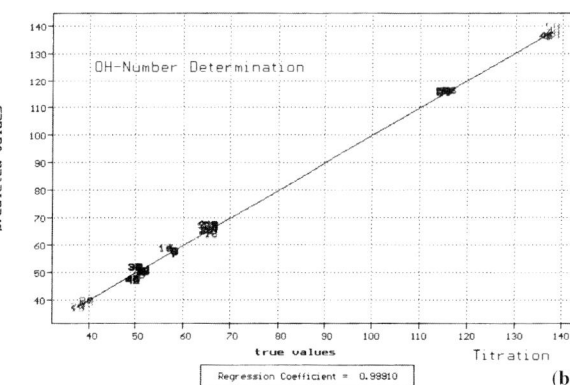

Fig. 4. Determination of OH-numbers for a group of different polyester products

"octane give-away" can be achieved. Figure 3 shows typical gasoline spectra obtained with an in-line ceramic transmission probe and a 50 m optical fibre.

In-process measurements of e.g. OH-numbers help to monitor polymerization reactions. Figure 4 illustrates an important feature of chemometric data reduction: with suitable learning sets, robust calibrations can be obtained even for a number of different products.

References

[1] D. A. Burns, E. W. Ciurczak (eds.), *Handbook of Near-Infrared Analysis*, Dekker, New York, 1992.
[2] H. W. Siesler, *NIR News* **1995**, *1*, 3.
[3] U. S. Patent No. 5, 157, 458.
[4] H. U. Gally, *Bühler Application Report*, 1992.
[5] A. Espinosa, M. Sanchez, S. Osta, C. Boniface, J. Gil, A. Martens, B. Deseales, D. Lambert, M. Valleur, *Oil Gas J.* 1994, *92*, No. 42, 49.

Mikrochim. Acta [Suppl.] 14, 773–775 (1997)
© Springer-Verlag 1997

Universal Near IR Rapid-Analyser Based on a High-Stability FT-IR Spectrometer for the Wavelength Range 750–2500 nm

Dmitry V. Matyunin* and **Marina V. Morozova**

Spectral Research Group, All-Russian Research Institute of Electrification of Agriculture, 2, 1st Vieshnyakovsky Proezd, 109456 Moscow, Russia

Abstract. The realization of quantitative multicomponent rapid analysis in the near infrared range (NIR), of various food and agricultural products, is very important. This method checks samples in vivo, in a short time without preliminary preparation. In this case the NIR-analyser registers a curve which is the sum of the individual spectra of the organic components of the real sample. On the other hand, the spectrum generated in the near-IR region consists entirely of overtones and combinations of overtones of fundamental bands within the mid-IR region. This curve, which may be obtained by the NIR-analyser in diffuse reflectance and diffuse transmittance modes is fairly sensitive to temperature and the signals are very low. For making an NIR rapid-analyser it is necessary to achieve long-term stability of the optical and mechanical units and electronics. NIR rapid-analysers are produced by some companies (Bran + Luebbe, Germany, NIR systems, U.S.A., and others). These instruments are either filter or high-energy grating spectrometers, for dedicated or universal application, respectively. Wavelength and photometric reproducibility remain the major problems of current NIR-analysers. The basic problem has been faced in making our rapid-analyser – the long-term stability of optical and mechanical units and electronics. The design of our instrument includes a highly stable Michelson interferometer with a precise drive for rapid scanning, and an integrating Ulbricht sphere with a unit for measuring channels alternately.

Key words: NIR- rapid-analyser, long-term stability, ceramic glass, superinvar, diffuse reflectance, noise level, photometric reproducibility.

The various commercial FT-IR instruments produced for the UV, VIS and NIR ranges still have some problems in connection with long-term stability. We have attempted to design a highly stable interferometer for the NIR-range, that would have an accuracy and reproducibility acceptable for multicomponent rapid analysis of organic compounds in various kinds of

products in vivo. All its optical, mechanical and electronic units remain constant in performance over a wide range of temperatures during continuous work (6–8 hours or even more). The analyser is universal, is it has to register diffuse reflection, diffuse transmission and transflectance over a wide spectral range (750–2500 nm), and it can be used in industry, medicine, pharmacy, etc. Norris [1] showed that high reproducibility and low noise level are major desiderata of NIR-analysers. The low noise level is demanded by the low values of the absorption in the NIR range and by the concentrations of the components to be determined. For example, a Cary 14 monochromator operating at a speed of 10 nm/s with 7.0 nm band-pass has a noise level less than 0.0005% O.D. (optical density) in the range 1000–2600 nm. These parameters were studied for a high-energy multiscan monochromator apparatus [2]. This instrument was based on a concave holographic grating optimized and fabricated specially for the apparatus by Jobin Yvon Optical Systems. Measurement of the noise level in the NIR range (1200–2400 nm) showed that it was 0.0004 O.D. (the full range scale was 4 O.D.). The parameters for one scientific spectrophotometer (Cary 5) [3] and two NIR-analysers (NIR-systems SY-2001 and InfraProver System) [4, 5] are presented in Table 1.

Among the commercial FT-IR spectrometers there are only two rapid analysers: the Buhler FT-NIR Universal Spectrometer and the Infra Prover System (they have the some optical design principle). These spectrometers do not have devices for integration of the infrared radiation from samples (integrating sphere

---

* To whom correspondence should be addressed

Table 1. Comparison of the technical parameters of the Cary 5 and two rapid analysers in the near infrared range

	Cary 5 NIR-spectrophotometer (Varian)	NIR-systems SY-2001 (NIR-system)	InfraProver system (Bran + Luebbe)
Wavelength range, nm	175–3300	1100–2500	1000–2500
Wavelength accuracy, nm	± 0.4	0.5	± 0.8 at 2000 nm ± 0.2 at 1000 nm
Wavelength reproducibility, nm	0.04	0.01–0.05	–
Photometric range, absorbance	3.0(5.5)	3.0(4.0)	–
Photometric noise, RMS at 0.0 absorbance at 1500 nm, 2 nm SBW	0.000040	$\leqslant 0.000020$	< 0.0005
Spectral band-width, nm (SBW)	0.04–20	10 ± 1	± 0.6 at 1000 nm

or special accessories for sample rotation). The application of fibre optics in the diffuse reflectance accessory the diameter of the cell is about 10 mm) is useful for medium values of the optical density (O.D.) As a result these analysers have limited application in quantitative analysis.

At present we do not have a standard method of metrological verification of the photometric scale for the reflectance mode in the near infrared range. There are only some diffuse reflectance calibration standards (Labsphere Inc., SRS-99–SRS-02) [6] for the range 0–2 O.D. There are still no special standard methods for calibration of NIR-analysers for diffuse reflectance/transmittance modes, but it is necessary to have standards of photometric range about 0–9 O.D. for the reflectance and transmittance modes in the near infrared range. The optical design for our NIR-analyser [7], is shown in Fig. 1, and consisting of an FT-IR and an integrating sphere, it may be optimal for an NIR rapid-analyser. It works as follows. The light from the halogen lamp (1), connected to the stabilized power, supply (2), falls on the focusing lens (3) and a parabolic mirror (4), which turns it through 90° and directly onto the beam-splitter (8). The two equivalent beams are then directed onto the mobile (10) and stationary (9) mirrors. The beams reflected from the mirrors interfere and fall on a focusing lens (13), which directs the light onto the unit (14) which switches the channels of measurement. The mirrors (14a) and (14b) fixed on the

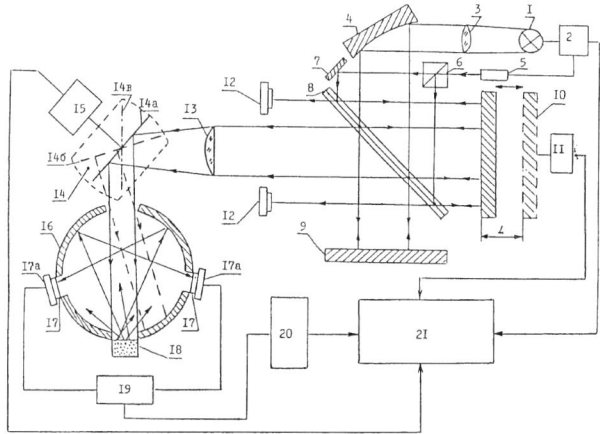

Fig. 1. Optical design of NIR-express-analyser on the basis of a high-stability FT-IR spectrometer for the wavelength range 750–2500 nm

unit serially direct the light beam onto a sample (18), placed on an internal wall of the integrating sphere (16). The shield (14c) reflects any escaping light back into the interferometer (the positions of the mirrors are shown by the dotted lines for each moment of measurement). Diffuse radiation reflected from the sample is registered by the detectors (17), connected with Peltier elements (17a). The signals from the detectors are amplified (19) and fall on the 18-bit A/D converter (20), which transmits them to a computer (21). All elements of the optical design are rigidly fixed on ceramic glass bases.

Fig. 2. NIR-express-analyser based on a high-stability FT-IR spectrometer for the wavelength range 750–2500 nm

The NIR rapid-analyser is based on a high-stability Michelson interferometer with a precise rapid-scan drive an integrating Ulbricht sphere, and a unit for measuring alternating channels, in an automatic mode (Fig. 2). To reach high long-term stability of the device we made the following technical decisions. Most of the constructional materials used in the interferometer should have the minimum thermal expansion coefficient (for example, its base and constructional parts are made of ceramic glass, the distance rings for the beam-splitters and other connections are made in superinvar, the beam-splitter and compensator are made of fused silica (for the IR-range), and special high-reflection and chemically resistant coatings are used on the mobile and stationary mirrors of the interferometer. A new design of precision unit for rapid scanning of the mobile mirror of the interferometer, with an optical path difference of 30 mm, has been developed. The drive of this rapid scanning device [8] includes a magnetic bearing (the system does not need compressed air). The integrating Ulbricht sphere coating is made from materials with minimum thermal expansion coef-

ficient. A high-sensitivity lead sulphide detector (PbS) is used with a Peltier element and optimized amplifier. A special high-stability halogen-lamp power supply is used with the fibre optic measurement system. Most of the rapid-analyser (about 75%) has now been constructed.

Experimental

Tests of separate units of the NIR rapid-analyser have shown the following results. The moving mirror provides an optical path difference of 30 mm. The angle of deflection of the mirror is less than 0.3″. Two devices, with optical paths of 30 mm and 100 mm, have been tested in a scientific FT-IR spectrometer for more than 10000 hours, and no changes in their technical parameters were observed. The following tolerances have been set for the various components.
– The beam-splitters and compensators to be plane-parallel to within 0.1 of an interference fringe.
– Non-parallelism of beam-splitter and compensator surfaces to be $\leqslant 0.05\,\mu m$.
– The thickness difference of the beam-splitters and compensators to be less than $0.1\,\mu m$.
  Measurement of the output of the stabilized power supply after heating for 30 min showed $< 0.2\%$ change from the initial value, and only 0.1% change after heating for 8 hr.

Acknowledgement. This work was supported by the Russian Fund of Technological Development under the Ministry of Science and Technical Policy of the Russian Federation.

References

[1] K. H. Norris, *Instrumentation for Near Infrared Spectroscopy NIR Conference*, Moscow, 1989.
[2] I. Landa, *Rev. Sci. Instrum.* **1979**, *50*, 34.
[3] Varian (Australia) Prospectus A.C.N. 004 559 540, Publication No. 85-100934-00, 1990.
[4] Perstorp Analytical Company, *NIR System*, Rapid Resin Analyzer, Sy-2001, 230 V AC model.
[5] Bran + Luebbe GmbH, *Analysentechnik* Catalogue No. 1.27, InfraProver System No. PC 46331 T1 (9104).
[6] Labsphere, Inc. Catalogue, 1994–1995.
[7] D. V. Matyunin, M. V. Morozova, *Technique in Agriculture* **1994**, *4*, 21.
[8] *Russian Patent* No. 1561645, 24 August 1987.

Mikrochim. Acta [Suppl.] 14, 777–779 (1997)

Extending the Spectral Range of an FT-VCD Spectrometer

Jennifer L. McCann*, **Dimiter Tsankov,** and **Hal Wieser**

Department of Chemistry, The University of Calgary, Calgary, Alberta, T2N 1N4, Canada

Abstract. Attempts were made to extend the spectral range of an FT-VCD spectrometer to lower wavenumbers than are currently practically possible. Experiments are described that test the measurement capabilities of a Bomem MB100 based spectrometer in which the normally used MCT type A detector and ZnSe photoelastic modulator are replaced by a selected MCT type B detector and a prototype CdTe photoelastic modulator, respectively. Realistic extension from the normal lower limit of $900\,cm^{-1}$ to $650\,cm^{-1}$ with an acceptable signal-to-noise ratio was achieved with the type B detector. While it could be demonstrated that the prototype CdTe photoelastic modulator is capable in principle of extending the range even further to $550\,cm^{-1}$, actual measurements could not be achieved because of very substantial signal reduction.

Key words: FT-VCD spectrometer, spectral range, extension to lower wavenumbers.

Vibrational circular dichroism (VCD) spectra can now be measured relatively routinely with an FT spectrometer between 1700 and $800\,cm^{-1}$. For the C–H and C=O stretching regions a dispersive instrument is usually preferred [1, 2]. In both cases, the circularly polarized light is produced by a photoelastic modulator (PEM). Though convenient, the PEM unfortunately also restricts the spectral range. The most commonly used PEM crystal, ZnSe, transmits light only down to $800\,cm^{-1}$. Practically achievable extensions to lower wavenumbers have been reported on two occasions. With an MCT type B detector in an FT interferometer the VCD spectra down to $600\,cm^{-1}$ were recorded for 3-methylcyclohexanone and α-pinene [3], although the reliability of the latter measurement is doubtful. For the same two compounds plus 3-bromocamphor the VCD spectra were measured to $650\,cm^{-1}$ also with a dispersive instrument [4]. Dependence on the optical transmissivity of the PEM could be avoided by replacing it, for example, with a polarizing beam-splitter as

the phase modulation element [5]. Though with this method it may be possible in principle to measure VCD spectra into the far-infrared [6], the practicality has yet to be proven. In this report we describe our recent attempts to extend the spectral range to $550\,cm^{-1}$ by using our Bomen MB100 based VCD spectrometer [7].

Experimental

The details of the spectrometer were published previously [7]. The advantageous feature differentiating this design from others is the direct use of the parallel light beam of 2.5 cm diameter emerging from the beam-splitter and passing through all optical elements up to the focusing lens before the detector. For the experiments described below, the ZnSe PEM (2 cm diameter) was replaced by a prototype CdTe PEM (1 cm diameter) on loan from Dr. G. Hoffmann [8], and the MCT type A detector by a specially selected MCT type B detector (designated below as detectors A and B, respectively, and both manufactured by Belov Technologies). The technical specifications of detectors A and B are, respectively, a D* of 5.4×10^{-10} with a bias current of 30 mA at 10 kHz and a long-wavelength cut-off (50%) of $760\,cm^{-1}$, and a D* of 2.7×10^{-10} with a bias current of 15 mA at 10 kHz and a long wavelength cut-off (50%) of $530\,cm^{-1}$. Both detectors are fitted with ZnSe windows. Using focusing lenses of ZnSe or KRS5 constituted another but minor variation. As originally described, the CdTe crystal is optically transparent from 3000 to $333\,cm^{-1}$ [8], and is driven at 70 kHz and 46 V peak-to-peak with one piezo device. In our hands, the maximum voltage applied to the piezoelectric element was 20 V peak-to-peak. We experienced problems with this CdTe PEM at those settings, including instability of the wavelength setting and overheating of the power transistors. As measures of the quality of operation we report below representative spectra of (−)-α-pinene (Fig. 1) and (−)-myretenal (Fig. 2), as well as calibration curves (Fig. 3).

Results and Discussion

The VCD spectra of (−)-α-pinene obtained with detector A down to $850\,cm^{-1}$ and the two photoelastic modulators are displayed in Fig. 1. They demonstrate the relative performance of the two modulators. It is

* To whom correspondence should be addressed

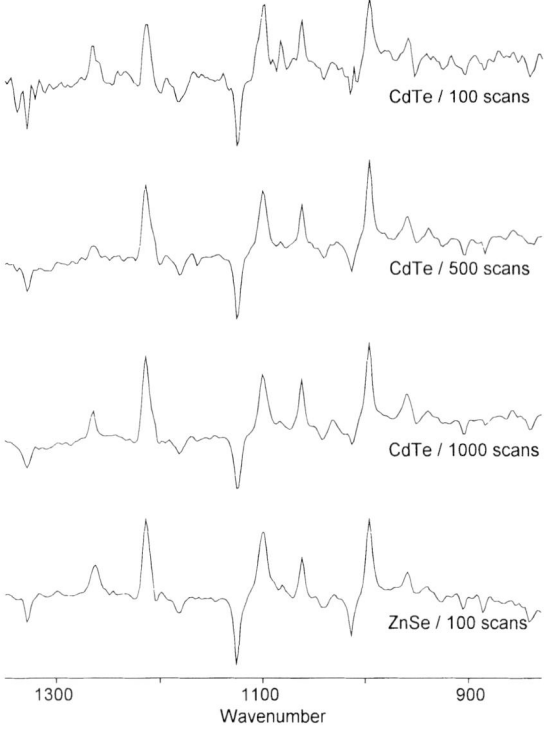

Fig. 1. VCD spectra of (−)-α-pinene enantiomer-corrected, 4 cm⁻¹
resolution

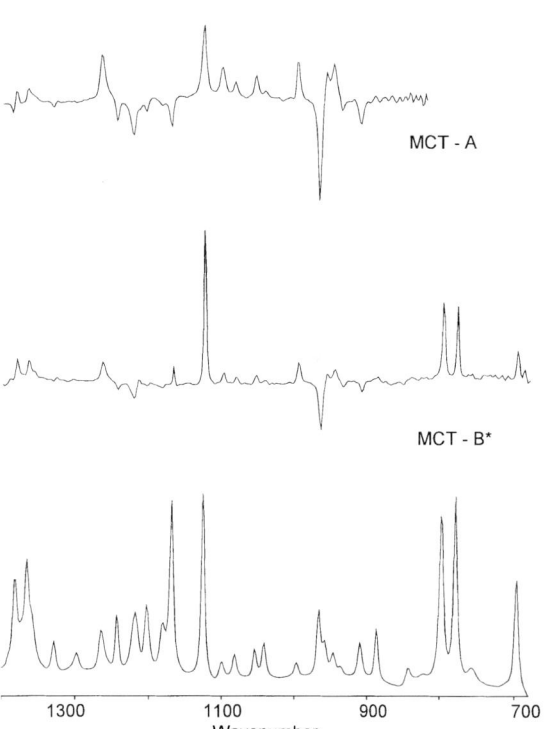

Fig. 2. Absorption (bottom) and VCD (top) spectra of (−)-myrtenal

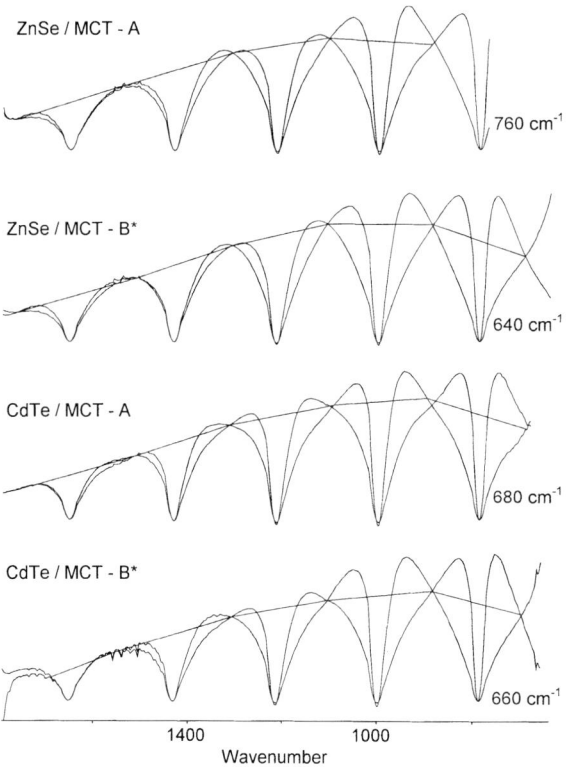

Fig. 3. Comparison of calibration spectra 50 scans, 4 cm⁻¹
resolution

seen that when recording with the CdTe PEM, about
10 times the number of scans are needed to obtain
comparable signal-to-noise ratios. In Fig. 2 the spectra
of (−)-myretenal obtained with ZnSe PEM are compared for the two detectors. The quality of the spectrum recorded with the B detector is clearly
comparable to that obtained with the A detector. The
comparison of the two VCD spectra also shows that
the former becomes saturated more quickly, as may be
seen from the relative intensities of the VCD bands
near 1110 and 800 cm⁻¹. These bands appear unduly
intense in the VCD spectrum in comparison with other
features, even though the corresponding absorbances
are only about 0.5 (scale not shown). To remedy this
distortion, closer optical filtering with the B detector
will be necessary. Figure 3 shows the quality of circular
polarization of the two modulators, as well as the lower
wavenumber limit which could optimally be achieved
with this arrangement. A wavelength extension by
about 120 cm⁻¹ can be realized with the B detector and
the ZnSe PEM without degrading the signal significantly. On the other hand, the prototype PEM fails to
produce reliable circular polarization at higher wavenumbers than its optical transmissivity would suggest.

We conclude, therefore, that an important extension in wavenumber response can be achieved by using a selected B detector, though closer optical filtering is required. While the CdTe PEM may be capable in principle of operating down to 333 cm^{-1}, this target is not achievable with this particular device.

References

[1] M. Diem, in: *Vibrational Spectra and Structure, Vol. 19* (J. R. Durig, ed.), Elsevier, Amsterdam, 1991.

[2] T. A. Keiderling, P. Panocoska, in: *Biomolecular Spectroscopy Part B, Advances in Spectroscopy, Vol. 21* (R. J. H. Clark, R. E. Hester, eds.), Wiley, Chichester, 1993, p. 267.

[3] P. L. Polavarapu, *Appl. Spectrosc.* **1984**, *38*, 26.

[4] F. Devlin, P. J. Stephens, *Appl. Spectrosc.* **1987**, *41*, 1142.

[5] N. Ragunathan, N. S. Lee, T. B. Freedman, L. A. Nafie, C. Tripp, H. Buijs, *Appl. Spectrosc.* **1990**, *44*, 5.

[6] P. L. Polavarapu, G. C. Chen, S. Weibel, *Appl. Spectrosc.* **1994**, *48*, 1224.

[7] D. Tsankov, T. Eggimann, H. Wieser, *Appl. Spectrosc.* **1995**, *49*, 132.

[8] G. G. Hoffmann, B. Schrader, G. Snatzke, *Rev. Sci. Instrum.* **1987**, *58*, 1675.

Mikrochim. Acta [Suppl.] 14, 781–783 (1997)
© Springer-Verlag 1997

A New Combined Spectroscopic Ellipsometer for the NIR and VIS-UV Range, using a Fourier-transform and a Grating Spectrometer

Uwe Richter*, **Georg Dittmar,** and **Uwe Wielsch**

SENTECH Instruments GmbH, Rudower Chaussee 6, D-12489 Berlin, Federal Republic of Germany

Abstract. Spectroscopic ellipsometry is a powerful tool in the UV/VIS and NIR. A new high-resolution NIR extension using a Fourier-transform spectrometer is added to a UV/VIS spectroscopic ellipsometer and a combined device has been developed, which is fully automated by using a high-precision mechanical fibre switch. The power of high-resolution NIR measurements is demonstrated with a directly bonded silicon on insulator (DBSOI) sample by measuring the thickness non-uniformity and other parameters without the necessity of a microspot.

Key words: spectroscopic ellipsometry, infrared, instrument, Fourier-transform spectroscopy, bonded silicon.

Spectroscopic ellipsometry is a well known and powerful tool for thin-film analysis. A wide spectrum of applications uses a broad spectral range from the UV to the far infrared. Many instruments have been developed to cover the UV/VIS, NIR, MIR and FIR. This paper introduces a new ellipsometer for the UV/VIS and NIR which is more powerful in the NIR than common instruments based on a grating spectrometer.

The special advantage of this combined ellipsometer is its automatic function through the whole spectral range, accompanied by high resolution and speed in the NIR. The new instrument allows precise analysis of non-ideal samples, for example a thickness variation within the measurement spot. Results are given for measurements on a directly bonded silicon-on-insulator (DBSOI) sample.

Instrument Description

The new ellipsometer, shown in Fig. 1, combines a conventional grating spectrometer with Xe-source with a Fourier-transform spectrometer which modulates light from a halogen lamp source. Both sources and detector systems are attached to a common goniometer carrying Polarizer-removable Compensator-Sample-Analyser (PCSA) optics.

The combination of two sources and two detector systems requires manual change of the optical fibre cables in other systems. In the instrument presented here a new high precision mechanical switch allows two fibres to be attached together with automatic switching between them by software. This has the advantage that the alignment of the measurement beam is fixed and the speed drastically increases because no user action is required. This set-up is required because no fibre cable is available with high throughput throughout the whole wavelength range from the UV to the NIR (quartz absorption bands at $1.4\,\mu m$ wavelength). Single-fibre set-ups would have weak intensities either in the UV or in the NIR. Our instrument has the throughput of systems with special fibres for UV and NIR, and the easy usage of single-fibre systems.

The second new feature of the combined ellipsometer is the NIR extension. Common solutions with grating spectrometers in the NIR have problems with the dynamic range when higher spectral resolutions are used.

In the UV/VIS, photomultiplier tubes with a dynamic range of 10^6 are used as detectors. In the NIR either InGaAs or Ge detectors are used, but their dynamic range (signal/noise ratio) is much smaller. For grating spectrometers the intensity decreases when the monochromator entrance slit is narrowed, but the spectral resolution increases. This behaviour and the

---

* To whom correspondence should be addressed

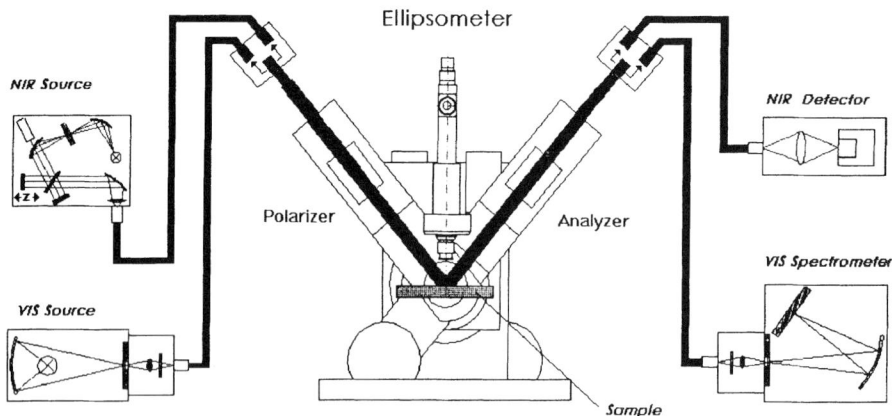

Fig. 1. Set-up of the new ellipsometer

demand for higher spectral resolution increases the measurement times and noise. So common NIR extensions require long measurement times for higher resolutions and unacceptable measurement times for obtaining the many data points which are necessary for analyzing non-ideal samples.

The new set-up uses a Fourier-transform spectrometer in the NIR to measure the single-beam spectra at different analyzer positions A_i. The Stokes vector [1] is calculated from these single-beam spectra by Fourier-transformation and the resulting ellipsometric angles follow the well known PCSA ellipsometer formulae [2].

The use of a Fourier-transform spectrometer allows high spectral resolution at high speed. A spectrum of 5000 data points is measured within a few seconds, whereas grating spectrometers require more than an hour for the same spectrum. Another improvement of this method is the modulation principle, allowing an increase in the signal/noise ratio. Thus the new NIR extension is more powerful than grating spectrometer solutions in three features: (1) higher measurement speed, (2) higher signal/noise ratio, (3) high spectral resolution for standard measurements.

These new features give excellent readings for ellipsometric spectra and allow the analysis of non-ideal samples. The following section demonstrates the application of this high resolution to a typical non-ideal sample.

Example: Thickness Non-uniformities

Many samples show non-ideal properties. A typical problem is the variation of the thickness with the location on the sample. This is often due to the deposi-tion or etching process. Many photoresists are spun onto a substrate, causing circularly oriented thickness gradients, sometimes accompanied by fringes. The aim of the following analysis is to demonstate the power of the new ellipsometer with its SpectraRay software to measure and model non-ideal samples. Etching can cause the thickness inhomogeneities, as is the case in the example discussed here.

A new technique allows the bonding of a silicon substrate onto a thermal oxide on a second silicon wafer. The bonded silicon wafer is etched on its other face and thinned down to a remaining thickness of a few microns. This process of etching about 500 microns is not ideal, and retains any non-parallelism of the wafer. After the etching process this causes non-uniformity of the remaining thickness, even if the etching process was excellently controlled.

The case of a directly bonded silicon on insulator (DBSOI) sample was analysed by El-Ghazzawi et al. [3] They found a good agreement between measurement and fit if the spot was reduced by using a microspot, and a slit for reducing the influence of focusing.

However, the use of a microspot lens is not possible in all ellipsometric applications (high requirements for sample alignment) and introduces new calibration constants for the change of state of polarization introduced by the microspot lenses and the focusing. In many cases it is not possible to measure with higher local resolution (this decreases the influence of the thickness-gradient) or to repeat the measurement with a microspot.

We have developed an averaging method, which takes the thickness non-uniformities into account and models them correctly. The starting point was the DBSOI sample described above, which was known to

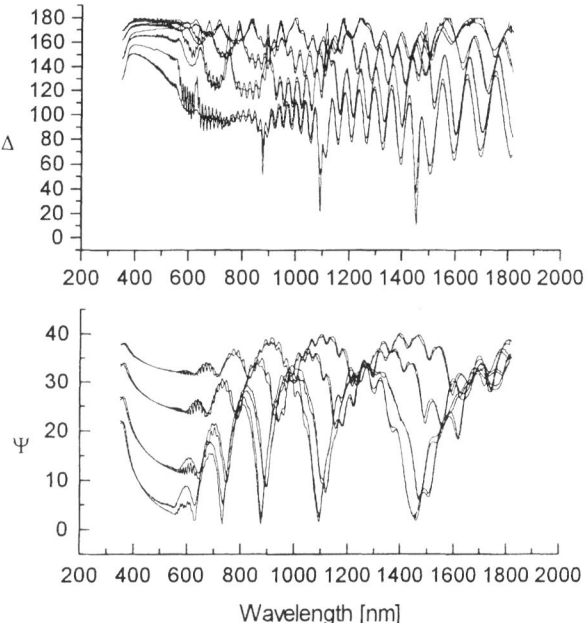

Fig. 2. Fit of multiple-angle measurement of the DBSOI sample

have such non-uniformity. A series of ellipsometric spectra at several angles of incidence was measured, using a large measurement spot (fully open aperture). The spot size changes from 4 to 8 mm with the angle of incidence.

Fits based on an ideal double-layer model showed bad performance. The thickness non-uniformity was introduced, assuming that the beam diameter of 4 mm at $0\,^\circ$ angle of incidence was changed to an ellipse for other angles. The averaging method for different thicknesses t was performed with intensities:

ideal stack $(\psi_i, \Delta_i) \rightarrow$

  ideal Stokes vectors $\vec{S}_i$

    $\int \mathbf{S} dt$

      $\rightarrow$ averaged Stokes vectors $\langle \vec{S} \rangle$

        $\rightarrow$ averaged ellipsometric angles $(\langle \psi \rangle, \langle \Delta \rangle)$ as a result

The fits gave good results in the whole spectral range and for all angles of incidence, as shown in Fig. 2. The results were: 2.2 nm oxide/3560 nm c-Si with 26 nm/mm gradient/2008 nm thermal oxide/c-Si.

A cross-reference measurement was performed with a SENTECH Film Thickness Probe FTP 500 for mapping purposes with high local resolution (10 µm spot). This gave the thickness variation of the top layer. A line scan through the measurement spot of the ellipsometer gave a thickness non-uniformity of 30 nm/mm, which agrees well with the results with the spectroscopic ellipsometer. The good agreement is achieved because the high spectral resolution creates no additional damping of interferences, as limited spectral resolution does. Otherwise two non-idealities have to be considered. In this case only the sample contributes to the non-idealities and this allows the resolution of thickness non-uniformities.

Conclusion

The new instrument is a fully automated ellipsometer for the UV/VIS and the NIR spectral range. It combines a high measurement speed and ease of use. The new type of NIR extension allows fast measurements at high spectral resolution (compared to common grating spectrometers).

The power of high spectral resolution in the NIR was demonstrated by measuring and modelling a DBSOI sample. This gave additional information about the thickness non-uniformity. A cross-reference with high local resolution showed good agreement between both methods.

References

[1] R. M. A. Azzam, N. M. Bashara, *Ellipsometry and Polarized Light*, North-Holland, Amsterdam, 1989, p. 55.
[2] R. M. A. Azzam, N. M. Bashara, *Ellipsometry and Polarized Light*, North-Holland, Amsterdam, 1989, p. 153.
[3] M. E. El-Ghazzawi, T. Saitoh, N. Hori, A. Sakai, T. Oka, *Thin Solid Films* **1993**, *233*, 218.

Mikrochim. Acta [Suppl.] 14, 785–787 (1997)
© Springer-Verlag 1997

Novel All-fibre-optic Fourier-transform Spectrometer with Thermally Scanned Interferometer

M. Stelzle[1,*,**], **J. Tuchtenhagen**[1], and **J. F. Rabolt**[2]

[1] Max-Planck-Institut für Kolloid- und Grenzflächenforschung, Rudower Chaussee 5, D-12489 Berlin, Federal Republic of Germany
[2] IBM Almaden Research Center, K13C/801, 650 Harry Road, San Jose, CA 95120, U.S.A.

Abstract. A novel Fourier-transform spectrometer based on optical fibres (FO-FTS) was constructed. The optical retardation between the arms of a Mach–Zehnder type interferometer made of a single-mode optical fibre is modulated by variation of the fibre temperature. Spectra of a multimode laser diode and a broad-band light source were recorded. Coherent averaging of multiple interferograms was successfully applied to enhance the signal-to-noise ratio of the interferograms. Application of the FO-FTS to chemically selective imaging of surfaces is proposed.

Key words: Fourier-transform spectroscopy fibre-optic interferometer.

The application of interferometers with air-bearings in common FTS designs makes these instruments sensitive towards vibrations. To date, no portable instruments or designs suited for a rough process-environment are commercially available. The FO-FTS we report here is based on an interferometer without any moving parts and is therefore inherently insensitive to environmental distortions. In addition the use of a fibre optic interferometer will allow for the design of very compact instruments. Earlier publications of fibre optical interferometers reported the use of the optical fibre only to guide the light [1, 2], the optical retardation being produced with an external moving mirror [3]. The use of a piezo fibre-stretching device was reported for the construction of an optical spectrum analyser by Krath et al. [4]. The approach chosen here, i.e. cycling the fibre temperature rather than mechanical stretching of the fibre, was previously proposed by Kersey et al. [5], but, no experimental data were presented.

---

* Present address: Natural and Medical Science Institute, NMI, Eberhardstrasse 29, D-72762 Reutlingen, Federal Republic of Germany.
** To whom correspondence should be addressed

Experimental

Figure 1 shows the schematic design of the FO-FTS. Light from the IR source and from an HeNe reference laser are coupled into the arms of a Mach–Zehnder type fibre-optic interferometer. The interferometer consists of a silica single-mode fibre with a cut-off wavelength of 630 nm and a total length of each arm of approximately 40 m. The fibre of one arm is wrapped around a cylinder which can be heated and cooled by means of thermoelectric elements. Thus, an optical retardation is produced since the refractive index, n, and therefore the optical path-length $L = ln$ (where l = geometric length of the fibre arms) will vary with temperature according to

$$L(v, \Delta T) = l \frac{dn}{dT}(v) \Delta T.$$

The maximum achievable optical retardation of our instrument was of the order of 2.2 cm, corresponding to a temperature span of one fibre arm of ± 20 K. The maximum achievable scan speed obviously depends on the available heating/cooling power and the heat capacity of the fibre/drum assembly. We achieved a scan speed of approximately 0.2 mm/s without specifically optimizing for maximum performance in this respect. A less massive cylinder than that employed in our set-up, temperature scanning of both arms with $180°$ phase shift, and the use of even longer fibres would allow for scan speeds of several cm/s.

At the output of the fibre coupler, the reference beam is reflected onto the reference detector by a dichroic mirror, while the IR light is transmitted and detected by an infrared detector. The ac-component of the reference signal is employed to create a trigger impulse for the AD converter board at each zero crossing. Thus, the infrared intensity is sampled at equidistant intervals $\lambda_{REF}/2$ (λ_{REF} = wavelength of the HeNe laser) of the optical retardation. According to the sampling theorem, the upper wavenumber limit for the FO-FTS is therefore given by $v_{max} = 1/\lambda_{REF}$.

Results and Discussion

In order to investigate the resolution of the FO-FTS an NIR multimode laser diode was used as light source. An interferogram and the spectrum calculated therefrom are displayed in Fig. 2. The areas appearing

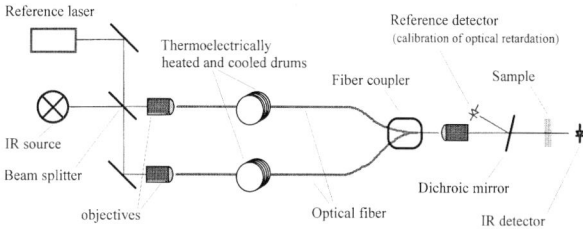

Fig. 1. Experimental set-up of the FO-FTS

shaded are parts of the interferogram with high fringe amplitude (Fig. 2a). The interferogram is displayed as acquired. The offset of the retardation is due to the fact that it is impossible to cut the fibre arms to exactly the same length within better than several mm. In addition, temperature differences between the fibre arms will

contribute to this offset of the optical retardation. The point of zero retardation is marked by an arrow in all interferograms. All amplitude spectra were calculated by complex Fast Fourier Transform (FFT) after zero-filling to the next power of 2. No special apodization function or additional zero-filling was applied. The high resolution and considerable signal/noise ratio achievable with the FO-FTS with only a single scan are evident. Clearly visible are about 15 peaks under an envelope with two maxima. This spectrum may be compared to a spectrum of the same laser diode recorded with a conventional grating instrument (Fig. 2b, inset) which also shows multiple peaks under an envelope with two maxima. It is interesting, however, to note the difference of about 600 cm^{-1} between the centre wavenumber in both spectra. This can be explained in terms of a frequency dispersion of the coefficient dn/dT [6]. In other words, dn/dT and therefore also the optical retardation produced by a certain change of the fibre temperature is smaller at 12200 cm^{-1} than it is at the reference frequency, i.e. 15800 cm^{-1}. In the case of the laser diode we find the ratio

$$(\mathrm{d}n/\mathrm{d}T)_{\nu = 12200\ \mathrm{cm}^{-1}}/(\mathrm{d}n/\mathrm{d}T)_{\nu = 15800\ \mathrm{cm}^{-1}}$$
$$= 11600\ \mathrm{cm}^{-1}/12200\ \mathrm{cm}^{-1} = 0.95$$

In turn, this type of interferometer would also allow for very precise measurements of dn/dT and its frequency dispersion.

In order to overcome the limitation with respect to the upper wavenumber limit, $\nu_{max} = 1/\lambda_{REF}$, described above, two prerequisites have to be met: first, the interferogram has to be oversampled by a factor of N in order to fulfil the sampling criteria for the high frequency limit, ν_{max}, of the spectrum, according to $\nu_{max} < N/\lambda_{REF}$. Secondly, a fibre with a cutoff wavelength, given by $\lambda_{cutoff} < 1/\nu_{max}$ must be employed in order to ensure pure single-mode transmission over the full spectral range that is to be covered with the FO-FTS. In order to fulfil the first criterion we devised an electronic circuit based on a phase-locked loop (PLL). The device produces trigger impulses at a multiple of the frequency of the reference signal which are phase-locked to the reference signal and equidistant. Figure 3 demonstrates the effectiveness of this approach. A spectrum was acquired, using multiplication of the trigger frequency by a factor of two. At high gain setting, the infrared detector picks up the residual HeNe light transmitted through the dichroic mirror. Along with the signal of the laser diode operated far below its laser threshold, this results in the spectrum

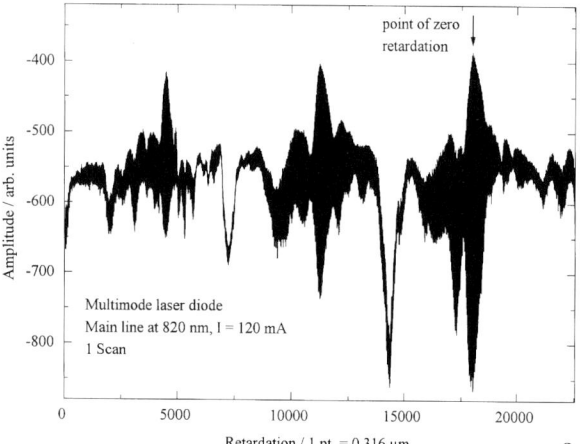

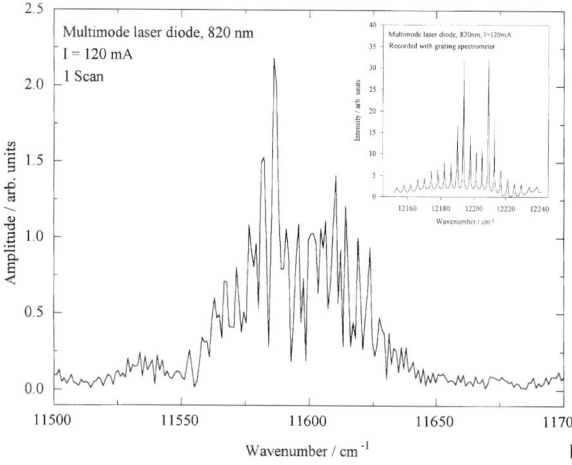

Fig. 2. NIR multimode laser diode as light source. **a** Interferogram of a single-scan measurement. The point of zero retardation is indicated by the arrow. **b** Spectrum calculated from the interferogram in (**a**). Inset: spectrum of the same laser diode, measured with a conventional grating spectrometer. Note the difference in peak wavenumber, which is due to the frequency dispersion of the coefficient dn/dT of the fibre

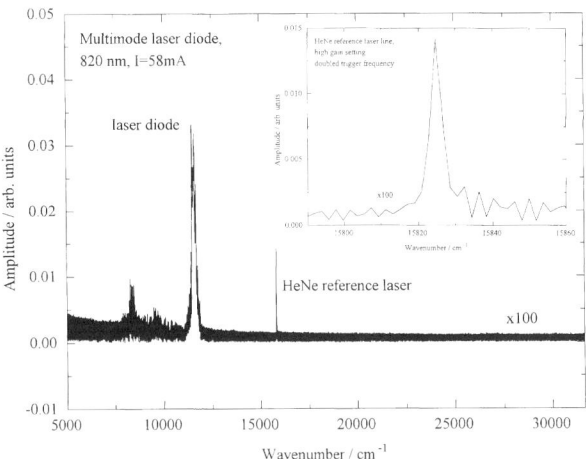

Fig. 3. Extension of the upper spectral limit of the FO-FTS by means of a phase-locked loop circuit. Spectrum of a low-intensity light source (laser diode operated at below laser threshold) acquired with frequency-doubled trigger signal. Note that the upper wavenumber limit imposed by the sampling criterion is now $31600 \, cm^{-1}$ even though an HeNe laser was used as reference light source. Inset: the resolution of the residual HeNe signal sampled by the IR detector approaches the theoretical limit for the applied optical retardation of 0.44 cm

displayed in Fig. 3. The HeNe peak is now located in the middle of the spectrum; the upper wavenumber limit is $31600 \, cm^{-1}$. The trace in the inset of Fig. 3 demonstrates the high resolution of the FO-FTS when operated in this mode, which approaches the theoretical limit (optical retardation was 0.44 cm). The use of a fibre with a cutoff wavelength at 300 nm, rather than 630 nm as in our set-up, would thus allow sampling of light sources with spectral components up to the near UV, while utilizing an HeNe laser as reference light source.

For very low intensity sources it is necessary to coherently average multiple interferograms in order to enhance the signal/noise ratio. This was successfully performed with multiple scans which had been previously acquired and stored on disk. A computer algorithm allowed detection of glitches of relative retardation between the interferograms and distortions of the interferograms due to erroneous trigger pulses. An example is shown in Fig. 4: the considerable gain of signal/noise ratio in the case of coherent averaging (upper trace) compared to a single-scan measurement (lower trace) is obvious. The constant amplitude of the fringes outside the centre-burst region is due to a residual transmission of the reference light (HeNe) onto the IR-detector. The fact that the fringe amplitude does not decrease with increasing retardation indicates that the interferograms were actually added in a coherent fashion.

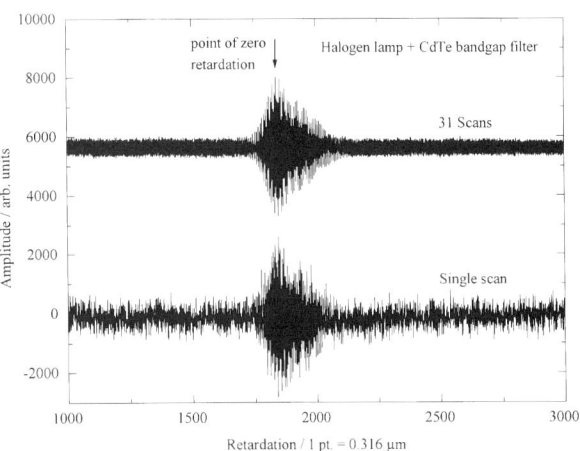

Fig. 4. Coherent averaging of multiple scans of a low-intensity broad-band light source. The upper trace resembles the average of 31 coherently co-added scans, while the lower trace is a single scan. The constant amplitude of the fringes outside the centre-burst region is due to a residual transmission of the reference light (HeNe) onto the IR-detector

Conclusions

The feasibility of an all-fibre-optic FTS based on a thermally scanned interferometer is demonstrated. High-resolution spectra can be acquired with only a single scan. A PLL circuit enables oversampling of the interferogram and allows for the acquisition of spectra containing frequency components beyond the reference laser frequency.

Chemically selective imaging of surface films is conceivable with the FO-FTS. The output terminal of the fibre coupler could serve as scanning probe. Lateral resolution superior to that of current FTIR microscopy seems achievable by employing near-field optical techniques such as are already commonly employed in scanning near-field optical microscopy (SNOM).

Acknowledgements. M. S. is grateful for support from the Deutsche Forschungsgemeinschaft (DFG) through grant Ste 639/1-1 for a postdoctoral fellowship. J. T. would like to thank the Max-Planck Society for a postdoctoral fellowship. Partial support by the NSF Centre on Polymer Interfaces and Macromolecular Assemblies is acknowledged.

References

[1] D. P. Hand, T. A. Carolan, J. S. Barton, J. D. C. Jones *Optics Commun.* **1993**, *97*, 295.
[2] E. L. Moore, O. G. Ramer, *Proc. SPIE* **1983**, *412*, 160.
[3] K. Takada, M. Kobayashi, J. Noda, *Appl. Optics* **1990**, *29*, 5170.
[4] K. J. Krath, B. Scholl, H. J. Schmitt, *IEEE Photonics Techn. Lett.* **1992**, *4*, 206.
[5] A. D. Kersey, A. Dandridge, A. B. Tveten, T. G. Giallorenzi, *Electron. Lett.* **1985**, *21*, 463.
[6] M. Stelzle, J. Tuchtenhagen, J. F. Rabolt, *Measurement Science and Technology*, accepted for publication.

Mikrochim. Acta [Suppl.] 14, 789–791 (1997)
© Springer-Verlag 1997

Fourier-transform Infrared Emission Spectroscopy of Kaolinite Dehydroxylation

Ray L. Frost[1,*] and **Anthony M. Vassallo**[2]

[1] Centre for Instrumental and Developmental Chemistry, Queensland University of Technology, 2 George Street, GPO Box 2434, Brisbane, Q 4001, Australia.
[2] CSIRO Division of Coal and Energy Technology, PO Box 136, North Ryde, NSW 2113, Australia.

Abstract. The dehydroxylation of kaolinite has been investigated by Fourier-transform *in situ* infrared emission spectroscopy from 100 to 800 °C at 5 degree intervals. The major advantage lies in the ability to obtain vibrational spectroscopic information in situ at the elevated temperature. Dehydroxylation was determined by the loss of intensity of the hydroxyl bands in the 3550–3750 cm^{-1} emission spectra. No clay phase changes occur until after dehydroxylation takes place. The kaolinite layers lose their outer and inner hydroxyl groups simultaneously. It is proposed that the kaolinite dehydroxylation process takes place homogeneously and involves two mechanisms.

Key words: FT-infrared, emission, hydroxyl, kaolinite, dehydroxylation.

Clay minerals, including kaolinite, have long been studied by using infrared spectroscopy [1–4] The technique of measurement of discrete vibrational frequencies emitted by thermally excited molecules, known as infrared emission spectroscopy (IES) [5] is not widely used. The major advantages of IES are that the samples are measured in situ at the elevated temperature and IES requires no sample treatment other than making the clay sample of submicron particle size. Further, the technique removes the difficulties of heating the sample to dehydroxylation temperatures and quenching it before measurement. IES measures the dehydroxylation process as it is actually taking place.

Experimental

FT-IR emission spectroscopy was performed with a Digilab FTS-60A spectrometer, which was modified by replacing the IR source with an emission cell. Approximately 0.2 mg of finely powdered clay was spread as a thin layer on a 6 mm diameter platinum surface and held in an inert atmosphere within a nitrogen-purged cell, with an additional flow of argon over the sample during heating. The time between scans (while the temperature was raised to the next hold point) was approximately 10 seconds. The spectra were acquired by co-addition of 256 scans for temperatures 100–300 °C (approximate scanning time 140 seconds) and 64 scans for temperatures 400–800 °C (approximate scanning time 45 seconds), with a nominal resolution of 4 cm^{-1}. The emittance spectrum at a particular temperature was calculated by subtraction of the single beam spectrum of the platinum support-plate from that of the platinum + sample, and the result ratioed to the single beam spectrum of an approximate black body (graphite). Emittance values vary from 0 to 1 with a scale equivalent to a transmission spectrum. The data were linearized with respect to concentration where required, by transforming to units of -log10 [1 − emittance (ν)].

Results and Discussion

The emission spectra of kaolinite (KGa-1) obtained at 5°C intervals from 200 to 600°C are shown in Figs. 1 and 2. Dehydroxylation, as evidenced by loss of the signal due to the OH stretching modes between 3600 and 3800 cm^{-1}, starts to occur at 425 °C. Four major bands are observed in the IES spectra, at 3684, 3664, 3649 and 3620 cm^{-1} at 175 °C; they correspond to the four infrared absorption bands found at 3707, 3674, 3655 and 3612 cm^{-1} for kaolinite. Thus there is excellent correspondence between the infrared emission spectra and the corresponding absorption bands, enabling the IES technique to be used for the *in situ* study of the kaolinite dehydroxylation. In a highly ordered kaolinite such as the Georgia kaolinite KGa-1 the four distinct bands are assigned as follows: the three higher frequency vibrations (v_1, v_2, v_3) are due to the three outer hydroxyl groups of the kaolinite layers, the v_4 band at about 3620 cm^{-1} is attributed to the inner hydroxyl group. All of the hydroxyl bands have been

---

* To whom correspondence should be addressed

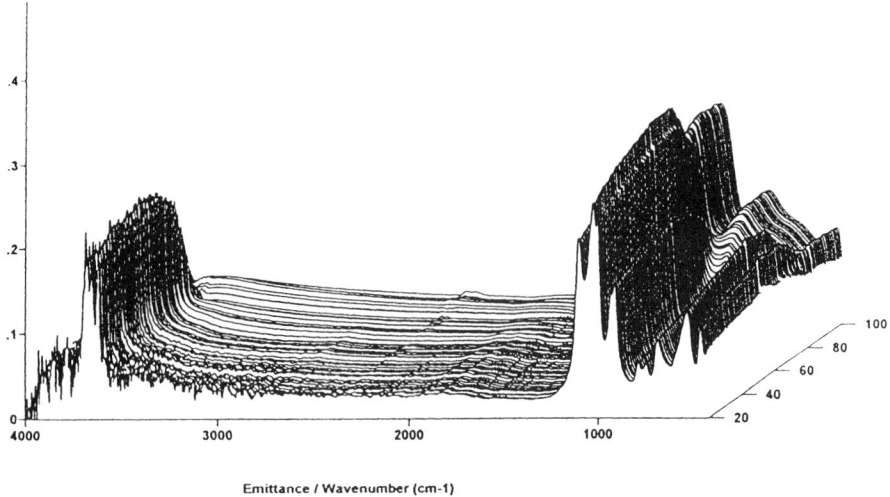

Fig. 1. The infrared emission spectra of kaolinite (KGa-1) over the 400 to 4000 cm^{-1} range and from 200 to 600 °C

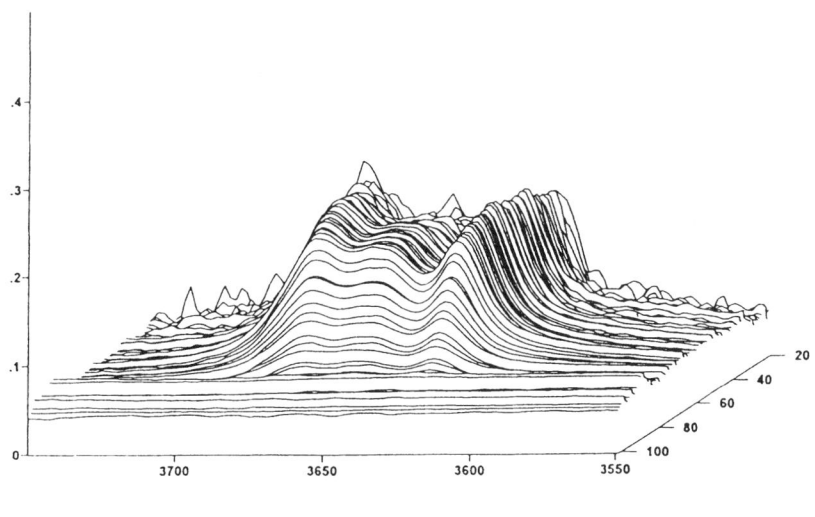

Fig. 2. The infrared emission spectra of kaolinite (KGa-1) over the hydroxyl stretching region 3550 to 3750 cm^{-1} and from 200 to 600 °C

found to be temperature-dependent. In this work the IES infrared peak of the inner hydroxyl group changed from 3620 to 3626 cm^{-1} over the temperature range from 175 to 520 °C at which point the hydroxyl was totally lost. In the IES, the peak for the outer hydroxyls is shifted from 3670 to 3655 cm^{-1} at 500 °C. Thus the band of the outer hydroxyls undergoes a red shift and the band attributed to the inner hydroxyl a blue shift, over this temperature range. One possible reason for these spectral shifts can be the hydrogen bonding capacity of the hydroxyl groups and their consequential acidity. The hydrogen bonding of the outer hy-

droxyls strengthens with increase in temperature as their protons become delocalized, shortening the hydrogen-bond lengths, whereas the hydrogen bonding of the inner hydroxyl weakens with increased temperature. The frequency shift of the stretching modes of the outer OH groups, located on the surface of the layers, is larger than for the internal groups: Δv for the 3670 cm^{-1} band is 16 cm^{-1} compared with Δv of 4 cm^{-1} for the 3620 cm^{-1} band for a temperature increase from 200 to 500 °C. There is no evidence to support the proposition that the different kaolinite hydroxyls are being removed at different temperatures or at different rates. The hydroxyls are progressively lost from 420 °C upwards and this loss is complete by 520 °C. It appears that the loss of the relative intensity of the 3620 cm^{-1} hydroxyl is linear over the temperature range from 420 to 480 °C. There is a change in the slope of the temperature–relative intensity graph at 480 °C until at 520 °C there is no intensity remaining. The change in the slope would support the hypothesis that two different mechanisms are in operation for the dehydroxylation of kaolinite.

References

[1] A. N. Lazarev, *Vibrational Spectra and Structure of Silicates*, Plenum, New York, 1972.
[2] V. C. Farmer, in: *The Infrared Spectra of Minerals* (V.C. Farmer ed.), Mineralogical Society, London, 1974, Chapter 15.
[3] G. C. Maiti, F. Freund, *Clay Minerals* **1981**, *16*, 395.
[4] J. J. Fripiat, F. Toussaint, *J. Phys. Chem.* **1963**, *67*, 30.
[5] A. M. Vassallo, P. A. Cole-Clarke, L. S. K. Pang, A. Palmisano, *Appl. Spectrosc.* **1992**, *46*, 73.
[6] C. T. Johnston, S. F. Agnew, D. L. Bish, *Clays & Clay Minerals* **1990**, *38*, 573.

Mikrochim. Acta [Suppl.] 14, 793–795 (1997)

Airborne FT-IR Emission Spectrometer for Trace Gas Monitoring

Peter Haschberger

Institut für Optoelektronik, Deutsche Forschungsanstalt für Luft- und Raumfahrt e.V., D-82230 Weßling, Federal Republic of Germany

Abstract. A new type of Michelson interferometer has been designed, based on use of a rotating retroreflector to change the path-length. The instrument is very insensitive to mechanical vibration and climatic conditions. It has been used for airborne measurement of air craft engine exhaust.

Key words: Michelson interferometer, rotating retroreflector, aircraft exhaust analysis, emission indices, trace gas analysis.

The DLR (German Aerospace Research Establishment) at Oberpfaffenhofen in Germany has recently developed a Fourier-transform infrared spectrometer (FTS) MIROR (Michelson interferometer with rotating retroreflector) based on a new type of Michelson interferometer that uses a continuously rotating retroreflector to generate the path-length alteration. This unique driving concept and the opto-mechanical lay-out make the instrument very insensitive to mechanical vibrations and climatic changes. Thus the system is suitable for mobile field and airborne operation.

With the MIROR spectrometer as the main part of a measurement system for emission spectroscopy, airborne measurements of aircraft jet engine exhaust were performed and quantitatively evaluated to calculate trace gas concentrations and emission indices.

Spectrometer

The opto-mechanical setup of the MIROR spectrometer (Fig. 1) is derived from the classical Michelson type. Instead of the plane mirror moving back and forth, a hollow corner cube retroreflector is used, overcoming tilt and shear problems. The optical path is varied in both interferometer arms by directing the beams of radiation to this eccentrically rotating retroreflector. As the angles of incidence with reference to the reflector's axis of symmetry have different signs ($\pm \alpha$) the path difference between the two arms changes sinusoidally during rotation. The fixed mirrors that reflect the beams the same way back through the reflector to the beamsplitter are designed as bore-hole mirrors to maximise the allowable beam diameter.

General advantages of this set-up in comparison with the conventional spectrometer lay-out (such as simplified drive and bearing concept, ruggedness, independence of spectral resolution and measurement rate) have been described in detail [1, 2]. In addition, it is an important feature for mobile and airborne operation that the reflector, as the one and only moving element, has no degree of mechanical freedom in the direction of the optical path alteration. The number of optical components is minimised and only first-surface reflectors are used, to reduce optical losses. With the exception of the corner cube and the drive unit, all parts of the interferometer are mounted on a thermally stabilized massive block of AlMg alloy. Therefore misalignment and performance losses due to vibrations of the components against each other or to thermal loads are reduced to a minimum. The compact and rugged set-up together with its spectroscopic performance (e.g. spectral resolution of down to $0.1\,cm^{-1}$, measurement rate of 2.5 interferograms per second, Table 1) make the spectrometer suitable for mobile operation under harsh environmental conditions.

Application

One application for the MIROR FTS is the quantitative analysis of aircraft exhaust emissions, not only on the ground (at test-rig sites) but also airborne. By means of this it is possible to determine the composition of the jet engine emissions under real flight condi-

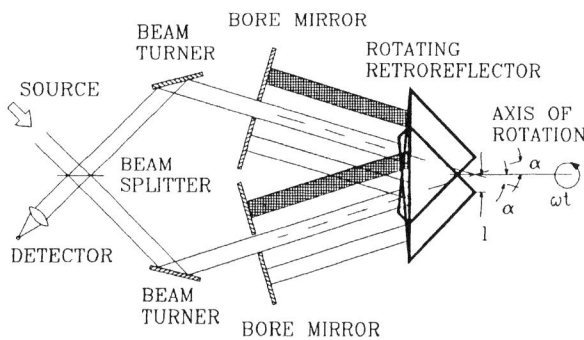

Fig. 1. MIROR, principle set-up: both extreme positions of the rotating reflector are shown

Table 1. MIROR spectrometer: opto-mechanical parameters and performance data

Opto-mechanical parameters	
Angle of incidence (α)	$16.5°$
Eccentricity of retroreflector (I)	$2.55\,cm$
Max. optical path difference	$11.4\,cm$
Max. spectral resolution (FWHH)	$0.12\,cm^{-1}$
Field of view (half angle)	$1\,mrad$
Aperture diameter (telescope)	$16\,cm$
Optical throughput	$2 \cdot 10^{-3}\,sr \cdot cm^2$
Beam splitter	Ge on KBr
Detector, spectral range	MCT, $800–2500\,cm^{-1}$
	InSb, $1800–3000\,cm^{-1}$
Dimensions, weight	$60 \times 40 \times 30\,cm$, $80\,kg$

Performance data	
Speed of rotation (retroreflector)	$1.3\,rps$
Measurement time	$0.25\,s$ per interferogram
NESR (with reference to one measurement)	MCT: $0.24\,\mu W/(cm^2 \cdot sr \cdot cm^{-1})$ @ $1000\,cm^{-1}$
	InSb: $0.08\,\mu W/(cm^2 \cdot sr \cdot cm^{-1})$ @ $2000\,cm^{-1}$
Acquisition system, ADC	16 bit, 10^6 samples/s

tions. This task is part of the German governmental joint programme "Schadstoffe in der Luftfahrt" (air traffic and environment) where more than 20 groups at universities, research institutes and private companies work together to estimate present and future effects of the increasing air traffic on the Earth's atmosphere and climate.

One important parameter for the global influence of aircraft exhaust constituents such as CO_2, H_2O, NO_x, CO, SO_2 is the total amount of each component emitted during flight time. Up to now these data have been calculated from kerosene consumption and engine test-rig runs where the so-called emission indices (grams of emitted gas species per kilogram of kerosene burnt) are measured. As these data are collected under ground level conditions (air temperature, pressure) various mathematical algorithms have been developed to estimate the true emission indices under different flight conditions.

By use of the airborne MIROR FTS it is now possible to measure gas concentrations and emission indices for all major trace gases in-flight to validate these algorithms and other in-situ measurement methods.

As a first platform the DLR research jet aircraft VFW 614 was chosen. With its two Rolls Royce engines mounted overwing the hot jets can be observed from inside the aircraft cabin in a range between the nozzle exit and some metres downstream from the nozzle. Figure 2 shows the measurement set-up in

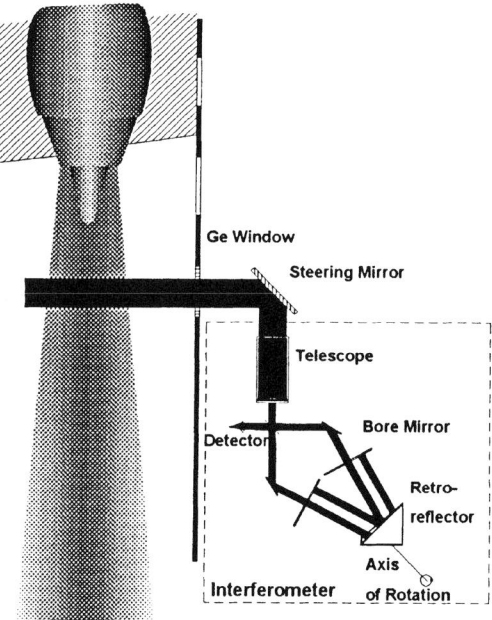

Fig. 2. MIROR on board VFW614

principle: the MIROR spectrometer is mounted in the cabin and focuses on the plume about one metre behind the nozzle. For this one of the cabin windows is replaced by an infrared transmitting germanium plate. The spectrometer works in the emission mode, analysing the spectral composition of the hot jet. The line of sight is controlled by a steering mirror. For in-flight radiometric calibration of the spectrometer this mirror

Table 2. Results of flight campaign: concentrations and emission indices (EI) of detected gas components, and gas temperatures for several measurements with identical flight parameters (altitude: 25000 ft, true airspeed: 145 m/s, fuel flow: 0.18 kg/s)

Time	[UTC]	9:54	9:56	9:59	10:03
NO	C [ppm V]	62.8	65.6	64.1	65.1
	EI [g/kg]	3.8	4.0	3.9	3.9
CO	C [ppm V]	153	156	156	159
	EI [g/kg]	8.7	8.8	8.8	8.9
CO_2	C [%]	3.6	3.5	3.6	3.6
	EI [g/kg]	3148	3132	3155	3176
H_2O	C [%]	2.2	2.3	2.3	2.3
	EI [g/kg]	813	829	830	829
Gas temperature [K]		636.5	633.5	633.2	633.9

is turned to direct successively to the instrument the radiation of three black body sources heated at different temperatures [3].

Analogue pre-processing such as amplification, filtering and signal conditioning is performed by a 19-in. rack mounted electronics; a PC containing a plug-in card with a fast 16-bit ADC controls the data acquisition and spectrometer operation.

To calculate the gas concentrations the line of sight of the spectrometer is divided into multiple layers, each of them characterized by a set of parameters (temperature, pressure, gas concentrations within the layer). These parameters are varied within a balancing algorithm (least-squares fit) to find an optimum match between the measured spectra and spectra calculated from the modelled scenario [4]. A simple model consists of only three layers (foreground, plume, background) whereas more complex models describe the jet core and bypass regions by a set of interdependent layers.

In addition, aircraft and flight parameters (i.e. altitude, speed, fuel flow, exit gas temperature, air temperature etc.) are recorded during the flight campaigns. From these data and the calculated gas concentrations the emission indices can be determined.

Results

In December 1994 the measurement system received its flight approval and a first test flight of the spectrometer on board the VFW 614 was performed. For the first time the exhaust species of an aircraft jet engine could be measured directly behind the nozzle exit at cruise altitudes. The results of this and several following campaigns demonstrate the excellent performance of the instrument (Fig. 3). Due to the special design mentioned above, the spectral resolution and NESR

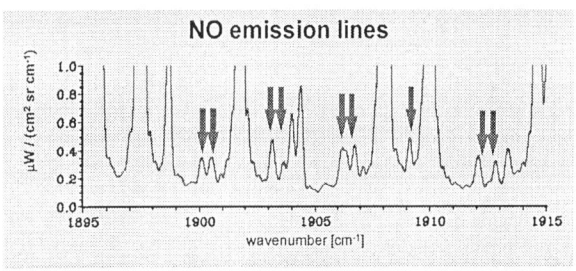

Fig. 3. Emission spectrum (NO range) of aircraft jet engine measured at cruise altitude (30000 ft); measurement time 75 s, spectral resolution $0.2\,cm^{-1}$

noise equivalent spectral radiance during flight are nearly identical to those of measurements in the laboratory. The species CO_2, H_2O, NO, and CO are detected and evaluated quantitatively (Table 2). Additionally, the temperature of the jet is determined with an uncertainty of only a few Kelvin (this value is an important input for models that simulate propagation and chemistry within the exhaust plume). In many cases discrepancies between the measured data and the extrapolated ground-based data were found, serving as starting points for further investigations and interdisciplinary discussions.

Operating the MIROR FTS on board the VFW 614 aircraft is only a first step to demonstrate the feasibility of the new instrument and the measurement set-up. Future projects plan to install the spectrometer in a more relevant, commercial, aircraft such as an Airbus A340 and perform similar flight campaigns.

References

[1] P. Haschberger, V. Tank, *J. Opt. Soc. Am. A* **1993**, *1*, 2338.
[2] P. Haschberger, *Appl. Spectrosc.* **1994**, *48*, 307.
[3] E. Lindermeir, P. Haschberger, V. Tank, H. Dietl, *Appl. Opt.* **1992**, *31*, 4527.
[4] E. Lindermeir, *Ann. Geophys.* **1994**, *12*, 417.

Mikrochim. Acta [Suppl.] 14, 797–799 (1997)
© Springer-Verlag 1997

Characterization of Gaseous Emission Sources by FT-IR Emission Spectrometry

Klaus Schäfer[1,*], **Rainer Haus**[2], and **Jörg Heland**[1]

Fraunhofer Institute of Atmospheric Environmental Research, [1] Kreuzeckbahnstrasse 19, D-82467 Garmisch-Partenkirchen, and [2] DLR Institute of Planetary Exploration, Rudower Chaussee 5, D-12484 Berlin, Federal Republic of Germany

Abstract. FT-IR emission spectrometry is used to detect the thermal radiation of warm exhausts, yielding in one measurement all information about the compounds present. Software units for the spectra retrieval has been developed, based on radiative transfer line-by-line calculations, a multi-layer plume model, and least-squares fit procedures. Measurements were performed with the mobile commercial K300 spectrometer (MCT- or InSb-detector). Temperature and concentrations of CO, CO_2, N_2O, NO, HCl, SO_2 and H_2O in smoke stack plumes from thermal power plants and municipal incinerators, of CO_2, H_2O, CO and CH_4 from flares, and of CO, CO_2, NO and H_2O in exhausts of aircraft engines were determined.

Key words: FT-IR, emission spectrometry, air pollution, emission inventory, environmental monitoring.

The use of FT-IR emission spectrometry in a remotely working analyzer unit for warm exhausts has been further developed and evaluated because this passive sensing enables monitoring of all IR-absorbing compounds in one measurement, emission source quantification up to distances of several 100 m (e.g. of smoke stack plumes), investigation of aircraft emissions under real flight conditions, as well as cost-effective inspection of aircraft engine and flare emissions. First investigations of greenhouse gas and other gaseous emissions with this method started 20 years ago. New regulatory requirements and environmental monitoring strategies need innovative and cost-effective measurement techniques. Hence FT-IR emission spectrometry was further developed for the determination of gaseous emission inventories and investigation of layered plumes, especially aircraft engine exhausts.

Theory and Experiment

To describe the radiation emission from distant warm sources, a radiative transfer simulation algorithm is used, which takes different atmospheric layers along the radiation path and their mutual influences into consideration [1, 2]. To measure smoke stack plume radiation emissions the telescope of the spectrometer is adjusted so that it receives the radiation from the slant path passing the plume directly above the stack top. The measured difference between this radiation and one without the plume in the field of view is simulated by the radiation transfer model. Thus information about the atmospheric background influencing the plume measurements is included in these measured quantities. Foreground parameters are obtained from active mode open-path FT-IR measurements made with an infrared light source.

Plume temperature and gas concentrations are then determined by a least-squares fit of the line-by-line simulation of the plume transmittances with the measured data.

The direct calculation of gaseous absorption coefficients on the basis of a line-by-line procedure from known data on molecular line parameters offers the possibility to include current atmospheric and geometrical parameters as well as spectroscopic features of the instrument (spectral resolution, instrumental line shape function). In this work the HITRAN92 [3] database was used.

To retrieve temperature and gas concentrations from aircraft exhaust measurements a multi-layer radiative transfer model has been developed. The model of engine plume is a homogeneous profile at exhaust

---

nozzle exit for the basic measurement. A parallel-layer model with symmetric distribution is used for downstream measurements: background, outer layer, inner layers, core layer around engine axis, inner layers, outer layer, and foreground.

Instrumentation and Data Analysis

The measurements were performed by using a van equipped with the commercial double pendulum interferometer K 300 of the Kayser-Threde Company [2, 4]. An InSb- and an MCT-detector were used. Spectra were measured with a spectral resolution of normally $0.2\,cm^{-1}$ ($0.08\,cm^{-1}$ maximum).

A multi-component program to interpret the measured spectra as described above has been developed and spectral windows for the retrieval of the temperature and the concentrations have been determined [2, 5, 6]. The detection limits of this remote sensing method are in the range from 1 to 10 ppm for CO_2, CO, N_2O, HCl and H_2O, 20 ppm for NO, NO_2, CH_4, and 100 ppm for SO_2 and depend on plume temperature. For the aircraft measurements, slightly higher detection limits apply [8].

Results

Numerous measurements and quantitative gas analyses of smoke stack emissions have been performed during the last few years at power plants with different types of firing (lignite, hard coal, oil, gas) to compare the FT-IR concentration results with data from CEM (continuous emission monitoring) sensors installed inside the stacks of some modern power plants [7]. The mean deviations between FT-IR and CEM data rarely exceed 25% and are normally 10%.

Measurements of methane emission from a test rig (heated air with controlled CH_4 release, no flame) were made, to define pure methane signatures for measurements of flare emissions [7]. As in the case of small-building smoke stacks, the gas quantification from flare plume measurements is very complicated because of strong variations in the narrow range plume location, owing to the influence of wind and a strongly variable plume diameter, but there are still enough data available, from which the plume temperature and the CH_4, H_2O, CO_2 and CO concentrations can be

derived. The emission and re-absorption of water vapour within the flare environment can not be simulated adequately with the one-layer plume model.

A multi-layer spectra retrieval program was developed and used to investigate gaseous emissions from different aircraft engine types with different fuels and at several thrust levels [8]. Plume temperature as well as CO_2, CO, H_2O and NO concentrations were retrieved for passenger aircraft engines such as JT8 and CFM56 and engines without bypass. The infrared radiation of the plume was measured at different distances from the nozzle exit. Emission indices of these compounds were determined by means of stoichiometric calculations.

Discussion

FT-IR remote measurements in the passive mode have been performed successfully to determine effluent concentrations in warm exhaust plumes. The main applications of this technique should be (a) to smoke stacks and flares for unannounced emission control, supervision of system breakdown, control of burning processes, or data sampling for stacks without CEM, and (b) to aircraft engines for investigation of emissions, for turbine development and engine-status control. The deviations mainly occur because of unknown wind influence on the plume diameter, which is one of the input parameters for the analysis software, and of the errors of the spectral line data. In some applications, optical methods are without a feasible alternative, as in the case of measurements of flare emissions and aircraft engine exhausts at different altitudes. Both the passive and active techniques can be used to complete the data base of regional emission inventories.

Acknowledgments. This work was supported financially by the German Federal Ministry of Research and Technology (BMFT), the German Science Foundation (DFG), and the Shell Research Center at Thornton, U. K. The measurement campaigns were performed with the technical assistance of Wilfried Bautzer.

References

[1] R. Haus, H. Goering, *Atmosphärenphysikalische Grundlagen der infrarotspektroskopischen Luftanalyse*, Heinrich-Hertz-Institute of the Academy of Sciences, Berlin, 1991, pp. 1–31.

[2] R. Haus, K. Schäfer, W. Bautzer, J. Heland, H. Mosebach, H. Bittner, T. Eisenmann, *Appl. Opt.* **1994**, *33*, 5682.

[3] L. S. Rothman, R. R. Gamache, R. H. Tipping, C. P. Rinsland, M. A. H. Smith, D. C. Benner, V. M. Devi, J. M. Flaud, C. Camy-Peyret, A. Perrin, A. Goldman, S. T. Massie, L. R. Brown, R. A. Toth, *J. Quant. Spectrosc. Radiat. Transfer* **1992**, *48*, 469.

[4] H. Mosebach, T. Eisenmann, Y. Schulz-Spahr, I. Neureither, H. Bittner, H. Rippel, K. Schäfer, D. Wehner, R. Haus, *Proc. SPIE* **1993**, *1717*, 149.

[5] R. Haus, K. Schäfer, H. Mosebach, *Umweltwiss. Schadst.– Forsch.* **1994**, *6*, 254.

[6] K. Schäfer, R. Haus, J. Heland, *Environ. Monit. Assess.* **1994**, *31*, 191.

[7] R. Haus, K. Schäfer, J. Hughes, J. Heland, W. Bautzer, *Proc. SPIE* **1995**, *2506*, 45.

[8] J. Heland, K. Schäfer, R. Haus, *Proc. SPIE* **1995**, *2506*, 62.

Mikrochim. Acta [Suppl.] 14, 801–802 (1997)
© Springer-Verlag 1997

Transient Infrared Spectroscopy for On-line Analysis of Solids and Viscous Liquids

Stephan J. Weeks[1,*], **John F. McClelland**[1,2], **Steven L. Wright**[3], and **Roger W. Jones**[1]

[1] Ames Laboratory, Ames, IA 50011, U.S.A.
[2] MTEC Photoacoustics, Inc., Ames, IA 50014, U.S.A.
[3] Pioneer Hi-Bred International, Inc., Johnston, IA 50131, U.S.A.

Abstract. Transient infrared spectroscopy (TIRS) is described as a developing technology that overcomes sample opacity, thickness and self-absorption problems to gain useful mid-IR spectral information from moving solids and viscous liquids for real-time, on-line process control. The utility of the technique is demonstrated by showing examples as diverse as monitoring the cure levels for acrylic coatings on polycarbonate sheets and monitoring the waste loading percentage for nuclear waste treatment by polymer microencapsulation.

Key words: FT-IR; on-line process control; transient infrared spectroscopy.

A unique attribute of the transient infrared spectroscopy (TIRS) technique is the ability to perform on-line infrared quantitative analyses on solids and viscous liquids in a moving stream. TIRS makes opaque materials yield their rich infrared spectral features by changing the thermal emission behaviour of a thin layer on the material's surface. This avoids the opacity and self-absorption problems that plague the more traditional IR techniques. TIRS is a pair of techniques in which a gas jet heats or cools a moving sample stream as it passes through the field of view of the FT-IR spectrometer. When a jet of hot gas impacts on the material's moving surface, a thin hot layer is produced that rapidly thickens as the heat diffuses into the material and is conveyed along in the process stream. The altered layer within the field of view of the FT-IR spectrometer, however, remains constant and thin. This heated surface layer then produces a transient infrared emission spectrum, or TIRES spectrum. Similarly, when a jet of cold gas is used, the blackbody radiation from the uncooled bulk material passes through the thin, cooled surface layer, producing a transient transmission spectrum, or TIRTS spectrum. These mid-IR spectra are used to determine chemical and physical properties of the process stream's material. The TIRS spectra can be used for reaction monitoring, compositional analysis, surface characterization, or monitoring any other material property reflected in the mid-IR spectrum. Applications of the TIRS technique to on-line, real-time monitoring of the cure level of acrylic coatings on polycarbonate sheets and of the waste loading percentage for low level mixed waste treatment by polymer microencapsulation are described.

Experimental

A Bomem MB 100 FT-IR spectrometer equipped with an external focusing mirror and a wide-band MCT detector cooled with liquid nitrogen was used to collect the spectral data. The normal infrared source was replaced by a cold surface. The spectrometer was operated with $8 \, cm^{-1}$ resolution. The gas jet was mounted to the spectrometer on a three-axis translator so that the jet could be aimed onto the moving material's surface within the spectrometer field of view. Detailed experimental descriptions are given in references [1] and [2]. For later work a syringe needle was used to deliver the gas jet stream onto the sample material's surface.

Results and Discussion

TIRS has moved out of the laboratory to perform tests on pilot and full scale process lines. TIRS has been successfully used to monitor the cure levels of coatings and uncoated bulk solid materials on process lines running at a wide range of line speeds [1]. The on-line transient infrared results proved to be as accurate as or more accurate than the off-line tests normally used by the manufacturers. Another process to which TIRS is

---

* To whom correspondence should be addressed

being applied is polyethylene encapsulation for the solidification of radioactive nitrate salts. In the encapsulation process, the salt waste is mixed with polyethylene pellets, heated, and extruded as a molten stream. TIRS has been successfully demonstrated as a real-time, on-line monitor for this process waste stream [1, 2]. In order to monitor the critical waste to polymer ratio, the TIRS system captured the real-time IR spectrum of the molten viscous stream as it left the extruder. Real-time monitoring for the treatment of sodium nitrate and sodium chloride surrogate waste streams, as well as a low-level mixed waste (LLMW) stream, has been accomplished. Typically, for a plot of percent waste loading (0–80%) *vs.* TIRS signal, a linear regression coefficient of ~ 0.99 with standard errors of prediction ~ 0.5–2% have been obtained. Work is continuing on using the data from the monitor to guide processing, minimize waste volume and certify the composition of the final waste form. Experimental parameters along with computer-aided computational techniques (e.g. principal component regression, partial least-squares) are being optimized to improve the accuracy and precision of the TIRS monitor.

Acknowledgements. Our thanks to Andrea Faucette and staff at Rocky Flats Environmental Technology Site and Paul Kalb and staff at Brookhaven National Laboratory for providing access to their waste processing extruder and their assistance in carrying out tests. Thanks also to the industrial staff that provided assistance with the other tests. Ames Laboratory is operated for the U.S. Department of Energy by Iowa State University under contract No. W-7405-ENG-82. This work is supported by the U.S. Department of Energy, Office of Science and Technology.

References

[1] R. W. Jones, J. F. McClelland, *Process Control Qual.* **1993**, *4*, 253.

[2] S. L. Wright, R. W. Jones, J. F. McClelland, P. D. Kalb, in: *Stabilization and Solidification of Hazardous Radioactive and Mixed Waste, 3rd Volume, ASTM STP 1240* (T. M. Gilliam, C. C. Wiles, eds.), American Society of Testing and Materials, West Conshohocken, PA, **1996**, pp. 198–208.

Mikrochim. Acta [Suppl.] 14, 803–805 (1997)
© Springer-Verlag 1997

Optical Design and Sampling Methods for the Measurement of Vibrational Circular Dichroism with a Nicolet Magna FT-IR Spectrometer

Laurence A. Nafie*, Xinhua Qu, Fujin Long, and **Teresa B. Freedman**

Department of Chemistry, Syracuse University, Syracuse, New York 13244-4100, U.S.A.

Abstract. A modified design for a Fourier-transform vibrational circular dichroism (FT-VCD) spectrometer is presented. The FT-VCD instrument is based on a Nicolet Magna-IR 550 equipped with an external bench for the measurement of VCD spectra. VCD spectra of high signal quality were obtained with 20 minutes of collection time at 4 cm^{-1} resolution in the mid-infrared region by conventional sequential collection of rapid-scan VCD and transmission interferograms, the simultaneous collection of these rapid-scan interferograms and step-scan collection. Of these three methods the simultaneous rapid-scan exhibited the lowest levels of noise.

Key words: FT-IR, vibrational circular dichroism (VCD), step-scan, polarization modulation.

A modified design for a Fourier-transform vibrational circular dichroism (FT-VCD) spectrometer is presented. VCD, the difference in absorption of left- and right-circularly polarized infrared radiation during vibrational excitation, has been measured with the double modulation technique by using modified commercial FT-IR instruments for more than a decade [1–4], but difficulties in removing baseline and absorption artifacts and in phase-correction remain. Our FT-VCD instrument is based on a Nicolet Magna-IR 550 equipped with an external bench for the measurement of VCD spectra. The design of the optical bench is based on an earlier design [1] in which focusing was accomplished by mirrors. In the new designs [2–4] the mirrors are replaced by infrared-transparent lenses to avoid baseline artifacts that arise from the reflection of circularly polarized radiation from the surfaces of the mirrors.

The most important optical element in the VCD optical bench is a ZnSe photoelastic modulator (PEM) which switches the polarization state of the radiation between the left and right circular states at 37 kHz. An optional ZnSe lens focuses the collimated beam from the interferometer, and an optical filter and a KRS-5 linear polarizer are placed between the lens and the PEM. The beam emerging from the sample, which is placed immediately after the PEM, is focused by a ZnSe lens onto a liquid-nitrogen cooled HgCdTe detector. The complete replacement of the mirrors used in previous versions of this instrument with lenses, enables us to obtain VCD spectra with high signal quality, minimized artifacts and flat baselines. VCD of single enantiomers can now be collected routinely for favorable samples in approximately 15 minutes. Several alternative optical arrangements have been explored to optimize the performance and fidelity of the FT-VCD instrument. An electro-optical diagram illustrating the experimental set-up for our most favorable FT-VCD optical arrangement is given in Fig. 1. The sample chamber features a 3-inch diameter 50-cm focal length ZnSe lens to provide mild focusing of the infrared beam on the sample, followed by a smaller fast-focusing ZnSe lens to form an image of the beam on the surface of the 2-mm square HgCdTe detector element.

In conjunction with Nicolet Instruments, we have tested both sequential and simultaneous collection of the AC spectrum (demodulated at both the PEM frequency and Fourier frequencies) and the DC spectrum (demodulated only at the Fourier frequencies), which are ratioed to obtain the raw VCD spectrum. A preliminary test of step-scan acquisition of the AC interferogram was also made. In Fig. 2, we compare the raw single-enantiomer VCD spectra of (−)-α-pinene

---

* To whom correspondence should be addressed

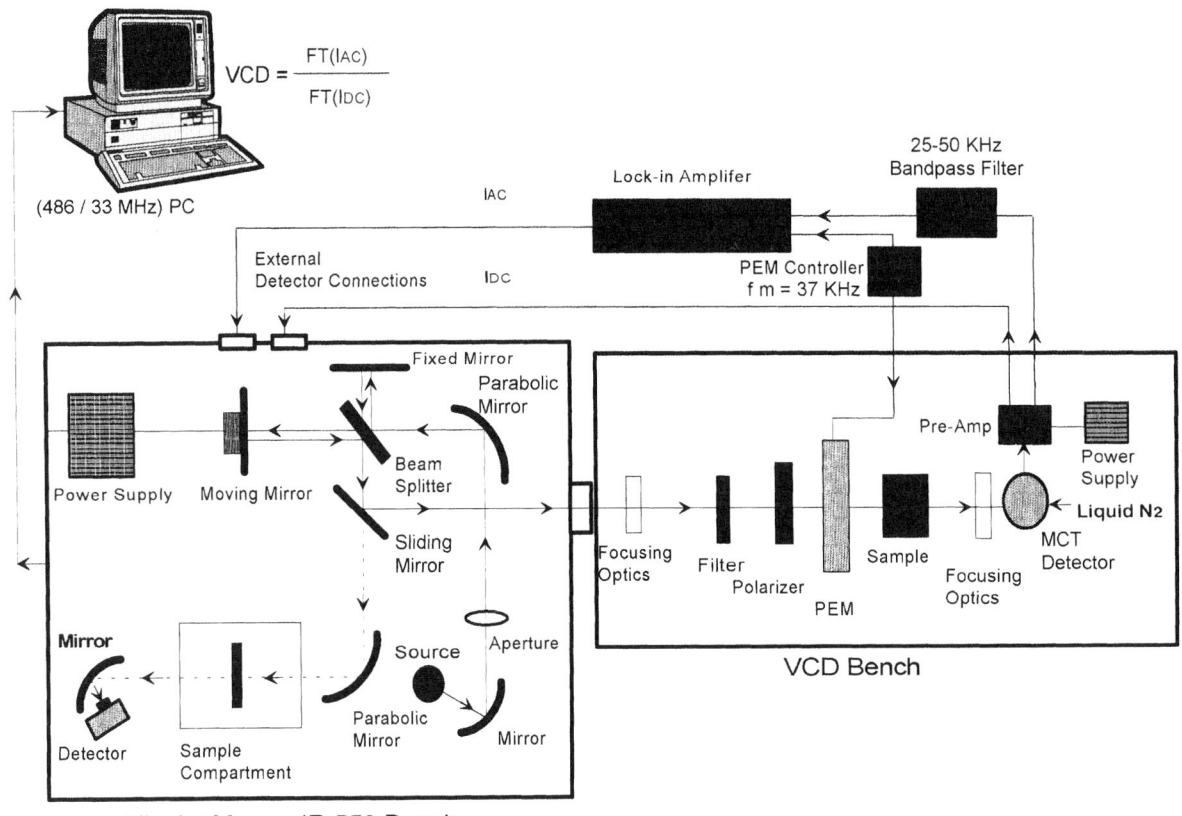

$$VCD = \frac{FT(I_{AC})}{FT(I_{DC})}$$

Fig. 1. Optical-electronic diagram of FT-VCD instrumental lay-out employed to collect the VCD spectra

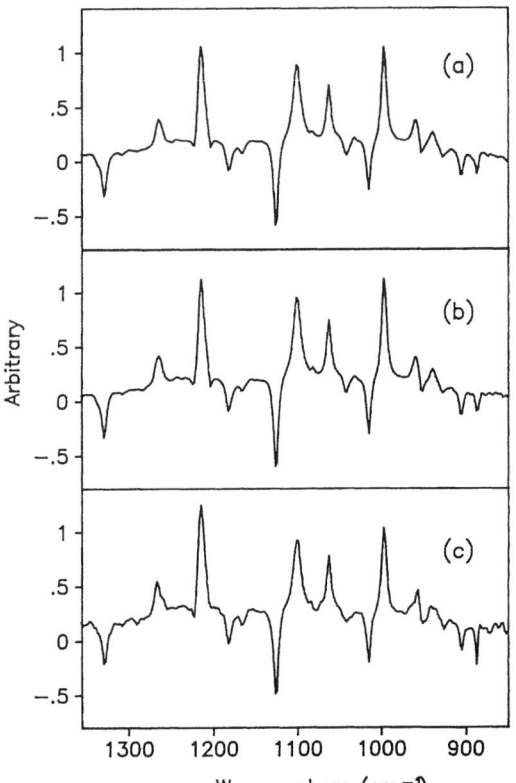

obtained under sequential, simultaneous and step-scan data acquisition conditions. The spectra were collected by using a 50 micron path-length BaF_2 cell, 4 cm^{-1} resolution, and a mirror velocity of 0.63 cm/s. We found that the signal-to-noise ratio is higher for the rapid-scan acquisition than for the step-scan acquisition, and of the two rapid-scan approaches, the simultaneous collection method held a slight advantage over the sequential scan method. The VCD step-scan results, while encouraging, should be regarded as preliminary since we did not explore various step-scan experimental parameters, as has been done recently in other laboratories [5, 6].

Finally, we have examined the effects of different phase corrections by processing the interferograms with an external program written with Mathcad

Fig. 2. Raw VCD spectra collected under the following conditions: **a** sequential rapid-scan collection of 1024 ac scans (~ 20 min) and 64 dc scans; **b** simultaneous rapid-scan collection of 1024 ac and 1024 dc scans; **c** step-scan collection for 15 min, ratioed by 64 dc rapid scans. In the step-scan mode, the detector signal is modulated only at the PEM frequency. Under the experimental conditions employed, the range of VCD absorbance values was approximately between $+5 \times 10^{-4}$ and -4×10^{-4}

(Mathsoft, Inc., Cambridge, MA). VCD intensity calibration curves were obtained by use of a CdSe quarter-wave plate and a second polarizer placed at the sample position. With these curves, the effects of different phase corrections on the Fourier-transformed spectra were investigated. Interferograms transformed with the de Haseth phase correction [7], which restricts the phase angle to $\pm \pi/2$, gave results that properly represent negative VCD intensities and yield calibration curves that are consistent with theoretical predictions.

Acknowledgements. The authors thank the Nicolet Instrument Corporation for the loan of an accessory to upgrade their Magna 550 to a Magna 850, enabling simultaneous and step-scan data acquisition, and Dr. R. Hapanowicz and Mr. R. Badeau for their assistance. This work was supported by NIH grant number GM-23567.

References

[1] E. D. Lipp, L. A. Nafie, *Appl. Spectrosc.* **1984**, *38*, 20.

[2] P. Malon, T. A. Keiderling, *Appl. Spectrosc.* **1988**, *42*, 32.

[3] G.-C. Chen, P. L. Polavarapu, S. Weibel, *Appl. Spectrosc.* **1994**, *48*, 1218.

[4] D. Tsankov, T. Eggimann, H. Wieser, *Appl. Spectrosc.* **1995**, *49*, 132.

[5] C. Marcott, A. E. Dowrey, I. Noda, *Appl. Spectrosc.* **1993**, *47*, 1324.

[6] B. Wang, T. A. Keiderling, *Appl. Spectrosc.* in press.

[7] C. A. McCoy, J. A. de Haseth, *Appl. Spectrosc.* **1988**, *42*, 336.

Mikrochim. Acta [Suppl.] 14, 807–808 (1997)
© Springer-Verlag 1997

Comparison of Fourier-Transform Vibrational Circular Dichroism and Multichannel-detected Raman Optical Activity

Laurence A. Nafie*, **Xinhua Qu**, **Eunah Lee**, **Gu-Sheng Yu**, and **Teresa B. Freedman**

Department of Chemistry, Syracuse University, Syracuse, New York 13244–4100, U.S.A.

Abstract. Infrared, vibrational circular dichroism, dual circular polarization Raman optical activity and Raman spectra have been used in conjunction for investigation of the vibrational modes of *trans*- and *cis*-pinane, and α- and β-pinene with a view of evaluating the usefulness of these techniques. It has been concluded that vibrational circular dichroism and Raman optical activity provide highly complementary and non-redundant information, of comparable spectral quality.

Key words: vibrational circular dichroism vibrational Raman optical activity, infrared, pinane, pinene.

Vibrational optical activity [1] occurs in nature in two principal forms: infrared vibrational circular dichroism (VCD) [2] and vibrational Raman optical activity (ROA) [3]. Both VCD and ROA arise from the differential response of a chiral molecule to its interaction with left and right circularly polarized radiation. In our laboratory, we have state-of-the-art instrumentation for both VCD and ROA [4]. In the case of VCD, this consists of an FT-IR spectrometer equipped with a linear polarizer, a ZnSe photoelastic modulator and an HgCdTe detector. The ROA instrument consists of an argon ion laser and two quartz zeroth-order quarter-wave plates for controlling the polarization states of incident and scattered light beams. The back-scattered Raman scattered light passes through a holographic filter and is detected with a single stage spectrograph and charge coupled device (CCD) detector. For a number of years, we have been interested in comparing VCD and ROA in various ways. These include the relative ease of obtaining spectra of a given level of quality and the signs and intensities of the VCD and ROA for a given vibrational mode.

VCD and back-scattered in-phase dual circular polarization (DCP$_I$) [1, 3, 4] ROA have been measured and compared quantitatively over the frequency range from 835 to 1345 cm^{-1} for *trans*-pinane, *cis*-pinane, α-pinene and β-pinene. For these molecules in this region of spectral overlap between VCD and ROA, the average ratio of VCD or ROA to its parent vibrational intensity is larger for ROA by approximately half an order of magnitude. Several vibrational modes in each molecule have been found to yield both large VCD and large ROA, whereas several other modes show little propensity toward significant VCD or ROA intensity. Beyond this general property of strongly chiral and strongly achiral vibrational modes, little additional correlation between VCD and ROA intensity was found. This quantitative compilation of VCD, infrared (IR), ROA and Raman intensities provides an experimental basis for computational intensity studies of VCD, ROA and their theoretical comparison.

As an example of these spectra, we show in Fig. 1 the IR, VCD, DCP$_I$ ROA and Raman spectra for (-)-*trans* pinane. The VCD and ROA show similar levels of spectral quality and resolution for collection times of approximately two hours each, and few correlations of sign or relative intensity. In comparing the VCD and ROA spectra for all four molecules, the signs of the VCD and ROA for corresponding vibrational modes were examined, the anisotropy ratios [2] and circular intensity differentials (CIDs) [3] were normalized to the largest measured value for each molecule, and the product of the normalized anisotropy ratio and normalized CID for each band was calculated. In order to obtain quantitative values for the IR, VCD, DCP$_I$ Raman and ROA, we curve-fitted all of our spectra. An example of the curve-fitting, employed to obtain integ-

---

* To whom correspondence should be addressed

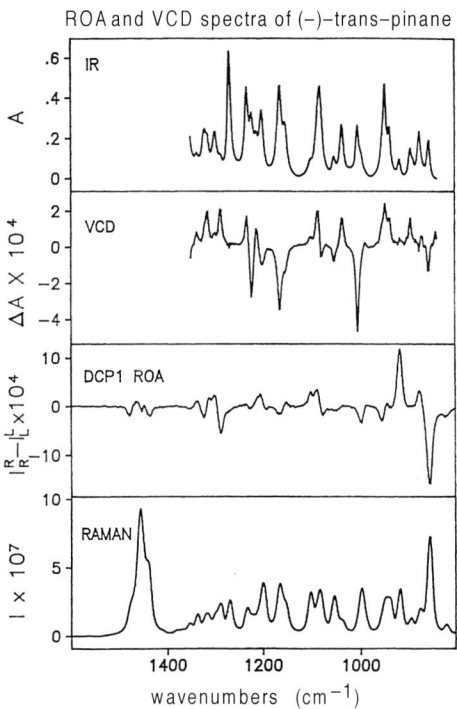

Fig. 1. Comparison of the infrared absorption, *A*, the VCD ($\Delta A = A_L - A_R$), back-scattering DCP$_1$ Raman, *I*, and ROA ($\Delta I = I_R^R - I_L^L$), of (-)-*trans*-pinane

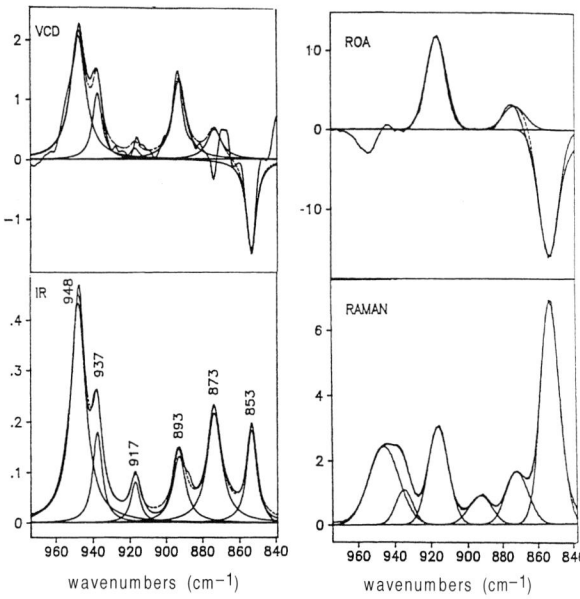

Fig. 2. Curve-fits (dotted) and experimental data (solid) for the IR, VCD, back-scattering DCP$_1$ Raman and ROA of (-)-*trans*-pinane in the 840–980 cm^{-1} region, showing individual component bands

rated band intensities in the 840–975 cm^{-1} region of (-)-*trans*-pinane, is shown in Fig. 2. For each spectrum, regions of 6–8 bands were curve-fitted separately with the curve-fitting routine in *SpectraCalc* (Galactic, Inc., Salem, NH), and the frequencies and band-widths obtained from the IR or Raman spectrum were used in the curve-fits for the corresponding VCD or ROA spectrum. IR and VCD band-shapes were Lorentzian, whereas Raman and ROA band-shapes were Gaussian.

Over the four molecules studied, we found a similarity in the number of VCD and ROA bands with the same or opposite signs (roughly equal), which indicates an absence of correlation between the electronic intensity mechanisms for VCD and ROA. Examination of the products of the normalized anisotropies and normalized CIDs reveal that there are a few bands that make very large contributions to both VCD and ROA

(indicative of modes that are highly chiral with large magnetic dipole contributions) and a few bands that make very small contributions in both spectra (indicative of achiral vibrational motion, probably localized distant from a chiral center). We conclude that ROA and VCD are highly complementary, non-redundant stereochemical techniques of comparable spectral quality.

Acknowledgement. This work was supported by NIH grant number GM-23567.

References

[1] L. A. Nafie, G.-S. Yu, X. Qu, T. B. Freedman, *Faraday Discuss.* **1994**, 99, 13.

[2] T. B. Freedman, L. A. Nafie, in: *Modern Nonlinear Optics, Part 3* (M. Evans, S. Kielich, eds.), Wiley Chichester, 1994, pp. 207–263.

[3] L. A. Nafie, D. Che, In: *Modern Nonlinear Optics, Part 3* (M. Evans, S. Kielich, eds.), Wiley, Chichester, 1994, pp. 105–149.

[4] L. A. Nafie, M. Citra, N. Ragunathan, D. Che, in: *Analytical Applications of Circular Dichroism* (N. Purdie, H. G. Brittain, eds.), Elsevier, Amsterdam, 1994, pp. 53–89.

Mikrochim. Acta [Suppl.] 14, 809–810 (1997)

Vibrational Circular Dichroism Spectra of Selected Chiral Polymers

Jennifer L. McCann[1], **Birgit Schulte**[2], **Dimiter Tsankov**[3], and **Hal Wieser**[1,*]

[1] Department of Chemistry, University of Calgary, Calgary, AB, T2N 1N4, Canada
[2] Department of Chemistry, Carl-von-Ossietzky-Universität, Oldenburg, Federal Republic of Germany
[3] Institute of Organic Chemistry, Bulgarian Academy of Sciences, BG-1113 Sofia, Bulgaria

Abstract. The vibrational circular dichroism (VCD) and absorption spectra of (−)-poly(menthyl methacrylate), (−)-poly(menthyl acrylate), and (−)-poly(menthyl vinyl ether) are reported and compared to those of (1R, 2S, 5R)-(−)-menthol in solution in CCl_4, from 830 to 1500 cm^{-1}. Many VCD peaks are common to both menthol and the corresponding polymers, but there is evidence of other peaks that are characteristic for each of the polymers.

Key words: vibrational circular dichroism, VCD, synthetic chiral polymers, menthol.

VCD is a spectroscopic technique which measures the differential absorption of left versus right circularly polarized light in the infrared region, and as such is very sensitive to details of molecular configuration and conformation. Though various biopolymers have been studied extensively by VCD in the past [1], only one report has dealt with a synthetic chiral polymer [2]. Synthetic polymers are ideal models for testing the applicability of VCD since they may be varied systematically, in both configuration and conformation. We decided some time ago to test VCD for probing synthetic chiral polymers in solution, with the hope that it would provide a wealth of new stereochemical information [3, 4]. In the present study we have used (−)-menthol as a chiral pendant for three different polymeric systems, each varying slightly in molecular structure.

Experimental

(1R,2S,5R)-(−)- and (1S,2R,5S)-(+)-menthol (Aldrich, 99%) were used without further purification. Menthyl methacrylate (MnMA)

and menthyl acrylate (MnA) were prepared as previously reported [5, 6] and subsequently subjected to free radical polymerization initiated with benzoyl peroxide to yield the corresponding polymers, PMnMA and PMnA respectively. Menthyl vinyl ether (MnVE) was prepared from ethyl vinyl ether and menthol [7], and polymerized with $EtAlCl_2$ in toluene at −78 °C to yield PMnVE.

The IR absorption and VCD spectra were recorded with a Bomem MB 100 spectrometer, equipped with standard VCD optics as described in detail elsewhere [8]. Both absorption and VCD spectra were recorded for CCl_4 solutions (0.031 g in 1 ml for each of the polymers, 0.39 M for menthol) in a 0.15 mm NaCl cell at 4 cm^{-1} resolution. The VCD spectra are a result of 5000 a.c. scans collected in a single block and ratioed with a block of 500 d.c. scans taken before the a.c. collection. The baseline was corrected for polarization artifacts by subtracting the VCD spectrum of the opposite enantiomer.

Results and Discussion

In each of the cases presented, (−)-menthol is appended to a different polymer backbone and is the primary source of chirality. The absorption and VCD spectra of (−)-menthol are shown in Fig. 1A.

In comparing PMnMA and PMnA we effectively remove the methyl group, yielding a less basic carbonyl group and greater flexibility of the polymer backbone. From the absorption spectra of PMnA (Fig. 1B) and PMnMA (Fig. 1C) it is evident that we have two separate molecules. However, with the exception of the region from 1250 to 1150 cm^{-1} the VCD spectra are remarkably similar even in regions where the absorption spectra differ. This similarity strongly suggests that the VCD signals are localized to the chiral centres, as we have removed a methyl group from the polymer backbone with little change in the VCD spectrum. The regions of the spectra that do change are the same regions which were affected in the spectra of PMnMA

_____

* To whom correspondence should be addressed

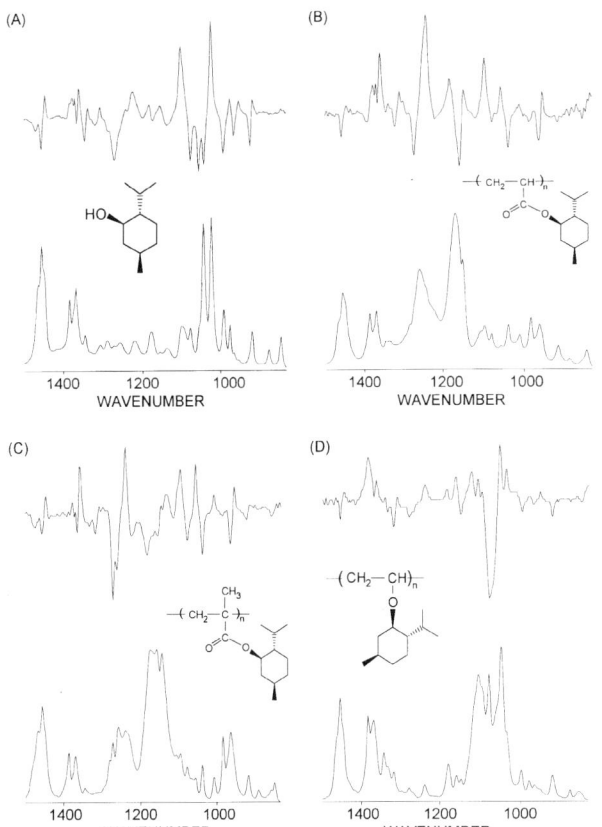

Fig. 1. Absorption (bottom) and VCD (top) spectra of **A** (−)-menthol, **B** PMnA, **C** PMnMA, and **D** PMnVE

when it was prepared with different tacticities [4], as well as when the spectra were taken for solutions in different solvents. This last point suggests that these regions are affected by a change in conformation either in the ester linkage, the polymer backbone, or both.

In shifting from PMnA to PMnVE we discriminate between the ester and the ether linkage. The asymmetric centre is closer to the polymer backbone in the ether than in the acrylate, and consequently it is expected that the chiral moiety will have a larger influence on the

inherent dissymmetry of the polymer. Conceivably, the chiral pendant may induce the polymer chain into short helical segments of a preferred screw sense.

The absorption and VCD spectra of PMnVE (Fig. 1D) may not be directly compared to those of the other polymers, but there are some obvious distinctions. Both the VCD spectra of PMnMA and PMnA are very detailed with numerous bisignate peaks throughout the displayed regions, many of which are common to the VCD spectrum of (−)-menthol. In contrast, the VCD spectrum of PMnVE is largely dominated by a large negative peak at $1077 \, cm^{-1}$. In fact, many of the bisignate peaks associated with the menthol moiety are weak in comparison. This large negative peak, assigned to C–O stretching, may aid in understanding the role of the chiral moiety within the polymer. With simpler models and appropriate computational methods we hope to further elucidate the underlying interactions.

In conclusion, this work clearly shows the potential use of VCD spectroscopy in extracting direct stereochemical information from optically active polymers in solution, in addition to providing insight into the interaction of the chiral moiety with the polymer backbone.

References

[1] T. A. Keiderling, P. Pancoska, in: *Advances in Spectroscopy Part B, Biomolecular Spectroscopy, Vol. 21* (R. J. H. Clark, R. Hester, eds.), Wiley, Chichester, 1993, p. 267.

[2] L. A. Nafie, T. A. Keiderling, P. J. Stephens, *J. Am. Chem. Soc.* **1976**, *98*, 2715.

[3] D. Tsankov, T. Eggimann, G. Liu, H. Wieser, *Proc. SPIE* **1993**, *2089*, 178.

[4] J. McCann, D. Tsankov, N. Hu, G. Liu, H. Wieser, *J. Mol. Struct.* **1994**, *349*, 309.

[5] M. M. Koton, T. A. Sokolova, M. H. Savitskaya, T. M. Kiseleva, *J. Gen. Chem.* (*U.S.S.R.*) **1958**, *28*, 409.

[6] R. L. Frank, H. R. Davis, Jr., S. S. Drake, J. J. McPherson, Jr., *J. Am. Chem. Soc.* **1944**, *66*, 1509.

[7] W. H. Watanabe, L. E. Conlon, *J. Am. Chem. Soc.* **1957**, *79*, 2828.

[8] D. Tsankov, T. Eggimann, H. Wieser, *Appl. Spectrosc.* **1995**, *49*, 132.

Mikrochim. Acta [Suppl.] 14, 811–813 (1997)

In Situ FT-IR Study of the Interfacial Oxide in the Anodic Dissolution of Silicon

Carlos da Fonseca, François Ozanam, and **Jean-Noël Chazalviel***

Laboratoire de Physique de la Matière Condensée, CNRS-École Polytechnique, Route de Saclay, F-91128 Palaiseau-Cedex, France

Abstract. The anodic dissolution of silicon in fluoride electrolyte has been investigated by using in situ infrared spectroscopy in the difference mode. The interfacial oxide layer has been found to thicken with increasing potential. Its thickness is little dependent upon electrolyte composition, but the oxide quality (defects, porosity), deduced from the shape of the νSiO spectrum, depends upon both the potential and the electrolyte composition. The current oscillation, which may be observed in the far anodic range, is associated with an oscillation of the oxide thickness and quality.

Key words: spectroelectrochemistry, electropolishing, semiconductors, polarization, vibrational.

The anodic dissolution of p-type silicon in fluoride media exhibits a very reproducible but rather complex behaviour, illustrated by current/potential curves of the type shown in Fig. 1a [1]: near the onset of anodic current flow, the current increases sharply and exhibits two peaks; the first peak, located near 0 V vs SCE, is associated with the transition between the regime of porous silicon formation and electropolishing. The presence of the second peak, at around 1–2 V vs SCE, may be ascribed to the existence of two distinct electropolishing regimes. In order to disentangle this behaviour, we have investigated the interface by using in situ infrared absorption as a function of electrode potential, using multiple total-internal reflection in the difference mode ("SNIFTIRS") and in the potential-modulation mode ("FTEMIRS"). Here we will limit our presentation to the differential measurements.

Figure 1b shows the kind of spectra which are obtained as a function of potential, in a given electrolyte. The reference potential was taken in the cathodic regime, where the Si surface is covered with hydrogen [2], which gives rise to the negative νSiH absorption near 2100 cm^{-1}. A νSiO absorption in the 1050–

1250 cm^{-1} range, characteristic of silicon oxide formation, is seen to develop when the potential increases above the value corresponding to the first current peak. The negative electrolyte absorption bands (νOH around 3400 cm^{-1}, νNH near 3200 cm^{-1}, δOH_2 near

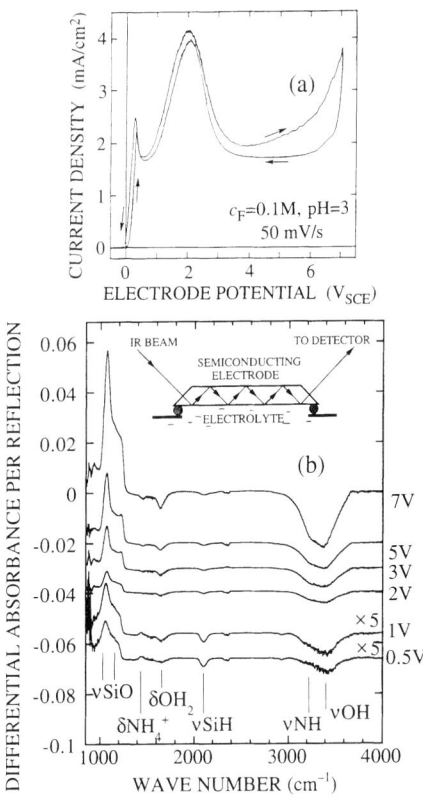

Fig. 1. a Current/potential curve of a typical p-Si/fluoride-electrolyte interface. Here the electrolyte is $1 M$ NH$_4$Cl + 0.05 M NH$_4$F + 0.05 M HF ($c_F = 0.1 M$, pH = 3). **b** Absorbance spectra, referred to -0.5 V vs SCE, for different values of the potential. Inset recalls the ATR geometry used here. No polarizer is used on the IR beam

---

* To whom correspondence should be addressed

$1650\,\mathrm{cm}^{-1}$ and $\delta\mathrm{NH}_4^+$ near $1450\,\mathrm{cm}^{-1}$) mainly arise from a shielding of electrolyte absorption in the internal-reflection geometry, due to formation of the oxide film, which replaces the electrolyte in the probed region.

A detailed study of the $\nu\mathrm{SiO}$ spectrum leads to the following conclusions.

(1) The thickness of the oxide is a roughly linear function of potential, and depends weakly upon electrolyte composition.

(2) The shape of the spectrum depends upon the polarization of the IR beam as expected for a thin layer, namely the LO and TO modes are clearly separated (see Fig. 2).

(3) The spectra can be represented as the superposition of five gaussian bands. Besides the two dominant contributions at 1070 and $1240\,\mathrm{cm}^{-1}$, these consist of two "disorder" modes at ~ 1150 and $1200\,\mathrm{cm}^{-1}$ [3], and a defect mode at $\sim 940\,\mathrm{cm}^{-1}$. Furthermore, the two main bands appear slightly asymmetric. The disorder modes are found to be weakly dependent upon polarization, in contrast to simple expectation [3], and the $940\,\mathrm{cm}^{-1}$ mode is insensitive to deuterium substitution, so its attribution to a SiOH mode [4] is probably to be rejected.

(4) The nature of the oxide appears to evolve considerably as a function of potential and electrolyte composition. Figure 3 shows, as a function of potential, the oxide thickness as deduced from TO-peak intensity [5], the mean LO/TO position

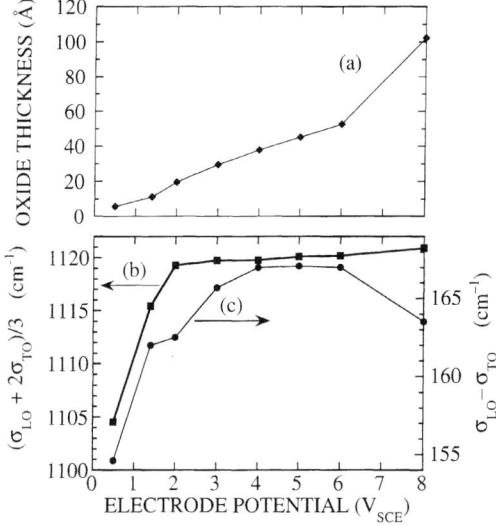

Fig. 3. Change of the oxide film as a function of potential (electrolyte $c_F = 0.05\,M$, pH $= 3$): *a* oxide thickness as deduced from the intensity of the $\nu\mathrm{SiO}$ TO peak: *b* average position $(\sigma_{\mathrm{LO}} + 2\sigma_{\mathrm{TO}})/3$ and *c* LO/TO splitting $\sigma_{\mathrm{LO}} - \sigma_{\mathrm{TO}}$

and the LO/TO splitting. These data provide evidence for deviations of the anodic oxide from ideality at thicknesses below $\sim 10\,\text{Å}$, and also for large potentials near the end of the second current plateau [6]. Furthermore, the area of the bands associated with defects (940, 1150, $1200\,\mathrm{cm}^{-1}$) appears to be minimum near the middle of the second-current plateau, and increases when going to electrolytes leading to larger anodic currents.

(5) In the second-plateau region, the interface exhibits a damped oscillating behaviour upon stepping the potential [7]. The infrared data show that this current oscillation is associated with an oscillation of the oxide thickness, of the order of $30\,\text{Å}$, and the concentration of defects is also found to be an oscillating function of time, which suggests that these defects may be connected with the mechanism giving rise to the oscillation. A quantitative theory for the mechanism of the oscillation is presently being worked out.

In conclusion, the anodic dissolution of silicon in dilute fluoride electrolytes offers a means for studying thin anodic oxides in conditions of very good reproducibility. Infrared absorption provides specific information on the oxide film as a function of experimental conditions, and especially information which may be valuable for the understanding of the oscillating behaviour observed in the high-potential region.

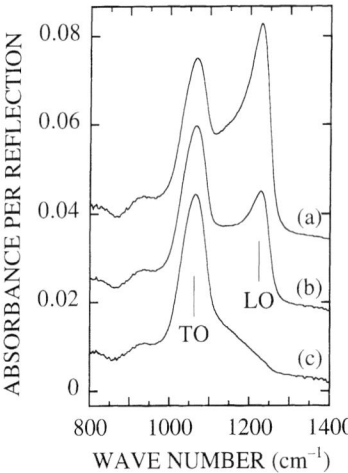

Fig. 2. $\nu\mathrm{SiO}$ spectrum for a p-polarized (*a*), unpolarized (*b*), and s-polarized (*c*) IR beam. The LO mode at $1240\,\mathrm{cm}^{-1}$ does not appear for s-polarization. Electrolyte $c_F = 0.1\,M$, pH $= 4.5$; potential $+8\,\mathrm{V}$ *vs* SCE

Acknowledgement. This work was partly supported by CEC through contract CI1 CT 93 0070.

References

[1] J.-N. Chazalviel, M. Etman, F. Ozanam, *J. Electroanal. Chem.* **1991**, *297*, 533.

[2] F. Ozanam, J.-N. Chazalviel, *J. Electron Spectrosc. Relat. Phenom.* **1993**, *64/65*, 395.

[3] C. T. Kirk, *Phys. Rev. B* **1988**, *38*, 1255.

[4] See e.g. K. Ramkumar, A. N. Saxena, *J. Electrochem. Soc.* **1992**, *139*, 1437.

[5] I. W. Boyd, J. I. B. Wilson, *J. Appl. Phys.* **1982**, *53*, 4166.

[6] C. da Fonseca, F. Ozanam, J.-N. Chazalviel. *Surface Sci.* **1996**, *365*, 1 (and to be published).

[7] F. Ozanam, J.-N. Chazalviel, A. Radi, M. Etman, *Ber. Bunsenges. Phys. Chem.* **1991**, *95*, 98.

Mikrochim. Acta [Suppl.] 14, 815–817 (1997)
© Springer-Verlag 1997

A Spectroelectrochemical in situ FT-IR Study on the Growth of Poly(3-octyl)thiophene from Dimer and from Monomer

Carita Kvarnström* and **Ari U. Ivaska**

Laboratory of Analytical Chemistry, Åbo Akademi University, FIN-20500 Åbo-Turku, Finland

Abstract. The electrochemical polymerization of poly(3-octyl-thiophene) was studied in situ by the external reflection FT-IR technique. Polymerization was made in 0.1 M LiBF$_4$–propylenecarbonate solution of the monomer, 3-octylthiophene, and of the dimer, 4,4′-dioctyl-2,2′-bithiophene. The growth of the film and the structure of the resulting polymer material from these solutions was compared. The effect of the current density on the polymer structure polymerized from the monomer was also studied.

Key words: in situ FT-IR, external reflection, conducting polymer poly(3-octylthiophene).

Poly-3-alkylthiophenes have been extensively studied by various spectroscopic techniques such as FT-IR, Raman and infrared reflection–absorption [1, 2]. There have been some studies on using the dimer instead of the monomer as starting material for poly(3-octylthiophene), P3T [3]. Louarn et al. reported that a more regular structure of poly(3-octylthiophene) was obtained when the dimer was used as starting material [4]. Results obtained from ^{1}H-NMR measurements show that monomer units are coupled "head-to-tail" whereas the dimer follows "head-to-head" or "tail-to-tail" coupling [5]. P3T grown from the dimer generally shows a better stereoregularity but the oxidation is retarded owing to the fact that the material has to undergo some structural rearrangement before oxidation can take place. The electrochemical oxidation potential of the dimer is approximately 300 mV lower than that of the monomer. Therefore the probability for side-reactions and for overoxidation of the formed film material is much lower in the polymerization of the dimer than of the monomer.

The aim of this work was to study in situ by FT-IR reflection spectroscopy the difference in the electrochemical polymerization route of 3-octylthiophene and of 4,4′-dioctyl-2,2′-bithiophene.

Experimental

A Mattson Cygnus 100 FT-IR spectrometer with an MCT-detector was used in the experiments. The in situ external reflection system was built from a variable-angle specular reflectance attachment from Specac. A three-electrode thin-layer spectroelectrochemical cell, modified from the one described by Bewick [6], was used in the experiments. All potentials reported are measured vs an aqueous Ag/AgCl/KCl (3 M) reference electrode. The working electrode was a 0.38 cm^2 Pt disk electrode. A detailed description of the procedure is given in [7]. Polymerization of 3-octylthiophene and of 4,4′-dioctyl-2,2′-bithiophene was performed in solutions of (0.1 M) LiBF$_4$ (Aldrich) in propylenecarbonate (Aldrich). The monomer concentrations used were 0.05–0.1 M and the dimer concentration was approximately 0.005 M. Nitrogen was passed through the electrolyte before every experiment. Galvanostatic polymerization was used with current densities of 0.26 mA/cm^2 or 0.8 mA/cm^2. Every reflection spectra is a Fourier-transform of 100 co-added interferograms and the resolution used is 4 cm^{-1}. Each spectrum represents the changes taking place at the electrode during approximately 50 seconds at the polymerization potential. The monomer 3-octylthiophene was received from Neste Oy, (Research Centre FIN-06850 Kulloo, Finland), 4,4′-dioctyl-2,2′-bithiophene was synthetized from the 3-octylthiophene according to the method of Lapkowski et al. [5].

Results and Discussion

The electrochemical polymerization was performed by using current densities that correspond to 1.4 V for the monomer and 1.1–1.2 V for the dimer. It is known that radical cations of monomeric/oligomeric species act as a major propagation intermediate for the polymer growth and that the oligomeric species have to reach a specific length before they will precipitate at the

---

* To whom correspondence should be addressed

electrode. Due to the fact that doping of the polymer takes place simultaneously with electrochemical polymerization the doping-induced IR-bands can be seen in the spectrum at the same time that oligomers with sufficient length have been formed. This can be seen in the series of spectra obtained both from monomer and from dimer polymerization. In those spectra the bands of the polymer and the doping-induced bands rise in the spectra approximately simultaneously, but after some time has elapsed from the initiation of the polymerization.

In Fig. 1 some selected spectra from a series of spectra taken in situ from the polymerization of 3-octylthiophene, Fig. (a) and (b), and from 4,4′-dioctyl-2,2′-thiophene, Fig. (c), can be seen. All IR-bands arising from the

polymer structure will not be treated here, but the focus will be rather on the vibrations that appear differently in polymerization of the monomer and of the dimer.

Monomer

In the monomer spectra taken during polymerization with a current density of $0.26 \, \text{mA/cm}^2$ practically all vibrations from both the polymer and from the doping-induced changes are visible already at the beginning of the polymerization. The change observed in the spectra during continuous polymerization is the increased intensity of the peaks from the conjugated polymer (simultaneous decreasing intensity of the peaks from

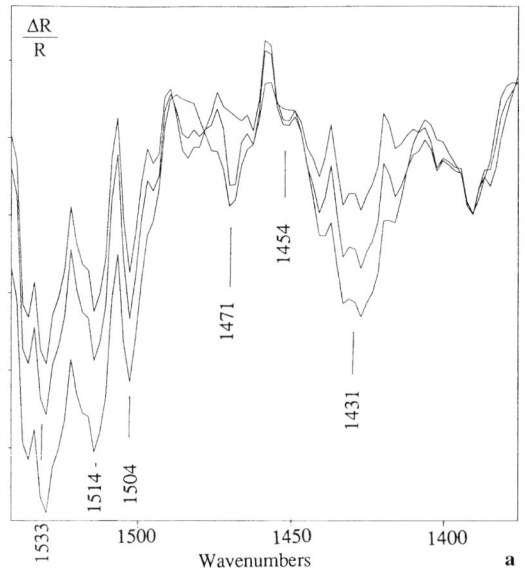

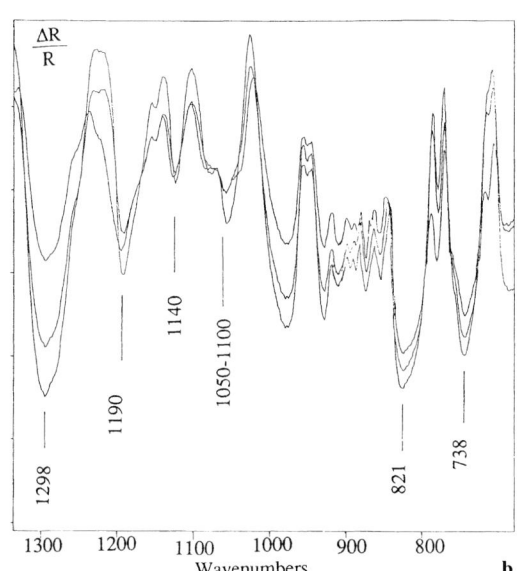

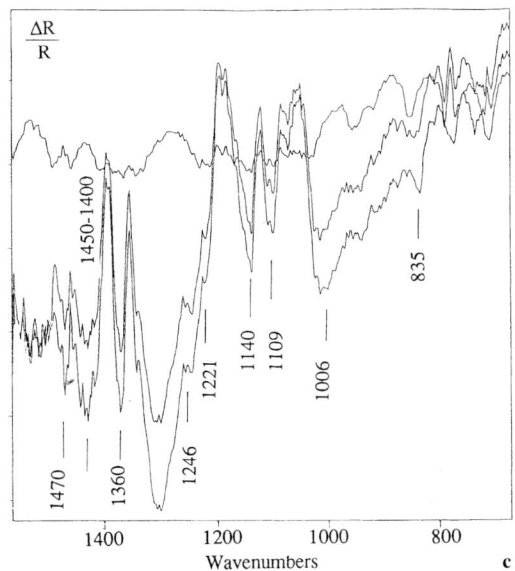

Fig. 1. External reflection FT-IR-spectra taken in situ during electrochemical polymerization of 3-octylthiophene **a** and **b** and of 4,4′-dioctyl-2,2′-bithiophene **c**; current density used was $0.26 \, \text{mA/cm}^2$. The in situ spectra, from top to bottom were taken after 6, 30 and 60 mC respectively, had been used for the polymerization. The peaks that are mentioned in the text are labelled

the monomer, salt and solvent). In the spectra of the polymer material made from the monomer the vibration at $821\,cm^{-1}$ from the $C_\beta-H$ out of plane band from three-substituted thiophene rings and the vibration at $3060\,cm^{-1}$ (not shown in Fig. 1) from $C_\beta-H$ stretching strongly indicates that linear polymer chains are built up. At 1431 (centre), 1454, 1504, 1514 (centre) and $1533\,cm^{-1}$, bands from the ring stretching can be seen. The last two bands are from the $C_\alpha=C_\beta$ stretch [1]. The only vibration in the monomer polymerization that appears first at a later state of polymerization is the one at $1471\,cm^{-1}$. This band originates from the thiophene ring stretching in longer chains. The conjugated backbone gives rise to vibrations at 738, 1140, 1190 and $1298\,cm^{-1}$. In the band at $738\,cm^{-1}$ there are also vibrations originating from the aliphatic substituents. The band at $1050-1100\,cm^{-1}$ is from the anion BF_4^-. When a higher current density, $0.8\,mA/cm^2$, is used for polymerization, the peaks at $3060\,cm^{-1}$ and at $821\,cm^{-1}$ do not appear in the spectra, indicating that the resulting film in this case has a more cross-linked structure.

Dimer

The lower polymerization potential, 1.1–1.2 V, of the dimer implies that the current density $(0.26\,mA/cm^2)$ is more efficiently used, owing to fewer side-reactions taking place than in the polymerization of the monomer. However, the low concentration of the dimer influences the rate of polymerization. This can be seen in Fig. 1(c) where hardly any vibrations from adsorbed material are visible in the dimer spectra after 6 mC has been consumed in the reaction. In the beginning of the polymerization the region $1000-500\,cm^{-1}$ is dominated by peaks from the solution species. Later, when the film is formed, a broad band between 1006 and $835\,cm^{-1}$ start to dominate this region. Stretching modes from the thiophene ring are visible between 1450 and $1400\,cm^{-1}$. In this region as well as at $1360\,cm^{-1}$, bands from the substituted octyl chain in the thiophene ring are visible. Due to the pseudo-centre of symmetry in the chain polymerized from dimer [4] the doping-induced bands are better resolved. Additionally to the band at $1290\,cm^{-1}$, bands at 1246, 1221, 1140 and $1109\,cm^{-1}$ from the conjugated backbone are visible. In order to show the growth of the film material both from the monomer and the dimer as a function of the charge consumed in the polymerization, the adsorption intensities of the vibra-

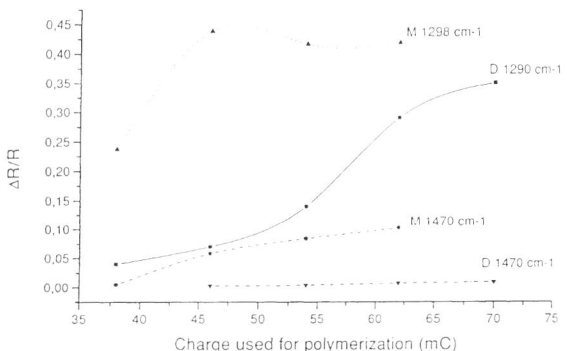

Fig. 2. The intensity of the band at $1290-1298\,cm^{-1}$ from the conjugation and at $1470\,cm^{-1}$ from the ring stretching is plotted *vs* charge used for polymerization of monomer and of dimer. The difference in growth rate of the adsorbed material $(1470\,cm^{-1})$ between the polymerizations made from monomer and from dimer is caused mainly by the differences in the concentrations of the monomer and the dimer. *M* indicates the monomer and *D* the dimer

tion at $1470\,cm^{-1}$ from ring stretching and of the band at $1290-1298\,cm^{-1}$ from conjugation are shown in Fig. 2. The steep but delayed increase of the intensity of the band from the conjugation at $1290\,cm^{-1}$ indicates that some rearrangements in the material obtained from the dimer have to take place before the polymer reaches the conjugated state.

To sum up the results in this work, it can be concluded that linear polymer chains can be obtained also with the monomer as starting material when low current density is used. The missing ring stretching modes of the $C_\alpha=C_\beta$ moiety at 1514 and $1533\,cm^{-1}$ in the spectra obtained with the dimer, confirm the different coupling mechanism from the mechanism of the monomer resembling the "head-to-tail" coupling for the former and the "head-to-head" coupling of the latter, as proposed by Lapkowski et al. [5]. Owing to the different coupling mechanisms the bands from the conjugation are more resolved in the polymer material that is obtained from the dimer.

References

[1] G. Louarn, J.-Y. Mevellec, J. Buisson, S. Lefrant, *Synth. Met.* **1993**, *55*, 587.
[2] X. Q. Yang, J. Chen, P. D. Hale, T. Inagaki, T. A. Skotheim, M. L. den Boer, *Synth. Met.* **1989**, *28*, C329.
[3] J. Bobacka, A. Ivaska, *Synth. Met.* **1991**, *43*, 3053.
[4] G. Louarn, J. Kruszka, S. Lefrant, M. Zagórska, I. Kulszewicz-Bayer, A. Proń, *Synth. Met.* **1993**, *61*, 233.
[5] M. Lapkowski, M. Zagórska, I. Kulszewicz-Bajer, K. Kozie, A. Proń, *J. Electroanal. Chem.* **1991**, *310*, 57.
[6] A. Bewick, K. Kunimatsu, B. S. Pons, J. W. Russell, *J. Electroanal. Chem.* **1984**, *160*, 47.
[7] C. Kvarnström, A. Ivaska, *Synth. Met.* **1994**, *62*, 125.

Mikrochim. Acta [Suppl.] 14, 819–822 (1997)

Isotopic Effect in Evolution of Structure and Optical Gap During Electrochromic Coloration of $WO_3 \cdot 1/3(H_2O)$ Films

Irina Shiyanovskaya*

Institute of Physics, Kiev 22, 252022 Ukraine

Abstract Evolution of the fundamental absorption edge and structure of $WO_3 \cdot 1/3(H_2O)$ films during electrochromic (EC) coloration with different cations has been studied in situ by optical and Fourier-transform Raman spectroscopy to clarify the role of inserted cations. With EC coloration, different high-energy shifts of the optical gap have been observed in the absorption spectra of the films coloured with H^+ and D^+ ions. The shift magnitude for coloration with H^+ ions is about twice that for coloration with D^+ ions at the same electron density. Analysis of the shifts of the vibration bands during EC coloration with H^+ and D^+ ions has shown that along with a lattice polarization induced by the injected electrons forming the polarons, a strong lattice distortion produced by inserted ions occurs as well. The large isotopic effect shows that the electronic structure of WO-octahedra is very sensitive to ion-insertion, and the electron rearrangement depends not only on the charge and radius of the inserted ions but also on their mass.

Key words: electrochromism, thin films, intercalation, FT-Raman spectroscopy, lattice polarization.

Electrochromic (EC) phenomena in transition metal oxides have been widely studied. A reversible colour change induced by an applied electric field or current has many potential applications.

It is now well known that EC coloration of WO_3 films results from a double injection of electrons from a conducting substrate and of small cations from an electrolyte into the films. The coloration is achieved by film polarizing in an electrolytic medium. Reverse polarization causes rejection of electrons and cations, and bleaching to the original colourless state.

Several models of EC coloration have been proposed [1–4]. The models of intervalence charge-transfer, by Faughnan et al. [3] and of a small polaron by Schrimer et al. [4] are the most widely accepted. The injected electron is trapped at a W^{6+} site, perturbing the surrounding lattice and forming a W^{5+} colour centre. Light absorption occurs through the charge transfer between two neighbouring tungsten sites, W^{5+}–W^{6+}. In these models, inserted cations play a passive role. They compensate for the negative charge of the injected electrons and are situated in the host matrix interstices.

The role of the mobile small cations injected from the electrolyte is not fully explained. It has been shown that different inserted ions, such as H^+ and Li^+, cause different high-energy shifts of the fundamental absorption edge, with coloration of amorphous WO_3 films, at the same electron density [5]. The magnitude of this shift is 0.05 eV at the colour centre concentration of $7.5 \times 10^{21}/cm^3$ when H^+ ions are used. If Li^+ ions are used, the magnitude of this shift is about three times larger. This may be connected with the fact that the ionic radius of Li^+ is about three times that of H^+. It allows the suggestion that inserted cations contribute more essentially in the coloration process and produce structural change in the host matrix. Moreover, the origin of the high-energy shift of the fundamental absorption edge during coloration may be caused not only by the Burstein–Moss shift of the Fermi level due to increased electron density in the conduction band during coloration, but also by structure rearrangement induced by inserted cations.

In the present work, in situ evolution of the fundamental absorption edge and vibration bands during coloration/bleaching of $WO_3 \cdot 1/3(H_2O)$ films has been studied, to clarify the role of cations inserted into the oxide matrix. The coloration has been performed in H^+ and D^+ acid electrolytes. The $WO_3 \cdot 1/3(H_2O)$ films were chosen as a suitable model EC system for simultaneous optical and structure studies.

* Present address: Department of Chemistry, State University of New York, College at Potsdam, 44 Pierrepont Av., Potsdam, NY 13676-2294, U.S.A.

Experimental

The crystalline hydrated films of $WO_3 \cdot 1/3(H_2O)$, about 500 nm thick, were obtained from amorphous WO_3 films deposited by thermal vacuum evaporation by a method described previously [6].

The optical measurements were performed with a KSVU-12 automatic spectrometer in the wavelength range 200–1200 nm in a electrochemical cell, for coloured and bleached films. Optical absorption spectra were obtained at room temperature. The FT-Raman spectra (in back-scattering geometry) of thin films in the coloured and bleached states were obtained with an IFS-88 spectrometer (Brucker) and a special FT-Raman module with an Nd: YAG laser (1064 nm), in the region 100–3500 cm^{-1}. Each FT-Raman spectrum is the average of 2000 scans with a spectral resolution of 4 cm^{-1}. A laser power of 40–60 mW was applied. A liquid-nitrogen cooled InGaAs detector was used.

Measurements were made with a classical three-electrode electrochemical cell. The H^+ and D^+ ions from 1% sulphuric acid solution in H_2O and D_2O, respectively, were used for the film coloration. The working electrode was a $WO_3 \cdot 1/3(H_2O)$ film on a transparent SnO_2-coated substrate. A platinum plate was used as a counter-electrode, in a quartz and glass cell. The reference electrode was a saturated calomel electrode. The potential of the working electrode with respect to the reference electrode was changed over the range from -0.6 to $+0.8$ V. The injected charge density was varied from 1 to 30 mC/cm^2.

Results and Discussion

The fundamental absorption of amorphous WO_3 and the disordered crystalline hydrate $WO_3 \cdot 1/3(H_2O)$ films is due to the non-direct transitions at an absorption coefficient higher than 10^4 cm^{-1} and follows the relation [7].

$$hv\alpha \sim (hv - E_g)^2 \qquad (1)$$

where hv is the photon energy, E_g the optical energy gap and α the absorption coefficient. Figure 1 shows the dependence of $(hv\alpha)^{1/2}$ on photon energy near the fundamental absorption edge for a colourless film (curve 1) and for films coloured by H^+ and D^+ ions (curves 2 and 3, respectively) at the same injected charge of 25 mC/cm^2. The value of E_g for the initial $WO_3 \cdot 1/3(H_2O)$ film was found to be 2.90 eV.

With EC coloration, the high-energy shift of the optical gap has been observed in spectra of the films coloured with either H^+ or D^+ ions. The shift magnitude for the coloration with H^+ ions is about twice that for the coloration with D^+ ions at the same electron density ($\Delta E_g = 0.02$ eV for D^+ ions and 0.04 eV for H^+ ions at a charge density of 25 mC/cm^2). Such a large isotopic effect shows that the electronic structure of the WO-octahedra is very sensitive to the ion insertion and that the electron rearrangement depends not only on the charge and radius of the intercalated ions but also on their mass. The larger delocalization of H^+ ions in

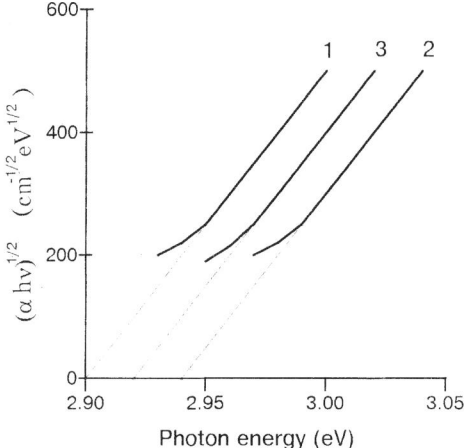

Fig. 1. Dependence of fundamental absorption edge on photon energy for $WO_3 \cdot 1/3(H_2O)$ film in the initial colourless state (curve 1) and for the film coloured with H^+ ions (curve 2) and D^+ ions (curve 3). Charge density is 25 mC/cm^2. Spectra are corrected for reflection

comparison with D^+ ions, because of their higher energy of zero vibrations and stronger penetration in classically forbidden regions, may be the origin of this isotopic effect.

To confirm the different electron rearrangements during the insertion of the H^+ and D^+ ions, the evolution of the vibration bands of the $WO_3 \cdot 1/3(H_2O)$ films during the coloration/bleaching process was studied. Fourier-transform Raman spectroscopy with near infrared excitation, the energy of which is too low to excite fluorescence and induce photochemical processes, gives a unique possibility to study strongly coloured films.

The structure of $WO_3 \cdot 1/3(H_2O)$ is similar to that of hexagonal WO_3. The structure units are tungsten–oxygen octahedra of two kinds [7]. The first contain a coordinated water molecule and an opposite terminal oxygen ion $W=O_t$. The second are WO_6 octahedra which share six bridge oxygen atoms, forming six-membered rings and -W$-O_b-$W-links. The crystalline hydrates $WO_3 \cdot 1/3(H_2O)$ are both covalent and molecular crystals with layered structure. The interaction between tungsten–oxygen octahedra inside the layers is covalent and the interaction between the layers is due to van der Waals bonds and hydrogen bonds of the coordinated water molecules.

The FT-Raman spectra below 1000 cm^{-1} are very sensitive to the film structure and reflect the features of the octahedra arrangement. Injection of cations and electrons into the film results in rearrangement of the

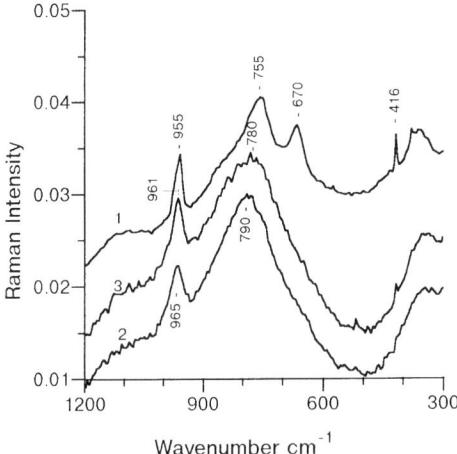

Fig. 2. FT-Raman spectra of $WO_3 \cdot 1/3(H_2O)$ film in initial colourless state (curve 1) and coloured states (curve 2 – coloration with H^+ ions, curve 3–coloration with D^+ ions). Charge density 25 mC/cm², resolution 4 cm⁻¹, 2000 scans

structure. In Fig. 2 the Raman spectra of a $WO_3 \cdot 1/3(H_2O)$ film in the initial colourless state (curve 1) and in the state coloured with H^+ and electrons (curve 2) at a charge density of 25 mC/cm² are shown. A small thermal and fluorescent background in the spectra is caused by the film's weak absorption of the near IR laser excitation.

The most obvious feature in the spectrum of the coloured films is the almost complete disappearance of the $v(W-O)$ stretching vibration band at 670 cm⁻¹ and $\delta(O-W-O)$ bending vibration band at 416 cm⁻¹, belonging to the vibrations in tungsten–oxygen octahedra with a coordinated water molecule [8]. The short-wave 35 cm⁻¹ shift of the band at 755 cm⁻¹ (corresponding to the stretching vibrations of O–W–O bonds in WO_6 octahedra) has also been observed. The disappearance of the vibration band associated with WO-octahedra with a coordinated water molecule may be attributed to predominant localization of the injected electrons and H^+ ions in the vicinity of WO-octahedra of this type.

Moreover, the 10 cm⁻¹ shift of the band at 950 cm⁻¹ (corresponding to the stretching vibration of the terminal bond $W=O_t$) is also observed. This shift is due to enhancement of this bond. What this means is that the H^+ ion very likely does not interact with the terminal oxygen ion, because if protonation of the terminal oxygen ions had occurred, the frequency $v(W=O_t)$ should have been shifted to the long-wave range, owing to the weakening of the $W=O_t$ bond.

The $W=O_t$ bond consists of a σ-hybrid bond and $p_\pi-d_\pi$ transfer of undivided electron pairs of oxygen to the 5d-orbital of W. The strong $p_\pi-d_\pi$ electron-transfer from terminal oxygen ions O_t brings about lowering of the electron density at the terminal oxygen ions O_t that is unfavourable for their protonation. The level of $p_\pi-d_\pi$ transfer depends on the bonds of the W ion with other bridge oxygen ions O_b. So it may be suggested that one of the orbitals of the bridge oxygen ions is involved in the formation of the bond with the proton, producing the weakening or even breaking of the $-W-O_b-W-$ bond and enhancement of $p_\pi-d_\pi$ transfer in $W=O_t$.

The film coloration induced by different cations is responsible for the different shifts of vibration bands. In Fig. 2 the Raman spectra of $WO_3 \cdot 1/3(H_2O)$ films coloured with H^+ ions (curve 2) and D^+ ions (curve 3) at the same injected charge density (25 mC/cm²) are shown. Coloration with D^+ ions leads to the smaller short-wave shift of the bands at 950 and 806 cm⁻¹ (6 and 25 cm⁻¹, respectively) in comparison with coloration by H^+. This testifies that along with the lattice polarization induced by the injected electrons forming the polarons, lattice distortion produced by the cations occurs as well.

It was found that the lattice distortion was reversible after ion-extraction during the bleaching process for injected charge density up to 30 mC/cm².

For higher coloration density (over 35 mC/cm²), strong enhancement of the Raman effect was observed. It may be caused by the sharp increase of the electron conductivity of the colured films during the injection of large charges.

This means that the cations do not passively compensate for the negative electron charge by remaining ionized in interstitial sites in the oxide matrix, but strongly interact with bridge oxygen ions, predominantly in WO-octahedra with coordinated water molecules.

Thus, the origin of the high-energy shift of the optical gap on coloration is caused not only by the Burstein–Moss shift of the Fermi level due to increased electron density in the conduction band, but also by the structure rearrangement caused by the interaction between the intercalated cations and the bridge oxygen atoms of tungsten–oxygen octahedra in the oxide matrix.

Acknowledgements. The author is grateful to Professor H. Ratajczak for financial support of this work and to Dr. J. Baran and Mr. M. Marchewka for expert technical assistance.

References

[1] S.K. Deb, *Phil. Mag.* **1973**, *27*, 801.

[2] S.K. Deb, *Solar Energy Mater Sol. Cells* **1992**, *25*, 327.

[3] B.W. Faughnan, R.S. Crandall P.M. Heyman, *RCA Rev.* **1975**, *36*, 177.

[4] O.F. Schirmer, V. Wittwer, B.Baur, G. Brandt, *J. Electrochem. Soc.* **1977**, *124*, 749.

[5] A. Nakamura, S. Yamada, *Appl. Phys.* **1981**, *24*, 55.

[6] I.V. Shiyanovskaya, T.A. Gavrilko, E.V. Gabrusenoks, *J. Mol. Struct.* **1993**, *293*, 295.

[7] B. Gerand, G. Nowogrocki, M. Figlarz, *J. Solid State Chem.* **1981**, *38*, 312.

[8] I.V. Shiyanovskaya, *Proc. SPIE* **1993**, *2089*, 394.

Mikrochim. Acta [Suppl.] 14, 823–825 (1997)
© Springer-Verlag 1997

FT-IR Analysis of Polyurethane Foams with the Use of Mid-Infrared Fibers

Sanmitra A. Bhat and **James A. de Haseth***

Department of Chemistry, University of Georgia, Athens, Georgia 30602–2556, U.S.A.

Abstract. Infrared fibers have found tremendous applications in the study of many systems. In this paper, mid-IR/FT-IR spectrometry was used to study the reaction kinetics of polyurethane foams. Polyurethane foams were produced with two different silicone surfactants, BF-2370 and B-8404. Data were collected for two hours on an FTS-60A/896 Bio-Rad Spectrometer. The temperature was monitored during the foam production. Infrared spectra at every 50 temperature drop were curve-fitted by use of Galactic GRAMS/386™ software. Based on the curve-fit results, percentages of free, monodentate and bidentate urea, and free and hydrogen-bonded urethane formed during the foam formation were determined. Relative reaction rates for the formation of urea and urethane were also calculated. For the BF-2370 surfactant foam, the initial rate of urethane formation was much greater than the urea formation. The rate of urea formation increased with time and was greater than urethane formation when it reached a steady state. Different results were obtained with the B-8404 foam, where both rates were in a steady state during the polymer formation. These results gave a better understanding of structural changes that occur with temperature in the formation of polyurethane formations.

Key words: infrared fiber, curve-fitting, polyurethane foams.

Among the many spectral treatment methods, curve-fitting is the one that is commonly used for the analysis of infrared spectra. Based on an original algorithm of non-linear peak fitting by Marquardt [1], it is sometimes referred as peak deconvolution or peak fitting. An infrared spectroscopist uses this method to obtain more information from the measured infrared spectra. In the present paper, it is used to fit the carbonyl region of the polyurethane foam. To study the relative amount of urea and urethane formed during the course of polyurethane formation, infrared spectra of the polymer were collected with the use of the mid-IR/FT-IR technique. The carbonyl region showed distinct peaks related to urea and urethane formation. Results of the curve-fitting of the carbonyl region are addressed.

---

* To whom correspondence should be addressed

Experimental

Polyurethane foams were produced with five ingredients: polyol, water, tin and amine catalyst, silicone surfactant and TDI (80:20 mixture of 2.6 and 2, 4-toluene di-isocyanate). An experimental set-up can be found elsewhere [2]. Except for TDI, all reactants were first mixed in a 4000ml plastic beaker for 2 min by use of a mixer at 2500 rpm. A known quantity of TDI was then added to the mixture, which was then mixed for 8 sec. The polymer solution was then immediately poured into a paper bucket that included a fiber probe, and spectrometer data collection was started. Data collections were made for 2 hr. Spectra were collected separately for the foams made with two different silicone surfactants: BF-2370 and B-8404. The concentrations of all other components were the same in the two foams, except for the B-8404 foam where the tin catalyst concentration was lowered to produce a stable foam. Temperatures and the corresponding times were recorded. The data collected were then transferred onto a PC to do the curve-fitting of the carbonyl region. Before the curve-fitting was done, the spectra were first normalized with respect to the CH stretching region (3000–2800 cm^{-1}). Curve-fitting was completed with the use of Galactic GRAMS/386™ software. For every 5 °C decrease in temperature, the corresponding infrared spectrum was curve-fitted. The height, width and area under each peak were determined.

Results and Discussion

The amide I region, 1800–1600 cm^{-1}, i.e., the carbonyl region, showed distinct peaks related to both free and hydrogen-bonded urea and urethane. The BF-2370 surfactant foam initially showed four peaks: 1735–1740 cm^{-1} for urethane, 1705–1710 cm^{-1} for hydrogen-bonded urethane, 1715–1720 cm^{-1} for urea and 1640 cm^{-1} for bidentate (doubly hydrogen-bonded carbonyl) urea. Towards the end of the reaction, it also showed a monodentate (singly hydrogen-bonded) peak at around 1662 cm^{-1}. The areas of the free, monodentate, and bidentate urea peaks were used to calculate total peak area for urea, and those of the free and hydrogen-bonded urethane peaks were used to calculate the total peak area for urethane. Percentages of

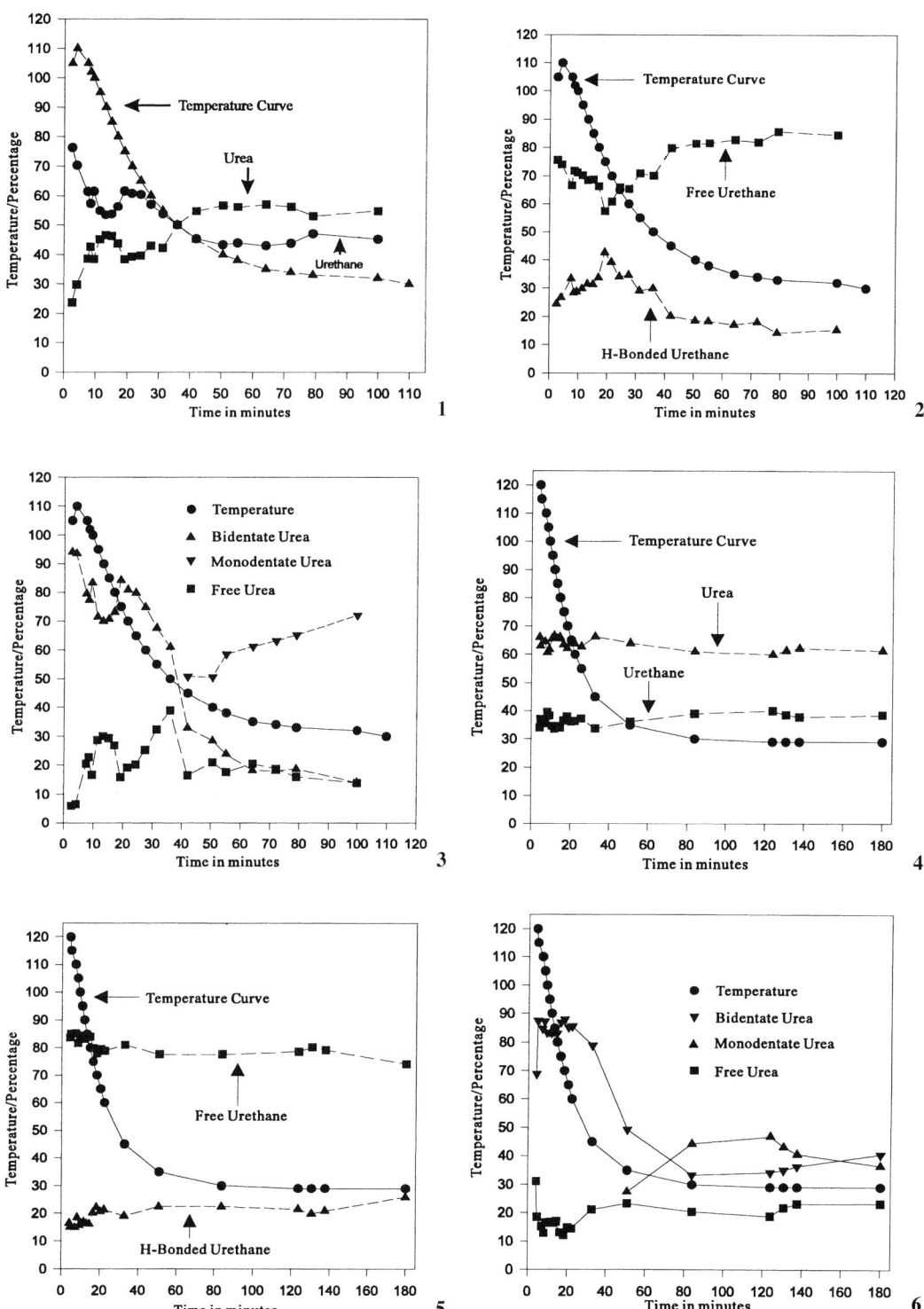

Fig. 1. Relative rates of urea and urethane formation for BF-2370 foam
Fig. 2. Relative rates within soft phase
Fig. 3. Relative rates within hard domain
Fig. 4. Time-temperature percentage relationship for B-8404 foam
Fig. 5. Formation rates within soft phase
Fig. 6. Formation rates within hard domain

urea and urethane were then computed for these peak total areas. In polyurethane foams, urethane is a soft or continuous phase and urea is the hard phase or domain. The percentage–temperature–time relationship is shown in Fig. 1. The rate of formation of urethane was much greater than that of urea. With time, there was a drop in urethane formation with a corresponding rise in urea formation. The rate of formation of urea was greater than that of urethane toward the end of polymer formation. After 40 min of polyurethane formation, the system reaches a steady state, where the rates of urethane and urea formation essentially remain constant. The data showed that 45% of urethane and 55% of urea are formed in BF-2370 foam. Within a soft segment, i.e. urethane, the rate of free urethane formation was much greater than that of hydrogen-bonded urethane. The percentage–temperature–time relationship is shown in Fig. 2. This rate free urethane rate initially decreased with time, but then slowly increased to reach a steady state. Nearly 85% of the polyurethane formed is free and only 15% is hydrogen bonded. For hard domains, i.e. urea, the rate of bidentate urea formation was greater than that of free urea. The rates increased concurrently and then decreased to reach a constant value. The rate of monodentate formation, formed later in the reaction, increased with time as shown in Fig. 3. For B-8404 surfactant foam, quite different results were obtained. The rate of urea formation was significantly greater than that of urethane, unlike BF-2370 where rate of urethane formation was greater than that of urea during the initial stages. These rates remained nearly constant during the polymer formation (Fig. 4). Within a soft phase, the rate of free urethane formation was greater (85%) during the initial stages of reaction. This rate decreased with time to reach a value of 74% (Fig. 5). In a hard domain, the rate of bidentate urea formation was greater than that of free urea. The rate of bidentate urea formation decreased with time, whereas that of the free urea increased only slightly during the polymer formation. The rate of monodentate urea formation, which was much later in the reaction, increased slowly with the polymer formation (Fig. 6).

Conclusion

Curve-fitting results showed that the rates of formation of urethane and urea are different in foams made with different silicone surfactants. At the end of polymerization, B-8404 and BF-2370 surfactant foams showed greater percentage of urea than urethane. Studies have shown that use of silicone surfactants aids the polymerization and stabilizes the cell wall of the polyurethane foams. It also affects the physical properties of the foams. In the present study, we did not compare the two silicone surfactants, but merely presented the effect that each silicone surfactant has on the relative rates of formation of urethane and urea. These relative rates would definitely affect the physical properties of the foam produced.

References

[1] D. W. Marquardt, *J. Soc. Ind. Appl. Math.* **1963**, *11*, 431.
[2] J. A. de Haseth, J. E. Andrews, J. V. McClusky, R. D. Priester, Jr., M. A. Harthcock, B. L. Davis, *Appl. Spectrosc* **1993**, *47*, 173.

Mikrochim. Acta [Suppl.] 14, 827–828 (1997)
© Springer-Verlag 1997

Evaluation of Microdialysis and FT-IR ATR-spectroscopy for in-vivo Blood Glucose Monitoring

A. Bittner[1], **H. M. Heise**[1], **Th. Koschinsky**[2], and **F. A. Gries**[2]

[1] Institut für Spektrochemie und Angewandte Spektroskopie, Bunsen-Kirchhoff-Str. 11, D-44139 Dortmund, Federal Republic of Germany
[2] Diabetes-Forschungsinstitut an der Universität Düsseldorf, Auf'm Hennekamp 65, D-40225 Düsseldorf, Federal Republic of Germany

Abstract. The development of a portable device for continuous glucose self-monitoring is desirable for diabetic patients dependent on insulin. Infrared spectrometry with attenuated total reflectance can be coupled to a microdialysis probe to reliably sense the low glucose concentrations in the dialysate. The results from multivariate calibration using partial least-squares are reported. Mean squared prediction errors achieved were 7.2 mg/dl (plasma dialysates) and 1.5 mg/dl (dialysates from glucose solutions).

Key words: IR-spectroscopy, microdialysis, blood glucose, attenuated total reflection.

Patients suffering from diabetes mellitus are advised to test their blood glucose concentrations regularly. Besides tests performed by non-invasive techniques based on near infrared spectrometry of body tissues [1], further approaches are described in the literature which apply the use of electrochemical enzyme biosensors implanted directly into the subcutaneous tissue [2], whereas ex vivo monitoring can be either enabled by open tissue perfusion [3] or in combination with a microdialysis probe [4]. The last of these techniques eases calibration problems, as the enzymatic glucose sensor based on glucose oxidase is markedly impaired when used in subcutaneous tissue and plasma. As the lifetime of the biosensors is usually limited, an alternative spectrometric glucose-sensing device would be more suitable. Infrared spectroscopy, by the attenuated total reflection technique, can be used for a reliable glucose determination in biotic fluids such as blood plasma, and recent progress based on partial least-squares analysis has shown its potential for reaching the lower concentration ranges necessary for coupling this technique to microdialysis [5]. Results on the coupling of a microdialysis probe to a micro-Circle cell in combination with FT-spectroscopy are reported.

Experimental

Plasma samples from seven patients of the clinical department of the Diabetes-Forschungsinstitut in Düsseldorf and microdialysates from plasma and glucose solutions were analysed in triplicate for glucose by a hexokinase/glucose-6-phosphate dehydrogenase (G6P-DH) assay programmed on a Hitachi 705 from Boehringer (Mannheim, Germany) for concentration determination. The original glucose concentration range for the plasma samples was between 100 and 480 mg/dl. A CMA/100 microinjection pump from CMA/Microdialysis (Stockholm, Sweden) was used for realizing perfusion rates with aqueous isotonic sodium chloride solution or distilled water of between 12 and 1.5 µl/min, in combination with CMA 10 microdialysis probes (polycarbonate membrane, outer radius 250 µm, length 16 mm, cut-off 20 kDa). Mid-infrared spectra were recorded with a Bruker IFS 66 FT-spectrometer equipped with an MCT detector from EC & G Judson (Montgomeryville, PA). A micro-Circle cell from Spectra-Tech Europe (Warrington, England) with a ZnSe crystal was employed for the ATR-measurements. The cell temperature was maintained at $37.00 \pm 0.02\ °C$ by a thermostat. For the spectra 500 interferograms providing a spectral resolution of $4\ cm^{-1}$ were co-added. A Blackman–Harris 3-term apodization function was applied before Fourier transformation.

Results and Discussion

Some spectra are presented in Fig. 1 to illustrate the performance of the spectrometer. ATR-spectroscopy has been proved to be suitable for the study of physiological fluids such as whole blood or plasma, but a post-measurement cleaning step must be used to eliminate the protein adsorbed onto the ATR-crystal, see e.g. [5]. The microdialysis helps to simplify the biotic matrix, avoiding this complication, but at the cost of dilution step. Different dilution factors are achieved by varying the flow-rate of the perfusion solution (see Table 1). Samples were drawn from 1.3 ml of blood plasma from each of several patients, or from aqueous glucose solutions of between 100 and

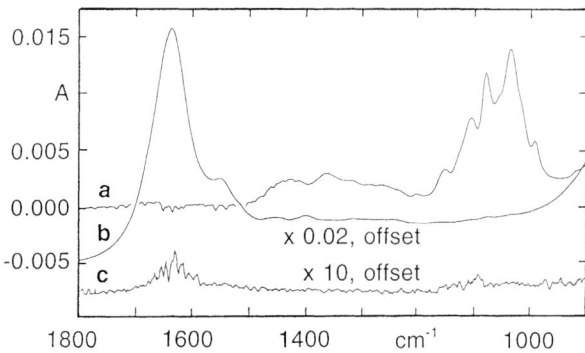

Fig. 1. Spectra recorded by using a micro-Circle cell: *a* aqueous glucose spectrum (water-compensated, concentration 500 mg/dl); *b* blood plasma spectrum; *c* noise level calculated from two subsequent single-beam spectra with water-filled cell

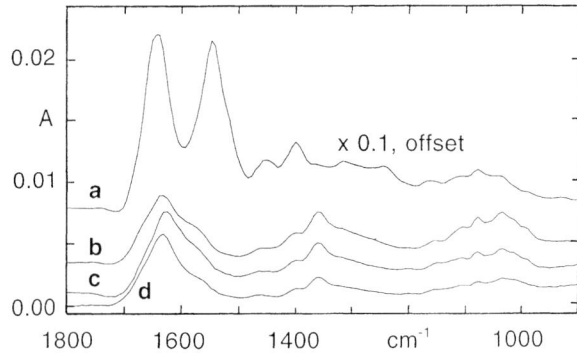

Fig. 2. Blood plasma spectrum with water compensation *a* and dialysates obtained by application of different perfusion rates: *b* 1.5 μl/min (glucose concentration 90 mg/dl); *c* 3 μl/min (60 mg/dl); *d* 6 μl/min (33 mg/dl)

Table 1. Extraction fractions (%) with different microdialysis probes for various perfusion rates (probe nos. 1 and 2 for glucose solutions only, others for blood plasma extractions)

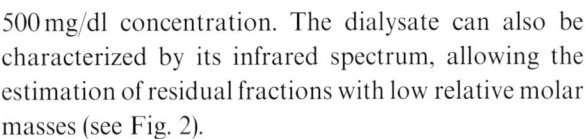

Probe no.	Perfusion flow-rate, μl/min			
	1.5	3.0	6.0	12.0
1	–	16.5	9.4	5.1
2	–	17.8	10.7	5.9
3	–	18.4	9.8	4.9
4	–	20.8	12.2	6.5
5	25.6	16.4	9.4	–
6	34.5	25.3	14.7	–

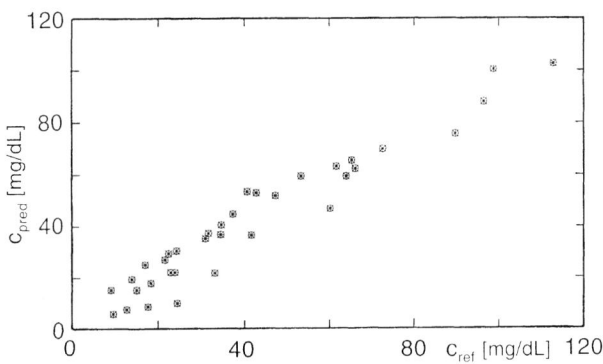

Fig. 3. Scatter plot of glucose concentrations predicted by a partial least-squares (PLS) model from 11 PLS factors versus reference values for microdialysates from blood plasma of seven patients

500 mg/dl concentration. The dialysate can also be characterized by its infrared spectrum, allowing the estimation of residual fractions with low relative molar masses (see Fig. 2).

As shown in Fig. 3, multivariate calibration by partial least-squares can yield adequate concentration predictions. Using the microdialysate spectra with data from between 1200 and 950 cm^{-1}, average prediction errors for glucose evaluated from a cross-validated calibration model are 7.2 mg/dl for plasma dialysates (spectral baseline corrected) and 1.5 mg/dl for aqueous glucose solutions (concentration range between 12 and 92 mg/dl with mean of 40 mg/dl). Even better results can be expected for one-person testing. The performance is satisfactory for ex vivo glucose monitoring by means of microdialysis, so continuous sensing by

a miniaturized IR-spectrometer is a feasible alternative to enzymatic biosensors.

Acknowledgements. Financial support by the Ministerium für Wissenschaft und Forschung des Landes Nordrhein-Westfalen, the Bundesminister für Bildung und Forschung and Axel Semrau GmbH (Sprockhövel, Germany) for the loan of microdialysis equipment is gratefully acknowledged.

References

[1] H. M. Heise, R. Marbach, Th. Koschinsky, F. A. Gries, *Artif. Organs* **1994**, *18*, 439.
[2] V. Poitout, D. Moatti-Sirat, G. Reach, Y. Zhang, G. S. Wilson, F. Lemonnier, J. C. Klein, *Diabetologia* **1993**, *36*, 658.
[3] F. Skrabal, Z. Trajanoski, H. Kontschieder, P. Kotanko, P. Wach, *Med. Biol. Eng. Comput.* **1995**, *33*, 116.
[4] C. Meyerhoff, F. J. Mennel, F. Bischof, F. Sternberg, E. F. Pfeiffer, *Horm. Metab. Res.* **1994**, *26*, 538.
[5] H. M. Heise, A. Bittner, *J. Mol. Struct.* **1995**, *348*, 21.

Mikrochim. Acta [Suppl.] 14, 829–831 (1997)
© Springer-Verlag 1997

The Use of Multivariate Calibration for Chlorinated Hydrocarbon Detection in Water with an MIR Fibre-Optic Sensor

R. Göbel*, M. Sengeis, R. Krska, and **R. Kellner**

Institute for Analytical Chemistry, Vienna University of Technology, Getreidemarkt 9/151, A-1060 Wien, Austria

Abstract. A detection system for chlorinated hydrocarbons (CHCs) in water has been based on a mid-infrared (MIR) fibre optic physicochemical sensor, the measurement principle being attenuated total reflection (ATR), with the IR-beam coupled into an ATR-element, an IR-transparent silver halide fibre. At each reflection the penetrating energy is attenuated by the IR-absorbing analytes present in the surrounding polymer matrix. The aim of the present study was to investigate multivariate calibration techniques for use of this sensor system to determine CHCs in water. Partial least-squares regression was compared with standard univariate calibration. The spectral range from 950 to 750 cm^{-1} was measured for different aqueous CHC solutions with the fibre-optic set-up coupled to a Fourier-transform infrared spectrometer. Mixed standards of tetrachloroethylene, trichloroethylene and chloroform in water were measured according to a central composite design.

Key words: IR, fibre optic, multivariate calibration, partial least-squares regression.

Why use multivariate calibration (MC) techniques in general? The major advantage of MC is that the need for complex sample preparation is greatly reduced. This is because selective input measurements are no longer needed – it is the output results that must be selective.

Why use MC techniques for a sensor for hydrocarbons in water? First we have to look at the result we get when measuring hydrocarbon-contaminated water by on-line infrared spectra of extractable organics as a function of time. In the first place the selectivity of the sensor depends on the separation of the environment from the sample measured. Any disturbances such as those arising from muddy or turbid samples, high salt concentrations or biological material, although they may influence the diffusion characteristics of the analytes, do not distrub the measurement at all. Time-resolved spectra contain all the necessary data about the actual concentration levels of the organic species in the polymer matrix of the ATR-element but we have to extract the information from the spectrum. Since one of the major advantages of the proposed sensor is its on-site and on-line detection capability, it is obvious that on-line data evaluation is necessary for this environmental monitoring tool.

One possibility is to use MC techniques for evaluation of the acquired spectral data, since this approach enables maintenance-free and rapid quantification of spectra. Most MC algorithms can also handle quantification of substances even when there are interferences or overlapping bands. The advantage of being able to quantify the target compounds on-line, however, is at the cost of flexibility, as a preceding calibration step is required. We have to decide whether we want to analyse (1) a completely unknown water sample, in which case the resulting spectra have to be interpreted manually, or (2) continuously and on-line a known water matrix by use of MC techniques.

A detailed description of all kinds of multivariate calibration techniques is given in a monograph by Martens and Næs [1].

Experimental

The detection system was that described earlier [2]. All measurements were performed with a Bruker IFS 88 FT-IR spectrometer with a polyisobutylene-coated (23 μm) silver halide fibre in a flow cell. One hundred co-added scans at 4 cm^{-1} resolution were acquired after analyte diffusion into the polymer layer for 5 min. After

---

* To whom correspondence should be addressed

Table 1. Composition of the standard mixtures for the multivariate calibration and the predicted residual error sum of squares calculated according to the "leave out one and predict" method

No.	TCE(μl/l) True	Predicted	TeCE(μl/l) True	Predicted	CF(μl/l) True	Predicted
1	32.2	32.4	8	7.9	18.1	18.6
2	67.8	66.4	8	7.4	18.1	17.9
3	32.2	30.8	17	17.3	18.1	17.1
4	67.8	66.8	17	18.5	18.1	18.7
5	32.2	31.9	8	8.7	41.9	46.2
6	67.8	67.7	8	7.1	41.9	43.4
7	32.2	33.6	17	15.6	41.9	42.3
8	67.8	68.1	17	17.7	41.9	41.6
9	20	21.9	12.5	13.7	30	29.5
10	80	82.1	12.5	11.9	30	29.4
11	50	51.7	5	5.9	30	28.2
12	50	47.3	20	16.0	30	26.8
13	50	52.3	12.5	13.7	10	13.8
14	50	48.6	12.5	13.8	50	47.2
15	50	48.5	12.5	12.3	30	29.3

Table 2. Statistical summary of the PLS1-calibration

Compound	σ [μl/l]	σ_{rel} [%]	R^2
TCE	2.09	4.2	99.2
TeCE	1.85	14.8	90
CF	2.57	8.6	96.9

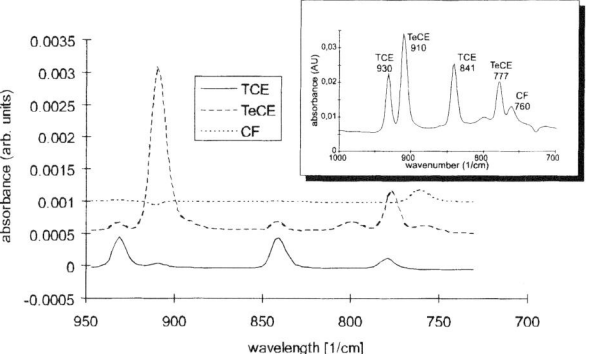

Fig. 1. Spectra of individual components for the K-matrix calibration of TCE, TeCE and CF

each measurement step the sensor was regenerated with clean water. The quantification software used was a partial least-squares algorithm, PLS1, implemented by OPUS™-measurement software.

Results and Discussion

First, the method has to be calibrated with a set of calibration solutions of known analyte concentration. Calibration sets of three CHCs, trichloroethylene (TCE), tetrachloroethylene (TeCE) and chloroform (CF), were prepared according to the central composite experimental design and measured. The next step was generating a "PRESS"-report (Predictive Residual Error Sum of Squares) in order to determine the optimal ranks for the PLS calibration, to calculate the true predictive error and to recognize bad calibration standards. The principle of these calculations is the "leave out one and predict" method where one calibration spectrum is removed from the calibration set and the

concentration values are predicted with the remaining data sets (Table 1). The statistical features are summarized in Table 2. It has to be noted that a good R^2 and good standard deviation of the calibration data do not guarantee reliable results. A possibility for looking behind the information within the data set is the so-called K-matrix. In the K-matrix the relevant information of a complete calibration set for each substance is extracted. That means, by looking at the K-matrix we should see the spectra of each single substance. Figure 1 shows the results for our calibration set. The small graph in the upper right hand corner is an example of a calibration spectrum. It is obvious that the extracted single-component spectra are rather good, except for that of TeCE, which is slightly influenced by TCE

Table 3. Example of PLS-prediction with two known CHC-mixtures

	TCE[μl/l] Real	Predicted	TeCE[μl/l] Real	Predicted	CF[μl/l] Real	Predicted
Sample 1	20	22.4	7	5.7	15	15.4
Deviation [%]	12.0		18.6		2.7	
Sample 2	35	37.0	3	3.8	38	34.2
Deviation [%]	5.7		26.7		10.0	

(at 937 and 842 cm^{-1}), indicating reliable calibration sets.

In order to test the quality of the calibration two "unknown" samples were measured and quantified. Table 3 compares the predicted analyte concentrations with the real ones. Summarizing, the relative deviations of the predicted values are typically in the 5–20% region, which is in the range of values obtained with univariate calibration methods. Predictions are slightly better for TCE and CF than for TeCE. One reason for that is the relatively good solubility of TCE and CF in water compared to TeCE. Standards for TCE and CF are much easier to prepare reproducibly than standards for TeCE.

Acknowledgements. Acknowledgement is given to the "Fonds zur Förderung der wissenschaftlichen Forschung" under Project No. 9289 TEC and to all members of the "GREAT-IR-TEAM".

References

[1] H. Martens, T. Næs, *Multivariate Calibration*, Wiley, New York, 1993.
[2] R. Krska, K. Taga, R. Kellner, *Appl. Spectrosc.* **1993**, *47*, 1484.

Mikrochim. Acta [Suppl.] 14, 833–835 (1997)

Application of Sapphire Fibres to IR Fibre-optic Evanescent Field Gas Sensors

R. Götz, B. Mizaikoff*, and **R. Kellner**

Institute for Analytical Chemistry, Vienna University of Technology, Getreidemarkt 9/151, A-1060 Vienna, Austria

Abstract. A fibre-optic gas sensor for hydrocarbons in the mid IR range (MIR) suitable for measurements at elevated temperatures, based on sapphire fibres and evanescent field spectroscopy, has been developed. This sensor takes advantage of the high thermal stability, mechanical ruggedness and the chemical resistivity of sapphire fibres. However, the spectral window of the fibres is confined to wavelengths below 4 μm. Thus, spectroscopic applications in the MIR are focused mainly on the exploitation of C–H, O–H and N–H stretch vibrations. An external fibre-optic set-up was developed, including a flow cell completely enclosed in an oven. A vacuum interface enabled direct coupling of the fibre to an MCT detector. Methane and butane were determined as analyte gases at various concentrations and temperatures. Calibration graphs were obtained for methane in a concentration range from 2 to 20% at temperatures from 20 to 300 °C and for butane in a concentration range from 1 to 10% at temperatures from 20 to 200 °C, respectively.

Key words: FT-IR, fibre-optic, sapphire fibre, sensor, gas analysis.

Fibre-optic evanescent wave sensors (FEWS) in the mid IR region have been successfully developed in our group both for measurements in aqueous environments [1, 2] and for gas analysis [3, 4]. Among the advantages of fibre-optic chemical sensors are flexibility, reversibility and the potential of continuous on-line monitoring, enabling applications in process control and measurements at remote locations. The emphasis of this work is the development of an IR fibre-optic sensor based on sapphire fibres for gas analysis at elevated temperatures up to 300 °C and suitable for measurements in a harsh environment.

Sapphire fibres are potentially useful for this purpose because of their superb mechanical and thermal (melting point 2053 °C) properties. Moreover, sapphire fibres are non-toxic, chemically inert and biocompat-

ible. Hence, besides spectroscopic applications they are also utilized for medical purposes [5, 6].

However, the transmission range of sapphire fibres in the MIR region is limited to the wavelength range below 4 μm. Hence, mainly C–H, N–H and O–H stretch vibrations can be detected in the fundamental region of the MIR. Since the spectral information accessible with sapphire fibre sensors is confined to this small spectral window, emphasis was placed on the development of a sensor for gaseous hydrocarbons. Although a number of spectroscopic applications of sapphire fibres can be found in the literature [7–9] the application to gas monitoring has not yet been reported.

Experimental

A sapphire fibre with a total length of 50 cm was used, but only a section of 18 cm served as the active sensor head, located in a flow cell. An overview of the most important properties of the sapphire fibres with respect to our application is given in Table 1 [10].

Table 1. Properties of sapphire optical fibres

General properties	
Composition	Al_2O_3
Melting point	2053 °C
Diameter	0.3 mm
Minimal bending radius	60 mm
Tensile strength (25 °C)	435 GPa
Hardness (Knoop)	2200 kg/mm^2

Optical properties	
Refractive index	1.7
Transmission range	0.25–4 μm
Attenuation losses (at 2.94 μm)	3 dB/m
Practical numerical aperture	0.14–0.34

---

* To whom correspondence should be addressed

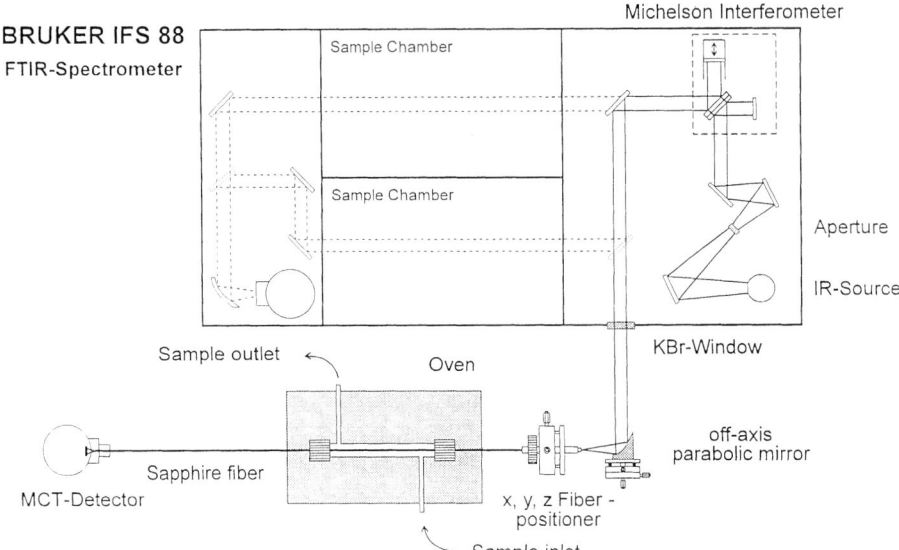

BRUKER IFS 88
FTIR-Spectrometer

Fig. 1. Fibre-optic sensor set-up based on a conventional Bruker FTIR-spectrometer with the fibre coupled externally to the spectrometer

All measurements were performed with a Bruker IFS 88 FT-IR spectrometer. In order to provide a flexible arrangement, the whole fibre-optic set-up was located outside the spectrometer (Fig. 1). After passing through the Michelson interferometer the IR beam is focused onto the tip of the fibre by an off-axis parabolic mirror. To optimize the energy throughput, the fibre is directly coupled to a broad-band MCT detector having maximum response in an adequate spectral region (at 3.9 µm) by a home-built vacuum interface. All spectra were obtained by co-adding 100 scans at a spectral resolution of $4 \, cm^{-1}$.

The flow cell had a length of 18 cm and a volume of about 2 ml. Temperature-resistant sealing was achieved by using Vespel® ferrules. The cell was completely enclosed in an oven equipped with a thermocouple, thus allowing isothermal measurements at elevated temperatures. Analyte gases were supplied from cylinders and diluted with nitrogen. The gas flow was adjusted by thermal mass-flow controllers, yielding a total flow of 200 ml/min. All measurements were performed at ambient pressure.

Results and Discussions

In order to determine the performance of the sapphire fibre-optic sensor, methane and butane were measured at various concentrations and temperatures. Spectra of the pure gases, measured at ambient temperature, are displayed in Fig. 2. Calibration graphs were obtained for methane in the concentration range from 2 to 20% v/v at temperatures from 20 to 300 °C and for butane in a concentration range from 1 to 10% v/v at temperatures from 20 to 200 °C (Fig. 3). Quantitative results were achieved either by evaluating the peak height (at $3017 \, cm^{-1}$ for methane) or the peak area (between 2980 and $2944 \, cm^{-1}$ for butane).

The significant decrease in absorbance at elevated temperature can partly be attributed to the lower gas

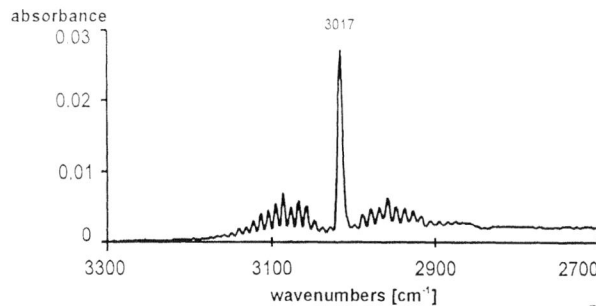

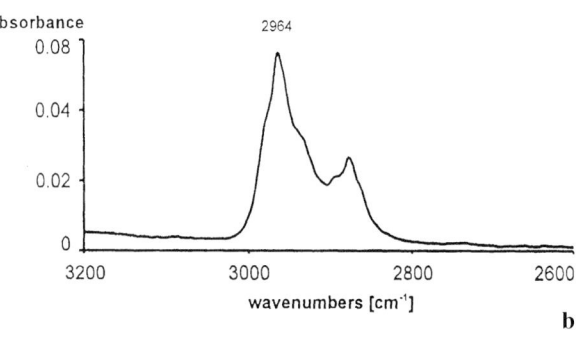

Fig. 2. Sapphire fibre ATR spectra methane (**a**) and butane (**b**) at 20 °C

density caused by heating the cell. Since all measurements were performed at ambient pressure, the gas density would consequently get smaller at higher temperature, according to the general gas equation.

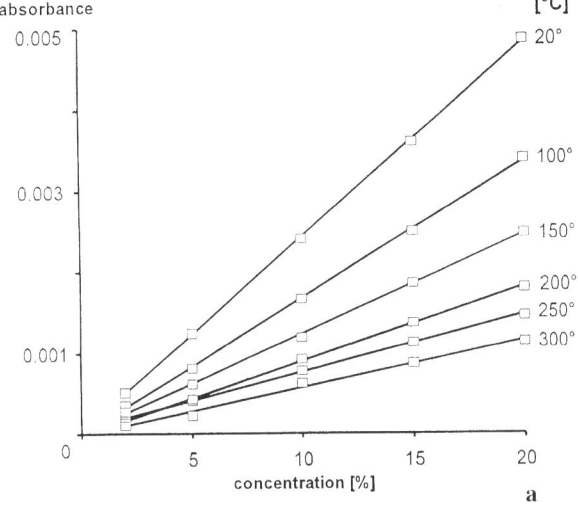

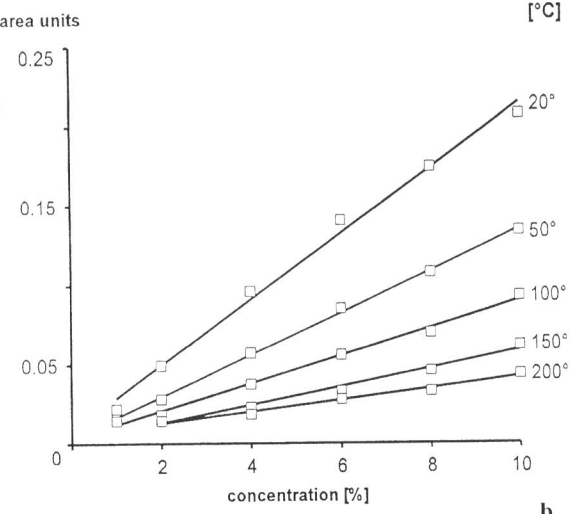

Fig. 3. Temperature-dependent calibration curves of methane (**a**) and butane (**b**) in N_2

Conclusions and Outlook

In this work the potential of measuring gaseous hydrocarbons in the low percentage concentration range with sapphire optical fibres at high temperature has been shown. Calibration curves at different temperatures have been obtained. Provided that the concept of the sapphire fibre-optic sensor can be extended to measurements in condensed phases, possible applications can be found in monitoring both gaseous and liquid process streams.

Acknowledgement. Acknowledgement is given to the "Fonds zur Förderung der wissenschaftlichen Forschung" for financial support under project no. 7898 CHE.

References

[1] R. Krska, E. Rosenberg, K. Taga, R. Kellner, A. Messica, A. Katzir, *Appl. Phys. Lett.* **1992**, *61*, 1778.

[2] R. Krska, K. Taga, R. Kellner, *Appl. Spectrosc.* **1993**, *47*, 1484.

[3] B. Mizaikoff, K. Taga, R. Kellner, *Vibr. Spectrosc.* **1995**, *8*, 103.

[4] B. Edl-Mizaikoff, R. Götz, R. Kellner, *Proc. SPIE*, **1995**, *2508*, 253.

[5] D. H. Jundt, M. M. Fejer, R. L. Byer, *Appl. Phys. Lett.* **1989**, *55*, 2170.

[6] G. Merberg, J. Harrington, *Proc. SPIE* **1992**, *1591*, 100.

[7] M. A. Druy, L. Elandjian, W. A. Stevenson, R. D. Driver, G. M. Leskovitz, L. E. Curtiss, *Proc. SPIE* **1989**, *1170*, 150.

[8] M. A. Serio, D. S. Pines, E. Kroo, K. S. Knight, P. R. Solomon, A. S. Bonanno, *Proc. SPIE* **1993**, *2069*, 20.

[9] M. A. Serio, H. Teng, K. S. Knight, S. C. Bates, S. Farquharson, A. S. Bonanno, P. R. Solomon, W. A. Stevenson, M. A. Druy, P. Glatkowski, J. A. Harrington, R. Nubling, Y. Ding, A. G. Comolli, *Proc. SPIE* **1993**, *2069*, 121.

[10] Saphikon, Inc., *Technical Information Bulletin.*

SpringerChemistry

D. Benoit, J.-F. Bresse, L. Van't dack,
H. Werner, J. Wernisch (eds.)

Microbeam and Nanobeam Analysis

1996. 394 partly coloured figures. XI, 643 pages.
Soft cover DM 280,–, öS 1960,–
Reduced price for subscribers to "Mikrochimica Acta":
Soft cover DM 252,–, öS 1764,–
ISBN 3-211-82874-5
Mikrochimica Acta / Supplement 13

The European Microanalysis Society held its Fourth Workshop in Saint Malo in
May 1995. This volume includes the revised presentations, 10 tutorial chapters and
50 brief articles, from leading experts in electron probe microanalysis, secondary mass
spectroscopy, analytical electron microscopy, and related fields.

Also available:

A. Boekestein, M. K. Pavićević (eds.)

Electron Microbeam Analysis

1992. 157 figures. IX, 278 pages.
Soft cover DM 218,–, öS 1525,–
Reduced price for subscribers to "Mikrochimica Acta":
Soft cover DM 196,20, öS 1372,50
ISBN 3-211-82359-X
Mikrochimica Acta / Supplement 12

 SpringerWienNewYork

Sachsenplatz 4 - 6, P.O.Box 89, A-1201 Wien • e-mail: springer@springer.co.at • Internet: http://www.springer.co.at
New York, NY 10010, 175 Fifth Avenue • Heidelberger Platz 3, D-14197 Berlin • Tokyo 113, 3-13, Hongo 3-chome, Bunkyo-ku

SpringerJournal

Edited by
K. M. Salikhov, Kazan (Managing Editor)
N. M. Atherton, Sheffield
G. Bodenhausen, Paris
P. Callaghan, Palmerston North
G. R. Eaton, Denver, CO
A. N. Garroway, Washington, DC
U. Haeberlen, Heidelberg
B. Maraviglia, Rome
J. Schmidt, Leiden
H. W. Spieß, Mainz
J. Stankowski, Poznan
D. Stehlik, Berlin
and an International Advisory Board

"Applied Magnetic Resonance" provides an international forum for the application of magnetic resonance in physics, chemistry, biology, medicine, geochemistry, ecology, engineering, and related fields.

"Applied Magnetic Resonance" publishes original articles with a strong emphasis on new applications of the technique and on new experimental methods. Routine applications to structural chemistry are outside the scope of the journal. Review articles are invited on methods and applications of NMR, NQR, EPR, Moessbauer spectroscopy, etc. Special issues on selected topics will be published under the guidance of guest editors.

A section is devoted to book reviews and Letters to the Editor. Information on conferences and new technical equipment are also accepted.

Applied Magnetic Resonance

Subscription Information:
1997. Vols. 12–13 (4 issues each):
DM 1.176,–, öS 8.232,–, plus carriage charges,
approx. US $ 869.00 incl. postage and handling
ISSN 0937-9347   Title No. 723

For customers in EU countries without VAT identification number

10 % VAT will be added to the subscription price

Sachsenplatz 4 - 6, P.O.Box 89, A-1201 Wien • e-mail: springer@springer.co.at • Internet: http://www.springer.co.at
New York, NY 10010, 175 Fifth Avenue • Heidelberger Platz 3, D-14197 Berlin • Tokyo 113, 3-13, Hongo 3-chome, Bunkyo-ku

Springer-Verlag
and the Environment